The Laws of Gravity and Electromagnetism

Arkady Poliakovsky

The Laws of Gravity and Electromagnetism

A Non-relativistic Model Invariant Under the Change of Inertial and Non-inertial Coordinate Systems

Springer

Arkady Poliakovsky
Department of Mathematics
Ben-Gurion University of the Negev
Be'er Sheva, Israel

ISBN 978-3-031-61406-4 ISBN 978-3-031-61407-1 (eBook)
https://doi.org/10.1007/978-3-031-61407-1

This Springer imprint is published by the registered company Springer Nature Switzerland AG
The registered company address is: Gewerbestrasse 11, 6330 Cham, Switzerland

Prologue

In [3], we showed how relativistic and non-relativistic approaches to space-time can be unified and proposed a new metrizable model of gravity in the frames of the idea of covariance (in tensor formulation) of the most fundamental Physical Laws under a change of general coordinate systems, given by arbitrary (both linear and smooth non-linear) transformations in the four-dimensional space-time. This approach is also one of the basic cornerstones of the classical Theory of General Relativity. However, as already was mentioned in [3] this idea do not really contradicts the non-relativistic considerations of space-time! For example, both the Galilean transformations and Lorentz's transformations can be considered with equal right as two different particular cases of general linear transformations of the space-time (Note that one can call a particular coordinate frame system by the word inertial if it is obtained by a general linear transformation from the other, commonly considered as inertial, coordinate frame). The existence of a global time scalar function in the four-dimensional space-time also implies no contradiction, since this scalar equals to the first coordinate of the space-time only in very special coordinate systems, while we assume the general covariance of the Physical Laws. The central goal of [3] was to show that both relativistic and non-relativistic approaches to space-time can be unified. In particular, the global universal time function can be concordantly introduced to relativistic theories. The particular type of coordinate systems where this global time equals to the first coordinate of the four-dimensional space-time and, moreover, the geometry of the remaining three-dimensional subspace is Euclidean, that were introduced in [3], are called Cartesian coordinate systems. Note that in many cases cartesian coordinate systems are often more convenient, than Lorentzian, since some Physical Laws have much simpler forms there. The basic advantage of Lorentzian coordinate systems is the very simple form of Maxwell Equations in the absence of genuine gravity. However, this advantage fades away in the presence of genuine gravity, that is the very common everyday practice! Moreover, as was showed in [3] and as we will demonstrate in the present work in more details, in the presence of genuine gravity, in particular in the frames of the most simple model of the Newtonian gravity, the Maxwell Equations have even more simple and convenient form rather in cartesian coordinate systems, than in Lorentzian ones.

In the present work, we deal mostly with cartesian coordinate systems, which are often called simply coordinate systems by default, and we develop the general forms of the most common physical laws in such coordinate systems, both in the classical and the quantum levels. Moreover, the introduction of cartesian coordinate systems, provides with possibility to formulate these laws without the tensor analysis and introducing of advanced Differential Geometry. Although, for the most generality, in several sections, we do introduce tensor and pseudo-metrical formulation and even more general coordinate systems, than cartesian ones, these section could be omitted by the readers who do not familiar with tensor analysis and advanced differential Geometry. One of the basic goals of this work is the general selfcontained and simple mathematical formulation of the most general Physical Laws, that is understandable to the reader, which is familiar with the classical mechanics, some basic elements of non-relativistic quantum mechanics and the basic calculus only.

As was shewed in [3] the change of coordinates from one coordinate system, which is both cartesian and inertial to another such a system is given, up to equivalence, by the classical Galilean transformations. As we called here Cartesian coordinate systems simply as coordinate systems by default, we also call coordinate systems, which are simultaneously cartesian and inertial, simply as inertial coordinate systems by default. In other words, for simplicity we just postulate here the classical space-time where the change of some inertial coordinate system $(*)$ to another inertial coordinate system $(**)$ is given (up to equivalence) by the Galilean transformation:

$$\begin{cases} \mathbf{x}' = \mathbf{x} + \mathbf{w}t, \\ t' = t, \end{cases} \tag{1}$$

and the change of some non-inertial coordinate system $(*)$ to another coordinate system $(**)$ is of the form:

$$\begin{cases} \mathbf{x}' = A(t) \cdot \mathbf{x} + \mathbf{z}(t), \\ t' = t, \end{cases} \tag{2}$$

where $A(t) \in SO(3)$ is a rotation, i.e. $A(t) \in \mathbb{R}^{3\times 3}$, $\det A(t) > 0$ and $A(t) \cdot A^T(t) = I$, where A^T is the transpose of the matrix A and I is the identity matrix. Then we show that the most common and general Physical Laws, both of the Classical and Quantum Physics can be formulated to be invariant under changes in coordinate systems both of the form (1) and of the form (2). Some important examples of such invariant Physical Laws are the Maxwell Equations, the Newtonian gravity as well as several more complicated models of gravity (including the simplified model presented in [3]) and many other Physical Laws including many Laws of Quantum Mechanics, Thermodynamics and Statistical Physics, Continuum Mechanics, Optics et al. Moreover, several basic Laws of Relativistic Physics, both in the classical and quantum levels can be still formulated to be invariant under the non-relativistic consideration of the space-time, like the Classical Relativistic Second Law of Newton and Quantum Dirac and Klein–Gordon equations for relativistic particles, including their interaction with the outer gravitational field! In particular, we introduce the Hamiltonian formulation of the Dirac equation, and moreover, we were able to formulate the Dirac equation for multiple particles, similarly to what was done for the Schrödinger equation of the non-relativistic quantum mechanics.

Be'er Sheva, Israel Arkady Poliakovsky

Contents

1 Introduction and Comprehensive Survey of the Main Results 1

1.1 A New Look to the Newtonian Gravity 1

1.1.1 Variational Principle for the Solution of the Cauchy-Problem for Hamilton–Jacobi Type Equation (1.20) 6

1.2 Non-relativistic Model of Electrodynamics 8

1.3 Local Gravitational Time and Maxwell Equations 13

1.4 Motion of Classical and Quantum Particles in the Gravitational and Electromagnetic Fields 15

1.5 Unified Gravitational-Electromagnetic Field and Conservation Laws 22

1.6 Transformations of General Scalar and Vector Fields Under the Change of Coordinate Systems 26

1.7 Genuine Gravity and Inertia 33

1.8 Lagrangian in the Case of Some Possible Alternative Model of the Gravity 35

1.9 Covariant Formulation of the Physical Laws in the Four-Dimensional Space-Time 40

1.9.1 Four-Vectors, Four-Covectors and Tensors in the Four-Dimensional Non-relativistic Space-Time 41

1.9.2 Pseudo-metric Tensors of the Four-Dimensional Space-Time 48

1.9.3 Maxwell Equations in Covariant Formulation 53

1.9.4 Covariant Formulation of Lagrangian of Motion of a Classical Charged Particle in the External Gravitational and Electromagnetic Fields 56

1.9.5 Kinematic Pseudo-metric Tensors of Inertia 59

1.9.6 Physical Laws in Curvilinear Coordinate Systems in the Non-relativistic Space-Time 64
1.9.7 Some General Covariant Identities in Non-Cartesian or Curvilinear Coordinate Systems 67
1.9.8 Certain Curvilinear Coordinate System in the Case of Stationary Radially Symmetric Gravitational Field and Relation to the Schwarzschild Metric 72
1.9.9 General Lagrangian of the Gravitational-Electromagnetic Field, Compatible with the General Lagrangian of the Motion in (1.280): The Case of the Newtonian-Type Gravity 76
1.9.10 General Lagrangian of the Gravitational-Electromagnetic Field, Compatible with the General Lagrangian of the Motion in (1.280): The Case of Some Possible Alternative Model of the Gravity 79
1.9.11 Covariant Formulation of the Laws of Gravity in Cartesian and Curvilinear Coordinate Systems: The Case of the Newtonian-Type Gravity 84
1.9.12 Covariant Formulation of the Laws of Gravity in Cartesian and Curvilinear Coordinate Systems: The Case of Some Alternative Model of the Gravity 93
1.10 Relativistic-Like Dirac Equation 99
1.11 Spinless Relativistic-Like Particles: The Klein–Gordon Equation 106
1.12 Thermodynamics of a Moving Continuum Medium 109
1.12.1 The Case of an Inviscid Fluid/Gas 112
1.12.2 The Case of the Simplest Viscous Fluid/Gas 113
1.12.3 Lagrangian Coordinates and the Simplest Models of Elastic Bodies 113
1.12.4 The Equations of the Motion of the General Continuum Medium in Lagrangian Coordinates 116
1.12.5 Propagation of Sound Wave in the Moving Inviscid Gas or Fluid 117
1.12.6 Propagation of Waves in the Moving Elastic Body 122

1.13 Macroscopic Electrodynamics in the Presence of Dielectric and/or Magnetic Mediums ... 128
1.13.1 Optical Dispersion in Moving Mediums ... 132
1.14 Some Further Consequences of Maxwell Equations ... 138
1.14.1 The Case of Quasistationary Electromagnetic Fields Inside a Slowly Moving Medium in a Weak Gravitational Field ... 145
1.14.2 Hertz's Notation for Quasistationary Electromagnetic Fields ... 149
1.15 Geometric Optics Inside a Moving Medium and/or in the Presence of Gravitational Field ... 156
1.15.1 Preliminary Calculations ... 156
1.15.2 Derivation of the Eikonal Equation ... 158
1.15.3 The Case of the Monochromatic Wave ... 162
1.15.4 The Laws of Reflection and Refraction ... 166
1.15.5 Sagnac Effect ... 169
1.15.6 Fizeau Experiment ... 170
1.15.7 The Case of the Non-monochromatic Wave or/and Moving Domains of Propagation of Light ... 172
1.15.8 Polarization of the Light Inside a Moving Medium and/or in the Presence of Gravitational Field ... 177
1.15.9 More Delicate Approximation of the Wave Equation ... 179
1.15.10 Wave Optics Inside a Moving Medium and/or in the Presence of Gravitational Field ... 183

2 Notations and Preliminaries ... 187

3 Transformations of Scalars and Vectors Under the Change of Cartesian Coordinate Systems ... 191

4 Non-Relativistic Gravity ... 225
4.1 Newtonian Gravity Revised ... 225
4.2 Variational Principle for Hamilton–Jacobi-Type Equation (4.46) ... 239
4.3 Genuine Gravity and Inertia ... 242

5 Maxwell Equations in Cartesian Coordinate Systems ... 245
5.1 Maxwell Equations Revised ... 245
5.2 Maxwell Equations in Non-inertial Cartesian Coordinate Systems ... 247
5.3 Scalar and Vectorial Electromagnetic Potentials ... 253
5.4 Lagrangian of the Electromagnetic Field ... 256
5.5 Alternative Lagrangian of the Electromagnetic Field ... 258
5.6 Local Gravitational Time and Maxwell Equations in a Non-rotating Coordinate System ... 260

6 Motion of Particles in External Gravitational-Electromagnetic Field ... 265
6.1 Lagrangian of a System of Classical Particles in a Gravitational-Electromagnetic Field ... 265
6.2 Hamiltonian of the Motion of a Finite System of Classical Particles in an Outer Gravitational-Electromagnetic Field ... 268
6.3 Classical Liouville's Equation ... 269
6.3.1 Thermodynamical Equilibrium; Canonical and Micro-canonical Ensembles ... 272
6.4 Shrödinger equation for a Finite System of Quantum Particles ... 279
6.5 Quantum Liouville's Equation for a Finite System of Quantum Particles ... 286
6.5.1 Thermodynamical Equilibrium; Canonical Ensemble ... 293
6.6 Shrödinger–Pauli Equation for a Spin-Half Quantum Particle ... 300
6.7 Shrödinger–Pauli Equation for a System of *n* Spin-Half Micro-particles ... 306
6.8 Quantum Liouville's Equation for a Finite System of Spin-Half Particles ... 312
6.8.1 Thermodynamical Equilibrium in Systems of Spin-Half Particles; Canonical Ensemble ... 319

7 Relation Between the Gravitational and Inertial Masses and Conservation Laws ... 329
7.1 Basic Assumptions and Their Consequences ... 329
7.2 Conservation of the Momentum, Angular Momentum and Energy ... 331

8 Lagrangian of the Unified Gravitational–Electromagnetic Field ... 339
8.1 Lagrangian in the Case of Newtonian Gravity ... 339
8.2 Lagrangian in the Case of Some Possible Alternative Model of the Gravity ... 341

9 Covariant Formulation of the Physical Laws in the Four-Dimensional Space-Time ... 355
9.1 Four Vectors, Four Covectors and Tensors in the Four-Dimensional Space-Time ... 355
9.2 Pseudo-metric Tensors of the Four-Dimensional Space-Time ... 365
9.3 Maxwell Equations in Covariant Formulation ... 373

9.4 Covariant Form of Lagrangian of Motion of a Classical Charged Particle 377
9.5 Kinematic Pseudo-Metric Tensors of Inertia 380
9.6 Physical Laws in Non-Cartesian or Curvilinear Coordinate Systems in the Space-Time 385
9.6.1 Some General Covariant Identities in Non-Cartesian or Curvilinear Coordinate Systems 388
9.7 Certain Coordinate System for Stationary Radial Field and the Schwarzschild Metric 393
9.8 General Lagrangian of the Gravitational Field, Related to the Lagrangian in (9.134) 401
9.8.1 The Case of the Newtonian-Type Gravity 401
9.8.2 The Case of Some Possible Alternative Model of the Gravity 404
9.9 Covariant Formulation of the Laws of Gravity in Curvilinear Coordinate Systems 417
9.9.1 The Case of the Newtonian-Type Gravity 417
9.9.2 The Case of Some Alternative Model of the Gravity 435

10 Relativistic-Like Dirac Equation 441
10.1 Classical Relativistic-Like Lagrangian and Hamiltonian of the Motion of Particles 441
10.2 Dirac Equation 443
10.3 Classical Relativistic-Like Lagrangian and Hamiltonian of a System of n Particles 450
10.4 Liouville's Equation for a System of n Relativistic-Like Classical Particles 451
10.4.1 Thermodynamical Equilibrium in the Relativistic-Like Case 454
10.5 Relativistic-Like Dirac Equation for a System of n Spin-Half Micro-particles 460
10.6 Quantum Liouville's Equation for a Finite System of Spin-Half Relativistic-Like Particles 470
10.6.1 Thermodynamical Equilibrium for Relativistic-Like Spin-Half Particles 477
10.7 Spinless Relativistic-Like Particles: The Klein–Gordon Equation 486

11 Thermodynamics of a Moving Continuum Medium 493
11.1 Some Special Cases of Continuum Mediums 497
11.1.1 The Case of an Inviscid Fluid/Gas 497
11.1.2 The Simplest Viscous Fluid/Gas 498

11.2 Lagrangian Coordinates and the Simplest Models of Elastic Bodies ... 499
11.2.1 Strain Tensor and Hooke's Law ... 502
11.2.2 The Equations of the Motion of the General Continuum Medium in Lagrangian Coordinates ... 503
11.3 Propagation of Sound Wave in the Moving Inviscid Gas or Fluid ... 505
11.4 Propagation of Waves in the Moving Elastic Body ... 514

12 Maxwell Equations in the Presence of Dielectrics and/or Magnetics ... 523
12.1 Polarization and Magnetization of a Totally Neutral System of Point Charges ... 523
12.2 General Setting of Maxwell Equations in the Medium ... 530
12.3 Change of Non-inertial Coordinate System ... 531
12.4 Case of Simplest Dielectrics/Magnetics ... 532
12.5 Case of Anisotropic Dielectrics/Magnetics ... 536
12.6 Ohm's Law in a Conducting Medium ... 536
12.7 Optical Dispersion in Moving Mediums ... 537
12.7.1 Optical Dispersion in Anisotropic Moving Mediums ... 547

13 Some Further Consequences of Maxwell Equations in a Medium or Vacuum ... 549
13.1 General Case ... 549
13.2 The Case of Quasistationary Electromagnetic Fields ... 558
13.2.1 Hertz's Notation for Quasistationary Electromagnetic Fields ... 562
13.3 Geometric Optics Inside a Moving Medium and/or in the Presence of Gravitational Field ... 572
13.3.1 Preliminary Calculations ... 572
13.3.2 Derivation of the Eikonal Equation ... 578
13.3.3 The Case of the Monochromatic Wave ... 582
13.3.4 The Laws of Reflection and Refraction ... 588
13.3.5 Sagnac Effect ... 591
13.3.6 Fizeau Experiment ... 592
13.3.7 Fermat Principle of Geometric Optics in the Case When We Cannot Neglect Effects of Order $O\left(\frac{|\bar{\mathbf{u}}|^2}{c_0^2}\right)$ ... 594
13.3.8 The Case of the Non-monochromatic Wave or/and Moving Domains of Propagation of Light ... 599
13.3.9 Polarization of the Light Inside a Moving Medium and/or in the Presence of Gravitational Field ... 614

13.3.10 More Delicate Approximation of the Wave Equation 619
13.3.11 Wave Optics Inside a Moving Medium and/or in the Presence of Gravitational Field 630

14 Appendix 637

References 641

Chapter 1
Introduction and Comprehensive Survey of the Main Results

1.1 A New Look to the Newtonian Gravity

Consider the classical space-time where the change of some Cartesian inertial coordinate system $(*)$ to another inertial coordinate system $(**)$ is given (up to equivalence) by the Galilean Transformation:

$$\begin{cases} \mathbf{x}' = \mathbf{x} + \mathbf{w}t, \\ t' = t \end{cases} \tag{1.1}$$

and the change of some non-inertial Cartesian coordinate system $(*)$ to another Cartesian coordinate system $(**)$ is of the form:

$$\begin{cases} \mathbf{x}' = A(t) \cdot \mathbf{x} + \mathbf{z}(t), \\ t' = t, \end{cases} \tag{1.2}$$

where $A(t) \in SO(3)$ is a rotation, i.e. $A(t) \in \mathbb{R}^{3\times 3}$, $\det A(t) > 0$ and $A(t) \cdot A^T(t) = I$, where A^T is the transpose of the matrix A and I is the identity matrix.

Similarly to the General Theory of Relativity, we assume that the most general laws of Classical Mechanics should be invariant in every non-inertial Cartesian coordinate system, i.e. they preserve their form under transformations of the form (1.2). Moreover, again as in the General Theory of Relativity, we assume that the fictitious forces in non-inertial coordinate systems and the forces of Newtonian gravitation have the same nature and represented by some field in somewhat similar to the Electromagnetic field.

We begin with some simple observations. Assume that we are away from essential gravitational masses. Then consider two Cartesian coordinate systems $(*)$ and $(**)$, such that the system $(**)$ is inertial and the change of coordinate system $(*)$ to coordinate system $(**)$ is given by (1.2). Then the fictitious-gravitational force in the

A. Poliakovsky, *The Laws of Gravity and Electromagnetism*,
https://doi.org/10.1007/978-3-031-61407-1_1

system (∗∗) is trivial $\mathbf{F}_0' = 0$. On the other hand, by (1.2) the fictitious-gravitational force in the system (∗), acting on the particle with inertial mass m and velocity $\mathbf{u}$, is given by

$$\mathbf{F}_0 = m\left(-2A^T(t)\cdot\frac{dA}{dt}(t)\cdot\mathbf{u} - A^T(t)\cdot\frac{d^2A}{dt^2}(t)\cdot\mathbf{x} - A^T(t)\cdot\frac{d^2\mathbf{z}}{dt^2}(t)\right). \tag{1.3}$$

Thus if we define a vector field $\mathbf{v} := \mathbf{v}(\mathbf{x}, t)$ by

$$\mathbf{v}(\mathbf{x}, t) := -A^T(t)\cdot\frac{dA}{dt}(t)\cdot\mathbf{x} - A^T(t)\cdot\frac{d\mathbf{z}}{dt}(t), \tag{1.4}$$

then, by straightforward calculations we rewrite (1.3) as

$$\mathbf{F}_0 = m\left(\frac{\partial\mathbf{v}}{\partial t} + \frac{1}{2}\nabla_{\mathbf{x}}\left(|\mathbf{v}|^2\right)\right) + m\mathbf{u}\times(-\mathrm{curl}_{\mathbf{x}}\mathbf{v}) \tag{1.5}$$

(see Sect. 4.1 for details).

Similarly, we assume that also in the general case of gravitational masses there exists a vector field $\mathbf{v} := \mathbf{v}(\mathbf{x}, t)$ such that in some inertial or non-inertial Cartesian coordinate system the fictitious-gravitational force is given by (1.5). Then we call the vector field $\mathbf{v}$ the vectorial gravitational potential. We see here the following analogy with Electrodynamics: denoting

$$\tilde{\mathbf{E}} := \partial_t\mathbf{v} + \nabla_{\mathbf{x}}\left(\frac{1}{2}|\mathbf{v}|^2\right) \quad\text{and}\quad \tilde{\mathbf{B}} := -c\,\mathrm{curl}_{\mathbf{x}}\mathbf{v},$$

we rewrite (1.5) as

$$\mathbf{F}_0 = m\left(\tilde{\mathbf{E}} + \frac{1}{c}\mathbf{u}\times\tilde{\mathbf{B}}\right),$$

where

$$\mathrm{curl}_{\mathbf{x}}\tilde{\mathbf{E}} + \frac{1}{c}\frac{\partial}{\partial t}\tilde{\mathbf{B}} = 0 \quad\text{and}\quad \mathrm{div}_{\mathbf{x}}\tilde{\mathbf{B}} = 0.$$

Next using (1.5) we rewrite the Second Law of Newton as

$$m\frac{d^2\mathbf{x}}{dt^2} = m\frac{d\mathbf{u}}{dt} = \mathbf{F}_0 + \mathbf{F} = m\left(\frac{\partial\mathbf{v}}{\partial t}(\mathbf{x}, t) + \frac{1}{2}\nabla_{\mathbf{x}}\left(|\mathbf{v}|^2\right)(\mathbf{x}, t)\right) + m\mathbf{u}\times(-\mathrm{curl}_{\mathbf{x}}\mathbf{v}(\mathbf{x}, t)) + \mathbf{F}, \tag{1.6}$$

where $\mathbf{x} := \mathbf{x}(t)$, $\mathbf{u} := \mathbf{u}(t) = \frac{d\mathbf{x}}{dt}(t)$ and m are the place, the velocity and the inertial mass of some given particle at the moment of time t, $\mathbf{v} := \mathbf{v}(\mathbf{x}, t)$ is the vectorial gravitational potential and $\mathbf{F}$ is the total non-gravitational force, acting on the given particle.

Once we considered the Second Law of Newton in the form (1.6) we show that this law is invariant under the change of inertial or non-inertial Cartesian coordinate system, provided that the law of transformation of the vectorial gravitational potential, under the change of coordinate system given by (1.2), is

$$\mathbf{v}' = A(t) \cdot \mathbf{v} + \frac{dA}{dt}(t) \cdot \mathbf{x} + \frac{d\mathbf{z}}{dt}(t), \tag{1.7}$$

i.e. it is the same as the transformation of a field of velocities. More precisely we have the following theorem (See Sect. 4.1 for the proof):

Theorem 1.1 *Consider that the change of some non-inertial Cartesian coordinate system* $(*)$ *to another Cartesian coordinate system* $(**)$ *is given by* (1.2). *Next, assume that in the coordinate system* $(**)$ *we observe a validity of the Second Law of Newton in the form:*

$$\frac{d\mathbf{u}'}{dt'} = -\mathbf{u}' \times curl_{\mathbf{x}'}\mathbf{v}' + \partial_{t'}\mathbf{v}' + \nabla_{\mathbf{x}'}\left(\frac{1}{2}|\mathbf{v}'|^2\right) + \frac{1}{m'}\mathbf{F}', \tag{1.8}$$

where $\mathbf{x}' := \mathbf{x}'(t')$, $\mathbf{u}' := \mathbf{u}'(t') = \frac{d\mathbf{x}'}{dt'}(t')$ *and* m' *are the place, the velocity and the inertial mass of some given particle at the moment of time* t', $\mathbf{v}' := \mathbf{v}'(\mathbf{x}', t')$ *is the vectorial gravitational potential and* $\mathbf{F}'$ *is a total non-gravitational force, acting on the given particle in the coordinate system* $(**)$. *Then in the coordinate system* $(*)$ *we have validity of the Second Law of Newton in the same as* (1.8) *form:*

$$\frac{d\mathbf{u}}{dt} = -\mathbf{u} \times curl_{\mathbf{x}}\mathbf{v} + \partial_t\mathbf{v} + \nabla_{\mathbf{x}}\left(\frac{1}{2}|\mathbf{v}|^2\right) + \frac{1}{m}\mathbf{F}, \tag{1.9}$$

provided that

$$\mathbf{v}' = A(t) \cdot \mathbf{v} + \frac{dA}{dt}(t) \cdot \mathbf{x} + \frac{d\mathbf{z}}{dt}(t) \tag{1.10}$$

$$\mathbf{F}' = A(t) \cdot \mathbf{F}, \tag{1.11}$$

$$m' = m, \tag{1.12}$$

$$\mathbf{u}' = A(t) \cdot \mathbf{u} + \frac{dA}{dt}(t) \cdot \mathbf{x} + \frac{d\mathbf{z}}{dt}(t). \tag{1.13}$$

We call a vector field that transforms by (1.7) under the change of Cartesian coordinate system, by the name speed-like vector field. Since the vectorial gravitational potential $\mathbf{v}$ is a speed-like vector field, i.e. under the changes of inertial or non-inertial coordinate system it behaves like a field of velocities of some continuum, we could introduce a fictitious continuum medium covering all the space, that we can call Aether, such that $\mathbf{v}(\mathbf{x}, t)$ is a fictitious velocity of this medium in the point $\mathbf{x}$ at the time t. Furthermore, if some particle with the place $\mathbf{r} := \mathbf{r}(t)$, the velocity $\mathbf{u} := \mathbf{u}(t) = \frac{d\mathbf{r}}{dt}(t)$ and the inertial mass m moves in the outer gravitational field with the vectorial gravitational potential $\mathbf{v} := \mathbf{v}(\mathbf{x}, t)$ in the absence of non-gravitational forces, then we can associate a Lagrangian with (1.6). Indeed, for this case we define a Lagrangian:

$$\mathcal{L}_0\left(\frac{d\mathbf{r}}{dt}, \mathbf{r}, t\right) := \frac{m}{2}\left|\frac{d\mathbf{r}}{dt} - \mathbf{v}(\mathbf{r}, t)\right|^2. \tag{1.14}$$

This Lagrangian is invariant under the change of non-inertial Cartesian coordinate systems, given by (1.2). Moreover, we can easily deduce that a trajectory $\mathbf{r}(t) : [0, T] \to \mathbb{R}^3$ is a critical point of the functional

$$I_0 = \int_0^T \mathcal{L}_0\left(\frac{d\mathbf{r}}{dt}(t), \mathbf{r}(t), t\right) dt \tag{1.15}$$

if and only if it satisfies

$$-m\frac{d^2\mathbf{r}}{dt^2} + m\left(\frac{\partial}{\partial t}\mathbf{v}(\mathbf{r}, t) + \nabla_{\mathbf{x}}\left(\frac{1}{2}|\mathbf{v}(\mathbf{r}, t)|^2\right) - \frac{d\mathbf{r}}{dt} \times \operatorname{curl}_{\mathbf{x}}\mathbf{v}(\mathbf{r}, t)\right) = 0, \tag{1.16}$$

consistently with (1.6) for the case $\mathbf{F} = 0$. Moreover, we would like to note that if in some inertial or non-inertial Cartesian coordinate system some material body with the place $\mathbf{r}(t)$ and velocity $\mathbf{u}(t) = \frac{d\mathbf{r}}{dt}(t)$ moves in the gravitational field, and, except of the force of gravity all other forces, acting on the body, are negligible then we can prove that the following equality for some instant of time t_0:

$$\mathbf{u}(t_0) := \frac{d\mathbf{r}}{dt}(t_0) = \mathbf{v}\left(\mathbf{r}(t_0), t_0\right)$$

implies

$$\mathbf{u}(t) := \frac{d\mathbf{r}}{dt}(t) = \mathbf{v}\left(\mathbf{r}(t), t\right),$$

for every instant of time. That is, if the velocity of the particle for some initial instant of time coincides with the local vectorial gravitational potential, then it will coincide with it at any instant of time and the trajectory of the motion will be tangent to the direction of the local vectorial gravitational potential.

Next, in order to fit the Second Law of Newton in the form (1.6) with the classical Second Law of Newton and the Newtonian Law of Gravity we consider that in inertial coordinate system (∗), at least in the first approximation, we should have

$$\begin{cases} \operatorname{curl}_{\mathbf{x}}\mathbf{v} = 0, \\ \frac{\partial \mathbf{v}}{\partial t} + \frac{1}{2}\nabla_{\mathbf{x}}\left(|\mathbf{v}|^2\right) = -\nabla_{\mathbf{x}}\Phi, \end{cases} \tag{1.17}$$

where Φ is a scalar Newtonian gravitational potential which satisfies

$$\Delta_{\mathbf{x}}\Phi = 4\pi G M, \tag{1.18}$$

where M is the gravitational mass density and G is the gravitational constant. Thus, since we require $\text{curl}_{\mathbf{x}}\mathbf{v} = 0$, (1.17) is equivalent to

$$\begin{cases} \text{curl}_{\mathbf{x}}\mathbf{v} = 0, \\ \frac{\partial \mathbf{v}}{\partial t} + d_{\mathbf{x}}\mathbf{v} \cdot \mathbf{v} = -\nabla_{\mathbf{x}}\Phi, \end{cases} \tag{1.19}$$

where $d_{\mathbf{x}}\mathbf{v}$ is the Jacobian matrix of the vector field $\mathbf{v}$. Clearly the law (1.19) is invariant under the change of inertial coordinate system, given by (1.1). Note also that, since in the system (∗) we have $\text{curl}_{\mathbf{x}}\mathbf{v} = 0$, we can write (1.17) as a Hamilton–Jacobi type equation:

$$\begin{cases} \mathbf{v} = \nabla_{\mathbf{x}} Z, \\ \frac{\partial Z}{\partial t} + \frac{1}{2}\left|\nabla_{\mathbf{x}} Z\right|^2 = -\Phi, \end{cases} \tag{1.20}$$

where Z is some scalar field. Next we introduce a law of gravity which is invariant in every non-inertial Cartesian coordinate system and is equivalent to (1.19) in every inertial coordinate system. This law has the form:

$$\begin{cases} \text{curl}_{\mathbf{x}}\left(\text{curl}_{\mathbf{x}}\mathbf{v}\right) = 0, \\ \frac{\partial}{\partial t}\left(\text{div}_{\mathbf{x}}\,\mathbf{v}\right) + \text{div}_{\mathbf{x}}\left\{\left(\text{div}_{\mathbf{x}}\,\mathbf{v}\right)\mathbf{v}\right\} + \frac{1}{4}\left|d_{\mathbf{x}}\mathbf{v} + \left\{d_{\mathbf{x}}\mathbf{v}\right\}^T\right|^2 - \left(\text{div}_{\mathbf{x}}\,\mathbf{v}\right)^2 = -4\pi GM, \end{cases} \tag{1.21}$$

together with some reasonable frame-independent assumptions for the asymptotic behavior of the vectorial gravitational potential $\mathbf{v}$ as $|\mathbf{x}| \to +\infty$ (see Sect. 4.1 for the details, in particular, see Proposition 4.2).

Next one can wonder: what should be the possible values of the vectorial gravitational potential $\mathbf{v}$ in the proximity of the Earth or another massive body? We attempt to answer this question in Remark 4.4. We obtain there that if we consider a non-rotating Cartesian coordinate system which center coincides with the center of the Earth, then in this system we should have either

$$\mathbf{v}(\mathbf{x}) = \frac{\sqrt{-2\Phi_1(|\mathbf{x}|)}}{|\mathbf{x}|}\mathbf{x}, \tag{1.22}$$

or

$$\mathbf{v}(\mathbf{x}) = -\frac{\sqrt{-2\Phi_1(|\mathbf{x}|)}}{|\mathbf{x}|}\mathbf{x}, \tag{1.23}$$

where Φ_1 is the usual Newtonian potential of the Earth, that satisfies $\Phi_1(r) = -\frac{Gm_0}{r}$ outside of the Earth. In particular, on the Earth surface we have

$$|\mathbf{v}| = \sqrt{\frac{2Gm_0}{r_0}}, \tag{1.24}$$

where r_0 is the Earth radius and m_0 is the Earth mass, i.e. the absolute value of the vectorial gravitational potential on the Earth surface approximately equals to the escape velocity and its direction is normal to the Earth, either downward or upward.

1.1.1 Variational Principle for the Solution of the Cauchy-Problem for Hamilton–Jacobi Type Equation (1.20)

Consider the Hamilton–Jacobi type equation (1.20) for the vectorial gravitational potential in the non-rotating coordinate system:

$$\begin{cases} \mathbf{v} = \nabla_{\mathbf{x}} Z, \\ \frac{\partial Z}{\partial t} + \frac{1}{2} \left|\nabla_{\mathbf{x}} Z\right|^2 = -\Phi, \end{cases} \tag{1.25}$$

where $Z := Z(\mathbf{x}, t)$ is some scalar field and $\Phi = \Phi(\mathbf{x}, t)$ is the scalar Newtonian gravitational potential which satisfies (1.18):

$$\Delta_{\mathbf{x}} \Phi = 4\pi G M, \tag{1.26}$$

where M is the gravitational mass density and G is the gravitational constant. Next consider the Cauchy problem for (1.25):

$$\begin{cases} \frac{\partial Z}{\partial t}(\mathbf{x}, t) + \frac{1}{2} \left|\nabla_{\mathbf{x}} Z(\mathbf{x}, t)\right|^2 = -\Phi(\mathbf{x}, t), \\ Z(\mathbf{x}, 0) = \varphi(\mathbf{x}), \end{cases} \tag{1.27}$$

where $\Phi(\mathbf{x}, t)$ and $\varphi(\mathbf{x})$ are prescribed smooth scalar functions. Furthermore, for every instant of time $t \geq 0$ consider the followed variational functional defined on trajectories $\mathbf{r}(s) : [0, t] \to \mathbb{R}^3$:

$$J_{gr}(\mathbf{r}) = \varphi(\mathbf{r}(0)) + \int_0^t \left(\frac{1}{2} \left| \frac{d\mathbf{r}}{ds}(s) \right|^2 - \Phi(\mathbf{r}(s), s) \right) ds. \tag{1.28}$$

Then, inserting the smooth solution Z of (1.27) into (1.28) and using the Chain Rule, in Sect. 4.2 we deduce that we can rewrite the definition of the functional J_{gr} in (1.28) as

$$\begin{aligned} J_{gr}(\mathbf{r}) &= Z(\mathbf{r}(t), t) + \frac{1}{2} \int_0^t \left| \frac{d\mathbf{r}}{ds}(s) - \nabla_{\mathbf{r}} Z(\mathbf{r}(s), s) \right|^2 ds \\ &= Z(\mathbf{r}(t), t) + \frac{1}{2} \int_0^t \left| \frac{d\mathbf{r}}{ds}(s) - \mathbf{v}(\mathbf{r}(s), s) \right|^2 ds, \end{aligned} \tag{1.29}$$

where consistently with (1.25) we consider $\mathbf{v}(\mathbf{x}, t) := \nabla_{\mathbf{x}} Z(\mathbf{x}, t)$. Therefore, if for every point $\mathbf{x} \in \mathbb{R}^3$ and every instant of time $t \geq 0$ we consider the function $g := g(\mathbf{x}, t) : \mathbb{R}^3 : [0, +\infty) \to \mathbb{R}$ defined as a minimum of the following variational problem:

$$g(\mathbf{x}, t) := \min \left\{ J_{gr}(\mathbf{r}) \;\middle|\; \mathbf{r}(s) : [0, t] \to \mathbb{R}^3,\ r(t) = \mathbf{x} \right\} =$$
$$\min \left\{ \varphi(\mathbf{r}(0)) + \int_0^t \left(\frac{1}{2} \left| \frac{d\mathbf{r}}{ds}(s) \right|^2 - \Phi(\mathbf{r}(s), s) \right) ds \;\middle|\; \mathbf{r}(s) : [0, t] \to \mathbb{R}^3,\ \mathbf{r}(t) = \mathbf{x} \right\}, \tag{1.30}$$

then by (1.29) we deduce that

$$g(\mathbf{x}, t) = Z(\mathbf{x}, t), \tag{1.31}$$

and this minimum is achieved on the unique trajectory $\mathbf{r}(s) : [0, t] \to \mathbb{R}^3$ that satisfies the following initial value problem for an ordinary differential equation:

$$\begin{cases} \frac{d\mathbf{r}}{ds}(s) = \nabla_{\mathbf{r}} Z(\mathbf{r}(s), s) = \mathbf{v}(\mathbf{r}(s), s) & \forall s \in [0, t], \\ \mathbf{r}(t) = \mathbf{x}. \end{cases} \tag{1.32}$$

So by inserting (1.31) into (1.30) we obtain the explicit formula for the solution Z of the Cauchy problem (1.27):

$$Z(\mathbf{x}, t) = \min \left\{ \varphi(\mathbf{r}(0)) + \int_0^t \left(\frac{1}{2} \left| \frac{d\mathbf{r}}{ds}(s) \right|^2 - \Phi(\mathbf{r}(s), s) \right) ds \;\middle|\; \mathbf{r}(s) : [0, t] \to \mathbb{R}^3,\ \mathbf{r}(t) = \mathbf{x} \right\}. \tag{1.33}$$

Furthermore, there exists a unique trajectory $\mathbf{r}(s) : [0, t] \to \mathbb{R}^3$ that minimizes the right-hand-side of (1.33) and moreover it satisfies (1.32). In particular for this trajectory we have

$$\mathbf{v}(\mathbf{x}, t) = \frac{d\mathbf{r}}{ds}(t). \tag{1.34}$$

On the other hand the minimizer of the right-hand-side of (1.33) clearly satisfies the following Euler–Lagrange equation:

$$\begin{cases} \frac{d^2\mathbf{r}}{ds^2}(s) = -\nabla_{\mathbf{r}} \Phi(\mathbf{r}(s), s) & \forall s \in [0, t], \\ \frac{d\mathbf{r}}{ds}(0) = \nabla_{\mathbf{r}} \varphi(\mathbf{r}(0)), \\ \mathbf{r}(t) = \mathbf{x}. \end{cases} \tag{1.35}$$

Equation (1.35) sometimes has an advantage with respect to (1.32) since the a priori unknown function Z dose not appear in (1.35).

So, in order to find $\mathbf{v}(\mathbf{x}, t)$ we need first to solve (1.35) and then use (1.34) with the solution of (1.35) that we just found.

1.2 Non-relativistic Model of Electrodynamics

Similarly to the General Theory of Relativity we assume that the electromagnetic field is influenced by the gravitational field. In Sect. 5.1 of this paper we propose the simple and natural quantitative relations of Electrodynamics, substituting (with minor changes) the classical Maxwell equations in the case of an arbitrarily vectorial gravitational potential, and invariant under Galilean Transformations. For this propose we appeal to the Maxwell equations in a medium. It is well known that the classical Maxwell equations in a medium have the following form in the Gaussian unit system:

$$\begin{cases} \text{curl}_{\mathbf{x}}\mathbf{H} = \frac{4\pi}{c}\mathbf{j} + \frac{1}{c}\frac{\partial \mathbf{D}}{\partial t}, \\ \text{div}_{\mathbf{x}}\,\mathbf{D} = 4\pi\rho, \\ \text{curl}_{\mathbf{x}}\mathbf{E} + \frac{1}{c}\frac{\partial \mathbf{B}}{\partial t} = 0, \\ \text{div}_{\mathbf{x}}\,\mathbf{B} = 0. \end{cases} \tag{1.36}$$

Here $\mathbf{x} \in \mathbb{R}^3$ and $t > 0$ are the place and the time, $\mathbf{E}$ is the electric field, $\mathbf{B}$ is the magnetic field, $\mathbf{D}$ is the electric displacement field, $\mathbf{H}$ is the $\mathbf{H}$-magnetic field, ρ is the charge density, $\mathbf{j}$ is the current density and c is the universal constant, called speed of light. It is assumed in the Classical Electrodynamics that for the vacuum we always have $\mathbf{D} = \mathbf{E}$ and $\mathbf{H} = \mathbf{B}$. We assume here that the Maxwell equations in the vacuum have the usual form of (1.36) in every inertial coordinate system, as in any other medium, however, we assume that, given some inertial coordinate system, the relations $\mathbf{D} = \mathbf{E}$ and $\mathbf{H} = \mathbf{B}$ in the vacuum are valid only for the parts of the space, where the vectorial gravitational potential is negligible.

So we assume that, given some inertial coordinate system, if in some point and at some instant the vectorial gravitational potential vanishes, then in this point and at this time we have $\mathbf{D} = \mathbf{E}$ and $\mathbf{H} = \mathbf{B}$. In order to obtain the relations $\mathbf{D} \sim \mathbf{E}$ and $\mathbf{H} \sim \mathbf{B}$ in the general case we assume that Eq. (1.36) and the Lorentz force

$$\mathbf{F} = \sigma\mathbf{E} + \frac{\sigma}{c}\mathbf{u} \times \mathbf{B} \tag{1.37}$$

(where σ is the charge of the test particle and $\mathbf{u}$ is its velocity) are invariant under the Galilean transformations, given by (1.1). Then the analysis of our assumptions, presented in Sect. 5.1, implies that the full system of Electrodynamics in the case of an arbitrarily vectorial gravitational potential $\mathbf{v} := \mathbf{v}(\mathbf{x}, t)$ has the following form:

$$\begin{cases} \operatorname{curl}_{\mathbf{x}}\mathbf{H} = \frac{4\pi}{c}\mathbf{j} + \frac{1}{c}\frac{\partial \mathbf{D}}{\partial t}, \\ \operatorname{div}_{\mathbf{x}}\mathbf{D} = 4\pi\rho, \\ \operatorname{curl}_{\mathbf{x}}\mathbf{E} + \frac{1}{c}\frac{\partial \mathbf{B}}{\partial t} = 0, \\ \operatorname{div}_{\mathbf{x}}\mathbf{B} = 0, \\ \mathbf{E} = \mathbf{D} - \frac{1}{c}\mathbf{v}\times\mathbf{B}, \\ \mathbf{H} = \mathbf{B} + \frac{1}{c}\mathbf{v}\times\mathbf{D}. \end{cases} \tag{1.38}$$

It can be easily checked that system (1.38) and the expression of the Lorentz force in (1.37) are invariant under the Galilean transformations (1.1), provided that

$$\begin{cases} \mathbf{D}' = \mathbf{D}, \\ \mathbf{B}' = \mathbf{B}, \\ \mathbf{E}' = \mathbf{E} - \frac{1}{c}\mathbf{w}\times\mathbf{B}, \\ \mathbf{H}' = \mathbf{H} + \frac{1}{c}\mathbf{w}\times\mathbf{D} \\ \mathbf{v}' = \mathbf{v} + \mathbf{w}. \end{cases} \tag{1.39}$$

In Sect. 5.2 we prove that the laws of Electrodynamics in the form (1.38) and the law of the Lorentz force (1.37), preserve their form also in non-inertial Cartesian coordinate systems. More precisely the following theorem is valid:

Theorem 1.2 *Consider that the change of some non-inertial Cartesian coordinate system* $(*)$ *to another Cartesian coordinate system* $(**)$ *is given by* (1.2)*. Next, assume that in the coordinate system* $(**)$ *we observe a validity of Maxwell Equations for the vacuum in the form:*

$$\begin{cases} \mathit{curl}_{\mathbf{x}'}\mathbf{H}' = \frac{4\pi}{c}\mathbf{j}' + \frac{1}{c}\frac{\partial \mathbf{D}'}{\partial t'}, \\ \operatorname{div}_{\mathbf{x}'}\mathbf{D}' = 4\pi\rho', \\ \mathit{curl}_{\mathbf{x}'}\mathbf{E}' + \frac{1}{c}\frac{\partial \mathbf{B}'}{\partial t'} = 0, \\ \operatorname{div}_{\mathbf{x}'}\mathbf{B}' = 0, \\ \mathbf{E}' = \mathbf{D}' - \frac{1}{c}\mathbf{v}'\times\mathbf{B}', \\ \mathbf{H}' = \mathbf{B}' + \frac{1}{c}\mathbf{v}'\times\mathbf{D}'. \end{cases} \tag{1.40}$$

Moreover, we assume that in coordinate system $(**)$ *we observe a validity of the expression for the Lorentz force in the form:*

$$\mathbf{F}' = \sigma'\mathbf{E}' + \frac{\sigma'}{c}\mathbf{u}'\times\mathbf{B}'. \tag{1.41}$$

Then in the coordinate system $(*)$ *we have the validity of Maxwell Equations for the vacuum in the same as* (1.40) *form:*

$$\begin{cases} curl_{\mathbf{x}}\mathbf{H} = \frac{4\pi}{c}\mathbf{j} + \frac{1}{c}\frac{\partial \mathbf{D}}{\partial t}, \\ \mathrm{div}_{\mathbf{x}}\,\mathbf{D} = 4\pi\rho, \\ curl_{\mathbf{x}}\mathbf{E} + \frac{1}{c}\frac{\partial \mathbf{B}}{\partial t} = 0, \\ \mathrm{div}_{\mathbf{x}}\,\mathbf{B} = 0, \\ \mathbf{E} = \mathbf{D} - \frac{1}{c}\mathbf{v} \times \mathbf{B}, \\ \mathbf{H} = \mathbf{B} + \frac{1}{c}\mathbf{v} \times \mathbf{D}, \end{cases} \tag{1.42}$$

and we have the validity of the expression for the Lorentz force in the same as (1.41) *form:*

$$\mathbf{F} = \sigma\mathbf{E} + \frac{\sigma}{c}\mathbf{u} \times \mathbf{B}, \tag{1.43}$$

provided that

$$\begin{cases} \mathbf{F}' = A(t)\cdot\mathbf{F}, \\ \sigma' = \sigma, \\ \mathbf{u}' = A(t)\cdot\mathbf{u} + \frac{dA}{dt}(t)\cdot\mathbf{x} + \frac{d\mathbf{z}}{dt}(t), \\ \rho' = \rho, \\ \mathbf{v}' = A(t)\cdot\mathbf{v} + \frac{dA}{dt}(t)\cdot\mathbf{x} + \frac{d\mathbf{z}}{dt}(t), \\ \mathbf{j}' = A(t)\cdot\mathbf{j} + \rho\,\frac{dA}{dt}(t)\cdot\mathbf{x} + \rho\,\frac{d\mathbf{z}}{dt}(t) \end{cases} \tag{1.44}$$

and

$$\begin{cases} \mathbf{D}' = A(t)\cdot\mathbf{D}, \\ \mathbf{B}' = A(t)\cdot\mathbf{B}, \\ \mathbf{E}' = A(t)\cdot\mathbf{E} - \frac{1}{c}\left(\frac{dA}{dt}(t)\cdot\mathbf{x} + \frac{d\mathbf{z}}{dt}(t)\right) \times (A(t)\cdot\mathbf{B}), \\ \mathbf{H}' = A(t)\cdot\mathbf{H} + \frac{1}{c}\left(\frac{dA}{dt}(t)\cdot\mathbf{x} + \frac{d\mathbf{z}}{dt}(t)\right) \times (A(t)\cdot\mathbf{D}). \end{cases} \tag{1.45}$$

Next we would like to note that, since as already mentioned before, the direction of the local vectorial gravitational potential is normal to the Earth surface, in the frames of our model, we provide a non-relativistic explanation of the classical Michelson–Morley experiment. Indeed in this experiment the axes of the apparatus are tangent to the Earth surface and thus the null result cannot be affected by the vectorial gravitational potential. Since, the value of the local vectorial gravitational potential equals to the escape velocity, if we consider the vertical Michelson–Morley experiment, where one of the axes of the apparatus is normal to the Earth surface, then in the frames of our model the expected result should be analogous to the positive result of Aether drift with the speed equal to the escape velocity. However, regarding the vertical Michelson–Morley experiment, i.e. the modification of Michelson–Morley experiment, where at least one of the axes of the apparatus is not tangent to the Earth surface, we found only very scarce and contradictory information.

Next, as in the classical electrodynamics, by the third and the fourth equations in (1.38) we can find a scalar field $\Psi := \Psi(\mathbf{x}, t)$ and a vector field $\mathbf{A} := \mathbf{A}(\mathbf{x}, t)$ such that

$$\begin{cases} \mathbf{B} = \operatorname{curl}_{\mathbf{x}} \mathbf{A}, \\ \mathbf{E} = -\nabla_{\mathbf{x}} \Psi - \frac{1}{c} \frac{\partial \mathbf{A}}{\partial t}. \end{cases} \tag{1.46}$$

We call Ψ and $\mathbf{A}$ the scalar and the vectorial electromagnetic potentials. Then by (1.46) and (1.38) we also have

$$\begin{cases} \mathbf{D} = -\nabla_{\mathbf{x}} \Psi - \frac{1}{c} \frac{\partial \mathbf{A}}{\partial t} + \frac{1}{c} \mathbf{v} \times \operatorname{curl}_{\mathbf{x}} \mathbf{A} \\ \mathbf{H} = \operatorname{curl}_{\mathbf{x}} \mathbf{A} + \frac{1}{c} \mathbf{v} \times \left(-\nabla_{\mathbf{x}} \Psi - \frac{1}{c} \frac{\partial \mathbf{A}}{\partial t} + \frac{1}{c} \mathbf{v} \times \operatorname{curl}_{\mathbf{x}} \mathbf{A}\right). \end{cases} \tag{1.47}$$

We also define the proper scalar electromagnetic potential $\Psi_0 := \Psi_0(\mathbf{x}, t)$ by

$$\Psi_0 := \Psi - \frac{1}{c} \mathbf{A} \cdot \mathbf{v}. \tag{1.48}$$

The name "proper scalar potential" is clarified below. The electromagnetic potentials are not uniquely defined and thus we need to choose a calibration. It is clear that if $(\tilde{\Psi}, \tilde{\Psi}_0, \tilde{\mathbf{A}})$ is another choice of electromagnetic potentials with a different calibration then there exists a scalar field $w := w(\mathbf{x}, t)$ such that we have

$$\begin{cases} \tilde{\Psi} = \Psi + \frac{1}{c} \frac{\partial w}{\partial t} \\ \tilde{\mathbf{A}} = \mathbf{A} - \nabla_{\mathbf{x}} w \\ \tilde{\Psi}_0 = \Psi_0 + \frac{1}{c} \left(\frac{\partial w}{\partial t} + \mathbf{v} \cdot \nabla_{\mathbf{x}} w\right). \end{cases} \tag{1.49}$$

For definiteness we can take $\mathbf{A}$ to satisfy

$$\operatorname{div}_{\mathbf{x}} \mathbf{A} = 0. \tag{1.50}$$

In Sect. 5.3 we show that, consistently with (1.45), under the change of non-inertial Cartesian coordinate system, given by (1.2), the electromagnetic potentials transform as

$$\begin{cases} \Psi' = \Psi + \frac{1}{c} \left(\frac{dA}{dt}(t) \cdot \mathbf{x} + \frac{dz}{dt}(t)\right) \cdot (A(t) \cdot \mathbf{A}) \\ \mathbf{A}' = A(t) \cdot \mathbf{A} \\ \Psi'_0 = \Psi_0. \end{cases} \tag{1.51}$$

The last equation in (1.51) clarifies the name "proper scalar potential". The equalities (1.51) are derived primarily under the choice of the calibration given by (1.50). However, as can be easily seen by (1.49), all the equalities in (1.51) still remain to hold, under any other choice of calibration scalar function w, provided that we have $w' = w$ under the transformation (1.2). In particular, under the Galilean transformations (1.1) the electromagnetic potentials transform as

$$\begin{cases} \Psi' = \Psi + \frac{1}{c}\mathbf{w} \cdot \mathbf{A} \\ \mathbf{A}' = \mathbf{A} \\ \Psi'_0 = \Psi_0. \end{cases} \tag{1.52}$$

Next we can associate a Lagrangian density related to electromagnetic field. Given known the charge distribution $\rho := \rho(\mathbf{x}, t)$, the current distribution $\mathbf{j} := \mathbf{j}(\mathbf{x}, t)$ and the vectorial gravitational potential $\mathbf{v} := \mathbf{v}(\mathbf{x}, t)$, consider a Lagrangian density L_1 defined by

$$L_1(\mathbf{A}, \Psi, \mathbf{x}, t) := \\ \frac{1}{8\pi}\left|-\nabla_{\mathbf{x}}\Psi - \frac{1}{c}\frac{\partial \mathbf{A}}{\partial t} + \frac{1}{c}\mathbf{v} \times \mathrm{curl}_{\mathbf{x}}\mathbf{A}\right|^2 - \frac{1}{8\pi}|\mathrm{curl}_{\mathbf{x}}\mathbf{A}|^2 - \left(\rho\Psi - \frac{1}{c}\mathbf{A}\cdot\mathbf{j}\right). \tag{1.53}$$

Using (1.51) we can deduce that Lagrangian L_1 is invariant, under the change of inertial or non-inertial Cartesian coordinate system, given by (1.2). Moreover, if, consistently with (1.46)–(1.48), we denote

$$\begin{cases} \mathbf{D} = -\nabla_{\mathbf{x}}\Psi - \frac{1}{c}\frac{\partial \mathbf{A}}{\partial t} + \frac{1}{c}\mathbf{v} \times \mathrm{curl}_{\mathbf{x}}\mathbf{A} \\ \mathbf{B} = \mathrm{curl}_{\mathbf{x}}\mathbf{A} \\ \mathbf{E} = -\nabla_{\mathbf{x}}\Psi - \frac{1}{c}\frac{\partial \mathbf{A}}{\partial t} \\ \mathbf{H} = \mathrm{curl}_{\mathbf{x}}\mathbf{A} + \frac{1}{c}\mathbf{v} \times \left(\nabla_{\mathbf{x}}\Psi - \frac{1}{c}\frac{\partial \mathbf{A}}{\partial t} + \frac{1}{c}\mathbf{v} \times \mathrm{curl}_{\mathbf{x}}\mathbf{A}\right) \\ \Psi_0 := \Psi - \frac{1}{c}\mathbf{A} \cdot \mathbf{v}, \end{cases} \tag{1.54}$$

then:

$$L_1(\mathbf{A}, \Psi, \mathbf{x}, t) = \\ \frac{1}{8\pi}|\mathbf{D}|^2 - \frac{1}{8\pi}|\mathbf{B}|^2 - \left(\rho\Psi - \frac{1}{c}\mathbf{A}\cdot\mathbf{j}\right) = \frac{1}{8\pi}|\mathbf{D}|^2 - \frac{1}{8\pi}|\mathbf{B}|^2 - \rho\Psi_0 + \frac{1}{c}\mathbf{A}\cdot(\mathbf{j} - \rho\mathbf{v}). \tag{1.55}$$

Then in Sect. 5.4 we obtain that a configuration $(\Psi, \mathbf{A})$ is a critical point of the functional

$$J_0 = \int_0^T \int_{\mathbb{R}^3} L_1(\mathbf{A}(\mathbf{x}, t), \Psi(\mathbf{x}, t), \mathbf{x}, t)\, d\mathbf{x}dt, \tag{1.56}$$

if and only if we have

$$\begin{cases} \operatorname{curl}_{\mathbf{x}}\mathbf{H} = \frac{4\pi}{c}\mathbf{j} + \frac{\partial \mathbf{D}}{\partial t} \\ \operatorname{div}_{\mathbf{x}}\mathbf{D} = 4\pi\rho \\ \operatorname{curl}_{\mathbf{x}}\mathbf{E} + \frac{1}{c}\frac{\partial \mathbf{B}}{\partial t} = 0 \\ \operatorname{div}_{\mathbf{x}}\mathbf{B} = 0 \\ \mathbf{E} = \mathbf{D} - \frac{1}{c}\mathbf{v}\times\mathbf{B} \\ \mathbf{H} = \mathbf{B} + \frac{1}{c}\mathbf{v}\times\mathbf{D}, \end{cases} \tag{1.57}$$

where $(\mathbf{D}, \mathbf{B}, \mathbf{E}, \mathbf{H})$ is given by (1.54). So we get a variational principle related to Maxwell equations in the form (1.38). See also Sect. 5.5 for the alternative Lagrangian of the electromagnetic field.

1.3 Local Gravitational Time and Maxwell Equations

Consider an inertial or more generally a non-rotating Cartesian coordinate system $(*)$. Then, as before, in this system we have

$$\mathbf{v}(\mathbf{x}, t) = \nabla_{\mathbf{x}} Z(\mathbf{x}, t), \tag{1.58}$$

where $\mathbf{v}$ is the vectorial gravitational potential and Z is a scalar field. Then define a scalar field $\tau := \tau(\mathbf{x}, t)$ by the following:

$$\tau(\mathbf{x}, t) = t + \frac{1}{c^2} Z(\mathbf{x}, t). \tag{1.59}$$

We call the quantity $\tau(\mathbf{x}, t)$ by the name local gravitational time. The name "local" and "gravitational" is quite clear, since τ depend on the space and time variables and derived by characteristic function Z of the gravitational field. The name "time" will be clarified below. Note also that, using (4.69) in Remark 4.3, one can easily deduce that under the change of inertial coordinate system $(*)$ to $(**)$ given by the Galilean Transformation (1.1) the local gravitational time τ transforms as

$$\tau' = \tau + \frac{1}{c^2}\mathbf{w}\cdot\mathbf{x} + \frac{|\mathbf{w}|^2}{2c^2}t \approx \tau + \frac{1}{c^2}\mathbf{w}\cdot\mathbf{x}, \tag{1.60}$$

where the last equality in (1.60) is valid if $\frac{|\mathbf{w}|^2}{c^2} \ll 1$.

Next consider the Maxwell equations in the vacuum of the form (1.38) and consider a curvilinear change of variables given by

$$\begin{cases} t' = \tau(\mathbf{x}, t) := t + \frac{Z(\mathbf{x},t)}{c^2} \\ \mathbf{x}' = \mathbf{x}. \end{cases} \tag{1.61}$$

Then, denoting

$$\begin{cases} \mathbf{E}^* := \mathbf{D} - \frac{1}{c}\,\mathbf{v} \times \mathbf{H} = \mathbf{E} - \frac{1}{c^2}\mathbf{v} \times (\mathbf{v} \times \mathbf{D}) \\ \mathbf{H}^* := \mathbf{B} + \frac{1}{c}\,\mathbf{v} \times \mathbf{E} = \mathbf{H} - \frac{1}{c^2}\mathbf{v} \times (\mathbf{v} \times \mathbf{B}) \end{cases} \tag{1.62}$$

by (1.38) we rewrite the Maxwell equations in the new curvilinear coordinates in the case of time independent $\mathbf{v}$ as

$$\begin{cases} \operatorname{curl}_{\mathbf{x}'}\mathbf{H} = \frac{4\pi}{c}\mathbf{j} + \frac{1}{c}\frac{\partial \mathbf{E}^*}{\partial t'}, \\ \operatorname{div}_{\mathbf{x}'}\mathbf{E}^* = 4\pi\left(\rho + \frac{1}{c^2}\,\mathbf{v}\cdot\mathbf{j}\right), \\ \operatorname{curl}_{\mathbf{x}'}\mathbf{E} + \frac{1}{c}\frac{\partial \mathbf{H}^*}{\partial t'} = 0, \\ \operatorname{div}_{\mathbf{x}'}\mathbf{H}^* = 0, \\ \mathbf{E}^* = \mathbf{E} - \frac{1}{c^2}\mathbf{v} \times (\mathbf{v} \times \mathbf{D}) \\ \mathbf{H}^* = \mathbf{H} - \frac{1}{c^2}\mathbf{v} \times (\mathbf{v} \times \mathbf{B}) \\ \mathbf{E} = \mathbf{D} - \frac{1}{c}\,\mathbf{v} \times \mathbf{B}, \\ \mathbf{H} = \mathbf{B} + \frac{1}{c}\,\mathbf{v} \times \mathbf{D}, \end{cases} \tag{1.63}$$

(See Sect. 5.6 for details). In particular, in the approximation, up to the order $\left(\frac{|\mathbf{v}|}{c}\right)^2 \ll 1$ we have $\mathbf{E}^* \approx \mathbf{E}$ and $\mathbf{H}^* \approx \mathbf{H}$ and then the approximate Maxwell equations have the form:

$$\begin{cases} \operatorname{curl}_{\mathbf{x}'}\mathbf{H} = \frac{4\pi}{c}\mathbf{j} + \frac{1}{c}\frac{\partial \mathbf{E}}{\partial t'}, \\ \operatorname{div}_{\mathbf{x}'}\mathbf{E} = 4\pi\left(\rho + \frac{1}{c^2}\,\mathbf{v}\cdot\mathbf{j}\right), \\ \operatorname{curl}_{\mathbf{x}'}\mathbf{E} + \frac{1}{c}\frac{\partial \mathbf{H}}{\partial t'} = 0, \\ \operatorname{div}_{\mathbf{x}'}\mathbf{H} = 0, \\ \mathbf{E} = \mathbf{D} - \frac{1}{c}\,\mathbf{v} \times \mathbf{B}, \\ \mathbf{H} = \mathbf{B} + \frac{1}{c}\,\mathbf{v} \times \mathbf{D}. \end{cases} \tag{1.64}$$

The first four equations in (1.64) form a following system of equation:

$$\begin{cases} \operatorname{curl}_{\mathbf{x}'}\mathbf{H} = \frac{4\pi}{c}\mathbf{j}^* + \frac{1}{c}\frac{\partial \mathbf{E}}{\partial t'}, \\ \operatorname{div}_{\mathbf{x}'}\mathbf{E} = 4\pi\rho^*, \\ \operatorname{curl}_{\mathbf{x}'}\mathbf{E} + \frac{1}{c}\frac{\partial \mathbf{H}}{\partial t'} = 0, \\ \operatorname{div}_{\mathbf{x}'}\mathbf{H} = 0, \end{cases} \tag{1.65}$$

where

$$\mathbf{j}^* := \mathbf{j} \quad \text{and} \quad \rho^* := \left(\rho + \frac{1}{c^2}\,\mathbf{v}\cdot\mathbf{j}\right) \tag{1.66}$$

The system (1.65) coincides with the classical Maxwell equations of the usual Electrodynamics. Therefore, given known $\mathbf{v}$, ρ and $\mathbf{j}$, (1.65) could be solved as easy as the usual wave equation, for example, by the method of retarded potentials. Then backward to (1.61) change of variables could be made in order to deduce the electromagnetic fields in coordinates $(\mathbf{x}, t)$. Next note that, since we defined $t' = \tau$, all the above clarifies the name "time" of the quantity τ. Finally we would like to note that if we have a motion of some material body with the place $\mathbf{r}(t)$ and the velocity $\mathbf{u}(t) := \frac{d\mathbf{r}}{dt}(t)$ and we associate the local gravitational time τ with this body then clearly

$$d\tau = \left(1 + \frac{1}{c^2}\mathbf{u}(t) \cdot \mathbf{v}\left(\mathbf{r}(t), t\right)\right) dt \approx dt, \tag{1.67}$$

where the last equality in (1.67) is valid if we have

$$\left(\frac{|\mathbf{v}|}{c}\right)^2 \ll 1 \quad \text{and} \quad \left(\frac{|\mathbf{u}(t)|}{c}\right)^2 \ll 1. \tag{1.68}$$

So we can use the local gravitational time τ in the approximate calculations instead of the true time t.

1.4 Motion of Classical and Quantum Particles in the Gravitational and Electromagnetic Fields

Given a classical particle with inertial mass m, charge σ, place $\mathbf{r}(t)$ and velocity $\mathbf{u}(t) = d\mathbf{r}/dt(t)$ in the outer gravitational field with the vectorial gravitational potential $\mathbf{v}(\mathbf{x}, t)$, the outer electromagnetic field with vectorial and scalar potentials $\mathbf{A}(\mathbf{x}, t)$ and $\Psi(\mathbf{x}, t)$, and additional conservative field with scalar potential $V(\mathbf{x}, t)$ we consider a Lagrangian:

$$L_0\left(\frac{d\mathbf{r}}{dt}, \mathbf{r}, t\right) := \frac{m}{2}\left|\frac{d\mathbf{r}}{dt} - \mathbf{v}(\mathbf{r}, t)\right|^2 - \sigma\left(\Psi(\mathbf{r}, t) - \frac{1}{c}\mathbf{A}(\mathbf{r}, t) \cdot \frac{d\mathbf{r}}{dt}\right) + V(\mathbf{r}, t). \tag{1.69}$$

Then this Lagrangian is invariant under the change of non-inertial coordinate system, given by (1.2). Moreover, we can show that a trajectory $\mathbf{r}(t) : [0, T] \to \mathbb{R}^3$ is a critical point of the functional

$$J_0 = \int_0^T L_0\left(\frac{d\mathbf{r}}{dt}(t), \mathbf{r}(t), t\right) dt \tag{1.70}$$

if and only if, consistently with (1.6) and (1.37), we have

$$m\frac{d^2\mathbf{r}}{dt^2} = m\left(\frac{\partial}{\partial t}\mathbf{v}(\mathbf{r},t) + \nabla_{\mathbf{x}}\left(\frac{1}{2}|\mathbf{v}(\mathbf{r},t)|^2\right) - \frac{d\mathbf{r}}{dt}\times \operatorname{curl}_{\mathbf{x}}\mathbf{v}(\mathbf{r},t)\right) + \nabla_{\mathbf{x}}V(\mathbf{r},t)$$
$$+\sigma\mathbf{E}(\mathbf{r},t) + \frac{\sigma}{c}\frac{d\mathbf{r}}{dt}\times\mathbf{B}(\mathbf{r},t), \quad (1.71)$$

where $\mathbf{E}$ and $\mathbf{B}$ are given by (1.46).

Next if we define the generalized momentum of the particle m by

$$\mathbf{P} := \nabla_{\mathbf{r}'}L_0\left(\mathbf{r}',\mathbf{r},t\right) = m\frac{d\mathbf{r}}{dt} - m\mathbf{v}(\mathbf{r},t) + \frac{\sigma}{c}\mathbf{A}(\mathbf{r},t), \quad (1.72)$$

and consider a Hamiltonian

$$H_0(\mathbf{P},\mathbf{r},t) := \mathbf{P}\cdot\frac{d\mathbf{r}}{dt} - L_0\left(\frac{d\mathbf{r}}{dt},\mathbf{r},t\right), \quad (1.73)$$

then we obtain

$$H_0(\mathbf{P},\mathbf{r},t) = \mathbf{P}\cdot\mathbf{v}(\mathbf{r},t) + \frac{1}{2m}\left|\mathbf{P} - \frac{\sigma}{c}\mathbf{A}(\mathbf{r},t)\right|^2 + \sigma\left(\Psi(\mathbf{r},t) - \frac{1}{c}\mathbf{A}(\mathbf{r},t)\cdot\mathbf{v}(\mathbf{r},t)\right) - V(\mathbf{r},t). \quad (1.74)$$

See Sects. 6.1 and 6.2 for the generalizations of the Lagrangian and Hamiltonian in the case of system of n classical particles. See also Sect. 6.3 for the invariance of the classical statistical Liouville's equation, arisen from this Hamiltonian, under the change of non-inertial coordinate system. Moreover, see Sect. 6.3.1 for the frame-independent formulations of the conditions of thermodynamical equilibrium and canonical and micro-canonical statistical ensembles.

Next, assume that our coordinate system is inertial. Then since by (1.20) we have the following Hamilton–Jacobi type equation

$$\begin{cases} \mathbf{v} = \nabla_{\mathbf{x}}Z, \\ \frac{\partial Z}{\partial t} + \frac{1}{2}|\nabla_{\mathbf{x}}Z|^2 = -\Phi, \end{cases} \quad (1.75)$$

where $Z := Z(\mathbf{x},t)$ is some scalar field and Φ is the Newtonian gravitational potential, we rewrite (1.69) as

$$L_0\left(\frac{d\mathbf{r}}{dt},\mathbf{r},t\right) = L_0'\left(\frac{d\mathbf{r}}{dt},\mathbf{r},t\right) - m\frac{d}{dt}\left\{Z\left(\mathbf{r}(t),t\right)\right\}. \quad (1.76)$$

where

$$L_0'\left(\frac{d\mathbf{r}}{dt},\mathbf{r},t\right) := \frac{m}{2}\left|\frac{d\mathbf{r}}{dt}\right|^2 - m\Phi(\mathbf{r},t) - \sigma\left(\Psi(\mathbf{r},t) - \frac{1}{c}\mathbf{A}(\mathbf{r},t)\cdot\frac{d\mathbf{r}}{dt}\right) + V(\mathbf{r},t)$$
$$(1.77)$$

(see Sect. 6.1 for details). Note that in the given inertial coordinate system L_0' coincides with the classical Lagrangian of motion in the gravitational and electromagnetic fields. Moreover, we rewrite (1.70) as

$$J_0 = \int_0^T L_0'\left(\frac{d\mathbf{r}}{dt}, \mathbf{r}, t\right) dt + m\left(Z\left(\mathbf{r}(0), 0\right) - Z\left(\mathbf{r}(T), T\right)\right). \tag{1.78}$$

Thus the stationary points of the functional J_0 will satisfy the same Euler–Lagrange equations as the stationary points of the functional

$$J_0' = \int_0^T L_0'\left(\frac{d\mathbf{r}}{dt}, \mathbf{r}, t\right) dt, \tag{1.79}$$

provided that the beginning and the ending points of the trajectory $\mathbf{r}(t)$ are fixed.

Next if we consider the motion of a quantum micro-particle with inertial mass m and charge σ in the outer gravitational field with the vectorial gravitational potential $\mathbf{v}(\mathbf{x}, t)$, the outer electromagnetic field with vectorial and scalar potentials $\mathbf{A}(\mathbf{x}, t)$ and $\Psi(\mathbf{x}, t)$, and additional conservative field with potential $V(\mathbf{x}, t)$, not taking into account the spin interaction, then the Shrödinger equation for this particle is

$$i\hbar\frac{\partial\psi}{\partial t} = \hat{H}_0 \cdot \psi, \tag{1.80}$$

where $\psi := \psi(\mathbf{x}, t) \in \mathbb{C}$ is a wave function and $\hat{H}_0$ is the Hamiltonian operator. Thus, since by (1.74) the Hermitian Hamiltonian operator has the form of

$$\begin{aligned}\hat{H}_0 \cdot \psi = &-\frac{i\hbar}{2}\operatorname{div}_{\mathbf{x}}\left\{\psi\mathbf{v}(\mathbf{x}, t)\right\} - \frac{i\hbar}{2}\mathbf{v}(\mathbf{x}, t)\cdot\nabla_{\mathbf{x}}\psi\\ &+\left\{\frac{1}{2m}\left(-i\hbar\nabla_{\mathbf{x}} - \frac{\sigma}{c}\mathbf{A}(\mathbf{x}, t)\right)\circ\left(-i\hbar\nabla_{\mathbf{x}} - \frac{\sigma}{c}\mathbf{A}(\mathbf{x}, t)\right)\right\}\cdot\psi\\ &+\sigma\left(\Psi(\mathbf{x}, t) - \frac{1}{c}\mathbf{A}(\mathbf{x}, t)\cdot\mathbf{v}(\mathbf{x}, t)\right)\cdot\psi - V(\mathbf{x}, t)\cdot\psi,\end{aligned} \tag{1.81}$$

we rewrite the corresponding Shrödinger equation as

$$\begin{aligned}&i\hbar\left(\frac{\partial\psi}{\partial t} + \mathbf{v}\cdot\nabla_{\mathbf{x}}\psi\right) + \frac{i\hbar}{2}\left(\operatorname{div}_{\mathbf{x}}\mathbf{v}\right)\psi =\\ &-\frac{\hbar^2}{2m}\Delta_{\mathbf{x}}\psi + \frac{i\hbar\sigma}{2mc}\operatorname{div}_{\mathbf{x}}\left\{\psi\mathbf{A}\right\} + \frac{i\hbar\sigma}{2mc}\mathbf{A}\cdot\nabla_{\mathbf{x}}\psi + \left(\sigma\Psi - \frac{\sigma}{c}\mathbf{A}\cdot\mathbf{v} + \frac{\sigma^2}{2mc^2}|\mathbf{A}|^2 - V\right)\psi.\end{aligned} \tag{1.82}$$

Then we can deduce that, under the change of non-inertial Cartesian coordinate system , given by (1.2), the Shrödinger equation of the form (1.82) stays invariant, provided that, under (1.2) we have

$$\begin{cases} \psi' = \psi \\ V' = V \\ \mathbf{v}' = A(t) \cdot \mathbf{v} + A'(t) \cdot \mathbf{x} + \frac{dz}{dt}(t) \\ \mathbf{A}' = A(t) \cdot \mathbf{A} \\ \Psi' - \frac{1}{c}\mathbf{A}' \cdot \mathbf{v}' = \Psi - \frac{1}{c}\mathbf{A} \cdot \mathbf{v}. \end{cases} \tag{1.83}$$

So the laws of Quantum Mechanics are also invariant in every non-inertial Cartesian coordinate system. Next, assume that we are in some inertial coordinate system and observe the Newtonian Law of Gravitation in the form of (1.19). Then, as a consequence, we have (1.20) for some scalar field Z and the scalar Newtonian gravitational potential Φ. Thus denoting

$$\psi_1 := e^{\frac{im}{\hbar}Z}\psi, \tag{1.84}$$

we rewrite (1.82) in the given inertial coordinate system as

$$i\hbar\frac{\partial \psi_1}{\partial t} = \\ -\frac{\hbar^2}{2m}\Delta_{\mathbf{x}}\psi_1 + \frac{i\hbar\sigma}{2mc}\operatorname{div}_{\mathbf{x}}\{\psi_1\mathbf{A}\} + \frac{i\hbar\sigma}{2mc}\mathbf{A}\cdot\nabla_{\mathbf{x}}\psi_1 + \left(\sigma\Psi + \frac{\sigma^2}{2mc^2}|\mathbf{A}|^2 - V + m\Phi\right)\psi_1,$$

which coincides with the classical Shrödinger equation for this case. Note also that by Remark 4.3, equality (1.84) implies that under the change of coordinate system given by the Galilean Transformation (1.1) the quantity ψ_1 transforms as

$$\psi_1' := e^{\frac{im}{\hbar}(\mathbf{w}\cdot\mathbf{x}+\frac{1}{2}|\mathbf{w}|^2 t)}\psi_1, \tag{1.85}$$

provided that $\psi' = \psi$. Moreover, (1.85) coincides with the classical law of transformation of the wave function, under the Galilean Transformation (see Sect. 17 in [2]).

Next, again consider the motion of a quantum micro-particle having the inertial mass m and the charges σ with the given gravitational and electromagnetical fields with potentials $\mathbf{v}(\mathbf{x}, t)$, $\mathbf{A}(\mathbf{x}, t)$ and $\Psi(\mathbf{x}, t)$ and additional conservative field with potential $V(\mathbf{x}, t)$, not taking into the account the spin interaction. Then consider a Lagrangian density L_0 defined by

$$L_0(\psi, \mathbf{x}, t) := \frac{i\hbar}{2}\left(\left(\frac{\partial\psi}{\partial t} + \mathbf{v}\cdot\nabla_{\mathbf{x}}\psi\right)\cdot\bar{\psi} - \psi\cdot\left(\frac{\partial\bar{\psi}}{\partial t} + \mathbf{v}\cdot\nabla_{\mathbf{x}}\bar{\psi}\right)\right) \\ -\frac{\hbar^2}{2m}\nabla_{\mathbf{x}}\psi\cdot\nabla_{\mathbf{x}}\bar{\psi} - \frac{\hbar\sigma i}{2mc}\left(\nabla_{\mathbf{x}}\psi\cdot\bar{\psi} - \psi\cdot\nabla_{\mathbf{x}}\bar{\psi}\right)\cdot\mathbf{A} - \frac{\sigma^2}{2mc^2}|\mathbf{A}|^2\psi\cdot\bar{\psi} \\ -\sigma\left(\Psi - \frac{1}{c}\mathbf{v}\cdot\mathbf{A}\right)\psi\cdot\bar{\psi} + V(\mathbf{x}, t)\psi\cdot\bar{\psi}, \tag{1.86}$$

where $\psi \in \mathbb{C}$ is a wave function. Then, as before, we can prove that L_0 is invariant under the change of inertial or non-inertial Cartesian coordinate system, given by (1.2), provided that we take into account (1.83). Moreover, if we consider a functional

$$J_0 = \int_0^T \int_{\mathbb{R}^3} L_0\,(\psi, \mathbf{x}, t)\, d\mathbf{x}dt. \tag{1.87}$$

Then, by (1.86) we get that the Euler–Lagrange equation for (1.87) coincides with the Shrödinger equation in the form of (1.82). Next we would like to note that the Lagrangian density L_0, defined by (1.86) obeys $U(1)$ local symmetry, i.e. for every scalar field $w := w(\mathbf{x}, t)$ one can easily deduce that L_0 in (1.86) is invariant under the transformation:

$$\begin{cases} \psi \to e^{-\frac{i\sigma w}{c\hbar}}\,\psi \\ \Psi \to \Psi + \frac{1}{c}\frac{\partial w}{\partial t} \\ \mathbf{A} \to \mathbf{A} - \nabla_{\mathbf{x}} w \\ \mathbf{v} \to \mathbf{v}. \end{cases} \tag{1.88}$$

See Sect. 6.4 for the generalizations of all mentioned above about the Shrödinger equation to the case of system of n quantum particles. Furthermore, see Sect. 6.5 for the invariance of the Quantum Liouville's equation, under the change of non-inertial coordinate system.

Next consider the motion of a spin-half quantum micro-particle with inertial mass m and the charge σ in the outer gravitational and electromagnetical field with potentials $\mathbf{v}(\mathbf{x}, t)$, $\mathbf{A}(\mathbf{x}, t)$ and $\Psi(\mathbf{x}, t)$ and additional conservative field with potential $V(\mathbf{x}, t)$. Since the Hamiltonian for a macro-particle has the form (1.74), we built the Hermitian Hamiltonian operator, taking into account the spin interaction as

$$\begin{aligned} \hat{H}_0 \cdot \psi = &-\frac{\hbar^2}{2m}\Delta_{\mathbf{x}}\psi + \frac{i\hbar\sigma}{2mc}\,\mathrm{div}_{\mathbf{x}}\,\{\psi\mathbf{A}(\mathbf{x}, t)\} + \frac{i\hbar\sigma}{2mc}\nabla_{\mathbf{x}}\psi \cdot \mathbf{A}(\mathbf{x}, t) + \frac{\sigma^2}{2mc^2}\,|\mathbf{A}(\mathbf{x}, t)|^2\,\psi \\ &+ \sigma\left(\Psi(\mathbf{x}, t) - \frac{1}{c}\mathbf{v}(\mathbf{x}, t) \cdot \mathbf{A}(\mathbf{x}, t)\right)\psi - V\,(\mathbf{x}, t)\,\psi - \frac{i\hbar}{2}\,\mathrm{div}_{\mathbf{x}}\,\{\psi\mathbf{v}(\mathbf{x}, t)\} - \frac{i\hbar}{2}\nabla_{\mathbf{x}}\psi \cdot \mathbf{v}(\mathbf{x}, t) \\ &- \frac{g\sigma\hbar}{2mc}\mathbf{S} \cdot (\mathrm{curl}_{\mathbf{x}}\mathbf{A}(\mathbf{x}, t)\,\psi) + \frac{\hbar}{4}\mathbf{S} \cdot (\mathrm{curl}_{\mathbf{x}}\mathbf{v}(\mathbf{x}, t)\,\psi), \end{aligned} \tag{1.89}$$

where $\psi(\mathbf{x}, t) = (\psi_1(\mathbf{x}, t), \psi_2(\mathbf{x}, t)) \in \mathbb{C}^2$ is a two-component wave function, $\hat{H}_0$ is the Hamiltonian operator, $\mathbf{S} := (S_1, S_2, S_3)$,

$$S_1 = \begin{pmatrix} 0 & 1 \\ 1 & 0 \end{pmatrix}, \quad S_2 = \begin{pmatrix} 0 & -i \\ i & 0 \end{pmatrix} \quad S_3 = \begin{pmatrix} 1 & 0 \\ 0 & -1 \end{pmatrix}$$

are Pauli matrices and g is a constant that depends on the type of the particle (for electron we have $g = 1$). Note that, in addition to the classical term of the spin-magnetic interaction, we added another term to the Hamiltonian, namely $\frac{\hbar}{4}\mathbf{S} \cdot (\mathrm{curl}_{\mathbf{x}}\mathbf{v}(\mathbf{x}, t)\,\psi)$. This term vanishes in every non-rotating and, in particular, in every inertial coordinate system, however, it provides the invariance of the Shrödinger–Pauli equation, under the change of non-inertial Cartesian coordinate system, as can be seen in the following Theorem 1.3. The Shrödinger–Pauli equation for this particle is

$$i\hbar\frac{\partial\psi}{\partial t} = \hat{H}_0 \cdot \psi. \tag{1.90}$$

That is,

$$\begin{aligned}
&i\hbar\left(\frac{\partial\psi}{\partial t} + \frac{1}{2}\,\mathrm{div}_{\mathbf{x}}\left\{\psi\mathbf{v}(\mathbf{x},t)\right\} + \frac{1}{2}\nabla_{\mathbf{x}}\psi\cdot\mathbf{v}(\mathbf{x},t)\right)\\
&= -\frac{\hbar^2}{2m}\Delta_{\mathbf{x}}\psi + \frac{i\hbar\sigma}{2mc}\,\mathrm{div}_{\mathbf{x}}\left\{\psi\mathbf{A}(\mathbf{x},t)\right\} + \frac{i\hbar\sigma}{2mc}\nabla_{\mathbf{x}}\psi\cdot\mathbf{A}(\mathbf{x},t) + \frac{\sigma^2}{2mc^2}\left|\mathbf{A}(\mathbf{x},t)\right|^2\psi\\
&\quad + \sigma\left(\Psi(\mathbf{x},t) - \frac{1}{c}\mathbf{v}(\mathbf{x},t)\cdot\mathbf{A}(\mathbf{x},t)\right)\psi - V(\mathbf{x},t)\,\psi\\
&\quad + \frac{\hbar}{2}\mathbf{S}\cdot\left(\left(\frac{1}{2}\mathrm{curl}_{\mathbf{x}}\mathbf{v}(\mathbf{x},t) - \frac{g\sigma}{mc}\mathrm{curl}_{\mathbf{x}}\mathbf{A}(\mathbf{x},t)\right)\psi\right).
\end{aligned} \tag{1.91}$$

In Sect. 6.6 we prove the following:

Theorem 1.3 *Consider that the change of some Cartesian coordinate system* $(*)$ *to another Cartesian coordinate system* $(**)$ *is given by* (1.2)*, where* $A(t) \in SO(3)$ *is a rotation. Next, assume that in the coordinate system* $(**)$ *we observe a validity of the Shrödinger–Pauli equation of the form:*

$$\begin{aligned}
&i\hbar\left(\frac{\partial\psi'}{\partial t'} + \frac{1}{2}\,\mathrm{div}_{\mathbf{x}'}\left\{\psi'\mathbf{v}'\right\} + \frac{1}{2}\nabla_{\mathbf{x}'}\psi'\cdot\mathbf{v}'\right) =\\
&\quad -\frac{\hbar^2}{2m'}\Delta_{\mathbf{x}'}\psi' + \frac{i\hbar\sigma'}{2m'c}\,\mathrm{div}_{\mathbf{x}'}\left\{\psi'\mathbf{A}'\right\} + \frac{i\hbar\sigma'}{2m'c}\nabla_{\mathbf{x}'}\psi'\cdot\mathbf{A}'\\
&\quad + \frac{(\sigma')^2}{2m'c^2}\left|\mathbf{A}'\right|^2\psi' + \sigma'\left(\Psi' - \frac{1}{c}\mathbf{v}'\cdot\mathbf{A}'\right)\psi' - V'\psi'\\
&\quad + \frac{\hbar}{2}\mathbf{S}\cdot\left(\left(\frac{1}{2}curl_{\mathbf{x}'}\mathbf{v}' - \frac{g'\sigma'}{m'c}curl_{\mathbf{x}'}\mathbf{A}'\right)\psi'\right),
\end{aligned} \tag{1.92}$$

where $\psi \in \mathbb{C}^2$*. Then in the coordinate system* $(*)$ *we have the validity of Shrödinger–Pauli equation of the same as* (1.92) *form:*

$$i\hbar\left(\frac{\partial\psi}{\partial t}+\frac{1}{2}\operatorname{div}_{\mathbf{x}}\{\psi\mathbf{v}\}+\frac{1}{2}\nabla_{\mathbf{x}}\psi\cdot\mathbf{v}\right)=$$
$$-\frac{\hbar^2}{2m}\Delta_{\mathbf{x}}\psi+\frac{i\hbar\sigma}{2mc}\operatorname{div}_{\mathbf{x}}\{\psi\mathbf{A}\}+\frac{i\hbar\sigma}{2mc}\nabla_{\mathbf{x}}\psi\cdot\mathbf{A}+\frac{\sigma^2}{2mc^2}|\mathbf{A}|^2\psi$$
$$+\sigma\left(\Psi-\frac{1}{c}\mathbf{v}\cdot\mathbf{A}\right)\psi-V\psi+\frac{\hbar}{2}\mathbf{S}\cdot\left(\left(\frac{1}{2}curl_{\mathbf{x}}\mathbf{v}-\frac{g\sigma}{mc}curl_{\mathbf{x}}\mathbf{A}\right)\psi\right) \quad (1.93)$$

provided that

$$\begin{cases} g'=g \\ V'=V, \\ \sigma'=\sigma, \\ m'=m, \\ \mathbf{v}'=A(t)\cdot\mathbf{v}+\frac{dA}{dt}(t)\cdot\mathbf{x}+\frac{d\mathbf{z}}{dt}(t), \\ \mathbf{A}'=A(t)\cdot\mathbf{A}, \\ \Psi'-\mathbf{v}'\cdot\mathbf{A}'=\Psi-\mathbf{v}\cdot\mathbf{A}, \\ \psi'=U(t)\cdot\psi, \end{cases} \quad (1.94)$$

where $U(t)\in SU(2)$ *is some special unitary* 2×2 *matrix, i.e.* $U(t)\in\mathbb{C}^{2\times 2}$, $\det U(t)=1$, $U(t)\cdot U^*(t)=I$ *where* $U^*(t)$ *is the Hermitian adjoint to* $U(t)$ *matrix:* $U^*(t):=\bar{U}(t)^T$ *and* I *is the identity* 2×2 *matrix. Moreover,* $U(t)$ *is characterized by the equality:*

$$U^*(t)\cdot\mathbf{S}\cdot U(t)=A(t)\cdot\mathbf{S}. \quad (1.95)$$

Next, again consider the motion of a quantum micro-particle with spin-half, inertial mass m and the charge σ with the given gravitational and electromagnetical fields with potentials $\mathbf{v}(\mathbf{x},t)$, $\mathbf{A}(\mathbf{x},t)$ and $\Psi(\mathbf{x},t)$ and additional conservative field with potential $V(\mathbf{x},t)$, taking into the account spin interaction. Then consider a Lagrangian density L defined by

$$L(\psi,\mathbf{x},t):=\frac{i\hbar}{2}\left(\left(\frac{\partial\psi}{\partial t}+\mathbf{v}\cdot\nabla_{\mathbf{x}}\psi\right)\cdot\bar{\psi}-\psi\cdot\left(\frac{\partial\bar{\psi}}{\partial t}+\mathbf{v}\cdot\nabla_{\mathbf{x}}\bar{\psi}\right)\right)-\frac{\hbar^2}{2m}\nabla_{\mathbf{x}}\psi\cdot\nabla_{\mathbf{x}}\bar{\psi}$$
$$-\frac{\hbar\sigma i}{2mc}\left(\nabla_{\mathbf{x}}\psi\cdot\bar{\psi}-\psi\cdot\nabla_{\mathbf{x}}\bar{\psi}\right)\cdot\mathbf{A}-\frac{\sigma^2}{2mc^2}|\mathbf{A}|^2\psi\cdot\bar{\psi}-\sigma\left(\Psi-\frac{1}{c}\mathbf{v}\cdot\mathbf{A}\right)\psi\cdot\bar{\psi}$$
$$-\frac{\hbar}{2}\left(\left(\mathbf{S}\cdot\left(\frac{1}{2}\mathrm{curl}_{\mathbf{x}}\mathbf{v}-\frac{g\sigma}{mc}\mathrm{curl}_{\mathbf{x}}\mathbf{A}\right)\right)\cdot\psi\right)\cdot\bar{\psi}+V(\mathbf{x},t)\,\psi\cdot\bar{\psi}, \quad (1.96)$$

where $\psi\in\mathbb{C}^2$ is a two-component wave function. Then similarly to the proof of Theorem 1.3 we can prove that L is invariant under the change of inertial or non-inertial Cartesian coordinate system, given by (1.2), provided that we take into account (1.94). Moreover, if we consider a functional

$$J = \int_0^T \int_{\mathbb{R}^3} L(\psi, \mathbf{x}, t)\, d\mathbf{x}dt, \tag{1.97}$$

then, by (1.96) we get that the Euler–Lagrange equation for (1.97) coincides with the Shrödinger–Pauli equation in the form of (1.91). Next we would like to note that, as before, the Lagrangian density L, defined by (1.96) obeys $U(1)$ local symmetry, i.e. for every scalar field $w := w(\mathbf{x}, t)$ one can easily deduce that L in (1.96) is invariant under the transformation:

$$\begin{cases} \psi \to e^{-\frac{i\sigma w}{c\hbar}}\psi \\ \Psi \to \Psi + \frac{1}{c}\frac{\partial w}{\partial t} \\ \mathbf{A} \to \mathbf{A} - \nabla_{\mathbf{x}} w \\ \mathbf{v} \to \mathbf{v}. \end{cases} \tag{1.98}$$

See Sect. 6.7 for the generalization of the Shrödinger–Pauli equation to the case of a system of n spin-half micro-particles. See also Sect. 6.8 for the Quantum Liouville's equation of a system of n spin-half particles.

1.5 Unified Gravitational-Electromagnetic Field and Conservation Laws

Similarly to our assumption that the electromagnetic field is influenced by gravitational field, we also can assume that the gravitational field is influenced by electromagnetic field. We remind that we assume that the first approximation of the law of gravitation is given by (1.21). However, till now we said nothing about the relation between the density of inertial and gravitational masses. If μ is the density of inertial masses and M is the density of gravitational masses, then consistently with the classical Newtonian theory of gravitation we assume that in the absence of essential electromagnetic fields we should have

$$M = \mu. \tag{1.99}$$

In order to satisfy the conservation laws of linear and angular momentums and energy, consider the following conserved scalar field Q, that we call "electromagnetical-gravitational" mass density, which is negligible in the absence of electromagnetic fields and satisfies the identity

$$\frac{\partial Q}{\partial t} + \operatorname{div}_{\mathbf{x}}\{Q\mathbf{v}\} = -\operatorname{div}_{\mathbf{x}}\left\{\frac{1}{4\pi c}\mathbf{D}\times\mathbf{B}\right\} \tag{1.100}$$

in the general case. Then, instead of (1.99), for the general case of gravitational-electromagnetic fields we consider the following relation between the gravitational and inertial mass densities

$$M = \mu + Q. \tag{1.101}$$

Then by (1.21) and (1.101) we have the following law of gravitation:

$$\begin{cases} \text{curl}_{\mathbf{x}} \left(\text{curl}_{\mathbf{x}} \mathbf{v}\right) = 0, \\ \frac{\partial}{\partial t} \left(\text{div}_{\mathbf{x}} \, \mathbf{v}\right) + \text{div}_{\mathbf{x}} \left\{\left(\text{div}_{\mathbf{x}} \, \mathbf{v}\right) \mathbf{v}\right\} + \frac{1}{4} \left| d_{\mathbf{x}} \mathbf{v} + \left\{d_{\mathbf{x}} \mathbf{v}\right\}^T \right|^2 - \left(\text{div}_{\mathbf{x}} \, \mathbf{v}\right)^2 = -4\pi G(\mu + Q). \end{cases} \tag{1.102}$$

The laws (1.100) and (1.102) are invariant under the change of non-inertial Cartesian coordinate system, given by (1.2), provided that, under (1.2) we have $Q' = Q$ and $\mu' = \mu$. In particular, in the inertial coordinate system $(*)$ we should have

$$\begin{cases} \text{curl}_{\mathbf{x}} \mathbf{v} = 0, \\ \frac{\partial \mathbf{v}}{\partial t} + d_{\mathbf{x}} \mathbf{v} \cdot \mathbf{v} = -\nabla_{\mathbf{x}} \Phi, \end{cases} \tag{1.103}$$

where Φ is the scalar gravitational potential which is a scalar field satisfying in every coordinate system:

$$\Delta_{\mathbf{x}} \Phi = 4\pi G(\mu + Q). \tag{1.104}$$

Remark 1.1 Lemma 14.1 from Appendix gives some insight that the "electromagnetical-gravitational" mass density Q in (1.100) should have the values of the same order as the quantity $\frac{1}{c^2}\left(|\mathbf{D}|^2 + |\mathbf{B}|^2\right)$ and therefore, in the usual circumstances is negligible with respect to the inertial mass density μ. Thus we can write $Q \approx 0$ in (1.102), i.e. the force of gravity in an inertial coordinate system approximately equals to the classical Newtonian force of gravity.

Next consider the Maxwell equation in the vacuum in the form (1.38) and consistently with (1.6), consider the second Law of Newton for the moving continuum with the inertial mass density μ and the field of velocities $\mathbf{u}$:

$$\mu \frac{\partial \mathbf{u}}{\partial t} + \mu d_{\mathbf{x}} \mathbf{u} \cdot \mathbf{u} = -\mu \mathbf{u} \times \text{curl}_{\mathbf{x}} \mathbf{v} + \mu \partial_t \mathbf{v} + \mu \nabla_{\mathbf{x}} \left(\frac{1}{2}|\mathbf{v}|^2\right) + \rho \mathbf{E} + \frac{1}{c} \mathbf{j} \times \mathbf{B} + \mathbf{G}, \tag{1.105}$$

where $\rho \mathbf{E} + \frac{1}{c} \mathbf{j} \times \mathbf{B}$ is the volume density of the Lorentz force and $\mathbf{G}$ is the total volume density of all non-gravitational and non-electromagnetic forces acting on the continuum with mass density μ. Then, in Chap. 7 we prove that in inertial coordinate systems we have conservation laws of the linear momentum, the angular momentum and the energy. More precisely, we have the following theorem:

Theorem 1.4 *Consider the Maxwell equation for the vacuum in the form* (1.38) *and the second Law of Newton for the moving continuum in the form* (1.105). *Next, assume that in some Cartesian coordinate system* $(*)$ *we observe the gravitational law in the form of* (1.103), (1.104) *and* (1.100). *Then in the system* $(*)$ *we have* *the following laws of conservation of the linear momentum, angular momentum and energy:*

$$
\frac{\partial}{\partial t}\left(\mu\mathbf{u}+Q\mathbf{v}+\frac{1}{4\pi c}\mathbf{D}\times\mathbf{B}\right)=
-\operatorname{div}_{\mathbf{x}}\left\{\mu\mathbf{u}\otimes\mathbf{u}+Q\mathbf{v}\otimes\mathbf{v}+\left(\frac{1}{4\pi c}\mathbf{D}\times\mathbf{B}\right)\otimes\mathbf{v}+\mathbf{v}\otimes\left(\frac{1}{4\pi c}\mathbf{D}\times\mathbf{B}\right)\right\}
+\frac{1}{4\pi}\operatorname{div}_{\mathbf{x}}\left\{\mathbf{D}\otimes\mathbf{D}+\mathbf{B}\otimes\mathbf{B}-\frac{1}{2}\left(|\mathbf{D}|^2+|\mathbf{B}|^2\right)I-\frac{1}{G}\nabla_{\mathbf{x}}\Phi\otimes\nabla_{\mathbf{x}}\Phi+\frac{1}{2G}|\nabla_{\mathbf{x}}\Phi|^2 I\right\}
+\mathbf{G}, \quad (1.106)
$$

$$
\frac{\partial}{\partial t}\left(\mathbf{x}\times(\mu\mathbf{u})+\mathbf{x}\times(Q\mathbf{v})+\mathbf{x}\times\left(\frac{1}{4\pi c}\mathbf{D}\times\mathbf{B}\right)\right)=
-\operatorname{div}_{\mathbf{x}}\left\{\mu(\mathbf{x}\times\mathbf{u})\otimes\mathbf{u}+Q(\mathbf{x}\times\mathbf{v})\otimes\mathbf{v}
+\left(\mathbf{x}\times\left(\frac{1}{4\pi c}\mathbf{D}\times\mathbf{B}\right)\right)\otimes\mathbf{v}+(\mathbf{x}\times\mathbf{v})\otimes\left(\frac{1}{4\pi c}\mathbf{D}\times\mathbf{B}\right)\right\}
+\frac{1}{4\pi}\operatorname{div}_{\mathbf{x}}\left\{(\mathbf{x}\times\mathbf{D})\otimes\mathbf{D}+(\mathbf{x}\times\mathbf{B})\otimes\mathbf{B}-\frac{1}{G}(\mathbf{x}\times\nabla_{\mathbf{x}}\Phi)\otimes\nabla_{\mathbf{x}}\Phi\right\}
+\frac{1}{8\pi}curl_{\mathbf{x}}\left\{\left(|\mathbf{D}|^2+|\mathbf{B}|^2-\frac{1}{G}|\nabla_{\mathbf{x}}\Phi|^2\right)\mathbf{x}\right\}+\mathbf{x}\times\mathbf{G}, \quad (1.107)
$$

and

$$
\frac{\partial}{\partial t}\left(\frac{1}{2}\mu|\mathbf{u}|^2+\frac{1}{2}Q|\mathbf{v}|^2+\frac{\mathbf{D}\cdot\mathbf{E}+\mathbf{B}\cdot\mathbf{H}}{8\pi}-\frac{1}{8\pi G}|\nabla_{\mathbf{x}}\Phi|^2\right)=
-\operatorname{div}_{\mathbf{x}}\left\{\left(\frac{\mu|\mathbf{u}|^2}{2}\right)\mathbf{u}+\left(\frac{Q|\mathbf{v}|^2}{2}\right)\mathbf{v}+\frac{1}{2}|\mathbf{v}|^2\left(\frac{1}{4\pi c}\mathbf{D}\times\mathbf{B}\right)+\left(\frac{\mathbf{D}\cdot\mathbf{E}+\mathbf{B}\cdot\mathbf{H}}{8\pi}\right)\mathbf{v}\right\}
+\frac{1}{4\pi}\operatorname{div}_{\mathbf{x}}\left\{(\mathbf{D}\otimes\mathbf{D}+\mathbf{B}\otimes\mathbf{B})\cdot\mathbf{v}-\frac{1}{2}\left(|\mathbf{D}|^2+|\mathbf{B}|^2\right)\mathbf{v}-c\mathbf{D}\times\mathbf{B}\right\}
-\operatorname{div}_{\mathbf{x}}\left\{\Phi\left(\mu\mathbf{u}+Q\mathbf{v}+\frac{1}{4\pi c}\mathbf{D}\times\mathbf{B}\right)\right\}-\frac{1}{4\pi G}\operatorname{div}_{\mathbf{x}}\left\{\Phi\frac{\partial}{\partial t}(\nabla_{\mathbf{x}}\Phi)\right\}+\mathbf{G}\cdot\mathbf{u}. \quad (1.108)
$$

Next given known the distribution of inertial mass density of some continuum medium $\mu := \mu(\mathbf{x}, t)$, the field of velocities of this medium $\mathbf{u} := \mathbf{u}(\mathbf{x}, t)$, the charge density $\rho := \rho(\mathbf{x}, t)$ and the current density $\mathbf{j} := \mathbf{j}(\mathbf{x}, t)$ consider a Lagrangian density L for the unified gravitational-electromagnetic field, defined by

$$
\begin{aligned}
&L(\mathbf{A}, \Psi, \mathbf{v}, \Phi, \mathbf{p}, \mathbf{x}, t) := \\
&\frac{1}{8\pi}\left|-\nabla_{\mathbf{x}}\Psi - \frac{1}{c}\frac{\partial \mathbf{A}}{\partial t} + \frac{1}{c}\mathbf{v}\times \mathrm{curl}_{\mathbf{x}}\mathbf{A}\right|^2 - \frac{1}{8\pi}\left|\mathrm{curl}_{\mathbf{x}}\mathbf{A}\right|^2 - \left(\rho\Psi - \frac{1}{c}\mathbf{A}\cdot\mathbf{j}\right) \\
&+\frac{\mu}{2}\left|\mathbf{u}-\mathbf{v}\right|^2 + \frac{1}{2}\left(d_{\mathbf{x}}\mathbf{v} + \{d_{\mathbf{x}}\mathbf{v}\}^T\right) : \left(d_{\mathbf{x}}\mathbf{p} + \{d_{\mathbf{x}}\mathbf{p}\}^T\right) - 2\left(\mathrm{div}_{\mathbf{x}}\,\mathbf{v}\right)\left(\mathrm{div}_{\mathbf{x}}\,\mathbf{p}\right) \\
&+\frac{1}{4\pi G}\left(\mathrm{div}_{\mathbf{x}}\,\mathbf{v}\right)\left(\frac{\partial\Phi}{\partial t} + \mathbf{v}\cdot\nabla_{\mathbf{x}}\Phi\right) + \frac{1}{4\pi G}\Phi\left(\mathrm{div}_{\mathbf{x}}\,\mathbf{v}\right)^2 - \frac{\Phi}{16\pi G}\left|d_{\mathbf{x}}\mathbf{v} + \{d_{\mathbf{x}}\mathbf{v}\}^T\right|^2 \\
&+\frac{1}{8\pi G}\left|\nabla_{\mathbf{x}}\Phi\right|^2,
\end{aligned}
\tag{1.109}
$$

where Φ is an ancillary proper scalar field and $\mathbf{p}$ is an ancillary proper vector field. Then, as before, we can show that L is invariant under the change of non-inertial Cartesian coordinate system given by (1.2), provided that, under (1.2) we have

$$
\begin{cases}
\mathbf{p}' = A(t)\cdot\mathbf{p}\\
\Phi' = \Phi\\
\mathbf{v}' = A(t)\cdot\mathbf{v} + A'(t)\cdot\mathbf{x} + \frac{d\mathbf{z}}{dt}(t)\\
\mathbf{A}' = A(t)\cdot\mathbf{A}\\
\Psi' - \frac{1}{c}\mathbf{A}'\cdot\mathbf{v}' = \Psi - \frac{1}{c}\mathbf{A}\cdot\mathbf{v}.
\end{cases}
\tag{1.110}
$$

Then in Sect. 8.1 we obtain that a configuration $(\mathbf{A}, \Psi, \mathbf{v}, \Phi, \mathbf{p})$ is a critical point of the functional

$$
J = \int_0^T\int_{\mathbb{R}^3} L(\mathbf{A}, \Psi, \mathbf{v}, \Phi, \mathbf{p}, \mathbf{x}, t)\,d\mathbf{x}dt,
\tag{1.111}
$$

if and only if it satisfies

$$
\begin{cases}
\mathrm{curl}_{\mathbf{x}}\mathbf{H} = \frac{4\pi}{c}\mathbf{j} + \frac{1}{c}\frac{\partial\mathbf{D}}{\partial t}\\
\mathrm{div}_{\mathbf{x}}\,\mathbf{D} = 4\pi\rho\\
\mathrm{curl}_{\mathbf{x}}\mathbf{E} + \frac{1}{c}\frac{\partial\mathbf{B}}{\partial t} = 0\\
\mathrm{div}_{\mathbf{x}}\,\mathbf{B} = 0\\
\mathbf{E} = \mathbf{D} - \frac{1}{c}\mathbf{v}\times\mathbf{B}\\
\mathbf{H} = \mathbf{B} + \frac{1}{c}\mathbf{v}\times\mathbf{D}\\
\mathrm{curl}_{\mathbf{x}}\left(\mathrm{curl}_{\mathbf{x}}\mathbf{v}\right) = 0\\
\frac{\partial}{\partial t}\left\{\mathrm{div}_{\mathbf{x}}\,\mathbf{v}\right\} + \mathbf{v}\cdot\nabla_{\mathbf{x}}\left(\mathrm{div}_{\mathbf{x}}\,\mathbf{v}\right) + \frac{1}{4}\left|d_{\mathbf{x}}\mathbf{v} + \{d_{\mathbf{x}}\mathbf{v}\}^T\right|^2 = -\Delta_{\mathbf{x}}\Phi\\
\left(\mu\mathbf{u} - \mu\mathbf{v} + \frac{1}{4\pi c}\mathbf{D}\times\mathbf{B}\right) =\\
\mathrm{curl}_{\mathbf{x}}\left(\mathrm{curl}_{\mathbf{x}}\mathbf{p}\right) - \frac{1}{4\pi G}\left(\frac{\partial}{\partial t}\left(\nabla_{\mathbf{x}}\Phi\right) - \mathrm{curl}_{\mathbf{x}}\left(\mathbf{v}\times\nabla_{\mathbf{x}}\Phi\right) + \left(\Delta_{\mathbf{x}}\Phi\right)\mathbf{v}\right),
\end{cases}
\tag{1.112}
$$

where, consistently with (1.54) we denote:

$$\begin{cases} \mathbf{D} := -\nabla_{\mathbf{x}}\Psi - \frac{1}{c}\frac{\partial \mathbf{A}}{\partial t} + \frac{1}{c}\mathbf{v} \times \mathrm{curl}_{\mathbf{x}}\mathbf{A} \\ \mathbf{B} := \mathrm{curl}_{\mathbf{x}}\mathbf{A} \\ \mathbf{E} := -\nabla_{\mathbf{x}}\Psi - \frac{1}{c}\frac{\partial \mathbf{A}}{\partial t} \\ \mathbf{H} := \mathrm{curl}_{\mathbf{x}}\mathbf{A} + \frac{1}{c}\mathbf{v} \times \left(-\nabla_{\mathbf{x}}\Psi - \frac{1}{c}\frac{\partial \mathbf{A}}{\partial t} + \frac{1}{c}\mathbf{v} \times \mathrm{curl}_{\mathbf{x}}\mathbf{A}\right). \end{cases} \tag{1.113}$$

In particular, using continuum equation $\partial_t \mu + \mathrm{div}_{\mathbf{x}}(\mu\mathbf{u}) = 0$ from the last equality in (1.112) we deduce

$$\frac{\partial}{\partial t}\left(\frac{1}{4\pi G}\Delta_{\mathbf{x}}\Phi - \mu\right) + \mathrm{div}_{\mathbf{x}}\left\{\left(\frac{1}{4\pi G}\Delta_{\mathbf{x}}\Phi - \mu\right)\mathbf{v}\right\} = -\,\mathrm{div}_{\mathbf{x}}\left\{\frac{1}{4\pi c}\mathbf{D} \times \mathbf{B}\right\}.$$

Thus denoting $Q = \Delta_{\mathbf{x}}\Phi/4\pi G - \mu$ we deduce the following system of equation for the gravitational-electromagnetic field, invariant under the change of non-inertial Cartesian coordinate system:

$$\begin{cases} \mathrm{curl}_{\mathbf{x}}\mathbf{H} = \frac{4\pi}{c}\mathbf{j} + \frac{1}{c}\frac{\partial \mathbf{D}}{\partial t} \\ \mathrm{div}_{\mathbf{x}}\,\mathbf{D} = 4\pi\rho \\ \mathrm{curl}_{\mathbf{x}}\mathbf{E} + \frac{1}{c}\frac{\partial \mathbf{B}}{\partial t} = 0 \\ \mathrm{div}_{\mathbf{x}}\,\mathbf{B} = 0 \\ \mathbf{E} = \mathbf{D} - \frac{1}{c}\mathbf{v} \times \mathbf{B} \\ \mathbf{H} = \mathbf{B} + \frac{1}{c}\mathbf{v} \times \mathbf{D} \\ \mathrm{curl}_{\mathbf{x}}\,(\mathrm{curl}_{\mathbf{x}}\mathbf{v}) = 0 \\ \frac{\partial}{\partial t}(\mathrm{div}_{\mathbf{x}}\,\mathbf{v}) + \mathrm{div}_{\mathbf{x}}\{(\mathrm{div}_{\mathbf{x}}\,\mathbf{v})\,\mathbf{v}\} + \frac{1}{4}\left|d_{\mathbf{x}}\mathbf{v} + \{d_{\mathbf{x}}\mathbf{v}\}^T\right|^2 - (\mathrm{div}_{\mathbf{x}}\,\mathbf{v})^2 = -4\pi G(\mu + Q) \\ \frac{\partial Q}{\partial t} + \mathrm{div}_{\mathbf{x}}\,(Q\mathbf{v}) = -\,\mathrm{div}_{\mathbf{x}}\left\{\frac{1}{4\pi c}\mathbf{D} \times \mathbf{B}\right\}, \end{cases} \tag{1.114}$$

which is consistent with (1.38), (1.102) and (1.100).

1.6 Transformations of General Scalar and Vector Fields Under the Change of Coordinate Systems

In order to get the above results we established some trivial calculus consequences about the behavior of scalar, vector and matrix fields, under the change of Cartesian coordinate system of the form (1.2). We combine them in the form of Proposition after the following definition:

Definition 1.1 Consider the change of some inertial or non-inertial Cartesian coordinate system $(*)$ to another Cartesian coordinate system $(**)$ of the form (1.2) where $A(t) \in SO(3)$ is a rotation.

- We say that a general scalar field $\psi := \psi(\mathbf{x}, t) : \mathbb{R}^3 \times [t_0, +\infty) \to \mathbb{R}$ is a proper scalar field if, under every change of coordinate system given by (1.2), this field transforms by the law:

$$\psi'(\mathbf{x}', t') = \psi(\mathbf{x}, t). \tag{1.115}$$

- We say that a general vector field $\mathbf{f} := \mathbf{f}(\mathbf{x}, t) : \mathbb{R}^3 \times [t_0, +\infty) \to \mathbb{R}^3$ is a proper vector field if, under every change of coordinate system given by (1.2), this field transforms by the law:
$$\mathbf{f}'(\mathbf{x}', t') = A(t) \cdot \mathbf{f}(\mathbf{x}, t). \tag{1.116}$$

- We say that a general vector field $\mathbf{v} := \mathbf{v}(\mathbf{x}, t) : \mathbb{R}^3 \times [t_0, +\infty) \to \mathbb{R}^3$ is a speed-like vector field if, under every change of coordinate system given by (1.2), this field transforms by the law:
$$\mathbf{v}'(\mathbf{x}', t') = A(t) \cdot \mathbf{v}(\mathbf{x}, t) + \frac{dA}{dt}(t) \cdot \mathbf{x} + \mathbf{w}(t), \tag{1.117}$$
where we set
$$\mathbf{w}(t) := \frac{d\mathbf{z}}{dt}(t) \qquad \forall t. \tag{1.118}$$

- We say that a general matrix-valued field $T := T(\mathbf{x}, t) : \mathbb{R}^3 \times [t_0, +\infty) \to \mathbb{R}^{3\times 3}$ is a proper matrix field if, under every change of coordinate system given by (1.2), this field transforms by the law:
$$T'(\mathbf{x}', t') = A(t) \cdot T(\mathbf{x}, t) \cdot A^T(t) = A(t) \cdot T(\mathbf{x}, t) \cdot \{A(t)\}^{-1}. \tag{1.119}$$

Proposition 1.1 *If $\psi : \mathbb{R}^3 \times [t_0, +\infty) \to \mathbb{R}$ is a proper scalar field, $\mathbf{f} : \mathbb{R}^3 \times [t_0, +\infty) \to \mathbb{R}^3$ and $\mathbf{g} : \mathbb{R}^3 \times [t_0, +\infty) \to \mathbb{R}^3$ are proper vector fields, $\mathbf{v} : \mathbb{R}^3 \times [t_0, +\infty) \to \mathbb{R}^3$ and $\mathbf{u} : \mathbb{R}^3 \times [t_0, +\infty) \to \mathbb{R}^3$ are speed-like vector fields and $T : \mathbb{R}^3 \times [t_0, +\infty) \to \mathbb{R}^{3\times 3}$ is a proper matrix field, then:*

(i) *scalar fields defined in every coordinate system as $\mathbf{f} \cdot \mathbf{g}$, $\operatorname{div}_{\mathbf{x}} \mathbf{f}$ and $\operatorname{div}_{\mathbf{x}} \mathbf{v}$ are proper scalar fields;*

(ii) *vector fields defined in every coordinate system as $\nabla_{\mathbf{x}}\psi$, $\operatorname{div}_{\mathbf{x}} T$, $curl_{\mathbf{x}}\mathbf{f}$, $\mathbf{f} \times \mathbf{g}$, $\operatorname{div}_{\mathbf{x}} \left(d_{\mathbf{x}}\mathbf{v} + \{d_{\mathbf{x}}\mathbf{v}\}^T\right)$, $\nabla_{\mathbf{x}} (\operatorname{div}_{\mathbf{x}} \mathbf{v})$, $\Delta_{\mathbf{x}}\mathbf{v}$, $curl_{\mathbf{x}} (curl_{\mathbf{x}}\mathbf{v})$ and $(\mathbf{u} - \mathbf{v})$ are proper vector fields;*

(iii) *matrix fields defined in every coordinate system as $d_{\mathbf{x}}\mathbf{f}$ and $\left(d_{\mathbf{x}}\mathbf{v} + \{d_{\mathbf{x}}\mathbf{v}\}^T\right)$ are proper matrix fields;*

(iv) *scalar fields $\xi : \mathbb{R}^3 \times [t_0, +\infty) \to \mathbb{R}$ and $\zeta : \mathbb{R}^3 \times [t_0, +\infty) \to \mathbb{R}$, defined in every coordinate system by*
$$\xi := \frac{\partial \psi}{\partial t} + \mathbf{v} \cdot \nabla_{\mathbf{x}}\psi \quad \text{and} \quad \zeta := \frac{\partial \psi}{\partial t} + \operatorname{div}_{\mathbf{x}} \{\psi \mathbf{v}\} \tag{1.120}$$
are proper scalar fields;

(v) *vector fields $\Theta : \mathbb{R}^3 \times [t_0, +\infty) \to \mathbb{R}^3$ and $\Xi : \mathbb{R}^3 \times [t_0, +\infty) \to \mathbb{R}^3$, defined in every coordinate system by*
$$\Theta := \frac{\partial \mathbf{f}}{\partial t} - curl_{\mathbf{x}} (\mathbf{v} \times \mathbf{f}) + (\operatorname{div}_{\mathbf{x}}\mathbf{f})\, \mathbf{v} \quad \text{and} \quad \Xi := \frac{\partial \mathbf{f}}{\partial t} - \mathbf{v} \times curl_{\mathbf{x}}\mathbf{f} + \nabla_{\mathbf{x}} (\mathbf{v} \cdot \mathbf{f}) \tag{1.121}$$

are proper vector fields and

$$\Xi = \Theta - (\mathrm{div}_{\mathbf{x}}\,\mathbf{v})\,\mathbf{f} + \left(d_{\mathbf{x}}\mathbf{v} + \{d_{\mathbf{x}}\mathbf{v}\}^T\right)\cdot\mathbf{f}. \tag{1.122}$$

We prove this Proposition in Chap. 3.

Next, we also prove the following Proposition in Chap. 3.

Proposition 1.2 *Consider some fixed inertial or non-inertial Cartesian coordinate system, denoted by the sign* ($\{0\}$), *and consider a vector field* $\mathbf{d} := \mathbf{d}(\mathbf{x}, t) : \mathbb{R}^3 \times [t_0, +\infty) \to \mathbb{R}^3$ *associated only with the chosen coordinate system* ($\{0\}$). *Then there exist a unique speed-like vector field* $\mathbf{v_d} : \mathbb{R}^3 \times [t_0, +\infty) \to \mathbb{R}^3$ *(associated with any inertial or non-inertial Cartesian coordinate system) such that in the particular system* ($\{0\}$) *we have*

$$\mathbf{v_d}(\mathbf{x}, t) = \mathbf{d}(\mathbf{x}, t) \quad \forall (\mathbf{x}, t) \in \mathbb{R}^3 \times [t_0, +\infty). \tag{1.123}$$

Moreover, there exist unique a proper vector field $\mathbf{h_d} : \mathbb{R}^3 \times [t_0, +\infty) \to \mathbb{R}^3$ *(associated with any inertial or non-inertial Cartesian coordinate system) such that in the particular system* ($\{0\}$) *we have*

$$\mathbf{h_d}(\mathbf{x}, t) = \mathbf{d}(\mathbf{x}, t) \quad \forall (\mathbf{x}, t) \in \mathbb{R}^3 \times [t_0, +\infty). \tag{1.124}$$

We also will need the following Definitions and Propositions and we prove these Propositions in Chap. 3.

Definition 1.2 Let $\mathbf{u} := \mathbf{u}(\mathbf{x}, t)$ be a speed-like vector field defined for $(\mathbf{x}, t) \in \Omega$ with domain of definition $\Omega \subset \mathbb{R}^3 \times \mathbb{R}$ and let $(*)$ be some inertial or non-inertial Cartesian coordinate system. Then $\mathbf{u}$ is called generally trivial speed-like field if there exists another Cartesian coordinate system $(**)$ such that under the change of coordinate system $(*)$ to another Cartesian coordinate system $(**)$ given by

$$\begin{cases} \mathbf{x}' = A(t)\cdot\mathbf{x} + \mathbf{z}(t), \\ t' = t, \end{cases} \tag{1.125}$$

where $A(t) \in SO(3)$ is a rotation, we have

$$A(t)\cdot\mathbf{u}(\mathbf{x}, t) + \frac{dA}{dt}(t)\cdot\mathbf{x} + \frac{d\mathbf{z}}{dt}(t) = \mathbf{u}'(\mathbf{x}', t') = 0. \tag{1.126}$$

Proposition 1.3 *Let* $\mathbf{u} := \mathbf{u}(\mathbf{x}, t)$ *be a smooth speed-like vector field defined for* $(\mathbf{x}, t) \in \Omega$ *with connected domain* $\Omega \subset \mathbb{R}^3 \times \mathbb{R}$, *such that for every instant of time* τ *the domain* $\Omega_\tau = \left\{\mathbf{x} \in \mathbb{R}^3 : (\mathbf{x}, \tau) \in \Omega\right\}$ *is also connected. Then* $\mathbf{u}$ *is a generally trivial speed-like field if and only if we have*

$$d_{\mathbf{x}}\mathbf{u} + \{d_{\mathbf{x}}\mathbf{u}\}^T = 0 \quad \forall (\mathbf{x}, t) \in \Omega. \tag{1.127}$$

Proposition 1.4 *Let* $\mathbf{v} := \mathbf{v}(\mathbf{x}, t) : \mathbb{R}^3 \times [t_0, +\infty) \to \mathbb{R}^3$ *be a speed-like vector field, such that there exists some Cartesian coordinate system* $(*)$, *where we have*

$$\lim_{|\mathbf{x}| \to +\infty} \mathbf{v}(\mathbf{x}, t) = 0 \quad \forall t \quad (\text{ in the given particular coordinate system } (*)). \tag{1.128}$$

Then there exist a uniquely defined <u>*generally trivial speed-like*</u> *vector field* $\mathbf{k} := \mathbf{k}(\mathbf{x}, t) : \mathbb{R}^3 \times [t_0, +\infty) \to \mathbb{R}^3$ *and a uniquely defined* <u>*proper*</u> *vector field* $\mathbf{h} := \mathbf{h}(\mathbf{x}, t) : \mathbb{R}^3 \times [t_0, +\infty) \to \mathbb{R}^3$, *so that in every Cartesian coordinate system we have*

$$\lim_{|\mathbf{x}| \to +\infty} \mathbf{h}(\mathbf{x}, t) = 0 \quad \forall t, \tag{1.129}$$

and in every Cartesian coordinate system we can decompose vector field $\mathbf{v}$ *as*

$$\mathbf{v}(\mathbf{x}, t) = \mathbf{h}(\mathbf{x}, t) + \mathbf{k}(\mathbf{x}, t) \qquad \forall (\mathbf{x}, t) \in \mathbb{R}^3 \times [t_0, +\infty). \tag{1.130}$$

Definition 1.3 We say that two (inertial or non-inertial) Cartesian coordinate systems $(*)$ and $(**)$ are <u>equivalent</u> if the change of coordinates from the system $(*)$ to the system $(**)$ is given by

$$\begin{cases} \mathbf{x}' = B \cdot \mathbf{x} + \mathbf{c}, \\ t' = t, \end{cases} \tag{1.131}$$

where $B \in SO(3)$ is a constant (independent on time) rotation and $\mathbf{c} \in \mathbb{R}^3$ is a constant (independent on time) vector. In other words, we obtain system $(**)$ from system $(*)$ just by a constant (independent on time) rotation in $\mathbb{R}^3$ and/or by a constant (independent on time) shift of the origin in $\mathbb{R}^3$.

Definition 1.4 Let $\mathbf{v} := \mathbf{v}(\mathbf{x}, t) : \mathbb{R}^3 \times [t_0, +\infty) \to \mathbb{R}^3$ be a speed-like smooth vector field and let $(*)$ be some (inertial or non-inertial) Cartesian coordinate system. We say that $\mathbf{v}$ is <u>asymptotically acceptable</u> in the coordinate system $(*)$, if there exist continuous mappings $B(t) : [t_0, +\infty) \to \mathbb{R}^{3\times 3}$ and $\mathbf{l}(t) : [t_0, +\infty) \to \mathbb{R}^3$, such that in the coordinate system $(*)$ we have

$$\begin{cases} \lim\limits_{|\mathbf{x}| \to +\infty} (d_{\mathbf{x}} \mathbf{v}(\mathbf{x}, t)) = B(t) \qquad \forall t \in [t_0, +\infty), \\ \lim\limits_{|\mathbf{x}| \to +\infty} \left(d_{\mathbf{x}} \mathbf{v}(\mathbf{x}, t) + \{d_{\mathbf{x}} \mathbf{v}(\mathbf{x}, t)\}^T \right) = 0 \qquad \forall t \in [t_0, +\infty), \\ \lim\limits_{|\mathbf{x}| \to +\infty} (\mathbf{v}(\mathbf{x}, t) - B(t) \cdot \mathbf{x}) = \mathbf{l}(t) \qquad \forall t \in [t_0, +\infty). \end{cases} \tag{1.132}$$

Proposition 1.5 *Let* $\mathbf{v} := \mathbf{v}(\mathbf{x}, t) : \mathbb{R}^3 \times [t_0, +\infty) \to \mathbb{R}^3$ *be a speed-like smooth vector field. Then the following three conditions are* <u>*equivalent:*</u>

(i) *There exists a (inertial or non-inertial) Cartesian coordinate system* $(*)$ *where the vector field* $\mathbf{v}$ *is asymptotically acceptable.*

(ii) *The vector field* $\mathbf{v}$ *is asymptotically acceptable in* <u>*every*</u> *(inertial or non-inertial) Cartesian coordinate system.*

(iii) *There exists a (inertial or non-inertial) Cartesian coordinate system* ({0}) *in which we have*

$$\lim_{|\mathbf{x}|\to+\infty} (d_{\mathbf{x}}\mathbf{v}(\mathbf{x}, t)) = 0 \qquad \forall t \in [t_0, +\infty)\,, \tag{1.133}$$

$$\lim_{|\mathbf{x}|\to+\infty} \mathbf{v}(\mathbf{x}, t) = 0 \qquad \forall t \in [t_0, +\infty)\,. \tag{1.134}$$

Moreover, in that case the Cartesian coordinate system where (1.134) *holds, is unique, up to an equivalence, in other words it is unique, up to a constant (independent on time) rotation in* $\mathbb{R}^3$ *and/or up to a constant (independent on time) shift of the origin in* $\mathbb{R}^3$.

As a direct consequence of both Propositions 1.4 and 1.5 we deduce the following Corollary:

Corollary 1.1 *Let* $\mathbf{v} := \mathbf{v}(\mathbf{x}, t) : \mathbb{R}^3 \times [t_0, +\infty) \to \mathbb{R}^3$ *be an asymptotically acceptable speed-like vector field. Then there exist a uniquely defined generally trivial speed-like vector field* $\mathbf{k} := \mathbf{k}(\mathbf{x}, t) : \mathbb{R}^3 \times [t_0, +\infty) \to \mathbb{R}^3$ *and a uniquely defined proper vector field* $\mathbf{h} := \mathbf{h}(\mathbf{x}, t) : \mathbb{R}^3 \times [t_0, +\infty) \to \mathbb{R}^3$, *so that in every Cartesian coordinate system we have*

$$\lim_{|\mathbf{x}|\to+\infty} \mathbf{h}(\mathbf{x}, t) = 0 \quad \forall t\,, \tag{1.135}$$

and in every Cartesian coordinate system we can decompose vector field $\mathbf{v}$ *as*

$$\mathbf{v}(\mathbf{x}, t) = \mathbf{h}(\mathbf{x}, t) + \mathbf{k}(\mathbf{x}, t) \qquad \forall (\mathbf{x}, t) \in \mathbb{R}^3 \times [t_0, +\infty). \tag{1.136}$$

Moreover, in addition, the proper vector field $\mathbf{h}$ *necessarily satisfies*

$$\lim_{|\mathbf{x}|\to+\infty} (d_{\mathbf{x}}\mathbf{h}(\mathbf{x}, t)) = 0 \quad \forall t\,. \tag{1.137}$$

Definition 1.5 Let $\mathcal{U}$ be a nonempty subset of the set of all inertial and non-inertial Cartesian coordinate systems. We say that $\mathcal{U}$ is an extended Galilean group if, given arbitrary Cartesian coordinate system $(*)$ inside the subset $\mathcal{U}$, the following statement holds:

- The arbitrary Cartesian coordinate system $(**)$ belongs to the subset $\mathcal{U}$ if and only if there exist a constant (independent on time) rotation $B \in SO(3)$ and constant (independent on time) vectors $\mathbf{c} \in \mathbb{R}^3$ and $\mathbf{w} \in \mathbb{R}^3$, so that the change of coordinates from the system $(*)$ to the system $(**)$ is given by

$$\begin{cases} \mathbf{x}' = B \cdot \mathbf{x} + \mathbf{c} + \mathbf{w}t, \\ t' = t. \end{cases} \tag{1.138}$$

In other words, the system (∗∗) belongs to $\mathcal{U}$ if and only if, up to equivalence of Cartesian coordinate systems, the system (∗∗) can be obtained from the system (∗) by the Galilean Transformation

$$\begin{cases} \mathbf{x}' = \mathbf{x} + \mathbf{w}t, \\ t' = t. \end{cases} \tag{1.139}$$

Definition 1.6 Let $\mathcal{V}$ be a nonempty subset of the set of all inertial and non-inertial Cartesian coordinate systems. We say that $\mathcal{V}$ is a non-rotational rigid motion group if, given arbitrary Cartesian coordinate system (∗) inside the subset $\mathcal{V}$, the following statement holds:

- The arbitrary Cartesian coordinate system (∗∗) belongs to the subset $\mathcal{V}$ if and only if there exist a constant (independent on time) rotation $B \in SO(3)$ and a smooth vector-valued function of the time variable $\mathbf{z}(t) : \mathbb{R} \to \mathbb{R}^3$, so that the change of coordinates from the system (∗) to the system (∗∗) is given by

$$\begin{cases} \mathbf{x}' = B \cdot \mathbf{x} + \mathbf{z}(t), \\ t' = t. \end{cases} \tag{1.140}$$

Definition 1.7 Let $\mathcal{W}$ be a certain nonempty subset of coordinate systems. We say that $\mathcal{W}$ is a general rigid motion group if, given arbitrary coordinate system (∗) inside the subset $\mathcal{W}$, the following statement holds:

- The arbitrary coordinate system (∗∗) belongs to the subset $\mathcal{W}$ if and only if there exist a smooth (time dependent) rotation $A(t) : \mathbb{R} \to SO(3)$ and a smooth vector-valued function of the time variable $\mathbf{z}(t) : \mathbb{R} \to \mathbb{R}^3$, so that the change of coordinates from the system (∗) to the system (∗∗) is given by

$$\begin{cases} \mathbf{x}' = A(t) \cdot \mathbf{x} + \mathbf{z}(t), \\ t' = t. \end{cases} \tag{1.141}$$

In particular, the set of all Cartesian coordinate systems is a general rigid motion group.

Definition 1.8 Let $\mathbf{v} := \mathbf{v}(\mathbf{x}, t) : \mathbb{R}^3 \times [t_0, +\infty) \to \mathbb{R}^3$ be an asymptotically acceptable speed-like smooth vector field (see Definition 1.4).

- We call that the Cartesian coordinate system ({0}) is preferable for the vector field $\mathbf{v}$ if in the system ({0}) we have

$$\lim_{|\mathbf{x}|\to+\infty} (d_{\mathbf{x}}\mathbf{v}(\mathbf{x}, t)) = 0 \qquad \forall t \in [t_0, +\infty), \tag{1.142}$$

$$\lim_{|\mathbf{x}|\to+\infty} \mathbf{v}(\mathbf{x}, t) = 0 \qquad \forall t \in [t_0, +\infty). \tag{1.143}$$

Note that by Proposition 1.5 the unique (up to equivalence) preferable for the vector field $\mathbf{v}$ Cartesian system ($\{0\}$) exists.

- We call that the Cartesian coordinate system $(*)$ is <u>inertial</u> with respect to the vector field $\mathbf{v}$ if in the system $(*)$ we have

$$\lim_{|\mathbf{x}|\to+\infty} (d_{\mathbf{x}}\mathbf{v}(\mathbf{x},t)) = 0 \qquad \forall t \in [t_0,+\infty)\,, \tag{1.144}$$

$$\lim_{|\mathbf{x}|\to+\infty} \mathbf{v}(\mathbf{x},t) = \mathbf{d} \qquad \forall t \in [t_0,+\infty)\,, \tag{1.145}$$

where $\mathbf{d} \in \mathbb{R}^3$ is a constant (independent on t) vector. Note that, in particular, by the definition, the preferable for $\mathbf{v}$ Cartesian system is also an inertial with respect to $\mathbf{v}$ coordinate system.

- We call that the Cartesian coordinate system $(*)$ is <u>non-rotating</u> with respect to the vector field $\mathbf{v}$ if in the system $(*)$ we have

$$\lim_{|\mathbf{x}|\to+\infty} (d_{\mathbf{x}}\mathbf{v}(\mathbf{x},t)) = 0 \qquad \forall t \in [t_0,+\infty)\,, \tag{1.146}$$

(without specifying $\lim_{|\mathbf{x}|\to+\infty} \mathbf{v}(\mathbf{x},t)$). Note that, in particular, by the definition, every inertial with respect to $\mathbf{v}$ Cartesian coordinate system is also a non-rotating with respect to $\mathbf{v}$ coordinate system.

Proposition 1.6 *Let* $\mathbf{v} := \mathbf{v}(\mathbf{x},t) : \mathbb{R}^3 \times [t_0,+\infty) \to \mathbb{R}^3$ *be an <u>asymptotically acceptable</u> speed-like smooth vector field (see Definition 1.4). Furthermore, let* $\mathcal{U}$ *be the set of all <u>inertial</u> with respect to* $\mathbf{v}$ *Cartesian coordinate systems and let* $\mathcal{V}$ *be the set of all <u>non-rotating</u> with respect to* $\mathbf{v}$ *Cartesian coordinate systems (see Definition 1.8). Then* $\mathcal{U}$ *is an <u>extended Galilean group</u> and* $\mathcal{V}$ *is a <u>non-rotational rigid motion group</u> (see Definitions 1.5 and 1.6).*

The next Proposition describes dynamics of the line, the surface and the volume integrals.

Proposition 1.7 **(i)** *Consider the moving continuum medium with velocity field* $\mathbf{u}(\mathbf{x},t)$ *and let* $\mathcal{S} := \mathcal{S}(t) \subset \mathbb{R}^3$ *be a two-dimensional surface oriented by the unit normal* $\mathbf{n} := \mathbf{n}(\mathbf{x},t)$ *and moving together with the given medium. Then, given a vector field* $\mathbf{f}(\mathbf{x},t)$ *we have*

$$\frac{d}{dt}\left(\iint \mathbf{f}\cdot\mathbf{n}\,d\mathcal{S}(t)\right) = \iint \left(\frac{\partial \mathbf{f}}{\partial t} - curl_{\mathbf{x}}\,(\mathbf{u}\times\mathbf{f}) + (\mathrm{div}_{\mathbf{x}}\mathbf{f})\,\mathbf{u}\right)\cdot\mathbf{n}\,d\mathcal{S}(t). \tag{1.147}$$

(ii) *Consider the moving continuum medium with velocity field* $\mathbf{u}(\mathbf{x},t)$ *and let* $\gamma := \gamma(t) \subset \mathbb{R}^3$ *be a one-dimensional curve oriented by the unit tangent vector* $\mathbf{t} := \mathbf{t}(\mathbf{x},t)$ *and moving together with the given medium. Then, given a vector field* $\mathbf{f}(\mathbf{x},t)$ *we have*

$$\frac{d}{dt}\left(\int \mathbf{f}\cdot\mathbf{t}\,d\gamma(t)\right) = \int\left(\frac{\partial\mathbf{f}}{\partial t} - \mathbf{u}\times curl_{\mathbf{x}}\mathbf{f} + \nabla_{\mathbf{x}}(\mathbf{u}\cdot\mathbf{f})\right)\cdot\mathbf{t}\,d\gamma(t). \quad (1.148)$$

(iii) *Consider the moving continuum medium with velocity field* $\mathbf{u}(\mathbf{x}, t)$ *and let* $\Omega := \Omega(t) \subset \mathbb{R}^3$ *be a three-dimensional domain moving together with the given medium. Then, given a scalar field* $\psi(\mathbf{x}, t)$ *we have*

$$\frac{d}{dt}\left(\iiint \psi\,d\Omega(t)\right) = \iiint\left(\frac{\partial\psi}{\partial t} + \mathrm{div}_{\mathbf{x}}\{\psi\mathbf{u}\}\right)d\Omega(t). \quad (1.149)$$

We prove this Proposition in Chap. 3. Note, however, that the result similar to Proposition 1.7 was obtained before by Cornille in Sect. 19.2 of [1].

1.7 Genuine Gravity and Inertia

Definition 1.9 In general, both in the case of the simplest model of the Newtonian gravity and in the case of any other alternative model of gravity, we always assume that the vectorial gravitational potential $\mathbf{v} := \mathbf{v}(\mathbf{x}, t)$ is an asymptotically acceptable speed-like vector field (see Definition 1.4 and Proposition 1.5). Therefore, Corollary 1.1 implies that there exist a uniquely defined generally trivial speed-like vector field $\mathbf{k} := \mathbf{k}(\mathbf{x}, t) \in \mathbb{R}^3$ (see Definition 1.2) and a uniquely defined proper vector field $\mathbf{h} := \mathbf{h}(\mathbf{x}, t) \in \mathbb{R}^3$, so that in every Cartesian coordinate system we have

$$\lim_{|\mathbf{x}|\to+\infty} \mathbf{h}(\mathbf{x}, t) = 0 \quad \forall t \quad \text{and} \quad \lim_{|\mathbf{x}|\to+\infty} (d_{\mathbf{x}}\mathbf{h}(\mathbf{x}, t)) = 0 \quad \forall t\,, \quad (1.150)$$

and in every Cartesian coordinate system we can decompose vector field $\mathbf{v}$ as

$$\mathbf{v}(\mathbf{x}, t) = \mathbf{h}(\mathbf{x}, t) + \mathbf{k}(\mathbf{x}, t) \qquad \forall(\mathbf{x}, t). \quad (1.151)$$

Then we call $\mathbf{h}$ vectorial potential of genuine gravity and $\mathbf{k}$ vectorial potential of inertia. In particular, it is clear that, given Cartesian coordinate system $(*)$, this system will be inertial if and only if the vectorial potential of inertia $\mathbf{k}$ is a constant (independent on $(\mathbf{x}, t)$) in the coordinate system $(*)$ (see Sect. 4.3 for the details).

The name of genuine gravity and inertia is clarified by the fact that since $\mathbf{k}$ is generally trivial speed-like vector field, by the definition, we can completely eliminate $\mathbf{k}$ (getting $\mathbf{k} = 0$) by the simple change of Cartesian coordinate system (it is fictitious). On the other hand, if $\mathbf{h}$ is nontrivial, since $\mathbf{h}$ is a proper vector field, we cannot make it trivial in any other coordinate system (it is genuine). Moreover, the vector of inertia $\mathbf{k}$ depends only on the coordinate system in the space and it is completely independent on the physical matter or physical fields filling this space. In contrast vectorial potential of genuine gravity **h** depends essentially on the surrounding physical matter (in the model of the Newtonian gravity through gravitational masses).

Next, assume the model of Newtonian gravity, given by (1.21), and assume that (∗) is some inertial Cartesian coordinate system. Then, by (1.17) in the system (∗) we have

$$\begin{cases} \mathrm{curl}_{\mathbf{x}}\mathbf{v} = 0, \\ \frac{\partial \mathbf{v}}{\partial t} + \frac{1}{2}\nabla_{\mathbf{x}}\left(|\mathbf{v}|^2\right) = -\nabla_{\mathbf{x}}\Phi, \end{cases} \tag{1.152}$$

where Φ is a scalar Newtonian gravitational potential which is assumed to be a proper scalar, which satisfies

$$\begin{cases} \Delta_{\mathbf{x}}\Phi = 4\pi G M \qquad \forall\,(\mathbf{x},t), \\ \lim\limits_{|\mathbf{x}|\to+\infty} \nabla_{\mathbf{x}}\Phi(\mathbf{x},t) = 0 \qquad \forall t, \end{cases} \tag{1.153}$$

with M being the gravitational mass density and G being the gravitational constant. Thus inserting (1.151) into (1.152) and using the fact that in the inertial system (∗) the vector **k** is a constant, we deduce

$$\begin{cases} \mathrm{curl}_{\mathbf{x}}\mathbf{h} = 0, \\ \frac{\partial \mathbf{h}}{\partial t} - \mathbf{v}\times \mathrm{curl}_{\mathbf{x}}\mathbf{h} + \nabla_{\mathbf{x}}\left(\mathbf{v}\cdot\mathbf{h}\right) - \nabla_{\mathbf{x}}\left(\frac{1}{2}|\mathbf{h}|^2\right) = -\nabla_{\mathbf{x}}\Phi, \end{cases} \tag{1.154}$$

that we can alternatively rewrite as

$$\begin{cases} \mathrm{curl}_{\mathbf{x}}\mathbf{h} = 0, \\ \frac{\partial \mathbf{h}}{\partial t} - \mathbf{k}\times \mathrm{curl}_{\mathbf{x}}\mathbf{h} + \nabla_{\mathbf{x}}\left(\mathbf{k}\cdot\mathbf{h}\right) + \nabla_{\mathbf{x}}\left(\frac{1}{2}|\mathbf{h}|^2\right) = -\nabla_{\mathbf{x}}\Phi. \end{cases} \tag{1.155}$$

However, since **h** is a proper vector field, by Proposition 1.1 we deduce, that both (1.154) and (1.155) preserve their form also in <u>non-inertial</u> Cartesian coordinate systems, provided Φ is a proper scalar, which is defined in <u>every</u> inertial or non-inertial Cartesian coordinate system by the following:

$$\begin{cases} \Delta_{\mathbf{x}}\Phi = 4\pi G M \qquad \forall\,(\mathbf{x},t), \\ \lim\limits_{|\mathbf{x}|\to+\infty} \nabla_{\mathbf{x}}\Phi(\mathbf{x},t) = 0 \qquad \forall t. \end{cases} \tag{1.156}$$

Finally, by Proposition 1.3 vector field **k** satisfies

$$d_{\mathbf{x}}\mathbf{k} + \{d_{\mathbf{x}}\mathbf{k}\}^T = 0 \qquad \forall\,(\mathbf{x},t), \tag{1.157}$$

in every inertial or non-inertial coordinate system. Therefore the Newtonian gravity in the form (1.21), can be rewritten in the terms of genuine gravity and inertia as (1.156) and either (1.154) or, alternatively, (1.155), that are complemented with (1.157) and (1.151). Moreover, note again that (1.156), (1.154), (1.155), (1.157), and (1.151) are invariant under the change of inertial or non-inertial Cartesian coordinate system.

Next note that, since we have $\operatorname{curl}_{\mathbf{x}}\mathbf{h} = 0$, we can write either (1.154) or (1.155) as the following Hamilton–Jacobi type equation:

$$\begin{cases} \mathbf{h} = \nabla_{\mathbf{x}} Y, \\ \frac{\partial Y}{\partial t} + \mathbf{v} \cdot \nabla_{\mathbf{x}} Y - \frac{1}{2}|\nabla_{\mathbf{x}} Y|^2 = \frac{\partial Y}{\partial t} + \mathbf{k} \cdot \nabla_{\mathbf{x}} Y + \frac{1}{2}|\nabla_{\mathbf{x}} Y|^2 = -\Phi, \end{cases} \tag{1.158}$$

where $Y := Y(\mathbf{x}, t)$ is some scalar field. Then, since $\mathbf{h}$ is a proper vector field and Φ is proper scalar field, by Proposition 1.1 we deduce, that (1.158) is invariant under the changes of inertial or non-inertial Cartesian coordinate systems, provided that we consider Y to be a proper scalar field. Note here that, in contrast to the fact that the quantity Z and Hamilton–Jacobi equation (1.20) were defined only in inertial coordinate systems, the quantity Y and Hamilton–Jacobi equation (1.158) are defined in every inertial or non-inertial coordinate system. Next in every inertial Cartesian coordinate system define the scalar field $X := X(\mathbf{x}, t)$ by

$$X(\mathbf{x}, t) = Z(\mathbf{x}, t) - Y(\mathbf{x}, t) \qquad \forall (\mathbf{x}, t)\,. \tag{1.159}$$

Then, by the first equation in (1.20) and the first equation in (1.158), together with (1.151), we deduce

$$\mathbf{k}(\mathbf{x}, t) = \nabla_{\mathbf{x}} X(\mathbf{x}, t) \qquad \forall (\mathbf{x}, t), \tag{1.160}$$

in every inertial Cartesian coordinate system.

1.8 Lagrangian in the Case of Some Possible Alternative Model of the Gravity

Consider $\mathbf{k}$ to be the vectorial potential of the inertia, which is a generally trivial speed-like vector field, assumed to be fixed in every fixed inertial or non-inertial Cartesian coordinate system (see Definition 1.9). Given known the distribution of inertial mass density of some continuum medium $\mu := \mu(\mathbf{x}, t)$, the field of velocities of this medium $\mathbf{u} := \mathbf{u}(\mathbf{x}, t)$, the charge density $\rho := \rho(\mathbf{x}, t)$ and the current density $\mathbf{j} := \mathbf{j}(\mathbf{x}, t)$ consider a Lagrangian density L defined by

$$L(\mathbf{A},\Psi,\mathbf{v},\Phi_0,\mathbf{h},\mathbf{x},t):=$$
$$\frac{1}{8\pi}\left|-\nabla_{\mathbf{x}}\Psi-\frac{1}{c}\frac{\partial\mathbf{A}}{\partial t}+\frac{1}{c}\mathbf{v}\times\operatorname{curl}_{\mathbf{x}}\mathbf{A}\right|^2-\frac{1}{8\pi}\left|\operatorname{curl}_{\mathbf{x}}\mathbf{A}\right|^2-\left(\rho\Psi-\frac{1}{c}\mathbf{A}\cdot\mathbf{j}\right)+\frac{\mu}{2}\left|\mathbf{u}-\mathbf{v}\right|^2$$
$$-\frac{c^2}{8\pi G}\left|-\nabla_{\mathbf{x}}\Phi_0-\frac{1}{c}\frac{\partial\mathbf{h}}{\partial t}+\frac{1}{c}\mathbf{v}\times\operatorname{curl}_{\mathbf{x}}\mathbf{h}-\frac{1}{c}\nabla_{\mathbf{x}}(\mathbf{h}\cdot\mathbf{v})\right|^2+\frac{c^2}{8\pi G}\left|\operatorname{curl}_{\mathbf{x}}\mathbf{h}\right|^2$$
$$+\frac{c^2\beta}{4\pi G}\left(\frac{1}{4}\left|d_{\mathbf{x}}\mathbf{v}+\{d_{\mathbf{x}}\mathbf{v}\}^T\right|^2-\left|\operatorname{div}_{\mathbf{x}}\mathbf{v}\right|^2\right), \qquad (1.161)$$

where

$$\mathbf{h}=\mathbf{v}-\mathbf{k}\quad\text{and}\quad\Phi_0=-\frac{1}{2c}\left|\mathbf{v}-\mathbf{k}\right|^2, \qquad (1.162)$$

and $\beta\in\mathbb{R}$ is some constant. In other words we have

$$L(\mathbf{A},\Psi,\mathbf{v},\Phi_0,\mathbf{h},\mathbf{x},t)=L_1(\mathbf{A},\Psi,\mathbf{v},\mathbf{x},t):=$$
$$\frac{1}{8\pi}\left|-\nabla_{\mathbf{x}}\Psi-\frac{1}{c}\frac{\partial\mathbf{A}}{\partial t}+\frac{1}{c}\mathbf{v}\times\operatorname{curl}_{\mathbf{x}}\mathbf{A}\right|^2-\frac{1}{8\pi}\left|\operatorname{curl}_{\mathbf{x}}\mathbf{A}\right|^2-\left(\rho\Psi-\frac{1}{c}\mathbf{A}\cdot\mathbf{j}\right)+\frac{\mu}{2}\left|\mathbf{u}-\mathbf{v}\right|^2$$
$$-\frac{c^2}{8\pi G}\left|\frac{1}{c}\nabla_{\mathbf{x}}\left(\frac{1}{2}\left|\mathbf{v}-\mathbf{k}\right|^2\right)-\frac{1}{c}\frac{\partial}{\partial t}(\mathbf{v}-\mathbf{k})+\frac{1}{c}\mathbf{v}\times\operatorname{curl}_{\mathbf{x}}(\mathbf{v}-\mathbf{k})-\frac{1}{c}\nabla_{\mathbf{x}}((\mathbf{v}-\mathbf{k})\cdot\mathbf{v})\right|^2$$
$$+\frac{c^2}{8\pi G}\left|\operatorname{curl}_{\mathbf{x}}(\mathbf{v}-\mathbf{k})\right|^2+\frac{c^2\beta}{4\pi G}\left(\frac{1}{4}\left|d_{\mathbf{x}}\mathbf{v}+\{d_{\mathbf{x}}\mathbf{v}\}^T\right|^2-\left|\operatorname{div}_{\mathbf{x}}\mathbf{v}\right|^2\right). \qquad (1.163)$$

In particular, in the inertial coordinate system where $d_{\mathbf{x}}\mathbf{k}=0$ and $\partial_t\mathbf{k}=0$ we have

$$L(\mathbf{A},\Psi,\mathbf{v},\Phi_0,\mathbf{h},\mathbf{x},t)=L_1(\mathbf{A},\Psi,\mathbf{v},\mathbf{x},t):=$$
$$\frac{1}{8\pi}\left|-\nabla_{\mathbf{x}}\Psi-\frac{1}{c}\frac{\partial\mathbf{A}}{\partial t}+\frac{1}{c}\mathbf{v}\times\operatorname{curl}_{\mathbf{x}}\mathbf{A}\right|^2-\frac{1}{8\pi}\left|\operatorname{curl}_{\mathbf{x}}\mathbf{A}\right|^2-\left(\rho\Psi-\frac{1}{c}\mathbf{A}\cdot\mathbf{j}\right)+\frac{\mu}{2}\left|\mathbf{u}-\mathbf{v}\right|^2$$
$$-\frac{c^2}{8\pi G}\left|-\frac{1}{c}\frac{\partial\mathbf{v}}{\partial t}+\frac{1}{c}\mathbf{v}\times\operatorname{curl}_{\mathbf{x}}\mathbf{v}-\frac{1}{c}\nabla_{\mathbf{x}}\left(\frac{1}{2}\left|\mathbf{v}\right|^2\right)\right|^2+\frac{c^2}{8\pi G}\left|\operatorname{curl}_{\mathbf{x}}\mathbf{v}\right|^2$$
$$+\frac{c^2\beta}{4\pi G}\left(\frac{1}{4}\left|d_{\mathbf{x}}\mathbf{v}+\{d_{\mathbf{x}}\mathbf{v}\}^T\right|^2-\left|\operatorname{div}_{\mathbf{x}}\mathbf{v}\right|^2\right). \qquad (1.164)$$

Note here the advantage of inertial coordinate systems, where L and L_1 are completely independent on the vectorial potential of the inertia $\mathbf{k}$. Furthermore, we rewrite (1.164) as

$$
\begin{aligned}
&L(\mathbf{A},\Psi,\mathbf{v},\Phi_0,\mathbf{h},\mathbf{x},t)=L_1(\mathbf{A},\Psi,\mathbf{v},\mathbf{x},t):=\\
&\frac{1}{8\pi}\left|-\nabla_{\mathbf{x}}\Psi-\frac{1}{c}\frac{\partial\mathbf{A}}{\partial t}+\frac{1}{c}\mathbf{v}\times\operatorname{curl}_{\mathbf{x}}\mathbf{A}\right|^2-\frac{1}{8\pi}|\operatorname{curl}_{\mathbf{x}}\mathbf{A}|^2-\left(\rho\Psi-\frac{1}{c}\mathbf{A}\cdot\mathbf{j}\right)\\
&\quad+\frac{\mu}{2}|\mathbf{u}-\mathbf{v}|^2-\frac{c^2}{8\pi G}\left|-\frac{1}{c}\left(\frac{\partial\mathbf{v}}{\partial t}+d_{\mathbf{x}}\mathbf{v}\cdot\mathbf{v}\right)\right|^2+\frac{c^2}{8\pi G}|\operatorname{curl}_{\mathbf{x}}\mathbf{v}|^2\\
&\qquad+\frac{c^2\beta}{4\pi G}\left(\frac{1}{4}\left|d_{\mathbf{x}}\mathbf{v}+\{d_{\mathbf{x}}\mathbf{v}\}^T\right|^2-|\operatorname{div}_{\mathbf{x}}\mathbf{v}|^2\right).
\end{aligned}\tag{1.165}
$$

Then, using Proposition 1.1 by (1.161)–(1.163) we deduce that L and L_1 are invariant under the change of inertial or non-inertial Cartesian coordinate system, provided that $\mathbf{h}$ and $\mathbf{A}$ are proper vector fields, $\mathbf{v}$ is a speed-like vector field and Φ_0 and $\Psi_0:=\Psi-\frac{1}{c}\mathbf{A}\cdot\mathbf{v}$ are proper scalar fields.

We investigate critical points of the functional

$$
J=\int_0^T\int_{\mathbb{R}^3}L_1(\mathbf{A},\Psi,\mathbf{v},\mathbf{x},t)\,d\mathbf{x}dt.\tag{1.166}
$$

We denote

$$
\begin{cases}
\Psi_0=\Psi-\frac{1}{c}\mathbf{A}\cdot\mathbf{v}\\
\mathbf{D}=-\nabla_{\mathbf{x}}\Psi-\frac{1}{c}\frac{\partial\mathbf{A}}{\partial t}+\frac{1}{c}\mathbf{v}\times\operatorname{curl}_{\mathbf{x}}\mathbf{A}\\
\quad=-\nabla_{\mathbf{x}}\Psi_0-\frac{1}{c}\frac{\partial\mathbf{A}}{\partial t}+\frac{1}{c}\mathbf{v}\times\operatorname{curl}_{\mathbf{x}}\mathbf{A}-\frac{1}{c}\nabla_{\mathbf{x}}(\mathbf{A}\cdot\mathbf{v})\\
\mathbf{B}=\operatorname{curl}_{\mathbf{x}}\mathbf{A}\\
\mathbf{E}=-\nabla_{\mathbf{x}}\Psi-\frac{1}{c}\frac{\partial\mathbf{A}}{\partial t}=\mathbf{D}-\frac{1}{c}\mathbf{v}\times\mathbf{B}\\
\mathbf{H}=\operatorname{curl}_{\mathbf{x}}\mathbf{A}+\frac{1}{c}\mathbf{v}\times\left(-\nabla_{\mathbf{x}}\Psi-\frac{1}{c}\frac{\partial\mathbf{A}}{\partial t}+\frac{1}{c}\mathbf{v}\times\operatorname{curl}_{\mathbf{x}}\mathbf{A}\right)=\mathbf{B}+\frac{1}{c}\mathbf{v}\times\mathbf{D}.
\end{cases}\tag{1.167}
$$

and

$$
\begin{cases}
\mathbf{R}=-\nabla_{\mathbf{x}}\Phi_0-\frac{1}{c}\frac{\partial\mathbf{h}}{\partial t}+\frac{1}{c}\mathbf{v}\times\operatorname{curl}_{\mathbf{x}}\mathbf{h}-\frac{1}{c}\nabla_{\mathbf{x}}(\mathbf{h}\cdot\mathbf{v})\\
\mathbf{Q}=\operatorname{curl}_{\mathbf{x}}\mathbf{h},
\end{cases}\tag{1.168}
$$

where $\mathbf{h}$ is a proper vector field and Φ_0 is a proper scalar field that are given by (1.162). In other words,

$$
\begin{cases}
\mathbf{R}=\frac{1}{c}\nabla_{\mathbf{x}}\left(\frac{1}{2}|\mathbf{v}-\mathbf{k}|^2\right)-\frac{1}{c}\frac{\partial}{\partial t}(\mathbf{v}-\mathbf{k})+\frac{1}{c}\mathbf{v}\times\operatorname{curl}_{\mathbf{x}}(\mathbf{v}-\mathbf{k})-\frac{1}{c}\nabla_{\mathbf{x}}((\mathbf{v}-\mathbf{k})\cdot\mathbf{v})\\
\mathbf{Q}=\operatorname{curl}_{\mathbf{x}}(\mathbf{v}-\mathbf{k}),
\end{cases}\tag{1.169}
$$

and in inertial coordinate system where $d_{\mathbf{x}}\mathbf{k}=0$ and $\partial_t\mathbf{k}=0$ we also have

$$
\begin{cases}
\mathbf{R}=-\frac{1}{c}\left(\frac{\partial\mathbf{v}}{\partial t}+d_{\mathbf{x}}\mathbf{v}\cdot\mathbf{v}\right)\\
\mathbf{Q}=\operatorname{curl}_{\mathbf{x}}\mathbf{v}.
\end{cases}\tag{1.170}
$$

As before, by (1.167) and (1.168) and Proposition 1.1 we infer that both $\mathbf{D}, \mathbf{B}$ and $\mathbf{R}, \mathbf{Q}$ are proper vector fields. Then in Sect. 8.2 we obtain that a configuration $(\mathbf{A}, \Psi, \mathbf{v}, \mathbf{x}, t)$ is a critical point of the functional

$$J = \int_0^T \int_{\mathbb{R}^3} L_1 (\mathbf{A}, \Psi, \mathbf{v}, \mathbf{x}, t)\, d\mathbf{x}dt, \tag{1.171}$$

if and only if it satisfies

$$\begin{cases} \text{curl}_{\mathbf{x}}\mathbf{B} = \frac{4\pi}{c}\mathbf{j} + \frac{1}{c}\frac{\partial \mathbf{D}}{\partial t} - \frac{1}{c}\text{curl}_{\mathbf{x}} (\mathbf{v} \times \mathbf{D}) \\ \text{div}_{\mathbf{x}}\, \mathbf{D} = 4\pi\rho \\ \text{curl}_{\mathbf{x}}\mathbf{D} + \frac{1}{c}\frac{\partial \mathbf{B}}{\partial t} - \frac{1}{c}\, \text{curl}_{\mathbf{x}} (\mathbf{v} \times \mathbf{B}) = 0 \\ \text{div}_{\mathbf{x}}\, \mathbf{B} = 0 \\ \text{curl}_{\mathbf{x}}\mathbf{R} + \frac{1}{c}\frac{\partial \mathbf{Q}}{\partial t} - \frac{1}{c}\text{curl}_{\mathbf{x}} (\mathbf{v} \times \mathbf{Q}) = 0 \\ \text{div}_{\mathbf{x}}\, \mathbf{Q} = 0 \\ \frac{1}{c}\left(\frac{\partial}{\partial t}(\text{div}_{\mathbf{x}}\, \mathbf{v}) + \mathbf{v}\cdot\nabla_{\mathbf{x}}(\text{div}_{\mathbf{x}}\, \mathbf{v}) + \frac{1}{4}\left|d_{\mathbf{x}}\mathbf{v} + \{d_{\mathbf{x}}\mathbf{v}\}^T\right|^2\right) = \frac{1}{2c}\,|\mathbf{Q}|^2 - \text{div}_{\mathbf{x}}\, \mathbf{R} \\ \text{curl}_{\mathbf{x}}(\text{curl}_{\mathbf{x}}\mathbf{v}) = \text{curl}_{\mathbf{x}}\mathbf{Q} \\ \frac{4\pi G}{c^2}\left(\mu\mathbf{u} - \mu\mathbf{v} + \frac{1}{4\pi c}\mathbf{D}\times\mathbf{B} - \frac{c}{4\pi G}\mathbf{R}\times\mathbf{Q}\right) = \\ (1+\beta)\,\text{curl}_{\mathbf{x}}\mathbf{Q} - \frac{1}{c}\left(\frac{\partial \mathbf{R}}{\partial t} - \text{curl}_{\mathbf{x}}\{\mathbf{v}\times\mathbf{R}\} + (\text{div}_{\mathbf{x}}\, \mathbf{R})\,\mathbf{v}\right), \end{cases} \tag{1.172}$$

and by (1.170) in the inertial frame we have

$$\begin{cases} \text{curl}_{\mathbf{x}}\mathbf{B} = \frac{4\pi}{c}\mathbf{j} + \frac{1}{c}\frac{\partial \mathbf{D}}{\partial t} - \frac{1}{c}\text{curl}_{\mathbf{x}} (\mathbf{v} \times \mathbf{D}) \\ \text{div}_{\mathbf{x}}\, \mathbf{D} = 4\pi\rho \\ \text{curl}_{\mathbf{x}}\mathbf{D} + \frac{1}{c}\frac{\partial \mathbf{B}}{\partial t} - \frac{1}{c}\, \text{curl}_{\mathbf{x}} (\mathbf{v} \times \mathbf{B}) = 0 \\ \text{div}_{\mathbf{x}}\, \mathbf{B} = 0 \\ \mathbf{R} = -\frac{1}{c}\left(\frac{\partial \mathbf{v}}{\partial t} + d_{\mathbf{x}}\mathbf{v}\cdot\mathbf{v}\right) \\ \mathbf{Q} = \text{curl}_{\mathbf{x}}\mathbf{v} \\ \frac{4\pi G}{c^2}\left(\mu\mathbf{u} - \mu\mathbf{v} + \frac{1}{4\pi c}\mathbf{D}\times\mathbf{B} - \frac{c}{4\pi G}\mathbf{R}\times\mathbf{Q}\right) = \\ (1+\beta)\,\text{curl}_{\mathbf{x}}\mathbf{Q} - \frac{1}{c}\left(\frac{\partial \mathbf{R}}{\partial t} - \text{curl}_{\mathbf{x}}\{\mathbf{v}\times\mathbf{R}\} + (\text{div}_{\mathbf{x}}\, \mathbf{R})\,\mathbf{v}\right). \end{cases} \tag{1.173}$$

Furthermore, taking $\text{div}_{\mathbf{x}}$ of the both sides of the last equality in (1.172) and using continuum equation $\partial_t\mu + \text{div}_{\mathbf{x}}\,(\mu\mathbf{u}) = 0$ we deduce

$$-\left(\partial_t\mu+\mathrm{div}_{\mathbf{x}}\,(\mu\mathbf{v})\right)+\mathrm{div}_{\mathbf{x}}\left\{\frac{1}{4\pi c}\mathbf{D}\times\mathbf{B}-\frac{c}{4\pi G}\mathbf{R}\times\mathbf{Q}\right\}=$$

$$\mathrm{div}_{\mathbf{x}}\left(\mu\mathbf{u}-\mu\mathbf{v}+\frac{1}{4\pi c}\mathbf{D}\times\mathbf{B}-\frac{c}{4\pi G}\mathbf{R}\times\mathbf{Q}\right)=$$

$$-\frac{c}{4\pi G}\left(\frac{\partial}{\partial t}\left(\mathrm{div}_{\mathbf{x}}\,\mathbf{R}\right)+\mathrm{div}_{\mathbf{x}}\left\{\left(\mathrm{div}_{\mathbf{x}}\,\mathbf{R}\right)\mathbf{v}\right\}\right). \quad (1.174)$$

Therefore, considering the proper scalar quantity Q_0, that we call the field mass, which satisfies

$$Q_0:=-\mu+\frac{c}{4\pi G}\,\mathrm{div}_{\mathbf{x}}\,\mathbf{R}, \quad (1.175)$$

by (1.174) we deduce

$$\frac{\partial Q_0}{\partial t}+\mathrm{div}_{\mathbf{x}}\,(Q_0\mathbf{v})=-\,\mathrm{div}_{\mathbf{x}}\left\{\frac{1}{4\pi c}\mathbf{D}\times\mathbf{B}-\frac{c}{4\pi G}\mathbf{R}\times\mathbf{Q}\right\}. \quad (1.176)$$

Thus, we rewrite (1.172) as

$$\begin{cases}
\mathrm{curl}_{\mathbf{x}}\mathbf{B}=\frac{4\pi}{c}\left(\mathbf{j}-\rho\mathbf{v}\right)+\frac{1}{c}\frac{\partial\mathbf{D}}{\partial t}-\frac{1}{c}\mathrm{curl}_{\mathbf{x}}\left(\mathbf{v}\times\mathbf{D}\right)+\frac{1}{c}\left(\mathrm{div}_{\mathbf{x}}\,\mathbf{D}\right)\mathbf{v},\\
\mathrm{div}_{\mathbf{x}}\,\mathbf{D}=4\pi\rho,\\
\mathrm{curl}_{\mathbf{x}}\mathbf{D}+\frac{1}{c}\frac{\partial\mathbf{B}}{\partial t}-\frac{1}{c}\,\mathrm{curl}_{\mathbf{x}}\left(\mathbf{v}\times\mathbf{B}\right)=0,\\
\mathrm{div}_{\mathbf{x}}\,\mathbf{B}=0,\\
\mathrm{curl}_{\mathbf{x}}\mathbf{R}+\frac{1}{c}\frac{\partial\mathbf{Q}}{\partial t}-\frac{1}{c}\mathrm{curl}_{\mathbf{x}}\left(\mathbf{v}\times\mathbf{Q}\right)=0,\\
\mathrm{div}_{\mathbf{x}}\,\mathbf{Q}=0,\\
\frac{4\pi G}{c^2}\left(\mu\mathbf{u}-\mu\mathbf{v}+\frac{1}{4\pi c}\mathbf{D}\times\mathbf{B}-\frac{c}{4\pi G}\mathbf{R}\times\mathbf{Q}\right)=\\
(1+\beta)\,\mathrm{curl}_{\mathbf{x}}\mathbf{Q}-\frac{1}{c}\left(\frac{\partial\mathbf{R}}{\partial t}-\mathrm{curl}_{\mathbf{x}}\left\{\mathbf{v}\times\mathbf{R}\right\}+\left(\mathrm{div}_{\mathbf{x}}\,\mathbf{R}\right)\mathbf{v}\right),\\
\mathrm{div}_{\mathbf{x}}\,\mathbf{R}=\frac{4\pi G}{c}\left(\mu+Q_0\right),\\
\frac{1}{c}\left(\frac{\partial}{\partial t}\left(\mathrm{div}_{\mathbf{x}}\,\mathbf{v}\right)+\mathbf{v}\cdot\nabla_{\mathbf{x}}\left(\mathrm{div}_{\mathbf{x}}\,\mathbf{v}\right)+\frac{1}{4}\left|d_{\mathbf{x}}\mathbf{v}+\left\{d_{\mathbf{x}}\mathbf{v}\right\}^T\right|^2\right)=\frac{1}{2c}\left|\mathbf{Q}\right|^2-\mathrm{div}_{\mathbf{x}}\,\mathbf{R},\\
\mathrm{curl}_{\mathbf{x}}(\mathrm{curl}_{\mathbf{x}}\mathbf{v})=\mathrm{curl}_{\mathbf{x}}\mathbf{Q},\\
\frac{\partial Q_0}{\partial t}+\mathrm{div}_{\mathbf{x}}\,(Q_0\mathbf{v})=-\,\mathrm{div}_{\mathbf{x}}\left\{\frac{1}{4\pi c}\mathbf{D}\times\mathbf{B}-\frac{c}{4\pi G}\mathbf{R}\times\mathbf{Q}\right\},
\end{cases} \quad (1.177)$$

and we rewrite (1.173) in the inertial frame as

$$\begin{cases} \mathrm{curl}_{\mathbf{x}}\mathbf{B} = \frac{4\pi}{c}\mathbf{j} + \frac{1}{c}\frac{\partial \mathbf{D}}{\partial t} - \frac{1}{c}\mathrm{curl}_{\mathbf{x}}\left(\mathbf{v}\times\mathbf{D}\right), \\ \mathrm{div}_{\mathbf{x}}\,\mathbf{D} = 4\pi\rho, \\ \mathrm{curl}_{\mathbf{x}}\mathbf{D} + \frac{1}{c}\frac{\partial \mathbf{B}}{\partial t} - \frac{1}{c}\,\mathrm{curl}_{\mathbf{x}}\left(\mathbf{v}\times\mathbf{B}\right) = 0, \\ \mathrm{div}_{\mathbf{x}}\,\mathbf{B} = 0, \\ \mathbf{R} = -\frac{1}{c}\left(\frac{\partial \mathbf{v}}{\partial t} + d_{\mathbf{x}}\mathbf{v}\cdot\mathbf{v}\right), \\ \mathbf{Q} = \mathrm{curl}_{\mathbf{x}}\mathbf{v}, \\ \frac{4\pi G}{c^2}\left(\mu\mathbf{u} - \mu\mathbf{v} + \frac{1}{4\pi c}\mathbf{D}\times\mathbf{B} - \frac{c}{4\pi G}\mathbf{R}\times\mathbf{Q}\right) = \\ (1+\beta)\,\mathrm{curl}_{\mathbf{x}}\mathbf{Q} - \frac{1}{c}\left(\frac{\partial \mathbf{R}}{\partial t} - \mathrm{curl}_{\mathbf{x}}\left\{\mathbf{v}\times\mathbf{R}\right\} + \left(\mathrm{div}_{\mathbf{x}}\,\mathbf{R}\right)\mathbf{v}\right), \\ \mathrm{div}_{\mathbf{x}}\,\mathbf{R} = \frac{4\pi G}{c}\left(\mu + Q_0\right), \\ \frac{\partial Q_0}{\partial t} + \mathrm{div}_{\mathbf{x}}\left(Q_0\mathbf{v}\right) = -\,\mathrm{div}_{\mathbf{x}}\left\{\frac{1}{4\pi c}\mathbf{D}\times\mathbf{B} - \frac{c}{4\pi G}\mathbf{R}\times\mathbf{Q}\right\}. \end{cases} \tag{1.178}$$

As before, by Proposition 1.1 we deduce that (1.177) is invariant under the change of inertial or non-inertial Cartesian coordinate systems. Moreover, (1.178) is invariant under the change of inertial Cartesian coordinate systems. Note also that, both, (1.177) in an arbitrary inertial or non-inertial Cartesian coordinate system and (1.178) in an arbitrary inertial Cartesian coordinate system, are completely independent on the vectorial potential of the inertia $\mathbf{k}$.

Finally, note that in the case of large constant $|\beta| \gg 1$ we have $\mathbf{Q} \to 0$ in (1.177) and thus, the gravity Eq. (1.177) reduce to the equations of the Newtonian-type Gravity in the form of (1.114). In that case the gravity field propagates with the infinite speed. On the other hand, in the case of vanishing constant $\beta = 0$ the form of equations for $\mathbf{R}$ and $\mathbf{Q}$ in (1.177) is completely the same as the form of the Maxwell equations for $\mathbf{D}$ and $\mathbf{B}$ in (1.177), except of the different meaning of "charges" and "currents" in these two sets of equations. In that case the electromagnetic and the gravity fields propagate with the same speed. However, in the mixed case of constant $\beta \sim 1$ the electromagnetic and the gravity fields propagate with different finite speeds.

1.9 Covariant Formulation of the Physical Laws in the Four-Dimensional Space-Time

In Chap. 9 we present the covariant (tensor) formulations of the Maxwell Equations and the Lagrangian density of the electromagnetic field and the covariant form of the Lagrangian of motion of charged particles in the outer gravitational and electromagnetic fields.

1.9.1 Four-Vectors, Four-Covectors and Tensors in the Four-Dimensional Non-relativistic Space-Time

First of all we would like to remind the definitions of the vectors, covectors and covariant and contravariant tensors of second order in $\mathbb{R}^4$.

Definition 1.10 Given $\mathcal{S}$, that is a certain subgroup of the group of all smooth non-degenerate invertible transformations from $\mathbb{R}^4$ onto $\mathbb{R}^4$ having the form

$$\begin{cases} x'^0 = f^{(0)}(x^0, x^1, x^2, x^3), \\ x'^1 = f^{(1)}(x^0, x^1, x^2, x^3), \\ x'^2 = f^{(2)}(x^0, x^1, x^2, x^3), \\ x'^3 = f^{(3)}(x^0, x^1, x^2, x^3), \end{cases} \tag{1.179}$$

we say that a one-component field $a := a(x^0, x^1, x^2, x^3)$ is a scalar field on the group $\mathcal{S}$, if under the coordinate transformation in the group $\mathcal{S}$ of the form (1.179) this field transforms as

$$a' = a. \tag{1.180}$$

Next we say that a four-component field (a^0, a^1, a^2, a^3) is a four-vector field on the group $\mathcal{S}$, if under the coordinate transformation in the group $\mathcal{S}$ of the form (1.179) every of four components of this field transforms as

$$a'^j = \sum_{k=0}^{3} \frac{\partial f^{(j)}}{\partial x^k} a^k \qquad \forall j = 0, 1, 2, 3. \tag{1.181}$$

Next we say that a four-component field (a_0, a_1, a_2, a_3) is a four-covector field on the group $\mathcal{S}$, if under the coordinate transformation in the group $\mathcal{S}$ of the form (1.179) every of four components of this field transforms as

$$a_j = \sum_{k=0}^{3} \frac{\partial f^{(k)}}{\partial x^j} a'_k \qquad \forall j = 0, 1, 2, 3. \tag{1.182}$$

Furthermore, we say that a 16-component field $\{a_{mn}\}_{m,n=0,1,2,3}$ is a two times covariant tensor field on the group $\mathcal{S}$, if under the coordinate transformation in the group $\mathcal{S}$ of the form (1.179) every of 16 components of this field transforms as

$$a_{mn} = \sum_{j=0}^{3} \sum_{k=0}^{3} \frac{\partial f^{(k)}}{\partial x^m} \frac{\partial f^{(j)}}{\partial x^n} a'_{kj} \qquad \forall m, n = 0, 1, 2, 3. \tag{1.183}$$

Next we say that a 16-component field $\{a^{mn}\}_{m,n=0,1,2,3}$ is a two times contravariant tensor field on the group $\mathcal{S}$, if under the coordinate transformation in the group $\mathcal{S}$ of the form (1.179) every of 16 components of this field transforms as

$$a'^{mn} = \sum_{j=0}^{3}\sum_{k=0}^{3} \frac{\partial f^{(m)}}{\partial x^k}\frac{\partial f^{(n)}}{\partial x^j} a^{kj} \quad \forall m, n = 0, 1, 2, 3. \tag{1.184}$$

Next consider the four-dimensional space-time $\mathbb{R}^4$, such that for every point in space $\mathbf{x} = (x_1, x_2, x_3) \in \mathbb{R}^3$ and every instant of time t we correspond the point $(x^0, x^1, x^2, x^3) \in \mathbb{R}^4$ that has the form:

$$(x^0, x^1, x^2, x^3) := (ct, x_1, x_2, x_3) = (ct, \mathbf{x}), \tag{1.185}$$

where c is the universal constant in Maxwell equations for vacuum. In this space we denote by $\mathcal{S}_0$, the subgroup of the group of smooth non-degenerate invertible mappings, containing transformations of the form

$$\begin{cases} x'^0 = x^0 \\ x'^j = \sum\limits_{k=1}^{3} A_{jk}\left(\frac{x^0}{c}\right) x_k + z_j\left(\frac{x^0}{c}\right) \quad \forall j = 1, 2, 3, \end{cases} \tag{1.186}$$

where

$$\left\{A_{jk}(t)\right\}_{j,k=1,2,3} = A(t) : \mathbb{R} \to SO(3)$$

is a rotation, smoothly dependent on t and

$$(z_1(t), z_2(t), z_3(t)) = \mathbf{z}(t) : \mathbb{R} \to \mathbb{R}^3$$

also smoothly dependent on t. Then in the terms of time t and three-dimensional space we rewrite (1.186) as (1.2), i.e.:

$$\begin{cases} \mathbf{x}' = A(t) \cdot \mathbf{x} + \mathbf{z}(t), \\ t' = t, \end{cases} \tag{1.187}$$

where $A(t) \in SO(3)$ is a rotation. That is, the group $\mathcal{S}_0$ represents all transformations of Cartesian non-inertial coordinate systems in the non-relativistic space-time. It can be easily checked by trivial calculations that $\mathcal{S}_0$ is indeed a group, i.e. for every two transformations $f, g \in \mathcal{S}_0$ the composition $g \circ f$ and the inverse transformation $f^{(-1)}$ are also contained in $\mathcal{S}_0$, mean that they also have a form of (1.186). Next assume that a four-covector (a_0, a_1, a_2, a_3) and a four-vector (b^0, b^1, b^2, b^3) on the group $\mathcal{S}_0$ are given. Then, by inserting (1.186) into (1.181) and (1.182) in Chap. 9 we obtained the following laws of transformation of four-covectors and four-vectors on the group $\mathcal{S}_0$, i.e. under the change of non-inertial Cartesian coordinate systems:

$$\begin{cases} a'_0 = a_0 - \sum_{k=1}^{3} \frac{1}{c} \left(\sum_{j=1}^{3} \frac{dA_{kj}}{dt} \left(\frac{x^0}{c}\right) x_j + \frac{dz_k}{dt} \left(\frac{x^0}{c}\right) \right) \left(\sum_{j=1}^{3} A_{kj} \left(\frac{x^0}{c}\right) a_j \right) \\ a'_k = \sum_{j=1}^{3} A_{kj} \left(\frac{x^0}{c}\right) a_j \quad \forall k = 1, 2, 3, \end{cases} \tag{1.188}$$

and

$$\begin{cases} b'^0 = b^0 \\ b'^j = \frac{1}{c} \left(\sum_{k=1}^{3} \frac{dA_{jk}}{dt} \left(\frac{x^0}{c}\right) x_k + \frac{dz_j}{dt} \left(\frac{x^0}{c}\right) \right) b^0 + \sum_{k=1}^{3} A_{jk} \left(\frac{x^0}{c}\right) b^k \quad \forall j = 1, 2, 3. \end{cases} \tag{1.189}$$

Therefore, if we denote the four-vector (b^0, b^1, b^2, b^3) and the four-covector (a_0, a_1, a_2, a_3) on the group $\mathcal{S}_0$ as

$$\begin{cases} (b^0, b^1, b^2, b^3) = \left(\sigma, \frac{1}{c}\mathbf{b}\right) \quad \text{where} \quad \sigma := b^0 \quad \text{and} \quad \mathbf{b} := c(b^1, b^2, b^3) \in \mathbb{R}^3, \\ (a_0, a_1, a_2, a_3) = (\psi, -\mathbf{a}) \quad \text{where} \quad \psi := a_0 \quad \text{and} \quad \mathbf{a} := -(a_1, a_2, a_3) \in \mathbb{R}^3, \end{cases} \tag{1.190}$$

then by (1.188) and (1.189) in the terms of time t and three-dimensional space $\mathbf{x}$, we obtain the following laws of transformations of σ, $\mathbf{b}$, ψ and $\mathbf{a}$ under the change of non-inertial Cartesian coordinate system:

$$\begin{cases} \sigma' = \sigma \\ \mathbf{b}' = A(t) \cdot \mathbf{b} + \left(\frac{dA}{dt}(t) \cdot \mathbf{x} + \frac{d\mathbf{z}}{dt}(t) \right) \sigma, \end{cases} \tag{1.191}$$

and

$$\begin{cases} \psi' = \psi + \frac{1}{c} \left(\frac{dA}{dt}(t) \cdot \mathbf{x} + \frac{d\mathbf{z}}{dt}(t) \right) \cdot (A(t) \cdot \mathbf{a}) \\ \mathbf{a}' = A(t) \cdot \mathbf{a}. \end{cases} \tag{1.192}$$

In particular, if $\sigma := b^0$ is the first coordinate of an arbitrary four-vector (b^0, b^1, b^2, b^3) on the group $\mathcal{S}_0$, then σ is a proper scalar field in the frames of Definition 1.1. Moreover, if $\mathbf{a} := -(a_1, a_2, a_3)$, where a_1, a_2, a_3 are the last three coordinates of an arbitrary four-covector (a_0, a_1, a_2, a_3) on the group $\mathcal{S}_0$, then $\mathbf{a}$ is a proper vector field in the frames of Definition 1.1.

Next, since by Definition 1.1 every three-dimensional speed-like vector field, $\mathbf{u}$ transforms under the change of non-inertial Cartesian coordinate system as

$$\mathbf{u}' = A(t) \cdot \mathbf{u} + \frac{dA}{dt}(t) \cdot \mathbf{x} + \frac{d\mathbf{z}}{dt}(t), \tag{1.193}$$

by comparing (1.193) with (1.191) we deduce that for every speed-like vector field $\mathbf{u}$ the four-component field (u^0, u^1, u^2, u^3) defined by

$$(u^0, u^1, u^2, u^3) := \left(1, \frac{1}{c}\mathbf{u}\right) \quad \text{where} \quad u^0 = 1 \quad \text{and} \quad (u^1, u^2, u^3) = \frac{1}{c}\mathbf{u} \in \mathbb{R}^3 \tag{1.194}$$

is a four-vector field on the group $\mathcal{S}_0$. We call such four-vectors by the name vectors of type 1. In particular, if $\mathbf{u}$ is the velocity field, then the quantity defined by (1.194) is a a four-vector field on the group $\mathcal{S}_0$ that we call the four-dimensional speed. Thus, in particular, if $\mathbf{r}(t) = (r_1(t), r_2(t), r_3(t))$ is a three-dimensional trajectory of the motion of some particle, parameterized by the global time t, then if we consider a curve $\frac{1}{c}(ct, r_1(t), r_2(t), r_3(t))$ in $\mathbb{R}^4$, parameterized by the global time t, then the four-component field:

$$\left(1, \frac{1}{c}\frac{d\mathbf{r}}{dt}(t)\right) := \left(1, \frac{1}{c}\frac{dr_1}{dt}(t), \frac{1}{c}\frac{dr_2}{dt}(t), \frac{1}{c}\frac{dr_3}{dt}(t)\right) \tag{1.195}$$

is a four-vector field on the group $\mathcal{S}_0$.

Similarly, if $\mathbf{v}$ is the vectorial gravitational potential, then since $\mathbf{v}$ is a speed-like vector field, the four-component field (v^0, v^1, v^2, v^3) defined by

$$(v^0, v^1, v^2, v^3) := \left(1, \frac{1}{c}\mathbf{v}\right) \quad \text{where} \quad v^0 = 1 \quad \text{and} \quad (v^1, v^2, v^3) = \frac{1}{c}\mathbf{v} \tag{1.196}$$

is also a four-vector field on the group $\mathcal{S}_0$ that we call the four-dimensional gravitational potential. In the same way, if $\mathbf{k}$ is the vectorial potential of inertia, defined by Definition 1.9, then since $\mathbf{k}$ is a speed-like vector field, the four-component field (k^0, k^1, k^2, k^3) defined by

$$(k^0, k^1, k^2, k^3) := \left(1, \frac{1}{c}\mathbf{k}\right) \quad \text{where} \quad k^0 = 1 \quad \text{and} \quad (k^1, k^2, k^3) = \frac{1}{c}\mathbf{k} \tag{1.197}$$

is also a four-vector field on the group $\mathcal{S}_0$ that we call the four-dimensional potential of inertia.

Moreover, by (1.191), if we consider the field of four-dimensional moment of a particle (p^0, p^1, p^2, p^3) defined by

$$(p^0, p^1, p^2, p^3) := \left(m, \frac{1}{c}(m\mathbf{u})\right) \quad \text{where} \quad p^0 = m \quad \text{and} \quad (p^1, p^2, p^3) = \frac{1}{c}(m\mathbf{u}), \tag{1.198}$$

where m is the mass of the particle and $\mathbf{u}$ is the velocity of the particle, then (p^0, p^1, p^2, p^3) is also a four-vector on the group $\mathcal{S}_0$. Moreover, by comparing (1.44) with (1.191) we deduce that if we consider the field of four-dimensional electric current (j^0, j^1, j^2, j^3) defined by

$$(j^0, j^1, j^2, j^3) := \left(\rho, \frac{1}{c}\mathbf{j}\right) \quad \text{where} \quad j^0 = \rho \quad \text{and} \quad (j^1, j^2, j^3) = \frac{1}{c}\mathbf{j}, \tag{1.199}$$

where ρ is the electric charge density and $\mathbf{j}$ is the electric current density, then (j^0, j^1, j^2, j^3) is also a four-vector on the group $\mathcal{S}_0$.

On the other hand, for every proper three-dimensional vector field $\mathbf{G}$ that satisfies due to Definition 1.1:

$$\mathbf{G}' = A(t) \cdot \mathbf{G}, \tag{1.200}$$

by comparing (1.200) with (1.191) we deduce that the four-component field (G^0, G^1, G^2, G^3) defined by

$$(G^0, G^1, G^2, G^3) := (0, \mathbf{G}) \quad \text{where} \quad G^0 = 0 \text{ and } (G^1, G^2, G^3) = \mathbf{G} \tag{1.201}$$

is also a four-vector field on the group $\mathcal{S}_0$. We call such four-vectors by the name vectors of type 0.

Next, since by (1.51) the scalar electromagnetic potential Ψ and the vector electromagnetic potential $\mathbf{A}$, under the change of non-inertial Cartesian coordinate system transform as

$$\begin{cases} \Psi' = \Psi + \frac{1}{c}\left(\frac{dA}{dt}(t) \cdot \mathbf{x} + \frac{dz}{dt}(t)\right) \cdot (A(t) \cdot \mathbf{A}) \\ \mathbf{A}' = A(t) \cdot \mathbf{A}, \end{cases} \tag{1.202}$$

by comparing (1.202) with (1.192) we deduce that the four-component field (A_0, A_1, A_2, A_3) defined as

$$(A_0, A_1, A_2, A_3) = (\Psi, -\mathbf{A}) \quad \text{where} \quad A_0 = \Psi \text{ and } (A_1, A_2, A_3) = -\mathbf{A} \tag{1.203}$$

is a four-covector field on the group $\mathcal{S}_0$. We call this four-covector field by the name four-dimensional electromagnetic potential. Next, since (A_0, A_1, A_2, A_3) is a four-covector field on the group $\mathcal{S}_0$, then it is well known from the tensor analysis that the 16-component field $\{F_{ij}\}_{0 \le i,j \le 3}$ defined in every non-inertial Cartesian coordinate system by

$$F_{ij} := \frac{\partial A_j}{\partial x^i} - \frac{\partial A_i}{\partial x^j} \qquad \forall i, j = 0, 1, 2, 3 \tag{1.204}$$

is an antisymmetric two times covariant tensor field on the group $\mathcal{S}_0$, which we call the covariant tensor of the electromagnetic field. In particular, by inserting (1.203) and (1.185) into (1.204) and denoting:

$$\begin{cases} (B_1, B_2, B_3) = \mathbf{B} := \text{curl}_{\mathbf{x}} \mathbf{A}, \\ (E_1, E_2, E_3) = \mathbf{E} := -\nabla_{\mathbf{x}} \Psi - \frac{1}{c}\frac{\partial \mathbf{A}}{\partial t}, \end{cases} \tag{1.205}$$

we deduce

$$\begin{cases} F_{00} = 0 \\ F_{0j} = -F_{j0} = E_j \quad \forall j = 1, 2, 3 \\ F_{jj} = 0 \quad \forall j = 1, 2, 3 \\ F_{12} = -F_{21} = -B_3 \\ F_{13} = -F_{31} = B_2 \\ F_{23} = -F_{32} = -B_1 . \end{cases} \tag{1.206}$$

Next assume that $T := \left\{T_{ij}\right\}_{i,j=1,2,3} \in \mathbb{R}^{3\times 3}$ is a 9-component proper matrix-valued field, which, being a proper matrix field, by Definition 1.1 satisfies

$$T' = A(t) \cdot T \cdot A^T(t) = A(t) \cdot T \cdot \{A(t)\}^{-1} . \tag{1.207}$$

Next consider a 16-component field $\{\mathcal{T}^{ij}\}_{0\leq i,j\leq 3}$ defined in every non-inertial Cartesian coordinate system by

$$\begin{cases} \mathcal{T}^{00} = 0 \\ \mathcal{T}^{0j} = \mathcal{T}^{j0} = 0 \quad \forall j = 1, 2, 3 \\ \mathcal{T}^{ij} := T_{ij} \quad \forall i, j = 1, 2, 3 , \end{cases} \tag{1.208}$$

Then, by inserting (1.186) and (1.207) into (1.184) in Chap. 9 we prove that the field $\{\mathcal{T}^{ij}\}_{0\leq i,j\leq 3}$ defined by (1.208) is a two times contravariant tensor field on the group $\mathcal{S}_0$.

In particular, if we consider the 9-component matrix field I that defined in every Cartesian coordinate system as $I := \left\{\delta_{ij}\right\}_{1,j=1,2,3} \in \mathbb{R}^{3\times 3}$, where

$$\delta_{ij} := \begin{cases} 1 & \text{if} \quad i = j \\ 0 & \text{if} \quad i \neq j, \end{cases} \tag{1.209}$$

which is a proper matrix field, since

$$I = A(t) \cdot I \cdot \{A(t)\}^{-1} , \tag{1.210}$$

then the 16-component field $\{\Theta^{ij}\}_{0\leq i,j\leq 3}$ defined in every non-inertial Cartesian coordinate system by

$$\begin{cases} \Theta^{00} = 0 \\ \Theta^{0j} = \Theta^{j0} = 0 \quad \forall j = 1, 2, 3 \\ \Theta^{ij} := \delta_{ij} \quad \forall i, j = 1, 2, 3 \end{cases} \tag{1.211}$$

is a two times contravariant tensor field on the group $\mathcal{S}_0$ and moreover, this tensor is symmetric. We call $\{\Theta^{ij}\}_{0\leq i,j\leq 3}$ the contravariant tensor of the three-dimensional geometry.

Next, the scalar field $\tau := \tau(x^0, x^1, x^2, x^3)$, defined in every Cartesian coordinate system as

$$\tau := \frac{x^0}{c} = t \tag{1.212}$$

is a scalar on the group $\mathcal{S}_0$. Here t is the global non-relativistic time. Moreover, by (1.212), the four-component field (v_0, v_1, v_2, v_3) defined as a gradient of the global time by

$$v_0 := c\,\frac{\partial \tau}{\partial x^0}(x^0, x^1, x^2, x^3) = 1 \quad \text{and} \quad v_j := c\,\frac{\partial \tau}{\partial x^j}(x^0, x^1, x^2, x^3) = 0 \qquad \forall j = 1, 2, 3 \tag{1.213}$$

is a four-covector field on the group $\mathcal{S}_0$.

Finally, consider a motion of a classical particle with inertial mass m, charge σ, place $\mathbf{r}(t)$ and velocity $\mathbf{u}(t) = d\mathbf{r}/dt(t)$ in the outer gravitational field with the vectorial gravitational potential $\mathbf{v}(\mathbf{x}, t)$, the outer electromagnetic field with vectorial and scalar potentials $\mathbf{A}(\mathbf{x}, t)$ and $\Psi(\mathbf{x}, t)$, and additional conservative field with scalar potential $V(\mathbf{x}, t)$ ruled by a Lagrangian (1.69):

$$L_0\left(\frac{d\mathbf{r}}{dt}, \mathbf{r}, t\right) := \frac{m}{2}\left|\frac{d\mathbf{r}}{dt} - \mathbf{v}(\mathbf{r}, t)\right|^2 - \sigma\left(\Psi(\mathbf{r}, t) - \frac{1}{c}\mathbf{A}(\mathbf{r}, t)\cdot\frac{d\mathbf{r}}{dt}\right) + V(\mathbf{r}, t). \tag{1.214}$$

Then L_0 is a scalar on the group $\mathcal{S}_0$. Moreover, consider the generalized momentum of the particle m by (1.72):

$$\mathbf{P} := \nabla_{\mathbf{r}'} L_0\left(\mathbf{r}', \mathbf{r}, t\right) = m\frac{d\mathbf{r}}{dt} - m\mathbf{v}(\mathbf{r}, t) + \frac{\sigma}{c}\mathbf{A}(\mathbf{r}, t) \tag{1.215}$$

consider a Hamiltonian

$$H_0\left(\mathbf{P}, \mathbf{r}, t\right) := \mathbf{P}\cdot\frac{d\mathbf{r}}{dt} - L_0\left(\frac{d\mathbf{r}}{dt}, \mathbf{r}, t\right), \tag{1.216}$$

which by (1.74) satisfies

$$H_0\left(\mathbf{P}, \mathbf{r}, t\right) = \mathbf{P}\cdot\mathbf{v}(\mathbf{r}, t) + \frac{1}{2m}\left|\mathbf{P} - \frac{\sigma}{c}\mathbf{A}(\mathbf{r}, t)\right|^2 + \sigma\left(\Psi(\mathbf{r}, t) - \frac{1}{c}\mathbf{A}(\mathbf{r}, t)\cdot\mathbf{v}(\mathbf{r}, t)\right) - V\left(\mathbf{r}, t\right), \tag{1.217}$$

and furthermore, define the four-dimensional generalized momentum (P_0, P_1, P_2, P_3) as

$$(P_0, P_1, P_2, P_3) := \left(\frac{1}{c}H_0, -\mathbf{P}\right) \quad \text{where} \quad P_0 = \frac{1}{c}H_0 \text{ and } (P_1, P_2, P_3) = -\mathbf{P}. \tag{1.218}$$

Then, since by (1.217) and (1.215), under the change of non-inertial Cartesian coordinate system H_0 and $\mathbf{P}$ transform as

$$\begin{cases} H'_0 = H_0 + \left(\frac{dA}{dt}(t) \cdot \mathbf{x} + \frac{d\mathbf{z}}{dt}(t)\right) \cdot (A(t) \cdot \mathbf{P}) \\ \mathbf{P}' = A(t) \cdot \mathbf{P}, \end{cases} \tag{1.219}$$

by comparing (1.219) with (1.192) we deduce that the four-dimensional momentum (P_0, P_1, P_2, P_3) is a four-covector on the group $\mathcal{S}_0$.

1.9.2 Pseudo-metric Tensors of the Four-Dimensional Space-Time

Consider $\{g^{ij}\}_{0\le i,j\le 3}$ to be a two times contravariant tensor field on the group $\mathcal{S}_0$, defined by

$$g^{ij} := v^i v^j - \Theta^{ij} \qquad \forall i, j = 0, 1, 2, 3, \tag{1.220}$$

where $\{\Theta^{ij}\}_{0\le i,j\le 3}$ is the contravariant tensor of the three-dimensional geometry, defined by (1.211) and being two times contravariant tensor, and (v^0, v^1, v^2, v^3) is the four-dimensional gravitational potential, defined by (1.196) and being a four-vector. Then, in Chap. 9 we obtain that $\{g^{ij}\}_{0\le i,j\le 3}$ is indeed two times contravariant tensor field on the group $\mathcal{S}_0$ and moreover, this tensor is symmetric. Moreover, by (1.211) and (1.196) we have

$$\begin{cases} g^{00} = 1 \\ g^{ij} = -\delta_{ij} + \frac{v^i v^j}{c^2} \quad \forall 1 \le i, j \le 3 \\ g^{0j} = g^{j0} = \frac{v^j}{c} \quad \forall 1 \le j \le 3, \end{cases} \tag{1.221}$$

where $\mathbf{v} = (v^1, v^2, v^3)$ is the three-dimensional vectorial gravitational potential. We call the tensor $\{g^{ij}\}_{0\le i,j\le 3}$ the contravariant pseudo-metric tensor of the four-dimensional space-time. Next consider a 16-component field $\{g_{ij}\}_{0\le i,j\le 3}$ defined by

$$\begin{cases} g_{00} = 1 - \frac{|\mathbf{v}|^2}{c^2} \\ g_{ij} = -\delta_{ij} \quad \forall 1 \le i, j \le 3 \\ g_{0j} = g_{j0} = \frac{v^j}{c} \quad \forall 1 \le j \le 3. \end{cases} \tag{1.222}$$

Then in Chap. 9 we deduce

$$\sum_{k=0}^{3} g^{ik} g_{kj} = \begin{cases} 1 & \text{if } i = j \\ 0 & \text{if } i \neq j \end{cases} \qquad \forall i, j = 0, 1, 2, 3. \tag{1.223}$$

Therefore, we obtain that $\{g_{ij}\}_{i,j=0,1,2,3}$ is a two times covariant tensor on the group $\mathcal{S}_0$, and moreover, this tensor is symmetric. We call the tensor $\{g_{ij}\}_{0\leq i,j\leq 3}$ covariant pseudo-metric tensor of the four-dimensional space-time. Using (1.223) we also obtain that the pseudo-metric tensors $\{g_{ij}\}_{i,j=0,1,2,3}$ and $\{g^{ij}\}_{0\leq i,j\leq 3}$ are non-degenerate. Moreover, it can be easily calculated that if we consider the 4×4-matrix:

$$G = \{g_{ij}\}_{0\leq i,j\leq 3}, \tag{1.224}$$

then

$$\det G = -1. \tag{1.225}$$

Thus, with the covariant and contravariant pseudo-metric tensors we can lower and lift indexes of arbitrary tensors. In particular given a four-covector (a_0, a_1, a_2, a_3) and a four-vector (b^0, b^1, b^2, b^3) on the group $\mathcal{S}_0$ we can define the corresponding lifted four-vector (a^0, a^1, a^2, a^3) and the corresponded lowered four-covector (b_0, b_1, b_2, b_3) by

$$(a^0, a^1, a^2, a^3) := \left\{\sum_{k=0}^{3} g^{mk} a_k\right\}_{m=0,1,2,3}$$

$$\text{and} \quad (b_0, b_1, b_2, b_3) := \left\{\sum_{k=0}^{3} g_{mk} b^k\right\}_{m=0,1,2,3} \tag{1.226}$$

Then by (1.221), (1.222) and (1.226) we have

$$a^0 = a_0 + \sum_{k=1}^{3} \frac{1}{c} v^k a_k \quad \text{and} \quad a^m = -a_m + \frac{1}{c} a^0 v^m \qquad \forall m = 1, 2, 3, \tag{1.227}$$

and

$$b_0 = b^0 - \sum_{k=1}^{3} \frac{1}{c} v^k b_k \quad \text{and} \quad b_m = -b^m + \frac{1}{c} b^0 v^m \qquad \forall m = 1, 2, 3. \tag{1.228}$$

In particular, we have

$$b^0 a_0 + \sum_{k=1}^{3} b^k a_k = b^0 a^0 - \sum_{k=1}^{3} b_k a_k. \tag{1.229}$$

Next, if for every speed-like vector field $\mathbf{u}$ we consider the four-vector field (u^0, u^1, u^2, u^3) defined by (1.194) as

$$(u^0, u^1, u^2, u^3) := \left(1, \frac{1}{c}\mathbf{u}\right) \quad \text{where} \quad u^0 = 1 \text{ and } (u^1, u^2, u^3) = \frac{1}{c}\mathbf{u} \in \mathbb{R}^3, \tag{1.230}$$

then, by (1.228) the corresponding lowered four-covector field (u_0, u_1, u_2, u_3) satisfies

$$(u_0, u_1, u_2, u_3) := \left(1 + \frac{1}{c^2}(\mathbf{u}-\mathbf{v})\cdot\mathbf{v}, -\frac{1}{c}(\mathbf{u}-\mathbf{v})\right) \quad \text{where}$$
$$u_0 = 1 + \frac{1}{c^2}(\mathbf{u}-\mathbf{v})\cdot\mathbf{v} \text{ and } (u_1, u_2, u_3) = -\frac{1}{c}(\mathbf{u}-\mathbf{v}) \in \mathbb{R}^3. \tag{1.231}$$

Moreover, in the case where (u^0, u^1, u^2, u^3) is a four-dimensional speed, we call the corresponding lowered four-covector field (u_0, u_1, u_2, u_3) by the name four-dimensional cospeed. In particular, if we consider the four-dimensional gravitational potential (v^0, v^1, v^2, v^3) defined by (1.196):

$$(v^0, v^1, v^2, v^3) := \left(1, \frac{1}{c}\mathbf{v}\right), \tag{1.232}$$

then by (1.231) we obtain that the corresponding lowered four-covector field (v_0, v_1, v_2, v_3), that we call the four-covector of gravitational potential, satisfies

$$(v_0, v_1, v_2, v_3) := (1, 0, 0, 0). \tag{1.233}$$

Note that the four-covector of gravitational potential, defined by (1.233) coincides with the four-covector defined by (1.213) as the gradient of the scalar of global time. Next, by (1.232) and (1.233) we clearly have

$$c^2\left(\sum_{j=0}^{3}\sum_{k=0}^{3} g^{jk}\frac{\partial \tau}{\partial x^j}\frac{\partial \tau}{\partial x^k}\right) = \sum_{j=0}^{3}\sum_{k=0}^{3} g^{jk} v_j v_k = \sum_{j=0}^{3}\sum_{k=0}^{3} g_{jk} v^j v^k = \sum_{j=0}^{3} v^j v_j = 1, \tag{1.234}$$

where τ is the scalar of the global time on the group $\mathcal{S}_0$, defined by (1.212). Finally, we clearly have

$$\sum_{k=0}^{3} \Theta^{mk}\frac{\partial \tau}{\partial x^k} = \sum_{k=0}^{3} \Theta^{mk} v_k = 0 \qquad \forall m = 0, 1, 2, 3, \tag{1.235}$$

where Θ^{ij} is the contravariant tensor of the three-dimensional geometry, defined by (1.211).

Moreover, if we consider the field of four-vector of the moment of a particle (p^0, p^1, p^2, p^3) defined by (1.198) as

$$(p^0, p^1, p^2, p^3) := \left(m, \frac{1}{c}(m\mathbf{u})\right) \quad \text{where} \quad p^0 = m \quad \text{and} \quad (p^1, p^2, p^3) = \frac{1}{c}(m\mathbf{u}), \tag{1.236}$$

where m is the mass of the particle and $\mathbf{u}$ is the velocity of the particle, then the corresponding lowered four-covector field (p_0, p_1, p_2, p_3), which we call the four-covector of momentum, satisfies

$$(p_0, p_1, p_2, p_3) := \left(m\left(1 + \frac{1}{c^2}(\mathbf{u} - \mathbf{v}) \cdot \mathbf{v}\right), -\frac{m}{c}(\mathbf{u} - \mathbf{v})\right) \quad \text{where}$$

$$p_0 = m\left(1 + \frac{1}{c^2}(\mathbf{u} - \mathbf{v}) \cdot \mathbf{v}\right) \quad \text{and} \quad (p_1, p_2, p_3) = -\frac{m}{c}(\mathbf{u} - \mathbf{v}). \tag{1.237}$$

In particular, by (1.229) we have

$$-\frac{c^2}{2m}\left(p^0 p_0 + \sum_{k=1}^{3} p^k p_k\right) = \frac{mc^2}{2}\left(\frac{1}{c^2}|\mathbf{u} - \mathbf{v}|^2 - 1\right) = \frac{m}{2}|\mathbf{u} - \mathbf{v}|^2 - \frac{mc^2}{2}. \tag{1.238}$$

Moreover, if we consider the four-dimensional electric current (j^0, j^1, j^2, j^3) defined by (1.199) as

$$(j^0, j^1, j^2, j^3) := \left(\rho, \frac{1}{c}\mathbf{j}\right) \quad \text{where} \quad j^0 = \rho \quad \text{and} \quad (j^1, j^2, j^3) = \frac{1}{c}\mathbf{j}, \tag{1.239}$$

where ρ is the electric charge density and $\mathbf{j}$ is the electric current density, then the corresponding lowered four-covector field (j_0, j_1, j_2, j_3), which we call the four-covector of current, satisfies

$$(j_0, j_1, j_2, j_3) := \left(\rho + \frac{1}{c^2}(\mathbf{j} - \rho\mathbf{v}) \cdot \mathbf{v}, -\frac{1}{c}(\mathbf{j} - \rho\mathbf{v})\right) \quad \text{where}$$

$$j_0 = \rho + \frac{1}{c^2}(\mathbf{j} - \rho\mathbf{v}) \cdot \mathbf{v} \quad \text{and} \quad (j_1, j_2, j_3) = -\frac{1}{c}(\mathbf{j} - \rho\mathbf{v}). \tag{1.240}$$

Finally, if Ψ is the scalar electromagnetic potential and $\mathbf{A}$ is the vector electromagnetic potential and we consider the four-<u>covector</u> field of four-dimensional electromagnetic potential (A_0, A_1, A_2, A_3), defined by (1.203) as

$$(A_0, A_1, A_2, A_3) = (\Psi, -\mathbf{A}) \quad \text{where} \quad A_0 = \Psi \quad \text{and} \quad (A_1, A_2, A_3) = -\mathbf{A}, \tag{1.241}$$

then by inserting (1.241) into (1.227) we deduce that the corresponding lifted four-vector field (A^0, A^1, A^2, A^3), which we call the four-vector of electromagnetic potential, satisfies

$$A^0 = \Psi - \frac{1}{c}\mathbf{v} \cdot \mathbf{A} \quad \text{and} \quad (A^1, A^2, A^3) = \mathbf{A} + \frac{1}{c}\left(\Psi - \frac{1}{c}\mathbf{v} \cdot \mathbf{A}\right)\mathbf{v}. \tag{1.242}$$

On the other hand, the proper scalar electromagnetic potential Ψ_0 was defined by (1.48) as

$$\Psi_0 := \Psi - \frac{1}{c}\mathbf{A} \cdot \mathbf{v}. \tag{1.243}$$

Thus we rewrite (1.242) as

$$A^0 = \Psi_0 \quad \text{and} \quad (A^1, A^2, A^3) = \mathbf{A} + \frac{1}{c}\Psi_0 \mathbf{v}. \tag{1.244}$$

Next given a two times covariant tensor $\{c_{mn}\}_{m,n=0,1,2,3}$ on the group $\mathcal{S}_0$ we consider two times contravariant lifted tensor on $\mathcal{S}_0$: $\{c^{mn}\}_{m,n=0,1,2,3}$ defined by

$$c^{mn} := \sum_{k=0}^{3}\sum_{j=0}^{3} g^{mj} g^{nk} c_{jk} \qquad \forall m, n = 0, 1, 2, 3. \tag{1.245}$$

In particular, if $\{F_{ij}\}_{0 \le i, j \le 3}$ is the antisymmetric two times covariant tensor field of the electromagnetic field on the group $\mathcal{S}_0$, which by (1.206) satisfies

$$\begin{cases} F_{00} = 0 \\ F_{0j} = -F_{j0} = E_j \qquad \forall j = 1, 2, 3 \\ F_{jj} = 0 \quad \forall j = 1, 2, 3 \\ F_{12} = -F_{21} = -B_3 \\ F_{13} = -F_{31} = B_2 \\ F_{23} = -F_{32} = -B_1 \,, \end{cases} \tag{1.246}$$

then by inserting (1.246) into (1.245), using (1.221) and denoting:

$$\begin{cases} (D_1, D_2, D_3) = \mathbf{D} := \mathbf{E} + \frac{1}{c}\mathbf{v} \times \mathbf{B} \\ (H_1, H_2, H_3) = \mathbf{H} := \mathbf{B} + \frac{1}{c}\mathbf{v} \times \mathbf{D}, \end{cases} \tag{1.247}$$

we deduce

$$\begin{cases} F^{00} = 0 \\ F^{0j} = -F^{j0} = -D_j \quad \forall\, j = 1, 2, 3, \\ F^{jj} = 0 \quad \forall\, j = 1, 2, 3, \\ F^{12} = -F^{21} = -H_3 \\ F^{13} = -F^{31} = H_2 \\ F^{23} = -F^{32} = -H_1. \end{cases} \tag{1.248}$$

In particular, by (1.246) and (1.248), using (1.247) we deduce that

$$-\sum_{j=0}^{3}\sum_{k=0}^{3}\frac{1}{4}F^{jk}F_{jk} = \frac{1}{2}|\mathbf{D}|^2 - \frac{1}{2}|\mathbf{B}|^2. \tag{1.249}$$

1.9.3 Maxwell Equations in Covariant Formulation

In Chap. 9 we prove that, since the lifted contravariant tensor of the electromagnetic field $\{F^{ij}\}_{0\le i,j\le 3}$ on the group $\mathcal{S}_0$, considered in (1.248) is antisymmetric, then the following four-component field:

$$\left\{\sum_{j=0}^{3}\frac{\partial F^{kj}}{\partial x^j} + \sum_{j=0}^{3}\frac{F^{kj}}{\sqrt{|\det G|}}\frac{\partial}{\partial x^j}\left(\sqrt{|\det G|}\right)\right\}_{0\le k\le 3} \tag{1.250}$$

is a four-vector on the group $\mathcal{S}_0$, where the 4×4-matrix G is defined as $G := \{g_{ij}\}_{0\le i,j\le 3}$. Then, since the matrix G satisfies $\det G = -1$ in every Cartesian coordinate system, then

$$\left\{\sum_{j=0}^{3}\frac{\partial F^{kj}}{\partial x^j} + \sum_{j=0}^{3}\frac{F^{kj}}{\sqrt{|\det G|}}\frac{\partial}{\partial x^j}\left(\sqrt{|\det G|}\right)\right\}_{0\le k\le 3} = \left\{\sum_{j=0}^{3}\frac{\partial F^{kj}}{\partial x^j}\right\}_{0\le k\le 3}. \tag{1.251}$$

Note here that we denoted the matrix $G = \{g_{ij}\}_{0\le i,j\le 3}$ by the same letter as the Gravitational Constant G. However, there is no ambiguity, since in the second case G is a constant scalar and in the first case G is a matrix. Moreover, we will use the matrix notation $G = \{g_{ij}\}_{0\le i,j\le 3}$ only in the expressions containing term $\det G$. Then, by (1.248), denoting $(x^0, x^1, x^2, x^3) := (ct, x_1, x_2, x_3) = (ct, \mathbf{x})$, we deduce

$$\left(\sum_{j=0}^{3}\frac{\partial F^{0j}}{\partial x^j}, \sum_{j=0}^{3}\frac{\partial F^{1j}}{\partial x^j}, \sum_{j=0}^{3}\frac{\partial F^{2j}}{\partial x^j}, \sum_{j=0}^{3}\frac{\partial F^{3j}}{\partial x^j}\right) = \left(-\operatorname{div}_{\mathbf{x}}\mathbf{D}, \left(\frac{1}{c}\frac{\partial \mathbf{D}}{\partial t} - \operatorname{curl}_{\mathbf{x}}\mathbf{H}\right)\right). \tag{1.252}$$

Therefore, by (1.252), the first pair of Maxwell Equations in (1.38):

$$\begin{cases} \text{curl}_{\mathbf{x}}\mathbf{H} = \frac{4\pi}{c}\mathbf{j} + \frac{1}{c}\frac{\partial \mathbf{D}}{\partial t} \\ \text{div}_{\mathbf{x}}\,\mathbf{D} = 4\pi\rho, \end{cases} \tag{1.253}$$

is equivalent to the following equations:

$$\left(\sum_{j=0}^{3}\frac{\partial F^{0j}}{\partial x^j}, \sum_{j=0}^{3}\frac{\partial F^{1j}}{\partial x^j}, \sum_{j=0}^{3}\frac{\partial F^{2j}}{\partial x^j}, \sum_{j=0}^{3}\frac{\partial F^{3j}}{\partial x^j}\right) = -4\pi(j^0, j^1, j^2, j^3), \tag{1.254}$$

where (j^0, j^1, j^2, j^3) is the four-vector of electric current on the group $\mathcal{S}_0$ defined by (1.199) as

$$(j^0, j^1, j^2, j^3) := \left(\rho, \frac{1}{c}\mathbf{j}\right) \tag{1.255}$$

Note that in both sides of the equation (1.254) we have four-vectors and thus (1.254) is a covariant form of (1.253). On the other hand, the second pair of Maxwell Equations in (1.38):

$$\begin{cases} \text{curl}_{\mathbf{x}}\mathbf{E} + \frac{1}{c}\frac{\partial \mathbf{B}}{\partial t} = 0 \\ \text{div}_{\mathbf{x}}\,\mathbf{B} = 0, \end{cases} \tag{1.256}$$

is equivalent to (1.205), i.e. to the following:

$$\begin{cases} \mathbf{B} = \text{curl}_{\mathbf{x}}\mathbf{A}, \\ \mathbf{E} = -\nabla_{\mathbf{x}}\Psi - \frac{1}{c}\frac{\partial \mathbf{A}}{\partial t}, \end{cases} \tag{1.257}$$

On the other hand, as before, by (1.206) we can rewrite (1.257) in the form of (1.204):

$$F_{ij} = \frac{\partial A_j}{\partial x^i} - \frac{\partial A_i}{\partial x^j} \qquad \forall i, j = 0, 1, 2, 3\,, \tag{1.258}$$

where (A_0, A_1, A_2, A_3) is the four-covector of the electromagnetic potential on the group $\mathcal{S}_0$ defined by (1.203) as

$$(A_0, A_1, A_2, A_3) = (\Psi, -\mathbf{A}). \tag{1.259}$$

Note that in both sides of the equation (1.258) we have two-time covariant tensors, and thus (1.258) is a covariant form of (1.256). Finally, the relations between $(\mathbf{E}, \mathbf{B})$ and $(\mathbf{D}, \mathbf{H})$ in (1.38):

$$\begin{cases} \mathbf{D} = \mathbf{E} + \frac{1}{c}\mathbf{v} \times \mathbf{B} \\ \mathbf{H} = \mathbf{B} + \frac{1}{c}\mathbf{v} \times \mathbf{D}, \end{cases} \tag{1.260}$$

are equivalent to the following covariant equations:

$$F^{mn} := \sum_{k=0}^{3}\sum_{j=0}^{3} g^{mj} g^{nk} F_{jk} \quad \forall m, n = 0, 1, 2, 3. \tag{1.261}$$

Thus by (1.258), (1.261) and (1.254) together, we deduce that the full system of Maxwell Equations in (1.38):

$$\begin{cases} \operatorname{curl}_{\mathbf{x}} \mathbf{H} = \frac{4\pi}{c}\mathbf{j} + \frac{1}{c}\frac{\partial \mathbf{D}}{\partial t} \\ \operatorname{div}_{\mathbf{x}} \mathbf{D} = 4\pi\rho \\ \operatorname{curl}_{\mathbf{x}} \mathbf{E} + \frac{1}{c}\frac{\partial \mathbf{B}}{\partial t} = 0 \\ \operatorname{div}_{\mathbf{x}} \mathbf{B} = 0 \\ \mathbf{E} = \mathbf{D} - \frac{1}{c}\mathbf{v} \times \mathbf{B} \\ \mathbf{H} = \mathbf{B} + \frac{1}{c}\mathbf{v} \times \mathbf{D}, \end{cases} \tag{1.262}$$

is equivalent to the following covariant equations:

$$\sum_{j=0}^{3} \frac{\partial}{\partial x^j}\left(\sum_{m=0}^{3}\sum_{n=0}^{3} g^{km} g^{jn}\left(\frac{\partial A_n}{\partial x^m} - \frac{\partial A_m}{\partial x^n}\right)\right) = -4\pi j^k \quad \forall k = 0, 1, 2, 3. \tag{1.263}$$

Note that equations (1.263) are fully analogous to the covariant formulation of Maxwell equations in Special Relativity and the only difference is the choice of the pseudo-metric tensor $\{g^{ij}\}_{0\leq i,j\leq 3}$ (Note that for the Special Relativity case we also have $\det G = -1$). As for the cases of the General relativity, the covariant formulation of Maxwell equations is still similar to (1.263), however, in addition to the different choices of the pseudo-metric tensor $\{g^{ij}\}_{0\leq i,j\leq 3}$ we also have $\det G \neq Const.$ and thus for the full analogy equations (1.263) should be rewritten in the enlarged form, due to (1.250):

$$\sum_{j=0}^{3} \frac{\partial}{\partial x^j}\left(\sum_{m=0}^{3}\sum_{n=0}^{3} g^{km} g^{jn}\left(\frac{\partial A_n}{\partial x^m} - \frac{\partial A_m}{\partial x^n}\right)\right) + \\ \sum_{j=0}^{3} \frac{1}{\sqrt{|\det G|}}\frac{\partial}{\partial x^j}\left(\sqrt{|\det G|}\right)\left(\sum_{m=0}^{3}\sum_{n=0}^{3} g^{km} g^{jn}\left(\frac{\partial A_n}{\partial x^m} - \frac{\partial A_m}{\partial x^n}\right)\right) = -4\pi j^k \\ \forall k = 0, 1, 2, 3. \tag{1.264}$$

Note also that we can rewrite (1.264) as

$$\sum_{j=0}^{3}\frac{\partial}{\partial x^j}\left(\sum_{m=0}^{3}\sum_{n=0}^{3}\sqrt{|\det G|}\,g^{km}g^{jn}\left(\frac{\partial A_n}{\partial x^m}-\frac{\partial A_m}{\partial x^n}\right)\right)=-4\pi\sqrt{|\det G|}\,j^k \quad \forall k=0,1,2,3. \tag{1.265}$$

Next by (1.249) we have

$$\frac{1}{2}|\mathbf{D}|^2-\frac{1}{2}|\mathbf{B}|^2=-\sum_{j=0}^{3}\sum_{k=0}^{3}\frac{1}{4}F^{jk}F_{jk}. \tag{1.266}$$

Therefore, by (1.255), (1.259) and (1.266), we can rewrite the density of the Lagrangian of the electromagnetic field, defined in (1.53) as

$$L_1(\mathbf{A},\Psi,\mathbf{x},t):=\frac{1}{4\pi}\left(\frac{1}{2}|\mathbf{D}|^2-\frac{1}{2}|\mathbf{B}|^2-4\pi\left(\rho\Psi-\frac{1}{c}\mathbf{A}\cdot\mathbf{j}\right)\right), \tag{1.267}$$

in the equivalent covariant form:

$$L_1=\frac{1}{4\pi}\left(-\sum_{n=0}^{3}\sum_{k=0}^{3}\frac{1}{4}F^{nk}F_{nk}-\sum_{k=0}^{3}4\pi j^kA_k\right)=$$
$$\frac{1}{4\pi}\left(-\sum_{n=0}^{3}\sum_{k=0}^{3}\sum_{m=0}^{3}\sum_{p=0}^{3}\frac{1}{4}g^{mn}g^{pk}\left(\frac{\partial A_p}{\partial x^m}-\frac{\partial A_m}{\partial x^p}\right)\left(\frac{\partial A_k}{\partial x^n}-\frac{\partial A_n}{\partial x^k}\right)-\sum_{k=0}^{3}4\pi j^kA_k\right). \tag{1.268}$$

The density of Lagrangian in (1.268) is also fully analogous to the covariant formulation of the Lagrangian density of the electromagnetic field in Special and General Relativity and the only difference is the choice of the pseudo-metric tensor $\{g^{ij}\}_{0\le i,j\le 3}$.

1.9.4 Covariant Formulation of Lagrangian of Motion of a Classical Charged Particle in the External Gravitational and Electromagnetic Fields

Given a classical charged particle with inertial mass m, charge σ, three-dimensional place $\mathbf{r}(t)$ and three-dimensional velocity $\frac{d\mathbf{r}}{dt}$ in the outer gravitational field with three-dimensional vectorial potential $\mathbf{v}(\mathbf{x},t)$, the outer electromagnetical field with three-dimensional vectorial potential $\mathbf{A}(\mathbf{x},t)$ and scalar potential $\Psi(\mathbf{x},t)$, consider a usual Lagrangian that is a particular case of (1.69):

$$L_0\left(\frac{d\mathbf{r}}{dt}, t\right) := \left\{\frac{m}{2}\left|\frac{d\mathbf{r}}{dt} - \mathbf{v}(\mathbf{r}, t)\right|^2 - \sigma\left(\Psi(\mathbf{r}, t) - \frac{1}{c}\mathbf{A}(\mathbf{r}, t) \cdot \frac{d\mathbf{r}}{dt}\right)\right\}. \quad (1.269)$$

Then, since we are interested in critical points of the functional

$$J_0 = \int_0^T L_0\left(\frac{d\mathbf{r}}{dt}, \mathbf{r}, t\right) dt, \quad (1.270)$$

adding a constant does not change the physical meaning of the Lagrangian and we can rewrite (1.269) as

$$L_0'\left(\frac{d\mathbf{r}}{dt}, t\right) := \left\{\left(\frac{m}{2}\left|\frac{d\mathbf{r}}{dt} - \mathbf{v}(\mathbf{r}, t)\right|^2 - \frac{mc^2}{2}\right) - \sigma\left(\Psi(\mathbf{r}, t) - \frac{1}{c}\mathbf{A}(\mathbf{r}, t) \cdot \frac{d\mathbf{r}}{dt}\right)\right\}. \quad (1.271)$$

and (1.270) as

$$J_0' := J_0 - \frac{Tmc^2}{2} = \int_0^T L_0'\left(\frac{d\mathbf{r}}{dt}, \mathbf{r}, t\right) dt =$$
$$\int_0^T \left\{\left(\frac{m}{2}\left|\frac{d\mathbf{r}}{dt} - \mathbf{v}(\mathbf{r}, t)\right|^2 - \frac{mc^2}{2}\right) - \sigma\left(\Psi(\mathbf{r}, t) - \frac{1}{c}\mathbf{A}(\mathbf{r}, t) \cdot \frac{d\mathbf{r}}{dt}\right)\right\} dt. \quad (1.272)$$

Next consider the four-vector field of the momentum on the group $\mathcal{S}_0$: $\left(p^0(t), p^1(t), p^2(t), p^3(t)\right)$, defined by (1.195) and (1.198) as

$$\left(p^0(t), p^1(t), p^2(t), p^3(t)\right) := \left(m, \frac{m}{c}\frac{d\mathbf{r}}{dt}(t)\right) = \left(m, \frac{m}{c}\frac{dr_1}{dt}(t), \frac{m}{c}\frac{dr_2}{dt}(t), \frac{m}{c}\frac{dr_3}{dt}(t)\right) \quad (1.273)$$

Then by (1.238) we have

$$\frac{mc^2}{2}\left(\frac{1}{c^2}\left|\frac{d\mathbf{r}}{dt} - \mathbf{v}(\mathbf{r}, t)\right|^2 - 1\right) = \frac{m}{2}\left|\frac{d\mathbf{r}}{dt} - \mathbf{v}(\mathbf{r}, t)\right|^2 - \frac{mc^2}{2}$$
$$= -\frac{c^2}{2m}\left(\sum_{k=0}^{3} p^k p_k\right) = -\frac{mc^2}{2}\left(\sum_{j=0}^{3}\sum_{k=0}^{3} g_{jk}(\mathbf{r}, t)\frac{p^j}{m}\frac{p^k}{m}\right). \quad (1.274)$$

On the other hand if we consider the four-covector of the electromagnetic potential on the group $\mathcal{S}_0$: (A_0, A_1, A_2, A_3), defined by (1.203) as

$$(A_0, A_1, A_2, A_3) = (\Psi, -\mathbf{A}), \quad (1.275)$$

then we can write

$$\sigma\left(\Psi(\mathbf{r},t)-\frac{1}{c}\mathbf{A}(\mathbf{r},t)\cdot\frac{d\mathbf{r}}{dt}\right)=\sum_{k=0}^{3}\sigma A_k(\mathbf{r},t)\,\frac{p^k}{m}. \tag{1.276}$$

Thus by (1.274) and (1.276) we rewrite (1.272) in a covariant form:

$$J_0'=\int_0^T L_0'\left(\frac{d\mathbf{r}}{dt},\mathbf{r},t\right)dt=$$
$$\int_0^T\left\{-\frac{mc^2}{2}\left(\sum_{j=0}^{3}\sum_{k=0}^{3}g_{jk}(\mathbf{r},t)\,\frac{p^j}{m}\frac{p^k}{m}\right)-\sum_{k=0}^{3}\sigma A_k(\mathbf{r},t)\,\frac{p^k}{m}\right\}dt. \tag{1.277}$$

Thus if we consider the four-dimensional space-time trajectory of the particle:

$$\left(\chi^0(t),\chi^1(t),\chi^2(t),\chi^3(t)\right)=\left(t,\frac{1}{c}r_1(t),\frac{1}{c}r_2(t),\frac{1}{c}r_3(t)\right), \tag{1.278}$$

then we rewrite (1.277) as

$$J_0'=\int_0^T\left\{-\frac{mc^2}{2}\left(\sum_{j=0}^{3}\sum_{k=0}^{3}g_{jk}\left(\chi(t)\right)\,\frac{d\chi^j}{dt}\frac{d\chi^k}{dt}\right)-\sum_{k=0}^{3}\sigma A_k\left(\chi(t)\right)\,\frac{d\chi^k}{dt}\right\}dt. \tag{1.279}$$

Moreover, $\left(\frac{d\chi^0}{dt},\frac{d\chi^1}{dt},\frac{d\chi^2}{dt},\frac{d\chi^3}{dt}\right)$ is a four-vector on the group $\mathcal{S}_0$ and the global non-relativistic time t is the scalar on the group $\mathcal{S}_0$.

Next we also can consider a more general Lagrangian than (1.279): given a function $\mathcal{G}(\tau):\mathbb{R}\to\mathbb{R}$ define

$$J_{\mathcal{G}}(\chi)=\int_0^T\left\{-mc^2\,\mathcal{G}\left(\sum_{j=0}^{3}\sum_{k=0}^{3}g_{jk}\left(\chi(t)\right)\,\frac{d\chi^j}{dt}\frac{d\chi^k}{dt}\right)-\sum_{k=0}^{3}\sigma A_k\left(\chi(t)\right)\,\frac{d\chi^k}{dt}\right\}dt. \tag{1.280}$$

Clearly, (1.280) is written in covariant form, and in particular, (1.280) is invariant under the change of non-inertial Cartesian coordinate systems. In particular, for $\mathcal{G}(\tau):=\frac{1}{2}\tau$ we obtain (1.279).

Another important particular case is the following choice: $\mathcal{G}(\tau):=\sqrt{\tau}$. Then we deduce

$$J_{rl}(\chi)=\int_0^T\left\{-mc^2\sqrt{\left(\sum_{j=0}^{3}\sum_{k=0}^{3}g_{jk}\left(\chi(t)\right)\,\frac{d\chi^j}{dt}\frac{d\chi^k}{dt}\right)}-\sum_{k=0}^{3}\sigma A_k\left(\chi(t)\right)\,\frac{d\chi^k}{dt}\right\}dt, \tag{1.281}$$

that is in somewhat analogous to the relativistic Lagrangian of the motion of charged particle. Due to (1.278) we rewrite (1.281) in a three-dimensional form as

$$J_{rl}(\mathbf{r}) = \int_0^T \left\{ -mc^2 \sqrt{1 - \frac{1}{c^2} \left| \frac{d\mathbf{r}}{dt} - \mathbf{v}(\mathbf{r}, t) \right|^2} - \sigma \left(\Psi(\mathbf{r}, t) - \frac{1}{c} \mathbf{A}(\mathbf{r}, t) \cdot \frac{d\mathbf{r}}{dt} \right) \right\} dt. \tag{1.282}$$

Thus in the case

$$\frac{1}{c^2} \left| \frac{d\mathbf{r}}{dt} - \mathbf{v}(\mathbf{r}, t) \right|^2 \ll 1,$$

up to additive constant, (1.282) becomes to be (1.270), where L_0 is given by (1.269). Note that the Lagrangian in (1.281) has the following advantage with respect to (1.279): if we parameterize the curve in (1.278) by some arbitrary parameter s with increasing mapping $t \leftrightarrow s$, which is however can differ from the global time t, then changing variables of integration in (1.281) from t to s gives

$$J_{rl}(\chi) = \int_a^b \left\{ -mc^2 \sqrt{\left(\sum_{j=0}^{3} \sum_{k=0}^{3} g_{jk}(\chi(s)) \frac{d\chi^j}{ds} \frac{d\chi^k}{ds} \right)} - \sum_{k=0}^{3} \sigma A_k(\chi(s)) \frac{d\chi^k}{ds} \right\} ds, \tag{1.283}$$

that has exactly the same form as (1.281), however s in (1.283) can be arbitrary parameter of the curve with increasing mapping $t \leftrightarrow s$.

Finally, we would like to note that if the motion of some particle is ruled by the relativistic-like Lagrangian in (1.282), then, although the absolute value of the velocity of the particle $\left|\frac{d\mathbf{r}}{dt}\right|$ can be arbitrary large, the absolute value of the difference between the velocity of the particle and the local gravitational potential cannot exceed the value c, i.e.:

$$|\mathbf{u}(t) - \mathbf{v}(\mathbf{r}, t)| := \left| \frac{d\mathbf{r}}{dt} - \mathbf{v}(\mathbf{r}, t) \right| < c \qquad \forall t, \tag{1.284}$$

provided that (1.284) is satisfied in some initial instant of time. Note also that the quantity in the left-hand side of (1.284) is invariant under the change of inertial or non-inertial Cartesian coordinate system.

1.9.5 Kinematic Pseudo-metric Tensors of Inertia

Consider $\{J^{ij}\}_{0 \le i,j \le 3}$ to be a two times contravariant tensor field on the group $\mathcal{S}_0$, defined by

$$J^{ij} := k^i k^j - \Theta^{ij} \qquad \forall i, j = 0, 1, 2, 3, \tag{1.285}$$

where $\{\Theta^{ij}\}_{0\leq i,j\leq 3}$ is the contravariant tensor of the three-dimensional geometry, defined by (1.211) and being two times contravariant tensor, and (k^0, k^1, k^2, k^3) is the four-dimensional potential of inertia, defined by (1.197) and being a four-vector. Then in Sect. 9.5 we obtain that $\{J^{ij}\}_{0\leq i,j\leq 3}$ is indeed a two times contravariant tensor field on the group $\mathcal{S}_0$ and moreover, this tensor is symmetric. Moreover, by (1.211) and (1.197) we have

$$\begin{cases} J^{00} = 1 \\ J^{ij} = -\delta_{ij} + \frac{k^i k^j}{c^2} \quad \forall 1 \leq i, j \leq 3 \\ J^{0j} = J^{j0} = \frac{k^j}{c} \quad \forall 1 \leq j \leq 3, \end{cases} \tag{1.286}$$

where $\mathbf{k} = (k^1, k^2, k^3)$ is the three-dimensional vectorial potential of inertia. We call the tensor $\{J^{ij}\}_{0\leq i,j\leq 3}$ the contravariant kinematic pseudo-metric tensor of inertia. Next consider a 16-component field $\{J_{ij}\}_{0\leq i,j\leq 3}$ defined by

$$\begin{cases} J_{00} = 1 - \frac{|\mathbf{k}|^2}{c^2} \\ J_{ij} = -\delta_{ij} \quad \forall 1 \leq i, j \leq 3 \\ J_{0j} = J_{j0} = \frac{k^j}{c} \quad \forall 1 \leq j \leq 3. \end{cases} \tag{1.287}$$

Then, as before in Sect. 1.9.5, we deduce

$$\sum_{m=0}^{3} J^{im} J_{mj} = \begin{cases} 1 & \text{if } i = j \\ 0 & \text{if } i \neq j \end{cases} \qquad \forall i, j = 0, 1, 2, 3. \tag{1.288}$$

Therefore, we obtain that $\{J_{ij}\}_{i,j=0,1,2,3}$ is a two times covariant tensor on the group $\mathcal{S}_0$, and moreover, this tensor is symmetric. We call the tensor $\{J_{ij}\}_{0\leq i,j\leq 3}$ covariant kinematic pseudo-metric tensors of inertia. Using (1.288) we also obtain that the pseudo-metric tensors $\{J_{ij}\}_{i,j=0,1,2,3}$ and $\{J^{ij}\}_{0\leq i,j\leq 3}$ are non-degenerate and they are inverse of each other. Moreover, as before, it can be easily calculated that if we consider the 4×4-matrix:

$$J = \{J_{ij}\}_{0\leq i,j\leq 3}, \tag{1.289}$$

then

$$\det J = -1. \tag{1.290}$$

In particular, by (1.290) and (1.225) we deduce

$$\det G = \det J, \tag{1.291}$$

where 4×4-matrix G is given by

$$G = \{g_{ij}\}_{0\leq i,j\leq 3}, \tag{1.292}$$

with the covariant pseudo-metric tensor of the four-dimensional space-time $\{g_{ij}\}_{0\leq i,j\leq 3}$, given by (1.222). Next, obviously we have

$$g^{ij} = J^{ij} + \left(v^i v^j - k^i k^j\right) \qquad \forall i, j = 0, 1, 2, 3, \tag{1.293}$$

where J^{ij} the is the contravariant kinematic pseudo-metric tensor of inertia, given by (1.285), and g^{ij} is the contravariant pseudo-metric tensor of the four-dimensional space-time, given by (1.220). In particular, in the case of the absence of the genuine gravity where $\mathbf{v} = \mathbf{k}$ we have

$$g^{ij} = J^{ij} \qquad \forall i, j = 0, 1, 2, 3, \tag{1.294}$$

or equivalently,

$$g_{ij} = J_{ij} \qquad \forall i, j = 0, 1, 2, 3. \tag{1.295}$$

Furthermore, by (1.286) we have

$$\sum_{m=0}^{3} J^{jm} \left(c\,\frac{\partial \tau}{\partial x^m}\right) = k^j \qquad \forall j = 0, 1, 2, 3, \tag{1.296}$$

where (k^0, k^1, k^2, k^3) is the four-dimensional potential of inertia, defined by (1.197) and τ is the scalar of the global time on the group $\mathcal{S}_0$, defined by (1.212). Thus since,

$$\sum_{j=0}^{3} k^j \left(c\,\frac{\partial \tau}{\partial x^j}\right) = 1, \tag{1.297}$$

we deduce the following eikonal type equation

$$c^2 \left(\sum_{j=0}^{3}\sum_{m=0}^{3} J^{jm}\,\frac{\partial \tau}{\partial x^j}\,\frac{\partial \tau}{\partial x^m}\right) = 1, \tag{1.298}$$

which is similar to (1.234), despite of the different pseudo-metrics. Moreover, by (1.296) and (1.288) we infer

$$\sum_{m=0}^{3} J_{jm}\, k^m = c\,\frac{\partial \tau}{\partial x^j} \qquad \forall j = 0, 1, 2, 3. \tag{1.299}$$

Next, as before, note that the Kinematic pseudo-metric tensors of inertia $\{J^{ij}\}_{0\leq i,j\leq 3}$ and $\{J_{ij}\}_{0\leq i,j\leq 3}$ depend only on the coordinate system in the space-time and are completely independent on the physical matter or physical fields filling this space. In contrast the pseudo-metric tensors of the four-dimensional space-time $\{g^{ij}\}_{0\leq i,j\leq 3}$ and $\{g_{ij}\}_{0\leq i,j\leq 3}$, given by (1.221) and (1.222), depend essentially on the surrounding physical matter (in the model of the Newtonian gravity through gravitational masses). Furthermore, note that in the absence of the genuine gravity $\{g^{ij}\}_{0\leq i,j\leq 3}$ and $\{g_{ij}\}_{0\leq i,j\leq 3}$ coincide with $\{J^{ij}\}_{0\leq i,j\leq 3}$ and $\{J_{ij}\}_{0\leq i,j\leq 3}$ respectively. In the model of the Newtonian gravity this happens away of essential gravitational masses. In particular, $\{g^{ij}\}_{0\leq i,j\leq 3}$ and $\{g_{ij}\}_{0\leq i,j\leq 3}$ tend to $\{J^{ij}\}_{0\leq i,j\leq 3}$ and $\{J_{ij}\}_{0\leq i,j\leq 3}$ respectively, as $|\mathbf{x}| \to +\infty$. Finally note that since in every inertial Cartesian coordinate system $\mathbf{k}$ is a constant we deduce that in every such a system $\{J^{ij}\}_{0\leq i,j\leq 3}$ and $\{J_{ij}\}_{0\leq i,j\leq 3}$ are constant in the four-dimensional space-time. So, in contrast to $\{g^{ij}\}_{0\leq i,j\leq 3}$ and $\{g_{ij}\}_{0\leq i,j\leq 3}$, the pseudo-metrics $\{J^{ij}\}_{0\leq i,j\leq 3}$ and $\{J_{ij}\}_{0\leq i,j\leq 3}$ are completely flat. In particular, if we consider the Christoffel Symbols of the Kinematic pseudo-metric tensors of inertia $\{J_{ij}\}_{0\leq i,j\leq 3}$ defined by

$$\begin{cases}\left\{\Gamma_{i,mn}\right\}_J := \frac{1}{2}\left(\frac{\partial J_{im}}{\partial x_n}+\frac{\partial J_{in}}{\partial x_m}-\frac{\partial J_{mn}}{\partial x_i}\right)\\ \left\{\Gamma^i_{mn}\right\}_J := \sum_{j=0}^{3} J^{ij}\left\{\Gamma_{j,mn}\right\}_J\end{cases} \quad \forall i,m,n=0,1,2,3, \tag{1.300}$$

then, since in every inertial Cartesian coordinate system $\{J_{ij}\}_{0\leq i,j\leq 3}$ is constant in the four-dimensional space-time, we deduce that in every such coordinate system we have

$$\left\{\Gamma^i_{mn}\right\}_J = \left\{\Gamma_{i,mn}\right\}_J = 0 \quad \forall i,m,n=0,1,2,3. \tag{1.301}$$

Thus, if we consider the two times covariant tensor of the covariant derivative to the covector of the gradient to the scalar of the global time τ from (1.212), denoted by $\left\{\delta_j\left(\frac{\partial\tau}{\partial x_i}\right)\right\}_J$, and defined by

$$\begin{aligned}\left\{\delta_j\left(\frac{\partial\tau}{\partial x_i}\right)\right\}_J = \left\{\delta_i\left(\frac{\partial\tau}{\partial x_j}\right)\right\}_J &:= \frac{\partial^2\tau}{\partial x_i\partial x_j} - \sum_{m=0}^{3}\left\{\Gamma^m_{ij}\right\}_J\frac{\partial\tau}{\partial x_m}\\ &= \frac{\partial^2\tau}{\partial x_i\partial x_j} - \sum_{m=0}^{3}\left\{\Gamma_{m,ij}\right\}_J\frac{k^m}{c} \quad \forall i,j=0,1,2,3,\end{aligned} \tag{1.302}$$

then by (1.213) and (1.301) we prove the following identity, first in every inertial Cartesian coordinate system and then, by the covariance of this identity, also in every non-inertial Cartesian coordinate system:

$$\left\{\delta_j\left(\frac{\partial\tau}{\partial x_i}\right)\right\}_J = 0 \quad \forall i,j=0,1,2,3. \tag{1.303}$$

Note here that we put brackets $\{\cdot\}_J$ for the Christoffel Symbols and covariant derivative with respect to Kinematical pseudo-metrics $\{J^{ij}\}_{0\leq i,j\leq 3}$ and $\{J_{ij}\}_{0\leq i,j\leq 3}$ in order to distinguish them from Christoffel Symbols and covariant derivative with respect to Dynamical pseudo-metrics $\{g^{ij}\}_{0\leq i,j\leq 3}$ and $\{g_{ij}\}_{0\leq i,j\leq 3}$, that we will write without any brackets.

Furthermore, since $\mathbf{k}$ is generally trivial speed-like vector field, by the definition, there exists a unique, up to equivalence, "preferable" inertial Cartesian coordinate system ($\{0\}$) where we have

$$\mathbf{k} = 0, \tag{1.304}$$

and inserting it into (1.286) and (1.287) we deduce that in system ($\{0\}$) we have

$$\begin{cases} J^{00} = 1 \\ J^{ij} = -\delta_{ij} \quad \forall 1 \le i, j \le 3 \\ J^{0j} = J^{j0} = 0 \quad \forall 1 \le j \le 3, \end{cases} \tag{1.305}$$

and

$$\begin{cases} J_{00} = 1 \\ J_{ij} = -\delta_{ij} \quad \forall 1 \le i, j \le 3 \\ J_{0j} = J_{j0} = 0 \quad \forall 1 \le j \le 3, \end{cases} \tag{1.306}$$

which is the same as the basic pseudo-metrics in the Special Relativity.

Next, we define the Dynamical four-covector of genuine gravity (s_0, s_1, s_2, s_3) by the following:

$$s_j = \frac{1}{2}\left(\sum_{m=0}^{3} g_{jm}k^m - \sum_{m=0}^{3} J_{jm}v^m\right) \quad \forall j = 0, 1, 2, 3, \tag{1.307}$$

where (k^0, k^1, k^2, k^3) is the four-dimensional potential of inertia, defined by (1.197), (v^0, v^1, v^2, v^3) is the four-dimensional gravitational potential, defined by (1.196), $\{g_{ij}\}_{0\le i,j\le 3}$ defined by (1.222) and $\{J_{ij}\}_{0\le i,j\le 3}$ defined by (1.287). Then, by inserting (1.197), (1.196), (1.222) and (1.287) into (1.307) we obtain

$$(s_0, s_1, s_2, s_3) = \left(-\frac{1}{2c^2}|\mathbf{h}|^2 - \frac{1}{c^2}\mathbf{k}\cdot\mathbf{h}\,,\ \frac{1}{c}\mathbf{h}\right) = \left(\frac{1}{2c^2}|\mathbf{h}|^2 - \frac{1}{c^2}\mathbf{v}\cdot\mathbf{h}\,,\ \frac{1}{c}\mathbf{h}\right) \quad \text{where}$$
$$s_0 = -\frac{1}{2c^2}|\mathbf{h}|^2 - \frac{1}{c^2}\mathbf{k}\cdot\mathbf{h} = \frac{1}{2c^2}|\mathbf{h}|^2 - \frac{1}{c^2}\mathbf{v}\cdot\mathbf{h} \quad \text{and} \quad (s_1, s_2, s_3) = \frac{1}{c}\mathbf{h}, \tag{1.308}$$

where $\mathbf{h}$ is the three-dimensional proper vector field of the vectorial potential of genuine gravity defined as in (1.151) by

$$\mathbf{h} := \mathbf{v} - \mathbf{k}, \tag{1.309}$$

(see Sect. 9.5 for details). In particular, by (1.222), (1.287), (1.308) and (1.213) together we deduce

$$g_{ij} = J_{ij} + \left(s_i\left(c\frac{\partial\tau}{\partial x^j}\right) + \left(c\frac{\partial\tau}{\partial x^i}\right)s_j\right) \quad \forall i, j = 0, 1, 2, 3. \tag{1.310}$$

Moreover, by (1.308) (1.211) and (1.285) we obtain

$$v^j - k^j = \sum_{m=0}^{3} \Theta^{jm} s_m = \sum_{m=0}^{3} \left(k^j k^m - J^{jm}\right) s_m \quad \forall j = 0, 1, 2, 3. \tag{1.311}$$

1.9.6 Physical Laws in Curvilinear Coordinate Systems in the Non-relativistic Space-Time

Let $\mathcal{S}$ be the group of all smooth non-degenerate invertible transformations from $\mathbb{R}^4$ onto $\mathbb{R}^4$ having the form (1.179):

$$\begin{cases} x'^0 = f^{(0)}(x^0, x^1, x^2, x^3), \\ x'^1 = f^{(1)}(x^0, x^1, x^2, x^3), \\ x'^2 = f^{(2)}(x^0, x^1, x^2, x^3), \\ x'^3 = f^{(3)}(x^0, x^1, x^2, x^3), \end{cases} \tag{1.312}$$

and let $\mathcal{S}_0$ be a subgroup of transformations of the form (1.186). Then, it is clear, that given any object that is a scalar, four-vector, four-covector, two times covariant tensor or two times contravariant tensor on the group $\mathcal{S}_0$, defined in every Cartesian non-inertial coordinate system, we can uniquely extend the definition of this object, in such a way that it will be defined also in every curvilinear coordinate systems in $\mathbb{R}^4$ and will be respectively a scalar, four-vector, four-covector, two times covariant tensor or two times contravariant tensor on the wider group $\mathcal{S}$. Thus all the physical laws that have a covariant form preserve their form also in transformations of the form (1.312), i.e. in curvilinear coordinate systems. In particular, the Maxwell equations in every curvilinear coordinate system have the form of (1.264) or equivalently of (1.265):

$$\sum_{j=0}^{3} \frac{\partial}{\partial x^j}\left(\sum_{m=0}^{3}\sum_{n=0}^{3} g^{km} g^{jn}\left(\frac{\partial A_n}{\partial x^m} - \frac{\partial A_m}{\partial x^n}\right)\right) +$$
$$\sum_{j=0}^{3} \frac{1}{\sqrt{|\det G|}} \frac{\partial}{\partial x^j}\left(\sqrt{|\det G|}\right)\left(\sum_{m=0}^{3}\sum_{n=0}^{3} g^{km} g^{jn}\left(\frac{\partial A_n}{\partial x^m} - \frac{\partial A_m}{\partial x^n}\right)\right) = -4\pi j^k$$
$$\forall k = 0, 1, 2, 3, \tag{1.313}$$

or equivalently:

$$\sum_{j-0}^{3}\frac{\partial}{\partial x^j}\left(\sum_{m=0}^{3}\sum_{n=0}^{3}\sqrt{|\det G|}\,g^{km}g^{jn}\left(\frac{\partial A_n}{\partial x^m}-\frac{\partial A_m}{\partial x^n}\right)\right)=-4\pi\sqrt{|\det G|}\,j^k$$

$$\forall k=0,1,2,3. \quad (1.314)$$

Here $\{A_k\}_{k=0,1,2,3}$ is the four-covector of the electromagnetic potential, $\{j^k\}_{k=0,1,2,3}$ is the four-vector of the current and $G := \{g_{kj}\}_{k,j=0,1,2,3}$, $\{g^{kj}\}_{k,j=0,1,2,3}$ are pseudo-metric covariant and contravariant tensors. Note, that in curvilinear coordinate system we can have $\det G \neq Const$ and thus we need to consider the enlarged form (1.264) instead of (1.263). Moreover, the density of the Lagrangian of the electromagnetic field in every curvilinear coordinate system in $\mathbb{R}^4$ also has a form of (1.268):

$$L_1=\frac{1}{4\pi}\left(-\sum_{n=0}^{3}\sum_{k=0}^{3}\frac{1}{4}F^{nk}F_{nk}-\sum_{k=0}^{3}4\pi j^k A_k\right)=$$
$$\frac{1}{4\pi}\left(-\sum_{n=0}^{3}\sum_{k=0}^{3}\sum_{m=0}^{3}\sum_{p=0}^{3}\frac{1}{4}g^{mn}g^{pk}\left(\frac{\partial A_p}{\partial x^m}-\frac{\partial A_m}{\partial x^p}\right)\left(\frac{\partial A_k}{\partial x^n}-\frac{\partial A_n}{\partial x^k}\right)-\sum_{k=0}^{3}4\pi j^k A_k\right), \quad (1.315)$$

where

$$F_{ij} := \frac{\partial A_j}{\partial x^i}-\frac{\partial A_i}{\partial x^j} \qquad \forall i,j=0,1,2,3\,. \quad (1.316)$$

Next the general Lagrangian of motion of the charged particle in the gravitational and electromagnetic field (1.280) preserve its form in every curvilinear coordinate system:

$$J_{\mathcal{G}}(\chi)=\int_0^T\left\{-mc^2\,\mathcal{G}\left(\sum_{j=0}^{3}\sum_{k=0}^{3}g_{jk}(\chi(t))\,\frac{d\chi^j}{dt}\frac{d\chi^k}{dt}\right)-\sum_{k=0}^{3}\sigma A_k(\chi(t))\,\frac{d\chi^k}{dt}\right\}dt. \quad (1.317)$$

where t is the global time, which is a scalar on the group $\mathcal{S}$,

$$\left(\chi^0(t),\chi^1(t),\chi^2(t),\chi^3(t)\right) := \left(\frac{1}{c}x^0(t),\frac{1}{c}x_1(t),\frac{1}{c}x_2(t),\frac{1}{c}x_3(t)\right), \quad (1.318)$$

and $\left(x^0(t),x^1(t),x^2(t),x^3(t)\right)\in\mathbb{R}^4$ is a four-dimensional space-time trajectory of the particle, parameterized by the global time.

Note that if we denote by t the scalar of global time, then in a general curvilinear coordinate system the coordinate x^0 can differ from ct, and the equality $x^0=ct$ valid, in general, only in Cartesian inertial or non-inertial coordinate systems. However, since the equality in (1.234) has a covariant form, the scalar of the global time t satisfies the following Eikonal-type equation in every curvilinear coordinate system:

$$\sum_{j=0}^{3}\sum_{k=0}^{3} g^{jk}\frac{\partial t}{\partial x^j}\frac{\partial t}{\partial x^k} = \frac{1}{c^2}. \tag{1.319}$$

Moreover, since the equality in (1.234) also has a covariant form, the following identity is valid in every curvilinear coordinate system:

$$\sum_{k=0}^{3}\Theta^{mk}\frac{\partial t}{\partial x^k} = 0 \quad \forall m = 0, 1, 2, 3, \tag{1.320}$$

where Θ^{ij} is the contravariant tensor of the three-dimensional geometry that has the form (1.211) only in Cartesian inertial or non-inertial coordinate systems.

Next, in the particular case of the relativistic-like Lagrangian where $\mathcal{G}(\tau) := \sqrt{\tau}$, the Lagrangian in (1.283) also preserve their form in every curvilinear coordinate system:

$$J_{rl}(\chi) = \int_a^b \left\{ -mc^2\sqrt{\left(\sum_{j=0}^{3}\sum_{k=0}^{3} g_{jk}\left(\chi(s)\right)\frac{d\chi^j}{ds}\frac{d\chi^k}{ds}\right)} - \sum_{k=0}^{3}\sigma A_k\left(\chi(s)\right)\frac{d\chi^k}{ds}\right\} ds, \tag{1.321}$$

where s is the arbitrary parameter of the trajectory with increasing mapping $t \leftrightarrow s$:

$$\left(\chi^0(s), \chi^1(s), \chi^2(s), \chi^3(s)\right) := \left(\frac{1}{c}x^0(s), \frac{1}{c}x_1(s), \frac{1}{c}x_2(s), \frac{1}{c}x_3(s)\right). \tag{1.322}$$

In particular, in the case $\frac{\partial \chi^0}{\partial t} > 0$ we can take $s := \chi^0$ in (1.321).

Finally, we would like to note the following fact: since in the absence of essential gravitational masses we have $g_{ij} = J_{ij}$, there exists a unique, up to equivalence, "preferable" inertial coordinate system where $\mathbf{v} = \mathbf{k} = 0$ everywhere. In this particular system by (1.306) we have

$$\begin{cases} g_{00} = 1 \\ g_{ij} = -\delta_{ij} \quad \forall 1 \le i, j \le 3 \\ g_{0j} = g_{j0} = 0 \quad \forall 1 \le j \le 3. \end{cases} \tag{1.323}$$

and thus the Maxwell equations are the same as in the Special Relativity. Moreover, in this system the Lagrangian of the motion of the particle of the form (1.321) is also the same as in the Special Relativity. Thus, since Maxwell equations (1.313) and the Lagrangian of the motion of particles (1.321) preserve their form in every Cartesian, non-Cartesian or curvilinear coordinate system of the group S, they stay the same as in Special Relativity also in the case of every Cartesian, non-Cartesian or curvilinear coordinate system. Thus in the particular case of $\mathcal{G}(\tau) := \sqrt{\tau}$ in (1.317) and in the absence of essential gravitational masses, the unique formal mathematical difference between our model and the Special Relativity is that in the frames of our model we consider the Galilean Transformations as transformations of the change of inertial Cartesian coordinate systems, up to equivalence, and (1.2) as transformations of the change of non-inertial Cartesian coordinate systems, however the Lorenz transformations lead to inertial non-Cartesian coordinate systems (see the following Definition 1.12). In contrast, in the Special Relativity the fundamental role of the Lorenz transformations, i.e. the transformations that preserve the form (1.323) of the pseudo-metric tensor, is postulated as the role of transformations of the change of inertial Cartesian coordinate systems, and at the same time the Galilean Transformations lead to inertial non-Cartesian coordinate systems, and transformations (1.2) lead to non-inertial non-Cartesian coordinate systems.

1.9.7 Some General Covariant Identities in Non-Cartesian or Curvilinear Coordinate Systems

Consider the contravariant kinematic pseudo-metric tensor of inertia $\{J^{ij}\}_{0\leq i,j\leq 3}$ on the group S, defined by

$$J^{ij} := k^i k^j - \Theta^{ij} \quad \forall i, j = 0, 1, 2, 3, \tag{1.324}$$

where $\{\Theta^{ij}\}_{0\leq i,j\leq 3}$ is the contravariant tensor of the three-dimensional geometry on the group S (the same as in (1.320)) defined in Cartesian coordinate systems by (1.211), and (k^0, k^1, k^2, k^3) is the four-vector of the potential of inertia on the group S, defined in Cartesian coordinate systems by (1.197). Then, as before, the reverse to $\{J^{ij}\}_{0\leq i,j\leq 3}$ covariant tensor on the group S is denoted as $\{J_{ij}\}_{0\leq i,j\leq 3}$ and satisfies

$$\sum_{m=0}^{3} J^{im} J_{mj} = \begin{cases} 1 & \text{if } i = j \\ 0 & \text{if } i \neq j \end{cases} \quad \forall i, j = 0, 1, 2, 3 \tag{1.325}$$

(as in (1.288)).

Definition 1.11 We say that a given general coordinate system $(*)$ is Cartesian if the contravariant tensor of the three-dimensional geometry $\{\Theta^{ij}\}_{0\leq i,j\leq 3}$ has the following simple form in the system $(*)$ (the same as in (1.211)):

$$\begin{cases} \Theta^{00} = 0 \\ \Theta^{0j} = \Theta^{j0} = 0 \quad \forall j = 1, 2, 3 \\ \Theta^{ij} := \delta_{ij} \quad \forall i, j = 1, 2, 3, \end{cases} \tag{1.326}$$

and at the same time in the system $(*)$ we have

$$c\frac{\partial t}{\partial x^0}(x^0, x^1, x^2, x^3) = 1, \tag{1.327}$$

where t is the scalar of global time (the same as in (1.320)). In particular, by (1.320), (1.326) and (1.327) all together in the system $(*)$ we have

$$c\left(\frac{\partial t}{\partial x^0}(x^0,x^1,x^2,x^3), \frac{\partial t}{\partial x^1}(x^0,x^1,x^2,x^3), \frac{\partial t}{\partial x^2}(x^0,x^1,x^2,x^3), \frac{\partial t}{\partial x^3}(x^0,x^1,x^2,x^3)\right) = (1,0,0,0). \quad (1.328)$$

Then, it easily can be shown that given an Cartesian coordinate system $(*)$ and a general coordinate system $(**)$, the system $(**)$ is Cartesian if and only if, up to a constant shift of the time, the change of coordinates from system $(*)$ to system $(**)$ is given by (1.186) or equivalently by (1.187):

$$\begin{cases} \mathbf{x}' = A\left(\frac{x_0}{c}\right)\cdot \mathbf{x} + \mathbf{z}\left(\frac{x_0}{c}\right) = A(t)\cdot\mathbf{x} + \mathbf{z}(t), \\ x_0' = x_0 = t, \end{cases} \quad (1.329)$$

where $A(t) \in SO(3)$ is a rotation.

Definition 1.12 Similarly to Cartesian coordinate systems, we say that a given non-Cartesian or curvilinear coordinate system $(*)$ is inertial if the four-vector of the potential of inertia (k^0, k^1, k^2, k^3), defined in Cartesian systems by (1.197) is constant in $\mathbb{R}^4$. Then, it easily can be shown that given an inertial coordinate system $(*)$ and a general coordinate system $(**)$, the system $(**)$ is inertial if and only if the change of coordinates from system $(*)$ to system $(**)$ is given by a linear transformation in $\mathbb{R}^4$. In particular, a general coordinate system $(***)$ is inertial if and only if the tensors $\{J^{ij}\}_{0\leq i,j\leq 3}$ and $\{J_{ij}\}_{0\leq i,j\leq 3}$ are constant in the four-dimensional space-time in the given system $(***)$. Moreover, given an inertial Cartesian coordinate system $(*)$ and a general coordinate system $(**)$, the system $(**)$ is inertial and at the same time is Cartesian if and only if, up to a constant shift of time and, up to equivalence of coordinate systems, the change of coordinates from system $(*)$ to system $(**)$ is given by the Galilean transformation.

In the general non-Cartesian or curvilinear coordinate system we obviously have the following list of general covariant identities:

•

$$\det G = \det J \quad (1.330)$$

(see (1.291)), where 4×4-matrix J is given by

$$J = \{J_{ij}\}_{0\leq i,j\leq 3}, \quad (1.331)$$

and 4×4-matrix G is given by

$$G = \{g_{ij}\}_{0\leq i,j\leq 3}, \quad (1.332)$$

with the covariant pseudo-metric tensor of the four-dimensional space-time $\{g_{ij}\}_{0\leq i,j\leq 3}$.

- $$\left\{\delta_j\left(\frac{\partial t}{\partial x_i}\right)\right\}_J = 0 \quad \forall i, j = 0, 1, 2, 3 \tag{1.333}$$

(see (1.303)), where

$$\left\{\delta_j\left(\frac{\partial t}{\partial x_i}\right)\right\}_J = \left\{\delta_i\left(\frac{\partial t}{\partial x_j}\right)\right\}_J := \frac{\partial^2 t}{\partial x^i \partial x^j} - \sum_{m=0}^{3}\left\{\Gamma^m_{ij}\right\}_J \frac{\partial t}{\partial x^m} \quad \forall i, j = 0, 1, 2, 3, \tag{1.334}$$

with

$$\begin{cases}\left\{\Gamma_{i,mn}\right\}_J := \frac{1}{2}\left(\frac{\partial J_{im}}{\partial x_n} + \frac{\partial J_{in}}{\partial x_m} - \frac{\partial J_{mn}}{\partial x_i}\right) \\ \left\{\Gamma^i_{mn}\right\}_J := \sum_{j=0}^{3} J^{ij}\left\{\Gamma_{j,mn}\right\}_J\end{cases} \quad \forall i, m, n = 0, 1, 2, 3. \tag{1.335}$$

- $$\sum_{m=0}^{3} J^{jm}\left(c\frac{\partial t}{\partial x^m}\right) = k^j \quad \forall j = 0, 1, 2, 3, \tag{1.336}$$

and

$$\sum_{m=0}^{3} g^{jm}\left(c\frac{\partial t}{\partial x^m}\right) = v^j \quad \forall j = 0, 1, 2, 3. \tag{1.337}$$

- $$\sum_{j=0}^{3} k^j\left(c\frac{\partial t}{\partial x^j}\right) = 1 = \sum_{j=0}^{3} v^j\left(c\frac{\partial t}{\partial x^j}\right), \tag{1.338}$$

- $$c^2\left(\sum_{j=0}^{3}\sum_{m=0}^{3} J^{jm}\frac{\partial t}{\partial x^j}\frac{\partial t}{\partial x^m}\right) = 1 = c^2\left(\sum_{j=0}^{3}\sum_{m=0}^{3} g^{jm}\frac{\partial t}{\partial x^j}\frac{\partial t}{\partial x^m}\right). \tag{1.339}$$

- $$\sum_{m=0}^{3} J_{jm}k^m = c\frac{\partial t}{\partial x^j} = \sum_{m=0}^{3} g_{jm}v^m \quad \forall j = 0, 1, 2, 3. \tag{1.340}$$

- $$\sum_{m=0}^{3} \Theta^{jm}\frac{\partial t}{\partial x^m} = 0 \quad \forall j = 0, 1, 2, 3. \tag{1.341}$$

•

$$J^{ij} = k^i k^j - \Theta^{ij} \qquad \forall i, j = 0, 1, 2, 3, \tag{1.342}$$

and

$$g^{ij} = v^i v^j - \Theta^{ij} \qquad \forall i, j = 0, 1, 2, 3, \tag{1.343}$$

that implies, in particular,

$$g^{ij} = \left(v^i v^j - k^i k^j\right) + J^{ij} \qquad \forall i, j = 0, 1, 2, 3. \tag{1.344}$$

•

$$\sum_{m=0}^{3} J^{im} J_{mj} = \sum_{m=0}^{3} g^{im} g_{mj} = \begin{cases} 1 & \text{if } i = j \\ 0 & \text{if } i \neq j \end{cases} \qquad \forall i, j = 0, 1, 2, 3, \tag{1.345}$$

that implies together with (1.340) and (1.339) the following:

$$\sum_{j=0}^{3} \sum_{m=0}^{3} J_{jm} k^j k^m = 1 = \sum_{j=0}^{3} \sum_{m=0}^{3} g_{jm} v^j v^m. \tag{1.346}$$

Moreover, in the general non-Cartesian or curvilinear coordinate system, as before, we can define the Dynamical four-covector of genuine gravity (s_0, s_1, s_2, s_3) by

$$s_j = \frac{1}{2}\left(\sum_{m=0}^{3} g_{jm} k^m - \sum_{m=0}^{3} J_{jm} v^m\right) \qquad \forall j = 0, 1, 2, 3 \tag{1.347}$$

(see (1.307)). Then we have the following covariant identities

$$g_{ij} = J_{ij} + \left(s_i \left(c \frac{\partial t}{\partial x^j}\right) + \left(c \frac{\partial t}{\partial x^i}\right) s_j\right) \qquad \forall i, j = 0, 1, 2, 3 \tag{1.348}$$

(see (1.310)), and

$$v^j - k^j = \sum_{m=0}^{3} \Theta^{jm} s_m = \sum_{m=0}^{3} \left(k^j k^m - J^{jm}\right) s_m \qquad \forall j = 0, 1, 2, 3 \tag{1.349}$$

(see (1.311)).

Definition 1.13 We say that a given general coordinate system $(*)$ is Lorentzian if in system $(*)$ kinematic pseudo-metric tensors of inertia $\{J^{ij}\}_{0 \leq i, j \leq 3}$ and $\{J_{ij}\}_{0 \leq i, j \leq 3}$ have the following simple form (the same as in (1.305) and (1.306)):

$$\begin{cases} J^{00} = 1 \\ J^{ij} = -\delta_{ij} \quad \forall 1 \le i, j \le 3 \\ J^{0j} = J^{j0} = 0 \quad \forall 1 \le j \le 3, \end{cases} \tag{1.350}$$

and

$$\begin{cases} J_{00} = 1 \\ J_{ij} = -\delta_{ij} \quad \forall 1 \le i, j \le 3 \\ J_{0j} = J_{j0} = 0 \quad \forall 1 \le j \le 3. \end{cases} \tag{1.351}$$

Thus, since in every Lorentzian coordinate system $\{J^{ij}\}_{0\le i,j\le 3}$ and $\{J_{ij}\}_{0\le i,j\le 3}$ are constant in $\mathbb{R}^4$, as it was explained in the end of Definition 1.12 every Lorentzian coordinate system is necessary inertial, however it is not necessary Cartesian. In particular, a unique, up to equivalence, "preferable" inertial Cartesian coordinate system ({0}) where the three-dimensional vector potential of inertia satisfies $\mathbf{k} = 0$, is a unique, up to equivalence, coordinate system which is Cartesian and Lorentzian simultaneously. On the other hand, there exists a plenty of Lorentzian coordinate systems. In particular, it is well known that, given a Lorentzian coordinate system (∗) and a general coordinate system (∗∗), the system (∗∗) is also Lorentzian, if and only if the change of coordinates from system (∗) to system (∗∗) is given by the usual Lorentz transformation, up to certain trivial transformations. In every Lorentzian coordinate system the four-vector of the potential of inertia (k^1, k^2, k^3, k^4) is constant which is by (1.346) and (1.351) always satisfies

$$\left(k^0\right)^2 - \left(\left(k^1\right)^2 + \left(k^2\right)^2 + \left(k^3\right)^2\right) = 1. \tag{1.352}$$

In fact in a general Lorentzian coordinate system the four-vector (k^1, k^2, k^3, k^4) can be arbitrary constant four-vector, satisfying (1.352). We remind that in the unique Cartesian Lorentzian coordinate system we have $(k^1, k^2, k^3, k^4) = (1, 0, 0, 0)$. Next by (1.340) together with (1.350) we have

$$\left(\frac{\partial t}{\partial x^0}, \frac{\partial t}{\partial x^1}, \frac{\partial t}{\partial x^2}, \frac{\partial t}{\partial x^3}\right) = \frac{1}{c}\left(k^0, -k^1, -k^2, -k^3\right), \tag{1.353}$$

and so, up to a constant shift in time, in every Lorentzian coordinate system we have

$$t = \frac{1}{c} k^0 x^0 - \frac{1}{c}\left(k^1 x^1 + k^2 x^2 + k^3 x^3\right). \tag{1.354}$$

Note that in the unique Cartesian Lorentzian coordinate system we have $t = \frac{x^0}{c}$.

1.9.8 Certain Curvilinear Coordinate System in the Case of Stationary Radially Symmetric Gravitational Field and Relation to the Schwarzschild Metric

Assume that for a given part of the space in some inertial or non-inertial Cartesian coordinate system $(*)$ the gravitational field is stationary and radially symmetric that means that the vectorial gravitational potential $\mathbf{v} = (v^1, v^2, v^3)$ is independent on time variable t and having the form

$$\mathbf{v}(\mathbf{x}) = g\left(|\mathbf{x}|\right)\frac{\mathbf{x}}{|\mathbf{x}|} \qquad \forall\mathbf{x}, \tag{1.355}$$

for some scalar function $g(s) : \mathbb{R} \to \mathbb{R}$. Next let $\Theta(\mathbf{x}) : \mathbb{R}^3 \to \mathbb{R}$ be defined as

$$\Theta(\mathbf{x}) = \xi\left(|\mathbf{x}|\right) \quad \forall\mathbf{x}, \qquad \text{where} \quad \frac{d\xi}{ds}(s) = \frac{g(s)}{1 - \frac{g^2(s)}{c^2}} \qquad \forall s. \tag{1.356}$$

Then, consider the change of variables in the four-dimensional space-time $\mathbb{R}^4$:

$$\begin{cases} x'^0 = x^0 + \frac{\Theta((x^1,x^2,x^3))}{c} \\ x'^j = x^j \qquad \forall j = 1, 2, 3. \end{cases} \tag{1.357}$$

that transforms the Cartesian coordinate system $(*)$ to the curvilinear coordinate system $(**)$ in the four-dimensional space-time $\mathbb{R}^4$. Then in the terms of the three-dimensional space and one-dimensional time:

$$(x^0, x^1, x^2, x^3) := (ct, x_1, x_2, x_3) = (ct, \mathbf{x}), \tag{1.358}$$

we rewrite (1.357) as

$$\begin{cases} t' = t + \frac{\Theta(\mathbf{x})}{c^2} \\ \mathbf{x}' = \mathbf{x}. \end{cases} \tag{1.359}$$

Note again, that since the new coordinate system $(**)$ in $\mathbb{R}^4$ is <u>curvilinear</u>, the time-like coordinate t' in coordinate system $(**)$ <u>differ</u> from the proper scalar of the global time.

Next consider the contravariant pseudo-metric tensor of the four-dimensional space-time $\{g^{ij}\}_{0\leq i,j\leq 3}$ that due to (1.221) has the form of

$$\begin{cases} g^{00} = 1 \\ g^{ij} = -\delta_{ij} + \frac{v^i v^j}{c^2} \quad \forall 1 \leq i, j \leq 3 \\ g^{0j} = g^{j0} = \frac{v^j}{c} \quad \forall 1 \leq j \leq 3, \end{cases} \tag{1.360}$$

in the Cartesian coordinate system (∗). We would like to find the form $\{g'^{ij}\}_{0\le i,j\le 3}$ of this tensor in the curvilinear coordinate system (∗∗). Then by (1.184) we have

$$g'^{mn} = \sum_{i=0}^{3}\sum_{j=0}^{3} \frac{\partial x'^m}{\partial x^i}\frac{\partial x'^n}{\partial x^j} g^{ij} \quad \forall 0 \le m, n \le 3. \tag{1.361}$$

Then straightforward calculations presented in Sect. 9.7 give that $\{g'^{ij}\}_{0\le i,j\le 3}$ has the following form in the system (∗∗):

$$\begin{cases} g'^{00} = \left(1 - \frac{|\mathbf{v}|^2}{c^2}\right)^{-1}, \\ g'^{0n} = g'^{n0} = 0 \quad \forall 1 \le n \le 3, \\ g'^{mn} = \frac{v^m}{c}\frac{v^n}{c} - \delta_{mn} \quad \forall 1 \le m, n \le 3. \end{cases} \tag{1.362}$$

Next we find that the <u>covariant</u> pseudo-metric tensor $\{g'_{ij}\}_{0\le i,j\le 3}$ in the curvilinear coordinate system (∗∗) has the following form:

$$\begin{cases} g'_{00} = \left(1 - \frac{|\mathbf{v}|^2}{c^2}\right), \\ g'_{0n} = g'_{n0} = 0 \quad \forall 1 \le n \le 3, \\ g'_{mn} = -\left(\left(1 - \frac{|\mathbf{v}|^2}{c^2}\right)^{-1} \frac{v^m}{c}\frac{v^n}{c} + \delta_{mn}\right) \quad \forall 1 \le m, n \le 3. \end{cases} \tag{1.363}$$

In particular, taking into account (1.359) and (1.355) we deduce that the quadratic form, induced by the covariant form of the pseudo-metric tensor $\{g'_{ij}\}_{0\le i,j\le 3}$ in the curvilinear coordinate system (∗∗), that defined on the tangent vectors $\left(dx'^0, dx'^1, dx'^2, dx'^3\right) \in \mathbb{R}^4$ where $d\mathbf{x}' := (dx'^1, dx'^2, dx'^3)$ has the following form:

$$\sum_{i=0}^{3}\sum_{j=0}^{3} g'_{ij} dx'^i dx'^j =$$

$$\left(1 - \frac{|\mathbf{v}(\mathbf{x}')|^2}{c^2}\right) dx_0'^2 - \left(\left(1 - \frac{|\mathbf{v}(\mathbf{x}')|^2}{c^2}\right)^{-1} \left|\frac{\mathbf{x}'}{|\mathbf{x}'|}\cdot d\mathbf{x}'\right|^2 + \left(|d\mathbf{x}'|^2 - \left|\frac{\mathbf{x}'}{|\mathbf{x}'|}\cdot d\mathbf{x}'\right|^2\right)\right). \tag{1.364}$$

Next, up to the end of this section, assume that our Cartesian coordinate system (∗) is non-rotating and our gravitational field is formed by the spherical symmetric massive body of mass m_0 and radius R_0 like the Earth, the Sun et al. with the center at the point 0. Then, as we get either in (1.22) and (1.23) (see Remark 4.4) in the case of the Newtonian gravity, or in Sect. 8.2 in the case of the more general gravity model, given by (1.177) with arbitrary constant β, we have either

$$\mathbf{v}(\mathbf{x}) = \frac{\sqrt{-2\Phi_1(|\mathbf{x}|)}}{|\mathbf{x}|}\mathbf{x}, \tag{1.365}$$

or

$$\mathbf{v}(\mathbf{x}) = -\frac{\sqrt{-2\Phi_1(|\mathbf{x}|)}}{|\mathbf{x}|}\mathbf{x}, \tag{1.366}$$

where Φ_1 is the classical Newtonian potential of our massive body m_0 that satisfies

$$\Phi_1(\mathbf{x}) = -\frac{Gm_0}{|\mathbf{x}|} \tag{1.367}$$

outside of the body surface. Both (1.365) and (1.366) are particular cases of (1.355), with

$$g(s) = \pm\sqrt{-2\Phi_1(s)}, \tag{1.368}$$

and in particular, outside of the massive body surface we have

$$g(|x|) = \pm\sqrt{\frac{2Gm_0}{|\mathbf{x}|}}. \tag{1.369}$$

Thus defining the function $\Theta(\mathbf{x})$ as in (1.356), that always can be done in the case $\frac{2Gm_0}{R_0} < c^2$, we can define the change of variables from coordinate system $(*)$ to the curvilinear coordinate system $(**)$ in the four-dimensional space-time $\mathbb{R}^4$ as in (1.359):

$$\begin{cases} t' = t + \frac{\Theta(\mathbf{x})}{c^2} \\ \mathbf{x}' = \mathbf{x}. \end{cases} \tag{1.370}$$

Then by inserting (1.365) or (1.366) into (1.363) we deduce the form of the covariant pseudo-metric tensor in the curvilinear coordinate system $(**)$:

$$\begin{cases} g'_{00} = \left(1 + \frac{2\Phi_1(|\mathbf{x}'|)}{c^2}\right), \\ g'_{0n} = g'_{n0} = 0 \quad \forall 1 \le n \le 3, \\ g'_{mn} = \left(\left(1 + \frac{2\Phi_1(|\mathbf{x}'|)}{c^2}\right)^{-1} \frac{2\Phi_1(|\mathbf{x}'|)}{c^2}\frac{x'_m}{|\mathbf{x}'|}\frac{x'_n}{|\mathbf{x}'|} - \delta_{mn}\right) \quad \forall 1 \le m, n \le 3. \end{cases} \tag{1.371}$$

Moreover, by (1.364) we have

$$\sum_{i=0}^{3}\sum_{j=0}^{3} g'_{ij}dx'^i dx'^j = \left(1 + \frac{2\Phi_1(|\mathbf{x}'|)}{c^2}\right)dx_0'^2$$
$$-\left(\left(1 + \frac{2\Phi_1(|\mathbf{x}'|)}{c^2}\right)^{-1}\left|\frac{\mathbf{x}'}{|\mathbf{x}'|}\cdot d\mathbf{x}'\right|^2 + \left(|d\mathbf{x}'|^2 - \left|\frac{\mathbf{x}'}{|\mathbf{x}'|}\cdot d\mathbf{x}'\right|^2\right)\right). \tag{1.372}$$

In particular, outside of the massive body surface, i.e. when $|x'| > R_0$ we rewrite (1.371) and (1.372) as

$$\begin{cases} g'_{00} = \left(1 - \frac{2Gm_0}{c^2|\mathbf{x}'|}\right), \\ g'_{0n} = g'_{n0} = 0 \quad \forall 1 \le n \le 3, \\ g'_{mn} = -\left(\left(1 - \frac{2Gm_0}{c^2|\mathbf{x}'|}\right)^{-1} \frac{2Gm_0}{c^2|\mathbf{x}'|} \frac{x'_m}{|\mathbf{x}'|} \frac{x'_n}{|\mathbf{x}'|} + \delta_{mn}\right) \quad \forall 1 \le m, n \le 3, \end{cases} \tag{1.373}$$

and

$$\sum_{i=0}^{3}\sum_{j=0}^{3} g'_{ij} dx'^i dx'^j =$$
$$\left(1 - \frac{2Gm_0}{c^2|\mathbf{x}'|}\right) dx_0'^2 - \left(\left(1 - \frac{2Gm_0}{c^2|\mathbf{x}'|}\right)^{-1} \left|\frac{\mathbf{x}'}{|\mathbf{x}'|} \cdot d\mathbf{x}'\right|^2 + \left(|d\mathbf{x}'|^2 - \left|\frac{\mathbf{x}'}{|\mathbf{x}'|} \cdot d\mathbf{x}'\right|^2\right)\right). \tag{1.374}$$

Therefore, we get that in coordinate system (∗∗), outside of the massive body, the covariant pseudo-metric tensor in (1.373) and (1.374) exactly the same as the well-known Schwarzschild metric from the General Relativity. Indeed in the spherical coordinates in $\mathbb{R}^3$ we rewrite (1.374) as

$$\sum_{i=0}^{3}\sum_{j=0}^{3} g'_{ij} dx'^i dx'^j =$$
$$\left(1 - \frac{2Gm_0}{c^2 r'}\right) dx_0'^2 - \left(\left(1 - \frac{2Gm_0}{c^2 r'}\right)^{-1} (dr')^2 + (r')^2 \left((d\theta')^2 + \sin^2(\theta')(d\varphi')^2\right)\right), \tag{1.375}$$

and this is exactly the classical Schwarzschild metric.

In particular, if we consider the monochromatic electromagnetic wave of frequency ω of the form $e^{i\omega t} U(\mathbf{x})$ in the coordinate system (∗), then by (1.370) in the coordinate system (∗∗) the form of this light is $e^{i\omega t'} U'(\mathbf{x}')$ where $U'(\mathbf{x}') = U(\mathbf{x}') e^{-i\omega \frac{\Theta(\mathbf{x}')}{c^2}}$, i.e. the electromagnetic wave in the coordinate system (∗∗) is also monochromatic of the same frequency ω. Thus all the optical effects that we find in the frames of our model coincides with the effects considered in the frames of General Relativity for the Schwarzschild metric. In particular, the Michelson–Morley experiment and all Sagnac-type effects will lead to the same result in the frame of our model like in the case of the General relativity. Moreover, since the Maxwell equations in both models have the same tensor form, all the electromagnetic effects, where the time does not appear explicitly will be the same. Similarly, the curvature of the light path in the Sun's gravitational field will be the same in both models. Finally, in the particular case of $\mathcal{G}(\tau) = \sqrt{\tau}$ in (1.317), i.e. in the case of the relativistic-like Lagrangian of the motion in (1.282) all the mechanical effects will be the same in the frame of our model like in the case of the General relativity for the Schwarzschild metric, provided that the time does not appear explicitly in this effects. In particular, the movement of the Mercury planet in the Sun's gravitational field will be the same in both models, provided we take into account the relativistic-like Lagrangian of the motion as in (1.282).

Finally, note that the similar, solution as in (1.374) or (1.375) is valid also for the general laws of the gravity, given by either (1.384) or (1.403), where in all places we take M_0 instead of m_0 and $\Phi_1(\mathbf{x}) = -\frac{GM_0}{|\mathbf{x}|}$ instead of (1.367), with M_0 being the total effective gravitational mass of the Earth (see Sect. 9.8.2 for the details).

1.9.9 General Lagrangian of the Gravitational-Electromagnetic Field, Compatible with the General Lagrangian of the Motion in (1.280)*: The Case of the Newtonian-Type Gravity*

Given known the distribution of inertial mass density of some continuum medium $\mu := \mu(\mathbf{x}, t)$, the field of velocities of this medium $\mathbf{u} := \mathbf{u}(\mathbf{x}, t)$, the charge density $\rho := \rho(\mathbf{x}, t)$ and the current density $\mathbf{j} := \mathbf{j}(\mathbf{x}, t)$ in a Cartesian coordinate system, consider a general Lagrangian density L of the unified gravitational-electromagnetic field that generalizes the Lagrangian density defined by (1.109) and is consistent with the Lagrangian of the motion of particles of the general form (1.280):

$$\begin{aligned} L\left(\mathbf{A}, \Psi, \mathbf{v}, \Phi, \mathbf{p}, \mathbf{x}, t\right) &:= \\ &\frac{1}{8\pi}\left|-\nabla_{\mathbf{x}}\Psi - \frac{1}{c}\frac{\partial \mathbf{A}}{\partial t} + \frac{1}{c}\mathbf{v}\times \operatorname{curl}_{\mathbf{x}}\mathbf{A}\right|^2 - \frac{1}{8\pi}\left|\operatorname{curl}_{\mathbf{x}}\mathbf{A}\right|^2 - \left(\rho\Psi - \frac{1}{c}\mathbf{A}\cdot\mathbf{j}\right) \\ &- \mu c^2 \mathcal{G}\left(1 - \frac{1}{c^2}\left|\mathbf{u}-\mathbf{v}\right|^2\right) + \frac{1}{2}\left(d_{\mathbf{x}}\mathbf{v} + \{d_{\mathbf{x}}\mathbf{v}\}^T\right) : \left(d_{\mathbf{x}}\mathbf{p} + \{d_{\mathbf{x}}\mathbf{p}\}^T\right) - 2\left(\operatorname{div}_{\mathbf{x}}\mathbf{v}\right)\left(\operatorname{div}_{\mathbf{x}}\mathbf{p}\right) \\ &+ \frac{1}{4\pi G}\left(\operatorname{div}_{\mathbf{x}}\mathbf{v}\right)\left(\frac{\partial\Phi}{\partial t} + \mathbf{v}\cdot\nabla_{\mathbf{x}}\Phi\right) + \frac{1}{4\pi G}\Phi\left(\operatorname{div}_{\mathbf{x}}\mathbf{v}\right)^2 - \frac{\Phi}{16\pi G}\left|d_{\mathbf{x}}\mathbf{v} + \{d_{\mathbf{x}}\mathbf{v}\}^T\right|^2 \\ &+ \frac{1}{8\pi G}\left|\nabla_{\mathbf{x}}\Phi\right|^2, \end{aligned} \tag{1.376}$$

where $\mathcal{G}(s) : \mathbb{R} \to \mathbb{R}$ is a given function, Φ is some ancillary proper scalar field and $\mathbf{p}$ is some ancillary proper vector field. Then L is invariant under the change of inertial or non-inertial Cartesian coordinate system of the form (9.16). Then denoting the function

$$g(s) := -c^2\mathcal{G}\left(1 - \frac{2s}{c^2}\right) \qquad \forall s \tag{1.377}$$

we rewrite (1.376) as

$$L\left(\mathbf{A},\Psi,\mathbf{v},\Phi,\mathbf{p},\mathbf{x},t\right):=$$
$$\frac{1}{8\pi}\left|-\nabla_{\mathbf{x}}\Psi-\frac{1}{c}\frac{\partial\mathbf{A}}{\partial t}+\frac{1}{c}\mathbf{v}\times\operatorname{curl}_{\mathbf{x}}\mathbf{A}\right|^2-\frac{1}{8\pi}\left|\operatorname{curl}_{\mathbf{x}}\mathbf{A}\right|^2-\left(\rho\Psi-\frac{1}{c}\mathbf{A}\cdot\mathbf{j}\right)$$
$$+\mu g\left(\frac{1}{2}\left|\mathbf{u}-\mathbf{v}\right|^2\right)+\frac{1}{2}\left(d_{\mathbf{x}}\mathbf{v}+\{d_{\mathbf{x}}\mathbf{v}\}^T\right):\left(d_{\mathbf{x}}\mathbf{p}+\{d_{\mathbf{x}}\mathbf{p}\}^T\right)-2\left(\operatorname{div}_{\mathbf{x}}\mathbf{v}\right)\left(\operatorname{div}_{\mathbf{x}}\mathbf{p}\right)$$
$$+\frac{1}{4\pi G}\left(\operatorname{div}_{\mathbf{x}}\mathbf{v}\right)\left(\frac{\partial\Phi}{\partial t}+\mathbf{v}\cdot\nabla_{\mathbf{x}}\Phi\right)+\frac{1}{4\pi G}\Phi\left(\operatorname{div}_{\mathbf{x}}\mathbf{v}\right)^2-\frac{\Phi}{16\pi G}\left|d_{\mathbf{x}}\mathbf{v}+\{d_{\mathbf{x}}\mathbf{v}\}^T\right|^2$$
$$+\frac{1}{8\pi G}\left|\nabla_{\mathbf{x}}\Phi\right|^2. \quad (1.378)$$

We point out two of the most important choices of function $\mathcal{G}(s)$: fully non-relativistic choice $\mathcal{G}(s)=\frac{s}{2}$ and correspondingly $g(s)=\left(s-\frac{c^2}{2}\right)$; and relativistic-like choice $\mathcal{G}(s)=\sqrt{s}$ and correspondingly $g(s):=-c^2\sqrt{1-\frac{2s}{c^2}}$. Note also that in the first case we have $\frac{dg}{ds}(s)=1$ and in the second case $\frac{dg}{ds}(s)=\left(1-\frac{2s}{c^2}\right)^{-\frac{1}{2}}\approx 1$, where the last equation is valid in the case where $2s\ll c^2$.

Then, as before, in Sect. 9.8.1 we obtain that a configuration $(\mathbf{A},\Psi,\mathbf{v},\Phi,\mathbf{p})$ is a critical point of the functional

$$J=\int_0^T\int_{\mathbb{R}^3}L\left(\mathbf{A},\Psi,\mathbf{v},\Phi,\mathbf{p},\mathbf{x},t\right)d\mathbf{x}dt \quad (1.379)$$

if and only if it satisfies

$$\begin{cases}\operatorname{curl}_{\mathbf{x}}\mathbf{H}=\frac{4\pi}{c}\mathbf{j}+\frac{1}{c}\frac{\partial\mathbf{D}}{\partial t}\\ \operatorname{div}_{\mathbf{x}}\mathbf{D}=4\pi\rho\\ \operatorname{curl}_{\mathbf{x}}\mathbf{E}+\frac{1}{c}\frac{\partial\mathbf{B}}{\partial t}=0\\ \operatorname{div}_{\mathbf{x}}\mathbf{B}=0\\ \mathbf{E}=\mathbf{D}-\frac{1}{c}\mathbf{v}\times\mathbf{B}\\ \mathbf{H}=\mathbf{B}+\frac{1}{c}\mathbf{v}\times\mathbf{D}\\ \operatorname{curl}_{\mathbf{x}}\left(\operatorname{curl}_{\mathbf{x}}\mathbf{v}\right)=0\\ \frac{\partial}{\partial t}\{\operatorname{div}_{\mathbf{x}}\mathbf{v}\}+\mathbf{v}\cdot\nabla_{\mathbf{x}}\left(\operatorname{div}_{\mathbf{x}}\mathbf{v}\right)+\frac{1}{4}\left|d_{\mathbf{x}}\mathbf{v}+\{d_{\mathbf{x}}\mathbf{v}\}^T\right|^2=-\Delta_{\mathbf{x}}\Phi\\ \left(\mu g'\left(\frac{1}{2}\left|\mathbf{u}-\mathbf{v}\right|^2\right)\left(\mathbf{u}-\mathbf{v}\right)+\frac{1}{4\pi c}\mathbf{D}\times\mathbf{B}\right)\\ =\operatorname{curl}_{\mathbf{x}}\left(\operatorname{curl}_{\mathbf{x}}\mathbf{p}\right)-\frac{1}{4\pi G}\left(\frac{\partial}{\partial t}\left(\nabla_{\mathbf{x}}\Phi\right)-\operatorname{curl}_{\mathbf{x}}\left(\mathbf{v}\times\nabla_{\mathbf{x}}\Phi\right)+\left(\Delta_{\mathbf{x}}\Phi\right)\mathbf{v}\right).\end{cases} \quad (1.380)$$

where, consistently with (1.54) we denote

$$\begin{cases} \mathbf{D} := -\nabla_{\mathbf{x}}\Psi - \frac{1}{c}\frac{\partial \mathbf{A}}{\partial t} + \frac{1}{c}\mathbf{v} \times \mathrm{curl}_{\mathbf{x}}\mathbf{A} \\ \mathbf{B} := \mathrm{curl}_{\mathbf{x}}\mathbf{A} \\ \mathbf{E} := -\nabla_{\mathbf{x}}\Psi - \frac{1}{c}\frac{\partial \mathbf{A}}{\partial t} \\ \mathbf{H} := \mathrm{curl}_{\mathbf{x}}\mathbf{A} + \frac{1}{c}\mathbf{v} \times \left(-\nabla_{\mathbf{x}}\Psi - \frac{1}{c}\frac{\partial \mathbf{A}}{\partial t} + \frac{1}{c}\mathbf{v} \times \mathrm{curl}_{\mathbf{x}}\mathbf{A}\right). \end{cases} \tag{1.381}$$

Next consider the equations of the gravitational-electromagnetic field in the form (1.380). Then, as before, defining the gravitational mass

$$M := \frac{1}{4\pi G}\Delta_{\mathbf{x}}\Phi \tag{1.382}$$

and using the continuum equation

$$\frac{\partial \mu}{\partial t} + \mathrm{div}_{\mathbf{x}}\,(\mu\mathbf{u}) = 0. \tag{1.383}$$

we rewrite (1.380) as

$$\begin{cases} \mathrm{curl}_{\mathbf{x}}\mathbf{H} = \frac{4\pi}{c}\mathbf{j} + \frac{1}{c}\frac{\partial \mathbf{D}}{\partial t}, \\ \mathrm{div}_{\mathbf{x}}\,\mathbf{D} = 4\pi\rho, \\ \mathrm{curl}_{\mathbf{x}}\mathbf{E} + \frac{1}{c}\frac{\partial \mathbf{B}}{\partial t} = 0, \\ \mathrm{div}_{\mathbf{x}}\,\mathbf{B} = 0, \\ \mathbf{E} = \mathbf{D} - \frac{1}{c}\mathbf{v} \times \mathbf{B}, \\ \mathbf{H} = \mathbf{B} + \frac{1}{c}\mathbf{v} \times \mathbf{D}, \\ \frac{\partial}{\partial t}\{\mathrm{div}_{\mathbf{x}}\,\mathbf{v}\} + \mathbf{v}\cdot\nabla_{\mathbf{x}}\,(\mathrm{div}_{\mathbf{x}}\,\mathbf{v}) + \frac{1}{4}\left|d_{\mathbf{x}}\mathbf{v} + \{d_{\mathbf{x}}\mathbf{v}\}^T\right|^2 = -4\pi G M, \\ \mathrm{curl}_{\mathbf{x}}\,(\mathrm{curl}_{\mathbf{x}}\mathbf{v}) = 0, \\ \frac{\partial}{\partial t}(M - \mu) + \mathrm{div}_{\mathbf{x}}\,\{(M - \mu)\,\mathbf{v}\} = \\ -\,\mathrm{div}_{\mathbf{x}}\left\{\mu\left(g'\left(\frac{1}{2}\,|\mathbf{u} - \mathbf{v}|^2\right) - 1\right)(\mathbf{u} - \mathbf{v}) + \frac{1}{4\pi c}\mathbf{D} \times \mathbf{B}\right\}. \end{cases} \tag{1.384}$$

Note again for the last equation in (1.384) that: in the fully non-relativistic case we have $g'(s) = 1$ and in the relativistic-like case we have $g'(s) = \left(1 - \frac{2s}{c^2}\right)^{-\frac{1}{2}} \approx 1$, where the last equation is valid in the case where $2s \ll c^2$.

1.9.10 General Lagrangian of the Gravitational-Electromagnetic Field, Compatible with the General Lagrangian of the Motion in (1.280): The Case of Some Possible Alternative Model of the Gravity

Consider $\mathbf{k}$ to be the vectorial potential of the inertia, which is a generally trivial speed-like vector field, assumed to be fixed in every fixed inertial or non-inertial Cartesian coordinate system (see Definition 1.9). Given known the distribution of inertial mass density of some continuum medium $\mu := \mu(\mathbf{x}, t)$, the field of velocities of this medium $\mathbf{u} := \mathbf{u}(\mathbf{x}, t)$, the charge density $\rho := \rho(\mathbf{x}, t)$ and the current density $\mathbf{j} := \mathbf{j}(\mathbf{x}, t)$ consider a Lagrangian density L defined by

$$\begin{aligned} L(\mathbf{A}, \Psi, \mathbf{v}, \Phi_0, \mathbf{h}, \mathbf{x}, t) := \frac{1}{8\pi}\left|-\nabla_{\mathbf{x}}\Psi - \frac{1}{c}\frac{\partial \mathbf{A}}{\partial t} + \frac{1}{c}\mathbf{v}\times \operatorname{curl}_{\mathbf{x}}\mathbf{A}\right|^2 - \frac{1}{8\pi}\left|\operatorname{curl}_{\mathbf{x}}\mathbf{A}\right|^2 \\ - \left(\rho\Psi - \frac{1}{c}\mathbf{A}\cdot\mathbf{j}\right) - \mu c^2 \mathcal{G}\left(1 - \frac{1}{c^2}\left|\mathbf{u}-\mathbf{v}\right|^2\right) \\ - \frac{c^2}{8\pi G}\left|-\nabla_{\mathbf{x}}\Phi_0 - \frac{1}{c}\frac{\partial \mathbf{h}}{\partial t} + \frac{1}{c}\mathbf{v}\times\operatorname{curl}_{\mathbf{x}}\mathbf{h} - \frac{1}{c}\nabla_{\mathbf{x}}(\mathbf{h}\cdot\mathbf{v})\right|^2 + \frac{c^2}{8\pi G}\left|\operatorname{curl}_{\mathbf{x}}\mathbf{h}\right|^2 \\ + \frac{c^2\beta}{4\pi G}\left(\frac{1}{4}\left|d_{\mathbf{x}}\mathbf{v} + \{d_{\mathbf{x}}\mathbf{v}\}^T\right|^2 - \left|\operatorname{div}_{\mathbf{x}}\mathbf{v}\right|^2\right), \end{aligned} \tag{1.385}$$

where

$$\mathbf{h} = \mathbf{v} - \mathbf{k} \quad \text{and} \quad \Phi_0 = -\frac{1}{2c}\left|\mathbf{v}-\mathbf{k}\right|^2, \tag{1.386}$$

$\mathcal{G}(s) : \mathbb{R} \to \mathbb{R}$ is a given function, and $\beta \in \mathbb{R}$ is some constant. Then, as before, denoting the function

$$g(s) := -c^2\mathcal{G}\left(1 - \frac{2s}{c^2}\right) \qquad \forall s \tag{1.387}$$

we rewrite (1.385) as

$$\begin{aligned} L(\mathbf{A}, \Psi, \mathbf{v}, \Phi_0, \mathbf{h}, \mathbf{x}, t) := \\ \frac{1}{8\pi}\left|-\nabla_{\mathbf{x}}\Psi - \frac{1}{c}\frac{\partial \mathbf{A}}{\partial t} + \frac{1}{c}\mathbf{v}\times \operatorname{curl}_{\mathbf{x}}\mathbf{A}\right|^2 - \frac{1}{8\pi}\left|\operatorname{curl}_{\mathbf{x}}\mathbf{A}\right|^2 - \left(\rho\Psi - \frac{1}{c}\mathbf{A}\cdot\mathbf{j}\right) + \mu g\left(\frac{1}{2}\left|\mathbf{u}-\mathbf{v}\right|^2\right) \\ - \frac{c^2}{8\pi G}\left|-\nabla_{\mathbf{x}}\Phi_0 - \frac{1}{c}\frac{\partial \mathbf{h}}{\partial t} + \frac{1}{c}\mathbf{v}\times\operatorname{curl}_{\mathbf{x}}\mathbf{h} - \frac{1}{c}\nabla_{\mathbf{x}}(\mathbf{h}\cdot\mathbf{v})\right|^2 + \frac{c^2}{8\pi G}\left|\operatorname{curl}_{\mathbf{x}}\mathbf{h}\right|^2 \\ + \frac{c^2\beta}{4\pi G}\left(\frac{1}{4}\left|d_{\mathbf{x}}\mathbf{v} + \{d_{\mathbf{x}}\mathbf{v}\}^T\right|^2 - \left|\operatorname{div}_{\mathbf{x}}\mathbf{v}\right|^2\right). \end{aligned} \tag{1.388}$$

In other words we have

$$
\begin{aligned}
& L\left(\mathbf{A}, \Psi, \mathbf{v}, \Phi_0, \mathbf{h}, \mathbf{x}, t\right)=L_1\left(\mathbf{A}, \Psi, \mathbf{v}, \mathbf{x}, t\right):= \\
& \frac{1}{8\pi}\left|-\nabla_{\mathbf{x}}\Psi-\frac{1}{c}\frac{\partial \mathbf{A}}{\partial t}+\frac{1}{c}\mathbf{v}\times\operatorname{curl}_{\mathbf{x}}\mathbf{A}\right|^2-\frac{1}{8\pi}\left|\operatorname{curl}_{\mathbf{x}}\mathbf{A}\right|^2-\left(\rho\Psi-\frac{1}{c}\mathbf{A}\cdot\mathbf{j}\right)+\mu g\left(\frac{1}{2}|\mathbf{u}-\mathbf{v}|^2\right) \\
& -\frac{c^2}{8\pi G}\left|\frac{1}{c}\nabla_{\mathbf{x}}\left(\frac{1}{2}|\mathbf{v}-\mathbf{k}|^2\right)-\frac{1}{c}\frac{\partial}{\partial t}(\mathbf{v}-\mathbf{k})+\frac{1}{c}\mathbf{v}\times\operatorname{curl}_{\mathbf{x}}(\mathbf{v}-\mathbf{k})-\frac{1}{c}\nabla_{\mathbf{x}}((\mathbf{v}-\mathbf{k})\cdot\mathbf{v})\right|^2 \\
& +\frac{c^2}{8\pi G}\left|\operatorname{curl}_{\mathbf{x}}(\mathbf{v}-\mathbf{k})\right|^2+\frac{c^2\beta}{4\pi G}\left(\frac{1}{4}\left|d_{\mathbf{x}}\mathbf{v}+\{d_{\mathbf{x}}\mathbf{v}\}^T\right|^2-\left|\operatorname{div}_{\mathbf{x}}\mathbf{v}\right|^2\right). \qquad (1.389)
\end{aligned}
$$

In particular, in the inertial coordinate system where $d_{\mathbf{x}}\mathbf{k}=0$ and $\partial_t\mathbf{k}=0$ we have

$$
\begin{aligned}
& L\left(\mathbf{A}, \Psi, \mathbf{v}, \Phi_0, \mathbf{h}, \mathbf{x}, t\right)=L_1\left(\mathbf{A}, \Psi, \mathbf{v}, \mathbf{x}, t\right):= \\
& \frac{1}{8\pi}\left|-\nabla_{\mathbf{x}}\Psi-\frac{1}{c}\frac{\partial \mathbf{A}}{\partial t}+\frac{1}{c}\mathbf{v}\times\operatorname{curl}_{\mathbf{x}}\mathbf{A}\right|^2-\frac{1}{8\pi}\left|\operatorname{curl}_{\mathbf{x}}\mathbf{A}\right|^2-\left(\rho\Psi-\frac{1}{c}\mathbf{A}\cdot\mathbf{j}\right)+\mu g\left(\frac{1}{2}|\mathbf{u}-\mathbf{v}|^2\right) \\
& -\frac{c^2}{8\pi G}\left|-\frac{1}{c}\frac{\partial \mathbf{v}}{\partial t}+\frac{1}{c}\mathbf{v}\times\operatorname{curl}_{\mathbf{x}}\mathbf{v}-\frac{1}{c}\nabla_{\mathbf{x}}\left(\frac{1}{2}|\mathbf{v}|^2\right)\right|^2+\frac{c^2}{8\pi G}\left|\operatorname{curl}_{\mathbf{x}}\mathbf{v}\right|^2 \\
& +\frac{c^2\beta}{4\pi G}\left(\frac{1}{4}\left|d_{\mathbf{x}}\mathbf{v}+\{d_{\mathbf{x}}\mathbf{v}\}^T\right|^2-\left|\operatorname{div}_{\mathbf{x}}\mathbf{v}\right|^2\right). \qquad (1.390)
\end{aligned}
$$

Note here the advantage of inertial coordinate systems, where L and L_1 are completely independent on the vectorial potential of the inertia $\mathbf{k}$. Furthermore, we rewrite (1.390) as

$$
\begin{aligned}
& L\left(\mathbf{A}, \Psi, \mathbf{v}, \Phi_0, \mathbf{h}, \mathbf{x}, t\right)=L_1\left(\mathbf{A}, \Psi, \mathbf{v}, \mathbf{x}, t\right):= \\
& \quad \frac{1}{8\pi}\left|-\nabla_{\mathbf{x}}\Psi-\frac{1}{c}\frac{\partial \mathbf{A}}{\partial t}+\frac{1}{c}\mathbf{v}\times\operatorname{curl}_{\mathbf{x}}\mathbf{A}\right|^2-\frac{1}{8\pi}\left|\operatorname{curl}_{\mathbf{x}}\mathbf{A}\right|^2 \\
& -\left(\rho\Psi-\frac{1}{c}\mathbf{A}\cdot\mathbf{j}\right)+\mu g\left(\frac{1}{2}|\mathbf{u}-\mathbf{v}|^2\right)-\frac{c^2}{8\pi G}\left|-\frac{1}{c}\left(\frac{\partial \mathbf{v}}{\partial t}+d_{\mathbf{x}}\mathbf{v}\cdot\mathbf{v}\right)\right|^2 \\
& +\frac{c^2}{8\pi G}\left|\operatorname{curl}_{\mathbf{x}}\mathbf{v}\right|^2+\frac{c^2\beta}{4\pi G}\left(\frac{1}{4}\left|d_{\mathbf{x}}\mathbf{v}+\{d_{\mathbf{x}}\mathbf{v}\}^T\right|^2-\left|\operatorname{div}_{\mathbf{x}}\mathbf{v}\right|^2\right). \qquad (1.391)
\end{aligned}
$$

Then, using Proposition 1.1 by (1.388), (1.386) and (1.389) we deduce that L and L_1 are invariant under the change of inertial or non-inertial Cartesian coordinate system, provided that $\mathbf{h}$ and $\mathbf{A}$ are proper vector fields, $\mathbf{v}$ is a speed-like vector field and Φ_0 and $\Psi_0 := \Psi - \frac{1}{c}\mathbf{A}\cdot\mathbf{v}$ are proper scalar fields.

We point out, again, two the most important choices of function $\mathcal{G}(s)$ in (1.385): fully non-relativistic choice $\mathcal{G}(s)=\frac{s}{2}$ and correspondingly $g(s)=\left(s-\frac{c^2}{2}\right)$; and relativistic-like choice $\mathcal{G}(s)=\sqrt{s}$ and correspondingly $g(s):=-c^2\sqrt{1-\frac{2s}{c^2}}$. Note also that in the first case we have $\frac{dg}{ds}(s) = 1$ and in the second case $\frac{dg}{ds}(s) = \left(1 - \frac{2s}{c^2}\right)^{-\frac{1}{2}} \approx 1$, where the last equation is valid in the case where $2s \ll c^2$.

We investigate critical points of the functional

$$J = \int_0^T \int_{\mathbb{R}^3} L_1(\mathbf{A}, \Psi, \mathbf{v}, \mathbf{x}, t)\, d\mathbf{x}dt. \tag{1.392}$$

We denote

$$\begin{cases} \Psi_0 = \Psi - \frac{1}{c}\mathbf{A}\cdot\mathbf{v} \\ \mathbf{D} = -\nabla_{\mathbf{x}}\Psi - \frac{1}{c}\frac{\partial\mathbf{A}}{\partial t} + \frac{1}{c}\mathbf{v}\times \operatorname{curl}_{\mathbf{x}}\mathbf{A} \\ = -\nabla_{\mathbf{x}}\Psi_0 - \frac{1}{c}\frac{\partial\mathbf{A}}{\partial t} + \frac{1}{c}\mathbf{v}\times \operatorname{curl}_{\mathbf{x}}\mathbf{A} - \frac{1}{c}\nabla_{\mathbf{x}}(\mathbf{A}\cdot\mathbf{v}) \\ \mathbf{B} = \operatorname{curl}_{\mathbf{x}}\mathbf{A} \\ \mathbf{E} = -\nabla_{\mathbf{x}}\Psi - \frac{1}{c}\frac{\partial\mathbf{A}}{\partial t} = \mathbf{D} - \frac{1}{c}\mathbf{v}\times\mathbf{B} \\ \mathbf{H} = \operatorname{curl}_{\mathbf{x}}\mathbf{A} + \frac{1}{c}\mathbf{v}\times\left(-\nabla_{\mathbf{x}}\Psi - \frac{1}{c}\frac{\partial\mathbf{A}}{\partial t} + \frac{1}{c}\mathbf{v}\times \operatorname{curl}_{\mathbf{x}}\mathbf{A}\right) = \mathbf{B} + \frac{1}{c}\mathbf{v}\times\mathbf{D}. \end{cases} \tag{1.393}$$

and

$$\begin{cases} \mathbf{R} = -\nabla_{\mathbf{x}}\Phi_0 - \frac{1}{c}\frac{\partial\mathbf{h}}{\partial t} + \frac{1}{c}\mathbf{v}\times \operatorname{curl}_{\mathbf{x}}\mathbf{h} - \frac{1}{c}\nabla_{\mathbf{x}}(\mathbf{h}\cdot\mathbf{v}) \\ \mathbf{Q} = \operatorname{curl}_{\mathbf{x}}\mathbf{h}, \end{cases} \tag{1.394}$$

where $\mathbf{h}$ is a proper vector field and Φ_0 is a proper scalar field that are given by (1.386). In other words,

$$\begin{cases} \mathbf{R} = \frac{1}{c}\nabla_{\mathbf{x}}\left(\frac{1}{2}|\mathbf{v}-\mathbf{k}|^2\right) - \frac{1}{c}\frac{\partial}{\partial t}(\mathbf{v}-\mathbf{k}) + \frac{1}{c}\mathbf{v}\times \operatorname{curl}_{\mathbf{x}}(\mathbf{v}-\mathbf{k}) - \frac{1}{c}\nabla_{\mathbf{x}}((\mathbf{v}-\mathbf{k})\cdot\mathbf{v}) \\ \mathbf{Q} = \operatorname{curl}_{\mathbf{x}}(\mathbf{v}-\mathbf{k}), \end{cases} \tag{1.395}$$

and in inertial coordinate system where $d_{\mathbf{x}}\mathbf{k} = 0$ and $\partial_t\mathbf{k} = 0$ we also have

$$\begin{cases} \mathbf{R} = -\frac{1}{c}\left(\frac{\partial\mathbf{v}}{\partial t} + d_{\mathbf{x}}\mathbf{v}\cdot\mathbf{v}\right) \\ \mathbf{Q} = \operatorname{curl}_{\mathbf{x}}\mathbf{v}. \end{cases} \tag{1.396}$$

Then in Sect. 9.8.2 we obtain that a configuration $(\mathbf{A}, \Psi, \mathbf{v}, \mathbf{x}, t)$ is a critical point of the functional

$$J = \int_0^T \int_{\mathbb{R}^3} L_1(\mathbf{A}, \Psi, \mathbf{v}, \mathbf{x}, t)\, d\mathbf{x}dt, \tag{1.397}$$

if and only if it satisfies

$$\begin{cases} \mathrm{curl}_{\mathbf{x}}\mathbf{B} = \frac{4\pi}{c}\mathbf{j} + \frac{1}{c}\frac{\partial \mathbf{D}}{\partial t} - \frac{1}{c}\mathrm{curl}_{\mathbf{x}}(\mathbf{v}\times\mathbf{D}) \\ \mathrm{div}_{\mathbf{x}}\,\mathbf{D} = 4\pi\rho \\ \mathrm{curl}_{\mathbf{x}}\mathbf{D} + \frac{1}{c}\frac{\partial \mathbf{B}}{\partial t} - \frac{1}{c}\,\mathrm{curl}_{\mathbf{x}}(\mathbf{v}\times\mathbf{B}) = 0 \\ \mathrm{div}_{\mathbf{x}}\,\mathbf{B} = 0 \\ \mathrm{curl}_{\mathbf{x}}\mathbf{R} + \frac{1}{c}\frac{\partial \mathbf{Q}}{\partial t} - \frac{1}{c}\mathrm{curl}_{\mathbf{x}}(\mathbf{v}\times\mathbf{Q}) = 0 \\ \mathrm{div}_{\mathbf{x}}\,\mathbf{Q} = 0 \\ \frac{1}{c}\left(\frac{\partial}{\partial t}(\mathrm{div}_{\mathbf{x}}\,\mathbf{v}) + \mathbf{v}\cdot\nabla_{\mathbf{x}}(\mathrm{div}_{\mathbf{x}}\,\mathbf{v}) + \frac{1}{4}\left|d_{\mathbf{x}}\mathbf{v} + \{d_{\mathbf{x}}\mathbf{v}\}^T\right|^2\right) = \frac{1}{2c}|\mathbf{Q}|^2 - \mathrm{div}_{\mathbf{x}}\,\mathbf{R} \\ \mathrm{curl}_{\mathbf{x}}(\mathrm{curl}_{\mathbf{x}}\mathbf{v}) = \mathrm{curl}_{\mathbf{x}}\mathbf{Q} \\ \frac{4\pi G}{c^2}\left(\mu g'\left(\frac{1}{2}|\mathbf{u}-\mathbf{v}|^2\right)(\mathbf{u}-\mathbf{v}) + \frac{1}{4\pi c}\mathbf{D}\times\mathbf{B} - \frac{c}{4\pi G}\mathbf{R}\times\mathbf{Q}\right) = \\ (1+\beta)\,\mathrm{curl}_{\mathbf{x}}\mathbf{Q} - \frac{1}{c}\left(\frac{\partial \mathbf{R}}{\partial t} - \mathrm{curl}_{\mathbf{x}}\{\mathbf{v}\times\mathbf{R}\} + (\mathrm{div}_{\mathbf{x}}\,\mathbf{R})\,\mathbf{v}\right), \end{cases} \tag{1.398}$$

and by (1.396) in the inertial frame we have

$$\begin{cases} \mathrm{curl}_{\mathbf{x}}\mathbf{B} = \frac{4\pi}{c}\mathbf{j} + \frac{1}{c}\frac{\partial \mathbf{D}}{\partial t} - \frac{1}{c}\mathrm{curl}_{\mathbf{x}}(\mathbf{v}\times\mathbf{D}) \\ \mathrm{div}_{\mathbf{x}}\,\mathbf{D} = 4\pi\rho \\ \mathrm{curl}_{\mathbf{x}}\mathbf{D} + \frac{1}{c}\frac{\partial \mathbf{B}}{\partial t} - \frac{1}{c}\,\mathrm{curl}_{\mathbf{x}}(\mathbf{v}\times\mathbf{B}) = 0 \\ \mathrm{div}_{\mathbf{x}}\,\mathbf{B} = 0 \\ \mathbf{R} = -\frac{1}{c}\left(\frac{\partial \mathbf{v}}{\partial t} + d_{\mathbf{x}}\mathbf{v}\cdot\mathbf{v}\right) \\ \mathbf{Q} = \mathrm{curl}_{\mathbf{x}}\mathbf{v} \\ \frac{4\pi G}{c^2}\left(\mu g'\left(\frac{1}{2}|\mathbf{u}-\mathbf{v}|^2\right)(\mathbf{u}-\mathbf{v}) + \frac{1}{4\pi c}\mathbf{D}\times\mathbf{B} - \frac{c}{4\pi G}\mathbf{R}\times\mathbf{Q}\right) = \\ (1+\beta)\,\mathrm{curl}_{\mathbf{x}}\mathbf{Q} - \frac{1}{c}\left(\frac{\partial \mathbf{R}}{\partial t} - \mathrm{curl}_{\mathbf{x}}\{\mathbf{v}\times\mathbf{R}\} + (\mathrm{div}_{\mathbf{x}}\,\mathbf{R})\,\mathbf{v}\right). \end{cases} \tag{1.399}$$

Furthermore, taking $\mathrm{div}_{\mathbf{x}}$ of the both sides of the last equality in (1.398) and using continuum equation $\partial_t\mu + \mathrm{div}_{\mathbf{x}}(\mu\mathbf{u}) = 0$ we deduce

$$\begin{aligned} &-(\partial_t\mu + \mathrm{div}_{\mathbf{x}}(\mu\mathbf{v})) \\ &+\mathrm{div}_{\mathbf{x}}\left\{\mu\left(g'\left(\frac{1}{2}|\mathbf{u}-\mathbf{v}|^2\right) - 1\right)(\mathbf{u}-\mathbf{v}) + \frac{1}{4\pi c}\mathbf{D}\times\mathbf{B} - \frac{c}{4\pi G}\mathbf{R}\times\mathbf{Q}\right\} \\ &\quad= \mathrm{div}_{\mathbf{x}}\left(\mu g'\left(\frac{1}{2}|\mathbf{u}-\mathbf{v}|^2\right)(\mathbf{u}-\mathbf{v}) + \frac{1}{4\pi c}\mathbf{D}\times\mathbf{B} - \frac{c}{4\pi G}\mathbf{R}\times\mathbf{Q}\right) = \\ &\qquad -\frac{c}{4\pi G}\left(\frac{\partial}{\partial t}(\mathrm{div}_{\mathbf{x}}\,\mathbf{R}) + \mathrm{div}_{\mathbf{x}}\{(\mathrm{div}_{\mathbf{x}}\,\mathbf{R})\,\mathbf{v}\}\right). \end{aligned} \tag{1.400}$$

Therefore, considering the proper scalar quantity Q_0, that we call the field mass, which satisfies

$$Q_0 := -\mu + \frac{c}{4\pi G}\,\mathrm{div}_{\mathbf{x}}\,\mathbf{R}, \tag{1.401}$$

by (1.400) we deduce

$$\frac{\partial Q_0}{\partial t} + \operatorname{div}_{\mathbf{x}}(Q_0\mathbf{v}) = \\ -\operatorname{div}_{\mathbf{x}}\left\{\mu\left(g'\left(\frac{1}{2}|\mathbf{u}-\mathbf{v}|^2\right)-1\right)(\mathbf{u}-\mathbf{v}) + \frac{1}{4\pi c}\mathbf{D}\times\mathbf{B} - \frac{c}{4\pi G}\mathbf{R}\times\mathbf{Q}\right\}. \tag{1.402}$$

Thus, we rewrite (1.398) as

$$\begin{cases} \operatorname{curl}_{\mathbf{x}}\mathbf{B} = \frac{4\pi}{c}(\mathbf{j}-\rho\mathbf{v}) + \frac{1}{c}\frac{\partial\mathbf{D}}{\partial t} - \frac{1}{c}\operatorname{curl}_{\mathbf{x}}(\mathbf{v}\times\mathbf{D}) + \frac{1}{c}(\operatorname{div}_{\mathbf{x}}\mathbf{D})\,\mathbf{v}, \\ \operatorname{div}_{\mathbf{x}}\mathbf{D} = 4\pi\rho, \\ \operatorname{curl}_{\mathbf{x}}\mathbf{D} + \frac{1}{c}\frac{\partial\mathbf{B}}{\partial t} - \frac{1}{c}\operatorname{curl}_{\mathbf{x}}(\mathbf{v}\times\mathbf{B}) = 0, \\ \operatorname{div}_{\mathbf{x}}\mathbf{B} = 0, \\ \operatorname{curl}_{\mathbf{x}}\mathbf{R} + \frac{1}{c}\frac{\partial\mathbf{Q}}{\partial t} - \frac{1}{c}\operatorname{curl}_{\mathbf{x}}(\mathbf{v}\times\mathbf{Q}) = 0, \\ \operatorname{div}_{\mathbf{x}}\mathbf{Q} = 0, \\ \frac{4\pi G}{c^2}\left(\mu g'\left(\frac{1}{2}|\mathbf{u}-\mathbf{v}|^2\right)(\mathbf{u}-\mathbf{v}) + \frac{1}{4\pi c}\mathbf{D}\times\mathbf{B} - \frac{c}{4\pi G}\mathbf{R}\times\mathbf{Q}\right) = \\ (1+\beta)\operatorname{curl}_{\mathbf{x}}\mathbf{Q} - \frac{1}{c}\left(\frac{\partial\mathbf{R}}{\partial t} - \operatorname{curl}_{\mathbf{x}}\{\mathbf{v}\times\mathbf{R}\} + (\operatorname{div}_{\mathbf{x}}\mathbf{R})\,\mathbf{v}\right), \\ \operatorname{div}_{\mathbf{x}}\mathbf{R} = \frac{4\pi G}{c}(\mu+Q_0), \\ \frac{1}{c}\left(\frac{\partial}{\partial t}(\operatorname{div}_{\mathbf{x}}\mathbf{v}) + \mathbf{v}\cdot\nabla_{\mathbf{x}}(\operatorname{div}_{\mathbf{x}}\mathbf{v}) + \frac{1}{4}\left|d_{\mathbf{x}}\mathbf{v} + \{d_{\mathbf{x}}\mathbf{v}\}^T\right|^2\right) = \frac{1}{2c}|\mathbf{Q}|^2 - \operatorname{div}_{\mathbf{x}}\mathbf{R}, \\ \operatorname{curl}_{\mathbf{x}}(\operatorname{curl}_{\mathbf{x}}\mathbf{v}) = \operatorname{curl}_{\mathbf{x}}\mathbf{Q}, \\ \frac{\partial Q_0}{\partial t} + \operatorname{div}_{\mathbf{x}}(Q_0\mathbf{v}) = \\ -\operatorname{div}_{\mathbf{x}}\left\{\mu\left(g'\left(\frac{1}{2}|\mathbf{u}-\mathbf{v}|^2\right)-1\right)(\mathbf{u}-\mathbf{v}) + \frac{1}{4\pi c}\mathbf{D}\times\mathbf{B} - \frac{c}{4\pi G}\mathbf{R}\times\mathbf{Q}\right\}, \end{cases} \tag{1.403}$$

and we rewrite (1.399) in the inertial frame as

$$\begin{cases} \operatorname{curl}_{\mathbf{x}}\mathbf{B} = \frac{4\pi}{c}\mathbf{j} + \frac{1}{c}\frac{\partial\mathbf{D}}{\partial t} - \frac{1}{c}\operatorname{curl}_{\mathbf{x}}(\mathbf{v}\times\mathbf{D}), \\ \operatorname{div}_{\mathbf{x}}\mathbf{D} = 4\pi\rho, \\ \operatorname{curl}_{\mathbf{x}}\mathbf{D} + \frac{1}{c}\frac{\partial\mathbf{B}}{\partial t} - \frac{1}{c}\operatorname{curl}_{\mathbf{x}}(\mathbf{v}\times\mathbf{B}) = 0, \\ \operatorname{div}_{\mathbf{x}}\mathbf{B} = 0, \\ \mathbf{R} = -\frac{1}{c}\left(\frac{\partial\mathbf{v}}{\partial t} + d_{\mathbf{x}}\mathbf{v}\cdot\mathbf{v}\right), \\ \mathbf{Q} = \operatorname{curl}_{\mathbf{x}}\mathbf{v}, \\ \frac{4\pi G}{c^2}\left(\mu g'\left(\frac{1}{2}|\mathbf{u}-\mathbf{v}|^2\right)(\mathbf{u}-\mathbf{v}) + \frac{1}{4\pi c}\mathbf{D}\times\mathbf{B} - \frac{c}{4\pi G}\mathbf{R}\times\mathbf{Q}\right) = \\ (1+\beta)\operatorname{curl}_{\mathbf{x}}\mathbf{Q} - \frac{1}{c}\left(\frac{\partial\mathbf{R}}{\partial t} - \operatorname{curl}_{\mathbf{x}}\{\mathbf{v}\times\mathbf{R}\} + (\operatorname{div}_{\mathbf{x}}\mathbf{R})\,\mathbf{v}\right), \\ \operatorname{div}_{\mathbf{x}}\mathbf{R} = \frac{4\pi G}{c}(\mu+Q_0), \\ \frac{\partial Q_0}{\partial t} + \operatorname{div}_{\mathbf{x}}(Q_0\mathbf{v}) = \\ -\operatorname{div}_{\mathbf{x}}\left\{\mu\left(g'\left(\frac{1}{2}|\mathbf{u}-\mathbf{v}|^2\right)-1\right)(\mathbf{u}-\mathbf{v}) + \frac{1}{4\pi c}\mathbf{D}\times\mathbf{B} - \frac{c}{4\pi G}\mathbf{R}\times\mathbf{Q}\right\}. \end{cases} \tag{1.404}$$

As before, by Proposition 1.1 we deduce that (1.403) is invariant under the change of inertial or non-inertial Cartesian coordinate systems. Moreover, (1.404) is invariant under the change of inertial Cartesian coordinate systems. Note also that, both, (1.403) in an arbitrary inertial or non-inertial Cartesian coordinate system and (1.404) in an arbitrary inertial Cartesian coordinate system, are completely independent on the vectorial potential of the inertia **k**. Note again for the last equation in (1.403) or (1.404) that: in the fully non-relativistic case we have $g'(s) = 1$ and in the relativistic-like case we have $g'(s) = \left(1 - \frac{2s}{c^2}\right)^{-\frac{1}{2}} \approx 1$, where the last equation is valid in the case where $2s \ll c^2$.

Finally, as before, note that in the case of large constant $|\beta| \gg 1$ we have $\mathbf{Q} \to 0$ in (1.403) and thus, the gravity equations (1.403) reduce to the equations of the Newtonian-type Gravity in the form of (1.384). In that case the gravity field propagates with the infinite speed. On the other hand, in the case of vanishing constant $\beta = 0$ the form of equations for **R** and **Q** in (1.403) is completely the same as the form of the Maxwell equations for **D** and **B** in (1.403), except of the different meaning of "charges" and "currents" in these two sets of equations. In that case the electromagnetic and the gravity fields propagate with the same speed. However, in the mixed case of constant $\beta \sim 1$ the electromagnetic and the gravity fields propagate with different finite speeds.

1.9.11 Covariant Formulation of the Laws of Gravity in Cartesian and Curvilinear Coordinate Systems: The Case of the Newtonian-Type Gravity

Next our purpose is to make the equivalent form of the Lagrangian density of the Newtonian-type gravity in (1.378) to be covariant and valid in every curvilinear coordinate system.

Assume first that our coordinate system is inertial or more generally non-inertial and Cartesian. Then consider a three-dimensional vectorial gravitational potential $\mathbf{v} = (v^1, v^2, v^3)$ and consider the covariant pseudo-metric tensor $G = \{g_{ij}\}_{0 \le i,j \le 3}$ defined by (1.222):

$$\begin{cases} g_{00} = 1 - \frac{|\mathbf{v}|^2}{c^2} \\ g_{ij} = -\delta_{ij} \quad \forall 1 \le i, j \le 3 \\ g_{0j} = g_{j0} = \frac{v^j}{c} \quad \forall 1 \le j \le 3. \end{cases} \tag{1.405}$$

Next consider the contravariant pseudo-metric tensor $\tilde{G} = \{g^{ij}\}_{0 \le i,j \le 3}$ defined by (1.221):

$$\begin{cases} g^{00} = 1 \\ g^{ij} = -\delta_{ij} + \frac{v^i v^j}{c^2} \quad \forall 1 \le i, j \le 3 \\ g^{0j} = g^{j0} = \frac{v^j}{c} \quad \forall 1 \le j \le 3. \end{cases} \tag{1.406}$$

Next consider the Christoffel Symbols:

$$\begin{cases}\Gamma_{i,kn} := \frac{1}{2}\left(\frac{\partial g_{ik}}{\partial x_n} + \frac{\partial g_{in}}{\partial x_k} - \frac{\partial g_{kn}}{\partial x_i}\right) \\ \Gamma^i_{kn} := \sum_{j=0}^{3} g^{ij}\Gamma_{j,kn}\end{cases} \quad \forall i,k,n = 0,1,2,3, \tag{1.407}$$

where $\mathbf{x} = (x_1, x_2, x_3)$, $x_0 = ct$ and the point in the four dimensional space-time is denoted as $(x^0, x^1, x^2, x^3) := (ct, \mathbf{x}) = (x_0, x_1, x_2, x_3)$. In particular, by (1.405) and by the first equation in (1.407) we obtain

$$\begin{cases}\Gamma_{0,00} = -\frac{1}{2c^3}\frac{\partial(|\mathbf{v}|^2)}{\partial t} \\ \Gamma_{0,k0} = \Gamma_{0,0k} = -\frac{1}{2c^2}\frac{\partial(|\mathbf{v}|^2)}{\partial x_k} \quad \forall k = 1,2,3, \\ \Gamma_{0,kn} = \frac{1}{2c}\left(\frac{\partial v^k}{\partial x_n} + \frac{\partial v^n}{\partial x_k}\right) \quad \forall k,n = 1,2,3 \\ \Gamma_{i,00} = \frac{1}{c^2}\left(\frac{\partial v^i}{\partial t} + \frac{1}{2}\frac{\partial(|\mathbf{v}|^2)}{\partial x_i}\right) \quad \forall i = 1,2,3, \\ \Gamma_{i,k0} = \Gamma_{i,0k} = \frac{1}{2c}\left(\frac{\partial v^i}{\partial x_k} - \frac{\partial v^k}{\partial x_i}\right) \quad \forall i,k = 1,2,3, \\ \Gamma_{i,kn} = 0 \quad \forall i,k,n = 1,2,3.\end{cases} \tag{1.408}$$

Next consider the four-dimensional gravitational potential (v^0, v^1, v^2, v^3) defined by (1.196) as

$$(v^0, v^1, v^2, v^3) := \left(1, \frac{1}{c}\mathbf{v}\right), \tag{1.409}$$

and the corresponding lowered four-covector field (v_0, v_1, v_2, v_3), that we called the four-covector of gravitational potential:

$$(v_0, v_1, v_2, v_3) := (1, 0, 0, 0)\,. \tag{1.410}$$

Note again that by (1.410) and (1.213) the four-covector of gravitational potential is a gradient of the global time multiplied by the constant c:

$$(v_0, v_1, v_2, v_3) := (1, 0, 0, 0) = c\left(\frac{\partial t}{\partial x^0}, \frac{\partial t}{\partial x^1}, \frac{\partial t}{\partial x^2}, \frac{\partial t}{\partial x^1}\right). \tag{1.411}$$

Furthermore, let $\{\Theta^{ij}\}_{0\le i,j\le 3}$ be the contravariant tensor of the three-dimensional geometry that satisfies (1.211) in every non-inertial Cartesian coordinate system:

$$\begin{cases}\Theta^{00} = 0 \\ \Theta^{0j} = \Theta^{j0} = 0 \quad \forall j = 1,2,3 \\ \Theta^{ij} := \delta_{ij} \quad \forall i,j = 1,2,3.\end{cases} \tag{1.412}$$

Moreover, by (1.220) we have

$$g^{ij} := v^i v^j - \Theta^{ij} \qquad \forall i, j = 0, 1, 2, 3. \tag{1.413}$$

Then in Sect. 9.9 we prove that we can write that given a three-dimensional vectorial gravitational potential $\mathbf{v}$, a proper scalar Φ and a proper three-dimensional vector $\mathbf{p}$ we have

$$\begin{aligned}
&\frac{1}{2}\left(d_{\mathbf{x}}\mathbf{v} + \{d_{\mathbf{x}}\mathbf{v}\}^T\right) : \left(d_{\mathbf{x}}\mathbf{p} + \{d_{\mathbf{x}}\mathbf{p}\}^T\right) - 2\,(\text{div}_{\mathbf{x}}\,\mathbf{v})\,(\text{div}_{\mathbf{x}}\,\mathbf{p}) \\
&\quad + \frac{1}{4\pi G}\,(\text{div}_{\mathbf{x}}\,\mathbf{v})\left(\frac{\partial \Phi}{\partial t} + \mathbf{v}\cdot\nabla_{\mathbf{x}}\Phi\right) + \frac{1}{4\pi G}\Phi\,(\text{div}_{\mathbf{x}}\,\mathbf{v})^2 - \frac{\Phi}{16\pi G}\left|d_{\mathbf{x}}\mathbf{v} + \{d_{\mathbf{x}}\mathbf{v}\}^T\right|^2 \\
&\quad + \frac{1}{8\pi G}\,|\nabla_{\mathbf{x}}\Phi|^2 = \sum_{m=0}^{3}\sum_{n=0}^{3}\frac{32\pi G}{c^4}\,\Theta^{mn}\,\frac{\partial}{\partial x^m}\left(\sum_{j=0}^{3}\sum_{k=0}^{3} g^{kj} v_k S_j\right)\frac{\partial}{\partial x^n}\left(\sum_{j=0}^{3}\sum_{k=0}^{3} g^{kj} v_k S_j\right) \\
&\quad + \left(\sum_{i=0}^{3}\sum_{j=0}^{3} g^{ij}\left(\delta_j v_i + \delta_i v_j\right)\right)\left(\sum_{i=0}^{3}\sum_{j=0}^{3} g^{ij}\left(\delta_j S_i + \delta_i S_j\right)\right) \\
&\quad - \sum_{k=0}^{3}\sum_{j=0}^{3}\sum_{m=0}^{3}\sum_{n=0}^{3} g^{km} g^{jn}\left(\delta_j S_k + \delta_k S_j\right)\left(\delta_m v_n + \delta_n v_m\right),
\end{aligned} \tag{1.414}$$

where the four-covector field (S_0, S_1, S_2, S_3) on the group $\mathcal{S}_0$ and the corresponding lifted four-covector field (S^0, S^1, S^2, S^3) are given by

$$\begin{cases} S_0 = \frac{c^2}{16\pi G}\Phi - \frac{1}{2}\mathbf{v}\cdot\mathbf{p} \\ S_j = \frac{c}{2}p_j \qquad \forall 1 \le j \le 3 \qquad \text{where} \quad (p_1, p_2, p_3) := \mathbf{p}, \end{cases} \tag{1.415}$$

and

$$\begin{cases} S^0 = \frac{c^2}{16\pi G}\Phi \\ S^j = \frac{c^2}{16\pi G}\Phi\frac{v^j}{c} - \frac{c}{2}p_j \qquad \forall 1 \le j \le 3, \end{cases} \tag{1.416}$$

and where $\delta_j S_i$ and $\delta_j v_i$ mean the covariant derivatives of the four-covectors (S_0, S_1, S_2, S_3) and (v_0, v_1, v_2, v_3), which are known from the Tensor Analysis to be two times covariant tensors and defined by the following:

$$\begin{cases} \delta_j S_i := \frac{\partial S_i}{\partial x_j} - \sum_{k=0}^{3}\Gamma_{ij}^k S_k \qquad \forall 0 \le i, j \le 3 \\ \delta_j v_i := \frac{\partial v_i}{\partial x_j} - \sum_{k=0}^{3}\Gamma_{ij}^k v_k \qquad \forall 0 \le i, j \le 3. \end{cases} \tag{1.417}$$

Note here that the right-hand side of (1.414) is written in a covariant form which is valid also in every curvilinear coordinate system.

Next, given a system of n particles with inertial masses $m_1, \ldots, m_n$, charges $\sigma_1, \ldots, \sigma_n$, places $\mathbf{r}_1(t), \ldots, \mathbf{r}_n(t)$ and velocities $\frac{d\mathbf{r}_1}{dt}(t), \ldots, \frac{d\mathbf{r}_n}{dt}(t)$ in the Cartesian coordinate system, the usual definitions of the charge density, current density and the mass density of this system are the following:

$$\begin{cases} \rho(\mathbf{x},t) := \sum_{j=1}^{n} \sigma_j\, \delta^{(3)}\left(\mathbf{x} - \mathbf{r}_j(t)\right), \\ \mathbf{j}(\mathbf{x},t) := \sum_{j=1}^{n} \sigma_j \frac{d\mathbf{r}_j}{dt}(t)\, \delta^{(3)}\left(\mathbf{x} - \mathbf{r}_j(t)\right), \\ \mu(\mathbf{x},t) := \sum_{j=1}^{n} m_j\, \delta^{(3)}\left(\mathbf{x} - \mathbf{r}_j(t)\right), \end{cases} \tag{1.418}$$

where $\delta^{(3)}$ is the usual Dirac-delta distribution (generalized function) in $\mathbb{R}^3$. Then denoting

$$\left(x^0, x^1, x^2, x^3\right) := \left(t, \frac{1}{c}\mathbf{x}\right), \tag{1.419}$$

denoting $\left(\chi_j^0(t), \chi_j^1(t), \chi_j^2(t), \chi_j^3(t)\right) \in \mathbb{R}^4$ to be a four-dimensional space-time trajectory of the j-th particle, parameterized by the global time, which is in Cartesian system defined by the following:

$$\left(\chi_j^0(t), \chi_j^1(t), \chi_j^2(t), \chi_j^3(t)\right) := \left(t, \frac{1}{c}\mathbf{r}_j(t)\right), \tag{1.420}$$

denoting by G the 4×4-matrix $G := \{g_{ij}\}_{0 \le i,j \le 3}$, which satisfies $\det G = -1$ in every Cartesian coordinate system, and denoting by $\left(j^0, j^1, j^2, j^3\right)$ the four vector of the current which is in Cartesian system defined by the following:

$$\left(j^0, j^1, j^2, j^3\right) := \left(\rho, \frac{1}{c}\mathbf{j}\right)(\mathbf{x}, t), \tag{1.421}$$

in Sect. 9.9 we prove, that similarly to the General Relativity, in every curvilinear coordinate system, where $\frac{\partial t}{\partial x^0} > 0$, we have

$$\left(j^0, j^1, j^2, j^3\right) := \frac{1}{\sqrt{|\det G|}}\left(\hat{\rho}, \frac{1}{c}\hat{\rho}\,\hat{\mathbf{u}}_{x^0}\right) \quad \text{where}$$

$$\hat{\rho} := \sum_{j=1}^{n} \sigma_j\, \delta^{(3)}\left(\hat{\mathbf{x}} - c\left(\chi_j^1, \chi_j^2, \chi_j^3\right)\left(\chi_j^0\right)\right)$$

is the local charge density, calculated in the curvilinear coordinate system. (1.422)

Here $\hat{\mathbf{x}} := (cx^1, cx^2, cx^3)$ and $\hat{\mathbf{u}}_{x^0}$ is the field of velocities of the system, calculated in a given curvilinear coordinate system by the differentiation of the last three coordinates of the particle: $\hat{\mathbf{r}} := (c\chi^1, c\chi^2, c\chi^3)$ by the coordinate χ^0 that can be considered as the local time instead of global time t. Moreover, the quantity in (1.422) equals to a four-vector, under the change of curvilinear coordinate systems.

Remark 1.2 Note here that we denoted the matrix $G = \{g_{ij}\}_{0\leq i,j\leq 3}$ by the same letter as the Gravitational Constant G. However, there is no ambiguity, since in the second case G is a constant scalar and in the first case G is a matrix. Moreover, we will use the matrix notation $G = \{g_{ij}\}_{0\leq i,j\leq 3}$ only in the expressions containing term $\det G$.

Similarly to (1.422), the following quantities, defined in every coordinate system where $\frac{\partial t}{\partial x^0} > 0$, equals to a four-vector and a covariant scalar respectively, under the change of curvilinear coordinate systems:

$$\frac{1}{\sqrt{|\det G|}}\left(\hat{\mu},\, \frac{1}{c}\hat{\mu}\,\hat{\mathbf{u}}_{x^0}\right) \quad\text{and}\quad \frac{\hat{\mu}}{\sqrt{|\det G|}}\sqrt{\left(\sum_{m=0}^{3}\sum_{k=0}^{3} g_{mk}\,\hat{u}^m_{x^0}\,\hat{u}^k_{x^0}\right)} \qquad \text{where}$$

$$\hat{\mu} := \sum_{j=1}^{n} m_j\,\delta^{(3)}\left(\hat{\mathbf{x}} - c\left(\chi^1_j, \chi^2_j, \chi^3_j\right)\left(\chi^0_j\right)\right)$$

is the local mass density, calculated in the curvilinear coordinate system, (1.423)

and $\left(\hat{u}^0, \hat{u}^1, \hat{u}^2, \hat{u}^3\right)_{x^0} = \left(1, \frac{1}{c}\hat{\mathbf{u}}_{x^0}\right)$ is the field of four-dimensional velocities of the system, calculated in a given curvilinear coordinate system by the differentiation of the four-dimensional coordinates of the particles by the first coordinate χ^0. Note here that although the quantities $\hat{\rho}$ and $\hat{\mu}$ are not covariant scalars and $\left(\hat{u}^0, \hat{u}^1, \hat{u}^2, \hat{u}^3\right)_{x^0}$ is not a four-vector, the first quantity in (1.422) equals to a four-vector and the two first quantities in (1.423) equal to a four-vector and a covariant scalar, under the change of curvilinear coordinate systems. Moreover, clearly the four dimensional speed $\left(u^0, u^1, u^2, u^3\right)_t$, obtained in curvilinear coordinate system by the differentiation of the global time t, instead of the first local coordinate χ_0, indeed forms a four-vector and therefore, the quantity

$$\frac{\hat{\mu}}{\sqrt{|\det G|}}\left(\sum_{m=0}^{3}\sum_{k=0}^{3} g_{mk}\,\hat{u}^m_{x^0}\,u^k_t\right)$$

$$\text{where}\quad \hat{\mu} = \sum_{j=1}^{n} m_j\,\delta^{(3)}\left(\hat{\mathbf{x}} - c\left(\chi^1_j, \chi^2_j, \chi^3_j\right)\left(\chi^0_j\right)\right) \qquad (1.424)$$

equals to a covariant scalar, under the change of curvilinear coordinate systems.

Next we can write the density of the Lagrangian of the electromagnetic field, defined in (1.55) in the equivalent form (1.268), where the right-hand side is written in a covariant form which is valid for every curvilinear coordinate system:

$$\frac{1}{4\pi}\left(\frac{1}{2}|\mathbf{D}|^2-\frac{1}{2}|\mathbf{B}|^2-4\pi\left(\rho\Psi-\frac{1}{c}\mathbf{A}\cdot\mathbf{j}\right)\right)=$$
$$\frac{1}{4\pi}\left(-\sum_{n=0}^{3}\sum_{k=0}^{3}\sum_{m=0}^{3}\sum_{p=0}^{3}\frac{1}{4}g^{mn}g^{pk}\left(\frac{\partial A_p}{\partial x^m}-\frac{\partial A_m}{\partial x^p}\right)\left(\frac{\partial A_k}{\partial x^n}-\frac{\partial A_n}{\partial x^k}\right)-\sum_{k=0}^{3}4\pi j^k A_k\right), \tag{1.425}$$

where (j^0, j^1, j^2, j^3) is the four-vector of the current that satisfies (1.422) in every curvilinear coordinate system where $\frac{\partial t}{\partial x^0} > 0$, and (A_0, A_1, A_2, A_3) is the four-covector of the electromagnetic potential. Then by (1.425), (1.414) and (1.424) we rewrite the Lagrangian density in (1.378) in the Cartesian coordinate system as

$$\frac{1}{8\pi}\left|-\nabla_{\mathbf{x}}\Psi-\frac{1}{c}\frac{\partial\mathbf{A}}{\partial t}+\frac{1}{c}\mathbf{v}\times\mathrm{curl}_{\mathbf{x}}\mathbf{A}\right|^2-\frac{1}{8\pi}|\mathrm{curl}_{\mathbf{x}}\mathbf{A}|^2-\left(\rho\Psi-\frac{1}{c}\mathbf{A}\cdot\mathbf{j}\right)$$
$$+\mu g\left(\frac{1}{2}|\mathbf{u}-\mathbf{v}|^2\right)+\frac{1}{2}\left(d_{\mathbf{x}}\mathbf{v}+\{d_{\mathbf{x}}\mathbf{v}\}^T\right):\left(d_{\mathbf{x}}\mathbf{p}+\{d_{\mathbf{x}}\mathbf{p}\}^T\right)-2\left(\mathrm{div}_{\mathbf{x}}\mathbf{v}\right)\left(\mathrm{div}_{\mathbf{x}}\mathbf{p}\right)$$
$$+\frac{1}{4\pi G}\left(\mathrm{div}_{\mathbf{x}}\mathbf{v}\right)\left(\frac{\partial\Phi}{\partial t}+\mathbf{v}\cdot\nabla_{\mathbf{x}}\Phi\right)+\frac{1}{4\pi G}\Phi\left(\mathrm{div}_{\mathbf{x}}\mathbf{v}\right)^2-\frac{\Phi}{16\pi G}\left|d_{\mathbf{x}}\mathbf{v}+\{d_{\mathbf{x}}\mathbf{v}\}^T\right|^2$$
$$+\frac{1}{8\pi G}|\nabla_{\mathbf{x}}\Phi|^2=L_{ge}=$$
$$\frac{1}{4\pi}\left(-\sum_{n=0}^{3}\sum_{k=0}^{3}\sum_{m=0}^{3}\sum_{p=0}^{3}\frac{1}{4}g^{mn}g^{pk}\left(\frac{\partial A_p}{\partial x^m}-\frac{\partial A_m}{\partial x^p}\right)\left(\frac{\partial A_k}{\partial x^n}-\frac{\partial A_n}{\partial x^k}\right)-\sum_{k=0}^{3}4\pi j^k A_k\right)$$
$$+\frac{\hat{\mu}}{\sqrt{|\det G|}}\left(\sum_{m=0}^{3}\sum_{k=0}^{3}g_{mk}\hat{u}^m_{x0}u^k_t\right)\left(\sum_{m=0}^{3}\sum_{k=0}^{3}g_{mk}u^m_t u^k_t\right)^{-1}g\left(\frac{c^2}{2}-\sum_{m=0}^{3}\sum_{k=0}^{3}\frac{c^2}{2}g_{mk}u^m_t u^k_t\right)$$
$$+\sum_{m=0}^{3}\sum_{n=0}^{3}\frac{32\pi G}{c^4}\Theta^{mn}\frac{\partial}{\partial x^m}\left(\sum_{j=0}^{3}\sum_{k=0}^{3}g^{kj}v_k S_j\right)\frac{\partial}{\partial x^n}\left(\sum_{j=0}^{3}\sum_{k=0}^{3}g^{kj}v_k S_j\right)$$
$$+\left(\sum_{i=0}^{3}\sum_{j=0}^{3}g^{ij}\left(\delta_j v_i+\delta_i v_j\right)\right)\left(\sum_{i=0}^{3}\sum_{j=0}^{3}g^{ij}\left(\delta_j S_i+\delta_i S_j\right)\right)$$
$$-\sum_{k=0}^{3}\sum_{j=0}^{3}\sum_{m=0}^{3}\sum_{n=0}^{3}g^{km}g^{jn}\left(\delta_j S_k+\delta_k S_j\right)\left(\delta_m v_n+\delta_n v_m\right), \tag{1.426}$$

where the right-hand side is written in the covariant form and the second equality is valid in every curvilinear coordinate system with $\frac{\partial t}{\partial x^0} > 0$, (j^0, j^1, j^2, j^3) is the four-vector of the current, that satisfies (1.422), $\hat{\mu}$ is given by (1.424) and in a Cartesian coordinate system we have

$$\begin{cases} S_0=\frac{c^2}{16\pi G}\Phi-\frac{1}{2}\mathbf{v}\cdot\mathbf{p} \\ S_j=\frac{c}{2}p_j \qquad \forall 1\le j\le 3. \end{cases} \tag{1.427}$$

and

$$\begin{cases} S^0 = \frac{c^2}{16\pi G}\Phi \\ S^j = \frac{c^2}{16\pi G}\Phi\frac{v^j}{c} - \frac{c}{2}p_j \qquad \forall 1 \le j \le 3. \end{cases} \tag{1.428}$$

Moreover in the particular case of the relativistic-like choice $g(s) := -c^2\sqrt{1-\frac{2s}{c^2}}$, by (1.423) we can write an alternative to (1.426) as

$$\begin{aligned}
&\frac{1}{8\pi}\left|-\nabla_{\mathbf{x}}\Psi - \frac{1}{c}\frac{\partial \mathbf{A}}{\partial t} + \frac{1}{c}\mathbf{v}\times \mathrm{curl}_{\mathbf{x}}\mathbf{A}\right|^2 - \frac{1}{8\pi}\left|\mathrm{curl}_{\mathbf{x}}\mathbf{A}\right|^2 - \left(\rho\Psi - \frac{1}{c}\mathbf{A}\cdot\mathbf{j}\right) \\
&+\mu g\left(\frac{1}{2}|\mathbf{u}-\mathbf{v}|^2\right) + \frac{1}{2}\left(d_{\mathbf{x}}\mathbf{v} + \{d_{\mathbf{x}}\mathbf{v}\}^T\right) : \left(d_{\mathbf{x}}\mathbf{p} + \{d_{\mathbf{x}}\mathbf{p}\}^T\right) - 2\left(\mathrm{div}_{\mathbf{x}}\,\mathbf{v}\right)\left(\mathrm{div}_{\mathbf{x}}\,\mathbf{p}\right) \\
&+\frac{1}{4\pi G}\left(\mathrm{div}_{\mathbf{x}}\,\mathbf{v}\right)\left(\frac{\partial\Phi}{\partial t} + \mathbf{v}\cdot\nabla_{\mathbf{x}}\Phi\right) + \frac{1}{4\pi G}\Phi\left(\mathrm{div}_{\mathbf{x}}\,\mathbf{v}\right)^2 - \frac{\Phi}{16\pi G}\left|d_{\mathbf{x}}\mathbf{v} + \{d_{\mathbf{x}}\mathbf{v}\}^T\right|^2 \\
&\qquad + \frac{1}{8\pi G}\left|\nabla_{\mathbf{x}}\Phi\right|^2 = L_{ge} = \\
&\frac{1}{4\pi}\left(-\sum_{n=0}^{3}\sum_{k=0}^{3}\sum_{m=0}^{3}\sum_{p=0}^{3}\frac{1}{4}g^{mn}g^{pk}\left(\frac{\partial A_p}{\partial x^m} - \frac{\partial A_m}{\partial x^p}\right)\left(\frac{\partial A_k}{\partial x^n} - \frac{\partial A_n}{\partial x^k}\right) - \sum_{k=0}^{3}4\pi j^k A_k\right) \\
&\qquad - \frac{c^2\hat{\mu}}{\sqrt{|\det G|}}\sqrt{\left(\sum_{m=0}^{3}\sum_{k=0}^{3}g_{mk}\,\hat{u}^m_{x^0}\,\hat{u}^k_{x^0}\right)} \\
&+\sum_{m=0}^{3}\sum_{n=0}^{3}\frac{32\pi G}{c^4}\Theta^{mn}\frac{\partial}{\partial x^m}\left(\sum_{j=0}^{3}\sum_{k=0}^{3}g^{kj}v_k S_j\right)\frac{\partial}{\partial x^n}\left(\sum_{j=0}^{3}\sum_{k=0}^{3}g^{kj}v_k S_j\right) \\
&\quad + \left(\sum_{i=0}^{3}\sum_{j=0}^{3}g^{ij}\left(\delta_j v_i + \delta_i v_j\right)\right)\left(\sum_{i=0}^{3}\sum_{j=0}^{3}g^{ij}\left(\delta_j S_i + \delta_i S_j\right)\right) \\
&\qquad - \sum_{k=0}^{3}\sum_{j=0}^{3}\sum_{m=0}^{3}\sum_{n=0}^{3}g^{km}g^{jn}\left(\delta_j S_k + \delta_k S_j\right)\left(\delta_m v_n + \delta_n v_m\right),
\end{aligned} \tag{1.429}$$

where the right-hand side is written in the covariant form and the second equality is valid in every curvilinear coordinate system with $\frac{\partial t}{\partial x^0} > 0$. On the other hand, in the case of fully non-relativistic Lagrangian, where $g(s) = \left(s - \frac{c^2}{2}\right)$, we can write an alternative to (1.426) as

$$
\begin{aligned}
&\frac{1}{8\pi}\left|-\nabla_{\mathbf{x}}\Psi-\frac{1}{c}\frac{\partial \mathbf{A}}{\partial t}+\frac{1}{c}\mathbf{v}\times \operatorname{curl}_{\mathbf{x}}\mathbf{A}\right|^2-\frac{1}{8\pi}\left|\operatorname{curl}_{\mathbf{x}}\mathbf{A}\right|^2-\left(\rho\Psi-\frac{1}{c}\mathbf{A}\cdot\mathbf{j}\right)\\
&+\mu g\left(\frac{1}{2}|\mathbf{u}-\mathbf{v}|^2\right)+\frac{1}{2}\left(d_{\mathbf{x}}\mathbf{v}+\{d_{\mathbf{x}}\mathbf{v}\}^T\right):\left(d_{\mathbf{x}}\mathbf{p}+\{d_{\mathbf{x}}\mathbf{p}\}^T\right)-2\left(\operatorname{div}_{\mathbf{x}}\mathbf{v}\right)\left(\operatorname{div}_{\mathbf{x}}\mathbf{p}\right)\\
&+\frac{1}{4\pi G}\left(\operatorname{div}_{\mathbf{x}}\mathbf{v}\right)\left(\frac{\partial\Phi}{\partial t}+\mathbf{v}\cdot\nabla_{\mathbf{x}}\Phi\right)+\frac{1}{4\pi G}\Phi\left(\operatorname{div}_{\mathbf{x}}\mathbf{v}\right)^2-\frac{\Phi}{16\pi G}\left|d_{\mathbf{x}}\mathbf{v}+\{d_{\mathbf{x}}\mathbf{v}\}^T\right|^2\\
&+\frac{1}{8\pi G}|\nabla_{\mathbf{x}}\Phi|^2=L_{ge}=\\
&\frac{1}{4\pi}\left(-\sum_{n=0}^{3}\sum_{k=0}^{3}\sum_{m=0}^{3}\sum_{p=0}^{3}\frac{1}{4}g^{mn}g^{pk}\left(\frac{\partial A_p}{\partial x^m}-\frac{\partial A_m}{\partial x^p}\right)\left(\frac{\partial A_k}{\partial x^n}-\frac{\partial A_n}{\partial x^k}\right)-\sum_{k=0}^{3}4\pi j^k A_k\right)\\
&-\frac{c^2\hat{\mu}}{\sqrt{|\det G|}}\left(\sum_{m=0}^{3}\sum_{k=0}^{3}g_{mk}\,\hat{u}^m_{x^0}\,u^k_t\right)\\
&+\sum_{m=0}^{3}\sum_{n=0}^{3}\frac{32\pi G}{c^4}\Theta^{mn}\frac{\partial}{\partial x^m}\left(\sum_{j=0}^{3}\sum_{k=0}^{3}g^{kj}v_kS_j\right)\frac{\partial}{\partial x^n}\left(\sum_{j=0}^{3}\sum_{k=0}^{3}g^{kj}v_kS_j\right)\\
&+\left(\sum_{i=0}^{3}\sum_{j=0}^{3}g^{ij}\left(\delta_j v_i+\delta_i v_j\right)\right)\left(\sum_{i=0}^{3}\sum_{j=0}^{3}g^{ij}\left(\delta_j S_i+\delta_i S_j\right)\right)\\
&-\sum_{k=0}^{3}\sum_{j=0}^{3}\sum_{m=0}^{3}\sum_{n=0}^{3}g^{km}g^{jn}\left(\delta_j S_k+\delta_k S_j\right)\left(\delta_m v_n+\delta_n v_m\right), \qquad (1.430)
\end{aligned}
$$

where again the right-hand side is written in the covariant form and the second equality is valid in every curvilinear coordinate system with $\frac{\partial t}{\partial x^0}>0$.

Once we wrote the Lagrangian density $L_{ge}:=L\left(S^k,v^k,A^k,x^k\right)_{k=0,\dots 4}$ as a covariant scalar, under the changes of curvilinear coordinate systems, we can write a covariant Lagrangian as

$$
J_{ge}(S^k,v^k,A^k):=\int_{(x^0,x^1,x^2,x^3)}L_{ge}\left(S^k,v^k,A^k,x^k\right)\sqrt{|\det G|}\,dx^0dx^1dx^2dx^3. \qquad (1.431)
$$

Although we need a term $\sqrt{|\det G|}$ for the covariance of the Lagrangian in curvilinear coordinate systems, in the Cartesian coordinate systems we always have $\sqrt{|\det G|}=1$.

Next note that the contravariant tensor of the three-dimensional geometry Θ^{ij} which satisfies (1.412) in non-inertial Cartesian coordinate systems and the scalar of the global time t are dependent on the geometry of the space-time only and are fully determined in a given curvilinear coordinate system by change of variables rules. In particular, the four-covector of the gravitational potential (v_0,v_1,v_2,v_3) is fully determined in the given curvilinear coordinate system, since we have

$$v_k = c\frac{\partial t}{\partial x^k} \qquad \forall k = 0, 1, 2, 3. \tag{1.432}$$

Moreover, by (1.319) and (1.320) we have the following covariant identities which are valid in every curvilinear coordinate system:

$$\begin{cases} \sum_{k=0}^{3} \Theta^{mk} v_k = \sum_{k=0}^{3} c\, \Theta^{mk} \frac{\partial t}{\partial x^k} = 0 & \forall m = 0, 1, 2, 3 \\ \sum_{k=0}^{3} \sum_{j=0}^{3} g^{kj} v_k v_j = \sum_{k=0}^{3} \sum_{j=0}^{3} c^2 g^{kj} \frac{\partial t}{\partial x^k} \frac{\partial t}{\partial x^j} = 1. \end{cases} \tag{1.433}$$

However the four-vector of the gravitational potential (v^0, v^1, v^2, v^3), the contravariant pseudo-metric tensor $g^{mn} = v^m v^n - \Theta^{mn}$ and thus also the covariant pseudo-metric tensor g_{mn} depend also on the physical properties of the matter. Without knowing the physical properties of the matter the four-vector of the gravitational potential can be an arbitrary vector (v^0, v^1, v^2, v^3) that satisfies

$$\sum_{k=0}^{3} v_k v^k = \sum_{k=0}^{3} c\frac{\partial t}{\partial x^k} v^k = 1. \tag{1.434}$$

Indeed for an arbitrary four-vector (v^0, v^1, v^2, v^3) that satisfies (1.434), denoting $g^{mn} := v^m v^n - \Theta^{mn}$, using (1.433) and (1.434) we clearly obtain the following consistency:

$$\sum_{j=0}^{3} g^{kj} v_j = \sum_{j=0}^{3} \left(v^k v^j - \Theta^{kj}\right) v_j = v^k \left(\sum_{j=0}^{3} v^j v_j\right) - \sum_{j=0}^{3} \Theta^{kj} v_j = v^k \qquad \forall k = 0, 1, 2, 3. \tag{1.435}$$

Thus we obtained that the four-vector of the gravitational potential can be arbitrary four-vector in (1.431) that satisfies the linear constraint (1.434) where the four-covector v_k is prescribed. So the four-vector (v^0, v^1, v^2, v^3) actually contains three independent scalar functions similarly as in Cartesian coordinate systems where we have $v^0 = 1$. On the other hand, the four-vector S^k contains four independent scalar functions. Thus the Lagrangian in (1.431) depends on seven independent scalar functions characterizing the gravitational field and the four-vector of electromagnetic potential, exactly as in Cartesian coordinate systems where we have four independent scalar functions that characterize the electromagnetic field: scalar Ψ and three-dimensional vector $\mathbf{A}$ and seven independent scalar functions that characterize the gravitational field: three are contained in the three-dimensional vectorial gravitational potential $\mathbf{v}$ and other four are the ancillary scalar field Φ and the ancillary three-dimensional vector field $\mathbf{p}$.

Finally note that since our model of the Newtonian-type gravity in the case of fully non-relativistic choice (1.430) and in the absence of electromagnetic fields coincides with the classical Newtonian gravitation, as a particular case, we obtained a covariant formulation of the classical Newtonian gravity in curvilinear coordinate systems.

1.9.12 Covariant Formulation of the Laws of Gravity in Cartesian and Curvilinear Coordinate Systems: The Case of Some Alternative Model of the Gravity

Consider $\mathbf{k}$ to be the vectorial potential of the inertia, which is a generally trivial speed-like vector field, assumed to be fixed in every fixed inertial or non-inertial Cartesian coordinate system (see Definition 1.9). In this section we find an equivalent form of the Lagrangian density of the gravitational-electromagnetic field in the case of the alternative model of the gravity having the general form (1.388) or (1.389):

$$
\begin{aligned}
&L\left(\mathbf{A}, \Psi, \mathbf{v}, \Phi_0, \mathbf{h}, \mathbf{x}, t\right) := \\
&\frac{1}{8\pi}\left|-\nabla_{\mathbf{x}}\Psi - \frac{1}{c}\frac{\partial \mathbf{A}}{\partial t} + \frac{1}{c}\mathbf{v}\times \operatorname{curl}_{\mathbf{x}}\mathbf{A}\right|^2 - \frac{1}{8\pi}\left|\operatorname{curl}_{\mathbf{x}}\mathbf{A}\right|^2 - \left(\rho\Psi - \frac{1}{c}\mathbf{A}\cdot\mathbf{j}\right) + \mu g\left(\frac{1}{2}\left|\mathbf{u}-\mathbf{v}\right|^2\right) \\
&\quad - \frac{c^2}{8\pi G}\left|-\nabla_{\mathbf{x}}\Phi_0 - \frac{1}{c}\frac{\partial \mathbf{h}}{\partial t} + \frac{1}{c}\mathbf{v}\times\operatorname{curl}_{\mathbf{x}}\mathbf{h} - \frac{1}{c}\nabla_{\mathbf{x}}\left(\mathbf{h}\cdot\mathbf{v}\right)\right|^2 + \frac{c^2}{8\pi G}\left|\operatorname{curl}_{\mathbf{x}}\mathbf{h}\right|^2 \\
&\quad + \frac{c^2\beta}{4\pi G}\left(\frac{1}{4}\left|d_{\mathbf{x}}\mathbf{v} + \{d_{\mathbf{x}}\mathbf{v}\}^T\right|^2 - \left|\operatorname{div}_{\mathbf{x}}\mathbf{v}\right|^2\right), \qquad (1.436)
\end{aligned}
$$

where

$$
\mathbf{h} = \mathbf{v} - \mathbf{k} \quad \text{and} \quad \Phi_0 = -\frac{1}{2c}\left|\mathbf{v}-\mathbf{k}\right|^2, \qquad (1.437)
$$

so that we have

$$
\begin{aligned}
&L\left(\mathbf{A}, \Psi, \mathbf{v}, \Phi_0, \mathbf{h}, \mathbf{x}, t\right) = L_1\left(\mathbf{A}, \Psi, \mathbf{v}, \mathbf{x}, t\right) := \\
&\frac{1}{8\pi}\left|-\nabla_{\mathbf{x}}\Psi - \frac{1}{c}\frac{\partial \mathbf{A}}{\partial t} + \frac{1}{c}\mathbf{v}\times \operatorname{curl}_{\mathbf{x}}\mathbf{A}\right|^2 - \frac{1}{8\pi}\left|\operatorname{curl}_{\mathbf{x}}\mathbf{A}\right|^2 - \left(\rho\Psi - \frac{1}{c}\mathbf{A}\cdot\mathbf{j}\right) + \mu g\left(\frac{1}{2}\left|\mathbf{u}-\mathbf{v}\right|^2\right) \\
&\quad - \frac{c^2}{8\pi G}\left|\frac{1}{c}\nabla_{\mathbf{x}}\left(\frac{1}{2}\left|\mathbf{v}-\mathbf{k}\right|^2\right) - \frac{1}{c}\frac{\partial}{\partial t}\left(\mathbf{v}-\mathbf{k}\right) + \frac{1}{c}\mathbf{v}\times\operatorname{curl}_{\mathbf{x}}\left(\mathbf{v}-\mathbf{k}\right) - \frac{1}{c}\nabla_{\mathbf{x}}\left(\left(\mathbf{v}-\mathbf{k}\right)\cdot\mathbf{v}\right)\right|^2 \\
&\quad + \frac{c^2}{8\pi G}\left|\operatorname{curl}_{\mathbf{x}}\left(\mathbf{v}-\mathbf{k}\right)\right|^2 + \frac{c^2\beta}{4\pi G}\left(\frac{1}{4}\left|d_{\mathbf{x}}\mathbf{v} + \{d_{\mathbf{x}}\mathbf{v}\}^T\right|^2 - \left|\operatorname{div}_{\mathbf{x}}\mathbf{v}\right|^2\right). \qquad (1.438)
\end{aligned}
$$

Our purpose is to make the equivalent form of this Lagrangian density to be covariant and valid in every curvilinear coordinate system. Assume first, that we deal with a Cartesian inertial or non-inertial coordinate system. Consider again the three-dimensional vectorial gravitational potential $\mathbf{v} = (v^1, v^2, v^3)$ and consider the covariant pseudo-metric tensor $G = \{g_{ij}\}_{0\leq i,j\leq 3}$ defined by (1.405) and the contravariant pseudo-metric tensor $\tilde{G} = \{g^{ij}\}_{0\leq i,j\leq 3}$ defined by (1.406). Next, as before, we consider the four-dimensional gravitational potential (v^0, v^1, v^2, v^3) defined by (1.409) and the corresponding lowered four-covector field (v_0, v_1, v_2, v_3), that we called the four-covector of gravitational potential, defined by (1.410). Then, we have

$$\left((\mathrm{div}_{\mathbf{x}}\,\mathbf{v})^2-\frac{1}{4}\left|d_{\mathbf{x}}\mathbf{v}+\{d_{\mathbf{x}}\mathbf{v}\}^T\right|^2\right)=\frac{c^2}{4}\left(\left(\sum_{i=0}^{3}\sum_{j=0}^{3}g^{ij}\left(\delta_j v_i+\delta_i v_j\right)\right)^2\right.$$
$$\left.-\sum_{k=0}^{3}\sum_{j=0}^{3}\sum_{m=0}^{3}\sum_{n=0}^{3}\left(\delta_j v_k+\delta_k v_j\right)g^{km}g^{jn}\left(\delta_m v_n+\delta_n v_m\right)\right), \quad (1.439)$$

where δ_j means the covariant derivative with respect to the dynamical pseudo-metrics $G=\{g_{ij}\}_{0\leq i,j\leq 3}$ (see Sect. 9.9.2 for the details). Furthermore, consider the Dynamical four-covector of genuine gravity (s_0, s_1, s_2, s_3), defined as in (1.308) by

$$(s_0,s_1,s_2,s_3)=-\frac{1}{c}\left(\left(\Phi_0+\frac{1}{c}\mathbf{v}\cdot\mathbf{h}\right),-\mathbf{h}\right)=\left(\frac{1}{2c^2}|\mathbf{h}|^2-\frac{1}{c^2}\mathbf{v}\cdot\mathbf{h},\ \frac{1}{c}\mathbf{h}\right) \quad \text{where}$$
$$s_0=-\frac{1}{c}\left(\Phi_0+\frac{1}{c}\mathbf{v}\cdot\mathbf{h}\right)=\frac{1}{2c^2}|\mathbf{h}|^2-\frac{1}{c^2}\mathbf{v}\cdot\mathbf{h} \quad \text{and} \quad (s_1,s_2,s_3)=\frac{1}{c}\mathbf{h}, \quad (1.440)$$

where $\mathbf{h}$ and Φ_0 are given by (1.437). Next, as in (1.266) and (1.268) we have the following:

$$\frac{1}{4\pi}\left(\frac{1}{2}\left|-\nabla_{\mathbf{x}}\Psi-\frac{1}{c}\frac{\partial\mathbf{A}}{\partial t}+\frac{1}{c}\mathbf{v}\times\mathrm{curl}_{\mathbf{x}}\mathbf{A}\right|^2-\frac{1}{2}\left|\mathrm{curl}_{\mathbf{x}}\mathbf{A}\right|^2-4\pi\left(\rho\Psi-\frac{1}{c}\mathbf{A}\cdot\mathbf{j}\right)\right)=$$
$$\frac{1}{4\pi}\left(-\sum_{n=0}^{3}\sum_{k=0}^{3}\sum_{m=0}^{3}\sum_{p=0}^{3}\frac{1}{4}g^{mn}g^{pk}\left(\frac{\partial A_p}{\partial x^m}-\frac{\partial A_m}{\partial x^p}\right)\left(\frac{\partial A_k}{\partial x^n}-\frac{\partial A_n}{\partial x^k}\right)-\sum_{k=0}^{3}4\pi j^k A_k\right), \quad (1.441)$$

where (A_0, A_1, A_2, A_3) is the four-covector of the four dimensional electromagnetic potential, defined as in (1.203) by

$$(A_0,A_1,A_2,A_3)=(\Psi,-\mathbf{A}) \quad \text{where} \quad A_0=\Psi \quad \text{and} \quad (A_1,A_2,A_3)=-\mathbf{A}, \quad (1.442)$$

and (j^0, j^1, j^2, j^3) is the four-vector of the current, given by (1.422). Then, completely analogously, as we get in (1.266) and (1.268) the following part of (1.441):

$$\left(\frac{1}{2}\left|-\nabla_{\mathbf{x}}\Psi-\frac{1}{c}\frac{\partial\mathbf{A}}{\partial t}+\frac{1}{c}\mathbf{v}\times\mathrm{curl}_{\mathbf{x}}\mathbf{A}\right|^2-\frac{1}{2}\left|\mathrm{curl}_{\mathbf{x}}\mathbf{A}\right|^2\right)=$$
$$-\sum_{n=0}^{3}\sum_{k=0}^{3}\sum_{m=0}^{3}\sum_{p=0}^{3}\frac{1}{4}g^{mn}g^{pk}\left(\frac{\partial A_p}{\partial x^m}-\frac{\partial A_m}{\partial x^p}\right)\left(\frac{\partial A_k}{\partial x^n}-\frac{\partial A_n}{\partial x^k}\right), \quad (1.443)$$

we can get also:

$$\left(\frac{1}{2}\left|-\nabla_{\mathbf{x}}\Phi_0-\frac{1}{c}\frac{\partial \mathbf{h}}{\partial t}+\frac{1}{c}\mathbf{v}\times\text{curl}_{\mathbf{x}}\mathbf{h}-\frac{1}{c}\nabla_{\mathbf{x}}\left(\mathbf{h}\cdot\mathbf{v}\right)\right|^2-\frac{1}{2}\left|\text{curl}_{\mathbf{x}}\mathbf{h}\right|^2\right)=$$
$$\left(\frac{1}{2}\left|-\nabla_{\mathbf{x}}\left(\Phi_0+\frac{1}{c}\mathbf{v}\cdot\mathbf{h}\right)-\frac{1}{c}\frac{\partial \mathbf{h}}{\partial t}+\frac{1}{c}\mathbf{v}\times\text{curl}_{\mathbf{x}}\mathbf{h}\right|^2-\frac{1}{2}\left|\text{curl}_{\mathbf{x}}\mathbf{h}\right|^2\right)=$$
$$-\sum_{n=0}^{3}\sum_{k=0}^{3}\sum_{m=0}^{3}\sum_{p=0}^{3}\frac{c^2}{4}g^{mn}g^{pk}\left(\frac{\partial s_p}{\partial x^m}-\frac{\partial s_m}{\partial x^p}\right)\left(\frac{\partial s_k}{\partial x^n}-\frac{\partial s_n}{\partial x^k}\right). \tag{1.444}$$

Then, similarly as it was done in (1.426), by (1.441), (1.444) and (1.439) we rewrite the Lagrangian density in (1.436) or (1.438) in the Cartesian coordinate system as

$$\frac{1}{8\pi}\left|-\nabla_{\mathbf{x}}\Psi-\frac{1}{c}\frac{\partial \mathbf{A}}{\partial t}+\frac{1}{c}\mathbf{v}\times\text{curl}_{\mathbf{x}}\mathbf{A}\right|^2-\frac{1}{8\pi}\left|\text{curl}_{\mathbf{x}}\mathbf{A}\right|^2$$
$$-\left(\rho\Psi-\frac{1}{c}\mathbf{A}\cdot\mathbf{j}\right)+\mu g\left(\frac{1}{2}\left|\mathbf{u}-\mathbf{v}\right|^2\right)$$
$$-\frac{c^2}{8\pi G}\left|-\nabla_{\mathbf{x}}\Phi_0-\frac{1}{c}\frac{\partial \mathbf{h}}{\partial t}+\frac{1}{c}\mathbf{v}\times\text{curl}_{\mathbf{x}}\mathbf{h}-\frac{1}{c}\nabla_{\mathbf{x}}\left(\mathbf{h}\cdot\mathbf{v}\right)\right|^2+\frac{c^2}{8\pi G}\left|\text{curl}_{\mathbf{x}}\mathbf{h}\right|^2$$
$$+\frac{c^2\beta}{4\pi G}\left(\frac{1}{4}\left|d_{\mathbf{x}}\mathbf{v}+\{d_{\mathbf{x}}\mathbf{v}\}^T\right|^2-\left|\text{div}_{\mathbf{x}}\mathbf{v}\right|^2\right)=L_{ge}=$$
$$\frac{1}{4\pi}\left(-\sum_{n=0}^{3}\sum_{k=0}^{3}\sum_{m=0}^{3}\sum_{p=0}^{3}\frac{1}{4}g^{mn}g^{pk}\left(\frac{\partial A_p}{\partial x^m}-\frac{\partial A_m}{\partial x^p}\right)\left(\frac{\partial A_k}{\partial x^n}-\frac{\partial A_n}{\partial x^k}\right)-\sum_{k=0}^{3}4\pi j^k A_k\right)$$
$$+\frac{\hat{\mu}}{\sqrt{|\det G|}}\left(\sum_{m=0}^{3}\sum_{k=0}^{3}g_{mk}\hat{u}^m_{x^0}u^k_t\right)\left(\sum_{m=0}^{3}\sum_{k=0}^{3}g_{mk}u^m_t u^k_t\right)^{-1}g\left(\frac{c^2}{2}-\sum_{m=0}^{3}\sum_{k=0}^{3}\frac{c^2}{2}g_{mk}u^m_t u^k_t\right)$$
$$+\frac{c^4}{4\pi G}\left(\sum_{n=0}^{3}\sum_{k=0}^{3}\sum_{m=0}^{3}\sum_{p=0}^{3}\frac{1}{4}g^{mn}g^{pk}\left(\frac{\partial s_p}{\partial x^m}-\frac{\partial s_m}{\partial x^p}\right)\left(\frac{\partial s_k}{\partial x^n}-\frac{\partial s_n}{\partial x^k}\right)\right)$$
$$-\frac{c^4\beta}{16\pi G}\left(\left(\sum_{i=0}^{3}\sum_{j=0}^{3}g^{ij}\left(\delta_j v_i+\delta_i v_j\right)\right)^2\right.$$
$$\left.-\sum_{k=0}^{3}\sum_{j=0}^{3}\sum_{m=0}^{3}\sum_{n=0}^{3}\left(\delta_j v_k+\delta_k v_j\right)g^{km}g^{jn}\left(\delta_m v_n+\delta_n v_m\right)\right), \tag{1.445}$$

where the right-hand side is written in the covariant form and the second equality is valid in every curvilinear coordinate system with $\frac{\partial t}{\partial x^0}>0$, (j^0, j^1, j^2, j^3) is the four-vector of the current, $\hat{\mu}$, u^j_t and $u^j_{x^0}$ are the same as in (1.424), and $\mathbf{h}$ and Φ_0 are given by

$$\mathbf{h} = \mathbf{v} - \mathbf{k} \quad \text{and} \quad \Phi_0 = -\frac{1}{2c}\left|\mathbf{v} - \mathbf{k}\right|^2. \tag{1.446}$$

Moreover in the particular case of the relativistic-like choice $g(s) := -c^2\sqrt{1 - \frac{2s}{c^2}}$, by (1.423) we can write an alternative to (1.445) as

$$
\begin{aligned}
&\frac{1}{8\pi}\left|-\nabla_{\mathbf{x}}\Psi - \frac{1}{c}\frac{\partial \mathbf{A}}{\partial t} + \frac{1}{c}\mathbf{v} \times \mathrm{curl}_{\mathbf{x}}\mathbf{A}\right|^2 \\
&\qquad - \frac{1}{8\pi}\left|\mathrm{curl}_{\mathbf{x}}\mathbf{A}\right|^2 - \left(\rho\Psi - \frac{1}{c}\mathbf{A}\cdot\mathbf{j}\right) + \mu g\left(\frac{1}{2}\left|\mathbf{u} - \mathbf{v}\right|^2\right) \\
&\qquad - \frac{c^2}{8\pi G}\left|-\nabla_{\mathbf{x}}\Phi_0 - \frac{1}{c}\frac{\partial \mathbf{h}}{\partial t} + \frac{1}{c}\mathbf{v} \times \mathrm{curl}_{\mathbf{x}}\mathbf{h} - \frac{1}{c}\nabla_{\mathbf{x}}\left(\mathbf{h}\cdot\mathbf{v}\right)\right|^2 + \frac{c^2}{8\pi G}\left|\mathrm{curl}_{\mathbf{x}}\mathbf{h}\right|^2 \\
&\qquad + \frac{c^2\beta}{4\pi G}\left(\frac{1}{4}\left|d_{\mathbf{x}}\mathbf{v} + \{d_{\mathbf{x}}\mathbf{v}\}^T\right|^2 - \left|\mathrm{div}_{\mathbf{x}}\,\mathbf{v}\right|^2\right) = L_{ge} = \\
&\frac{1}{4\pi}\left(-\sum_{n=0}^{3}\sum_{k=0}^{3}\sum_{m=0}^{3}\sum_{p=0}^{3}\frac{1}{4}g^{mn}g^{pk}\left(\frac{\partial A_p}{\partial x^m} - \frac{\partial A_m}{\partial x^p}\right)\left(\frac{\partial A_k}{\partial x^n} - \frac{\partial A_n}{\partial x^k}\right) - \sum_{k=0}^{3}4\pi j^k A_k\right) \\
&\qquad - \frac{c^2\hat{\mu}}{\sqrt{|\det G|}}\sqrt{\left(\sum_{m=0}^{3}\sum_{k=0}^{3}g_{mk}\,\hat{u}^m_{x^0}\,\hat{u}^k_{x^0}\right)} \\
&\qquad + \frac{c^4}{4\pi G}\left(\sum_{n=0}^{3}\sum_{k=0}^{3}\sum_{m=0}^{3}\sum_{p=0}^{3}\frac{1}{4}g^{mn}g^{pk}\left(\frac{\partial s_p}{\partial x^m} - \frac{\partial s_m}{\partial x^p}\right)\left(\frac{\partial s_k}{\partial x^n} - \frac{\partial s_n}{\partial x^k}\right)\right) \\
&\qquad - \frac{c^4\beta}{16\pi G}\left(\left(\sum_{i=0}^{3}\sum_{j=0}^{3}g^{ij}\left(\delta_j v_i + \delta_i v_j\right)\right)^2\right. \\
&\qquad \left. - \sum_{k=0}^{3}\sum_{j=0}^{3}\sum_{m=0}^{3}\sum_{n=0}^{3}\left(\delta_j v_k + \delta_k v_j\right)g^{km}g^{jn}\left(\delta_m v_n + \delta_n v_m\right)\right),
\end{aligned}
\tag{1.447}
$$

where the right-hand side is written in the covariant form and the second equality is valid in every curvilinear coordinate system with $\frac{\partial t}{\partial x^0} > 0$. On the other hand, in the case of fully non-relativistic Lagrangian, where $g(s) = \left(s - \frac{c^2}{2}\right)$, we can write an alternative to (1.445) as

$$
\begin{aligned}
&\frac{1}{8\pi}\left|-\nabla_{\mathbf{x}}\Psi-\frac{1}{c}\frac{\partial \mathbf{A}}{\partial t}+\frac{1}{c}\mathbf{v}\times \mathrm{curl}_{\mathbf{x}}\mathbf{A}\right|^2-\frac{1}{8\pi}\left|\mathrm{curl}_{\mathbf{x}}\mathbf{A}\right|^2\\
&\qquad-\left(\rho\Psi-\frac{1}{c}\mathbf{A}\cdot\mathbf{j}\right)+\mu g\left(\frac{1}{2}\left|\mathbf{u}-\mathbf{v}\right|^2\right)\\
&-\frac{c^2}{8\pi G}\left|-\nabla_{\mathbf{x}}\Phi_0-\frac{1}{c}\frac{\partial \mathbf{h}}{\partial t}+\frac{1}{c}\mathbf{v}\times \mathrm{curl}_{\mathbf{x}}\mathbf{h}-\frac{1}{c}\nabla_{\mathbf{x}}\left(\mathbf{h}\cdot\mathbf{v}\right)\right|^2+\frac{c^2}{8\pi G}\left|\mathrm{curl}_{\mathbf{x}}\mathbf{h}\right|^2\\
&\qquad+\frac{c^2\beta}{4\pi G}\left(\frac{1}{4}\left|d_{\mathbf{x}}\mathbf{v}+\{d_{\mathbf{x}}\mathbf{v}\}^T\right|^2-\left|\mathrm{div}_{\mathbf{x}}\mathbf{v}\right|^2\right)=L_{ge}=\\
&\frac{1}{4\pi}\left(-\sum_{n=0}^{3}\sum_{k=0}^{3}\sum_{m=0}^{3}\sum_{p=0}^{3}\frac{1}{4}g^{mn}g^{pk}\left(\frac{\partial A_p}{\partial x^m}-\frac{\partial A_m}{\partial x^p}\right)\left(\frac{\partial A_k}{\partial x^n}-\frac{\partial A_n}{\partial x^k}\right)-\sum_{k=0}^{3}4\pi j^k A_k\right)\\
&\qquad-\frac{c^2\hat{\mu}}{\sqrt{|\det G|}}\left(\sum_{m=0}^{3}\sum_{k=0}^{3}g_{mk}\,\hat{u}^m_{x^0}\,u^k_t\right)\\
&+\frac{c^4}{4\pi G}\left(\sum_{n=0}^{3}\sum_{k=0}^{3}\sum_{m=0}^{3}\sum_{p=0}^{3}\frac{1}{4}g^{mn}g^{pk}\left(\frac{\partial s_p}{\partial x^m}-\frac{\partial s_m}{\partial x^p}\right)\left(\frac{\partial s_k}{\partial x^n}-\frac{\partial s_n}{\partial x^k}\right)\right)\\
&\qquad-\frac{c^4\beta}{16\pi G}\left(\left(\sum_{i=0}^{3}\sum_{j=0}^{3}g^{ij}\left(\delta_j v_i+\delta_i v_j\right)\right)^2\right.\\
&\qquad\left.-\sum_{k=0}^{3}\sum_{j=0}^{3}\sum_{m=0}^{3}\sum_{n=0}^{3}\left(\delta_j v_k+\delta_k v_j\right)g^{km}g^{jn}\left(\delta_m v_n+\delta_n v_m\right)\right),
\end{aligned}
\tag{1.448}
$$

where again the right-hand side is written in the covariant form and the second equality is valid in every curvilinear coordinate system with $\frac{\partial t}{\partial x^0}>0$.

Note that the four-vector of the gravitational potential (v^0, v^1, v^2, v^3) and the four-covector of the genuine gravity (s_0, s_1, s_2, s_3) are <u>not</u> independent arguments of L_{ge}. Moreover, in the general non-Cartesian or curvilinear coordinate system, knowing (v^0, v^1, v^2, v^3) and thus also knowing $\{g_{ij}\}_{0\le i,j\le 3}$, as before in (1.347), we can find (s_0, s_1, s_2, s_3) by

$$
s_j=\frac{1}{2}\left(\sum_{m=0}^{3}g_{jm}k^m-\sum_{m=0}^{3}J_{jm}v^m\right)\quad\forall j=0,1,2,3,
\tag{1.449}
$$

where (k^0, k^1, k^2, k^3) is the four-vector of the potential of inertia and $\{J_{ij}\}_{0\le i,j\le 3}$ is the covariant kinematic pseudo-metric tensor of inertia, that are prescribed in every Cartesian, non-Cartesian or curvilinear coordinate system. On the other hand, knowing (s_0, s_1, s_2, s_3), we can find $\{g_{ij}\}_{0\le i,j\le 3}$ and (v^0, v^1, v^2, v^3) by the following covariant identities

$$
g_{ij}=J_{ij}+\left(s_i\left(c\frac{\partial t}{\partial x^j}\right)+\left(c\frac{\partial t}{\partial x^i}\right)s_j\right)\quad\forall i,j=0,1,2,3
\tag{1.450}
$$

(see (1.348)), and

$$v^j - k^j = \sum_{m=0}^{3} \Theta^{jm} s_m = \sum_{m=0}^{3} \left(k^j k^m - J^{jm}\right) s_m \quad \forall j = 0, 1, 2, 3 \tag{1.451}$$

(see (1.349)).

Once we wrote the Lagrangian density $L_{ge} := L_{ge1}\left(v^k, A^k, x^k\right)_{k=0,\dots 4} = L_{ge2}\left(s^k, A^k, x^k\right)_{k=0,\dots 4}$ as a covariant scalar, under the changes of curvilinear coordinate systems, we can write a covariant Lagrangian as

$$\begin{aligned} J_{ge1}(v^k, A^k) = J_{ge1}(s^k, A^k) :=& \\ \int_{(x^0,x^1,x^2,x^3)} & L_{ge1}\left(v^k, A^k, x^k\right) \sqrt{|\det G|}\, dx^0 dx^1 dx^2 dx^3 \\ = \int_{(x^0,x^1,x^2,x^3)} & L_{ge2}\left(s^k, A^k, x^k\right) \sqrt{|\det G|}\, dx^0 dx^1 dx^2 dx^3. \end{aligned} \tag{1.452}$$

Although we need a term $\sqrt{|\det G|}$ for the covariance of the Lagrangian in curvilinear coordinate systems, in the Cartesian coordinate systems we always have $\sqrt{|\det G|} = 1$.

Finally, note again, that the four-vector of the gravitational potential, as an independent argument, is restricted by (1.434):

$$\sum_{k=0}^{3} c \frac{\partial t}{\partial x^k} v^k = 1. \tag{1.453}$$

So the four-vector (v^0, v^1, v^2, v^3) actually contains three independent scalar functions similarly as in Cartesian coordinate systems where we have $v^0 = 1$. On the other hand if we consider (s_0, s_1, s_2, s_3) instead of (v^0, v^1, v^2, v^3) , as an independent argument, then by (1.291) and (1.450) it is restricted by the following identity

$$\det\left(\left\{J_{ij} + s_i \left(c\frac{\partial t}{\partial x^j}\right) + \left(c\frac{\partial t}{\partial x^i}\right) s_j\right\}_{0\le i,j\le 3}\right) = \det\left(\{J_{ij}\}_{0\le i,j\le 3}\right), \tag{1.454}$$

which is valid in every Cartesian, non-Cartesian, or curvilinear coordinate system. So the four-covector (s_0, s_1, s_2, s_3) also contains three independent scalar functions.

1.10 Relativistic-Like Dirac Equation

As in (1.282) consider the relativistic-like Lagrangian of the motion of the particle with mass m and charge σ in the outer gravitational and electromagnetic fields and additional field with potential $V(\mathbf{x}, t)$:

$$J_{rl}(\mathbf{r}) = \int_0^T L_0\left(\frac{d\mathbf{r}}{dt}, \mathbf{r}, t\right) dt :=$$
$$\int_0^T \left\{ -mc^2\sqrt{1 - \frac{1}{c^2}\left|\frac{d\mathbf{r}}{dt} - \mathbf{v}(\mathbf{r}, t)\right|^2} - \sigma\left(\Psi(\mathbf{r}, t) - \frac{1}{c}\mathbf{A}(\mathbf{r}, t) \cdot \frac{d\mathbf{r}}{dt}\right) + V(\mathbf{r}, t) \right\} dt. \tag{1.455}$$

Next define the generalized momentum of the particle by

$$\mathbf{P} := \nabla_{\mathbf{r}'} L_0(\mathbf{r}', \mathbf{r}, t) = m\left(1 - \frac{1}{c^2}\left|\frac{d\mathbf{r}}{dt} - \mathbf{v}(\mathbf{r}, t)\right|^2\right)^{-\frac{1}{2}} \left(\frac{d\mathbf{r}}{dt} - \mathbf{v}(\mathbf{r}, t)\right) + \frac{\sigma}{c}\mathbf{A}(\mathbf{r}, t). \tag{1.456}$$

Then

$$\frac{d\mathbf{r}}{dt} = \left(1 + \frac{1}{c^2}\left|\frac{1}{m}\mathbf{P} - \frac{\sigma}{mc}\mathbf{A}(\mathbf{r}, t)\right|^2\right)^{-\frac{1}{2}} \left(\frac{1}{m}\mathbf{P} - \frac{\sigma}{mc}\mathbf{A}(\mathbf{r}, t)\right) + \mathbf{v}(\mathbf{r}, t). \tag{1.457}$$

Thus, if we consider a Hamiltonian

$$H_0(\mathbf{P}, \mathbf{r}, t) := \mathbf{P} \cdot \frac{d\mathbf{r}}{dt} - L_0\left(\frac{d\mathbf{r}}{dt}, \mathbf{r}, t\right), \tag{1.458}$$

then we deduce, that the relativistic-like Hamiltonian for macro-particles has the form:

$$H_0(\mathbf{P}, \mathbf{r}, t) = mc^2\left(1 + \frac{1}{c^2}\left|\frac{1}{m}\mathbf{P} - \frac{\sigma}{mc}\mathbf{A}(\mathbf{r}, t)\right|^2\right)^{\frac{1}{2}} + \sigma\left(\Psi(\mathbf{r}, t) - \frac{1}{c}\mathbf{v}(\mathbf{r}, t) \cdot \mathbf{A}(\mathbf{r}, t)\right)$$
$$- V(\mathbf{r}, t) + \mathbf{v}(\mathbf{r}, t) \cdot \mathbf{P} \tag{1.459}$$

(see Chap. 10 for the details). In particular, if similarly to (1.218) we define the four-dimensional generalized momentum (P_0, P_1, P_2, P_3) as

$$(P_0, P_1, P_2, P_3) := \left(\frac{1}{c}H_0, -\mathbf{P}\right) \quad \text{where} \quad P_0 = \frac{1}{c}H_0 \quad \text{and} \quad (P_1, P_2, P_3) = -\mathbf{P}. \tag{1.460}$$

Then, since by (1.456) and (1.459), under the change of non-inertial Cartesian coordinate system H_0 and $\mathbf{P}$ transform as

$$\begin{cases} H_0' = H_0 + \left(\frac{dA}{dt}(t) \cdot \mathbf{x} + \frac{d\mathbf{z}}{dt}(t)\right) \cdot (A(t) \cdot \mathbf{P}) \\ \mathbf{P}' = A(t) \cdot \mathbf{P}, \end{cases} \tag{1.461}$$

by comparing (1.461) with (1.192) we deduce that the four-dimensional momentum (P_0, P_1, P_2, P_3) is a four-covector on the group $\mathcal{S}_0$ that is the group of changes of Cartesian non-inertial coordinate systems.

See Sect. 10.3 for the Relativistic-like Lagrangian and Hamiltonian of the motion of the system of n particles. See also Sect. 10.4 for Liouville's equation for a system of n relativistic-like classical particles.

Next consider the motion of a spin-half quantum relativistic-like micro-particle with inertial mass m and the charge σ in the outer gravitational and electromagnetical field with characteristics $\mathbf{v}(\mathbf{x}, t)$, $\mathbf{A}(\mathbf{x}, t)$ and $\Psi(\mathbf{x}, t)$ and additional conservative field with potential $V(\mathbf{y}, t)$. The evolution equation for this particle is

$$i\hbar\frac{\partial \psi}{\partial t} = \hat{H}_0 \cdot \psi, \tag{1.462}$$

where $\psi(\mathbf{x}, t) = (\psi_1(\mathbf{x}, t), \psi_2(\mathbf{x}, t)) \in \mathbb{C}^2 \times \mathbb{C}^2$ is a four-component wave function and $\hat{H}_0$ is the Hamiltonian operator. Since the relativistic-like Hamiltonian for macro-particles has the form (1.459), analogously to the usual Dirac Hamiltonian operator, we built the Hermitian Hamiltonian operator as $\hat{H}_0 \cdot \psi = \left(\hat{H}_1 \cdot \psi, \hat{H}_2 \cdot \psi\right)$, where

$$\begin{aligned} \hat{H}_1 \cdot \psi = {} & mc^2\psi_1 - c\,\mathbf{S} \cdot \left(i\hbar\nabla_{\mathbf{x}}\psi_2 + \frac{\sigma}{c}\mathbf{A}(\mathbf{x}, t)\psi_2\right) + \sigma\left(\Psi(\mathbf{x}, t) - \frac{1}{c}\mathbf{v}(\mathbf{x}, t) \cdot \mathbf{A}(\mathbf{x}_j, t)\right)\psi_1 \\ & - V(\mathbf{x}, t)\,\psi_1 - \frac{i\hbar}{2}\operatorname{div}_{\mathbf{x}}\{\psi_1\mathbf{v}(\mathbf{x}, t)\} - \frac{i\hbar}{2}\nabla_{\mathbf{x}}\psi_1 \cdot \mathbf{v}(\mathbf{x}, t) + \frac{\hbar}{4}\mathbf{S} \cdot (\operatorname{curl}_{\mathbf{x}}\mathbf{v}(\mathbf{x}, t)\,\psi_1) \\ & - \frac{(g-1)\sigma\hbar}{2mc}\mathbf{S} \cdot (\operatorname{curl}_{\mathbf{x}}\mathbf{A}(\mathbf{x}, t)\,\psi_1), \end{aligned} \tag{1.463}$$

and

$$\begin{aligned} \hat{H}_2 \cdot \psi = {} & -mc^2\psi_2 - c\,\mathbf{S} \cdot \left(i\hbar\nabla_{\mathbf{x}}\psi_1 + \frac{\sigma}{c}\mathbf{A}(\mathbf{x}, t)\psi_1\right) + \sigma\left(\Psi(\mathbf{x}, t) - \frac{1}{c}\mathbf{v}(\mathbf{x}, t) \cdot \mathbf{A}(\mathbf{x}, t)\right)\psi_2 \\ & - V(\mathbf{x}, t)\,\psi_2 - \frac{i\hbar}{2}\operatorname{div}_{\mathbf{x}}\{\psi_2\mathbf{v}(\mathbf{x}, t)\} - \frac{i\hbar}{2}\nabla_{\mathbf{x}}\psi_2 \cdot \mathbf{v}(\mathbf{x}, t) + \frac{\hbar}{4}\mathbf{S} \cdot (\operatorname{curl}_{\mathbf{x}}\mathbf{v}(\mathbf{x}, t)\,\psi_2) \\ & - \frac{(g-1)\sigma\hbar}{2mc}\mathbf{S} \cdot (\operatorname{curl}_{\mathbf{x}}\mathbf{A}(\mathbf{x}, t)\,\psi_2), \end{aligned} \tag{1.464}$$

where $\mathbf{S} := (S_1, S_2, S_3)$ and

$$S_1 = \begin{pmatrix} 0 & 1 \\ 1 & 0 \end{pmatrix}, \quad S_2 = \begin{pmatrix} 0 & -i \\ i & 0 \end{pmatrix} \quad S_3 = \begin{pmatrix} 1 & 0 \\ 0 & -1 \end{pmatrix}$$

are Pauli matrices and g is a constant that depends on the type of the particle (for electron we have $g = 1$). As before for the Schrödinger–Pauli equation, we added an additional term to the Hamiltonian, namely $\frac{\hbar}{4}\mathbf{S} \cdot (\text{curl}_{\mathbf{x}}\mathbf{v}(\mathbf{x}, t)\, \psi)$. Although this term vanishes in inertial coordinate systems, it provides however invariance of our Dirac-type equation, under the change of non-inertial coordinate systems as we will see below. Thus, we have the following two evolution equations that we call together the Dirac system of equations:

$$
\begin{aligned}
i\hbar\frac{\partial \psi_1}{\partial t} = mc^2\psi_1 - c\,\mathbf{S} \cdot \left(i\hbar\nabla_{\mathbf{x}}\psi_2 + \frac{\sigma}{c}\mathbf{A}(\mathbf{x}, t)\psi_2\right) + \sigma\left(\Psi(\mathbf{x}, t) - \frac{1}{c}\mathbf{v}(\mathbf{x}, t) \cdot \mathbf{A}(\mathbf{x}_j, t)\right)\psi_1 \\
- V(\mathbf{x}, t)\,\psi_1 - \frac{i\hbar}{2}\,\text{div}_{\mathbf{x}}\left\{\psi_1\mathbf{v}(\mathbf{x}, t)\right\} - \frac{i\hbar}{2}\nabla_{\mathbf{x}}\psi_1 \cdot \mathbf{v}(\mathbf{x}, t) + \frac{\hbar}{4}\mathbf{S} \cdot (\text{curl}_{\mathbf{x}}\mathbf{v}(\mathbf{x}, t)\,\psi_1) \\
- \frac{(g-1)\sigma\hbar}{2mc}\mathbf{S} \cdot (\text{curl}_{\mathbf{x}}\mathbf{A}(\mathbf{x}, t)\,\psi_1)\,, \qquad (1.465)
\end{aligned}
$$

and

$$
\begin{aligned}
i\hbar\frac{\partial \psi_2}{\partial t} = \\
- mc^2\psi_2 - c\,\mathbf{S} \cdot \left(i\hbar\nabla_{\mathbf{x}}\psi_1 + \frac{\sigma}{c}\mathbf{A}(\mathbf{x}, t)\psi_1\right) + \sigma\left(\Psi(\mathbf{x}, t) - \frac{1}{c}\mathbf{v}(\mathbf{x}, t) \cdot \mathbf{A}(\mathbf{x}, t)\right)\psi_2 \\
- V(\mathbf{x}, t)\,\psi_2 - \frac{i\hbar}{2}\,\text{div}_{\mathbf{x}}\left\{\psi_2\mathbf{v}(\mathbf{x}, t)\right\} - \frac{i\hbar}{2}\nabla_{\mathbf{x}}\psi_2 \cdot \mathbf{v}(\mathbf{x}, t) + \frac{\hbar}{4}\mathbf{S} \cdot (\text{curl}_{\mathbf{x}}\mathbf{v}(\mathbf{x}, t)\,\psi_2) \\
- \frac{(g-1)\sigma\hbar}{2mc}\mathbf{S} \cdot (\text{curl}_{\mathbf{x}}\mathbf{A}(\mathbf{x}, t)\,\psi_2)\,. \qquad (1.466)
\end{aligned}
$$

Then we can rewrite Dirac equations as

$$
\begin{aligned}
i\hbar\left(\frac{\partial \psi_1}{\partial t} + \frac{1}{2}\,\text{div}_{\mathbf{x}}\left\{\psi_1\mathbf{v}(\mathbf{x}, t)\right\} + \frac{1}{2}\nabla_{\mathbf{x}}\psi_1 \cdot \mathbf{v}(\mathbf{x}, t)\right) \\
= mc^2\psi_1 - c\,\mathbf{S} \cdot \left(i\hbar\nabla_{\mathbf{x}}\psi_2 + \frac{\sigma}{c}\mathbf{A}(\mathbf{x}, t)\psi_2\right) \\
+ \sigma\left(\Psi(\mathbf{x}, t) - \frac{1}{c}\mathbf{v}(\mathbf{x}, t) \cdot \mathbf{A}(\mathbf{x}, t)\right)\psi_1 - V(\mathbf{x}, t)\,\psi_1 + \frac{\hbar}{4}\mathbf{S} \cdot (\text{curl}_{\mathbf{x}}\mathbf{v}(\mathbf{x}, t)\,\psi_1) \\
- \frac{(g-1)\sigma\hbar}{2mc}\mathbf{S} \cdot (\text{curl}_{\mathbf{x}}\mathbf{A}(\mathbf{x}, t)\,\psi_1)\,, \qquad (1.467)
\end{aligned}
$$

and

$$
\begin{aligned}
i\hbar\left(\frac{\partial\psi_2}{\partial t}+\frac{1}{2}\,\mathrm{div}_{\mathbf{x}}\left\{\psi_2\mathbf{v}(\mathbf{x},t)\right\}+\frac{1}{2}\nabla_{\mathbf{x}}\psi_2\cdot\mathbf{v}(\mathbf{x},t)\right)\\
=-mc^2\psi_2-c\,\mathbf{S}\cdot\left(i\hbar\nabla_{\mathbf{x}}\psi_1+\frac{\sigma}{c}\mathbf{A}(\mathbf{x},t)\psi_1\right)\\
+\sigma\left(\Psi(\mathbf{x},t)-\frac{1}{c}\mathbf{v}(\mathbf{x},t)\cdot\mathbf{A}(\mathbf{x},t)\right)\psi_2-V\left(\mathbf{x},t\right)\psi_2+\frac{\hbar}{4}\mathbf{S}\cdot\left(\mathrm{curl}_{\mathbf{x}}\mathbf{v}(\mathbf{x},t)\,\psi_2\right)\\
-\frac{(g-1)\sigma\hbar}{2mc}\mathbf{S}\cdot\left(\mathrm{curl}_{\mathbf{x}}\mathbf{A}(\mathbf{x},t)\,\psi_2\right).\quad(1.468)
\end{aligned}
$$

Then similarly to the proof of Theorem 1.3 about the invariance of Shrödinger–Pauli equation we can prove the following Theorem for Dirac equations:

Theorem 1.5 *Consider that the change of some Cartesian coordinate system (∗) to another Cartesian coordinate system (∗∗) is given by*

$$
\begin{cases}\mathbf{x}'=A(t)\cdot\mathbf{x}+\mathbf{z}(t),\\ t'=t,\end{cases}\quad(1.469)
$$

where $A(t)\in SO(3)$ is a rotation. Next, assume that in the coordinate system (∗∗) we observe a validity of the Dirac equations of the form:

$$
\begin{aligned}
i\hbar\left(\frac{\partial\psi_1'}{\partial t'}+\frac{1}{2}\,\mathrm{div}_{\mathbf{x}'}\left\{\psi_1'\mathbf{v}'(\mathbf{x}',t')\right\}+\frac{1}{2}\nabla_{\mathbf{x}'}\psi_1'\cdot\mathbf{v}'(\mathbf{x}',t')\right)\\
=m'c^2\psi_1'-c\,\mathbf{S}\cdot\left(i\hbar\nabla_{\mathbf{x}'}\psi_2'+\frac{\sigma'}{c}\mathbf{A}'(\mathbf{x}',t)\psi_2'\right)\\
+\sigma'\left(\Psi'(\mathbf{x}',t')-\frac{1}{c}\mathbf{v}'(\mathbf{x}',t')\cdot\mathbf{A}'(\mathbf{x}',t')\right)\psi_1'-V'\left(\mathbf{x}',t'\right)\psi_1'+\frac{\hbar}{4}\mathbf{S}\cdot\left(curl_{\mathbf{x}'}\mathbf{v}'(\mathbf{x}',t')\,\psi_1'\right)\\
-\frac{(g'-1)\sigma'\hbar}{2\,m'c}\mathbf{S}\cdot\left(curl_{\mathbf{x}'}\mathbf{A}'(\mathbf{x}',t')\,\psi_1'\right),\quad(1.470)
\end{aligned}
$$

and

$$
\begin{aligned}
i\hbar\left(\frac{\partial\psi_2'}{\partial t'}+\frac{1}{2}\,\mathrm{div}_{\mathbf{x}'}\left\{\psi_2'\mathbf{v}'(\mathbf{x}',t')\right\}+\frac{1}{2}\nabla_{\mathbf{x}'}\psi_2'\cdot\mathbf{v}'(\mathbf{x}',t')\right)=-m'c^2\psi_2'\\
-c\,\mathbf{S}\cdot\left(i\hbar\nabla_{\mathbf{x}'}\psi_1'+\frac{\sigma'}{c}\mathbf{A}'(\mathbf{x}',t')\psi_1'\right)+\sigma'\left(\Psi'(\mathbf{x}',t')-\frac{1}{c}\mathbf{v}'(\mathbf{x}',t')\cdot\mathbf{A}'(\mathbf{x}',t')\right)\psi_2'\\
-V'\left(\mathbf{x}',t'\right)\psi_2'+\frac{\hbar}{4}\mathbf{S}\cdot\left(curl_{\mathbf{x}'}\mathbf{v}'(\mathbf{x}',t')\,\psi_2'\right)\\
-\frac{(g'-1)\sigma'\hbar}{2\,m'c}\mathbf{S}\cdot\left(curl_{\mathbf{x}'}\mathbf{A}'(\mathbf{x}',t')\,\psi_2'\right),\quad(1.471)
\end{aligned}
$$

where $\psi = (\psi_1, \psi_2) \in \mathbb{C}^2 \times \mathbb{C}^2$ *is a four-component wave function. Then in the coordinate system* $(*)$ *we have the validity of Dirac equations of the same as* (1.470) *and* (1.471) *form:*

$$\begin{aligned} i\hbar\left(\frac{\partial \psi_1}{\partial t} + \frac{1}{2}\,\mathrm{div}_{\mathbf{x}}\,\{\psi_1 \mathbf{v}(\mathbf{x},t)\} + \frac{1}{2}\nabla_{\mathbf{x}}\psi_1 \cdot \mathbf{v}(\mathbf{x},t)\right) \\ = mc^2\psi_1 - c\,\mathbf{S}\cdot\left(i\hbar\nabla_{\mathbf{x}}\psi_2 + \frac{\sigma}{c}\mathbf{A}(\mathbf{x},t)\psi_2\right) \\ +\sigma\left(\Psi(\mathbf{x},t) - \frac{1}{c}\mathbf{v}(\mathbf{x},t)\cdot\mathbf{A}(\mathbf{x},t)\right)\psi_1 - V\,(\mathbf{x},t)\,\psi_1 + \frac{\hbar}{4}\mathbf{S}\cdot(curl_{\mathbf{x}}\mathbf{v}(\mathbf{x},t)\,\psi_1) \\ - \frac{(g-1)\sigma\hbar}{2mc}\mathbf{S}\cdot(curl_{\mathbf{x}}\mathbf{A}(\mathbf{x},t)\,\psi_1)\,, \end{aligned} \tag{1.472}$$

and

$$\begin{aligned} i\hbar\left(\frac{\partial \psi_2}{\partial t} + \frac{1}{2}\,\mathrm{div}_{\mathbf{x}}\,\{\psi_2 \mathbf{v}(\mathbf{x},t)\} + \frac{1}{2}\nabla_{\mathbf{x}}\psi_2 \cdot \mathbf{v}(\mathbf{x},t)\right) = \\ - mc^2\psi_2 - c\,\mathbf{S}\cdot\left(i\hbar\nabla_{\mathbf{x}}\psi_1 + \frac{\sigma}{c}\mathbf{A}(\mathbf{x},t)\psi_1\right) \\ +\sigma\left(\Psi(\mathbf{x},t) - \frac{1}{c}\mathbf{v}(\mathbf{x},t)\cdot\mathbf{A}(\mathbf{x},t)\right)\psi_2 - V\,(\mathbf{x},t)\,\psi_2 + \frac{\hbar}{4}\mathbf{S}\cdot(curl_{\mathbf{x}}\mathbf{v}(\mathbf{x},t)\,\psi_2) \\ - \frac{(g-1)\sigma\hbar}{2mc}\mathbf{S}\cdot(curl_{\mathbf{x}}\mathbf{A}(\mathbf{x},t)\,\psi_2)\,, \end{aligned} \tag{1.473}$$

provided that

$$\begin{cases} V' = V, \\ \sigma' = \sigma, \\ g' = g, \\ m' = m, \\ \mathbf{v}' = A(t)\cdot\mathbf{v} + \frac{dA}{dt}(t)\cdot\mathbf{x} + \frac{d\mathbf{z}}{dt}(t), \\ \mathbf{A}' = A(t)\cdot\mathbf{A}, \\ \Psi' - \mathbf{v}'\cdot\mathbf{A}' = \Psi - \mathbf{v}\cdot\mathbf{A}, \\ \psi_1' = U(t)\cdot\psi_1, \\ \psi_2' = U(t)\cdot\psi_2, \end{cases} \tag{1.474}$$

where, as before, $U(t) \in SU(2)$ *is characterized by*

$$U^*(t)\cdot\mathbf{S}\cdot U(t) = A(t)\cdot\mathbf{S}, \tag{1.475}$$

that means

$$\Big(U^*(t)\cdot S_1\cdot U(t), U^*(t)\cdot S_2\cdot U(t), U^*(t)\cdot S_3\cdot U(t)\Big) =$$
$$\Big(a_{11}(t)S_1 + a_{12}(t)S_2 + a_{13}(t)S_3\,,\; a_{21}(t)S_1 + a_{22}(t)S_2 + a_{23}(t)S_3\,,$$
$$a_{31}(t)S_1 + a_{32}(t)S_2 + a_{33}(t)S_3\Big),$$

where $A(t) = \{a_{mk}(t)\}_{\{1\le m,k\le 3\}}$.

Next, in the case that our particle has a positive energy, define

$$(\phi_1, \phi_2) = \left(e^{-\frac{ic^2mt}{\hbar}}\psi_1, e^{-\frac{ic^2mt}{\hbar}}\psi_2\right).$$

Then, as we show in Chap. 10, we rewrite (1.465) and (1.466) in the non-relativistic limit as

$$\phi_2 \approx -\frac{1}{2cm}\mathbf{S}\cdot\left(i\hbar\nabla_{\mathbf{x}}\phi_1 + \frac{\sigma}{c}\mathbf{A}(\mathbf{x},t)\phi_1\right). \tag{1.476}$$

and:

$$\begin{aligned}
i\hbar\frac{\partial\phi_1}{\partial t} \approx\\
-\frac{\hbar^2}{2m}\Delta_{\mathbf{x}}\phi_1 + \frac{i\hbar\sigma}{2mc}\operatorname{div}_{\mathbf{x}}\{\phi_1\mathbf{A}(\mathbf{x},t)\} + \frac{i\hbar\sigma}{2mc}\nabla_{\mathbf{x}}\phi_1\cdot\mathbf{A}(\mathbf{x},t) + \frac{\sigma^2}{2mc^2}\left|\mathbf{A}(\mathbf{x},t)\right|^2\phi_1\\
-\frac{g\sigma\hbar}{2mc}\mathbf{S}\cdot(\operatorname{curl}_{\mathbf{x}}\mathbf{A}(\mathbf{x},t)\,\phi_1) + \sigma\left(\Psi(\mathbf{x},t) - \frac{1}{c}\mathbf{v}(\mathbf{x},t)\cdot\mathbf{A}(\mathbf{x},t)\right)\phi_1 - V(\mathbf{x},t)\,\phi_1\\
-\frac{i\hbar}{2}\operatorname{div}_{\mathbf{x}}\{\phi_1\mathbf{v}(\mathbf{x},t)\} - \frac{i\hbar}{2}\nabla_{\mathbf{x}}\phi_1\cdot\mathbf{v}(\mathbf{x},t) + \frac{\hbar}{4}\mathbf{S}\cdot(\operatorname{curl}_{\mathbf{x}}\mathbf{v}(\mathbf{x},t)\,\phi_1).
\end{aligned} \tag{1.477}$$

The last equation coincides with the non-relativistic Shrödinger–Pauli equation, that we studied above.

Next, consider a Lagrangian density L associated with the motion of a spin-half quantum relativistic-like micro-particle with inertial mass m and the charge σ in the outer gravitational and electromagnetical field with characteristics $\mathbf{v}(\mathbf{x},t)$, $\mathbf{A}(\mathbf{x},t)$ and $\Psi(\mathbf{x},t)$ and additional conservative field with potential $V(\mathbf{y},t)$:

$$
\begin{aligned}
L(\psi, \mathbf{x}, t) := \frac{i\hbar}{2}\left(\left(\frac{\partial \psi_1}{\partial t} + \mathbf{v}\cdot\nabla_{\mathbf{x}}\psi_1\right)\cdot\bar{\psi}_1 - \psi_1\cdot\left(\frac{\partial \bar{\psi}_1}{\partial t} + \mathbf{v}\cdot\nabla_{\mathbf{x}}\bar{\psi}_1\right)\right) \\
+\frac{c}{2}\left(\left(\mathbf{S}\cdot\left(i\hbar\nabla_{\mathbf{x}}\psi_2 + \frac{\sigma}{c}\mathbf{A}(\mathbf{x},t)\psi_2\right)\right)\cdot\bar{\psi}_1 - \psi_1\cdot\left(\mathbf{S}\cdot\left(i\hbar\nabla_{\mathbf{x}}\bar{\psi}_2 - \frac{\sigma}{c}\mathbf{A}(\mathbf{x},t)\bar{\psi}_2\right)\right)\right) \\
+\frac{i\hbar}{2}\left(\left(\frac{\partial \psi_2}{\partial t} + \mathbf{v}\cdot\nabla_{\mathbf{x}}\psi_2\right)\cdot\bar{\psi}_2 - \psi_2\cdot\left(\frac{\partial \bar{\psi}_2}{\partial t} + \mathbf{v}\cdot\nabla_{\mathbf{x}}\bar{\psi}_2\right)\right) \\
+\frac{c}{2}\left(\left(\mathbf{S}\cdot\left(i\hbar\nabla_{\mathbf{x}}\psi_1 + \frac{\sigma}{c}\mathbf{A}(\mathbf{x},t)\psi_1\right)\right)\cdot\bar{\psi}_2 - \psi_2\cdot\left(\mathbf{S}\cdot\left(i\hbar\nabla_{\mathbf{x}}\bar{\psi}_1 - \frac{\sigma}{c}\mathbf{A}(\mathbf{x},t)\bar{\psi}_1\right)\right)\right) \\
- mc^2\left(\psi_1\cdot\bar{\psi}_1 - \psi_2\cdot\bar{\psi}_2\right) - \sigma\left(\Psi - \frac{1}{c}\mathbf{v}\cdot\mathbf{A}\right)\left(\psi_1\cdot\bar{\psi}_1 + \psi_2\cdot\bar{\psi}_2\right) \\
+ V(\mathbf{x},t)\left(\psi_1\cdot\bar{\psi}_1 + \psi_2\cdot\bar{\psi}_2\right) - \frac{\hbar}{4}\left(\mathbf{S}\cdot(\mathrm{curl}_{\mathbf{x}}\mathbf{v})\,\psi_1\right)\cdot\bar{\psi}_1 - \frac{\hbar}{4}\left(\mathbf{S}\cdot(\mathrm{curl}_{\mathbf{x}}\mathbf{v})\,\psi_2\right)\cdot\bar{\psi}_2 \\
+\frac{(g-1)\sigma\hbar}{2mc}\left(\mathbf{S}\cdot(\mathrm{curl}_{\mathbf{x}}\mathbf{A})\,\psi_1\right)\cdot\bar{\psi}_1 + \frac{(g-1)\sigma\hbar}{2mc}\left(\mathbf{S}\cdot(\mathrm{curl}_{\mathbf{x}}\mathbf{A})\,\psi_2\right)\cdot\bar{\psi}_2, \quad (1.478)
\end{aligned}
$$

where $\psi = (\psi_1, \psi_2) \in \mathbb{C}^2 \times \mathbb{C}^2$ is a four-component wave function. Then similarly to the proof of Theorem 1.5 we can prove that L is invariant under the change of inertial or non-inertial Cartesian coordinate system, given by (1.469), provided that we take into account (1.474). Moreover, if we consider a functional

$$
J = \int_0^T \int_{\mathbb{R}^3} L(\psi, \mathbf{x}, t)\, d\mathbf{x} dt, \quad (1.479)
$$

then, by (1.478) we deduce the Euler–Lagrange equation for (1.479) coincide with Dirac equations in the form of (1.467) and (1.468). Next, as before, we would like to note that, as before, the Lagrangian density L, defined by (1.478) obeys $U(1)$ local symmetry, i.e. for every scalar field $w := w(\mathbf{x}, t)$ one can easily deduce that L in (1.478) is invariant under the transformation:

$$
\begin{cases}
\psi_1 \to e^{-\frac{i\sigma w}{c\hbar}}\psi_1 \\
\psi_2 \to e^{-\frac{i\sigma w}{c\hbar}}\psi_2 \\
\Psi \to \Psi + \frac{1}{c}\frac{\partial w}{\partial t} \\
\mathbf{A} \to \mathbf{A} - \nabla_{\mathbf{x}} w \\
\mathbf{v} \to \mathbf{v}.
\end{cases} \quad (1.480)
$$

See also Sect. 10.5 for generalizations of Dirac equation for a system of n spin-half stable particles. See also Sect. 10.6 for the Quantum Liouville's equation for a finite system of spin-half relativistic-like stable particles.

1.11 Spinless Relativistic-Like Particles: The Klein–Gordon Equation

Consider the motion of a relativistic-like spinless quantum micro-particle with inertial masses m and the charge σ in the outer gravitational and electromagnetical field with characteristics $\mathbf{v}(\mathbf{x}, t)$, $\mathbf{A}(\mathbf{x}, t)$ and $\Psi(\mathbf{x}, t)$ and additional conservative field with potential $V(\mathbf{x}, t)$. The Klein–Gordon equation for this system of particles is the following:

$$\begin{aligned}
&\frac{\hbar^2}{2c^2 m}\frac{\partial}{\partial t}\left(\frac{\partial\psi}{\partial t}+\mathbf{v}(\mathbf{x},t)\cdot\nabla_{\mathbf{x}}\psi+\frac{i\sigma}{\hbar}\Psi_0(\mathbf{x},t)\psi\right)\\
&\quad+\frac{\hbar^2}{2c^2 m}\operatorname{div}_{\mathbf{x}}\left\{\mathbf{v}(\mathbf{x},t)\left(\frac{\partial\psi}{\partial t}+\mathbf{v}(\mathbf{x},t)\cdot\nabla_{\mathbf{x}}\psi+\frac{i\sigma}{\hbar}\Psi_0(\mathbf{x},t)\psi\right)\right\}\\
&\quad+\frac{\hbar^2}{2c^2 m}\left(\frac{i\sigma}{\hbar}\Psi_0(\mathbf{x},t)\right)\left(\frac{\partial\psi}{\partial t}+\mathbf{v}(\mathbf{x},t)\cdot\nabla_{\mathbf{x}}\psi+\frac{i\sigma}{\hbar}\Psi_0(\mathbf{x},t)\psi\right)\\
&\qquad+\frac{c^2 m}{2}\psi-\frac{\hbar^2}{2m}\Delta_{\mathbf{x}}\psi-V(\mathbf{x},t)\,\psi\\
&+\frac{i\hbar\sigma}{2mc}\operatorname{div}_{\mathbf{x}}\{\psi\mathbf{A}(\mathbf{x},t)\}+\frac{i\hbar\sigma}{2mc}\mathbf{A}(\mathbf{x},t)\cdot\nabla_{\mathbf{x}}\psi+\left(\frac{\sigma^2}{2mc^2}|\mathbf{A}(\mathbf{x},t)|^2\right)\psi=0,
\end{aligned} \tag{1.481}$$

where, as before,

$$\Psi_0(\mathbf{x},t):=\Psi(\mathbf{x},t)-\frac{1}{c}\mathbf{v}(\mathbf{x},t)\cdot\mathbf{A}(\mathbf{x},t) \tag{1.482}$$

is the proper electrical potential and $\psi=\psi(\mathbf{x},t)\in\mathbb{C}$ is the scalar wave function. Next consider a change of some non-inertial Cartesian coordinate system $(*)$ to another Cartesian coordinate system $(**)$ of the form:

$$\begin{cases}\mathbf{x}'=A(t)\cdot\mathbf{x}+\mathbf{z}(t),\\ t'=t,\end{cases} \tag{1.483}$$

where $A(t)\in SO(3)$ is a rotation. Then, we deduce that the Klein–Gordon equation of the form (1.481) is invariant under the change of non-inertial Cartesian coordinate system, provided that under (1.483) we have

$$\begin{cases}\psi'=\psi\\ V'=V\\ \mathbf{A}'=A(t)\cdot\mathbf{A}\\ \Psi_0':=\Psi'-\frac{1}{c}\mathbf{A}'\cdot\mathbf{v}'=\Psi-\frac{1}{c}\mathbf{A}\cdot\mathbf{v}:=\Psi_0.\end{cases} \tag{1.484}$$

Next defining

$$\psi_1(\mathbf{x}, t) := e^{\frac{ic^2 mt}{\hbar}} \psi(\mathbf{x}, t) \tag{1.485}$$

we rewrite (1.481) as

$$\begin{aligned}
&\frac{\hbar^2}{2c^2 m}\frac{\partial}{\partial t}\left(\frac{\partial \psi_1}{\partial t} + \mathbf{v}(\mathbf{x}, t)\cdot\nabla_{\mathbf{x}}\psi_1 + \frac{i\sigma}{\hbar}\Psi_0(\mathbf{x}, t)\psi_1\right) \\
&\quad + \frac{\hbar^2}{2c^2 m}\operatorname{div}_{\mathbf{x}}\left\{\mathbf{v}(\mathbf{x}, t)\left(\frac{\partial \psi_1}{\partial t} + \mathbf{v}(\mathbf{x}, t)\cdot\nabla_{\mathbf{x}}\psi_1 + \frac{i\sigma}{\hbar}\Psi_0(\mathbf{x}, t)\psi_1\right)\right\} \\
&\quad + \frac{\hbar^2}{2c^2 m}\left(\frac{i\sigma}{\hbar}\Psi_0(\mathbf{x}, t)\right)\left(\frac{\partial \psi_1}{\partial t} + \mathbf{v}(\mathbf{x}, t)\cdot\nabla_{\mathbf{x}}\psi_1 + \frac{i\sigma}{\hbar}\Psi_0(\mathbf{x}, t)\psi_1\right) \\
&\quad - i\hbar\left(\frac{\partial \psi_1}{\partial t} + \frac{1}{2}\mathbf{v}(\mathbf{x}, t)\cdot\nabla_{\mathbf{x}}\psi_1 + \frac{1}{2}\operatorname{div}_{\mathbf{x}}\{\psi_1\mathbf{v}(\mathbf{x}, t)\} + \frac{i\sigma}{\hbar}\Psi_0(\mathbf{x}, t)\psi_1\right) - \frac{\hbar^2}{2m}\Delta_{\mathbf{x}}\psi_1 \\
&- V(\mathbf{x}, t)\psi_1 + \frac{i\hbar\sigma}{2mc}\operatorname{div}_{\mathbf{x}}\{\psi_1\mathbf{A}(\mathbf{x}, t)\} + \frac{i\hbar\sigma}{2mc}\mathbf{A}(\mathbf{x}, t)\cdot\nabla_{\mathbf{x}}\psi_1 + \left(\frac{\sigma^2}{2mc^2}|\mathbf{A}(\mathbf{x}, t)|^2\right)\psi_1 = 0,
\end{aligned} \tag{1.486}$$

see Sect. 10.7 for the details. Thus by (1.486) in the non-relativistic limit we obtain

$$\begin{aligned}
&i\hbar\left(\frac{\partial \psi_1}{\partial t} + \mathbf{v}(\mathbf{x}, t)\cdot\nabla_{\mathbf{x}}\psi_1\right) + \frac{i\hbar}{2}\left(\operatorname{div}_{\mathbf{x}}\mathbf{v}(\mathbf{x}, t)\right)\psi_1 \\
&\qquad \approx -\frac{\hbar^2}{2m}\Delta_{\mathbf{x}}\psi_1 - V(\mathbf{x}, t)\psi_1 + \frac{i\hbar\sigma}{2mc}\mathbf{A}(\mathbf{x}, t)\cdot\nabla_{\mathbf{x}}\psi_1 \\
&\quad + \frac{i\hbar\sigma}{2mc}\operatorname{div}_{\mathbf{x}}\{\psi_1\mathbf{A}(\mathbf{x}, t)\} + \left(\sigma\Psi(\mathbf{x}, t) - \frac{\sigma}{c}\mathbf{A}(\mathbf{x}, t)\cdot\mathbf{v}(\mathbf{x}, t) + \frac{\sigma^2}{2mc^2}|\mathbf{A}(\mathbf{x}, t)|^2\right)\psi_1,
\end{aligned} \tag{1.487}$$

that coincides with the Schrödinger equation of the form (1.82).

Next, again consider the motion of a quantum micro-particle having inertial masses m and the charge σ with the known gravitational and electromagnetical field with potentials $\mathbf{v}(\mathbf{x}, t)$, $\mathbf{A}(\mathbf{x}, t)$ and $\Psi(\mathbf{x}, t)$ and additional conservative field with potential $V(\mathbf{x}, t)$. Then consider a Lagrangian density L defined by

$$\begin{aligned}
&L_2(\psi, \mathbf{A}, \Psi, \mathbf{v}, \mathbf{x}, t) := -\frac{c^2 m}{2}\psi\cdot\bar{\psi} - \frac{\hbar^2}{2m}\nabla_{\mathbf{x}}\psi\cdot\nabla_{\mathbf{x}}\bar{\psi} \\
&+ \frac{\hbar^2}{2c^2 m}\left(\frac{\partial \psi}{\partial t} + \mathbf{v}(\mathbf{x}, t)\cdot\nabla_{\mathbf{x}}\psi + \frac{i\sigma}{\hbar}\Psi_0(\mathbf{x}, t)\psi\right)\left(\frac{\partial \bar{\psi}}{\partial t} + \mathbf{v}(\mathbf{x}, t)\cdot\nabla_{\mathbf{x}}\bar{\psi} - \frac{i\sigma}{\hbar}\Psi_0(\mathbf{x}, t)\bar{\psi}\right) \\
&- \frac{\hbar\sigma i}{2mc}\left(\nabla_{\mathbf{x}}\psi\cdot\bar{\psi} - \psi\cdot\nabla_{\mathbf{x}}\bar{\psi}\right)\cdot\mathbf{A}(\mathbf{x}, t) - \frac{\sigma^2}{2mc^2}|\mathbf{A}(\mathbf{x}, t)|^2\psi\cdot\bar{\psi} + V(\mathbf{x}, t)\psi\cdot\bar{\psi},
\end{aligned} \tag{1.488}$$

where $\psi(\mathbf{x}, t) \in \mathbb{C}$ is a wave function of the system. Then, as before, it can be proven that L is invariant under the change of inertial or non-inertial Cartesian coordinate systems of the form (1.483), provided that $\psi' = \psi$. We investigate stationary points of the functional

$$J = \int_0^T \int_{\mathbb{R}^3} L(\psi, \mathbf{A}, \Psi, \mathbf{v}, \mathbf{x}, t)\, d\mathbf{x}dt. \tag{1.489}$$

Then,

$$\begin{aligned}
0 = \frac{\delta L_2}{\delta(\bar{\psi})} = &-\frac{\hbar^2}{2c^2 m}\frac{\partial}{\partial t}\left(\frac{\partial \psi}{\partial t} + \mathbf{v}(\mathbf{x},t)\cdot\nabla_{\mathbf{x}}\psi + \frac{i\sigma}{\hbar}\Psi_0(\mathbf{x},t)\psi\right)\\
&-\frac{\hbar^2}{2c^2 m}\operatorname{div}_{\mathbf{x}}\left\{\mathbf{v}(\mathbf{x},t)\left(\frac{\partial \psi}{\partial t} + \mathbf{v}(\mathbf{x},t)\cdot\nabla_{\mathbf{x}}\psi + \frac{i\sigma}{\hbar}\Psi_0(\mathbf{x},t)\psi\right)\right\}\\
&-\frac{\hbar^2}{2c^2 m}\left(\frac{i\sigma}{\hbar}\Psi_0(\mathbf{x},t)\right)\left(\frac{\partial \psi}{\partial t} + \mathbf{v}(\mathbf{x},t)\cdot\nabla_{\mathbf{x}}\psi + \frac{i\sigma}{\hbar}\Psi_0(\mathbf{x},t)\psi\right) - \frac{c^2 m}{2}\psi + \frac{\hbar^2}{2m}\Delta_{\mathbf{x}}\psi\\
&-\frac{\hbar\sigma i}{2mc}\left(\mathbf{A}(\mathbf{x},t)\cdot\nabla_{\mathbf{x}}\psi + \operatorname{div}_{\mathbf{x}}\{\psi\mathbf{A}(\mathbf{x},t)\}\right) - \frac{\sigma^2}{2mc^2}|\mathbf{A}(\mathbf{x},t)|^2\psi + V(\mathbf{x},t)\psi,
\end{aligned} \tag{1.490}$$

and

$$\begin{aligned}
0 = \frac{\delta L_2}{\delta(\psi)} = &-\frac{\hbar^2}{2c^2 m}\frac{\partial}{\partial t}\left(\frac{\partial \bar{\psi}}{\partial t} + \mathbf{v}(\mathbf{x},t)\cdot\nabla_{\mathbf{x}}\bar{\psi} + \frac{(\bar{i})\sigma}{\hbar}\Psi_0(\mathbf{x},t)\bar{\psi}\right)\\
&-\frac{\hbar^2}{2c^2 m}\operatorname{div}_{\mathbf{x}}\left\{\mathbf{v}(\mathbf{x},t)\left(\frac{\partial \bar{\psi}}{\partial t} + \mathbf{v}(\mathbf{x},t)\cdot\nabla_{\mathbf{x}}\bar{\psi} + \frac{(\bar{i})\sigma}{\hbar}\Psi_0(\mathbf{x},t)\bar{\psi}\right)\right\}\\
&-\frac{\hbar^2}{2c^2 m}\left(\frac{(\bar{i})\sigma}{\hbar}\Psi_0(\mathbf{x},t)\right)\left(\frac{\partial \bar{\psi}}{\partial t} + \mathbf{v}(\mathbf{x},t)\cdot\nabla_{\mathbf{x}}\bar{\psi} + \frac{(\bar{i})\sigma}{\hbar}\Psi_0(\mathbf{x},t)\bar{\psi}\right) - \frac{c^2 m}{2}\bar{\psi} + \frac{\hbar^2}{2m}\Delta_{\mathbf{x}}\bar{\psi}\\
&-\frac{\hbar\sigma(\bar{i})}{2mc}\left(\mathbf{A}(\mathbf{x},t)\cdot\nabla_{\mathbf{x}}\bar{\psi} + \operatorname{div}_{\mathbf{x}}\{\bar{\psi}\mathbf{A}(\mathbf{x},t)\}\right) - \frac{\sigma^2}{2mc^2}|\mathbf{A}(\mathbf{x},t)|^2\bar{\psi} + V(\mathbf{x},t)\bar{\psi},
\end{aligned} \tag{1.491}$$

where the last equality is just the complex conjugate of (1.490). So we get that the Euler–Lagrange equation for (1.489) coincides with the Klein–Gordon equation of the form (1.481). Next we would like to note that the Lagrangian density L_2, defined by (1.488) obeys $U(1)$ local symmetry, i.e. for every scalar field $w := w(\mathbf{x}, t)$ one can easily deduce that L_2 in (1.488) is invariant under the transformation:

$$\begin{cases}
\psi \to e^{-\frac{i\sigma w}{c\hbar}}\psi\\
\Psi \to \Psi + \frac{1}{c}\frac{\partial w}{\partial t}\\
\mathbf{A} \to \mathbf{A} - \nabla_{\mathbf{x}} w\\
\mathbf{v} \to \mathbf{v}\\
\Psi_0 \to \Psi_0 + \frac{1}{c}\left(\frac{\partial w}{\partial t} + \mathbf{v}\cdot\nabla_{\mathbf{x}} w\right).
\end{cases} \tag{1.492}$$

See Sect. 10.7 for the generalization of all the above to the system of n relativistic-like spinless particles.

1.12 Thermodynamics of a Moving Continuum Medium

Again, consistently with (1.6), consider in some Cartesian coordinate system (*) the Second Law of Newton for the moving continuum medium with the inertial mass density μ, the field of average (macroscopic) velocities $\mathbf{u}$, the charge density ρ and the electric current density $\mathbf{j}$:

$$\mu\left(\frac{\partial \mathbf{u}}{\partial t}+d_{\mathbf{x}}\mathbf{u}\cdot\mathbf{u}\right)=\frac{\partial(\mu\mathbf{u})}{\partial t}+\operatorname{div}_{\mathbf{x}}\{\mu\mathbf{u}\otimes\mathbf{u}\}=$$
$$-\mu\mathbf{u}\times\operatorname{curl}_{\mathbf{x}}\mathbf{v}+\mu\left(\partial_t\mathbf{v}+\nabla_{\mathbf{x}}\frac{1}{2}|\mathbf{v}|^2\right)+\rho\mathbf{E}+\frac{1}{c}\mathbf{j}\times\mathbf{B}+\operatorname{div}_{\mathbf{x}}\mathcal{T}=$$
$$=-\mu(\mathbf{u}-\mathbf{v})\times\operatorname{curl}_{\mathbf{x}}\mathbf{v}+\mu\left(\partial_t\mathbf{v}+d_{\mathbf{x}}\mathbf{v}\cdot\mathbf{v}\right)+\rho\mathbf{E}+\frac{1}{c}\mathbf{j}\times\mathbf{B}+\operatorname{div}_{\mathbf{x}}\mathcal{T}. \quad (1.493)$$

Here $\rho\mathbf{E}+\frac{1}{c}\mathbf{j}\times\mathbf{B}$ is the volume density of the Lorentz force where $\mathbf{E}$ and $\mathbf{B}$ are outer electric and magnetic fields, assumed to be changing smoothly and almost constant in the microscopic level, $\mathbf{v}$ is a vectorial gravitational potential also assumed to be changing smoothly and almost constant in the microscopic level, and $\mathcal{T}\in\mathbb{R}^{3\times 3}$ is the symmetric Cauchy stress tensor of the continuum medium. Moreover, the mass density μ, clearly satisfies the continuum equation:

$$\frac{\partial\mu}{\partial t}+\operatorname{div}_{\mathbf{x}}(\mu\mathbf{u})=0. \quad (1.494)$$

In particular, multiplying (11.1) by $\mathbf{u}$ and using (11.2) we deduce the equality of the balance of the kinetic energy:

$$\frac{\partial}{\partial t}\left(\frac{\mu}{2}|\mathbf{u}|^2\right)+\operatorname{div}_{\mathbf{x}}\left\{\left(\frac{\mu}{2}|\mathbf{u}|^2\right)\mathbf{u}\right\}=\mu\left(\frac{\partial}{\partial t}\left(\frac{1}{2}|\mathbf{u}|^2\right)+\mathbf{u}\cdot\nabla_{\mathbf{x}}\left(\frac{1}{2}|\mathbf{u}|^2\right)\right)=$$
$$\mu\left(\partial_t\mathbf{v}+\nabla_{\mathbf{x}}\frac{1}{2}|\mathbf{v}|^2\right)\cdot\mathbf{u}+\left(\rho\mathbf{E}+\frac{1}{c}\mathbf{j}\times\mathbf{B}\right)\cdot\mathbf{u}+(\operatorname{div}_{\mathbf{x}}\mathcal{T})\cdot\mathbf{u}=$$
$$\mu\left(\partial_t\mathbf{v}+\nabla_{\mathbf{x}}\frac{1}{2}|\mathbf{v}|^2\right)\cdot\mathbf{u}+\rho\mathbf{E}\cdot\mathbf{u}-\frac{1}{c}(\mathbf{u}\times\mathbf{B})\cdot\mathbf{j}+(\operatorname{div}_{\mathbf{x}}\mathcal{T})\cdot\mathbf{u}. \quad (1.495)$$

Next, it is well known (see [4]), that the First Law of Thermodynamics of this moving medium has the following form:

$$\frac{\partial E}{\partial t}+\operatorname{div}_{\mathbf{x}}\{E\mathbf{u}\}=\mu\left(\frac{\partial}{\partial t}\left(\frac{E}{\mu}\right)+\mathbf{u}\cdot\nabla_{\mathbf{x}}\left(\frac{E}{\mu}\right)\right)$$
$$=\frac{1}{2}\left(d_{\mathbf{x}}\mathbf{u}+\{d_{\mathbf{x}}\mathbf{u}\}^T\right):\mathcal{T}-\operatorname{div}_{\mathbf{x}}\mathbf{q}+(\mathbf{j}-\rho\mathbf{u})\cdot\left(\mathbf{E}+\frac{1}{c}\mathbf{u}\times\mathbf{B}\right). \quad (1.496)$$

Here E is the volume density of the internal energy (energy per unit volume) and consistently $\frac{E}{\mu}$ is the internal energy per unit mass, $\mathbf{q}$ is the heat flux and

$$(\mathbf{j} - \rho\mathbf{u}) \cdot \left(\mathbf{E} + \frac{1}{c}\mathbf{u} \times \mathbf{B}\right)$$

is the Joules heat term. In particular, adding (1.496) with (1.495) and using the symmetry of $\mathcal{T}$, we deduce the following equality of the balance of the energy:

$$\begin{aligned}
&\frac{\partial}{\partial t}\left(\frac{\mu}{2}|\mathbf{u}|^2 + E\right) + \operatorname{div}_{\mathbf{x}}\left\{\left(\frac{\mu}{2}|\mathbf{u}|^2 + E\right)\mathbf{u}\right\} = \\
&\qquad \mu\left(\frac{\partial}{\partial t}\left(\frac{1}{2}|\mathbf{u}|^2 + \frac{E}{\mu}\right) + \mathbf{u} \cdot \nabla_{\mathbf{x}}\left(\frac{1}{2}|\mathbf{u}|^2 + \frac{E}{\mu}\right)\right) \\
&\qquad = \operatorname{div}_{\mathbf{x}}(\mathcal{T} \cdot \mathbf{u}) - \operatorname{div}_{\mathbf{x}} \mathbf{q} + \mu\left(\partial_t \mathbf{v} + \nabla_{\mathbf{x}}\frac{1}{2}|\mathbf{v}|^2\right) \cdot \mathbf{u} + \mathbf{E} \cdot \mathbf{j}. \qquad (1.497)
\end{aligned}$$

Next the Second Law of Thermodynamics states that

$$T\left(\frac{\partial S}{\partial t} + \operatorname{div}_{\mathbf{x}}\{S\mathbf{u}\}\right) = T\mu\left(\frac{\partial}{\partial t}\left(\frac{S}{\mu}\right) + \mathbf{u} \cdot \nabla_{\mathbf{x}}\left(\frac{S}{\mu}\right)\right) \geq -\operatorname{div}_{\mathbf{x}} \mathbf{q}. \qquad (1.498)$$

Here $T := T(\mathbf{x}, t)$ is the Kelvin temperature field and S is the volume density of the entropy (entropy per unit volume) and consistently $\frac{S}{\mu}$ is the entropy per unit mass. Moreover, we have the equality in (1.498) in the case of reversible or quasi-reversible processes. In the latter case we rewrite the First Law (1.496) and the Second Law (1.498) together as

$$\begin{aligned}
&\frac{\partial E}{\partial t} + \operatorname{div}_{\mathbf{x}}\{E\mathbf{u}\} = \\
&\quad T\left(\frac{\partial S}{\partial t} + \operatorname{div}_{\mathbf{x}}\{S\mathbf{u}\}\right) + \frac{1}{2}\left(d_{\mathbf{x}}\mathbf{u} + \{d_{\mathbf{x}}\mathbf{u}\}^T\right) : \mathcal{T} + (\mathbf{j} - \rho\mathbf{u}) \cdot \left(\mathbf{E} + \frac{1}{c}\mathbf{u} \times \mathbf{B}\right) \\
&= T\mu\left(\frac{\partial}{\partial t}\left(\frac{S}{\mu}\right) + \mathbf{u} \cdot \nabla_{\mathbf{x}}\left(\frac{S}{\mu}\right)\right) + \frac{1}{2}\left(d_{\mathbf{x}}\mathbf{u} + \{d_{\mathbf{x}}\mathbf{u}\}^T\right) : \mathcal{T} + (\mathbf{j} - \rho\mathbf{u}) \cdot \left(\mathbf{E} + \frac{1}{c}\mathbf{u} \times \mathbf{B}\right) \\
&\qquad = \mu\left(\frac{\partial}{\partial t}\left(\frac{E}{\mu}\right) + \mathbf{u} \cdot \nabla_{\mathbf{x}}\left(\frac{E}{\mu}\right)\right). \qquad (1.499)
\end{aligned}$$

In particular, if the stress tensor has the following particular form

$$\mathcal{T} = -p\,I, \qquad (1.500)$$

where p is the scalar pressure and $I := Id \in \mathbb{R}^{3\times 3}$ is the identity matrix, then we rewrite the First Law of Thermodynamics (1.496) as

$$\mu\left(\frac{\partial}{\partial t}\left(\frac{E}{\mu}\right)+\mathbf{u}\cdot\nabla_{\mathbf{x}}\left(\frac{E}{\mu}\right)\right)=-\left(\operatorname{div}_{\mathbf{x}}\mathbf{u}\right)p-\operatorname{div}_{\mathbf{x}}\mathbf{q}+(\mathbf{j}-\rho\mathbf{u})\cdot\left(\mathbf{E}+\frac{1}{c}\mathbf{u}\times\mathbf{B}\right), \tag{1.501}$$

or equivalently as

$$\begin{aligned}&\frac{\partial}{\partial t}\left(\frac{E}{\mu}\right)+\mathbf{u}\cdot\nabla_{\mathbf{x}}\left(\frac{E}{\mu}\right)=\\&-p\left(\frac{\partial}{\partial t}\left(\frac{1}{\mu}\right)+\mathbf{u}\cdot\nabla_{\mathbf{x}}\left(\frac{1}{\mu}\right)\right)-\frac{1}{\mu}\operatorname{div}_{\mathbf{x}}\mathbf{q}+\frac{1}{\mu}(\mathbf{j}-\rho\mathbf{u})\cdot\left(\mathbf{E}+\frac{1}{c}\mathbf{u}\times\mathbf{B}\right),\end{aligned} \tag{1.502}$$

and in the case of quasi-reversible process we rewrite (1.499) as

$$\begin{aligned}\frac{\partial}{\partial t}\left(\frac{E}{\mu}\right)+\mathbf{u}\cdot\nabla_{\mathbf{x}}\left(\frac{E}{\mu}\right)=-p\left(\frac{\partial}{\partial t}\left(\frac{1}{\mu}\right)+\mathbf{u}\cdot\nabla_{\mathbf{x}}\left(\frac{1}{\mu}\right)\right)+T\left(\frac{\partial}{\partial t}\left(\frac{S}{\mu}\right)+\mathbf{u}\cdot\nabla_{\mathbf{x}}\left(\frac{S}{\mu}\right)\right)\\+\frac{1}{\mu}(\mathbf{j}-\rho\mathbf{u})\cdot\left(\mathbf{E}+\frac{1}{c}\mathbf{u}\times\mathbf{B}\right),\end{aligned} \tag{1.503}$$

where clearly $\frac{1}{\mu}$ is the volume per unit mass. Moreover, the following inequality always holds:

$$\mathbf{q}\cdot\nabla_{\mathbf{x}}T\leq 0. \tag{1.504}$$

In particular, by inserting (1.504) into (1.498) we deduce

$$\frac{\partial S}{\partial t}+\operatorname{div}_{\mathbf{x}}\{S\mathbf{u}\}=\mu\left(\frac{\partial}{\partial t}\left(\frac{S}{\mu}\right)+\mathbf{u}\cdot\nabla_{\mathbf{x}}\left(\frac{S}{\mu}\right)\right)\geq-\operatorname{div}_{\mathbf{x}}\left(\frac{1}{T}\mathbf{q}\right). \tag{1.505}$$

Note that by (1.505), the total entropy of an arbitrary moving thermally isolated system is a non-decreasing function of time. Finally, we remind the approximate Fourier's law (which is consistent with (1.504)):

$$\mathbf{q}=-\chi\nabla_{\mathbf{x}}T, \tag{1.506}$$

where χ is some positive material coefficient (not necessary a constant).

Next consider the change of certain non-inertial Cartesian coordinate system $(*)$ to another Cartesian coordinate system $(**)$ of the form (1.2). Then by Proposition 1.1 we easily deduce that the Laws in (1.496), (1.498), (1.499), (1.500), (1.502), (1.503), (1.504), (1.505) and (1.506) are invariant under the change of a non-inertial Cartesian coordinate system given by (1.2), provided that under (1.2) we have

$$\begin{cases} \mu' = \mu, \\ E' = E, \\ \mathcal{S}' = \mathcal{S}, \\ T' = T, \\ \mathbf{q}' = A(t) \cdot \mathbf{q}, \\ \mathcal{T}' = A(t) \cdot \mathcal{T} \cdot A^T(t) \\ p' = p, \\ \chi' = \chi, \\ \rho' = \rho, \\ \mathbf{u}' = A(t) \cdot \mathbf{u} + \frac{dA}{dt}(t) \cdot \mathbf{x} + \frac{d\mathbf{z}}{dt}(t), \\ \mathbf{v}' = A(t) \cdot \mathbf{v} + \frac{dA}{dt}(t) \cdot \mathbf{x} + \frac{d\mathbf{z}}{dt}(t), \\ \mathbf{j}' = A(t) \cdot \mathbf{j} + \rho \frac{dA}{dt}(t) \cdot \mathbf{x} + \rho \frac{d\mathbf{z}}{dt}(t). \end{cases} \tag{1.507}$$

See also Chap. 11 for the generalization of Fourier's law (1.506) to the case of anisotropic mediums.

1.12.1 The Case of an Inviscid Fluid/Gas

In the case of an inviscid fluid or gas equality (1.500) indeed holds. As a consequence, equality (1.502) holds, and moreover, in the case of quasi-reversible process equality (1.503) also holds. Moreover, in the case of a classical ideal gas the following state equality is well known:

$$p = \frac{\mu}{m_0} k T, \tag{1.508}$$

where m_0 is the mass of the single molecule of the given gas and k is the Boltzmann constant. Finally, for the ideal gas, we have the following expression for the volume density of the internal energy E:

$$E = \frac{\mu}{m_0} C_0 k T, \tag{1.509}$$

where $C_0 > 0$ is a constant that depends on the kind of the gas (for the monatomic gas we have $C_0 = \frac{3}{2}$). On the other hand, in the case of incompressible fluid we have

$$\mathrm{div}_{\mathbf{x}} \mathbf{u} = 0, \tag{1.510}$$

and the pressure p is unspecified.

Then, as before, we easily deduce that the Laws in (1.508), (1.509) and (1.510) are invariant under the change of a non-inertial Cartesian coordinate system given by (1.2), provided that under (1.2) we have (1.507).

1.12.2 The Case of the Simplest Viscous Fluid/Gas

In the case of the simplest viscous fluid or gas we have the following equality, that substitutes (1.500):

$$\mathcal{T} = -p\, I + \left(\alpha \left(d_{\mathbf{x}}\mathbf{u} + \{d_{\mathbf{x}}\mathbf{u}\}^T\right) + \beta\, (\mathrm{div}_{\mathbf{x}}\, \mathbf{u})\, I\right), \tag{1.511}$$

where $\alpha \geq 0$ and β are some material coefficients. Then, as in (1.502) and (1.503), by (1.511) we rewrite the First Law of Thermodynamics (1.496) as

$$\begin{aligned}\frac{\partial}{\partial t}\left(\frac{E}{\mu}\right) + \mathbf{u}\cdot\nabla_{\mathbf{x}}\left(\frac{E}{\mu}\right) &= \frac{\alpha}{2}\left|d_{\mathbf{x}}\mathbf{u} + \{d_{\mathbf{x}}\mathbf{u}\}^T\right|^2 + \frac{\beta}{2}\left|\mathrm{div}_{\mathbf{x}}\,\mathbf{u}\right|^2 \\ &- p\left(\frac{\partial}{\partial t}\left(\frac{1}{\mu}\right) + \mathbf{u}\cdot\nabla_{\mathbf{x}}\left(\frac{1}{\mu}\right)\right) - \frac{1}{\mu}\,\mathrm{div}_{\mathbf{x}}\,\mathbf{q} + \frac{1}{\mu}\,(\mathbf{j} - \rho\mathbf{u})\cdot\left(\mathbf{E} + \frac{1}{c}\mathbf{u}\times\mathbf{B}\right),\end{aligned} \tag{1.512}$$

and in the case of quasi-reversible process we rewrite (1.499) as

$$\begin{aligned}\frac{\partial}{\partial t}\left(\frac{E}{\mu}\right) + \mathbf{u}\cdot\nabla_{\mathbf{x}}\left(\frac{E}{\mu}\right) &= \frac{\alpha}{2}\left|d_{\mathbf{x}}\mathbf{u} + \{d_{\mathbf{x}}\mathbf{u}\}^T\right|^2 + \frac{\beta}{2}\left|\mathrm{div}_{\mathbf{x}}\,\mathbf{u}\right|^2 \\ &- p\left(\frac{\partial}{\partial t}\left(\frac{1}{\mu}\right) + \mathbf{u}\cdot\nabla_{\mathbf{x}}\left(\frac{1}{\mu}\right)\right) + T\left(\frac{\partial}{\partial t}\left(\frac{S}{\mu}\right) + \mathbf{u}\cdot\nabla_{\mathbf{x}}\left(\frac{S}{\mu}\right)\right) \\ &+ \frac{1}{\mu}\,(\mathbf{j} - \rho\mathbf{u})\cdot\left(\mathbf{E} + \frac{1}{c}\mathbf{u}\times\mathbf{B}\right).\end{aligned} \tag{1.513}$$

Moreover, in the case of a classical ideal gas the equalities (1.508) and (1.509) are still valid in the case of viscous flow. On the other hand, in the case of incompressible viscous fluid we have (1.510) and the pressure p is also unspecified.

Then, as before, by Proposition 1.1 we easily deduce that the Laws in (1.511), (1.512), (1.513), (1.508), (1.509) and (1.510) are invariant under the change of a non-inertial Cartesian coordinate system given by (1.2), provided that under (1.2) we have (1.507).

See also the end of Sect. 11.1.2 for the generalization of the viscosity law (1.511) to the case of an anisotropic Newtonian fluid.

1.12.3 Lagrangian Coordinates and the Simplest Models of Elastic Bodies

In some Cartesian coordinate system consider a motion of some continuum medium occupying a region $\Omega \subset \mathbb{R}^3$ at some fixed instant of time $t = t_0$ and having the velocity field $\mathbf{u} := \mathbf{u}(\mathbf{x}, t)$. Next let $\mathbf{r}(t, \mathbf{y}) : \mathbb{R} \times \Omega \to \mathbb{R}^3$ be a solution of the following initial value problem for an ordinary differential equation:

$$\begin{cases} \frac{\partial \mathbf{r}}{\partial t}(t, \mathbf{y}) = \mathbf{u}\left(\mathbf{r}(t, \mathbf{y}), t\right) \quad \forall t \in \mathbb{R}, \ \forall \mathbf{y} \in \Omega \\ \mathbf{r}(t_0, \mathbf{y}) = \mathbf{y} \quad \forall \mathbf{y} \in \Omega. \end{cases} \tag{1.514}$$

Then, clearly $\mathbf{x} = \mathbf{r}(t, \mathbf{y})$ stands for the spatial coordinates at the instant of time t of the parcel of continuum, having initial coordinates $\mathbf{y}$. Moreover, we can deduce that $det\left\{d_{\mathbf{y}}\mathbf{r}(t, \mathbf{y})\right\} \neq 0$ for every instant of time t and so, for the given instant of time t the mapping $\mathbf{x} = \mathbf{r}(t, \mathbf{y})$ is locally invertible, i.e. the equation $\mathbf{x} = \mathbf{r}(t, \mathbf{y})$ can be resolved in $\mathbf{y}$ (see Sect. 11.2 for details). Thus there exists a regular mapping $\mathbf{y} = \mathbf{f}(\mathbf{x}, t)$, such that

$$\mathbf{f}\left(\mathbf{r}(t, \mathbf{y}), t\right) = \mathbf{y} \quad \forall \mathbf{y} \in \Omega \qquad \text{and} \qquad \mathbf{r}\left(t, \mathbf{f}(\mathbf{x}, t)\right) = \mathbf{x}. \tag{1.515}$$

Then clearly $\mathbf{y} = \mathbf{f}(\mathbf{x}, t)$ stands for the initial coordinates of the parcel of continuum, having coordinates $\mathbf{x}$ at the instant of time t. Next differentiating the first equation in (1.515) by t due to the chain rule and using (1.514) we deduce

$$\begin{cases} \frac{\partial \mathbf{f}}{\partial t}\left(\mathbf{x}, t\right) + d_{\mathbf{x}}\mathbf{f}\left(\mathbf{x}, t\right) \cdot \mathbf{u}\left(\mathbf{x}, t\right) = 0 \\ \mathbf{f}(\mathbf{x}, t_0) = \mathbf{x}. \end{cases} \tag{1.516}$$

The Cartesian coordinates $(\mathbf{x}, t)$ are called the Eulerian coordinates of the continuum medium. In contrast, change of variables

$$\begin{cases} t = t \\ \mathbf{y} = \mathbf{f}\left(\mathbf{x}, t\right) \end{cases} \tag{1.517}$$

to the new coordinates $(\mathbf{y}, t)$ of the space-time leads to generally non-Cartesian curvilinear coordinates that called the Lagrangian or the reference coordinates of the continuum medium. Next note that in the case of a given scalar field in the Eulerian coordinates $\Theta = \Theta(\mathbf{x}, t)$ and the corresponding field in Lagrangian coordinates $\Theta_1(\mathbf{y}, t) := \Theta\left(\mathbf{r}(t, \mathbf{y}), t\right)$, due to the chain rule and using (1.514) we obtain that for $\frac{\partial \Theta_1}{\partial t}(\mathbf{y}, t)$ in the Lagrangian coordinates corresponds to the following expression in the Eulerian coordinates:

$$\frac{\partial \Theta_1}{\partial t}(\mathbf{y}, t) = \frac{\partial \Theta}{\partial t}\left(\mathbf{x}, t\right) + \mathbf{u}\left(\mathbf{x}, t\right) \cdot \nabla_{\mathbf{x}}\Theta\left(\mathbf{x}, t\right). \tag{1.518}$$

Thus by (1.518) we also obtain that in the case of a given vector field in the Eulerian coordinates $\mathbf{g} = \mathbf{g}(\mathbf{x}, t)$ and the corresponding field in Lagrangian coordinates $\mathbf{g}_1(\mathbf{y}, t) := \mathbf{g}\left(\mathbf{x}, t\right) = \mathbf{g}\left(\mathbf{r}(t, \mathbf{y}), t\right)$ for $\frac{\partial \mathbf{g}_1}{\partial t}(\mathbf{y}, t)$ in the Lagrangian coordinates corresponds to the following expression in the Eulerian coordinates:

$$\frac{\partial \mathbf{g}_1}{\partial t}(\mathbf{y},t) = \frac{\partial \mathbf{g}}{\partial t}(\mathbf{x},t) + d_{\mathbf{x}}\mathbf{g}(\mathbf{x},t)\cdot \mathbf{u}(\mathbf{x},t)\,. \tag{1.519}$$

Next, as before, assume that the change of some non-inertial Cartesian system (∗) of Eulerian coordinates to another Cartesian system (∗∗) of Eulerian coordinates is of the form:

$$\begin{cases} \mathbf{x}' = A(t)\cdot \mathbf{x} + \mathbf{z}(t), \\ t' = t, \end{cases} \tag{1.520}$$

where $A(t) \in SO(3)$. Then in Sect. 11.2 we deduce that the law of transformation of the Lagrangian coordinates $(\mathbf{y},t) \to (\mathbf{y}',t')$, consistent with (1.520), is the following:

$$\begin{cases} \mathbf{y}' = A(t_0)\cdot \mathbf{y} + \mathbf{z}(t_0), \\ t' = t, \end{cases} \tag{1.521}$$

i.e. if we define $\mathbf{r}'(t',\mathbf{y}') : \mathbb{R}\times\mathbb{R}^3 \to \mathbb{R}^3$ as

$$\mathbf{r}'(t',\mathbf{y}') = \mathbf{r}'\left(t, A(t_0)\cdot \mathbf{y} + \mathbf{z}(t_0)\right) := A(t)\cdot \mathbf{r}(t,\mathbf{y}) + \mathbf{z}(t), \tag{1.522}$$

then consistently with (1.514) we have

$$\begin{cases} \frac{\partial \mathbf{r}'}{\partial t'}(t',\mathbf{y}') = \mathbf{u}'\left(\mathbf{r}'(t',\mathbf{y}'),t'\right) \\ \mathbf{r}'(t_0,\mathbf{y}') = \mathbf{y}'. \end{cases} \tag{1.523}$$

Moreover, for inverse mappings we have

$$\mathbf{f}'(\mathbf{x}',t') = \mathbf{f}'\left(A(t)\cdot \mathbf{x} + \mathbf{z}(t),t\right) = A(t_0)\cdot \mathbf{f}(\mathbf{x},t) + \mathbf{z}(t_0). \tag{1.524}$$

Next it is well known that the finite strain tensor $\mathcal{E}(\mathbf{x},t) \in \mathbb{R}^{3\times 3}$ of an elastic continuum medium in the Cartesian Eulerian coordinates $(\mathbf{x},t)$ has the form:

$$\mathcal{E}(\mathbf{x},t) := \frac{1}{2}\left(I - \{d_{\mathbf{x}}\mathbf{f}(\mathbf{x},t)\}^T \cdot d_{\mathbf{x}}\mathbf{f}(\mathbf{x},t)\right), \tag{1.525}$$

where the mapping $\mathbf{f}(\mathbf{x},t)$ is given by (1.515) and satisfies (1.516). Then by (1.520) and (1.524) we deduce that under (1.520) the matrix-valued field $\mathcal{E}$ transforms as

$$\mathcal{E}'(\mathbf{x}',t) = A(t)\cdot \mathcal{E}(\mathbf{x},t)\cdot A^T(t), \tag{1.526}$$

i.e. $\mathcal{E}$ is a proper matrix field as it was defined in Definition 1.1.

Remark 1.3 It is quite clear that the finite strain tensor, defined by (1.525) depends essentially on the choice of initial instant of time t_0. Thus, the initial time t_0 for (1.525) is always chosen (possibly fictitiously) in such a way that our elastic body is relaxed at time t_0.

Next, in the case of the simplest elastic body we have the following Hooke's law that is similar to (1.511):

$$\mathcal{T}(\mathbf{x}, t) = (\alpha \mathcal{E}(\mathbf{x}, t) + \beta\,(\mathrm{tr}\,\{\mathcal{E}(\mathbf{x}, t)\})\, I)\,, \tag{1.527}$$

where α and β are some material coefficients, which are not necessarily constant. Here $\mathcal{T}(\mathbf{x}, t)$ is the Cauchy stress tensor appearing in the equations of the motion of the medium (1.493) in the Eulerian coordinates $(\mathbf{x}, t)$ and $\mathcal{E}(\mathbf{x}, t)$ is the finite strain tensor defined by (1.525). Then by (1.526) we easily deduce that Hooke's law (1.527) is invariant under the change of a non-inertial Cartesian coordinate system given by (1.520), provided that, as usual, under (1.520) we have

$$\begin{cases} \alpha' = \alpha, \\ \beta' = \beta, \\ \mathcal{T}' = A(t) \cdot \mathcal{T} \cdot A^T(t). \end{cases} \tag{1.528}$$

See the end of Sect. 11.2.1 for the generalization of Hooke's law (1.527) to the case of anisotropic elastic bodies.

1.12.4 *The Equations of the Motion of the General Continuum Medium in Lagrangian Coordinates*

The equation of the motion of the continuum in the Cartesian Eulerian coordinates $(\mathbf{x}, t)$ has the form (1.493), i.e.:

$$\mu\left(\frac{\partial \mathbf{u}}{\partial t} + d_{\mathbf{x}}\mathbf{u} \cdot \mathbf{u}\right) = -\mu \mathbf{u} \times \mathrm{curl}_{\mathbf{x}}\mathbf{v} + \mu\left(\partial_t \mathbf{v} + \nabla_{\mathbf{x}}\frac{1}{2}|\mathbf{v}|^2\right) + \rho \mathbf{E} + \frac{1}{c}\mathbf{j} \times \mathbf{B} + \mathrm{div}_{\mathbf{x}}\,\mathcal{T} =$$
$$= -\mu(\mathbf{u} - \mathbf{v}) \times \mathrm{curl}_{\mathbf{x}}\mathbf{v} + \mu\,(\partial_t \mathbf{v} + d_{\mathbf{x}}\mathbf{v} \cdot \mathbf{v}) + \rho \mathbf{E} + \frac{1}{c}\mathbf{j} \times \mathbf{B} + \mathrm{div}_{\mathbf{x}}\,\mathcal{T}, \tag{1.529}$$

and the continuum equation has the form (1.494), i.e.:

$$\frac{\partial \mu}{\partial t} + \mathrm{div}_{\mathbf{x}}\,(\mu \mathbf{u}) = 0. \tag{1.530}$$

We would like to get the analogs of these equations in the Lagrangian coordinates $(\mathbf{y}, t)$. First of all we can obtain

$$\mu\,(\mathbf{r}(t, \mathbf{y}), t) = \mu(\mathbf{y}, t_0)\,\left|det\{d_{\mathbf{y}}\mathbf{r}(t, \mathbf{y})\}\right|^{-1} \quad \forall t,\ \forall \mathbf{y} \in \Omega. \tag{1.531}$$

Then equality (1.531) substitutes the continuum equation (1.530) in the Lagrangian coordinates $(\mathbf{y}, t)$ (see Section 11.2 for details). In particular, by (1.515), (1.531) in Eulerian coordinates reeds as

$$\mu(\mathbf{x}, t) = \mu(\mathbf{x}, t_0)\, |det\{d_{\mathbf{x}}\mathbf{f}(\mathbf{x}, t)\}|\,. \tag{1.532}$$

Next, the following Calculus fact is well known in Continuum Mechanics: if, given some matrix-valued field

$$\mathcal{T}(\mathbf{x}, t) \in \mathbb{R}^{3\times 3}, \tag{1.533}$$

we denote

$$F(\mathbf{y}, t) := d_{\mathbf{y}}\mathbf{r}(t, \mathbf{y}) \in \mathbb{R}^{3\times 3}, \tag{1.534}$$

and

$$\mathcal{R}(\mathbf{y}, t) := \mathcal{T}(\mathbf{r}(t, \mathbf{y}), t) \cdot \{F^{-1}(\mathbf{y}, t)\}^T\, (\det F(\mathbf{y}, t)), \tag{1.535}$$

then we must have

$$\operatorname{div}_{\mathbf{y}} \mathcal{R}(\mathbf{y}, t) = (\operatorname{div}_{\mathbf{x}} \mathcal{T}(\mathbf{r}(t, \mathbf{y}), t))\, (\det F(\mathbf{y}, t))\,. \tag{1.536}$$

In particular, if $\mathcal{T}(\mathbf{x}, t)$ is the Cauchy stress tensor, then $\mathcal{R}(\mathbf{y}, t)$ defined by (1.535) is called the first Piola–Kirchhoff stress tensor. Then by (1.536), (1.531), (1.514) and (1.519) we finally rewrite (1.529) in Lagrangian coordinates as

$$\begin{aligned}
&\mu(\mathbf{y}, t_0)\frac{\partial^2 \mathbf{r}}{\partial t^2}(t, \mathbf{y}) = \operatorname{div}_{\mathbf{y}} \mathcal{R}(\mathbf{y}, t)\\
&- \mu(\mathbf{y}, t_0)\frac{\partial \mathbf{r}}{\partial t}(t, \mathbf{y}) \times \operatorname{curl}_{\mathbf{x}}\mathbf{v}(\mathbf{r}(t, \mathbf{y}), t) + \mu(\mathbf{y}, t_0)\left(\partial_t \mathbf{v}(\mathbf{r}(t, \mathbf{y}), t) + \nabla_{\mathbf{x}}\frac{1}{2}\,|\mathbf{v}(\mathbf{r}(t, \mathbf{y}), t)|^2\right)\\
&+ \frac{\mu(\mathbf{y}, t_0)}{\mu(\mathbf{r}(t, \mathbf{y}), t)}\left(\rho(\mathbf{r}(t, \mathbf{y}), t)\,\mathbf{E}(\mathbf{r}(t, \mathbf{y}), t) + \frac{1}{c}\mathbf{j}(\mathbf{r}(t, \mathbf{y}), t) \times \mathbf{B}(\mathbf{r}(t, \mathbf{y}), t)\right).
\end{aligned} \tag{1.537}$$

1.12.5 *Propagation of Sound Wave in the Moving Inviscid Gas or Fluid*

Since in the case of inviscid fluid or gas (1.500) holds:

$$\mathcal{T} = -pI, \tag{1.538}$$

consistently with (1.493), we rewrite the second Law of Newton for the our inviscid continuum medium with the inertial mass density μ, the field of average (macroscopic) velocities $\mathbf{u}$, the charge density ρ and the electric current density $\mathbf{j}$:

$$\mu\left(\frac{\partial \mathbf{u}}{\partial t}+d_{\mathbf{x}}\mathbf{u}\cdot\mathbf{u}\right)=\frac{\partial}{\partial t}(\mu\mathbf{u})+\operatorname{div}_{\mathbf{x}}\{\mu\mathbf{u}\otimes\mathbf{u}\}=$$
$$-\mu\mathbf{u}\times\operatorname{curl}_{\mathbf{x}}\mathbf{v}+\mu\left(\partial_t\mathbf{v}+\nabla_{\mathbf{x}}\frac{1}{2}|\mathbf{v}|^2\right)+\rho\mathbf{E}+\frac{1}{c}\mathbf{j}\times\mathbf{B}-\nabla_{\mathbf{x}}p=$$
$$=-\mu(\mathbf{u}-\mathbf{v})\times\operatorname{curl}_{\mathbf{x}}\mathbf{v}+\mu\left(\partial_t\mathbf{v}+d_{\mathbf{x}}\mathbf{v}\cdot\mathbf{v}\right)+\rho\mathbf{E}+\frac{1}{c}\mathbf{j}\times\mathbf{B}-\nabla_{\mathbf{x}}p. \quad (1.539)$$

Here, as before, $\rho\mathbf{E}+\frac{1}{c}\mathbf{j}\times\mathbf{B}$ is the volume density of the Lorentz force where $\mathbf{E}$ and $\mathbf{B}$ are outer electric and magnetic fields, assumed to be changing smoothly and almost constant in the microscopic level, $\mathbf{v}$ is a vectorial gravitational potential also assumed to be changing smoothly and almost constant in the microscopic level, and p is the pressure. Moreover, the mass density μ, clearly satisfies the continuum equation (1.494):

$$\frac{\partial\mu}{\partial t}+\operatorname{div}_{\mathbf{x}}(\mu\mathbf{u})=0. \quad (1.540)$$

Next, by (1.501) the First Law of Thermodynamics of this moving medium has the following form:

$$\frac{\partial E}{\partial t}+\operatorname{div}_{\mathbf{x}}\{E\mathbf{u}\}=\mu\left(\frac{\partial}{\partial t}\left(\frac{E}{\mu}\right)+\mathbf{u}\cdot\nabla_{\mathbf{x}}\left(\frac{E}{\mu}\right)\right)$$
$$=-\left(\operatorname{div}_{\mathbf{x}}\mathbf{u}\right)p-\operatorname{div}_{\mathbf{x}}\mathbf{q}+(\mathbf{j}-\rho\mathbf{u})\cdot\left(\mathbf{E}+\frac{1}{c}\mathbf{u}\times\mathbf{B}\right). \quad (1.541)$$

Here E is the volume density of the internal energy (energy per unit volume) and consistently $\frac{E}{\mu}$ is the internal energy per unit mass and $\mathbf{q}$ is the heat flux. Moreover, the internal energy per unit mass is a function of the density μ, pressure p and Kelvin temperature T:

$$\frac{E}{\mu}=U(\mu,p,T). \quad (1.542)$$

There is also a state equation of the form:

$$T=g(\mu,p). \quad (1.543)$$

We remind that in the case of the simplest ideal gas (1.542) takes particular form of (1.509):

$$\frac{E}{\mu}=\frac{C_0}{m_0}kT, \quad (1.544)$$

and (1.543) takes particular form of (1.508):

$$T=\frac{m_0}{k}\frac{p}{\mu}, \quad (1.545)$$

where m_0 is the mass of the single molecule of the given gas and k is the Boltzmann constant and $C_0 > 0$ is a constant that depends on the kind of the gas (for the monatomic gas we have $C_0 = \frac{3}{2}$). So, by inserting (1.543) into (1.542) we have

$$\frac{E}{\mu} = U\left(\mu, p, g\left(\mu, p\right)\right) := F\left(\mu, p\right). \tag{1.546}$$

In particular, in the case where (1.544) and (1.545) are valid we have

$$\frac{E}{\mu} = F\left(\mu, p\right) := C_0 \frac{p}{\mu}. \tag{1.547}$$

Next we assume that

$$\mathbf{u} = \mathbf{u}_0 + \mathbf{u}_1, \quad \mu = \mu_0 + \mu_1, \quad p = p_0 + p_1, \tag{1.548}$$

where $\mathbf{u}_0(\mathbf{x}, t)$, $\mu_0(\mathbf{x}, t)$, $p_0(\mathbf{x}, t)$ are the averages of $\mathbf{u}(\mathbf{x}, t)$, $\mu(\mathbf{x}, t)$, $p(\mathbf{x}, t)$ on small spatial and temporal intervals, surrounding the point $(\mathbf{x}, t)$. Although we assume these intervals of space and time to be very small, we also assume them to be quite macroscopic. We call $\mathbf{u}_1, \mu_1, p_1$ the oscillating parts of $\mathbf{u}, \mu, p$ (they characterize the sound wave) and we assume that they are small with respect to the averages $\mathbf{u}_0, \mu_0, p_0$, i.e. we have

$$|\mathbf{u}_1| \ll |\mathbf{u}_0|, \quad |\mu_1| \ll |\mu_0|, \quad |p_1| \ll |p_0|. \tag{1.549}$$

However, we assume that $\mathbf{u}_1, \mu_1, p_1$ are highly oscillate and thus they change spatially and temporarily much faster than the averages $\mathbf{u}_0, \mu_0, p_0$ and the fields $\mathbf{v}, \mathbf{E}, \mathbf{B}, \rho, \mathbf{j}$, i.e. we have

$$\begin{gathered}
\frac{\left|d_{\mathbf{x}}\left(\rho\mathbf{E} + \frac{1}{c}\mathbf{j}\times\mathbf{B}\right)\right|}{\left|\rho\mathbf{E} + \frac{1}{c}\mathbf{j}\times\mathbf{B}\right|} + \frac{|d_{\mathbf{x}}\mathbf{v}|}{|\mathbf{v}|} + \frac{|d_{\mathbf{x}}\mu_0|}{|\mu_0|} + \frac{|d_{\mathbf{x}}p_0|}{|p_0|} + \frac{|d_{\mathbf{x}}\mathbf{u}_0|}{|\mathbf{u}_0|} \\
\ll \min\left\{\frac{|d_{\mathbf{x}}\mathbf{u}_1|}{|\mathbf{u}_1|}, \frac{|d_{\mathbf{x}}\mu_1|}{|\mu_1|}, \frac{|d_{\mathbf{x}}p_1|}{|p_1|}\right\}, \\
\frac{\left|\partial_t\left(\rho\mathbf{E} + \frac{1}{c}\mathbf{j}\times\mathbf{B}\right)\right|}{\left|\rho\mathbf{E} + \frac{1}{c}\mathbf{j}\times\mathbf{B}\right|} + \frac{|\partial_t\mathbf{v}|}{|\mathbf{v}|} + \frac{|\partial_t\mu_0|}{|\mu_0|} + \frac{|\partial_t p_0|}{|p_0|} + \frac{|\partial_t\mathbf{u}_0|}{|\mathbf{u}_0|} \\
\ll \min\left\{\frac{|\partial_t\mathbf{u}_1|}{|\mathbf{u}_1|}, \frac{|\partial_t\mu_1|}{|\mu_1|}, \frac{|\partial_t p_1|}{|p_1|}\right\}, \\
\frac{|\mathbf{u}_1|}{|\mathbf{u}_0|} \ll \min\left\{\frac{|d_{\mathbf{x}}\mathbf{u}_1|}{|d_{\mathbf{x}}\mathbf{u}_0|}, \frac{|d_{\mathbf{x}}\mathbf{u}_1|}{|d_{\mathbf{x}}\mathbf{v}|}\right\}, \quad \frac{|\mu_1|}{|\mu_0|} + \frac{|p_1|}{|p_0|} \ll \min\left\{\frac{|d_{\mathbf{x}}p_1|}{|d_{\mathbf{x}}p_0|}, \frac{|d_{\mathbf{x}}\mu_1|}{|d_{\mathbf{x}}\mu_0|}\right\} \\
\text{and} \quad \frac{|\mu_1|}{|\mu_0|} \ll \frac{|d_{\mathbf{x}}\mathbf{u}_1||\mathbf{u}_0||\mu_0|}{\left|\rho\mathbf{E} + \frac{1}{c}\mathbf{j}\times\mathbf{B}\right|}.
\end{gathered} \tag{1.550}$$

Finally, we assume that the fields $\mathbf{v}$, $\mathbf{E}$, $\mathbf{B}$, ρ, $\mathbf{j}$ and $(\operatorname{div}_{\mathbf{x}} \mathbf{q})$ change slowly with respect to the oscillations of $\mathbf{u}_1$, μ_1, p_1 and thus we assume that $\mathbf{v}$, $\mathbf{E}$, $\mathbf{B}$, ρ, $\mathbf{j}$ and $(\operatorname{div}_{\mathbf{x}} \mathbf{q})$ can be replaced by their spatial and temporal averages. Note that μ, p, μ_0, p_0, μ_1, p_1 behave like proper scalar fields and $\mathbf{u}$, $\mathbf{u}_0$ behave like speed-like vector fields under the change of Cartesian coordinate systems. Thus, since $\mathbf{u}_1 = \mathbf{u} - \mathbf{u}_0$, we deduce that $\mathbf{u}_1$ behaves like a proper vector field under the change of Cartesian coordinate systems. Finally, note that obviously the averages of $\mathbf{u}_1$, μ_1, p_1 vanish.

Then in Sect. 11.3 we deduce

$$\frac{\partial}{\partial t}\left(\frac{\partial \mu_1}{\partial t} + \mathbf{u}_0 \cdot \nabla_{\mathbf{x}} \mu_1\right) + \operatorname{div}_{\mathbf{x}}\left\{\left(\frac{\partial \mu_1}{\partial t} + \mathbf{u}_0 \cdot \nabla_{\mathbf{x}} \mu_1\right) \mathbf{u}_0\right\} \approx \Delta_{\mathbf{x}} p_1, \tag{1.551}$$

and

$$\mu_0\left(\frac{\partial \mathbf{u}_1}{\partial t} - \operatorname{curl}_{\mathbf{x}}\{\mathbf{u}_0 \times \mathbf{u}_1\} + (\operatorname{div}_{\mathbf{x}} \mathbf{u}_1)\,\mathbf{u}_0\right) \approx -\nabla_{\mathbf{x}} p_1. \tag{1.552}$$

Note that, as before, it can be easily proved that (1.551) and (1.552) are invariant under the change of inertial or non-inertial Cartesian coordinate system. Thus, (1.551) and (1.552) are still valid if (1.549) and (1.550) hold only in some specific (possibly artificial) inertial or non-inertial Cartesian coordinate system.

Next we proceed in two cases:

(i) In the case of the simple models of inviscid gas.
(ii) In the case of inviscid barotropic fluid.

In the case of inviscid gas, in Sect. 11.3 we prove that

$$\mu_1 \approx \left(\frac{p_0}{\mu_0} - \mu_0 \frac{\partial F}{\partial \mu}(\mu_0, p_0)\right)^{-1} \mu_0 \frac{\partial F}{\partial p}(\mu_0, p_0)\; p_1. \tag{1.553}$$

Therefore, inserting (1.553) into (1.551) and using again (1.550) we finally obtain the wave equation for the oscillating part of the pressure p_1 of the form:

$$\frac{\partial}{\partial t}\left\{\frac{1}{c_0^2}\left(\frac{\partial p_1}{\partial t} + \mathbf{u}_0 \cdot \nabla_{\mathbf{x}} p_1\right)\right\} + \operatorname{div}_{\mathbf{x}}\left\{\frac{1}{c_0^2}\left(\frac{\partial p_1}{\partial t} + \mathbf{u}_0 \cdot \nabla_{\mathbf{x}} p_1\right) \mathbf{u}_0\right\} \approx \Delta_{\mathbf{x}} p_1, \tag{1.554}$$

where we denote

$$c_0 := \sqrt{\frac{\left(\frac{p_0}{\mu_0} - \mu_0 \frac{\partial F}{\partial \mu}(\mu_0, p_0)\right)}{\mu_0 \frac{\partial F}{\partial p}(\mu_0, p_0)}}. \tag{1.555}$$

Moreover, the oscillating parts of the density μ_1 and the velocity $\mathbf{u}_1$ can be found from (1.553) and (1.552) respectively, i.e. we have

$$\mu_1 \approx \left(\frac{p_0}{\mu_0} - \mu_0 \frac{\partial F}{\partial \mu}(\mu_0, p_0)\right)^{-1} \mu_0 \frac{\partial F}{\partial p}(\mu_0, p_0)\; p_1, \tag{1.556}$$

and

$$\mu_0 \left(\frac{\partial \mathbf{u}_1}{\partial t} - \operatorname{curl}_{\mathbf{x}} \{\mathbf{u}_0 \times \mathbf{u}_1\} + (\operatorname{div}_{\mathbf{x}} \mathbf{u}_1)\, \mathbf{u}_0 \right) \approx -\nabla_{\mathbf{x}} p_1. \tag{1.557}$$

Note that, as before, it can be easily proved that (1.554), (1.556) and (1.557) are invariant under the change of inertial or non-inertial Cartesian coordinate system. Thus, as before, (1.554), (1.556) and (1.557) are still valid if (1.549) and (1.550) hold only in some specific (possibly artificial) inertial or non-inertial Cartesian coordinate system.

In particular, in the case of in the case of the simplest ideal gas where (1.544) and (1.545) are valid, by (1.547) we have

$$\frac{E}{\mu} = F(\mu, p) := C_0 \frac{p}{\mu}, \tag{1.558}$$

and therefore, inserting (1.558) into (1.555) gives

$$c_0 := \sqrt{\frac{(1 + C_0)\, p_0}{C_0 \mu_0}}. \tag{1.559}$$

Moreover, in this case (1.553) reeds as

$$\mu_1 \approx \frac{C_0}{1 + C_0} \frac{\mu_0}{p_0} p_1. \tag{1.560}$$

On the other hand, in the case barotropic fluid the pressure p is a function of the density μ only, i.e.

$$p = \mathcal{R}(\mu). \tag{1.561}$$

Thus, inserting (1.548) into (1.561) and using (1.549) we deduce

$$p_1 \approx \frac{d\mathcal{R}}{d\mu}(\mu_0)\, \mu_1. \tag{1.562}$$

Therefore, inserting (1.562) into (1.551) and using again (1.550) we finally obtain the analogous to (1.554) wave equation for the oscillating part of the pressure p_1 of the form:

$$\frac{\partial}{\partial t} \left\{ \frac{1}{c_0^2} \left(\frac{\partial p_1}{\partial t} + \mathbf{u}_0 \cdot \nabla_{\mathbf{x}} p_1 \right) \right\} + \operatorname{div}_{\mathbf{x}} \left\{ \frac{1}{c_0^2} \left(\frac{\partial p_1}{\partial t} + \mathbf{u}_0 \cdot \nabla_{\mathbf{x}} p_1 \right) \mathbf{u}_0 \right\} \approx \Delta_{\mathbf{x}} p_1, \tag{1.563}$$

where we denote

$$c_0 := \sqrt{\frac{d\mathcal{R}}{d\mu}(\mu_0)}. \tag{1.564}$$

Moreover, the oscillating parts of the density μ_1 and the velocity $\mathbf{u}_1$ can be found from (1.562) and (1.552) respectively, i.e. we have

$$p_1 \approx \frac{d\mathcal{R}}{d\mu}(\mu_0)\,\mu_1, \tag{1.565}$$

and

$$\mu_0\left(\frac{\partial \mathbf{u}_1}{\partial t} - \operatorname{curl}_{\mathbf{x}}\{\mathbf{u}_0 \times \mathbf{u}_1\} + (\operatorname{div}_{\mathbf{x}} \mathbf{u}_1)\,\mathbf{u}_0\right) \approx -\nabla_{\mathbf{x}} p_1. \tag{1.566}$$

Again note that (1.563), (1.565) and (1.566) are invariant under the change of inertial or non-inertial Cartesian coordinate system. Thus, as before, (1.563), (1.565) and (1.566) are still valid if (1.549) and (1.550) hold only in some specific (possibly artificial) inertial or non-inertial Cartesian coordinate system.

1.12.6 Propagation of Waves in the Moving Elastic Body

Consistently with the general equation of motion of a continuum medium (1.493), consider the motion of an elastic body:

$$\mu\left(\frac{\partial \mathbf{u}}{\partial t} + d_{\mathbf{x}}\mathbf{u}\cdot\mathbf{u}\right) = \\ -\mu\mathbf{u}\times\operatorname{curl}_{\mathbf{x}}\mathbf{v} + \mu\left(\partial_t\mathbf{v} + \nabla_{\mathbf{x}}\frac{1}{2}|\mathbf{v}|^2\right) + \rho\mathbf{E} + \frac{1}{c}\mathbf{j}\times\mathbf{B} + \operatorname{div}_{\mathbf{x}}\left\{\alpha\mathcal{E}(\mathbf{x},t) + \beta\left(\operatorname{tr}\{\mathcal{E}(\mathbf{x},t)\}\right)I\right\}, \tag{1.567}$$

where

$$\mathcal{E}(\mathbf{x},t) := \frac{1}{2}\left(I - \{d_{\mathbf{x}}\mathbf{f}(\mathbf{x},t)\}^T\cdot d_{\mathbf{x}}\mathbf{f}(\mathbf{x},t)\right), \tag{1.568}$$

consistently with (1.525) and (1.527), with

$$\begin{cases}\frac{\partial \mathbf{f}}{\partial t}(\mathbf{x},t) + d_{\mathbf{x}}\mathbf{f}(\mathbf{x},t)\cdot\mathbf{u}(\mathbf{x},t) = 0\\ \mathbf{f}(\mathbf{x},t_0) = \mathbf{x},\end{cases} \tag{1.569}$$

consistently with (1.516), where the initial instant of time t_0 is chosen (possibly fictitiously) in such a way that our elastic body is relaxed at time t_0. Moreover, by (1.532) we have

$$\mu(\mathbf{x},t) = \mu(\mathbf{x},t_0)\,|det\{d_{\mathbf{x}}\mathbf{f}(\mathbf{x},t)\}|\,. \tag{1.570}$$

Next we assume that

$$\mathbf{u}(\mathbf{x},t) = \mathbf{u}_0(\mathbf{x},t) + \mathbf{u}_1(\mathbf{x},t), \quad \mu(\mathbf{x},t) = \mu_0(\mathbf{x},t) + \mu_1(\mathbf{x},t), \\ \mathbf{f}(\mathbf{x},t) = \mathbf{f}_0(\mathbf{x},t) + \mathbf{f}_1(\mathbf{x},t) \text{ and } \mathcal{E}(\mathbf{x},t) = \mathcal{E}_0(\mathbf{x},t) + \mathcal{E}_1(\mathbf{x},t), \tag{1.571}$$

where $\mathbf{u}_0(\mathbf{x}, t)$, $\mu_0(\mathbf{x}, t)$, $\mathbf{f}_0(\mathbf{x}, t)$, $\mathcal{E}_0(\mathbf{x}, t)$ are the averages of $\mathbf{u}(\mathbf{x}, t)$, $\mu(\mathbf{x}, t)$, $\mathbf{f}(\mathbf{x}, t)$, $\mathcal{E}(\mathbf{x}, t)$ on small spatial and temporal intervals, surrounding the point $(\mathbf{x}, t)$. Although we assume these intervals of space and time to be very small, we also assume them to be quite macroscopic. We call $\mathbf{u}_1, \mu_1, \mathbf{f}_1, \mathcal{E}_1$ the oscillating parts of $\mathbf{u}, \mu, \mathbf{f}, \mathcal{E}$ and we assume that they are small with respect to the averages $\mathbf{u}_0$, μ_0, $\mathbf{f}_0$, $\mathcal{E}_0$, i.e. we have

$$|\mathbf{u}_1| \ll |\mathbf{u}_0|, \quad |\mu_1| \ll |\mu_0|, \quad |\mathbf{f}_1| \ll |\mathbf{f}_0|. \tag{1.572}$$

However, we assume that $\mathbf{u}_1, \mu_1, \mathbf{f}_1$ are highly oscillate and thus they change spatially and temporarily much faster than the averages $\mathbf{u}_0$, μ_0, $\mathbf{f}_0$ and the fields $\mathbf{v}$, $\mathbf{E}$, $\mathbf{B}$, ρ, $\mathbf{j}$, i.e. we have

$$\begin{aligned}
&\frac{|d_{\mathbf{x}}\alpha|}{|\alpha|} + \frac{|d_{\mathbf{x}}\beta|}{|\beta|} + \frac{\left|d_{\mathbf{x}}\left(\rho\mathbf{E} + \frac{1}{c}\mathbf{j}\times\mathbf{B}\right)\right|}{\left|\rho\mathbf{E} + \frac{1}{c}\mathbf{j}\times\mathbf{B}\right|} + \frac{|d_{\mathbf{x}}\mathbf{v}|}{|\mathbf{v}|} + \frac{|d_{\mathbf{x}}\mu_0|}{|\mu_0|} + \frac{|d_{\mathbf{x}}\mathbf{f}_0|}{|\mathbf{f}_0|} + \frac{|d_{\mathbf{x}}\mathbf{u}_0|}{|\mathbf{u}_0|} \\
&\qquad\qquad \ll \min\left\{\frac{|d_{\mathbf{x}}\mathbf{u}_1|}{|\mathbf{u}_1|}, \frac{|d_{\mathbf{x}}\mu_1|}{|\mu_1|}, \frac{|d_{\mathbf{x}}\mathbf{f}_1|}{|\mathbf{f}_1|}\right\}, \\
&\frac{\left|\partial_t\left(\rho\mathbf{E} + \frac{1}{c}\mathbf{j}\times\mathbf{B}\right)\right|}{\left|\rho\mathbf{E} + \frac{1}{c}\mathbf{j}\times\mathbf{B}\right|} + \frac{|\partial_t\mathbf{v}|}{|\mathbf{v}|} + \frac{|\partial_t\mu_0|}{|\mu_0|} + \frac{|\partial_t\mathbf{f}_0|}{|\mathbf{f}_0|} + \frac{|\partial_t\mathbf{u}_0|}{|\mathbf{u}_0|} \\
&\qquad\qquad \ll \min\left\{\frac{|\partial_t\mathbf{u}_1|}{|\mathbf{u}_1|}, \frac{|\partial_t\mu_1|}{|\mu_1|}, \frac{|\partial_t\mathbf{f}_1|}{|\mathbf{f}_1|}\right\}, \\
&\frac{|d_{\mathbf{x}}^2\mathbf{f}_0|}{|d_{\mathbf{x}}\mathbf{f}_0|} \ll \frac{|d_{\mathbf{x}}^2\mathbf{f}_1|}{|d_{\mathbf{x}}\mathbf{f}_1|} \qquad \frac{|\mathbf{u}_1|}{|\mathbf{u}_0|} \ll \min\left\{\frac{|d_{\mathbf{x}}\mathbf{u}_1|}{|d_{\mathbf{x}}\mathbf{u}_0|}, \frac{|d_{\mathbf{x}}\mathbf{u}_1|}{|d_{\mathbf{x}}\mathbf{v}|}\right\}, \\
&\qquad\qquad \text{and} \quad \frac{|\mu_1|}{|\mu_0|} \ll \frac{|d_{\mathbf{x}}\mathbf{u}_1||\mathbf{u}_0||\mu_0|}{\left|\rho\mathbf{E} + \frac{1}{c}\mathbf{j}\times\mathbf{B}\right|}.
\end{aligned} \tag{1.573}$$

Finally, we assume that the fields $\mathbf{v}$, $\mathbf{E}$, $\mathbf{B}$, ρ, $\mathbf{j}$ change slowly with respect to the oscillations of $\mathbf{u}_1, \mu_1, \mathbf{f}_1$ and thus we assume that $\mathbf{v}$, $\mathbf{E}$, $\mathbf{B}$, ρ, $\mathbf{j}$ can be replaced by their spatial and temporal averages. Note that μ, μ_0, μ_1 behave like proper scalar fields, $\mathcal{E}$, $\mathcal{E}_0$, $\mathcal{E}_1$ behave like proper matrix fields and $\mathbf{u}$, $\mathbf{u}_0$ behave like speed-like vector fields under the change of Cartesian coordinate systems. Thus, since $\mathbf{u}_1 = \mathbf{u} - \mathbf{u}_0$, we deduce that $\mathbf{u}_1$ behaves like a proper vector field under the change of Cartesian coordinate systems. Furthermore, using (1.524) we deduce that the vector field $\mathbf{g}_1(\mathbf{x}, t)$ defined by the following:

$$\mathbf{g}_1(\mathbf{x}, t) := \left(d_{\mathbf{x}}\mathbf{f}_0(\mathbf{x}, t)\right)^{-1} \cdot \mathbf{f}_1(\mathbf{x}, t) \tag{1.574}$$

behaves like a proper vector field. Finally, note that obviously the averages of $\mathbf{u}_1, \mu_1, \mathbf{f}_1, \mathcal{E}_1$ vanish.

Then in Sect. 11.4 we deduce

$$\mu_0 \left(\frac{\partial}{\partial t} \left(\frac{\partial \mathbf{g}_1}{\partial t} + d_\mathbf{x}\mathbf{g}_1 \cdot \mathbf{u}_0 \right) + d_\mathbf{x} \left(\frac{\partial \mathbf{g}_1}{\partial t} + d_\mathbf{x}\mathbf{g}_1 \cdot \mathbf{u}_0 \right) \cdot \mathbf{u}_0 \right) \approx$$
$$- \alpha \operatorname{div}_\mathbf{x} \{\mathcal{E}_0 \cdot d_\mathbf{x}\mathbf{g}_1\} - \alpha \operatorname{div}_\mathbf{x} \left\{\{d_\mathbf{x}\mathbf{g}_1\}^T \cdot \mathcal{E}_0\right\}$$
$$- \beta \nabla_\mathbf{x} (\operatorname{tr} (\mathcal{E}_0 \cdot d_\mathbf{x}\mathbf{g}_1)) - \beta \nabla_\mathbf{x} \left(\operatorname{tr} \left(\{d_\mathbf{x}\mathbf{g}_1\}^T \cdot \mathcal{E}_0\right)\right)$$
$$+ \frac{\alpha}{2} \Delta_\mathbf{x}\mathbf{g}_1 + \left(\frac{\alpha}{2} + \beta\right) \nabla_\mathbf{x} (\operatorname{div}_\mathbf{x} \mathbf{g}_1), \quad (1.575)$$

and

$$\frac{\partial \mathbf{g}_1}{\partial t} - \operatorname{curl}_\mathbf{x} \{\mathbf{u}_0 \times \mathbf{g}_1\} + (\operatorname{div}_\mathbf{x} \mathbf{g}_1)\, \mathbf{u}_0 \approx -\mathbf{u}_1, \quad (1.576)$$

where $\mathbf{g}_1(\mathbf{x}, t)$ is given by (1.574). Furthermore, we also derive

$$\mathcal{E}_0(\mathbf{x}, t) \approx \frac{1}{2} \left(I - \{d_\mathbf{x}\mathbf{f}_0(\mathbf{x}, t)\}^T \cdot d_\mathbf{x}\mathbf{f}_0(\mathbf{x}, t)\right), \quad (1.577)$$

and

$$\mathcal{E}_1(\mathbf{x}, t) \approx$$
$$\mathcal{E}_0(\mathbf{x}, t) \cdot d_\mathbf{x}\mathbf{g}_1(\mathbf{x}, t) + \{d_\mathbf{x}\mathbf{g}_1(\mathbf{x}, t)\}^T \cdot \mathcal{E}_0(\mathbf{x}, t) - \frac{1}{2} \left(d_\mathbf{x}\mathbf{g}_1(\mathbf{x}, t) + \{d_\mathbf{x}\mathbf{g}_1(\mathbf{x}, t)\}^T\right). \quad (1.578)$$

Moreover, we can derive the approximation of (1.575) as

$$\mu_0 \frac{\partial}{\partial t} \left(\frac{\partial \mathbf{g}_1}{\partial t} - \operatorname{curl}_\mathbf{x} \{\mathbf{u}_0 \times \mathbf{g}_1\} + (\operatorname{div}_\mathbf{x} \mathbf{g}_1)\, \mathbf{u}_0 \right)$$
$$- \mu_0 \operatorname{curl}_\mathbf{x} \left\{ \mathbf{u}_0 \times \left(\frac{\partial \mathbf{g}_1}{\partial t} - \operatorname{curl}_\mathbf{x} \{\mathbf{u}_0 \times \mathbf{g}_1\} + (\operatorname{div}_\mathbf{x} \mathbf{g}_1)\, \mathbf{u}_0 \right) \right\}$$
$$+ \left(\operatorname{div}_\mathbf{x} \left\{ \frac{\partial \mathbf{g}_1}{\partial t} - \operatorname{curl}_\mathbf{x} \{\mathbf{u}_0 \times \mathbf{g}_1\} + (\operatorname{div}_\mathbf{x} \mathbf{g}_1)\, \mathbf{u}_0 \right\} \right) \mathbf{u}_0 \approx \frac{\alpha}{2} \Delta_\mathbf{x}\mathbf{g}_1 + \left(\frac{\alpha}{2} + \beta\right) \nabla_\mathbf{x} (\operatorname{div}_\mathbf{x} \mathbf{g}_1)$$
$$- \alpha \operatorname{div}_\mathbf{x} \{\mathcal{E}_0 \cdot d_\mathbf{x}\mathbf{g}_1\} - \alpha \operatorname{div}_\mathbf{x} \left\{\{d_\mathbf{x}\mathbf{g}_1\}^T \cdot \mathcal{E}_0\right\}$$
$$- \beta \nabla_\mathbf{x} (\operatorname{tr} (\mathcal{E}_0 \cdot d_\mathbf{x}\mathbf{g}_1)) - \beta \nabla_\mathbf{x} \left(\operatorname{tr} \left(\{d_\mathbf{x}\mathbf{g}_1\}^T \cdot \mathcal{E}_0\right)\right). \quad (1.579)$$

In, particular, if our elastic body is nearly rigid we have

$$\mathcal{E}_0 \approx 0, \quad (1.580)$$

and we simplify (1.575) as

$$\mu_0\left(\frac{\partial}{\partial t}\left(\frac{\partial \mathbf{g}_1}{\partial t}+d_{\mathbf{x}}\mathbf{g}_1\cdot\mathbf{u}_0\right)+d_{\mathbf{x}}\left(\frac{\partial \mathbf{g}_1}{\partial t}+d_{\mathbf{x}}\mathbf{g}_1\cdot\mathbf{u}_0\right)\cdot\mathbf{u}_0\right)$$
$$\approx\frac{\alpha}{2}\,\Delta_{\mathbf{x}}\mathbf{g}_1+\left(\frac{\alpha}{2}+\beta\right)\nabla_{\mathbf{x}}\left(\mathrm{div}_{\mathbf{x}}\,\mathbf{g}_1\right), \quad (1.581)$$

and (1.578) as

$$\mathcal{E}_1\approx-\frac{1}{2}\left(d_{\mathbf{x}}\mathbf{g}_1+\{d_{\mathbf{x}}\mathbf{g}_1\}^T\right). \quad (1.582)$$

Moreover, we rewrite (1.579) as

$$\mu_0\frac{\partial}{\partial t}\left(\frac{\partial \mathbf{g}_1}{\partial t}-\mathrm{curl}_{\mathbf{x}}\,\{\mathbf{u}_0\times\mathbf{g}_1\}+(\mathrm{div}_{\mathbf{x}}\,\mathbf{g}_1)\,\mathbf{u}_0\right)$$
$$-\mu_0\,\mathrm{curl}_{\mathbf{x}}\left\{\mathbf{u}_0\times\left(\frac{\partial \mathbf{g}_1}{\partial t}-\mathrm{curl}_{\mathbf{x}}\,\{\mathbf{u}_0\times\mathbf{g}_1\}+(\mathrm{div}_{\mathbf{x}}\,\mathbf{g}_1)\,\mathbf{u}_0\right)\right\}$$
$$+\mu_0\left(\mathrm{div}_{\mathbf{x}}\left\{\frac{\partial \mathbf{g}_1}{\partial t}-\mathrm{curl}_{\mathbf{x}}\,\{\mathbf{u}_0\times\mathbf{g}_1\}+(\mathrm{div}_{\mathbf{x}}\,\mathbf{g}_1)\,\mathbf{u}_0\right\}\right)\mathbf{u}_0$$
$$\approx\frac{\alpha}{2}\,\Delta_{\mathbf{x}}\mathbf{g}_1+\left(\frac{\alpha}{2}+\beta\right)\nabla_{\mathbf{x}}\left(\mathrm{div}_{\mathbf{x}}\,\mathbf{g}_1\right). \quad (1.583)$$

Next, note that, since μ, μ_0, μ_1 are proper scalar fields, $\mathcal{E}$, $\mathcal{E}_0$, $\mathcal{E}_1$ are proper matrix fields, $\mathbf{u}$, $\mathbf{u}_0$ are speed-like vector fields, $\mathbf{u}_1$ is a proper vector field and the vector field $\mathbf{g}_1(\mathbf{x},t)$ defined by (1.574) is a proper vector field, as before, it can be easily proved that (1.579), (1.576), (1.577), (1.578), (1.583) (1.580) and (1.582) are invariant under the change of inertial or non-inertial Cartesian coordinate system. Thus, (1.579), (1.576), (1.577), (1.578) and in the case of nearly rigid elastic body also (1.583) (1.580) and (1.582) are still valid if (1.572) and (1.573) hold only in some specific (possibly artificial) inertial or non-inertial Cartesian coordinate system.

Next, taking the divergence of both sides of (1.583) and using (1.573) gives

$$\frac{\partial}{\partial t}\left\{\frac{1}{c_{01}^2}\left(\frac{\partial h_1}{\partial t}+\mathbf{u}_0\cdot\nabla_{\mathbf{x}}h_1\right)\right\}+\mathrm{div}_{\mathbf{x}}\left\{\frac{1}{c_{01}^2}\left(\frac{\partial h_1}{\partial t}+\mathbf{u}_0\cdot\nabla_{\mathbf{x}}h_1\right)\mathbf{u}_0\right\}\approx\Delta_{\mathbf{x}}h_1, \quad (1.584)$$

where

$$h_1(\mathbf{x},t):=\mathrm{div}_{\mathbf{x}}\,\mathbf{g}_1(\mathbf{x},t)\quad\text{and}\quad c_{01}:=\sqrt{\frac{(\alpha+\beta)}{\mu_0}}. \quad (1.585)$$

On the other hand taking the curl of both sides of (1.581) and using (1.573) gives

$$\frac{\partial}{\partial t}\left\{\frac{1}{c_{02}^2}\left(\frac{\partial \mathbf{h}_2}{\partial t}+d_{\mathbf{x}}\mathbf{h}_2\cdot\mathbf{u}_0\right)\right\}+\mathrm{div}_{\mathbf{x}}\left\{\frac{1}{c_{02}^2}\left(\frac{\partial \mathbf{h}_2}{\partial t}+d_{\mathbf{x}}\mathbf{h}_2\cdot\mathbf{u}_0\right)\otimes\mathbf{u}_0\right\}\approx\Delta_{\mathbf{x}}\mathbf{h}_2, \quad (1.586)$$

where

$$\mathbf{h}_2(\mathbf{x},t) := \text{curl}_{\mathbf{x}}\, \mathbf{g}_1(\mathbf{x},t) \quad \text{and} \quad c_{02} := \sqrt{\frac{\alpha}{2\mu_0}}. \tag{1.587}$$

Again, using (1.573) we can rewrite (1.586) as

$$\begin{aligned}\frac{1}{c_{02}^2}\frac{\partial}{\partial t}\left(\frac{\partial \mathbf{h}_2}{\partial t} - \mathbf{u}_0 \times \text{curl}_{\mathbf{x}}\mathbf{h}_2 + \nabla_{\mathbf{x}}\left(\mathbf{h}_2 \cdot \mathbf{u}_0\right)\right)\\ - \frac{1}{c_{02}^2}\text{curl}_{\mathbf{x}}\left\{\mathbf{u}_0 \times \left(\frac{\partial \mathbf{h}_2}{\partial t} - \mathbf{u}_0 \times \text{curl}_{\mathbf{x}}\mathbf{h}_2 + \nabla_{\mathbf{x}}\left(\mathbf{h}_2 \cdot \mathbf{u}_0\right)\right)\right\}\\ + \frac{1}{c_{02}^2}\left(\text{div}_{\mathbf{x}}\left(\frac{\partial \mathbf{h}_2}{\partial t} - \mathbf{u}_0 \times \text{curl}_{\mathbf{x}}\mathbf{h}_2 + \nabla_{\mathbf{x}}\left(\mathbf{h}_2 \cdot \mathbf{u}_0\right)\right)\right)\mathbf{u}_0 \approx \Delta_{\mathbf{x}}\mathbf{h}_2.\end{aligned} \tag{1.588}$$

Next we deduce that equations (1.584) and (1.588) are invariant under the change of inertial or non-inertial Cartesian coordinate system, provided that $\mathbf{h}_2$ is a proper vector field and h_1 is a proper scalar field. Thus, in the case of nearly rigid elastic body (1.584) and (1.588) are still valid if (1.572) and (1.573) hold only in some specific (possibly artificial) inertial or non-inertial Cartesian coordinate system. Finally note that, although the equation (1.586) is invariant under the Galilean Transformation, it is not invariant under the more general change of non-inertial Cartesian coordinate system. However, (1.586) is more convenient than (1.588), since every of the three scalar components of the vector field $\mathbf{h}_2$ in (1.586) satisfies three decoupled wave equations of the same type. On the other hand, if we consider some three proper vector fields $\mathbf{e}_1 := \mathbf{e}_1(\mathbf{x},t), \mathbf{e}_2 := \mathbf{e}_2(\mathbf{x},t)$, and $\mathbf{e}_3 := \mathbf{e}_3(\mathbf{x},t)$, which are mutually orthogonal to each other and satisfy the following approximation analogous to (1.573):

$$\frac{|d_{\mathbf{x}}\mathbf{e}_1| + c_{02}|\partial_t \mathbf{e}_1|}{|\mathbf{e}_1|} + \frac{|d_{\mathbf{x}}\mathbf{e}_2| + c_{02}|\partial_t \mathbf{e}_2|}{|\mathbf{e}_2|} + \frac{|d_{\mathbf{x}}\mathbf{e}_3| + c_{02}|\partial_t \mathbf{e}_3|}{|\mathbf{e}_3|} \ll \frac{|d_{\mathbf{x}}\mathbf{h}_2| + c_{02}|\partial_t \mathbf{h}_2|}{|\mathbf{h}_2|}, \tag{1.589}$$

i.e. the field $\mathbf{e}_k$ vary in space and time much weaker than $\mathbf{h}_2$, then we may write the alternative to (1.586) approximate equations in the form of three decoupled scalar wave equations of the same type:

$$\begin{aligned}\frac{\partial}{\partial t}\left\{\frac{1}{c_{02}^2}\left(\frac{\partial}{\partial t}\left(\mathbf{e}_k \cdot \mathbf{h}_2\right) + \mathbf{u}_0 \cdot \nabla_{\mathbf{x}}\left(\mathbf{e}_k \cdot \mathbf{h}_2\right)\right)\right\}\\ + \text{div}_{\mathbf{x}}\left\{\frac{1}{c_{02}^2}\left(\frac{\partial}{\partial t}\left(\mathbf{e}_k \cdot \mathbf{h}_2\right) + \mathbf{u}_0 \cdot \nabla_{\mathbf{x}}\left(\mathbf{e}_k \cdot \mathbf{h}_2\right)\right)\mathbf{u}_0\right\}\\ \approx \Delta_{\mathbf{x}}\left(\mathbf{e}_k \cdot \mathbf{h}_2\right) \qquad \forall k = 1,2,3.\end{aligned} \tag{1.590}$$

Then, clearly, the new alternative approximate equations (1.590) together with (1.584) are indeed invariant under the more general change of non-inertial Cartesian coordinate system.

As a final corollary of the above in the case of nearly rigid body we have (1.590) and (1.584), i.e.:

$$\frac{\partial}{\partial t}\left\{\frac{1}{c_{01}^2}\left(\frac{\partial h_1}{\partial t}+\mathbf{u}_0\cdot\nabla_{\mathbf{x}}h_1\right)\right\}+\operatorname{div}_{\mathbf{x}}\left\{\frac{1}{c_{01}^2}\left(\frac{\partial h_1}{\partial t}+\mathbf{u}_0\cdot\nabla_{\mathbf{x}}h_1\right)\mathbf{u}_0\right\}\approx\Delta_{\mathbf{x}}h_1, \tag{1.591}$$

and

$$\frac{\partial}{\partial t}\left\{\frac{1}{c_{02}^2}\left(\frac{\partial}{\partial t}(\mathbf{e}_k\cdot\mathbf{h}_2)+\mathbf{u}_0\cdot\nabla_{\mathbf{x}}(\mathbf{e}_k\cdot\mathbf{h}_2)\right)\right\}$$
$$+\operatorname{div}_{\mathbf{x}}\left\{\frac{1}{c_{02}^2}\left(\frac{\partial}{\partial t}(\mathbf{e}_k\cdot\mathbf{h}_2)+\mathbf{u}_0\cdot\nabla_{\mathbf{x}}(\mathbf{e}_k\cdot\mathbf{h}_2)\right)\mathbf{u}_0\right\}$$
$$\approx\Delta_{\mathbf{x}}(\mathbf{e}_k\cdot\mathbf{h}_2)\qquad\forall k=1,2,3, \tag{1.592}$$

where

$$h_1(\mathbf{x},t):=\operatorname{div}_{\mathbf{x}}\mathbf{g}_1(\mathbf{x},t)\qquad\text{and}\qquad c_{01}:=\sqrt{\frac{(\alpha+\beta)}{\mu_0}} \tag{1.593}$$

and

$$\mathbf{h}_2(\mathbf{x},t):=\operatorname{curl}_{\mathbf{x}}\mathbf{g}_1(\mathbf{x},t)\qquad\text{and}\qquad c_{02}:=\sqrt{\frac{\alpha}{2\mu_0}}. \tag{1.594}$$

Moreover, we have (1.582), i.e.:

$$\mathcal{E}_1\approx-\frac{1}{2}\left(d_{\mathbf{x}}\mathbf{g}_1+\{d_{\mathbf{x}}\mathbf{g}_1\}^T\right), \tag{1.595}$$

and (1.576), i.e.:

$$\frac{\partial\mathbf{g}_1}{\partial t}-\operatorname{curl}_{\mathbf{x}}\{\mathbf{u}_0\times\mathbf{g}_1\}+(\operatorname{div}_{\mathbf{x}}\mathbf{g}_1)\,\mathbf{u}_0\approx-\mathbf{u}_1, \tag{1.596}$$

and the above six equations are invariant under the change of inertial or non-inertial Cartesian coordinate system.

As we can easily see, the divergence of $\mathbf{g}_1$ and any of the scalar components of the curl of $\mathbf{g}_1$ satisfy the invariant wave equation of the same type as (1.554) or (1.563). However, as we can see from (1.593) and (1.594) the characteristic parameter c_{01} in the wave equation for the divergence part of $\mathbf{g}_1$ (1.591) differ from the characteristic parameter c_{02} in the wave equation for the curl part of $\mathbf{g}_1$ (1.592), i.e. the divergence part and the curl parts of $\mathbf{g}_1$ propagate as two different waves with the different speeds. As it can be easily seen, in the case of the flat waves in the resting body the divergence part of $\mathbf{g}_1$ (which is curl-free) propagate as a longitudinal wave, similarly to the sound in fluid or gas, and at the same time the curl part of $\mathbf{g}_1$ (which is divergence-free) propagate as a transverse wave. Moreover, the longitudinal and transverse waves in an elastic body propagate with two different speeds.

1.13 Macroscopic Electrodynamics in the Presence of Dielectric and/or Magnetic Mediums

Consider system (1.38) in some inertial or non-inertial Cartesian coordinate systems inside a dielectric and/or magnetic medium:

$$\begin{cases} \operatorname{curl}_{\mathbf{x}} \mathbf{H}_0 = \frac{4\pi}{c}\left(\mathbf{j} + \mathbf{j}_{mp}\right) + \frac{1}{c}\frac{\partial \mathbf{D}_0}{\partial t} \\ \operatorname{div}_{\mathbf{x}} \mathbf{D}_0 = 4\pi\left(\rho + \rho_p\right) \\ \operatorname{curl}_{\mathbf{x}} \mathbf{E} + \frac{1}{c}\frac{\partial \mathbf{B}}{\partial t} = 0 \\ \operatorname{div}_{\mathbf{x}} \mathbf{B} = 0, \end{cases} \tag{1.597}$$

where $\mathbf{E}$ is the electric field, $\mathbf{B}$ is the magnetic field, $\mathbf{v} := \mathbf{v}(\mathbf{x}, t)$ is the vectorial gravitational potential, ρ is the average (macroscopic) charge density, ρ_p is the density of the charge of polarization, $\mathbf{j}$ is the average (macroscopic) current density, $\mathbf{j}_{mp}$ is the density of the current of polarization and magnetization and

$$\mathbf{D}_0 := \mathbf{E} + \frac{1}{c}\mathbf{v} \times \mathbf{B} \quad \text{and} \quad \mathbf{H}_0 := \mathbf{B} + \frac{1}{c}\mathbf{v} \times \mathbf{D}_0. \tag{1.598}$$

Consistently with Sect. 12.1, it is well known from the Lorentz theory that in the case of a moving dielectric/magnetic medium

$$\rho_p = -\operatorname{div}_{\mathbf{x}} \mathbf{P} \quad \text{and} \quad \mathbf{j}_{mp} = \frac{\partial \mathbf{P}}{\partial t} + c \operatorname{curl}_{\mathbf{x}} \mathbf{M}, \tag{1.599}$$

where $\mathbf{P} : \mathbb{R}^3 \times [0, +\infty) \to \mathbb{R}^3$ is the field of polarization, $\mathbf{M} : \mathbb{R}^3 \times [0, +\infty) \to \mathbb{R}^3$ is the field of magnetization and $\mathbf{u} := \mathbf{u}(\mathbf{x}, t)$ is the field of velocities of the dielectric medium, see Proposition 12.1 for the proof of (1.599). Thus, if we consider

$$\mathbf{D} := \mathbf{D}_0 + 4\pi\mathbf{P} = \mathbf{E} + \frac{1}{c}\mathbf{v} \times \mathbf{B} + 4\pi\mathbf{P}, \tag{1.600}$$

and

$$\mathbf{H} := \mathbf{H}_0 - 4\pi\mathbf{M} = \mathbf{B} + \frac{1}{c}\mathbf{v} \times \mathbf{E} + \frac{1}{c}\mathbf{v} \times \left(\frac{1}{c}\mathbf{v} \times \mathbf{B}\right) - 4\pi\mathbf{M}, \tag{1.601}$$

we obtain the usual Maxwell equations of the form:

$$\begin{cases} \operatorname{curl}_{\mathbf{x}} \mathbf{H} = \frac{4\pi}{c}\mathbf{j} + \frac{1}{c}\frac{\partial \mathbf{D}}{\partial t} \\ \operatorname{div}_{\mathbf{x}} \mathbf{D} = 4\pi\rho \\ \operatorname{curl}_{\mathbf{x}} \mathbf{E} + \frac{1}{c}\frac{\partial \mathbf{B}}{\partial t} = 0 \\ \operatorname{div}_{\mathbf{x}} \mathbf{B} = 0. \end{cases} \tag{1.602}$$

We call **D** by the electric displacement field and **H** by the **H**-magnetic field in a medium.

Next, in Chap. 12 we prove that the laws of transformation of electromagnetic fields in dielectric/magnetic medium, under the change of non-inertial Cartesian coordinate system of the form (1.2), are exactly the same as (1.45) in the vacuum, i.e. having the form of

$$\begin{cases} \mathbf{D}' = A(t) \cdot \mathbf{D} \\ \mathbf{B}' = A(t) \cdot \mathbf{B} \\ \mathbf{E}' = A(t) \cdot \mathbf{E} - \frac{1}{c}\left(\frac{dA}{dt}(t) \cdot \mathbf{x} + \frac{d\mathbf{z}}{dt}(t)\right) \times (A(t) \cdot \mathbf{B}) \\ \mathbf{H}' = A(t) \cdot \mathbf{H} + \frac{1}{c}\left(\frac{dA}{dt}(t) \cdot \mathbf{x} + \frac{d\mathbf{z}}{dt}(t)\right) \times (A(t) \cdot \mathbf{D}), \end{cases} \tag{1.603}$$

provided that

$$\begin{cases} \mathbf{u}' = A(t) \cdot \mathbf{u} + \frac{dA}{dt}(t) \cdot \mathbf{x} + \frac{d\mathbf{z}}{dt}(t) \\ \mathbf{v}' = A(t) \cdot \mathbf{v} + \frac{dA}{dt}(t) \cdot \mathbf{x} + \frac{d\mathbf{z}}{dt}(t) \\ \mathbf{P}' = A(t) \cdot \mathbf{P}, \\ \mathbf{M}' = A(t) \cdot \mathbf{M} - \frac{1}{c}\left(\frac{dA}{dt}(t) \cdot \mathbf{x} + \frac{d\mathbf{z}}{dt}(t)\right) \times (A(t) \cdot \mathbf{P}), \end{cases} \tag{1.604}$$

(see Sect. 12.1 for the justification of the last two equalities in (1.604)). Moreover, as before, the Maxwell equations in the medium of the form (1.602) stays invariant under the change of inertial or non-inertial Cartesian coordinate system, provided we have (1.603) and

$$\mathbf{j}' = A(t) \cdot \mathbf{j} + \rho\frac{dA}{dt}(t) \cdot \mathbf{x} + \rho\frac{d\mathbf{z}}{dt}(t). \tag{1.605}$$

Next, one can show that in the case of simplest moving homogenous isotropic nonmagnetic dielectric medium we have

$$\begin{cases} \mathbf{P} = \frac{n^2-1}{4\pi}\left(\mathbf{E} + \frac{1}{c}\mathbf{u} \times \mathbf{B}\right), \\ \mathbf{M} = -\frac{1}{c}\frac{n^2-1}{4\pi}\mathbf{u} \times \left(\mathbf{E} + \frac{1}{c}\mathbf{u} \times \mathbf{B}\right) = -\frac{1}{c}\mathbf{u} \times \mathbf{P}, \end{cases} \tag{1.606}$$

where n is a material coefficient (not necessarily constant), called refraction index and **u** is the velocity of the medium. Moreover, in the case of the simplest homogenous isotropic dielectric medium with certain magnetic properties one generalizes (1.606) by the following:

$$\begin{cases} \mathbf{P} = \frac{n^2-1}{4\pi}\left(\mathbf{E} + \frac{1}{c}\mathbf{u} \times \mathbf{B}\right) - \kappa\mathbf{D}, \\ \mathbf{M} = -\frac{1}{c}\frac{n^2-1}{4\pi}\mathbf{u} \times \left(\mathbf{E} + \frac{1}{c}\mathbf{u} \times \mathbf{B}\right) + \kappa\mathbf{H} = -\frac{1}{c}\mathbf{u} \times \mathbf{P} + \kappa\left(\mathbf{H} - \frac{1}{c}\mathbf{u} \times \mathbf{D}\right), \end{cases} \tag{1.607}$$

where n is the refraction index and κ is an additional material coefficient (not necessarily constant), see Section 12.4 for the details. In the case of nonmagnetic dielectric we just put $\kappa = 0$ and (1.607) becomes to be the same as (1.606). Then, using (1.604) and (1.603), it can be easily seen that the laws in (1.607) are invariant under the changes of inertial or non-inertial Cartesian coordinate system. Next, denoting

$$\kappa_0 = \frac{1}{1+4\pi\kappa} \quad \text{and} \quad \gamma_0 = \frac{1+4\pi\kappa}{n^2} = \frac{1}{\kappa_0 n^2} \quad \text{so that} \quad n = \frac{1}{\sqrt{\gamma_0\kappa_0}} \tag{1.608}$$

and defining the speed-like vector field

$$\tilde{\mathbf{u}} := \left(\frac{1}{n^2}\mathbf{v} + \left(1 - \frac{1}{n^2}\right)\mathbf{u}\right) = \left(\gamma_0\kappa_0\mathbf{v} + (1-\gamma_0\kappa_0)\,\mathbf{u}\right), \tag{1.609}$$

by plugging (1.607) into (1.600) and (1.601) we deduce

$$\mathbf{E} = \gamma_0\mathbf{D} - \frac{1}{c}\,\tilde{\mathbf{u}} \times \mathbf{B}, \tag{1.610}$$

and

$$\mathbf{H} = \kappa_0\mathbf{B} + \frac{1}{c}\,\tilde{\mathbf{u}} \times \mathbf{D} + \frac{\kappa_0(1-\kappa_0\gamma_0)}{c^2}\,(\mathbf{u}-\mathbf{v}) \times ((\mathbf{u}-\mathbf{v}) \times \mathbf{B}), \tag{1.611}$$

where we call γ_0 and κ_0 dielectric and magnetic permeability of the medium, see Sect. 12.4 for the details. Thus by (1.602), (1.609), (1.610) and (1.611) we have

$$\begin{cases} \mathrm{curl}_{\mathbf{x}}\mathbf{H} = \frac{4\pi}{c}\mathbf{j} + \frac{1}{c}\frac{\partial\mathbf{D}}{\partial t}, \\ \mathrm{div}_{\mathbf{x}}\,\mathbf{D} = 4\pi\rho, \\ \mathrm{curl}_{\mathbf{x}}\mathbf{E} + \frac{1}{c}\frac{\partial\mathbf{B}}{\partial t} = 0, \\ \mathrm{div}_{\mathbf{x}}\,\mathbf{B} = 0, \\ \mathbf{E} = \gamma_0\mathbf{D} - \frac{1}{c}\,\tilde{\mathbf{u}} \times \mathbf{B}, \\ \mathbf{H} = \kappa_0\mathbf{B} + \frac{1}{c}\,\tilde{\mathbf{u}} \times \mathbf{D} + \frac{\kappa_0(1-\kappa_0\gamma_0)}{c^2}\,(\mathbf{u}-\mathbf{v}) \times ((\mathbf{u}-\mathbf{v}) \times \mathbf{B}), \\ \tilde{\mathbf{u}} := (\kappa_0\gamma_0\mathbf{v} + (1-\kappa_0\gamma_0)\mathbf{u}), \end{cases} \tag{1.612}$$

where $\tilde{\mathbf{u}}$ is a speed-like vector field that we call the optical displacement of the moving medium. Note that for the case $\gamma_0 = 1$ and $\kappa_0 = 1$, the system (1.612) is exactly the same as the corresponding system in the vacuum. The equations in (1.612) take much simpler forms in the case where the quantity

$$\frac{|\kappa_0||1-\kappa_0\gamma_0| \cdot |\mathbf{u}-\mathbf{v}|^2}{c^2} \ll 1 \tag{1.613}$$

is negligible, that happens if either the absolute value of the difference between the medium velocity and vectorial gravitational potential is much less then the constant

c or $\gamma_0\kappa_0 = \frac{1}{n^2}$ is close to the value 1. Indeed, in this case, instead of (1.610) and (1.611) we obtain the following relations:

$$\mathbf{E} = \gamma_0\mathbf{D} - \frac{1}{c}\tilde{\mathbf{u}} \times \mathbf{B}, \tag{1.614}$$

$$\mathbf{H} = \kappa_0\mathbf{B} + \frac{1}{c}\tilde{\mathbf{u}} \times \mathbf{D}. \tag{1.615}$$

As a consequence we obtain the full system of Maxwell equations in the medium:

$$\begin{cases} \text{curl}_{\mathbf{x}}\mathbf{H} = \frac{4\pi}{c}\mathbf{j} + \frac{1}{c}\frac{\partial \mathbf{D}}{\partial t}, \\ \text{div}_{\mathbf{x}}\,\mathbf{D} = 4\pi\rho, \\ \text{curl}_{\mathbf{x}}\mathbf{E} + \frac{1}{c}\frac{\partial \mathbf{B}}{\partial t} = 0, \\ \text{div}_{\mathbf{x}}\,\mathbf{B} = 0, \\ \mathbf{E} = \gamma_0\mathbf{D} - \frac{1}{c}\tilde{\mathbf{u}} \times \mathbf{B}, \\ \mathbf{H} = \kappa_0\mathbf{B} + \frac{1}{c}\tilde{\mathbf{u}} \times \mathbf{D}, \\ \tilde{\mathbf{u}} = (\kappa_0\gamma_0\mathbf{v} + (1-\kappa_0\gamma_0)\mathbf{u}), \end{cases} \tag{1.616}$$

where $\tilde{\mathbf{u}}$ is the speed-like vector field and γ_0 and κ_0 are dielectric and magnetic permeability of the medium. Note that (1.616) is analogous to the system of Maxwell equations in the vacuum and it is also invariant under the change of inertial or non-inertial Cartesian coordinate system, provided that under this transformation we have (1.604).

Next, in the case of the simplest anisotropic dielectric and/or magnetic, it is well known that we have the following generalization of (1.607):

$$\begin{cases} \mathbf{P} = \Gamma \cdot \left(\mathbf{E} + \frac{1}{c}\mathbf{u} \times \mathbf{B}\right) - \Upsilon \cdot \mathbf{D}, \\ \mathbf{M} = -\frac{1}{c}\mathbf{u} \times \mathbf{P} + \Upsilon \cdot \left(\mathbf{H} - \frac{1}{c}\mathbf{u} \times \mathbf{D}\right), \end{cases} \tag{1.617}$$

where $\Gamma \in \mathbb{R}^{3\times 3}$ and $\Upsilon \in \mathbb{R}^{3\times 3}$ are matrix-valued fields. Then, using (1.604) and (1.603) it can be easily seen that the laws in (1.617) are invariant under the changes of inertial or non-inertial Cartesian coordinate system, provided that Γ and Υ are proper matrix fields (see Definition 1.1).

Next, it is well known that Ohm's Law in a conducting medium has the form

$$\mathbf{j} - \rho\,\mathbf{u} = \varepsilon\left(\mathbf{E} + \frac{1}{c}\mathbf{u} \times \mathbf{B}\right), \tag{1.618}$$

where $\mathbf{u}$ is the velocity of the medium and ε is a material coefficient. As before, using (1.604) and (1.603), it can be easily seen that Ohm's Law is invariant under the changes of inertial or non-inertial Cartesian coordinate system.

Furthermore, it is well known that in the case of the strong magnetic field the modification of Ohm's Law including the Hall effect has the following form:

$$\mathbf{j}-\rho\,\mathbf{u}=\varepsilon\left(\mathbf{E}+\frac{1}{c}\mathbf{u}\times\mathbf{B}\right)-\varsigma\,(\mathbf{j}-\rho\,\mathbf{u})\times\mathbf{B}, \tag{1.619}$$

where ς is a material coefficient. Then, as before, using (1.604) and (1.603), it can be easily seen that the generalized Ohm's Law (1.619), including the Hall effect, is invariant under the changes of inertial or non-inertial Cartesian coordinate system.

Next, it is well known that, in the case of the anisotropic conducting medium, Ohm's Law has the following form, generalizing (1.618):

$$\mathbf{j}-\rho\,\mathbf{u}=\Xi\cdot\left(\mathbf{E}+\frac{1}{c}\mathbf{u}\times\mathbf{B}\right), \tag{1.620}$$

where $\Xi\in\mathbb{R}^{3\times3}$ is a matrix-valued field. As before, using (1.604) and (1.603), it can be easily seen that (1.620) is invariant under the changes of inertial or non-inertial Cartesian coordinate system, provided that Ξ is a proper matrix field.

Finally, the generalization of (1.619) to anisotropic mediums is the following:

$$\mathbf{j}-\rho\,\mathbf{u}=\Xi\cdot\left(\mathbf{E}+\frac{1}{c}\mathbf{u}\times\mathbf{B}\right)-\Pi\cdot\left((\mathbf{j}-\rho\,\mathbf{u})\times\mathbf{B}\right), \tag{1.621}$$

where $\Xi\in\mathbb{R}^{3\times3}$ and $\Pi\in\mathbb{R}^{3\times3}$ are matrix-valued fields. As before, using (1.604) and (1.603), it can be easily seen that (1.621) is invariant under the changes of inertial or non-inertial Cartesian coordinate system, provided that Ξ and Π are proper matrix fields.

1.13.1 Optical Dispersion in Moving Mediums

Remind that if U is a real-valued field, then we can write the field U as a Fourier's Transform on the time variable:

$$U(\mathbf{x},t)=Re\left\{2\int_0^{+\infty}\hat{U}(\mathbf{x},\omega)e^{i\omega t}d\omega\right\}$$
$$\text{where}\quad \hat{U}(\mathbf{x},\omega):=\frac{1}{2\pi}\int_{-\infty}^{+\infty}U(\mathbf{x},t)e^{-i\omega t}dt\,. \tag{1.622}$$

Thus, since **E**, **B**, **D** and **H** are real vectors, we can write them as Fourier's Transform on the time variable:

$$\mathbf{E}(\mathbf{x},t)=Re\left\{2\int_0^{+\infty}\hat{\mathbf{E}}(\omega,\mathbf{x})e^{i\omega t}d\omega\right\}$$
$$\text{where}\quad \hat{\mathbf{E}}(\omega,\mathbf{x}):=\frac{1}{2\pi}\int_{-\infty}^{+\infty}\mathbf{E}(\mathbf{x},t)e^{-i\omega t}dt, \tag{1.623}$$

$$\mathbf{B}(\mathbf{x},t) = Re\left\{2\int_0^{+\infty} \hat{\mathbf{B}}(\omega,\mathbf{x})e^{i\omega t}d\omega\right\}$$
$$\text{where } \hat{\mathbf{B}}(\omega,\mathbf{x}) := \frac{1}{2\pi}\int_{-\infty}^{+\infty} \mathbf{B}(\mathbf{x},t)e^{-i\omega t}dt, \quad (1.624)$$

$$\mathbf{D}(\mathbf{x},t) = Re\left\{2\int_0^{+\infty} \hat{\mathbf{D}}(\omega,\mathbf{x})e^{i\omega t}d\omega\right\}$$
$$\text{where } \hat{\mathbf{D}}(\omega,\mathbf{x}) := \frac{1}{2\pi}\int_{-\infty}^{+\infty} \mathbf{D}(\mathbf{x},t)e^{-i\omega t}dt, \quad (1.625)$$

and

$$\mathbf{H}(\mathbf{x},t) = Re\left\{2\int_0^{+\infty} \hat{\mathbf{H}}(\omega,\mathbf{x})e^{i\omega t}d\omega\right\}$$
$$\text{where } \hat{\mathbf{H}}(\omega,\mathbf{x}) := \frac{1}{2\pi}\int_{-\infty}^{+\infty} \mathbf{H}(\mathbf{x},t)e^{-i\omega t}dt, \quad (1.626)$$

where given $\mathbf{a} = (a_1, a_2, a_3) \in \mathbb{C}^3$ we denote $Re\{\mathbf{a}\} = (Re\{a_1\}, Re\{a_2\}, Re\{a_3\}) \in \mathbb{R}^3$. Then, it is well known that in the case of the simplest optical dispersion in the resting medium with $\mathbf{u} = 0$ we have the following generalization of (1.607):

$$\begin{cases} \mathbf{P}(\mathbf{x},t) = Re\left\{2\int_0^{+\infty} \frac{n^2(\omega,\mathbf{x},t)-1}{4\pi}\hat{\mathbf{E}}(\omega,\mathbf{x})e^{i\omega t}d\omega\right\} \\ -Re\left\{2\int_0^{+\infty} \kappa(\omega,\mathbf{x},t)\hat{\mathbf{D}}(\omega,\mathbf{x})e^{i\omega t}d\omega\right\}, \\ \mathbf{M}(\mathbf{x},t) = Re\left\{2\int_0^{+\infty} \kappa(\omega,\mathbf{x},t)\hat{\mathbf{H}}(\omega,\mathbf{x})e^{i\omega t}d\omega\right\}, \end{cases} \quad (1.627)$$

where $\hat{\mathbf{E}}(\omega,\mathbf{x})$, $\hat{\mathbf{B}}(\omega,\mathbf{x})$, $\hat{\mathbf{D}}(\omega,\mathbf{x})$ and $\hat{\mathbf{H}}(\omega,\mathbf{x})$ are Fourier's Transforms by time of $\mathbf{E}(\mathbf{x},t)$, $\mathbf{B}(\mathbf{x},t)$, $\mathbf{D}(\mathbf{x},t)$ and $\mathbf{H}(\mathbf{x},t)$, given by the second equalities of (1.623), (1.624), (1.625) and (1.626) and the quantities $n(\omega,\mathbf{x},t)$ and $\kappa(\omega,\mathbf{x},t)$ in (1.627) are assumed to be complex in general and assumed to depend on ω in addition to the dependence on $(\mathbf{x},t)$.

We would like to obtain the law being an analog of (1.627) in the case of moving medium (i.e. $\mathbf{u} \neq 0$), which is invariant under the change of inertial or non-inertial Cartesian coordinate system. In addition to the invariance under the change of Cartesian coordinate systems and the equivalence to (1.627) in the particular case $\mathbf{u} = 0$, we also need to assume that in the case of an arbitrarily moving transparent medium without dispersion our law is equivalent to (1.607).

Then, in some Cartesian coordinate systems consider a motion of some continuum medium occupying a region $\Omega \subset \mathbb{R}^3$ at some fixed instant of time $t = t_0$ and having the velocity field $\mathbf{u} := \mathbf{u}(\mathbf{x},t)$. Next, as before, let $\mathbf{r}(t,\mathbf{y}) : \mathbb{R} \times \Omega \to \mathbb{R}^3$ be a solution of (1.514), i.e. it satisfies the following initial value problem for an ordinary differential equation:

$$\begin{cases} \frac{\partial \mathbf{r}}{\partial t}(t,\mathbf{y}) = \mathbf{u}\left(\mathbf{r}(t,\mathbf{y}),t\right) & \forall t \in \mathbb{R},\ \forall \mathbf{y} \in \Omega \\ \mathbf{r}(t_0,\mathbf{y}) = \mathbf{y} \quad \forall \mathbf{y} \in \Omega. \end{cases} \tag{1.628}$$

Then, clearly $\mathbf{x} = \mathbf{r}(t,\mathbf{y})$ stands for the spatial coordinates at the instant of time t of the parcel of continuum, having initial coordinates $\mathbf{y}$. Thus, as before, we deduce that $det\left\{d_{\mathbf{y}}\mathbf{r}(t,\mathbf{y})\right\} \neq 0$ for every instant of time t and so, for the given instant of time t the mapping $\mathbf{x} = \mathbf{r}(t,\mathbf{y})$ is locally invertible, i.e. the equation $\mathbf{x} = \mathbf{r}(t,\mathbf{y})$ can be resolved in $\mathbf{y}$. Thus there exists a regular mapping $\mathbf{y} = \mathbf{f}(\mathbf{x},t)$, such that

$$\mathbf{f}\left(\mathbf{r}(t,\mathbf{y}),t\right) = \mathbf{y} \quad \forall \mathbf{y} \in \Omega \qquad \text{and} \qquad \mathbf{r}\left(t,\mathbf{f}(\mathbf{x},t)\right) = \mathbf{x}. \tag{1.629}$$

Then, clearly $\mathbf{y} = \mathbf{f}(\mathbf{x},t)$ stands for the initial coordinates at the time t_0 of the parcel of continuum, having coordinates $\mathbf{x}$ at the instant of time t. Next $\mathbf{f}(\mathbf{x},t)$ satisfies

$$\begin{cases} \frac{\partial \mathbf{f}}{\partial t}(\mathbf{x},t) + d_{\mathbf{x}}\mathbf{f}(\mathbf{x},t) \cdot \mathbf{u}(\mathbf{x},t) = 0 \\ \mathbf{f}(\mathbf{x},t_0) = \mathbf{x}. \end{cases} \tag{1.630}$$

Next consider,

$$\begin{cases} \mathbf{E}^*(t,\mathbf{y}) := \left\{d_{\mathbf{y}}\mathbf{r}(t,\mathbf{y})\right\}^T \cdot \left(\mathbf{E}\left(\mathbf{r}(t,\mathbf{y}),t\right) + \frac{1}{c}\mathbf{u}\left(\mathbf{r}(t,\mathbf{y}),t\right) \times \mathbf{B}\left(\mathbf{r}(t,\mathbf{y}),t\right)\right) \\ \quad \forall t \in \mathbb{R},\ \forall \mathbf{y} \in \Omega, \\ \mathbf{B}^*(t,\mathbf{y}) := \left\{d_{\mathbf{y}}\mathbf{r}(t,\mathbf{y})\right\}^T \cdot \mathbf{B}\left(\mathbf{r}(t,\mathbf{y}),t\right) \quad \forall t \in \mathbb{R},\ \forall \mathbf{y} \in \Omega, \\ \mathbf{D}^*(t,\mathbf{y}) := \left\{d_{\mathbf{y}}\mathbf{r}(t,\mathbf{y})\right\}^T \cdot \mathbf{D}\left(\mathbf{r}(t,\mathbf{y}),t\right) \quad \forall t \in \mathbb{R},\ \forall \mathbf{y} \in \Omega, \\ \mathbf{H}^*(t,\mathbf{y}) := \left\{d_{\mathbf{y}}\mathbf{r}(t,\mathbf{y})\right\}^T \cdot \left(\mathbf{H}\left(\mathbf{r}(t,\mathbf{y}),t\right) - \frac{1}{c}\mathbf{u}\left(\mathbf{r}(t,\mathbf{y}),t\right) \times \mathbf{D}\left(\mathbf{r}(t,\mathbf{y}),t\right)\right) \\ \quad \forall t \in \mathbb{R},\ \forall \mathbf{y} \in \Omega, \end{cases} \tag{1.631}$$

where $\mathbf{r}(t,\mathbf{y})$ is given by (12.68). Then, since $\mathbf{E}^*$, $\mathbf{B}^*$, $\mathbf{D}^*$ and $\mathbf{H}^*$ are real vectors, we can write them as Fourier's Transform on the time variable:

$$\mathbf{E}^*(t,\mathbf{y}) = Re\left\{2\int_0^{+\infty} \hat{\mathbf{E}}^*(\tilde{\omega},\mathbf{y})e^{i\tilde{\omega}t}d\tilde{\omega}\right\}$$
$$\text{where} \quad \hat{\mathbf{E}}^*(\tilde{\omega},\mathbf{y}) := \frac{1}{2\pi}\int_{-\infty}^{+\infty} \mathbf{E}^*(t,\mathbf{y})e^{-i\tilde{\omega}t}dt\,, \tag{1.632}$$

$$\mathbf{B}^*(t,\mathbf{y}) = Re\left\{2\int_0^{+\infty}\hat{\mathbf{B}}^*(\tilde{\omega},\mathbf{y})e^{i\tilde{\omega}t}d\tilde{\omega}\right\}$$
$$\text{where } \hat{\mathbf{B}}^*(\tilde{\omega},\mathbf{y}) := \frac{1}{2\pi}\int_{-\infty}^{+\infty}\mathbf{B}^*(t,\mathbf{y})e^{-i\tilde{\omega}t}dt\,, \quad (1.633)$$

$$\mathbf{D}^*(t,\mathbf{y}) = Re\left\{2\int_0^{+\infty}\hat{\mathbf{D}}^*(\tilde{\omega},\mathbf{y})e^{i\tilde{\omega}t}d\tilde{\omega}\right\}$$
$$\text{where } \hat{\mathbf{D}}^*(\tilde{\omega},\mathbf{y}) := \frac{1}{2\pi}\int_{-\infty}^{+\infty}\mathbf{D}^*(t,\mathbf{y})e^{-i\tilde{\omega}t}dt\,, \quad (1.634)$$

$$\mathbf{H}^*(t,\mathbf{y}) = Re\left\{2\int_0^{+\infty}\hat{\mathbf{H}}^*(\tilde{\omega},\mathbf{y})e^{i\tilde{\omega}t}d\tilde{\omega}\right\}$$
$$\text{where } \hat{\mathbf{H}}^*(\tilde{\omega},\mathbf{y}) := \frac{1}{2\pi}\int_{-\infty}^{+\infty}\mathbf{H}^*(t,\mathbf{y})e^{-i\tilde{\omega}t}dt\,. \quad (1.635)$$

Then we write the generalization of (1.627) as

$$\mathbf{P}\left(\mathbf{r}(t,\mathbf{y}),t\right) =$$
$$Re\left\{2\int_0^{+\infty}\frac{n^2\left(\tilde{\omega},\mathbf{r}(t,\mathbf{y}),t\right)-1}{4\pi}\left(\left(\left\{d_{\mathbf{y}}\mathbf{r}\left(t,\mathbf{y}\right)\right\}^T\right)^{-1}\cdot\hat{\mathbf{E}}^*\left(\omega,\mathbf{x}\right)\right)e^{i\omega t}d\omega\right\}$$
$$-\,Re\left\{2\int_0^{+\infty}\kappa\left(\tilde{\omega},\mathbf{r}(t,\mathbf{y}),t\right)\left(\left(\left\{d_{\mathbf{y}}\mathbf{r}\left(t,\mathbf{y}\right)\right\}^T\right)^{-1}\cdot\hat{\mathbf{D}}^*\left(\omega,\mathbf{x}\right)\right)e^{i\omega t}d\omega\right\}$$
$$\text{and}\qquad \mathbf{M}\left(\mathbf{r}(t,\mathbf{y}),t\right)+\frac{1}{c}\mathbf{u}\left(\mathbf{r}(t,\mathbf{y}),t\right)\times\mathbf{P}\left(\mathbf{r}(t,\mathbf{y}),t\right) =$$
$$Re\left\{2\int_0^{+\infty}\kappa\left(\tilde{\omega},\mathbf{r}(t,\mathbf{y}),t\right)\left(\left(\left\{d_{\mathbf{y}}\mathbf{r}\left(t,\mathbf{y}\right)\right\}^T\right)^{-1}\cdot\hat{\mathbf{H}}^*(\tilde{\omega},\mathbf{y})\right)e^{i\tilde{\omega}t}d\tilde{\omega}\right\}, \quad (1.636)$$

i.e.

$$\mathbf{P}\left(\mathbf{x},t\right) = Re\left\{2\int_0^{+\infty}\frac{n^2\left(\tilde{\omega},\mathbf{x},t\right)-1}{4\pi}\left(\left\{d_{\mathbf{x}}\mathbf{f}\left(\mathbf{x},t\right)\right\}^T\cdot\hat{\mathbf{E}}^*\left(\tilde{\omega},\mathbf{f}\left(\mathbf{x},t\right)\right)\right)e^{i\tilde{\omega}t}d\tilde{\omega}\right\}$$
$$-\,Re\left\{2\int_0^{+\infty}\kappa\left(\tilde{\omega},\mathbf{x},t\right)\left(\left\{d_{\mathbf{x}}\mathbf{f}\left(\mathbf{x},t\right)\right\}^T\cdot\hat{\mathbf{D}}^*\left(\tilde{\omega},\mathbf{f}\left(\mathbf{x},t\right)\right)\right)e^{i\tilde{\omega}t}d\tilde{\omega}\right\}\qquad\text{and}$$
$$\mathbf{M}\left(\mathbf{x},t\right)+\frac{1}{c}\mathbf{u}(\mathbf{x},t)\times\mathbf{P}(\mathbf{x},t) =$$
$$Re\left\{2\int_0^{+\infty}\kappa\left(\tilde{\omega},\mathbf{x},t\right)\left(\left\{d_{\mathbf{x}}\mathbf{f}\left(\mathbf{x},t\right)\right\}^T\cdot\hat{\mathbf{H}}^*\left(\tilde{\omega},\mathbf{f}\left(\mathbf{x},t\right)\right)\right)e^{i\tilde{\omega}t}d\tilde{\omega}\right\}. \quad (1.637)$$

The quantities $n(\tilde{\omega}, \mathbf{x}, t)$ and $\kappa(\tilde{\omega}, \mathbf{x}, t)$ in (1.636) and (1.637) are assumed to be complex in general and assumed to depend on $\tilde{\omega}$ in addition to the dependence on $(\mathbf{x}, t)$.

In particular, in the case $\mathbf{u} = 0$ we clearly have $\mathbf{r}(t, \mathbf{y}) = \mathbf{y}$, $\mathbf{f}(\mathbf{x}, t) = \mathbf{x}$ and $\mathbf{E}^*(t, \mathbf{x}) = \mathbf{E}(\mathbf{x}, t)$, $\mathbf{B}^*(t, \mathbf{x}) = \mathbf{B}(\mathbf{x}, t)$, $\mathbf{D}^*(t, \mathbf{x}) = \mathbf{D}(\mathbf{x}, t)$, $\mathbf{H}^*(t, \mathbf{x}) = \mathbf{H}(\mathbf{x}, t)$ and therefore, (1.636) and (1.637) coincide with (1.627).

On the other hand, in Sect. 12.7 we prove that in the case of an arbitrarily moving transparent medium without dispersion, i.e. in the case where we assume that $n := n(\mathbf{x}, t)$ and $\kappa := \kappa(\mathbf{x}, t)$ are independent on the argument $\tilde{\omega}$ and moreover we assume them to be real, we can rewrite (1.637) in the simpler form:

$$\begin{cases} \mathbf{P}(\mathbf{x}, t) = \frac{n^2(\mathbf{x},t)-1}{4\pi} \left(\mathbf{E}(\mathbf{x}, t) + \frac{1}{c}\mathbf{u}(\mathbf{x}, t) \times \mathbf{B}(\mathbf{x}, t)\right) - \kappa(\mathbf{x}, t)\,\mathbf{D}(\mathbf{x}, t), \\ \mathbf{M}(\mathbf{x}, t) = \kappa(\mathbf{x}, t) \left(\mathbf{H}(\mathbf{x}, t) - \frac{1}{c}\mathbf{u}(\mathbf{x}, t) \times \mathbf{D}(\mathbf{x}, t)\right) - \frac{1}{c}\mathbf{u}(\mathbf{x}, t) \times \mathbf{P}(\mathbf{x}, t), \end{cases} \tag{1.638}$$

consistently with (1.607).

Next, as before, assume that the change of some non-inertial Cartesian system $(*)$ of coordinates to another Cartesian system $(**)$ of coordinates is of the form:

$$\begin{cases} \mathbf{x}' = A(t) \cdot \mathbf{x} + \mathbf{z}(t), \\ t' = t, \end{cases}$$

where $A(t) \in SO(3)$ is a rotation. Then, in Sect. 12.7 we prove that the dispersion law, in the forms of (1.636) or equivalently (1.637), is invariant under the change of inertial or non-inertial Cartesian coordinate system, provided we have

$$\tilde{\omega}' = \tilde{\omega}, \quad n'\left(\tilde{\omega}', \mathbf{x}', t'\right) = n(\tilde{\omega}, \mathbf{x}, t) \quad \text{and} \quad \kappa'\left(\tilde{\omega}', \mathbf{x}', t'\right) = \kappa(\tilde{\omega}, \mathbf{x}, t). \tag{1.639}$$

Next, in Sect. 12.7 we prove the right-hand side of (1.636) or (1.637) is independent on the initial instant of time t_0 in (1.628).

Finally, we can write the following possible alternatives to the dispersion law in the form of (1.631) and (1.636) or (1.637). We could consider,

$$\begin{cases} \mathbf{E}^*(t, \mathbf{y}) := \left\{d_{\mathbf{y}}\mathbf{r}(t, \mathbf{y})\right\}^{-1} \cdot \left(\mathbf{E}(\mathbf{r}(t, \mathbf{y}), t) + \frac{1}{c}\mathbf{u}(\mathbf{r}(t, \mathbf{y}), t) \times \mathbf{B}(\mathbf{r}(t, \mathbf{y}), t)\right) \\ \qquad \forall t \in \mathbb{R}, \ \forall \mathbf{y} \in \Omega, \\ \mathbf{B}^*(t, \mathbf{y}) := \left\{d_{\mathbf{y}}\mathbf{r}(t, \mathbf{y})\right\}^{-1} \cdot \mathbf{B}(\mathbf{r}(t, \mathbf{y}), t) \qquad \forall t \in \mathbb{R}, \ \forall \mathbf{y} \in \Omega, \\ \mathbf{D}^*(t, \mathbf{y}) := \left\{d_{\mathbf{y}}\mathbf{r}(t, \mathbf{y})\right\}^{-1} \cdot \mathbf{D}(\mathbf{r}(t, \mathbf{y}), t) \qquad \forall t \in \mathbb{R}, \ \forall \mathbf{y} \in \Omega, \\ \mathbf{H}^*(t, \mathbf{y}) := \left\{d_{\mathbf{y}}\mathbf{r}(t, \mathbf{y})\right\}^{-1} \cdot \left(\mathbf{H}(\mathbf{r}(t, \mathbf{y}), t) - \frac{1}{c}\mathbf{u}(\mathbf{r}(t, \mathbf{y}), t) \times \mathbf{D}(\mathbf{r}(t, \mathbf{y}), t)\right) \\ \qquad \forall t \in \mathbb{R}, \ \forall \mathbf{y} \in \Omega, \end{cases} \tag{1.640}$$

where $\mathbf{r}(t, \mathbf{y})$ is given by (12.68). Then, since $\mathbf{E}^*$, $\mathbf{D}^*$, $\mathbf{B}^*$ and $\mathbf{H}^*$ are real vectors, we could write them as Fourier's Transform on the time variable:

$$\mathbf{E}^*(t,\mathbf{y}) = Re\left\{2\int_0^{+\infty} \hat{\mathbf{E}}^*(\tilde{\omega},\mathbf{y})e^{i\tilde{\omega}t}d\tilde{\omega}\right\}$$
$$\text{where}\;\; \hat{\mathbf{E}}^*(\tilde{\omega},\mathbf{y}) := \frac{1}{2\pi}\int_{-\infty}^{+\infty} \mathbf{E}^*(t,\mathbf{y})e^{-i\tilde{\omega}t}dt\,, \quad (1.641)$$

$$\mathbf{B}^*(t,\mathbf{y}) = Re\left\{2\int_0^{+\infty} \hat{\mathbf{B}}^*(\tilde{\omega},\mathbf{y})e^{i\tilde{\omega}t}d\tilde{\omega}\right\}$$
$$\text{where}\;\; \hat{\mathbf{B}}^*(\tilde{\omega},\mathbf{y}) := \frac{1}{2\pi}\int_{-\infty}^{+\infty} \mathbf{B}^*(t,\mathbf{y})e^{-i\tilde{\omega}t}dt\,, \quad (1.642)$$

$$\mathbf{D}^*(t,\mathbf{y}) = Re\left\{2\int_0^{+\infty} \hat{\mathbf{D}}^*(\tilde{\omega},\mathbf{y})e^{i\tilde{\omega}t}d\tilde{\omega}\right\}$$
$$\text{where}\;\; \hat{\mathbf{D}}^*(\tilde{\omega},\mathbf{y}) := \frac{1}{2\pi}\int_{-\infty}^{+\infty} \mathbf{D}^*(t,\mathbf{y})e^{-i\tilde{\omega}t}dt\,, \quad (1.643)$$

$$\mathbf{H}^*(t,\mathbf{y}) = Re\left\{2\int_0^{+\infty} \hat{\mathbf{H}}^*(\tilde{\omega},\mathbf{y})e^{i\tilde{\omega}t}d\tilde{\omega}\right\}$$
$$\text{where}\;\; \hat{\mathbf{H}}^*(\tilde{\omega},\mathbf{y}) := \frac{1}{2\pi}\int_{-\infty}^{+\infty} \mathbf{H}^*(t,\mathbf{y})e^{-i\tilde{\omega}t}dt\,. \quad (1.644)$$

Then, alternatively to (1.631) and (1.636) we could consider

$$\mathbf{P}\left(\mathbf{r}(t,\mathbf{y}),t\right) = Re\left\{2\int_0^{+\infty} \frac{n^2\left(\tilde{\omega},\mathbf{r}(t,\mathbf{y}),t\right)-1}{4\pi}\left(d_{\mathbf{y}}\mathbf{r}\left(t,\mathbf{y}\right)\cdot\hat{\mathbf{E}}^*\left(\omega,\mathbf{x}\right)\right)e^{i\omega t}d\omega\right\}$$
$$- Re\left\{2\int_0^{+\infty} \kappa\left(\tilde{\omega},\mathbf{r}(t,\mathbf{y}),t\right)\left(d_{\mathbf{y}}\mathbf{r}\left(t,\mathbf{y}\right)\cdot\hat{\mathbf{D}}^*\left(\omega,\mathbf{x}\right)\right)e^{i\omega t}d\omega\right\}$$
$$\text{and}\qquad \mathbf{M}\left(\mathbf{r}(t,\mathbf{y}),t\right) + \frac{1}{c}\mathbf{u}\left(\mathbf{r}(t,\mathbf{y}),t\right)\times\mathbf{P}\left(\mathbf{r}(t,\mathbf{y}),t\right) =$$
$$Re\left\{2\int_0^{+\infty} \kappa\left(\tilde{\omega},\mathbf{r}(t,\mathbf{y}),t\right)\left(d_{\mathbf{y}}\mathbf{r}\left(t,\mathbf{y}\right)\cdot\hat{\mathbf{H}}^*\left(\tilde{\omega},\mathbf{y}\right)\right)e^{i\tilde{\omega}t}d\tilde{\omega}\right\}, \quad (1.645)$$

i.e.

$$\mathbf{P}(\mathbf{x},t) = Re\left\{2\int_0^{+\infty}\frac{n^2(\tilde{\omega},\mathbf{x},t)-1}{4\pi}\left(\{d_\mathbf{x}\mathbf{f}(\mathbf{x},t)\}^{-1}\cdot\hat{\mathbf{E}}^*(\tilde{\omega},\mathbf{f}(\mathbf{x},t))\right)e^{i\tilde{\omega}t}d\tilde{\omega}\right\}$$
$$-Re\left\{2\int_0^{+\infty}\kappa(\tilde{\omega},\mathbf{x},t)\left(\{d_\mathbf{x}\mathbf{f}(\mathbf{x},t)\}^{-1}\cdot\hat{\mathbf{D}}^*(\tilde{\omega},\mathbf{f}(\mathbf{x},t))\right)e^{i\tilde{\omega}t}d\tilde{\omega}\right\} \quad \text{and}$$
$$\mathbf{M}(\mathbf{x},t)+\frac{1}{c}\mathbf{u}(\mathbf{x},t)\times\mathbf{P}(\mathbf{x},t)=$$
$$Re\left\{2\int_0^{+\infty}\kappa(\tilde{\omega},\mathbf{x},t)\left(\{d_\mathbf{x}\mathbf{f}(\mathbf{x},t)\}^{-1}\cdot\hat{\mathbf{H}}^*(\tilde{\omega},\mathbf{f}(\mathbf{x},t))\right)e^{i\tilde{\omega}t}d\tilde{\omega}\right\}. \tag{1.646}$$

Then, exactly in the same way we already proved for (1.631) and (1.636) or (1.637) we can also prove that the above alternative (1.640) and (1.645) or (1.646) is invariant under the change of inertial or non-inertial Cartesian coordinate system. Furthermore, as before, the above alternative is equivalent to (1.627) in the particular case $\mathbf{u} = 0$. Moreover, in the case of an arbitrary moving transparent medium without dispersion the alternative law is equivalent to (1.607). Finally, as before, (1.640) and (1.645) or (1.646) is also independent on the initial instant of time t_0.

We are unable to see any clear advantage of the dispersion law in the form of (1.631) and (1.636) or (1.637) with respect to the law (1.640) and (1.645) or (1.646). We just note that the above two forms of the dispersion law coincide in the case where our medium moves as a rigid body.

See also Sect. 12.7.1 to the generalization of the dispersion laws to the case of anisotropic mediums.

1.14 Some Further Consequences of Maxwell Equations

Again consider the system of Maxwell equations in the vacuum or in a medium, without dispersion, having the form (1.616):

$$\begin{cases}\text{curl}_\mathbf{x}\mathbf{H} = \frac{4\pi}{c}\mathbf{j}+\frac{1}{c}\frac{\partial\mathbf{D}}{\partial t},\\ \text{div}_\mathbf{x}\mathbf{D} = 4\pi\rho,\\ \text{curl}_\mathbf{x}\mathbf{E}+\frac{1}{c}\frac{\partial\mathbf{B}}{\partial t}=0,\\ \text{div}_\mathbf{x}\mathbf{B}=0\\ \mathbf{E}=\gamma_0\mathbf{D}-\frac{1}{c}\tilde{\mathbf{u}}\times\mathbf{B}\\ \mathbf{H}=\kappa_0\mathbf{B}+\frac{1}{c}\tilde{\mathbf{u}}\times\mathbf{D},\\ \tilde{\mathbf{u}}=(\gamma_0\kappa_0\mathbf{v}+(1-\gamma_0\kappa_0)\mathbf{u}),\end{cases} \tag{1.647}$$

where $\gamma_0 \neq 0$ and $\kappa_0 \neq 0$ are material coefficients, $\mathbf{v}$ is the vectorial gravitational potential $\mathbf{u}$ is the medium velocity and $\tilde{\mathbf{u}} = (\gamma_0\kappa_0\mathbf{v} + (1-\gamma_0\kappa_0)\mathbf{u})$ is the speed-like vector field. Remind that in the case of the vacuum we have $\gamma_0 = \kappa_0 = 1$, $\tilde{\mathbf{u}} = \mathbf{v}$ and

Eq. (1.647) are precise (in the frames of our model). Otherwise, in the case $\gamma_0\kappa_0 \neq 1$ Eq. (1.647) are just an approximation that is good enough for the case:

$$\frac{|\kappa_0||1-\gamma_0\kappa_0|\cdot|\mathbf{u}-\mathbf{v}|^2}{c^2} \ll 1. \tag{1.648}$$

Throughout this section we study equation (1.647) in domains where we assume that the coefficients $\gamma_0 \neq 0$ and $\kappa_0 \neq 0$ vary sufficiently slow on the place and time and thus their spatial and temporal derivatives are negligible with respect to other terms. Next again by the third and the fourth equations in (1.647) we can write

$$\begin{cases} \mathbf{B} = \operatorname{curl}_{\mathbf{x}}\mathbf{A}, \\ \mathbf{E} = -\nabla_{\mathbf{x}}\Psi - \frac{1}{c}\frac{\partial \mathbf{A}}{\partial t}, \end{cases} \tag{1.649}$$

where Ψ and $\mathbf{A}$ are the usual scalar and the vectorial electromagnetic potentials. Then by (1.649) and (1.647) we have

$$\begin{cases} \mathbf{B} = \operatorname{curl}_{\mathbf{x}}\mathbf{A} \\ \mathbf{E} = -\nabla_{\mathbf{x}}\Psi - \frac{1}{c}\frac{\partial \mathbf{A}}{\partial t} \\ \mathbf{D} = -\frac{1}{\gamma_0}\nabla_{\mathbf{x}}\Psi - \frac{1}{\gamma_0 c}\frac{\partial \mathbf{A}}{\partial t} + \frac{1}{c\gamma_0}\tilde{\mathbf{u}} \times \operatorname{curl}_{\mathbf{x}}\mathbf{A} \\ \mathbf{H} = \kappa_0 \operatorname{curl}_{\mathbf{x}}\mathbf{A} + \frac{1}{c}\tilde{\mathbf{u}} \times \left(-\frac{1}{\gamma_0}\nabla_{\mathbf{x}}\Psi - \frac{1}{\gamma_0 c}\frac{\partial \mathbf{A}}{\partial t} + \frac{1}{\gamma_0 c}\tilde{\mathbf{u}} \times \operatorname{curl}_{\mathbf{x}}\mathbf{A}\right). \end{cases} \tag{1.650}$$

Next we remind the definition of the proper scalar electromagnetic potential:

$$\Psi_0 := \Psi - \frac{1}{c}\mathbf{A}\cdot\mathbf{v} \tag{1.651}$$

and remind also that $\mathbf{A}$ is a proper vector field and Ψ_0 is a proper scalar field. Then in the case of the medium we also define an additional scalar electromagnetic potential:

$$\Psi_1 := \Psi - \frac{1}{c}\mathbf{A}\cdot\tilde{\mathbf{u}}. \tag{1.652}$$

Then, since $\mathbf{A}$ is a proper vector field, we deduce that Ψ_1 is also a proper scalar field. Moreover, in the case of the vacuum or more generally in the case where $\gamma_0\kappa_0 \approx 1$ we have $\Psi_1 = \Psi_0$. Thus by (1.652) we rewrite (1.650) as

$$\begin{cases} \mathbf{B} = \operatorname{curl}_{\mathbf{x}}\mathbf{A} \\ \mathbf{E} = -\nabla_{\mathbf{x}}\Psi_1 - \frac{1}{c}\frac{\partial \mathbf{A}}{\partial t} - \frac{1}{c}\nabla_{\mathbf{x}}\left(\mathbf{A}\cdot\tilde{\mathbf{u}}\right) \\ \mathbf{D} = -\frac{1}{\gamma_0}\nabla_{\mathbf{x}}\Psi_1 - \frac{1}{\gamma_0 c}\left(\frac{\partial \mathbf{A}}{\partial t} - \tilde{\mathbf{u}} \times \operatorname{curl}_{\mathbf{x}}\mathbf{A} + \nabla_{\mathbf{x}}\left(\mathbf{A}\cdot\tilde{\mathbf{u}}\right)\right) \\ \mathbf{H} = \kappa_0 \operatorname{curl}_{\mathbf{x}}\mathbf{A} - \frac{1}{c}\tilde{\mathbf{u}} \times \left(\frac{1}{\gamma_0}\nabla_{\mathbf{x}}\Psi_1 + \frac{1}{\gamma_0 c}\left(\frac{\partial \mathbf{A}}{\partial t} - \tilde{\mathbf{u}} \times \operatorname{curl}_{\mathbf{x}}\mathbf{A} + \nabla_{\mathbf{x}}\left(\mathbf{A}\cdot\tilde{\mathbf{u}}\right)\right)\right). \end{cases} \tag{1.653}$$

Then by inserting (1.653) into (1.647) straightforward calculations presented in Sect. 13.1 lead to the following equations:

$$\left(\frac{\partial}{\partial t}\left\{\frac{1}{c}\operatorname{div}_{\mathbf{x}}\mathbf{A}\right\}+\operatorname{div}_{\mathbf{x}}\left\{\frac{1}{c}\left(\operatorname{div}_{\mathbf{x}}\mathbf{A}\right)\tilde{\mathbf{u}}\right\}\right)+\Delta_{\mathbf{x}}\Psi_1$$
$$+\operatorname{div}_{\mathbf{x}}\left\{\frac{1}{c}\left(d_{\mathbf{x}}\tilde{\mathbf{u}}+\{d_{\mathbf{x}}\tilde{\mathbf{u}}\}^T\right)\cdot\mathbf{A}-\frac{1}{c}\left(\operatorname{div}_{\mathbf{x}}\tilde{\mathbf{u}}\right)\mathbf{A}\right\}=-4\pi\gamma_0\rho \quad (1.654)$$

and

$$-\Delta_{\mathbf{x}}\mathbf{A}=\frac{4\pi}{\kappa_0 c}\left(\mathbf{j}-\rho\tilde{\mathbf{u}}\right)-\nabla_{\mathbf{x}}\left(\frac{1}{\kappa_0\gamma_0 c}\left(\frac{\partial\Psi_1}{\partial t}+\tilde{\mathbf{u}}\cdot\nabla_{\mathbf{x}}\Psi_1\right)+\left(\operatorname{div}_{\mathbf{x}}\mathbf{A}\right)\right)$$
$$+\left(\left(d_{\mathbf{x}}\tilde{\mathbf{u}}+\{d_{\mathbf{x}}\tilde{\mathbf{u}}\}^T\right)\cdot\left\{\frac{1}{\kappa_0\gamma_0 c}\nabla_{\mathbf{x}}\Psi_1\right\}-\left(\operatorname{div}_{\mathbf{x}}\tilde{\mathbf{u}}\right)\left\{\frac{1}{\kappa_0\gamma_0 c}\nabla_{\mathbf{x}}\Psi_1\right\}\right)$$
$$-\frac{\partial}{\partial t}\left\{\frac{1}{\gamma_0\kappa_0 c^2}\left(\frac{\partial\mathbf{A}}{\partial t}+d_{\mathbf{x}}\mathbf{A}\cdot\tilde{\mathbf{u}}+\{d_{\mathbf{x}}\tilde{\mathbf{u}}\}^T\cdot\mathbf{A}\right)\right\}$$
$$-\operatorname{div}_{\mathbf{x}}\left\{\left(\frac{1}{\gamma_0\kappa_0 c^2}\left(\frac{\partial\mathbf{A}}{\partial t}+d_{\mathbf{x}}\mathbf{A}\cdot\tilde{\mathbf{u}}+\{d_{\mathbf{x}}\tilde{\mathbf{u}}\}^T\cdot\mathbf{A}\right)\right)\otimes\tilde{\mathbf{u}}\right\}$$
$$+\left(d_{\mathbf{x}}\tilde{\mathbf{u}}\cdot\left\{\frac{1}{\gamma_0\kappa_0 c^2}\left(\frac{\partial\mathbf{A}}{\partial t}+d_{\mathbf{x}}\mathbf{A}\cdot\tilde{\mathbf{u}}+\{d_{\mathbf{x}}\tilde{\mathbf{u}}\}^T\cdot\mathbf{A}\right)\right\}\right). \quad (1.655)$$

Alternatively, we rewrite (1.655) as

$$-\Delta_{\mathbf{x}}\mathbf{A}=\frac{4\pi}{\kappa_0 c}\left(\mathbf{j}-\rho\tilde{\mathbf{u}}\right)-\nabla_{\mathbf{x}}\left(\operatorname{div}_{\mathbf{x}}\mathbf{A}\right)$$
$$-\frac{1}{\gamma_0\kappa_0 c}\left(\frac{\partial}{\partial t}\{\nabla_{\mathbf{x}}\Psi_1\}-\operatorname{curl}_{\mathbf{x}}\{\tilde{\mathbf{u}}\times\nabla_{\mathbf{x}}\Psi_1\}+\left(\Delta_{\mathbf{x}}\Psi_1\right)\tilde{\mathbf{u}}\right)$$
$$-\frac{\partial}{\partial t}\left\{\frac{1}{\gamma_0\kappa_0 c^2}\left(\frac{\partial\mathbf{A}}{\partial t}+d_{\mathbf{x}}\mathbf{A}\cdot\tilde{\mathbf{u}}+\{d_{\mathbf{x}}\tilde{\mathbf{u}}\}^T\cdot\mathbf{A}\right)\right\}$$
$$-\operatorname{div}_{\mathbf{x}}\left\{\left(\frac{1}{\gamma_0\kappa_0 c^2}\left(\frac{\partial\mathbf{A}}{\partial t}+d_{\mathbf{x}}\mathbf{A}\cdot\tilde{\mathbf{u}}+\{d_{\mathbf{x}}\tilde{\mathbf{u}}\}^T\cdot\mathbf{A}\right)\right)\otimes\tilde{\mathbf{u}}\right\}$$
$$+\left(d_{\mathbf{x}}\tilde{\mathbf{u}}\cdot\left\{\frac{1}{\gamma_0\kappa_0 c^2}\left(\frac{\partial\mathbf{A}}{\partial t}+d_{\mathbf{x}}\mathbf{A}\cdot\tilde{\mathbf{u}}+\{d_{\mathbf{x}}\tilde{\mathbf{u}}\}^T\cdot\mathbf{A}\right)\right\}\right). \quad (1.656)$$

Next if we assume the following calibration of the potentials:

$$\operatorname{div}_{\mathbf{x}}\mathbf{A}=0, \quad (1.657)$$

then by (1.657), (1.654) and (1.656) we have

$$-\Delta_{\mathbf{x}}\Psi_1=4\pi\gamma_0\rho+\operatorname{div}_{\mathbf{x}}\left\{\frac{1}{c}\left(d_{\mathbf{x}}\tilde{\mathbf{u}}+\{d_{\mathbf{x}}\tilde{\mathbf{u}}\}^T\right)\cdot\mathbf{A}-\frac{1}{c}\left(\operatorname{div}_{\mathbf{x}}\tilde{\mathbf{u}}\right)\mathbf{A}\right\}, \quad (1.658)$$

and

$$-\Delta_{\mathbf{x}}\mathbf{A} = \frac{4\pi}{\kappa_0 c}\left(\mathbf{j} - \frac{1}{4\pi\gamma_0}\left(\frac{\partial}{\partial t}\left\{\nabla_{\mathbf{x}}\Psi_1\right\} - \operatorname{curl}_{\mathbf{x}}\left\{\tilde{\mathbf{u}} \times \nabla_{\mathbf{x}}\Psi_1\right\}\right)\right)$$
$$+ \frac{1}{\gamma_0\kappa_0 c^2}\left(\operatorname{div}_{\mathbf{x}}\left\{\left(d_{\mathbf{x}}\tilde{\mathbf{u}} + \{d_{\mathbf{x}}\tilde{\mathbf{u}}\}^T\right)\cdot\mathbf{A} - (\operatorname{div}_{\mathbf{x}}\tilde{\mathbf{u}})\,\mathbf{A}\right\}\right)\tilde{\mathbf{u}}$$
$$- \frac{\partial}{\partial t}\left\{\frac{1}{\gamma_0\kappa_0 c^2}\left(\frac{\partial \mathbf{A}}{\partial t} + d_{\mathbf{x}}\mathbf{A}\cdot\tilde{\mathbf{u}} + \{d_{\mathbf{x}}\tilde{\mathbf{u}}\}^T\cdot\mathbf{A}\right)\right\}$$
$$- \operatorname{div}_{\mathbf{x}}\left\{\left(\frac{1}{\gamma_0\kappa_0 c^2}\left(\frac{\partial \mathbf{A}}{\partial t} + d_{\mathbf{x}}\mathbf{A}\cdot\tilde{\mathbf{u}} + \{d_{\mathbf{x}}\tilde{\mathbf{u}}\}^T\cdot\mathbf{A}\right)\right)\otimes\tilde{\mathbf{u}}\right\}$$
$$+ \left(d_{\mathbf{x}}\tilde{\mathbf{u}}\cdot\left\{\frac{1}{\gamma_0\kappa_0 c^2}\left(\frac{\partial \mathbf{A}}{\partial t} + d_{\mathbf{x}}\mathbf{A}\cdot\tilde{\mathbf{u}} + \{d_{\mathbf{x}}\tilde{\mathbf{u}}\}^T\cdot\mathbf{A}\right)\right\}\right), \tag{1.659}$$

(See Sect. 13.1 for details).

On the other hand, if we assume the following alternative calibration of the potentials:

$$\frac{1}{\kappa_0\gamma_0 c}\left(\frac{\partial \Psi_1}{\partial t} + \tilde{\mathbf{u}}\cdot\nabla_{\mathbf{x}}\Psi_1\right) + \operatorname{div}_{\mathbf{x}}\mathbf{A} = 0, \tag{1.660}$$

then by (1.654), (1.655) we have

$$\left(\frac{\partial}{\partial t}\left\{\frac{1}{\kappa_0\gamma_0 c^2}\left(\frac{\partial \Psi_1}{\partial t} + \tilde{\mathbf{u}}\cdot\nabla_{\mathbf{x}}\Psi_1\right)\right\} + \operatorname{div}_{\mathbf{x}}\left\{\frac{1}{\kappa_0\gamma_0 c^2}\left(\frac{\partial \Psi_1}{\partial t} + \tilde{\mathbf{u}}\cdot\nabla_{\mathbf{x}}\Psi_1\right)\tilde{\mathbf{u}}\right\}\right) - \Delta_{\mathbf{x}}\Psi_1$$
$$= 4\pi\gamma_0\rho + \operatorname{div}_{\mathbf{x}}\left\{\frac{1}{c}\left(d_{\mathbf{x}}\tilde{\mathbf{u}} + \{d_{\mathbf{x}}\tilde{\mathbf{u}}\}^T\right)\cdot\mathbf{A} - \frac{1}{c}(\operatorname{div}_{\mathbf{x}}\tilde{\mathbf{u}})\,\mathbf{A}\right\}, \tag{1.661}$$

and

$$-\Delta_{\mathbf{x}}\mathbf{A} =$$
$$\frac{4\pi}{\kappa_0 c}(\mathbf{j} - \rho\tilde{\mathbf{u}}) + \left(\left(d_{\mathbf{x}}\tilde{\mathbf{u}} + \{d_{\mathbf{x}}\tilde{\mathbf{u}}\}^T\right)\cdot\left\{\frac{1}{\kappa_0\gamma_0 c}\nabla_{\mathbf{x}}\Psi_1\right\} - (\operatorname{div}_{\mathbf{x}}\tilde{\mathbf{u}})\left\{\frac{1}{\kappa_0\gamma_0 c}\nabla_{\mathbf{x}}\Psi_1\right\}\right)$$
$$- \frac{\partial}{\partial t}\left\{\frac{1}{\gamma_0\kappa_0 c^2}\left(\frac{\partial \mathbf{A}}{\partial t} + d_{\mathbf{x}}\mathbf{A}\cdot\tilde{\mathbf{u}} + \{d_{\mathbf{x}}\tilde{\mathbf{u}}\}^T\cdot\mathbf{A}\right)\right\}$$
$$- \operatorname{div}_{\mathbf{x}}\left\{\left(\frac{1}{\gamma_0\kappa_0 c^2}\left(\frac{\partial \mathbf{A}}{\partial t} + d_{\mathbf{x}}\mathbf{A}\cdot\tilde{\mathbf{u}} + \{d_{\mathbf{x}}\tilde{u}\}^T\cdot\mathbf{A}\right)\right)\otimes\tilde{\mathbf{u}}\right\}$$
$$+ \left(d_{\mathbf{x}}\tilde{\mathbf{u}}\cdot\left\{\frac{1}{\gamma_0\kappa_0 c^2}\left(\frac{\partial \mathbf{A}}{\partial t} + d_{\mathbf{x}}\mathbf{A}\cdot\tilde{\mathbf{u}} + \{d_{\mathbf{x}}\tilde{\mathbf{u}}\}^T\cdot\mathbf{A}\right)\right\}\right). \tag{1.662}$$

Next, from now we assume that $\tilde{\mathbf{u}}$ varies sufficiently slowly in space and time variables, so that the following approximation is valid:

$$\frac{\left|d_{\mathbf{x}}\tilde{\mathbf{u}}+\{d_{\mathbf{x}}\tilde{\mathbf{u}}\}^T\right|}{c} \ll \frac{|\nabla_{\mathbf{x}}\Psi_1|}{|\mathbf{A}|+|\Psi_1|}+\frac{|\gamma_0\kappa_0||\Delta_{\mathbf{x}}\mathbf{A}|}{|d_{\mathbf{x}}\mathbf{A}|+|\nabla_{\mathbf{x}}\Psi_1|}. \tag{1.663}$$

In particular, if in some Cartesian coordinate system we have both

$$\frac{\left|d_{\mathbf{x}}\tilde{\mathbf{u}}+\{d_{\mathbf{x}}\tilde{\mathbf{u}}\}^T\right|^2}{|\tilde{\mathbf{u}}|^2} \ll \left(\frac{|\nabla_{\mathbf{x}}\Psi_1|}{|\mathbf{A}|+|\Psi_1|}\right)^2+\left(\frac{|\gamma_0\kappa_0||\Delta_{\mathbf{x}}\mathbf{A}|}{|d_{\mathbf{x}}\mathbf{A}|+|\nabla_{\mathbf{x}}\Psi_1|}\right)^2, \tag{1.664}$$

and

$$\frac{|\tilde{\mathbf{u}}|^2}{c^2} \ll 1, \tag{1.665}$$

then (1.663) indeed holds! Furthermore, taking into the account (1.663), under the calibration (1.657), we rewrite (1.658) and (1.659) as

$$-\Delta_{\mathbf{x}}\Psi_1 \approx 4\pi\gamma_0\rho, \tag{1.666}$$

and

$$\begin{aligned}
-\Delta_{\mathbf{x}}\mathbf{A} \approx \frac{4\pi}{\kappa_0 c}\Bigg(\mathbf{j}-\frac{1}{4\pi\gamma_0}\left(\frac{\partial}{\partial t}\{\nabla_{\mathbf{x}}\Psi_1\}-\operatorname{curl}_{\mathbf{x}}\{\tilde{\mathbf{u}}\times\nabla_{\mathbf{x}}\Psi_1\}\right)\Bigg)\\
-\frac{\partial}{\partial t}\left\{\frac{1}{\gamma_0\kappa_0 c^2}\left(\frac{\partial\mathbf{A}}{\partial t}+d_{\mathbf{x}}\mathbf{A}\cdot\tilde{\mathbf{u}}+\{d_{\mathbf{x}}\tilde{\mathbf{u}}\}^T\cdot\mathbf{A}\right)\right\}\\
-\operatorname{div}_{\mathbf{x}}\left\{\left(\frac{1}{\gamma_0\kappa_0 c^2}\left(\frac{\partial\mathbf{A}}{\partial t}+d_{\mathbf{x}}\mathbf{A}\cdot\tilde{\mathbf{u}}+\{d_{\mathbf{x}}\tilde{\mathbf{u}}\}^T\cdot\mathbf{A}\right)\right)\otimes\tilde{\mathbf{u}}\right\}\\
+\left(d_{\mathbf{x}}\tilde{\mathbf{u}}\cdot\left\{\frac{1}{\gamma_0\kappa_0 c^2}\left(\frac{\partial\mathbf{A}}{\partial t}+d_{\mathbf{x}}\mathbf{A}\cdot\tilde{\mathbf{u}}+\{d_{\mathbf{x}}\tilde{\mathbf{u}}\}^T\cdot\mathbf{A}\right)\right\}\right).
\end{aligned} \tag{1.667}$$

Note that, using Proposition 1.1 we deduce that the approximate equations (1.666) and (1.667) are still invariant under the change of inertial or non-inertial Cartesian coordinate system, provided that $\mathbf{A}$ is a proper vector field and Ψ_1 is a proper scalar field.

On the other hand, taking into the account (1.663), under the calibration (1.660), we rewrite (1.661) and (1.662) as

$$\frac{1}{\kappa_0\gamma_0 c^2}\left(\frac{\partial}{\partial t}\left(\frac{\partial\Psi_1}{\partial t}+\tilde{\mathbf{u}}\cdot\nabla_{\mathbf{x}}\Psi_1\right)+\operatorname{div}_{\mathbf{x}}\left\{\left(\frac{\partial\Psi_1}{\partial t}+\tilde{\mathbf{u}}\cdot\nabla_{\mathbf{x}}\Psi_1\right)\tilde{\mathbf{u}}\right\}\right)-\Delta_{\mathbf{x}}\Psi_1\approx 4\pi\gamma_0\rho \tag{1.668}$$

and

$$-\Delta_{\mathbf{x}}\mathbf{A} \approx \frac{4\pi}{\kappa_0 c}\left(\mathbf{j}-\rho\tilde{\mathbf{u}}\right) - \frac{\partial}{\partial t}\left\{\frac{1}{\gamma_0\kappa_0 c^2}\left(\frac{\partial \mathbf{A}}{\partial t} + d_{\mathbf{x}}\mathbf{A}\cdot\tilde{\mathbf{u}} + \{d_{\mathbf{x}}\tilde{\mathbf{u}}\}^T\cdot\mathbf{A}\right)\right\}$$
$$-\operatorname{div}_{\mathbf{x}}\left\{\left(\frac{1}{\gamma_0\kappa_0 c^2}\left(\frac{\partial \mathbf{A}}{\partial t} + d_{\mathbf{x}}\mathbf{A}\cdot\tilde{\mathbf{u}} + \{d_{\mathbf{x}}\tilde{\mathbf{u}}\}^T\cdot\mathbf{A}\right)\right)\otimes\tilde{\mathbf{u}}\right\}$$
$$+\left(d_{\mathbf{x}}\tilde{\mathbf{u}}\cdot\left\{\frac{1}{\gamma_0\kappa_0 c^2}\left(\frac{\partial \mathbf{A}}{\partial t} + d_{\mathbf{x}}\mathbf{A}\cdot\tilde{\mathbf{u}} + \{d_{\mathbf{x}}\tilde{\mathbf{u}}\}^T\cdot\mathbf{A}\right)\right\}\right). \qquad (1.669)$$

Again note that, using Proposition 1.1 we deduce that the approximate Eqs. (1.668) and (1.669) are still invariant under the change of inertial or non-inertial Cartesian coordinate system, provided that $\mathbf{A}$ is a proper vector field and Ψ_1 is a proper scalar field.

In particular, assume that in some Cartesian coordinate system (∗) we have the following stronger than (1.663) approximation:

$$\frac{|d_{\mathbf{x}}\tilde{\mathbf{u}}|}{c} \ll \frac{|\gamma_0\kappa_0||\Delta_{\mathbf{x}}\mathbf{A}| + \frac{1}{c}\left|d_{\mathbf{x}}\left\{\frac{\partial \mathbf{A}}{\partial t}\right\}\right| + \frac{1}{c^2}\left|\frac{\partial^2 \mathbf{A}}{\partial t^2}\right| + |\gamma_0\kappa_0||\nabla^2_{\mathbf{x}}\Psi_1| + \frac{1}{c}\left|\frac{\partial}{\partial t}\left(\nabla_{\mathbf{x}}\Psi_1\right)\right| + \frac{1}{c^2}\left|\frac{\partial^2\Psi_1}{\partial t^2}\right|}{|d_{\mathbf{x}}\mathbf{A}| + \frac{1}{c}\left|\frac{\partial \mathbf{A}}{\partial t}\right| + |\nabla_{\mathbf{x}}\Psi_1| + \frac{1}{c}\left|\frac{\partial \Psi_1}{\partial t}\right|}, \qquad (1.670)$$

i.e. the field $\tilde{\mathbf{u}}$ changes in space much slower than $(\Psi_1, \mathbf{A})$. Estimation (1.670) holds especially good for the electromagnetic waves of high frequency, for example, for the visible light. However, (1.670) is still well for almost every electromagnetic field we meet in the common life, except probably the magnetic field of the Earth. Then, by (1.668), (1.669) and (1.670) we can write the further approximating equations:

$$\left(\frac{\partial}{\partial t}\left\{\frac{1}{c_0^2}\left(\frac{\partial \Psi_1}{\partial t} + \tilde{\mathbf{u}}\cdot\nabla_{\mathbf{x}}\Psi_1\right)\right\} + \operatorname{div}_{\mathbf{x}}\left\{\frac{1}{c_0^2}\left(\frac{\partial \Psi_1}{\partial t} + \tilde{\mathbf{u}}\cdot\nabla_{\mathbf{x}}\Psi_1\right)\tilde{\mathbf{u}}\right\}\right) - \Delta_{\mathbf{x}}\Psi_1$$
$$\approx 4\pi\gamma_0\rho, \qquad (1.671)$$

and

$$\left(\frac{\partial}{\partial t}\left\{\frac{1}{c_0^2}\left(\frac{\partial \mathbf{A}}{\partial t} + d_{\mathbf{x}}\mathbf{A}\cdot\tilde{\mathbf{u}}\right)\right\} + \operatorname{div}_{\mathbf{x}}\left\{\frac{1}{c_0^2}\left(\frac{\partial \mathbf{A}}{\partial t} + d_{\mathbf{x}}\mathbf{A}\cdot\tilde{\mathbf{u}}\right)\otimes\tilde{\mathbf{u}}\right\}\right) - \Delta_{\mathbf{x}}\mathbf{A}$$
$$\approx \frac{4\pi}{\kappa_0 c}\left(\mathbf{j}-\rho\tilde{\mathbf{u}}\right), \qquad (1.672)$$

where the scalar quantity c_0, defined by

$$c_0 = c\sqrt{\kappa_0\gamma_0} \qquad (1.673)$$

is called speed of light in the medium. Note that, although the approximate Eq. (1.671) and (1.672) are invariant under the Galilean Transformation, they are not invariant under the more general change of non-inertial Cartesian coordinate system. However, (1.671) and (1.672) are more convenient than (1.668) and (1.669), since the scalar potential Ψ_1 and every of the three scalar components of the vector potential $\mathbf{A}$ in (1.671) and (1.672) satisfies four decoupled equations of the same type, that differ only by the right parts. On the other hand, if we consider some three proper vector fields $\mathbf{e}_1 := \mathbf{e}_1(\mathbf{x}, t)$, $\mathbf{e}_2 := \mathbf{e}_2(\mathbf{x}, t)$, and $\mathbf{e}_3 := \mathbf{e}_3(\mathbf{x}, t)$, which are mutually orthogonal to each other and satisfy the following approximation:

$$\frac{|d_{\mathbf{x}}\mathbf{e}_1| + c_0|\partial_t \mathbf{e}_1|}{|\mathbf{e}_1|} + \frac{|d_{\mathbf{x}}\mathbf{e}_2| + c_0|\partial_t \mathbf{e}_2|}{|\mathbf{e}_2|} + \frac{|d_{\mathbf{x}}\mathbf{e}_3| + c_0|\partial_t \mathbf{e}_3|}{|\mathbf{e}_3|} \ll \frac{|\nabla_{\mathbf{x}}\Psi_1| + c_0|\partial_t \Psi_1| + |d_{\mathbf{x}}\mathbf{A}| + c_0|\partial_t \mathbf{A}|}{|\Psi_1| + |\mathbf{A}|} \tag{1.674}$$

in the given before coordinate system $(*)$, i.e. the field $\mathbf{e}_k$ vary in space and time much weaker than $(\Psi_1, \mathbf{A})$, then we may write the alternative to (1.672) and (1.671) approximate equations in the form of four decoupled scalar invariant wave equations of the same type:

$$\frac{\partial}{\partial t}\left\{\frac{1}{c_0^2}\left(\frac{\partial}{\partial t}(\mathbf{e}_k \cdot \mathbf{A}) + \tilde{\mathbf{u}} \cdot \nabla_{\mathbf{x}}(\mathbf{e}_k \cdot \mathbf{A})\right)\right\} + \operatorname{div}_{\mathbf{x}}\left\{\frac{1}{c_0^2}\left(\frac{\partial}{\partial t}(\mathbf{e}_k \cdot \mathbf{A}) + \tilde{\mathbf{u}} \cdot \nabla_{\mathbf{x}}(\mathbf{e}_k \cdot \mathbf{A})\right)\tilde{\mathbf{u}}\right\} - \Delta_{\mathbf{x}}(\mathbf{e}_k \cdot \mathbf{A}) \approx \frac{4\pi}{\kappa_0 c}(\mathbf{j} - \rho\tilde{\mathbf{u}}) \cdot \mathbf{e}_k \quad \forall k = 1, 2, 3, \tag{1.675}$$

and

$$\left(\frac{\partial}{\partial t}\left\{\frac{1}{c_0^2}\left(\frac{\partial \Psi_1}{\partial t} + \tilde{\mathbf{u}} \cdot \nabla_{\mathbf{x}}\Psi_1\right)\right\} + \operatorname{div}_{\mathbf{x}}\left\{\frac{1}{c_0^2}\left(\frac{\partial \Psi_1}{\partial t} + \tilde{\mathbf{u}} \cdot \nabla_{\mathbf{x}}\Psi_1\right)\tilde{\mathbf{u}}\right\}\right) - \Delta_{\mathbf{x}}\Psi_1 \approx 4\pi\gamma_0\rho. \tag{1.676}$$

Then, clearly, the new alternative approximate Eqs. (1.675), (1.676) are indeed invariant under the more general change of non-inertial Cartesian coordinate system. So we can use approximate Eqs. (1.675) and (1.676) in the arbitrary Cartesian coordinate system $(*)$ even if (1.670) and (1.674) are not satisfied in the system $(*)$, provided that (1.670) and (1.674) are satisfied in another Cartesian system $(**)$.

In the absence of charges and currents (for example, for electromagnetic waves) Eqs. (1.671) and (1.672) become

$$\left(\frac{\partial}{\partial t}\left\{\frac{1}{c_0^2}\left(\frac{\partial \Psi_1}{\partial t} + \tilde{\mathbf{u}} \cdot \nabla_{\mathbf{x}}\Psi_1\right)\right\} + \operatorname{div}_{\mathbf{x}}\left\{\frac{1}{c_0^2}\left(\frac{\partial \Psi_1}{\partial t} + \tilde{\mathbf{u}} \cdot \nabla_{\mathbf{x}}\Psi_1\right)\tilde{\mathbf{u}}\right\}\right) - \Delta_{\mathbf{x}}\Psi_1 = 0, \tag{1.677}$$

and

$$\left(\left\{\frac{1}{c_0^2}\frac{\partial}{\partial t}\left(\frac{\partial \mathbf{A}}{\partial t}+d_{\mathbf{x}}\mathbf{A}\cdot\tilde{\mathbf{u}}\right)\right\}+\operatorname{div}_{\mathbf{x}}\left\{\frac{1}{c_0^2}\left(\frac{\partial \mathbf{A}}{\partial t}+d_{\mathbf{x}}\mathbf{A}\cdot\tilde{\mathbf{u}}\right)\otimes\tilde{\mathbf{u}}\right\}\right)-\Delta_{\mathbf{x}}\mathbf{A}=0, \tag{1.678}$$

and Eqs. (1.675), (1.676) become

$$\begin{aligned}\frac{\partial}{\partial t}\left\{\frac{1}{c_0^2}\left(\frac{\partial}{\partial t}\left(\mathbf{e}_k\cdot\mathbf{A}\right)+\tilde{\mathbf{u}}\cdot\nabla_{\mathbf{x}}\left(\mathbf{e}_k\cdot\mathbf{A}\right)\right)\right\}\\+\operatorname{div}_{\mathbf{x}}\left\{\frac{1}{c_0^2}\left(\frac{\partial}{\partial t}\left(\mathbf{e}_k\cdot\mathbf{A}\right)+\tilde{\mathbf{u}}\cdot\nabla_{\mathbf{x}}\left(\mathbf{e}_k\cdot\mathbf{A}\right)\right)\tilde{\mathbf{u}}\right\}\\-\Delta_{\mathbf{x}}\left(\mathbf{e}_k\cdot\mathbf{A}\right)\approx 0\quad\forall k=1,2,3,\end{aligned} \tag{1.679}$$

and

$$\left(\frac{\partial}{\partial t}\left\{\frac{1}{c_0^2}\left(\frac{\partial \Psi_1}{\partial t}+\tilde{\mathbf{u}}\cdot\nabla_{\mathbf{x}}\Psi_1\right)\right\}+\operatorname{div}_{\mathbf{x}}\left\{\frac{1}{c_0^2}\left(\frac{\partial \Psi_1}{\partial t}+\tilde{\mathbf{u}}\cdot\nabla_{\mathbf{x}}\Psi_1\right)\tilde{\mathbf{u}}\right\}\right)-\Delta_{\mathbf{x}}\Psi_1\approx 0. \tag{1.680}$$

Therefore, by (1.653), differentiating (1.677) and (1.678) or (1.679) and (1.680) and further usage of (1.670) and (1.674) gives that if the scalar field $U := U(\mathbf{x}, t)$ is one of any three scalar components of every of the fields **E**, **B**, **D** or **H**, then U satisfies the following approximate scalar equation of the wave type:

$$\left(\frac{\partial}{\partial t}\left\{\frac{1}{c_0^2}\left(\frac{\partial U}{\partial t}+\tilde{\mathbf{u}}\cdot\nabla_{\mathbf{x}}U\right)\right\}+\operatorname{div}_{\mathbf{x}}\left\{\frac{1}{c_0^2}\left(\frac{\partial U}{\partial t}+\tilde{\mathbf{u}}\cdot\nabla_{\mathbf{x}}U\right)\tilde{\mathbf{u}}\right\}\right)-\Delta_{\mathbf{x}}U\approx 0, \tag{1.681}$$

where

$$\tilde{\mathbf{u}}=\left(\gamma_0\kappa_0\mathbf{v}+\left(1-\gamma_0\kappa_0\right)\mathbf{u}\right). \tag{1.682}$$

Note that the wave equation (1.681) is of the same type as (1.554), (1.563), (1.591) or (1.592), with the only difference that the field $\tilde{\mathbf{u}}$ appear in (1.681) instead of the field $\mathbf{u}_0$ in (1.554), (1.563). (1.591) or (1.592).

1.14.1 The Case of Quasistationary Electromagnetic Fields Inside a Slowly Moving Medium in a Weak Gravitational Field

Assume that in the given inertial or non-inertial Cartesian coordinate system (∗) the field $\tilde{\mathbf{u}}$ is weak, meaning that at any instant on every point:

$$\frac{1}{\kappa_0\gamma_0}\frac{|\tilde{\mathbf{u}}|^2}{c^2}\ll 1. \tag{1.683}$$

Here $\tilde{\mathbf{u}} = (\gamma_0\kappa_0\mathbf{v} + (1 - \gamma_0\kappa_0)\mathbf{u})$ is the speed-like vector field, where $\mathbf{v}$ is a vectorial gravitational potential in the system $(*)$ and $\mathbf{u}$ is the medium velocity. Furthermore, consider quasistationary electromagnetic fields. This means the following: assume that the changes in time of the physical characteristics of the electromagnetic fields become essential after certain interval of time T_e and the changes in space of the physical characteristics of the fields become essential in the spatial landscape L_e. Then we assume that

$$(\kappa_0\gamma_0)\frac{c^2T_e^2}{L_e^2} \gg 1. \tag{1.684}$$

Next assume that we are under the calibration (1.657). Then by (1.683) and (1.684) we rewrite (1.658) and (1.659) as

$$-\Delta_{\mathbf{x}}\Psi_1 = 4\pi\gamma_0\rho + \frac{1}{c}\,\mathrm{div}_{\mathbf{x}}\left\{\left(d_{\mathbf{x}}\tilde{\mathbf{u}} + \{d_{\mathbf{x}}\tilde{\mathbf{u}}\}^T\right)\cdot\mathbf{A} - (\mathrm{div}_{\mathbf{x}}\,\tilde{\mathbf{u}})\,\mathbf{A}\right\}, \tag{1.685}$$

and

$$-\Delta_{\mathbf{x}}\mathbf{A} \approx \frac{4\pi}{\kappa_0 c}(\mathbf{j} - \rho\tilde{\mathbf{u}}) - \frac{1}{\kappa_0\gamma_0 c}\left(\frac{\partial}{\partial t}(\nabla_{\mathbf{x}}\Psi_1) - \mathrm{curl}_{\mathbf{x}}(\tilde{\mathbf{u}}\times\nabla_{\mathbf{x}}\Psi_1) + (\Delta_{\mathbf{x}}\Psi_1)\,\tilde{\mathbf{u}}\right). \tag{1.686}$$

Moreover, by (1.683) and (1.684) we can perform further approximation of (1.686) and we get

$$-\Delta_{\mathbf{x}}\mathbf{A} \approx \frac{4\pi}{\kappa_0 c}\mathbf{j} - \frac{1}{\kappa_0 c}\left(\frac{\partial}{\partial t}(\nabla_{\mathbf{x}}\psi_0) - \mathrm{curl}_{\mathbf{x}}(\mathbf{v}\times\nabla_{\mathbf{x}}\psi_0)\right), \tag{1.687}$$

where $\psi_0(\mathbf{x}, t)$ is the classical Coulomb's potential which satisfies

$$-\Delta_{\mathbf{x}}\psi_0 = 4\pi\rho. \tag{1.688}$$

So we rewrite (1.685) and (1.687) as

$$\begin{cases} -\Delta_{\mathbf{x}}\mathbf{A} \approx \frac{4\pi}{\kappa_0 c}\tilde{\mathbf{j}}, \\ -\Delta_{\mathbf{x}}\Psi_1 = 4\pi\gamma_0\rho + \frac{1}{c}\,\mathrm{div}_{\mathbf{x}}\left\{\left(d_{\mathbf{x}}\tilde{\mathbf{u}} + \{d_{\mathbf{x}}\tilde{\mathbf{u}}\}^T\right)\cdot\mathbf{A} - (\mathrm{div}_{\mathbf{x}}\,\tilde{\mathbf{u}})\,\mathbf{A}\right\}, \end{cases} \tag{1.689}$$

where we set the reduced current:

$$\begin{cases} \tilde{\mathbf{j}} := \mathbf{j} - \frac{1}{4\pi}\frac{\partial}{\partial t}(\nabla_{\mathbf{x}}\psi_0) + \frac{1}{4\pi}\mathrm{curl}_{\mathbf{x}}(\tilde{\mathbf{u}}\times\nabla_{\mathbf{x}}\psi_0), \\ -\Delta_{\mathbf{x}}\psi_0 = 4\pi\rho. \end{cases} \tag{1.690}$$

Note that by the Continuum Equation of the Conservation of Charges:

$$\frac{\partial\rho}{\partial t} + \mathrm{div}_{\mathbf{x}}\,\mathbf{j} = 0, \tag{1.691}$$

the reduced current clearly satisfies

$$\operatorname{div}_{\mathbf{x}} \widetilde{\mathbf{j}} = 0. \tag{1.692}$$

Moreover, using Proposition 1.1 we can easily deduce that $\widetilde{\mathbf{j}}$ is a proper vector field (see Sect. 13.2 for details). Finally, the approximate vectorial electromagnetic potential $\mathbf{A}$ from (1.689) clearly satisfies

$$\operatorname{div}_{\mathbf{x}} \mathbf{A} = 0. \tag{1.693}$$

Next, since by (1.652) we have

$$\Psi_1 := \Psi - \frac{1}{c}\mathbf{A} \cdot \tilde{\mathbf{u}}, \tag{1.694}$$

we rewrite (1.689) as

$$\begin{cases} -\Delta_{\mathbf{x}} \mathbf{A} \approx \frac{4\pi}{\kappa_0 c} \widetilde{\mathbf{j}}, \\ -\Delta_{\mathbf{x}} \Psi = 4\pi \gamma_0 \rho - \frac{1}{c} \operatorname{div}_{\mathbf{x}} \left(\tilde{\mathbf{u}} \times \operatorname{curl}_{\mathbf{x}} \mathbf{A}\right). \end{cases} \tag{1.695}$$

where

$$\begin{cases} \widetilde{\mathbf{j}} := \mathbf{j} - \frac{1}{4\pi} \frac{\partial}{\partial t} \left(\nabla_{\mathbf{x}} \psi_0\right) + \frac{1}{4\pi} \operatorname{curl}_{\mathbf{x}} \left(\tilde{\mathbf{u}} \times \nabla_{\mathbf{x}} \psi_0\right), \\ -\Delta_{\mathbf{x}} \psi_0 = 4\pi \rho, \end{cases} \tag{1.696}$$

(see Sect. 13.2 for details). So in order to find the scalar and the vectorial electromagnetic potentials we just need to solve Laplace equations. Knowing the approximate electromagnetic potentials by (1.650) we can find the approximations of the electromagnetic fields:

$$\begin{cases} \mathbf{B} = \operatorname{curl}_{\mathbf{x}} \mathbf{A} \\ \mathbf{E} = -\nabla_{\mathbf{x}} \Psi - \frac{1}{c} \frac{\partial \mathbf{A}}{\partial t} \\ \mathbf{D} = -\frac{1}{\gamma_0} \nabla_{\mathbf{x}} \Psi - \frac{1}{\gamma_0 c} \frac{\partial \mathbf{A}}{\partial t} + \frac{1}{c \gamma_0} \tilde{\mathbf{u}} \times \operatorname{curl}_{\mathbf{x}} \mathbf{A} \\ \mathbf{H} = \kappa_0 \operatorname{curl}_{\mathbf{x}} \mathbf{A} + \frac{1}{c} \tilde{\mathbf{u}} \times \left(-\frac{1}{\gamma_0} \nabla_{\mathbf{x}} \Psi - \frac{1}{\gamma_0 c} \frac{\partial \mathbf{A}}{\partial t} + \frac{1}{\gamma_0 c} \tilde{\mathbf{u}} \times \operatorname{curl}_{\mathbf{x}} \mathbf{A}\right), \end{cases} \tag{1.697}$$

where Ψ and $\mathbf{A}$ are given by (1.695). Note also that, since $\widetilde{\mathbf{j}}$ is a proper vector field, by Proposition 1.1 we deduce that equations (1.689) and thus also equations (1.695) are invariant under the change of non-inertial Cartesian coordinate system, provided that $\mathbf{A}$ is a proper vector field and $\Psi_1 = \Psi - \frac{1}{c}\mathbf{A} \cdot \tilde{\mathbf{u}}$ is a proper scalar field. So the approximate solutions in the case of quasistationary fields in a weak gravitational field satisfy the same transformation as the exact solutions of Maxwell Equations. Therefore, if in coordinate system $(*)$ we can use the approximate equations, given by (1.695) and (1.697), then we can use the similar approximation also in coordinate system $(**)$, even in the case when in system $(**)$ (1.683) or (1.684) are not satisfied.

Remark 1.4 The solutions of (1.695) and (1.697) satisfy the following equations:

$$\begin{cases} \operatorname{curl}_{\mathbf{x}}\left(\kappa_0\mathbf{B} + \frac{1}{c}\,\tilde{\mathbf{u}} \times (-\nabla_{\mathbf{x}}\psi_0)\right) = \frac{4\pi}{c}\mathbf{j} + \frac{1}{c}\frac{\partial(-\nabla_{\mathbf{x}}\psi_0)}{\partial t}, \\ \operatorname{div}_{\mathbf{x}}\mathbf{D} = 4\pi\rho, \\ \operatorname{curl}_{\mathbf{x}}\mathbf{E} + \frac{1}{c}\frac{\partial \mathbf{B}}{\partial t} = 0, \\ \operatorname{div}_{\mathbf{x}}\mathbf{B} = 0 \\ \mathbf{E} = \gamma_0\mathbf{D} - \frac{1}{c}\,\tilde{\mathbf{u}} \times \mathbf{B} \\ \mathbf{H} = \kappa_0\mathbf{B} + \frac{1}{c}\,\tilde{\mathbf{u}} \times \mathbf{D}, \\ \tilde{\mathbf{u}} = (\gamma_0\kappa_0\mathbf{v} + (1-\gamma_0\kappa_0)\mathbf{u}), \end{cases} \tag{1.698}$$

where ψ_0 was defined by (1.688). Equations (1.698) differ from the original Maxwell Eq. (1.647) only by neglecting the divergence-free part of the vector field $\mathbf{D}$ on the first equation.

Next, assume that, in addition to the validity of the approximation (1.683) and (1.684), the approximation (1.670) also holds. Then we further approximate (1.689) as

$$\begin{cases} -\Delta_{\mathbf{x}}\Psi_1 = 4\pi\gamma_0\rho, \\ -\Delta_{\mathbf{x}}\mathbf{A} \approx \frac{4\pi}{\kappa_0 c}\mathbf{j} - \frac{1}{\kappa_0\gamma_0 c}\left(\frac{\partial}{\partial t}(\nabla_{\mathbf{x}}\Psi_1) - \operatorname{curl}_{\mathbf{x}}(\tilde{\mathbf{u}} \times \nabla_{\mathbf{x}}\Psi_1)\right) \\ \Psi = \Psi_1 + \frac{1}{c}\mathbf{A}\cdot\tilde{\mathbf{u}}. \end{cases} \tag{1.699}$$

Moreover, as before, we deduce that equations (1.699) are also invariant under the change of non-inertial Cartesian coordinate system. Therefore, as before, if in the coordinate system (∗) we can use the approximation equations, given by (1.699) then we can use the similar equations also in coordinate system (∗∗), even in the case when in system (∗∗) (1.683), (1.684) or (1.670) are not satisfied.

Finally, assume that we are under the alternative calibration (1.660). Then by (1.683) and (1.684) we rewrite (1.661) and (1.662) as

$$-\Delta_{\mathbf{x}}\Psi_1 \approx 4\pi\gamma_0\rho + \frac{1}{c}\operatorname{div}_{\mathbf{x}}\left\{\left(d_{\mathbf{x}}\tilde{\mathbf{u}} + \{d_{\mathbf{x}}\tilde{\mathbf{u}}\}^T\right)\cdot\mathbf{A} - (\operatorname{div}_{\mathbf{x}}\tilde{\mathbf{u}})\,\mathbf{A}\right\}, \tag{1.700}$$

and

$$-\Delta_{\mathbf{x}}\mathbf{A} \approx \frac{4\pi}{\kappa_0 c}(\mathbf{j} - \rho\tilde{\mathbf{u}}) + \frac{1}{\kappa_0\gamma_0 c}\left(\left(d_{\mathbf{x}}\tilde{\mathbf{u}} + \{d_{\mathbf{x}}\tilde{\mathbf{u}}\}^T\right)\cdot\nabla_{\mathbf{x}}\Psi_1 - (\operatorname{div}_{\mathbf{x}}\tilde{\mathbf{u}})\,\nabla_{\mathbf{x}}\Psi_1\right). \tag{1.701}$$

Thus if we assume that in addition to the approximation (1.683) and (1.684) the approximation (1.670) also holds, we further approximate (1.700) and (1.701) as

$$\begin{cases} -\Delta_{\mathbf{x}}\Psi_1 \approx 4\pi\gamma_0\rho, \\ -\Delta_{\mathbf{x}}\mathbf{A} \approx \frac{4\pi}{\kappa_0 c}(\mathbf{j} - \rho\tilde{\mathbf{u}}) \\ \Psi = \Psi_1 + \frac{1}{c}\mathbf{A}\cdot\tilde{\mathbf{u}}. \end{cases} \tag{1.702}$$

Moreover, as before, we deduce that equations (1.702) are also invariant under the change of non-inertial Cartesian coordinate system. Therefore, as before, if in coordinate system $(*)$ we can use the approximation equations, given by (1.702) then we can use the similar equations also in coordinate system $(**)$, even in the case when in system $(**)$ (1.683), (1.684) or (1.670) are not satisfied.

1.14.2 Hertz's Notation for Quasistationary Electromagnetic Fields

Again consider in some domain the full system of Maxwell equations in the vacuum or in a medium, without dispersion, having the form (1.647):

$$\begin{cases} \text{curl}_{\mathbf{x}}\mathbf{H} = \frac{4\pi}{c}\mathbf{j} + \frac{1}{c}\frac{\partial \mathbf{D}}{\partial t}, \\ \text{div}_{\mathbf{x}}\,\mathbf{D} = 4\pi\rho, \\ \text{curl}_{\mathbf{x}}\mathbf{E} + \frac{1}{c}\frac{\partial \mathbf{B}}{\partial t} = 0, \\ \text{div}_{\mathbf{x}}\,\mathbf{B} = 0 \\ \mathbf{E} = \gamma_0\mathbf{D} - \frac{1}{c}\,\tilde{\mathbf{u}} \times \mathbf{B} \\ \mathbf{H} = \kappa_0\mathbf{B} + \frac{1}{c}\,\tilde{\mathbf{u}} \times \mathbf{D}, \\ \tilde{\mathbf{u}} = (\gamma_0\kappa_0\mathbf{v} + (1 - \gamma_0\kappa_0)\mathbf{u})\,. \end{cases} \tag{1.703}$$

Next define the electromagnetic fields in Hertz's notation:

$$\begin{cases} \tilde{\mathbf{E}} := \mathbf{E} + \frac{1}{c}\,\mathbf{u} \times \mathbf{B}, \\ \tilde{\mathbf{D}} := \frac{1}{\gamma_0}\tilde{\mathbf{E}} = \frac{1}{\gamma_0}\left(\mathbf{E} + \frac{1}{c}\,\mathbf{u} \times \mathbf{B}\right), \\ \tilde{\mathbf{B}} := \frac{1}{\kappa_0}\left(\mathbf{H} - \frac{1}{c}\,\mathbf{u} \times \mathbf{D}\right), \\ \tilde{\mathbf{H}} := \kappa_0\tilde{\mathbf{B}} = \left(\mathbf{H} - \frac{1}{c}\,\mathbf{u} \times \mathbf{D}\right). \end{cases} \tag{1.704}$$

Note here that $\mathbf{u}$ is the velocity field in the medium, thus since this field is absent in vacuum we use the convention that for the vacuum we have $\mathbf{u} := \mathbf{v}$ (and in addition $\tilde{\mathbf{u}} = \mathbf{v}$). In particular, by (1.704) and (1.703) we have

$$\begin{cases} \tilde{\mathbf{E}} = \gamma_0\mathbf{D} + \frac{1}{c}\,(\mathbf{u} - \tilde{\mathbf{u}}) \times \mathbf{B} = \gamma_0\left(\mathbf{D} + \frac{\kappa_0}{c}\,(\mathbf{u} - \mathbf{v}) \times \mathbf{B}\right), \\ \tilde{\mathbf{D}} = \mathbf{D} + \frac{\kappa_0}{c}\,(\mathbf{u} - \mathbf{v}) \times \mathbf{B}, \\ \tilde{\mathbf{B}} = \frac{1}{\kappa_0}\left(\kappa_0\mathbf{B} - \frac{1}{c}\,(\mathbf{u} - \tilde{\mathbf{u}}) \times \mathbf{D}\right) = \mathbf{B} - \frac{\gamma_0}{c}\,(\mathbf{u} - \mathbf{v}) \times \mathbf{D}, \\ \tilde{\mathbf{H}} = \kappa_0\mathbf{B} - \frac{\gamma_0\kappa_0}{c}\,(\mathbf{u} - \mathbf{v}) \times \mathbf{D}. \end{cases} \tag{1.705}$$

Therefore, since $\mathbf{D}$, $\mathbf{B}$, $(\mathbf{u} - \mathbf{v})$ are all proper vector fields, we deduce that $\tilde{\mathbf{E}}$, $\tilde{\mathbf{D}}$, $\tilde{\mathbf{B}}$, $\tilde{\mathbf{H}}$ are all proper vector fields. Next, analogously with (1.683), assume that in the given inertial or non-inertial Cartesian coordinate system $(*)$ the fields $\mathbf{u}$, $\mathbf{v}$, $\tilde{\mathbf{u}}$ are weak, meaning that at any instant on every point:

$$\frac{1}{\kappa_0\gamma_0}\left(\frac{|\mathbf{u}|^2}{c^2}+\frac{|\mathbf{v}|^2}{c^2}+\frac{|\tilde{\mathbf{u}}|^2}{c^2}\right)\ll 1, \tag{1.706}$$

and consider quasistationary electromagnetic fields, meaning the following: assume that the changes in time of the physical characteristics of the electromagnetic fields become essential after certain interval of time T_e and the changes in space of the physical characteristics of the fields become essential in the spatial landscape L_e. Then, analogously with (1.684), we assume that

$$(\kappa_0\gamma_0)\frac{c^2T_e^2}{L_e^2}\gg 1. \tag{1.707}$$

Moreover, assume that we have the following approximation analogous to (1.670): if the changes in space of the physical characteristics of the electromagnetic fields become essential in the spatial landscape L_e and the changes in space of the fields $\mathbf{u}$, $\mathbf{v}$, $\tilde{\mathbf{u}}$ become essential in the spatial landscape L_u, then we assume

$$L_e\ll L_u, \tag{1.708}$$

i.e. the fields $\mathbf{u}$, $\mathbf{v}$, $\tilde{\mathbf{u}}$ change in space much slower than the electromagnetic fields.

Then, denoting

$$\begin{cases}\tilde{\rho}^* := \rho-\frac{1}{c^2}\left(\mathbf{u}-\mathbf{v}\right)\cdot\left(\mathbf{j}-\rho\mathbf{u}\right),\\ \tilde{\mathbf{j}}^* = \left(\mathbf{j}-\rho\mathbf{u}\right)+\tilde{\rho}^*\mathbf{u},\end{cases} \tag{1.709}$$

by (1.706), (1.707) and (1.708), using the equality of the conservation of the charge $\frac{\partial\rho}{\partial t}+\operatorname{div}_{\mathbf{x}}\mathbf{j}=0$ we deduce

$$\begin{cases}\mathbf{j}\approx\tilde{\mathbf{j}}^*,\\ \frac{\partial\tilde{\rho}^*}{\partial t}+\operatorname{div}_{\mathbf{x}}\tilde{\mathbf{j}}^*\approx 0.\end{cases} \tag{1.710}$$

Moreover, by (1.709) we easily obtain that $\tilde{\rho}^*$ is a proper scalar and $\left(\tilde{\mathbf{j}}^*-\tilde{\rho}^*\mathbf{u}\right)$ is a proper vector. Finally, by (1.706), (1.707), (1.708) and (1.709), (1.710), (1.704) we deduce from the original Eq. (1.703) the following approximate Maxwell equations in Hertz's form and write them as the following two sets of equations:

$$\begin{cases}\operatorname{curl}_{\mathbf{x}}\left(\kappa_0\tilde{\mathbf{B}}\right)\approx\frac{4\pi}{c}\tilde{\mathbf{j}}^*+\frac{1}{c}\frac{\partial}{\partial t}\left(\frac{1}{\gamma_0}\tilde{\mathbf{E}}\right)-\frac{1}{c}\operatorname{curl}_{\mathbf{x}}\left(\mathbf{u}\times\frac{1}{\gamma_0}\tilde{\mathbf{E}}\right),\\ \operatorname{div}_{\mathbf{x}}\left(\frac{1}{\gamma_0}\tilde{\mathbf{E}}\right)\approx 4\pi\tilde{\rho}^*,\\ \operatorname{curl}_{\mathbf{x}}\tilde{\mathbf{E}}\approx-\left(\frac{1}{c}\frac{\partial\tilde{\mathbf{B}}}{\partial t}-\frac{1}{c}\operatorname{curl}_{\mathbf{x}}\left(\mathbf{u}\times\tilde{\mathbf{B}}\right)\right),\\ \operatorname{div}_{\mathbf{x}}\tilde{\mathbf{B}}\approx 0,\\ \frac{\partial\tilde{\rho}^*}{\partial t}+\operatorname{div}_{\mathbf{x}}\tilde{\mathbf{j}}^*\approx 0,\end{cases} \tag{1.711}$$

and

$$\begin{cases} \rho = \tilde{\rho}^* + \frac{1}{c^2}(\mathbf{u} - \mathbf{v}) \cdot \left(\tilde{\mathbf{j}}^* - \tilde{\rho}^* \mathbf{u}\right), \\ \mathbf{j} = \left(\tilde{\mathbf{j}}^* - \tilde{\rho}^* \mathbf{u}\right) + \rho \mathbf{u}, \\ \mathbf{B} \approx \tilde{\mathbf{B}} + \frac{1}{c}(\mathbf{u} - \mathbf{v}) \times \tilde{\mathbf{E}} \\ \mathbf{E} = \tilde{\mathbf{E}} - \frac{1}{c}\mathbf{u} \times \mathbf{B}, \end{cases} \tag{1.712}$$

(see Sect. 13.2.1 for the details) that are valid, however, only for quasistationary electromagnetic fields. Next, since $\tilde{\mathbf{E}}, \tilde{\mathbf{D}}, \tilde{\mathbf{B}}, \tilde{\mathbf{H}}, \mathbf{B}, \left(\mathbf{E} + \frac{1}{c}\mathbf{u} \times \mathbf{B}\right)$ and $\left(\tilde{\mathbf{j}}^* - \tilde{\rho}^* \mathbf{u}\right)$, $(\mathbf{j} - \rho \mathbf{u})$, $(\mathbf{u} - \mathbf{v})$ are all proper vector fields and $\tilde{\rho}^*$, ρ are proper scalars, as before we deduce that all equations in (1.711) and (1.712) are invariant under the change of inertial or non-inertial Cartesian coordinate system. Therefore, if in coordinate system $(*)$ we can use the approximate equations, given by (1.711) and (1.712), then we can use the similar approximation also in coordinate system $(**)$, even in the case when in system $(**)$ (1.706), (1.707) or (1.708) are not satisfied. Next, by (1.618) we adjoint (1.711) and (1.712) with Ohm's Law that in our notations has the following simple form:

$$\tilde{\mathbf{j}}^* - \tilde{\rho}^* \mathbf{u} = \varepsilon \tilde{\mathbf{E}}, \tag{1.713}$$

and the Joules heat term, appearing in the First Law of Thermodynamics (1.496), in our notations has the following simple form

$$(\mathbf{j} - \rho \mathbf{u}) \cdot \left(\mathbf{E} + \frac{1}{c}\mathbf{u} \times \mathbf{B}\right) = \left(\tilde{\mathbf{j}}^* - \tilde{\rho}^* \mathbf{u}\right) \cdot \tilde{\mathbf{E}}. \tag{1.714}$$

Finally, note that the system (1.711) is completely analogous to the system (1.703), where the field $\frac{1}{\gamma_0}\tilde{\mathbf{E}}$ substitutes the field $\mathbf{D}$, the field $\tilde{\mathbf{B}}$ substitutes the field $\mathbf{B}$, $\tilde{\rho}^*, \tilde{\mathbf{j}}^*$ substitute $\rho, \mathbf{j}$, and with the real velocity of the medium $\mathbf{u}$ instead of the field $\tilde{\mathbf{u}} = (\gamma_0 \mathbf{v} + (1 - \gamma_0)\mathbf{u})$. On the other hand, in the case of the real medium (not vacuum) the equations in (1.711) are completely independent on the vectorial gravitation potential $\mathbf{v}$. Since the systems (1.711) and (1.703) are similar they both obeys wave-type solutions and thus one can wonder here: is it possible that (1.711) is approximately equivalent to (1.703) also for highly oscillating in time electromagnetic fields and/or for wave-type solutions in the absence of charges and currents, for example in the case of the visible light? Unfortunately, the answer for the last question should be negative, indeed as it can be seen, the term $(1 - \gamma_0)\mathbf{u}$ (consistently with the equality $\tilde{\mathbf{u}} = (\gamma_0 \mathbf{v} + (1 - \gamma_0)\mathbf{u})$) appear in the Fizeau experiment for the light (see Subsect. 1.15.6) rather than the term $\mathbf{u}$. The main reason for it is that (1.707) is not valid for highly oscillating in time electromagnetic fields. Moreover, in the case of the complete absence of charges and currents in the whole space (1.707) is not valid also for any nontrivial wave-type solution, even with moderate frequency. So (1.711) and (1.712) are valid, only for quasistationary electromagnetic fields, generated by charges and currents moderately changing in time, for example in the case of estimation of DC or AC chains.

Next, as before, again by the third and the fourth equations in (1.711) we can write

$$\begin{cases} \tilde{\mathbf{B}} \approx \operatorname{curl}_{\mathbf{x}} \tilde{\mathbf{A}}^*, \\ \tilde{\mathbf{E}} \approx -\nabla_{\mathbf{x}} \tilde{\Psi}^* - \frac{1}{c}\frac{\partial \tilde{\mathbf{A}}^*}{\partial t} + \frac{1}{c}\mathbf{u} \times \left(\operatorname{curl}_{\mathbf{x}} \tilde{\mathbf{A}}^*\right), \end{cases} \tag{1.715}$$

where $\tilde{\Psi}^*$ and $\tilde{\mathbf{A}}^*$ are the scalar and the vectorial electromagnetic potentials in Hertz's notation. Next, analogously to (1.652) we define the proper scalar electromagnetic potential in Hertz's notation:

$$\tilde{\Psi}_1^* := \tilde{\Psi}^* - \frac{1}{c}\tilde{\mathbf{A}}^* \cdot \mathbf{u}. \tag{1.716}$$

As before, the electromagnetic potentials in Hertz's notation are not uniquely defined and thus we need to choose a calibration. For definiteness we can take $\tilde{\mathbf{A}}^*$ to satisfy

$$\operatorname{div}_{\mathbf{x}} \tilde{\mathbf{A}}^* = 0. \tag{1.717}$$

Then, as before, $\tilde{\mathbf{A}}^*$ is a proper vector field and $\tilde{\Psi}_1^*$ is a proper scalar field. It is clear that any other choice of electromagnetic potentials with a different calibration, as before, can be obtained by

$$\begin{cases} \tilde{\Psi}^* \;\to\; \tilde{\Psi}^* + \frac{1}{c}\frac{\partial w}{\partial t} \\ \tilde{\mathbf{A}}^* \;\to\; \tilde{\mathbf{A}}^* - \nabla_{\mathbf{x}} w \\ \tilde{\Psi}_1^* \;\to\; \tilde{\Psi}_1^* + \frac{1}{c}\left(\frac{\partial w}{\partial t} + \mathbf{u} \cdot \nabla_{\mathbf{x}} w\right). \end{cases} \tag{1.718}$$

where $w := w(\mathbf{x}, t)$ is an arbitrary proper scalar field. Moreover, as before, under such a change of calibration, $\tilde{\mathbf{A}}^*$ is still a proper vector field and $\tilde{\Psi}_1^*$ is still a proper scalar field provided the scalar field w is proper. In particular the following alternative to (1.717) calibration of the potentials can be chosen:

$$\frac{1}{\kappa_0 \gamma_0 c}\left(\frac{\partial \tilde{\Psi}_1^*}{\partial t} + \mathbf{u} \cdot \nabla_{\mathbf{x}} \tilde{\Psi}_1^*\right) + \operatorname{div}_{\mathbf{x}} \tilde{\mathbf{A}}^* = 0. \tag{1.719}$$

Next, inserting (1.716) into (1.715) gives

$$\begin{cases} \tilde{\mathbf{B}} \approx \operatorname{curl}_{\mathbf{x}} \tilde{\mathbf{A}}^*, \\ \tilde{\mathbf{E}} \approx -\nabla_{\mathbf{x}} \tilde{\Psi}_1^* - \frac{1}{c}\left(\frac{\partial \tilde{\mathbf{A}}^*}{\partial t} - \mathbf{u} \times \left(\operatorname{curl}_{\mathbf{x}} \tilde{\mathbf{A}}^*\right) + \nabla_{\mathbf{x}}\left(\tilde{\mathbf{A}}^* \cdot \mathbf{u}\right)\right). \end{cases} \tag{1.720}$$

Furthermore, inserting (1.720) into (1.711) and using (1.706), (1.707) and (1.708) we deduce

$$\begin{cases}\operatorname{curl}_{\mathbf{x}}\left(\kappa_0\tilde{\mathbf{B}}\right)\approx\frac{4\pi}{c}\left(\tilde{\mathbf{j}}^*-\tilde{\rho}^*\mathbf{u}\right)\\ -\frac{1}{c}\left(\frac{\partial}{\partial t}\left(\frac{1}{\gamma_0}\nabla_{\mathbf{x}}\tilde{\Psi}_1^*\right)-\operatorname{curl}_{\mathbf{x}}\left(\mathbf{u}\times\frac{1}{\gamma_0}\nabla_{\mathbf{x}}\tilde{\Psi}_1^*\right)+\left(\operatorname{div}_{\mathbf{x}}\left\{\frac{1}{\gamma_0}\nabla_{\mathbf{x}}\tilde{\Psi}_1^*\right\}\right)\mathbf{u}\right),\\ \operatorname{div}_{\mathbf{x}}\left(\frac{1}{\gamma_0}\tilde{\mathbf{E}}\right)\approx 4\pi\,\tilde{\rho}^*,\\ \frac{\partial\tilde{\rho}^*}{\partial t}+\operatorname{div}_{\mathbf{x}}\tilde{\mathbf{j}}^*\approx 0,\\ \tilde{\mathbf{E}}\approx-\nabla_{\mathbf{x}}\tilde{\Psi}_1^*-\frac{1}{c}\left(\frac{\partial\tilde{\mathbf{A}}^*}{\partial t}-\operatorname{curl}_{\mathbf{x}}\left(\mathbf{u}\times\tilde{\mathbf{A}}^*\right)+\left(\operatorname{div}_{\mathbf{x}}\tilde{\mathbf{A}}^*\right)\mathbf{u}\right),\\ \tilde{\mathbf{B}}\approx\operatorname{curl}_{\mathbf{x}}\tilde{\mathbf{A}}^*,\end{cases}\tag{1.721}$$

(see Sect. 13.2.1 for the details). Then again all equations in the new system (1.721) are invariant under the change of inertial or non-inertial Cartesian coordinate system.

Furthermore, if we assume that in the domain of the study the coefficients $\gamma_0\neq 0$ and $\kappa_0\neq 0$ vary sufficiently slow on the place and time and thus their spatial and temporal derivatives are negligible then by (1.706), (1.707) and (1.708), (1.721) can be approximately rewritten as

$$\begin{cases}-\Delta_{\mathbf{x}}\tilde{\mathbf{A}}^*\approx\frac{4\pi}{\kappa_0 c}\left(\tilde{\mathbf{j}}^*-\tilde{\rho}^*\mathbf{u}\right)-\nabla_{\mathbf{x}}\left\{\operatorname{div}_{\mathbf{x}}\tilde{\mathbf{A}}^*+\frac{1}{c}\frac{1}{\gamma_0\kappa_0}\left(\frac{\partial\tilde{\Psi}_1^*}{\partial t}+\mathbf{u}\cdot\nabla_{\mathbf{x}}\tilde{\Psi}_1^*\right)\right\},\\ -\Delta_{\mathbf{x}}\tilde{\Psi}_1\approx 4\pi\gamma_0\tilde{\rho}^*+\frac{1}{c}\left(\frac{\partial}{\partial t}\left(\operatorname{div}_{\mathbf{x}}\tilde{\mathbf{A}}^*\right)+\operatorname{div}_{\mathbf{x}}\left\{\left(\operatorname{div}_{\mathbf{x}}\tilde{\mathbf{A}}^*\right)\mathbf{u}\right\}\right),\\ \frac{\partial\tilde{\rho}^*}{\partial t}+\operatorname{div}_{\mathbf{x}}\tilde{\mathbf{j}}^*\approx 0,\\ \tilde{\mathbf{E}}\approx-\nabla_{\mathbf{x}}\tilde{\Psi}_1^*-\frac{1}{c}\left(\frac{\partial\tilde{\mathbf{A}}^*}{\partial t}-\mathbf{u}\times\left(\operatorname{curl}_{\mathbf{x}}\tilde{\mathbf{A}}^*\right)+\nabla_{\mathbf{x}}\left(\tilde{\mathbf{A}}^*\cdot\mathbf{u}\right)\right),\\ \tilde{\mathbf{B}}\approx\operatorname{curl}_{\mathbf{x}}\tilde{\mathbf{A}}^*,\end{cases}\tag{1.722}$$

(again see Sect. 13.2.1 for the details). Furthermore, defining the conduction currents:

$$\begin{cases}\mathbf{j}_{cond}:=\mathbf{j}-\rho\mathbf{u},\\ \tilde{\mathbf{j}}^*{}_{cond}:=\tilde{\mathbf{j}}^*-\tilde{\rho}^*\mathbf{u},\end{cases}\tag{1.723}$$

we obviously have

$$\begin{cases}\mathbf{j}_{cond}=\tilde{\mathbf{j}}^*{}_{cond},\\ \frac{\partial\rho}{\partial t}=\operatorname{div}_{\mathbf{x}}\{\rho\mathbf{u}\}=-\operatorname{div}_{\mathbf{x}}\{\mathbf{j}_{cond}\},\\ \frac{\partial\tilde{\rho}^*}{\partial t}+\operatorname{div}_{\mathbf{x}}\{\tilde{\rho}^*\mathbf{u}\}\approx-\operatorname{div}_{\mathbf{x}}\left\{\tilde{\mathbf{j}}^*{}_{cond}\right\},\end{cases}\tag{1.724}$$

and moreover, obviously $\mathbf{j}_{cond}=\tilde{\mathbf{j}}^*{}_{cond}$ is a proper vector field. Thus, in the case of calibration (1.717) by (1.121) and (1.122) in Proposition 1.1, using (1.708), (1.722) can be rewritten as

$$\begin{cases} -\Delta_{\mathbf{x}}\tilde{\Psi}_1 \approx 4\pi\gamma_0\tilde{\rho}^*, \\ -\Delta_{\mathbf{x}}\tilde{\mathbf{A}}^* \approx \frac{4\pi}{\kappa_0 c}\tilde{\mathbf{j}}^*{}_{cond} - \frac{1}{c}\frac{1}{\gamma_0\kappa_0}\left(\frac{\partial}{\partial t}\left(\nabla_{\mathbf{x}}\tilde{\Psi}_1^*\right) - \text{curl}_{\mathbf{x}}\left(\mathbf{u}\times\nabla_{\mathbf{x}}\tilde{\Psi}_1^*\right) + \left(\Delta_{\mathbf{x}}\tilde{\Psi}_1^*\right)\mathbf{u}\right), \\ \frac{\partial\tilde{\rho}^*}{\partial t} + \text{div}_{\mathbf{x}}\left\{\tilde{\rho}^*\mathbf{u}\right\} \approx -\text{div}_{\mathbf{x}}\left\{\tilde{\mathbf{j}}^*{}_{cond}\right\}, \\ \tilde{\mathbf{E}} \approx -\nabla_{\mathbf{x}}\tilde{\Psi}_1^* - \frac{1}{c}\left(\frac{\partial\tilde{\mathbf{A}}^*}{\partial t} - \mathbf{u}\times\left(\text{curl}_{\mathbf{x}}\tilde{\mathbf{A}}^*\right) + \nabla_{\mathbf{x}}\left(\tilde{\mathbf{A}}^*\cdot\mathbf{u}\right)\right), \\ \tilde{\mathbf{B}} \approx \text{curl}_{\mathbf{x}}\tilde{\mathbf{A}}^*. \end{cases} \tag{1.725}$$

On the other hand, under the alternative calibration (1.719), again using (1.706), (1.707) and (1.708) in the second equation of (1.722), we approximately rewrite (1.722) as

$$\begin{cases} -\Delta_{\mathbf{x}}\tilde{\mathbf{A}}^* \approx \frac{4\pi}{\kappa_0 c}\tilde{\mathbf{j}}^*{}_{cond}, \\ -\Delta_{\mathbf{x}}\tilde{\Psi}_1 \approx 4\pi\gamma_0\tilde{\rho}^*, \\ \frac{\partial\tilde{\rho}^*}{\partial t} + \text{div}_{\mathbf{x}}\left\{\tilde{\rho}^*\mathbf{u}\right\} \approx -\text{div}_{\mathbf{x}}\left\{\tilde{\mathbf{j}}^*{}_{cond}\right\}, \\ \tilde{\mathbf{E}} \approx -\nabla_{\mathbf{x}}\tilde{\Psi}_1^* - \frac{1}{c}\left(\frac{\partial\tilde{\mathbf{A}}^*}{\partial t} - \mathbf{u}\times\left(\text{curl}_{\mathbf{x}}\tilde{\mathbf{A}}^*\right) + \nabla_{\mathbf{x}}\left(\tilde{\mathbf{A}}^*\cdot\mathbf{u}\right)\right), \\ \tilde{\mathbf{B}} \approx \text{curl}_{\mathbf{x}}\tilde{\mathbf{A}}^*. \end{cases} \tag{1.726}$$

In both given cases of calibration in (1.725) or (1.726) we just need to solve two Laplace equations in order to find $\tilde{\Psi}_1$ and $\tilde{\mathbf{A}}^*$ and then we find $\tilde{\mathbf{E}}$ and $\tilde{\mathbf{B}}$ by differentiation. Moreover, in either the case of (1.725) or the case of (1.726) we have

$$\text{div}_{\mathbf{x}}\left\{\tilde{\mathbf{E}} + \nabla_{\mathbf{x}}\tilde{\Psi}_1^*\right\} \approx 0 \tag{1.727}$$

and Ohm's Law in the form of (1.713):

$$\tilde{\mathbf{j}}^*{}_{cond} = \varepsilon\tilde{\mathbf{E}}. \tag{1.728}$$

Furthermore, the Joules heat term has the following simple form

$$\tilde{\mathbf{j}}^*{}_{cond}\cdot\tilde{\mathbf{E}}. \tag{1.729}$$

Moreover, in both cases of (1.725) or (1.726), by the last two equations there we obviously have

$$\begin{cases} \text{curl}_{\mathbf{x}}\tilde{\mathbf{E}} \approx -\left(\frac{1}{c}\frac{\partial\tilde{\mathbf{B}}}{\partial t} - \frac{1}{c}\,\text{curl}_{\mathbf{x}}\left(\mathbf{u}\times\tilde{\mathbf{B}}\right)\right), \\ \text{div}_{\mathbf{x}}\,\tilde{\mathbf{B}} \approx 0. \end{cases} \tag{1.730}$$

Next, in order to find the real electromagnetic fields in the usual notation we have (1.712), i.e. the following:

$$\begin{cases} \mathbf{B} \approx \tilde{\mathbf{B}} + \frac{1}{c}(\mathbf{u} - \mathbf{v}) \times \tilde{\mathbf{E}}, \\ \mathbf{E} = \tilde{\mathbf{E}} - \frac{1}{c}\mathbf{u} \times \mathbf{B}, \\ \rho = \tilde{\rho}^* + \frac{1}{c^2}(\mathbf{u} - \mathbf{v}) \cdot \tilde{\mathbf{j}^*}_{cond}, \\ \mathbf{j}_{cond} = \tilde{\mathbf{j}^*}_{cond}, \\ \mathbf{j} = \tilde{\mathbf{j}^*}_{cond} + \rho\mathbf{u}. \end{cases} \tag{1.731}$$

Finally, as before, we deduce that all equations in either (1.725) or (1.726), and, in addition, (1.727), (1.728), (1.730) and (1.731) are invariant under the change of inertial or non-inertial Cartesian coordinate system. Therefore, if in coordinate system ($*$) we can use the approximate equations, given by either (1.725) or (1.726), and (1.727), (1.728), (1.730) and (1.731), then we can use the similar approximation also in coordinate system ($**$), even in the case when in system ($**$) (1.706), (1.707) or (1.708) are not satisfied.

Next, note that either (1.725) or (1.726), and (1.727), (1.728) and (1.730) are completely independent on the vectorial gravitational potential $\mathbf{v}$. Moreover, note that the first two equations in (1.726) are completely independent also on the velocity field of the medium $\mathbf{u}$. Furthermore, if $\Omega := \Omega(t) \subset \mathbb{R}^3$ is a three-dimensional domain, moving with velocity field $\mathbf{u}$ together with the given medium, then, by the third equation of either (1.725) or (1.726), using part **(iii)** of Proposition 1.7 and the Divergence Integral Theorem of the Calculus we deduce

$$\begin{aligned} \frac{d}{dt}\left(\iiint \rho \, d\Omega(t)\right) &= -\iint \mathbf{j}_{cond} \cdot \mathbf{n} \, d\left(\partial\Omega(t)\right) \\ &= -\iint \tilde{\mathbf{j}^*}_{cond} \cdot \mathbf{n} \, d\left(\partial\Omega(t)\right) \approx \frac{d}{dt}\left(\iiint \tilde{\rho}^* \, d\Omega(t)\right). \end{aligned} \tag{1.732}$$

Thus, denoting

$$\begin{cases} Q_{\Omega(t)}(t) := \iiint \rho \, d\Omega(t), \quad \tilde{Q}^*_{\Omega(t)}(t) := \iiint \tilde{\rho}^* \, d\Omega(t) \quad \text{and} \\ I_{\partial\Omega(t)}(t) := \iint \mathbf{j}_{cond} \cdot \mathbf{n} = \iint \tilde{\mathbf{j}^*}_{cond} \cdot \mathbf{n} \, d\left(\partial\Omega(t)\right) := \tilde{I}^*_{\partial\Omega(t)}(t), \end{cases} \tag{1.733}$$

which are the total charge inside the domain $\Omega(t)$, the total charge* inside the domain $\Omega(t)$ and the corresponding total current outward the boundary of $\Omega(t)$, by (1.732) we have

$$\frac{d}{dt}\left(Q_{\Omega(t)}(t)\right) = -I_{\partial\Omega(t)}(t) = -\tilde{I}^*_{\partial\Omega(t)}(t) \approx \frac{d}{dt}\left(\tilde{Q}^*_{\Omega(t)}(t)\right). \tag{1.734}$$

On the other hand, if $\gamma := \gamma(t) \subset \mathbb{R}^3$ is a one-dimensional curve oriented by the unit tangent vector $\mathbf{t} := \mathbf{t}(\mathbf{x}, t)$, having the starting and the ending points $\mathbf{r}_{begin}(t)$, $\mathbf{r}_{end}(t)$ and moving with velocity field $\mathbf{u}$ together with the given medium, then, by the fourth equation in either (1.725) or (1.726) using part **(ii)** of Proposition 1.7 we obtain

$$\int \tilde{\mathbf{E}} \cdot \mathbf{t} \, d\gamma(t) \approx \tilde{\Psi}_1^* \left(\mathbf{r}_{begin}(t), t \right) - \tilde{\Psi}_1^* \left(\mathbf{r}_{end}(t), t \right) - \frac{1}{c} \frac{d}{dt} \left(\int \tilde{\mathbf{A}}^* \cdot \mathbf{t} \, d\gamma(t) \right). \tag{1.735}$$

In particular, in the case where $\gamma(t) := \partial \mathcal{S}(t)$ is a boundary of a two-dimensional surface $\mathcal{S} := \mathcal{S}(t) \subset \mathbb{R}^3$, oriented by the unit normal $\mathbf{n} := \mathbf{n}(\mathbf{x}, t)$, since in the later case we have $\mathbf{r}_{begin}(t) = \mathbf{r}_{end}(t)$, using the last equation in either (1.725) or (1.726) and the Stokes Theorem, we rewrite (1.735) as

$$\int \tilde{\mathbf{E}} \cdot \mathbf{t} \, d\gamma(t) \approx -\frac{1}{c} \frac{d}{dt} \left(\iint \tilde{\mathbf{B}} \cdot \mathbf{n} \, d\mathcal{S}(t) \right) = -\frac{1}{c} \frac{d}{dt} \left(\iint \operatorname{curl}_{\mathbf{x}} \tilde{\mathbf{A}}^* \cdot \mathbf{n} \, d\mathcal{S}(t) \right). \tag{1.736}$$

Finally, the obtained relations in (1.727), (1.728), (1.729), (1.734) with (1.733), (1.735), (1.736) and the first two equations in (1.726) are completely independent on the velocity field $\mathbf{u}$ (and on the vectorial gravitational potential $\mathbf{v}$). However, the mentioned relations in (1.727), (1.728), (1.734) with (1.733), (1.735), (1.736) and the first two equations in (1.726) are together sufficient for the estimation of DC or AC linear chains and, in particular, they are sufficient to obtain Kirchhoff's current and voltage laws, Faraday's law of induction and the law of capacity. Thus, the estimation of DC or quasistationary AC currents in the case of moving electrical chains and/or in the case of nontrivial gravitation is completely analogous to that estimation for the resting chains without influence of any gravitation!

1.15 Geometric Optics Inside a Moving Medium and/or in the Presence of Gravitational Field

1.15.1 Preliminary Calculations

Assume that in some inertial or non-inertial Cartesian coordinate system a scalar real-valued field $U := U(\mathbf{x}, t)$, characterizing some wave, satisfies the following wave equation

$$\frac{\partial}{\partial t} \left\{ \frac{1}{c_0^2} \left(\frac{\partial U}{\partial t} + \tilde{\mathbf{u}} \cdot \nabla_{\mathbf{x}} U \right) \right\} + \operatorname{div}_{\mathbf{x}} \left\{ \frac{1}{c_0^2} \left(\frac{\partial U}{\partial t} + \tilde{\mathbf{u}} \cdot \nabla_{\mathbf{x}} U \right) \tilde{\mathbf{u}} \right\} - \Delta_{\mathbf{x}} U = 0, \tag{1.737}$$

where $\tilde{\mathbf{u}} := \tilde{\mathbf{u}}(\mathbf{x}, t)$ is some moderately changing (in space and in time) speed-like vector field and $c_0 := c_0(\mathbf{x}, t) > 0$ is a moderately changing (in space and in time) scalar quantity, that we call wave propagation speed. Note that (1.737) coincides with (1.681) and thus, in particular, U can represent one of any scalar components of the electromagnetic field in the medium without dispersion, i.e. when (1.607) is valid. In this case $\tilde{\mathbf{u}}$ is the linear combination of the velocity of the medium $\mathbf{u}$ and the vectorial gravitational potential $\mathbf{v}$. Note also that (1.554) or (1.563) also coincide with (1.737) and thus, in particular, U could represent the oscillating part of the pressure p_1 in the sound wave. However, in the later case $\tilde{\mathbf{u}}$ is equal to the averaged (macroscopical) velocity of the fluid/gas $\mathbf{u}_0$. Moreover, (1.591) and (1.592) also coincide with (1.737) and thus, in particular, U could represent either longitudinal or transverse wave in an elastic body. In the later case $\tilde{\mathbf{u}}$ is also equal to the averaged (macroscopical) velocity of the elastic medium $\mathbf{u}_0$.

Next if we assume that the fields $\tilde{\mathbf{u}}$ and c_0 are independent on the time variable, since U is a real-valued field, then we can write the field U as Fourier's Transform on the time variable:

$$U(\mathbf{x},t) = Re\left\{2\int_0^{+\infty} \hat{U}(\mathbf{x},\omega)e^{i\omega t}d\omega\right\}$$
$$\text{where}\quad \hat{U}(\mathbf{x},\omega) := \frac{1}{2\pi}\int_{-\infty}^{+\infty} U(\mathbf{x},t)e^{-i\omega t}dt\,. \tag{1.738}$$

Moreover, by (1.737) we obtain that Fourier's Transform $\hat{U}(\mathbf{x},\omega)$ satisfies

$$\frac{i\omega}{c_0^2}\left(i\omega\hat{U} + \tilde{\mathbf{u}}\cdot\nabla_{\mathbf{x}}\hat{U}\right) + \text{div}_{\mathbf{x}}\left\{\frac{1}{c_0^2}\left(i\omega\hat{U} + \tilde{\mathbf{u}}\cdot\nabla_{\mathbf{x}}\hat{U}\right)\tilde{\mathbf{u}}\right\} - \Delta_{\mathbf{x}}\hat{U} = 0. \tag{1.739}$$

Thus by (1.739), for every given ω the monochromatic wave-type function

$$U_\omega(\mathbf{x},t) := \hat{U}(\mathbf{x},\omega)e^{i\omega t} \tag{1.740}$$

is a complex solution of

$$\frac{\partial}{\partial t}\left\{\frac{1}{c_0^2}\left(\frac{\partial U_\omega}{\partial t} + \tilde{\mathbf{u}}\cdot\nabla_{\mathbf{x}}U_\omega\right)\right\} + \text{div}_{\mathbf{x}}\left\{\frac{1}{c_0^2}\left(\frac{\partial U_\omega}{\partial t} + \tilde{\mathbf{u}}\cdot\nabla_{\mathbf{x}}U_\omega\right)\tilde{\mathbf{u}}\right\} - \Delta_{\mathbf{x}}U_\omega = 0. \tag{1.741}$$

Note that Eq. (1.741) coincides with (1.737). Moreover, by (1.738) a general solution of (1.737) can be represented as a superposition of monochromatic waves of type $U_\omega = f(\mathbf{x},\omega)e^{i\omega t}$ that satisfy (1.741) for every ω. Finally if we consider a complex-valued function $\mathcal{U}(\mathbf{x},t)$, defined by

$$\mathcal{U}(\mathbf{x},t) := 2\int_0^{+\infty} \hat{U}(\mathbf{x},\omega)e^{i\omega t}d\omega, \tag{1.742}$$

then $Re\,\mathcal{U} = U$ and $\mathcal{U}$ is a complex solution of (1.737) (i.e. not only the real part of $\mathcal{U}$ but also the imaginary part solve (1.737)).

Next assume that a scalar complex field $U := U(\mathbf{x},t)$ satisfies (1.737). In particular, U can be a monochromatic solution of (1.741).

Furthermore, we represent the complex field U as

$$U(\mathbf{x},t) = A(\mathbf{x},t)e^{iT(\mathbf{x},t)}, \tag{1.743}$$

where $A := A(\mathbf{x},t)$ and $T := T(\mathbf{x},t)$ are real scalar fields. Then define

$$\omega := \left\langle \left| \frac{\partial T}{\partial t} \right| \right\rangle, \tag{1.744}$$

where the sign $\langle \cdot \rangle$ means the spatial and temporal averaging. Next define k_0 and a scalar field $S := S(\mathbf{x}, t)$ by

$$k_0 := \frac{\omega}{c} \quad \text{and} \quad S(\mathbf{x}, t) = \frac{1}{k_0} T(\mathbf{x}, t), \tag{1.745}$$

where c is a constant in the Maxwell equations for the vacuum. So we clearly have

$$U(\mathbf{x}, t) = A(\mathbf{x}, t) e^{ik_0 S(\mathbf{x}, t)} . \tag{1.746}$$

We in position to insert it into the wave equation (1.737) of the form

$$\left(\frac{\partial}{\partial t} \left\{ \frac{1}{c_0^2} \left(\frac{\partial U}{\partial t} + \tilde{\mathbf{u}} \cdot \nabla_{\mathbf{x}} U \right) \right\} + \operatorname{div}_{\mathbf{x}} \left\{ \frac{1}{c_0^2} \left(\frac{\partial U}{\partial t} + \tilde{\mathbf{u}} \cdot \nabla_{\mathbf{x}} U \right) \tilde{\mathbf{u}} \right\} \right) - \Delta_{\mathbf{x}} U = 0. \tag{1.747}$$

Then, inserting (1.746) into (1.747) we deduce

$$\begin{aligned} &k_0^2 \left(|\nabla_{\mathbf{x}} S|^2 - \frac{1}{c_0^2} \left(\frac{\partial S}{\partial t} + \tilde{\mathbf{u}} \cdot \nabla_{\mathbf{x}} S \right)^2 \right) A \\ &+ \left(\frac{\partial}{\partial t} \left\{ \frac{1}{c_0^2} \left(\frac{\partial A}{\partial t} + \tilde{\mathbf{u}} \cdot \nabla_{\mathbf{x}} A \right) \right\} + \operatorname{div}_{\mathbf{x}} \left\{ \frac{1}{c_0^2} \left(\frac{\partial A}{\partial t} + \tilde{\mathbf{u}} \cdot \nabla_{\mathbf{x}} A \right) \tilde{\mathbf{u}} \right\} \right) - \Delta_{\mathbf{x}} A \\ &+ i\kappa_0 A \left(\frac{\partial}{\partial t} \left\{ \frac{1}{c_0^2} \left(\frac{\partial S}{\partial t} + \tilde{\mathbf{u}} \cdot \nabla_{\mathbf{x}} S \right) \right\} + \operatorname{div}_{\mathbf{x}} \left\{ \frac{1}{c_0^2} \left(\frac{\partial S}{\partial t} + \tilde{\mathbf{u}} \cdot \nabla_{\mathbf{x}} S \right) \tilde{\mathbf{u}} \right\} - \Delta_{\mathbf{x}} S \right) \\ &+ 2i\kappa_0 \left(\frac{1}{c_0^2} \left(\frac{\partial S}{\partial t} + \tilde{\mathbf{u}} \cdot \nabla_{\mathbf{x}} S \right) \left(\frac{\partial A}{\partial t} + \tilde{\mathbf{u}} \cdot \nabla_{\mathbf{x}} A \right) - \nabla_{\mathbf{x}} A \cdot \nabla_{\mathbf{x}} S \right) = 0 \end{aligned} \tag{1.748}$$

(see Sect. 13.3 for details).

1.15.2 *Derivation of the Eikonal Equation*

Comparing both real and imaginary parts of (1.748) to zero we obtain two equations:

$$\begin{aligned} &k_0^2 \left(|\nabla_{\mathbf{x}} S|^2 - \frac{1}{c_0^2} \left(\frac{\partial S}{\partial t} + \tilde{\mathbf{u}} \cdot \nabla_{\mathbf{x}} S \right)^2 \right) A \\ &+ \left(\frac{\partial}{\partial t} \left\{ \frac{1}{c_0^2} \left(\frac{\partial A}{\partial t} + \tilde{\mathbf{u}} \cdot \nabla_{\mathbf{x}} A \right) \right\} + \operatorname{div}_{\mathbf{x}} \left\{ \frac{1}{c_0^2} \left(\frac{\partial A}{\partial t} + \tilde{\mathbf{u}} \cdot \nabla_{\mathbf{x}} A \right) \tilde{\mathbf{u}} \right\} \right) - \Delta_{\mathbf{x}} A = 0, \end{aligned} \tag{1.749}$$

and

$$A\left(\frac{\partial}{\partial t}\left\{\frac{1}{c_0^2}\left(\frac{\partial S}{\partial t}+\tilde{\mathbf{u}}\cdot\nabla_{\mathbf{x}}S\right)\right\}+\operatorname{div}_{\mathbf{x}}\left\{\frac{1}{c_0^2}\left(\frac{\partial S}{\partial t}+\tilde{\mathbf{u}}\cdot\nabla_{\mathbf{x}}S\right)\tilde{\mathbf{u}}\right\}-\Delta_{\mathbf{x}}S\right)$$
$$+\frac{2}{c_0^2}\left(\frac{\partial S}{\partial t}+\tilde{\mathbf{u}}\cdot\nabla_{\mathbf{x}}S\right)\left(\frac{\partial A}{\partial t}+\tilde{\mathbf{u}}\cdot\nabla_{\mathbf{x}}A\right)-2\nabla_{\mathbf{x}}A\cdot\nabla_{\mathbf{x}}S=0 \quad (1.750)$$

Next assume the Geometric Optics approximation that is good for the electromagnetic wave of high frequency, for example, for the visible light. The delicate Geometric Optics approximation means the following: assume that the changes in time of c_0, $\tilde{\mathbf{u}}$, A and S become essential after certain interval of time T_e and the changes in space of c_0, $\tilde{\mathbf{u}}$, A and S become essential in the spatial landscape L_e. Then we assume that

$$k_0^2c_0^2T_e^2 \gg 1 \quad \text{and} \quad k_0^2L_e^2 \gg 1. \quad (1.751)$$

On the other hand the rough Geometric Optics approximation (stronger than (1.751)) means

$$k_0c_0T_e \gg 1 \quad \text{and} \quad k_0L_e \gg 1. \quad (1.752)$$

Thus, by (1.751) we approximate (1.749) as the Eikonal-type equation:

$$\frac{1}{c_0^2}\left(\frac{\partial S}{\partial t}+\tilde{\mathbf{u}}\cdot\nabla_{\mathbf{x}}S\right)^2=|\nabla_{\mathbf{x}}S|^2, \quad (1.753)$$

and, without any use in (1.751), we rewrite (1.750) as

$$A\left(\frac{\partial}{\partial t}\left\{\frac{1}{c_0^2}\left(\frac{\partial S}{\partial t}+\tilde{\mathbf{u}}\cdot\nabla_{\mathbf{x}}S\right)\right\}+\operatorname{div}_{\mathbf{x}}\left\{\frac{1}{c_0^2}\left(\frac{\partial S}{\partial t}+\tilde{\mathbf{u}}\cdot\nabla_{\mathbf{x}}S\right)\tilde{\mathbf{u}}\right\}-\Delta_{\mathbf{x}}S\right)$$
$$+\frac{2}{c_0^2}\left(\frac{\partial S}{\partial t}+\tilde{\mathbf{u}}\cdot\nabla_{\mathbf{x}}S\right)\left(\frac{\partial A}{\partial t}+\tilde{\mathbf{u}}\cdot\nabla_{\mathbf{x}}A\right)-2\nabla_{\mathbf{x}}A\cdot\nabla_{\mathbf{x}}S=0. \quad (1.754)$$

Equality (1.753) is called the Eikonal equation and equality (1.754) is called the equation of the beam propagation. Then, as before, we deduce that equation (1.753) is invariant under the change of non-inertial Cartesian coordinate system, provided that under such change we have $S' = S$. Moreover, (1.754) is also invariant under the change of non-inertial Cartesian coordinate system, in the case that under such change we have $A' = A$, provided that $S' = S$. So if the approximation (1.751) is valid in some Cartesian coordinate system (∗), then we can use (1.753) and (1.754) also in any other inertial or non-inertial Cartesian coordinate system (∗∗) even in the case when (1.751) is not valid in the system (∗∗), provided that under the change of coordinate system we have $A' = A$ and $S' = S$. Furthermore, note that although we established the Eikonal equation (1.753) in the delicate Geometric Optics approximation (1.751), since by (1.746) we have

$$U(\mathbf{x},t)=A(\mathbf{x},t)e^{ik_0 S(\mathbf{x},t)}\,, \tag{1.755}$$

then, considering the approximations of A and S as in (1.754), (1.753) and putting them into (1.755) we actually establish $U(\mathbf{x},t)$ only in the rough Geometric Optics approximation (1.752). Finally, denoting

$$\mathbf{h}(\mathbf{x},t):=\tilde{\mathbf{u}}(\mathbf{x},t)-c_0^2(\mathbf{x},t)\left(\frac{\partial S}{\partial t}(\mathbf{x},t)+\tilde{\mathbf{u}}(\mathbf{x},t)\cdot\nabla_{\mathbf{x}}S(\mathbf{x},t)\right)^{-1}\nabla_{\mathbf{x}}S(\mathbf{x},t), \tag{1.756}$$

and

$$G\left(\mathbf{x},t\right):=\frac{c_0^2}{2\left(\frac{\partial S}{\partial t}+\tilde{\mathbf{u}}\cdot\nabla_{\mathbf{x}}S\right)}\Bigg(\Delta_{\mathbf{x}}S-\frac{\partial}{\partial t}\left\{\frac{1}{c_0^2}\left(\frac{\partial S}{\partial t}+\tilde{\mathbf{u}}\cdot\nabla_{\mathbf{x}}S\right)\right\}-\operatorname{div}_{\mathbf{x}}\left\{\frac{1}{c_0^2}\left(\frac{\partial S}{\partial t}+\tilde{\mathbf{u}}\cdot\nabla_{\mathbf{x}}S\right)\tilde{\mathbf{u}}\right\}\Bigg), \tag{1.757}$$

we rewrite (1.754) and (1.753) as

$$\frac{\partial A}{\partial t}(\mathbf{x},t)+\mathbf{h}(\mathbf{x},t)\cdot\nabla_{\mathbf{x}}A(\mathbf{x},t)-G(\mathbf{x},t)A(\mathbf{x},t)=0\,, \tag{1.758}$$

and

$$|\mathbf{h}(\mathbf{x},t)-\tilde{\mathbf{u}}(\mathbf{x},t)|^2=c_0^2(\mathbf{x},t)\,. \tag{1.759}$$

Next consider the curve $\mathbf{r}(t):[t_0,b]\to\mathbb{R}^3$, parameterized by the time variable t, defined as a solution of ordinary differential equation:

$$\begin{cases}\frac{d\mathbf{r}}{dt}(t)=\mathbf{h}\left(\mathbf{r}(t),t\right)\\ \mathbf{r}(t_0)=\mathbf{x}_0,\end{cases} \tag{1.760}$$

where $\mathbf{h}$ was defined in (1.756), then, by (1.758) and the Chain rule we have

$$\frac{d}{dt}\left(A\left(\mathbf{r}(t),t\right)\right)=\nabla_{\mathbf{x}}A\left(\mathbf{r}(t),t\right)\cdot\frac{d\mathbf{r}}{dt}(t)+\frac{\partial A}{\partial t}\left(\mathbf{r}(t),t\right)=G\left(\mathbf{r}(t),t\right)A\left(\mathbf{r}(t),t\right), \tag{1.761}$$

where G was defined in (1.757). Then (1.761) implies

$$A\left(\mathbf{r}(t),t\right)=A\left(\mathbf{x}_0,t_0\right)e^{\int_{t_0}^{t}G(\mathbf{r}(\tau),\tau)d\tau}\qquad\forall t\in[t_0,b]. \tag{1.762}$$

In particular,

$$A\left(\mathbf{x}_0,t_0\right)=0\;\text{ implies }\;A\left(\mathbf{r}(t),t\right)=0\quad\forall t\in[t_0,b],\quad\text{and }\;A\left(\mathbf{x}_0,t_0\right)\neq 0\;\text{ implies }\;A\left(\mathbf{r}(t),t\right)\neq 0\quad\forall t\in[t_0,b]. \tag{1.763}$$

Therefore, by (1.763) we deduce that the curve that satisfies (1.760) coincides with the ray of light that passes through the point $\mathbf{x}_0$ at the instant of time t_0. So, equality (1.760) is the equation of a ray and the vector field $\mathbf{h}$ defined for every $\mathbf{x}$ by (1.756) is the direction of the propagation of the ray that passes through point $\mathbf{x}$ at the instant of time t. On the other hand, as before, we can easily prove that the vector field defined in every inertial or non-inertial coordinate system by (1.756) is a speed-like vector field. Moreover, by (1.759) the following implication holds:

$$\tilde{\mathbf{u}} = 0 \quad \text{implies} \quad |\mathbf{h}| = c_0. \tag{1.764}$$

Finally, by chain rule, for the curve that satisfies (13.167) we have

$$\begin{aligned}\frac{d}{dt}\left(S\left(\mathbf{r}(t), t\right)\right) &= \frac{\partial S}{\partial t}\left(\mathbf{r}(t), t\right) + \nabla_{\mathbf{x}} S\left(\mathbf{r}(t), t\right) \cdot \frac{d\mathbf{r}}{dt}(t) \\ &= \frac{\partial S}{\partial t}\left(\mathbf{r}(t), t\right) + \nabla_{\mathbf{x}} S\left(\mathbf{r}(t), t\right) \cdot \mathbf{h}\left(\mathbf{r}(t), t\right).\end{aligned} \tag{1.765}$$

Thus, by (1.756), (1.759) and (1.765) we deduce

$$\frac{d}{dt}\left(S\left(\mathbf{r}(t), t\right)\right) = 0 \qquad \forall t \geq t_0, \tag{1.766}$$

and so

$$S\left(\mathbf{r}(t), t\right) = S\left(\mathbf{x}_0, t_0\right) \qquad \forall t \geq t_0. \tag{1.767}$$

By all these facts, vector field $\mathbf{h}(\mathbf{x}, t)$ defined by (13.157) can be considered as the vector of the velocity (speed) of the wave at the point $\mathbf{x}$ at the instant of time t. Moreover, by (1.759) we have

$$|\mathbf{h} - \tilde{\mathbf{u}}|^2 = c_0^2. \tag{1.768}$$

Remark 1.5 In contrast to the proof of (1.753), we do not use any of the Geometric Optics approximations (1.751) or (1.752) in the proof of (1.754) and (1.758). So, (1.758) is still valid without assumptions of the Geometric Optics approximation (1.751) or (1.752) and thus, the vector field $\mathbf{h}$, defined by (1.756), is the direction of the propagation of the ray that passes through point $\mathbf{x}$ also in the general case. However, without assumptions of the Geometric Optics approximation we cannot derive (1.753), (1.764), (1.767) and (1.768) anymore.

1.15.3 The Case of the Monochromatic Wave

Next, up to the end of this section, consider the case of monochromatic wave of the constant frequency $\nu = \frac{\omega}{2\pi}$ where the fields $\tilde{\mathbf{u}}$ and c_0 are independent on the time variable, i.e. the case of (1.743) where we have

$$\begin{cases} \frac{\partial T}{\partial t} = \omega \\ \frac{\partial A}{\partial t} = 0 \\ \frac{\partial \tilde{\mathbf{u}}}{\partial t} = 0 \\ \frac{\partial c_0}{\partial t} = 0. \end{cases} \tag{1.769}$$

We also assume that either our medium has no dispersion (i.e. in the case of an electromagnetic wave (1.607) is valid), or the velocity of the medium is negligible, i.e. $\mathbf{u} = 0$ (and so in the case of an electromagnetic wave we have $\tilde{\mathbf{u}} = \mathbf{v}$) and our medium is transparent for the given frequency $\nu = \frac{\omega}{2\pi}$. Then, by (1.744) and (1.745) we rewrite (1.769) as

$$\begin{cases} \frac{\partial S}{\partial t} = c \\ \frac{\partial A}{\partial t} = 0. \end{cases} \tag{1.770}$$

Thus $\nabla_{\mathbf{x}} S$ is independent on t and moreover, by (1.770) we rewrite (1.753) as

$$\frac{c^2}{c_0^2}\left(1 + \frac{1}{c}\tilde{\mathbf{u}} \cdot \nabla_{\mathbf{x}} S\right)^2 = |\nabla_{\mathbf{x}} S|^2, \tag{1.771}$$

and, using (1.770) we rewrite (1.754) as

$$2\left(\nabla_{\mathbf{x}} S - \frac{c}{c_0}\left(1 + \frac{1}{c}\left(\tilde{\mathbf{u}} \cdot \nabla_{\mathbf{x}} S\right)\right)\frac{\tilde{\mathbf{u}}}{c_0}\right) \cdot \nabla_{\mathbf{x}} A = \left(\operatorname{div}_{\mathbf{x}}\left\{\frac{1}{c_0^2}\left(c + \tilde{\mathbf{u}} \cdot \nabla_{\mathbf{x}} S\right)\tilde{\mathbf{u}}\right\} - \left(\Delta_{\mathbf{x}} S\right)\right) A. \tag{1.772}$$

In particular, in the case of the region of the space where the following approximation is valid:

$$\frac{|\tilde{\mathbf{u}}|^2}{c_0^2} \ll 1, \tag{1.773}$$

up to order $O\left(\frac{|\tilde{\mathbf{u}}|^2}{c_0^2}\right)$, we rewrite (1.771) as

$$\left|\frac{c\tilde{\mathbf{u}}}{c_0^2} - \nabla_{\mathbf{x}} S\right|^2 = \frac{c^2}{c_0^2}, \tag{1.774}$$

and (1.772) as

$$\left(\frac{c\tilde{\mathbf{u}}}{c_0^2} - \nabla_{\mathbf{x}} S\right) \cdot \nabla_{\mathbf{x}} A + \frac{1}{2}\left(\operatorname{div}_{\mathbf{x}}\left\{\frac{c}{c_0^2}\tilde{\mathbf{u}}\right\} - \Delta_{\mathbf{x}} S\right) A = 0. \tag{1.775}$$

The Eikonal equation (1.774) and equation of the beam propagation (1.775) are two basic equations of propagation of monochromatic light in the Geometric Optics approximation inside a moving medium or/and in the presence of nontrivial gravitational field, provided that the field $\tilde{\mathbf{u}}$ satisfies (1.773).

Next if we consider an arbitrary characteristic curve $\mathbf{r}(s) : [a, b] \to \mathbb{R}^3$ of equation (1.775) defined as a solution of ordinary differential equation

$$\begin{cases} \frac{d\mathbf{r}}{ds}(s) = \frac{c}{c_0^2(\mathbf{r}(s))}\tilde{\mathbf{u}}(\mathbf{r}(s)) - \nabla_{\mathbf{x}} S(\mathbf{r}(s)) \\ \mathbf{r}(a) = \mathbf{x}_0, \end{cases} \tag{1.776}$$

then, as before, by (1.775) and (1.776) we have

$$\begin{aligned} \frac{d}{ds}(A(\mathbf{r}(s))) &= \nabla_{\mathbf{x}} A(\mathbf{r}(s)) \cdot \frac{d\mathbf{r}}{ds}(s) \\ &= \frac{1}{2}\left(\Delta_{\mathbf{x}} S(\mathbf{r}(s)) - \operatorname{div}_{\mathbf{x}}\left\{\frac{c}{c_0^2(\mathbf{r}(s))}\tilde{\mathbf{u}}(\mathbf{r}(s))\right\}\right) A(\mathbf{r}(s)) \end{aligned} \tag{1.777}$$

that implies

$$A(\mathbf{r}(s)) = A(\mathbf{x}_0)\, e^{\frac{1}{2}\int_a^s \left(\Delta_{\mathbf{x}} S(\mathbf{r}(\tau)) - \operatorname{div}_{\mathbf{x}}\left\{\frac{c}{c_0^2(\mathbf{r}(\tau))}\tilde{\mathbf{u}}(\mathbf{r}(\tau))\right\}\right) d\tau} \quad \forall s \in [a, b]. \tag{1.778}$$

In particular,

$$\begin{aligned} &A(\mathbf{x}_0) = 0 \text{ implies } A(\mathbf{r}(s)) = 0 \quad \forall s \in [a, b], \\ &\qquad \text{and } A(\mathbf{x}_0) \neq 0 \text{ implies } A(\mathbf{r}(s)) \neq 0 \quad \forall s \in [a, b]. \end{aligned} \tag{1.779}$$

Therefore, by (1.779) we deduce that in the case of (1.773) the curve that satisfies (1.776) coincides with the ray of light that passes through the point $\mathbf{x}_0$. So in the case of (1.773), equality (1.776) is the equation of a ray and the vector field $\mathbf{h}_0$ defined for every $\mathbf{x}$ by

$$\mathbf{h}_0(\mathbf{x}) := \frac{c}{c_0^2(\mathbf{x})}\tilde{\mathbf{u}}(\mathbf{x}) - \nabla_{\mathbf{x}} S(\mathbf{x}) \approx \frac{c}{c_0^2(\mathbf{x})}\mathbf{h}(\mathbf{x}) \tag{1.780}$$

is the direction of the propagation of the ray that passes through point $\mathbf{x}$. Moreover, by (1.774) $\mathbf{h}_0$ satisfies

$$|\mathbf{h}_0|^2 = \frac{c^2}{c_0^2}. \tag{1.781}$$

Remark 1.6 In contrast to the proof of (1.753), (1.771) or (1.774), we do not use the Geometric Optics approximations (1.751) or (1.752) in the proof of (1.750), (1.754), (1.772) and (1.775). We just need the estimation (1.773) for the proof of (1.775). So, (1.779) is still valid without assumptions of the Geometric Optics approximation (1.751) or (1.752) and thus, the vector field $\mathbf{h}_0$, defined by (1.780), is the direction of the propagation of the ray that passes through point $\mathbf{x}$ also in the general case, provided the estimation (1.773) holds. However, without assumption of the Geometric Optics approximation we cannot derive (1.781) anymore.

Next by (1.774) and (1.775) in Sect. 13.3 we prove the following Fermat Principle:

Proposition 1.8 *Assume Geometric Optics approximation together with* (1.773). *Then the light that travels from point N to point M chooses the path $\mathbf{r}(s) : [a, b] \to \mathbb{R}^3$ with endpoints $\mathbf{r}(a) = N$ and $\mathbf{r}(b) = M$ which minimizes the quantity:*

$$J(\mathbf{r}(\cdot)) := \int_a^b n(\mathbf{r}(s)) \left|\mathbf{r}'(s)\right| ds - \int_a^b \frac{1}{c} n^2 (\mathbf{r}(s))\, \tilde{\mathbf{u}}(\mathbf{r}(s)) \cdot \mathbf{r}'(s) ds, \tag{1.782}$$

where we set the refraction index:

$$n(\mathbf{x}) := \frac{c}{c_0(\mathbf{x})}. \tag{1.783}$$

Moreover, if the path $\mathbf{r}(s) : [a, b] \to \mathbb{R}^3$ with endpoints $\mathbf{r}(a) = N$ and $\mathbf{r}(b) = M$ is the real path of the light, then:

$$(-S(M)) - (-S(N)) = \int_a^b n(\mathbf{r}(s)) \left|\mathbf{r}'(s)\right| ds - \int_a^b \frac{1}{c} n^2 (\mathbf{r}(s))\, \tilde{\mathbf{u}}(\mathbf{r}(s)) \cdot \mathbf{r}'(s) ds. \tag{1.784}$$

See also Sect. 13.3 for the generalization of the Fermat Principle to the case where we cannot take (1.773) into account.

In particular, by Proposition 1.8 the path of travel of the light satisfies the Euler–Lagrange equation for the functional $J(\mathbf{r}(\cdot))$, that is the differential equation of the path of light:

$$\frac{d}{d\lambda}\left(n(\mathbf{r}) \frac{d\mathbf{r}}{d\lambda}\right) = \frac{1}{c} n^2(\mathbf{r}) \left(\operatorname{curl}_{\mathbf{x}} \tilde{\mathbf{u}}(\mathbf{r})\right) \times \frac{d\mathbf{r}}{d\lambda}$$
$$+ \nabla_{\mathbf{x}} n(\mathbf{r}) + \frac{2}{c} n(\mathbf{r}) \left\{\tilde{\mathbf{u}}(\mathbf{r}) \otimes \nabla_{\mathbf{x}} n(\mathbf{r}) - \nabla_{\mathbf{x}} n(\mathbf{r}) \otimes \tilde{\mathbf{u}}(\mathbf{r})\right\} \cdot \frac{d\mathbf{r}}{d\lambda}, \tag{1.785}$$

where

$$\lambda := \int_a^s \left|\mathbf{r}'(\tau)\right| d\tau \tag{1.786}$$

is the natural parameter of the curve (see Sect. 13.3 for details).

Next, assume that the wave we consider has an electromagnetic nature. Then by (1.673) and (1.682) we have

$$c_0 = c\sqrt{\kappa_0\gamma_0} \quad \text{and} \quad \tilde{\mathbf{u}} = (\gamma_0\kappa_0\mathbf{v} + (1 - \gamma_0\kappa_0)\mathbf{u})\,, \tag{1.787}$$

where $\mathbf{u}$ is the medium velocity and $\mathbf{v}$ is the local vectorial gravitational potential. Moreover, assume that we consider light traveling in some regions either filled with the resting medium of constant dielectric permeability γ_0 and magnetic permeability κ_0 or in the vacuum. Then by (1.787) and (1.783) we have

$$n = \frac{1}{\sqrt{\kappa_0\gamma_0}} \quad \text{is a constant,} \quad \text{and} \quad \tilde{\mathbf{u}} = \gamma_0\kappa_0\mathbf{v}. \tag{1.788}$$

Then by (1.788) we rewrite (1.785) as

$$\frac{d^2\mathbf{r}}{d\lambda^2} = \frac{1}{c}\sqrt{\gamma_0\kappa_0}\,\left(\text{curl}_{\mathbf{x}}\mathbf{v}\,(\mathbf{r})\right) \times \frac{d\mathbf{r}}{d\lambda}. \tag{1.789}$$

In particular, if our coordinate system is inertial, or more generally non-rotating, then $\text{curl}_{\mathbf{x}}\mathbf{v} = 0$ and we deduce that the path of the light from the point N to the point M is the direct line connecting these points, provided we take in the account estimation (1.773).

On the other hand, if our system is rotating, then, since $\mathbf{v}$ is a speed-like vector field, we clearly deduce

$$\text{curl}_{\mathbf{x}}\mathbf{v} = -2\mathbf{w}, \tag{1.790}$$

where $\mathbf{w}$ is the vector of the angular speed of rotation of our coordinate system. Thus by inserting (1.790) into (1.789) we deduce

$$\frac{d^2\mathbf{r}}{d\lambda^2} = -\frac{2}{c}\sqrt{\gamma_0\kappa_0}\,\mathbf{w} \times \frac{d\mathbf{r}}{d\lambda}. \tag{1.791}$$

The solution of (1.791) is the following:

$$\begin{cases} x(\lambda) = C_1\frac{c\sqrt{\kappa_0\gamma_0}}{2w}\left(\cos\left(\frac{2w}{c}\sqrt{\kappa_0\gamma_0}\lambda\right) - 1\right) + C_2\frac{c\sqrt{\kappa_0\gamma_0}}{2w}\sin\left(\frac{2w}{c}\sqrt{\kappa_0\gamma_0}\lambda\right) + D_1 \\ y(\lambda) = -C_1\frac{c\sqrt{\kappa_0\gamma_0}}{2w}\sin\left(\frac{2w}{c}\sqrt{\kappa_0\gamma_0}\lambda\right) + C_2\frac{c\sqrt{\kappa_0\gamma_0}}{2w}\left(\cos\left(\frac{2w}{c}\sqrt{\kappa_0\gamma_0}\lambda\right) - 1\right) + D_2 \\ z(\lambda) = C_3\lambda + D_3, \end{cases} \tag{1.792}$$

where, since λ is a natural parameter of the curve, we have

$$C_1^2 + C_2^2 + C_3^2 = 1. \tag{1.793}$$

So, the curve in (1.792) is the trajectory of the light in the rotating coordinate system, provided we assume (1.773). In particular, by (1.792) we have

$$\begin{cases} x(0) = D_1, \quad y(0) = D_2, \quad z(0) = D_3, \\ \frac{dx}{d\lambda}(0) = C_2, \quad \frac{dy}{d\lambda}(0) = -C_1, \quad \frac{dz}{d\lambda}(0) = C_3. \end{cases} \tag{1.794}$$

The constants $C_1, C_2, C_3, D_1, D_2, D_3$ can be determined either by the initial data (1.794) or by the beginning and the ending points N and M of the curve.

1.15.4 The Laws of Reflection and Refraction

Next consider a monochromatic wave of the frequency $\nu = \omega/(2\pi)$ characterized by

$$U(\mathbf{x}, t) = A(\mathbf{x})e^{ik_0 S(\mathbf{x},t)}, \quad \text{where} \quad k_0 = \frac{\omega}{c} \quad \text{and} \quad \frac{\partial S}{\partial t} = c\,, \tag{1.795}$$

and, consistently with (1.780) consider a direction field:

$$\mathbf{h}_0(\mathbf{x}) = \frac{c}{c_0^2(\mathbf{x})}\tilde{\mathbf{u}}(\mathbf{x}) - \nabla_{\mathbf{x}} S(\mathbf{x}) \approx \frac{c}{c_0^2(\mathbf{x})}\mathbf{h}(\mathbf{x}). \tag{1.796}$$

Furthermore, assume that this wave undergoes reflection and/or refraction on the stationary (time independent) surface $\mathcal{T}$ with the outcoming unit normal $\mathbf{n}$, separating two regions characterized respectively by $c_0 = c_0^{(1)}$ and $\tilde{\mathbf{u}} = \tilde{\mathbf{u}}_1$ and by $c_0^{(2)}$ and $\tilde{\mathbf{u}}_2$, with the formation of the reflected wave (of the same frequency), characterized by

$$U_1(\mathbf{x}, t) = A_1(\mathbf{x})e^{ik_0 S_1(\mathbf{x},t)}, \quad \text{where} \quad \frac{\partial S_1}{\partial t} = c\,, \tag{1.797}$$

and by a direction field:

$$\mathbf{h}_1(\mathbf{x}) = \frac{c}{c_0^2(\mathbf{x})}\tilde{\mathbf{u}}(\mathbf{x}) - \nabla_{\mathbf{x}} S_1(\mathbf{x}), \tag{1.798}$$

and formation of the refracted wave (of the same frequency), characterized by

$$U_2(\mathbf{x}, t) = A_2(\mathbf{x})e^{ik_0 S_2(\mathbf{x},t)}, \quad \text{where} \quad \frac{\partial S_2}{\partial t} = c \tag{1.799}$$

and by a direction field:

$$\mathbf{h}_2(\mathbf{x}) = \frac{c}{\left(c_0^{(2)}(\mathbf{x})\right)^2}\tilde{\mathbf{u}_2}(\mathbf{x}) - \nabla_{\mathbf{x}} S_2(\mathbf{x}). \tag{1.800}$$

Then the boundary conditions of U, U_1 and U_2 depend on the physical meaning of these fields. However, one of the necessary conditions should be that

$$S_1(\mathbf{x}, t) = S_2(\mathbf{x}, t) + C_2 = S(\mathbf{x}, t) \qquad \forall \mathbf{x} \in \mathcal{T}, \tag{1.801}$$

where C_2 is a real constant. In particular (1.801) implies:

$$\nabla_{\mathbf{x}} S_1 - (\mathbf{n} \cdot \nabla_{\mathbf{x}} S_1)\,\mathbf{n} = \nabla_{\mathbf{x}} S_2 - (\mathbf{n} \cdot \nabla_{\mathbf{x}} S_2)\,\mathbf{n} = \nabla_{\mathbf{x}} S - (\mathbf{n} \cdot \nabla_{\mathbf{x}} S)\,\mathbf{n} \qquad \forall \mathbf{x} \in \mathcal{T}. \tag{1.802}$$

In particular, for every point on the surface $\mathcal{T}$ vectors $\nabla_{\mathbf{x}} S_1$ and $\nabla_{\mathbf{x}} S_2$ lie in the plane formed by vectors $\mathbf{n}$ and $\nabla_{\mathbf{x}} S$. Moreover, by (1.796), (1.798) and (1.802) we have

$$\mathbf{h}_1 - (\mathbf{n} \cdot \mathbf{h}_1)\,\mathbf{n} = \mathbf{h}_0 - (\mathbf{n} \cdot \mathbf{h}_0)\,\mathbf{n} \qquad \forall \mathbf{x} \in \mathcal{T}, \tag{1.803}$$

and in particular, for every point on the surface $\mathcal{T}$ vector $\mathbf{h}_1$ lies in the plane formed by vectors $\mathbf{n}$ and $\mathbf{h}_0$. Next, assume that the approximate equations in (1.774) and (1.775) are valid in every of two regions on the both sides of $\mathcal{T}$. Then by (1.781) we have

$$|\mathbf{h}_1| = |\mathbf{h}_0| = \frac{c}{c_0}. \tag{1.804}$$

Then, since $\mathbf{h}_1 \neq \mathbf{h}_0$, by (1.803) and (1.804) we deduce

$$\mathbf{n} \cdot \mathbf{h}_1 = -\mathbf{n} \cdot \mathbf{h}_0 \qquad \forall \mathbf{x} \in \mathcal{T}. \tag{1.805}$$

So, by (1.804) and (1.805) we obtain the law of reflection: vector $\mathbf{h}_1$ lies in the plane formed by vectors $\mathbf{n}$ and $\mathbf{h}_0$, and we have

$$\theta\,(\mathbf{h}_0, -\mathbf{n}) = \theta_1\,(\mathbf{h}_1, \mathbf{n}) \tag{1.806}$$

where $\theta\,(\mathbf{h}_0, -\mathbf{n})$ is the angle between the incoming ray direction $\mathbf{h}_0$ and the incoming normal to the surface $-\mathbf{n}$ and $\theta_1\,(\mathbf{h}_1, \mathbf{n})$ is the angle between the reflected ray direction $\mathbf{h}_1$ and the outcoming normal $\mathbf{n}$.

Next assume that the wave we consider in (1.795) has an electromagnetic nature. Then by (1.787) we have

$$c_0 = c\sqrt{\kappa_0 \gamma_0} \quad \text{and} \quad \tilde{\mathbf{u}} = (\gamma_0 \kappa_0 \mathbf{v} + (1 - \gamma_0 \kappa_0)\mathbf{u}), \tag{1.807}$$

where, $\mathbf{u}$ is the medium velocity and $\mathbf{v}$ is the local vectorial gravitational potential. Similarly, on the second side of surface $\mathcal{T}$ we have

$$c_0^{(2)} = c\sqrt{\kappa_0^{(2)} \gamma_0^{(2)}} \quad \text{and} \quad \tilde{\mathbf{u}}_2 = \left(\gamma_0^{(2)} \kappa_0^{(2)} \mathbf{v} + (1 - \gamma_0^{(2)} \kappa_0^{(2)})\mathbf{u}^{(2)}\right), \tag{1.808}$$

where, $\mathbf{u}^{(2)}$ is the medium velocity on the second side of surface $\mathcal{T}$. Furthermore, assume that the medium rests on the both sides of surface $\mathcal{T}$. Then in this particular case we rewrite (1.807) and (1.808) as

$$c_0 = c\sqrt{\kappa_0\gamma_0} \quad \text{and} \quad \tilde{\mathbf{u}} = \gamma_0\kappa_0\mathbf{v}, \tag{1.809}$$

and

$$c_0^{(2)} = c\sqrt{\kappa_0^{(2)}\gamma_0^{(2)}} \quad \text{and} \quad \tilde{\mathbf{u}}_2 = \gamma_0^{(2)}\kappa_0^{(2)}\mathbf{v}. \tag{1.810}$$

Then in particular, by (1.809) and (1.810) we deduce

$$\frac{c}{\left(c_0^{(2)}\right)^2}\tilde{\mathbf{u}}_2 = \frac{c}{c_0^2}\tilde{\mathbf{u}} = \frac{1}{c}\,\mathbf{v}. \tag{1.811}$$

Thus, by inserting (1.796) and (1.811) into (1.802), we deduce

$$\mathbf{h}_2 - (\mathbf{n}\cdot\mathbf{h}_2)\,\mathbf{n} = \mathbf{h}_0 - (\mathbf{n}\cdot\mathbf{h}_0)\,\mathbf{n} \qquad \forall \mathbf{x} \in \mathcal{T}, \tag{1.812}$$

and in particular, for every point on the surface $\mathcal{T}$ vector $\mathbf{h}_2$ lies in the plane formed by vectors $\mathbf{n}$ and $\mathbf{h}_0$. On the other hand by (1.781) we have

$$|\mathbf{h}_0| = \frac{c}{c_0} \quad \text{and} \quad |\mathbf{h}_2| = \frac{c}{c_0^{(2)}}. \tag{1.813}$$

So, by (1.812) and (1.813), we have Snell's law of refraction: vector $\mathbf{h}_2$ lies in the plane formed by vectors $\mathbf{n}$ and $\mathbf{h}_0$, and we have

$$n\sin\left(\theta\left(\mathbf{h}_0,\mathbf{n}\right)\right) = n_2\sin\left(\theta_2\left(\mathbf{h}_2,\mathbf{n}\right)\right) \tag{1.814}$$

where $\theta\left(\mathbf{h}_0,\mathbf{n}\right)$ is the angle between the incoming ray direction $\mathbf{h}_0$ and the normal to the surface $\mathbf{n}$, $\theta_2\left(\mathbf{h}_2,\mathbf{n}\right)$ is the angle between the refracted ray direction $\mathbf{h}_2$ and the normal $\mathbf{n}$ and as in (1.783) we set refraction indexes:

$$n := \frac{c}{c_0} \quad \text{and} \quad n_2 := \frac{c}{c_0^{(2)}}. \tag{1.815}$$

1.15.5 Sagnac Effect

Assume again the monochromatic electromagnetic wave of the frequency $\nu = \omega/(2\pi)$ characterized by

$$U(\mathbf{x}, t) = A(\mathbf{x}, t)e^{iT(\mathbf{x},t)} = A(\mathbf{x}, t)e^{ik_0 S(\mathbf{x},t)}, \quad \text{where} \quad k_0 = \frac{\omega}{c} \quad \text{and} \quad \frac{\partial S}{\partial t} = c\,. \tag{1.816}$$

Then by (1.787) we have

$$c_0 = c\sqrt{\kappa_0\gamma_0} \quad \text{and} \quad \tilde{\mathbf{u}} = (\gamma_0\kappa_0\mathbf{v} + (1 - \gamma_0\kappa_0)\mathbf{u})\,, \tag{1.817}$$

where, $\mathbf{u}$ is the medium velocity and $\mathbf{v}$ is the local vectorial gravitational potential. Moreover, assume again that we consider light traveling in some regions either filled with the resting medium of constant dielectric permeability γ_0 and magnetic permeability κ_0 or in the vacuum. Then by (1.817) and (1.783) we have

$$n = \frac{1}{\sqrt{\kappa_0\gamma_0}} \quad \text{is a constant,} \quad \text{and} \quad \tilde{\mathbf{u}} = \gamma_0\kappa_0\mathbf{v}. \tag{1.818}$$

Next, assume that the light travels from point N to point M across the curve $\mathbf{r}(s) : [a, b] \to \mathbb{R}^3$ with endpoints $\mathbf{r}(a) = N$ and $\mathbf{r}(b) = M$ undergoing possibly certain number of reflections from mirrors during its travel. Then by (1.784), (1.818) and (1.801) we have

$$\delta(-S) := \left(-S(M^-)\right) - \left(-S(N^+)\right) = \frac{1}{\sqrt{\kappa_0\gamma_0}} \int_a^b \left|\mathbf{r}'(s)\right| ds - \frac{1}{c}\int_a^b \mathbf{v}\left(\mathbf{r}(s)\right) \cdot \mathbf{r}'(s) ds. \tag{1.819}$$

In particular, if we assume that $M = N$, i.e. our curve is closed and moreover, our curve is the boundary of some surface $\mathcal{S}_0$, then by Stokes Theorem we have

$$\begin{aligned}\delta(-S) = \left(-S(M^-)\right) - \left(-S(M^+)\right) &= \frac{1}{\sqrt{\kappa_0\gamma_0}} \int_a^b \left|\mathbf{r}'(s)\right| ds - \frac{1}{c}\iint (\text{curl}_{\mathbf{x}}\mathbf{v}) \cdot \mathbf{n}\, d\mathcal{S}_0 \\ &= \frac{1}{\sqrt{\kappa_0\gamma_0}} \left|\partial\mathcal{S}_0\right| - \frac{1}{c}\iint (\text{curl}_{\mathbf{x}}\mathbf{v}) \cdot \mathbf{n}\, d\mathcal{S}_0,\end{aligned} \tag{1.820}$$

where $\mathbf{n}$ is the unit normal to the surface. In particular, if our coordinate system is inertial, or more generally non-rotating, then $\text{curl}_{\mathbf{x}}\mathbf{v} = 0$ and by (1.820) we deduce

$$\delta(-S) = \frac{1}{\sqrt{\kappa_0\gamma_0}} \left|\partial\mathcal{S}_0\right|. \tag{1.821}$$

On the other hand, if our system is rotating, then as in (1.790) we clearly deduce

$$\text{curl}_{\mathbf{x}}\mathbf{v} = -2\mathbf{w}, \tag{1.822}$$

where **w** is the vector of the angular speed of rotation of our coordinate system. Then by (1.822) and (1.820) we deduce

$$\delta(-S) = \frac{1}{\sqrt{\kappa_0\gamma_0}}\,|\partial\mathcal{S}_0| + \frac{2}{c}\iint \mathbf{w}\cdot\mathbf{n}\,d\mathcal{S}_0. \tag{1.823}$$

In particular, if the surface $\mathcal{S}_0$ is a part of some plain then we rewrite (1.823) as

$$\delta(-S) = \frac{1}{\sqrt{\kappa_0\gamma_0}}\,|\partial\mathcal{S}_0| + \frac{2}{c}\,(\mathbf{w}\cdot\mathbf{n})\,|\mathcal{S}_0|\,. \tag{1.824}$$

On the other hand, if the light travels across the same curve in the opposite direction, then we must have

$$\delta(-S^-) = \frac{1}{\sqrt{\kappa_0\gamma_0}}\,|\partial\mathcal{S}_0| - \frac{2}{c}\,(\mathbf{w}\cdot\mathbf{n})\,|\mathcal{S}_0|\,. \tag{1.825}$$

Thus, by taking the difference in two cases and using (1.816), we deduce

$$\big(\delta(-T) - \delta(-T^-)\big) = k_0\,\big(\delta(-S) - \delta(-S^-)\big) = \frac{4\omega}{c^2}\cdot(\mathbf{w}\cdot\mathbf{n})\,|\mathcal{S}_0|\,. \tag{1.826}$$

Here, γ_0 and κ_0 are the dielectric and the magnetic permeability of the medium, T is given in (1.816), $|\mathcal{S}_0|$ is the area of the flat surface bounded by the closed path of the light, $\mathbf{n}$ is the unit normal to the surface, ω is the frequency of the light and $\mathbf{w}$ is the angular speed vector of the rotation of our coordinate system.

1.15.6 *Fizeau Experiment*

Assume again the monochromatic electromagnetic wave of the frequency $\nu = \omega/(2\pi)$ characterized by

$$U(\mathbf{x},t) = A(\mathbf{x},t)e^{iT(\mathbf{x},t)} = A(\mathbf{x},t)e^{ik_0 S(\mathbf{x},t)}, \quad \text{where } k_0 = \frac{\omega}{c} \text{ and } \frac{\partial S}{\partial t} = c\,. \tag{1.827}$$

Then by (1.787) we have

$$c_0 = c\sqrt{\kappa_0\gamma_0} \text{ and } \tilde{\mathbf{u}} = (\gamma_0\kappa_0\mathbf{v} + (1-\gamma_0\kappa_0)\mathbf{u})\,, \tag{1.828}$$

where, $\mathbf{u}$ is the medium velocity and $\mathbf{v}$ is the local vectorial gravitational potential. Moreover, assume that we consider light traveling in some regions filled with the moving medium of constant dielectric permeability γ_0 and magnetic permeability κ_0. Then by (1.828) and (1.783) we have

$$n=\frac{c}{c_0}=\frac{1}{\sqrt{\kappa_0\gamma_0}} \quad \text{is a constant,} \quad \text{and} \quad \tilde{\mathbf{u}}=\frac{1}{n^2}\mathbf{v}+\left(1-\frac{1}{n^2}\right)\mathbf{u}. \tag{1.829}$$

Next, assume that the light travels from point N to point M across the curve $\mathbf{r}(s) : [a,b] \to \mathbb{R}^3$ with endpoints $\mathbf{r}(a) = N$ and $\mathbf{r}(b) = M$ undergoing possibly certain number of reflections from mirrors during its travel. Then, as before, by (1.784), (1.829) and (1.801) we have

$$\delta(-S) := \left(-S(M^-)\right) - \left(-S(N^+)\right) =$$
$$n\int_a^b \left|\mathbf{r}'(s)\right| ds - \frac{1}{c}\int_a^b \mathbf{v}\left(\mathbf{r}(s)\right)\cdot\mathbf{r}'(s)ds - \frac{n^2}{c}\left(1-\frac{1}{n^2}\right)\int_a^b \mathbf{u}\left(\mathbf{r}(s)\right)\cdot\mathbf{r}'(s)ds. \tag{1.830}$$

Next assume that, either our curve is perpendicular to the direction of the vectorial gravitational potential $\mathbf{v}$, that happens, for example, if our path of the light is tangent to the Earth surface, or assume that our curve is closed, i.e. $M = N$ and moreover, our coordinate system is inertial, or more generally non-rotating. In particular, if we assume that $M = N$, i.e. our curve is closed and moreover, our coordinate system is inertial, or more generally non-rotating, then, as before, by Stokes Theorem we have

$$\int_a^b \mathbf{v}\left(\mathbf{r}(s)\right)\cdot\mathbf{r}'(s)ds = 0. \tag{1.831}$$

On the other hand in the case that our curve is perpendicular to the direction of the vectorial gravitational potential $\mathbf{v}$, (1.831) also trivially follows. Therefore, by inserting (1.831) into (1.830) in both cases we obtain

$$\begin{aligned}\delta(-S) &= \left(-S(M^-)\right) - \left(-S(N^+)\right)\\ &= n\int_a^b \left|\mathbf{r}'(s)\right| ds - \frac{n^2}{c}\left(1-\frac{1}{n^2}\right)\int_a^b \mathbf{u}\left(\mathbf{r}(s)\right)\cdot\mathbf{r}'(s)ds.\end{aligned} \tag{1.832}$$

Then by (1.832) and (1.827) we deduce

$$\begin{aligned}\delta(-T) &:= \left(-T(M^-)\right) - \left(-T(N^+)\right) = k_0\delta(-S)\\ &= \frac{n\omega}{c}\int_a^b \left|\mathbf{r}'(s)\right| ds - \frac{n^2\omega}{c^2}\left(1-\frac{1}{n^2}\right)\int_a^b \mathbf{u}\left(\mathbf{r}(s)\right)\cdot\mathbf{r}'(s)ds\\ &\frac{n^2\omega}{c^2}\left(c_0\int_a^b \left|\mathbf{r}'(s)\right| ds - \left(1-\frac{1}{n^2}\right)\int_a^b \mathbf{u}\left(\mathbf{r}(s)\right)\cdot\mathbf{r}'(s)ds\right).\end{aligned} \tag{1.833}$$

In particular, if the absolute value $|\mathbf{u}\,(\mathbf{r}(s))|$ is a constant across the curve and if the angle between $r'(s)$ and $\mathbf{u}\,(\mathbf{r}(s))$ is a constant across the curve and equals to the value θ then denoting the length of the path by L:

$$L := \int_a^b \left|\mathbf{r}'(s)\right| ds, \tag{1.834}$$

by (1.833) we deduce

$$\delta(-T) = k_0\delta(-S) = \frac{\omega L n^2}{c^2}\left(c_0 - \left(1 - \frac{1}{n^2}\right)|\mathbf{u}|\cos(\theta)\right). \tag{1.835}$$

Thus, if the direction of $\mathbf{u}$ coincides with the direction of the light, i.e. $\theta = 0$ then

$$\delta(-T) = k_0\delta(-S) = \frac{\omega L n^2}{c^2}\left(c_0 - \left(1 - \frac{1}{n^2}\right)|\mathbf{u}|\right) \approx \frac{\omega L}{\left(c_0 + \left(1 - \frac{1}{n^2}\right)|\mathbf{u}|\right)}. \tag{1.836}$$

On the other hand, if the direction of $\mathbf{u}$ is opposite to the direction of the light, i.e. $\theta = \pi$ then

$$\delta(-T) = k_0\delta(-S) = \frac{\omega L n^2}{c^2}\left(c_0 + \left(1 - \frac{1}{n^2}\right)|\mathbf{u}|\right) \approx \frac{\omega L}{\left(c_0 - \left(1 - \frac{1}{n^2}\right)|\mathbf{u}|\right)}. \tag{1.837}$$

So, in the frames of our model we explain the results of the Fizeau experiment.

1.15.7 The Case of the Non-monochromatic Wave or/and Moving Domains of Propagation of Light

Next, assume that the wave is not monochromatic or/and the fields $\tilde{\mathbf{u}}$ and c_0 depend on the time variable or/and we consider the case of moving domains of propagation of light (in particular moving surfaces of reflection/refraction). In other words we can not assume (1.769) or (1.770) anymore. However, we do the Geometric Optics approximation (1.751). We also assume that our medium has no dispersion. Then, due to (1.746) we have

$$U(\mathbf{x}, t) = A(\mathbf{x}, t)e^{ik_0 S(\mathbf{x},t)}, \tag{1.838}$$

and by (1.744) and (1.745) we have

$$\left\langle \left|\frac{\partial S}{\partial t}\right| \right\rangle = c, \tag{1.839}$$

where the sign $\langle\cdot\rangle$ means the spatial and temporal averaging. Then, due to (1.753) we have the Eikonal-type equation:

$$\frac{1}{c_0^2}\left(\frac{\partial S}{\partial t}+\tilde{\mathbf{u}}\cdot\nabla_{\mathbf{x}}S\right)^2=|\nabla_{\mathbf{x}}S|^2, \tag{1.840}$$

and we rewrite the equation of the propagation of the beam (1.754) as

$$\frac{\partial A}{\partial t}(\mathbf{x},t)+\mathbf{h}(\mathbf{x},t)\cdot\nabla_{\mathbf{x}}A(\mathbf{x},t)=G(\mathbf{x},t)A(\mathbf{x},t), \tag{1.841}$$

where we denote

$$\mathbf{h}(\mathbf{x},t):=\tilde{\mathbf{u}}(\mathbf{x},t)-c_0^2(\mathbf{x},t)\left(\frac{\partial S}{\partial t}(\mathbf{x},t)+\tilde{\mathbf{u}}(\mathbf{x},t)\cdot\nabla_{\mathbf{x}}S(\mathbf{x},t)\right)^{-1}\nabla_{\mathbf{x}}S(\mathbf{x},t), \tag{1.842}$$

and

$$G(\mathbf{x},t):=\frac{c_0^2}{2\left(\frac{\partial S}{\partial t}+\tilde{\mathbf{u}}\cdot\nabla_{\mathbf{x}}S\right)}\Bigg(\Delta_{\mathbf{x}}S$$
$$-\frac{\partial}{\partial t}\left\{\frac{1}{c_0^2}\left(\frac{\partial S}{\partial t}+\tilde{\mathbf{u}}\cdot\nabla_{\mathbf{x}}S\right)\right\}-\operatorname{div}_{\mathbf{x}}\left\{\frac{1}{c_0^2}\left(\frac{\partial S}{\partial t}+\tilde{\mathbf{u}}\cdot\nabla_{\mathbf{x}}S\right)\tilde{\mathbf{u}}\right\}\Bigg). \tag{1.843}$$

Next, consider a curve $\mathbf{r}(t):[t_0,b]\to\mathbb{R}^3$ parameterized by the time variable t, defined as a solution of ordinary differential equation:

$$\begin{cases}\frac{d\mathbf{r}}{dt}(t)=\mathbf{h}\left(\mathbf{r}(t),t\right)\\ \mathbf{r}(t_0)=\mathbf{x}_0,\end{cases} \tag{1.844}$$

where $\mathbf{h}$ was defined in (1.842). Note that the equations of the ray propagation of the form (1.844) are not always convenient since $\mathbf{h}$ in the right-hand side of (1.844) depend on the partial derivatives of the quantity S, which we do not know apriory unless we solve (1.840). In the following proposition, that we prove in subsection 13.3.8, we present the alternative form for the equations of the ray propagation being the second-order ordinary differential equations which dose not contain the quantity S or its partial derivatives.

Proposition 1.9 *Consider a smooth solution of* (1.840). *Next let* $\mathbf{r}(t):[t_0,b]\to\mathbb{R}^3$ *be any curve parameterized by the time variable* t*, defined as a solution* (1.844). *Then,* $\mathbf{r}(t)$ *satisfies the following second-order ordinary differential equations of the Ray Propagation:*

$$\frac{d}{dt}\left(\frac{1}{c_0\left(\mathbf{r}(t),t\right)}\left(\frac{d\mathbf{r}}{dt}(t)-\tilde{\mathbf{u}}\left(\mathbf{r}(t),t\right)\right)\right)=$$
$$-\nabla_{\mathbf{x}}c_0(\mathbf{r},t)-\frac{1}{c_0(\mathbf{r},t)}\left\{d_{\mathbf{x}}\tilde{\mathbf{u}}(\mathbf{r},t)\right\}^T\cdot\left(\frac{d\mathbf{r}}{dt}-\tilde{\mathbf{u}}(\mathbf{r},t)\right)$$
$$+\frac{1}{c_0^2(\mathbf{r},t)}\left(\left(\frac{d\mathbf{r}}{dt}-\tilde{\mathbf{u}}(\mathbf{r},t)\right)\cdot\nabla_{\mathbf{x}}c_0(\mathbf{r},t)\right)\left(\frac{d\mathbf{r}}{dt}-\tilde{\mathbf{u}}(\mathbf{r},t)\right)$$
$$+\frac{1}{2c_0^3(\mathbf{r},t)}\left(\left(\left(\left\{d_{\mathbf{x}}\tilde{\mathbf{u}}(\mathbf{r},t)\right\}^T\right.\right.\right.$$
$$\left.\left.\left.+d_{\mathbf{x}}\tilde{\mathbf{u}}(\mathbf{r},t)\right)\cdot\left(\frac{d\mathbf{r}}{dt}-\tilde{\mathbf{u}}(\mathbf{r},t)\right)\right)\cdot\left(\frac{d\mathbf{r}}{dt}-\tilde{\mathbf{u}}(\mathbf{r},t)\right)\right)\left(\frac{d\mathbf{r}}{dt}-\tilde{\mathbf{u}}(\mathbf{r},t)\right). \quad (1.845)$$

Moreover, the proper scalar quantity $\tilde{\omega}(t):[t_0,b]\to\mathbb{R}$, defined by

$$\tilde{\omega}(t)=\frac{\partial S}{\partial t}\left(\mathbf{r}(t),t\right)+\tilde{\mathbf{u}}\left(\mathbf{r}(t),t\right)\cdot\nabla_{\mathbf{x}}S\left(\mathbf{r}(t),t\right) \quad (1.846)$$

satisfies the following ordinary differential equation:

$$\frac{d\tilde{\omega}}{dt}(t)=\frac{\tilde{\omega}(t)}{c_0(\mathbf{r},t)}\left(\frac{\partial c_0}{\partial t}(\mathbf{r},t)+\tilde{\mathbf{u}}(\mathbf{r},t)\cdot\nabla_{\mathbf{x}}c_0(\mathbf{r},t)\right)$$
$$-\frac{\tilde{\omega}(t)}{2c_0^2(\mathbf{r},t)}\left(\left(\left\{d_{\mathbf{x}}\tilde{\mathbf{u}}(\mathbf{r},t)\right\}^T+d_{\mathbf{x}}\tilde{\mathbf{u}}(\mathbf{r},t)\right)\cdot\left(\frac{d\mathbf{r}}{dt}(t)-\tilde{\mathbf{u}}(\mathbf{r},t)\right)\right)\cdot\left(\frac{d\mathbf{r}}{dt}(t)-\tilde{\mathbf{u}}(\mathbf{r},t)\right). \quad (1.847)$$

Note that, as we deduce in Sect. 13.3.8, if $\mathbf{r}(t)$ satisfies (1.845) and moreover it satisfies the following equality for the initial instant of time t_0:

$$\frac{1}{c_0^2\left(\mathbf{r}(t_0),t_0\right)}\left|\frac{d\mathbf{r}}{dt}(t_0)-\tilde{\mathbf{u}}\left(\mathbf{r}(t_0),t_0\right)\right|^2=1,$$

then we have

$$\frac{1}{c_0^2\left(\mathbf{r}(t),t\right)}\left|\frac{d\mathbf{r}}{dt}(t)-\tilde{\mathbf{u}}\left(\mathbf{r}(t),t\right)\right|^2=1$$

for every instant of time. So we have the following consistency with (1.768) and (1.844):

$$\left|\frac{d\mathbf{r}}{dt}(t)-\tilde{\mathbf{u}}(\mathbf{r},t)\right|^2=c_0^2(\mathbf{r},t). \quad (1.848)$$

Next note that, as before, we can easily deduce that equations of the ray propagation either of the form (1.844) or of the form (1.845) are invariant under the change of non-inertial Cartesian coordinate system. Moreover, (1.847) is also invariant under the change of non-inertial Cartesian coordinate system.

Next consider a wave characterized by

$$U(\mathbf{x}, t) = A(\mathbf{x}, t)e^{ik_0 S(\mathbf{x},t)}, \tag{1.849}$$

and, consistently with (1.842) consider a velocity field of the wave:

$$\mathbf{h}(\mathbf{x}, t) := \tilde{\mathbf{u}}(\mathbf{x}, t) - c_0^2(\mathbf{x}, t)\left(\frac{\partial S}{\partial t}(\mathbf{x}, t) + \tilde{\mathbf{u}}(\mathbf{x}, t)\cdot\nabla_{\mathbf{x}} S(\mathbf{x}, t)\right)^{-1} \nabla_{\mathbf{x}} S(\mathbf{x}, t). \tag{1.850}$$

Furthermore, assume that the wave we consider undergoes reflection and/or refraction on the time-dependent surface $\mathcal{T}$ having the outcoming three-dimensional unit normal $\mathbf{n}(\mathbf{x}, t)$ and the motion velocity field $\mathbf{w}_{\mathcal{T}}(\mathbf{x}, t)$, separating two regions characterized respectively by $c_0 = c_0^{(1)}$ and $\tilde{\mathbf{u}} = \tilde{\mathbf{u}}_1$ and by $c_0^{(2)}$ and $\tilde{\mathbf{u}}_2$, with the formation of the reflected wave, characterized by

$$U_1(\mathbf{x}, t) = A_1(\mathbf{x}, t)e^{ik_0 S_1(\mathbf{x},t)}, \tag{1.851}$$

and by the velocity field:

$$\mathbf{h}_1(\mathbf{x}) = \tilde{\mathbf{u}}_1(\mathbf{x}, t) - c_0^2(\mathbf{x}, t)\left(\frac{\partial S_1}{\partial t}(\mathbf{x}, t) + \tilde{\mathbf{u}}_1(\mathbf{x}, t)\cdot\nabla_{\mathbf{x}} S_1(\mathbf{x}, t)\right)^{-1} \nabla_{\mathbf{x}} S_1(\mathbf{x}, t), \tag{1.852}$$

and formation of the refracted wave characterized by

$$U_2(\mathbf{x}, t) = A_2(\mathbf{x}, t)e^{ik_0 S_2(\mathbf{x},t)}, \tag{1.853}$$

and by the velocity field:

$$\mathbf{h}_2(\mathbf{x}) = \tilde{\mathbf{u}}_2(\mathbf{x}, t) - (c_0^{(2)})^2(\mathbf{x}, t)\left(\frac{\partial S_2}{\partial t}(\mathbf{x}, t) + \tilde{\mathbf{u}}_2(\mathbf{x}, t)\cdot\nabla_{\mathbf{x}} S_2(\mathbf{x}, t)\right)^{-1} \nabla_{\mathbf{x}} S_2(\mathbf{x}, t). \tag{1.854}$$

Then the boundary conditions of U, U_1 and U_2 depend on the physical meaning of these fields. However, one of the necessary conditions should be that

$$S_1(\mathbf{x}, t) = S_2(\mathbf{x}, t) + C_2 = S(\mathbf{x}, t) \quad \forall \mathbf{x} \in \mathcal{T},\ \forall t, \tag{1.855}$$

where C_2 is a real constant. In particular (1.855) implies

$$\nabla_{\mathbf{x}} S_1 - (\mathbf{n}\cdot\nabla_{\mathbf{x}} S_1)\,\mathbf{n} = \nabla_{\mathbf{x}} S_2 - (\mathbf{n}\cdot\nabla_{\mathbf{x}} S_2)\,\mathbf{n} = \nabla_{\mathbf{x}} S - (\mathbf{n}\cdot\nabla_{\mathbf{x}} S)\,\mathbf{n} \quad \forall \mathbf{x} \in \mathcal{T},\ \forall t. \tag{1.856}$$

In particular, for every point on the surface $\mathcal{T}$ vectors $\nabla_{\mathbf{x}} S_1$ and $\nabla_{\mathbf{x}} S_2$ lie in the plane formed by vectors $\mathbf{n}$ and $\nabla_{\mathbf{x}} S$. Moreover, by (1.855) we also have

$$\frac{\partial S}{\partial t} + \mathbf{w}_{\mathcal{T}} \cdot \nabla_{\mathbf{x}} S = \frac{\partial S_1}{\partial t} + \mathbf{w}_{\mathcal{T}} \cdot \nabla_{\mathbf{x}} S_1 = \frac{\partial S_2}{\partial t} + \mathbf{w}_{\mathcal{T}} \cdot \nabla_{\mathbf{x}} S_2 \qquad \forall \mathbf{x} \in \mathcal{T},\ \forall t. \tag{1.857}$$

Finally, by (1.768) we have

$$|\mathbf{h} - \tilde{\mathbf{u}}|^2 = c_0^2, \quad |\mathbf{h}_1 - \tilde{\mathbf{u}}|^2 = c_0^2 \ \text{ and } \ |\mathbf{h}_2 - \tilde{\mathbf{u}}_2|^2 = (c_0^{(2)})^2. \tag{1.858}$$

In Sect. 13.3.8 we prove that if the following approximation is valid on both sides of the surface $\mathcal{T}$:

$$\frac{|\mathbf{w}_{\mathcal{T}}|^2}{c_0^2 + (c_0^{(2)})^2} \ll 1, \quad \frac{|\tilde{\mathbf{u}}|^2}{c_0^2} \ll 1 \ \text{ and } \ \frac{|\tilde{\mathbf{u}}_2|^2}{(c_0^{(2)})^2} \ll 1, \tag{1.859}$$

then, we have the following law of reflection: vector $(\mathbf{h}_1 - \mathbf{w}_{\mathcal{T}})$ lies in the plane formed by vectors $\mathbf{n}$ and $(\mathbf{h} - \mathbf{w}_{\mathcal{T}})$, and we have

$$\theta\left((\mathbf{h} - \mathbf{w}_{\mathcal{T}}), -\mathbf{n}\right) = \theta_1\left((\mathbf{h}_1 - \mathbf{w}_{\mathcal{T}}), \mathbf{n}\right) \tag{1.860}$$

where $\theta\left((\mathbf{h} - \mathbf{w}_{\mathcal{T}}), -\mathbf{n}\right)$ is the angle between the vector of the relative velocity of the incoming ray, relative to the surface of reflection, $(\mathbf{h} - \mathbf{w}_{\mathcal{T}})$ and the incoming normal to the surface $-\mathbf{n}$ and $\theta_1\left((\mathbf{h}_1 - \mathbf{w}_{\mathcal{T}}), \mathbf{n}\right)$ is the angle between the vector of the relative velocity of the reflected ray, relative to the surface of reflection, $(\mathbf{h}_1 - \mathbf{w}_{\mathcal{T}})$ and the outcoming normal $\mathbf{n}$.

Moreover, if we assume that the wave we consider in (1.849) has an electromagnetic nature. Then by (1.787) we have

$$c_0 = c\sqrt{\kappa_0 \gamma_0} \ \text{ and } \ \tilde{\mathbf{u}} = \left(\gamma_0 \kappa_0 \mathbf{v} + (1 - \gamma_0 \kappa_0)\mathbf{u}\right), \tag{1.861}$$

where, $\mathbf{u}$ is the medium velocity and $\mathbf{v}$ is the local vectorial gravitational potential. Similarly, on the second side of surface $\mathcal{T}$ we have

$$c_0^{(2)} = c\sqrt{\kappa_0^{(2)} \gamma_0^{(2)}} \ \text{ and } \ \tilde{\mathbf{u}}_2 = \left(\gamma_0^{(2)} \kappa_0^{(2)} \mathbf{v} + (1 - \gamma_0^{(2)} \kappa_0^{(2)})\mathbf{u}_2\right), \tag{1.862}$$

where, $\mathbf{u}_2$ is the medium velocity on the second side of surface $\mathcal{T}$. Then in Sect. 13.3.8 we also prove that we have the following Snell's law of refraction: $(\mathbf{h}_2 - \mathbf{w}_{\mathcal{T}})$ lies in the plane formed by vectors $\mathbf{n}$ and $(\mathbf{h} - \mathbf{w}_{\mathcal{T}})$ and we have

$$n \sin\left(\theta\left((\mathbf{h} - \mathbf{w}_{\mathcal{T}}), \mathbf{n}\right)\right) = n_2 \sin\left(\theta_2\left((\mathbf{h}_2 - \mathbf{w}_{\mathcal{T}}), \mathbf{n}\right)\right) \tag{1.863}$$

where $\theta\left((\mathbf{h} - \mathbf{w}_{\mathcal{T}}), \mathbf{n}\right)$ is the vector of the relative velocity of the incoming ray, relative to the surface of refraction, $(\mathbf{h} - \mathbf{w}_{\mathcal{T}})$ and the normal to the surface $\mathbf{n}$, $\theta_2\left((\mathbf{h}_2 - \mathbf{w}_{\mathcal{T}}), \mathbf{n}\right)$ is the vector of the relative velocity of the refracted ray, relative to the surface of refraction, $(\mathbf{h}_2 - \mathbf{w}_{\mathcal{T}})$ and the normal $\mathbf{n}$ and as in (1.783) we set refraction indexes:

$$n := \frac{c}{c_0} \quad \text{and} \quad n_2 := \frac{c}{c_0^{(2)}}. \tag{1.864}$$

Note that, since $\mathbf{h}$ and $\mathbf{w}_{\mathcal{T}}$ are both speed like vector fields then $(\mathbf{h} - \mathbf{w}_{\mathcal{T}})$ is a proper vector field and thus the above law of reflection together with (1.860) and Snell's law together with (1.863) are invariant under the change of inertial or non-inertial Cartesian coordinate systems. In particular, if (1.859) holds for some Cartesian coordinate system, then we can use this laws also in other coordinate systems where (1.859) does not hold. Therefore, for the validity of the above laws of reflection and refraction we may assume the following relation instead of (1.859):

$$\frac{|\tilde{\mathbf{u}} - \mathbf{w}_{\mathcal{T}}|^2}{c_0^2} \ll 1 \quad \text{and} \quad \frac{|\tilde{\mathbf{u}}_2 - \mathbf{w}_{\mathcal{T}}|^2}{(c_0^{(2)})^2} \ll 1. \tag{1.865}$$

1.15.8 Polarization of the Light Inside a Moving Medium and/or in the Presence of Gravitational Field

Again consider the system of Maxwell equations in the vacuum or in a medium of the form (1.616) in the absence of macroscopic charges and/or currents:

$$\begin{cases} \text{curl}_{\mathbf{x}}\mathbf{H} = \frac{1}{c}\frac{\partial \mathbf{D}}{\partial t}, \\ \text{div}_{\mathbf{x}}\,\mathbf{D} = 0, \\ \text{curl}_{\mathbf{x}}\mathbf{E} + \frac{1}{c}\frac{\partial \mathbf{B}}{\partial t} = 0, \\ \text{div}_{\mathbf{x}}\,\mathbf{B} = 0, \\ \mathbf{E} = \gamma_0\mathbf{D} - \frac{1}{c}\tilde{\mathbf{u}} \times \mathbf{B}, \\ \mathbf{H} = \kappa_0\mathbf{B} + \frac{1}{c}\tilde{\mathbf{u}} \times \mathbf{D}, \\ \tilde{\mathbf{u}} = (\gamma_0\kappa_0\mathbf{v} + (1 - \gamma_0\kappa_0)\mathbf{u}), \end{cases} \tag{1.866}$$

and consider the case of a monochromatic wave of the constant frequency $\nu = \frac{\omega}{2\pi}$ where the fields $\tilde{\mathbf{u}}$, γ_0 and κ_0 are independent on the time variable. We also assume that either our medium has no dispersion, i.e. (1.607) is valid, or the velocity of the medium is negligible, i.e. $\mathbf{u} = 0$ (and so $\tilde{\mathbf{u}} = \mathbf{v}$) and our medium is transparent for the given frequency $\nu = \frac{\omega}{2\pi}$.

Next assume the rough Geometric Optics approximation (1.752) (stronger than (1.751)) that means the following: assume that the changes in time of the basic characteristics of the electromagnetic field become essential after certain interval of time T_e and the changes in space of the basic characteristics of the electromagnetic field become essential in the spatial landscape L_e. Then we assume that

$$k_0 c_0 T_e \gg 1 \quad \text{and} \quad k_0 L_e \gg 1, \tag{1.867}$$

where

$$k_0 = \frac{\omega}{c}. \tag{1.868}$$

We also assume (1.773):

$$\frac{|\tilde{\mathbf{u}}|^2}{c_0^2} \ll 1. \tag{1.869}$$

Then consistently with (1.746) we can write

$$\begin{cases} \mathbf{D}(\mathbf{x},t) = \Xi_1 \cdot \mathbf{D}_a(\mathbf{x})e^{ik_0 S(\mathbf{x},t)} \\ \mathbf{B}(\mathbf{x},t) = \Xi_2 \cdot \mathbf{B}_a(\mathbf{x})e^{ik_0 S(\mathbf{x},t)} \\ \mathbf{E}(\mathbf{x},t) = \Xi_3 \cdot \mathbf{E}_a(\mathbf{x})e^{ik_0 S(\mathbf{x},t)} \\ \mathbf{H}(\mathbf{x},t) = \Xi_4 \cdot \mathbf{H}_a(\mathbf{x})e^{ik_0 S(\mathbf{x},t)} \\ \mathbf{E} = \gamma_0 \mathbf{D} - \frac{1}{c}\tilde{\mathbf{u}} \times \mathbf{B} \\ \mathbf{H} = \kappa_0 \mathbf{B} + \frac{1}{c}\tilde{\mathbf{u}} \times \mathbf{D}. \end{cases} \tag{1.870}$$

Here $\mathbf{D}_a(\mathbf{x}), \mathbf{B}_a(\mathbf{x}), \mathbf{E}_a(\mathbf{x}), \mathbf{H}_a(\mathbf{x}) : \mathbb{R}^3 \to \mathbb{R}^3$ are real vector fields, independent on the time variable, $S(\mathbf{x},t)$ is a real function such that $\frac{\partial S}{\partial t} = c$ and $\Xi_1, \Xi_2, \Xi_3, \Xi_4 \in \mathbb{C}^{3\times 3}$ are constant complex diagonal matrices of the form:

$$\Xi_k = \begin{pmatrix} e^{i\theta_{1k}} & 0 & 0 \\ 0 & e^{i\theta_{2k}} & 0 \\ 0 & 0 & e^{i\theta_{3k}} \end{pmatrix} \qquad \forall k = 1,2,3,4, \tag{1.871}$$

where $\theta_{1k}, \theta_{2k}, \theta_{3k}$ are real constants.

Then, consistently with (1.774), S satisfies the Eikonal equation:

$$\left| \frac{c\tilde{\mathbf{u}}}{c_0^2} - \nabla_{\mathbf{x}} S \right|^2 = \frac{c^2}{c_0^2}, \tag{1.872}$$

and consistently with (1.775) if $\mathbf{A}_1$ denotes one of the vectors $\mathbf{D}_a, \mathbf{B}_a, \mathbf{E}_a, \mathbf{H}_a$ then:

$$\{d_{\mathbf{x}}\mathbf{A}_1\}^T \cdot \left(\frac{c\tilde{\mathbf{u}}}{c_0^2} - \nabla_{\mathbf{x}} S \right) + \frac{1}{2}(-\Delta_{\mathbf{x}} S)\,\mathbf{A}_1 = 0. \tag{1.873}$$

Moreover, consistently with (1.769), (1.770) and (1.787) we have

$$\begin{cases} c_0 = c\sqrt{\kappa_0 \gamma_0} \\ \frac{\partial S}{\partial t} = c \\ \frac{\partial \mathbf{A}_1}{\partial t} = 0, \\ \frac{\partial \tilde{\mathbf{u}}}{\partial t} = 0 \\ \frac{\partial c_0}{\partial t} = 0, \end{cases} \tag{1.874}$$

and, consistently with (1.780), the vector field $\mathbf{h}_0$ defined for every $\mathbf{x}$ by

$$\mathbf{h}_0(\mathbf{x}) := \frac{c}{c_0^2(\mathbf{x})}\tilde{\mathbf{u}}(\mathbf{x}) - \nabla_{\mathbf{x}} S(\mathbf{x}) \tag{1.875}$$

is the direction of the propagation of the ray that passes through point $\mathbf{x}$. Then in Sect. 13.3.9 we deduce that

$$\begin{cases} \mathbf{B} \approx \frac{\gamma_0}{(1+\frac{1}{c}\tilde{\mathbf{u}}\cdot\nabla_{\mathbf{x}}S)} (-\nabla_{\mathbf{x}} S) \times \mathbf{D} \\ \mathbf{D} \approx -\frac{\kappa_0}{(1+\frac{1}{c}\tilde{\mathbf{u}}\cdot\nabla_{\mathbf{x}}S)} (-\nabla_{\mathbf{x}} S) \times \mathbf{B} \\ (-\nabla_{\mathbf{x}} S) \cdot \mathbf{D} \approx 0 \\ (-\nabla_{\mathbf{x}} S) \cdot \mathbf{B} \approx 0. \end{cases} \tag{1.876}$$

So the vectors $(-\nabla_{\mathbf{x}} S)$, $\mathbf{D}$ and $\mathbf{B}$ form together rightly orientated orthogonal system of vectors. Moreover, in Sect. 13.3.9 we also deduce that

$$\begin{cases} \mathbf{H} \approx \kappa_0 \mathbf{h}_0 \times \mathbf{E} \\ \mathbf{E} \approx -\gamma_0 \mathbf{h}_0 \times \mathbf{H} \\ \mathbf{h}_0 \cdot \mathbf{E} \approx 0 \\ \mathbf{h}_0 \cdot \mathbf{H} \approx 0. \end{cases} \tag{1.877}$$

So the vectors $\mathbf{h}_0$, $\mathbf{E}$ and $\mathbf{H}$ form together rightly orientated orthogonal system of vectors. We remind here again that $\mathbf{h}_0$ is the direction of the propagation of the ray.

1.15.9 More Delicate Approximation of the Wave Equation

Again consider the scalar wave equation of the form

$$\left(\frac{\partial}{\partial t}\left\{\frac{1}{c_0^2}\left(\frac{\partial U}{\partial t} + \tilde{\mathbf{u}} \cdot \nabla_{\mathbf{x}} U\right)\right\} + \operatorname{div}_{\mathbf{x}}\left\{\frac{1}{c_0^2}\left(\frac{\partial U}{\partial t} + \tilde{\mathbf{u}} \cdot \nabla_{\mathbf{x}} U\right)\tilde{\mathbf{u}}\right\}\right) - \Delta_{\mathbf{x}} U = 0. \tag{1.878}$$

Above, we established $U(\mathbf{x}, t)$ only in the rough Geometric Optics approximation (1.752). Since we have (1.755) in the rough Geometric Optics approximation and $S(\mathbf{x}, t)$ satisfies the following Eikonal equation

$$\frac{1}{c_0^2}\left(\frac{\partial S}{\partial t} + \tilde{\mathbf{u}} \cdot \nabla_{\mathbf{x}} S\right)^2 = |\nabla_{\mathbf{x}} S|^2 \tag{1.879}$$

in the case of more precise delicate Geometric Optics approximation, in order to get better approximation for $U(\mathbf{x}, t)$ we can use the following alternative consideration, slightly different than we did above: from now we consider $U(\mathbf{x}, t)$ be of the form

$$U(\mathbf{x},t) = A(\mathbf{x},t)e^{ik_0 S(\mathbf{x},t)}, \tag{1.880}$$

similarly to (1.755). However, we consider that S in (1.880) satisfied (1.879) precisely, although we consider the amplitude A to be complex rather than real! So we include the correction term of $\kappa_0 S$ into the complex amplitude A! Thus either (1.754) or (1.758) is not valid anymore, since it was derived under the assumption of real amplitude A. However, either (1.753) or (1.759) still holds. That is, we have (1.879), which assumed to hold precisely. Moreover, by (1.748) together with (1.879) we have

$$\begin{aligned}
&\left(\frac{\partial}{\partial t}\left\{\frac{1}{c_0^2}\left(\frac{\partial A}{\partial t} + \tilde{\mathbf{u}}\cdot\nabla_{\mathbf{x}}A\right)\right\} + \operatorname{div}_{\mathbf{x}}\left\{\frac{1}{c_0^2}\left(\frac{\partial A}{\partial t} + \tilde{\mathbf{u}}\cdot\nabla_{\mathbf{x}}A\right)\tilde{\mathbf{u}}\right\}\right) - \Delta_{\mathbf{x}}A \\
&+ i\kappa_0 A\left(\frac{\partial}{\partial t}\left\{\frac{1}{c_0^2}\left(\frac{\partial S}{\partial t} + \tilde{\mathbf{u}}\cdot\nabla_{\mathbf{x}}S\right)\right\} + \operatorname{div}_{\mathbf{x}}\left\{\frac{1}{c_0^2}\left(\frac{\partial S}{\partial t} + \tilde{\mathbf{u}}\cdot\nabla_{\mathbf{x}}S\right)\tilde{\mathbf{u}}\right\} - \Delta_{\mathbf{x}}S\right) \\
&+ 2i\kappa_0\left(\frac{1}{c_0^2}\left(\frac{\partial S}{\partial t} + \tilde{\mathbf{u}}\cdot\nabla_{\mathbf{x}}S\right)\left(\frac{\partial A}{\partial t} + \tilde{\mathbf{u}}\cdot\nabla_{\mathbf{x}}A\right) - \nabla_{\mathbf{x}}A\cdot\nabla_{\mathbf{x}}S\right) = 0.
\end{aligned} \tag{1.881}$$

Then denoting, the proper scalar field:

$$\tilde{\omega}(\mathbf{x},t) := \kappa_0\left(\frac{\partial S}{\partial t}(\mathbf{x},t) + \tilde{\mathbf{u}}(\mathbf{x},t)\cdot\nabla_{\mathbf{x}}S(\mathbf{x},t)\right), \tag{1.882}$$

the proper vector field:

$$\mathbf{k}(\mathbf{x},t) := c_0(\mathbf{x},t)\left(\frac{\partial S}{\partial t}(\mathbf{x},t) + \tilde{\mathbf{u}}(\mathbf{x},t)\cdot\nabla_{\mathbf{x}}S(\mathbf{x},t)\right)^{-1}\nabla_{\mathbf{x}}S(\mathbf{x},t), \tag{1.883}$$

the speed-like vector field:

$$\begin{aligned}
\mathbf{h}(\mathbf{x},t) &:= \tilde{\mathbf{u}}(\mathbf{x},t) - c_0^2(\mathbf{x},t)\left(\frac{\partial S}{\partial t}(\mathbf{x},t) + \tilde{\mathbf{u}}(\mathbf{x},t)\cdot\nabla_{\mathbf{x}}S(\mathbf{x},t)\right)^{-1}\nabla_{\mathbf{x}}S(\mathbf{x},t) \\
&= \tilde{\mathbf{u}}(\mathbf{x},t) - c_0(\mathbf{x},t)\mathbf{k}(\mathbf{x},t),
\end{aligned} \tag{1.884}$$

and the proper scalar field:

$$\begin{aligned}
G(\mathbf{x},t) := \Bigg(&\frac{c_0^2}{2\left(\frac{\partial S}{\partial t} + \tilde{\mathbf{u}}\cdot\nabla_{\mathbf{x}}S\right)}\Bigg(\Delta_{\mathbf{x}}S \\
&- \frac{\partial}{\partial t}\left\{\frac{1}{c_0^2}\left(\frac{\partial S}{\partial t} + \tilde{\mathbf{u}}\cdot\nabla_{\mathbf{x}}S\right)\right\} - \operatorname{div}_{\mathbf{x}}\left\{\frac{1}{c_0^2}\left(\frac{\partial S}{\partial t} + \tilde{\mathbf{u}}\cdot\nabla_{\mathbf{x}}S\right)\tilde{\mathbf{u}}\right\}\Bigg)\Bigg)(\mathbf{x},t),
\end{aligned} \tag{1.885}$$

we clearly rewrite (1.881) as

$$\frac{c_0^2}{2\tilde{\omega}}\left\{\frac{\partial}{\partial t}\left\{\frac{1}{c_0^2}\left(\frac{\partial A}{\partial t}+\tilde{\mathbf{u}}\cdot\nabla_{\mathbf{x}}A\right)\right\}+\operatorname{div}_{\mathbf{x}}\left\{\frac{1}{c_0^2}\left(\frac{\partial A}{\partial t}+\tilde{\mathbf{u}}\cdot\nabla_{\mathbf{x}}A\right)\tilde{\mathbf{u}}\right\}-\Delta_{\mathbf{x}}A\right\} + i\left(\frac{\partial A}{\partial t}+\mathbf{h}\cdot\nabla_{\mathbf{x}}A-G\,A\right)=0, \quad (1.886)$$

and (1.879), as

$$|\mathbf{k}(\mathbf{x},t)|^2=1 \quad \text{or equivalently} \quad |\mathbf{h}(\mathbf{x},t)-\tilde{\mathbf{u}}(\mathbf{x},t)|^2=c_0^2(\mathbf{x},t)\,. \quad (1.887)$$

Furthermore, from now assume the case of delicate Geometric Optics approximation. Remind that the delicate Geometric Optics approximation means the following: assume that the changes in time of c_0, $\tilde{\mathbf{u}}$, A and S become essential after certain interval of time T_e and the changes in space of c_0, A and S become essential in the spatial landscape L_e. Then we assume that

$$k_0^2c_0^2T_e^2 \gg 1 \ \text{ and } \ k_0^2L_e^2 \gg 1. \quad (1.888)$$

Moreover, as before, we assume that the order of c_0 is less or equal to the order of c. Then, using estimation (1.888) we rewrite (1.886) as

$$-\frac{c_0}{2\tilde{\omega}}\left(\nabla_{\mathbf{x}}A-(\mathbf{k}\cdot\nabla_{\mathbf{x}}A)\,\mathbf{k}\right)\cdot\left(\left(d_{\mathbf{x}}\tilde{\mathbf{u}}+\{d_{\mathbf{x}}\tilde{\mathbf{u}}\}^T\right)\cdot\mathbf{k}-\left(\left(\mathbf{k}\cdot\left(\{d_{\mathbf{x}}\tilde{\mathbf{u}}\}^T+d_{\mathbf{x}}\tilde{\mathbf{u}}\right)\right)\cdot\mathbf{k}\right)\mathbf{k}\right)$$
$$+\frac{c_0^2}{2\tilde{\omega}}A\left(\frac{\partial}{\partial t}\left\{\frac{1}{c_0^2}G\right\}+\operatorname{div}_{\mathbf{x}}\left\{\frac{1}{c_0^2}G\tilde{\mathbf{u}}\right\}+\frac{1}{c_0^2}(c_0\mathbf{k})\cdot\nabla_{\mathbf{x}}G+\frac{1}{c_0^2}G^2\right)$$
$$-\frac{c_0^2}{2}\left(\operatorname{div}_{\mathbf{x}}\left\{\frac{1}{\tilde{\omega}}\left(\nabla_{\mathbf{x}}A-(\mathbf{k}\cdot\nabla_{\mathbf{x}}A)\,\mathbf{k}\right)\right\}\right)\approx -i\left\{\frac{\partial A}{\partial t}+\mathbf{h}\cdot\nabla_{\mathbf{x}}A+\frac{1}{2}\left(\operatorname{div}_{\mathbf{x}}\mathbf{h}\right)A\right\}$$
$$-iA\frac{1}{2c_0}(c_0\mathbf{k}\cdot\nabla_{\mathbf{x}}c_0)+iA\frac{1}{2c_0}\left(\frac{\partial c_0}{\partial t}+\mathbf{h}\cdot\nabla_{\mathbf{x}}c_0\right)+\frac{iA}{4}\left(\mathbf{k}\cdot\left(\{d_{\mathbf{x}}\tilde{\mathbf{u}}\}^T+d_{\mathbf{x}}\tilde{\mathbf{u}}\right)\right)\cdot\mathbf{k} \quad (1.889)$$

(see Sect. 13.3.10 for the details). Moreover, as before, we can easily deduce that (1.889) invariant under the change of non-inertial Cartesian coordinate system, in the case that under such change we have $A'=A$ and $S'=S$.

In particular, if in some Cartesian coordinate system we assume time independent settings of the problem so that $\frac{\partial S}{\partial t}=c$, $\frac{\partial c_0}{\partial t}=0$ and $\frac{\partial \tilde{\mathbf{u}}}{\partial t}=0$, then we also have $\frac{\partial \mathbf{k}}{\partial t}=\frac{\partial \mathbf{h}}{\partial t}=0$ and $\frac{\partial \tilde{\omega}}{\partial t}=\frac{\partial G}{\partial t}=0$, and therefore, time-independent solutions $A:=A(\mathbf{x})$, of (1.889) are admitted and they satisfy

$$-\frac{c_0}{2\tilde{\omega}}\left(\nabla_{\mathbf{x}}A-(\mathbf{k}\cdot\nabla_{\mathbf{x}}A)\,\mathbf{k}\right)\cdot\left(\left(d_{\mathbf{x}}\tilde{\mathbf{u}}+\{d_{\mathbf{x}}\tilde{\mathbf{u}}\}^T\right)\cdot\mathbf{k}-\left(\left(\mathbf{k}\cdot\left(\{d_{\mathbf{x}}\tilde{\mathbf{u}}\}^T+d_{\mathbf{x}}\tilde{\mathbf{u}}\right)\right)\cdot\mathbf{k}\right)\mathbf{k}\right)$$
$$+\frac{c_0^2}{2\tilde{\omega}}A\left(\operatorname{div}_{\mathbf{x}}\left\{\frac{1}{c_0^2}G\tilde{\mathbf{u}}\right\}+\frac{1}{c_0^2}(c_0\mathbf{k})\cdot\nabla_{\mathbf{x}}G+\frac{1}{c_0^2}G^2\right)$$
$$-\frac{c_0^2}{2}\left(\operatorname{div}_{\mathbf{x}}\left\{\frac{1}{\tilde{\omega}}\left(\nabla_{\mathbf{x}}A-(\mathbf{k}\cdot\nabla_{\mathbf{x}}A)\,\mathbf{k}\right)\right\}\right)\approx -i\left\{\mathbf{h}\cdot\nabla_{\mathbf{x}}A+\frac{1}{2}\left(\operatorname{div}_{\mathbf{x}}\mathbf{h}\right)A\right\}$$
$$-iA\frac{1}{2c_0}\left(c_0\mathbf{k}\cdot\nabla_{\mathbf{x}}c_0\right)+iA\frac{1}{2c_0}\left(\mathbf{h}\cdot\nabla_{\mathbf{x}}c_0\right)+\frac{iA}{4}\left(\mathbf{k}\cdot\left(\{d_{\mathbf{x}}\tilde{\mathbf{u}}\}^T+d_{\mathbf{x}}\tilde{\mathbf{u}}\right)\right)\cdot\mathbf{k}. \qquad (1.890)$$

Finally, from now we assume that $\tilde{\mathbf{u}}$ and c_0 vary sufficiently slowly in space and time variables, so that the following approximations are valid:

$$\frac{|\nabla_{\mathbf{x}}c_0|+\frac{1}{c_0}\left|\frac{\partial c_0}{\partial t}+\tilde{\mathbf{u}}\cdot\nabla_{\mathbf{x}}c_0\right|}{c_0}+\frac{\left|d_{\mathbf{x}}\tilde{\mathbf{u}}+\{d_{\mathbf{x}}\tilde{\mathbf{u}}\}^T\right|}{c_0}\ll\frac{\left(|\nabla_{\mathbf{x}}A|+\frac{1}{c_0}\left|\frac{\partial A}{\partial t}+\tilde{\mathbf{u}}\cdot\nabla_{\mathbf{x}}A\right|\right)}{|A|}. \qquad (1.891)$$

In particular, if in some Cartesian coordinate system we have

$$\begin{cases}\frac{\left|d_{\mathbf{x}}\tilde{\mathbf{u}}+\{d_{\mathbf{x}}\tilde{\mathbf{u}}\}^T\right|^2}{|\tilde{\mathbf{u}}|^2}\ll\frac{\left(|\nabla_{\mathbf{x}}A|+\frac{1}{c_0}\left|\frac{\partial A}{\partial t}+\tilde{\mathbf{u}}\cdot\nabla_{\mathbf{x}}A\right|\right)^2}{|A|^2}\\ \frac{|\tilde{\mathbf{u}}|^2}{c_0^2}\ll 1\\ \frac{|\nabla_{\mathbf{x}}c_0|+\frac{1}{c_0}\left|\frac{\partial c_0}{\partial t}+\tilde{\mathbf{u}}\cdot\nabla_{\mathbf{x}}c_0\right|}{c_0}\ll\frac{\left(|\nabla_{\mathbf{x}}A|+\frac{1}{c_0}\left|\frac{\partial A}{\partial t}+\tilde{\mathbf{u}}\cdot\nabla_{\mathbf{x}}A\right|\right)}{|A|},\end{cases} \qquad (1.892)$$

then (1.891) indeed holds! Then, by (1.891) we rewrite and simplify (1.889) as

$$\frac{1}{2\tilde{\omega}}A\left(\frac{\partial G}{\partial t}+\operatorname{div}_{\mathbf{x}}\{G\tilde{\mathbf{u}}\}+(c_0\mathbf{k})\cdot\nabla_{\mathbf{x}}G+G^2\right)$$
$$-\frac{1}{2}\left(\operatorname{div}_{\mathbf{x}}\left\{\frac{c_0^2}{\tilde{\omega}}\left(\nabla_{\mathbf{x}}A-(\mathbf{k}\cdot\nabla_{\mathbf{x}}A)\,\mathbf{k}\right)\right\}\right)\approx -i\left\{\frac{\partial A}{\partial t}+\mathbf{h}\cdot\nabla_{\mathbf{x}}A+\frac{1}{2}\left(\operatorname{div}_{\mathbf{x}}\mathbf{h}\right)A\right\} \qquad (1.893)$$

(see Sect. 13.3.10 for the details). Moreover, as before, we can easily deduce that (1.893) invariant under the change of non-inertial Cartesian coordinate system, in the case that under such change we have $A' = A$ and $S' = S$.

Furthermore, in time-independent settings we rewrite (1.890) as

$$\frac{1}{2\tilde{\omega}}A\left(\operatorname{div}_{\mathbf{x}}\{G\tilde{\mathbf{u}}\}+(c_0\mathbf{k})\cdot\nabla_{\mathbf{x}}G+G^2\right)$$
$$-\frac{1}{2}\left(\operatorname{div}_{\mathbf{x}}\left\{\frac{c_0^2}{\tilde{\omega}}\left(\nabla_{\mathbf{x}}A-(\mathbf{k}\cdot\nabla_{\mathbf{x}}A)\,\mathbf{k}\right)\right\}\right)\approx -i\left\{\mathbf{h}\cdot\nabla_{\mathbf{x}}A+\frac{1}{2}\left(\operatorname{div}_{\mathbf{x}}\mathbf{h}\right)A\right\}. \qquad (1.894)$$

In particular, in the case $\tilde{\mathbf{u}} = 0$ and constant $\mathbf{k}$, c_0 and $\tilde{\omega}$, (1.894) simplifies as the paraxial approximation of the Helmholtz equation, with respect to direction $(-\mathbf{k})$, which is common in Optics:

$$\frac{c_0}{2\tilde{\omega}} \left(\operatorname{div}_{\mathbf{x}} \{(\nabla_{\mathbf{x}} A - (\mathbf{h} \cdot \nabla_{\mathbf{x}} A)\, \mathbf{h})\}\right) \approx -i\mathbf{k} \cdot \nabla_{\mathbf{x}} A. \tag{1.895}$$

Next, note that, the time-dependent equation (1.893) is very similar to the Schrödinger equation and similarly can be written in the form

$$i\hbar \frac{\partial A}{\partial t} = \hat{H} \cdot A, \tag{1.896}$$

where

$$\hat{H} \cdot A := \frac{\hbar}{2} \left(\operatorname{div}_{\mathbf{x}} \left\{ \frac{c_0^2}{\tilde{\omega}} (\nabla_{\mathbf{x}} A - (\mathbf{k} \cdot \nabla_{\mathbf{x}} A)\, \mathbf{k}) \right\}\right) - i\hbar \left\{ \mathbf{h} \cdot \nabla_{\mathbf{x}} A + \frac{1}{2} (\operatorname{div}_{\mathbf{x}} \mathbf{h})\, A \right\}$$
$$- \frac{\hbar}{2\tilde{\omega}} \left(\frac{\partial G}{\partial t} + \operatorname{div}_{\mathbf{x}} \{G\tilde{\mathbf{u}}\} + (c_0 \mathbf{k}) \cdot \nabla_{\mathbf{x}} G + G^2 \right) A \tag{1.897}$$

is a self-adjoint linear operator on the complex Hilbert space.

Finally, solving either (1.889) or (1.893) and inserting the solution into (1.880) we established $U(\mathbf{x}, t)$ in (1.878) in the delicate Geometric Optics approximation (1.751).

1.15.10 Wave Optics Inside a Moving Medium and/or in the Presence of Gravitational Field

Assume that in some inertial or non-inertial Cartesian coordinate system a complex valued scalar field $U := U(\mathbf{x}, t)$, characterizing some wave, satisfies the wave equation (1.737):

$$\frac{\partial}{\partial t} \left\{ \frac{1}{c_0^2} \left(\frac{\partial U}{\partial t} + \tilde{\mathbf{u}} \cdot \nabla_{\mathbf{x}} U \right) \right\} + \operatorname{div}_{\mathbf{x}} \left\{ \frac{1}{c_0^2} \left(\frac{\partial U}{\partial t} + \tilde{\mathbf{u}} \cdot \nabla_{\mathbf{x}} U \right) \tilde{\mathbf{u}} \right\} - \Delta_{\mathbf{x}} U = 0, \tag{1.898}$$

where, as before, $\tilde{\mathbf{u}} := \tilde{\mathbf{u}}(\mathbf{x}, t)$ is some moderately changing (in space and in time) speed-like vector field and $c_0 := c_0(\mathbf{x}, t) > 0$ is a moderately changing scalar quantity, that we call wave propagation speed. In particular, if we assume that the fields $\tilde{\mathbf{u}}$ and c_0 are independent on the time variable, and we consider the case of a monochromatic wave of the constant frequency $\nu = \frac{\omega}{2\pi}$:

$$U(\mathbf{x}, t) := \mathcal{U}(\mathbf{x}) e^{i\omega t} \qquad \forall (\mathbf{x}, t), \tag{1.899}$$

where $\mathcal{U} := \mathcal{U}(\mathbf{x})$ is a complex-valued scalar field, independent on time, then inserting (1.899) into (1.898) gives the following time-independent equation:

$$\frac{i\omega}{c_0^2(\mathbf{x})}\left(i\omega\mathcal{U}(\mathbf{x}) + \tilde{\mathbf{u}}(\mathbf{x}) \cdot \nabla_{\mathbf{x}}\mathcal{U}(\mathbf{x})\right) + \operatorname{div}_{\mathbf{x}}\left\{\frac{1}{c_0^2(\mathbf{x})}\left(i\omega\mathcal{U}(\mathbf{x}) + \tilde{\mathbf{u}}(\mathbf{x}) \cdot \nabla_{\mathbf{x}}\mathcal{U}(\mathbf{x})\right)\tilde{\mathbf{u}}(\mathbf{x})\right\}$$
$$- \Delta_{\mathbf{x}}\mathcal{U}(\mathbf{x}) = 0. \quad (1.900)$$

Next, suppose that, although we cannot consider the Geometric Optics approximation (1.751), we can consider, however, the following assumption:

$$\tilde{\mathbf{u}}(\mathbf{x}, t) = \nabla_{\mathbf{x}}\tilde{Z}(\mathbf{x}, t) \qquad \forall(\mathbf{x}, t), \tag{1.901}$$

where $\tilde{Z} := \tilde{Z}(\mathbf{x}, t)$ is some real-valued scalar function, which satisfies

$$\frac{1}{c_0^2}\left|\frac{\partial \tilde{Z}}{\partial t}\right| + \frac{1}{c_0^2}\left|\nabla_{\mathbf{x}}\tilde{Z}\right|^2 \ll 1. \tag{1.902}$$

Note that (1.901) is valid in the particular case of the electromagnetic waves, propagating in vacuum, in the <u>inertial</u> or more generally non-rotating Cartesian coordinate system (where we indeed have $\tilde{\mathbf{u}} = \mathbf{v}$ and $\operatorname{curl}_{\mathbf{x}}\mathbf{v} = 0$). On the other hand, (1.902) means that $\frac{|\tilde{\mathbf{u}}|^2}{c_0^2} \ll 1$ and $\tilde{\mathbf{u}}$ is either independent on time or depends on time slowly. Moreover, assume that c_0 is a constant in the given region.

Next we do the change of variables $(\mathbf{x}, t) \to (\mathbf{z}, \tau)$ as

$$\begin{cases} \mathbf{z} := \mathbf{x}, \\ \tau := t + \frac{1}{c_0^2}\tilde{Z}(\mathbf{x}, t), \end{cases} \tag{1.903}$$

so that, we can find a complex-valued function $V := V(\mathbf{z}, \tau)$ depending on $\mathbf{z} \in \mathbb{R}^3$ and a real τ, which satisfies

$$U(\mathbf{x}, t) := V\left(\mathbf{x}, t + \frac{1}{c_0^2}\tilde{Z}(\mathbf{x}, t)\right). \tag{1.904}$$

Then in Sect. 13.3.11 we deduce that the function $V(\mathbf{z}, \tau)$ approximately solves the standard type of the wave equation of the form:

$$\frac{1}{c_0^2}\frac{\partial^2 V}{\partial \tau^2}(\mathbf{z}, \tau) - \Delta_{\mathbf{z}}V(\mathbf{z}, \tau) = 0. \tag{1.905}$$

Fortunately, the plenty of tools for analytical resolution of the <u>standard</u> wave equation (1.905) is well known. In particular, if we assume that the fields $\tilde{\mathbf{u}}$ and c_0 are independent on the time variable, and we consider the case of a monochromatic wave of the constant frequency $\nu = \frac{\omega}{2\pi}$, as in (1.899), then, we can choose $\tilde{Z}$ to be independent on time and moreover, inserting (1.899) into (1.904) gives

$$V(\mathbf{z}, \tau) = \mathcal{V}(\mathbf{z})\, e^{i\omega\tau} \quad \text{and} \quad \mathcal{U}(\mathbf{x}) := \mathcal{V}(\mathbf{x})\, e^{\frac{i\omega}{c_0^2(\mathbf{x})}\tilde{Z}(\mathbf{x})}, \tag{1.906}$$

where $\mathcal{V} := \mathcal{V}(\mathbf{z})$ is a complex-valued scalar field, independent on time. Moreover, inserting (1.906) into (1.905) gives that $\mathcal{V}(\mathbf{x})$ solves the Helmholtz equation of the form:

$$\frac{\omega^2}{c_0^2}\mathcal{V}(\mathbf{x}) + \Delta_{\mathbf{x}}\mathcal{V}(\mathbf{x}) = 0, \tag{1.907}$$

and

$$U(\mathbf{x}, t) = \mathcal{V}(\mathbf{x})\, e^{i\omega\left(t + \frac{1}{c_0^2}\tilde{Z}(\mathbf{x})\right)}. \tag{1.908}$$

Thus, if, as before in (1.745), we denote $k_0 := \frac{\omega}{c}$, where c is a constant in the Maxwell equations for the vacuum, and express $U(\mathbf{x}, t)$ as in (1.746) by the following:

$$U(\mathbf{x}, t) = A(\mathbf{x}, t)e^{ik_0 S(\mathbf{x},t)}, \tag{1.909}$$

where $A := A(\mathbf{x}, t)$ and $S := S(\mathbf{x}, t)$ are real scalar fields, then by (1.908) we deduce

$$U(\mathbf{x}, t) = A(\mathbf{x}, t)e^{ik_0 S(\mathbf{x},t)}, \quad \text{where} \quad A(\mathbf{x}, t) = B(\mathbf{x}) \quad \text{and} \quad S(\mathbf{x}, t) = ct + P(\mathbf{x}) + \frac{c}{c_0^2}\tilde{Z}(\mathbf{x}), \tag{1.910}$$

with

$$B(\mathbf{x})e^{ik_0 P(\mathbf{x})} := \mathcal{V}(\mathbf{x}), \tag{1.911}$$

where $B := B(\mathbf{x})$ and $P := P(\mathbf{x})$ are real scalar fields and $\mathcal{V}(\mathbf{x})$ is a complex solution of (1.907). In particular, if the vector field $\mathbf{h}_0$ is the direction of the propagation of the ray that passes through point $\mathbf{x}$, defined for every $\mathbf{x}$, as in (1.780) by

$$\mathbf{h}_0(\mathbf{x}) := \frac{c}{c_0^2}\tilde{\mathbf{u}}(\mathbf{x}) - \nabla_{\mathbf{x}} S(\mathbf{x}), \tag{1.912}$$

then by inserting (1.910) and (1.901) into (1.912) gives

$$\mathbf{h}_0(\mathbf{x}) = -\nabla_{\mathbf{x}} P(\mathbf{x}). \tag{1.913}$$

As $P(\mathbf{x})$ being the phase of $\mathcal{V}(\mathbf{x})$ is independent on the field $\tilde{\mathbf{u}}$ we obtain that the direction of the ray $\mathbf{h}$, in the general case of nontrivial field $\tilde{\mathbf{u}}$, is the same as in the case $\tilde{\mathbf{u}}(\mathbf{x}) = 0$. So the nontrivial motion of the medium and the nontrivial vectorial gravitational potential does not affect the direction of the ray propagation $\mathbf{h}$, provided we assume (1.901) and (1.902). Moreover, if we denote by $U(\mathbf{x}, t) = A(\mathbf{x}, t)e^{ik_0 S(\mathbf{x},t)}$

the monochromatic wave solution of the problem in the general case of nontrivial field $\tilde{\mathbf{u}}$ and by $U_0(\mathbf{x}, t) = A_0(\mathbf{x}, t)e^{ik_0 S_0(\mathbf{x},t)}$ the monochromatic wave solution of the corresponding simpler problem in the case of the trivial field $\tilde{\mathbf{u}} = 0$ then we have

$$U(\mathbf{x}, t) = U_0(\mathbf{x}, t)e^{\frac{i\omega}{c_0^2}\tilde{Z}(\mathbf{x})}, \quad A(\mathbf{x}, t) = A_0(\mathbf{x}, t) \quad \text{and} \quad S(\mathbf{x}, t) = S_0(\mathbf{x}, t) + \frac{c}{c_0^2}\tilde{Z}(\mathbf{x}), \tag{1.914}$$

i.e. in the case, where (1.901) and (1.902) hold, the wave solution for the general case of nontrivial $\tilde{\mathbf{u}}$ can be obtained from the corresponding solution in the simplest trivial case by simply adding the term $\frac{\omega}{c_0^2}\tilde{Z}(\mathbf{x})$ to the phase. In particular, all the pictures of the interference or the diffraction for given nontrivial $\tilde{\mathbf{u}}$ will be the same as in the trivial case, provided we assume (1.901) and (1.902). Finally, by simply adding the phase $\frac{\omega}{c_0^2}\tilde{Z}(\mathbf{x})$ we obtain the particular monochromatic solutions of (1.898) from the following well known particular monochromatic solutions of the simplest wave equation (i.e. for the case $\tilde{\mathbf{u}} = 0$):

- plane wave,
- spherical wave,
- the approximate solution in the form of the Gaussian beam.

Chapter 2
Notations and Preliminaries

• By $\mathbb{R}^{p\times q}$, we denote the set of $p \times q$-matrixes with real coefficients.
• For a $p \times q$ matrix A with ijth entry a_{ij} and for a $q \times d$ matrix B with ijth entry b_{ij}, we denote by $AB := A \cdot B$ their product, i.e. the $p \times d$ matrix, with ijth entry $\sum_{k=1}^{q} a_{ik}b_{kj}$.
• We identify a vector $\mathbf{u} = (u_1, \ldots, u_q) \in \mathbb{R}^q$ with the $q \times 1$ matrix having $i1$th entry u_i, so that, for the $p \times q$ matrix A with ijth entry a_{ij} and for $\mathbf{v} = (v_1, v_2, \ldots, v_q) \in \mathbb{R}^q$, we denote by $A\,\mathbf{v} := A \cdot \mathbf{v}$ the p-dimensional vector $\mathbf{u} = (u_1, \ldots, u_p) \in \mathbb{R}^p$, given by $u_i = \sum_{k=1}^{q} a_{ik}v_k$ for every $1 \le i \le p$.
• For a $p \times q$ matrix A with ijth entry a_{ij}, we denote by A^T the transpose $q \times p$ matrix with ijth entry a_{ji}.
• For a $p \times p$ matrix A with ijth entry a_{ij}, we denote $\mathrm{tr}(A) := \sum_{k=1}^{p} a_{kk}$ (the trace of the matrix A).
• For a $p \times q$ real matrices A and B with ijth entries a_{ij} and b_{ij} denote by $A : B$ their scalar product $A : B := \mathrm{tr}(A \cdot B^T) := \sum_{j=1}^{p} \sum_{k=1}^{q} a_{jk}b_{jk}$
• For a $p \times p$ real matrix A with ijth entry a_{ij} denote by $|A|$ its standard norm $|A| := \sqrt{A : A} = \sqrt{\sum_{j=1}^{p} \sum_{k=1}^{p} a_{jk}^2}$.
• For $\mathbf{u} = (u_1, \ldots, u_p) \in \mathbb{R}^p$ and $\mathbf{v} = (v_1, \ldots, v_p) \in \mathbb{R}^p$, we denote by $\mathbf{u}\mathbf{v} := \mathbf{u} \cdot \mathbf{v} := \sum_{k=1}^{p} u_k v_k$ the standard scalar product. We also note that $\mathbf{u}\mathbf{v} = \mathbf{u}^T\mathbf{v} = \mathbf{v}^T\mathbf{u}$ as products of matrices.
• For $\mathbf{u} = (u_1, u_2, u_3) \in \mathbb{R}^3$ and $\mathbf{v} = (v_1, v_2, v_3) \in \mathbb{R}^3$, we denote

$$\mathbf{u} \times \mathbf{v} := (u_2v_3 - u_3v_2, u_3v_1 - u_1v_3, u_1v_2 - u_2v_1) \in \mathbb{R}^3.$$

• For $\mathbf{u} = (u_1, \ldots, u_p) \in \mathbb{R}^p$ and $\mathbf{v} = (v_1, \ldots, v_q) \in \mathbb{R}^q$, we denote by $\mathbf{u} \otimes \mathbf{v}$ the $p \times q$ matrix with ijth entry $u_i v_j$ (i.e. $\mathbf{u} \otimes \mathbf{v} = \mathbf{u}\,\mathbf{v}^T$ as a product of matrices).

A. Poliakovsky, *The Laws of Gravity and Electromagnetism*,
https://doi.org/10.1007/978-3-031-61407-1_2

• Given a vector-valued function $\mathbf{f}(\mathbf{x}) = (f_1(\mathbf{x}), \ldots, f_k(\mathbf{x})) : G \to \mathbb{R}^k$ $(G \subset \mathbb{R}^N)$, we denote by $D\mathbf{f}$ the $k \times N$ matrix with ijth entry $\frac{\partial f_i}{\partial x_j}$. In the case of a scalar-valued function $\psi(\mathbf{x}) : G \subset \mathbb{R}^N \to \mathbb{R}$ we associate with $D\psi$ (which, by definition, belongs to $\mathbb{R}^{1\times N}$) the corresponding vector $\nabla\psi := \left(\frac{\partial\psi}{\partial x_1}, \ldots, \frac{\partial\psi}{\partial x_N}\right)$.

• Given a matrix-valued function $F(\mathbf{x}) := \{F_{ij}(\mathbf{x})\} : G \subset \mathbb{R}^N \to \mathbb{R}^{k\times N}$, we denote by div F the $\mathbb{R}^k$-valued vector field defined by div $F(\mathbf{x}) := (l_1, \ldots, l_k)(\mathbf{x})$ where $l_i(\mathbf{x}) = \sum_{j=1}^{N} \frac{\partial F_{ij}}{\partial x_j}(\mathbf{x})$. Given a vector-valued function

$\mathbf{f}(\mathbf{x}) := (f_1(\mathbf{x}), \ldots, f_N(\mathbf{x})) : G \subset \mathbb{R}^N \to \mathbb{R}^N$, we denote $\operatorname{div}\mathbf{f} := \sum_{j=1}^{N} \frac{\partial f_j}{\partial x_j}$.

• Given a scalar- or vector-valued function $\mathbf{f}(\mathbf{x}) : G \subset \mathbb{R}^N \to \mathbb{R}^k$ we denote by $\Delta\mathbf{f}$, the Laplacian of $\mathbf{f}$ defined by $\Delta\mathbf{f} := \sum_{j=1}^{N} \frac{\partial^2\mathbf{f}}{\partial x_j^2}$.

• Given a vector-valued function $\mathbf{f}(\mathbf{x}) = (f_1(\mathbf{x}), f_2(\mathbf{x}), f_3(\mathbf{x})) : G \subset \mathbb{R}^3 \to \mathbb{R}^3$, we denote

$$\operatorname{curl}\mathbf{f}(\mathbf{x}) := \left(\frac{\partial f_3}{\partial x_2} - \frac{\partial f_2}{\partial x_3}, \frac{\partial f_1}{\partial x_3} - \frac{\partial f_3}{\partial x_1}, \frac{\partial f_2}{\partial x_1} - \frac{\partial f_1}{\partial x_2}\right)(\mathbf{x}).$$

We have the following trivial identities:

$$\mathbf{a}\times\mathbf{b} = -\mathbf{b}\times\mathbf{a} \quad \text{and} \quad \mathbf{a}\cdot(\mathbf{b}\times\mathbf{c}) = (\mathbf{a}\times\mathbf{b})\cdot\mathbf{c} \qquad \forall\,\mathbf{a},\mathbf{b},\mathbf{c}\in\mathbb{R}^3, \tag{2.1}$$

$$\mathbf{a}\times(\mathbf{b}\times\mathbf{c}) = (\mathbf{a}\cdot\mathbf{c})\,\mathbf{b} - (\mathbf{a}\cdot\mathbf{b})\,\mathbf{c} \qquad \forall\,\mathbf{a},\mathbf{b},\mathbf{c}\in\mathbb{R}^3, \tag{2.2}$$

$$(A\cdot\mathbf{b})\times\mathbf{c} - (A\cdot\mathbf{c})\times\mathbf{b} = \operatorname{tr}(A)\,(\mathbf{b}\times\mathbf{c}) - A^T\cdot(\mathbf{b}\times\mathbf{c}) \qquad \forall A\in\mathbb{R}^{3\times3},\ \forall\,\mathbf{b},\mathbf{c}\in\mathbb{R}^3, \tag{2.3}$$

$$A^T\cdot((A\cdot\mathbf{b})\times(A\cdot\mathbf{c})) = (\det A)\,(\mathbf{b}\times\mathbf{c}) \qquad \forall A\in\mathbb{R}^{3\times3},\ \forall\,\mathbf{b},\mathbf{c}\in\mathbb{R}^3, \tag{2.4}$$

$$\operatorname{div}(\mathbf{f}\times\mathbf{g}) = \mathbf{g}\cdot\operatorname{curl}\mathbf{f} - \mathbf{f}\cdot\operatorname{curl}\mathbf{g} \qquad \forall\,\mathbf{f},\mathbf{g} : G\subset\mathbb{R}^3\to\mathbb{R}^3, \tag{2.5}$$

$$\operatorname{div}(\psi\mathbf{f}) = \psi\operatorname{div}\mathbf{f} + \nabla\psi\cdot\mathbf{f} \qquad \forall\,\psi : G\subset\mathbb{R}^3\to\mathbb{R},\ \forall\,\mathbf{f} : G\subset\mathbb{R}^3\to\mathbb{R}^3, \tag{2.6}$$

$$\operatorname{curl}(\psi\mathbf{f}) = \psi\operatorname{curl}\mathbf{f} + \nabla\psi\times\mathbf{f} \qquad \forall\,\psi : G\subset\mathbb{R}^3\to\mathbb{R},\ \forall\,\mathbf{f} : G\subset\mathbb{R}^3\to\mathbb{R}^3, \tag{2.7}$$

$$\operatorname{div}(\operatorname{curl}\mathbf{f}) = 0 \qquad \forall\,\mathbf{f} : G\subset\mathbb{R}^3\to\mathbb{R}^3, \tag{2.8}$$

$$\operatorname{curl}(\nabla\psi) = 0 \qquad \forall\,\psi : G\subset\mathbb{R}^3\to\mathbb{R}, \tag{2.9}$$

$$\operatorname{curl}(\operatorname{curl}\mathbf{f}) = \nabla(\operatorname{div}\mathbf{f}) - \Delta\mathbf{f} \qquad \forall\,\mathbf{f} : G\subset\mathbb{R}^3\to\mathbb{R}^3, \tag{2.10}$$

$$\operatorname{curl}(\mathbf{f} \times \mathbf{g}) = (\operatorname{div} \mathbf{g})\, \mathbf{f} - (\operatorname{div} \mathbf{f})\, \mathbf{g} + (D\mathbf{f}) \cdot \mathbf{g} - (D\mathbf{g}) \cdot \mathbf{f} \quad \forall \mathbf{f}, \mathbf{g} : G \subset \mathbb{R}^3 \to \mathbb{R}^3, \tag{2.11}$$

$$\operatorname{curl}(\mathbf{f} \times \mathbf{g}) = \operatorname{div}(\mathbf{f} \otimes \mathbf{g} - \mathbf{g} \otimes \mathbf{f}) \quad \forall \mathbf{f}, \mathbf{g} : G \subset \mathbb{R}^3 \to \mathbb{R}^3, \tag{2.12}$$

$$\operatorname{div}(\mathbf{f} \otimes \mathbf{g}) = (D\mathbf{f}) \cdot \mathbf{g} + (\operatorname{div} \mathbf{g})\, \mathbf{f} \quad \forall \mathbf{f}, \mathbf{g} : G \subset \mathbb{R}^3 \to \mathbb{R}^3, \tag{2.13}$$

$$\nabla(\mathbf{f} \cdot \mathbf{g}) = (D\mathbf{f})^T \cdot \mathbf{g} + (D\mathbf{g})^T \cdot \mathbf{f} \quad \forall \mathbf{f}, \mathbf{g} : G \subset \mathbb{R}^3 \to \mathbb{R}^3, \tag{2.14}$$

$$\mathbf{f} \times (\operatorname{curl} \mathbf{g}) = (D\mathbf{g})^T \cdot \mathbf{f} - (D\mathbf{g}) \cdot \mathbf{f} \quad \forall \mathbf{f}, \mathbf{g} : G \subset \mathbb{R}^3 \to \mathbb{R}^3, \tag{2.15}$$

$$\nabla(\mathbf{f} \cdot \mathbf{g}) = \mathbf{f} \times (\operatorname{curl} \mathbf{g}) + \mathbf{g} \times (\operatorname{curl} \mathbf{f}) + (D\mathbf{f}) \cdot \mathbf{g} + (D\mathbf{g}) \cdot \mathbf{f} \quad \forall \mathbf{f}, \mathbf{g} : G \subset \mathbb{R}^3 \to \mathbb{R}^3, \tag{2.16}$$

where we mean, by $A \cdot \mathbf{l}$, the usual product of matrix $A \in \mathbb{R}^{3\times 3}$ and vector $\mathbf{l} \in \mathbb{R}^3$ and, by A^T, we mean the transpose of matrix A.

Chapter 3
Transformations of Scalars and Vectors Under the Change of Cartesian Coordinate Systems

Definition 3.1 Consider the change of some inertial or non-inertial Cartesian coordinate system (∗) to another Cartesian coordinate system (∗∗) of the form:

$$\begin{cases} \mathbf{x}' = A(t) \cdot \mathbf{x} + \mathbf{z}(t), \\ t' = t, \end{cases} \tag{3.1}$$

where $A(t) \in SO(3)$ is a rotation, i.e. $A(t) \in \mathbb{R}^{3\times 3}$, $\det A(t) > 0$ and $A(t) \cdot A^T(t) = I$.

- We say that the scalar field $\psi := \psi(\mathbf{x}, t) : \mathbb{R}^3 \times [t_0, +\infty) \to \mathbb{R}$ is a proper scalar field if, under every change of coordinate system given by (3.1), this field transforms by the law:
$$\psi'(\mathbf{x}', t') = \psi(\mathbf{x}, t). \tag{3.2}$$
- We say that the vector field $\mathbf{f} := \mathbf{f}(\mathbf{x}, t) : \mathbb{R}^3 \times [t_0, +\infty) \to \mathbb{R}^3$ is a proper vector field if, under every change of coordinate system given by (3.1), this field transforms by the law:
$$\mathbf{f}'(\mathbf{x}', t') = A(t) \cdot \mathbf{f}(\mathbf{x}, t). \tag{3.3}$$
- We say that the vector field $\mathbf{v} := \mathbf{v}(\mathbf{x}, t) : \mathbb{R}^3 \times [t_0, +\infty) \to \mathbb{R}^3$ is a speed-like vector field if, under every change of coordinate system given by (3.1), this field transforms by the law:
$$\mathbf{v}'(\mathbf{x}', t') = A(t) \cdot \mathbf{v}(\mathbf{x}, t) + \frac{dA}{dt}(t) \cdot \mathbf{x} + \mathbf{w}(t), \tag{3.4}$$
where we set
$$\mathbf{w}(t) := \frac{d\mathbf{z}}{dt}(t) \qquad \forall t. \tag{3.5}$$

A. Poliakovsky, *The Laws of Gravity and Electromagnetism*,
https://doi.org/10.1007/978-3-031-61407-1_3

- We say that the matrix-valued field $T := T(\mathbf{x}, t) : \mathbb{R}^3 \times [t_0, +\infty) \to \mathbb{R}^{3\times 3}$ is a proper matrix field if, under every change of coordinate system given by (3.1), this field transforms by the law:

$$T'(\mathbf{x}', t') = A(t) \cdot T(\mathbf{x}, t) \cdot A^T(t) = A(t) \cdot T(\mathbf{x}, t) \cdot \{A(t)\}^{-1}. \tag{3.6}$$

Proposition 3.1 *If $\psi : \mathbb{R}^3 \times [t_0, +\infty) \to \mathbb{R}$ is a proper scalar field, $\mathbf{f} : \mathbb{R}^3 \times [t_0, +\infty) \to \mathbb{R}^3$ and $\mathbf{g} : \mathbb{R}^3 \times [t_0, +\infty) \to \mathbb{R}^3$ are proper vector fields, $\mathbf{v} : \mathbb{R}^3 \times [t_0, +\infty) \to \mathbb{R}^3$ and $\mathbf{u} : \mathbb{R}^3 \times [t_0, +\infty) \to \mathbb{R}^3$ are speed-like vector fields and $T : \mathbb{R}^3 \times [t_0, +\infty) \to \mathbb{R}^{3\times 3}$ is a proper matrix field, then:*

(i) *scalar fields defined in every coordinate system as $\mathbf{f} \cdot \mathbf{g}$, $\mathrm{div}_{\mathbf{x}}\, \mathbf{f}$ and $\mathrm{div}_{\mathbf{x}}\, \mathbf{v}$ are proper scalar fields;*

(ii) *vector fields defined in every coordinate system as $\nabla_{\mathbf{x}}\psi$, $\mathrm{div}_{\mathbf{x}}\, T$, $curl_{\mathbf{x}}\mathbf{f}$, $\mathbf{f} \times \mathbf{g}$, $\mathrm{div}_{\mathbf{x}} \left(d_{\mathbf{x}}\mathbf{v} + \{d_{\mathbf{x}}\mathbf{v}\}^T\right)$, $\nabla_{\mathbf{x}} (\mathrm{div}_{\mathbf{x}}\, \mathbf{v})$, $\Delta_{\mathbf{x}}\mathbf{v}$, $curl_{\mathbf{x}} (curl_{\mathbf{x}}\mathbf{v})$ and $(\mathbf{u} - \mathbf{v})$ are proper vector fields;*

(iii) *matrix fields defined in every coordinate system as $d_{\mathbf{x}}\mathbf{f}$ and $\left(d_{\mathbf{x}}\mathbf{v} + \{d_{\mathbf{x}}\mathbf{v}\}^T\right)$ are proper matrix fields;*

(iv) *scalar fields $\xi : \mathbb{R}^3 \times [t_0, +\infty) \to \mathbb{R}$ and $\zeta : \mathbb{R}^3 \times [t_0, +\infty) \to \mathbb{R}$, defined in every coordinate system by*

$$\xi := \frac{\partial \psi}{\partial t} + \mathbf{v} \cdot \nabla_{\mathbf{x}}\psi \quad and \quad \zeta := \frac{\partial \psi}{\partial t} + \mathrm{div}_{\mathbf{x}} \{\psi \mathbf{v}\}, \tag{3.7}$$

are proper scalar fields;

(v) *vector fields $\Theta : \mathbb{R}^3 \times [t_0, +\infty) \to \mathbb{R}^3$ and $\Xi : \mathbb{R}^3 \times [t_0, +\infty) \to \mathbb{R}^3$, defined in every coordinate system by*

$$\Theta := \frac{\partial \mathbf{f}}{\partial t} - curl_{\mathbf{x}} (\mathbf{v} \times \mathbf{f}) + (\mathrm{div}_{\mathbf{x}}\mathbf{f})\, \mathbf{v} \quad and \quad \Xi := \frac{\partial \mathbf{f}}{\partial t} - \mathbf{v} \times curl_{\mathbf{x}}\mathbf{f} + \nabla_{\mathbf{x}} (\mathbf{v} \cdot \mathbf{f}), \tag{3.8}$$

are proper vector fields and

$$\Xi = \Theta - (\mathrm{div}_{\mathbf{x}}\, \mathbf{v})\, \mathbf{f} + \left(d_{\mathbf{x}}\mathbf{v} + \{d_{\mathbf{x}}\mathbf{v}\}^T\right) \cdot \mathbf{f}. \tag{3.9}$$

Proof By (3.1) and the chain rule for every vector fields $\Gamma : \mathbb{R}^3 \times [t_0, +\infty) \to \mathbb{R}^3$ and $\Lambda : \mathbb{R}^3 \times [t_0, +\infty) \to \mathbb{R}^3$ we have

$$\begin{cases} (A(t) \cdot \Gamma) \cdot (A(t) \cdot \Lambda) = \Gamma \cdot \Lambda \\ (A(t) \cdot \Gamma) \times (A(t) \cdot \Lambda) = A(t) \cdot (\Gamma \times \Lambda) \\ d_{\mathbf{x}'}\Gamma = (d_{\mathbf{x}}\Gamma) \cdot A^{-1}(t) \\ \mathrm{curl}_{\mathbf{x}'} (A(t) \cdot \Gamma) = A(t) \cdot \mathrm{curl}_{\mathbf{x}}\Gamma \\ \mathrm{div}_{\mathbf{x}'} (A(t) \cdot \Gamma) = \mathrm{div}_{\mathbf{x}}\, \Gamma. \end{cases} \tag{3.10}$$

Thus, in particular, by (3.10) and (3.3) we have

$$\mathbf{f}' \cdot \mathbf{g}' = \mathbf{f} \cdot \mathbf{g}, \qquad \mathbf{f}' \times \mathbf{g}' = A(t)\,(\mathbf{f} \times \mathbf{g})\,, \tag{3.11}$$

and

$$\operatorname{div}_{\mathbf{x}'} \mathbf{f}' = \operatorname{div}_{\mathbf{x}'} (A(t) \cdot \mathbf{f}) = \operatorname{div}_{\mathbf{x}} \mathbf{f}, \tag{3.12}$$

and by (3.10) and (3.4) we have

$$\begin{aligned} \operatorname{div}_{\mathbf{x}'} \mathbf{v}' &= \operatorname{div}_{\mathbf{x}'} \left\{A(t) \cdot \mathbf{v} + A'(t) \cdot \mathbf{x} + \mathbf{w}(t)\right\} \\ &= \operatorname{div}_{\mathbf{x}} \left\{\mathbf{v} + A^{-1}(t) \cdot A'(t) \cdot \mathbf{x} + A^{-1}(t) \cdot \mathbf{w}(t)\right\} = \operatorname{div}_{\mathbf{x}} \mathbf{v} + \operatorname{tr}\left(A^{-1}(t) \cdot A'(t)\right), \end{aligned} \tag{3.13}$$

where $\operatorname{tr}\left(A^{-1}(t) \cdot A'(t)\right)$ is the trace of the matrix $A^{-1}(t) \cdot A'(t)$ (sum of diagonal elements). However, since $A^T(t) \cdot A(t) = I$ we have $A^{-1}(t) = A^T(t)$ and $A^{-1}(t) \cdot A'(t) = S(t)$, where $S^T(t) = -S(t)$. In particular $tr\ S(t) = 0$ and thus

$$tr\ \left(A^{-1}(t) \cdot A'(t)\right) = 0. \tag{3.14}$$

Thus by (3.13) and (3.14) we have

$$\operatorname{div}_{\mathbf{x}'} \mathbf{v}' = \operatorname{div}_{\mathbf{x}} \mathbf{v}. \tag{3.15}$$

So by (3.11), (3.12) and (3.15) we proved **(i)**.

Next by (3.10) and (3.3) we have

$$d_{\mathbf{x}'}\mathbf{f}' = d_{\mathbf{x}'} (A(t) \cdot \mathbf{f}) = A(t) \cdot d_{\mathbf{x}'}\mathbf{f} = A(t) \cdot (d_{\mathbf{x}}\mathbf{f}) \cdot A^{-1}(t) = A(t) \cdot (d_{\mathbf{x}}\mathbf{f}) \cdot A^T(t), \tag{3.16}$$

and by (3.10) and (3.4) we have

$$\begin{aligned} d_{\mathbf{x}'}\mathbf{v}' &= d_{\mathbf{x}'} \left(A(t) \cdot \mathbf{v} + A'(t) \cdot \mathbf{x} + \mathbf{w}(t)\right) = A(t) \cdot d_{\mathbf{x}'}\mathbf{v} + d_{\mathbf{x}} \left(A'(t) \cdot \mathbf{x}\right) \cdot A^{-1}(t) \\ &= A(t) \cdot (d_{\mathbf{x}}\mathbf{v}) \cdot A^{-1}(t) + A'(t)A^{-1}(t) = A(t) \cdot (d_{\mathbf{x}}\mathbf{v}) \cdot A^T(t) + A'(t) \cdot A^T(t). \end{aligned} \tag{3.17}$$

Then taking the transpose of the both sides of (3.17) we infer

$$\left\{d_{\mathbf{x}'}\mathbf{v}'\right\}^T = A(t) \cdot \{d_{\mathbf{x}}\mathbf{v}\}^T \cdot A^T(t) + A(t) \cdot \left\{A'(t)\right\}^T. \tag{3.18}$$

However, as before, since $A(t) \cdot A^T(t) = I$ we have $A'(t) \cdot A^T(t) + A(t) \cdot \left\{A'(t)\right\}^T = 0$, by (3.17) and (3.18), we have

$$\left(d_{\mathbf{x}'}\mathbf{v}' + \left\{d_{\mathbf{x}'}\mathbf{v}'\right\}^T\right) = A(t) \cdot \left(d_{\mathbf{x}}\mathbf{v} + \{d_{\mathbf{x}}\mathbf{v}\}^T\right) \cdot A^T(t). \tag{3.19}$$

So by (3.16) and (3.19) we proved **(iii)**.

Next by the chain rule and (3.2), we obtain

$$\nabla_{\mathbf{x}'}\psi' = \nabla_{\mathbf{x}'}\psi = \left\{A^{-1}(t)\right\}^T \cdot \nabla_{\mathbf{x}}\psi = A(t)\cdot\nabla_{\mathbf{x}}\psi, \tag{3.20}$$

by (3.3) and (3.10) we obtain

$$\text{curl}_{\mathbf{x}'}\mathbf{f}' = \text{curl}_{\mathbf{x}'}\left(A(t)\cdot\mathbf{f}\right) = A(t)\cdot\text{curl}_{\mathbf{x}}\mathbf{f}, \tag{3.21}$$

and by the chain rule and (3.6) we have

$$\text{div}_{\mathbf{x}'}\, T' = \text{div}_{\mathbf{x}'}\left(A(t)\cdot T\cdot A^T(t)\right) = A(t)\cdot\left(\text{div}_{\mathbf{x}}\, T\right). \tag{3.22}$$

Thus by (3.22) and (3.19) we have

$$\text{div}_{\mathbf{x}'}\left(d_{\mathbf{x}'}\mathbf{v}' + \left\{d_{\mathbf{x}'}\mathbf{v}'\right\}^T\right) = A(t)\cdot\left\{\text{div}_{\mathbf{x}}\left(d_{\mathbf{x}}\mathbf{v} + \{d_{\mathbf{x}}\mathbf{v}\}^T\right)\right\}. \tag{3.23}$$

On the other hand, by (3.12) and (3.20) we have

$$\nabla_{\mathbf{x}'}\left(\text{div}_{\mathbf{x}'}\,\mathbf{v}'\right) = A(t)\cdot\nabla_{\mathbf{x}}\left(\text{div}_{\mathbf{x}}\,\mathbf{v}\right). \tag{3.24}$$

Therefore, by (3.23) and (3.24), using (2.10) we deduce

$$\Delta_{\mathbf{x}'}\mathbf{v}' = A(t)\cdot\Delta_{\mathbf{x}}\mathbf{v} \quad \text{and} \quad \text{curl}_{\mathbf{x}'}\left(\text{curl}_{\mathbf{x}'}\mathbf{v}'\right) = A(t)\cdot\text{curl}_{\mathbf{x}}\left(\text{curl}_{\mathbf{x}}\mathbf{v}\right). \tag{3.25}$$

Next by (3.4) we deduce

$$(\mathbf{u}' - \mathbf{v}') = A(t)\cdot(\mathbf{u} - \mathbf{v}). \tag{3.26}$$

So by (3.11), (3.20), (3.21), (3.22), (3.23), (3.24), (3.25) and (3.26) we deduce **(ii)**.

Furthermore, by the chain rule for every scalar field $\gamma : \mathbb{R}^3 \times [t_0, +\infty) \to \mathbb{R}$ and for every vector field $\Gamma : \mathbb{R}^3 \times [t_0, +\infty) \to \mathbb{R}^3$ we obtain

$$\frac{\partial\gamma}{\partial t} = \frac{\partial\gamma}{\partial t'} + \left(A'(t)\cdot\mathbf{x} + \mathbf{w}(t)\right)\cdot\nabla_{\mathbf{x}'}\gamma \tag{3.27}$$

and

$$\frac{\partial\Gamma}{\partial t} = \frac{\partial\Gamma}{\partial t'} + (d_{\mathbf{x}'}\Gamma)\cdot\left(A'(t)\cdot\mathbf{x} + \mathbf{w}(t)\right). \tag{3.28}$$

Therefore, by (3.28) and (3.10)

$$\frac{\partial\Gamma}{\partial t'} = \frac{\partial\Gamma}{\partial t} - (d_{\mathbf{x}}\Gamma)\cdot\left(A^{-1}(t)\cdot A'(t)\cdot\mathbf{x} + A^{-1}(t)\cdot\mathbf{w}(t)\right), \tag{3.29}$$

and by (3.10) (3.20) and (3.27)

$$\frac{\partial \gamma}{\partial t'} + \big(A(t)\cdot\Gamma + A'(t)\cdot \mathbf{x} + \mathbf{w}(t)\big)\cdot \nabla_{\mathbf{x}'}\gamma = \frac{\partial \gamma}{\partial t} + \Gamma\cdot\nabla_{\mathbf{x}}\gamma. \tag{3.30}$$

In particular, by (3.2), (3.4) and (3.30), we have

$$\frac{\partial \psi}{\partial t'} + \mathbf{v}'\cdot\nabla_{\mathbf{x}'}\psi = \frac{\partial \psi}{\partial t} + \mathbf{v}\cdot\nabla_{\mathbf{x}}\psi \tag{3.31}$$

and then since

$$\frac{\partial \psi}{\partial t} + \operatorname{div}_{\mathbf{x}}\{\psi\mathbf{v}\} = \frac{\partial \psi}{\partial t} + \mathbf{v}\cdot\nabla_{\mathbf{x}}\psi + \psi\,(\operatorname{div}_{\mathbf{x}}\mathbf{v}), \tag{3.32}$$

by (3.31), (3.2) and (3.15) we infer **(iv)**. On the other hand, by (3.10), (3.29) and (3.4) for every vector field $\Gamma : \mathbb{R}^3 \times [t_0, +\infty) \to \mathbb{R}^3$ we get

$$\begin{aligned}
&\frac{\partial\,(A(t)\cdot\Gamma)}{\partial t'} - \operatorname{curl}_{\mathbf{x}'}\big(\mathbf{v}'\times(A(t)\cdot\Gamma)\big) + (\operatorname{div}_{\mathbf{x}'}(A(t)\cdot\Gamma))\,\mathbf{v}' = \\
&\left(A(t)\cdot\frac{\partial\Gamma}{\partial t} + A'(t)\cdot\Gamma - A(t)\cdot(d_{\mathbf{x}}\Gamma)\cdot\big(A^{-1}(t)\cdot A'(t)\cdot\mathbf{x} + A^{-1}(t)\cdot\mathbf{w}(t)\big)\right) \\
&\quad - A(t)\cdot\operatorname{curl}_{\mathbf{x}}\big(\big(\mathbf{v} + A^{-1}(t)\cdot A'(t)\cdot\mathbf{x} + A^{-1}(t)\cdot\mathbf{w}(t)\big)\times\Gamma\big) \\
&\qquad + (\operatorname{div}_{\mathbf{x}}\Gamma)\big(A(t)\cdot\mathbf{v} + A'(t)\cdot\mathbf{x} + \mathbf{w}(t)\big) \\
&\qquad = A(t)\cdot\left(\frac{\partial\Gamma}{\partial t} - \operatorname{curl}_{\mathbf{x}}(\mathbf{v}\times\Gamma) + (\operatorname{div}_{\mathbf{x}}\Gamma)\,\mathbf{v}\right) \\
&\qquad + A(t)\cdot\big(d_{\mathbf{x}}\big(A^{-1}(t)\cdot A'(t)\cdot\mathbf{x} + A^{-1}(t)\cdot\mathbf{w}(t)\big)\big)\cdot\Gamma \\
&\qquad - A(t)\cdot(d_{\mathbf{x}}\Gamma)\cdot\big(A^{-1}(t)\cdot A'(t)\cdot\mathbf{x} + A^{-1}(t)\cdot\mathbf{w}(t)\big) \\
&\qquad + A(t)\cdot\big((\operatorname{div}_{\mathbf{x}}\Gamma)\big(A^{-1}(t)\cdot A'(t)\cdot\mathbf{x} + A^{-1}(t)\cdot\mathbf{w}(t)\big)\big) \\
&\qquad - A(t)\cdot\operatorname{curl}_{\mathbf{x}}\big(\big(A^{-1}(t)\cdot A'(t)\cdot\mathbf{x} + A^{-1}(t)\cdot\mathbf{w}(t)\big)\times\Gamma\big).
\end{aligned} \tag{3.33}$$

On the other hand, by (2.11), we have

$$\begin{aligned}
&\big(d_{\mathbf{x}}\big(A^{-1}(t)\cdot A'(t)\cdot\mathbf{x} + A^{-1}(t)\cdot\mathbf{w}(t)\big)\big)\cdot\Gamma \\
&\qquad - (d_{\mathbf{x}}\Gamma)\cdot\big(A^{-1}(t)\cdot A'(t)\cdot\mathbf{x} + A^{-1}(t)\cdot\mathbf{w}(t)\big) \\
&\qquad + (\operatorname{div}_{\mathbf{x}}\Gamma)\big(A^{-1}(t)\cdot A'(t)\cdot\mathbf{x} + A^{-1}(t)\cdot\mathbf{w}(t)\big) \\
&\qquad - \operatorname{curl}_{\mathbf{x}}\big(\big(A^{-1}(t)\cdot A'(t)\cdot\mathbf{x} + A^{-1}(t)\cdot\mathbf{w}(t)\big)\times\Gamma\big) \\
&\qquad\qquad = \big(\operatorname{div}_{\mathbf{x}}\big(A^{-1}(t)\cdot A'(t)\cdot\mathbf{x} + A^{-1}(t)\cdot\mathbf{w}(t)\big)\big)\,\Gamma.
\end{aligned} \tag{3.34}$$

Therefore, by (3.33) and (3.34) we deduce

$$\frac{\partial\left(A(t)\cdot\Gamma\right)}{\partial t'}-\operatorname{curl}_{\mathbf{x}'}\left(\mathbf{v}'\times\left(A(t)\cdot\Gamma\right)\right)+\left(\operatorname{div}_{\mathbf{x}'}\left(A(t)\cdot\Gamma\right)\right)\mathbf{v}'=$$
$$A(t)\cdot\left(\frac{\partial\Gamma}{\partial t}-\operatorname{curl}_{\mathbf{x}}\left(\mathbf{v}\times\Gamma\right)+\left(\operatorname{div}_{\mathbf{x}}\Gamma\right)\mathbf{v}\right)$$
$$+A(t)\cdot\left(\left(\operatorname{div}_{\mathbf{x}}\left(A^{-1}(t)\cdot A'(t)\cdot\mathbf{x}+A^{-1}(t)\cdot\mathbf{w}(t)\right)\right)\Gamma\right)$$
$$=A(t)\cdot\left(\frac{\partial\Gamma}{\partial t}-\operatorname{curl}_{\mathbf{x}}\left(\mathbf{v}\times\Gamma\right)+\left(\operatorname{div}_{\mathbf{x}}\Gamma\right)\mathbf{v}\right)+\left(\operatorname{tr}\left(A^{-1}(t)\cdot A'(t)\right)\right)A(t)\cdot\Gamma, \tag{3.35}$$

where $\operatorname{tr}\left(A^{-1}(t)\cdot A'(t)\right)$ is the trace of the matrix $A^{-1}(t)\cdot A'(t)$. Therefore, by (3.35) and (3.14) for every vector field $\Gamma:\mathbb{R}^3\times[t_0,+\infty)\to\mathbb{R}^3$ we have

$$\frac{\partial\left(A(t)\cdot\Gamma\right)}{\partial t'}-\operatorname{curl}_{\mathbf{x}'}\left(\mathbf{v}'\times\left(A(t)\cdot\Gamma\right)\right)+\left(\operatorname{div}_{\mathbf{x}'}\left(A(t)\cdot\Gamma\right)\right)\mathbf{v}'$$
$$=A(t)\cdot\left(\frac{\partial\Gamma}{\partial t}-\operatorname{curl}_{\mathbf{x}}\left(\mathbf{v}\times\Gamma\right)+\left(\operatorname{div}_{\mathbf{x}}\Gamma\right)\mathbf{v}\right). \tag{3.36}$$

Thus, by (3.36) and (3.3) we infer

$$\frac{\partial\mathbf{f}'}{\partial t'}-\operatorname{curl}_{\mathbf{x}'}\left(\mathbf{v}'\times\mathbf{f}'\right)+\left(\operatorname{div}_{\mathbf{x}'}\mathbf{f}'\right)\mathbf{v}'=A(t)\cdot\left(\frac{\partial\mathbf{f}}{\partial t}-\operatorname{curl}_{\mathbf{x}}\left(\mathbf{v}\times\mathbf{f}\right)+\left(\operatorname{div}_{\mathbf{x}}\mathbf{f}\right)\mathbf{v}\right). \tag{3.37}$$

Finally, by (2.15), (2.14) and (2.11) we deduce

$$\frac{\partial\mathbf{f}}{\partial t}-\mathbf{v}\times\operatorname{curl}_{\mathbf{x}}\mathbf{f}+\nabla_{\mathbf{x}}\left(\mathbf{f}\cdot\mathbf{v}\right)=\nabla_{\mathbf{x}}\left(\mathbf{f}\cdot\mathbf{v}\right)+\frac{\partial\mathbf{f}}{\partial t}+d_{\mathbf{x}}\mathbf{f}\cdot\mathbf{v}-\left\{d_{\mathbf{x}}\mathbf{f}\right\}^T\cdot\mathbf{v}$$
$$=\frac{\partial\mathbf{f}}{\partial t}+d_{\mathbf{x}}\mathbf{f}\cdot\mathbf{v}+\left\{d_{\mathbf{x}}\mathbf{v}\right\}^T\cdot\mathbf{f}=\frac{\partial\mathbf{f}}{\partial t}+d_{\mathbf{x}}\mathbf{f}\cdot\mathbf{v}-d_{\mathbf{x}}\mathbf{v}\cdot\mathbf{f}+\left(d_{\mathbf{x}}\mathbf{v}+\left\{d_{\mathbf{x}}\mathbf{v}\right\}^T\right)\cdot\mathbf{f}$$
$$=\left(\frac{\partial\mathbf{f}}{\partial t}-\operatorname{curl}_{\mathbf{x}}\left(\mathbf{v}\times\mathbf{f}\right)+\left(\operatorname{div}_{\mathbf{x}}\mathbf{f}\right)\mathbf{v}\right)-\left(\operatorname{div}_{\mathbf{x}}\mathbf{v}\right)\mathbf{f}+\left(d_{\mathbf{x}}\mathbf{v}+\left\{d_{\mathbf{x}}\mathbf{v}\right\}^T\right)\cdot\mathbf{f}. \tag{3.38}$$

So we get (3.9). Moreover, by (3.3), (3.15), (3.19), (3.38) and (3.37) we infer

$$\frac{\partial\mathbf{f}'}{\partial t'}-\mathbf{v}'\times\operatorname{curl}_{\mathbf{x}'}\mathbf{f}'+\nabla_{\mathbf{x}'}\left(\mathbf{f}'\cdot\mathbf{v}'\right)=A(t)\cdot\left(\frac{\partial\mathbf{f}}{\partial t}-\mathbf{v}\times\operatorname{curl}_{\mathbf{x}}\mathbf{f}+\nabla_{\mathbf{x}}\left(\mathbf{f}\cdot\mathbf{v}\right)\right). \tag{3.39}$$

So by (3.37) and (3.39) we finally obtain **(v)**. □

Proposition 3.2 *Consider some fixed inertial or non-inertial Cartesian coordinate system, denoted by the sign* ({0})*, and consider a vector field* $\mathbf{d}:=\mathbf{d}(\mathbf{x},t):\mathbb{R}^3\times[t_0,+\infty)\to\mathbb{R}^3$ *associated only with the chosen coordinate system* ({0})*. Then there exist a unique speed-like vector field* $\mathbf{v}_{\mathbf{d}}:\mathbb{R}^3\times[t_0,+\infty)\to\mathbb{R}^3$ *(associated with* *any inertial or non-inertial Cartesian coordinate system) such that in the particular system ({0}) we have*

$$\mathbf{v_d}(\mathbf{x}, t) = \mathbf{d}(\mathbf{x}, t) \quad \forall (\mathbf{x}, t) \in \mathbb{R}^3 \times [t_0, +\infty) . \tag{3.40}$$

Moreover, there exist unique a proper vector field $\mathbf{h_d} : \mathbb{R}^3 \times [t_0, +\infty) \to \mathbb{R}^3$ *(associated with any inertial or non-inertial Cartesian coordinate system) such that in the particular system ({0}) we have*

$$\mathbf{h_d}(\mathbf{x}, t) = \mathbf{d}(\mathbf{x}, t) \quad \forall (\mathbf{x}, t) \in \mathbb{R}^3 \times [t_0, +\infty) . \tag{3.41}$$

Proof Since uniqueness is trivial, we concentrate ourselves on existence. Similar to (3.1), given an arbitrary inertial or non-inertial Cartesian coordinate system, denoted by the sign ({1}), the change of coordinate system ({0}) to the system ({1}) can be written as

$$\begin{cases} \mathbf{x}^{(1)} = B_1(t) \cdot \mathbf{x} + \mathbf{z}_1(t), \\ t^{(1)} = t, \end{cases} \tag{3.42}$$

where $B_1(t) \in SO(3)$ is a rotation, i.e. $B_1(t) \in \mathbb{R}^{3\times 3}$, $\det B_1(t) > 0$ and $B_1(t) \cdot B_1^T(t) = I$. Similarly, given another arbitrary inertial or non-inertial Cartesian coordinate system, denoted by the sign ({2}), the change of coordinate system ({0}) to the system ({2}) can be written as

$$\begin{cases} \mathbf{x}^{(2)} = B_2(t) \cdot \mathbf{x} + \mathbf{z}_2(t), \\ t^{(2)} = t, \end{cases} \tag{3.43}$$

where $B_2(t) \in SO(3)$ is another rotation. On the other hand, the change coordinate system ({1}) to the system ({2}) can be written as

$$\begin{cases} \mathbf{x}^{(2)} = A(t^{(1)}) \cdot \mathbf{x}^{(1)} + \mathbf{z}(t^{(1)}), \\ t^{(2)} = t^{(1)}, \end{cases} \tag{3.44}$$

where $A(t) \in SO(3)$ is a rotation. In particular, inserting (3.42) into (3.44) gives

$$\mathbf{x}^{(2)} = A(t) \cdot (B_1(t) \cdot \mathbf{x} + \mathbf{z}_1(t)) + \mathbf{z}(t) = (A(t) \cdot B_1(t)) \cdot \mathbf{x} + (A(t) \cdot \mathbf{z}_1(t) + \mathbf{z}(t)) . \tag{3.45}$$

Thus, comparing (3.45) with (3.43) we easily deduce

$$B_2(t) = (A(t) \cdot B_1(t)) \quad \text{and} \quad \mathbf{z}_2(t) = (A(t) \cdot \mathbf{z}_1(t) + \mathbf{z}(t)) \quad \forall t. \tag{3.46}$$

Next, for the system ({1}) define

$$\mathbf{v}_\mathbf{d}^{(1)}(\mathbf{x}^{(1)}, t^{(1)}) = B_1(t) \cdot \mathbf{d}(\mathbf{x}, t) + \frac{dB_1}{dt}(t) \cdot \mathbf{x} + \frac{d\mathbf{z}_1}{dt}(t), \tag{3.47}$$

and

$$\mathbf{h}_{\mathbf{d}}^{(1)}(\mathbf{x}^{(1)}, t^{(1)}) = B_1(t) \cdot \mathbf{d}(\mathbf{x}, t). \tag{3.48}$$

Correspondingly, for the system ({2}) we have

$$\mathbf{v}_{\mathbf{d}}^{(2)}(\mathbf{x}^{(2)}, t^{(2)}) = B_2(t) \cdot \mathbf{d}(\mathbf{x}, t) + \frac{dB_2}{dt}(t) \cdot \mathbf{x} + \frac{d\mathbf{z}_2}{dt}(t), \tag{3.49}$$

and

$$\mathbf{h}_{\mathbf{d}}^{(2)}(\mathbf{x}^{(2)}, t^{(2)}) = B_2(t) \cdot \mathbf{d}(\mathbf{x}, t). \tag{3.50}$$

Thus, in the particular system ({0}) we have

$$\mathbf{v}_{\mathbf{d}}(\mathbf{x}, t) = \mathbf{h}_{\mathbf{d}}(\mathbf{x}, t) = \mathbf{d}(\mathbf{x}, t) \qquad \forall (\mathbf{x}, t) \in \mathbb{R}^3 \times [t_0, +\infty). \tag{3.51}$$

Thus, in order to complete the proof we need to show that the vector field, defined by (3.47) and (3.49), is a speed-like vector field and the vector field, defined by (3.48) and (3.50), is a proper vector field. Indeed, by (3.48) and (3.50) together with (3.46) and (3.42) we easily deduce

$$\mathbf{h}_{\mathbf{d}}^{(2)}(\mathbf{x}^{(2)}, t^{(2)}) = (A(t) \cdot B_1(t)) \cdot \mathbf{d}(\mathbf{x}, t) = A(t) \cdot (B_1(t) \cdot \mathbf{d}(\mathbf{x}, t)) = A(t^{(1)}) \cdot \mathbf{h}_{\mathbf{d}}^{(1)}(\mathbf{x}^{(1)}, t^{(1)}), \tag{3.52}$$

and so, comparing (3.52) with (3.3) in Definition 3.1, we deduce that $\mathbf{h}_{\mathbf{d}}$ is a proper vector field. Similarly, by (3.47) and (3.49) together with (3.46) and (3.42) we deduce

$$\begin{aligned}
\mathbf{v}_{\mathbf{d}}^{(2)}(\mathbf{x}^{(2)}, t^{(2)}) &= B_2(t) \cdot \mathbf{d}(\mathbf{x}, t) + \frac{dB_2}{dt}(t) \cdot \mathbf{x} + \frac{d\mathbf{z}_2}{dt}(t) = \\
&(A(t) \cdot B_1(t)) \cdot \mathbf{d}(\mathbf{x}, t) + \left(\frac{d}{dt}\{A(t) \cdot B_1(t)\}\right) \cdot \mathbf{x} + \frac{d}{dt}\{A(t) \cdot \mathbf{z}_1(t) + \mathbf{z}(t)\} = \\
&A(t) \cdot (B_1(t) \cdot \mathbf{d}(\mathbf{x}, t)) + \left(A(t) \cdot \frac{dB_1}{dt}(t) + \frac{dA}{dt}(t) \cdot B_1(t)\right) \cdot \mathbf{x} \\
&+ \left\{A(t) \cdot \frac{d\mathbf{z}_1}{dt}(t) + \frac{dA}{dt}(t) \cdot \mathbf{z}_1(t) + \frac{d\mathbf{z}}{dt}(t)\right\} \\
= A(t) \cdot \left(B_1(t) \cdot \mathbf{d}(\mathbf{x}, t) + \frac{dB_1}{dt}(t) \cdot \mathbf{x} + \frac{d\mathbf{z}_1}{dt}(t)\right) &+ \frac{dA}{dt}(t) \cdot (B_1(t) \cdot \mathbf{x} + \mathbf{z}_1(t)) + \frac{d\mathbf{z}}{dt}(t) \\
&= \mathbf{A}(t^{(1)}) \cdot \mathbf{v}_{\mathbf{d}}^{(1)}(\mathbf{x}^{(1)}, t^{(1)}) + \frac{dA}{dt^{(1)}}(t^{(1)}) \cdot \mathbf{x}^{(1)} + \frac{d\mathbf{z}}{dt^{(1)}}(t^{(1)}),
\end{aligned} \tag{3.53}$$

and so, comparing (3.53) with (3.4) in Definition 3.1, we deduce that $\mathbf{v}_{\mathbf{d}}$ is a speed-like vector field. This completes the proof. □

Definition 3.2 Let $\mathbf{u} := \mathbf{u}(\mathbf{x}, t)$ be a speed-like vector field defined for $(\mathbf{x}, t) \in \Omega$ with domain of definition $\Omega \subset \mathbb{R}^3 \times \mathbb{R}$ and let $(*)$ be some inertial or non-inertial Cartesian coordinate system. Then $\mathbf{u}$ is called generally trivial speed-like field if there exists another Cartesian coordinate system $(**)$ such that under the change of coordinate system $(*)$ to another Cartesian coordinate system $(**)$ given by

$$\begin{cases} \mathbf{x}' = A(t) \cdot \mathbf{x} + \mathbf{z}(t), \\ t' = t, \end{cases} \tag{3.54}$$

where $A(t) \in SO(3)$ is a rotation, we have

$$A(t) \cdot \mathbf{u}(\mathbf{x}, t) + \frac{dA}{dt}(t) \cdot \mathbf{x} + \frac{d\mathbf{z}}{dt}(t) = \mathbf{u}'(\mathbf{x}', t') = 0. \tag{3.55}$$

Proposition 3.3 *Let* $\mathbf{u} := \mathbf{u}(\mathbf{x}, t)$ *be a smooth speed-like vector field defined for* $(\mathbf{x}, t) \in \Omega$ *with connected domain* $\Omega \subset \mathbb{R}^3 \times \mathbb{R}$, *such that for every instant of time* τ *the domain* $\Omega_\tau = \{\mathbf{x} \in \mathbb{R}^3 : (\mathbf{x}, \tau) \in \Omega\}$ *is also connected. Then* $\mathbf{u}$ *is a generally trivial speed-like field if and only if we have*

$$d_{\mathbf{x}}\mathbf{u} + \{d_{\mathbf{x}}\mathbf{u}\}^T = 0 \quad \forall (\mathbf{x}, t) \in \Omega. \tag{3.56}$$

Proof Assume that (3.56) is satisfied and denotes $(u_1, u_2, u_3) = \mathbf{u} \in \mathbb{R}^3$ and $(x_1, x_2, x_3) = \mathbf{x} \in \mathbb{R}^3$. Then (3.56) is equivalent to

$$\frac{\partial u_m}{\partial x_n} + \frac{\partial u_n}{\partial x_m} = 0 \quad \forall (\mathbf{x}, t) \in \Omega, \ \forall m, n = 1, 2, 3. \tag{3.57}$$

One more time differentiating of (3.57) gives

$$\frac{\partial^2 u_m}{\partial x_n \partial x_k} + \frac{\partial^2 u_n}{\partial x_m \partial x_k} = 0 \quad \forall (\mathbf{x}, t) \in \Omega, \ \forall m, n, k = 1, 2, 3. \tag{3.58}$$

Therefore,

$$\begin{aligned} 2\frac{\partial^2 u_n}{\partial x_m \partial x_k} &= \left(\frac{\partial^2 u_m}{\partial x_n \partial x_k} + \frac{\partial^2 u_n}{\partial x_m \partial x_k}\right) - \left(\frac{\partial^2 u_m}{\partial x_k \partial x_n} + \frac{\partial^2 u_k}{\partial x_m \partial x_n}\right) \\ &+ \left(\frac{\partial^2 u_k}{\partial x_n \partial x_m} + \frac{\partial^2 u_n}{\partial x_k \partial x_m}\right) = 0 \quad \forall (\mathbf{x}, t) \in \Omega, \ \forall m, n, k = 1, 2, 3. \end{aligned} \tag{3.59}$$

So,

$$\frac{\partial}{\partial x_k}\left(\frac{\partial u_n}{\partial x_m}\right) = \frac{\partial^2 u_n}{\partial x_m \partial x_k} = 0 \quad \forall (\mathbf{x}, t) \in \Omega, \ \forall m, n, k = 1, 2, 3. \tag{3.60}$$

Thus since, for every instant of time t the domain $\Omega_t \subset \mathbb{R}^3$ is connected, by (3.60) we deduce that $d_{\mathbf{x}}\mathbf{u}$ is independent on the spatial variable $\mathbf{x}$ and can depend only on the time variable t. So by (3.60) and (3.57) together we deduce that there exists an antisymmetric matrix $B(t) \in \mathbb{R}^{3\times 3}$, depending on the time variable t only, such that

$$d_{\mathbf{x}}\mathbf{u} = B(t) \ \ \forall (\mathbf{x}, t) \in \Omega \quad \text{where} \quad B^T(t) = -B(t). \tag{3.61}$$

Thus there exists a regular mapping $\mathbf{d}(t) : \mathbb{R} \to \mathbb{R}^3$ such that

$$\mathbf{u}(\mathbf{x}, t) = B(t) \cdot \mathbf{x} + \mathbf{d}(t) \quad \forall (\mathbf{x}, t) \in \Omega \qquad \text{where} \quad B^T(t) = -B(t). \tag{3.62}$$

Next let $\mathbf{h}(t) \in \mathbb{R}^3$ be a solution of

$$\begin{cases} \frac{d\mathbf{h}}{dt}(t) = B(t) \cdot \mathbf{h}(t) + \mathbf{d}(t) & \forall t \\ \mathbf{h}(0) = 0, \end{cases} \tag{3.63}$$

and $R(t) \in \mathbb{R}^{3\times 3}$ be a matrix-valued solution of

$$\begin{cases} \frac{dR}{dt}(t) = B(t) \cdot R(t) & \forall t \\ R(0) = I, \end{cases} \tag{3.64}$$

where I is the identity matrix. In particular,

$$R^T(t) \cdot \frac{dR}{dt}(t) = R^T(t) \cdot B(t) \cdot R(t) \qquad \forall t. \tag{3.65}$$

On the other hand, by (3.64) and (3.62), we have

$$\frac{dR^T}{dt}(t) = R^T(t) \cdot B^T(t) = -R^T(t) \cdot B(t) \qquad \forall t, \tag{3.66}$$

and thus

$$\frac{dR^T}{dt}(t) \cdot R(t) = -R^T(t) \cdot B(t) \cdot R(t) \qquad \forall t. \tag{3.67}$$

Thus, by (3.65) and (3.67) together we obtain

$$\frac{d}{dt}\left(R^T(t) \cdot R(t)\right) = R^T(t) \cdot \frac{dR}{dt}(t) + \frac{dR^T}{dt}(t) \cdot R(t) = 0 \qquad \forall t. \tag{3.68}$$

Therefore, by (3.68) and (3.64) we obtain

$$R^T(t) \cdot R(t) = R^T(0) \cdot R(0) = I^T \cdot I = I \qquad \forall t. \tag{3.69}$$

In particular, we have $(\det R(t))^2 = 1$ and thus, since $\det R(0) = 1$, we deduce $\det R(t) = 1$. So we obtain

$$R(t) \in SO(3) \qquad \forall t. \tag{3.70}$$

Next consider $A(t) \in SO(3)$ and $\mathbf{z}(t) \in \mathbb{R}^3$ defined by

$$A(t) = R^T(t) = R^{-1}(t) \quad \text{and} \quad \mathbf{z}(t) = -R^T(t) \cdot \mathbf{h}(t) \qquad \forall t, \tag{3.71}$$

and consider the change of Cartesian coordinate system given by

$$\begin{cases} \mathbf{x}' = A(t)\cdot\mathbf{x}+\mathbf{z}(t) = R^T(t)\cdot\mathbf{x} - R^T(t)\cdot\mathbf{h}(t) \\ t' = t. \end{cases} \tag{3.72}$$

Then since the vector field **u** is a speed-like vector field, under the change of coordinate system, given by (3.72), **u** transforms as

$$\mathbf{u}'(\mathbf{x}',t') = A(t)\cdot\mathbf{u}(\mathbf{x},t) + \frac{dA}{dt}(t)\cdot\mathbf{x} + \frac{d\mathbf{z}}{dt}(t). \tag{3.73}$$

Therefore, inserting (3.62) into (3.73) gives

$$\mathbf{u}'(\mathbf{x}',t') = A(t)\cdot(B(t)\cdot\mathbf{x}+\mathbf{d}(t)) + \frac{dA}{dt}(t)\cdot\mathbf{x} + \frac{d\mathbf{z}}{dt}(t). \tag{3.74}$$

Then inserting (3.63), (3.64) and (3.71) into (3.74) together with (3.68) gives

$$\begin{aligned}
\mathbf{u}'(\mathbf{x}',t') &= R^T(t)\cdot\left(\frac{dR}{dt}(t)\cdot R^T(t)\cdot\mathbf{x}+\mathbf{d}(t)\right) + \frac{dR^T}{dt}(t)\cdot\mathbf{x} - \frac{d}{dt}\left(R^T(t)\cdot\mathbf{h}(t)\right) = \\
R^T(t)&\cdot\left(\frac{dR}{dt}(t)\cdot R^T(t)+R(t)\cdot\frac{dR^T}{dt}(t)\right)\cdot\mathbf{x} + R^T(t)\cdot\mathbf{d}(t) - \frac{dR^T}{dt}(t)\cdot\mathbf{h}(t) - R^T(t)\cdot\frac{d\mathbf{h}}{dt}(t) \\
&= 0 + R^T(t)\cdot\mathbf{d}(t) - \frac{dR^T}{dt}(t)\cdot\mathbf{h}(t) - R^T(t)\cdot(B(t)\cdot\mathbf{h}(t)+\mathbf{d}(t)) \\
&= R^T(t)\cdot\mathbf{d}(t) - \frac{dR^T}{dt}(t)\cdot\mathbf{h}(t) - R^T(t)\cdot\left(\frac{dR}{dt}(t)\cdot R^T(t)\cdot\mathbf{h}(t)+\mathbf{d}(t)\right) \\
&= 0 - R^T(t)\cdot\left(R(t)\cdot\frac{dR^T}{dt}(t)+\frac{dR}{dt}(t)\cdot R^T(t)\right)\cdot\mathbf{h}(t) = 0.
\end{aligned} \tag{3.75}$$

So, starting from equality (3.56) we finally deduce in (3.75) that

$$\mathbf{u}'(\mathbf{x}',t') = 0 \quad \forall\,(\mathbf{x}',t') \tag{3.76}$$

in some Cartesian coordinate system and thus by the definition **u** is a generally trivial speed-like field.

Conversely, assume that **u** is a generally trivial speed-like field and (∗) is some inertial or non-inertial Cartesian coordinate system. Then by the Definition 3.2 there exists another Cartesian coordinate system (∗∗) such that under the change of coordinate system (∗) to another Cartesian coordinate system (∗∗) given by (3.54) we have (3.55). In particular, in system (∗∗) we have

$$d_{\mathbf{x}'}\mathbf{u}' + \left\{d_{\mathbf{x}'}\mathbf{u}'\right\}^T = 0. \tag{3.77}$$

Thus, since by Proposition 3.1, **(iii)**, the quantity $\left(d_{\mathbf{x}}\mathbf{u} + \{d_{\mathbf{x}}\mathbf{u}\}^T\right)$ is a proper matrix field, by (3.77) we infer that in system (∗) we also have

$$d_{\mathbf{x}}\mathbf{u} + \{d_{\mathbf{x}}\mathbf{u}\}^T = 0,$$

i.e. we obtain (3.56). □

Proposition 3.4 *Let* $\mathbf{v} := \mathbf{v}(\mathbf{x}, t) : \mathbb{R}^3 \times [t_0, +\infty) \to \mathbb{R}^3$ *be a speed-like vector field, such that there exists some Cartesian coordinate system* $(*)$*, where we have*

$$\lim_{|\mathbf{x}|\to+\infty} \mathbf{v}(\mathbf{x}, t) = 0 \quad \forall t \quad (\textit{ in the given particular coordinate system } (*)). \tag{3.78}$$

Then there exist a uniquely defined generally trivial speed-like vector field $\mathbf{k} := \mathbf{k}(\mathbf{x}, t) : \mathbb{R}^3 \times [t_0, +\infty) \to \mathbb{R}^3$ *and a uniquely defined proper vector field* $\mathbf{h} := \mathbf{h}(\mathbf{x}, t) : \mathbb{R}^3 \times [t_0, +\infty) \to \mathbb{R}^3$*, so that in every Cartesian coordinate system we have*

$$\lim_{|\mathbf{x}|\to+\infty} \mathbf{h}(\mathbf{x}, t) = 0 \quad \forall t\,, \tag{3.79}$$

and in every Cartesian coordinate system we can decompose vector field $\mathbf{v}$ *as*

$$\mathbf{v}(\mathbf{x}, t) = \mathbf{h}(\mathbf{x}, t) + \mathbf{k}(\mathbf{x}, t) \qquad \forall (\mathbf{x}, t) \in \mathbb{R}^3 \times [t_0, +\infty). \tag{3.80}$$

Proof Let $(*)$ be a Cartesian coordinate system, where (3.78) satisfied. Then, by Proposition 3.2 there exist a unique speed-like vector field $\mathbf{k} : \mathbb{R}^3 \times [t_0, +\infty) \to \mathbb{R}^3$, such that in the given particular system $(*)$ we have

$$\mathbf{k}(\mathbf{x}, t) = 0 \quad \forall (\mathbf{x}, t) \in \mathbb{R}^3 \times [t_0, +\infty)\,. \tag{3.81}$$

Obviously, by (3.81) we obtain that $\mathbf{k}$ is generally trivial speed-like vector field. Furthermore, by the same Proposition 3.2 there exist a unique proper vector field $\mathbf{h} : \mathbb{R}^3 \times [t_0, +\infty) \to \mathbb{R}^3$, such that in the given particular system $(*)$ we have

$$\mathbf{h}(\mathbf{x}, t) = \mathbf{v}(\mathbf{x}, t) \quad \forall (\mathbf{x}, t) \in \mathbb{R}^3 \times [t_0, +\infty)\,. \tag{3.82}$$

In particular, by (3.82) and (3.78) in the given particular system $(*)$ we have

$$\lim_{|\mathbf{x}|\to+\infty} \mathbf{h}(\mathbf{x}, t) = 0 \quad \forall t\,. \tag{3.83}$$

Moreover, by together (3.81) and (3.82) in the given particular system $(*)$ we also have

$$\mathbf{h}(\mathbf{x}, t) = \mathbf{v}(\mathbf{x}, t) - \mathbf{k}(\mathbf{x}, t) \quad \forall (\mathbf{x}, t) \in \mathbb{R}^3 \times [t_0, +\infty)\,. \tag{3.84}$$

However, since $\mathbf{v}$ and $\mathbf{k}$ are both speed-like vector fields, by **(i)** in Proposition 3.1 we deduce that $(\mathbf{v} - \mathbf{k})$ is a proper vector field. Thus, since $(\mathbf{v} - \mathbf{k})$ and $\mathbf{h}$ are both proper vector fields, we deduce that (3.83) and (3.84) hold in every Cartesian coordinate system. So we deduce (3.79) and (3.80).

Finally, assume that an arbitrary generally trivial speed-like vector field $\mathbf{k} := \mathbf{k}(\mathbf{x}, t) : \mathbb{R}^3 \times [t_0, +\infty) \to \mathbb{R}^3$ and an arbitrary proper vector field $\mathbf{h} := \mathbf{h}(\mathbf{x}, t) : \mathbb{R}^3 \times [t_0, +\infty) \to \mathbb{R}^3$ satisfy (3.79) and (3.80). Then, inserting (3.78) and (3.79) into (3.80) we deduce that

$$\lim_{|\mathbf{x}| \to +\infty} \mathbf{k}(\mathbf{x}, t) = 0 \quad \forall t \quad (\text{ in the given system } (*))\,. \tag{3.85}$$

However, since $\mathbf{k}$ is a generally trivial speed-like vector field, by transformation law (3.4) in Definition 3.1 we deduce that for every fixed instant of time t $\mathbf{k}$ depends on $\mathbf{x}$ linearly in every Cartesian coordinate system. Plugging this fact with (3.85) together gives

$$\mathbf{k}(\mathbf{x}, t) = 0 \quad \forall (\mathbf{x}, t) \in \mathbb{R}^3 \times [t_0, +\infty) \quad (\text{in the given system } (*))\,. \tag{3.86}$$

So $\mathbf{k}$ is defined uniquely (exactly as in (3.81)). Therefore, by (3.80) and (3.86) we obtain

$$\mathbf{h}(\mathbf{x}, t) = \mathbf{v}(\mathbf{x}, t) \quad \forall (\mathbf{x}, t) \in \mathbb{R}^3 \times [t_0, +\infty) \quad (\text{in the given system } (*))\,. \tag{3.87}$$

So $\mathbf{h}$ is also defined uniquely (exactly as in (3.82)). This completes the proof. □

Definition 3.3 We say that two (inertial or non-inertial) Cartesian coordinate systems $(*)$ and $(**)$ are equivalent if the change of coordinates from the system $(*)$ to the system $(**)$ is given by

$$\begin{cases} \mathbf{x}' = B \cdot \mathbf{x} + \mathbf{c}, \\ t' = t, \end{cases} \tag{3.88}$$

where $B \in SO(3)$ is a constant (independent of time) rotation and $\mathbf{c} \in \mathbb{R}^3$ is a constant (independent of time) vector. In other words, we obtain system $(**)$ from system $(*)$ just by a constant (independent of time) rotation in $\mathbb{R}^3$ and/or by a constant (independent of time) shift of the origin in $\mathbb{R}^3$.

Lemma 3.1 *Let $\mathbf{v} := \mathbf{v}(\mathbf{x}, t) : \mathbb{R}^3 \times [t_0, +\infty) \to \mathbb{R}^3$ be a speed-like vector field and let $\mathbf{h} := \mathbf{h}(\mathbf{x}, t) : \mathbb{R}^3 \times [t_0, +\infty) \to \mathbb{R}^3$ be a proper vector field. Moreover, let $(*)$ and $(**)$ be two equivalent Cartesian coordinate systems. Then we have*

$$\mathbf{v}(\mathbf{x}, t) = \mathbf{h}(\mathbf{x}, t) \quad \forall (\mathbf{x}, t) \in \mathbb{R}^3 \times [t_0, +\infty) \quad \textit{in the system } (*), \tag{3.89}$$

if and only if

$$\mathbf{v}'(\mathbf{x}', t') = \mathbf{h}'(\mathbf{x}', t') \quad \forall (\mathbf{x}', t') \in \mathbb{R}^3 \times [t_0, +\infty) \quad \textit{in the system } (**). \tag{3.90}$$

In particular,

$$\mathbf{v}(\mathbf{x}, t) = 0 \quad \forall (\mathbf{x}, t) \in \mathbb{R}^3 \times [t_0, +\infty) \quad \textit{in the system } (*), \tag{3.91}$$

if and only if

$$\mathbf{v}'(\mathbf{x}',t')=0 \quad \forall(\mathbf{x}',t')\in\mathbb{R}^3\times[t_0,+\infty) \qquad \textit{in the system } (**). \tag{3.92}$$

Proof Obviously, by Definition 3.3 the change of coordinates from system $(*)$ to system $(**)$ is given by

$$\begin{cases}\mathbf{x}'=A(t)\cdot\mathbf{x}+\mathbf{z}(t),\\ t'=t,\end{cases} \tag{3.93}$$

where $A(t):=B\in SO(3)$ is a constant (independent on time) rotation and $\mathbf{z}(t):=\mathbf{c}\in\mathbb{R}^3$ is a constant (independent on time) vector. In particular,

$$\begin{cases}\frac{dA}{dt}(t)=0 & \forall t,\\ \frac{d\mathbf{z}}{dt}(t)=0 & \forall t,\end{cases} \tag{3.94}$$

and so in this case the law of transformation of a speed-like vector field in (3.4) of Definition 3.1 coincides with the law of transformation of a proper vector field in (3.3) of Definition 3.1. □

Definition 3.4 Let $\mathbf{v}:=\mathbf{v}(\mathbf{x},t):\mathbb{R}^3\times[t_0,+\infty)\to\mathbb{R}^3$ be a speed-like smooth vector field and let $(*)$ be some (inertial or non-inertial) Cartesian coordinate system. We say that $\mathbf{v}$ is <u>asymptotically acceptable</u> in the coordinate system $(*)$, if there exist continuous mappings $B(t):[t_0,+\infty)\to\mathbb{R}^{3\times3}$ and $\mathbf{l}(t):[t_0,+\infty)\to\mathbb{R}^3$, such that in the coordinate system $(*)$ we have

$$\begin{cases}\lim\limits_{|\mathbf{x}|\to+\infty}\left(d_{\mathbf{x}}\mathbf{v}(\mathbf{x},t)\right)=B(t) & \forall t\in[t_0,+\infty),\\ \lim\limits_{|\mathbf{x}|\to+\infty}\left(d_{\mathbf{x}}\mathbf{v}(\mathbf{x},t)+\{d_{\mathbf{x}}\mathbf{v}(\mathbf{x},t)\}^T\right)=0 & \forall t\in[t_0,+\infty),\\ \lim\limits_{|\mathbf{x}|\to+\infty}\left(\mathbf{v}(\mathbf{x},t)-B(t)\cdot\mathbf{x}\right)=\mathbf{l}(t) & \forall t\in[t_0,+\infty).\end{cases} \tag{3.95}$$

Proposition 3.5 *Let $\mathbf{v}:=\mathbf{v}(\mathbf{x},t):\mathbb{R}^3\times[t_0,+\infty)\to\mathbb{R}^3$ be a speed-like smooth vector field. Then the following three conditions are <u>equivalent</u>:*

(i) *There exists a (inertial or non-inertial) Cartesian coordinate system $(*)$ where the vector field $\mathbf{v}$ is asymptotically acceptable.*
(ii) *The vector field $\mathbf{v}$ is asymptotically acceptable in <u>every</u> (inertial or non-inertial) Cartesian coordinate system.*
(iii) *There exists a (inertial or non-inertial) Cartesian coordinate system $(\{0\})$ in which we have*

$$\lim_{|\mathbf{x}|\to+\infty}\left(d_{\mathbf{x}}\mathbf{v}(\mathbf{x},t)\right)=0 \qquad \forall t\in[t_0,+\infty), \tag{3.96}$$

$$\lim_{|\mathbf{x}|\to+\infty}\mathbf{v}(\mathbf{x},t)=0 \qquad \forall t\in[t_0,+\infty). \tag{3.97}$$

Moreover, in that case the Cartesian coordinate system where (3.97) *holds is unique up to an equivalence, in other words, it is unique up to a constant (independent on time) rotation in* $\mathbb{R}^3$ *and/or up to a constant (independent on time) shift of the origin in* $\mathbb{R}^3$.

Proof Consider some Cartesian coordinate system $(*)$ where we have (3.95):

$$\begin{cases}\lim_{|\mathbf{x}|\to+\infty}\left(d_{\mathbf{x}}\mathbf{v}(\mathbf{x},t)\right)=B(t) & \forall t\in[t_0,+\infty)\,,\\ \lim_{|\mathbf{x}|\to+\infty}\left(d_{\mathbf{x}}\mathbf{v}(\mathbf{x},t)+\{d_{\mathbf{x}}\mathbf{v}(\mathbf{x},t)\}^T\right)=0 & \forall t\in[t_0,+\infty)\,,\\ \lim_{|\mathbf{x}|\to+\infty}\left(\mathbf{v}(\mathbf{x},t)-B(t)\cdot\mathbf{x}\right)=\mathbf{l}(t) & \forall t\in[t_0,+\infty)\,.\end{cases} \tag{3.98}$$

In particular, by the first and the second equations in (3.98) we easily deduce that

$$B^T(t)=-B(t)\qquad \forall t\in[t_0,+\infty). \tag{3.99}$$

Next, let $(**)$ be another Cartesian coordinate system, such that the change of the coordinate system $(*)$ to the Cartesian coordinate system $(**)$ is given by

$$\begin{cases}\mathbf{x}'=A(t)\cdot\mathbf{x}+\mathbf{z}(t),\\ t'=t,\end{cases} \tag{3.100}$$

where $A(t)\in SO(3)$ is a rotation. Then since $\mathbf{v}$ is a speed-like vector field, as before, we have

$$\mathbf{v}'(\mathbf{x}',t')=A(t)\cdot\mathbf{v}(\mathbf{x},t)+\frac{dA}{dt}(t)\cdot\mathbf{x}+\frac{d\mathbf{z}}{dt}(t). \tag{3.101}$$

In particular, as before, we obtain

$$\begin{aligned}d_{\mathbf{x}'}\mathbf{v}'(\mathbf{x}',t')&=A(t)\cdot\left(d_{\mathbf{x}}\mathbf{v}(\mathbf{x},t)\right)\cdot A^{-1}(t)+\frac{dA}{dt}(t)\cdot A^{-1}(t)\\ &=A(t)\cdot\left(d_{\mathbf{x}}\mathbf{v}(\mathbf{x},t)\right)\cdot A^{T}(t)+\frac{dA}{dt}(t)\cdot A^{T}(t).\end{aligned} \tag{3.102}$$

Therefore, by the first equation in (3.98) we deduce

$$\begin{aligned}&\lim_{|\mathbf{x}'|\to+\infty}\left(d_{\mathbf{x}'}\mathbf{v}'(\mathbf{x}',t')\right)=\lim_{|\mathbf{x}|\to+\infty}\left(A(t)\cdot\left(d_{\mathbf{x}}\mathbf{v}(\mathbf{x},t)\right)\cdot A^{T}(t)+\frac{dA}{dt}(t)\cdot A^{T}(t)\right)\\ &=A(t)\cdot B(t)\cdot A^{T}(t)+\frac{dA}{dt}(t)\cdot A^{T}(t)=B'(t')\qquad \forall t'=t\in[t_0,+\infty),\end{aligned} \tag{3.103}$$

where we denote

$$B'(t') := A(t) \cdot B(t) \cdot A^T(t) + \frac{dA}{dt}(t) \cdot A^T(t) \qquad \forall t' = t \in [t_0, +\infty). \quad (3.104)$$

Then, by the third equation in (3.98) together with (3.104), (3.101) and (3.100) we infer

$$\begin{aligned}
\lim_{|\mathbf{x}'| \to +\infty} \left(\mathbf{v}'(\mathbf{x}', t') - B'(t') \cdot \mathbf{x}'\right) &= \lim_{|\mathbf{x}| \to +\infty} \Bigg\{ \left(A(t) \cdot \mathbf{v}(\mathbf{x}, t) + \frac{dA}{dt}(t) \cdot \mathbf{x} + \frac{d\mathbf{z}}{dt}(t) \right) \\
&\quad - \left(A(t) \cdot B(t) \cdot A^T(t) + \frac{dA}{dt}(t) \cdot A^T(t) \right) \cdot (A(t) \cdot \mathbf{x} + \mathbf{z}(t)) \Bigg\} \\
&= \lim_{|\mathbf{x}| \to +\infty} \Bigg\{ \left(A(t) \cdot \mathbf{v}(\mathbf{x}, t) + \frac{dA}{dt}(t) \cdot \mathbf{x} + \frac{d\mathbf{z}}{dt}(t) \right) \\
&\quad - \left(A(t) \cdot B(t) \cdot A^{-1}(t) + \frac{dA}{dt}(t) \cdot A^{-1}(t) \right) \cdot (A(t) \cdot \mathbf{x}) \\
&\quad - \left(A(t) \cdot B(t) \cdot A^T(t) + \frac{dA}{dt}(t) \cdot A^T(t) \right) \cdot \mathbf{z}(t) \Bigg\} = \\
&\lim_{|\mathbf{x}| \to +\infty} \Bigg\{ \left(A(t) \cdot (\mathbf{v}(\mathbf{x}, t) - B(t) \cdot \mathbf{x}) + \frac{d\mathbf{z}}{dt}(t) \right) \\
&\quad - \left(A(t) \cdot B(t) \cdot A^T(t) + \frac{dA}{dt}(t) \cdot A^T(t) \right) \cdot \mathbf{z}(t) \Bigg\} \\
= \left(A(t) \cdot \mathbf{l}(t) + \frac{d\mathbf{z}}{dt}(t) \right) &- \left(A(t) \cdot B(t) \cdot A^T(t) + \frac{dA}{dt}(t) \cdot A^T(t) \right) \cdot \mathbf{z}(t) := \mathbf{l}'(t') \\
&\forall t' = t \in [t_0, +\infty), \quad (3.105)
\end{aligned}$$

where we denote

$$\mathbf{l}'(t') := \left(A(t) \cdot \mathbf{l}(t) + \frac{d\mathbf{z}}{dt}(t) \right) - \left(A(t) \cdot B(t) \cdot A^T(t) + \frac{dA}{dt}(t) \cdot A^T(t) \right) \cdot \mathbf{z}(t)$$
$$\forall t' = t \in [t_0, +\infty). \quad (3.106)$$

On the other hand, by **(iii)** in Proposition 3.1 together with the second equation in (3.98) we obtain

$$\begin{aligned}
&\lim_{|\mathbf{x}'| \to +\infty} \left(d_{\mathbf{x}'} \mathbf{v}'(\mathbf{x}', t') + \{ d_{\mathbf{x}'} \mathbf{v}'(\mathbf{x}', t') \}^T \right) \\
&= A(t) \cdot \left\{ \lim_{|\mathbf{x}| \to +\infty} \left(d_{\mathbf{x}} \mathbf{v}(\mathbf{x}, t) + \{ d_{\mathbf{x}} \mathbf{v}(\mathbf{x}, t) \}^T \right) \right\} \cdot A^T(t) = 0 \qquad \forall t = t' \in [t_0, +\infty).
\end{aligned}$$
$$(3.107)$$

So, by (3.103), (3.105) and (3.107) we obtain

$$\begin{cases} \lim_{|\mathbf{x}'|\to+\infty} \left(d_{\mathbf{x}'}\mathbf{v}'(\mathbf{x}',t')\right) = B'(t') \quad \forall t' \in [t_0,+\infty), \\ \lim_{|\mathbf{x}'|\to+\infty} \left(d_{\mathbf{x}'}\mathbf{v}'(\mathbf{x}',t') + \left\{d_{\mathbf{x}'}\mathbf{v}'(\mathbf{x}',t')\right\}^T\right) = 0 \quad \forall t' \in [t_0,+\infty), \\ \lim_{|\mathbf{x}'|\to+\infty} \left(\mathbf{v}'(\mathbf{x}',t') - B'(t')\cdot\mathbf{x}'\right) = \mathbf{l}'(t') \quad \forall t' \in [t_0,+\infty), \end{cases} \tag{3.108}$$

and therefore if $\mathbf{v}$ is asymptotically acceptable in the coordinate system (*) then by (3.108) $\mathbf{v}'$ is also asymptotically acceptable in the coordinate system (**) and implication **(i)** $\Longleftrightarrow$ **(ii)** follows. Moreover, by (3.103) and (3.105) we deduced

$$\lim_{|\mathbf{x}'|\to+\infty} \left(d_{\mathbf{x}'}\mathbf{v}'(\mathbf{x}',t')\right) = B'(t') := \left(\frac{dA}{dt}(t) + A(t)\cdot B(t)\right)\cdot A^T(t) \qquad \forall t' = t \in [t_0,+\infty), \tag{3.109}$$

and

$$\lim_{|\mathbf{x}'|\to+\infty} \left(\mathbf{v}'(\mathbf{x}',t') - B'(t')\cdot\mathbf{x}'\right) = \mathbf{l}'(t') :=$$
$$\frac{d\mathbf{z}}{dt}(t) - \left(A(t)\cdot B(t)\cdot A^T(t) + \frac{dA}{dt}(t)\cdot A^T(t)\right)\cdot\mathbf{z}(t) + A(t)\cdot\mathbf{l}(t) \quad \forall t' = t \in [t_0,+\infty), \tag{3.110}$$

where by (3.99) $B(t)$ satisfies

$$B^T(t) = -B(t) \qquad \forall t \in [t_0,+\infty). \tag{3.111}$$

Next, assume that $\tau_0 \in [t_0,+\infty)$ and smooth mappings $A_0(t) : [t_0,+\infty) \to \mathbb{R}^{3\times3}$ and $\mathbf{z}_0(t) : [t_0,+\infty) \to \mathbb{R}^3$ satisfy

$$\begin{cases} \frac{dA_0}{dt}(t) = -A_0(t)\cdot B(t) \qquad \forall t \in [t_0,+\infty) \\ A_0(\tau_0) = I, \end{cases} \tag{3.112}$$

and

$$\frac{d\mathbf{z}_0}{dt}(t) = \left(A(t)\cdot B(t)\cdot A^T(t) + \frac{dA}{dt}(t)\cdot A^T(t)\right)\cdot\mathbf{z}_0(t) - A(t)\cdot\mathbf{l}(t) \quad \forall t = t \in [t_0,+\infty), \tag{3.113}$$

where $I \in \mathbb{R}^{3\times3}$ is the identity matrix. By the Theory of Ordinary Differential Equations, it is well known that such mappings $A_0(t)$ and $\mathbf{z}_0(t)$ indeed exist. Then by (3.112) and (3.111), we deduce

$$\frac{d}{dt}\left\{A_0^T(t)\right\} = -B^T(t)\cdot A_0^T = B(t)\cdot\left\{A_0^T(t)\right\} \qquad \forall t \in [t_0,+\infty), \tag{3.114}$$

and so

$$\begin{cases} \frac{d}{dt}\left\{A_0^T(t)\right\} = B(t)\cdot\left\{A_0^T(t)\right\} & \forall t \in [t_0, +\infty) \\ A_0^T(\tau_0) = I. \end{cases} \tag{3.115}$$

In particular, again by the Theory of Ordinary Differential Equations using (3.115), we deduce that

$$\det A_0^T(t) > 0 \qquad \forall t \in [t_0, +\infty) \tag{3.116}$$

and then also

$$\det A_0(t) > 0 \qquad \forall t \in [t_0, +\infty). \tag{3.117}$$

So, $A_0(t)$ and $A_0^T(t)$ are invertible matrices. On the other hand, since differentiating the identity

$$A_0^{-1}(t)\cdot A_0(t) = I \qquad \forall t' = t \in [t_0, +\infty)$$

gives

$$\left(\frac{d}{dt}\left\{A_0^{-1}(t)\right\}\right)\cdot A_0(t) + A_0^{-1}(t)\cdot \frac{dA_0}{dt}(t) = 0 \qquad \forall t' = t \in [t_0, +\infty),$$

we deduce

$$\frac{d}{dt}\left\{A_0^{-1}(t)\right\} = -A_0^{-1}(t)\cdot\frac{dA_0}{dt}(t)\cdot A_0^{-1}(t) \qquad \forall t' = t \in [t_0, +\infty). \tag{3.118}$$

Thus, by (3.112) and (3.118) we deduce

$$\frac{d}{dt}\left\{A_0^{-1}(t)\right\} = -A_0^{-1}(t)\cdot(-A_0(t)\cdot B(t))\cdot A_0^{-1}(t) = B(t)\cdot A_0^{-1}(t) \qquad \forall t \in [t_0, +\infty), \tag{3.119}$$

and so

$$\begin{cases} \frac{d}{dt}\left\{A_0^{-1}(t)\right\} = B(t)\cdot\left\{A_0^{-1}(t)\right\} & \forall t \in [t_0, +\infty) \\ A_0^{-1}(\tau_0) = I. \end{cases} \tag{3.120}$$

Therefore, by Uniqueness Theorem for the Cauchy Problem for Ordinary Differential Equations, (3.115) and (3.120) together imply that

$$A_0^{-1}(t) = A_0^T(t) \qquad \forall t \in [t_0, +\infty). \tag{3.121}$$

So by (3.117) and (3.121) we deduce that $A_0(t)$, which solves (3.112), necessarily satisfies $A_0(t) \in SO(3)$. In other words $A_0(t)$ which solves (3.112) is a rotation.

Therefore, there exists some particular Cartesian coordinate system $(**)$, such that the change of the coordinate system $(*)$ to the Cartesian coordinate system $(**)$

is given by (3.100) with $A(t) = A_0(t)$ and $\mathbf{z}(t) = \mathbf{z}_0(t)$:

$$\begin{cases} \mathbf{x}' = A(t) \cdot \mathbf{x} + \mathbf{z}(t) = A_0(t) \cdot \mathbf{x} + \mathbf{z}_0(t), \\ t' = t, \end{cases} \tag{3.122}$$

and moreover by (3.112) for this system transformation we have

$$\frac{dA}{dt}(t) + A(t) \cdot B(t) = 0 \qquad \forall t \in [t_0, +\infty), \tag{3.123}$$

and by (3.113) we have

$$\frac{d\mathbf{z}}{dt}(t) - \left(A(t) \cdot B(t) \cdot A^T(t) + \frac{dA}{dt}(t) \cdot A^T(t) \right) \cdot \mathbf{z}(t) + A(t) \cdot \mathbf{l}(t) = 0, \qquad \forall t \in [t_0, +\infty). \tag{3.124}$$

Thus, inserting (3.123) into (3.109) we deduce

$$\lim_{|\mathbf{x}'| \to +\infty} \left(d_{\mathbf{x}'} \mathbf{v}'(\mathbf{x}', t') \right) = 0 \qquad \forall t' \in [t_0, +\infty) \qquad \text{(in the chosen system (**))}. \tag{3.125}$$

Moreover, inserting (3.124) into (3.110) we deduce

$$\lim_{|\mathbf{x}'| \to +\infty} \mathbf{v}'(\mathbf{x}', t') = \lim_{|\mathbf{x}'| \to +\infty} \left(\mathbf{v}'(\mathbf{x}', t') - B'(t') \cdot \mathbf{x}' \right) = 0$$
$$\forall t' \in [t_0, +\infty) \qquad \text{(in the chosen system (**))}, \tag{3.126}$$

and by (3.125) together with (3.126) implication **(i)** $\Longleftrightarrow$ **(iii)** also follows.

Finally, consider two Cartesian coordinate systems ({1}) and ({2}) so that the change of coordinate system ({1}) to the system ({2}) can be written as

$$\begin{cases} \mathbf{x}^{(2)} = A(t^{(1)}) \cdot \mathbf{x}^{(1)} + \mathbf{z}(t^{(1)}), \\ t^{(2)} = t^{(1)}, \end{cases} \tag{3.127}$$

where $A(t) \in SO(3)$ is a rotation. Moreover, assume that the field $\mathbf{v}$ in both these systems satisfies

$$\lim_{|\mathbf{x}^{(1)}| \to +\infty} \mathbf{v}^{(1)}(\mathbf{x}^{(1)}, t^{(1)}) = 0 \qquad \forall t^{(1)} = t \in [t_0, +\infty), \tag{3.128}$$

$$\lim_{|\mathbf{x}^{(2)}| \to +\infty} \mathbf{v}^{(2)}(\mathbf{x}^{(2)}, t^{(2)}) = 0 \qquad \forall t^{(2)} = t \in [t_0, +\infty). \tag{3.129}$$

Then, since $\mathbf{v}$ is a speed-like vector field, by (3.4) in Definition 3.1 we deduce that

$$\mathbf{v}^{(2)}(\mathbf{x}^{(2)}, t) = A(t) \cdot \mathbf{v}^{(1)}(\mathbf{x}^{(1)}, t) + \frac{dA}{dt}(t) \cdot \mathbf{x}^{(1)} + \frac{d\mathbf{z}}{dt}(t) \qquad \forall t \in [t_0, +\infty). \tag{3.130}$$

Thus, by (3.130) two equalities (3.128) and (3.129) can be satisfied simultaneously if and only if we have

$$\frac{dA}{dt}(t) = 0 \quad \text{and} \quad \frac{d\mathbf{z}}{dt}(t) = 0 \qquad \forall t \in [t_0, +\infty), \tag{3.131}$$

and so this happens if and only if A and $\mathbf{z}$ in (3.127) are constants (independent on time), and that means two systems ($\{1\}$) and ($\{2\}$) are equivalent. This completes the proof. □

Corollary 3.1 *Let* $\mathbf{v} := \mathbf{v}(\mathbf{x}, t) : \mathbb{R}^3 \times [t_0, +\infty) \to \mathbb{R}^3$ *be an asymptotically acceptable speed-like vector field. Then there exist a uniquely defined generally trivial speed-like vector field* $\mathbf{k} := \mathbf{k}(\mathbf{x}, t) : \mathbb{R}^3 \times [t_0, +\infty) \to \mathbb{R}^3$ *and a uniquely defined proper vector field* $\mathbf{h} := \mathbf{h}(\mathbf{x}, t) : \mathbb{R}^3 \times [t_0, +\infty) \to \mathbb{R}^3$*, so that in every Cartesian coordinate system we have*

$$\lim_{|\mathbf{x}| \to +\infty} \mathbf{h}(\mathbf{x}, t) = 0 \quad \forall t\,, \tag{3.132}$$

and in every Cartesian coordinate system we can decompose vector field $\mathbf{v}$ *as*

$$\mathbf{v}(\mathbf{x}, t) = \mathbf{h}(\mathbf{x}, t) + \mathbf{k}(\mathbf{x}, t) \qquad \forall (\mathbf{x}, t) \in \mathbb{R}^3 \times [t_0, +\infty). \tag{3.133}$$

Moreover, in addition, the proper vector field $\mathbf{h}$ *necessarily satisfies*

$$\lim_{|\mathbf{x}| \to +\infty} (d_{\mathbf{x}} \mathbf{h}(\mathbf{x}, t)) = 0 \quad \forall t \qquad \textit{(in every Cartesian coordinate system)}. \tag{3.134}$$

Proof The result is a direct consequence of both Propositions 3.4 and 3.5, where only identity (3.134) requires to be proved. In order to prove (3.134) consider the system $(*)$ where we have

$$\begin{cases} \lim\limits_{|\mathbf{x}| \to +\infty} (d_{\mathbf{x}} \mathbf{v}(\mathbf{x}, t)) = 0 \qquad \forall t \in [t_0, +\infty)\,, \\ \lim\limits_{|\mathbf{x}| \to +\infty} \mathbf{v}(\mathbf{x}, t) = 0 \qquad \forall t \in [t_0, +\infty)\,, \end{cases} \tag{3.135}$$

and consider the system $(**)$ where we have

$$\mathbf{k}'(\mathbf{x}', t') = 0 \qquad \forall (\mathbf{x}', t') \in \mathbb{R}^3 \times [t_0, +\infty). \tag{3.136}$$

On the other hand, as in (3.132) and (3.133) in the system $(**)$ we also have

$$\lim_{|\mathbf{x}'| \to +\infty} \mathbf{h}'(\mathbf{x}', t') = 0 \quad \forall t'\,, \tag{3.137}$$

and

$$\mathbf{v}'(\mathbf{x}', t') = \mathbf{h}'(\mathbf{x}', t') + \mathbf{k}'(\mathbf{x}', t') \qquad \forall (\mathbf{x}', t') \in \mathbb{R}^3 \times [t_0, +\infty). \tag{3.138}$$

In particular, by (3.138), (3.136) and (3.137) together we deduce that in system $(**)$ we have

$$\mathbf{v}'(\mathbf{x}', t') = \mathbf{h}'(\mathbf{x}', t') \qquad \forall (\mathbf{x}', t') \in \mathbb{R}^3 \times [t_0, +\infty), \tag{3.139}$$

and

$$\lim_{|\mathbf{x}'| \to +\infty} \mathbf{v}'(\mathbf{x}', t') = 0 \quad \forall t'. \tag{3.140}$$

Therefore, again using Proposition 3.5, by the second equality in (3.135) together with (3.140) we obtain that systems $(*)$ and $(**)$ are equivalent. By this equivalence and by Lemma 3.1, using (3.139) we deduce that in system $(*)$ we also have

$$\mathbf{v}(\mathbf{x}, t) = \mathbf{h}(\mathbf{x}, t) \qquad \forall (\mathbf{x}, t) \in \mathbb{R}^3 \times [t_0, +\infty). \tag{3.141}$$

Thus, by (3.141) and by the first equality in (3.135) we finally deduce (3.134). □

Definition 3.5 Let $\mathcal{U}$ be a nonempty subset of the set of all inertial and non-inertial Cartesian coordinate systems. We say that $\mathcal{U}$ is an extended Galilean group if, given arbitrary Cartesian coordinate system $(*)$ inside the subset $\mathcal{U}$, the following statement holds:

- The arbitrary Cartesian coordinate system $(**)$ belongs to the subset $\mathcal{U}$ if and only if there exist a constant (independent on time) rotation $B \in SO(3)$ and constant (independent on time) vectors $\mathbf{c} \in \mathbb{R}^3$ and $\mathbf{w} \in \mathbb{R}^3$, so that the change of coordinates from the system $(*)$ to the system $(**)$ is given by

$$\begin{cases} \mathbf{x}' = B \cdot \mathbf{x} + \mathbf{c} + \mathbf{w}t, \\ t' = t. \end{cases} \tag{3.142}$$

In other words, the system $(**)$ belongs to $\mathcal{U}$ if and only if, up to equivalence of Cartesian coordinate systems, the system $(**)$ can be obtained from the system $(*)$ by the Galilean transformation

$$\begin{cases} \mathbf{x}' = \mathbf{x} + \mathbf{w}t, \\ t' = t. \end{cases} \tag{3.143}$$

Definition 3.6 Let $\mathcal{V}$ be a nonempty subset of the set of all inertial and non-inertial Cartesian coordinate systems. We say that $\mathcal{V}$ is a non-rotational rigid motion group if, given arbitrary Cartesian coordinate system $(*)$ inside the subset $\mathcal{V}$, the following statement holds:

- The arbitrary Cartesian coordinate system (∗∗) belongs to the subset $\mathcal{V}$ if and only if there exist a constant (independent on time) rotation $B \in SO(3)$ and a smooth vector-valued function of the time variable $\mathbf{z}(t) : \mathbb{R} \to \mathbb{R}^3$, so that the change of coordinates from the system (∗) to the system (∗∗) is given by

$$\begin{cases} \mathbf{x}' = B \cdot \mathbf{x} + \mathbf{z}(t), \\ t' = t. \end{cases} \tag{3.144}$$

Definition 3.7 Let $\mathcal{W}$ be a certain nonempty subset of coordinate systems. We say that $\mathcal{W}$ is a general rigid motion group if, given arbitrary coordinate system (∗) inside the subset $\mathcal{W}$, the following statement holds:

- The arbitrary coordinate system (∗∗) belongs to the subset $\mathcal{W}$ if and only if there exist a smooth (time dependent) rotation $A(t) : \mathbb{R} \to SO(3)$ and a smooth vector-valued function of the time variable $\mathbf{z}(t) : \mathbb{R} \to \mathbb{R}^3$, so that the change of coordinates from the system (∗) to the system (∗∗) is given by

$$\begin{cases} \mathbf{x}' = A(t) \cdot \mathbf{x} + \mathbf{z}(t), \\ t' = t. \end{cases} \tag{3.145}$$

In particular, the set of all Cartesian coordinate systems is a general rigid motion group.

Definition 3.8 Let $\mathbf{v} := \mathbf{v}(\mathbf{x}, t) : \mathbb{R}^3 \times [t_0, +\infty) \to \mathbb{R}^3$ be an asymptotically acceptable speed-like smooth vector field (see Definition 3.4).

- We call that the Cartesian coordinate system ({0}) is preferable for the vector field $\mathbf{v}$ if in the system ({0}) we have

$$\lim_{|\mathbf{x}| \to +\infty} (d_{\mathbf{x}} \mathbf{v}(\mathbf{x}, t)) = 0 \qquad \forall t \in [t_0, +\infty)\,, \tag{3.146}$$

$$\lim_{|\mathbf{x}| \to +\infty} \mathbf{v}(\mathbf{x}, t) = 0 \qquad \forall t \in [t_0, +\infty)\,. \tag{3.147}$$

  Note that by Proposition 3.5 the unique (up to equivalence) preferable for the vector field $\mathbf{v}$ Cartesian system ({0}) exists.
- We call that the Cartesian coordinate system (∗) is inertial with respect to the vector field $\mathbf{v}$ if in the system (∗) we have

$$\lim_{|\mathbf{x}| \to +\infty} (d_{\mathbf{x}} \mathbf{v}(\mathbf{x}, t)) = 0 \qquad \forall t \in [t_0, +\infty)\,, \tag{3.148}$$

$$\lim_{|\mathbf{x}| \to +\infty} \mathbf{v}(\mathbf{x}, t) = \mathbf{d} \qquad \forall t \in [t_0, +\infty)\,, \tag{3.149}$$

  where $\mathbf{d} \in \mathbb{R}^3$ is a constant (independent on t) vector. Note that, in particular, by the definition, the preferable for $\mathbf{v}$ Cartesian system is also an inertial with respect to $\mathbf{v}$ coordinate system.

- We call that the Cartesian coordinate system $(*)$ is non-rotating with respect to the vector field $\mathbf{v}$ if in the system $(*)$ we have

$$\lim_{|\mathbf{x}|\to+\infty} \left(d_{\mathbf{x}}\mathbf{v}(\mathbf{x},t)\right) = 0 \qquad \forall t \in [t_0,+\infty)\,, \tag{3.150}$$

(without specifying $\lim_{|\mathbf{x}|\to+\infty} \mathbf{v}(\mathbf{x},t)$). Note that, in particular, by the definition, every inertial with respect to $\mathbf{v}$ Cartesian coordinate system is also a non-rotating with respect to $\mathbf{v}$ coordinate system.

Proposition 3.6 *Let* $\mathbf{v} := \mathbf{v}(\mathbf{x},t) : \mathbb{R}^3 \times [t_0,+\infty) \to \mathbb{R}^3$ *be an asymptotically acceptable speed-like smooth vector field (see Definition 3.4). Furthermore, let* $\mathcal{U}$ *be the set of all inertial with respect to* $\mathbf{v}$ *Cartesian coordinate systems and let* $\mathcal{V}$ *be the set of all non-rotating with respect to* $\mathbf{v}$ *Cartesian coordinate systems (see Definition 3.8). Then* $\mathcal{U}$ *is an extended Galilean group and* $\mathcal{V}$ *is a non-rotational rigid motion group (see Definitions 3.5 and 3.6).*

Proof Consider the change of some Cartesian coordinate system $(*)$ to another Cartesian coordinate system $(**)$ of the form:

$$\begin{cases} \mathbf{x}' = A(t)\cdot\mathbf{x} + \mathbf{z}(t), \\ t' = t, \end{cases} \tag{3.151}$$

where $A(t) \in SO(3)$ and correspondingly to (3.4) we have

$$\mathbf{v}'(\mathbf{x}',t') = A(t)\cdot\mathbf{v}(\mathbf{x},t) + \frac{dA}{dt}(t)\cdot\mathbf{x} + \frac{d\mathbf{z}}{dt}(t). \tag{3.152}$$

In particular, by (3.151), (3.152) and the chain rule we deduce

$$\begin{aligned}\left\{d_{\mathbf{x}'}\mathbf{v}'(\mathbf{x}',t')\right\}\cdot A(t) &= d_{\mathbf{x}}\left(A(t)\cdot\mathbf{v}(\mathbf{x},t) + \frac{dA}{dt}(t)\cdot\mathbf{x} + \frac{d\mathbf{z}}{dt}(t)\right) \\ &= A(t)\cdot\{d_{\mathbf{x}}\mathbf{v}(\mathbf{x},t)\} + \frac{dA}{dt}(t),\end{aligned} \tag{3.153}$$

and thus

$$\left(\lim_{|\mathbf{x}'|\to+\infty}\left\{d_{\mathbf{x}'}\mathbf{v}'(\mathbf{x}',t')\right\}\right)\cdot A(t) = A(t)\cdot\left(\lim_{|\mathbf{x}|\to+\infty}\{d_{\mathbf{x}}\mathbf{v}(\mathbf{x},t)\}\right) + \frac{dA}{dt}(t). \tag{3.154}$$

Therefore, by (3.154), in both systems $(*)$ and $(**)$ we have simultaneously

$$\lim_{|\mathbf{x}|\to+\infty} \left(d_{\mathbf{x}}\mathbf{v}(\mathbf{x},t)\right) = 0 \qquad \forall t \in [t_0,+\infty)\,, \tag{3.155}$$

and

$$\lim_{|\mathbf{x}'|\to+\infty} \left(d_{\mathbf{x}'}\mathbf{v}'(\mathbf{x}',t')\right) = 0 \qquad \forall t' \in [t_0, +\infty)\,, \tag{3.156}$$

if and only if

$$\frac{dA}{dt}(t) = 0 \qquad \forall t\,, \tag{3.157}$$

or equivalently

$$A(t) = B \qquad \forall t\,, \tag{3.158}$$

where $B \in SO(3)$ is a constant rotation, so that

$$\begin{cases} \mathbf{x}' = B \cdot \mathbf{x} + \mathbf{z}(t), \\ t' = t, \end{cases} \tag{3.159}$$

and

$$\mathbf{v}'(\mathbf{x}',t') = B \cdot \mathbf{v}(\mathbf{x},t) + \frac{d\mathbf{z}}{dt}(t). \tag{3.160}$$

In particular, by Definitions 3.6 and 3.8 $\mathcal{V}$ is indeed a non-rotational rigid motion group. Moreover, by (3.160) we deduce

$$\lim_{|\mathbf{x}'|\to+\infty} \mathbf{v}'(\mathbf{x}',t') = B \cdot \left(\lim_{|\mathbf{x}|\to+\infty} \mathbf{v}(\mathbf{x},t)\right) + \frac{d\mathbf{z}}{dt}(t), \tag{3.161}$$

and thus in both systems $(*)$ and $(**)$ the limits $\lim_{|\mathbf{x}'|\to+\infty} \mathbf{v}'(\mathbf{x}',t')$ and $\lim_{|\mathbf{x}|\to+\infty} \mathbf{v}(\mathbf{x},t)$ are independent on time simultaneously if and only if $\mathbf{w} := \frac{d\mathbf{z}}{dt}(t)$ is a constant (independent on time) vector. In that case there exists constant vectors $\mathbf{c} \in \mathbb{R}^3$ and $\mathbf{w} \in \mathbb{R}^3$ so that

$$\mathbf{z}(t) = \mathbf{c} + \mathbf{w}t \qquad \forall t. \tag{3.162}$$

Inserting (3.162) into (3.159) gives

$$\begin{cases} \mathbf{x}' = B \cdot \mathbf{x} + \mathbf{c} + \mathbf{w}t, \\ t' = t. \end{cases} \tag{3.163}$$

In particular, by Definitions 3.5 and 3.8 $\mathcal{U}$ is indeed an extended Galilean group. □

Definition 3.9 Given an extended Galilean group $\mathcal{U}$ (see Definition 3.5), we say that the scalar field $Z := Z(\mathbf{x},t) : \mathbb{R}^3 \times [t_0, +\infty) \to \mathbb{R}$, defined in every Cartesian system inside the set $\mathcal{U}$, is a speed-like field generator on $\mathcal{U}$ if, under every change of two coordinate systems $(*)$ and $(**)$ inside $\mathcal{U}$, given by (3.142), the quantity Z transforms as

$$Z'(\mathbf{x}', t') = Z(\mathbf{x}, t) + \mathbf{w} \cdot (B \cdot \mathbf{x}) + \frac{1}{2}|\mathbf{w}|^2 t + C_*^{**}, \tag{3.164}$$

where $C_*^{**} \in \mathbb{R}$ is an unspecified constant, independent on the space and time variables $(\mathbf{x}, t)$, which is, however, allowed to depend on both coordinate systems $(*)$ and $(**)$.

Lemma 3.2 *Let $\mathbf{v} := \mathbf{v}(\mathbf{x}, t) : \mathbb{R}^3 \times [t_0, +\infty) \to \mathbb{R}^3$ be a speed-like vector field. Furthermore, let $\mathcal{U}$ be an extended Galilean group and let $Z := Z(\mathbf{x}, t) : \mathbb{R}^3 \times [t_0, +\infty) \to \mathbb{R}$ be a speed-like field generator on $\mathcal{U}$, such that there exists Cartesian coordinate system $(*)$ inside $\mathcal{U}$, where we have*

$$\mathbf{v}(\mathbf{x}, t) = \nabla_{\mathbf{x}} Z(\mathbf{x}, t) \quad \forall (\mathbf{x}, t) \in \mathbb{R}^3 \times [t_0, +\infty) \quad \textit{in the given particular system } (*) \,. \tag{3.165}$$

Then, in every Cartesian coordinate system $(**)$ inside the set $\mathcal{U}$ we also have

$$\mathbf{v}'(\mathbf{x}', t') = \nabla_{\mathbf{x}'} Z'(\mathbf{x}', t') \quad \forall (\mathbf{x}', t') \in \mathbb{R}^3 \times [t_0, +\infty) \quad \textit{in the system } (**) \,. \tag{3.166}$$

Proof By (3.164) and (3.142) we have

$$\nabla_{\mathbf{x}'} Z'(\mathbf{x}', t') = \left(B^{-1}\right)^T \cdot \nabla_{\mathbf{x}} \left(Z(\mathbf{x}, t) + \mathbf{w} \cdot (B \cdot \mathbf{x}) + \frac{1}{2}|\mathbf{w}|^2 t + C_*^{**} \right) =$$
$$B \cdot \nabla_{\mathbf{x}} Z(\mathbf{x}, t) + B \cdot (B^T \cdot \mathbf{w}) = B \cdot \nabla_{\mathbf{x}} Z(\mathbf{x}, t) + \mathbf{w} \quad \forall (\mathbf{x}, t) \in \mathbb{R}^3 \times [t_0, +\infty). \tag{3.167}$$

On the other hand, the law of transformation of a speed-like vector field in (3.4) of Definition 3.1 with $A(t) := B$ and $\mathbf{z}(t) := (\mathbf{c} + \mathbf{w}t)$ in the case of (3.142) reads as

$$\mathbf{v}'(\mathbf{x}', t') = B \cdot \mathbf{v}(\mathbf{x}, t) + \mathbf{w} \quad \forall (\mathbf{x}, t) \in \mathbb{R}^3 \times [t_0, +\infty) \,. \tag{3.168}$$

Thus, by (3.167) and (3.168) together with (3.165) we finally deduce (3.166). □

Lemma 3.3 *Let $\mathcal{U}$ be an extended Galilean group and let $(\{0\})$ be some fixed inertial or non-inertial Cartesian coordinate system inside $\mathcal{U}$. Next consider a scalar field $\gamma := \gamma(\mathbf{x}, t) : \mathbb{R}^3 \times [t_0, +\infty) \to \mathbb{R}$ associated only with the chosen coordinate system $(\{0\})$. Then there exist a speed-like field generator on $\mathcal{U}$, denoted by $Z_\gamma : \mathbb{R}^3 \times [t_0, +\infty) \to \mathbb{R}$ (and associated with any inertial or non-inertial Cartesian coordinate system inside $\mathcal{U}$), such that in the particular system $(\{0\})$ we have*

$$Z_\gamma(\mathbf{x}, t) = \gamma(\mathbf{x}, t) \quad \forall (\mathbf{x}, t) \in \mathbb{R}^3 \times [t_0, +\infty) \,. \tag{3.169}$$

Proof Similar to (3.142), given an arbitrary inertial or non-inertial Cartesian coordinate system inside $\mathcal{U}$, denoted by the sign $(\{1\})$, the change of coordinate system $(\{0\})$ to the system $(\{1\})$ can be written as

$$\begin{cases} \mathbf{x}^{(1)} = B_1 \cdot \mathbf{x} + \mathbf{c}_1 + \mathbf{w}_1 t, \\ t^{(1)} = t, \end{cases} \tag{3.170}$$

where $B_1 \in SO(3)$ is a constant rotation and $\mathbf{c}_1, \mathbf{w}_1 \in \mathbb{R}^3$ are constant vectors. Similarly, given another arbitrary inertial or non-inertial Cartesian coordinate system inside $\mathcal{U}$, denoted by the sign ($\{2\}$), the change of coordinate system ($\{0\}$) to the system ($\{2\}$) can be written as

$$\begin{cases} \mathbf{x}^{(2)} = B_2 \cdot \mathbf{x} + \mathbf{c}_2 + \mathbf{w}_2 t, \\ t^{(2)} = t, \end{cases} \tag{3.171}$$

where $B_2 \in SO(3)$ is another constant rotation and $\mathbf{c}_2, \mathbf{w}_2 \in \mathbb{R}^3$ are other constant vectors. On the other hand, the change coordinate system ($\{1\}$) to the system ($\{2\}$) can be written as

$$\begin{cases} \mathbf{x}^{(2)} = B \cdot \mathbf{x}^{(1)} + \mathbf{c} + \mathbf{w} t^{(1)}, \\ t^{(2)} = t^{(1)}, \end{cases} \tag{3.172}$$

where $B \in SO(3)$ is a constant rotation and $\mathbf{c}, \mathbf{w} \in \mathbb{R}^3$ are constant vectors. In particular, inserting (3.170) into (3.172) gives

$$\mathbf{x}^{(2)} = B \cdot (B_1 \cdot \mathbf{x} + \mathbf{c}_1 + \mathbf{w}_1 t) + \mathbf{c} + \mathbf{w} t = (B \cdot B_1) \cdot \mathbf{x} + (B \cdot \mathbf{c}_1 + \mathbf{c}) + (B \cdot \mathbf{w}_1 + \mathbf{w})\, t. \tag{3.173}$$

Thus, comparing (3.173) with (3.171) we easily deduce

$$B_2 = (B \cdot B_1), \quad \mathbf{c}_2 = (B \cdot \mathbf{c}_1 + \mathbf{c}) \quad \text{and} \quad \mathbf{w}_2 = (B \cdot \mathbf{w}_1 + \mathbf{w}). \tag{3.174}$$

Next, for the system ($\{1\}$) define

$$Z_\gamma^{(1)}(\mathbf{x}^{(1)}, t^{(1)}) = \gamma(\mathbf{x}, t) + \mathbf{w}_1 \cdot (B_1 \cdot \mathbf{x}) + \frac{1}{2} |\mathbf{w}_1|^2 t. \tag{3.175}$$

Correspondingly, for the system ($\{2\}$) we have

$$Z_\gamma^{(2)}(\mathbf{x}^{(2)}, t^{(2)}) = \gamma(\mathbf{x}, t) + \mathbf{w}_2 \cdot (B_2 \cdot \mathbf{x}) + \frac{1}{2} |\mathbf{w}_2|^2 t. \tag{3.176}$$

Thus, in the particular system ($\{0\}$) we have

$$Z_\gamma(\mathbf{x}, t) = \gamma(\mathbf{x}, t) \qquad \forall (\mathbf{x}, t) \in \mathbb{R}^3 \times [t_0, +\infty). \tag{3.177}$$

Thus in order to complete the proof we need to show that the scalar field, defined by (3.175) and (3.176), is a speed-like field generator. Indeed, by (3.175) and (3.176) together with (3.174) and (3.170) we deduce

$$
\begin{aligned}
Z_\gamma^{(2)}(\mathbf{x}^{(2)}, t^{(2)}) &= \gamma(\mathbf{x}, t) + \mathbf{w}_2 \cdot (B_2 \cdot \mathbf{x}) + \frac{1}{2} |\mathbf{w}_2|^2 t \\
&= \gamma(\mathbf{x}, t) + (B \cdot \mathbf{w}_1 + \mathbf{w}) \cdot (B \cdot B_1 \cdot \mathbf{x}) + \frac{1}{2} |B \cdot \mathbf{w}_1 + \mathbf{w}|^2 t = \gamma(\mathbf{x}, t) \\
&+ (B \cdot \mathbf{w}_1) \cdot (B \cdot (B_1 \cdot \mathbf{x})) + \mathbf{w} \cdot (B \cdot (B_1 \cdot \mathbf{x})) + \frac{1}{2} |B \cdot \mathbf{w}_1|^2 t + \frac{1}{2} |\mathbf{w}|^2 t + (\mathbf{w} \cdot (B \cdot \mathbf{w}_1))\, t \\
&= \gamma(\mathbf{x}, t) + \mathbf{w}_1 \cdot (B_1 \cdot \mathbf{x}) + \mathbf{w} \cdot \left(B \cdot \left(\mathbf{x}^{(1)} - \mathbf{c}_1\right)\right) + \frac{1}{2} |\mathbf{w}_1|^2 t + \frac{1}{2} |\mathbf{w}|^2 t \\
&= \left\{\gamma(\mathbf{x}, t) + \mathbf{w}_1 \cdot (B_1 \cdot \mathbf{x}) + \frac{1}{2} |\mathbf{w}_1|^2 t\right\} + \mathbf{w} \cdot \left(B \cdot \mathbf{x}^{(1)}\right) + \frac{1}{2} |\mathbf{w}|^2 t - \mathbf{w} \cdot (B \cdot \mathbf{c}_1) \\
&= Z_\gamma^{(1)}(\mathbf{x}^{(1)}, t^{(1)}) + \mathbf{w} \cdot \left(B \cdot \mathbf{x}^{(1)}\right) + \frac{1}{2} |\mathbf{w}|^2 t^{(1)} - \mathbf{w} \cdot (B \cdot \mathbf{c}_1), \qquad (3.178)
\end{aligned}
$$

where we also used the identity $(B \cdot \mathbf{x}) \cdot (B \cdot \mathbf{y}) = \mathbf{x} \cdot \mathbf{y}$, which is valid for any $B \in SO(3)$ and any $\mathbf{x}, \mathbf{y} \in \mathbb{R}^3$. Thus, comparing (3.178) with (3.164) in Definition 3.9, we deduce that Z_γ is a speed-like field generator on $\mathcal{U}$. This completes the proof. □

Lemma 3.4 *Let $Z := Z(\mathbf{x}, t) : \mathbb{R}^3 \times [t_0, +\infty) \to \mathbb{R}$ be a speed-like field generator on an extended Galilean group $\mathcal{U}$ and let $(*)$ and $(**)$ be two Cartesian coordinate systems inside $\mathcal{U}$, so that the change of coordinates from the system $(*)$ to the system $(**)$ is given by*

$$
\begin{cases} \mathbf{x}' = B \cdot \mathbf{x} + \mathbf{c} + \mathbf{w}t, \\ t' = t, \end{cases} \qquad (3.179)
$$

with constant rotation $B \in SO(3)$ and constant vectors $\mathbf{c}, \mathbf{w} \in \mathbb{R}^3$, and correspondingly we have

$$
Z'(\mathbf{x}', t') = Z(\mathbf{x}, t) + \mathbf{w} \cdot (B \cdot \mathbf{x}) + \frac{1}{2}|\mathbf{w}|^2 t + C_*^{**}, \qquad (3.180)
$$

where $C_*^{**} \in \mathbb{R}$ is an unspecified constant. Then we have

$$
\frac{\partial Z'}{\partial t'}(\mathbf{x}', t') + \frac{1}{2} \left|\nabla_{\mathbf{x}'} Z'(\mathbf{x}', t')\right|^2 = \frac{\partial Z}{\partial t}(\mathbf{x}, t) + \frac{1}{2} |\nabla_{\mathbf{x}} Z(\mathbf{x}, t)|^2 \quad \forall (\mathbf{x}, t) \in \mathbb{R}^3 \times [t_0, +\infty). \qquad (3.181)
$$

Proof As before, in (3.167), we have

$$
\begin{aligned}
\nabla_{\mathbf{x}'} Z'(\mathbf{x}', t') &= \left(B^{-1}\right)^T \cdot \nabla_{\mathbf{x}} \left(Z(\mathbf{x}, t) + \mathbf{w} \cdot (B \cdot \mathbf{x}) + \frac{1}{2}|\mathbf{w}|^2 t + C_*^{**}\right) = \\
& B \cdot \nabla_{\mathbf{x}} Z(\mathbf{x}, t) + B \cdot (B^T \cdot \mathbf{w}) = B \cdot \nabla_{\mathbf{x}} Z(\mathbf{x}, t) + \mathbf{w} \quad \forall (\mathbf{x}, t) \in \mathbb{R}^3 \times [t_0, +\infty). \qquad (3.182)
\end{aligned}
$$

On the other hand, by the chain rule we have

$$\frac{\partial Z'}{\partial t'} + \mathbf{w} \cdot \nabla_{\mathbf{x}'} Z' \\ = \frac{\partial Z'}{\partial t} = \frac{\partial}{\partial t}\left(Z(\mathbf{x}, t) + \mathbf{w} \cdot (B \cdot \mathbf{x}) + \frac{1}{2}|\mathbf{w}|^2 t + C_*^{**} \right) = \frac{\partial Z}{\partial t} + \frac{1}{2}|\mathbf{w}|^2. \tag{3.183}$$

Therefore, by (3.182) and (3.183) we have

$$\frac{\partial Z'}{\partial t'} + \frac{1}{2}\left|\nabla_{\mathbf{x}'} Z'\right|^2 = \frac{\partial Z}{\partial t} + \frac{1}{2}|\mathbf{w}|^2 - \mathbf{w} \cdot \nabla_{\mathbf{x}'} Z' + \frac{1}{2}\left|\nabla_{\mathbf{x}'} Z'\right|^2 = \frac{\partial Z}{\partial t} + \frac{1}{2}\left|\nabla_{\mathbf{x}'} Z' - \mathbf{w}\right|^2 \\ = \frac{\partial Z}{\partial t} + \frac{1}{2}|B \cdot \nabla_{\mathbf{x}} Z|^2 = \frac{\partial Z}{\partial t} + \frac{1}{2}|\nabla_{\mathbf{x}} Z|^2. \tag{3.184}$$

This completes the proof. □

Proposition 3.7 **(i)** *Consider the moving continuum medium with velocity field* $\mathbf{u}(\mathbf{x}, t)$ *and let* $\mathcal{S} := \mathcal{S}(t) \subset \mathbb{R}^3$ *be a two-dimensional surface oriented by the unit normal* $\mathbf{n} := \mathbf{n}(\mathbf{x}, t)$ *and moving together with the given medium. Then, given a vector field* $\mathbf{f}(\mathbf{x}, t)$ *we have*

$$\frac{d}{dt}\left(\iint \mathbf{f} \cdot \mathbf{n}\, d\mathcal{S}(t) \right) = \iint \left(\frac{\partial \mathbf{f}}{\partial t} - curl_{\mathbf{x}}\left(\mathbf{u} \times \mathbf{f}\right) + \left(\mathrm{div}_{\mathbf{x}} \mathbf{f}\right) \mathbf{u} \right) \cdot \mathbf{n}\, d\mathcal{S}(t). \tag{3.185}$$

(ii) *Consider the moving continuum medium with velocity field* $\mathbf{u}(\mathbf{x}, t)$ *and let* $\gamma := \gamma(t) \subset \mathbb{R}^3$ *be a one-dimensional curve oriented by the unit tangent vector* $\mathbf{t} := \mathbf{t}(\mathbf{x}, t)$ *and moving together with the given medium. Then, given a vector field* $\mathbf{f}(\mathbf{x}, t)$ *we have*

$$\frac{d}{dt}\left(\int \mathbf{f} \cdot \mathbf{t}\, d\gamma(t) \right) = \int \left(\frac{\partial \mathbf{f}}{\partial t} - \mathbf{u} \times curl_{\mathbf{x}} \mathbf{f} + \nabla_{\mathbf{x}}\left(\mathbf{u} \cdot \mathbf{f}\right) \right) \cdot \mathbf{t}\, d\gamma(t). \tag{3.186}$$

(iii) *Consider the moving continuum medium with velocity field* $\mathbf{u}(\mathbf{x}, t)$ *and let* $\Omega := \Omega(t) \subset \mathbb{R}^3$ *be a three-dimensional domain moving together with the given medium. Then, given a scalar field* $\psi(\mathbf{x}, t)$ *we have*

$$\frac{d}{dt}\left(\iiint \psi\, d\Omega(t) \right) = \iiint \left(\frac{\partial \psi}{\partial t} + \mathrm{div}_{\mathbf{x}}\left\{\psi \mathbf{u}\right\} \right) d\Omega(t). \tag{3.187}$$

Proof Assume that $\mathcal{S}(t)$ is given by the following:

$$\mathcal{S}(t) := \left\{ \mathbf{x} \in \mathbb{R}^3 \;:\; \mathbf{x} = \mathbf{h}(u, v, t),\; (u, v) \in \mathcal{W} \right\}, \tag{3.188}$$

where $\mathcal{W} \subset \mathbb{R}^2$ is some flat domain (in the general case $\mathcal{S}(t)$ is just the union of several parts, where every part is as in (3.188)). Then, as it is well known from the calculus, we have

$$\iint \mathbf{f} \cdot \mathbf{n}\, d\mathcal{S}(t) = \iint\limits_{\mathcal{W}} \mathbf{f}\left(\mathbf{h}(u,v,t),t\right) \cdot \left(\frac{\partial \mathbf{h}}{\partial u}(u,v,t) \times \frac{\partial \mathbf{h}}{\partial v}(u,v,t)\right) du dv. \tag{3.189}$$

Therefore, differentiating (3.189), by chain rule we obtain

$$\begin{aligned}
&\frac{d}{dt}\left(\iint \mathbf{f} \cdot \mathbf{n}\, d\mathcal{S}(t)\right) = \\
&\qquad \iint\limits_{\mathcal{W}} \frac{d}{dt}\left(\mathbf{f}\left(\mathbf{h}(u,v,t),t\right)\right) \cdot \left(\frac{\partial \mathbf{h}}{\partial u}(u,v,t) \times \frac{\partial \mathbf{h}}{\partial v}(u,v,t)\right) du dv \\
&\qquad\quad + \iint\limits_{\mathcal{W}} \mathbf{f}\left(\mathbf{h}(u,v,t),t\right) \cdot \left(\frac{\partial^2 \mathbf{h}}{\partial u \partial t}(u,v,t) \times \frac{\partial \mathbf{h}}{\partial v}(u,v,t)\right. \\
&\quad \left. + \frac{\partial \mathbf{h}}{\partial u}(u,v,t) \times \frac{\partial^2 \mathbf{h}}{\partial v \partial t}(u,v,t)\right) du dv = \iint\limits_{\mathcal{W}} \left(\frac{\partial \mathbf{f}}{\partial t}\left(\mathbf{h}(u,v,t),t\right)\right. \\
&\quad \left. + d_{\mathbf{x}}\mathbf{f}\left(\mathbf{h}(u,v,t),t\right) \cdot \frac{\partial \mathbf{h}}{\partial t}(u,v,t)\right) \cdot \left(\frac{\partial \mathbf{h}}{\partial u}(u,v,t) \times \frac{\partial \mathbf{h}}{\partial v}(u,v,t)\right) du dv \\
&\qquad\quad + \iint\limits_{\mathcal{W}} \mathbf{f}\left(\mathbf{h}(u,v,t),t\right) \cdot \left(\frac{\partial^2 \mathbf{h}}{\partial u \partial t}(u,v,t) \times \frac{\partial \mathbf{h}}{\partial v}(u,v,t)\right. \\
&\qquad\qquad \left. - \frac{\partial^2 \mathbf{h}}{\partial v \partial t}(u,v,t) \times \frac{\partial \mathbf{h}}{\partial u}(u,v,t)\right) du dv.
\end{aligned} \tag{3.190}$$

On the other hand, since the surface $\mathcal{S}(t)$ moves together with the medium with velocity field $\mathbf{u}(\mathbf{x},t)$ we obviously have

$$\frac{\partial \mathbf{h}}{\partial t}(u,v,t) = \mathbf{u}\left(\mathbf{h}(u,v,t),t\right). \tag{3.191}$$

Moreover, differentiating (3.191) by u and v and using again the chain rule give

$$\begin{cases} \frac{\partial^2 \mathbf{h}}{\partial u \partial t}(u,v,t) = d_{\mathbf{x}}\mathbf{u}\left(\mathbf{h}(u,v,t),t\right) \cdot \frac{\partial \mathbf{h}}{\partial u}(u,v,t), \\ \frac{\partial^2 \mathbf{h}}{\partial v \partial t}(u,v,t) = d_{\mathbf{x}}\mathbf{u}\left(\mathbf{h}(u,v,t),t\right) \cdot \frac{\partial \mathbf{h}}{\partial v}(u,v,t). \end{cases} \tag{3.192}$$

Thus inserting (3.191) and (3.192) into (3.190) gives

$$\frac{d}{dt}\left(\iint \mathbf{f}\cdot\mathbf{n}\,d\mathcal{S}(t)\right)=\iint\limits_{\mathcal{W}}\left(\frac{\partial\mathbf{f}}{\partial t}\left(\mathbf{h}(u,v,t),t\right)\right.$$

$$\left.+\,d_{\mathbf{x}}\mathbf{f}\left(\mathbf{h}(u,v,t),t\right)\cdot\mathbf{u}\left(\mathbf{h}(u,v,t),t\right)\right)\cdot\left(\frac{\partial\mathbf{h}}{\partial u}(u,v,t)\times\frac{\partial\mathbf{h}}{\partial v}(u,v,t)\right)dudv$$

$$+\iint\limits_{\mathcal{W}}\mathbf{f}\left(\mathbf{h}(u,v,t),t\right)\cdot\left(\left(d_{\mathbf{x}}\mathbf{u}\left(\mathbf{h}(u,v,t),t\right)\cdot\frac{\partial\mathbf{h}}{\partial u}(u,v,t)\right)\times\frac{\partial\mathbf{h}}{\partial v}(u,v,t)\right)dudv$$

$$-\iint\limits_{\mathcal{W}}\mathbf{f}\left(\mathbf{h}(u,v,t),t\right)\cdot\left(\left(d_{\mathbf{x}}\mathbf{u}\left(\mathbf{h}(u,v,t),t\right)\cdot\frac{\partial\mathbf{h}}{\partial v}(u,v,t)\right)\times\frac{\partial\mathbf{h}}{\partial u}(u,v,t)\right)dudv. \tag{3.193}$$

On the other hand, using (2.3) we rewrite (3.193) as

$$\frac{d}{dt}\left(\iint \mathbf{f}\cdot\mathbf{n}\,d\mathcal{S}(t)\right)=\iint\limits_{\mathcal{W}}\left(\frac{\partial\mathbf{f}}{\partial t}\left(\mathbf{h}(u,v,t),t\right)\right.$$

$$\left.+\,d_{\mathbf{x}}\mathbf{f}\left(\mathbf{h}(u,v,t),t\right)\cdot\mathbf{u}\left(\mathbf{h}(u,v,t),t\right)\right)\cdot\left(\frac{\partial\mathbf{h}}{\partial u}(u,v,t)\times\frac{\partial\mathbf{h}}{\partial v}(u,v,t)\right)dudv$$

$$+\iint\limits_{\mathcal{W}}\left(\mathrm{tr}\left\{d_{\mathbf{x}}\mathbf{u}\left(\mathbf{h}(u,v,t),t\right)\right\}\right)\mathbf{f}\left(\mathbf{h}(u,v,t),t\right)\cdot\left(\frac{\partial\mathbf{h}}{\partial u}(u,v,t)\times\frac{\partial\mathbf{h}}{\partial v}(u,v,t)\right)dudv$$

$$-\iint\limits_{\mathcal{W}}\mathbf{f}\left(\mathbf{h}(u,v,t),t\right)\cdot\left(\left\{d_{\mathbf{x}}\mathbf{u}\left(\mathbf{h}(u,v,t),t\right)\right\}^{T}\cdot\left(\frac{\partial\mathbf{h}}{\partial u}(u,v,t)\times\frac{\partial\mathbf{h}}{\partial v}(u,v,t)\right)\right)dudv$$

$$=\iint\limits_{\mathcal{W}}\left(\frac{\partial\mathbf{f}}{\partial t}\left(\mathbf{h}(u,v,t),t\right)\right.$$

$$\left.d_{\mathbf{x}}\mathbf{f}\left(\mathbf{h}(u,v,t),t\right)\cdot\mathbf{u}\left(\mathbf{h}(u,v,t),t\right)\right)\cdot\left(\frac{\partial\mathbf{h}}{\partial u}(u,v,t)\times\frac{\partial\mathbf{h}}{\partial v}(u,v,t)\right)dudv$$

$$+\iint\limits_{\mathcal{W}}\left(\mathrm{div}_{\mathbf{x}}\,\mathbf{u}\left(\mathbf{h}(u,v,t),t\right)\right)\mathbf{f}\left(\mathbf{h}(u,v,t),t\right)\cdot\left(\frac{\partial\mathbf{h}}{\partial u}(u,v,t)\times\frac{\partial\mathbf{h}}{\partial v}(u,v,t)\right)dudv$$

$$-\iint\limits_{\mathcal{W}}\left(d_{\mathbf{x}}\mathbf{u}\left(\mathbf{h}(u,v,t),t\right)\cdot\mathbf{f}\left(\mathbf{h}(u,v,t),t\right)\right)\cdot\left(\frac{\partial\mathbf{h}}{\partial u}(u,v,t)\times\frac{\partial\mathbf{h}}{\partial v}(u,v,t)\right)dudv. \tag{3.194}$$

Thus, by (2.11) we rewrite (3.194) as

$$\frac{d}{dt}\left(\iint \mathbf{f}\cdot\mathbf{n}\,d\mathcal{S}(t)\right)=\iint\limits_{\mathcal{W}}\left(\frac{\partial\mathbf{f}}{\partial t}\left(\mathbf{h}(u,v,t),t\right)\right)\cdot\left(\frac{\partial\mathbf{h}}{\partial u}(u,v,t)\times\frac{\partial\mathbf{h}}{\partial v}(u,v,t)\right)dudv$$

$$-\iint\limits_{\mathcal{W}}\left(\mathrm{curl}_{\mathbf{x}}\left(\mathbf{u}\left(\mathbf{h}(u,v,t),t\right)\times\mathbf{f}\left(\mathbf{h}(u,v,t),t\right)\right)\right)\cdot\left(\frac{\partial\mathbf{h}}{\partial u}(u,v,t)\times\frac{\partial\mathbf{h}}{\partial v}(u,v,t)\right)dudv$$

$$+\iint\limits_{\mathcal{W}}\left(\left(\mathrm{div}_{\mathbf{x}}\mathbf{f}\left(\mathbf{h}(u,v,t),t\right)\right)\mathbf{u}\left(\mathbf{h}(u,v,t),t\right)\right)\cdot\left(\frac{\partial\mathbf{h}}{\partial u}(u,v,t)\times\frac{\partial\mathbf{h}}{\partial v}(u,v,t)\right)dudv. \tag{3.195}$$

Therefore, using the reverse direction of (3.189) in the right-hand side of (3.195) gives

$$\frac{d}{dt}\left(\iint \mathbf{f}\cdot\mathbf{n}\,dS(t)\right)=$$
$$\iint \frac{\partial \mathbf{f}}{\partial t}\cdot\mathbf{n}\,dS(t)-\iint (\mathrm{curl}_{\mathbf{x}}(\mathbf{u}\times\mathbf{f}))\cdot\mathbf{n}\,dS(t)+\iint (\mathrm{div}_{\mathbf{x}}\mathbf{f})\,\mathbf{u}\cdot\mathbf{n}\,dS(t), \tag{3.196}$$

and thus (3.185) follows. So we proved part **(i)** of the proposition.

Next assume that $\gamma(t)$ is given by the following:

$$\gamma(t):=\left\{\mathbf{x}\in\mathbb{R}^3\,:\,\mathbf{x}=\mathbf{h}(s,t),\ s\in[0,1]\right\}, \tag{3.197}$$

(in the general case $\gamma(t)$ is just the union of several parts, where every part is as in (3.197)). Then, as it is well known from the calculus, we have

$$\int \mathbf{f}\cdot\mathbf{t}\,d\gamma(t)=\int_0^1 \mathbf{f}\left(\mathbf{h}(s,t),t\right)\cdot\frac{\partial \mathbf{h}}{\partial s}(s,t)\,ds. \tag{3.198}$$

Therefore, differentiating (3.198), by chain rule we obtain

$$\frac{d}{dt}\left(\int \mathbf{f}\cdot\mathbf{t}\,d\gamma(t)\right)=$$
$$\int_0^1 \frac{d}{dt}\left(\mathbf{f}\left(\mathbf{h}(s,t),t\right)\right)\cdot\frac{\partial \mathbf{h}}{\partial s}(s,t)\,ds+\int_0^1 \mathbf{f}\left(\mathbf{h}(s,t),t\right)\cdot\frac{\partial^2 \mathbf{h}}{\partial s\partial t}(s,t)\,ds=$$
$$\int_0^1\left(\frac{\partial \mathbf{f}}{\partial t}\left(\mathbf{h}(s,t),t\right)+d_{\mathbf{x}}\mathbf{f}\left(\mathbf{h}(s,t),t\right)\cdot\frac{\partial \mathbf{h}}{\partial t}(s,t)\right)\cdot\frac{\partial \mathbf{h}}{\partial s}(s,t)\,ds$$
$$+\int_0^1 \mathbf{f}\left(\mathbf{h}(s,t),t\right)\cdot\frac{\partial^2 \mathbf{h}}{\partial s\partial t}(s,t)\,ds. \tag{3.199}$$

On the other hand, since the curve $\gamma(t)$ moves together with the medium with velocity field $\mathbf{u}(\mathbf{x},t)$ we obviously have

$$\frac{\partial \mathbf{h}}{\partial t}(s,t)=\mathbf{u}\left(\mathbf{h}(s,t),t\right). \tag{3.200}$$

Moreover, differentiating (3.200) by s and using again the chain rule gives:

$$\frac{\partial^2 \mathbf{h}}{\partial s\partial t}(s,t)=d_{\mathbf{x}}\mathbf{u}\left(\mathbf{h}(s,t),t\right)\cdot\frac{\partial \mathbf{h}}{\partial s}(s,t). \tag{3.201}$$

Thus inserting (3.200) and (3.201) into (3.199) gives

$$
\begin{aligned}
\frac{d}{dt}\left(\int \mathbf{f}\cdot\mathbf{t}\,d\gamma(t)\right) &= \int_0^1 \left(\frac{\partial \mathbf{f}}{\partial t}\left(\mathbf{h}(s,t),t\right) + d_{\mathbf{x}}\mathbf{f}\left(\mathbf{h}(s,t),t\right)\cdot\mathbf{u}\left(\mathbf{h}(s,t),t\right)\right)\cdot\frac{\partial \mathbf{h}}{\partial s}(s,t)\,ds \\
&\quad + \int_0^1 \mathbf{f}\left(\mathbf{h}(s,t),t\right)\cdot\left(d_{\mathbf{x}}\mathbf{u}\left(\mathbf{h}(s,t),t\right)\cdot\frac{\partial \mathbf{h}}{\partial s}(s,t)\right)\,ds \\
&= \int_0^1 \left(\frac{\partial \mathbf{f}}{\partial t}\left(\mathbf{h}(s,t),t\right) + d_{\mathbf{x}}\mathbf{f}\left(\mathbf{h}(s,t),t\right)\cdot\mathbf{u}\left(\mathbf{h}(s,t),t\right)\right)\cdot\frac{\partial \mathbf{h}}{\partial s}(s,t)\,ds \\
&\quad + \int_0^1 \left(\left\{d_{\mathbf{x}}\mathbf{u}\left(\mathbf{h}(s,t),t\right)\right\}^T\cdot\mathbf{f}\left(\mathbf{h}(s,t),t\right)\right)\cdot\frac{\partial \mathbf{h}}{\partial s}(s,t)\,ds. \qquad (3.202)
\end{aligned}
$$

On the other hand, using (2.15) we rewrite (3.202) as

$$
\begin{aligned}
&\frac{d}{dt}\left(\int \mathbf{f}\cdot\mathbf{t}\,d\gamma(t)\right) = \\
&\int_0^1 \left(\frac{\partial \mathbf{f}}{\partial t}\left(\mathbf{h}(s,t),t\right) - \mathbf{u}\left(\mathbf{h}(s,t),t\right)\times \mathrm{curl}_{\mathbf{x}}\mathbf{f}\left(\mathbf{h}(s,t),t\right)\right)\cdot\frac{\partial \mathbf{h}}{\partial s}(s,t)\,ds \\
&+ \int_0^1 \left(\left\{d_{\mathbf{x}}\mathbf{f}\left(\mathbf{h}(s,t),t\right)\right\}^T\cdot\mathbf{u}\left(\mathbf{h}(s,t),t\right) + \left\{d_{\mathbf{x}}\mathbf{u}\left(\mathbf{h}(s,t),t\right)\right\}^T\cdot\mathbf{f}\left(\mathbf{h}(s,t),t\right)\right)\cdot\frac{\partial \mathbf{h}}{\partial s}(s,t)\,ds \\
&= \int_0^1 \left(\frac{\partial \mathbf{f}}{\partial t}\left(\mathbf{h}(s,t),t\right) - \mathbf{u}\left(\mathbf{h}(s,t),t\right)\times \mathrm{curl}_{\mathbf{x}}\mathbf{f}\left(\mathbf{h}(s,t),t\right)\right)\cdot\frac{\partial \mathbf{h}}{\partial s}(s,t)\,ds \\
&\quad + \int_0^1 \nabla_{\mathbf{x}}\left(\mathbf{u}\left(\mathbf{h}(s,t),t\right)\cdot\mathbf{f}\left(\mathbf{h}(s,t),t\right)\right)\cdot\frac{\partial \mathbf{h}}{\partial s}(s,t)\,ds. \qquad (3.203)
\end{aligned}
$$

Thereforc, using the reverse direction of (3.198) in the right-hand side of (3.203) gives

$$
\frac{d}{dt}\left(\int \mathbf{f}\cdot\mathbf{t}\,d\gamma(t)\right) = \int\left(\frac{\partial \mathbf{f}}{\partial t} - \mathbf{u}\times\mathrm{curl}_{\mathbf{x}}\mathbf{f}\right)\cdot\mathbf{t}\,d\gamma(t) + \int \nabla_{\mathbf{x}}\left(\mathbf{u}\cdot\mathbf{f}\right)\cdot\mathbf{t}\,d\gamma(t), \tag{3.204}
$$

and thus (3.186) follows. So we proved part **(ii)** of the proposition.

Next assume that $\Omega(t)$ is given by the following:

$$
\Omega(t) := \left\{\mathbf{x}\in\mathbb{R}^3\ :\ \mathbf{x} = \mathbf{h}(\mathbf{y},t),\ \mathbf{y}\in G\right\}, \tag{3.205}
$$

where $G \subset \mathbb{R}^3$ is some three-dimensional domain. Then, as it is well known from the rule of the change of variables of integration from the calculus, we have

$$
\iiint \psi\,d\Omega(t) = \iiint_G \psi\left(\mathbf{h}(u,v,t),t\right)\left|det\left\{d_{\mathbf{y}}\mathbf{h}(\mathbf{y},t)\right\}\right|\,d\mathbf{y}. \tag{3.206}
$$

Therefore, differentiating (3.206), by chain rule we obtain

$$\frac{d}{dt}\left(\iiint \psi\, d\Omega(t)\right) = \iiint_G \frac{d}{dt}\left(\psi\left(\mathbf{h}(\mathbf{y},t),t\right)\right)\left|det\left\{d_{\mathbf{y}}\mathbf{h}(\mathbf{y},t)\right\}\right| d\mathbf{y}$$

$$+\iiint_G \psi\left(\mathbf{h}(\mathbf{y},t),t\right)\frac{\partial}{\partial t}\left\{\left|det\left\{d_{\mathbf{y}}\mathbf{h}(\mathbf{y},t)\right\}\right|\right\} d\mathbf{y} =$$

$$\iiint_G \left(\frac{\partial\psi}{\partial t}\left(\mathbf{h}(\mathbf{y},t),t\right) + \nabla_{\mathbf{x}}\psi\left(\mathbf{h}(\mathbf{y},t),t\right)\cdot\frac{\partial\mathbf{h}}{\partial t}(\mathbf{y},t)\right)\left|det\left\{d_{\mathbf{y}}\mathbf{h}(\mathbf{y},t)\right\}\right| d\mathbf{y}$$

$$+\iiint_G \psi\left(\mathbf{h}(\mathbf{y},t),t\right)\frac{\partial}{\partial t}\left\{\left|det\left\{d_{\mathbf{y}}\mathbf{h}(\mathbf{y},t)\right\}\right|\right\} d\mathbf{y}. \tag{3.207}$$

On the other hand, since the domain $\Omega(t)$ moves together with the medium with velocity field $\mathbf{u}(\mathbf{x}, t)$ we obviously have

$$\frac{\partial\mathbf{h}}{\partial t}(\mathbf{y},t) = \mathbf{u}\left(\mathbf{h}(\mathbf{y},t),t\right). \tag{3.208}$$

Moreover, differentiating (3.208) by $\mathbf{y}$ and using again the chain rule give

$$\frac{\partial}{\partial t}\left\{d_{\mathbf{y}}\mathbf{h}(\mathbf{y},t)\right\} = d_{\mathbf{x}}\mathbf{u}\left(\mathbf{h}(\mathbf{y},t),t\right)\cdot\left\{d_{\mathbf{y}}\mathbf{h}(\mathbf{y},t)\right\}. \tag{3.209}$$

In particular, using the Liouville theorem in the theory of linear ordinary differential systems by (3.209) we deduce that $det\left\{d_{\mathbf{y}}\mathbf{h}(\mathbf{y},t)\right\}$ satisfies

$$\frac{\partial}{\partial t}\left(det\left\{d_{\mathbf{y}}\mathbf{h}(\mathbf{y},t)\right\}\right) = \left(\mathrm{tr}\left\{d_{\mathbf{x}}\mathbf{u}\left(\mathbf{h}(\mathbf{y},t),t\right)\right\}\right)\left(det\left\{d_{\mathbf{y}}\mathbf{h}(\mathbf{y},t)\right\}\right)$$

$$= \left(\mathrm{div}_{\mathbf{x}}\,\mathbf{u}\left(\mathbf{h}(\mathbf{y},t),t\right)\right)\left(det\left\{d_{\mathbf{y}}\mathbf{h}(\mathbf{y},t)\right\}\right), \tag{3.210}$$

and thus

$$\frac{\partial}{\partial t}\left(\left|det\left\{d_{\mathbf{y}}\mathbf{h}(\mathbf{y},t)\right\}\right|\right) = \left(\mathrm{div}_{\mathbf{x}}\,\mathbf{u}\left(\mathbf{h}(\mathbf{y},t),t\right)\right)\left|det\left\{d_{\mathbf{y}}\mathbf{h}(\mathbf{y},t)\right\}\right|. \tag{3.211}$$

Thus inserting (3.208) and (3.211) into (3.207) gives

$$\frac{d}{dt}\left(\iiint \psi\, d\Omega(t)\right) =$$

$$\iiint_G \left(\frac{\partial\psi}{\partial t}\left(\mathbf{h}(\mathbf{y},t),t\right) + \mathbf{u}\left(\mathbf{h}(\mathbf{y},t),t\right)\cdot\nabla_{\mathbf{x}}\psi\left(\mathbf{h}(\mathbf{y},t),t\right)\right)\left|det\left\{d_{\mathbf{y}}\mathbf{h}(\mathbf{y},t)\right\}\right| d\mathbf{y}$$

$$+\iiint_G \psi\left(\mathbf{h}(\mathbf{y},t),t\right)\left(\mathrm{div}_{\mathbf{x}}\,\mathbf{u}\left(\mathbf{h}(\mathbf{y},t),t\right)\right)\left|det\left\{d_{\mathbf{y}}\mathbf{h}(\mathbf{y},t)\right\}\right| d\mathbf{y}. \tag{3.212}$$

Therefore, using the reverse direction of (3.206) in the right-hand side of (3.212) gives

$$\frac{d}{dt}\left(\iiint \psi\, d\Omega(t)\right) = \iiint \left(\frac{\partial \psi}{\partial t} + \mathbf{u}\cdot\nabla_{\mathbf{x}}\psi + (\mathrm{div}_{\mathbf{x}}\,\mathbf{u})\psi\right) d\Omega(t)$$
$$= \iiint \left(\frac{\partial \psi}{\partial t} + \mathrm{div}_{\mathbf{x}}\,\{\psi\mathbf{u}\}\right) d\Omega(t), \quad (3.213)$$

and thus (3.187) follows. So we proved part **(iii)** of the proposition. □

Lemma 3.5 *The matrix-valued field* $T := T(\mathbf{x}, t) : \mathbb{R}^3 \times [t_0, +\infty) \to \mathbb{R}^{3\times 3}$ *is a proper matrix field if and only if, under every change of coordinate system given by* (3.1), *this field transforms as*

$$T'(\mathbf{x}', t') = (A(t) \otimes A(t)) \cdot T(\mathbf{x}, t), \quad (3.214)$$

where the sign $\otimes$ *in* (3.214) *means the tensor product of the matrices, i.e. for given two linear operators (matrices)* $A \in \mathbb{R}^{3\times 3}$ *and* $B \in \mathbb{R}^{3\times 3}$ *their tensor product* $A \otimes B$ *is a linear operator from* $\mathbb{R}^{3\times 3}$ *to* $\mathbb{R}^{3\times 3}$, *defined by the identity:*

$$(A \otimes B)\cdot(\mathbf{a}\otimes\mathbf{b}) = (A\cdot\mathbf{a})\otimes(B\cdot\mathbf{b}) \qquad \forall\mathbf{a}\in\mathbb{R}^3,\ \forall\mathbf{b}\in\mathbb{R}^3. \quad (3.215)$$

Proof Since we identify a vector $\mathbf{c} \in \mathbb{R}^3$ with the corresponding 3×1 matrix, clearly we have

$$\mathbf{a}\otimes\mathbf{b} = \mathbf{a}\cdot\mathbf{b}^T \qquad \forall\mathbf{a}\in\mathbb{R}^3,\ \forall\mathbf{b}\in\mathbb{R}^3. \quad (3.216)$$

Thus, inserting (3.216) into (3.215) gives

$$(A\otimes A)\cdot(\mathbf{a}\otimes\mathbf{b}) = (A\cdot\mathbf{a})\otimes(A\cdot\mathbf{b}) = (A\cdot\mathbf{a})\cdot(A\cdot\mathbf{b})^T = (A\cdot\mathbf{a})\cdot(\mathbf{b}^T\cdot A^T) = A\cdot(\mathbf{a}\cdot\mathbf{b}^T)\cdot A^T$$
$$= A\cdot(\mathbf{a}\otimes\mathbf{b})\cdot A^T \qquad \forall\mathbf{a}\in\mathbb{R}^3,\ \forall\mathbf{b}\in\mathbb{R}^3. \quad (3.217)$$

Then, by the linearity, (3.217) gives

$$(A\otimes A)\cdot G = A\cdot G\cdot A^T \qquad \forall G\in\mathbb{R}^{3\times 3}. \quad (3.218)$$

Thus, by (3.218) we deduce that (3.214) is equivalent to (3.6). □

Chapter 4
Non-Relativistic Gravity

4.1 Newtonian Gravity Revised

Consider the classical space-time where the change of some inertial coordinate system $(*)$ to another inertial coordinate system $(**)$ is given (up to equivalence) by the Galilean transformation:

$$\begin{cases} \mathbf{x}' = \mathbf{x} + \mathbf{w}t, \\ t' = t, \end{cases} \tag{4.1}$$

and the change of some non-inertial Cartesian coordinate system $(*)$ to another Cartesian coordinate system $(**)$ is of the form:

$$\begin{cases} \mathbf{x}' = A(t) \cdot \mathbf{x} + \mathbf{z}(t), \\ t' = t, \end{cases} \tag{4.2}$$

where $A(t) \in SO(3)$ is a rotation, i.e. $A(t) \in \mathbb{R}^{3\times 3}$, $\det A(t) > 0$ and $A(t) \cdot A^T(t) = I$, where A^T is the transpose of the matrix A.

Similar to the General Theory of Relativity, we assume that the most general laws of classical mechanics should be invariant in every non-inertial Cartesian coordinate system, i.e. they preserve their form under transformations of the form (4.2). Moreover, again as in the General Theory of Relativity, we assume that the fictitious forces (inertial forces) in non-inertial coordinate systems and the forces of Newtonian gravitation have the same nature and represented by some field in somewhat similar to the electromagnetic field.

We begin with some simple observation. Assume that we are away from essential gravitational masses and strong electromagnetic fields. Then consider two Cartesian coordinate systems $(*)$ and $(**)$, such that the system $(**)$ is inertial and the change of coordinate system $(*)$ to coordinate system $(**)$ is given by (4.2). Then the fictitious-gravitational force in the system $(**)$ is trivial $\mathbf{F}'_0 = 0$. On the other hand, since under the change of coordinate system of the form (4.2) the velocity transforms as

A. Poliakovsky, *The Laws of Gravity and Electromagnetism*,
https://doi.org/10.1007/978-3-031-61407-1_4

$$\mathbf{u}' = A(t)\cdot\mathbf{u} + \frac{dA}{dt}(t)\cdot\mathbf{x} + \frac{d\mathbf{z}}{dt}(t) \tag{4.3}$$

and the acceleration transforms as

$$\mathbf{a}' = A(t)\cdot\mathbf{a} + 2\frac{dA}{dt}(t)\cdot\mathbf{u} + \frac{d^2A}{dt^2}(t)\cdot\mathbf{x} + \frac{d^2\mathbf{z}}{dt^2}(t) \tag{4.4}$$

the fictitious-gravitational force in the system $(*)$, acting on the particle with inertial mass m and velocity $\mathbf{u}$, is given by

$$\mathbf{F}_0 = m\left(-2A^T(t)\cdot\frac{dA}{dt}(t)\cdot\mathbf{u} - A^T(t)\cdot\frac{d^2A}{dt^2}(t)\cdot\mathbf{x} - A^T(t)\cdot\frac{d^2\mathbf{z}}{dt^2}(t)\right). \tag{4.5}$$

On the other hand, since $A(t)\cdot A^T(t) = I$ and thus $A^T(t)\cdot\frac{dA}{dt}(t) + \frac{dA^T}{dt}(t)\cdot A(t) = 0$, if we define a vector field

$$\mathbf{v}(\mathbf{x},t) := -A^T(t)\cdot\frac{dA}{dt}(t)\cdot\mathbf{x} - A^T(t)\cdot\frac{d\mathbf{z}}{dt}(t), \tag{4.6}$$

then we obviously have

$$\begin{cases} d_{\mathbf{x}}\mathbf{v} = -A^T(t)\cdot\frac{dA}{dt}(t) = \frac{dA^T}{dt}(t)\cdot A(t) \\ \{d_{\mathbf{x}}\mathbf{v}\}^T = -\frac{dA^T}{dt}(t)\cdot A(t) = A^T(t)\cdot\frac{dA}{dt}(t) \\ \frac{\partial\mathbf{v}}{\partial t} = -A^T(t)\cdot\left(\frac{d^2A}{dt^2}(t)\cdot\mathbf{x} + \frac{d^2\mathbf{z}}{dt^2}(t)\right) - \frac{dA^T}{dt}(t)\cdot\left(\frac{dA}{dt}(t)\cdot\mathbf{x} + \frac{d\mathbf{z}}{dt}(t)\right). \end{cases} \tag{4.7}$$

Thus by (4.6) and (4.7) we rewrite (4.5) as

$$\mathbf{F}_0 = m\left(-2A^T(t)\cdot\frac{dA}{dt}(t)\cdot\mathbf{u} + \frac{\partial\mathbf{v}}{\partial t} - \frac{dA^T}{dt}(t)\cdot A(t)\cdot\mathbf{v}\right). \tag{4.8}$$

Then using (2.15) and (4.7) we finally rewrite (4.8) as

$$\mathbf{F}_0 = m\left(\frac{\partial\mathbf{v}}{\partial t} + \frac{1}{2}\nabla_{\mathbf{x}}\left(|\mathbf{v}|^2\right)\right) + m\mathbf{u}\times(-\mathrm{curl}_{\mathbf{x}}\mathbf{v}). \tag{4.9}$$

Similarly assume that also in the general case of essential gravitational masses there exists a vector field $\mathbf{v}(\mathbf{x},t)$ such that in some inertial or non-inertial Cartesian coordinate system the fictitious-gravitational force is given by (4.9). Then we call the vector field $\mathbf{v}$ the vectorial gravitational potential. We see here the following analogy with electrodynamics, denoting

$$\tilde{\mathbf{E}} := \partial_t\mathbf{v} + \nabla_{\mathbf{x}}\left(\frac{1}{2}|\mathbf{v}|^2\right) \quad \text{and} \quad \tilde{\mathbf{B}} := -c\,\mathrm{curl}_{\mathbf{x}}\mathbf{v},$$

we rewrite (4.9) as

$$\mathbf{F}_0 = m\left(\tilde{\mathbf{E}} + \frac{1}{c}\mathbf{u}\times\tilde{\mathbf{B}}\right), \tag{4.10}$$

where

$$\text{curl}_{\mathbf{x}}\tilde{\mathbf{E}} + \frac{1}{c}\frac{\partial}{\partial t}\tilde{\mathbf{B}} = 0 \quad \text{and} \quad \text{div}_{\mathbf{x}}\,\tilde{\mathbf{B}} = 0.$$

Next using (4.9) we rewrite the Second Law of Newton as

$$m\frac{d^2\mathbf{x}}{dt^2} = m\frac{d\mathbf{u}}{dt} = \mathbf{F}_0 + \mathbf{F} = m\left(\frac{\partial \mathbf{v}}{\partial t}(\mathbf{x},t) + \frac{1}{2}\nabla_{\mathbf{x}}\left(|\mathbf{v}|^2\right)(\mathbf{x},t)\right) + m\mathbf{u}\times(-\text{curl}_{\mathbf{x}}\mathbf{v}(\mathbf{x},t)) + \mathbf{F}, \tag{4.11}$$

where $\mathbf{x} := \mathbf{x}(t)$, $\mathbf{u} := \mathbf{u}(t) = \frac{d\mathbf{x}}{dt}(t)$ and m are the place, the velocity and the inertial mass of some given particle at the moment of time t, $\mathbf{v} := \mathbf{v}(\mathbf{x},t)$ is the vectorial gravitational potential and $\mathbf{F}$ is the total non-gravitational force, acting on the given particle.

Once we considered the Second Law of Newton in the form

$$\frac{d\mathbf{u}}{dt} = -\mathbf{u}\times\text{curl}_{\mathbf{x}}\mathbf{v} + \partial_t\mathbf{v} + \nabla_{\mathbf{x}}\left(\frac{1}{2}|\mathbf{v}|^2\right) + \frac{1}{m}\mathbf{F}, \tag{4.12}$$

we still need to prove that this law is invariant under the change of inertial or non-inertial Cartesian coordinate system and to determine the law of transformation for the vectorial gravitational potential under the change of coordinate systems. As we will show above this is indeed the case and moreover the law of transformation of the vectorial gravitational potential under the change of coordinate system, given by (4.2), is

$$\mathbf{v}' = A(t)\cdot\mathbf{v} + \frac{dA}{dt}(t)\cdot\mathbf{x} + \frac{d\mathbf{z}}{dt}(t)$$

,i.e. it is the same as the transformation of a field of velocities. More precisely we have the following:

Proposition 4.1 *Consider the change of some non-inertial Cartesian coordinate system $(*)$ to another Cartesian coordinate system $(**)$ of the form:*

$$\begin{cases} \mathbf{x}' = A(t)\cdot\mathbf{x} + \mathbf{z}(t), \\ t' = t, \end{cases} \tag{4.13}$$

where $A(t) \in SO(3)$ is a rotation, i.e. $A(t) \in \mathbb{R}^{3\times 3}$, $\det A(t) > 0$ and $A(t)\cdot A^T(t) = I$. Next, assume that in the coordinate system $(**)$ we observe a validity of the Second Law of Newton in the form:

$$\frac{d\mathbf{u}'}{dt'} = -\mathbf{u}'\times curl_{\mathbf{x}'}\mathbf{v}' + \partial_{t'}\mathbf{v}' + \nabla_{\mathbf{x}'}\left(\frac{1}{2}|\mathbf{v}'|^2\right) + \frac{1}{m'}\mathbf{F}', \tag{4.14}$$

where $\mathbf{x}' := \mathbf{x}'(t')$, $\mathbf{u}' := \mathbf{u}'(t') = \frac{d\mathbf{x}'}{dt'}(t')$ *and* m' *are the place, the velocity and the mass of some given particle at the moment of time* t', $\mathbf{v}' := \mathbf{v}'(\mathbf{x}', t')$ *is the vectorial gravitational potential and* $\mathbf{F}'$ *is a total non-gravitational force, acting on the given particle in the coordinate system* $(**)$. *Then in the coordinate system* $(*)$ *we observe a validity of the Second Law of Newton in the (same as* (4.14)*) form:*

$$\frac{d\mathbf{u}}{dt} = -\mathbf{u} \times curl_{\mathbf{x}}\mathbf{v} + \partial_t \mathbf{v} + \nabla_{\mathbf{x}}\left(\frac{1}{2}|\mathbf{v}|^2\right) + \frac{1}{m}\mathbf{F}, \tag{4.15}$$

where

$$\mathbf{v}' = A(t) \cdot \mathbf{v} + \frac{dA}{dt}(t) \cdot \mathbf{x} + \frac{d\mathbf{z}}{dt}(t) \tag{4.16}$$

$$\mathbf{F}' = A(t) \cdot \mathbf{F}, \tag{4.17}$$

$$m' = m, \tag{4.18}$$

$$\mathbf{u}' = A(t) \cdot \mathbf{u} + \frac{dA}{dt}(t) \cdot \mathbf{x} + \frac{d\mathbf{z}}{dt}(t). \tag{4.19}$$

Proof Using (2.15) we rewrite (4.14) as

$$\frac{d\mathbf{u}'}{dt'} = -(\mathbf{u}' - \mathbf{v}') \times \mathrm{curl}_{\mathbf{x}'}\mathbf{v}' + \partial_{t'}\mathbf{v}' + d_{\mathbf{x}'}\mathbf{v}' \cdot \mathbf{v}' + \frac{1}{m'}\mathbf{F}'. \tag{4.20}$$

Next define the vector field $\mathbf{v}$ in the system $(*)$ in such a way that it will be related to $\mathbf{v}'$ in the system $(**)$ due to (4.16). I.e. $\mathbf{v}$ is given by

$$\mathbf{v} := A^T(t) \cdot \left(\mathbf{v}' - \frac{dA}{dt}(t) \cdot \mathbf{x} - \frac{d\mathbf{z}}{dt}(t)\right).$$

We are going to prove (4.15) in the system $(*)$ using the following relations between the physical characteristics in coordinate systems $(*)$ and $(**)$:

$$\mathbf{F}' = A(t) \cdot \mathbf{F}, \tag{4.21}$$

$$m' = m, \tag{4.22}$$

$$\mathbf{u}' = A(t) \cdot \mathbf{u} + A'(t) \cdot \mathbf{x} + \mathbf{w}(t), \tag{4.23}$$

$$\mathbf{v}' = A(t) \cdot \mathbf{v} + A'(t) \cdot \mathbf{x} + \mathbf{w}(t), \tag{4.24}$$

where $\mathbf{w}(t) := \frac{d\mathbf{z}}{dt}(t)$ and $A'(t) = \frac{dA}{dt}(t)$. Indeed, inserting these relations into (4.20) we obtain

$$\frac{d}{dt}\left(A(t) \cdot \mathbf{u}(t) + A'(t) \cdot \mathbf{x}(t) + \mathbf{w}(t)\right) =$$
$$- \left(A(t) \cdot (\mathbf{u} - \mathbf{v})\right) \times \mathrm{curl}_{\mathbf{x}'}\left(A(t) \cdot \mathbf{v} + A'(t) \cdot \mathbf{x} + \mathbf{w}(t)\right)$$
$$+ \partial_{t'}\left(A(t) \cdot \mathbf{v} + A'(t) \cdot \mathbf{x} + \mathbf{w}(t)\right)$$

$$+ d_{\mathbf{x}'}\left(A(t)\cdot\mathbf{v}+A'(t)\cdot\mathbf{x}+\mathbf{w}(t)\right)\cdot\left(A(t)\cdot\mathbf{v}+A'(t)\cdot\mathbf{x}+\mathbf{w}(t)\right)+\frac{1}{m}A(t)\cdot\mathbf{F}. \tag{4.25}$$

Next using the chain rule, we deduce

$$\partial_{t'}\left(A(t)\cdot\mathbf{v}+A'(t)\cdot\mathbf{x}+\mathbf{w}(t)\right)+d_{\mathbf{x}'}\left(A(t)\cdot\mathbf{v}+A'(t)\cdot\mathbf{x}+\mathbf{w}(t)\right)\cdot\left(A'(t)\cdot\mathbf{x}+\mathbf{w}(t)\right)= \partial_t\left(A(t)\cdot\mathbf{v}+A'(t)\cdot\mathbf{x}+\mathbf{w}(t)\right). \tag{4.26}$$

Inserting it into (4.25) we deduce

$$\frac{d}{dt}\left(A(t)\cdot\mathbf{u}(t)+A'(t)\cdot\mathbf{x}(t)+\mathbf{w}(t)\right)= -\left(A(t)\cdot(\mathbf{u}-\mathbf{v})\right)\times\operatorname{curl}_{\mathbf{x}'}\left(A(t)\cdot\mathbf{v}+A'(t)\cdot\mathbf{x}+\mathbf{w}(t)\right) + \partial_t\left(A(t)\cdot\mathbf{v}+A'(t)\cdot\mathbf{x}+\mathbf{w}(t)\right)+d_{\mathbf{x}}\left(A(t)\cdot\mathbf{v}+A'(t)\cdot\mathbf{x}+\mathbf{w}(t)\right)\cdot\mathbf{v}+\frac{1}{m}A(t)\cdot\mathbf{F}. \tag{4.27}$$

On the other hand, by (4.13) and by Proposition 3.1 we clearly have

$$\operatorname{curl}_{\mathbf{x}'}\left(\left(A(t)\cdot\mathbf{v}+A'(t)\cdot\mathbf{x}+\mathbf{w}(t)\right)\right) = A(t)\cdot\operatorname{curl}_{\mathbf{x}}\left(\mathbf{v}+A^{-1}(t)\cdot A'(t)\cdot\mathbf{x}+A^{-1}(t)\cdot\mathbf{w}(t)\right). \tag{4.28}$$

Inserting it into (4.27) we deduce

$$\frac{d}{dt}\left(A(t)\cdot\mathbf{u}(t)+A'(t)\cdot\mathbf{x}(t)+\mathbf{w}(t)\right)= -\left(A(t)\cdot(\mathbf{u}-\mathbf{v})\right)\times\left(A(t)\cdot\operatorname{curl}_{\mathbf{x}}\left(\mathbf{v}+A^{-1}(t)\cdot A'(t)\cdot\mathbf{x}+A^{-1}(t)\cdot w(t)\right)\right) + \partial_t\left(A(t)\cdot\mathbf{v}+A'(t)\cdot\mathbf{x}+\mathbf{w}(t)\right)+d_{\mathbf{x}}\left(A(t)\cdot\mathbf{v}+A'(t)\cdot\mathbf{x}+\mathbf{w}(t)\right)\cdot\mathbf{v}+\frac{1}{m}A(t)\cdot\mathbf{F}. \tag{4.29}$$

Thus by (4.29) and (2.4) we have

$$\frac{d}{dt}\left(A(t)\cdot\mathbf{u}(t)+A'(t)\cdot\mathbf{x}(t)+\mathbf{w}(t)\right)= -A(t)\cdot\left((\mathbf{u}-\mathbf{v})\times\operatorname{curl}_{\mathbf{x}}\left(\mathbf{v}+A^{-1}(t)\cdot A'(t)\cdot\mathbf{x}+A^{-1}(t)\cdot w(t)\right)\right) + \partial_t\left(A(t)\cdot\mathbf{v}+A'(t)\cdot\mathbf{x}+\mathbf{w}(t)\right)+d_{\mathbf{x}}\left(A(t)\cdot\mathbf{v}+A'(t)\cdot\mathbf{x}+\mathbf{w}(t)\right)\cdot\mathbf{v}+\frac{1}{m}A(t)\cdot\mathbf{F}. \tag{4.30}$$

On the other hand, clearly we have

$$\frac{d}{dt}\left(A(t)\cdot\mathbf{u}(t)+A'(t)\cdot\mathbf{x}(t)+\mathbf{w}(t)\right)=A(t)\cdot\frac{d\mathbf{u}}{dt}+2A'(t)\cdot\mathbf{u}+A''(t)\cdot\mathbf{x}(t)+\frac{d\mathbf{w}}{dt}(t).$$

Inserting it into (4.30) we deduce

$$\begin{aligned}
&A(t)\cdot\frac{d\mathbf{u}}{dt}+2A'(t)\cdot\mathbf{u}+A''(t)\cdot\mathbf{x}(t)+\frac{d\mathbf{w}}{dt}(t)=\\
&\qquad -A(t)\cdot\Big((\mathbf{u}-\mathbf{v})\times\operatorname{curl}_{\mathbf{x}}\Big(\mathbf{v}+A^{-1}(t)\cdot A'(t)\cdot\mathbf{x}+A^{-1}(t)\cdot w(t)\Big)\Big)\\
&\quad+\partial_t\left(A(t)\cdot\mathbf{v}+A'(t)\cdot\mathbf{x}+\mathbf{w}(t)\right)+d_{\mathbf{x}}\left(A(t)\cdot\mathbf{v}+A'(t)\cdot\mathbf{x}+\mathbf{w}(t)\right)\cdot\mathbf{v}+\frac{1}{m}A(t)\cdot\mathbf{F}\\
&\quad=-A(t)\cdot((\mathbf{u}-\mathbf{v})\times\operatorname{curl}_{\mathbf{x}}\mathbf{v})-A(t)\cdot\Big((\mathbf{u}-\mathbf{v})\times\operatorname{curl}_{\mathbf{x}}\left(A^{-1}(t)\cdot A'(t)\cdot\mathbf{x}\right)\Big)\\
&\quad+A(t)\cdot\partial_t\mathbf{v}+2A'(t)\cdot\mathbf{v}+A''(t)\cdot\mathbf{x}+\frac{d\mathbf{w}}{dt}(t)+A(t)\cdot d_{\mathbf{x}}\mathbf{v}\cdot\mathbf{v}+\frac{1}{m}A(t)\cdot\mathbf{F}. \qquad (4.31)
\end{aligned}$$

We rewrite (4.31) as

$$\begin{aligned}
\frac{d\mathbf{u}}{dt}&=-(\mathbf{u}-\mathbf{v})\times\operatorname{curl}_{\mathbf{x}}\left(A^{-1}(t)\cdot A'(t)\cdot\mathbf{x}\right)-2A^{-1}(t)\cdot A'(t)\cdot(\mathbf{u}-\mathbf{v})\\
&\quad-(\mathbf{u}-\mathbf{v})\times\operatorname{curl}_{\mathbf{x}}\mathbf{v}+\partial_t\mathbf{v}+d_{\mathbf{x}}\mathbf{v}\cdot\mathbf{v}+\frac{1}{m}\mathbf{F}. \qquad (4.32)
\end{aligned}$$

Thus by (2.15) and (4.32) we deduce

$$\begin{aligned}
&\frac{d\mathbf{u}}{dt}=\\
&d_{\mathbf{x}}\left(A^{-1}(t)\cdot A'(t)\cdot\mathbf{x}\right)\cdot(\mathbf{u}-\mathbf{v})-\left\{d_{\mathbf{x}}\left(A^{-1}(t)\cdot A'(t)\cdot\mathbf{x}\right)\right\}^T\cdot(\mathbf{u}-\mathbf{v})-2A^{-1}(t)\cdot A'(t)\cdot(\mathbf{u}-\mathbf{v})\\
&\qquad-(\mathbf{u}-\mathbf{v})\times\operatorname{curl}_{\mathbf{x}}\mathbf{v}+\partial_t\mathbf{v}+d_{\mathbf{x}}\mathbf{v}\cdot\mathbf{v}+\frac{1}{m}\mathbf{F}\\
&=\left(A^{-1}(t)\cdot A'(t)\right)\cdot(\mathbf{u}-\mathbf{v})-\left\{A^{-1}(t)\cdot A'(t)\right\}^T\cdot(\mathbf{u}-\mathbf{v})-2A^{-1}(t)\cdot A'(t)\cdot(\mathbf{u}-\mathbf{v})\\
&\qquad-(\mathbf{u}-\mathbf{v})\times\operatorname{curl}_{\mathbf{x}}\mathbf{v}+\partial_t\mathbf{v}+d_{\mathbf{x}}\mathbf{v}\cdot\mathbf{v}+\frac{1}{m}\mathbf{F}. \qquad (4.33)
\end{aligned}$$

On the other hand, the matrix $A^{-1}(t)\cdot A'(t)$ is antisymmetric and thus

$$\left\{A^{-1}(t)\cdot A'(t)\right\}^T=-\left(A^{-1}(t)\cdot A'(t)\right).$$

Inserting it into (4.33) we deduce

$$\frac{d\mathbf{u}}{dt}=-(\mathbf{u}-\mathbf{v})\times\operatorname{curl}_{\mathbf{x}}\mathbf{v}+\partial_t\mathbf{v}+d_{\mathbf{x}}\mathbf{v}\cdot\mathbf{v}+\frac{1}{m}\mathbf{F}. \qquad (4.34)$$

Thus again by (2.15) we finally rewrite (4.34) as

$$\frac{d\mathbf{u}}{dt}=-\mathbf{u}\times\operatorname{curl}_{\mathbf{x}}\mathbf{v}+\partial_t\mathbf{v}+\nabla_{\mathbf{x}}\left(\frac{1}{2}|\mathbf{v}|^2\right)+\frac{1}{m}\mathbf{F}. \qquad (4.35)$$

Therefore, in the coordinate system (∗), we observe a validity of Second Law of Newton in the same form as (4.14). □

Remark 4.1 Assume that in some inertial or non-inertial Cartesian coordinate system some particle with the place $\mathbf{r}(t)$ and velocity $\mathbf{u}(t) = \frac{d\mathbf{r}}{dt}(t)$ moves in the gravitational field, and all other forces, acting on the particle, except the gravitational forces are negligible. Then since, as before, by (4.12) with $\mathbf{F} = 0$ we have

$$\frac{d\mathbf{u}}{dt}(t) = -\left(\mathbf{u}(t) - \mathbf{v}\left(\mathbf{r}(t), t\right)\right) \times \mathrm{curl}_{\mathbf{x}}\mathbf{v}\left(\mathbf{r}(t), t\right) + \partial_t \mathbf{v}\left(\mathbf{r}(t), t\right) + d_{\mathbf{x}}\mathbf{v}\left(\mathbf{r}(t), t\right) \cdot \mathbf{v}\left(\mathbf{r}(t), t\right) =$$
$$\partial_t \mathbf{v}\left(\mathbf{r}(t), t\right) + d_{\mathbf{x}}\mathbf{v}\left(\mathbf{r}(t), t\right) \cdot \frac{d\mathbf{r}}{dt}(t) - d_{\mathbf{x}}\mathbf{v}\left(\mathbf{r}(t), t\right) \cdot \left(\mathbf{u}(t) - \mathbf{v}\left(\mathbf{r}(t), t\right)\right)$$
$$-\left(\mathbf{u}(t) - \mathbf{v}\left(\mathbf{r}(t), t\right)\right) \times \mathrm{curl}_{\mathbf{x}}\mathbf{v}\left(\mathbf{r}(t), t\right)$$
$$= \frac{d}{dt}\left\{\mathbf{v}\left(\mathbf{r}(t), t\right)\right\} - \left\{d_{\mathbf{x}}\mathbf{v}\left(\mathbf{r}(t), t\right)\right\}^T \cdot \left(\mathbf{u}(t) - \mathbf{v}\left(\mathbf{r}(t), t\right)\right), \quad (4.36)$$

we deduce that the vectorial quantity $\left(\frac{d\mathbf{r}}{dt}(t) - \mathbf{v}\left(\mathbf{r}(t), t\right)\right) = \left(\mathbf{u}(t) - \mathbf{v}\left(\mathbf{r}(t), t\right)\right)$ satisfies the following first-order homogenous vectorial linear ordinary differential equation:

$$\frac{d}{dt}\left\{u(t) - \mathbf{v}\left(\mathbf{r}(t), t\right)\right\} + \left\{d_{\mathbf{x}}\mathbf{v}\left(\mathbf{r}(t), t\right)\right\}^T \cdot \left\{\mathbf{u}(t) - \mathbf{v}\left(\mathbf{r}(t), t\right)\right\} = 0. \quad (4.37)$$

In particular if for some instant of time t_0 we have

$$\mathbf{u}(t_0) := \frac{d\mathbf{r}}{dt}(t_0) = \mathbf{v}\left(\mathbf{r}(t_0), t_0\right) \quad (4.38)$$

then by uniqueness theorem for ordinary differential equations, (4.37) and (4.38) together imply

$$\mathbf{u}(t) := \frac{d\mathbf{r}}{dt}(t) = \mathbf{v}\left(\mathbf{r}(t), t\right) \qquad \forall t, \quad (4.39)$$

for every instant of time. I.e. if the velocity of the particle for some initial instant of time coincides with the local vectorial gravitational potential, then it will coincide with it at any instant of time and the trajectory of motion will be tangent to the direction of the local vectorial gravitational potential.

Remark 4.2 Assume that some particle with the place $\mathbf{r} := \mathbf{r}(t)$, the velocity $\mathbf{u} := \mathbf{u}(t) = \frac{d\mathbf{r}}{dt}(t)$ and the inertial mass m moves in the outer gravitational field with the vectorial gravitational potential $\mathbf{v} := \mathbf{v}(\mathbf{x}, t)$ in the absence of non-gravitational forces. Then we can associate a Lagrangian with (4.12). Indeed, for this case, we define a Lagrangian:

$$\mathcal{L}_0\left(\frac{d\mathbf{r}}{dt}, \mathbf{r}, t\right) := \frac{m}{2}\left|\frac{d\mathbf{r}}{dt} - \mathbf{v}(\mathbf{r}, t)\right|^2. \quad (4.40)$$

This Lagrangian is invariant under the change of non-inertial Cartesian coordinate systems, given by (4.2). Moreover, we can easily deduce (see Sect. 6.1 for more details) that a trajectory $\mathbf{r}(t) : [0, T] \to \mathbb{R}^3$ is a critical point of the functional

$$I_0 = \int_0^T \mathcal{L}_0 \left(\frac{d\mathbf{r}}{dt}(t), \mathbf{r}(t), t \right) dt \tag{4.41}$$

if and only if it satisfies

$$-m\frac{d^2\mathbf{r}}{dt^2} + m\left(\frac{\partial}{\partial t}\mathbf{v}(\mathbf{r},t) + \nabla_{\mathbf{x}}\left(\frac{1}{2}|\mathbf{v}(\mathbf{r},t)|^2\right) - \frac{d\mathbf{r}}{dt} \times \text{curl}_{\mathbf{x}}\mathbf{v}(\mathbf{r},t)\right) = 0, \tag{4.42}$$

consistently with (4.12) for the case $\mathbf{F} = 0$.

Next, in order to fit the Second Law of Newton in the form (4.12) with the classical Second Law of Newton and the Newtonian Law of Gravity we consider that in inertial coordinate system (∗), at least in the first approximation, we should have

$$\begin{cases} \text{curl}_{\mathbf{x}}\mathbf{v} = 0, \\ \frac{\partial \mathbf{v}}{\partial t} + \frac{1}{2}\nabla_{\mathbf{x}}\left(|\mathbf{v}|^2\right) = -\nabla_{\mathbf{x}}\Phi, \end{cases} \tag{4.43}$$

where Φ is a scalar Newtonian gravitational potential which is assumed to be a proper scalar, which satisfies

$$\begin{cases} \Delta_{\mathbf{x}}\Phi = 4\pi G M & \forall(\mathbf{x},t), \\ \lim\limits_{|\mathbf{x}|\to+\infty} \nabla_{\mathbf{x}}\Phi(\mathbf{x},t) = 0 & \forall t, \end{cases} \tag{4.44}$$

where M is the gravitational mass density and G is the gravitational constant. Thus, since we require $\text{curl}_{\mathbf{x}}\mathbf{v} = 0$, (4.43) is equivalent to:

$$\begin{cases} \text{curl}_{\mathbf{x}}\mathbf{v} = 0, \\ \frac{\partial \mathbf{v}}{\partial t} + d_{\mathbf{x}}\mathbf{v}\cdot\mathbf{v} = -\nabla_{\mathbf{x}}\Phi. \end{cases} \tag{4.45}$$

Clearly the law (4.45) is invariant under the change of inertial coordinate system given by (4.1). Note also that, since in the system (∗) we have $\text{curl}_{\mathbf{x}}\mathbf{v} = 0$, we can write (4.43) as the following Hamilton–Jacobi-type equation:

$$\begin{cases} \mathbf{v} = \nabla_{\mathbf{x}} Z, \\ \frac{\partial Z}{\partial t} + \frac{1}{2}|\nabla_{\mathbf{x}} Z|^2 = -\Phi, \end{cases} \tag{4.46}$$

where $Z := Z(\mathbf{x}, t)$ is some scalar field. We would like to derive the law which is invariant in every non-inertial Cartesian coordinate system and is equivalent to (4.45) in every inertial coordinate system. Note that (4.45) and (4.44) imply

$$\begin{cases} \text{curl}_{\mathbf{x}}\left(\text{curl}_{\mathbf{x}}\mathbf{v}\right) = 0, \\ \text{div}_{\mathbf{x}}\left\{\frac{\partial \mathbf{v}}{\partial t} + d_{\mathbf{x}}\mathbf{v}\cdot\mathbf{v} + \frac{1}{2}\mathbf{v}\times\text{curl}_{\mathbf{x}}\mathbf{v}\right\} = -4\pi G M \end{cases} \tag{4.47}$$

that we rewrite using (2.5) as

$$\begin{cases} \mathrm{curl}_{\mathbf{x}}\left(\mathrm{curl}_{\mathbf{x}}\mathbf{v}\right) = 0, \\ \frac{\partial}{\partial t}\left(\mathrm{div}_{\mathbf{x}}\,\mathbf{v}\right) + \mathbf{v}\cdot\nabla_{\mathbf{x}}\left(\mathrm{div}_{\mathbf{x}}\,\mathbf{v}\right) + \frac{1}{4}\left|d_{\mathbf{x}}\mathbf{v} + \{d_{\mathbf{x}}\mathbf{v}\}^T\right|^2 = -4\pi G M, \end{cases} \tag{4.48}$$

or, equivalently, as

$$\begin{cases} \mathrm{curl}_{\mathbf{x}}\left(\mathrm{curl}_{\mathbf{x}}\mathbf{v}\right) = 0, \\ \frac{\partial}{\partial t}\left(\mathrm{div}_{\mathbf{x}}\,\mathbf{v}\right) + \mathrm{div}_{\mathbf{x}}\left\{\left(\mathrm{div}_{\mathbf{x}}\,\mathbf{v}\right)\mathbf{v}\right\} + \frac{1}{4}\left|d_{\mathbf{x}}\mathbf{v} + \{d_{\mathbf{x}}\mathbf{v}\}^T\right|^2 - \left(\mathrm{div}_{\mathbf{x}}\,\mathbf{v}\right)^2 = -4\pi G M. \end{cases} \tag{4.49}$$

Next observe that using Proposition 3.1 we deduce that the laws in (4.48) and (4.49) are invariant under the change of non-inertial Cartesian coordinate system, given by (4.2). So, instead of condition (4.45), we can consider the frame-independent condition (4.49) together with the frame-independent requirement that the vectorial gravitational potential $\mathbf{v}$ is an asymptotically acceptable speed-like vector field (see Definition 3.4 and Proposition 3.5). Then we can prove the following:

Proposition 4.2 *Let* $\mathbf{v} := \mathbf{v}(\mathbf{x}, t)$ *be the vectorial gravitational potential which is assumed to be an asymptotically acceptable speed-like vector field (see Definition 3.4 and Proposition 3.5) and let* $\Phi := \Phi(\mathbf{x}, t)$ *be a scalar Newtonian gravitational potential which is assumed to be a proper scalar, which satisfies*

$$\lim_{|\mathbf{x}|\to+\infty} \nabla_{\mathbf{x}}\Phi(\mathbf{x}, t) = 0 \qquad \forall t. \tag{4.50}$$

Furthermore, let $(*)$ *be some Cartesian coordinate system. Then:*

- *If the coordinate system* $(*)$ *is non-rotating with respect to the speed-like vector field* $\mathbf{v}$ *(see Definition 3.8) then in the system* $(*)$ *we have*

$$curl_{\mathbf{x}}\mathbf{v} = 0 \qquad \forall(\mathbf{x}, t), \tag{4.51}$$

  if and only if

$$curl_{\mathbf{x}}\left(curl_{\mathbf{x}}\mathbf{v}\right) = 0 \qquad \forall(\mathbf{x}, t). \tag{4.52}$$

- *If, in addition, the coordinate system* $(*)$ *is inertial with respect to the speed-like vector field* $\mathbf{v}$ *(see Definition 3.8) then in the system* $(*)$ *we have*

$$\begin{cases} curl_{\mathbf{x}}\mathbf{v} = 0 \qquad \forall(\mathbf{x}, t), \\ \frac{\partial \mathbf{v}}{\partial t} + d_{\mathbf{x}}\mathbf{v}\cdot\mathbf{v} = -\nabla_{\mathbf{x}}\Phi \qquad \forall(\mathbf{x}, t), \end{cases} \tag{4.53}$$

  if and only if

$$\begin{cases} curl_{\mathbf{x}}\,(curl_{\mathbf{x}}\mathbf{v}) = 0 \quad \forall(\mathbf{x}, t), \\ \frac{\partial}{\partial t}\,(\text{div}_{\mathbf{x}}\,\mathbf{v}) + \text{div}_{\mathbf{x}}\,\{(\text{div}_{\mathbf{x}}\,\mathbf{v})\,\mathbf{v}\} + \frac{1}{4}\left|d_{\mathbf{x}}\mathbf{v} + \{d_{\mathbf{x}}\mathbf{v}\}^T\right|^2 - (\text{div}_{\mathbf{x}}\,\mathbf{v})^2 = -\Delta_{\mathbf{x}}\Phi \\ \quad \forall(\mathbf{x}, t). \end{cases} \tag{4.54}$$

Proof Obviously, (4.51) implies (4.52). On the other hand, in the case that the coordinate system $(*)$ is non-rotating with respect to the speed-like vector field $\mathbf{v}$, by Definition 3.8 in the system $(*)$ we have

$$\lim_{|\mathbf{x}|\to+\infty} (d_{\mathbf{x}}\mathbf{v}(\mathbf{x}, t)) = 0 \qquad \forall t. \tag{4.55}$$

Therefore, if (4.52) holds, then by (4.52) and (4.55) in the coordinate system $(*)$ we deduce

$$\begin{cases} \Delta_{\mathbf{x}}\,(\text{curl}_{\mathbf{x}}\mathbf{v}(\mathbf{x}, t)) = 0 \qquad \forall\,(\mathbf{x}, t), \\ \lim_{|\mathbf{x}|\to+\infty} (\text{curl}_{\mathbf{x}}\mathbf{v}(\mathbf{x}, t)) = 0 \qquad \forall t. \end{cases} \tag{4.56}$$

However, by Liouville's theorem for the Laplace equation, (4.56) implies that in the coordinate system $(*)$ we also have

$$\text{curl}_{\mathbf{x}}\mathbf{v}(\mathbf{x}, t) = 0 \qquad \forall\,(\mathbf{x}, t). \tag{4.57}$$

So for that case (4.51) and (4.52) are indeed equivalent.

Next, if, in addition, the coordinate system $(*)$ is inertial with respect to the speed-like vector field $\mathbf{v}$, then by Definition 3.8 in the system $(*)$ we have

$$\lim_{|\mathbf{x}|\to+\infty} (d_{\mathbf{x}}\mathbf{v}(\mathbf{x}, t)) = 0 \qquad \forall t\,, \tag{4.58}$$

$$\lim_{|\mathbf{x}|\to+\infty} \mathbf{v}(\mathbf{x}, t) = \mathbf{d} \qquad \forall t\,, \tag{4.59}$$

where $\mathbf{d} \in \mathbb{R}^3$ is a constant (independent on t) vector. In particular, by (4.50), (4.58) and (4.59) in system $(*)$ we also have

$$\lim_{|\mathbf{x}|\to+\infty} \left(\frac{\partial \mathbf{v}}{\partial t} + d_{\mathbf{x}}\mathbf{v}\cdot\mathbf{v} + \nabla_{\mathbf{x}}\Phi\right) = 0 \qquad \forall t. \tag{4.60}$$

On the other hand, using (2.5) and (2.15) we deduce

$$\left(\frac{\partial}{\partial t}\,(\text{div}_{\mathbf{x}}\,\mathbf{v}) + \text{div}_{\mathbf{x}}\,\{(\text{div}_{\mathbf{x}}\,\mathbf{v})\,\mathbf{v}\} + \frac{1}{4}\left|d_{\mathbf{x}}\mathbf{v} + \{d_{\mathbf{x}}\mathbf{v}\}^T\right|^2 - (\text{div}_{\mathbf{x}}\,\mathbf{v})^2\right)$$
$$- \frac{1}{2}\mathbf{v}\cdot\text{curl}_{\mathbf{x}}\,(\text{curl}_{\mathbf{x}}\mathbf{v}) = \frac{\partial}{\partial t}\,(\text{div}_{\mathbf{x}}\,\mathbf{v}) + \mathbf{v}\cdot\nabla_{\mathbf{x}}\,(\text{div}_{\mathbf{x}}\,\mathbf{v})$$
$$+ \frac{1}{4}\left|d_{\mathbf{x}}\mathbf{v} + \{d_{\mathbf{x}}\mathbf{v}\}^T\right|^2 - \frac{1}{2}\mathbf{v}\cdot\text{curl}_{\mathbf{x}}\,(\text{curl}_{\mathbf{x}}\mathbf{v}) =$$

$$\begin{aligned}\text{div}_{\mathbf{x}}\left\{\frac{\partial \mathbf{v}}{\partial t}+d_{\mathbf{x}}\mathbf{v}\cdot\mathbf{v}\right\}+\frac{1}{4}\left|d_{\mathbf{x}}\mathbf{v}-\{d_{\mathbf{x}}\mathbf{v}\}^T\right|^2-\frac{1}{2}\mathbf{v}\cdot\text{curl}_{\mathbf{x}}\left(\text{curl}_{\mathbf{x}}\mathbf{v}\right)=\\
\text{div}_{\mathbf{x}}\left\{\frac{\partial \mathbf{v}}{\partial t}+d_{\mathbf{x}}\mathbf{v}\cdot\mathbf{v}\right\}+\frac{1}{2}\left|\text{curl}_{\mathbf{x}}\mathbf{v}\right|^2-\frac{1}{2}\mathbf{v}\cdot\text{curl}_{\mathbf{x}}\left(\text{curl}_{\mathbf{x}}\mathbf{v}\right)=\\
\text{div}_{\mathbf{x}}\left\{\frac{\partial \mathbf{v}}{\partial t}+d_{\mathbf{x}}\mathbf{v}\cdot\mathbf{v}+\frac{1}{2}\mathbf{v}\times\text{curl}_{\mathbf{x}}\mathbf{v}\right\}=\text{div}_{\mathbf{x}}\left\{\frac{\partial \mathbf{v}}{\partial t}+\nabla_{\mathbf{x}}\left(\frac{1}{2}|\mathbf{v}|^2\right)-\frac{1}{2}\mathbf{v}\times\text{curl}_{\mathbf{x}}\mathbf{v}\right\}.\end{aligned}\tag{4.61}$$

In particular, by (4.61) we deduce that (4.53) implies (4.54). On the other hand, if (4.54) satisfied, then by (4.61) and (4.57) we deduce

$$\begin{cases}\text{curl}_{\mathbf{x}}\mathbf{v}=0,\\ \text{div}_{\mathbf{x}}\left\{\frac{\partial \mathbf{v}}{\partial t}+\nabla_{\mathbf{x}}\left(\frac{1}{2}|\mathbf{v}|^2\right)\right\}=-\Delta_{\mathbf{x}}\Phi.\end{cases}\tag{4.62}$$

In particular, by (4.62) together with (2.10) we deduce

$$\begin{aligned}-\Delta_{\mathbf{x}}\left(\nabla_{\mathbf{x}}\Phi\right)=\nabla_{\mathbf{x}}\left(\text{div}_{\mathbf{x}}\left\{\frac{\partial \mathbf{v}}{\partial t}+\nabla_{\mathbf{x}}\left(\frac{1}{2}|\mathbf{v}|^2\right)\right\}\right)=\\
\nabla_{\mathbf{x}}\left(\text{div}_{\mathbf{x}}\left\{\frac{\partial \mathbf{v}}{\partial t}\right\}\right)+\Delta_{\mathbf{x}}\left(\nabla_{\mathbf{x}}\left\{\frac{1}{2}|\mathbf{v}|^2\right\}\right)=\\
\Delta_{\mathbf{x}}\left(\frac{\partial \mathbf{v}}{\partial t}+\nabla_{\mathbf{x}}\left(\frac{1}{2}|\mathbf{v}|^2\right)\right)=\Delta_{\mathbf{x}}\left(\frac{\partial \mathbf{v}}{\partial t}+d_{\mathbf{x}}\mathbf{v}\cdot\mathbf{v}\right).\end{aligned}\tag{4.63}$$

Therefore, by (4.63) together with asymptotic condition (4.60) we deduce

$$\begin{cases}\Delta_{\mathbf{x}}\left(\frac{\partial \mathbf{v}}{\partial t}+d_{\mathbf{x}}\mathbf{v}\cdot\mathbf{v}+\nabla_{\mathbf{x}}\Phi\right)=0. & \forall(\mathbf{x},t),\\ \lim\limits_{|\mathbf{x}|\to+\infty}\left(\frac{\partial \mathbf{v}}{\partial t}+d_{\mathbf{x}}\mathbf{v}\cdot\mathbf{v}+\nabla_{\mathbf{x}}\Phi\right)=0 & \forall t.\end{cases}\tag{4.64}$$

Thus, again by Liouville's theorem for the Laplace equation, (4.64) implies that in the coordinate system $(*)$ we have

$$\frac{\partial \mathbf{v}}{\partial t}+d_{\mathbf{x}}\mathbf{v}\cdot\mathbf{v}+\nabla_{\mathbf{x}}\Phi=0\qquad\forall(\mathbf{x},t).\tag{4.65}$$

Then in the system $(*)$ (4.65) and (4.57) indeed implies (4.53) and the result follows. □

Definition 4.1 We can call a Cartesian coordinate system, which is inertial with respect to the vectorial gravitational potential $\mathbf{v}$, by the usual name an inertial Cartesian coordinate system. Then, by Proposition 3.6 it is clear that every Cartesian coordinate system $(**)$ that we can get from such an inertial coordinate system $(*)$ by the usual Galilean transformations and all equivalent coordinate systems also will be inertial. Note also that by Proposition 4.2, in the case of the simplest Newtonian-like model of the gravity, given by (4.49), equalities in (4.45) are valid in every inertial Cartesian system. More generally, we call every Cartesian coordinate system, which is non-rotating with respect to the vectorial gravitational potential $\mathbf{v}$, by the usual name a non-rotating Cartesian coordinate system. Note again that by Proposition 4.2, in the case of the gravity given by (4.49), equality (4.51) holds in every non-rotating Cartesian system.

As a consequence of all mentioned above, the second law of Newton, invariant under the change of non-inertial Cartesian coordinate system, has the following form:

$$m\frac{d^2\mathbf{x}}{dt^2} = m\frac{d\mathbf{u}}{dt} = m\left(\frac{\partial \mathbf{v}}{\partial t}(\mathbf{x}, t) + \frac{1}{2}\nabla_{\mathbf{x}}\left(|\mathbf{v}|^2\right)(\mathbf{x}, t)\right) - m\mathbf{u} \times \mathrm{curl}_{\mathbf{x}}\mathbf{v}(\mathbf{x}, t) + \mathbf{F}, \tag{4.66}$$

and the first approximation of the law of gravity, which is equivalent to the Newtonian gravity and which is invariant under the change of non-inertial Cartesian coordinate system, has the following form:

$$\begin{cases} \mathrm{curl}_{\mathbf{x}}\left(\mathrm{curl}_{\mathbf{x}}\mathbf{v}\right) = 0, \\ \frac{\partial}{\partial t}\left(\mathrm{div}_{\mathbf{x}}\,\mathbf{v}\right) + \mathrm{div}_{\mathbf{x}}\left\{\left(\mathrm{div}_{\mathbf{x}}\,\mathbf{v}\right)\mathbf{v}\right\} + \frac{1}{4}\left|d_{\mathbf{x}}\mathbf{v} + \{d_{\mathbf{x}}\mathbf{v}\}^T\right|^2 - \left(\mathrm{div}_{\mathbf{x}}\,\mathbf{v}\right)^2 = -4\pi G M. \end{cases} \tag{4.67}$$

Here $\mathbf{x} := \mathbf{x}(t)$, $\mathbf{u} := \mathbf{u}(t) = \frac{d\mathbf{x}}{dt}(t)$ and m are the place, the velocity and the inertial mass of some given particle at the moment of time t; $\mathbf{v} := \mathbf{v}(\mathbf{x}, t)$ is the vectorial gravitational potential; M is the volume density of gravitational masses and $\mathbf{F}$ is the total non-gravitational force, acting on the given particle. Moreover, the vectorial gravitational potential $\mathbf{v}$ is a speed-like vector field, i.e. under the changes of inertial or non-inertial Cartesian coordinate system it behaves like a field of velocities of some continuum. Thus we could introduce the fictitious continuum medium covering all the space that we can call Aether, such that $\mathbf{v}(\mathbf{x}, t)$ is a fictitious velocity of this medium in the point $\mathbf{x}$ at the time t.

Remark 4.3 In the case of Newtonian-like gravity, given by (4.49), the quantity $Z(\mathbf{x}, t)$ in (4.46) is well defined (modulo additive constant) in every non-rotating Cartesian coordinate system and, in particular, it is well defined (modulo additive constant) in inertial coordinate systems. It can be easily checked by straightforward calculations that, if under the change of coordinate system $(*)$ to $(**)$ given by the Galilean transformation

$$\begin{cases} \mathbf{x}' = \mathbf{x} + \mathbf{w}t, \\ t' = t, \end{cases} \tag{4.68}$$

the quantity Z transforms as

$$Z'(\mathbf{x}', t') = Z(\mathbf{x}, t) + \mathbf{w}\cdot\mathbf{x} + \frac{1}{2}|\mathbf{w}|^2 t$$
$$\text{(modulo unspecified additive constant, independent on } (\mathbf{x}, t)), \tag{4.69}$$

then equalities

$$\begin{cases} \mathbf{v} + \mathbf{w} = \mathbf{v}' = \nabla_{\mathbf{x}'} Z', \\ \frac{\partial Z'}{\partial t'} + \frac{1}{2} \left|\nabla_{\mathbf{x}'} Z'\right|^2 = -\Phi' \end{cases} \tag{4.70}$$

in coordinate system $(**)$ imply the similar equalities

$$\begin{cases} \mathbf{v} = \nabla_{\mathbf{x}} Z, \\ \frac{\partial Z}{\partial t} + \frac{1}{2} \left|\nabla_{\mathbf{x}} Z\right|^2 = -\Phi \end{cases} \tag{4.71}$$

in coordinate system $(*)$, provided that $\Phi' = \Phi$ (see Lemma 3.4 for the details). More generally, by Lemma 3.2 we deduce that we can choose the quantity Z to be a speed-like field generator (see Definition 3.9) on the extended Galilean group of all inertial Cartesian coordinate systems. Moreover, again by Lemma 3.4, in that case the equations in (4.46) preserve their form in every inertial Cartesian coordinate system.

Remark 4.4 One can wonder: what should be possible values of the vectorial gravitational potential $\mathbf{v}$ in the proximity of the Earth or another massive body? In order to try to answer this question in the case of the Newtonian-type gravity, given by (4.49), consider two Cartesian coordinate systems: non-rotating system $(*)$ with the center that coincides with the center of masses of the Earth and inertial system $(**)$ related to some external cosmic bodies. Assume that the center of masses of the Earth has place $\mathbf{R}(t')$ and velocity $\mathbf{W}(t') := \frac{d\mathbf{R}}{dt'}(t')$ in the coordinate system $(**)$. Thus the change of coordinate system $(*)$ to coordinate system $(**)$ is given by

$$\begin{cases} \mathbf{x}' = \mathbf{x} + \mathbf{R}(t), \\ t' = t, \end{cases} \tag{4.72}$$

and the vectorial gravitational potential $\mathbf{v}$, being a speed like vector field, transforms as

$$\mathbf{v}' = \mathbf{v} + \mathbf{W}(t). \tag{4.73}$$

Next, since the system $(**)$ is inertial, consistently with (4.45) and (4.44) we have

$$\begin{cases} \operatorname{curl}_{\mathbf{x}'} \mathbf{v}' = 0, \\ \frac{\partial \mathbf{v}'}{\partial t'} + d_{\mathbf{x}'} \mathbf{v}' \cdot \mathbf{v}' = -\nabla_{\mathbf{x}'} \Phi_1' - \nabla_{\mathbf{x}'} \Phi_2', \end{cases} \tag{4.74}$$

with

$$\Delta_{\mathbf{x}'} \Phi_1' = 4\pi G M_1' \quad \text{and} \quad \Delta_{\mathbf{x}'} \Phi_2' = 4\pi G M_2', \tag{4.75}$$

where M_1 is the gravitational mass density of the Earth and M_2 is the gravitational mass density of all other external cosmic bodies like sun, etc. Moreover, again since the system $(**)$ is inertial, we clearly have

$$\frac{d\mathbf{W}}{\partial t}(t) = \frac{d\mathbf{W}}{\partial t'}(t') = -\nabla_{\mathbf{x}'}\Phi'_2\left(\mathbf{R}(t), t\right). \tag{4.76}$$

On the other hand, inserting (4.72) and (4.73) into (4.74) and using Proposition 3.1 we deduce

$$\mathrm{curl}_{\mathbf{x}}\mathbf{v} = 0, \tag{4.77}$$

and

$$\begin{aligned}\frac{d\mathbf{W}}{\partial t}(t) + \frac{\partial \mathbf{v}}{\partial t} + d_{\mathbf{x}}\mathbf{v}\cdot\mathbf{v} = \frac{d\mathbf{W}}{\partial t'}(t') + \frac{\partial \mathbf{v}}{\partial t'} + d_{\mathbf{x}'}\mathbf{v}\cdot\mathbf{W}(t') + d_{\mathbf{x}'}\mathbf{v}\cdot\mathbf{v} = -\nabla_{\mathbf{x}'}\Phi'_1 - \nabla_{\mathbf{x}'}\Phi'_2 \\ = -\nabla_{\mathbf{x}}\Phi_1 - \nabla_{\mathbf{x}'}\Phi'_2.\end{aligned} \tag{4.78}$$

On the other hand, the quantities $\nabla_{\mathbf{x}'}\Phi'_2 = \nabla_{\mathbf{x}}\Phi_2$ and $\frac{d\mathbf{W}}{\partial t}(t)$, being generated by the gravitational field from the far bodies, are insignificant, at the first approximation, with respect to the quantity $\nabla_{\mathbf{x}}\Phi_1$ in the scale compatible to the Earth size. Moreover, even if we wish to consider a finer approximation then we can observe that the quantity $\nabla_{\mathbf{x}'}\Phi'_2 = \nabla_{\mathbf{x}}\Phi_2$ varies insignificantly in the space variables in the scale compatible to the Earth size, and thus using (4.76) we get

$$\nabla_{\mathbf{x}}\Phi_2\left(\mathbf{x}, t\right) = \nabla_{\mathbf{x}'}\Phi'_2\left(\mathbf{x}', t'\right) \approx \nabla_{\mathbf{x}'}\Phi'_2\left(\mathbf{R}(t), t\right) = -\frac{d\mathbf{W}}{\partial t}(t). \tag{4.79}$$

Therefore, in both cases by (4.77), (4.78) and (4.79) we finally deduce

$$\begin{cases}\mathrm{curl}_{\mathbf{x}}\mathbf{v} = 0, \\ \frac{\partial \mathbf{v}}{\partial t} + \frac{1}{2}\nabla_{\mathbf{x}}\left(|\mathbf{v}|^2\right) = \frac{\partial \mathbf{v}}{\partial t} + d_{\mathbf{x}}\mathbf{v}\cdot\mathbf{v} \approx -\nabla_{\mathbf{x}}\Phi_1,\end{cases} \tag{4.80}$$

where

$$\Delta_{\mathbf{x}}\Phi_1 = 4\pi G M_1. \tag{4.81}$$

Being in the system $(*)$ which is stationary with respect to the center of the Earth we look for stationary (i.e. time independent) solutions of (4.80). Thus (4.80) implies

$$\begin{cases}\mathrm{curl}_{\mathbf{x}}\mathbf{v}(\mathbf{x}) = 0, \\ |\mathbf{v}(\mathbf{x})|^2 = -2\Phi_1(\mathbf{x}).\end{cases} \tag{4.82}$$

On the other hand, the scalar field Φ_1, being the Newtonian potential of the Earth, is radial and outside of the Earth surface it is known that $\Phi_1(\mathbf{x}) = -\frac{Gm_0}{|\mathbf{x}|}$, where m_0 is the Earth mass. Thus, since there exists a scalar field $Z_0(\mathbf{x})$ such that $\mathbf{v}(\mathbf{x}) = \nabla_{\mathbf{x}} Z_0(\mathbf{x})$ and since of symmetry considerations $Z_0(\mathbf{x}) = Z_0(|\mathbf{x}|)$ should be radial, by (4.82) we obtain

$$\left|\frac{dZ_0}{d(|\mathbf{x}|)}(|\mathbf{x}|)\right| = \sqrt{-2\Phi_1(\mathbf{x})} \tag{4.83}$$

that implies either

$$\mathbf{v}(\mathbf{x}) = \frac{\sqrt{-2\Phi_1(|\mathbf{x}|)}}{|\mathbf{x}|}\mathbf{x}, \tag{4.84}$$

or

$$\mathbf{v}(\mathbf{x}) = -\frac{\sqrt{-2\Phi_1(|\mathbf{x}|)}}{|\mathbf{x}|}\mathbf{x}. \tag{4.85}$$

In particular on the Earth surface we have

$$|\mathbf{v}| = \sqrt{\frac{2Gm_0}{r_0}}, \tag{4.86}$$

where r_0 is the Earth radius and m_0 is the Earth mass, i.e. the absolute value of the vectorial gravitational potential on the Earth surface approximately equals to the escape velocity and its direction is normal to the Earth, either downward or upward.

4.2 Variational Principle for Hamilton–Jacobi-Type Equation (4.46)

Consider the Hamilton–Jacobi-type equation (4.46) for the vectorial gravitational potential in the non-rotating coordinate system:

$$\begin{cases} \mathbf{v} = \nabla_{\mathbf{x}} Z, \\ \frac{\partial Z}{\partial t} + \frac{1}{2}\left|\nabla_{\mathbf{x}} Z\right|^2 = -\Phi, \end{cases} \tag{4.87}$$

where $Z := Z(\mathbf{x}, t)$ is some scalar field and $\Phi = \Phi(\mathbf{x}, t)$ is the scalar Newtonian gravitational potential which satisfies (4.44):

$$\Delta_{\mathbf{x}}\Phi = 4\pi G M, \tag{4.88}$$

where M is the gravitational mass density and G is the gravitational constant. Next consider the Cauchy problem for (4.87):

$$\begin{cases} \frac{\partial Z}{\partial t}(\mathbf{x}, t) + \frac{1}{2}\left|\nabla_{\mathbf{x}} Z(\mathbf{x}, t)\right|^2 = -\Phi(\mathbf{x}, t), \\ Z(\mathbf{x}, 0) = \varphi(\mathbf{x}), \end{cases} \tag{4.89}$$

where $\Phi(\mathbf{x}, t)$ and $\varphi(\mathbf{x})$ are prescribed smooth scalar functions. Furthermore, for every instant of time $t \geq 0$ consider the following variational functional defined on trajectories $\mathbf{r}(s) : [0, t] \to \mathbb{R}^3$:

$$J_{gr}(\mathbf{r}) = \varphi(\mathbf{r}(0)) + \int_0^t \left(\frac{1}{2} \left| \frac{d\mathbf{r}}{ds}(s) \right|^2 - \Phi(\mathbf{r}(s), s) \right) ds. \tag{4.90}$$

Then inserting the smooth solution Z of (4.89) into (4.90) and using the chain rule we deduce

$$J_{gr}(\mathbf{r}) = Z(\mathbf{r}(0), 0) + \int_0^t \left(\frac{1}{2} \left| \frac{d\mathbf{r}}{ds}(s) \right|^2 + \frac{\partial Z}{\partial s}(\mathbf{r}(s), s) + \frac{1}{2} |\nabla_{\mathbf{r}} Z(\mathbf{r}(s), s)|^2 \right) ds$$
$$= Z(\mathbf{r}(0), 0)$$
$$+ \int_0^t \left(\frac{1}{2} \left| \frac{d\mathbf{r}}{ds}(s) \right|^2 + \frac{d}{ds}(Z(\mathbf{r}(s), s)) - \nabla_{\mathbf{r}} Z(\mathbf{r}(s), s) \cdot \frac{d\mathbf{r}}{ds}(s) + \frac{1}{2} |\nabla_{\mathbf{r}} Z(\mathbf{r}(s), s)|^2 \right) ds$$
$$= Z(\mathbf{r}(0), 0) + \int_0^t \frac{d}{ds}(Z(\mathbf{r}(s), s))\, ds + \frac{1}{2} \int_0^t \left| \frac{d\mathbf{r}}{ds}(s) - \nabla_{\mathbf{r}} Z(\mathbf{r}(s), s) \right|^2 ds$$
$$= Z(\mathbf{r}(0), 0) + (Z(\mathbf{r}(t), t) - Z(\mathbf{r}(0), 0)) + \frac{1}{2} \int_0^t \left| \frac{d\mathbf{r}}{ds}(s) - \nabla_{\mathbf{r}} Z(\mathbf{r}(s), s) \right|^2 ds$$
$$= Z(\mathbf{r}(t), t) + \frac{1}{2} \int_0^t \left| \frac{d\mathbf{r}}{ds}(s) - \nabla_{\mathbf{r}} Z(\mathbf{r}(s), s) \right|^2 ds. \tag{4.91}$$

Thus we rewrite the definition of the functional J_{gr} in (4.90) as

$$J_{gr}(\mathbf{r}) = Z(\mathbf{r}(t), t) + \frac{1}{2} \int_0^t \left| \frac{d\mathbf{r}}{ds}(s) - \nabla_{\mathbf{r}} Z(\mathbf{r}(s), s) \right|^2 ds$$
$$= Z(\mathbf{r}(t), t) + \frac{1}{2} \int_0^t \left| \frac{d\mathbf{r}}{ds}(s) - \mathbf{v}(\mathbf{r}(s), s) \right|^2 ds, \tag{4.92}$$

where consistently with (4.87) we consider $\mathbf{v}(\mathbf{x}, t) := \nabla_{\mathbf{x}} Z(\mathbf{x}, t)$. Therefore, if for every point $\mathbf{x} \in \mathbb{R}^3$ and every instant of time $t \geq 0$ we consider the function $g := g(\mathbf{x}, t) : \mathbb{R}^3 : [0, +\infty) \to \mathbb{R}$ defined as a minimum of the following variational problem:

$$g(\mathbf{x}, t) := \min \left\{ J_{gr}(\mathbf{r}) \mid \mathbf{r}(s) : [0, t] \to \mathbb{R}^3,\ r(t) = \mathbf{x} \right\} =$$
$$\min \left\{ \varphi(\mathbf{r}(0)) + \int_0^t \left(\frac{1}{2} \left| \frac{d\mathbf{r}}{ds}(s) \right|^2 - \Phi(\mathbf{r}(s), s) \right) ds \quad | \right.$$
$$\left. \mathbf{r}(s) : [0, t] \to \mathbb{R}^3,\ \mathbf{r}(t) = \mathbf{x} \right\}, \tag{4.93}$$

then by (4.92) we deduce that

$$g(\mathbf{x}, t) = Z(\mathbf{x}, t), \tag{4.94}$$

and this minimum is achieved on the unique trajectory $\mathbf{r}(s) : [0, t] \to \mathbb{R}^3$ that satisfies the following initial value problem for an ordinary differential equation:

$$\begin{cases} \frac{d\mathbf{r}}{ds}(s) = \nabla_{\mathbf{r}} Z(\mathbf{r}(s), s) = \mathbf{v}(\mathbf{r}(s), s) & \forall s \in [0, t], \\ \mathbf{r}(t) = \mathbf{x}. \end{cases} \tag{4.95}$$

So by inserting (4.94) into (4.93) we obtain the explicit formula for the solution Z of the Cauchy problem (4.89):

$$Z(\mathbf{x}, t) = \min \left\{ \varphi(\mathbf{r}(0)) + \int_0^t \left(\frac{1}{2} \left| \frac{d\mathbf{r}}{ds}(s) \right|^2 - \Phi(\mathbf{r}(s), s) \right) ds \;\middle|\; \mathbf{r}(s) : [0, t] \to \mathbb{R}^3, \; \mathbf{r}(t) = \mathbf{x} \right\}. \tag{4.96}$$

Furthermore, there exists a unique trajectory $\mathbf{r}(s) : [0, t] \to \mathbb{R}^3$ that minimizes the right-hand side of (4.96) and moreover it satisfies

$$\begin{cases} \frac{d\mathbf{r}}{ds}(s) = \nabla_{\mathbf{r}} Z(\mathbf{r}(s), s) = \mathbf{v}(\mathbf{r}(s), s) & \forall s \in [0, t], \\ \mathbf{r}(t) = \mathbf{x}. \end{cases} \tag{4.97}$$

In particular for this trajectory we have

$$\mathbf{v}(\mathbf{x}, t) = \frac{d\mathbf{r}}{ds}(t). \tag{4.98}$$

On the other hand, the minimizer of the right-hand side of (4.96) clearly satisfies the following Euler–Lagrange equation:

$$\begin{cases} \frac{d^2\mathbf{r}}{ds^2}(s) = -\nabla_{\mathbf{r}} \Phi(\mathbf{r}(s), s) & \forall s \in [0, t], \\ \frac{d\mathbf{r}}{ds}(0) = \nabla_{\mathbf{r}} \varphi(\mathbf{r}(0)), \\ \mathbf{r}(t) = \mathbf{x}. \end{cases} \tag{4.99}$$

Equation (4.99) sometimes has an advantage with respect to (4.97) since the a priori unknown function Z does not appear in (4.99).

So, in order to find $\mathbf{v}(\mathbf{x}, t)$ we need first to solve (4.99) and then use (4.98) with the solution of (4.99) that we just found.

4.3 Genuine Gravity and Inertia

Definition 4.2 In general, both in the case of the simplest model of the Newtonian gravity and in the case of any other alternative model of gravity, we always assume that the vectorial gravitational potential $\mathbf{v} := \mathbf{v}(\mathbf{x}, t)$ is an asymptotically acceptable speed-like vector field (see Definition 3.4 and Proposition 3.5). Therefore, Corollary 3.1 implies that there exist a uniquely defined generally trivial speed-like vector field $\mathbf{k} := \mathbf{k}(\mathbf{x}, t) \in \mathbb{R}^3$ (see Definition 3.2) and a uniquely defined proper vector field $\mathbf{h} := \mathbf{h}(\mathbf{x}, t) \in \mathbb{R}^3$, so that in every Cartesian coordinate system we have

$$\lim_{|\mathbf{x}|\to+\infty} \mathbf{h}(\mathbf{x}, t) = 0 \quad \forall t \quad \text{and} \quad \lim_{|\mathbf{x}|\to+\infty} (d_{\mathbf{x}}\mathbf{h}(\mathbf{x}, t)) = 0 \quad \forall t\,, \tag{4.100}$$

and in every Cartesian coordinate system we can decompose vector field $\mathbf{v}$ as

$$\mathbf{v}(\mathbf{x}, t) = \mathbf{h}(\mathbf{x}, t) + \mathbf{k}(\mathbf{x}, t) \qquad \forall(\mathbf{x}, t). \tag{4.101}$$

Then we call $\mathbf{h}$ vectorial potential of genuine gravity and $\mathbf{k}$ vectorial potential of inertia. In particular, it is clear that, given Cartesian coordinate system $(*)$, this system will be inertial (in the frames of Definition 4.1) if and only if the vectorial potential of inertia $\mathbf{k}$ is a constant (independent on $(\mathbf{x}, t)$) in the coordinate system $(*)$.

The name of genuine gravity and inertia is clarified by the fact that since $\mathbf{k}$ is generally trivial speed-like vector field, by the definition, we can completely eliminate $\mathbf{k}$ (getting $\mathbf{k} = 0$) by the simple change of Cartesian coordinate system (it is fictitious). On the other hand, if $\mathbf{h}$ is nontrivial, since $\mathbf{h}$ is a proper vector field, we cannot make it trivial in any other coordinate system (it is genuine). Moreover, the vector of inertia $\mathbf{k}$ depends only on the coordinate system in the space and it is completely independent of the physical matter or physical fields filling this space. In contrast, vectorial potential of genuine gravity $\mathbf{h}$ depends essentially on the surrounding physical matter (in the model of the Newtonian gravity through gravitational masses).

Next, assume the model of Newtonian gravity, given by (4.49), and assume that $(*)$ is some inertial (in the frames of Definition 4.1) Cartesian coordinate system. Then by Proposition 4.2 in the system $(*)$ we have

$$\begin{cases} \operatorname{curl}_{\mathbf{x}}\mathbf{v} = 0, \\ \frac{\partial \mathbf{v}}{\partial t} + d_{\mathbf{x}}\mathbf{v} \cdot \mathbf{v} = -\nabla_{\mathbf{x}}\Phi, \end{cases} \tag{4.102}$$

where Φ is a scalar Newtonian gravitational potential which is assumed to be a proper scalar, which satisfies

$$\begin{cases} \Delta_{\mathbf{x}}\Phi = 4\pi G M \qquad \forall\,(\mathbf{x}, t), \\ \lim\limits_{|\mathbf{x}|\to+\infty} \nabla_{\mathbf{x}}\Phi(\mathbf{x}, t) = 0 \qquad \forall t, \end{cases} \tag{4.103}$$

with M being the gravitational mass density and G being the gravitational constant. Thus inserting (4.101) into (4.102) and using the fact that in the inertial system (*) the vector **k** is a constant, we deduce

$$\begin{cases} \mathrm{curl}_{\mathbf{x}}\mathbf{h} = 0, \\ \frac{\partial \mathbf{h}}{\partial t} + d_{\mathbf{x}}\mathbf{h} \cdot (\mathbf{k} + \mathbf{h}) = -\nabla_{\mathbf{x}}\Phi. \end{cases} \tag{4.104}$$

In particular, by (4.104), using the fact $\mathrm{curl}_{\mathbf{x}}\mathbf{h} = 0$ we deduce

$$\begin{cases} \mathrm{curl}_{\mathbf{x}}\mathbf{h} = 0, \\ \frac{\partial \mathbf{h}}{\partial t} - (\mathbf{k} + \mathbf{h}) \times \mathrm{curl}_{\mathbf{x}}\mathbf{h} + \{d_{\mathbf{x}}\mathbf{h}\}^T \cdot (\mathbf{k} + \mathbf{h}) = -\nabla_{\mathbf{x}}\Phi \end{cases} \tag{4.105}$$

that we finally rewrite as

$$\begin{cases} \mathrm{curl}_{\mathbf{x}}\mathbf{h} = 0, \\ \frac{\partial \mathbf{h}}{\partial t} - \mathbf{v} \times \mathrm{curl}_{\mathbf{x}}\mathbf{h} + \nabla_{\mathbf{x}}(\mathbf{v} \cdot \mathbf{h}) - \nabla_{\mathbf{x}}\left(\frac{1}{2}|\mathbf{h}|^2\right) = -\nabla_{\mathbf{x}}\Phi, \end{cases} \tag{4.106}$$

or alternatively as

$$\begin{cases} \mathrm{curl}_{\mathbf{x}}\mathbf{h} = 0, \\ \frac{\partial \mathbf{h}}{\partial t} - \mathbf{k} \times \mathrm{curl}_{\mathbf{x}}\mathbf{h} + \nabla_{\mathbf{x}}(\mathbf{k} \cdot \mathbf{h}) + \nabla_{\mathbf{x}}\left(\frac{1}{2}|\mathbf{h}|^2\right) = -\nabla_{\mathbf{x}}\Phi. \end{cases} \tag{4.107}$$

However, since **h** is a proper vector field, by Proposition 3.1 we deduce that both (4.106) and (4.107) preserve their form also in non-inertial Cartesian coordinate systems, provided Φ is a proper scalar, which is defined in every inertial or non-inertial Cartesian coordinate system by the following:

$$\begin{cases} \Delta_{\mathbf{x}}\Phi = 4\pi G M & \forall\,(\mathbf{x}, t), \\ \lim_{|\mathbf{x}|\to+\infty} \nabla_{\mathbf{x}}\Phi(\mathbf{x}, t) = 0 & \forall t. \end{cases} \tag{4.108}$$

Finally, by Proposition 3.3 vector field **k** satisfies

$$d_{\mathbf{x}}\mathbf{k} + \{d_{\mathbf{x}}\mathbf{k}\}^T = 0 \qquad \forall\,(\mathbf{x}, t) \tag{4.109}$$

in every inertial or non-inertial coordinate system. Therefore the Newtonian gravity in the form (4.49), can be rewritten in the terms of genuine gravity and inertia as (4.108) and either (4.106) or, alternatively, (4.107), that are complemented with (4.109) and (4.101). Moreover, note again that (4.108), (4.106), (4.107), (4.109), and (4.101) are invariant under the change of inertial or non-inertial Cartesian coordinate system.

Next note that, since we have $\mathrm{curl}_{\mathbf{x}}\mathbf{h} = 0$, we can write either (4.106) or (4.107) as the following Hamilton–Jacobi-type equation:

$$\begin{cases} \mathbf{h} = \nabla_{\mathbf{x}} Y, \\ \frac{\partial Y}{\partial t} + \mathbf{v} \cdot \nabla_{\mathbf{x}} Y - \frac{1}{2} |\nabla_{\mathbf{x}} Y|^2 = \frac{\partial Y}{\partial t} + \mathbf{k} \cdot \nabla_{\mathbf{x}} Y + \frac{1}{2} |\nabla_{\mathbf{x}} Y|^2 = -\Phi, \end{cases} \tag{4.110}$$

where $Y := Y(\mathbf{x}, t)$ is some scalar field. Then, since $\mathbf{h}$ is a proper vector field and Φ is proper scalar field, by Proposition 3.1 we deduce that (4.110) is invariant under the changes of inertial or non-inertial Cartesian coordinate systems, provided that we consider Y to be a proper scalar field. Note here that, in contrast to the fact that the quantity Z and Hamilton–Jacobi equation (4.46) were defined only in inertial coordinate systems, the quantity Y and Hamilton–Jacobi equation (4.110) are defined in every inertial or non-inertial coordinate system. Next in every inertial Cartesian coordinate system define the scalar field $X := X(\mathbf{x}, t)$ by

$$X(\mathbf{x}, t) = Z(\mathbf{x}, t) - Y(\mathbf{x}, t) \qquad \forall (\mathbf{x}, t)\,. \tag{4.111}$$

Then, by the first equation in (4.46) and the first equation in (4.110), together with (4.101), we deduce

$$\mathbf{k}(\mathbf{x}, t) = \nabla_{\mathbf{x}} X(\mathbf{x}, t) \qquad\qquad \forall (\mathbf{x}, t), \tag{4.112}$$

in every inertial Cartesian coordinate system. Moreover, since the quantity Y is a proper scalar and since by Remark 4.3 the quantity Z is a speed-like field generator (see Definition 3.9) on the extended Galilean group of all inertial Cartesian coordinate systems, by (4.111) we deduce that the quantity X in (4.112) is also a speed-like field generator on the extended Galilean group of all inertial Cartesian coordinate systems. In particular, under the change of coordinate system $(*)$ to $(**)$ given by the Galilean transformation

$$\begin{cases} \mathbf{x}' = \mathbf{x} + \mathbf{w}t, \\ t' = t, \end{cases} \tag{4.113}$$

the quantity X transforms as

$$X'(\mathbf{x}', t') = X(\mathbf{x}, t) + \mathbf{w} \cdot \mathbf{x} + \frac{1}{2} |\mathbf{w}|^2 t$$
$$\text{(modulo unspecified additive constant, independent on } (\mathbf{x}, t)). \tag{4.114}$$

Finally note that, since in every inertial Cartesian system $\mathbf{k}$ is a constant independent on $(\mathbf{x}, t)$, by (4.112) we deduce that the quantity $X(\mathbf{x}, t)$ is a linear function of $(\mathbf{x}, t)$.

Chapter 5
Maxwell Equations in Cartesian Coordinate Systems

5.1 Maxwell Equations Revised

We would like to make the laws of Electrodynamics in the vacuum to be invariant under the Galilean transformations. For this purpose we refer to the analogy with the Maxwell equations in a medium. It is well known that the classical Maxwell equations in a medium have the form of

$$\begin{cases} \text{curl}_{\mathbf{x}}\mathbf{H} = \frac{4\pi}{c}\mathbf{j} + \frac{1}{c}\frac{\partial \mathbf{D}}{\partial t}, \\ \text{div}_{\mathbf{x}}\,\mathbf{D} = 4\pi\rho, \\ \text{curl}_{\mathbf{x}}\mathbf{E} + \frac{1}{c}\frac{\partial \mathbf{B}}{\partial t} = 0, \\ \text{div}_{\mathbf{x}}\,\mathbf{B} = 0. \end{cases} \tag{5.1}$$

Here $\mathbf{E}$ is the electric field, $\mathbf{B}$ is the magnetic field, $\mathbf{D}$ is the electric displacement field, $\mathbf{H}$ is the $\mathbf{H}$-magnetic field, ρ is the charge density, $\mathbf{j}$ is the current density and c is the universal constant, called speed of light. It is assumed in the Classical Electrodynamics that for the vacuum we always have $\mathbf{D} = \mathbf{E}$ and $\mathbf{H} = \mathbf{B}$.

We assume that the Maxwell equations in the vacuum have the usual form (5.1), as in any other medium, however, similarly to the General Theory of Relativity we assume that the electromagnetic field is influenced by the gravitational field. Then, we assume that for a given inertial coordinate system we have $\mathbf{D}(\mathbf{x}, t) = \mathbf{E}(\mathbf{x}, t)$ and $\mathbf{H}(\mathbf{x}, t) = \mathbf{B}(\mathbf{x}, t)$ for the vacuum only in the case where the vectorial gravitational potential $\mathbf{v}(\mathbf{x}, t)$ on the point $\mathbf{x}$ at the time t equals to zero in the given coordinate system, i.e.

$$\text{If}\;\; \mathbf{v}(\mathbf{x}, t) = 0 \;\;\text{for some}\;\; (\mathbf{x}, t) \in \mathbb{R}^3 \times \mathbb{R} \;\;\text{then}\;\; \mathbf{D}(\mathbf{x}, t) = \mathbf{E}(\mathbf{x}, t) \;\;\text{and}\;\; \mathbf{H}(\mathbf{x}, t) = \mathbf{B}(\mathbf{x}, t), \tag{5.2}$$

A. Poliakovsky, *The Laws of Gravity and Electromagnetism*,
https://doi.org/10.1007/978-3-031-61407-1_5

where $\mathbf{v}(\mathbf{x}, t)$ is the same as in (4.66). In order to obtain the relations $\mathbf{D} \sim \mathbf{E}$ and $\mathbf{H} \sim \mathbf{B}$ in the general case we assume that Eq. (5.1) and the Lorentz force $\mathbf{F} := \sigma\mathbf{E} + \frac{\sigma}{c}\mathbf{u} \times \mathbf{B}$ (where σ is the charge of the test particle and $\mathbf{u}$ is its velocity) are invariant under the Galilean Transformations:

$$\begin{cases} \mathbf{x}' = \mathbf{x} + t\mathbf{w}, \\ t' = t. \end{cases} \tag{5.3}$$

First observe that if $\mathbf{u}$ is a velocity of the test particle then $\mathbf{u}' = \mathbf{u} + \mathbf{w}$. Thus, since we assumed that the Lorentz force $\mathbf{F} := \sigma\mathbf{E} + \frac{\sigma}{c}\mathbf{u} \times \mathbf{B}$ is invariant under Galilean transformation we infer

$$\sigma\mathbf{E}' + \frac{\sigma}{c}(\mathbf{u} + \mathbf{w}) \times \mathbf{B}' = \sigma\mathbf{E}' + \frac{\sigma}{c}\mathbf{u}' \times \mathbf{B}' = \mathbf{F}' = \mathbf{F} = \sigma\mathbf{E} + \frac{\sigma}{c}\mathbf{u} \times \mathbf{B}.$$

Therefore, we obtain the following identities:

$$\begin{cases} \mathbf{E}' = \mathbf{E} - \frac{1}{c}\mathbf{w} \times \mathbf{B}, \\ \mathbf{B}' = \mathbf{B}. \end{cases} \tag{5.4}$$

It is easy to check that, under transformations (5.3) and (5.4), the last two equations in (5.1) are invariant. Next observe that in the absence of currents and charges the first two equations in (5.1) for $\mathbf{H}$ and $\mathbf{D}$ will be the same as the last two for $\mathbf{E}$ and $\mathbf{B}$ if we will change the sign of the time there. Therefore, it can be assumed that the first two equations will stay invariant under the transformation:

$$\begin{cases} \mathbf{H}' = \mathbf{H} + \frac{1}{c}\mathbf{w} \times \mathbf{D}, \\ \mathbf{D}' = \mathbf{D}. \end{cases} \tag{5.5}$$

Indeed, since $\rho' = \rho$ and $\mathbf{j}' = \mathbf{j} + \rho\mathbf{w}$, it can be easily checked that under the transformations (5.3) and (5.5) the first two equations will stay invariant also in the case of charges and currents. Therefore, we obtained that all equations in (5.1) are invariant under the transformations (5.3) and

$$\begin{cases} \mathbf{D}' = \mathbf{D}, \\ \mathbf{B}' = \mathbf{B}, \\ \mathbf{E}' = \mathbf{E} - \frac{1}{c}\mathbf{w} \times \mathbf{B}, \\ \mathbf{H}' = \mathbf{H} + \frac{1}{c}\mathbf{w} \times \mathbf{D}. \end{cases} \tag{5.6}$$

Next fix some point $(\mathbf{x}_1, t_1) \in \mathbb{R}^3 \times \mathbb{R}$ and consider $\mathbf{w} := -\mathbf{v}(\mathbf{x}_1, t_1)$, where $\mathbf{v}$ is the vectorial gravitational potential. Then, since $\mathbf{v}' = \mathbf{v} + \mathbf{w}$ (speed-like vector field), we obtain that at the point $(\mathbf{x}_1', t_1')$ we have $\mathbf{v}' = 0$. Therefore, by the assumption (5.2) we must have $\mathbf{E}' = \mathbf{D}'$ and $\mathbf{H}' = \mathbf{B}'$ at this point. Plugging it into (5.6), for this point we obtain

$$\mathbf{E}(\mathbf{x}_1, t_1) + \frac{\mathbf{v}(\mathbf{x}_1, t_1)}{c} \times \mathbf{B}(\mathbf{x}_1, t_1) = \mathbf{E}(\mathbf{x}_1, t_1) - \frac{\mathbf{w}}{c} \times \mathbf{B}(\mathbf{x}_1, t_1)$$
$$= \mathbf{E}'(\mathbf{x}'_1, t'_1) = \mathbf{D}'(\mathbf{x}'_1, t'_1) = \mathbf{D}(\mathbf{x}_1, t_1), \quad (5.7)$$

and

$$\mathbf{H}(\mathbf{x}_1, t_1) - \frac{\mathbf{v}(\mathbf{x}_1, t_1)}{c} \times \mathbf{D}(\mathbf{x}_1, t_1) = \mathbf{H}(\mathbf{x}_1, t_1) + \frac{\mathbf{w}}{c} \times \mathbf{D}(\mathbf{x}_1, t_1)$$
$$= \mathbf{H}'(\mathbf{x}'_1, t'_1) = \mathbf{B}'(\mathbf{x}'_1, t'_1) = \mathbf{B}(\mathbf{x}_1, t_1). \quad (5.8)$$

Thus, since the point $(\mathbf{x}_1, t_1) \in \mathbb{R}^3 \times \mathbb{R}$ was arbitrarily chosen, by (5.7) and (5.8) we obtain the following relations

$$\begin{cases} \mathbf{E}(\mathbf{x}, t) = \mathbf{D}(\mathbf{x}, t) - \frac{1}{c}\mathbf{v}(\mathbf{x}, t) \times \mathbf{B}(\mathbf{x}, t) & \forall (\mathbf{x}, t) \in \mathbb{R}^3 \times \mathbb{R} \\ \mathbf{H}(\mathbf{x}, t) = \mathbf{B}(\mathbf{x}, t) + \frac{1}{c}\mathbf{v}(\mathbf{x}, t) \times \mathbf{D}(\mathbf{x}, t) & \forall (\mathbf{x}, t) \in \mathbb{R}^3 \times \mathbb{R}. \end{cases} \quad (5.9)$$

Plugging (5.9) into (5.1) we obtain the full system of Electrodynamics in the case of an arbitrarily vectorial gravitational potential:

$$\begin{cases} \mathrm{curl}_{\mathbf{x}}\mathbf{H} = \frac{4\pi}{c}\mathbf{j} + \frac{1}{c}\frac{\partial \mathbf{D}}{\partial t}, \\ \mathrm{div}_{\mathbf{x}}\,\mathbf{D} = 4\pi\rho, \\ \mathrm{curl}_{\mathbf{x}}\mathbf{E} + \frac{1}{c}\frac{\partial \mathbf{B}}{\partial t} = 0, \\ \mathrm{div}_{\mathbf{x}}\,\mathbf{B} = 0, \\ \mathbf{E} = \mathbf{D} - \frac{1}{c}\mathbf{v} \times \mathbf{B}, \\ \mathbf{H} = \mathbf{B} + \frac{1}{c}\mathbf{v} \times \mathbf{D}, \end{cases} \quad (5.10)$$

where $\mathbf{v}$ is the vectorial gravitational potential. It can be easily checked that system (5.10) and the Lorentz force $\mathbf{F} := \sigma(\mathbf{E} + \frac{\mathbf{u}}{c} \times \mathbf{B})$ are invariant under transformations (5.3) and (5.6). Note here that $\mathbf{D}$ and $\mathbf{B}$ are invariant under the change of inertial coordinate system. Moreover, we can write the Lorentz force as $\mathbf{F} := \sigma(\mathbf{D} + \frac{\mathbf{u}-\mathbf{v}}{c} \times \mathbf{B})$, where $(\mathbf{u} - \mathbf{v})$ is the relative velocity of the test particle with respect to the fictitious aether.

5.2 Maxwell Equations in Non-inertial Cartesian Coordinate Systems

Consider the change of certain non-inertial Cartesian coordinate system $(*)$ to another Cartesian coordinate system $(**)$:

$$\begin{cases} \mathbf{x}' = A(t) \cdot \mathbf{x} + \mathbf{z}(t), \\ t' = t, \end{cases} \tag{5.11}$$

where $A(t) \in SO(3)$ is a rotation, i.e. $A(t) \in \mathbb{R}^{3\times 3}$, $\det A(t) > 0$ and $A(t) \cdot A^T(t) = I$ (here A^T is the transpose matrix of A and I is the identity matrix). Next, assume that in coordinate system $(**)$ we observe a validity of Maxwell Equations for the vacuum in the form:

$$\begin{cases} \operatorname{curl}_{\mathbf{x}'} \mathbf{H}' = \frac{4\pi}{c}\mathbf{j}' + \frac{1}{c}\frac{\partial \mathbf{D}'}{\partial t'}, \\ \operatorname{div}_{\mathbf{x}'} \mathbf{D}' = 4\pi\rho', \\ \operatorname{curl}_{\mathbf{x}'} \mathbf{E}' + \frac{1}{c}\frac{\partial \mathbf{B}'}{\partial t'} = 0, \\ \operatorname{div}_{\mathbf{x}'} \mathbf{B}' = 0, \\ \mathbf{E}' = \mathbf{D}' - \frac{1}{c}\mathbf{v}' \times \mathbf{B}', \\ \mathbf{H}' = \mathbf{B}' + \frac{1}{c}\mathbf{v}' \times \mathbf{D}'. \end{cases} \tag{5.12}$$

Moreover, we assume that in coordinate system $(**)$ we observe a validity of expression for the Lorentz force

$$\mathbf{F}' := \sigma'\mathbf{E}' + \frac{\sigma'}{c}\mathbf{u}' \times \mathbf{B}' \tag{5.13}$$

(where σ' is the charge of the test particle and $\mathbf{u}'$ is its velocity in coordinate system $(**)$). All above happens, in particular, if coordinate system $(**)$ is inertial. Observe that if $\mathbf{F}$ is the force in coordinate system $(*)$ which corresponds to the Lorentz force $\mathbf{F}'$ in coordinate system $(**)$, then we must have $\mathbf{F}' = A(t) \cdot \mathbf{F}$. Moreover, denoting $\mathbf{w}(t) = \mathbf{z}'(t)$, we have the following obvious relations between the physical characteristics in coordinate systems $(*)$ and $(**)$:

$$\mathbf{F}' = A(t) \cdot \mathbf{F}, \tag{5.14}$$

$$\sigma' = \sigma, \tag{5.15}$$

$$\mathbf{u}' = A(t) \cdot \mathbf{u} + A'(t) \cdot \mathbf{x} + \mathbf{w}(t), \tag{5.16}$$

$$\rho' = \rho, \tag{5.17}$$

$$\mathbf{v}' = A(t) \cdot \mathbf{v} + A'(t) \cdot \mathbf{x} + \mathbf{w}(t), \tag{5.18}$$

$$\mathbf{j}' = A(t) \cdot \mathbf{j} + \rho A'(t) \cdot \mathbf{x} + \rho\mathbf{w}(t) \tag{5.19}$$

(where $A'(t)$ is a derivative of $A(t)$). We consider the fields $\mathbf{E}$ and $\mathbf{B}$ in the coordinate system $(*)$ to be defined by the expression of Lorentz force:

$$\mathbf{F} = \sigma\mathbf{E} + \frac{\sigma}{c}\mathbf{u} \times \mathbf{B}. \tag{5.20}$$

Plugging it into (5.13) and using (5.14), (5.15) and (5.16) we deduce

$$\sigma\left(\mathbf{E}'+\frac{1}{c}\left(A'(t)\cdot\mathbf{x}+\mathbf{w}(t)\right)\times\mathbf{B}'\right)+\frac{\sigma}{c}\left(A(t)\cdot\mathbf{u}\right)\times\mathbf{B}'$$
$$=\sigma\mathbf{E}'+\frac{\sigma}{c}\left(A(t)\cdot\mathbf{u}+A'(t)\cdot\mathbf{x}+\mathbf{w}(t)\right)\times\mathbf{B}'$$
$$=\sigma'\mathbf{E}'+\frac{\sigma'}{c}\mathbf{u}'\times\mathbf{B}'=\mathbf{F}'=A(t)\cdot\mathbf{F}=\sigma A(t)\cdot\mathbf{E}+\frac{\sigma}{c}A(t)\cdot(\mathbf{u}\times\mathbf{B}). \quad (5.21)$$

Thus using the trivial identity

$$A\cdot(\mathbf{a}\times\mathbf{b})=(A\cdot\mathbf{a})\times(A\cdot\mathbf{b})\quad\forall\,\mathbf{a}\in\mathbb{R}^3,\ \ \forall\,\mathbf{b}\in\mathbb{R}^3,\ \ \forall\,A\in SO(3), \quad (5.22)$$

by (5.21) we deduce

$$\sigma\left(\mathbf{E}'+\frac{1}{c}\left(A'(t)\cdot\mathbf{x}+\mathbf{w}(t)\right)\times\mathbf{B}'\right)+\frac{\sigma}{c}\left(A(t)\cdot\mathbf{u}\right)\times\mathbf{B}'$$
$$=\sigma A(t)\cdot\mathbf{E}+\frac{\sigma}{c}\left(A(t)\cdot\mathbf{u}\right)\times\left(A(t)\cdot\mathbf{B}\right). \quad (5.23)$$

Therefore, since (5.23) must be valid for arbitrary choices of $\mathbf{u}$ we deduce

$$\begin{cases}\mathbf{B}'=A(t)\cdot\mathbf{B},\\ \mathbf{E}'+\frac{1}{c}\left(A'(t)\cdot\mathbf{x}+\mathbf{w}(t)\right)\times\mathbf{B}'=A(t)\cdot\mathbf{E}.\end{cases}$$

Therefore,

$$\mathbf{E}'=A(t)\cdot\mathbf{E}-\frac{1}{c}\left(A'(t)\cdot\mathbf{x}+\mathbf{w}(t)\right)\times\mathbf{B}'$$
$$=A(t)\cdot\mathbf{E}-\frac{1}{c}\left(A'(t)\cdot\mathbf{x}+\mathbf{w}(t)\right)\times\left(A(t)\cdot\mathbf{B}\right).$$

So we obtained the following relations linking the fields $\mathbf{E}$, $\mathbf{B}$ in coordinate system $(*)$ and $\mathbf{E}'$, $\mathbf{B}'$ in coordinate system $(**)$:

$$\begin{cases}\mathbf{E}'=A(t)\cdot\mathbf{E}-\frac{1}{c}\left(A'(t)\cdot\mathbf{x}+\mathbf{w}(t)\right)\times\left(A(t)\cdot\mathbf{B}\right),\\ \mathbf{B}'=A(t)\cdot\mathbf{B}.\end{cases} \quad (5.24)$$

Next, by (5.12) in coordinate system $(**)$ we have the relations

$$\begin{cases}\mathbf{D}'=\mathbf{E}'+\frac{1}{c}\mathbf{v}'\times\mathbf{B}',\\ \mathbf{H}'=\mathbf{B}'+\frac{1}{c}\mathbf{v}'\times\mathbf{D}'.\end{cases}$$

Analogously we define $\mathbf{D}$ and $\mathbf{H}$ in coordinate system $(*)$ by the formulas

$$\begin{cases}\mathbf{D}=\mathbf{E}+\frac{1}{c}\mathbf{v}\times\mathbf{B},\\ \mathbf{H}=\mathbf{B}+\frac{1}{c}\mathbf{v}\times\mathbf{D}.\end{cases} \quad (5.25)$$

Then with the help of (5.24), (5.18) and (5.22) we deduce

$$\begin{aligned}
\mathbf{D}' =&\mathbf{E}' + \frac{1}{c}\mathbf{v}' \times \mathbf{B}' = \\
&A(t)\cdot\mathbf{E} - \frac{1}{c}\left(A'(t)\cdot\mathbf{x}+\mathbf{w}(t)\right)\times\left(A(t)\cdot\mathbf{B}\right) + \frac{1}{c}\mathbf{v}'\times\left(A(t)\cdot\mathbf{B}\right) = \\
&A(t)\cdot\mathbf{E} - \frac{1}{c}\left(A'(t)\cdot\mathbf{x}+\mathbf{w}(t)\right)\times\left(A(t)\cdot\mathbf{B}\right) \\
&\quad + \frac{1}{c}\left(A(t)\cdot\mathbf{v}+A'(t)\cdot\mathbf{x}+\mathbf{w}(t)\right)\times\left(A(t)\cdot\mathbf{B}\right) \\
&= A(t)\cdot\mathbf{E} + \frac{1}{c}\left(A(t)\cdot\mathbf{v}\right)\times\left(A(t)\cdot\mathbf{B}\right) = A(t)\cdot\left(\mathbf{E}+\frac{1}{c}\mathbf{v}\times\mathbf{B}\right) = A(t)\cdot\mathbf{D},
\end{aligned}$$

and thus

$$\begin{aligned}
\mathbf{H}' = \mathbf{B}' + \frac{1}{c}\mathbf{v}'\times\mathbf{D}' = A(t)\cdot\mathbf{B} + \frac{1}{c}\left(A(t)\cdot\mathbf{v}+A'(t)\cdot\mathbf{x}+\mathbf{w}(t)\right)\times\left(A(t)\cdot\mathbf{D}\right) = \\
A(t)\cdot\mathbf{B} + \frac{1}{c}\left(A(t)\cdot\mathbf{v}\right)\times\left(A(t)\cdot\mathbf{D}\right) + \frac{1}{c}\left(A'(t)\cdot\mathbf{x}+\mathbf{w}(t)\right)\times\left(A(t)\cdot\mathbf{D}\right) = \\
A(t)\cdot\left(\mathbf{B}+\frac{1}{c}\mathbf{v}\times\mathbf{D}\right) + \frac{1}{c}\left(A'(t)\cdot\mathbf{x}+\mathbf{w}(t)\right)\times\left(A(t)\cdot\mathbf{D}\right) \\
= A(t)\cdot\mathbf{H} + \frac{1}{c}\left(A'(t)\cdot\mathbf{x}+\mathbf{w}(t)\right)\times\left(A(t)\cdot\mathbf{D}\right).
\end{aligned}$$

That is, the following relations are valid:

$$\begin{cases}
\mathbf{D}' = A(t)\cdot\mathbf{D}, \\
\mathbf{B}' = A(t)\cdot\mathbf{B}, \\
\mathbf{E}' = A(t)\cdot\mathbf{E} - \frac{1}{c}\left(A'(t)\cdot\mathbf{x}+\mathbf{w}(t)\right)\times\left(A(t)\cdot\mathbf{B}\right), \\
\mathbf{H}' = A(t)\cdot\mathbf{H} + \frac{1}{c}\left(A'(t)\cdot\mathbf{x}+\mathbf{w}(t)\right)\times\left(A(t)\cdot\mathbf{D}\right).
\end{cases} \tag{5.26}$$

In particular vector fields **D** and **B** are proper vector fields.

Next, by (5.11) and by Proposition 3.1, for every vector field $\Gamma : \mathbb{R}^3 \times [t_0, +\infty) \to \mathbb{R}^3$ we have

$$\begin{cases}
d_{\mathbf{x}'}\Gamma = (d_{\mathbf{x}}\Gamma)\cdot A^{-1}(t) \\
\operatorname{curl}_{\mathbf{x}'}\left(A(t)\cdot\Gamma\right) = A(t)\cdot\operatorname{curl}_{\mathbf{x}}\Gamma \\
\operatorname{div}_{\mathbf{x}'}\left(A(t)\cdot\Gamma\right) = \operatorname{div}_{\mathbf{x}}\Gamma.
\end{cases} \tag{5.27}$$

Furthermore, by Proposition 3.1, for every vector field $\Gamma : \mathbb{R}^3 \times [t_0, +\infty) \to \mathbb{R}^3$ we have

$$\frac{\partial\left(A(t)\cdot\Gamma\right)}{\partial t'} - \operatorname{curl}_{\mathbf{x}'}\left(\mathbf{v}'\times\left(A(t)\cdot\Gamma\right)\right) + \left(\operatorname{div}_{\mathbf{x}'}\left(A(t)\cdot\Gamma\right)\right)\mathbf{v}'$$

$$= A(t) \cdot \left(\frac{\partial \Gamma}{\partial t} - \text{curl}_{\mathbf{x}} (\mathbf{v} \times \Gamma) + (\text{div}_{\mathbf{x}} \Gamma) \, \mathbf{v} \right). \quad (5.28)$$

On the other hand, by (5.12) we have

$$\text{curl}_{\mathbf{x}'} \mathbf{B}' - \frac{4\pi}{c} (\mathbf{j}' - \rho' \mathbf{v}') - \frac{1}{c} \left(\frac{\partial \mathbf{D}'}{\partial t'} - \text{curl}_{\mathbf{x}'} (\mathbf{v}' \times \mathbf{D}') + (\text{div}_{\mathbf{x}'} \mathbf{D}') \, \mathbf{v}' \right)$$
$$= \text{curl}_{\mathbf{x}'} \mathbf{H}' - \frac{4\pi}{c} \mathbf{j}' - \frac{1}{c} \frac{\partial \mathbf{D}'}{\partial t'} = 0 \quad (5.29)$$

and

$$\text{curl}_{\mathbf{x}'} \mathbf{D}' + \frac{1}{c} \left(\frac{\partial \mathbf{B}'}{\partial t'} - \text{curl}_{\mathbf{x}'} (\mathbf{v}' \times \mathbf{B}') + (\text{div}_{\mathbf{x}'} \mathbf{B}') \, \mathbf{v}' \right) = \text{curl}_{\mathbf{x}'} \mathbf{E}' + \frac{1}{c} \frac{\partial \mathbf{B}'}{\partial t'} = 0. \quad (5.30)$$

Thus plugging (5.29) and (5.30) into (5.28) and using (5.25), (5.17), (5.18), (5.19) and (5.27) gives

$$A(t) \cdot \left(\text{curl}_{\mathbf{x}} \mathbf{H} - \frac{4\pi}{c} \mathbf{j} - \frac{1}{c} \frac{\partial \mathbf{D}}{\partial t} + \frac{1}{c} (4\pi\rho - \text{div}_{\mathbf{x}} \mathbf{D}) \, \mathbf{v} \right) =$$
$$A(t) \cdot \left(\text{curl}_{\mathbf{x}} \mathbf{B} - \frac{4\pi}{c} (\mathbf{j} - \rho \mathbf{v}) - \frac{1}{c} \left(\frac{\partial \mathbf{D}}{\partial t} - \text{curl}_{\mathbf{x}} (\mathbf{v} \times \mathbf{D}) + (\text{div}_{\mathbf{x}} \mathbf{D}) \, \mathbf{v} \right) \right) =$$
$$\text{curl}_{\mathbf{x}'} \mathbf{B}' - \frac{4\pi}{c} (\mathbf{j}' - \rho' \mathbf{v}') - \frac{1}{c} \left(\frac{\partial \mathbf{D}'}{\partial t'} - \text{curl}_{\mathbf{x}'} (\mathbf{v}' \times \mathbf{D}') + (\text{div}_{\mathbf{x}'} \mathbf{D}) \, \mathbf{v}' \right) = 0. \quad (5.31)$$

Similarly

$$A(t) \cdot \left(\text{curl}_{\mathbf{x}} \mathbf{E} + \frac{1}{c} \frac{\partial \mathbf{B}}{\partial t} + \frac{1}{c} (\text{div}_{\mathbf{x}} \mathbf{B}) \, \mathbf{v} \right) =$$
$$A(t) \cdot \left(\text{curl}_{\mathbf{x}} \mathbf{D} + \frac{1}{c} \left(\frac{\partial \mathbf{B}}{\partial t} - \text{curl}_{\mathbf{x}} (\mathbf{v} \times \mathbf{B}) + (\text{div}_{\mathbf{x}} \mathbf{B}) \, \mathbf{v} \right) \right)$$
$$= \text{curl}_{\mathbf{x}'} \mathbf{D}' + \frac{1}{c} \left(\frac{\partial \mathbf{B}'}{\partial t'} - \text{curl}_{\mathbf{x}'} (\mathbf{v}' \times \mathbf{B}') + (\text{div}_{\mathbf{x}'} \mathbf{B}) \, \mathbf{v}' \right) = 0. \quad (5.32)$$

On the other hand, by (5.26), (5.12), (5.27) and (5.17) we obtain

$$4\pi\rho = 4\pi\rho' = \text{div}_{\mathbf{x}'} \mathbf{D}' = \text{div}_{\mathbf{x}} \mathbf{D} \quad \text{and} \quad 0 = \text{div}_{\mathbf{x}'} \mathbf{B}' = \text{div}_{\mathbf{x}} \mathbf{B}. \quad (5.33)$$

Thus plugging (5.31), (5.32) and (5.33) we obtain

$$\begin{cases} \text{curl}_{\mathbf{x}}\mathbf{H} = \frac{4\pi}{c}\mathbf{j} + \frac{1}{c}\frac{\partial \mathbf{D}}{\partial t}, \\ \text{div}_{\mathbf{x}}\,\mathbf{D} = 4\pi\rho, \\ \text{curl}_{\mathbf{x}}\mathbf{E} + \frac{1}{c}\frac{\partial \mathbf{B}}{\partial t} = 0, \\ \text{div}_{\mathbf{x}}\,\mathbf{B} = 0. \end{cases} \tag{5.34}$$

Then, plugging (5.34) into (5.25), we finally obtain that in coordinate system ($*$) the Maxwell equations have the same form as in system ($**$), i.e.

$$\begin{cases} \text{curl}_{\mathbf{x}}\mathbf{H} = \frac{4\pi}{c}\mathbf{j} + \frac{1}{c}\frac{\partial \mathbf{D}}{\partial t}, \\ \text{div}_{\mathbf{x}}\,\mathbf{D} = 4\pi\rho, \\ \text{curl}_{\mathbf{x}}\mathbf{E} + \frac{1}{c}\frac{\partial \mathbf{B}}{\partial t} = 0, \\ \text{div}_{\mathbf{x}}\,\mathbf{B} = 0, \\ \mathbf{E} = \mathbf{D} - \frac{1}{c}\mathbf{v} \times \mathbf{B}, \\ \mathbf{H} = \mathbf{B} + \frac{1}{c}\mathbf{v} \times \mathbf{D}. \end{cases} \tag{5.35}$$

Therefore, since the assumption, that coordinate system ($**$) is inertial, implies the relations of (5.12), we deduce that the expressions of Maxwell equations in the form (5.35) and of the Lorentz force in the form (5.20) are valid in every non-inertial Cartesian coordinate system. Moreover, under the change of the system, given by (5.11), the transformations of the electromagnetic fields are given by (5.26), i.e.

$$\begin{cases} \mathbf{D}' = A(t) \cdot \mathbf{D}, \\ \mathbf{B}' = A(t) \cdot \mathbf{B}, \\ \mathbf{E}' = A(t) \cdot \mathbf{E} - \frac{1}{c}\left(\frac{dA}{dt}(t) \cdot \mathbf{x} + \frac{dz}{dt}(t)\right) \times \left(A(t) \cdot \mathbf{B}\right), \\ \mathbf{H}' = A(t) \cdot \mathbf{H} + \frac{1}{c}\left(\frac{dA}{dt}(t) \cdot \mathbf{x} + \frac{dz}{dt}(t)\right) \times \left(A(t) \cdot \mathbf{D}\right). \end{cases} \tag{5.36}$$

So the laws of Electrodynamics are also invariant in non-inertial coordinate systems.

Remark 5.1 Since as already mentioned before, the direction of the local vectorial gravitational potential is normal to the Earth surface, in the frames of our model, we provide a non-relativistic explanation of the classical Michelson–Morley experiment. Indeed in this experiment the axes of the apparatus are tangent to the Earth surface and thus the null result cannot be affected by the vectorial gravitational potential. Since, the value of the local vectorial gravitational potential equals to the escape velocity, if we consider the vertical Michelson–Morley experiment, where one of the axes of the apparatus is normal to the Earth surface, then in the frames of our model the expected result should be analogous to the positive result of Aether drift with the speed equal to the escape velocity. However, regarding the vertical Michelson–Morley experiment, i.e. the modification of Michelson–Morley experiment, where at least one of the axes of the apparatus is not tangent to the Earth surface, we found only very scarce and contradictory information.

5.3 Scalar and Vectorial Electromagnetic Potentials

Consider the system of Maxwell equations in the vacuum of the form

$$\begin{cases} \mathrm{curl}_{\mathbf{x}}\mathbf{H} = \frac{4\pi}{c}\mathbf{j} + \frac{1}{c}\frac{\partial \mathbf{D}}{\partial t}, \\ \mathrm{div}_{\mathbf{x}}\,\mathbf{D} = 4\pi\rho, \\ \mathrm{curl}_{\mathbf{x}}\mathbf{E} + \frac{1}{c}\frac{\partial \mathbf{B}}{\partial t} = 0, \\ \mathrm{div}_{\mathbf{x}}\,\mathbf{B} = 0, \\ \mathbf{E} = \mathbf{D} - \frac{1}{c}\mathbf{v}\times\mathbf{B}, \\ \mathbf{H} = \mathbf{B} + \frac{1}{c}\mathbf{v}\times\mathbf{D}, \end{cases} \tag{5.37}$$

where $\mathbf{v}$ is the vectorial gravitational potential. Then by the third and the fourth equations in (5.37) we can write

$$\begin{cases} \mathbf{B} = \mathrm{curl}_{\mathbf{x}}\mathbf{A}, \\ \mathbf{E} = -\nabla_{\mathbf{x}}\Psi - \frac{1}{c}\frac{\partial \mathbf{A}}{\partial t}, \end{cases} \tag{5.38}$$

where we call Ψ and $\mathbf{A}$ the scalar and the vectorial electromagnetic potentials. Then by (5.38) and (5.37) we have

$$\begin{cases} \mathbf{B} = \mathrm{curl}_{\mathbf{x}}\mathbf{A} \\ \mathbf{E} = -\nabla_{\mathbf{x}}\Psi - \frac{1}{c}\frac{\partial \mathbf{A}}{\partial t} \\ \mathbf{D} = -\nabla_{\mathbf{x}}\Psi - \frac{1}{c}\frac{\partial \mathbf{A}}{\partial t} + \frac{1}{c}\mathbf{v}\times\mathrm{curl}_{\mathbf{x}}\mathbf{A} \\ \mathbf{H} = \mathrm{curl}_{\mathbf{x}}\mathbf{A} + \frac{1}{c}\mathbf{v}\times\left(-\nabla_{\mathbf{x}}\Psi - \frac{1}{c}\frac{\partial \mathbf{A}}{\partial t} + \frac{1}{c}\mathbf{v}\times\mathrm{curl}_{\mathbf{x}}\mathbf{A}\right). \end{cases} \tag{5.39}$$

We also define the proper scalar electromagnetic potential $\Psi_0 = \Psi_0(\mathbf{x}, t)$ by

$$\Psi_0 := \Psi - \frac{1}{c}\mathbf{A}\cdot\mathbf{v}. \tag{5.40}$$

The name "proper" will be clarified bellow. Then, by (5.39) and (5.40) we have

$$\begin{cases} \mathbf{B} = \mathrm{curl}_{\mathbf{x}}\mathbf{A} \\ \mathbf{E} = -\nabla_{\mathbf{x}}\Psi_0 - \frac{1}{c}\frac{\partial \mathbf{A}}{\partial t} - \frac{1}{c}\nabla_{\mathbf{x}}(\mathbf{A}\cdot\mathbf{v}) \\ \mathbf{D} = -\nabla_{\mathbf{x}}\Psi_0 - \frac{1}{c}\left(\frac{\partial \mathbf{A}}{\partial t} - \mathbf{v}\times\mathrm{curl}_{\mathbf{x}}\mathbf{A} + \nabla_{\mathbf{x}}(\mathbf{A}\cdot\mathbf{v})\right) \\ \mathbf{H} = \mathrm{curl}_{\mathbf{x}}\mathbf{A} - \frac{1}{c}\mathbf{v}\times\left(-\nabla_{\mathbf{x}}\Psi_0 + \frac{1}{c}\left(\frac{\partial \mathbf{A}}{\partial t} - \mathbf{v}\times\mathrm{curl}_{\mathbf{x}}\mathbf{A} + \nabla_{\mathbf{x}}(\mathbf{A}\cdot\mathbf{v})\right)\right). \end{cases} \tag{5.41}$$

The electromagnetic potentials are not uniquely defined and thus we need to choose a calibration. For definiteness we can take $\mathbf{A}$ to satisfy

$$\mathrm{div}_{\mathbf{x}}\,\mathbf{A} = 0. \tag{5.42}$$

It is clear that if $(\tilde{\Psi}, \tilde{\Psi}_0, \tilde{\mathbf{A}})$ is another choice of electromagnetic potentials with a different calibration then there exists a scalar field $w := w(\mathbf{x}, t)$ such that we have

$$\begin{cases} \tilde{\Psi} = \Psi + \frac{1}{c}\frac{\partial w}{\partial t} \\ \tilde{\mathbf{A}} = \mathbf{A} - \nabla_{\mathbf{x}} w \\ \tilde{\Psi}_0 = \Psi_0 + \frac{1}{c}\left(\frac{\partial w}{\partial t} + \mathbf{v}\cdot\nabla_{\mathbf{x}} w\right). \end{cases} \tag{5.43}$$

Next consider the change of certain non-inertial Cartesian coordinate system $(*)$ to another Cartesian coordinate system $(**)$:

$$\begin{cases} \mathbf{x}' = A(t)\cdot\mathbf{x} + \mathbf{z}(t), \\ t' = t, \end{cases} \tag{5.44}$$

where $A(t) \in SO(3)$ is a rotation, i.e. $A(t) \in \mathbb{R}^{3\times 3}$, $\det A(t) > 0$ and $A(t)\cdot A^T(t) = I$ (here A^T is the transpose matrix of A and I is the identity matrix). We are going to investigate, what are the transformations of $(\Psi, \Psi_0, \mathbf{A}) \sim (\Psi', \Psi_0', \mathbf{A}')$ under the change of coordinates, given by (5.44). Since, by (5.26) the following relations are valid

$$\begin{cases} \mathbf{D}' = A(t)\cdot\mathbf{D}, \\ \mathbf{B}' = A(t)\cdot\mathbf{B}, \\ \mathbf{E}' = A(t)\cdot\mathbf{E} - \frac{1}{c}\left(\frac{dA}{dt}(t)\cdot\mathbf{x} + \frac{d\mathbf{z}}{dt}(t)\right)\times\left(A(t)\cdot\mathbf{B}\right), \\ \mathbf{H}' = A(t)\cdot\mathbf{H} + \frac{1}{c}\left(\frac{dA}{dt}(t)\cdot\mathbf{x} + \frac{d\mathbf{z}}{dt}(t)\right)\times\left(A(t)\cdot\mathbf{D}\right), \end{cases} \tag{5.45}$$

by the second equality in (5.45), the first equality in (5.38) and (5.42) we deduce

$$\mathbf{A}' = A(t)\cdot\mathbf{A}, \tag{5.46}$$

i.e. if $\mathbf{A}$ satisfies calibration (5.42) then it is a proper vector field. On the other hand, by (5.41) we have

$$\nabla_{\mathbf{x}}\Psi_0 = -\mathbf{D} - \frac{1}{c}\left(\frac{\partial\mathbf{A}}{\partial t} - \mathbf{v}\times\mathrm{curl}_{\mathbf{x}}\mathbf{A} + \nabla_{\mathbf{x}}\left(\mathbf{A}\cdot\mathbf{v}\right)\right). \tag{5.47}$$

Thus by (5.46) and (5.45), using Proposition 3.1 we deduce that $\nabla_{\mathbf{x}}\Psi_0$ is a proper vector field, i.e.

$$\nabla_{\mathbf{x}'}\Psi_0' = A(t)\cdot\nabla_{\mathbf{x}}\Psi_0. \tag{5.48}$$

So

$$\Psi_0' = \Psi_0, \tag{5.49}$$

i.e. Ψ_0 is a is proper scalar field, invariant under the change of non-inertial Cartesian coordinate systems. This explains why we called Ψ_0 the proper scalar electromagnetic potential. Then by (5.49) and (5.40) we deduce

$$\left(\frac{1}{c}\mathbf{A}'\cdot\mathbf{v}'-\Psi'\right)=\left(\frac{1}{c}\mathbf{A}\cdot\mathbf{v}-\Psi\right). \tag{5.50}$$

Therefore, by (5.50), (5.46) and the fact that

$$\mathbf{v}'=A(t)\cdot\mathbf{v}+\frac{dA}{dt}(t)\cdot\mathbf{x}+\frac{d\mathbf{z}}{dt}(t), \tag{5.51}$$

we deduce

$$\frac{1}{c}\mathbf{A}\cdot\left(\mathbf{v}+A^T(t)\cdot\frac{dA}{dt}(t)\cdot\mathbf{x}+A^T(t)\cdot\frac{d\mathbf{z}}{dt}(t)\right)-\Psi'=\frac{1}{c}\mathbf{A}\cdot\mathbf{v}-\Psi. \tag{5.52}$$

So

$$\begin{aligned}\Psi'&=\Psi+\frac{1}{c}\mathbf{A}\cdot\left(A^T(t)\cdot\frac{dA}{dt}(t)\cdot\mathbf{x}+A^T(t)\cdot\frac{d\mathbf{z}}{dt}(t)\right)\\&=\Psi+\frac{1}{c}\left(A(t)\cdot\mathbf{A}\right)\cdot\left(\frac{dA}{dt}(t)\cdot\mathbf{x}+\frac{d\mathbf{z}}{dt}(t)\right).\end{aligned} \tag{5.53}$$

Therefore, under the change of some non-inertial Cartesian coordinate system (∗) to another Cartesian coordinate system (∗∗), given by (5.44), the electromagnetic potentials transform as

$$\begin{cases}\Psi'=\Psi+\frac{1}{c}\left(\frac{dA}{dt}(t)\cdot\mathbf{x}+\frac{d\mathbf{z}}{dt}(t)\right)\cdot\left(A(t)\cdot\mathbf{A}\right)\\ \mathbf{A}'=A(t)\cdot\mathbf{A}\\ \Psi_0':=\left(\Psi'-\frac{1}{c}\mathbf{A}'\cdot\mathbf{v}'\right)=\Psi_0:=\left(\Psi-\frac{1}{c}\mathbf{A}\cdot\mathbf{v}\right).\end{cases} \tag{5.54}$$

In particular, under the Galilean transformations (4.1) the electromagnetic potentials transform as

$$\begin{cases}\Psi'=\Psi+\frac{1}{c}\mathbf{w}\cdot\mathbf{A}\\ \mathbf{A}'=\mathbf{A}\\ \Psi_0'=\Psi_0.\end{cases} \tag{5.55}$$

In the proof of (5.54) we used equality (5.42) only for proof of equality (5.46). Thus relations (5.54) are still valid for every choice of calibration of $(\Psi,\Psi_0,\mathbf{A})$, which implies (5.46). In particular if w is a proper scalar field, i.e. $w'=w$ and if $(\tilde{\Psi},\tilde{\Psi}_0,\tilde{\mathbf{A}})$ is another choice of electromagnetic potentials defined by

$$\begin{cases}\tilde{\Psi}=\Psi+\frac{1}{c}\frac{\partial w}{\partial t}\\ \tilde{\mathbf{A}}=\mathbf{A}-\nabla_{\mathbf{x}}w\\ \tilde{\Psi}_0=\Psi_0+\frac{1}{c}\left(\frac{\partial w}{\partial t}+\mathbf{v}\cdot\nabla_{\mathbf{x}}w\right),\end{cases} \tag{5.56}$$

then, by Proposition 3.1 we have

$$\begin{cases} \tilde{\Psi}' = \tilde{\Psi} + \frac{1}{c}\left(\frac{dA}{dt}(t)\cdot \mathbf{x} + \frac{d\mathbf{z}}{dt}(t)\right)\cdot \left(A(t)\cdot \tilde{\mathbf{A}}\right) \\ \tilde{\mathbf{A}}' = A(t)\cdot \tilde{\mathbf{A}} \\ \tilde{\Psi}_0' = \tilde{\Psi}_0. \end{cases} \tag{5.57}$$

On the other hand, we always can find a proper scalar field w for calibration to illuminate $\tilde{\Psi}_0$ in (5.56). Then we have $\tilde{\Psi}_0 = 0$ and the electromagnetic fields are fully represented by the vectorial electromagnetic potential $\tilde{\mathbf{A}}$ analogously as the vectorial gravitational potential represents the gravitational field. For this case, we rewrite (5.41) as

$$\begin{cases} \tilde{\Psi}_0 = 0 \\ -\frac{1}{c}\operatorname{div}_{\mathbf{x}}\left\{\frac{\partial \tilde{\mathbf{A}}}{\partial t} - \mathbf{v}\times \operatorname{curl}_{\mathbf{x}}\tilde{\mathbf{A}} + \nabla_{\mathbf{x}}\left(\tilde{\mathbf{A}}\cdot \mathbf{v}\right)\right\} = 4\pi\rho \\ \mathbf{B} = \operatorname{curl}_{\mathbf{x}}\tilde{\mathbf{A}} \\ \mathbf{E} = -\frac{1}{c}\frac{\partial \tilde{\mathbf{A}}}{\partial t} - \frac{1}{c}\nabla_{\mathbf{x}}\left(\tilde{\mathbf{A}}\cdot \mathbf{v}\right) \\ \mathbf{D} = -\frac{1}{c}\left(\frac{\partial \tilde{\mathbf{A}}}{\partial t} - \mathbf{v}\times \operatorname{curl}_{\mathbf{x}}\tilde{\mathbf{A}} + \nabla_{\mathbf{x}}\left(\tilde{\mathbf{A}}\cdot \mathbf{v}\right)\right) \\ \mathbf{H} = \operatorname{curl}_{\mathbf{x}}\tilde{\mathbf{A}} - \frac{1}{c}\mathbf{v}\times\left(\frac{1}{c}\left(\frac{\partial \tilde{\mathbf{A}}}{\partial t} - \mathbf{v}\times \operatorname{curl}_{\mathbf{x}}\tilde{\mathbf{A}} + \nabla_{\mathbf{x}}\left(\tilde{\mathbf{A}}\cdot \mathbf{v}\right)\right)\right). \end{cases} \tag{5.58}$$

Moreover, in this case (5.57) is satisfied.

5.4 Lagrangian of the Electromagnetic Field

We would like to present a Lagrangian and a variational principle for the electromagnetic field and to obtain the Maxwell equations in the form (5.37) from this principle. Given known the charge distribution $\rho := \rho(\mathbf{x}, t)$, the current distribution $\mathbf{j} := \mathbf{j}(\mathbf{x}, t)$ and the vectorial gravitational potential $\mathbf{v} := \mathbf{v}(\mathbf{x}, t)$, consider a Lagrangian density L_1 defined by

$$L_1(\mathbf{A}, \Psi, \mathbf{x}, t) :=$$

$$\frac{1}{8\pi}\left|-\nabla_{\mathbf{x}}\Psi - \frac{1}{c}\frac{\partial \mathbf{A}}{\partial t} + \frac{1}{c}\mathbf{v}\times \operatorname{curl}_{\mathbf{x}}\mathbf{A}\right|^2 - \frac{1}{8\pi}|\operatorname{curl}_{\mathbf{x}}\mathbf{A}|^2 - \left(\rho\Psi - \frac{1}{c}\mathbf{A}\cdot\mathbf{j}\right). \tag{5.59}$$

We investigate stationary points of the functional

$$J = \int_0^T \int_{\mathbb{R}^3} L_1(\mathbf{A}, \Psi, \mathbf{x}, t)\, d\mathbf{x}dt. \tag{5.60}$$

We denote

$$\begin{cases} \mathbf{D} = -\nabla_{\mathbf{x}}\Psi - \frac{1}{c}\frac{\partial \mathbf{A}}{\partial t} + \frac{1}{c}\mathbf{v} \times \operatorname{curl}_{\mathbf{x}}\mathbf{A} \\ \mathbf{B} = \operatorname{curl}_{\mathbf{x}}\mathbf{A} \\ \mathbf{E} = -\nabla_{\mathbf{x}}\Psi - \frac{1}{c}\frac{\partial \mathbf{A}}{\partial t} = \mathbf{D} - \frac{1}{c}\mathbf{v} \times \mathbf{B} \\ \mathbf{H} = \operatorname{curl}_{\mathbf{x}}\mathbf{A} + \frac{1}{c}\mathbf{v} \times \left(-\nabla_{\mathbf{x}}\Psi - \frac{1}{c}\frac{\partial \mathbf{A}}{\partial t} + \frac{1}{c}\mathbf{v} \times \operatorname{curl}_{\mathbf{x}}\mathbf{A}\right) = \mathbf{B} + \frac{1}{c}\mathbf{v} \times \mathbf{D} \\ \Psi_0 := \Psi - \frac{1}{c}\mathbf{A} \cdot \mathbf{v}. \end{cases} \tag{5.61}$$

So we can write

$$\begin{aligned} L_1(\mathbf{A}, \Psi, \mathbf{x}, t) &:= \frac{1}{8\pi}|\mathbf{D}|^2 - \frac{1}{8\pi}|\mathbf{B}|^2 - \left(\rho\Psi - \frac{1}{c}\mathbf{A} \cdot \mathbf{j}\right) \\ &= \frac{1}{8\pi}|\mathbf{D}|^2 - \frac{1}{8\pi}|\mathbf{B}|^2 - \rho\Psi_0 + \frac{1}{c}\mathbf{A} \cdot (\mathbf{j} - \rho\mathbf{v}), \end{aligned} \tag{5.62}$$

and by (5.61) we have

$$\begin{cases} \operatorname{curl}_{\mathbf{x}}\mathbf{E} + \frac{1}{c}\frac{\partial \mathbf{B}}{\partial t} = 0 \\ \operatorname{div}_{\mathbf{x}}\mathbf{B} = 0. \end{cases} \tag{5.63}$$

Moreover by (5.59) and (2.5) we have

$$0 = \frac{\delta L_1}{\delta \Psi} = \frac{1}{4\pi}\operatorname{div}_{\mathbf{x}}\mathbf{D} - \rho, \tag{5.64}$$

and

$$\begin{aligned} 0 = \frac{\delta L_1}{\delta \mathbf{A}} &= \frac{1}{c}\mathbf{j} + \frac{1}{4\pi c}\frac{\partial \mathbf{D}}{\partial t} - \frac{1}{4\pi}\operatorname{curl}_{\mathbf{x}}\mathbf{B} - \frac{1}{4\pi c}\operatorname{curl}_{\mathbf{x}}(\mathbf{v} \times \mathbf{D}) \\ &= \frac{1}{c}\mathbf{j} + \frac{1}{4\pi c}\frac{\partial \mathbf{D}}{\partial t} - \frac{1}{4\pi}\operatorname{curl}_{\mathbf{x}}\mathbf{H}. \end{aligned} \tag{5.65}$$

So by (5.64), (5.65), (5.61) and (5.63) we obtain the Maxwell equations in the form:

$$\begin{cases} \operatorname{curl}_{\mathbf{x}}\mathbf{H} = \frac{4\pi}{c}\mathbf{j} + \frac{\partial \mathbf{D}}{\partial t} \\ \operatorname{div}_{\mathbf{x}}\mathbf{D} = 4\pi\rho \\ \operatorname{curl}_{\mathbf{x}}\mathbf{E} + \frac{1}{c}\frac{\partial \mathbf{B}}{\partial t} = 0 \\ \operatorname{div}_{\mathbf{x}}\mathbf{B} = 0 \\ \mathbf{E} = \mathbf{D} - \frac{1}{c}\mathbf{v} \times \mathbf{B} \\ \mathbf{H} = \mathbf{B} + \frac{1}{c}\mathbf{v} \times \mathbf{D}. \end{cases} \tag{5.66}$$

Note also that, using (5.62), by (5.54) and (5.45) the Lagrangian L_1 is invariant, under the change of inertial or non-inertial coordinate system, given by (5.44), i.e. for this change we have

$$L'_1\left(\mathbf{A}', \Psi', \mathbf{x}', t'\right) = L_1(\mathbf{A}, \Psi, \mathbf{x}, t). \tag{5.67}$$

5.5 Alternative Lagrangian of the Electromagnetic Field

Next we can associate an alternative Lagrangian density related to electromagnetic field. Given known the charge distribution $\rho := \rho(\mathbf{x}, t)$, the current distribution $\mathbf{j} := \mathbf{j}(\mathbf{x}, t)$ and the vectorial gravitational potential $\mathbf{v} := \mathbf{v}(\mathbf{x}, t)$, again consider $\mathbf{D}, \mathbf{B}, \mathbf{E}, \mathbf{H}$ be as in (5.61). Then, by the elementary calculus there exist proper vector fields $\mathbf{A}$, $\mathbf{C}$ and a proper scalar field ψ_0 such that

$$\begin{cases} \mathbf{D} = \mathrm{curl}_{\mathbf{x}}\mathbf{C} - \nabla_{\mathbf{x}}\psi_0 \\ \mathbf{B} = \mathrm{curl}_{\mathbf{x}}\mathbf{A}\,, \end{cases} \tag{5.68}$$

so that together with the identities

$$\begin{cases} \mathbf{E} = \mathbf{D} - \frac{1}{c}\mathbf{v} \times \mathbf{B} \\ \mathbf{H} = \mathbf{B} + \frac{1}{c}\mathbf{v} \times \mathbf{D}, \end{cases} \tag{5.69}$$

by (5.68) we deduce

$$\begin{cases} \mathbf{D} = \mathrm{curl}_{\mathbf{x}}\mathbf{C} - \nabla_{\mathbf{x}}\psi_0 \\ \mathbf{B} = \mathrm{curl}_{\mathbf{x}}\mathbf{A} \\ \mathbf{E} = (\mathrm{curl}_{\mathbf{x}}\mathbf{C} - \nabla_{\mathbf{x}}\psi_0) - \frac{1}{c}\mathbf{v} \times (\mathrm{curl}_{\mathbf{x}}\mathbf{A}) \\ \mathbf{H} = \mathrm{curl}_{\mathbf{x}}\mathbf{A} + \frac{1}{c}\mathbf{v} \times (\mathrm{curl}_{\mathbf{x}}\mathbf{C} - \nabla_{\mathbf{x}}\psi_0)\ . \end{cases} \tag{5.70}$$

Then Maxwell equations (5.66) in the form:

$$\begin{cases} \mathrm{curl}_{\mathbf{x}}\mathbf{H} = \frac{4\pi}{c}\mathbf{j} + \frac{1}{c}\frac{\partial \mathbf{D}}{\partial t} \\ \mathrm{div}_{\mathbf{x}}\,\mathbf{D} = 4\pi\rho \\ \mathrm{curl}_{\mathbf{x}}\mathbf{E} + \frac{1}{c}\frac{\partial \mathbf{B}}{\partial t} = 0 \\ \mathrm{div}_{\mathbf{x}}\,\mathbf{B} = 0 \\ \mathbf{E} = \mathbf{D} - \frac{1}{c}\mathbf{v} \times \mathbf{B} \\ \mathbf{H} = \mathbf{B} + \frac{1}{c}\mathbf{v} \times \mathbf{D}, \end{cases} \tag{5.71}$$

are equivalent to

$$\begin{cases} \mathrm{curl}_{\mathbf{x}}\left\{\mathrm{curl}_{\mathbf{x}}\mathbf{A} + \frac{1}{c}\mathbf{v} \times (\mathrm{curl}_{\mathbf{x}}\mathbf{C} - \nabla_{\mathbf{x}}\psi_0)\right\} = \frac{4\pi}{c}\mathbf{j} + \frac{1}{c}\frac{\partial}{\partial t}\left\{\mathrm{curl}_{\mathbf{x}}\mathbf{C} - \nabla_{\mathbf{x}}\psi_0\right\} \\ -\Delta_{\mathbf{x}}\psi_0 = 4\pi\rho \\ \mathrm{curl}_{\mathbf{x}}\left\{(\mathrm{curl}_{\mathbf{x}}\mathbf{C} - \nabla_{\mathbf{x}}\psi_0) - \frac{1}{c}\mathbf{v} \times (\mathrm{curl}_{\mathbf{x}}\mathbf{A})\right\} + \frac{1}{c}\frac{\partial}{\partial t}(\mathrm{curl}_{\mathbf{x}}\mathbf{A}) = 0. \end{cases} \tag{5.72}$$

that we rewrite as

$$
\begin{cases}
-\Delta_{\mathbf{x}}\psi_0 = 4\pi\rho \\
\mathrm{curl}_{\mathbf{x}}\left\{\mathrm{curl}_{\mathbf{x}}\mathbf{A}\right\} - \frac{4\pi}{c}\widetilde{\mathbf{j}} = \frac{1}{c}\frac{\partial}{\partial t}\left\{\mathrm{curl}_{\mathbf{x}}\mathbf{C}\right\} - \mathrm{curl}_{\mathbf{x}}\left\{\frac{1}{c}\mathbf{v}\times(\mathrm{curl}_{\mathbf{x}}\mathbf{C})\right\} \\
\mathrm{curl}_{\mathbf{x}}\left(\mathrm{curl}_{\mathbf{x}}\mathbf{C}\right) = -\frac{1}{c}\frac{\partial}{\partial t}\left(\mathrm{curl}_{\mathbf{x}}\mathbf{A}\right) + \mathrm{curl}_{\mathbf{x}}\left\{\frac{1}{c}\mathbf{v}\times(\mathrm{curl}_{\mathbf{x}}\mathbf{A})\right\}.
\end{cases}
\tag{5.73}
$$

where we set the reduced current:

$$
\begin{cases}
\widetilde{\mathbf{j}} := \mathbf{j} - \frac{1}{4\pi}\frac{\partial}{\partial t}\left(\nabla_{\mathbf{x}}\psi_0\right) + \frac{1}{4\pi}\mathrm{curl}_{\mathbf{x}}\left(\mathbf{v}\times\nabla_{\mathbf{x}}\psi_0\right), \\
-\Delta_{\mathbf{x}}\psi_0 = 4\pi\rho.
\end{cases}
\tag{5.74}
$$

that clearly satisfies

$$
\mathrm{div}_{\mathbf{x}}\,\widetilde{\mathbf{j}} = 0 \tag{5.75}
$$

and

$$
\widetilde{\mathbf{j}} := (\mathbf{j} - \rho\mathbf{v}) - \frac{1}{4\pi}\left(\frac{\partial}{\partial t}\left(\nabla_{\mathbf{x}}\psi_0\right) - \mathrm{curl}_{\mathbf{x}}\left(\mathbf{v}\times\nabla_{\mathbf{x}}\psi_0\right) + \left(\mathrm{div}_{\mathbf{x}}\left\{\nabla_{\mathbf{x}}\psi_0\right\}\right)\mathbf{v}\right). \tag{5.76}
$$

In particular since by (5.76) it can be easily deduced that the reduced current $\widetilde{\mathbf{j}}$ is a proper vector field, and as before, we deduce that (5.73) is invariant under the change of inertial or non-inertial Cartesian coordinate systems. Next, defining the following alternative Lagrangian density:

$$
\begin{aligned}
\tilde{L}_1\left(\mathbf{A},\mathbf{C},\mathbf{x},t\right) := {} & \frac{1}{8\pi}\left(-\nabla_{\mathbf{x}}\left(\frac{1}{c}\mathbf{v}\cdot\mathbf{A}\right) - \frac{1}{c}\frac{\partial\mathbf{A}}{\partial t} + \frac{1}{c}\mathbf{v}\times\mathrm{curl}_{\mathbf{x}}\mathbf{A}\right)\cdot\left(\mathrm{curl}_{\mathbf{x}}\mathbf{C}\right) \\
& - \frac{1}{8\pi}\left(-\nabla_{\mathbf{x}}\left(\frac{1}{c}\mathbf{v}\cdot\mathbf{C}\right) - \frac{1}{c}\frac{\partial\mathbf{C}}{\partial t} + \frac{1}{c}\mathbf{v}\times\mathrm{curl}_{\mathbf{x}}\mathbf{C}\right)\cdot\left(\mathrm{curl}_{\mathbf{x}}\mathbf{A}\right) \\
& - \frac{1}{8\pi}\left|\mathrm{curl}_{\mathbf{x}}\mathbf{A}\right|^2 - \frac{1}{8\pi}\left|\mathrm{curl}_{\mathbf{x}}\mathbf{C}\right|^2 + \frac{1}{c}\mathbf{A}\cdot\widetilde{\mathbf{j}}
\end{aligned}
\tag{5.77}
$$

and considering the functional

$$
\tilde{J} = \int_0^T\int_{\mathbb{R}^3} L_1\left(\mathbf{A},\mathbf{C},\mathbf{x},t\right)d\mathbf{x}dt, \tag{5.78}
$$

we can easily deduce that $\tilde{J}$ admits the last two equations in (5.73) as the Euler–Lagrange identities. Here ψ_0 is assumed to be fixed and given by the first equation in (5.73). Moreover, as before, we deduce that the Lagrangian density $\tilde{L}_1$ in (5.77) is also invariant under the change of inertial or non-inertial Cartesian coordinate systems.

Next denoting complex proper vector fields:

$$
\mathbf{T} := \mathbf{A} - i\mathbf{C} \quad \text{ana} \quad \overline{\mathbf{T}} := \mathbf{A} + i\mathbf{C}, \tag{5.79}
$$

we rewrite (5.77) as

$$\begin{aligned}\tilde{L}_1(\mathbf{T},\mathbf{x},t) := \frac{i}{16\pi}\left(\nabla_{\mathbf{x}}\left(\frac{1}{c}\mathbf{v}\cdot\mathbf{T}\right)+\frac{1}{c}\frac{\partial\mathbf{T}}{\partial t}-\frac{1}{c}\mathbf{v}\times\mathrm{curl}_{\mathbf{x}}\mathbf{T}\right)\cdot\left(\mathrm{curl}_{\mathbf{x}}\overline{\mathbf{T}}\right)\\ -\frac{i}{16\pi}\left(\nabla_{\mathbf{x}}\left(\frac{1}{c}\mathbf{v}\cdot\overline{\mathbf{T}}\right)+\frac{1}{c}\frac{\partial\overline{\mathbf{T}}}{\partial t}-\frac{1}{c}\mathbf{v}\times\mathrm{curl}_{\mathbf{x}}\overline{\mathbf{T}}\right)\cdot\left(\mathrm{curl}_{\mathbf{x}}\mathbf{T}\right)\\ -\frac{1}{8\pi}\left(\mathrm{curl}_{\mathbf{x}}\mathbf{T}\right)\cdot\left(\mathrm{curl}_{\mathbf{x}}\overline{\mathbf{T}}\right)+\frac{1}{2c}\left(\mathbf{T}+\overline{\mathbf{T}}\right)\cdot\tilde{\mathbf{j}}\end{aligned} \quad (5.80)$$

and the last two equations in (5.73) together as

$$\begin{aligned}&\frac{i}{c}\frac{\partial}{\partial t}\{\mathrm{curl}_{\mathbf{x}}\mathbf{T}\}-\frac{i}{c}\mathrm{curl}_{\mathbf{x}}\{\mathbf{v}\times\mathrm{curl}_{\mathbf{x}}\mathbf{T}\}\\ &=i\,\mathrm{curl}_{\mathbf{x}}\left\{\nabla_{\mathbf{x}}\left(\frac{1}{c}\mathbf{v}\cdot\mathbf{T}\right)+\frac{1}{c}\frac{\partial\mathbf{T}}{\partial t}-\frac{1}{c}\mathbf{v}\times\mathrm{curl}_{\mathbf{x}}\mathbf{T}\right)=\mathrm{curl}\{\mathrm{curl}_{\mathbf{x}}\mathbf{T}\}-\frac{4\pi}{c}\tilde{\mathbf{j}}.\end{aligned} \quad (5.81)$$

5.6 Local Gravitational Time and Maxwell Equations in a Non-rotating Coordinate System

Throughout this section considers an inertial or more generally a non-rotating Cartesian coordinate system (∗). Then, as before, in this system we have

$$\mathbf{v}(\mathbf{x},t)=\nabla_{\mathbf{x}}Z(\mathbf{x},t), \quad (5.82)$$

where $\mathbf{v}$ is the vectorial gravitational potential and Z is a scalar field. Then define a scalar field $\tau := \tau(\mathbf{x}, t)$ by the following:

$$\tau(\mathbf{x},t)=t+\frac{1}{c^2}Z(\mathbf{x},t). \quad (5.83)$$

We call the quantity $\tau(\mathbf{x}, t)$ by the name local gravitational time. The name "local" and "gravitational" is quite clear, since τ depend on the space and time variables and derived by characteristic function Z of the gravitational field. The name "time" will be clarified bellow. Note also that, using (4.69) in Remark 4.3, one can easily deduce that under the change of inertial coordinate system (∗) to (∗∗) given by the Galilean Transformation

$$\begin{cases}\mathbf{x}'=\mathbf{x}+\mathbf{w}t,\\ t'=t,\end{cases} \quad (5.84)$$

the local gravitational time τ transforms as

$$\tau'(\mathbf{x}',t') := t' + \frac{1}{c^2}Z'(\mathbf{x}',t') = \left(1+\frac{|\mathbf{w}|^2}{2c^2}\right)t + \frac{1}{c^2}Z(\mathbf{x},t) + \frac{1}{c^2}\mathbf{w}\cdot\mathbf{x}$$
$$= \tau(\mathbf{x},t) + \frac{1}{c^2}\mathbf{w}\cdot\mathbf{x} + \frac{|\mathbf{w}|^2}{2c^2}t \approx \tau(\mathbf{x},t) + \frac{1}{c^2}\mathbf{w}\cdot\mathbf{x}, \qquad (5.85)$$

where the last equality in (5.85) is valid if $\frac{|\mathbf{w}|^2}{c^2} \ll 1$. So, under (5.84) we have

$$\tau' = \tau + \frac{1}{c^2}\mathbf{w}\cdot\mathbf{x} + \frac{|\mathbf{w}|^2}{2c^2}t \approx \tau + \frac{1}{c^2}\mathbf{w}\cdot\mathbf{x}. \qquad (5.86)$$

Next consider the Maxwell equations in the vacuum of the form:

$$\begin{cases} \mathrm{curl}_{\mathbf{x}}\mathbf{H} = \frac{4\pi}{c}\mathbf{j} + \frac{1}{c}\frac{\partial \mathbf{D}}{\partial t}, \\ \mathrm{div}_{\mathbf{x}}\,\mathbf{D} = 4\pi\rho, \\ \mathrm{curl}_{\mathbf{x}}\mathbf{E} + \frac{1}{c}\frac{\partial \mathbf{B}}{\partial t} = 0, \\ \mathrm{div}_{\mathbf{x}}\,\mathbf{B} = 0, \\ \mathbf{E} = \mathbf{D} - \frac{1}{c}\mathbf{v}\times\mathbf{B}, \\ \mathbf{H} = \mathbf{B} + \frac{1}{c}\mathbf{v}\times\mathbf{D}. \end{cases} \qquad (5.87)$$

where $\mathbf{E}$ is the electric field, $\mathbf{B}$ is the magnetic field, $\mathbf{D}$ is the electric displacement field and $\mathbf{H}$ is the $\mathbf{H}$-magnetic field, $\mathbf{v} := \mathbf{v}(\mathbf{x},t) = \nabla_{\mathbf{x}} Z(\mathbf{x},t)$ is the vectorial gravitational potential, ρ is the charge density and $\mathbf{j}$ is the current density. Next consider a curvilinear change of variables given by

$$\begin{cases} t' = \tau(\mathbf{x},t) := t + \frac{Z(\mathbf{x},t)}{c^2} \\ \mathbf{x}' = \mathbf{x}. \end{cases} \qquad (5.88)$$

Then by the chain rule, for every vector field $\mathbf{F}$ we have

$$\begin{cases} \frac{\partial \mathbf{F}}{\partial t} = \frac{\partial \mathbf{F}}{\partial t'}\left(1 + \frac{1}{c^2}\frac{\partial Z}{\partial t}\right), \\ d_{\mathbf{x}}\mathbf{F} = d_{\mathbf{x}'}\mathbf{F} + \frac{1}{c^2}\frac{\partial \mathbf{F}}{\partial t'} \otimes \nabla_{\mathbf{x}} Z, \\ \mathrm{div}_{\mathbf{x}}\,\mathbf{F} = \mathrm{div}_{\mathbf{x}'}\,\mathbf{F} + \frac{1}{c^2}\frac{\partial \mathbf{F}}{\partial t'}\cdot\nabla_{\mathbf{x}} Z, \\ \mathrm{curl}_{\mathbf{x}}\mathbf{F} = \mathrm{curl}_{\mathbf{x}'}\mathbf{F} + \frac{1}{c^2}\nabla_{\mathbf{x}} Z \times \frac{\partial \mathbf{F}}{\partial t'}. \end{cases} \qquad (5.89)$$

Thus inserting (5.89) into (5.87), since $\mathbf{v} = \nabla_{\mathbf{x}} Z$, we deduce

$$\begin{cases} \operatorname{curl}_{\mathbf{x}'} \mathbf{H} + \frac{1}{c^2} \mathbf{v} \times \frac{\partial \mathbf{H}}{\partial t'} = \frac{4\pi}{c} \mathbf{j} + \frac{1}{c} \frac{\partial \mathbf{D}}{\partial t'} \left(1 + \frac{1}{c^2} \frac{\partial Z}{\partial t}\right), \\ \operatorname{div}_{\mathbf{x}'} \mathbf{D} + \frac{1}{c^2} \frac{\partial \mathbf{D}}{\partial t'} \cdot \mathbf{v} = 4\pi\rho, \\ \operatorname{curl}_{\mathbf{x}'} \mathbf{E} + \frac{1}{c^2} \mathbf{v} \times \frac{\partial \mathbf{E}}{\partial t'} + \frac{1}{c} \frac{\partial \mathbf{B}}{\partial t'} \left(1 + \frac{1}{c^2} \frac{\partial Z}{\partial t}\right) = 0, \\ \operatorname{div}_{\mathbf{x}'} \mathbf{B} + \frac{1}{c^2} \frac{\partial \mathbf{B}}{\partial t'} \cdot \mathbf{v} = 0, \\ \mathbf{E} = \mathbf{D} - \frac{1}{c} \mathbf{v} \times \mathbf{B}, \\ \mathbf{H} = \mathbf{B} + \frac{1}{c} \mathbf{v} \times \mathbf{D}. \end{cases} \tag{5.90}$$

In particular if Z is independent of t or quasistatic then we rewrite (5.90) as

$$\begin{cases} \operatorname{curl}_{\mathbf{x}'} \mathbf{H} = \frac{4\pi}{c} \mathbf{j} + \frac{1}{c} \frac{\partial}{\partial t'} \left(\mathbf{D} - \frac{1}{c} \mathbf{v} \times \mathbf{H}\right), \\ \operatorname{div}_{\mathbf{x}'} \mathbf{D} + \frac{1}{c} \mathbf{v} \cdot \operatorname{curl}_{\mathbf{x}'} \mathbf{H} = 4\pi \left(\rho + \frac{1}{c^2} \mathbf{v} \cdot \mathbf{j}\right), \\ \operatorname{curl}_{\mathbf{x}'} \mathbf{E} + \frac{1}{c} \frac{\partial}{\partial t'} \left(\mathbf{B} + \frac{1}{c} \mathbf{v} \times \mathbf{E}\right) = 0, \\ \operatorname{div}_{\mathbf{x}'} \mathbf{B} - \frac{1}{c} \mathbf{v} \cdot \operatorname{curl}_{\mathbf{x}'} \mathbf{E} = 0, \\ \mathbf{E} = \mathbf{D} - \frac{1}{c} \mathbf{v} \times \mathbf{B}, \\ \mathbf{H} = \mathbf{B} + \frac{1}{c} \mathbf{v} \times \mathbf{D}. \end{cases} \tag{5.91}$$

That is,

$$\begin{cases} \operatorname{curl}_{\mathbf{x}'} \mathbf{H} = \frac{4\pi}{c} \mathbf{j} + \frac{1}{c} \frac{\partial}{\partial t'} \left(\mathbf{D} - \frac{1}{c} \mathbf{v} \times \mathbf{H}\right), \\ \operatorname{div}_{\mathbf{x}'} \left(\mathbf{D} - \frac{1}{c} \mathbf{v} \times \mathbf{H}\right) = 4\pi \left(\rho + \frac{1}{c^2} \mathbf{v} \cdot \mathbf{j}\right), \\ \operatorname{curl}_{\mathbf{x}'} \mathbf{E} + \frac{1}{c} \frac{\partial}{\partial t'} \left(\mathbf{B} + \frac{1}{c} \mathbf{v} \times \mathbf{E}\right) = 0, \\ \operatorname{div}_{\mathbf{x}'} \left(\mathbf{B} + \frac{1}{c} \mathbf{v} \times \mathbf{E}\right) = 0, \\ \mathbf{E} = \mathbf{D} - \frac{1}{c} \mathbf{v} \times \left(\mathbf{H} - \frac{1}{c} \mathbf{v} \times \mathbf{D}\right), \\ \mathbf{H} = \mathbf{B} + \frac{1}{c} \mathbf{v} \times \left(\mathbf{E} + \frac{1}{c} \mathbf{v} \times \mathbf{B}\right) \\ \mathbf{E} = \mathbf{D} - \frac{1}{c} \mathbf{v} \times \mathbf{B}, \\ \mathbf{H} = \mathbf{B} + \frac{1}{c} \mathbf{v} \times \mathbf{D}. \end{cases} \tag{5.92}$$

In particular, denoting

$$\begin{cases} \mathbf{E}^* := \mathbf{D} - \frac{1}{c} \mathbf{v} \times \mathbf{H} = \mathbf{E} - \frac{1}{c^2} \mathbf{v} \times (\mathbf{v} \times \mathbf{D}) \\ \mathbf{H}^* := \mathbf{B} + \frac{1}{c} \mathbf{v} \times \mathbf{E} = \mathbf{H} - \frac{1}{c^2} \mathbf{v} \times (\mathbf{v} \times \mathbf{B}), \end{cases} \tag{5.93}$$

by (5.92) we rewrite the Maxwell equations in the new curvilinear coordinates in the case of time independent $\mathbf{v}$ as

$$\begin{cases} \text{curl}_{\mathbf{x}'}\mathbf{H} = \frac{4\pi}{c}\mathbf{j} + \frac{1}{c}\frac{\partial \mathbf{E}^*}{\partial t'}, \\ \text{div}_{\mathbf{x}'}\,\mathbf{E}^* = 4\pi\left(\rho + \frac{1}{c^2}\,\mathbf{v}\cdot\mathbf{j}\right), \\ \text{curl}_{\mathbf{x}'}\mathbf{E} + \frac{1}{c}\frac{\partial \mathbf{H}^*}{\partial t'} = 0, \\ \text{div}_{\mathbf{x}'}\,\mathbf{H}^* = 0, \\ \mathbf{E}^* = \mathbf{E} - \frac{1}{c^2}\mathbf{v}\times(\mathbf{v}\times\mathbf{D}) \\ \mathbf{H}^* = \mathbf{H} - \frac{1}{c^2}\mathbf{v}\times(\mathbf{v}\times\mathbf{B}) \\ \mathbf{E} = \mathbf{D} - \frac{1}{c}\,\mathbf{v}\times\mathbf{B}, \\ \mathbf{H} = \mathbf{B} + \frac{1}{c}\,\mathbf{v}\times\mathbf{D}. \end{cases} \tag{5.94}$$

In particular, in the approximation, up to the order $\left(\frac{|\mathbf{v}|}{c}\right)^2 \ll 1$ we have $\mathbf{E}^* \approx \mathbf{E}$ and $\mathbf{H}^* \approx \mathbf{H}$ and then the approximate Maxwell equations have the form:

$$\begin{cases} \text{curl}_{\mathbf{x}'}\mathbf{H} = \frac{4\pi}{c}\mathbf{j} + \frac{1}{c}\frac{\partial \mathbf{E}}{\partial t'}, \\ \text{div}_{\mathbf{x}'}\,\mathbf{E} = 4\pi\left(\rho + \frac{1}{c^2}\,\mathbf{v}\cdot\mathbf{j}\right), \\ \text{curl}_{\mathbf{x}'}\mathbf{E} + \frac{1}{c}\frac{\partial \mathbf{H}}{\partial t'} = 0, \\ \text{div}_{\mathbf{x}'}\,\mathbf{H} = 0, \\ \mathbf{E} = \mathbf{D} - \frac{1}{c}\,\mathbf{v}\times\mathbf{B}, \\ \mathbf{H} = \mathbf{B} + \frac{1}{c}\,\mathbf{v}\times\mathbf{D}. \end{cases} \tag{5.95}$$

The first four equations in (5.95) form the following system of equation:

$$\begin{cases} \text{curl}_{\mathbf{x}'}\mathbf{H} = \frac{4\pi}{c}\mathbf{j}^* + \frac{1}{c}\frac{\partial \mathbf{E}}{\partial t'}, \\ \text{div}_{\mathbf{x}'}\,\mathbf{E} = 4\pi\rho^*, \\ \text{curl}_{\mathbf{x}'}\mathbf{E} + \frac{1}{c}\frac{\partial \mathbf{H}}{\partial t'} = 0, \\ \text{div}_{\mathbf{x}'}\,\mathbf{H} = 0, \end{cases} \tag{5.96}$$

where

$$\mathbf{j}^* := \mathbf{j} \quad \text{and} \quad \rho^* := \left(\rho + \frac{1}{c^2}\,\mathbf{v}\cdot\mathbf{j}\right) \tag{5.97}$$

The system (5.96) coincides with the classical Maxwell equations of the usual Electrodynamics and is similar to (5.87) for the case $\mathbf{v} = 0$. Therefore, given known $\mathbf{v}$, ρ and $\mathbf{j}$, (5.96) could be solved as easy as the usual wave equation, for example, by the method of retarded potentials. Then backward to (5.88) change of variables could be made in order to deduce the electromagnetic fields in coordinates $(\mathbf{x}, t)$. Next note that, since we defined $t' = \tau$ all the above clarifies the name "time" of the quantity τ. Finally, we would like to note that if we have a motion of some material body with the place $\mathbf{r}(t)$ and the velocity $\mathbf{u}(t) := \frac{d\mathbf{r}}{dt}(t)$ and we associate the local gravitational time τ with this body then clearly

$$d\tau = \left(1 + \frac{1}{c^2}\mathbf{u}(t) \cdot \mathbf{v}\left(\mathbf{r}(t), t\right)\right) dt \approx dt, \tag{5.98}$$

where the last equality in (5.98) is valid if we have

$$\left(\frac{|\mathbf{v}|}{c}\right)^2 \ll 1 \quad \text{and} \quad \left(\frac{|\mathbf{u}(t)|}{c}\right)^2 \ll 1. \tag{5.99}$$

So we can use the local gravitational time τ in the approximate calculations instead of the true time t.

Chapter 6
Motion of Particles in External Gravitational-Electromagnetic Field

6.1 Lagrangian of a System of Classical Particles in a Gravitational-Electromagnetic Field

Given a system of n particles with inertial masses $m_1, \ldots, m_n$, charges $\sigma_1, \ldots, \sigma_n$, places $\mathbf{r}_1(t), \ldots, \mathbf{r}_n(t)$ and velocities $\mathbf{r}'_1(t), \ldots, \mathbf{r}'_n(t)$ in the outer gravitational field with vectorial potential $\mathbf{v}(\mathbf{x}, t)$, the outer electromagnetical fields with potentials $\mathbf{A}(\mathbf{x}, t)$ and $\Psi(\mathbf{x}, t)$ and additional conservative field with the classical scalar potential $V(\mathbf{y}_1, \ldots, \mathbf{y}_n, t)$, consider a Lagrangian:

$$L_0\left(\frac{d\mathbf{r}_1}{dt}, \ldots, \frac{d\mathbf{r}_n}{dt}, \mathbf{r}_1, \ldots, \mathbf{r}_n, t\right) :=$$
$$\sum_{j=1}^{n}\left\{\frac{m_j}{2}\left|\frac{d\mathbf{r}_j}{dt} - \mathbf{v}(\mathbf{r}_j, t)\right|^2 - \sigma_j\left(\Psi(\mathbf{r}_j, t) - \frac{1}{c}\mathbf{A}(\mathbf{r}_j, t)\cdot\frac{d\mathbf{r}_j}{dt}\right)\right\} + V(\mathbf{r}_1, \ldots, \mathbf{r}_n, t). \tag{6.1}$$

This Lagrangian is invariant under the change of inertial and non-inertial Cartesian coordinate systems. We investigate stationary points of the functional

$$J_0 = \int_0^T L_0\left(\frac{d\mathbf{r}_1}{dt}, \ldots, \frac{d\mathbf{r}_n}{dt}, , \mathbf{r}_1, \ldots, \mathbf{r}_n, t\right) dt. \tag{6.2}$$

Then for every $j = 1, \ldots, n$ we have

A. Poliakovsky, *The Laws of Gravity and Electromagnetism*,
https://doi.org/10.1007/978-3-031-61407-1_6

$$
\begin{aligned}
\frac{\delta L_0}{\delta \mathbf{r}_j} = -m_j \frac{d}{dt}\left(\frac{d\mathbf{r}_j}{dt} - \mathbf{v}(\mathbf{r}_j,t)\right) - \frac{\sigma_j}{c}\frac{d}{dt}\left(\mathbf{A}(\mathbf{r}_j,t)\right) \\
- m_j \left\{\nabla_{\mathbf{x}}\mathbf{v}(\mathbf{r}_j,t)\right\}^T \cdot \left(\frac{d\mathbf{r}_j}{dt} - \mathbf{v}(\mathbf{r}_j,t)\right) \\
- \sigma_j \left(\nabla_{\mathbf{x}}\Psi(\mathbf{r}_j,t) - \frac{1}{c}\left\{d_{\mathbf{x}}\mathbf{A}(\mathbf{r}_j,t)\right\}^T \cdot \frac{d\mathbf{r}_j}{dt}\right) + \nabla_{\mathbf{y}_j} V(\mathbf{r}_1,\dots,\mathbf{r}_n,t) = \\
- m_j \frac{d^2\mathbf{r}_j}{dt^2} + m_j\left(\frac{\partial}{\partial t}\mathbf{v}(\mathbf{r}_j,t) + \nabla_{\mathbf{x}}\left(\frac{1}{2}\left|\mathbf{v}(\mathbf{r}_j,t)\right|^2\right) - \frac{1}{c}\frac{d\mathbf{r}_j}{dt} \times \mathrm{curl}_{\mathbf{x}}\mathbf{v}(\mathbf{r}_j,t)\right) \\
+ \sigma_j\left(-\nabla_{\mathbf{x}}\Psi(\mathbf{r}_j,t) - \frac{1}{c}\frac{\partial}{\partial t}\left(\mathbf{A}(\mathbf{r}_j,t)\right) + \frac{1}{c}\frac{d\mathbf{r}_j}{dt} \times \mathrm{curl}_{\mathbf{x}}\mathbf{A}(\mathbf{r}_j,t)\right) \\
+ \nabla_{\mathbf{y}_j} V(\mathbf{r}_1,\dots,\mathbf{r}_n,t) = 0. \quad (6.3)
\end{aligned}
$$

So denoting

$$
\begin{cases}
\mathbf{E} = -\nabla_{\mathbf{x}}\Psi - \frac{1}{c}\frac{\partial \mathbf{A}}{\partial t} \\
\mathbf{B} = \mathrm{curl}_{\mathbf{x}}\mathbf{A}
\end{cases}
\quad (6.4)
$$

we rewrite (6.3) as

$$
\begin{aligned}
m_j\frac{d^2\mathbf{r}_j}{dt^2} = m_j\left(\frac{\partial}{\partial t}\mathbf{v}(\mathbf{r}_j,t) + \nabla_{\mathbf{x}}\left(\frac{1}{2}\left|\mathbf{v}(\mathbf{r}_j,t)\right|^2\right) - \frac{d\mathbf{r}_j}{dt} \times \mathrm{curl}_{\mathbf{x}}\mathbf{v}(\mathbf{r}_j,t)\right) \\
+ \sigma_j\mathbf{E}(\mathbf{r}_j,t) + \frac{\sigma_j}{c}\frac{d\mathbf{r}_j}{dt} \times \mathbf{B}(\mathbf{r}_j,t) + \nabla_{\mathbf{y}_j} V(\mathbf{r}_1,\dots,\mathbf{r}_n,t) \\
= \sigma_j\mathbf{E}(\mathbf{r}_j,t) + \frac{\sigma_j}{c}\frac{d\mathbf{r}_j}{dt} \times \mathbf{B}(\mathbf{r}_j,t) + \nabla_{\mathbf{y}_j} V(\mathbf{r}_1,\dots,\mathbf{r}_n,t) + \\
m_j\left(\frac{\partial}{\partial t}\mathbf{v}(\mathbf{r}_j,t) + d_{\mathbf{x}}\mathbf{v}(\mathbf{r}_j,t)\cdot\mathbf{v}(\mathbf{r}_j,t) - \left(\frac{d\mathbf{r}_j}{dt} - \mathbf{v}(\mathbf{r}_j,t)\right) \times \mathrm{curl}_{\mathbf{x}}\mathbf{v}(\mathbf{r}_j,t)\right).
\end{aligned}
\quad (6.5)
$$

So for each particle we get the second law of Newton, consistent with (4.11), including the gravitational and the Lorentz force.

Next, assume that our coordinate system is inertial. Then since by (4.46) we have the following Hamilton–Jacobi type equation

$$
\begin{cases}
\mathbf{v} = \nabla_{\mathbf{x}} Z, \\
\frac{\partial Z}{\partial t} + \frac{1}{2}\left|\nabla_{\mathbf{x}} Z\right|^2 = -\Phi,
\end{cases}
\quad (6.6)
$$

where $Z := Z(\mathbf{x}, t)$ is some scalar field and Φ is the Newtonian gravitational potential, using (6.1) we deduce

$$L_0\left(\frac{d\mathbf{r}_1}{dt},\dots,\frac{d\mathbf{r}_n}{dt},\mathbf{r}_1,\dots,\mathbf{r}_n,t\right)=$$
$$\sum_{j=1}^{n}\left\{\frac{m_j}{2}\left|\frac{d\mathbf{r}_j}{dt}-\nabla_{\mathbf{x}}Z(\mathbf{r}_j,t)\right|^2-\sigma_j\left(\Psi(\mathbf{r}_j,t)-\frac{1}{c}\mathbf{A}(\mathbf{r}_j,t)\cdot\frac{d\mathbf{r}_j}{dt}\right)\right\}+V(\mathbf{r}_1,\dots,\mathbf{r}_n,t)$$
$$=\sum_{j=1}^{n}m_j\left(\frac{1}{2}\left|\nabla_{\mathbf{x}}Z(\mathbf{r}_j,t)\right|^2-\nabla_{\mathbf{x}}Z(\mathbf{r}_j,t)\cdot\frac{d\mathbf{r}_j}{dt}\right)$$
$$+\sum_{j=1}^{n}\left\{\frac{m_j}{2}\left|\frac{d\mathbf{r}_j}{dt}\right|^2-\sigma_j\left(\Psi(\mathbf{r}_j,t)-\frac{1}{c}\mathbf{A}(\mathbf{r}_j,t)\cdot\frac{d\mathbf{r}_j}{dt}\right)\right\}+V(\mathbf{r}_1,\dots,\mathbf{r}_n,t)$$
$$=\sum_{j=1}^{n}m_j\left(\frac{\partial Z}{\partial t}(\mathbf{r}_j,t)+\frac{1}{2}\left|\nabla_{\mathbf{x}}Z(\mathbf{r}_j,t)\right|^2\right)-\sum_{j=1}^{n}m_j\frac{d}{dt}\left\{Z\left(\mathbf{r}_j(t),t\right)\right\}$$
$$+\sum_{j=1}^{n}\left\{\frac{m_j}{2}\left|\frac{d\mathbf{r}_j}{dt}\right|^2-\sigma_j\left(\Psi(\mathbf{r}_j,t)-\frac{1}{c}\mathbf{A}(\mathbf{r}_j,t)\cdot\frac{d\mathbf{r}_j}{dt}\right)\right\}+V(\mathbf{r}_1,\dots,\mathbf{r}_n,t)=$$
$$\sum_{j=1}^{n}\left\{\frac{m_j}{2}\left|\frac{d\mathbf{r}_j}{dt}\right|^2-m_j\Phi(\mathbf{r}_j,t)-\sigma_j\left(\Psi(\mathbf{r}_j,t)-\frac{1}{c}\mathbf{A}(\mathbf{r}_j,t)\cdot\frac{d\mathbf{r}_j}{dt}\right)\right\}+V(\mathbf{r}_1,\dots,\mathbf{r}_n,t)$$
$$-\frac{d}{dt}\left\{\sum_{j=1}^{n}m_jZ\left(\mathbf{r}_j(t),t\right)\right\}. \tag{6.7}$$

So we rewrite (6.1) as

$$L_0\left(\frac{d\mathbf{r}_1}{dt},\dots,\frac{d\mathbf{r}_n}{dt},\mathbf{r}_1,\dots,\mathbf{r}_n,t\right)=$$
$$L_0'\left(\frac{d\mathbf{r}_1}{dt},\dots,\frac{d\mathbf{r}_n}{dt},\mathbf{r}_1,\dots,\mathbf{r}_n,t\right)-\frac{d}{dt}\left\{\sum_{j=1}^{n}m_jZ\left(\mathbf{r}_j(t),t\right)\right\}, \tag{6.8}$$

where

$$L_0'\left(\frac{d\mathbf{r}_1}{dt},\dots,\frac{d\mathbf{r}_n}{dt},\mathbf{r}_1,\dots,\mathbf{r}_n,t\right):=$$
$$\sum_{j=1}^{n}\left\{\frac{m_j}{2}\left|\frac{d\mathbf{r}_j}{dt}\right|^2-m_j\Phi(\mathbf{r}_j,t)-\sigma_j\left(\Psi(\mathbf{r}_j,t)-\frac{1}{c}\mathbf{A}(\mathbf{r}_j,t)\cdot\frac{d\mathbf{r}_j}{dt}\right)\right\}$$
$$+V(\mathbf{r}_1,\dots,\mathbf{r}_n,t). \tag{6.9}$$

Note that in the given inertial coordinate system L_0' coincides with the classical Lagrangian of motion in the gravitational and electromagnetic fields. Moreover, we rewrite (6.2) as

$$J_0 = \int_0^T L_0'\left(\frac{d\mathbf{r}_1}{dt}, \ldots, \frac{d\mathbf{r}_n}{dt}, \mathbf{r}_1, \ldots, \mathbf{r}_n, t\right) dt + \sum_{j=1}^n m_j \left(Z\left(\mathbf{r}_j(0), 0\right) - Z\left(\mathbf{r}_j(T), T\right)\right). \tag{6.10}$$

Thus the stationary points of the functional J_0 will satisfy the same Euler–Lagrange equations as the stationary points of the functional

$$J_0' = \int_0^T L_0'\left(\frac{d\mathbf{r}_1}{dt}, \ldots, \frac{d\mathbf{r}_n}{dt}, \mathbf{r}_1, \ldots, \mathbf{r}_n, t\right) dt, \tag{6.11}$$

provided that the beginning and the ending points of trajectories $\mathbf{r}_j(t)$ are fixed.

6.2 Hamiltonian of the Motion of a Finite System of Classical Particles in an Outer Gravitational-Electromagnetic Field

For every $j = 1, \ldots, n$ define the generalized momentum of the particle m_j by

$$\mathbf{P}_j := \nabla_{\mathbf{r}_j'} L_0\left(\mathbf{r}_1', \ldots, \mathbf{r}_n', \mathbf{r}_1, \ldots, \mathbf{r}_n, t\right) = m_j \frac{d\mathbf{r}_j}{dt} - m_j \mathbf{v}(\mathbf{r}_j, t) + \frac{\sigma_j}{c}\mathbf{A}(\mathbf{r}_j, t), \tag{6.12}$$

where L_0 is given by (6.1). Then

$$\frac{d\mathbf{r}_j}{dt} = \frac{1}{m_j}\mathbf{P}_j + \mathbf{v}(\mathbf{r}_j, t) - \frac{\sigma_j}{m_j c}\mathbf{A}(\mathbf{r}_j, t). \tag{6.13}$$

Thus if we consider a Hamiltonian

$$H_0\left(\mathbf{P}_1, \ldots, \mathbf{P}_n, \mathbf{r}_1, \ldots, \mathbf{r}_n, t\right) := \sum_{j=1}^n \mathbf{P}_j \cdot \frac{d\mathbf{r}_j}{dt} - L_0\left(\frac{d\mathbf{r}_1}{dt}, \ldots, \frac{d\mathbf{r}_n}{dt}, \mathbf{r}_1, \ldots, \mathbf{r}_n, t\right) \tag{6.14}$$

then by (6.1), (6.14) and (6.13) we have

$$H_0\left(\mathbf{P}_1,\ldots,\mathbf{P}_n,\mathbf{r}_1,\ldots,\mathbf{r}_n,t\right)=-V\left(\mathbf{r}_1,\ldots,\mathbf{r}_n,t\right)+\sum_{j=1}^{n}\mathbf{P}_j\cdot\frac{d\mathbf{r}_j}{dt}$$

$$-\sum_{j=1}^{n}\left(\frac{m_j}{2}\left|\frac{d\mathbf{r}_j}{dt}-\mathbf{v}(\mathbf{r}_j,t)\right|^2-\sigma_j\left(\Psi(\mathbf{r}_j,t)-\frac{1}{c}\mathbf{A}(\mathbf{r}_j,t)\cdot\frac{d\mathbf{r}_j}{dt}\right)\right)=$$

$$\sum_{j=1}^{n}\mathbf{P}_j\cdot\left(\frac{1}{m_j}\mathbf{P}_j+\mathbf{v}(\mathbf{r}_j,t)-\frac{\sigma_j}{m_jc}\mathbf{A}(\mathbf{r}_j,t)\right)-\sum_{j=1}^{n}\frac{m_j}{2}\left|\frac{1}{m_j}\mathbf{P}_j-\frac{\sigma_j}{m_jc}\mathbf{A}(\mathbf{r}_j,t)\right|^2$$

$$+\sum_{j=1}^{n}\sigma_j\left(\Psi(\mathbf{r}_j,t)-\frac{1}{c}\mathbf{A}(\mathbf{r}_j,t)\cdot\left(\frac{1}{m_j}\mathbf{P}_j+\mathbf{v}(\mathbf{r}_j,t)-\frac{\sigma_j}{m_jc}\mathbf{A}(\mathbf{r}_j,t)\right)\right)-V\left(\mathbf{r}_1,\ldots,\mathbf{r}_n,t\right)$$

$$=\sum_{j=1}^{n}\mathbf{P}_j\cdot\mathbf{v}(\mathbf{r}_j,t)+\sum_{j=1}^{n}\frac{1}{2m_j}\left|\mathbf{P}_j-\frac{\sigma_j}{c}\mathbf{A}(\mathbf{r}_j,t)\right|^2$$

$$+\sum_{j=1}^{n}\sigma_j\left(\Psi(\mathbf{r}_j,t)-\frac{1}{c}\mathbf{A}(\mathbf{r}_j,t)\cdot\mathbf{v}(\mathbf{r}_j,t)\right)-V\left(\mathbf{r}_1,\ldots,\mathbf{r}_n,t\right). \quad (6.15)$$

6.3 Classical Liouville's Equation

Assume that the number of particles n in the system, ruled by the Hamiltonian (6.15), is large and we describe this system statistically. Then let $w\left(\mathbf{P}_1,\ldots,\mathbf{P}_n,\mathbf{r}_1,\ldots,\mathbf{r}_n,t\right)\to[0,+\infty)$ be the probability density of the system which satisfies the well known classical Liouville's equation of the form:

$$\frac{\partial w}{\partial t}+\sum_{j=1}^{n}\left(\operatorname{div}_{\mathbf{r}_j}\left\{w\,\nabla_{\mathbf{P}_j}H_0\right\}-\operatorname{div}_{\mathbf{P}_j}\left\{w\,\nabla_{\mathbf{r}_j}H_0\right\}\right)=$$

$$\frac{\partial w}{\partial t}+\sum_{j=1}^{n}\left(\nabla_{\mathbf{P}_j}H_0\cdot\nabla_{\mathbf{r}_j}w-\nabla_{\mathbf{r}_j}H_0\cdot\nabla_{\mathbf{P}_j}w\right)=0. \quad (6.16)$$

Then since by (6.15) we have

$$H_0\left(\mathbf{P}_1,\ldots,\mathbf{P}_n,\mathbf{r}_1,\ldots,\mathbf{r}_n,t\right)=$$

$$\sum_{j=1}^{n}\mathbf{P}_j\cdot\mathbf{v}(\mathbf{r}_j,t)+\sum_{j=1}^{n}\frac{1}{2m_j}\left|\mathbf{P}_j-\frac{\sigma_j}{c}\mathbf{A}(\mathbf{r}_j,t)\right|^2$$

$$+\sum_{j=1}^{n}\sigma_j\left(\Psi(\mathbf{r}_j,t)-\frac{1}{c}\mathbf{A}(\mathbf{r}_j,t)\cdot\mathbf{v}(\mathbf{r}_j,t)\right)-V\left(\mathbf{r}_1,\ldots,\mathbf{r}_n,t\right), \quad (6.17)$$

and in particular,

$$\nabla_{\mathbf{P}_j} H_0 = \frac{1}{m_j}\left(\mathbf{P}_j - \frac{\sigma_j}{c}\mathbf{A}(\mathbf{r}_j, t)\right) + \mathbf{v}(\mathbf{r}_j, t) \quad \text{and}$$

$$\nabla_{\mathbf{r}_j} H_0 = \left\{d_{\mathbf{x}}\mathbf{v}(\mathbf{r}_j, t)\right\}^T \cdot \mathbf{P}_j - \frac{\sigma_j}{cm_j}\left\{d_{\mathbf{x}}\mathbf{A}(\mathbf{r}_j, t)\right\}^T \cdot \left(\mathbf{P}_j - \frac{\sigma_j}{c}\mathbf{A}(\mathbf{r}_j, t)\right)$$
$$+\sigma_j \nabla_{\mathbf{x}}\left(\Psi(\mathbf{r}_j, t) - \frac{1}{c}\mathbf{A}(\mathbf{r}_j, t)\cdot\mathbf{v}(\mathbf{r}_j, t)\right) - \nabla_{\mathbf{y}_j} V(\mathbf{r}_1, \ldots, \mathbf{r}_n, t) \qquad \forall j = 1, \ldots n, \tag{6.18}$$

inserting (6.18) into (6.16) gives

$$0 = \frac{\partial w}{\partial t} + \sum_{j=1}^{n}\nabla_{\mathbf{P}_j} H_0 \cdot \nabla_{\mathbf{r}_j} w - \sum_{j=1}^{n}\nabla_{\mathbf{r}_j} H_0 \cdot \nabla_{\mathbf{P}_j} w = \frac{\partial w}{\partial t} + \sum_{j=1}^{n}\mathbf{v}(\mathbf{r}_j, t)\cdot\nabla_{\mathbf{r}_j} w$$
$$+\sum_{j=1}^{n}\frac{1}{m_j}\left(\mathbf{P}_j - \frac{\sigma_j}{c}\mathbf{A}(\mathbf{r}_j, t)\right)\cdot\nabla_{\mathbf{r}_j} w - \sum_{j=1}^{n}\left(\left\{d_{\mathbf{x}}\mathbf{v}(\mathbf{r}_j, t)\right\}^T\cdot\mathbf{P}_j\right)\cdot\nabla_{\mathbf{P}_j} w$$
$$+\sum_{j=1}^{n}\left(\frac{\sigma_j}{cm_j}\left\{d_{\mathbf{x}}\mathbf{A}(\mathbf{r}_j, t)\right\}^T\cdot\left(\mathbf{P}_j - \frac{\sigma_j}{c}\mathbf{A}(\mathbf{r}_j, t)\right)\right)\cdot\nabla_{\mathbf{P}_j} w$$
$$-\sum_{j=1}^{n}\left(\sigma_j\nabla_{\mathbf{x}}\left(\Psi(\mathbf{r}_j, t) - \frac{1}{c}\mathbf{A}(\mathbf{r}_j, t)\cdot\mathbf{v}(\mathbf{r}_j, t)\right) - \nabla_{\mathbf{y}_j} V(\mathbf{r}_1, \ldots, \mathbf{r}_n, t)\right)\cdot\nabla_{\mathbf{P}_j} w =$$
$$\frac{\partial w}{\partial t} + \sum_{j=1}^{n}\mathbf{v}(\mathbf{r}_j, t)\cdot\nabla_{\mathbf{r}_j} w + \sum_{j=1}^{n}\frac{1}{m_j}\left(\mathbf{P}_j - \frac{\sigma_j}{c}\mathbf{A}(\mathbf{r}_j, t)\right)\cdot\nabla_{\mathbf{r}_j} w$$
$$+\sum_{j=1}^{n}\frac{1}{2}\left(\left(d_{\mathbf{x}}\mathbf{v}(\mathbf{r}_j, t) - \left\{d_{\mathbf{x}}\mathbf{v}(\mathbf{r}_j, t)\right\}^T\right)\cdot\mathbf{P}_j\right)\cdot\nabla_{\mathbf{P}_j} w$$
$$-\sum_{j=1}^{n}\frac{1}{2}\left(\left(d_{\mathbf{x}}\mathbf{v}(\mathbf{r}_j, t) + \left\{d_{\mathbf{x}}\mathbf{v}(\mathbf{r}_j, t)\right\}^T\right)\cdot\mathbf{P}_j\right)\cdot\nabla_{\mathbf{P}_j} w$$
$$+\sum_{j=1}^{n}\left(\frac{\sigma_j}{cm_j}\left\{d_{\mathbf{x}}\mathbf{A}(\mathbf{r}_j, t)\right\}^T\cdot\left(\mathbf{P}_j - \frac{\sigma_j}{c}\mathbf{A}(\mathbf{r}_j, t)\right)\right)\cdot\nabla_{\mathbf{P}_j} w$$
$$-\sum_{j=1}^{n}\left(\sigma_j\nabla_{\mathbf{x}}\left(\Psi(\mathbf{r}_j, t) - \frac{1}{c}\mathbf{A}(\mathbf{r}_j, t)\cdot\mathbf{v}(\mathbf{r}_j, t)\right) - \nabla_{\mathbf{y}_j} V(\mathbf{r}_1, \ldots, \mathbf{r}_n, t)\right)\cdot\nabla_{\mathbf{P}_j} w. \tag{6.19}$$

Thus, by (6.19), using (2.15), we rewrite Liouville's equation as

$$
\begin{aligned}
&\frac{\partial w}{\partial t}+\sum_{j=1}^{n}\mathbf{v}(\mathbf{r}_j,t)\cdot\nabla_{\mathbf{r}_j}w+\sum_{j=1}^{n}\frac{1}{2}\left(\left(\mathrm{curl}_{\mathbf{x}}\mathbf{v}(\mathbf{r}_j,t)\right)\times\mathbf{P}_j\right)\cdot\nabla_{\mathbf{P}_j}w\\
&+\sum_{j=1}^{n}\frac{1}{m_j}\left(\mathbf{P}_j-\frac{\sigma_j}{c}\mathbf{A}(\mathbf{r}_j,t)\right)\cdot\nabla_{\mathbf{r}_j}w-\sum_{j=1}^{n}\frac{1}{2}\left(\left(d_{\mathbf{x}}\mathbf{v}(\mathbf{r}_j,t)+\left\{d_{\mathbf{x}}\mathbf{v}(\mathbf{r}_j,t)\right\}^T\right)\cdot\mathbf{P}_j\right)\cdot\nabla_{\mathbf{P}_j}w\\
&\qquad+\sum_{j=1}^{n}\left(\frac{\sigma_j}{cm_j}\left\{d_{\mathbf{x}}\mathbf{A}(\mathbf{r}_j,t)\right\}^T\cdot\left(\mathbf{P}_j-\frac{\sigma_j}{c}\mathbf{A}(\mathbf{r}_j,t)\right)\right)\cdot\nabla_{\mathbf{P}_j}w\\
&-\sum_{j=1}^{n}\left(\sigma_j\nabla_{\mathbf{x}}\left(\Psi(\mathbf{r}_j,t)-\frac{1}{c}\mathbf{A}(\mathbf{r}_j,t)\cdot\mathbf{v}(\mathbf{r}_j,t)\right)-\nabla_{\mathbf{y}_j}V\left(\mathbf{r}_1,\ldots,\mathbf{r}_n,t\right)\right)\cdot\nabla_{\mathbf{P}_j}w=0.
\end{aligned}
\tag{6.20}
$$

Next if the change of some non-inertial Cartesian coordinate system $(*)$ to another Cartesian coordinate system $(**)$ is of the form (4.2):

$$
\begin{cases}\mathbf{x}'=A(t)\cdot\mathbf{x}+\mathbf{z}(t),\\ t'=t,\end{cases}
\tag{6.21}
$$

where $A(t)\in SO(3)$ is a rotation, then consistently with (6.21) and (6.12) we have the following change of variables

$$
(\mathbf{P}_1,\ldots,\mathbf{P}_n,\mathbf{x}_1,\ldots,\mathbf{x}_n,t)\to(\mathbf{P}'_1,\ldots,\mathbf{P}'_n,\mathbf{x}'_1,\ldots,\mathbf{x}'_n,t'):
$$

$$
\begin{cases}t'=t,\\ \mathbf{x}'_k=A(t)\cdot\mathbf{x}_k+\mathbf{z}(t)\qquad\forall k=1,\ldots,n,\\ \mathbf{P}'_k=A(t)\cdot\mathbf{P}_k\qquad\forall k=1,\ldots,n.\end{cases}
\tag{6.22}
$$

Thus, since consistently with (6.21) we have

$$
\begin{cases}V'\left(\mathbf{x}'_1,\ldots,\mathbf{x}'_n,t'\right)=V\left(\mathbf{x}_1,\ldots,\mathbf{x}_n,t\right),\\ \sigma'_j=\sigma_j,\\ m'_j=m_j,\\ \mathbf{v}'(\mathbf{x}',t)=A(t)\cdot\mathbf{v}(\mathbf{x},t)+\frac{dA}{dt}(t)\cdot\mathbf{x}+\frac{d\mathbf{z}}{dt}(t),\\ \mathbf{A}'(\mathbf{x}',t)=A(t)\cdot\mathbf{A}(\mathbf{x},t),\\ \Psi'(\mathbf{x}',t)-\mathbf{v}'(\mathbf{x}',t)\cdot\mathbf{A}'(\mathbf{x}',t)=\Psi(\mathbf{x},t)-\mathbf{v}(\mathbf{x},t)\cdot\mathbf{A}(\mathbf{x},t),\end{cases}
\tag{6.23}
$$

by (6.22) and (6.23) we deduce that the Liouville equation (6.20) is invariant under the change of non-inertial Cartesian coordinate system of the form (4.2).

6.3.1 Thermodynamical Equilibrium; Canonical and Micro-canonical Ensembles

Next let $w(\mathbf{P}_1, \ldots, \mathbf{P}_n, \mathbf{r}_1, \ldots, \mathbf{r}_n, t) \to [0, +\infty)$ be the probability density of the system, ruled by the Hamiltonian H_0 from (6.17), in the case of thermodynamical equilibrium. Then in the case that $\frac{\partial H_0}{\partial t} = 0$ and the given equilibrium system rests macroscopically, i.e. it has the macroscopical velocity field zero: $\mathbf{u}(\mathbf{x}, t) = 0$ it is well known that

$$w(\mathbf{P}_1, \ldots, \mathbf{P}_n, \mathbf{r}_1, \ldots, \mathbf{r}_n, t) := \frac{1}{K(f, H_0)} f\left(H_0(\mathbf{P}_1, \ldots, \mathbf{P}_n, \mathbf{r}_1, \ldots, \mathbf{r}_n, t)\right), \tag{6.24}$$

with

$$K(f, H_0) := \int f\left(H_0(\mathbf{P}_1, \ldots, \mathbf{P}_n, \mathbf{r}_1, \ldots, \mathbf{r}_n, t)\right) d\mathbf{P}_1 \ldots d\mathbf{P}_n d\mathbf{r}_1 \ldots d\mathbf{r}_n, \tag{6.25}$$

and,

- in the case of a system in thermostat, having the Kelvin temperature T, we have the canonical ensemble: $f(s) = e^{-\frac{s}{kT}}\ \forall s$, where k is the Boltzmann constant,
- in the case of a thermally isolated system with the average internal energy E we have the micro-canonical ensemble: $f(s) = \delta(s - E)\ \forall s$.

We would like to find alternative forms of the above laws of thermodynamical equilibrium, which are invariant under the change of inertial or non-inertial Cartesian coordinate system. Then it is clear that if the given system rests in the old coordinate system, then it obviously has non-trivial macroscopical velocity field $\mathbf{u}(\mathbf{x}, t)$ in the new one. Moreover, $\mathbf{u}(\mathbf{x}, t)$ can depend on $\mathbf{x}$ and t, as it indeed happens in the case of a rotation of the new coordinate system with respect to the old one. On the other hand the concept of thermodynamical equilibrium is clearly independent on the coordinate system.

In order to find the forms of the laws of thermodynamical equilibrium, which are indeed invariant under the change of inertial or non-inertial Cartesian coordinate system we follow the steps as below. By (6.17) the Hamiltonian H_0 of our system has the following form:

$$\begin{aligned} &H_0(\mathbf{P}_1, \ldots, \mathbf{P}_n, \mathbf{r}_1, \ldots, \mathbf{r}_n, t) = \\ &\qquad \sum_{j=1}^{n} \mathbf{P}_j \cdot \mathbf{v}(\mathbf{r}_j, t) + \sum_{j=1}^{n} \frac{1}{2m_j} \left| \mathbf{P}_j - \frac{\sigma_j}{c} \mathbf{A}(\mathbf{r}_j, t) \right|^2 \\ &\qquad + \sum_{j=1}^{n} \sigma_j \left(\Psi(\mathbf{r}_j, t) - \frac{1}{c} \mathbf{A}(\mathbf{r}_j, t) \cdot \mathbf{v}(\mathbf{r}_j, t) \right) - V(\mathbf{r}_1, \ldots, \mathbf{r}_n, t). \end{aligned} \tag{6.26}$$

Next define an invariant quantity $H_{\mathbf{u}}$ as

$$H_{\mathbf{u}}\left(\mathbf{P}_1,\ldots,\mathbf{P}_n,\mathbf{r}_1,\ldots,\mathbf{r}_n,t\right):=$$

$$H_0\left(\mathbf{P}_1,\ldots,\mathbf{P}_n,\mathbf{r}_1,\ldots,\mathbf{r}_n,t\right)-\sum_{j=1}^{n}\mathbf{P}_j\cdot\mathbf{u}(\mathbf{r}_j,t)=\sum_{j=1}^{n}\mathbf{P}_j\cdot\left(\mathbf{v}(\mathbf{r}_j,t)-\mathbf{u}(\mathbf{r}_j,t)\right)$$

$$+\sum_{j=1}^{n}\frac{1}{2m_j}\left|\mathbf{P}_j-\frac{\sigma_j}{c}\mathbf{A}(\mathbf{r}_j,t)\right|^2+\sum_{j=1}^{n}\sigma_j\left(\Psi(\mathbf{r}_j,t)-\frac{1}{c}\mathbf{A}(\mathbf{r}_j,t)\cdot\mathbf{v}(\mathbf{r}_j,t)\right)$$

$$-V\left(\mathbf{r}_1,\ldots,\mathbf{r}_n,t\right), \quad (6.27)$$

where $\mathbf{u}(\mathbf{x},t)$ is the macroscopical velocity field of the given system of particles. Then, as before, it can be easily deduced that under the change of some non-inertial Cartesian coordinate system $(*)$ to another Cartesian coordinate system $(**)$ of the form (6.21) the quantity $H_{\mathbf{u}}$ transforms as

$$H'_{\mathbf{u}'}=H_{\mathbf{u}}, \quad (6.28)$$

provided that $\mathbf{u}$ is the speed-like vector field, i.e. under the above change we have

$$\mathbf{u}'(\mathbf{x}',t)\;=\;A(t)\cdot\mathbf{u}(\mathbf{x},t)+\frac{dA}{dt}(t)\cdot\mathbf{x}+\frac{d\mathbf{z}}{dt}(t), \quad (6.29)$$

and we have (6.22) and (6.23). Next consider the canonical and micro-canonical ensembles as

$$w\left(\mathbf{P}_1,\ldots,\mathbf{P}_n,\mathbf{r}_1,\ldots,\mathbf{r}_n,t\right):=\frac{1}{K(f,H_{\mathbf{u}})}f\left(H_{\mathbf{u}}\left(\mathbf{P}_1,\ldots,\mathbf{P}_n,\mathbf{r}_1,\ldots,\mathbf{r}_n,t\right)\right), \quad (6.30)$$

where as before,

$$K(f,H_{\mathbf{u}}):=\int f\left(H_{\mathbf{u}}\left(\mathbf{P}_1,\ldots,\mathbf{P}_n,\mathbf{r}_1,\ldots,\mathbf{r}_n,t\right)\right)d\mathbf{P}_1\ldots d\mathbf{P}_n d\mathbf{r}_1\ldots d\mathbf{r}_n, \quad (6.31)$$

and,

- in the case of a system in thermostat, having the Kelvin temperature T, we have $f(s)=e^{-\frac{s}{kT}}\ \forall s$,
- in the case of a thermally isolated system with the average internal energy E we have $f(s)=\delta(s-E)\ \forall s$.

Then, by (6.28) we deduce that under the change of some non-inertial Cartesian coordinate system $(*)$ to another Cartesian coordinate system $(**)$ of the form (6.21) the quantity w transforms as

$$w'=w. \quad (6.32)$$

Moreover, in the case $\mathbf{u} = 0$ (6.30) coincides with (6.24). However, we still need to derive the restrictions on the field $\mathbf{u}$ and the Hamiltonian H_0, providing that our system can indeed be found in the state of thermodynamical equilibrium. We remind that in the case $\mathbf{u} = 0$ the appropriate restriction is $\frac{\partial H_0}{\partial t} = 0$. In order to get these restrictions in the general case, we need to insert w in (6.30) into Liouville's equation in (6.16) having the form:

$$\frac{\partial w}{\partial t} + \sum_{j=1}^{n} \left(\nabla_{\mathbf{P}_j} H_0 \cdot \nabla_{\mathbf{r}_j} w - \nabla_{\mathbf{r}_j} H_0 \cdot \nabla_{\mathbf{P}_j} w\right) = 0. \tag{6.33}$$

By (6.27) we have

$$\nabla_{\mathbf{P}_j} H_{\mathbf{u}} = \nabla_{\mathbf{P}_j} H_0 - \mathbf{u}(\mathbf{r}_j, t)$$
$$\text{and} \quad \nabla_{\mathbf{r}_j} H_{\mathbf{u}} = \nabla_{\mathbf{r}_j} H_0 - \left\{d_{\mathbf{x}} \mathbf{u}(\mathbf{r}_j, t)\right\}^T \cdot \mathbf{P}_j \quad \forall j = 1, \ldots n. \tag{6.34}$$

Next inserting (6.30) into (6.33) we deuce

$$\frac{\partial H_{\mathbf{u}}}{\partial t} + \sum_{j=1}^{n} \left(\nabla_{\mathbf{P}_j} H_0 \cdot \nabla_{\mathbf{r}_j} H_{\mathbf{u}} - \nabla_{\mathbf{r}_j} H_0 \cdot \nabla_{\mathbf{P}_j} H_{\mathbf{u}}\right) = 0. \tag{6.35}$$

Thus inserting (6.34) into (6.35) we obtain

$$0 =$$
$$\frac{\partial H_{\mathbf{u}}}{\partial t} + \sum_{j=1}^{n} \left(\left(\nabla_{\mathbf{P}_j} H_{\mathbf{u}} + \mathbf{u}(\mathbf{r}_j, t)\right) \cdot \nabla_{\mathbf{r}_j} H_{\mathbf{u}} - \left(\nabla_{\mathbf{r}_j} H_{\mathbf{u}} + \left\{d_{\mathbf{x}} \mathbf{u}(\mathbf{r}_j, t)\right\}^T \cdot \mathbf{P}_j\right) \cdot \nabla_{\mathbf{P}_j} H_{\mathbf{u}}\right)$$
$$= \frac{\partial H_{\mathbf{u}}}{\partial t} + \sum_{j=1}^{n} \mathbf{u}(\mathbf{r}_j, t) \cdot \nabla_{\mathbf{r}_j} H_{\mathbf{u}} - \sum_{j=1}^{n} \left(\left\{d_{\mathbf{x}} \mathbf{u}(\mathbf{r}_j, t)\right\}^T \cdot \mathbf{P}_j\right) \cdot \nabla_{\mathbf{P}_j} H_{\mathbf{u}} =$$
$$\frac{\partial H_{\mathbf{u}}}{\partial t} + \sum_{j=1}^{n} \mathbf{u}(\mathbf{r}_j, t) \cdot \nabla_{\mathbf{r}_j} H_{\mathbf{u}} + \sum_{j=1}^{n} \frac{1}{2} \left(\left(d_{\mathbf{x}} \mathbf{u}(\mathbf{r}_j, t) - \left\{d_{\mathbf{x}} \mathbf{u}(\mathbf{r}_j, t)\right\}^T\right) \cdot \mathbf{P}_j\right) \cdot \nabla_{\mathbf{P}_j} H_{\mathbf{u}}$$
$$- \sum_{j=1}^{n} \frac{1}{2} \left(\left(d_{\mathbf{x}} \mathbf{u}(\mathbf{r}_j, t) + \left\{d_{\mathbf{x}} \mathbf{u}(\mathbf{r}_j, t)\right\}^T\right) \cdot \mathbf{P}_j\right) \cdot \nabla_{\mathbf{P}_j} H_{\mathbf{u}}. \tag{6.36}$$

Thus, by (2.15), we rewrite (6.36) as

$$\frac{\partial H_{\mathbf{u}}}{\partial t} + \sum_{j=1}^{n} \mathbf{u}(\mathbf{r}_j, t) \cdot \nabla_{\mathbf{r}_j} H_{\mathbf{u}} + \sum_{j=1}^{n} \frac{1}{2} \left(\left(\operatorname{curl}_{\mathbf{x}} \mathbf{u}(\mathbf{r}_j, t)\right) \times \mathbf{P}_j\right) \cdot \nabla_{\mathbf{P}_j} H_{\mathbf{u}}$$
$$- \sum_{j=1}^{n} \frac{1}{2} \left(\left(d_{\mathbf{x}} \mathbf{u}(\mathbf{r}_j, t) + \left\{d_{\mathbf{x}} \mathbf{u}(\mathbf{r}_j, t)\right\}^T\right) \cdot \mathbf{P}_j\right) \cdot \nabla_{\mathbf{P}_j} H_{\mathbf{u}} = 0. \tag{6.37}$$

Equality (6.37) is the required restriction on the field $\mathbf{u}$ and the Hamiltonian H_0, providing that our system can indeed be found in state of thermodynamical equilibrium. In particular, if $\frac{\partial H_0}{\partial t} = 0$ and $\mathbf{u} = 0$ then (6.37) indeed holds. Moreover, as before in the case of Liouville's equation, we deduce that the equation (6.37) is invariant under the change of non-inertial Cartesian coordinate system of the form (6.21), provided that we have (6.29), (6.22) and (6.23).

Next, we still need to prove that if w is given by (6.30) then the vector field $\mathbf{u}(\mathbf{x}, t)$ is indeed the macroscopic (average) velocity field of every particle that we can find near the point $\mathbf{x}$ at the instant of time t. Indeed, we need to prove the following:

$$\mathbf{u}(\mathbf{x}, t) := \frac{1}{\mu_j(\mathbf{x}, t)}\mathbf{p}_j(\mathbf{x}, t) + \mathbf{v}(\mathbf{x}, t) - \frac{\sigma_j}{cm_j}\mathbf{A}(\mathbf{x}, t) \qquad \forall j \in \{1, \ldots n\}, \tag{6.38}$$

where

$$\mu_j(\mathbf{x}, t) := \int m_j\, w\left(\mathbf{P}_1, \ldots, \mathbf{P}_n, \mathbf{r}_1, \ldots, \mathbf{r}_n, t\right) \delta(\mathbf{x} - \mathbf{r}_j)\, d\mathbf{P}_1, \ldots, d\mathbf{P}_n, d\mathbf{r}_1, \ldots, d\mathbf{r}_n, \tag{6.39}$$

and

$$\mathbf{p}_j(\mathbf{x}, t) := \int \mathbf{P}_j\, w\left(\mathbf{P}_1, \ldots, \mathbf{P}_n, \mathbf{r}_1, \ldots, \mathbf{r}_n, t\right) \delta(\mathbf{x} - \mathbf{r}_j)\, d\mathbf{P}_1, \ldots, d\mathbf{P}_n, d\mathbf{r}_1, \ldots, d\mathbf{r}_n. \tag{6.40}$$

On the other hand, by (6.27) we have

$$\begin{aligned}
H_{\mathbf{u}}\left(\mathbf{P}_1, \ldots, \mathbf{P}_n, \mathbf{r}_1, \ldots, \mathbf{r}_n, t\right) &= \sum_{j=1}^{n} \frac{1}{2m_j}\left|\mathbf{P}_j - \frac{\sigma_j}{c}\mathbf{A}(\mathbf{r}_j, t)\right|^2 \\
&+ \sum_{j=1}^{n}\left(\mathbf{P}_j - \frac{\sigma_j}{c}\mathbf{A}(\mathbf{r}_j, t)\right)\cdot\left(\mathbf{v}(\mathbf{r}_j, t) - \mathbf{u}(\mathbf{r}_j, t)\right) + \sum_{j=1}^{n}\frac{\sigma_j}{c}\mathbf{A}(\mathbf{r}_j, t)\cdot\left(\mathbf{v}(\mathbf{r}_j, t) - \mathbf{u}(\mathbf{r}_j, t)\right) \\
&+ \sum_{j=1}^{n}\sigma_j\left(\Psi(\mathbf{r}_j, t) - \frac{1}{c}\mathbf{A}(\mathbf{r}_j, t)\cdot\mathbf{v}(\mathbf{r}_j, t)\right) - V\left(\mathbf{r}_1, \ldots, \mathbf{r}_n, t\right) \\
&= \sum_{j=1}^{n}\frac{1}{2m_j}\left|\mathbf{P}_j + m_j\mathbf{v}(\mathbf{r}_j, t) - \frac{\sigma_j}{c}\mathbf{A}(\mathbf{r}_j, t) - m_j\mathbf{u}(\mathbf{r}_j, t)\right|^2 - \sum_{j=1}^{n}\frac{m_j}{2}\left|\mathbf{v}(\mathbf{r}_j, t) - \mathbf{u}(\mathbf{r}_j, t)\right|^2 \\
&+ \sum_{j=1}^{n}\frac{\sigma_j}{c}\mathbf{A}(\mathbf{r}_j, t)\cdot\left(\mathbf{v}(\mathbf{r}_j, t) - \mathbf{u}(\mathbf{r}_j, t)\right) + \sum_{j=1}^{n}\sigma_j\left(\Psi(\mathbf{r}_j, t) - \frac{1}{c}\mathbf{A}(\mathbf{r}_j, t)\cdot\mathbf{v}(\mathbf{r}_j, t)\right) \\
&- V\left(\mathbf{r}_1, \ldots, \mathbf{r}_n, t\right).
\end{aligned} \tag{6.41}$$

Thus by inserting (6.41) into (6.30) we deduce that (6.38) indeed holds.

Finally, by (6.27) we have

$$H_{\mathbf{u}}\left(\mathbf{P}_1,\ldots,\mathbf{P}_n,\mathbf{r}_1,\ldots,\mathbf{r}_n,t\right)= \sum_{j=1}^{n}\frac{1}{2m_j}\left|\mathbf{P}_j\right|^2-\sum_{j=1}^{n}\mathbf{P}_j\cdot\left(\mathbf{u}(\mathbf{r}_j,t)-\mathbf{v}(\mathbf{r}_j,t)+\frac{\sigma_j}{m_jc}\mathbf{A}(\mathbf{r}_j,t)\right)$$
$$+\sum_{j=1}^{n}\left(\frac{\sigma_j^2}{2m_jc^2}\left|\mathbf{A}(\mathbf{r}_j,t)\right|^2+\sigma_j\Psi(\mathbf{r}_j,t)-\frac{\sigma_j}{c}\mathbf{A}(\mathbf{r}_j,t)\cdot\mathbf{v}(\mathbf{r}_j,t)\right)-V\left(\mathbf{r}_1,\ldots,\mathbf{r}_n,t\right) \tag{6.42}$$

that we rewrite as

$$H_{\mathbf{u}}\left(\mathbf{P}_1,\ldots,\mathbf{P}_n,\mathbf{r}_1,\ldots,\mathbf{r}_n,t\right)=\sum_{j=1}^{n}\frac{1}{2m_j}\left|\mathbf{P}_j\right|^2-\sum_{j=1}^{n}\mathbf{P}_j\cdot\mathbf{h}_j(\mathbf{r}_j,t)+U\left(\mathbf{r}_1,\ldots,\mathbf{r}_n,t\right), \tag{6.43}$$

where by $\mathbf{h}_j(\mathbf{x},t)$ we denote a proper vector field, defined as

$$\mathbf{h}_j(\mathbf{x},t):=\left(\mathbf{u}(\mathbf{x},t)-\mathbf{v}(\mathbf{x},t)+\frac{\sigma_j}{m_jc}\mathbf{A}(\mathbf{x},t)\right), \tag{6.44}$$

and by U we denote a proper scalar, defined as

$$U\left(\mathbf{x}_1,\ldots,\mathbf{x}_n,t\right):= \sum_{j=1}^{n}\left(\frac{\sigma_j^2}{2m_jc^2}\left|\mathbf{A}(\mathbf{x}_j,t)\right|^2+\sigma_j\Psi(\mathbf{x}_j,t)-\frac{\sigma_j}{c}\mathbf{A}(\mathbf{x}_j,t)\cdot\mathbf{v}(\mathbf{x}_j,t)\right)-V\left(\mathbf{x}_1,\ldots,\mathbf{x}_n,t\right). \tag{6.45}$$

Thus,

$$\frac{\partial H_{\mathbf{u}}}{\partial t}=-\sum_{j=1}^{n}\mathbf{P}_j\cdot\frac{\partial\mathbf{h}_j}{\partial t}(\mathbf{r}_j,t)+\frac{\partial U}{\partial t}\left(\mathbf{r}_1,\ldots,\mathbf{r}_n,t\right), \tag{6.46}$$

$$\nabla_{\mathbf{r}_j}H_{\mathbf{u}}=-\left\{d_{\mathbf{x}}\mathbf{h}_j(\mathbf{r}_j,t)\right\}^T\cdot\mathbf{P}_j+\nabla_{\mathbf{r}_j}U\left(\mathbf{r}_1,\ldots,\mathbf{r}_n,t\right), \tag{6.47}$$

and

$$\nabla_{\mathbf{P}_j}H_{\mathbf{u}}=\frac{1}{m_j}\mathbf{P}_j-\mathbf{h}_j(\mathbf{r}_j,t). \tag{6.48}$$

On the other hand, by (6.36) we have

$$\frac{\partial H_{\mathbf{u}}}{\partial t}+\sum_{j=1}^{n}\mathbf{u}(\mathbf{r}_j,t)\cdot\nabla_{\mathbf{r}_j}H_{\mathbf{u}}-\sum_{j=1}^{n}\left(\left\{d_{\mathbf{x}}\mathbf{u}(\mathbf{r}_j,t)\right\}^T\cdot\mathbf{P}_j\right)\cdot\nabla_{\mathbf{P}_j}H_{\mathbf{u}}=0. \tag{6.49}$$

Thus inserting (6.46), (6.47) and (6.48) into (6.49) gives

$$
\begin{aligned}
& -\sum_{j=1}^{n} \mathbf{P}_j \cdot \frac{\partial \mathbf{h}_j}{\partial t}(\mathbf{r}_j, t) + \frac{\partial U}{\partial t}(\mathbf{r}_1, \ldots, \mathbf{r}_n, t) \\
& \quad + \sum_{j=1}^{n} \mathbf{u}(\mathbf{r}_j, t) \cdot \left(-\left\{ d_{\mathbf{x}} \mathbf{h}_j(\mathbf{r}_j, t) \right\}^T \cdot \mathbf{P}_j + \nabla_{\mathbf{r}_j} U(\mathbf{r}_1, \ldots, \mathbf{r}_n, t) \right) \\
& \quad - \sum_{j=1}^{n} \left(\left\{ d_{\mathbf{x}} \mathbf{u}(\mathbf{r}_j, t) \right\}^T \cdot \mathbf{P}_j \right) \cdot \left(\frac{1}{m_j} \mathbf{P}_j - \mathbf{h}_j(\mathbf{r}_j, t) \right) = 0. \qquad (6.50)
\end{aligned}
$$

That is,

$$
\begin{aligned}
& \left(\frac{\partial U}{\partial t}(\mathbf{r}_1, \ldots, \mathbf{r}_n, t) + \sum_{j=1}^{n} \mathbf{u}(\mathbf{r}_j, t) \cdot \nabla_{\mathbf{r}_j} U(\mathbf{r}_1, \ldots, \mathbf{r}_n, t) \right) \\
& \quad - \sum_{j=1}^{n} \mathbf{P}_j \cdot \left(\frac{\partial \mathbf{h}_j}{\partial t}(\mathbf{r}_j, t) + d_{\mathbf{x}} \mathbf{h}_j(\mathbf{r}_j, t) \cdot \mathbf{u}(\mathbf{r}_j, t) - d_{\mathbf{x}} \mathbf{u}(\mathbf{r}_j, t) \cdot \mathbf{h}_j(\mathbf{r}_j, t) \right) \\
& \quad - \sum_{j=1}^{n} \frac{1}{2m_j} \left(\left(d_{\mathbf{x}} \mathbf{u}(\mathbf{r}_j, t) + \left\{ d_{\mathbf{x}} \mathbf{u}(\mathbf{r}_j, t) \right\}^T \right) \cdot \mathbf{P}_j \right) \cdot \mathbf{P}_j = 0. \qquad (6.51)
\end{aligned}
$$

Thus by (2.11) we rewrite (6.51) as

$$
\begin{aligned}
& \left(\frac{\partial U}{\partial t}(\mathbf{r}_1, \ldots, \mathbf{r}_n, t) + \sum_{j=1}^{n} \mathbf{u}(\mathbf{r}_j, t) \cdot \nabla_{\mathbf{r}_j} U(\mathbf{r}_1, \ldots, \mathbf{r}_n, t) \right) \\
& - \sum_{j=1}^{n} \mathbf{P}_j \cdot \left(\frac{\partial \mathbf{h}_j}{\partial t}(\mathbf{r}_j, t) - \operatorname{curl}_{\mathbf{x}} \left(\mathbf{u}(\mathbf{r}_j, t) \times \mathbf{h}_j(\mathbf{r}_j, t) \right) + \left(\operatorname{div}_{\mathbf{x}} \mathbf{h}_j(\mathbf{r}_j, t) \right) \mathbf{u}(\mathbf{r}_j, t) \right. \\
& \left. - \left(\operatorname{div}_{\mathbf{x}} \mathbf{u}(\mathbf{r}_j, t) \right) \mathbf{h}_j(\mathbf{r}_j, t) \right) - \sum_{j=1}^{n} \frac{1}{2m_j} \left(\left(d_{\mathbf{x}} \mathbf{u}(\mathbf{r}_j, t) + \left\{ d_{\mathbf{x}} \mathbf{u}(\mathbf{r}_j, t) \right\}^T \right) \cdot \mathbf{P}_j \right) \cdot \mathbf{P}_j = 0
\end{aligned}
\qquad (6.52)
$$

that is equivalent to the following

$$
\begin{cases}
\frac{\partial U}{\partial t}(\mathbf{x}_1, \ldots, \mathbf{x}_n, t) + \sum_{j=1}^{n} \mathbf{u}(\mathbf{x}_j, t) \cdot \nabla_{\mathbf{x}_j} U(\mathbf{x}_1, \ldots, \mathbf{x}_n, t) = 0 \\
\frac{\partial \mathbf{h}_j}{\partial t}(\mathbf{x}, t) - \operatorname{curl}_{\mathbf{x}} \left(\mathbf{u}(\mathbf{x}, t) \times \mathbf{h}_j(\mathbf{x}, t) \right) \\
+ \left(\operatorname{div}_{\mathbf{x}} \mathbf{h}_j(\mathbf{x}, t) \right) \mathbf{u}(\mathbf{x}, t) = 0 \qquad \forall j = 1, \ldots, n, \\
d_{\mathbf{x}} \mathbf{u}(\mathbf{x}, t) + \{ d_{\mathbf{x}} \mathbf{u}(\mathbf{x}, t) \}^T = 0.
\end{cases}
\qquad (6.53)
$$

Next, using (6.44) we rewrite (6.53) as

$$\begin{cases} \frac{\partial U}{\partial t}(\mathbf{x}_1, \ldots, \mathbf{x}_n, t) + \sum_{j=1}^{n} \mathbf{u}(\mathbf{x}_j, t) \cdot \nabla_{\mathbf{x}_j} U(\mathbf{x}_1, \ldots, \mathbf{x}_n, t) = 0, \\ \frac{\partial}{\partial t}\left(\mathbf{u}(\mathbf{x}, t) - \mathbf{v}(\mathbf{x}, t) + \frac{\sigma_j}{m_j c}\mathbf{A}(\mathbf{x}, t)\right) \\ -\mathrm{curl}_{\mathbf{x}}\left(\mathbf{u}(\mathbf{x}, t) \times \left(\mathbf{u}(\mathbf{x}, t) - \mathbf{v}(\mathbf{x}, t) + \frac{\sigma_j}{m_j c}\mathbf{A}(\mathbf{x}, t)\right)\right) \\ +\left(\mathrm{div}_{\mathbf{x}}\left(\mathbf{u}(\mathbf{x}, t) - \mathbf{v}(\mathbf{x}, t) + \frac{\sigma_j}{m_j c}\mathbf{A}(\mathbf{x}, t)\right)\right)\mathbf{u}(\mathbf{x}, t) = 0 \qquad \forall j = 1, \ldots, n, \\ d_{\mathbf{x}}\mathbf{u}(\mathbf{x}, t) + \{d_{\mathbf{x}}\mathbf{u}(\mathbf{x}, t)\}^T = 0. \end{cases} \tag{6.54}$$

Thus since the second equation in (6.54) must be valid for every $j = 1, \ldots, n$ by (6.54) we deduce

$$\begin{cases} \frac{\partial U}{\partial t}(\mathbf{x}_1, \ldots, \mathbf{x}_n, t) + \sum_{j=1}^{n} \mathbf{u}(\mathbf{x}_j, t) \cdot \nabla_{\mathbf{x}_j} U(\mathbf{x}_1, \ldots, \mathbf{x}_n, t) = 0, \\ \frac{\partial}{\partial t}(\mathbf{u}(\mathbf{x}, t) - \mathbf{v}(\mathbf{x}, t)) - \mathrm{curl}_{\mathbf{x}}(\mathbf{u}(\mathbf{x}, t) \times (\mathbf{u}(\mathbf{x}, t) - \mathbf{v}(\mathbf{x}, t))) \\ + (\mathrm{div}_{\mathbf{x}}(\mathbf{u}(\mathbf{x}, t) - \mathbf{v}(\mathbf{x}, t)))\mathbf{u}(\mathbf{x}, t) = 0, \\ \frac{\partial \mathbf{A}}{\partial t}(\mathbf{x}, t) - \mathrm{curl}_{\mathbf{x}}(\mathbf{u}(\mathbf{x}, t) \times \mathbf{A}(\mathbf{x}, t)) + (\mathrm{div}_{\mathbf{x}}\mathbf{A}(\mathbf{x}, t))\mathbf{u}(\mathbf{x}, t) = 0, \\ d_{\mathbf{x}}\mathbf{u}(\mathbf{x}, t) + \{d_{\mathbf{x}}\mathbf{u}(\mathbf{x}, t)\}^T = 0, \end{cases} \tag{6.55}$$

where

$$U(\mathbf{x}_1, \ldots, \mathbf{x}_n, t) := \sum_{j=1}^{n}\left(\frac{\sigma_j^2}{2m_j c^2}|\mathbf{A}(\mathbf{x}_j, t)|^2 + \sigma_j \Psi(\mathbf{x}_j, t) - \frac{\sigma_j}{c}\mathbf{A}(\mathbf{x}_j, t) \cdot \mathbf{v}(\mathbf{x}_j, t)\right) - V(\mathbf{x}_1, \ldots, \mathbf{x}_n, t). \tag{6.56}$$

Then, clearly equations (6.54) and (6.55), (6.56) are invariant under the change of inertial or non-inertial Cartesian coordinate systems. Moreover, (6.55), (6.56) are equivalent to (6.36) and then represent the necessary conditions on the field **u** and the system for the existence of a possibility to achieve the Thermodynamical equilibrium in the given system.

Next note that by (6.55) and Proposition 3.3 there exists another Cartesian coordinate system $(**)$ such that under the change of coordinate system $(*)$ to another Cartesian coordinate system $(**)$, given by (3.54), we have

$$A(t) \cdot \mathbf{u}(\mathbf{x}, t) + \frac{dA}{dt}(t) \cdot \mathbf{x} + \frac{d\mathbf{z}}{dt}(t) = \mathbf{u}'(\mathbf{x}', t') = 0. \tag{6.57}$$

Thus, since (6.55), (6.56) are invariant under the change of inertial or non-inertial Cartesian coordinate systems, in system $(**)$ we have

$$\begin{cases} \frac{\partial U'}{\partial t'}\left(\mathbf{x}'_1, \ldots, \mathbf{x}'_n, t'\right) = 0, \\ \frac{\partial \mathbf{v}'}{\partial t'}(\mathbf{x}', t') = 0, \\ \frac{\partial \mathbf{A}'}{\partial t'}(\mathbf{x}', t') = 0 \\ \mathbf{u}'(\mathbf{x}', t') = 0, \end{cases} \tag{6.58}$$

where

$$U'\left(\mathbf{x}'_1, \ldots, \mathbf{x}'_n, t'\right) := \sum_{j=1}^{n} \left(\frac{(\sigma'_j)^2}{2\, m'_j c^2} \left|\mathbf{A}'(\mathbf{x}'_j, t')\right|^2 + \sigma'_j \Psi'(\mathbf{x}'_j, t') - \frac{\sigma'_j}{c} \mathbf{A}'(\mathbf{x}'_j, t') \cdot \mathbf{v}'(\mathbf{x}'_j, t') \right) - V'\left(\mathbf{x}'_1, \ldots, \mathbf{x}'_n, t'\right). \tag{6.59}$$

On the other hand, (6.58) is equivalent to

$$\frac{\partial H'_0}{\partial t'}\left(\mathbf{P}'_1, \ldots, \mathbf{P}'_n, \mathbf{r}'_1, \ldots, \mathbf{r}'_n, t'\right) = 0 \quad \text{and} \quad \mathbf{u}'(\mathbf{x}', t') = 0. \tag{6.60}$$

Thus we obtain that the condition for existing the possibility to achieve the Thermodynamical equilibrium in the given system, ruled by the Hamiltonian $H_0\left(\mathbf{P}_1, \ldots, \mathbf{P}_n, \mathbf{r}_1, \ldots, \mathbf{r}_n, t\right)$, of the form (6.49), is equivalent to the existence of a Cartesian coordinate system (**), where the average velocity of every particle vanishes and at the same time in the system (**) the Hamiltonian is independent on the time variable explicitly.

6.4 Shrödinger equation for a Finite System of Quantum Particles

Consider the motion of a system of n quantum micro-particles with inertial masses $m_1, \ldots, m_n$ and the charges $\sigma_1, \ldots, \sigma_n$ in the outer gravitational and electromagnetical field with characteristics $\mathbf{v}(\mathbf{x}, t)$, $\mathbf{A}(\mathbf{x}, t)$ and $\Psi(\mathbf{x}, t)$ and additional conservative field with potential $V(\mathbf{y}_1, \ldots, \mathbf{y}_n, t)$, not taking into account the spin interaction. The Shrödinger equation for this system of particles is

$$i\hbar \frac{\partial \psi}{\partial t} = \hat{H}_0 \cdot \psi, \tag{6.61}$$

where $\psi(\mathbf{x}_1, \ldots, \mathbf{x}_n, t) \in \mathbb{C}$ is a wave function and $\hat{H}_0$ is the Hamiltonian operator. Since by (6.15) the Hamiltonian for a macro-particles has the form

$$
\begin{aligned}
&H_{\text{macro}}\left(\mathbf{P}_1,\ldots,\mathbf{P}_n,\mathbf{r}_1,\ldots,\mathbf{r}_n,t\right) = \\
&\qquad - V\left(\mathbf{r}_1,\ldots,\mathbf{r}_n,t\right) + \sum_{j=1}^{n}\frac{1}{2}\mathbf{P}_j\cdot\mathbf{v}(\mathbf{r}_j,t) + \sum_{j=1}^{n}\frac{1}{2}\mathbf{v}(\mathbf{r}_j,t)\cdot\mathbf{P}_j \\
&+\sum_{j=1}^{n}\frac{1}{2m_j}\left|\mathbf{P}_j - \frac{\sigma_j}{c}\mathbf{A}(\mathbf{r}_j,t)\right|^2 + \sum_{j=1}^{n}\sigma_j\left(\Psi(\mathbf{r}_j,t) - \frac{1}{c}\mathbf{A}(\mathbf{r}_j,t)\cdot\mathbf{v}(\mathbf{r}_j,t)\right),
\end{aligned}
\tag{6.62}
$$

we built the Hermitian Hamiltonian operator as

$$
\begin{aligned}
\hat{H}_0\cdot\psi &= -\sum_{j=1}^{n}\frac{i\hbar}{2}\operatorname{div}_{\mathbf{x}_j}\left\{\psi\mathbf{v}(\mathbf{x}_j,t)\right\} - \sum_{j=1}^{n}\frac{i\hbar}{2}\mathbf{v}(\mathbf{x}_j,t)\cdot\nabla_{\mathbf{x}_j}\psi \\
&+\sum_{j=1}^{n}\left\{\frac{1}{2m_j}\left(-i\hbar\nabla_{\mathbf{x}_j} - \frac{\sigma_j}{c}\mathbf{A}(\mathbf{x}_j,t)\right)\circ\left(-i\hbar\nabla_{\mathbf{x}_j} - \frac{\sigma_j}{c}\mathbf{A}(\mathbf{x}_j,t)\right)\right\}\cdot\psi \\
&+\sum_{j=1}^{n}\sigma_j\left(\Psi(\mathbf{x}_j,t) - \frac{1}{c}\mathbf{A}(\mathbf{x}_j,t)\cdot\mathbf{v}(\mathbf{x}_j,t)\right)\cdot\psi - V\left(\mathbf{x}_1,\ldots,\mathbf{x}_n,t\right)\cdot\psi \\
&= -\sum_{j=1}^{n}\frac{\hbar^2}{2m_j}\Delta_{\mathbf{x}_j}\psi - \sum_{j=1}^{n}\frac{i\hbar}{2}\operatorname{div}_{\mathbf{x}_j}\left\{\psi\mathbf{v}(\mathbf{x}_j,t)\right\} \\
&-\sum_{j=1}^{n}\frac{i\hbar}{2}\mathbf{v}(\mathbf{x}_j,t)\cdot\nabla_{\mathbf{x}_j}\psi + \sum_{j=1}^{n}\frac{i\hbar\sigma_j}{2m_jc}\operatorname{div}_{\mathbf{x}_j}\left\{\psi\mathbf{A}(\mathbf{x}_j,t)\right\} + \sum_{j=1}^{n}\frac{i\hbar\sigma_j}{2m_jc}\mathbf{A}(\mathbf{x}_j,t)\cdot\nabla_{\mathbf{x}_j}\psi \\
&+\sum_{j=1}^{n}\left(\sigma_j\Psi(\mathbf{x}_j,t) - \frac{\sigma_j}{c}\mathbf{A}(\mathbf{x}_j,t)\cdot\mathbf{v}(\mathbf{x}_j,t) + \frac{\sigma_j^2}{2m_jc^2}\left|\mathbf{A}(\mathbf{x}_j,t)\right|^2\right)\psi - V\left(\mathbf{x}_1,\ldots,\mathbf{x}_n,t\right)\psi.
\end{aligned}
\tag{6.63}
$$

Thus the corresponding Shrödinger equation will be

$$
\begin{aligned}
i\hbar\frac{\partial\psi}{\partial t} &= \hat{H}_0\cdot\psi = -\sum_{j=1}^{n}\frac{\hbar^2}{2m_j}\Delta_{\mathbf{x}_j}\psi - \sum_{j=1}^{n}\frac{i\hbar}{2}\operatorname{div}_{\mathbf{x}_j}\left\{\psi\mathbf{v}(\mathbf{x}_j,t)\right\} - \sum_{j=1}^{n}\frac{i\hbar}{2}\mathbf{v}(\mathbf{x}_j,t)\cdot\nabla_{\mathbf{x}_j}\psi \\
&+\sum_{j=1}^{n}\frac{i\hbar\sigma_j}{2m_jc}\operatorname{div}_{\mathbf{x}_j}\left\{\psi\mathbf{A}(\mathbf{x}_j,t)\right\} + \sum_{j=1}^{n}\frac{i\hbar\sigma_j}{2m_jc}\mathbf{A}(\mathbf{x}_j,t)\cdot\nabla_{\mathbf{x}_j}\psi \\
&+\sum_{j=1}^{n}\left(\sigma_j\Psi(\mathbf{x}_j,t) - \frac{\sigma_j}{c}\mathbf{A}(\mathbf{x}_j,t)\cdot\mathbf{v}(\mathbf{x}_j,t) + \frac{\sigma_j^2}{2m_jc^2}\left|\mathbf{A}(\mathbf{x}_j,t)\right|^2\right)\psi \\
&\qquad - V\left(\mathbf{x}_1,\ldots,\mathbf{x}_n,t\right)\psi.
\end{aligned}
\tag{6.64}
$$

So

$$i\hbar\left(\frac{\partial\psi}{\partial t}+\sum_{j=1}^{n}\mathbf{v}(\mathbf{x}_j,t)\cdot\nabla_{\mathbf{x}_j}\psi\right)+\sum_{j=1}^{n}\frac{i\hbar}{2}\left(\operatorname{div}_{\mathbf{x}_j}\mathbf{v}(\mathbf{x}_j,t)\right)\psi$$
$$=-\sum_{j=1}^{n}\frac{\hbar^2}{2m_j}\Delta_{\mathbf{x}_j}\psi-V\left(\mathbf{x}_1,\dots,\mathbf{x}_n,t\right)\psi$$
$$+\sum_{j=1}^{n}\frac{i\hbar\sigma_j}{2m_jc}\operatorname{div}_{\mathbf{x}_j}\left\{\psi\mathbf{A}(\mathbf{x}_j,t)\right\}+\sum_{j=1}^{n}\frac{i\hbar\sigma_j}{2m_jc}\mathbf{A}(\mathbf{x}_j,t)\cdot\nabla_{\mathbf{x}_j}\psi$$
$$+\sum_{j=1}^{n}\left(\sigma_j\Psi(\mathbf{x}_j,t)-\frac{\sigma_j}{c}\mathbf{A}(\mathbf{x}_j,t)\cdot\mathbf{v}(\mathbf{x}_j,t)+\frac{\sigma_j^2}{2m_jc^2}\left|\mathbf{A}(\mathbf{x}_j,t)\right|^2\right)\psi. \quad (6.65)$$

Next consider a change of some non-inertial Cartesian coordinate system $(*)$ to another Cartesian coordinate system $(**)$ of the form:

$$\begin{cases}\mathbf{x}'=A(t)\cdot\mathbf{x}+\mathbf{z}(t),\\ t'=t,\end{cases} \quad (6.66)$$

where $A(t)\in SO(3)$ is a rotation, i.e. $A(t)\in\mathbb{R}^{3\times3}$, $\det A(t)>0$ and $A(t)\cdot A^T(t)=I$. Then, since

$$\begin{cases}\psi'=\psi\\ V'=V\\ \mathbf{A}'=A(t)\cdot\mathbf{A}\\ \Psi'-\frac{1}{c}\mathbf{A}'\cdot\mathbf{v}'=\Psi-\frac{1}{c}\mathbf{A}\cdot\mathbf{v},\end{cases} \quad (6.67)$$

we deduce that the Shrödinger equation of the form (6.65) is invariant under the change of non-inertial Cartesian coordinate system. So the quantum mechanical laws are also invariant in every non-inertial Cartesian coordinate system.

Next, assume that in inertial coordinate system $(*)$ we have

$$\begin{cases}\operatorname{curl}_{\mathbf{x}}\mathbf{v}=0,\\ \frac{\partial\mathbf{v}}{\partial t}+d_{\mathbf{x}}\mathbf{v}\cdot\mathbf{v}=-\nabla_{\mathbf{x}}\Phi,\end{cases} \quad (6.68)$$

where Φ is the scalar gravitational potential. Since in the system $(*)$ we have $\operatorname{curl}_{\mathbf{x}}\mathbf{v}=0$ we can rewrite (6.68) as

$$\begin{cases}\mathbf{v}=\nabla_{\mathbf{x}}Z,\\ \frac{\partial Z}{\partial t}+\frac{1}{2}\left|\nabla_{\mathbf{x}}Z\right|^2=-\Phi.\end{cases} \quad (6.69)$$

Thus by (6.69) we rewrite (6.65) as

$$
\begin{aligned}
i\hbar\frac{\partial\psi}{\partial t} &+ \sum_{j=1}^{n} i\hbar\nabla_{\mathbf{x}_j} Z(\mathbf{x}_j,t)\cdot\nabla_{\mathbf{x}_j}\psi + \sum_{j=1}^{n}\frac{i\hbar}{2}\left(\Delta_{\mathbf{x}_j} Z(\mathbf{x}_j,t)\right)\psi + \sum_{j=1}^{n}\frac{\hbar^2}{2m_j}\Delta_{\mathbf{x}_j}\psi = \\
&\sum_{j=1}^{n}\frac{i\hbar\sigma_j}{2m_jc}\left(\operatorname{div}_{\mathbf{x}_j}\mathbf{A}(\mathbf{x}_j,t)\right)\psi + \sum_{j=1}^{n}\frac{i\hbar\sigma_j}{m_jc}\mathbf{A}(\mathbf{x}_j,t)\cdot\nabla_{\mathbf{x}_j}\psi \\
&-\sum_{j=1}^{n}\frac{\sigma_j}{c}\left(\mathbf{A}(\mathbf{x}_j,t)\cdot\nabla_{\mathbf{x}_j} Z(\mathbf{x}_j,t)\right)\psi + \sum_{j=1}^{n}\left(\sigma_j\Psi(\mathbf{x}_j,t) + \frac{\sigma_j^2}{2m_jc^2}\left|\mathbf{A}(\mathbf{x}_j,t)\right|^2\right)\psi - V\psi.
\end{aligned}
\tag{6.70}
$$

Then multiplying (6.70) by factor $e^{\sum_{j=1}^{n}\frac{im_j}{\hbar}Z(\mathbf{x}_j,t)}$ gives

$$
\begin{aligned}
i\hbar\frac{\partial\psi}{\partial t}e^{\sum_{j=1}^{n}\frac{im_j}{\hbar}Z(\mathbf{x}_j,t)} &+ \sum_{k=1}^{n} i\hbar\left(\nabla_{\mathbf{x}_k} Z(\mathbf{x}_k,t)\cdot\nabla_{\mathbf{x}_k}\psi\right)e^{\sum_{j=1}^{n}\frac{im_j}{\hbar}Z(\mathbf{x}_j,t)} \\
+\sum_{k=1}^{n}\frac{i\hbar}{2}\left(\Delta_{\mathbf{x}_k} Z(\mathbf{x}_k,t)\right)e^{\sum_{j=1}^{n}\frac{im_j}{\hbar}Z(\mathbf{x}_j,t)}\psi &+ \sum_{k=1}^{n}\frac{\hbar^2}{2m_k}\left(\Delta_{\mathbf{x}_k}\psi\right)e^{\sum_{j=1}^{n}\frac{im_j}{\hbar}Z(\mathbf{x}_j,t)} = \\
&\sum_{k=1}^{n}\frac{i\hbar\sigma_k}{2m_kc}\left(\operatorname{div}_{\mathbf{x}_k}\mathbf{A}(\mathbf{x}_k,t)\right)e^{\sum_{j=1}^{n}\frac{im_j}{\hbar}Z(\mathbf{x}_j,t)}\psi \\
&+\sum_{k=1}^{n}\frac{i\hbar\sigma_k}{m_kc}\left(\mathbf{A}(\mathbf{x}_k,t)\cdot\nabla_{\mathbf{x}_k}\psi\right)e^{\sum_{j=1}^{n}\frac{im_j}{\hbar}Z(\mathbf{x}_j,t)} \\
-\sum_{k=1}^{n}\frac{\sigma_k}{c}\left(\mathbf{A}(\mathbf{x}_k,t)\cdot\nabla_{\mathbf{x}_k} Z(\mathbf{x}_k,t)\right)&e^{\sum_{j=1}^{n}\frac{im_j}{\hbar}Z(\mathbf{x}_j,t)}\psi - V\left(e^{\sum_{j=1}^{n}\frac{im_j}{\hbar}Z(\mathbf{x}_j,t)}\psi\right) \\
&+\sum_{k=1}^{n}\left(\sigma_k\Psi(\mathbf{x}_k,t) + \frac{\sigma_k^2}{2m_kc^2}\left|\mathbf{A}(\mathbf{x}_k,t)\right|^2\right)\left(e^{\sum_{j=1}^{n}\frac{im_j}{\hbar}Z(\mathbf{x}_j,t)}\psi\right).
\end{aligned}
\tag{6.71}
$$

We rewrite (6.71) as

$$
\begin{aligned}
i\hbar\frac{\partial}{\partial t}\left(e^{\sum_{j=1}^{n}\frac{im_j}{\hbar}Z(\mathbf{x}_j,t)}\psi\right) &+ \sum_{k=1}^{n}\frac{\hbar^2}{2m_k}\Delta_{\mathbf{x}_k}\left(e^{\sum_{j=1}^{n}\frac{im_j}{\hbar}Z(\mathbf{x}_j,t)}\psi\right) = \\
&\sum_{k=1}^{n}\frac{i\hbar\sigma_k}{2m_kc}\left(\operatorname{div}_{\mathbf{x}_k}\mathbf{A}(\mathbf{x}_k,t)\right)e^{\sum_{j=1}^{n}\frac{im_j}{\hbar}Z(\mathbf{x}_j,t)}\psi \\
&+\sum_{k=1}^{n}\frac{i\hbar\sigma_k}{m_kc}\mathbf{A}(\mathbf{x}_k,t)\cdot\nabla_{\mathbf{x}_k}\left(e^{\sum_{j=1}^{n}\frac{im_j}{\hbar}Z(\mathbf{x}_j,t)}\psi\right) \\
+\sum_{k=1}^{n}\left(\sigma_k\Psi(\mathbf{x}_k,t) + \frac{\sigma_k^2}{2m_kc^2}\left|\mathbf{A}(\mathbf{x}_k,t)\right|^2\right)&\left(e^{\sum_{j=1}^{n}\frac{im_j}{\hbar}Z(\mathbf{x}_j,t)}\psi\right) - V\left(e^{\sum_{j=1}^{n}\frac{im_j}{\hbar}Z(\mathbf{x}_j,t)}\psi\right) \\
&-\sum_{k=1}^{n}m_k\left(\frac{\partial Z}{\partial t}(\mathbf{x}_k,t) + \frac{1}{2}\left|\nabla_{\mathbf{x}_k} Z(\mathbf{x}_k,t)\right|^2\right)\left(e^{\sum_{j=1}^{n}\frac{im_j}{\hbar}Z(\mathbf{x}_j,t)}\psi\right).
\end{aligned}
\tag{6.72}
$$

Therefore, inserting (6.69) into (6.72) gives

$$
\begin{aligned}
i\hbar\frac{\partial}{\partial t}\left(e^{\sum_{j=1}^{n}\frac{im_j}{\hbar}Z(\mathbf{x}_j,t)}\psi\right) &= -\sum_{k=1}^{n}\frac{\hbar^2}{2m_k}\Delta_{\mathbf{x}}\left(e^{\sum_{j=1}^{n}\frac{im_j}{\hbar}Z(\mathbf{x}_j,t)}\psi\right)\\
&-V\left(e^{\sum_{j=1}^{n}\frac{im_j}{\hbar}Z(\mathbf{x}_j,t)}\psi\right)+\sum_{k=1}^{n}\frac{i\hbar\sigma_k}{2m_kc}\left(\operatorname{div}_{\mathbf{x}_k}\mathbf{A}(\mathbf{x}_k,t)\right)e^{\sum_{j=1}^{n}\frac{im_j}{\hbar}Z(\mathbf{x}_j,t)}\psi\\
&+\sum_{k=1}^{n}\frac{i\hbar\sigma_k}{m_kc}\mathbf{A}(\mathbf{x}_k,t)\cdot\nabla_{\mathbf{x}_k}\left(e^{\sum_{j=1}^{n}\frac{im_j}{\hbar}Z(\mathbf{x}_j,t)}\psi\right)\\
+\sum_{k=1}^{n}&\left(\sigma_k\Psi(\mathbf{x}_k,t)+\frac{\sigma_k^2}{2m_kc^2}|\mathbf{A}(\mathbf{x}_k,t)|^2+m_k\Phi(\mathbf{x}_k,t)\right)\left(e^{\sum_{j=1}^{n}\frac{im_j}{\hbar}Z(\mathbf{x}_j,t)}\psi\right).
\end{aligned}
\tag{6.73}
$$

Then denoting

$$
\psi_1 := e^{\sum_{j=1}^{n}\frac{im_j}{\hbar}Z(\mathbf{x}_j,t)}\psi, \tag{6.74}
$$

we obtain in the coordinate system (∗) the Shrödinger equation in the form

$$
\begin{aligned}
i\hbar\frac{\partial\psi_1}{\partial t} &= -\sum_{j=1}^{n}\frac{\hbar^2}{2m_j}\Delta_{\mathbf{x}_j}\psi_1+\sum_{j=1}^{n}\frac{i\hbar\sigma_j}{2m_jc}\operatorname{div}_{\mathbf{x}_j}\left\{\psi_1\mathbf{A}(\mathbf{x}_j,t)\right\}+\sum_{j=1}^{n}\frac{i\hbar\sigma_j}{2m_jc}\mathbf{A}(\mathbf{x}_j,t)\cdot\nabla_{\mathbf{x}_j}\psi_1\\
&+\sum_{j=1}^{n}\left(\sigma_j\Psi(\mathbf{x}_j,t)+\frac{\sigma_j^2}{2m_jc^2}\left|\mathbf{A}(\mathbf{x}_j,t)\right|^2+m_j\Phi(\mathbf{x}_j,t)\right)\psi_1-V\psi_1,
\end{aligned}
\tag{6.75}
$$

which coincides with the classical Shrödinger equation for this case.

Remark 6.1 Note that by (4.69) in Remark 4.3, equality (6.74) implies that under the change of coordinate system given by the Galilean Transformation

$$
\begin{cases}\mathbf{x}'=\mathbf{x}+\mathbf{w}t,\\ t'=t,\end{cases} \tag{6.76}
$$

the quantity ψ_1 transforms as

$$
\psi_1' := e^{\sum_{j=1}^{n}\frac{im_j}{\hbar}\left(\mathbf{w}\cdot\mathbf{x}_j+\frac{1}{2}|\mathbf{w}|^2t\right)}\psi_1, \tag{6.77}
$$

provided that $\psi'=\psi$. Moreover, (6.77) coincides with the classical law of transformation of the wave function, under the Galilean Transformation (see Sect. 17 in [2], the end of the section).

Next, again consider the motion and interaction of system of n quantum microparticles having inertial masses $m_1,\ldots,m_n$ and the charges $\sigma_1,\ldots,\sigma_n$ with the known gravitational and electromagnetical field with potentials $\mathbf{v}(\mathbf{x},t)$, $\mathbf{A}(\mathbf{x},t)$

and $\Psi(\mathbf{x}, t)$ and additional conservative field with potential $V(\mathbf{y}_1, \ldots, \mathbf{y}_n, t)$. Then consider a Lagrangian density L defined by

$$
\begin{aligned}
&L_2(\psi, \mathbf{A}, \Psi, \mathbf{v}, \mathbf{x}_1, \ldots, \mathbf{x}_n, t) := \\
&\frac{i\hbar}{2}\left(\left(\frac{\partial\psi}{\partial t} + \sum_{k=1}^{n} \mathbf{v}(\mathbf{x}_k, t) \cdot \nabla_{\mathbf{x}_k}\psi\right) \cdot \bar{\psi} - \psi \cdot \left(\frac{\partial\bar{\psi}}{\partial t} + \sum_{k=1}^{n} \mathbf{v}(\mathbf{x}_k, t) \cdot \nabla_{\mathbf{x}_k}\bar{\psi}\right)\right) \\
&\qquad - \sum_{k=1}^{n} \frac{\hbar^2}{2m_k} \nabla_{\mathbf{x}_k}\psi \cdot \nabla_{\mathbf{x}_k}\bar{\psi} + V(\mathbf{x}_1, \ldots, \mathbf{x}_n, t)\, \psi \cdot \bar{\psi} \\
&- \sum_{k=1}^{n} \frac{\hbar\sigma_k i}{2m_k c}\left(\nabla_{\mathbf{x}_k}\psi \cdot \bar{\psi} - \psi \cdot \nabla_{\mathbf{x}_k}\bar{\psi}\right) \cdot \mathbf{A}(\mathbf{x}_k, t) - \sum_{k=1}^{n} \frac{\sigma_k^2}{2m_k c^2} |\mathbf{A}(\mathbf{x}_k, t)|^2 \psi \cdot \bar{\psi} \\
&\qquad - \sum_{k=1}^{n} \sigma_k \left(\Psi(\mathbf{x}_k, t) - \frac{1}{c}\mathbf{v}(\mathbf{x}_k, t) \cdot \mathbf{A}(\mathbf{x}_k, t)\right) \psi \cdot \bar{\psi}, \qquad (6.78)
\end{aligned}
$$

where $\psi(\mathbf{x}_1, \ldots, \mathbf{x}_n, t) \in \mathbb{C}$ is a wave function of the system. Then, as before, it can be proven that L is invariant under the change of inertial or non-inertial Cartesian coordinate systems of the form

$$
\begin{cases}
t' = t \\
\mathbf{x}'_k = A(t) \cdot \mathbf{x}_k + \mathbf{z}(t) \qquad \forall k = 1, \ldots, n,
\end{cases}
$$

provided that $\psi' = \psi$. We investigate stationary points of the functional

$$
J = \int_0^T \int_{(\mathbb{R}^3)^n} L(\psi, \mathbf{A}, \Psi, \mathbf{v}, \mathbf{x}_1, \ldots, \mathbf{x}_n, t)\, d\mathbf{x}_1 \ldots, d\mathbf{x}_n dt. \qquad (6.79)
$$

Then,

$$
\begin{aligned}
0 = \frac{\delta L_2}{\delta(\bar{\psi})} = i\hbar&\left(\frac{\partial\psi}{\partial t} + \sum_{k=1}^{n} \frac{1}{2}\mathbf{v}(\mathbf{x}_k, t) \cdot \nabla_{\mathbf{x}_k}\psi + \sum_{k=1}^{n} \frac{1}{2}\operatorname{div}_{\mathbf{x}_k}\{\psi\mathbf{v}(\mathbf{x}_k, t)\}\right) \\
&+ \sum_{k=1}^{n} \frac{\hbar^2}{2m_k} \Delta_{\mathbf{x}_k}\psi + V(\mathbf{x}_1, \ldots, \mathbf{x}_n, t)\, \psi \\
- \sum_{k=1}^{n} \frac{\hbar\sigma_k i}{2m_k c}&\left(\mathbf{A}(\mathbf{x}_k, t) \cdot \nabla_{\mathbf{x}_k}\psi + \operatorname{div}_{\mathbf{x}_k}\{\psi\mathbf{A}(\mathbf{x}_k, t)\}\right) - \sum_{k=1}^{n} \frac{\sigma_k^2}{2m_k c^2} |\mathbf{A}(\mathbf{x}_k, t)|^2 \psi \\
&- \sum_{k=1}^{n} \sigma_k \left(\Psi(\mathbf{x}_k, t) - \frac{1}{c}\mathbf{v}(\mathbf{x}_k, t) \cdot \mathbf{A}(\mathbf{x}_k, t)\right) \psi, \qquad (6.80)
\end{aligned}
$$

and

$$0 = \frac{\delta L_2}{\delta(\psi)} = (\bar{i})\hbar\left(\frac{\partial\bar{\psi}}{\partial t} + \sum_{k=1}^{n}\frac{1}{2}\mathbf{v}(\mathbf{x}_k,t)\cdot\nabla_{\mathbf{x}_k}\bar{\psi} + \sum_{k=1}^{n}\frac{1}{2}\operatorname{div}_{\mathbf{x}_k}\left\{\bar{\psi}\mathbf{v}(\mathbf{x}_k,t)\right\}\right)$$

$$+\sum_{k=1}^{n}\frac{\hbar^2}{2m_k}\Delta_{\mathbf{x}_k}\bar{\psi} + V(\mathbf{x}_1,\ldots,\mathbf{x}_n,t)\,\bar{\psi}$$

$$-\sum_{k=1}^{n}\frac{\hbar\sigma_k(\bar{i})}{2m_k c}\left(\mathbf{A}(\mathbf{x}_k,t)\cdot\nabla_{\mathbf{x}_k}\bar{\psi} + \operatorname{div}_{\mathbf{x}_k}\left\{\bar{\psi}\mathbf{A}(\mathbf{x}_k,t)\right\}\right) - \sum_{k=1}^{n}\frac{\sigma_k^2}{2m_k c^2}\left|\mathbf{A}(\mathbf{x}_k,t)\right|^2\bar{\psi}$$

$$-\sum_{k=1}^{n}\sigma_k\left(\Psi(\mathbf{x}_k,t) - \frac{1}{c}\mathbf{v}(\mathbf{x}_k,t)\cdot\mathbf{A}(\mathbf{x}_k,t)\right)\bar{\psi}, \qquad (6.81)$$

where the last equality is just the complex conjugate of (6.80). So we get that the Euler–Lagrange equation for (6.79) coincides with the Schrödinger equation of the form (6.65).

Finally, assume that the first and the second particles have the same mass $m_1 = m_2$ and the same charge $\sigma_1 = \sigma_2$ and moreover, assume that we have

$$V(\mathbf{x}_1,\mathbf{x}_2,\mathbf{x}_3,\ldots,\mathbf{x}_n,t) = V(\mathbf{x}_2,\mathbf{x}_1,\mathbf{x}_3,\ldots,\mathbf{x}_n,t) \qquad \forall\mathbf{x}_1,\mathbf{x}_2,\mathbf{x}_3,\ldots,\mathbf{x}_n \in \mathbb{R}^3, \quad \forall t.$$

In this case it can be easily deduced that if $\psi(\mathbf{x}_1,\mathbf{x}_2,\mathbf{x}_3,\ldots,\mathbf{x}_n,t)$ is a solution of (6.65), then $\psi(\mathbf{x}_2,\mathbf{x}_1,\mathbf{x}_3,\ldots,\mathbf{x}_n,t)$ as also a solutions of (6.65), Therefore, if $\psi(\mathbf{x}_1,\mathbf{x}_2,\mathbf{x}_3,\ldots,\mathbf{x}_n,t)$ is a solution of (6.65), then for every $t \geq 0$ we will have either

$$\psi(\mathbf{x}_2,\mathbf{x}_1,\mathbf{x}_3,\ldots,\mathbf{x}_n,t) = -\psi(\mathbf{x}_1,\mathbf{x}_2,\mathbf{x}_3,\ldots,\mathbf{x}_n,t) \qquad \forall\mathbf{x}_1,\mathbf{x}_2,\mathbf{x}_3,\ldots,\mathbf{x}_n \in \mathbb{R}^3, \qquad (6.82)$$

or

$$\psi(\mathbf{x}_2,\mathbf{x}_1,\mathbf{x}_3,\ldots,\mathbf{x}_n,t) = \psi(\mathbf{x}_1,\mathbf{x}_2,\mathbf{x}_3,\ldots,\mathbf{x}_n,t) \qquad \forall\mathbf{x}_1,\mathbf{x}_2,\mathbf{x}_3,\ldots,\mathbf{x}_n \in \mathbb{R}^3, \qquad (6.83)$$

provided that either (6.82) or (6.83) respectively holds for the initial instant of time $t = 0$. So we have a consistency with the principles of identity for two or more identical fermions or bosons, in the cases where we do not take into account the spin interaction.

6.5 Quantum Liouville's Equation for a Finite System of Quantum Particles

Consider the statistical description of the motion of a system of n quantum micro-particles with inertial masses $m_1, \dots, m_n$ and the charges $\sigma_1, \dots, \sigma_n$ in the outer gravitational and electromagnetical field with characteristics $\mathbf{v}(\mathbf{x}, t)$, $\mathbf{A}(\mathbf{x}, t)$ and $\Psi(\mathbf{x}, t)$ and additional conservative field with potential $V(\mathbf{y}_1, \dots, \mathbf{y}_n, t)$, not taking into account the spin interaction. Then it is well known that the Quantum Liouville's equation for this system of particles has the following form:

$$i\hbar\frac{\partial\xi}{\partial t}(\mathbf{x}_1, \dots, \mathbf{x}_n, \mathbf{y}_1, \dots, \mathbf{y}_n, t) = \hat{H}_0(\mathbf{x}_1, \dots, \mathbf{x}_n, t)\cdot\xi(\mathbf{x}_1, \dots, \mathbf{x}_n, \mathbf{y}_1, \dots, \mathbf{y}_n, t)$$
$$- \hat{H}_0^*(\mathbf{y}_1, \dots, \mathbf{y}_n, t)\cdot\xi(\mathbf{x}_1, \dots, \mathbf{x}_n, \mathbf{y}_1, \dots, \mathbf{y}_n, t), \quad (6.84)$$

where $\xi(\mathbf{x}_1, \dots, \mathbf{x}_n, \mathbf{y}_1, \dots, \mathbf{y}_n, t) \in \mathbb{C}$ is a density matrix function and $\hat{H}_0(\mathbf{x}_1, \dots, \mathbf{x}_n, t)$ is the Hamiltonian operator acting on the variables $(\mathbf{x}_1, \dots, \mathbf{x}_n)$ and $\hat{H}_0^*(\mathbf{y}_1, \dots, \mathbf{y}_n, t)$ is the complex conjugate (not the Hermitian adjoint) to the Hamiltonian operator acting on the variables $(\mathbf{y}_1, \dots, \mathbf{y}_n)$. Since by (6.15) the Hamiltonian for a macro-particles has the form

$$H_{\text{macro}}(\mathbf{P}_1, \dots, \mathbf{P}_n, \mathbf{r}_1, \dots, \mathbf{r}_n, t) =$$
$$- V(\mathbf{r}_1, \dots, \mathbf{r}_n, t) + \sum_{j=1}^{n}\frac{1}{2}\mathbf{P}_j\cdot\mathbf{v}(\mathbf{r}_j, t) + \sum_{j=1}^{n}\frac{1}{2}\mathbf{v}(\mathbf{r}_j, t)\cdot\mathbf{P}_j$$
$$+ \sum_{j=1}^{n}\frac{1}{2m_j}\left|\mathbf{P}_j - \frac{\sigma_j}{c}\mathbf{A}(\mathbf{r}_j, t)\right|^2 + \sum_{j=1}^{n}\sigma_j\left(\Psi(\mathbf{r}_j, t) - \frac{1}{c}\mathbf{A}(\mathbf{r}_j, t)\cdot\mathbf{v}(\mathbf{r}_j, t)\right), \quad (6.85)$$

as before in (6.63), we built the Hamiltonian operator as

$$\hat{H}_0(\mathbf{x}_1, \dots, \mathbf{x}_n, t)\cdot\psi(\mathbf{x}_1, \dots, \mathbf{x}_n, t) = -\sum_{j=1}^{n}\frac{\hbar^2}{2m_j}\Delta_{\mathbf{x}_j}\psi - \sum_{j=1}^{n}\frac{i\hbar}{2}\,\mathrm{div}_{\mathbf{x}_j}\{\psi\mathbf{v}(\mathbf{x}_j, t)\}$$
$$- \sum_{j=1}^{n}\frac{i\hbar}{2}\mathbf{v}(\mathbf{x}_j, t)\cdot\nabla_{\mathbf{x}_j}\psi + \sum_{j=1}^{n}\frac{i\hbar\sigma_j}{2m_jc}\,\mathrm{div}_{\mathbf{x}_j}\{\psi\mathbf{A}(\mathbf{x}_j, t)\} + \sum_{j=1}^{n}\frac{i\hbar\sigma_j}{2m_jc}\mathbf{A}(\mathbf{x}_j, t)\cdot\nabla_{\mathbf{x}_j}\psi$$
$$+ \sum_{j=1}^{n}\left(\sigma_j\Psi(\mathbf{x}_j, t) - \frac{\sigma_j}{c}\mathbf{A}(\mathbf{x}_j, t)\cdot\mathbf{v}(\mathbf{x}_j, t) + \frac{\sigma_j^2}{2m_jc^2}|\mathbf{A}(\mathbf{x}_j, t)|^2\right)\psi - V(\mathbf{x}_1, \dots, \mathbf{x}_n, t)\,\psi, \quad (6.86)$$

and consistently with (6.86):

$$\hat{H}_0^* \left(\mathbf{y}_1, \ldots, \mathbf{y}_n, t\right) \cdot \psi \left(\mathbf{y}_1, \ldots, \mathbf{y}_n, t\right) = -\sum_{j=1}^{n} \frac{\hbar^2}{2m_j} \Delta_{\mathbf{y}_j} \psi + \sum_{j=1}^{n} \frac{i\hbar}{2} \operatorname{div}_{\mathbf{y}_j} \left\{\psi \mathbf{v}(\mathbf{y}_j, t)\right\}$$
$$+ \sum_{j=1}^{n} \frac{i\hbar}{2} \mathbf{v}(\mathbf{y}_j, t) \cdot \nabla_{\mathbf{y}_j} \psi - \sum_{j=1}^{n} \frac{i\hbar\sigma_j}{2m_j c} \operatorname{div}_{\mathbf{y}_j} \left\{\psi \mathbf{A}(\mathbf{y}_j, t)\right\} - \sum_{j=1}^{n} \frac{i\hbar\sigma_j}{2m_j c} \mathbf{A}(\mathbf{y}_j, t) \cdot \nabla_{\mathbf{y}_j} \psi$$
$$+ \sum_{j=1}^{n} \left(\sigma_j \Psi(\mathbf{y}_j, t) - \frac{\sigma_j}{c} \mathbf{A}(\mathbf{y}_j, t) \cdot \mathbf{v}(\mathbf{y}_j, t) + \frac{\sigma_j^2}{2m_j c^2} \left|\mathbf{A}(\mathbf{y}_j, t)\right|^2\right) \psi$$
$$- V \left(\mathbf{y}_1, \ldots, \mathbf{y}_n, t\right) \psi. \quad (6.87)$$

Thus we rewrite the corresponding Quantum Liouville's equation (6.84) as

$$i\hbar \left(\frac{\partial \xi}{\partial t} + \sum_{j=1}^{n} \mathbf{v}(\mathbf{x}_j, t) \cdot \nabla_{\mathbf{x}_j} \xi + \sum_{j=1}^{n} \mathbf{v}(\mathbf{y}_j, t) \cdot \nabla_{\mathbf{y}_j} \xi\right)$$
$$+ \sum_{j=1}^{n} \frac{i\hbar}{2} \left(\operatorname{div}_{\mathbf{x}_j} \mathbf{v}(\mathbf{x}_j, t) + \operatorname{div}_{\mathbf{y}_j} \mathbf{v}(\mathbf{y}_j, t)\right) \xi = -\sum_{j=1}^{n} \frac{\hbar^2}{2m_j} \left(\Delta_{\mathbf{x}_j} \xi - \Delta_{\mathbf{y}_j} \xi\right)$$
$$- \left(V \left(\mathbf{x}_1, \ldots, \mathbf{x}_n, t\right) - V \left(\mathbf{y}_1, \ldots, \mathbf{y}_n, t\right)\right) \xi$$
$$+ \sum_{j=1}^{n} \frac{i\hbar\sigma_j}{2m_j c} \left(\operatorname{div}_{\mathbf{x}_j} \left\{\xi \mathbf{A}(\mathbf{x}_j, t)\right\} + \operatorname{div}_{\mathbf{y}_j} \left\{\xi \mathbf{A}(\mathbf{y}_j, t)\right\}\right)$$
$$+ \sum_{j=1}^{n} \frac{i\hbar\sigma_j}{2m_j c} \left(\mathbf{A}(\mathbf{x}_j, t) \cdot \nabla_{\mathbf{x}_j} \xi + \mathbf{A}(\mathbf{y}_j, t) \cdot \nabla_{\mathbf{y}_j} \xi\right)$$
$$+ \sum_{j=1}^{n} \left(\sigma_j \Psi(\mathbf{x}_j, t) - \frac{\sigma_j}{c} \mathbf{A}(\mathbf{x}_j, t) \cdot \mathbf{v}(\mathbf{x}_j, t) + \frac{\sigma_j^2}{2m_j c^2} \left|\mathbf{A}(\mathbf{x}_j, t)\right|^2\right) \xi$$
$$- \sum_{j=1}^{n} \left(\sigma_j \Psi(\mathbf{y}_j, t) - \frac{\sigma_j}{c} \mathbf{A}(\mathbf{y}_j, t) \cdot \mathbf{v}(\mathbf{y}_j, t) + \frac{\sigma_j^2}{2m_j c^2} \left|\mathbf{A}(\mathbf{y}_j, t)\right|^2\right) \xi. \quad (6.88)$$

Next consider a change of some non-inertial Cartesian coordinate system (∗) to another Cartesian coordinate system (∗∗) of the form (4.2):

$$\begin{cases} \mathbf{x}' = A(t) \cdot \mathbf{x} + \mathbf{z}(t), \\ t' = t, \end{cases} \quad (6.89)$$

where $A(t) \in SO(3)$ is a rotation. Then, since

$$\begin{cases} \mathbf{x}'_j = A(t) \cdot \mathbf{x}_j + \mathbf{z}(t) \quad \forall j = 1, \ldots, n \\ \mathbf{y}'_j = A(t) \cdot \mathbf{y}_j + \mathbf{z}(t) \quad \forall j = 1, \ldots, n \\ \xi'\left(\mathbf{x}'_1, \ldots, \mathbf{x}'_n, \mathbf{y}'_1, \ldots, \mathbf{y}'_n, t\right) = \xi\left(\mathbf{x}_1, \ldots, \mathbf{x}_n, \mathbf{y}_1, \ldots, \mathbf{y}_n, t\right) \\ V'\left(\mathbf{y}'_1, \ldots, \mathbf{y}'_n, t\right) = V\left(\mathbf{y}_1, \ldots, \mathbf{y}_n, t\right) \\ \mathbf{A}' = A(t) \cdot \mathbf{A} \\ \Psi' - \frac{1}{c}\mathbf{A}' \cdot \mathbf{v}' = \Psi - \frac{1}{c}\mathbf{A} \cdot \mathbf{v}, \end{cases} \tag{6.90}$$

we deduce that the Quantum Liouville's equation of the form (6.88) is invariant under the change of non-inertial Cartesian coordinate system, provided we have $\xi' = \xi$.

Next assume that $\hat{A}\left(\mathbf{x}_1, \ldots, \mathbf{x}_n, t\right)$ is some Hermitian operator acting on the functions $\psi\left(\mathbf{x}_1, \ldots, \mathbf{x}_n, t\right) \in \mathbb{C}$ with respect to spatial variables $(\mathbf{x}_1, \ldots, \mathbf{x}_n)$. Then the average $\tilde{A}(t)$ of $\hat{A}$ on the density matrix ξ is defined by

$$\tilde{A}(t) = \frac{\int \left(\hat{A}\left(\mathbf{x}_1, \ldots, \mathbf{x}_n, t\right) \cdot \xi\right)\left(\mathbf{z}_1, \ldots, \mathbf{z}_n, \mathbf{z}_1, \ldots, \mathbf{z}_n, t\right) d\mathbf{z}_1 \ldots d\mathbf{z}_n}{\int \xi\left(\mathbf{z}_1, \ldots, \mathbf{z}_n, \mathbf{z}_1, \ldots, \mathbf{z}_n, t\right) d\mathbf{z}_1 \ldots d\mathbf{z}_n}. \tag{6.91}$$

Thus, using the fact that $\hat{A}$ is Hermitian, it can be easily deduced that in addition to (6.91) we have the following identity:

$$\int \left(\hat{A}\left(\mathbf{x}_1, \ldots, \mathbf{x}_n, t\right) \cdot \xi\right)\left(\mathbf{z}_1, \ldots, \mathbf{z}_n, \mathbf{z}_1, \ldots, \mathbf{z}_n, t\right) d\mathbf{z}_1 \ldots d\mathbf{z}_n = \\ \int \left(\hat{A}^*\left(\mathbf{y}_1, \ldots, \mathbf{y}_n, t\right) \cdot \xi\right)\left(\mathbf{z}_1, \ldots, \mathbf{z}_n, \mathbf{z}_1, \ldots, \mathbf{z}_n, t\right) d\mathbf{z}_1 \ldots d\mathbf{z}_n, \tag{6.92}$$

where $\hat{A}^*\left(\mathbf{y}_1, \ldots, \mathbf{y}_n, t\right)$ is the complex conjugate (not the Hermitian adjoint) to the operator $\hat{A}$ acting on the variables $(\mathbf{y}_1, \ldots, \mathbf{y}_n)$ and so,

$$\tilde{A}(t) = \frac{\int \left(\hat{A}^*\left(\mathbf{y}_1, \ldots, \mathbf{y}_n, t\right) \cdot \xi\right)\left(\mathbf{z}_1, \ldots, \mathbf{z}_n, \mathbf{z}_1, \ldots, \mathbf{z}_n, t\right) d\mathbf{z}_1 \ldots d\mathbf{z}_n}{\int \xi\left(\mathbf{z}_1, \ldots, \mathbf{z}_n, \mathbf{z}_1, \ldots, \mathbf{z}_n, t\right) d\mathbf{z}_1 \ldots d\mathbf{z}_n}. \tag{6.93}$$

In particular, by inserting (6.92) in the particular case $\hat{A} = \hat{H}_0$ into (6.84) we deduce

$$i\hbar\frac{\partial}{\partial t} \int \xi\left(\mathbf{z}_1, \ldots, \mathbf{z}_n, \mathbf{z}_1, \ldots, \mathbf{z}_n, t\right) d\mathbf{z}_1 \ldots d\mathbf{z}_n = 0, \tag{6.94}$$

and so,

$$\int \xi\left(\mathbf{z}_1, \ldots, \mathbf{z}_n, \mathbf{z}_1, \ldots, \mathbf{z}_n, t\right) d\mathbf{z}_1 \ldots d\mathbf{z}_n = \int \xi\left(\mathbf{z}_1, \ldots, \mathbf{z}_n, \mathbf{z}_1, \ldots, \mathbf{z}_n, 0\right) d\mathbf{z}_1 \ldots d\mathbf{z}_n. \quad (6.95)$$

Next, we remind that by (6.84) we have

$$i\hbar \frac{\partial \xi}{\partial t}\left(\mathbf{x}_1, \ldots, \mathbf{x}_n, \mathbf{y}_1, \ldots, \mathbf{y}_n, t\right) = \hat{H}_0\left(\mathbf{x}_1, \ldots, \mathbf{x}_n, t\right) \cdot \xi\left(\mathbf{x}_1, \ldots, \mathbf{x}_n, \mathbf{y}_1, \ldots, \mathbf{y}_n, t\right) - \hat{H}_0^*\left(\mathbf{y}_1, \ldots, \mathbf{y}_n, t\right) \cdot \xi\left(\mathbf{x}_1, \ldots, \mathbf{x}_n, \mathbf{y}_1, \ldots, \mathbf{y}_n, t\right). \quad (6.96)$$

Thus if we denote,

$$\xi_1\left(\mathbf{x}_1, \ldots, \mathbf{x}_n, \mathbf{y}_1, \ldots, \mathbf{y}_n, t\right) := \bar{\xi}\left(\mathbf{y}_1, \ldots, \mathbf{y}_n, \mathbf{x}_1, \ldots, \mathbf{x}_n, t\right), \quad (6.97)$$

and

$$\zeta\left(\mathbf{x}_1, \ldots, \mathbf{x}_n, \mathbf{y}_1, \ldots, \mathbf{y}_n, t\right) := \xi\left(\mathbf{x}_1, \ldots, \mathbf{x}_n, \mathbf{y}_1, \ldots, \mathbf{y}_n, t\right) - \bar{\xi}\left(\mathbf{y}_1, \ldots, \mathbf{y}_n, \mathbf{x}_1, \ldots, \mathbf{x}_n, t\right), \quad (6.98)$$

then

$$-i\hbar \frac{\partial \bar{\xi}}{\partial t}\left(\mathbf{y}_1, \ldots, \mathbf{y}_n, \mathbf{x}_1, \ldots, \mathbf{x}_n, t\right) = \hat{H}_0^*\left(\mathbf{y}_1, \ldots, \mathbf{y}_n, t\right) \cdot \bar{\xi}\left(\mathbf{y}_1, \ldots, \mathbf{y}_n, \mathbf{x}_1, \ldots, \mathbf{x}_n, t\right) - \hat{H}_0\left(\mathbf{x}_1, \ldots, \mathbf{x}_n, t\right) \cdot \bar{\xi}\left(\mathbf{y}_1, \ldots, \mathbf{y}_n, \mathbf{x}_1, \ldots, \mathbf{x}_n, t\right), \quad (6.99)$$

and then implies

$$i\hbar \frac{\partial \xi_1}{\partial t}\left(\mathbf{x}_1, \ldots, \mathbf{x}_n, \mathbf{y}_1, \ldots, \mathbf{y}_n, t\right) = \hat{H}_0\left(\mathbf{x}_1, \ldots, \mathbf{x}_n, t\right) \cdot \xi_1\left(\mathbf{x}_1, \ldots, \mathbf{x}_n, \mathbf{y}_1, \ldots, \mathbf{y}_n, t\right) - \hat{H}_0^*\left(\mathbf{y}_1, \ldots, \mathbf{y}_n, t\right) \cdot \xi_1\left(\mathbf{x}_1, \ldots, \mathbf{x}_n, \mathbf{y}_1, \ldots, \mathbf{y}_n, t\right). \quad (6.100)$$

Therefore,

$$i\hbar \frac{\partial \zeta}{\partial t}\left(\mathbf{x}_1, \ldots, \mathbf{x}_n, \mathbf{y}_1, \ldots, \mathbf{y}_n, t\right) = \hat{H}_0\left(\mathbf{x}_1, \ldots, \mathbf{x}_n, t\right) \cdot \zeta\left(\mathbf{x}_1, \ldots, \mathbf{x}_n, \mathbf{y}_1, \ldots, \mathbf{y}_n, t\right) - \hat{H}_0^*\left(\mathbf{y}_1, \ldots, \mathbf{y}_n, t\right) \cdot \zeta\left(\mathbf{x}_1, \ldots, \mathbf{x}_n, \mathbf{y}_1, \ldots, \mathbf{y}_n, t\right). \quad (6.101)$$

Thus, if ζ satisfies initial condition $\zeta\left(\mathbf{x}_1, \ldots, \mathbf{x}_n, \mathbf{y}_1, \ldots, \mathbf{y}_n, 0\right) = 0$, then $\zeta\left(\mathbf{x}_1, \ldots, \mathbf{x}_n, \mathbf{y}_1, \ldots, \mathbf{y}_n, t\right) = 0$ for every $t > 0$. So

$$\xi\left(\mathbf{x}_1,\ldots,\mathbf{x}_n,\mathbf{y}_1,\ldots,\mathbf{y}_n,0\right)=\bar{\xi}\left(\mathbf{y}_1,\ldots,\mathbf{y}_n,\mathbf{x}_1,\ldots,\mathbf{x}_n,0\right) \quad \text{implies}$$
$$\xi\left(\mathbf{x}_1,\ldots,\mathbf{x}_n,\mathbf{y}_1,\ldots,\mathbf{y}_n,t\right)=\bar{\xi}\left(\mathbf{y}_1,\ldots,\mathbf{y}_n,\mathbf{x}_1,\ldots,\mathbf{x}_n,t\right) \quad \forall t\geq 0. \tag{6.102}$$

Finally, assume that the first and the second particles have the same mass $m_1 = m_2$ and the same charge $\sigma_1 = \sigma_2$ and moreover, assume that we have

$$V\left(\mathbf{x}_1,\mathbf{x}_2,\mathbf{x}_3,\ldots,\mathbf{x}_n,t\right)=V\left(\mathbf{x}_2,\mathbf{x}_1,\mathbf{x}_3,\ldots,\mathbf{x}_n,t\right) \qquad \forall\mathbf{x}_1,\mathbf{x}_2,\mathbf{x}_3,\ldots,\mathbf{x}_n\in\mathbb{R}^3, \quad \forall t.$$

In this case it can be easily deduced that if $\xi\left(\mathbf{x}_1,\mathbf{x}_2,\mathbf{x}_3,\ldots,\mathbf{x}_n,\mathbf{y}_1,\mathbf{y}_2,\mathbf{y}_3,\ldots,\mathbf{y}_n,t\right)$ is a solution of (6.96), then $\xi(\mathbf{x}_2,\mathbf{x}_1,\mathbf{x}_3,\ldots,\mathbf{x}_n,\mathbf{y}_2,\mathbf{y}_1,\mathbf{y}_3,\ldots,\mathbf{y}_n,t)$ is also a solution of (6.96). Therefore, if $\xi\left(\mathbf{x}_1,\mathbf{x}_2,\mathbf{x}_3,\ldots,\mathbf{x}_n,\mathbf{y}_1,\mathbf{y}_2,\mathbf{y}_3,\ldots,\mathbf{y}_n,t\right)$ is a solution of (6.96), then for every $t\geq 0$ we will have

$$\xi(\mathbf{x}_2,\mathbf{x}_1,\mathbf{x}_3,\ldots,\mathbf{x}_n,\mathbf{y}_2,\mathbf{y}_1,\mathbf{y}_3,\ldots,\mathbf{y}_n,t)=\xi\left(\mathbf{x}_1,\mathbf{x}_2,\mathbf{x}_3,\ldots,\mathbf{x}_n,\mathbf{y}_1,\mathbf{y}_2,\mathbf{y}_3,\ldots,\mathbf{y}_n,t\right)$$
$$\forall\mathbf{x}_1,\mathbf{x}_2,\mathbf{x}_3,\ldots,\mathbf{x}_n,\mathbf{y}_1,\mathbf{y}_2,\mathbf{y}_3,\ldots,\mathbf{y}_n\in\mathbb{R}^3, \tag{6.103}$$

provided that (6.103) holds for the initial instant of time $t = 0$. So we have a consistency with the principles of identity for two or more identical fermions or bosons, in the cases where we do not take into account the spin interaction.

Next given $\xi\left(\mathbf{x}_1,\ldots,\mathbf{x}_n,\mathbf{y}_1,\ldots,\mathbf{y}_n,t\right)$ that satisfies

$$\xi\left(\mathbf{x}_1,\ldots,\mathbf{x}_n,\mathbf{y}_1,\ldots,\mathbf{y}_n,t\right)=\bar{\xi}\left(\mathbf{y}_1,\ldots,\mathbf{y}_n,\mathbf{x}_1,\ldots,\mathbf{x}_n,t\right) \quad \forall t\geq 0 \tag{6.104}$$

define the operator $\hat{R}_\xi$ by

$$\hat{R}_\xi\left(\mathbf{x}_1,\ldots,\mathbf{x}_n,t\right)\cdot\psi\left(\mathbf{x}_1,\ldots,\mathbf{x}_n,t\right)$$
$$=\int\xi\left(\mathbf{x}_1,\ldots,\mathbf{x}_n,\mathbf{y}_1,\ldots,\mathbf{y}_n,t\right)\psi\left(\mathbf{y}_1,\ldots,\mathbf{y}_n,t\right)d\mathbf{y}_1\ldots d\mathbf{y}_n. \tag{6.105}$$

Then the mapping $\xi\to\hat{R}_\xi$ is one-to-one. Moreover, by (6.104) $\hat{R}_\xi$ is an Hermitian operator. Finally by (6.96) we have

$$i\hbar\frac{\partial\hat{R}_\xi}{\partial t}\left(\mathbf{x}_1,\ldots,\mathbf{x}_n,t\right)=$$
$$\hat{H}_0\left(\mathbf{x}_1,\ldots,\mathbf{x}_n,t\right)\cdot\hat{R}_\xi\left(\mathbf{x}_1,\ldots,\mathbf{x}_n,t\right)-\hat{R}_\xi\left(\mathbf{x}_1,\ldots,\mathbf{x}_n,t\right)\cdot\hat{H}_0\left(\mathbf{x}_1,\ldots,\mathbf{x}_n,t\right). \tag{6.106}$$

Equation (6.106) is equivalent to (6.96).

Next, clearly ξ satisfies (6.103) if and only if we have either the following:

$$\psi(\mathbf{x}_2, \mathbf{x}_1, \mathbf{x}_3, \ldots, \mathbf{x}_n, t) = -\psi(\mathbf{x}_1, \mathbf{x}_2, \mathbf{x}_3, \ldots, \mathbf{x}_n, t) \;\; \forall \mathbf{x}_1, \ldots, \mathbf{x}_n \in \mathbb{R}^3$$
$$\text{and} \quad \phi = \hat{R}_\xi \cdot \psi$$
$$\text{implies} \quad \phi(\mathbf{x}_2, \mathbf{x}_1, \mathbf{x}_3, \ldots, \mathbf{x}_n, t) = -\phi(\mathbf{x}_1, \mathbf{x}_2, \mathbf{x}_3, \ldots, \mathbf{x}_n, t) \;\; \forall \mathbf{x}_1, \ldots, \mathbf{x}_n \in \mathbb{R}^3, \tag{6.107}$$

or the following:

$$\psi(\mathbf{x}_2, \mathbf{x}_1, \mathbf{x}_3, \ldots, \mathbf{x}_n, t) = \psi(\mathbf{x}_1, \mathbf{x}_2, \mathbf{x}_3, \ldots, \mathbf{x}_n, t) \;\; \forall \mathbf{x}_1, \ldots, \mathbf{x}_n \in \mathbb{R}^3$$
$$\text{and} \quad \phi = \hat{R}_\xi \cdot \psi$$
$$\text{implies} \quad \phi(\mathbf{x}_2, \mathbf{x}_1, \mathbf{x}_3, \ldots, \mathbf{x}_n, t) = \phi(\mathbf{x}_1, \mathbf{x}_2, \mathbf{x}_3, \ldots, \mathbf{x}_n, t) \;\; \forall \mathbf{x}_1, \ldots, \mathbf{x}_n \in \mathbb{R}^3. \tag{6.108}$$

Finally, assume that $\hat{A}(\mathbf{x}_1, \ldots, \mathbf{x}_n, t)$ is some Hermitian operator acting on the functions $\psi(\mathbf{x}_1, \ldots, \mathbf{x}_n, t) \in \mathbb{C}$ with respect to spatial variables $(\mathbf{x}_1, \ldots, \mathbf{x}_n)$. Then we remind that the average $\tilde{A}(t)$ of $\hat{A}$ on the density matrix ξ, is defined by (6.91):

$$\tilde{A}(t) = \frac{\int \left(\hat{A}(\mathbf{x}_1, \ldots, \mathbf{x}_n, t) \cdot \xi\right)(\mathbf{z}_1, \ldots, \mathbf{z}_n, \mathbf{z}_1, \ldots, \mathbf{z}_n, t)\, d\mathbf{z}_1 \ldots d\mathbf{z}_n}{\int \xi(\mathbf{z}_1, \ldots, \mathbf{z}_n, \mathbf{z}_1, \ldots, \mathbf{z}_n, t)\, d\mathbf{z}_1 \ldots d\mathbf{z}_n}. \tag{6.109}$$

Thus,

$$trace\left(\hat{A} \circ \hat{R}_\xi\right) = \int \left(\hat{A}(\mathbf{x}_1, \ldots, \mathbf{x}_n, t) \cdot \xi\right)(\mathbf{z}_1, \ldots, \mathbf{z}_n, \mathbf{z}_1, \ldots, \mathbf{z}_n, t)\, d\mathbf{z}_1 \ldots d\mathbf{z}_n, \tag{6.110}$$

and so,

$$\tilde{A}(t) = \frac{trace\left(\hat{A} \circ \hat{R}_\xi\right)}{trace\left(\hat{R}_\xi\right)} = \frac{trace\left(\hat{R}_\xi \circ \hat{A}\right)}{trace\left(\hat{R}_\xi\right)}, \tag{6.111}$$

where by the trace we mean the trace of an operator on a Hilbert space. Moreover, by (6.95) and (6.110) we have

$$trace\left(\hat{R}_\xi\right)(t) = trace\left(\hat{R}_\xi\right)(0). \tag{6.112}$$

We also prove the following:

Proposition 6.1 *Let $\hat{A} := \hat{A}(t)$ and $\hat{H}_0 := \hat{H}_0(t)$ be two time-dependent Hermitian operators on a given abstract Hilbert space, such that*

$$i\hbar \frac{\partial \hat{A}}{\partial t}(t) = \hat{H}_0(t) \cdot \hat{A}(t) - \hat{A}(t) \cdot \hat{H}_0(t) \qquad \forall t. \tag{6.113}$$

If, given the holomorphic function $f(z) : \mathbb{C} \to \mathbb{C}$ defined as a sum of the power series

$$f(z) := \sum_{m=0}^{+\infty} a_m z^m, \tag{6.114}$$

with $a_m \in \mathbb{C}$, we define the operator $(f \circ \hat{A}) := (f \circ \hat{A})(t)$ as

$$(f \circ \hat{A})(t) := \sum_{m=0}^{+\infty} a_m \hat{A}^m(t), \tag{6.115}$$

then:

$$i\hbar \frac{\partial}{\partial t}\left(f \circ \hat{A}\right)(t) = \hat{H}_0(t) \cdot \left(f \circ \hat{A}\right)(t) - \left(f \circ \hat{A}\right)(t) \cdot \hat{H}_0(t). \tag{6.116}$$

Proof We will prove now that for every $m = 0, 1, 2, 3, \ldots$ we have

$$i\hbar \frac{\partial \hat{A}^m}{\partial t} = \hat{H}_0 \cdot \hat{A}^m - \hat{A}^m \cdot \hat{H}_0. \tag{6.117}$$

Indeed, for $m = 0$ we have

$$i\hbar \frac{\partial \hat{A}^0}{\partial t} = i\hbar \frac{\partial(\hat{I}d)}{\partial t} = 0 = \hat{H}_0 \cdot \hat{I}d - \hat{I}d \cdot \hat{H}_0 = \hat{H}_0 \cdot \hat{A}^0 - \hat{A}^0 \cdot \hat{H}_0. \tag{6.118}$$

Next, if for some $m = k$ we have

$$i\hbar \frac{\partial \hat{A}^k}{\partial t} = \hat{H}_0 \cdot \hat{A}^k - \hat{A}^k \cdot \hat{H}_0, \tag{6.119}$$

then by (6.113) and (6.119) we obtain

$$\begin{aligned} i\hbar \frac{\partial \hat{A}^{k+1}}{\partial t} &= \hat{A} \cdot \left(i\hbar \frac{\partial \hat{A}^k}{\partial t}\right) + \left(i\hbar \frac{\partial \hat{A}}{\partial t}\right) \cdot \hat{A}^k \\ &= \hat{A} \cdot \left(\hat{H}_0 \cdot \hat{A}^k - \hat{A}^k \cdot \hat{H}_0\right) + \left(\hat{H}_0 \cdot \hat{A} - \hat{A} \cdot \hat{H}_0\right) \cdot \hat{A}^k = \hat{H}_0 \cdot \hat{A}^{k+1} - \hat{A}^{k+1} \cdot \hat{H}_0. \end{aligned} \tag{6.120}$$

So we prove (6.117) for $m = k + 1$ and thus for every m.

Therefore, given the holomorphic function $f(z) : \mathbb{C} \to \mathbb{C}$ defined as a sum of the power series

$$f(z) := \sum_{m=0}^{+\infty} a_m z^m, \tag{6.121}$$

with $a_m \in \mathbb{C}$, if we define the operator $f \circ \hat{A}$ as

$$f \circ \hat{A} := \sum_{m=0}^{+\infty} a_m \hat{A}^m, \tag{6.122}$$

by (6.117) and (6.122) we finally deduce (6.116).

6.5.1 Thermodynamical Equilibrium; Canonical Ensemble

Let $\xi\,(\mathbf{x}_1, \ldots, \mathbf{x}_n, \mathbf{y}_1, \ldots, \mathbf{y}_n, t)$ be the density matrix of the quantum system, ruled by the Hamiltonian operator $\hat{H}_0$ from (6.86), in the case of thermodynamical equilibrium and $\hat{R}_\xi$ is given by (6.105) for this ξ. Then in the case that $\frac{\partial \hat{H}_0}{\partial t} = 0$ and the given equilibrium system rests macroscopically, i.e. it has the macroscopical velocity field zero: $\mathbf{u}(\mathbf{x}, t) = 0$ it is well known that

$$\hat{R}_\xi = \frac{1}{trace\left(f \circ \hat{H}_0\right)} f \circ \hat{H}_0, \tag{6.123}$$

and in the case of a system in thermostat, having the Kelvin temperature T, we have the canonical ensemble:

$$f(s) = e^{-\frac{s}{kT}} = \sum_{m=0}^{+\infty} \frac{1}{m!}\left(-\frac{s}{kT}\right)^m \quad \forall s \in \mathbb{C}, \tag{6.124}$$

where k is the Boltzmann constant and by $f \circ \hat{H}_0$ we denote the following operator

$$f \circ \hat{H}_0 = \sum_{m=0}^{+\infty} \frac{1}{m!}\left(-\frac{1}{kT}\hat{H}_0\right)^m \quad \forall s \in \mathbb{C}. \tag{6.125}$$

We would like to find an alternative form of the above law of thermodynamical equilibrium, which is invariant under the change of inertial or non-inertial Cartesian coordinate system. Then it is clear that if the given system rests in the old coordinate system, then it obviously has non-trivial macroscopical velocity field $\mathbf{u}(\mathbf{x}, t)$ in the new one. Moreover, $\mathbf{u}(\mathbf{x}, t)$ can depend on $\mathbf{x}$ and t, as it indeed happens in the case of a rotation of the new coordinate system with respect to the old one. On the other hand the concept of thermodynamical equilibrium is clearly independent on the coordinate system.

In order to find the forms of the law of thermodynamical equilibrium, which are indeed invariant under the change of inertial or non-inertial Cartesian coordinate system we follow the steps as below. Given a speed-like vector field $\mathbf{u} = \mathbf{u}(\mathbf{x}, t)$, define

$$\hat{H}_{\mathbf{u}}\left(\mathbf{x}_1, \ldots, \mathbf{x}_n, t\right) := \hat{H}_0\left(\mathbf{x}_1, \ldots, \mathbf{x}_n, t\right) - \frac{1}{2}\left(\sum_{j=1}^{n} \hat{\mathbf{P}}_j \cdot \mathbf{u}(\mathbf{x}_j, t) + \mathbf{u}(\mathbf{x}_j, t) \cdot \hat{\mathbf{P}}_j\right), \tag{6.126}$$

where $\hat{\mathbf{P}}_j$ is the operator defined as

$$\hat{\mathbf{P}}_j \cdot \psi\left(\mathbf{x}_1, \ldots, \mathbf{x}_n\right) = -i\hbar\nabla_{\mathbf{x}_j}\psi\left(\mathbf{x}_1, \ldots, \mathbf{x}_n\right). \tag{6.127}$$

Then by (6.86) we have

$$\begin{aligned}
\hat{H}_{\mathbf{u}}\left(\mathbf{x}_1, \ldots, \mathbf{x}_n, t\right) \cdot \psi\left(\mathbf{x}_1, \ldots, \mathbf{x}_n, t\right) &= -\sum_{j=1}^{n} \frac{\hbar^2}{2m_j}\Delta_{\mathbf{x}_j}\psi \\
-\sum_{j=1}^{n} \frac{i\hbar}{2}\operatorname{div}_{\mathbf{x}_j}\left\{\psi\left(\mathbf{v}(\mathbf{x}_j, t) - \mathbf{u}(\mathbf{x}_j, t)\right)\right\} &- \sum_{j=1}^{n} \frac{i\hbar}{2}\left(\mathbf{v}(\mathbf{x}_j, t) - \mathbf{u}(\mathbf{x}_j, t)\right) \cdot \nabla_{\mathbf{x}_j}\psi \\
+\sum_{j=1}^{n} \frac{i\hbar\sigma_j}{2m_j c}\operatorname{div}_{\mathbf{x}_j}\left\{\psi\mathbf{A}(\mathbf{x}_j, t)\right\} &+ \sum_{j=1}^{n} \frac{i\hbar\sigma_j}{2m_j c}\mathbf{A}(\mathbf{x}_j, t) \cdot \nabla_{\mathbf{x}_j}\psi \\
+\sum_{j=1}^{n}\left(\sigma_j\Psi(\mathbf{x}_j, t) - \frac{\sigma_j}{c}\mathbf{A}(\mathbf{x}_j, t) \cdot \mathbf{v}(\mathbf{x}_j, t) + \frac{\sigma_j^2}{2m_j c^2}\left|\mathbf{A}(\mathbf{x}_j, t)\right|^2\right)\psi \\
- V\left(\mathbf{x}_1, \ldots, \mathbf{x}_n, t\right)\psi.
\end{aligned} \tag{6.128}$$

Thus, as before, we can prove that (6.128) is invariant under the change of inertial or non-inertial Cartesian coordinate systems, i.e.

$$\begin{aligned}
\hat{H}'_{\mathbf{u}'}\left(\mathbf{x}'_1, \ldots, \mathbf{x}'_n, t'\right) \cdot \psi'\left(\mathbf{x}'_1, \ldots, \mathbf{x}'_n, t'\right) = \phi'\left(\mathbf{x}'_1, \ldots, \mathbf{x}'_n, t'\right) \quad \text{if and only if} \\
\hat{H}_{\mathbf{u}}\left(\mathbf{x}_1, \ldots, \mathbf{x}_n, t\right) \cdot \psi\left(\mathbf{x}_1, \ldots, \mathbf{x}_n, t\right) = \phi\left(\mathbf{x}_1, \ldots, \mathbf{x}_n, t\right),
\end{aligned} \tag{6.129}$$

provided that, as before, we have

$$\begin{cases}
V'\left(\mathbf{x}'_1, \ldots, \mathbf{x}'_n, t'\right) = V\left(\mathbf{x}_1, \ldots, \mathbf{x}_n, t\right), \\
\sigma'_j = \sigma_j, \\
m'_j = m_j, \\
\mathbf{v}'(\mathbf{x}', t) = A(t) \cdot \mathbf{v}(\mathbf{x}, t) + \frac{dA}{dt}(t) \cdot \mathbf{x} + \frac{d\mathbf{z}}{dt}(t), \\
\mathbf{u}'(\mathbf{x}', t) = A(t) \cdot \mathbf{u}(\mathbf{x}, t) + \frac{dA}{dt}(t) \cdot \mathbf{x} + \frac{d\mathbf{z}}{dt}(t), \\
\mathbf{A}'(\mathbf{x}', t) = A(t) \cdot \mathbf{A}(\mathbf{x}, t), \\
\Psi'(\mathbf{x}', t) - \mathbf{v}'(\mathbf{x}', t) \cdot \mathbf{A}'(\mathbf{x}', t) = \Psi(\mathbf{x}, t) - \mathbf{v}(\mathbf{x}, t) \cdot \mathbf{A}(\mathbf{x}, t), \\
\psi'\left(\mathbf{x}'_1, \ldots, \mathbf{x}'_n, t'\right) = \psi\left(\mathbf{x}_1, \ldots, \mathbf{x}_n, t\right), \\
\phi'\left(\mathbf{x}'_1, \ldots, \mathbf{x}'_n, t'\right) = \phi\left(\mathbf{x}_1, \ldots, \mathbf{x}_n, t\right).
\end{cases} \tag{6.130}$$

Next consider

$$\hat{R}_{\xi} = \frac{1}{trace\left(f \circ \hat{H}_{\mathbf{u}}\right)} f \circ \hat{H}_{\mathbf{u}}, \tag{6.131}$$

where, as before, in the case of a system in thermostat, having the Kelvin temperature T, we have the canonical ensemble:

$$f(s) = e^{-\frac{s}{kT}} = \sum_{m=0}^{+\infty} \frac{1}{m!}\left(-\frac{s}{kT}\right)^m \quad \forall s \in \mathbb{C} \tag{6.132}$$

with k being the Boltzmann constant and $f \circ \hat{H}_{\mathbf{u}}$ is the following operator

$$f \circ \hat{H}_{\mathbf{u}} = \sum_{m=0}^{+\infty} \frac{1}{m!}\left(-\frac{1}{kT}\hat{H}_{\mathbf{u}}\right)^m \quad \forall s \in \mathbb{C}. \tag{6.133}$$

We would like to note here that since the operator $\hat{H}_{\mathbf{u}}$ is bounded from below and $f(s)$ decays rapidly as $s \to +\infty$, there exists a density matrix ξ_1, such that the operator $\hat{R}_{\xi_1}$, given by (6.105), equals to the operator $f \circ \hat{H}_{\mathbf{u}}$ and thus (6.131) indeed has sense. Next by (6.129) and (6.133) the operator $\hat{R}_{\xi}$ in the left-hand side of (6.131) is invariant under the change of inertial or non-inertial Cartesian coordinate systems, i.e.

$$\hat{R}_{\xi'}\left(\mathbf{x}'_1, \ldots, \mathbf{x}'_n, t'\right) \cdot \psi'\left(\mathbf{x}'_1, \ldots, \mathbf{x}'_n, t'\right) = \phi'\left(\mathbf{x}'_1, \ldots, \mathbf{x}'_n, t'\right) \quad \text{if and only if}$$
$$\hat{R}_{\xi}\left(\mathbf{x}_1, \ldots, \mathbf{x}_n, t\right) \cdot \psi\left(\mathbf{x}_1, \ldots, \mathbf{x}_n, t\right) = \phi\left(\mathbf{x}_1, \ldots, \mathbf{x}_n, t\right), \tag{6.134}$$

provided that, as before, we have (6.130) and $\xi' = \xi$. Moreover, in the case $\mathbf{u} = 0$ (6.131) coincides with (6.123). However, we still need to derive the restrictions on the field $\mathbf{u}$ and the Hamiltonian operator $\hat{H}_0$, providing that our system can indeed be found in the state of thermodynamical equilibrium. We remind that in the case $\mathbf{u} = 0$ the appropriate restriction is $\frac{\partial \hat{H}_0}{\partial t} = 0$. In order to get these restrictions in the general case, we need to insert $\hat{R}_{\xi}$ in (6.131) into the equation in (6.106) which is equivalent to the quantum Liouville equation.

Therefore, assume that the vector field $\mathbf{u}$ satisfies (6.55) with (6.56), i.e.

$$\begin{cases} \frac{\partial U}{\partial t}\left(\mathbf{x}_1, \ldots, \mathbf{x}_n, t\right) + \sum_{j=1}^{n} \mathbf{u}(\mathbf{x}_j, t) \cdot \nabla_{\mathbf{x}_j} U\left(\mathbf{x}_1, \ldots, \mathbf{x}_n, t\right) = 0, \\ \frac{\partial}{\partial t}\left(\mathbf{u}(\mathbf{x}, t) - \mathbf{v}(\mathbf{x}, t)\right) - \operatorname{curl}_{\mathbf{x}}\left(\mathbf{u}(\mathbf{x}, t) \times \left(\mathbf{u}(\mathbf{x}, t) - \mathbf{v}(\mathbf{x}, t)\right)\right) \\ + \left(\operatorname{div}_{\mathbf{x}}\left(\mathbf{u}(\mathbf{x}, t) - \mathbf{v}(\mathbf{x}, t)\right)\right)\mathbf{u}(\mathbf{x}, t) = 0, \\ \frac{\partial \mathbf{A}}{\partial t}(\mathbf{x}, t) - \operatorname{curl}_{\mathbf{x}}\left(\mathbf{u}(\mathbf{x}, t) \times \mathbf{A}(\mathbf{x}, t)\right) + \left(\operatorname{div}_{\mathbf{x}} \mathbf{A}(\mathbf{x}, t)\right)\mathbf{u}(\mathbf{x}, t) = 0, \\ d_{\mathbf{x}}\mathbf{u}(\mathbf{x}, t) + \left\{d_{\mathbf{x}}\mathbf{u}(\mathbf{x}, t)\right\}^T = 0, \end{cases} \tag{6.135}$$

where

$$U(\mathbf{x}_1,\ldots,\mathbf{x}_n,t):=\\ \sum_{j=1}^{n}\left(\frac{\sigma_j^2}{2m_jc^2}\left|\mathbf{A}(\mathbf{x}_j,t)\right|^2+\sigma_j\Psi(\mathbf{x}_j,t)-\frac{\sigma_j}{c}\mathbf{A}(\mathbf{x}_j,t)\cdot\mathbf{v}(\mathbf{x}_j,t)\right)-V(\mathbf{x}_1,\ldots,\mathbf{x}_n,t). \tag{6.136}$$

Then, by Proposition 6.2 we have

$$i\hbar\frac{\partial}{\partial t}\left(f\circ\hat{H}_{\mathbf{u}}\right)=\hat{H}_0\cdot\left(f\circ\hat{H}_{\mathbf{u}}\right)-\left(f\circ\hat{H}_{\mathbf{u}}\right)\cdot\hat{H}_0. \tag{6.137}$$

Then by (6.137) together with (6.112) we deduce that

$$i\hbar\frac{\partial}{\partial t}\left(\hat{R}_\xi\right)=\hat{H}_0\cdot\left(\hat{R}_\xi\right)-\left(\hat{R}_\xi\right)\cdot\hat{H}_0, \tag{6.138}$$

where $\hat{R}_\xi$ is the operator in the left-hand side of (6.131). So we indeed get (6.106) in the case of (6.135), (6.136). Moreover, (6.135), (6.136) are invariant under the change of inertial or non-inertial coordinate system.

Proposition 6.2 *Assume that the speed-like vector field* $\mathbf{u}$ *satisfies* (6.135) *and* (6.136). *Next, assume that the holomorphic function* $f(z):\mathbb{C}\to\mathbb{C}$ *defined as a sum of the power series*

$$f(z):=\sum_{m=0}^{+\infty}a_m z^m, \tag{6.139}$$

with $a_m\in\mathbb{C}$, *is such that for the operator* $f\circ\hat{H}_{\mathbf{u}}$, *given by*

$$f\circ\hat{H}_{\mathbf{u}}:=\sum_{m=0}^{+\infty}a_m\hat{H}_{\mathbf{u}}^{m}, \tag{6.140}$$

where the operator $\hat{H}_{\mathbf{u}}$ *is given by* (6.128), *there exists a density matrix* $\xi(\mathbf{x}_1,\ldots,\mathbf{x}_n,\mathbf{y}_1,\ldots,\mathbf{y}_n,t)$, *such that*

$$\left(f\circ\hat{H}_{\mathbf{u}}\right)(\mathbf{x}_1,\ldots,\mathbf{x}_n,t)=\hat{R}_\xi(\mathbf{x}_1,\ldots,\mathbf{x}_n,t), \tag{6.141}$$

where $\hat{R}_\xi$ *is given by* (6.105). *Then the operator* $f\circ\hat{H}_{\mathbf{u}}$ *satisfies*

$$i\hbar\frac{\partial}{\partial t}\left(f\circ\hat{H}_{\mathbf{u}}\right)=\hat{H}_0\cdot\left(f\circ\hat{H}_{\mathbf{u}}\right)-\left(f\circ\hat{H}_{\mathbf{u}}\right)\cdot\hat{H}_0, \tag{6.142}$$

or equivalently

$$i\hbar\frac{\partial \xi}{\partial t}(\mathbf{x}_1,\dots,\mathbf{x}_n,\mathbf{y}_1,\dots,\mathbf{y}_n,t) = \hat{H}_0(\mathbf{x}_1,\dots,\mathbf{x}_n,t)\cdot\xi(\mathbf{x}_1,\dots,\mathbf{x}_n,\mathbf{y}_1,\dots,\mathbf{y}_n,t)$$
$$-\hat{H}_0^*(\mathbf{y}_1,\dots,\mathbf{y}_n,t)\cdot\xi(\mathbf{x}_1,\dots,\mathbf{x}_n,\mathbf{y}_1,\dots,\mathbf{y}_n,t). \quad (6.143)$$

Moreover, $f \circ \hat{H}_{\mathbf{u}}$ is invariant under the change of inertial or non-inertial Cartesian coordinate systems, i.e.

$$\left(f\circ\hat{H}'_{\mathbf{u}'}\right)(\mathbf{x}'_1,\dots,\mathbf{x}'_n,t')\cdot\psi'(\mathbf{x}'_1,\dots,\mathbf{x}'_n,t') = \phi'(\mathbf{x}'_1,\dots,\mathbf{x}'_n,t') \quad \textit{if and only if}$$
$$\left(f\circ\hat{H}_{\mathbf{u}}\right)(\mathbf{x}_1,\dots,\mathbf{x}_n,t)\cdot\psi(\mathbf{x}_1,\dots,\mathbf{x}_n,t) = \phi(\mathbf{x}_1,\dots,\mathbf{x}_n,t), \quad (6.144)$$

provided that we have (6.130). Finally, if we assume that the first and the second particles have the same mass $m_1 = m_2$ and the same charge $\sigma_1 = \sigma_2$ and moreover, if we assume that

$$V(\mathbf{x}_1,\mathbf{x}_2,\mathbf{x}_3,\dots,\mathbf{x}_n,t) = V(\mathbf{x}_2,\mathbf{x}_1,\mathbf{x}_3,\dots,\mathbf{x}_n,t) \quad \forall\mathbf{x}_1,\mathbf{x}_2,\mathbf{x}_3,\dots,\mathbf{x}_n\in\mathbb{R}^3, \ \ \forall t,$$

then

$$\psi(\mathbf{x}_2,\mathbf{x}_1,\mathbf{x}_3,\dots,\mathbf{x}_n,t) = -\psi(\mathbf{x}_1,\mathbf{x}_2,\mathbf{x}_3,\dots,\mathbf{x}_n,t) \ \forall\mathbf{x}_1,\dots,\mathbf{x}_n\in\mathbb{R}^3$$
$$\textit{and} \ \ \phi = \left(f\circ\hat{H}_{\mathbf{u}}\right)\cdot\psi$$
$$\textit{implies} \ \ \phi(\mathbf{x}_2,\mathbf{x}_1,\mathbf{x}_3,\dots,\mathbf{x}_n,t) = -\phi(\mathbf{x}_1,\mathbf{x}_2,\mathbf{x}_3,\dots,\mathbf{x}_n,t) \ \forall\mathbf{x}_1,\dots,\mathbf{x}_n\in\mathbb{R}^3, \quad (6.145)$$

and

$$\psi(\mathbf{x}_2,\mathbf{x}_1,\mathbf{x}_3,\dots,\mathbf{x}_n,t) = \psi(\mathbf{x}_1,\mathbf{x}_2,\mathbf{x}_3,\dots,\mathbf{x}_n,t) \ \forall\mathbf{x}_1,\dots,\mathbf{x}_n\in\mathbb{R}^3$$
$$\textit{and} \ \ \phi = \left(f\circ\hat{H}_{\mathbf{u}}\right)\cdot\psi$$
$$\textit{implies} \ \ \phi(\mathbf{x}_2,\mathbf{x}_1,\mathbf{x}_3,\dots,\mathbf{x}_n,t) = \phi(\mathbf{x}_1,\mathbf{x}_2,\mathbf{x}_3,\dots,\mathbf{x}_n,t) \ \forall\mathbf{x}_1,\dots,\mathbf{x}_n\in\mathbb{R}^3. \quad (6.146)$$

Proof Again by (6.135) and Proposition 3.3 there exists another Cartesian coordinate system $(**)$ such that under the change of coordinate system $(*)$ to another Cartesian coordinate system $(**)$, given by (3.54), we have

$$A(t)\cdot\mathbf{u}(\mathbf{x},t) + \frac{dA}{dt}(t)\cdot\mathbf{x} + \frac{d\mathbf{z}}{dt}(t) = \mathbf{u}'(\mathbf{x}',t') = 0. \quad (6.147)$$

Thus, since (6.135), (6.136) are invariant under the change of inertial or non-inertial Cartesian coordinate systems, as before, in system $(**)$ we have

$$\begin{cases} \frac{\partial U'}{\partial t'}\left(\mathbf{x}'_1, \ldots, \mathbf{x}'_n, t'\right) = 0, \\ \frac{\partial \mathbf{v}'}{\partial t'}(\mathbf{x}', t') = 0, \\ \frac{\partial \mathbf{A}'}{\partial t'}(\mathbf{x}', t') = 0 \\ \mathbf{u}'(\mathbf{x}', t') = 0, \end{cases} \tag{6.148}$$

where

$$U'\left(\mathbf{x}'_1, \ldots, \mathbf{x}'_n, t'\right) := \sum_{j=1}^{n} \left(\frac{(\sigma'_j)^2}{2\, m'_j c^2} \left|\mathbf{A}'(\mathbf{x}'_j, t')\right|^2 + \sigma'_j \Psi'(\mathbf{x}'_j, t') - \frac{\sigma'_j}{c} \mathbf{A}'(\mathbf{x}'_j, t') \cdot \mathbf{v}'(\mathbf{x}'_j, t') \right) - V'\left(\mathbf{x}'_1, \ldots, \mathbf{x}'_n, t'\right). \tag{6.149}$$

On the other hand, (6.148) is equivalent to

$$\frac{\partial \hat{H}'_0}{\partial t'}\left(\mathbf{x}'_1, \ldots, \mathbf{x}'_n, t'\right) = 0 \quad \text{and} \quad \mathbf{u}'(\mathbf{x}', t') = 0, \tag{6.150}$$

in system $(**)$. Thus, by (6.150), in system $(**)$ we deduce

$$\hat{H}'_0 \cdot \hat{H}'_{\mathbf{u}'} - \hat{H}'_{\mathbf{u}'} \cdot \hat{H}'_0 = \hat{H}'_0 \cdot \hat{H}'_0 - \hat{H}'_0 \cdot \hat{H}'_0 = 0 = i\hbar \frac{\partial \hat{H}'_0}{\partial t'} = i\hbar \frac{\partial \hat{H}'_{\mathbf{u}'}}{\partial t'}. \tag{6.151}$$

So we get

$$i\hbar \frac{\partial \hat{H}'_{\mathbf{u}'}}{\partial t'} = \hat{H}'_0 \cdot \hat{H}'_{\mathbf{u}'} - \hat{H}'_{\mathbf{u}'} \cdot \hat{H}'_0. \tag{6.152}$$

Therefore, given the holomorphic function $f(z) : \mathbb{C} \to \mathbb{C}$ defined as a sum of the power series

$$f(z) := \sum_{m=0}^{+\infty} a_m z^m, \tag{6.153}$$

with $a_m \in \mathbb{C}$, if we define the operator $f \circ \hat{H}_{\mathbf{u}}$ as

$$f \circ \hat{H}_{\mathbf{u}} := \sum_{m=0}^{+\infty} a_m \hat{H}_{\mathbf{u}}^m, \tag{6.154}$$

by (6.152) and Proposition 6.1 we deduce

$$i\hbar \frac{\partial}{\partial t'}\left(f \circ \hat{H}'_{\mathbf{u}'}\right) = \hat{H}'_0 \cdot \left(f \circ \hat{H}'_{\mathbf{u}'}\right) - \left(f \circ \hat{H}'_{\mathbf{u}'}\right) \cdot \hat{H}'_0. \tag{6.155}$$

Moreover, by (6.128) and (6.154), we can easily prove that $f \circ \hat{H}_{\mathbf{u}}$ is invariant under the change of inertial or non-inertial Cartesian coordinate systems, i.e.

$$\left(f \circ \hat{H}'_{\mathbf{u}'}\right)\left(\mathbf{x}'_1, \ldots, \mathbf{x}'_n, t'\right) \cdot \psi'\left(\mathbf{x}'_1, \ldots, \mathbf{x}'_n, t'\right) = \phi'\left(\mathbf{x}'_1, \ldots, \mathbf{x}'_n, t'\right) \quad \text{if and only if}$$
$$\left(f \circ \hat{H}_{\mathbf{u}}\right)(\mathbf{x}_1, \ldots, \mathbf{x}_n, t) \cdot \psi(\mathbf{x}_1, \ldots, \mathbf{x}_n, t) = \phi(\mathbf{x}_1, \ldots, \mathbf{x}_n, t), \tag{6.156}$$

provided that, as before, we have (6.130). Next assume that the holomorphic function f in (6.153) is such that there exists a density matrix $\xi(\mathbf{x}_1, \ldots, \mathbf{x}_n, \mathbf{y}_1, \ldots, \mathbf{y}_n, t)$ satisfying

$$\left(f \circ \hat{H}_{\mathbf{u}}\right)(\mathbf{x}_1, \ldots, \mathbf{x}_n, t) = \hat{R}_{\xi}(\mathbf{x}_1, \ldots, \mathbf{x}_n, t). \tag{6.157}$$

Then, by (6.156), for the density matrix $\xi'\left(\mathbf{x}'_1, \ldots, \mathbf{x}'_n, \mathbf{y}'_1, \ldots, \mathbf{y}'_n, t'\right)$ we have

$$\left(f \circ \hat{H}'_{\mathbf{u}'}\right)\left(\mathbf{x}'_1, \ldots, \mathbf{x}'_n, t'\right) = \hat{R}_{\xi'}\left(\mathbf{x}'_1, \ldots, \mathbf{x}'_n, t'\right), \tag{6.158}$$

provided that $\xi' = \xi$. Therefore, since we obtained before, that equation (6.155) is equivalent to the primed version of (6.96) and at the same time equation (6.96) is invariant under the change of non-inertial Cartesian coordinate system, provided we have $\xi' = \xi$, with the help of (6.157), as before, we deduce

$$i\hbar\frac{\partial}{\partial t}\left(f \circ \hat{H}_{\mathbf{u}}\right) = \hat{H}_0 \cdot \left(f \circ \hat{H}_{\mathbf{u}}\right) - \left(f \circ \hat{H}_{\mathbf{u}}\right) \cdot \hat{H}_0 \tag{6.159}$$

in an arbitrary coordinate system. So we obtain (6.142) or equivalently (6.143).

Finally, assume that the first and the second particles have the same mass $m_1 = m_2$ and the same charge $\sigma_1 = \sigma_2$ and moreover, assume that we have

$$V(\mathbf{x}_1, \mathbf{x}_2, \mathbf{x}_3, \ldots, \mathbf{x}_n, t) = V(\mathbf{x}_2, \mathbf{x}_1, \mathbf{x}_3, \ldots, \mathbf{x}_n, t) \qquad \forall \mathbf{x}_1, \mathbf{x}_2, \mathbf{x}_3, \ldots, \mathbf{x}_n \in \mathbb{R}^3, \quad \forall t.$$

In this case clearly by (6.128) we deduce the following relations:

$$\psi(\mathbf{x}_2, \mathbf{x}_1, \mathbf{x}_3, \ldots, \mathbf{x}_n, t) = -\psi(\mathbf{x}_1, \mathbf{x}_2, \mathbf{x}_3, \ldots, \mathbf{x}_n, t) \;\; \forall \mathbf{x}_1, \ldots, \mathbf{x}_n \in \mathbb{R}^3$$
$$\text{and} \quad \phi = \hat{H}_{\mathbf{u}} \cdot \psi$$
$$\text{implies} \quad \phi(\mathbf{x}_2, \mathbf{x}_1, \mathbf{x}_3, \ldots, \mathbf{x}_n, t) = -\phi(\mathbf{x}_1, \mathbf{x}_2, \mathbf{x}_3, \ldots, \mathbf{x}_n, t) \;\; \forall \mathbf{x}_1, \ldots, \mathbf{x}_n \in \mathbb{R}^3, \tag{6.160}$$

and

$$\psi(\mathbf{x}_2, \mathbf{x}_1, \mathbf{x}_3, \ldots, \mathbf{x}_n, t) = \psi(\mathbf{x}_1, \mathbf{x}_2, \mathbf{x}_3, \ldots, \mathbf{x}_n, t) \;\; \forall \mathbf{x}_1, \ldots, \mathbf{x}_n \in \mathbb{R}^3$$
$$\text{and} \quad \phi = \hat{H}_{\mathbf{u}} \cdot \psi$$
$$\text{implies} \quad \phi(\mathbf{x}_2, \mathbf{x}_1, \mathbf{x}_3, \ldots, \mathbf{x}_n, t) = \phi(\mathbf{x}_1, \mathbf{x}_2, \mathbf{x}_3, \ldots, \mathbf{x}_n, t) \;\; \forall \mathbf{x}_1, \ldots, \mathbf{x}_n \in \mathbb{R}^3. \tag{6.161}$$

Therefore, by (6.154) and (6.160), (6.161) we obtain

$$\psi(\mathbf{x}_2, \mathbf{x}_1, \mathbf{x}_3, \ldots, \mathbf{x}_n, t) = -\psi(\mathbf{x}_1, \mathbf{x}_2, \mathbf{x}_3, \ldots, \mathbf{x}_n, t) \;\; \forall \mathbf{x}_1, \ldots, \mathbf{x}_n \in \mathbb{R}^3$$
$$\text{and} \;\; \phi = \left(f \circ \hat{H}_{\mathbf{u}}\right) \cdot \psi$$
$$\text{implies} \;\; \phi(\mathbf{x}_2, \mathbf{x}_1, \mathbf{x}_3, \ldots, \mathbf{x}_n, t) = -\phi(\mathbf{x}_1, \mathbf{x}_2, \mathbf{x}_3, \ldots, \mathbf{x}_n, t) \;\; \forall \mathbf{x}_1, \ldots, \mathbf{x}_n \in \mathbb{R}^3, \tag{6.162}$$

and

$$\psi(\mathbf{x}_2, \mathbf{x}_1, \mathbf{x}_3, \ldots, \mathbf{x}_n, t) = \psi(\mathbf{x}_1, \mathbf{x}_2, \mathbf{x}_3, \ldots, \mathbf{x}_n, t) \;\; \forall \mathbf{x}_1, \ldots, \mathbf{x}_n \in \mathbb{R}^3$$
$$\text{and} \;\; \phi = \left(f \circ \hat{H}_{\mathbf{u}}\right) \cdot \psi$$
$$\text{implies} \;\; \phi(\mathbf{x}_2, \mathbf{x}_1, \mathbf{x}_3, \ldots, \mathbf{x}_n, t) = \phi(\mathbf{x}_1, \mathbf{x}_2, \mathbf{x}_3, \ldots, \mathbf{x}_n, t) \;\; \forall \mathbf{x}_1, \ldots, \mathbf{x}_n \in \mathbb{R}^3, \tag{6.163}$$

consistently with (6.107) and (6.108).

6.6 Shrödinger–Pauli Equation for a Spin-Half Quantum Particle

Consider the motion of a spin-half quantum micro-particle with inertial mass m and the charge σ in the outer gravitational and electromagnetical field with characteristics $\mathbf{v}(\mathbf{x}, t)$, $\mathbf{A}(\mathbf{x}, t)$ and $\Psi(\mathbf{x}, t)$ and additional conservative field with potential $V(\mathbf{y}, t)$. Since the Hamiltonian for a macro-particle has the form

$$H_{\text{macro}}(\mathbf{P}, \mathbf{r}, t) =$$
$$\frac{m}{2}\left|\mathbf{P} - \frac{\sigma}{c}\mathbf{A}(\mathbf{r}, t)\right|^2 + \sigma\left(\Psi(\mathbf{r}, t) - \frac{1}{c}\mathbf{v}(\mathbf{r}, t)\cdot\mathbf{A}(\mathbf{r}, t)\right) - V(\mathbf{r}, t) + \frac{1}{2}\mathbf{v}(\mathbf{r}, t)\cdot\mathbf{P} + \frac{1}{2}\mathbf{P}\cdot\mathbf{v}(\mathbf{r}, t), \tag{6.164}$$

we built the Hermitian Hamiltonian operator, taking into account the spin interaction as

$$\hat{H}_0 \cdot \psi = -\frac{\hbar^2}{2m}\Delta_{\mathbf{x}}\psi + \frac{i\hbar\sigma}{2mc}\operatorname{div}_{\mathbf{x}}\{\psi\mathbf{A}(\mathbf{x}, t)\} + \frac{i\hbar\sigma}{2mc}\nabla_{\mathbf{x}}\psi\cdot\mathbf{A}(\mathbf{x}, t) + \frac{\sigma^2}{2mc^2}|\mathbf{A}(\mathbf{x}, t)|^2\psi$$
$$+\sigma\left(\Psi(\mathbf{x}, t) - \frac{1}{c}\mathbf{v}(\mathbf{x}, t)\cdot\mathbf{A}(\mathbf{x}, t)\right)\psi - V(\mathbf{x}, t)\psi - \frac{i\hbar}{2}\operatorname{div}_{\mathbf{x}}\{\psi\mathbf{v}(\mathbf{x}, t)\} - \frac{i\hbar}{2}\nabla_{\mathbf{x}}\psi\cdot\mathbf{v}(\mathbf{x}, t)$$
$$-\frac{g\sigma\hbar}{2mc}\mathbf{S}\cdot\left(\operatorname{curl}_{\mathbf{x}}\mathbf{A}(\mathbf{x}, t)\,\psi\right) + \frac{\hbar}{4}\mathbf{S}\cdot\left(\operatorname{curl}_{\mathbf{x}}\mathbf{v}(\mathbf{x}, t)\,\psi\right), \tag{6.165}$$

where $\psi(\mathbf{x}, t) = (\psi_1(\mathbf{x}, t), \psi_2(\mathbf{x}, t)) \in \mathbb{C}^2$ is a two-component wave function, $\hat{H}_0$ is the Hamiltonian operator, $\mathbf{S} := (S_1, S_2, S_3)$,

$$S_1 = \begin{pmatrix} 0 & 1 \\ 1 & 0 \end{pmatrix}, \quad S_2 = \begin{pmatrix} 0 & -i \\ i & 0 \end{pmatrix} \quad S_3 = \begin{pmatrix} 1 & 0 \\ 0 & -1 \end{pmatrix}$$

are Pauli matrices and g is a constant that depends on the type of the particle (for electron we have $g = 1$). Note that, in addition to the classical term of the spin-magnetic interaction, we added another term to the Hamiltonian, namely $\frac{\hbar}{4}\mathbf{S} \cdot (\mathrm{curl}_{\mathbf{x}}\mathbf{v}(\mathbf{x}, t)\, \psi)$. In the case of the Newtonian-type gravity, this term vanishes in every non-rotating and, in particular, in every inertial coordinate system, however, it provides the invariance of the Shrödinger–Pauli equation, under the change of non-inertial Cartesian coordinate system, as can be seen in the following Theorem 6.1. The Shrödinger–Pauli equation for this particle is

$$i\hbar\frac{\partial \psi}{\partial t} = \hat{H}_0 \cdot \psi. \tag{6.166}$$

Thus,

$$\begin{aligned} i\hbar\frac{\partial \psi}{\partial t} &= -\frac{\hbar^2}{2m}\Delta_{\mathbf{x}}\psi + \frac{i\hbar\sigma}{2mc}\,\mathrm{div}_{\mathbf{x}}\{\psi\mathbf{A}(\mathbf{x}, t)\} + \frac{i\hbar\sigma}{2mc}\nabla_{\mathbf{x}}\psi \cdot \mathbf{A}(\mathbf{x}, t) + \frac{\sigma^2}{2mc^2}\left|\mathbf{A}(\mathbf{x}, t)\right|^2\psi \\ &+\sigma\left(\Psi(\mathbf{x}, t) - \frac{1}{c}\mathbf{v}(\mathbf{x}, t) \cdot \mathbf{A}(\mathbf{x}, t)\right)\psi - V(\mathbf{x}, t)\,\psi - \frac{i\hbar}{2}\,\mathrm{div}_{\mathbf{x}}\{\psi\mathbf{v}(\mathbf{x}, t)\} - \frac{i\hbar}{2}\nabla_{\mathbf{x}}\psi \cdot \mathbf{v}(\mathbf{x}, t) \\ &+ \frac{\hbar}{2}\mathbf{S} \cdot \left(\left(\frac{1}{2}\mathrm{curl}_{\mathbf{x}}\mathbf{v}(\mathbf{x}, t) - \frac{g\sigma}{mc}\mathrm{curl}_{\mathbf{x}}\mathbf{A}(\mathbf{x}, t)\right)\psi\right). \end{aligned} \tag{6.167}$$

That is,

$$\begin{aligned} &i\hbar\left(\frac{\partial \psi}{\partial t} + \frac{1}{2}\,\mathrm{div}_{\mathbf{x}}\{\psi\mathbf{v}(\mathbf{x}, t)\} + \frac{1}{2}\nabla_{\mathbf{x}}\psi \cdot \mathbf{v}(\mathbf{x}, t)\right) \\ &= -\frac{\hbar^2}{2m}\Delta_{\mathbf{x}}\psi + \frac{i\hbar\sigma}{2mc}\,\mathrm{div}_{\mathbf{x}}\{\psi\mathbf{A}(\mathbf{x}, t)\} + \frac{i\hbar\sigma}{2mc}\nabla_{\mathbf{x}}\psi \cdot \mathbf{A}(\mathbf{x}, t) + \frac{\sigma^2}{2mc^2}\left|\mathbf{A}(\mathbf{x}, t)\right|^2\psi \\ &\quad +\sigma\left(\Psi(\mathbf{x}, t) - \frac{1}{c}\mathbf{v}(\mathbf{x}, t) \cdot \mathbf{A}(\mathbf{x}, t)\right)\psi - V(\mathbf{x}, t)\,\psi \\ &\quad + \frac{\hbar}{2}\mathbf{S} \cdot \left(\left(\frac{1}{2}\mathrm{curl}_{\mathbf{x}}\mathbf{v}(\mathbf{x}, t) - \frac{g\sigma}{mc}\mathrm{curl}_{\mathbf{x}}\mathbf{A}(\mathbf{x}, t)\right)\psi\right). \end{aligned} \tag{6.168}$$

Theorem 6.1 *Consider that the change of some Cartesian coordinate system* $(*)$ *to another Cartesian coordinate system* $(**)$ *is given by*

$$\begin{cases} \mathbf{x}' = A(t) \cdot \mathbf{x} + \mathbf{z}(t), \\ t' = t, \end{cases} \tag{6.169}$$

where $A(t) \in SO(3)$ is a rotation. Next, assume that in the coordinate system (**) we observe a validity of the Shrödinger–Pauli equation of the form:

$$
\begin{aligned}
i\hbar\left(\frac{\partial\psi'}{\partial t'}+\frac{1}{2}\,\mathrm{div}_{\mathbf{x}'}\left\{\psi'\mathbf{v}'\right\}+\frac{1}{2}\nabla_{\mathbf{x}'}\psi'\cdot\mathbf{v}'\right)\\
=-\frac{\hbar^2}{2\,m'}\Delta_{\mathbf{x}'}\psi'+\frac{i\hbar\sigma'}{2\,m'c}\,\mathrm{div}_{\mathbf{x}'}\left\{\psi'\mathbf{A}'\right\}+\frac{i\hbar\sigma'}{2\,m'c}\nabla_{\mathbf{x}'}\psi'\cdot\mathbf{A}'\\
+\frac{(\sigma')^2}{2\,m'c^2}\left|\mathbf{A}'\right|^2\psi'+\sigma'\left(\Psi'-\frac{1}{c}\mathbf{v}'\cdot\mathbf{A}'\right)\psi'-V'\psi'\\
+\frac{\hbar}{2}\mathbf{S}\cdot\left(\left(\frac{1}{2}curl_{\mathbf{x}'}\mathbf{v}'-\frac{g'\sigma'}{m'c}curl_{\mathbf{x}'}\mathbf{A}'\right)\psi'\right),
\end{aligned}
\tag{6.170}
$$

where $\psi \in \mathbb{C}^2$. Then in the coordinate system (*) we have the validity of Shrödinger–Pauli equation of the same as (6.170) form:

$$
\begin{aligned}
i\hbar\left(\frac{\partial\psi}{\partial t}+\frac{1}{2}\,\mathrm{div}_{\mathbf{x}}\left\{\psi\mathbf{v}\right\}+\frac{1}{2}\nabla_{\mathbf{x}}\psi\cdot\mathbf{v}\right)=-\frac{\hbar^2}{2\,m}\Delta_{\mathbf{x}}\psi+\frac{i\hbar\sigma}{2mc}\,\mathrm{div}_{\mathbf{x}}\left\{\psi\mathbf{A}\right\}+\frac{i\hbar\sigma}{2mc}\nabla_{\mathbf{x}}\psi\cdot\mathbf{A}\\
+\frac{\sigma^2}{2mc^2}\left|\mathbf{A}\right|^2\psi+\sigma\left(\Psi-\frac{1}{c}\mathbf{v}\cdot\mathbf{A}\right)\psi-V\psi+\frac{\hbar}{2}\mathbf{S}\cdot\left(\left(\frac{1}{2}curl_{\mathbf{x}}\mathbf{v}-\frac{g\sigma}{mc}curl_{\mathbf{x}}\mathbf{A}\right)\psi\right),
\end{aligned}
\tag{6.171}
$$

provided that

$$
\begin{cases}
g'=g\\
V'=V,\\
\sigma'=\sigma,\\
m'=m,\\
\mathbf{v}'=A(t)\cdot\mathbf{v}+\frac{dA}{dt}(t)\cdot\mathbf{x}+\frac{d\mathbf{z}}{dt}(t),\\
\mathbf{A}'=A(t)\cdot\mathbf{A},\\
\Psi'-\mathbf{v}'\cdot\mathbf{A}'=\Psi-\mathbf{v}\cdot\mathbf{A},\\
\psi'=U(t)\cdot\psi,
\end{cases}
\tag{6.172}
$$

where $U(t) \in SU(2)$ is some special unitary 2×2 matrix, i.e. $U(t) \in \mathbb{C}^{2\times 2}$, $\det U(t) = 1$, $U(t)\cdot U^*(t) = I$ *where $U^*(t)$ is the Hermitian adjoint to $U(t)$ matrix: $U^*(t) := \bar{U}(t)^T$ and I is the identity 2×2 matrix. Moreover, $U(t)$ is characterized by*

$$
U^*(t)\cdot\mathbf{S}\cdot U(t)=A(t)\cdot\mathbf{S}
\tag{6.173}
$$

that means

$$\Big(U^*(t)\cdot S_1\cdot U(t), U^*(t)\cdot S_2\cdot U(t), U^*(t)\cdot S_3\cdot U(t)\Big) =$$
$$\Big(a_{11}(t)S_1 + a_{12}(t)S_2 + a_{13}(t)S_3\,,\; a_{21}(t)S_1 + a_{22}(t)S_2 + a_{23}(t)S_3\,,$$
$$a_{31}(t)S_1 + a_{32}(t)S_2 + a_{33}(t)S_3\Big),$$

where $A(t) = \{a_{mk}(t)\}_{\{1\le m,k\le 3\}}$.

Proof It is well known that we have

$$\begin{cases} S_1^2 = S_2^2 = S_3^2 = I, \\ S_1\cdot S_2 = -S_2\cdot S_1 = iS_3, \quad S_2\cdot S_3 = -S_3\cdot S_2 = iS_1, \quad S_3\cdot S_1 = -S_1\cdot S_3 = iS_2. \end{cases} \tag{6.174}$$

Next, it is well known that $SO(3)$ is smoothly double covered by $SU(2)$ and the cover mapping is regular and locally one-to-one. In particular, for every t there exists $U(t)\in SU(2)$ such that (6.173) is satisfied (note that the second choice is $(-U(t))$). Moreover, by Implicit Function Theorem we deduce that if $A(t)$ is differentiable by t then $U(t)$ is also differentiable by t. Thus if $\psi' = U(t)\cdot\psi$ in (6.170), then by (6.170), (6.172) and proposition 3.1 we have

$$\begin{aligned} i\hbar U(t)\cdot\left(U^{-1}(t)\cdot\frac{dU}{dt}(t)\cdot\psi + \frac{\partial\psi}{\partial t} + \frac{1}{2}\operatorname{div}_{\mathbf{x}}\{\psi\mathbf{v}\} + \frac{1}{2}\nabla_{\mathbf{x}}\psi\cdot\mathbf{v}\right) = \\ U(t)\cdot\Bigg(-\frac{\hbar^2}{2m}\Delta_{\mathbf{x}}\psi + \frac{i\hbar\sigma}{2mc}\operatorname{div}_{\mathbf{x}}\{\psi\mathbf{A}\} + \frac{i\hbar\sigma}{2mc}\nabla_{\mathbf{x}}\psi\cdot\mathbf{A} \\ + \frac{\sigma^2}{2mc^2}|\mathbf{A}|^2\psi + \sigma\left(\Psi - \frac{1}{c}\mathbf{v}\cdot\mathbf{A}\right)\psi - V\psi\Bigg) \\ + \frac{\hbar}{2}\mathbf{S}\cdot\left(A(t)\cdot\left(\frac{1}{2}\operatorname{curl}_{\mathbf{x}}\left(A^{-1}(t)\cdot\frac{dA}{dt}(t)\cdot\mathbf{x}\right) + \frac{1}{2}\operatorname{curl}_{\mathbf{x}}\mathbf{v} - \frac{g\sigma}{mc}\operatorname{curl}_{\mathbf{x}}\mathbf{A}\right)U(t)\cdot\psi\right). \end{aligned} \tag{6.175}$$

Thus, since $U(t)$ is unitary and then $U^{-1}(t) = U^*(t)$ and $A^{-1}(t) = A^T(t)$, by (6.175) we have

$$\begin{aligned} \frac{\hbar}{4}\left(4iU^*(t)\cdot\frac{dU}{dt}(t)\cdot\psi - U^*(t)\cdot\mathbf{S}\cdot U(t)\cdot\left(A(t)\cdot\left(\operatorname{curl}_{\mathbf{x}}\left(A^T(t)\cdot\frac{dA}{dt}(t)\cdot\mathbf{x}\right)\right)\psi\right)\right) \\ + i\hbar\left(\frac{\partial\psi}{\partial t} + \frac{1}{2}\operatorname{div}_{\mathbf{x}}\{\psi\mathbf{v}\} + \frac{1}{2}\nabla_{\mathbf{x}}\psi\cdot\mathbf{v}\right) = \\ -\frac{\hbar^2}{2m}\Delta_{\mathbf{x}}\psi + \frac{i\hbar\sigma}{2mc}\operatorname{div}_{\mathbf{x}}\{\psi\mathbf{A}\} + \frac{i\hbar\sigma}{2mc}\nabla_{\mathbf{x}}\psi\cdot\mathbf{A} + \frac{\sigma^2}{2mc^2}|\mathbf{A}|^2\psi + \sigma\left(\Psi - \frac{1}{c}\mathbf{v}\cdot\mathbf{A}\right)\psi - V\psi \\ + \frac{\hbar}{2}U^*(t)\cdot\mathbf{S}\cdot U(t)\cdot\left(A(t)\cdot\left(\frac{1}{2}\operatorname{curl}_{\mathbf{x}}\mathbf{v} - \frac{g\sigma}{mc}\operatorname{curl}_{\mathbf{x}}\mathbf{A}\right)\psi\right). \end{aligned} \tag{6.176}$$

Thus by (6.176) and (6.173) we deduce

$$\frac{\hbar}{4}\left(4iU^*(t)\cdot\frac{dU}{dt}(t)\cdot\psi-\mathbf{S}\cdot\left(\operatorname{curl}_{\mathbf{x}}\left(A^T(t)\cdot\frac{dA}{dt}(t)\cdot\mathbf{x}\right)\right)\psi\right)$$
$$+i\hbar\left(\frac{\partial\psi}{\partial t}+\frac{1}{2}\operatorname{div}_{\mathbf{x}}\{\psi\mathbf{v}\}+\frac{1}{2}\nabla_{\mathbf{x}}\psi\cdot\mathbf{v}\right)=$$
$$-\frac{\hbar^2}{2m}\Delta_{\mathbf{x}}\psi+\frac{i\hbar\sigma}{2mc}\operatorname{div}_{\mathbf{x}}\{\psi\mathbf{A}\}+\frac{i\hbar\sigma}{2mc}\nabla_{\mathbf{x}}\psi\cdot\mathbf{A}+\frac{\sigma^2}{2mc^2}|\mathbf{A}|^2\psi+\sigma\left(\Psi-\frac{1}{c}\mathbf{v}\cdot\mathbf{A}\right)\psi-V\psi$$
$$+\frac{\hbar}{2}\mathbf{S}\cdot\left(\left(\frac{1}{2}\operatorname{curl}_{\mathbf{x}}\mathbf{v}-\frac{g\sigma}{mc}\operatorname{curl}_{\mathbf{x}}\mathbf{A}\right)\psi\right). \quad (6.177)$$

On the other hand differentiating the identities $A^T(t)\cdot A(t)=I$ and $U^*(t)\cdot U(t)=I$ by t we deduce that the real 3×3 matrix $A^T(t)\cdot\frac{dA}{dt}(t)$ is antisymmetric and the complex 2×2 matrix $iU^*(t)\cdot\frac{dU}{dt}(t)$ is Hermitian self-adjoint. Moreover, differentiating the identity $\det U(t)=1$ and using that $U^*(t)=U^{-1}(t)$ we deduce that the matrix $iU^*(t)\cdot\frac{dU}{dt}(t)$ is traceless. In particular, since $A^T(t)\cdot\frac{dA}{dt}(t)$ is antisymmetric, there exists $\mathbf{w}(t)=(w_1(t),w_2(t),w_3(t))\in\mathbb{R}^3$ such that

$$A^T(t)\cdot\frac{dA}{dt}(t)\cdot\mathbf{x}=\mathbf{w}(t)\times\mathbf{x}\qquad\forall\mathbf{x}\in\mathbb{R}^3, \quad (6.178)$$

and then,

$$\operatorname{curl}_{\mathbf{x}}\left(A^T(t)\cdot\frac{dA}{dt}(t)\cdot\mathbf{x}\right)=2\mathbf{w}(t). \quad (6.179)$$

On the other hand, since the matrix $iU^*(t)\cdot\frac{dU}{dt}(t)$ is Hermitian self-adjoint and traceless, clearly there exist $\mathbf{q}(t)=(q_1(t),q_2(t),q_3(t))\in\mathbb{R}^3$ such that

$$iU^*(t)\cdot\frac{dU}{dt}(t)=\mathbf{q}(t)\cdot\mathbf{S}:=q_1(t)S_1+q_2(t)S_2+q_3(t)S_3. \quad (6.180)$$

Finally differentiating the identity (6.173) by t we deduce

$$\frac{dU^*}{dt}(t)\cdot\mathbf{S}\cdot U(t)+U^*(t)\cdot\mathbf{S}\cdot\frac{dU}{dt}(t)=\frac{dA}{dt}(t)\cdot\mathbf{S}, \quad (6.181)$$

and then again inserting (6.173) into (6.181) and using the antisymmetry of $A^T(t)\cdot\frac{dA}{dt}(t)$ and the Hermitian property of $iU^*(t)\cdot\frac{dU}{dt}(t)$ we obtain

$$-i\left((A(t)\cdot\mathbf{S})\cdot\left(iU^*(t)\cdot\frac{dU}{dt}(t)\right)-\left(i\cdot U^*(t)\cdot\frac{dU}{dt}(t)\right)\cdot(A(t)\cdot\mathbf{S})\right)$$
$$=A(t)\cdot\left(A^T(t)\cdot\frac{dA}{dt}(t)\right)\cdot\mathbf{S}, \quad (6.182)$$

that implies

$$-i\left(\mathbf{S}\cdot\left(iU^*(t)\cdot\frac{dU}{dt}(t)\right)-\left(i\cdot U^*(t)\cdot\frac{dU}{dt}(t)\right)\cdot\mathbf{S}\right)=\left(A^T(t)\cdot\frac{dA}{dt}(t)\right)\cdot\mathbf{S}. \tag{6.183}$$

Then inserting (6.178) and (6.180) into (6.183) we deduce

$$-i\left(\mathbf{S}\cdot(q_1(t)S_1+q_2(t)S_2+q_3(t)S_3)-(q_1(t)S_1+q_2(t)S_2+q_3(t)S_3)\cdot\mathbf{S}\right)=\mathbf{w}(t)\times\mathbf{S}. \tag{6.184}$$

Thus by (6.174) and (6.184) we get

$$2\mathbf{q}(t)=\mathbf{w}(t). \tag{6.185}$$

Therefore, (6.185), (6.180) and (6.179) together imply

$$4iU^*(t)\cdot\frac{dU}{dt}(t)=\mathbf{S}\cdot\left(\operatorname{curl}_{\mathbf{x}}\left(A^T(t)\cdot\frac{dA}{dt}(t)\cdot\mathbf{x}\right)\right), \tag{6.186}$$

and inserting (6.186) into (6.177) we finally conclude (6.171).

Next, again consider the motion of a quantum micro-particle with spin-half, inertial mass m and the charges σ with the known gravitational and electromagnetical field with characteristics $\mathbf{v}(\mathbf{x},t)$, $\mathbf{A}(\mathbf{x},t)$ and $\Psi(\mathbf{x},t)$ and additional conservative field with potential $V(\mathbf{y},t)$, taking into the account spin interaction. Then consider a Lagrangian density L defined by

$$\begin{aligned}L(\psi,\mathbf{x},t):=\frac{i\hbar}{2}\left(\left(\frac{\partial\psi}{\partial t}+\mathbf{v}\cdot\nabla_{\mathbf{x}}\psi\right)\cdot\bar{\psi}-\psi\cdot\left(\frac{\partial\bar{\psi}}{\partial t}+\mathbf{v}\cdot\nabla_{\mathbf{x}}\bar{\psi}\right)\right)\\ -\frac{\hbar^2}{2m}\nabla_{\mathbf{x}}\psi\cdot\nabla_{\mathbf{x}}\bar{\psi}+V(\mathbf{x},t)\,\psi\cdot\bar{\psi}\\ -\frac{\hbar\sigma i}{2mc}\left(\nabla_{\mathbf{x}}\psi\cdot\bar{\psi}-\psi\cdot\nabla_{\mathbf{x}}\bar{\psi}\right)\cdot\mathbf{A}-\frac{\sigma^2}{2mc^2}|\mathbf{A}|^2\,\psi\cdot\bar{\psi}-\sigma\left(\Psi-\frac{1}{c}\mathbf{v}\cdot\mathbf{A}\right)\psi\cdot\bar{\psi}\\ -\frac{\hbar}{2}\left(\left(\mathbf{S}\cdot\left(\frac{1}{2}\operatorname{curl}_{\mathbf{x}}\mathbf{v}-\frac{g\sigma}{mc}\operatorname{curl}_{\mathbf{x}}\mathbf{A}\right)\right)\cdot\psi\right)\cdot\bar{\psi},\end{aligned} \tag{6.187}$$

where $\psi\in\mathbb{C}^2$ is a two-component wave function. Then similarly to the proof of Theorem 6.1 we can prove that L is invariant under the change of inertial or non-inertial Cartesian coordinate system, given by (6.169), provided that we take into account (6.172). We investigate stationary points of the functional

$$J=\int_0^T\int_{\mathbb{R}^3}L(\psi,\mathbf{x},t)\,d\mathbf{x}dt. \tag{6.188}$$

Then, by (6.187) we have

$$0 = \frac{\delta L}{\delta(\bar{\psi})} = i\hbar\left(\frac{\partial\psi}{\partial t} + \frac{1}{2}\mathbf{v}\cdot\nabla_{\mathbf{x}}\psi + \frac{1}{2}\operatorname{div}_{\mathbf{x}}\{\psi\mathbf{v}\}\right) + \frac{\hbar^2}{2m}\Delta_{\mathbf{x}}\psi + V(\mathbf{x},t)\,\psi$$
$$-\frac{\hbar\sigma i}{2mc}\left(\mathbf{A}\cdot\nabla_{\mathbf{x}}\psi + \operatorname{div}_{\mathbf{x}}\{\psi\mathbf{A}\}\right) - \frac{\sigma^2}{2mc^2}|\mathbf{A}|^2\,\psi - \sigma\left(\Psi - \frac{1}{c}\mathbf{v}\cdot\mathbf{A}\right)\psi$$
$$-\frac{\hbar}{2}\left(\mathbf{S}\cdot\left(\frac{1}{2}\operatorname{curl}_{\mathbf{x}}\mathbf{v} - \frac{g\sigma}{mc}\operatorname{curl}_{\mathbf{x}}\mathbf{A}\right)\right)\cdot\psi, \tag{6.189}$$

and

$$0 = \frac{\delta L}{\delta(\psi)} = (\bar{i})\hbar\left(\frac{\partial\bar{\psi}}{\partial t} + \frac{1}{2}\mathbf{v}\cdot\nabla_{\mathbf{x}}\bar{\psi} + \frac{1}{2}\operatorname{div}_{\mathbf{x}}\left\{\bar{\psi}\mathbf{v}\right\}\right) + \frac{\hbar^2}{2m}\Delta_{\mathbf{x}}\bar{\psi} + V(\mathbf{x},t)\,\bar{\psi}$$
$$-\frac{\hbar\sigma(\bar{i})}{2mc}\left(\mathbf{A}\cdot\nabla_{\mathbf{x}}\bar{\psi} + \operatorname{div}_{\mathbf{x}}\left\{\bar{\psi}\mathbf{A}\right\}\right) - \frac{\sigma^2}{2mc^2}|\mathbf{A}|^2\,\bar{\psi} - \sigma\left(\Psi - \frac{1}{c}\mathbf{v}\cdot\mathbf{A}\right)\bar{\psi}$$
$$-\frac{\hbar}{2}\left(\mathbf{S}^T\cdot\left(\frac{1}{2}\operatorname{curl}_{\mathbf{x}}\mathbf{v} - \frac{g\sigma}{mc}\operatorname{curl}_{\mathbf{x}}\mathbf{A}\right)\right)\cdot\bar{\psi}. \tag{6.190}$$

Note that the last equality is just the complex conjugate of equality (6.189). So we get that the Euler–Lagrange equation for (6.187) coincides with the Shrödinger–Pauli equation in the form of (6.168).

6.7 Shrödinger–Pauli Equation for a System of *n* Spin-Half Micro-particles

Consider the motion of a system of n spin-half quantum micro-particles with inertial masses $m_1, \ldots, m_n$ and the charges $\sigma_1, \ldots, \sigma_n$ in the outer gravitational and electromagnetical field with characteristics $\mathbf{v}(\mathbf{x}, t)$, $\mathbf{A}(\mathbf{x}, t)$ and $\Psi(\mathbf{x}, t)$ and additional conservative field with potential $V(\mathbf{y}_1, \ldots, \mathbf{y}_n, t)$, taking into account the spin interaction. Then the system is characterized by 2^n-component complex wave function $\psi(\mathbf{x}_1, \ldots, \mathbf{x}_n, t) \in \mathbb{C}^{2^n}$ where by $\mathbb{C}^{2^n}$ we denote the tensor product of n copies of the space $\mathbb{C}^2$:

$$\mathbb{C}^{2^n} := \left(\mathbb{C}^2\right) \otimes_1 \left(\mathbb{C}^2\right) \otimes_2 \left(\mathbb{C}^2\right) \ldots \otimes_{(n-1)} \left(\mathbb{C}^2\right) \tag{6.191}$$

Then we built the Hermitian Hamiltonian operator as

$$\hat{H}_0 \cdot \psi = -\sum_{j=1}^{n} \frac{i\hbar}{2} \operatorname{div}_{\mathbf{x}_j} \{\psi \mathbf{v}(\mathbf{x}_j, t)\} - \sum_{j=1}^{n} \frac{i\hbar}{2} \mathbf{v}(\mathbf{x}_j, t) \cdot \nabla_{\mathbf{x}_j} \psi$$
$$+ \sum_{j=1}^{n} \left\{ \frac{1}{2m_j} \left(-i\hbar\nabla_{\mathbf{x}_j} - \frac{\sigma_j}{c} \mathbf{A}(\mathbf{x}_j, t) \right) \circ \left(-i\hbar\nabla_{\mathbf{x}_j} - \frac{\sigma_j}{c} \mathbf{A}(\mathbf{x}_j, t) \right) \right\} \cdot \psi$$
$$+ \sum_{j=1}^{n} \sigma_j \left(\Psi(\mathbf{x}_j, t) - \frac{1}{c} \mathbf{A}(\mathbf{x}_j, t) \cdot \mathbf{v}(\mathbf{x}_j, t) \right) \cdot \psi - V(\mathbf{x}_1, \ldots, \mathbf{x}_n, t) \cdot \psi$$
$$- \sum_{j=1}^{n} \frac{g_j \sigma_j \hbar}{2m_j c} \mathbf{S}_j \cdot (\operatorname{curl}_{\mathbf{x}_j} \mathbf{A}(\mathbf{x}_j, t)\, \psi) + \sum_{j=1}^{n} \frac{\hbar}{4} \mathbf{S}_j \cdot (\operatorname{curl}_{\mathbf{x}_j} \mathbf{v}(\mathbf{x}_j, t)\, \psi) = -\sum_{j=1}^{n} \frac{\hbar^2}{2m_j} \Delta_{\mathbf{x}_j} \psi$$
$$- \sum_{j=1}^{n} \frac{i\hbar}{2} \operatorname{div}_{\mathbf{x}_j} \{\psi \mathbf{v}(\mathbf{x}_j, t)\} - \sum_{j=1}^{n} \frac{i\hbar}{2} \mathbf{v}(\mathbf{x}_j, t) \cdot \nabla_{\mathbf{x}_j} \psi$$
$$+ \sum_{j=1}^{n} \frac{i\hbar\sigma_j}{2m_j c} \operatorname{div}_{\mathbf{x}_j} \{\psi \mathbf{A}(\mathbf{x}_j, t)\} + \sum_{j=1}^{n} \frac{i\hbar\sigma_j}{2m_j c} \mathbf{A}(\mathbf{x}_j, t) \cdot \nabla_{\mathbf{x}_j} \psi$$
$$+ \sum_{j=1}^{n} \left(\sigma_j \Psi(\mathbf{x}_j, t) - \frac{\sigma_j}{c} \mathbf{A}(\mathbf{x}_j, t) \cdot \mathbf{v}(\mathbf{x}_j, t) + \frac{\sigma_j^2}{2m_j c^2} \left| \mathbf{A}(\mathbf{x}_j, t) \right|^2 \right) \psi - V(\mathbf{x}_1, \ldots, \mathbf{x}_n, t)\, \psi$$
$$- \sum_{j=1}^{n} \frac{g_j \sigma_j \hbar}{2m_j c} \mathbf{S}_j \cdot (\operatorname{curl}_{\mathbf{x}_j} \mathbf{A}(\mathbf{x}_j, t)\, \psi) + \sum_{j=1}^{n} \frac{\hbar}{4} \mathbf{S}_j \cdot (\operatorname{curl}_{\mathbf{x}_j} \mathbf{v}(\mathbf{x}_j, t)\, \psi), \tag{6.192}$$

where $\hat{H}_0$ is the Hamiltonian operator, for every $j = 1, \ldots, n$ g_j is a constant that depends on the type of the particle (for electron we have $g_j = 1$), and for every $j = 1, \ldots, n$ we denote

$$\mathbf{S}_j := (S_1^j, S_2^j, S_3^j) \qquad \forall\, j = 1, 2, \ldots, n, \tag{6.193}$$

where for every $k = 1, 2, 3$ and every $j = 1, 2, \ldots, n$: $S_k^j : \mathbb{C}^{2^n} \to \mathbb{C}^{2^n}$ is a linear operator on $\mathbb{C}^{2^n}$ (i.e. it is a $2^n \times 2^n$-complex matrix) defined by the following identities:

$$S_k^1 := (S_k) \otimes_1 \left(I^{2\times 2}\right) \otimes_2 \left(I^{2\times 2}\right) \ldots \otimes_{(n-1)} \left(I^{2\times 2}\right), \quad \ldots$$
$$S_k^j := \left(I^{2\times 2}\right) \otimes_1 \left(I^{2\times 2}\right) \otimes_2 \left(I^{2\times 2}\right) \ldots \otimes_{(j-1)} (S_k) \otimes_j \left(I^{2\times 2}\right) \otimes_{(j+1)} \left(I^{2\times 2}\right) \ldots \otimes_{(n-1)} \left(I^{2\times 2}\right),$$
$$\ldots \quad \text{and} \quad S_k^n := \left(I^{2\times 2}\right) \otimes_1 \left(I^{2\times 2}\right) \otimes_2 \left(I^{2\times 2}\right) \ldots \otimes_{(n-1)} (S_k). \tag{6.194}$$

Here S_k for $k = 1, 2, 3$ are Pauli matrixes defined as

$$S_1 = \begin{pmatrix} 0 & 1 \\ 1 & 0 \end{pmatrix}, \quad S_2 = \begin{pmatrix} 0 & -i \\ i & 0 \end{pmatrix} \quad S_3 = \begin{pmatrix} 1 & 0 \\ 0 & -1 \end{pmatrix}$$

and the sign $\otimes$ in (6.194) means the tensor product of the matrices, i.e. for given two linear operators $A : \mathbb{C}^p \to \mathbb{C}^p$ and $B : \mathbb{C}^q \to \mathbb{C}^q$ their tensor product $A \otimes B$ is a linear operator from $\mathbb{C}^p \otimes \mathbb{C}^q$ to $\mathbb{C}^p \otimes \mathbb{C}^q$, defined by the identity:

$$(A \otimes B) \cdot (a \otimes b) = (A \cdot a) \otimes (B \cdot b) \qquad \forall a \in \mathbb{C}^p,\ \forall b \in \mathbb{C}^q. \tag{6.195}$$

Note that, in addition to the classical term of the spin-magnetic interaction, we added another term to the Hamiltonian, namely $\sum_{j=1}^{n} \frac{\hbar}{4}\mathbf{S}_j \cdot \left(\text{curl}_{\mathbf{x}_j} \mathbf{v}(\mathbf{x}_j, t)\, \psi\right)$. In the case of the Newtonian-type gravity, this term vanishes in every non-rotating and, in particular, in every inertial coordinate system, however, it provides the invariance of the Shrödinger–Pauli equation, under the change of non-inertial Cartesian coordinate system, as can be seen in the following Theorem 6.2. Thus the corresponding Shrödinger–Pauli equation will be the following:

$$\begin{aligned}
i\hbar\frac{\partial \psi}{\partial t} = \hat{H}_0 \cdot \psi = &-\sum_{j=1}^{n} \frac{\hbar^2}{2m_j}\Delta_{\mathbf{x}_j}\psi - \sum_{j=1}^{n} \frac{i\hbar}{2}\,\text{div}_{\mathbf{x}_j}\left\{\psi \mathbf{v}(\mathbf{x}_j, t)\right\} - \sum_{j=1}^{n} \frac{i\hbar}{2}\mathbf{v}(\mathbf{x}_j, t) \cdot \nabla_{\mathbf{x}_j}\psi \\
&+ \sum_{j=1}^{n} \frac{i\hbar\sigma_j}{2m_j c}\,\text{div}_{\mathbf{x}_j}\left\{\psi \mathbf{A}(\mathbf{x}_j, t)\right\} + \sum_{j=1}^{n} \frac{i\hbar\sigma_j}{2m_j c}\mathbf{A}(\mathbf{x}_j, t) \cdot \nabla_{\mathbf{x}_j}\psi \\
&+ \sum_{j=1}^{n}\left(\sigma_j \Psi(\mathbf{x}_j, t) - \frac{\sigma_j}{c}\mathbf{A}(\mathbf{x}_j, t) \cdot \mathbf{v}(\mathbf{x}_j, t) + \frac{\sigma_j^2}{2m_j c^2}\left|\mathbf{A}(\mathbf{x}_j, t)\right|^2\right)\psi - V\left(\mathbf{x}_1, \ldots, \mathbf{x}_n, t\right)\psi \\
&- \sum_{j=1}^{n} \frac{g_j \sigma_j \hbar}{2m_j c}\mathbf{S}_j \cdot \left(\text{curl}_{\mathbf{x}_j} \mathbf{A}(\mathbf{x}_j, t)\, \psi\right) + \sum_{j=1}^{n} \frac{\hbar}{4}\mathbf{S}_j \cdot \left(\text{curl}_{\mathbf{x}_j} \mathbf{v}(\mathbf{x}_j, t)\, \psi\right).
\end{aligned} \tag{6.196}$$

So

$$\begin{aligned}
i\hbar&\left(\frac{\partial \psi}{\partial t} + \sum_{j=1}^{n} \mathbf{v}(\mathbf{x}_j, t) \cdot \nabla_{\mathbf{x}_j}\psi\right) + \sum_{j=1}^{n} \frac{i\hbar}{2}\left(\text{div}_{\mathbf{x}_j} \mathbf{v}(\mathbf{x}_j, t)\right)\psi \\
&= -\sum_{j=1}^{n} \frac{\hbar^2}{2m_j}\Delta_{\mathbf{x}_j}\psi - V\left(\mathbf{x}_1, \ldots, \mathbf{x}_n, t\right)\psi \\
&+ \sum_{j=1}^{n} \frac{i\hbar\sigma_j}{2m_j c}\,\text{div}_{\mathbf{x}_j}\left\{\psi \mathbf{A}(\mathbf{x}_j, t)\right\} + \sum_{j=1}^{n} \frac{i\hbar\sigma_j}{2m_j c}\mathbf{A}(\mathbf{x}_j, t) \cdot \nabla_{\mathbf{x}_j}\psi \\
&+ \sum_{j=1}^{n}\left(\sigma_j \Psi(\mathbf{x}_j, t) - \frac{\sigma_j}{c}\mathbf{A}(\mathbf{x}_j, t) \cdot \mathbf{v}(\mathbf{x}_j, t) + \frac{\sigma_j^2}{2m_j c^2}\left|\mathbf{A}(\mathbf{x}_j, t)\right|^2\right)\psi \\
&- \sum_{j=1}^{n} \frac{g_j \sigma_j \hbar}{2m_j c}\mathbf{S}_j \cdot \left(\text{curl}_{\mathbf{x}_j} \mathbf{A}(\mathbf{x}_j, t)\, \psi\right) + \sum_{j=1}^{n} \frac{\hbar}{4}\mathbf{S}_j \cdot \left(\text{curl}_{\mathbf{x}_j} \mathbf{v}(\mathbf{x}_j, t)\, \psi\right).
\end{aligned} \tag{6.197}$$

Then in the similar way as the proof of Theorem 6.1 we can prove the following more general Theorem about the invariance of the Shrödinger–Pauli equation (6.197) under the change of inertial or non-inertial Cartesian coordinate system:

Theorem 6.2 *Consider that the change of some Cartesian coordinate system* $(*)$ *to another Cartesian coordinate system* $(**)$ *is given by*

$$\begin{cases} t' = t, \\ \mathbf{x}' = A(t)\cdot\mathbf{x} + \mathbf{z}(t), \\ \mathbf{x}'_k = A(t)\cdot\mathbf{x}_k + \mathbf{z}(t) \quad \forall k = 1,\ldots,n, \end{cases} \tag{6.198}$$

where $A(t) \in SO(3)$ *is a rotation. Next, assume that in the coordinate system* $(**)$ *we observe a validity of the Shrödinger–Pauli equation of the form:*

$$\begin{aligned} &i\hbar\left(\frac{\partial\psi'}{\partial t'} + \sum_{j=1}^{n}\mathbf{v}'(\mathbf{x}'_j,t')\cdot\nabla_{\mathbf{x}'_j}\psi'\right) + \sum_{j=1}^{n}\frac{i\hbar}{2}\left(\operatorname{div}_{\mathbf{x}'_j}\mathbf{v}'(\mathbf{x}'_j,t')\right)\psi' = -\sum_{j=1}^{n}\frac{\hbar^2}{2m'_j}\Delta_{\mathbf{x}'_j}\psi' \\ &- V'\left(\mathbf{x}'_1,\ldots,\mathbf{x}'_n,t'\right)\psi' + \sum_{j=1}^{n}\frac{i\hbar\sigma'_j}{2m'_jc}\operatorname{div}_{\mathbf{x}'_j}\left\{\psi'\mathbf{A}'(\mathbf{x}'_j,t')\right\} + \sum_{j=1}^{n}\frac{i\hbar\sigma'_j}{2m'_jc}\mathbf{A}'(\mathbf{x}'_j,t')\cdot\nabla_{\mathbf{x}'_j}\psi' \\ &+\sum_{j=1}^{n}\left(\sigma'_j\Psi'(\mathbf{x}'_j,t') - \frac{\sigma'_j}{c}\mathbf{A}'(\mathbf{x}'_j,t')\cdot\mathbf{v}'(\mathbf{x}'_j,t') + \frac{(\sigma')_j^2}{2m'_jc^2}\left|\mathbf{A}'(\mathbf{x}'_j,t')\right|^2\right)\psi' \\ &-\sum_{j=1}^{n}\frac{g'_j\sigma'_j\hbar}{2m'_jc}\mathbf{S}_j\cdot\left(curl_{\mathbf{x}'_j}\mathbf{A}'(\mathbf{x}'_j,t')\,\psi'\right) + \sum_{j=1}^{n}\frac{\hbar}{4}\mathbf{S}_j\cdot\left(curl_{\mathbf{x}'_j}\mathbf{v}'(\mathbf{x}'_j,t')\,\psi'\right) \end{aligned} \tag{6.199}$$

where $\psi(\mathbf{x}_1,\ldots,\mathbf{x}_n,t) \in \mathbb{C}^{2^n}$ *is a* 2^n*-component complex wave function defined above. Then in the coordinate system* $(*)$ *we have the validity of Shrödinger–Pauli equation of the same as* (6.199) *form:*

$$\begin{aligned} &i\hbar\left(\frac{\partial\psi}{\partial t} + \sum_{j=1}^{n}\mathbf{v}(\mathbf{x}_j,t)\cdot\nabla_{\mathbf{x}_j}\psi\right) + \sum_{j=1}^{n}\frac{i\hbar}{2}\left(\operatorname{div}_{\mathbf{x}_j}\mathbf{v}(\mathbf{x}_j,t)\right)\psi = -\sum_{j=1}^{n}\frac{\hbar^2}{2m_j}\Delta_{\mathbf{x}_j}\psi \\ &- V\left(\mathbf{x}_1,\ldots,\mathbf{x}_n,t\right)\psi + \sum_{j=1}^{n}\frac{i\hbar\sigma_j}{2m_jc}\operatorname{div}_{\mathbf{x}_j}\left\{\psi\mathbf{A}(\mathbf{x}_j,t)\right\} + \sum_{j=1}^{n}\frac{i\hbar\sigma_j}{2m_jc}\mathbf{A}(\mathbf{x}_j,t)\cdot\nabla_{\mathbf{x}_j}\psi \\ &+\sum_{j=1}^{n}\left(\sigma_j\Psi(\mathbf{x}_j,t) - \frac{\sigma_j}{c}\mathbf{A}(\mathbf{x}_j,t)\cdot\mathbf{v}(\mathbf{x}_j,t) + \frac{\sigma_j^2}{2m_jc^2}\left|\mathbf{A}(\mathbf{x}_j,t)\right|^2\right)\psi \\ &-\sum_{j=1}^{n}\frac{g_j\sigma_j\hbar}{2m_jc}\mathbf{S}_j\cdot\left(curl_{\mathbf{x}_j}\mathbf{A}(\mathbf{x}_j,t)\,\psi\right) + \sum_{j=1}^{n}\frac{\hbar}{4}\mathbf{S}_j\cdot\left(curl_{\mathbf{x}_j}\mathbf{v}(\mathbf{x}_j,t)\,\psi\right), \end{aligned} \tag{6.200}$$

provided that we have

$$
\begin{cases}
g'_j = g_j \\
V'\left(\mathbf{x}'_1,\ldots,\mathbf{x}'_n,t'\right) = V\left(\mathbf{x}_1,\ldots,\mathbf{x}_n,t\right), \\
\sigma'_j = \sigma_j, \\
m'_j = m_j, \\
\mathbf{v}'(\mathbf{x}',t) = A(t)\cdot\mathbf{v}(\mathbf{x},t) + \frac{dA}{dt}(t)\cdot\mathbf{x} + \frac{d\mathbf{z}}{dt}(t), \\
\mathbf{A}'(\mathbf{x}',t) = A(t)\cdot\mathbf{A}(\mathbf{x},t), \\
\Psi'(\mathbf{x}',t) - \mathbf{v}'(\mathbf{x}',t)\cdot\mathbf{A}'(\mathbf{x}',t) = \Psi(\mathbf{x},t) - \mathbf{v}(\mathbf{x},t)\cdot\mathbf{A}(\mathbf{x},t), \\
\psi'\left(\mathbf{x}'_1,\ldots,\mathbf{x}'_n,t'\right) = \left(U(t)\otimes_1 U(t)\otimes_2 U(t)\ldots\otimes_{(n-1)} U(t)\right)\cdot\psi\left(\mathbf{x}_1,\ldots,\mathbf{x}_n,t\right),
\end{cases}
\tag{6.201}
$$

where, as before, $U(t) \in SU(2)$ is characterized by

$$
U^*(t)\cdot\mathbf{S}\cdot U(t) = A(t)\cdot\mathbf{S}, \tag{6.202}
$$

where $\mathbf{S} := (S_1, S_2, S_3)$, that means

$$
\Big(U^*(t)\cdot S_1\cdot U(t), U^*(t)\cdot S_2\cdot U(t), U^*(t)\cdot S_3\cdot U(t)\Big) =
\Big(a_{11}(t)S_1 + a_{12}(t)S_2 + a_{13}(t)S_3\,,\; a_{21}(t)S_1 + a_{22}(t)S_2 + a_{23}(t)S_3\,,\; a_{31}(t)S_1 + a_{32}(t)S_2 + a_{33}(t)S_3\Big),
$$

where $A(t) = \{a_{mk}(t)\}_{\{1\le m,k\le 3\}}$.

Next, consider the Lagrangian of the motion of system of n spin-half quantum micro-particles with inertial masses $m_1,\ldots,m_n$ and the charges $\sigma_1,\ldots,\sigma_n$ in the outer gravitational and electromagnetical fields with characteristics $\mathbf{v}(\mathbf{x},t)$, $\mathbf{A}(\mathbf{x},t)$ and $\Psi(\mathbf{x},t)$ and additional conservative field with potential $V(\mathbf{y}_1,\ldots,\mathbf{y}_n,t)$. Then consider a Lagrangian density L_0 defined by

$$
\begin{aligned}
L_0\left(\psi,\mathbf{A},\Psi,\mathbf{v},\mathbf{x}_1,\ldots,\mathbf{x}_n,t\right) :=\;&
\frac{i\hbar}{2}\left(\left(\frac{\partial\psi}{\partial t} + \sum_{k=1}^{n}\mathbf{v}(\mathbf{x}_k,t)\cdot\nabla_{\mathbf{x}_k}\psi\right)\cdot\bar{\psi} - \psi\cdot\left(\frac{\partial\bar{\psi}}{\partial t} + \sum_{k=1}^{n}\mathbf{v}(\mathbf{x}_k,t)\cdot\nabla_{\mathbf{x}_k}\bar{\psi}\right)\right) \\
&- \sum_{k=1}^{n}\frac{\hbar^2}{2m_k}\nabla_{\mathbf{x}_k}\psi\cdot\nabla_{\mathbf{x}_k}\bar{\psi} + V\left(\mathbf{x}_1,\ldots,\mathbf{x}_n,t\right)\psi\cdot\bar{\psi} \\
&- \sum_{k=1}^{n}\frac{\hbar\sigma_k i}{2m_k c}\left(\nabla_{\mathbf{x}_k}\psi\cdot\bar{\psi} - \psi\cdot\nabla_{\mathbf{x}_k}\bar{\psi}\right)\cdot\mathbf{A}(\mathbf{x}_k,t) - \sum_{k=1}^{n}\frac{\sigma_k^2}{2m_k c^2}\left|\mathbf{A}(\mathbf{x}_k,t)\right|^2\psi\cdot\bar{\psi} \\
&- \sum_{k=1}^{n}\sigma_k\left(\Psi(\mathbf{x}_k,t) - \frac{1}{c}\mathbf{v}(\mathbf{x}_k,t)\cdot\mathbf{A}(\mathbf{x}_k,t)\right)\psi\cdot\bar{\psi} \\
&- \sum_{k=1}^{n}\frac{\hbar}{2}\left(\left(\mathbf{S}_k\cdot\left(\frac{1}{2}\mathrm{curl}_{\mathbf{x}_k}\mathbf{v}(\mathbf{x}_k,t) - \frac{g_k\sigma_k}{m_k c}\mathrm{curl}_{\mathbf{x}_k}\mathbf{A}(\mathbf{x}_k,t)\right)\right)\cdot\psi\right)\cdot\bar{\psi},
\end{aligned}
\tag{6.203}
$$

where $\psi := \psi(\mathbf{x}_1, \dots \mathbf{x}_n, t) \in \mathbb{C}^{2^n}$ is a wave function of the system. Then, as before, we can get that L_0 is invariant under the change of inertial or non-inertial Cartesian coordinate systems of the form

$$\begin{cases} t' = t \\ \mathbf{x}' = A(t) \cdot \mathbf{x} + \mathbf{z}(t) \\ \mathbf{x}'_k = A(t) \cdot \mathbf{x}_k + \mathbf{z}(t) \qquad \forall k = 1, \dots, n, \end{cases}$$

provided, we take (6.201) into account. Next we investigate stationary points of the functional

$$J(\psi) = \int_0^T \int_{(\mathbb{R}^3)^n} L_0(\psi, \mathbf{A}, \Psi, \mathbf{v}, \mathbf{x}_1, \dots, \mathbf{x}_n, t)\, d\mathbf{x}_1 \dots, d\mathbf{x}_n dt. \tag{6.204}$$

Then,

$$\begin{aligned} 0 = \frac{\delta L_0}{\delta(\bar{\psi})} = \\ i\hbar \left(\frac{\partial \psi}{\partial t} + \sum_{k=1}^n \frac{1}{2} \mathbf{v}(\mathbf{x}_k, t) \cdot \nabla_{\mathbf{x}_k} \psi + \sum_{k=1}^n \frac{1}{2} \operatorname{div}_{\mathbf{x}_k} \{\psi \mathbf{v}(\mathbf{x}_k, t)\} \right) + \sum_{k=1}^n \frac{\hbar^2}{2m_k} \Delta_{\mathbf{x}_k} \psi \\ + V(\mathbf{x}_1, \dots, \mathbf{x}_n, t)\, \psi - \frac{\hbar \sigma_k i}{2m_k c} \left(\sum_{k=1}^n \mathbf{A}(\mathbf{x}_k, t) \cdot \nabla_{\mathbf{x}_k} \psi + \sum_{k=1}^n \operatorname{div}_{\mathbf{x}_k} \{\psi \mathbf{A}(\mathbf{x}_k, t)\} \right) \\ - \sum_{k=1}^n \frac{\sigma_k^2}{2m_k c^2} |\mathbf{A}(\mathbf{x}_k, t)|^2 \psi - \sum_{k=1}^n \sigma_k \left(\Psi(\mathbf{x}_k, t) - \frac{1}{c} \mathbf{v}(\mathbf{x}_k, t) \cdot \mathbf{A}(\mathbf{x}_k, t) \right) \psi \\ - \sum_{k=1}^n \frac{\hbar}{2} \left(\mathbf{S}_k \cdot \left(\frac{1}{2} \operatorname{curl}_{\mathbf{x}_k} \mathbf{v}(\mathbf{x}_k, t) - \frac{g_k \sigma_k}{m_k c} \operatorname{curl}_{\mathbf{x}_k} \mathbf{A}(\mathbf{x}_k, t) \right) \right) \cdot \psi, \end{aligned} \tag{6.205}$$

and

$$\begin{aligned} 0 = \frac{\delta L_0}{\delta(\psi)} = \\ (\bar{i})\hbar \left(\frac{\partial \bar{\psi}}{\partial t} + \sum_{k=1}^n \frac{1}{2} \mathbf{v}(\mathbf{x}_k, t) \cdot \nabla_{\mathbf{x}_k} \bar{\psi} + \sum_{k=1}^n \frac{1}{2} \operatorname{div}_{\mathbf{x}_k} \left\{\bar{\psi} \mathbf{v}(\mathbf{x}_k, t)\right\} \right) + \sum_{k=1}^n \frac{\hbar^2}{2m_k} \Delta_{\mathbf{x}_k} \bar{\psi} \\ + V(\mathbf{x}_1, \dots, \mathbf{x}_n, t)\, \bar{\psi} - \sum_{k=1}^n \frac{\hbar \sigma_k (\bar{i})}{2m_k c} \left(\mathbf{A}(\mathbf{x}_k, t) \cdot \nabla_{\mathbf{x}_k} \bar{\psi} + \operatorname{div}_{\mathbf{x}_k} \left\{\bar{\psi} \mathbf{A}(\mathbf{x}_k, t)\right\} \right) \\ - \sum_{k=1}^n \frac{\sigma_k^2}{2m_k c^2} |\mathbf{A}(\mathbf{x}_k, t)|^2 \bar{\psi} - \sum_{k=1}^n \sigma_k \left(\Psi(\mathbf{x}_k, t) - \frac{1}{c} \mathbf{v}(\mathbf{x}_k, t) \cdot \mathbf{A}(\mathbf{x}_k, t) \right) \bar{\psi} \\ - \sum_{k=1}^n \frac{\hbar}{2} \left(\mathbf{S}_k^T \cdot \left(\frac{1}{2} \operatorname{curl}_{\mathbf{x}_k} \mathbf{v}(\mathbf{x}_k, t) - \frac{g_k \sigma_k}{m_k c} \operatorname{curl}_{\mathbf{x}_k} \mathbf{A}(\mathbf{x}_k, t) \right) \right) \cdot \bar{\psi}. \end{aligned} \tag{6.206}$$

Equation (6.206) is just a complex conjugate of equation (6.205). Thus the Euler–Lagrange for (6.204) coincides with the Shrödinger–Pauli equation (6.197).

Finally, assume that the first and the second particles have the same mass $m_1 = m_2$ and the same charge $\sigma_1 = \sigma_2$ and moreover, assume that we have

$$V(\mathbf{x}_1, \mathbf{x}_2, \mathbf{x}_3, \ldots, \mathbf{x}_n, t) = V(\mathbf{x}_2, \mathbf{x}_1, \mathbf{x}_3, \ldots, \mathbf{x}_n, t) \qquad \forall \mathbf{x}_1, \mathbf{x}_2, \mathbf{x}_3, \ldots, \mathbf{x}_n \in \mathbb{R}^3, \quad \forall t.$$

In this case it can be easily deduced that if $\psi(\mathbf{x}_1, \mathbf{x}_2, \mathbf{x}_3, \ldots, \mathbf{x}_n, t) \in \mathbb{C}^{2^n}$ is a solution of (6.197), then $A_{1,2} \cdot \psi(\mathbf{x}_2, \mathbf{x}_1, \mathbf{x}_3, \ldots, \mathbf{x}_n, t)$ is also a solution of (6.197), where by $A_{1,2} : \mathbb{C}^{2^n} \to \mathbb{C}^{2^n}$ we denote the linear operator (matrix) defined as

$$A_{1,2} \cdot (a_1 \otimes a_2 \otimes a_3 \otimes \ldots \otimes a_n) = (a_2 \otimes a_1 \otimes a_3 \otimes \ldots \otimes a_n) \qquad \forall a_1, \ldots a_n \in \mathbb{C}^2. \tag{6.207}$$

Therefore, if $\psi(\mathbf{x}_1, \mathbf{x}_2, \mathbf{x}_3, \ldots, \mathbf{x}_n, t) \in \mathbb{C}^{2^n}$ is a solution of (6.197) then for every $t \geq 0$ we will have

$$A_{1,2} \cdot \psi(\mathbf{x}_2, \mathbf{x}_1, \mathbf{x}_3, \ldots, \mathbf{x}_n, t) = -\psi(\mathbf{x}_1, \mathbf{x}_2, \mathbf{x}_3, \ldots, \mathbf{x}_n, t) \qquad \forall \mathbf{x}_1, \mathbf{x}_2, \mathbf{x}_3, \ldots, \mathbf{x}_n \in \mathbb{R}^3, \tag{6.208}$$

provided that (6.208) holds for the initial instant of time $t = 0$. So we have a consistency with the Pauli Exclusion Principle for two or more identical fermions.

6.8 Quantum Liouville's Equation for a Finite System of Spin-Half Particles

Consider the statistical description of the motion of a system of n spin-half microparticles with inertial masses $m_1, \ldots, m_n$ and the charges $\sigma_1, \ldots, \sigma_n$ in the outer gravitational and electromagnetical field with characteristics $\mathbf{v}(\mathbf{x}, t)$, $\mathbf{A}(\mathbf{x}, t)$ and $\Psi(\mathbf{x}, t)$ and additional conservative field with potential $V(\mathbf{y}_1, \ldots, \mathbf{y}_n, t)$, taking into account the spin interaction. Then, as before, the Quantum Liouville's equation for this system of particles has the following form:

$$\begin{aligned} i\hbar\frac{\partial \xi}{\partial t}(\mathbf{x}_1, \ldots, \mathbf{x}_n, \mathbf{y}_1, \ldots, \mathbf{y}_n, t) = \\ \left(\hat{H}_0(\mathbf{x}_1, \ldots, \mathbf{x}_n, t) \otimes I\right) \cdot \xi(\mathbf{x}_1, \ldots, \mathbf{x}_n, \mathbf{y}_1, \ldots, \mathbf{y}_n, t) \\ - \left(I \otimes \hat{H}_0^*(\mathbf{y}_1, \ldots, \mathbf{y}_n, t)\right) \cdot \xi(\mathbf{x}_1, \ldots, \mathbf{x}_n, \mathbf{y}_1, \ldots, \mathbf{y}_n, t), \end{aligned} \tag{6.209}$$

where $\xi(\mathbf{x}_1, \ldots, \mathbf{x}_n, \mathbf{y}_1, \ldots, \mathbf{y}_n, t) \in \mathbb{C}^{2^n} \otimes \mathbb{C}^{2^n}$ is a density matrix function, $\hat{H}_0(\mathbf{x}_1, \ldots, \mathbf{x}_n, t)$ is the Hamiltonian operator acting on the variables $(\mathbf{x}_1, \ldots, \mathbf{x}_n)$, $\hat{H}_0^*(\mathbf{y}_1, \ldots, \mathbf{y}_n, t)$ is the complex conjugate (not the Hermitian adjoint) to the Hamiltonian operator acting on the variables $(\mathbf{y}_1, \ldots, \mathbf{y}_n)$ and $\hat{H}_0(\mathbf{x}_1, \ldots, \mathbf{x}_n, t) \otimes I$ and

$I \otimes \hat{H}_0^* (\mathbf{y}_1, \ldots, \mathbf{y}_n, t)$ are linear operators acting on functions $\xi (\mathbf{x}_1, \ldots, \mathbf{x}_n, \mathbf{y}_1, \ldots, \mathbf{y}_n, t) \in \mathbb{C}^{2^n} \otimes \mathbb{C}^{2^n}$, defined by

$$\left(\hat{H}_0 (\mathbf{x}_1, \ldots, \mathbf{x}_n, t) \otimes I\right) \cdot (\psi_1 (\mathbf{x}_1, \ldots, \mathbf{x}_n, t) \otimes \psi_2 (\mathbf{y}_1, \ldots, \mathbf{y}_n, t)) =$$
$$\left(\hat{H}_0 (\mathbf{x}_1, \ldots, \mathbf{x}_n, t) \cdot \psi_1 (\mathbf{x}_1, \ldots, \mathbf{x}_n, t)\right) \otimes \psi_2 (\mathbf{y}_1, \ldots, \mathbf{y}_n, t) \quad \text{and}$$
$$\left(I \otimes \hat{H}_0^* (\mathbf{y}_1, \ldots, \mathbf{y}_n, t)\right) \cdot (\psi_1 (\mathbf{x}_1, \ldots, \mathbf{x}_n, t) \otimes \psi_2 (\mathbf{y}_1, \ldots, \mathbf{y}_n, t)) =$$
$$\psi_1 (\mathbf{x}_1, \ldots, \mathbf{x}_n, t) \otimes \left(\hat{H}_0^* (\mathbf{y}_1, \ldots, \mathbf{y}_n, t) \cdot \psi_2 (\mathbf{y}_1, \ldots, \mathbf{y}_n, t)\right) \quad \forall \psi_1, \psi_2 \in \mathbb{C}^{2^n}. \tag{6.210}$$

Since by (6.15) the Hamiltonian operator $\hat{H}_0$ has the forms:

$$\hat{H}_0 (\mathbf{x}_1, \ldots, \mathbf{x}_n, t) \cdot \psi (\mathbf{x}_1, \ldots, \mathbf{x}_n, t) = -\sum_{j=1}^{n} \frac{\hbar^2}{2m_j} \Delta_{\mathbf{x}_j} \psi - \sum_{j=1}^{n} \frac{i\hbar}{2} \operatorname{div}_{\mathbf{x}_j} \{\psi \mathbf{v}(\mathbf{x}_j, t)\}$$
$$-\sum_{j=1}^{n} \frac{i\hbar}{2} \mathbf{v}(\mathbf{x}_j, t) \cdot \nabla_{\mathbf{x}_j} \psi + \sum_{j=1}^{n} \frac{i\hbar\sigma_j}{2m_j c} \operatorname{div}_{\mathbf{x}_j} \{\psi \mathbf{A}(\mathbf{x}_j, t)\} + \sum_{j=1}^{n} \frac{i\hbar\sigma_j}{2m_j c} \mathbf{A}(\mathbf{x}_j, t) \cdot \nabla_{\mathbf{x}_j} \psi$$
$$+\sum_{j=1}^{n} \left(\sigma_j \Psi(\mathbf{x}_j, t) - \frac{\sigma_j}{c} \mathbf{A}(\mathbf{x}_j, t) \cdot \mathbf{v}(\mathbf{x}_j, t) + \frac{\sigma_j^2}{2m_j c^2} |\mathbf{A}(\mathbf{x}_j, t)|^2\right) \psi - V (\mathbf{x}_1, \ldots, \mathbf{x}_n, t) \psi$$
$$-\sum_{j=1}^{n} \frac{g_j \sigma_j \hbar}{2m_j c} \mathbf{S}_j \cdot \left(\operatorname{curl}_{\mathbf{x}_j} \mathbf{A}(\mathbf{x}_j, t) \psi\right) + \sum_{j=1}^{n} \frac{\hbar}{4} \mathbf{S}_j \cdot \left(\operatorname{curl}_{\mathbf{x}_j} \mathbf{v}(\mathbf{x}_j, t) \psi\right), \tag{6.211}$$

and consistently with (6.211) we write

$$\hat{H}_0^* (\mathbf{y}_1, \ldots, \mathbf{y}_n, t) \cdot \psi (\mathbf{y}_1, \ldots, \mathbf{y}_n, t) = -\sum_{j=1}^{n} \frac{\hbar^2}{2m_j} \Delta_{\mathbf{y}_j} \psi + \sum_{j=1}^{n} \frac{i\hbar}{2} \operatorname{div}_{\mathbf{y}_j} \{\psi \mathbf{v}(\mathbf{y}_j, t)\}$$
$$+\sum_{j=1}^{n} \frac{i\hbar}{2} \mathbf{v}(\mathbf{y}_j, t) \cdot \nabla_{\mathbf{y}_j} \psi - \sum_{j=1}^{n} \frac{i\hbar\sigma_j}{2m_j c} \operatorname{div}_{\mathbf{y}_j} \{\psi \mathbf{A}(\mathbf{y}_j, t)\} - \sum_{j=1}^{n} \frac{i\hbar\sigma_j}{2m_j c} \mathbf{A}(\mathbf{y}_j, t) \cdot \nabla_{\mathbf{y}_j} \psi$$
$$+\sum_{j=1}^{n} \left(\sigma_j \Psi(\mathbf{y}_j, t) - \frac{\sigma_j}{c} \mathbf{A}(\mathbf{y}_j, t) \cdot \mathbf{v}(\mathbf{y}_j, t) + \frac{\sigma_j^2}{2m_j c^2} |\mathbf{A}(\mathbf{y}_j, t)|^2\right) \psi - V (\mathbf{y}_1, \ldots, \mathbf{y}_n, t) \psi$$
$$-\sum_{j=1}^{n} \frac{g_j \sigma_j \hbar}{2m_j c} \bar{\mathbf{S}}_j \cdot \left(\operatorname{curl}_{\mathbf{y}_j} \mathbf{A}(\mathbf{y}_j, t) \psi\right) + \sum_{j=1}^{n} \frac{\hbar}{4} \bar{\mathbf{S}}_j \cdot \left(\operatorname{curl}_{\mathbf{y}_j} \mathbf{v}(\mathbf{y}_j, t) \psi\right), \tag{6.212}$$

where $\bar{\mathbf{S}}_j$ is the complex conjugate (not the Hermitian adjoint) of the operator $\mathbf{S}_j$. Thus we rewrite the corresponding Quantum Liouville's equation (6.209) as

$$
\begin{aligned}
&i\hbar\left(\frac{\partial \xi}{\partial t}+\sum_{j=1}^{n}\mathbf{v}(\mathbf{x}_j,t)\cdot\nabla_{\mathbf{x}_j}\xi+\sum_{j=1}^{n}\mathbf{v}(\mathbf{y}_j,t)\cdot\nabla_{\mathbf{y}_j}\xi\right)\\
&\qquad+\sum_{j=1}^{n}\frac{i\hbar}{2}\Big(\mathrm{div}_{\mathbf{x}_j}\,\mathbf{v}(\mathbf{x}_j,t)+\mathrm{div}_{\mathbf{y}_j}\,\mathbf{v}(\mathbf{y}_j,t)\Big)\xi=\\
&\sum_{j=1}^{n}\frac{\hbar}{4}(\mathbf{S}_j\otimes I)\cdot\Big(\mathrm{curl}_{\mathbf{x}_j}\mathbf{v}(\mathbf{x}_j,t)\,\xi\Big)-\sum_{j=1}^{n}\frac{\hbar}{4}(I\otimes\bar{\mathbf{S}}_j)\cdot\Big(\mathrm{curl}_{\mathbf{y}_j}\mathbf{v}(\mathbf{y}_j,t)\,\xi\Big)\\
&-\sum_{j=1}^{n}\frac{\hbar^2}{2m_j}\Big(\Delta_{\mathbf{x}_j}\xi-\Delta_{\mathbf{y}_j}\xi\Big)-\left(V\left(\mathbf{x}_1,\ldots,\mathbf{x}_n,t\right)-V\left(\mathbf{y}_1,\ldots,\mathbf{y}_n,t\right)\right)\xi\\
&\qquad+\sum_{j=1}^{n}\frac{i\hbar\sigma_j}{2m_jc}\Big(\mathrm{div}_{\mathbf{x}_j}\left\{\xi\mathbf{A}(\mathbf{x}_j,t)\right\}+\mathrm{div}_{\mathbf{y}_j}\left\{\xi\mathbf{A}(\mathbf{y}_j,t)\right\}\Big)\\
&\qquad+\sum_{j=1}^{n}\frac{i\hbar\sigma_j}{2m_jc}\Big(\mathbf{A}(\mathbf{x}_j,t)\cdot\nabla_{\mathbf{x}_j}\xi+\mathbf{A}(\mathbf{y}_j,t)\cdot\nabla_{\mathbf{y}_j}\xi\Big)\\
&+\sum_{j=1}^{n}\left(\sigma_j\Psi(\mathbf{x}_j,t)-\frac{\sigma_j}{c}\mathbf{A}(\mathbf{x}_j,t)\cdot\mathbf{v}(\mathbf{x}_j,t)+\frac{\sigma_j^2}{2m_jc^2}\left|\mathbf{A}(\mathbf{x}_j,t)\right|^2\right)\xi\\
&-\sum_{j=1}^{n}\left(\sigma_j\Psi(\mathbf{y}_j,t)-\frac{\sigma_j}{c}\mathbf{A}(\mathbf{y}_j,t)\cdot\mathbf{v}(\mathbf{y}_j,t)+\frac{\sigma_j^2}{2m_jc^2}\left|\mathbf{A}(\mathbf{y}_j,t)\right|^2\right)\xi\\
&-\sum_{j=1}^{n}\frac{g_j\sigma_j\hbar}{2m_jc}(\mathbf{S}_j\otimes I)\cdot\Big(\mathrm{curl}_{\mathbf{x}_j}\mathbf{A}(\mathbf{x}_j,t)\,\xi\Big)+\sum_{j=1}^{n}\frac{g_j\sigma_j\hbar}{2m_jc}(I\otimes\bar{\mathbf{S}}_j)\cdot\Big(\mathrm{curl}_{\mathbf{y}_j}\mathbf{A}(\mathbf{y}_j,t)\,\xi\Big),
\end{aligned}
\tag{6.213}
$$

where, as before,

$$
\begin{aligned}
&(\mathbf{S}_j\otimes I)\cdot(a\otimes b)=(\mathbf{S}_j\cdot a)\otimes b \quad\text{and}\\
&(I\otimes\bar{\mathbf{S}}_j)\cdot(a\otimes b)=a\otimes(\bar{\mathbf{S}}_j\cdot b)\quad\forall a\in\mathbb{C}^{2^n},\ \forall b\in\mathbb{C}^{2^n}.
\end{aligned}
\tag{6.214}
$$

Next consider a change of some non-inertial Cartesian coordinate system $(*)$ to another Cartesian coordinate system $(**)$ of the form (4.2):

$$
\begin{cases}
\mathbf{x}'=A(t)\cdot\mathbf{x}+\mathbf{z}(t),\\
t'=t,
\end{cases}
\tag{6.215}
$$

where $A(t)\in SO(3)$ is a rotation. Then, as before in Theorem 6.2, we deduce that the Quantum Liouville's equation of the form (6.213) is invariant under the change of non-inertial Cartesian coordinate system, provided we have

$$\begin{cases}\mathbf{x}'_j = A(t)\cdot\mathbf{x}_j + \mathbf{z}(t) \quad \forall j = 1,\dots,n\\ \mathbf{y}'_j = A(t)\cdot\mathbf{y}_j + \mathbf{z}(t) \quad \forall j = 1,\dots,n\\ \xi'\left(\mathbf{x}'_1,\dots,\mathbf{x}'_n,\mathbf{y}'_1,\dots,\mathbf{y}'_n,t'\right) = \Big(\left(U(t)\otimes_1 U(t)\dots\otimes_{(n-1)} U(t)\right)\otimes\\ \left(\bar{U}(t)\otimes_1\bar{U}(t)\dots\otimes_{(n-1)}\bar{U}(t)\right)\Big)\cdot\xi\left(\mathbf{x}_1,\dots,\mathbf{x}_n,\mathbf{y}_1,\dots,\mathbf{y}_n,t\right)\\ V'\left(\mathbf{y}'_1,\dots,\mathbf{y}'_n,t'\right) = V\left(\mathbf{y}_1,\dots,\mathbf{y}_n,t\right)\\ \mathbf{A}' = A(t)\cdot\mathbf{A}\\ \Psi' - \frac{1}{c}\mathbf{A}'\cdot\mathbf{v}' = \Psi - \frac{1}{c}\mathbf{A}\cdot\mathbf{v},\end{cases} \tag{6.216}$$

where, as before, $U(t)\in SU(2)$ is characterized by

$$U^*(t)\cdot\mathbf{S}\cdot U(t) = A(t)\cdot\mathbf{S}, \tag{6.217}$$

and $\bar{U}(t)$ is the complex conjugate (not the Hermitian adjoint) of the matrix $U(t)$.

Next assume that $\hat{A}\left(\mathbf{x}_1,\dots,\mathbf{x}_n,t\right)$ is some Hermitian operator acting on the functions $\psi\left(\mathbf{x}_1,\dots,\mathbf{x}_n,t\right)\in\mathbb{C}^{2^n}$ with respect to spatial variables $(\mathbf{x}_1,\dots,\mathbf{x}_n)$. Then the average $\tilde{A}(t)$ of $\hat{A}$ on the density matrix ξ is defined by

$$\tilde{A}(t) = \frac{\int trace\left(\left(\hat{A}\left(\mathbf{x}_1,\dots,\mathbf{x}_n,t\right)\otimes I\right)\cdot\xi\right)\left(\mathbf{z}_1,\dots,\mathbf{z}_n,\mathbf{z}_1,\dots,\mathbf{z}_n,t\right)d\mathbf{z}_1\dots d\mathbf{z}_n}{\int trace\left(\xi\right)\left(\mathbf{z}_1,\dots,\mathbf{z}_n,\mathbf{z}_1,\dots,\mathbf{z}_n,t\right)d\mathbf{z}_1\dots d\mathbf{z}_n}, \tag{6.218}$$

where for given $R\in\mathbb{C}^p\otimes\mathbb{C}^p$ $trace(R)$ is a linear functional from $\mathbb{C}^p\otimes\mathbb{C}^p$ to $\mathbb{C}$ defined by

$$trace(a\otimes b) = \sum_{k=1}^{p} a_k b_k \quad \forall a = (a_1,\dots a_p)\in\mathbb{C}^p,\ \forall b = (b_1,\dots b_p)\in\mathbb{C}^p. \tag{6.219}$$

On the other hand, using the fact that $\hat{A}$ is Hermitian, it can be easily deduced that in addition to (6.218) we have the following identity:

$$\int trace\left(\left(\hat{A}\left(\mathbf{x}_1,\dots,\mathbf{x}_n,t\right)\otimes I\right)\cdot\xi\right)\left(\mathbf{z}_1,\dots,\mathbf{z}_n,\mathbf{z}_1,\dots,\mathbf{z}_n,t\right)d\mathbf{z}_1\dots d\mathbf{z}_n =$$
$$\int trace\left(\left(I\otimes\hat{A}^*\left(\mathbf{y}_1,\dots,\mathbf{y}_n,t\right)\right)\cdot\xi\right)\left(\mathbf{z}_1,\dots,\mathbf{z}_n,\mathbf{z}_1,\dots,\mathbf{z}_n,t\right)d\mathbf{z}_1\dots d\mathbf{z}_n, \tag{6.220}$$

where $\hat{A}^*\left(\mathbf{y}_1,\dots,\mathbf{y}_n,t\right)$ is the complex conjugate (not the Hermitian adjoint) to the operator $\hat{A}$ acting on the variables $(\mathbf{y}_1,\dots,\mathbf{y}_n)$ and so,

$$\tilde{A}(t) = \frac{\int trace\left(\left(I\otimes\hat{A}^*\left(\mathbf{y}_1,\dots,\mathbf{y}_n,t\right)\right)\cdot\xi\right)\left(\mathbf{z}_1,\dots,\mathbf{z}_n,\mathbf{z}_1,\dots,\mathbf{z}_n,t\right)d\mathbf{z}_1\dots d\mathbf{z}_n}{\int trace\left(\xi\right)\left(\mathbf{z}_1,\dots,\mathbf{z}_n,\mathbf{z}_1,\dots,\mathbf{z}_n,t\right)d\mathbf{z}_1\dots d\mathbf{z}_n}. \tag{6.221}$$

In particular, by inserting (6.220) in the particular case $\hat{A} = \hat{H}_0$ into (6.209) we deduce

$$i\hbar\frac{\partial}{\partial t}\int trace\,(\xi)\,(\mathbf{z}_1,\ldots,\mathbf{z}_n,\mathbf{z}_1,\ldots,\mathbf{z}_n,t)\,d\mathbf{z}_1\ldots d\mathbf{z}_n = 0, \tag{6.222}$$

and so,

$$\int trace\,(\xi)\,(\mathbf{z}_1,\ldots,\mathbf{z}_n,\mathbf{z}_1,\ldots,\mathbf{z}_n,t)\,d\mathbf{z}_1\ldots d\mathbf{z}_n$$
$$= \int trace\,(\xi)\,(\mathbf{z}_1,\ldots,\mathbf{z}_n,\mathbf{z}_1,\ldots,\mathbf{z}_n,0)\,d\mathbf{z}_1\ldots d\mathbf{z}_n. \tag{6.223}$$

Next, by (6.209) we have

$$i\hbar\frac{\partial\xi}{\partial t}(\mathbf{x}_1,\ldots,\mathbf{x}_n,\mathbf{y}_1,\ldots,\mathbf{y}_n,t) =$$
$$\left(\hat{H}_0\,(\mathbf{x}_1,\ldots,\mathbf{x}_n,t)\otimes I\right)\cdot\xi\,(\mathbf{x}_1,\ldots,\mathbf{x}_n,\mathbf{y}_1,\ldots,\mathbf{y}_n,t)$$
$$-\left(I\otimes\hat{H}_0^*\,(\mathbf{y}_1,\ldots,\mathbf{y}_n,t)\right)\cdot\xi\,(\mathbf{x}_1,\ldots,\mathbf{x}_n,\mathbf{y}_1,\ldots,\mathbf{y}_n,t). \tag{6.224}$$

Then, denote

$$\xi_1\,(\mathbf{x}_1,\ldots,\mathbf{x}_n,\mathbf{y}_1,\ldots,\mathbf{y}_n,t) := M\cdot\bar{\xi}\,(\mathbf{y}_1,\ldots,\mathbf{y}_n,\mathbf{x}_1,\ldots,\mathbf{x}_n,t)\,, \tag{6.225}$$

where M is a linear operator (matrix), acting on vectors in $\mathbb{C}^{2^n}\otimes\mathbb{C}^{2^n}$ and satisfying:

$$M\cdot(a\otimes b) = b\otimes a \quad \forall a\in\mathbb{C}^{2^n},\ \forall b\in\mathbb{C}^{2^n} \tag{6.226}$$

and consider

$$\zeta\,(\mathbf{x}_1,\ldots,\mathbf{x}_n,\mathbf{y}_1,\ldots,\mathbf{y}_n,t) :=$$
$$\xi\,(\mathbf{x}_1,\ldots,\mathbf{x}_n,\mathbf{y}_1,\ldots,\mathbf{y}_n,t) - \xi_1\,(\mathbf{x}_1,\ldots,\mathbf{x}_n,\mathbf{y}_1,\ldots,\mathbf{y}_n,t). \tag{6.227}$$

Thus,

$$-i\hbar M\cdot\frac{\partial\bar{\xi}}{\partial t}(\mathbf{y}_1,\ldots,\mathbf{y}_n,\mathbf{x}_1,\ldots,\mathbf{x}_n,t) =$$
$$\left(I\otimes\hat{H}_0^*\,(\mathbf{y}_1,\ldots,\mathbf{y}_n,t)\right)\cdot\left(M\cdot\bar{\xi}\,(\mathbf{y}_1,\ldots,\mathbf{y}_n,\mathbf{x}_1,\ldots,\mathbf{x}_n,t)\right)$$
$$-\left(\hat{H}_0\,(\mathbf{x}_1,\ldots,\mathbf{x}_n,t)\otimes I\right)\cdot\left(M\cdot\bar{\xi}\,(\mathbf{y}_1,\ldots,\mathbf{y}_n,\mathbf{x}_1,\ldots,\mathbf{x}_n,t)\right), \tag{6.228}$$

and then

$$i\hbar\frac{\partial\xi_1}{\partial t}\left(\mathbf{x}_1,\ldots,\mathbf{x}_n,\mathbf{y}_1,\ldots,\mathbf{y}_n,t\right)=\\ \left(\hat{H}_0\left(\mathbf{x}_1,\ldots,\mathbf{x}_n,t\right)\otimes I\right)\cdot\xi_1\left(\mathbf{x}_1,\ldots,\mathbf{x}_n,\mathbf{y}_1,\ldots,\mathbf{y}_n,t\right)\\ -\left(I\otimes\hat{H}_0^*\left(\mathbf{y}_1,\ldots,\mathbf{y}_n,t\right)\right)\cdot\xi_1\left(\mathbf{x}_1,\ldots,\mathbf{x}_n,\mathbf{y}_1,\ldots,\mathbf{y}_n,t\right). \quad (6.229)$$

Therefore,

$$i\hbar\frac{\partial\zeta}{\partial t}\left(\mathbf{x}_1,\ldots,\mathbf{x}_n,\mathbf{y}_1,\ldots,\mathbf{y}_n,t\right)=\\ \left(\hat{H}_0\left(\mathbf{x}_1,\ldots,\mathbf{x}_n,t\right)\otimes I\right)\cdot\zeta\left(\mathbf{x}_1,\ldots,\mathbf{x}_n,\mathbf{y}_1,\ldots,\mathbf{y}_n,t\right)\\ -\left(I\otimes\hat{H}_0^*\left(\mathbf{y}_1,\ldots,\mathbf{y}_n,t\right)\right)\cdot\zeta\left(\mathbf{x}_1,\ldots,\mathbf{x}_n,\mathbf{y}_1,\ldots,\mathbf{y}_n,t\right). \quad (6.230)$$

Thus if ζ satisfies initial condition $\zeta\left(\mathbf{x}_1,\ldots,\mathbf{x}_n,\mathbf{y}_1,\ldots,\mathbf{y}_n,0\right)=0$ then $\zeta\left(\mathbf{x}_1,\ldots,\mathbf{x}_n,\mathbf{y}_1,\ldots,\mathbf{y}_n,t\right)=0$ for every $t>0$. So

$$\xi\left(\mathbf{x}_1,\ldots,\mathbf{x}_n,\mathbf{y}_1,\ldots,\mathbf{y}_n,0\right)=M\cdot\bar{\xi}\left(\mathbf{y}_1,\ldots,\mathbf{y}_n,\mathbf{x}_1,\ldots,\mathbf{x}_n,0\right)\quad\text{implies}\\ \xi\left(\mathbf{x}_1,\ldots,\mathbf{x}_n,\mathbf{y}_1,\ldots,\mathbf{y}_n,t\right)=M\cdot\bar{\xi}\left(\mathbf{y}_1,\ldots,\mathbf{y}_n,\mathbf{x}_1,\ldots,\mathbf{x}_n,t\right)\quad\forall t\geq 0. \quad (6.231)$$

Finally, assume that the first and the second particles have the same mass $m_1=m_2$ and the same charge $\sigma_1=\sigma_2$ and moreover, assume that we have

$$V\left(\mathbf{x}_1,\mathbf{x}_2,\mathbf{x}_3,\ldots,\mathbf{x}_n,t\right)=V\left(\mathbf{x}_2,\mathbf{x}_1,\mathbf{x}_3,\ldots,\mathbf{x}_n,t\right)\qquad\forall\mathbf{x}_1,\mathbf{x}_2,\mathbf{x}_3,\ldots,\mathbf{x}_n\in\mathbb{R}^3,\quad\forall t.$$

In this case it can be easily deduced that if $\xi\left(\mathbf{x}_1,\mathbf{x}_2,\mathbf{x}_3,\ldots,\mathbf{x}_n,\mathbf{y}_1,\mathbf{y}_2,\mathbf{y}_3,\ldots,\mathbf{y}_n,t\right)\in\mathbb{C}^{2^n}\otimes\mathbb{C}^{2^n}$ is a solution of (6.224), then $B_{1,2}\cdot\xi(\mathbf{x}_2,\mathbf{x}_1,\mathbf{x}_3,\ldots,\mathbf{x}_n,\mathbf{y}_2,\mathbf{y}_1,\mathbf{y}_3,\ldots,\mathbf{y}_n,t)$ is also a solution of (6.224), where by $B_{1,2}:\mathbb{C}^{2^n}\otimes\mathbb{C}^{2^n}\to\mathbb{C}^{2^n}\otimes\mathbb{C}^{2^n}$ we denote the linear operator (matrix) defined as

$$B_{1,2}\cdot\left(\left(a_1\otimes a_2\otimes a_3\otimes\ldots\otimes a_n\right)\otimes\left(b_1\otimes b_2\otimes b_3\otimes\ldots\otimes b_n\right)\right)=\\ \left(\left(a_2\otimes a_1\otimes a_3\otimes\ldots\otimes a_n\right)\otimes\left(b_2\otimes b_1\otimes b_3\otimes\ldots\otimes b_n\right)\right)\\ \forall a_1,\ldots a_n,b_1,\ldots,b_n\in\mathbb{C}^2. \quad (6.232)$$

Therefore, if $\xi\left(\mathbf{x}_1,\mathbf{x}_2,\mathbf{x}_3,\ldots,\mathbf{x}_n,\mathbf{y}_1,\mathbf{y}_2,\mathbf{y}_3,\ldots,\mathbf{y}_n,t\right)\in\mathbb{C}^{2^n}\otimes\mathbb{C}^{2^n}$ is a solution of (6.224) then for every $t\geq 0$ we will have

$$B_{1,2} \cdot \xi(\mathbf{x}_2, \mathbf{x}_1, \mathbf{x}_3, \ldots, \mathbf{x}_n, \mathbf{y}_2, \mathbf{y}_1, \mathbf{y}_3, \ldots, \mathbf{y}_n, t) \\ = \xi(\mathbf{x}_1, \mathbf{x}_2, \mathbf{x}_3, \ldots, \mathbf{x}_n, \mathbf{y}_1, \mathbf{y}_2, \mathbf{y}_3, \ldots, \mathbf{y}_n, t) \\ \forall \mathbf{x}_1, \mathbf{x}_2, \mathbf{x}_3, \ldots, \mathbf{x}_n, \mathbf{y}_1, \mathbf{y}_2, \mathbf{y}_3, \ldots, \mathbf{y}_n \in \mathbb{R}^3, \quad (6.233)$$

provided that (6.233) holds for the initial instant of time $t = 0$. So we have a consistency with the Pauli Exclusion Principle for two or more identical fermions.

Next given $\xi(\mathbf{x}_1, \ldots, \mathbf{x}_n, \mathbf{y}_1, \ldots, \mathbf{y}_n, t)$ that satisfies

$$\xi(\mathbf{x}_1, \ldots, \mathbf{x}_n, \mathbf{y}_1, \ldots, \mathbf{y}_n, t) = M \cdot \bar{\xi}(\mathbf{y}_1, \ldots, \mathbf{y}_n, \mathbf{x}_1, \ldots, \mathbf{x}_n, t) \quad \forall t \geq 0 \quad (6.234)$$

define the operator $\hat{R}_\xi$ by

$$\hat{R}_\xi(\mathbf{x}_1, \ldots, \mathbf{x}_n, t) \cdot \psi(\mathbf{x}_1, \ldots, \mathbf{x}_n, t) = \\ \int A\left(\xi(\mathbf{x}_1, \ldots, \mathbf{x}_n, \mathbf{y}_1, \ldots, \mathbf{y}_n, t), \psi(\mathbf{y}_1, \ldots, \mathbf{y}_n, t)\right) d\mathbf{y}_1 \ldots d\mathbf{y}_n, \quad (6.235)$$

where A is a bilinear mapping, acting on the pairs of vectors in $\left(\mathbb{C}^{2^n} \otimes \mathbb{C}^{2^n}\right) \times \mathbb{C}^{2^n}$, taking the values in $\mathbb{C}^{2^n}$ and having the form

$$A((a \otimes b), c) = (b \cdot c)a \quad \forall a \in \mathbb{C}^{2^n}, \; \forall b \in \mathbb{C}^{2^n}, \; \forall c \in \mathbb{C}^{2^n}. \quad (6.236)$$

Then the mapping $\xi \to \hat{R}_\xi$ is one-to-one. Moreover, by (6.234) $\hat{R}_\xi$ is an Hermitian operator. Finally by (6.224) we have

$$i\hbar \frac{\partial \hat{R}_\xi}{\partial t}(\mathbf{x}_1, \ldots, \mathbf{x}_n, t) = \\ \hat{H}_0(\mathbf{x}_1, \ldots, \mathbf{x}_n, t) \cdot \hat{R}_\xi(\mathbf{x}_1, \ldots, \mathbf{x}_n, t) - \hat{R}_\xi(\mathbf{x}_1, \ldots, \mathbf{x}_n, t) \cdot \hat{H}_0(\mathbf{x}_1, \ldots, \mathbf{x}_n, t). \\ (6.237)$$

Equation (6.237) is equivalent to (6.224).

Next, clearly, ξ satisfies (6.233) if and only if we have the following relation:

$$A_{1,2} \cdot \psi(\mathbf{x}_2, \mathbf{x}_1, \mathbf{x}_3, \ldots, \mathbf{x}_n, t) = -\psi(\mathbf{x}_1, \mathbf{x}_2, \mathbf{x}_3, \ldots, \mathbf{x}_n, t) \;\; \forall \mathbf{x}_1, \ldots, \mathbf{x}_n \in \mathbb{R}^3 \\ \text{and} \quad \phi = \hat{R}_\xi \cdot \psi \\ \text{implies} \quad A_{1,2} \cdot \phi(\mathbf{x}_2, \mathbf{x}_1, \mathbf{x}_3, \ldots, \mathbf{x}_n, t) = -\phi(\mathbf{x}_1, \mathbf{x}_2, \mathbf{x}_3, \ldots, \mathbf{x}_n, t) \\ \forall \mathbf{x}_1, \ldots, \mathbf{x}_n \in \mathbb{R}^3, \quad (6.238)$$

where by $A_{1,2} : \mathbb{C}^{2^n} \to \mathbb{C}^{2^n}$ we denote the linear operator (matrix) defined as in (6.207) by the following:

$$A_{1,2} \cdot (a_1 \otimes a_2 \otimes a_3 \otimes \ldots \otimes a_n) = (a_2 \otimes a_1 \otimes a_3 \otimes \ldots \otimes a_n) \qquad \forall a_1, \ldots a_n \in \mathbb{C}^2. \tag{6.239}$$

Finally, assume that $\hat{A}(\mathbf{x}_1, \ldots, \mathbf{x}_n, t)$ is some Hermitian operator acting on the functions $\psi(\mathbf{x}_1, \ldots, \mathbf{x}_n, t) \in \mathbb{C}^{2^n}$ with respect to spatial variables $(\mathbf{x}_1, \ldots, \mathbf{x}_n)$. Then, we remind that the average $\tilde{A}(t)$ of $\hat{A}$ on the density matrix ξ is defined by (6.218):

$$\tilde{A}(t) = \frac{\int trace\left(\left(\hat{A}(\mathbf{x}_1, \ldots, \mathbf{x}_n, t) \otimes I\right) \cdot \xi\right)(\mathbf{z}_1, \ldots, \mathbf{z}_n, \mathbf{z}_1, \ldots, \mathbf{z}_n, t)\, d\mathbf{z}_1 \ldots d\mathbf{z}_n}{\int trace(\xi)(\mathbf{z}_1, \ldots, \mathbf{z}_n, \mathbf{z}_1, \ldots, \mathbf{z}_n, t)\, d\mathbf{z}_1 \ldots d\mathbf{z}_n}. \tag{6.240}$$

Thus,

$$trace\left(\hat{A} \circ \hat{R}_\xi\right) = \int trace\left(\left(\hat{A}(\mathbf{x}_1, \ldots, \mathbf{x}_n, t) \otimes I\right) \cdot \xi\right)(\mathbf{z}_1, \ldots, \mathbf{z}_n, \mathbf{z}_1, \ldots, \mathbf{z}_n, t)\, d\mathbf{z}_1 \ldots d\mathbf{z}_n, \tag{6.241}$$

and so,

$$\tilde{A}(t) = \frac{trace\left(\hat{A} \circ \hat{R}_\xi\right)}{trace\left(\hat{R}_\xi\right)} = \frac{trace\left(\hat{R}_\xi \circ \hat{A}\right)}{trace\left(\hat{R}_\xi\right)}, \tag{6.242}$$

where by the trace we mean the trace of an operator on a Hilbert space. Moreover, by (6.223) and (6.241) we have

$$trace\left(\hat{R}_\xi\right)(t) = trace\left(\hat{R}_\xi\right)(0). \tag{6.243}$$

6.8.1 *Thermodynamical Equilibrium in Systems of Spin-Half Particles; Canonical Ensemble*

Let $\xi(\mathbf{x}_1, \ldots, \mathbf{x}_n, \mathbf{y}_1, \ldots, \mathbf{y}_n, t)$ be the density matrix of the quantum system, ruled by the Hamiltonian operator $\hat{H}_0$ from (6.211), in the case of thermodynamical equilibrium and $\hat{R}_\xi$ is given by (6.235) for this ξ. Then in the case that $\frac{\partial \hat{H}_0}{\partial t} = 0$ and the given equilibrium system rests macroscopically, i.e. it has the macroscopical velocity field zero: $\mathbf{u}(\mathbf{x}, t) = 0$ it is well known that

$$\hat{R}_{\xi} = \frac{1}{trace\left(f \circ \hat{H}_0\right)} f \circ \hat{H}_0, \tag{6.244}$$

and in the case of a system in thermostat, having the Kelvin temperature T, we have the canonical ensemble:

$$f(s) = e^{-\frac{s}{kT}} = \sum_{m=0}^{+\infty} \frac{1}{m!} \left(-\frac{s}{kT}\right)^m \quad \forall s \in \mathbb{C}, \tag{6.245}$$

where k is the Boltzmann constant and by $f \circ \hat{H}_0$ we denote the following operator

$$f \circ \hat{H}_0 = \sum_{m=0}^{+\infty} \frac{1}{m!} \left(-\frac{1}{kT} \hat{H}_0\right)^m \quad \forall s \in \mathbb{C}. \tag{6.246}$$

We would like to find alternative form of the above law of thermodynamical equilibrium, which is invariant under the change of inertial or non-inertial Cartesian coordinate system. Then it is clear that if the given system rests in the old coordinate system, then it obviously has non-trivial macroscopical velocity field $\mathbf{u}(\mathbf{x}, t)$ in the new one. Moreover, $\mathbf{u}(\mathbf{x}, t)$ can depend on $\mathbf{x}$ and t, as it indeed happens in the case of a rotation of the new coordinate system with respect to the old one. On the other hand the concept of thermodynamical equilibrium is clearly independent on the coordinate system.

In order to find the forms of the law of thermodynamical equilibrium, which are indeed invariant under the change of inertial or non-inertial Cartesian coordinate system we follow the steps as below. Given a speed-like vector field $\mathbf{u} = \mathbf{u}(\mathbf{x}, t)$, define

$$\hat{H}_{\mathbf{u}}(\mathbf{x}_1, \ldots, \mathbf{x}_n, t) :=$$
$$\hat{H}_0(\mathbf{x}_1, \ldots, \mathbf{x}_n, t) - \frac{1}{2}\left(\sum_{j=1}^{n} \hat{\mathbf{P}}_j \cdot \mathbf{u}(\mathbf{x}_j, t) + \mathbf{u}(\mathbf{x}_j, t) \cdot \hat{\mathbf{P}}_j\right) - \sum_{j=1}^{n} \frac{\hbar}{4} \mathbf{S}_j \cdot \left(\mathrm{curl}_{\mathbf{x}_j} \mathbf{u}(\mathbf{x}_j, t)\right), \tag{6.247}$$

where $\hat{\mathbf{P}}_j$ is operator defined as

$$\hat{\mathbf{P}}_j \cdot \psi(\mathbf{x}_1, \ldots, \mathbf{x}_n) = -i\hbar \nabla_{\mathbf{x}_j} \psi(\mathbf{x}_1, \ldots, \mathbf{x}_n). \tag{6.248}$$

Then by (6.211) we have

$$\hat{H}_{\mathbf{u}}(\mathbf{x}_1,\ldots,\mathbf{x}_n,t)\cdot\psi(\mathbf{x}_1,\ldots,\mathbf{x}_n,t) = -\sum_{j=1}^{n}\frac{\hbar^2}{2m_j}\Delta_{\mathbf{x}_j}\psi$$
$$-\sum_{j=1}^{n}\frac{i\hbar}{2}\operatorname{div}_{\mathbf{x}_j}\left\{\psi\left(\mathbf{v}(\mathbf{x}_j,t)-\mathbf{u}(\mathbf{x}_j,t)\right)\right\} - \sum_{j=1}^{n}\frac{i\hbar}{2}\left(\mathbf{v}(\mathbf{x}_j,t)-\mathbf{u}(\mathbf{x}_j,t)\right)\cdot\nabla_{\mathbf{x}_j}\psi$$
$$+\sum_{j=1}^{n}\frac{i\hbar\sigma_j}{2m_jc}\operatorname{div}_{\mathbf{x}_j}\left\{\psi\mathbf{A}(\mathbf{x}_j,t)\right\} + \sum_{j=1}^{n}\frac{i\hbar\sigma_j}{2m_jc}\mathbf{A}(\mathbf{x}_j,t)\cdot\nabla_{\mathbf{x}_j}\psi$$
$$+\sum_{j=1}^{n}\left(\sigma_j\Psi(\mathbf{x}_j,t)-\frac{\sigma_j}{c}\mathbf{A}(\mathbf{x}_j,t)\cdot\mathbf{v}(\mathbf{x}_j,t)+\frac{\sigma_j^2}{2m_jc^2}\left|\mathbf{A}(\mathbf{x}_j,t)\right|^2\right)\psi - V(\mathbf{x}_1,\ldots,\mathbf{x}_n,t)\psi$$
$$-\sum_{j=1}^{n}\frac{g_j\sigma_j\hbar}{2m_jc}\mathbf{S}_j\cdot\left(\operatorname{curl}_{\mathbf{x}_j}\mathbf{A}(\mathbf{x}_j,t)\,\psi\right) + \sum_{j=1}^{n}\frac{\hbar}{4}\mathbf{S}_j\cdot\left(\operatorname{curl}_{\mathbf{x}_j}\left(\mathbf{v}(\mathbf{x}_j,t)-\mathbf{u}(\mathbf{x}_j,t)\right)\psi\right). \quad (6.249)$$

Thus, as before, we can prove that (6.249) is invariant under the change of inertial or non-inertial Cartesian coordinate systems, i.e.

$$\hat{H}'_{\mathbf{u}'}\left(\mathbf{x}'_1,\ldots,\mathbf{x}'_n,t'\right)\cdot\psi'\left(\mathbf{x}'_1,\ldots,\mathbf{x}'_n,t'\right) = \phi'\left(\mathbf{x}'_1,\ldots,\mathbf{x}'_n,t'\right) \quad \text{if and only if}$$
$$\hat{H}_{\mathbf{u}}(\mathbf{x}_1,\ldots,\mathbf{x}_n,t)\cdot\psi(\mathbf{x}_1,\ldots,\mathbf{x}_n,t) = \phi(\mathbf{x}_1,\ldots,\mathbf{x}_n,t), \quad (6.250)$$

provided that, as before in (6.201), we have

$$\begin{cases} V'\left(\mathbf{x}'_1,\ldots,\mathbf{x}'_n,t'\right) = V(\mathbf{x}_1,\ldots,\mathbf{x}_n,t),\\ \sigma'_j = \sigma_j,\\ m'_j = m_j,\\ g'_j = g_j,\\ \mathbf{v}'(\mathbf{x}',t) = A(t)\cdot\mathbf{v}(\mathbf{x},t)+\frac{dA}{dt}(t)\cdot\mathbf{x}+\frac{d\mathbf{z}}{dt}(t),\\ \mathbf{u}'(\mathbf{x}',t) = A(t)\cdot\mathbf{u}(\mathbf{x},t)+\frac{dA}{dt}(t)\cdot\mathbf{x}+\frac{d\mathbf{z}}{dt}(t),\\ \mathbf{A}'(\mathbf{x}',t) = A(t)\cdot\mathbf{A}(\mathbf{x},t),\\ \Psi'(\mathbf{x}',t)-\mathbf{v}'(\mathbf{x}',t)\cdot\mathbf{A}'(\mathbf{x}',t) = \Psi(\mathbf{x},t)-\mathbf{v}(\mathbf{x},t)\cdot\mathbf{A}(\mathbf{x},t),\\ \psi'\left(\mathbf{x}'_1,\ldots,\mathbf{x}'_n,t'\right) = \left(U(t)\otimes_1 U(t)\otimes_2 U(t)\ldots\otimes_{(n-1)}U(t)\right)\cdot\psi(\mathbf{x}_1,\ldots,\mathbf{x}_n,t),\\ \phi'\left(\mathbf{x}'_1,\ldots,\mathbf{x}'_n,t'\right) = \left(U(t)\otimes_1 U(t)\otimes_2 U(t)\ldots\otimes_{(n-1)}U(t)\right)\cdot\phi(\mathbf{x}_1,\ldots,\mathbf{x}_n,t). \end{cases} \quad (6.251)$$

Next consider

$$\hat{R}_{\xi} = \frac{1}{trace\left(f\circ\hat{H}_{\mathbf{u}}\right)}f\circ\hat{H}_{\mathbf{u}}, \quad (6.252)$$

where, as before, in the case of a system in thermostat, having the Kelvin temperature T, we have the canonical ensemble:

$$f(s) = e^{-\frac{s}{kT}} = \sum_{m=0}^{+\infty} \frac{1}{m!} \left(-\frac{s}{kT}\right)^m \quad \forall s \in \mathbb{C} \tag{6.253}$$

with k being the Boltzmann constant and $f \circ \hat{H}_{\mathbf{u}}$ is the following operator

$$f \circ \hat{H}_{\mathbf{u}} = \sum_{m=0}^{+\infty} \frac{1}{m!} \left(-\frac{1}{kT} \hat{H}_{\mathbf{u}}\right)^m \quad \forall s \in \mathbb{C}. \tag{6.254}$$

We would like to note here that since the operator $\hat{H}_{\mathbf{u}}$ is bounded from below and $f(s)$ decays rapidly as $s \to +\infty$, there exists a density matrix ξ_1, such that the operator $\hat{R}_{\xi_1}$, given by (6.235), equals to the operator $f \circ \hat{H}_{\mathbf{u}}$ and thus (6.252) indeed has sense. Next by (6.250) and (6.254) the operator $\hat{R}_\xi$ in the left-hand side of (6.252) is invariant under the change of inertial or non-inertial Cartesian coordinate systems, i.e.

$$\hat{R}_{\xi'} \left(\mathbf{x}'_1, \ldots, \mathbf{x}'_n, t'\right) \cdot \psi' \left(\mathbf{x}'_1, \ldots, \mathbf{x}'_n, t'\right) = \phi' \left(\mathbf{x}'_1, \ldots, \mathbf{x}'_n, t'\right) \quad \text{if and only if}$$
$$\hat{R}_\xi \left(\mathbf{x}_1, \ldots, \mathbf{x}_n, t\right) \cdot \psi \left(\mathbf{x}_1, \ldots, \mathbf{x}_n, t\right) = \phi \left(\mathbf{x}_1, \ldots, \mathbf{x}_n, t\right), \tag{6.255}$$

provided that, as before, we have (6.251) and

$$\xi' \left(\mathbf{x}'_1, \ldots, \mathbf{x}'_n, \mathbf{y}'_1, \ldots, \mathbf{y}'_n, t'\right) = \Big(\left(U(t) \otimes_1 U(t) \ldots \otimes_{(n-1)} U(t)\right) \otimes$$
$$\left(\bar{U}(t) \otimes_1 \bar{U}(t) \ldots \otimes_{(n-1)} \bar{U}(t)\right) \Big) \cdot \xi \left(\mathbf{x}_1, \ldots, \mathbf{x}_n, \mathbf{y}_1, \ldots, \mathbf{y}_n, t\right). \tag{6.256}$$

Moreover, in the case $\mathbf{u} = 0$ (6.252) coincides with (6.244). However, we still need to derive the restrictions on the field $\mathbf{u}$ and the Hamiltonian operator $\hat{H}_0$, providing that our system can indeed be found in the state of thermodynamical equilibrium. We remind that in the case $\mathbf{u} = 0$ the appropriate restriction is $\frac{\partial \hat{H}_0}{\partial t} = 0$. In order to get these restrictions in the general case, we need to insert $\hat{R}_\xi$ in (6.252) into the equation in (6.237) which is equivalent to the quantum Liouville equation.

Therefore, assume that the vector field $\mathbf{u}$ satisfies (6.55) with (6.56), i.e.

$$\begin{cases} \frac{\partial U}{\partial t} \left(\mathbf{x}_1, \ldots, \mathbf{x}_n, t\right) + \sum_{j=1}^n \mathbf{u}(\mathbf{x}_j, t) \cdot \nabla_{\mathbf{x}_j} U \left(\mathbf{x}_1, \ldots, \mathbf{x}_n, t\right) = 0, \\ \frac{\partial}{\partial t} \left(\mathbf{u}(\mathbf{x}, t) - \mathbf{v}(\mathbf{x}, t)\right) - \mathrm{curl}_{\mathbf{x}} \left(\mathbf{u}(\mathbf{x}, t) \times \left(\mathbf{u}(\mathbf{x}, t) - \mathbf{v}(\mathbf{x}, t)\right)\right) \\ + \left(\mathrm{div}_{\mathbf{x}} \left(\mathbf{u}(\mathbf{x}, t) - \mathbf{v}(\mathbf{x}, t)\right)\right) \mathbf{u}(\mathbf{x}, t) = 0, \\ \frac{\partial \mathbf{A}}{\partial t}(\mathbf{x}, t) - \mathrm{curl}_{\mathbf{x}} \left(\mathbf{u}(\mathbf{x}, t) \times \mathbf{A}(\mathbf{x}, t)\right) + \left(\mathrm{div}_{\mathbf{x}} \mathbf{A}(\mathbf{x}, t)\right) \mathbf{u}(\mathbf{x}, t) = 0, \\ d_{\mathbf{x}} \mathbf{u}(\mathbf{x}, t) + \{d_{\mathbf{x}} \mathbf{u}(\mathbf{x}, t)\}^T = 0, \end{cases} \tag{6.257}$$

where

$$U(\mathbf{x}_1, \ldots, \mathbf{x}_n, t) :=$$
$$\sum_{j=1}^{n}\left(\frac{\sigma_j^2}{2m_j c^2}\left|\mathbf{A}(\mathbf{x}_j, t)\right|^2 + \sigma_j \Psi(\mathbf{x}_j, t) - \frac{\sigma_j}{c}\mathbf{A}(\mathbf{x}_j, t)\cdot \mathbf{v}(\mathbf{x}_j, t)\right) - V(\mathbf{x}_1, \ldots, \mathbf{x}_n, t). \tag{6.258}$$

Then, by Proposition 6.3 we have

$$i\hbar\frac{\partial}{\partial t}\left(f \circ \hat{H}_{\mathbf{u}}\right) = \hat{H}_0 \cdot \left(f \circ \hat{H}_{\mathbf{u}}\right) - \left(f \circ \hat{H}_{\mathbf{u}}\right) \cdot \hat{H}_0. \tag{6.259}$$

Then by (6.259) together with (6.243) we deduce that

$$i\hbar\frac{\partial}{\partial t}\left(\hat{R}_\xi\right) = \hat{H}_0 \cdot \left(\hat{R}_\xi\right) - \left(\hat{R}_\xi\right) \cdot \hat{H}_0, \tag{6.260}$$

where $\hat{R}_\xi$ is the operator in the left-hand side of (6.252). So we indeed get (6.237) in the case of (6.257), (6.258). Moreover, (6.257), (6.258) are invariant under the change of inertial or non-inertial coordinate system.

Proposition 6.3 *Assume that the speed-like vector field* $\mathbf{u}$ *satisfies* (6.257) *and* (6.258). *Next assume that the holomorphic function* $f(z) : \mathbb{C} \to \mathbb{C}$ *defined as a sum of the power series*

$$f(z) := \sum_{m=0}^{+\infty} a_m z^m, \tag{6.261}$$

with $a_m \in \mathbb{C}$, *is such that for the operator* $f \circ \hat{H}_{\mathbf{u}}$, *given by*

$$f \circ \hat{H}_{\mathbf{u}} := \sum_{m=0}^{+\infty} a_m \hat{H}_{\mathbf{u}}^m, \tag{6.262}$$

where the operator $\hat{H}_{\mathbf{u}}$ *is given by* (6.249), *there exists a density matrix* $\xi(\mathbf{x}_1, \ldots, \mathbf{x}_n, \mathbf{y}_1, \ldots, \mathbf{y}_n, t)$, *such that*

$$\left(f \circ \hat{H}_{\mathbf{u}}\right)(\mathbf{x}_1, \ldots, \mathbf{x}_n, t) = \hat{R}_\xi(\mathbf{x}_1, \ldots, \mathbf{x}_n, t), \tag{6.263}$$

where $\hat{R}_\xi$ *is given by* (6.235). *Then the operator* $f \circ \hat{H}_{\mathbf{u}}$ *satisfies*

$$i\hbar\frac{\partial}{\partial t}\left(f \circ \hat{H}_{\mathbf{u}}\right) = \hat{H}_0 \cdot \left(f \circ \hat{H}_{\mathbf{u}}\right) - \left(f \circ \hat{H}_{\mathbf{u}}\right) \cdot \hat{H}_0, \tag{6.264}$$

or equivalently

$$i\hbar\frac{\partial\xi}{\partial t}(\mathbf{x}_1,\ldots,\mathbf{x}_n,\mathbf{y}_1,\ldots,\mathbf{y}_n,t) = \left(\hat{H}_0(\mathbf{x}_1,\ldots,\mathbf{x}_n,t)\otimes I\right)\cdot\xi(\mathbf{x}_1,\ldots,\mathbf{x}_n,\mathbf{y}_1,\ldots,\mathbf{y}_n,t) - \left(I\otimes\hat{H}_0^*(\mathbf{y}_1,\ldots,\mathbf{y}_n,t)\right)\cdot\xi(\mathbf{x}_1,\ldots,\mathbf{x}_n,\mathbf{y}_1,\ldots,\mathbf{y}_n,t). \quad (6.265)$$

Moreover, $f\circ\hat{H}_{\mathbf{u}}$ is invariant under the change of inertial or non-inertial Cartesian coordinate systems, i.e.

$$\left(f\circ\hat{H}'_{\mathbf{u}'}\right)(\mathbf{x}'_1,\ldots,\mathbf{x}'_n,t')\cdot\psi'(\mathbf{x}'_1,\ldots,\mathbf{x}'_n,t') = \phi'(\mathbf{x}'_1,\ldots,\mathbf{x}'_n,t') \quad \textit{if and only if} \quad \left(f\circ\hat{H}_{\mathbf{u}}\right)(\mathbf{x}_1,\ldots,\mathbf{x}_n,t)\cdot\psi(\mathbf{x}_1,\ldots,\mathbf{x}_n,t) = \phi(\mathbf{x}_1,\ldots,\mathbf{x}_n,t), \quad (6.266)$$

provided that we have (6.251). *Finally, if we assume that the first and the second particles have the same mass $m_1 = m_2$ and the same charge $\sigma_1 = \sigma_2$ and moreover, if we assume that*

$$V(\mathbf{x}_1,\mathbf{x}_2,\mathbf{x}_3,\ldots,\mathbf{x}_n,t) = V(\mathbf{x}_2,\mathbf{x}_1,\mathbf{x}_3,\ldots,\mathbf{x}_n,t) \qquad \forall\mathbf{x}_1,\mathbf{x}_2,\mathbf{x}_3,\ldots,\mathbf{x}_n\in\mathbb{R}^3, \quad \forall t,$$

then

$$A_{1,2}\cdot\psi(\mathbf{x}_2,\mathbf{x}_1,\mathbf{x}_3,\ldots,\mathbf{x}_n,t) = -\psi(\mathbf{x}_1,\mathbf{x}_2,\mathbf{x}_3,\ldots,\mathbf{x}_n,t)\ \forall\mathbf{x}_1,\ldots,\mathbf{x}_n\in\mathbb{R}^3 \quad \textit{and} \quad \phi = \left(f\circ\hat{H}_{\mathbf{u}}\right)\cdot\psi \quad \textit{implies} \quad A_{1,2}\cdot\phi(\mathbf{x}_2,\mathbf{x}_1,\mathbf{x}_3,\ldots,\mathbf{x}_n,t) = -\phi(\mathbf{x}_1,\mathbf{x}_2,\mathbf{x}_3,\ldots,\mathbf{x}_n,t) \quad \forall\mathbf{x}_1,\ldots,\mathbf{x}_n\in\mathbb{R}^3, \quad (6.267)$$

where, as before, by $A_{1,2}:\mathbb{C}^{2^n}\to\mathbb{C}^{2^n}$ we denote the linear operator (matrix) defined as in (6.207) *by the following:*

$$A_{1,2}\cdot(a_1\otimes a_2\otimes a_3\otimes\ldots\otimes a_n) = (a_2\otimes a_1\otimes a_3\otimes\ldots\otimes a_n) \qquad \forall a_1,\ldots a_n\in\mathbb{C}^2. \quad (6.268)$$

Proof Again by (6.257) and Proposition 3.3 there exists another Cartesian coordinate system $(**)$ such that under the change of coordinate system $(*)$ to another Cartesian coordinate system $(**)$, given by (3.54), we have

$$A(t)\cdot\mathbf{u}(\mathbf{x},t) + \frac{dA}{dt}(t)\cdot\mathbf{x} + \frac{d\mathbf{z}}{dt}(t) = \mathbf{u}'(\mathbf{x}',t') = 0. \quad (6.269)$$

Thus, since (6.257), (6.258) are invariant under the change of inertial or non-inertial Cartesian coordinate systems, as before, in system $(**)$ we have

$$\begin{cases} \frac{\partial U'}{\partial t'}\left(\mathbf{x}_1', \ldots, \mathbf{x}_n', t'\right) = 0, \\ \frac{\partial \mathbf{v}'}{\partial t'}(\mathbf{x}', t') = 0, \\ \frac{\partial \mathbf{A}'}{\partial t'}(\mathbf{x}', t') = 0, \\ \mathbf{u}'(\mathbf{x}', t') = 0, \end{cases} \tag{6.270}$$

where

$$U'\left(\mathbf{x}_1', \ldots, \mathbf{x}_n', t'\right) := \sum_{j=1}^{n} \left(\frac{(\sigma_j')^2}{2\, m_j' c^2} \left|\mathbf{A}'(\mathbf{x}_j', t')\right|^2 + \sigma_j' \Psi'(\mathbf{x}_j', t') - \frac{\sigma_j'}{c} \mathbf{A}'(\mathbf{x}_j', t') \cdot \mathbf{v}'(\mathbf{x}_j', t') \right) - V'\left(\mathbf{x}_1', \ldots, \mathbf{x}_n', t'\right). \tag{6.271}$$

On the other hand, (6.270) is equivalent to

$$\frac{\partial \hat{H}_0'}{\partial t'}\left(\mathbf{x}_1', \ldots, \mathbf{x}_n', t'\right) = 0 \quad \text{and} \quad \mathbf{u}'(\mathbf{x}', t') = 0, \tag{6.272}$$

in system (∗∗). Thus, by (6.272), in system (∗∗) we deduce

$$\hat{H}_0' \cdot \hat{H}_{\mathbf{u}'}' - \hat{H}_{\mathbf{u}'}' \cdot \hat{H}_0' = \hat{H}_0' \cdot \hat{H}_0' - \hat{H}_0' \cdot \hat{H}_0' = 0 = i\hbar \frac{\partial \hat{H}_0'}{\partial t'} = i\hbar \frac{\partial \hat{H}_{\mathbf{u}'}'}{\partial t'}. \tag{6.273}$$

So we get

$$i\hbar \frac{\partial \hat{H}_{\mathbf{u}'}'}{\partial t'} = \hat{H}_0' \cdot \hat{H}_{\mathbf{u}'}' - \hat{H}_{\mathbf{u}'}' \cdot \hat{H}_0'. \tag{6.274}$$

Therefore, given the holomorphic function $f(z) : \mathbb{C} \to \mathbb{C}$ defined as a sum of the power series

$$f(z) := \sum_{m=0}^{+\infty} a_m z^m, \tag{6.275}$$

with $a_m \in \mathbb{C}$, if we define the operator $f \circ \hat{H}_{\mathbf{u}}$ as

$$f \circ \hat{H}_{\mathbf{u}} := \sum_{m=0}^{+\infty} a_m \hat{H}_{\mathbf{u}}^m, \tag{6.276}$$

by (6.274) and Proposition 6.1 we deduce

$$i\hbar \frac{\partial}{\partial t'}\left(f \circ \hat{H}_{\mathbf{u}'}'\right) = \hat{H}_0' \cdot \left(f \circ \hat{H}_{\mathbf{u}'}'\right) - \left(f \circ \hat{H}_{\mathbf{u}'}'\right) \cdot \hat{H}_0'. \tag{6.277}$$

Moreover, by (6.250) and (6.276), we can easily prove that $f \circ \hat{H}_{\mathbf{u}}$ is invariant under the change of inertial or non-inertial Cartesian coordinate systems, i.e.

$$\left(f \circ \hat{H}'_{\mathbf{u}'}\right)\left(\mathbf{x}'_1, \ldots, \mathbf{x}'_n, t'\right) \cdot \psi'\left(\mathbf{x}'_1, \ldots, \mathbf{x}'_n, t'\right) = \phi'\left(\mathbf{x}'_1, \ldots, \mathbf{x}'_n, t'\right) \quad \text{if and only if}$$
$$\left(f \circ \hat{H}_{\mathbf{u}}\right)(\mathbf{x}_1, \ldots, \mathbf{x}_n, t) \cdot \psi(\mathbf{x}_1, \ldots, \mathbf{x}_n, t) = \phi(\mathbf{x}_1, \ldots, \mathbf{x}_n, t), \tag{6.278}$$

provided that, as before, we have (6.251). Next assume that the holomorphic function f in (6.275) is such that there exists a density matrix $\xi(\mathbf{x}_1, \ldots, \mathbf{x}_n, \mathbf{y}_1, \ldots, \mathbf{y}_n, t)$ satisfying

$$\left(f \circ \hat{H}_{\mathbf{u}}\right)(\mathbf{x}_1, \ldots, \mathbf{x}_n, t) = \hat{R}_{\xi}(\mathbf{x}_1, \ldots, \mathbf{x}_n, t), \tag{6.279}$$

where $\hat{R}_{\xi}$ is given by (6.235). Then, by (6.278) for the density matrix $\xi'\left(\mathbf{x}'_1, \ldots, \mathbf{x}'_n, \mathbf{y}'_1, \ldots, \mathbf{y}'_n, t'\right)$ we have

$$\left(f \circ \hat{H}'_{\mathbf{u}'}\right)\left(\mathbf{x}'_1, \ldots, \mathbf{x}'_n, t'\right) = \hat{R}_{\xi'}\left(\mathbf{x}'_1, \ldots, \mathbf{x}'_n, t'\right), \tag{6.280}$$

provided that

$$\xi'\left(\mathbf{x}'_1, \ldots, \mathbf{x}'_n, \mathbf{y}'_1, \ldots, \mathbf{y}'_n, t'\right) = \Big(\left(U(t) \otimes_1 U(t) \ldots \otimes_{(n-1)} U(t)\right) \otimes$$
$$\left(\bar{U}(t) \otimes_1 \bar{U}(t) \ldots \otimes_{(n-1)} \bar{U}(t)\right)\Big) \cdot \xi(\mathbf{x}_1, \ldots, \mathbf{x}_n, \mathbf{y}_1, \ldots, \mathbf{y}_n, t). \tag{6.281}$$

Therefore, since we obtained before, that equation (6.277) is equivalent to the primed version of (6.224) and at the same time equation (6.224) is invariant under the change of non-inertial Cartesian coordinate system, provided we have (6.281), with the help of (6.279), as before, we deduce

$$i\hbar\frac{\partial}{\partial t}\left(f \circ \hat{H}_{\mathbf{u}}\right) = \hat{H}_0 \cdot \left(f \circ \hat{H}_{\mathbf{u}}\right) - \left(f \circ \hat{H}_{\mathbf{u}}\right) \cdot \hat{H}_0 \tag{6.282}$$

in an arbitrary coordinate system. So we get (6.264) or equivalently (6.265).

Finally, assume that the first and the second particles have the same mass $m_1 = m_2$ and the same charge $\sigma_1 = \sigma_2$ and moreover, assume that we have

$$V(\mathbf{x}_1, \mathbf{x}_2, \mathbf{x}_3, \ldots, \mathbf{x}_n, t) = V(\mathbf{x}_2, \mathbf{x}_1, \mathbf{x}_3, \ldots, \mathbf{x}_n, t) \qquad \forall \mathbf{x}_1, \mathbf{x}_2, \mathbf{x}_3, \ldots, \mathbf{x}_n \in \mathbb{R}^3, \quad \forall t.$$

In this case clearly by (6.249) we deduce the following relation:

$$A_{1,2} \cdot \psi(\mathbf{x}_2, \mathbf{x}_1, \mathbf{x}_3, \ldots, \mathbf{x}_n, t) = -\psi(\mathbf{x}_1, \mathbf{x}_2, \mathbf{x}_3, \ldots, \mathbf{x}_n, t) \;\; \forall \mathbf{x}_1, \ldots, \mathbf{x}_n \in \mathbb{R}^3$$
$$\text{and} \quad \phi = \hat{H}_{\mathbf{u}} \cdot \psi$$
$$\text{implies} \quad A_{1,2} \cdot \phi(\mathbf{x}_2, \mathbf{x}_1, \mathbf{x}_3, \ldots, \mathbf{x}_n, t) = -\phi(\mathbf{x}_1, \mathbf{x}_2, \mathbf{x}_3, \ldots, \mathbf{x}_n, t)$$
$$\forall \mathbf{x}_1, \ldots, \mathbf{x}_n \in \mathbb{R}^3, \qquad (6.283)$$

where, as before, by $A_{1,2} : \mathbb{C}^{2^n} \to \mathbb{C}^{2^n}$ we denote the linear operator (matrix) defined as in (6.207) by the following:

$$A_{1,2} \cdot (a_1 \otimes a_2 \otimes a_3 \otimes \ldots \otimes a_n) = (a_2 \otimes a_1 \otimes a_3 \otimes \ldots \otimes a_n) \qquad \forall a_1, \ldots a_n \in \mathbb{C}^2. \qquad (6.284)$$

Therefore, by (6.276) and (6.283) we obtain

$$A_{1,2} \cdot \psi(\mathbf{x}_2, \mathbf{x}_1, \mathbf{x}_3, \ldots, \mathbf{x}_n, t) = -\psi(\mathbf{x}_1, \mathbf{x}_2, \mathbf{x}_3, \ldots, \mathbf{x}_n, t) \;\; \forall \mathbf{x}_1, \ldots, \mathbf{x}_n \in \mathbb{R}^3$$
$$\text{and} \quad \phi = \left(f \circ \hat{H}_{\mathbf{u}}\right) \cdot \psi$$
$$\text{implies} \quad A_{1,2} \cdot \phi(\mathbf{x}_2, \mathbf{x}_1, \mathbf{x}_3, \ldots, \mathbf{x}_n, t) = -\phi(\mathbf{x}_1, \mathbf{x}_2, \mathbf{x}_3, \ldots, \mathbf{x}_n, t)$$
$$\forall \mathbf{x}_1, \ldots, \mathbf{x}_n \in \mathbb{R}^3, \qquad (6.285)$$

consistently with (6.238).

Chapter 7
Relation Between the Gravitational and Inertial Masses and Conservation Laws

7.1 Basic Assumptions and Their Consequences

We assumed before that the electromagnetic field is influenced by the gravitational field. We also can assume that the gravitational field is influenced by the electromagnetic field. We remind that we assume the first approximation of the law of gravitation in the form of (4.67), i.e.

$$\begin{cases} \operatorname{curl}_{\mathbf{x}}\left(\operatorname{curl}_{\mathbf{x}}\mathbf{v}\right) = 0, \\ \frac{\partial}{\partial t}\left(\operatorname{div}_{\mathbf{x}}\mathbf{v}\right) + \operatorname{div}_{\mathbf{x}}\left\{\left(\operatorname{div}_{\mathbf{x}}\mathbf{v}\right)\mathbf{v}\right\} + \frac{1}{4}\left|d_{\mathbf{x}}\mathbf{v} + \left\{d_{\mathbf{x}}\mathbf{v}\right\}^T\right|^2 - \left(\operatorname{div}_{\mathbf{x}}\mathbf{v}\right)^2 = -4\pi G M, \end{cases} \tag{7.1}$$

where M is the density of gravitational masses. However, till now, we said nothing about the relation between the density of inertial and gravitational masses. If μ is the density of inertial masses, then consistently with the classical Newtonian theory of gravitation, we assume that, in the absence of essential electromagnetic fields, we should have

$$M = \mu. \tag{7.2}$$

In order to satisfy the laws of conservation of the linear and angular momentums and energy, consider the following conserved proper scalar field Q, that we call "electromagnetical–gravitational" mass density, which is negligible in the absence of electromagnetic fields and satisfies the identity

$$\frac{\partial Q}{\partial t} + \operatorname{div}_{\mathbf{x}}\left\{Q\mathbf{v}\right\} = -\operatorname{div}_{\mathbf{x}}\left\{\frac{1}{4\pi c}\mathbf{D}\times\mathbf{B}\right\} \tag{7.3}$$

in the general case. Then, instead of (7.2), for the general case of gravitational–electromagnetic fields, we consider the following relation between gravitational and inertial mass densities:

$$M = \mu + Q. \tag{7.4}$$

A. Poliakovsky, *The Laws of Gravity and Electromagnetism*,
https://doi.org/10.1007/978-3-031-61407-1_7

Then, by (7.1) and (7.4), we have the following law of gravitation:

$$\begin{cases} \operatorname{curl}_{\mathbf{x}} (\operatorname{curl}_{\mathbf{x}} \mathbf{v}) = 0, \\ \frac{\partial}{\partial t} (\operatorname{div}_{\mathbf{x}} \mathbf{v}) + \operatorname{div}_{\mathbf{x}} \{(\operatorname{div}_{\mathbf{x}} \mathbf{v})\, \mathbf{v}\} + \frac{1}{4} \left|d_{\mathbf{x}}\mathbf{v} + \{d_{\mathbf{x}}\mathbf{v}\}^T\right|^2 - (\operatorname{div}_{\mathbf{x}} \mathbf{v})^2 = -4\pi G(\mu + Q). \end{cases} \tag{7.5}$$

Then as before, we deduce that the laws (7.3) and (7.5) are invariant under the change of non-inertial Cartesian coordinate system, provided that $Q' = Q$. We can rewrite (7.5) as

$$\begin{cases} \operatorname{curl}_{\mathbf{x}} (\operatorname{curl}_{\mathbf{x}} \mathbf{v}) = 0, \\ \operatorname{div}_{\mathbf{x}} \left\{ \frac{\partial \mathbf{v}}{\partial t} + d_{\mathbf{x}}\mathbf{v} \cdot \mathbf{v} + \frac{1}{2} \mathbf{v} \times \operatorname{curl}_{\mathbf{x}} \mathbf{v} \right\} = -4\pi G(\mu + Q). \end{cases} \tag{7.6}$$

In particular, in the inertial coordinate system (∗), we have

$$\begin{cases} \operatorname{curl}_{\mathbf{x}} \mathbf{v} = 0, \\ \operatorname{div}_{\mathbf{x}} \left\{ \frac{\partial \mathbf{v}}{\partial t} + d_{\mathbf{x}}\mathbf{v} \cdot \mathbf{v} \right\} = -4\pi G(\mu + Q), \end{cases} \tag{7.7}$$

that we can rewrite as

$$\begin{cases} \operatorname{curl}_{\mathbf{x}} \mathbf{v} = 0, \\ \frac{\partial \mathbf{v}}{\partial t} + d_{\mathbf{x}}\mathbf{v} \cdot \mathbf{v} = -\nabla_{\mathbf{x}} \Phi, \end{cases} \tag{7.8}$$

where Φ is the scalar gravitational potential: a proper scalar field which satisfies in every coordinate system:

$$\Delta_{\mathbf{x}} \Phi = 4\pi G(\mu + Q). \tag{7.9}$$

Since in the system (∗) we have $\operatorname{curl}_{\mathbf{x}} \mathbf{v} = 0$, we can write

$$\begin{cases} \mathbf{v} = \nabla_{\mathbf{x}} Z, \\ \frac{\partial Z}{\partial t} + \frac{1}{2} |\nabla_{\mathbf{x}} Z|^2 = -\Phi. \end{cases} \tag{7.10}$$

Remark 7.1 Lemma 14.1 from appendix gives some insight that the "electromagnetic–gravitational" mass density Q in (7.3) should have the values of the same order as the quantity $\frac{1}{c^2} \left(|\mathbf{D}|^2 + |\mathbf{B}|^2\right)$ and therefore, in the usual circumstances is negligible with respect to the inertial mass density μ. Thus, we can write $Q \approx 0$ in (7.5), i.e. the force of gravity in an inertial coordinate system approximately equals to the classical Newtonian force of gravity.

7.2 Conservation of the Momentum, Angular Momentum and Energy

Consider the Maxwell equation in the vacuum in some Cartesian coordinate system (∗):

$$\begin{cases} \mathrm{curl}_{\mathbf{x}}\mathbf{H} = \frac{4\pi}{c}\mathbf{j} + \frac{1}{c}\frac{\partial \mathbf{D}}{\partial t}, \\ \mathrm{div}_{\mathbf{x}}\,\mathbf{D} = 4\pi\rho, \\ \mathrm{curl}_{\mathbf{x}}\mathbf{E} + \frac{1}{c}\frac{\partial \mathbf{B}}{\partial t} = 0, \\ \mathrm{div}_{\mathbf{x}}\,\mathbf{B} = 0, \\ \mathbf{E} = \mathbf{D} - \frac{1}{c}\,\mathbf{v}\times\mathbf{B}, \\ \mathbf{H} = \mathbf{B} + \frac{1}{c}\,\mathbf{v}\times\mathbf{D}, \end{cases} \tag{7.11}$$

and consistently with (4.11), consider in the system (∗) the second Law of Newton for the moving continuum with the inertial mass density μ and the field of velocities $\mathbf{u}$:

$$\frac{\partial \mathbf{u}}{\partial t} + d_{\mathbf{x}}\mathbf{u}\cdot\mathbf{u} = -(\mathbf{u}-\mathbf{v})\times \mathrm{curl}_{\mathbf{x}}\mathbf{v} + \partial_t\mathbf{v} + d_{\mathbf{x}}\mathbf{v}\cdot\mathbf{v} + \frac{1}{\mu}\tilde{\mathbf{F}}, \tag{7.12}$$

where $\tilde{\mathbf{F}}$ is the total volume density of all non-gravitational forces acting on the continuum with mass density μ. Thus, again by (2.15), we rewrite (7.12) as

$$\mu\left(\frac{\partial \mathbf{u}}{\partial t} + d_{\mathbf{x}}\mathbf{u}\cdot\mathbf{u}\right) = \frac{\partial(\mu\mathbf{u})}{\partial t} + \mathrm{div}_{\mathbf{x}}\left\{\mu\mathbf{u}\otimes\mathbf{u}\right\} =$$
$$-\mu\mathbf{u}\times \mathrm{curl}_{\mathbf{x}}\mathbf{v} + \mu\partial_t\mathbf{v} + \mu\nabla_{\mathbf{x}}\left(\frac{1}{2}|\mathbf{v}|^2\right) + \rho\mathbf{E} + \frac{1}{c}\mathbf{j}\times\mathbf{B} + \mathbf{F} =$$
$$= -\mu(\mathbf{u}-\mathbf{v})\times \mathrm{curl}_{\mathbf{x}}\mathbf{v} + \mu\partial_t\mathbf{v} + d_{\mathbf{x}}\mathbf{v}\cdot(\mu\mathbf{v}) + \rho\mathbf{E} + \frac{1}{c}\mathbf{j}\times\mathbf{B} + \mathbf{F} \tag{7.13}$$

where $\rho\mathbf{E} + \frac{1}{c}\mathbf{j}\times\mathbf{B}$ is the volume density of the Lorentz force and $\mathbf{F}$ is the total volume density of all non-gravitational and non-electromagnetic forces acting on the continuum with mass density μ, which satisfies the continuum equation:

$$\frac{\partial \mu}{\partial t} + \mathrm{div}_{\mathbf{x}}\left(\mu\mathbf{u}\right) = 0. \tag{7.14}$$

Then, again by (2.15), we rewrite (7.13) as

$$\mu\frac{\partial}{\partial t}\left\{(\mathbf{u}-\mathbf{v})\right\} + \mu\, d_{\mathbf{x}}\left\{(\mathbf{u}-\mathbf{v})\right\}\cdot\mathbf{u} + \mu\left\{d_{\mathbf{x}}\mathbf{v}\right\}^T\cdot(\mathbf{u}-\mathbf{v}) = \rho\mathbf{E} + \frac{1}{c}\mathbf{j}\times\mathbf{B} + \mathbf{F}. \tag{7.15}$$

Thus, by (7.15) and (7.14), we obtain

$$\frac{\partial}{\partial t}\{\mu(\mathbf{u}-\mathbf{v})\}+\operatorname{div}_{\mathbf{x}}\{\mu(\mathbf{u}-\mathbf{v})\otimes\mathbf{u}\}+\mu\{d_{\mathbf{x}}\mathbf{v}\}^T\cdot(\mathbf{u}-\mathbf{v})=\rho\mathbf{E}+\frac{1}{c}\mathbf{j}\times\mathbf{B}+\mathbf{F}. \tag{7.16}$$

Moreover, multiplying (7.15) by $(\mathbf{u}-\mathbf{v})$ and using (7.14), we have

$$\frac{\partial}{\partial t}\left\{\frac{\mu}{2}|\mathbf{u}-\mathbf{v}|^2\right\}+\operatorname{div}_{\mathbf{x}}\left\{\frac{\mu}{2}|\mathbf{u}-\mathbf{v}|^2\mathbf{u}\right\}=$$
$$-\frac{1}{2}\left(d_{\mathbf{x}}\mathbf{v}+\{d_{\mathbf{x}}\mathbf{v}\}^T\right)^T:\{\mu(\mathbf{u}-\mathbf{v})\otimes(\mathbf{u}-\mathbf{v})\}+\mathbf{j}\cdot\mathbf{E}-\mathbf{v}\cdot\left(\rho\mathbf{E}+\frac{1}{c}\mathbf{j}\times\mathbf{B}\right)+(\mathbf{u}-\mathbf{v})\cdot\mathbf{F}. \tag{7.17}$$

On the other hand, by Lemmas 14.2 and 14.1 in the appendix, we have

$$\frac{\partial}{\partial t}\left(\frac{1}{4\pi c}\mathbf{D}\times\mathbf{B}\right)+\operatorname{div}_{\mathbf{x}}\left\{\left(\frac{1}{4\pi c}\mathbf{D}\times\mathbf{B}\right)\otimes\mathbf{v}\right\}=-(d_{\mathbf{x}}\mathbf{v})^T\cdot\left(\frac{1}{4\pi c}\mathbf{D}\times\mathbf{B}\right)$$
$$+\frac{1}{4\pi}\operatorname{div}_{\mathbf{x}}\left\{\mathbf{D}\otimes\mathbf{D}+\mathbf{B}\otimes\mathbf{B}-\frac{1}{2}\left(|\mathbf{D}|^2+|\mathbf{B}|^2\right)I\right\}-\left(\rho\mathbf{E}+\frac{1}{c}\mathbf{j}\times\mathbf{B}\right) \tag{7.18}$$

and

$$\frac{\partial}{\partial t}\left(\frac{|\mathbf{D}|^2+|\mathbf{B}|^2}{8\pi}\right)+\operatorname{div}_{\mathbf{x}}\left\{\left(\frac{|\mathbf{D}|^2+|\mathbf{B}|^2}{8\pi}\right)\mathbf{v}\right\}=$$
$$\frac{1}{4\pi}\operatorname{div}_{\mathbf{x}}\left\{(\mathbf{D}\otimes\mathbf{D}+\mathbf{B}\otimes\mathbf{B})\cdot\mathbf{v}-\frac{1}{2}\left(|\mathbf{D}|^2+|\mathbf{B}|^2\right)\mathbf{v}-c\mathbf{D}\times\mathbf{B}\right\}$$
$$-\left\{\frac{1}{4\pi}\left(\operatorname{div}_{\mathbf{x}}\left\{\mathbf{D}\otimes\mathbf{D}+\mathbf{B}\otimes\mathbf{B}-\frac{1}{2}\left(|\mathbf{D}|^2+|\mathbf{B}|^2\right)I\right\}\right)-\left(\rho\mathbf{E}+\frac{1}{c}\mathbf{j}\times\mathbf{B}\right)\right\}\cdot\mathbf{v}-\mathbf{j}\cdot\mathbf{E}$$
$$=-\frac{c}{4\pi}\operatorname{div}_{\mathbf{x}}\{\mathbf{D}\times\mathbf{B}\}+\left(\rho\mathbf{E}+\frac{1}{c}\mathbf{j}\times\mathbf{B}\right)\cdot\mathbf{v}-\mathbf{j}\cdot\mathbf{E}$$
$$+\frac{1}{8\pi}\left(d_{\mathbf{x}}\mathbf{v}+\{d_{\mathbf{x}}\mathbf{v}\}^T\right):\left\{\mathbf{D}\otimes\mathbf{D}+\mathbf{B}\otimes\mathbf{B}-\frac{1}{2}\left(|\mathbf{D}|^2+|\mathbf{B}|^2\right)I\right\}. \tag{7.19}$$

Thus, by (7.16) and (7.18), we have

$$\frac{\partial}{\partial t}\left\{\mu(\mathbf{u}-\mathbf{v})+\frac{1}{4\pi c}\mathbf{D}\times\mathbf{B}\right\}+\operatorname{div}_{\mathbf{x}}\left\{\left(\mu(\mathbf{u}-\mathbf{v})+\frac{1}{4\pi c}\mathbf{D}\times\mathbf{B}\right)\otimes\mathbf{v}\right\}$$
$$+\{d_{\mathbf{x}}\mathbf{v}\}^T\cdot\left\{\mu(\mathbf{u}-\mathbf{v})+\frac{1}{4\pi c}\mathbf{D}\times\mathbf{B}\right\}=$$
$$\frac{1}{4\pi}\operatorname{div}_{\mathbf{x}}\left\{\mathbf{D}\otimes\mathbf{D}+\mathbf{B}\otimes\mathbf{B}-\frac{1}{2}\left(|\mathbf{D}|^2+|\mathbf{B}|^2\right)I-4\pi\mu(\mathbf{u}-\mathbf{v})\otimes(\mathbf{u}-\mathbf{v})\right\}+\mathbf{F}, \tag{7.20}$$

and by (7.17) and (7.19), we have

$$\frac{\partial}{\partial t}\left\{\frac{\mu}{2}|\mathbf{u}-\mathbf{v}|^2+\frac{|\mathbf{D}|^2+|\mathbf{B}|^2}{8\pi}\right\}+\operatorname{div}_{\mathbf{x}}\left\{\left(\frac{\mu}{2}|\mathbf{u}-\mathbf{v}|^2+\frac{|\mathbf{D}|^2+|\mathbf{B}|^2}{8\pi}\right)\mathbf{v}\right\}$$
$$+\operatorname{div}_{\mathbf{x}}\left\{\frac{\mu}{2}|\mathbf{u}-\mathbf{v}|^2(\mathbf{u}-\mathbf{v})+\frac{c}{4\pi}\mathbf{D}\times\mathbf{B}\right\}$$
$$=-\frac{1}{2}\left(d_{\mathbf{x}}\mathbf{v}+\{d_{\mathbf{x}}\mathbf{v}\}^T\right):\{\mu(\mathbf{u}-\mathbf{v})\otimes(\mathbf{u}-\mathbf{v})\}$$
$$+\frac{1}{8\pi}\left(d_{\mathbf{x}}\mathbf{v}+\{d_{\mathbf{x}}\mathbf{v}\}^T\right):\left\{\mathbf{D}\otimes\mathbf{D}+\mathbf{B}\otimes\mathbf{B}-\frac{1}{2}\left(|\mathbf{D}|^2+|\mathbf{B}|^2\right)I\right\}+(\mathbf{u}-\mathbf{v})\cdot\mathbf{F}. \tag{7.21}$$

In particular, by (7.21) and (7.20), we have

$$\frac{\partial}{\partial t}\left\{\frac{\mu}{2}|\mathbf{u}-\mathbf{v}|^2+\frac{|\mathbf{D}|^2+|\mathbf{B}|^2}{8\pi}\right\}+\operatorname{div}_{\mathbf{x}}\left\{\left(\frac{\mu}{2}|\mathbf{u}-\mathbf{v}|^2+\left(\frac{|\mathbf{D}|^2+|\mathbf{B}|^2}{8\pi}\right)\right)\mathbf{v}\right\}$$
$$+\operatorname{div}_{\mathbf{x}}\left\{\frac{\mu}{2}|\mathbf{u}-\mathbf{v}|^2(\mathbf{u}-\mathbf{v})+\frac{c}{4\pi}\mathbf{D}\times\mathbf{B}\right\}=-\mathbf{v}\cdot\frac{\partial}{\partial t}\left\{\mu(\mathbf{u}-\mathbf{v})+\frac{1}{4\pi c}\mathbf{D}\times\mathbf{B}\right\}$$
$$-\operatorname{div}_{\mathbf{x}}\left\{\left(\left(\mu(\mathbf{u}-\mathbf{v})+\frac{1}{4\pi c}\mathbf{D}\times\mathbf{B}\right)\cdot\mathbf{v}\right)\mathbf{v}\right\}-\operatorname{div}_{\mathbf{x}}\{\mu((\mathbf{u}-\mathbf{v})\cdot\mathbf{v})(\mathbf{u}-\mathbf{v})\}$$
$$+\frac{1}{4\pi}\operatorname{div}_{\mathbf{x}}\left\{\left(\mathbf{D}\otimes\mathbf{D}+\mathbf{B}\otimes\mathbf{B}-\frac{1}{2}\left(|\mathbf{D}|^2+|\mathbf{B}|^2\right)I\right)\cdot\mathbf{v}\right\}+\mathbf{u}\cdot\mathbf{F}, \tag{7.22}$$

and thus

$$\frac{\partial}{\partial t}\left\{\frac{\mu}{2}|\mathbf{u}-\mathbf{v}|^2+\frac{|\mathbf{D}|^2+|\mathbf{B}|^2}{8\pi}+\mathbf{v}\cdot\left(\mu(\mathbf{u}-\mathbf{v})+\frac{1}{4\pi c}\mathbf{D}\times\mathbf{B}\right)\right\}$$
$$+\operatorname{div}_{\mathbf{x}}\left\{\left(\frac{\mu}{2}|\mathbf{u}-\mathbf{v}|^2+\left(\frac{|\mathbf{D}|^2+|\mathbf{B}|^2}{8\pi}\right)\right)\mathbf{v}\right\}+\operatorname{div}_{\mathbf{x}}\left\{\frac{\mu}{2}|\mathbf{u}-\mathbf{v}|^2(\mathbf{u}-\mathbf{v})+\frac{c}{4\pi}\mathbf{D}\times\mathbf{B}\right\}$$
$$=\frac{\partial\mathbf{v}}{\partial t}\cdot\left(\mu(\mathbf{u}-\mathbf{v})+\frac{1}{4\pi c}\mathbf{D}\times\mathbf{B}\right)-\operatorname{div}_{\mathbf{x}}\left\{\left(\left(\mu(\mathbf{u}-\mathbf{v})+\frac{1}{4\pi c}\mathbf{D}\times\mathbf{B}\right)\cdot\mathbf{v}\right)\mathbf{v}\right\}$$
$$-\operatorname{div}_{\mathbf{x}}\{\mu((\mathbf{u}-\mathbf{v})\cdot\mathbf{v})(\mathbf{u}-\mathbf{v})\}+\frac{1}{4\pi}\operatorname{div}_{\mathbf{x}}\left\{\left(\mathbf{D}\otimes\mathbf{D}+\mathbf{B}\otimes\mathbf{B}-\frac{1}{2}\left(|\mathbf{D}|^2+|\mathbf{B}|^2\right)I\right)\cdot\mathbf{v}\right\}$$
$$+\mathbf{u}\cdot\mathbf{F}. \tag{7.23}$$

Moreover, by (7.20) and (2.15), we have

$$\frac{\partial}{\partial t}\left\{\mu(\mathbf{u}-\mathbf{v})+\frac{1}{4\pi c}\mathbf{D}\times\mathbf{B}\right\}$$
$$+\operatorname{div}_{\mathbf{x}}\left\{\left(\mu(\mathbf{u}-\mathbf{v})+\frac{1}{4\pi c}\mathbf{D}\times\mathbf{B}\right)\otimes\mathbf{v}+\mathbf{v}\otimes\left(\mu(\mathbf{u}-\mathbf{v})+\frac{1}{4\pi c}\mathbf{D}\times\mathbf{B}\right)\right\}$$
$$+\left(\mu(\mathbf{u}-\mathbf{v})+\frac{1}{4\pi c}\mathbf{D}\times\mathbf{B}\right)\times\operatorname{curl}_{\mathbf{x}}\mathbf{v}-\left(\operatorname{div}_{\mathbf{x}}\left\{\mu(\mathbf{u}-\mathbf{v})+\frac{1}{4\pi c}\mathbf{D}\times\mathbf{B}\right\}\right)\mathbf{v}=$$
$$\frac{1}{4\pi}\operatorname{div}_{\mathbf{x}}\left\{\mathbf{D}\otimes\mathbf{D}+\mathbf{B}\otimes\mathbf{B}-\frac{1}{2}\left(|\mathbf{D}|^2+|\mathbf{B}|^2\right)I-4\pi\mu(\mathbf{u}-\mathbf{v})\otimes(\mathbf{u}-\mathbf{v})\right\}+\mathbf{F},\tag{7.24}$$

and by (7.20) and (2.12), we have

$$\frac{\partial}{\partial t}\left\{\mu(\mathbf{u}-\mathbf{v})+\frac{1}{4\pi c}\mathbf{D}\times\mathbf{B}\right\}-\operatorname{curl}_{\mathbf{x}}\left\{\mathbf{v}\times\left(\mu(\mathbf{u}-\mathbf{v})+\frac{1}{4\pi c}\mathbf{D}\times\mathbf{B}\right)\right\}$$
$$+\left(\operatorname{div}_{\mathbf{x}}\left\{\mu(\mathbf{u}-\mathbf{v})+\frac{1}{4\pi c}\mathbf{D}\times\mathbf{B}\right\}\right)\mathbf{v}+\left(d_{\mathbf{x}}\mathbf{v}+\{d_{\mathbf{x}}\mathbf{v}\}^T\right)\cdot\left\{\mu(\mathbf{u}-\mathbf{v})+\frac{1}{4\pi c}\mathbf{D}\times\mathbf{B}\right\}$$
$$=\frac{1}{4\pi}\operatorname{div}_{\mathbf{x}}\left\{\mathbf{D}\otimes\mathbf{D}+\mathbf{B}\otimes\mathbf{B}-\frac{1}{2}\left(|\mathbf{D}|^2+|\mathbf{B}|^2\right)I-4\pi\mu(\mathbf{u}-\mathbf{v})\otimes(\mathbf{u}-\mathbf{v})\right\}+\mathbf{F}.\tag{7.25}$$

On the other hand, for every vector fields $\Gamma:\mathbb{R}^3\to\mathbb{R}^3$ and $\Lambda:\mathbb{R}^3\to\mathbb{R}^3$ and every scalar field $P:\mathbb{R}^3\to\mathbb{R}$, we have

$$\mathbf{x}\times\operatorname{div}_{\mathbf{x}}\{\Gamma\otimes\Lambda+\Lambda\otimes\Gamma\}=\operatorname{div}_{\mathbf{x}}\{(\mathbf{x}\times\Gamma)\otimes\Lambda+(\mathbf{x}\times\Lambda)\otimes\Gamma\},$$
$$\mathbf{x}\times\operatorname{div}_{\mathbf{x}}\{P\Gamma\otimes\Gamma\}=\operatorname{div}_{\mathbf{x}}\{P(\mathbf{x}\times\Gamma)\otimes\Gamma\}\quad\text{and}\quad\mathbf{x}\times\nabla_{\mathbf{x}}P=-\operatorname{curl}_{\mathbf{x}}\{P\,\mathbf{x}\}.\tag{7.26}$$

Thus inserting (7.26) into (7.24), we obtain

$$\frac{\partial}{\partial t}\left\{\mathbf{x}\times\left(\mu(\mathbf{u}-\mathbf{v})+\frac{1}{4\pi c}\mathbf{D}\times\mathbf{B}\right)\right\}+\operatorname{div}_{\mathbf{x}}\left\{\mu\left(\mathbf{x}\times(\mathbf{u}-\mathbf{v})\right)\otimes(\mathbf{u}-\mathbf{v})\right\}$$
$$+\operatorname{div}_{\mathbf{x}}\left\{(\mathbf{x}\times\mathbf{v})\otimes\left(\mu(\mathbf{u}-\mathbf{v})+\frac{1}{4\pi c}\mathbf{D}\times\mathbf{B}\right)+\left(\mathbf{x}\times\left(\mu(\mathbf{u}-\mathbf{v})+\frac{1}{4\pi c}\mathbf{D}\times\mathbf{B}\right)\right)\otimes\mathbf{v}\right\}$$
$$-\mathbf{x}\times\left\{\left(\operatorname{div}_{\mathbf{x}}\left\{\mu(\mathbf{u}-\mathbf{v})+\frac{1}{4\pi c}\mathbf{D}\times\mathbf{B}\right\}\right)\mathbf{v}-\left(\mu(\mathbf{u}-\mathbf{v})+\frac{1}{4\pi c}\mathbf{D}\times\mathbf{B}\right)\times\operatorname{curl}_{\mathbf{x}}\mathbf{v}\right\}$$
$$=\frac{1}{4\pi}\operatorname{div}_{\mathbf{x}}\left\{(\mathbf{x}\times\mathbf{D})\otimes\mathbf{D}+(\mathbf{x}\times\mathbf{B})\otimes\mathbf{B}\right\}+\operatorname{curl}_{\mathbf{x}}\left\{\frac{1}{2}\left(|\mathbf{D}|^2+|\mathbf{B}|^2\right)\mathbf{x}\right\}+\mathbf{x}\times\mathbf{F}.\tag{7.27}$$

Next assume that the system $(*)$ is inertial. Then, since, by (7.3) and (7.14), we have

$$\frac{\partial}{\partial t}(\mu+Q)+\operatorname{div}_{\mathbf{x}}\{(\mu+Q)\,\mathbf{v}\}=-\operatorname{div}_{\mathbf{x}}\left\{\mu(\mathbf{u}-\mathbf{v})+\frac{1}{4\pi c}\mathbf{D}\times\mathbf{B}\right\},\tag{7.28}$$

by (7.28), (7.8) and (7.9), we have

$$-\left(\mathrm{div}_{\mathbf{x}}\left\{\mu\,(\mathbf{u}-\mathbf{v})+\frac{1}{4\pi c}\,\mathbf{D}\times\mathbf{B}\right\}\right)\mathbf{v}=\left(\frac{\partial}{\partial t}\,(\mu+Q)\right)\mathbf{v}+\left(\mathrm{div}_{\mathbf{x}}\,\{(\mu+Q)\,\mathbf{v}\}\right)\mathbf{v}$$
$$=\frac{\partial}{\partial t}\,((\mu+Q)\,\mathbf{v})+\mathrm{div}_{\mathbf{x}}\,\{(\mu+Q)\,\mathbf{v}\otimes\mathbf{v}\}-(\mu+Q)\left(\frac{\partial \mathbf{v}}{\partial t}+d_{\mathbf{x}}\mathbf{v}\cdot\mathbf{v}\right)$$
$$=\frac{\partial}{\partial t}\,((\mu+Q)\,\mathbf{v})+\mathrm{div}_{\mathbf{x}}\,\{(\mu+Q)\,\mathbf{v}\otimes\mathbf{v}\}+(\mu+Q)\,\nabla_{\mathbf{x}}\Phi=$$
$$\frac{\partial}{\partial t}\,((\mu+Q)\,\mathbf{v})+\mathrm{div}_{\mathbf{x}}\,\{(\mu+Q)\,\mathbf{v}\otimes\mathbf{v}\}+\frac{1}{4\pi G}\,(\Delta_{\mathbf{x}}\Phi)\,\nabla_{\mathbf{x}}\Phi=$$
$$\frac{\partial}{\partial t}\,((\mu+Q)\,\mathbf{v})+\mathrm{div}_{\mathbf{x}}\,\{(\mu+Q)\,\mathbf{v}\otimes\mathbf{v}\}+\frac{1}{4\pi G}\,\mathrm{div}_{\mathbf{x}}\left\{\nabla_{\mathbf{x}}\Phi\otimes\nabla_{\mathbf{x}}\Phi-\frac{1}{2}\,|\nabla_{\mathbf{x}}\Phi|^2\,I\right\}. \qquad (7.29)$$

Moreover, by (7.26) and (7.29), we have

$$-\left(\mathrm{div}_{\mathbf{x}}\left\{\mu\,(\mathbf{u}-\mathbf{v})+\frac{1}{4\pi c}\,\mathbf{D}\times\mathbf{B}\right\}\right)\mathbf{x}\times\mathbf{v}=$$
$$\frac{\partial}{\partial t}\,((\mu+Q)\,\mathbf{x}\times\mathbf{v})+\mathrm{div}_{\mathbf{x}}\,\{(\mu+Q)\,(\mathbf{x}\times\mathbf{v})\otimes\mathbf{v}\}$$
$$+\frac{1}{4\pi G}\,\mathrm{div}_{\mathbf{x}}\,\{(\mathbf{x}\times\nabla_{\mathbf{x}}\Phi)\otimes\nabla_{\mathbf{x}}\Phi\}+\frac{1}{8\pi G}\mathrm{curl}_{\mathbf{x}}\left\{|\nabla_{\mathbf{x}}\Phi|^2\,\mathbf{x}\right\}. \qquad (7.30)$$

Therefore, by inserting (7.29) into (7.24) and using (7.8), we deduce the following conservation of the momentum:

$$\frac{\partial}{\partial t}\left\{\mu\mathbf{u}+Q\mathbf{v}+\frac{1}{4\pi c}\,\mathbf{D}\times\mathbf{B}\right\}$$
$$+\,\mathrm{div}_{\mathbf{x}}\left\{\mu\mathbf{u}\otimes\mathbf{u}+Q\mathbf{v}\otimes\mathbf{v}+\left(\frac{1}{4\pi c}\mathbf{D}\times\mathbf{B}\right)\otimes\mathbf{v}+\mathbf{v}\otimes\left(\frac{1}{4\pi c}\mathbf{D}\times\mathbf{B}\right)\right\}$$
$$+\frac{1}{4\pi G}\,\mathrm{div}_{\mathbf{x}}\left\{\nabla_{\mathbf{x}}\Phi\otimes\nabla_{\mathbf{x}}\Phi-\frac{1}{2}\,|\nabla_{\mathbf{x}}\Phi|^2\,I\right\}-\mathrm{div}_{\mathbf{x}}\,\{\mu\,(\mathbf{u}-\mathbf{v})\otimes(\mathbf{u}-\mathbf{v})\}=$$
$$\frac{\partial}{\partial t}\left\{\mu\,(\mathbf{u}-\mathbf{v})+\frac{1}{4\pi c}\,\mathbf{D}\times\mathbf{B}\right\}$$
$$+\,\mathrm{div}_{\mathbf{x}}\left\{\left(\mu\,(\mathbf{u}-\mathbf{v})+\frac{1}{4\pi c}\mathbf{D}\times\mathbf{B}\right)\otimes\mathbf{v}+\mathbf{v}\otimes\left(\mu\,(\mathbf{u}-\mathbf{v})+\frac{1}{4\pi c}\mathbf{D}\times\mathbf{B}\right)\right\}$$
$$+\,\frac{\partial}{\partial t}\,((\mu+Q)\,\mathbf{v})+\mathrm{div}_{\mathbf{x}}\,\{(\mu+Q)\,\mathbf{v}\otimes\mathbf{v}\}+\frac{1}{4\pi G}\,\mathrm{div}_{\mathbf{x}}\left\{\nabla_{\mathbf{x}}\Phi\otimes\nabla_{\mathbf{x}}\Phi-\frac{1}{2}\,|\nabla_{\mathbf{x}}\Phi|^2\,I\right\}$$
$$=\frac{1}{4\pi}\,\mathrm{div}_{\mathbf{x}}\left\{\mathbf{D}\otimes\mathbf{D}+\mathbf{B}\otimes\mathbf{B}-\frac{1}{2}\,(|\mathbf{D}|^2+|\mathbf{B}|^2)\,I-4\pi\mu\,(\mathbf{u}-\mathbf{v})\otimes(\mathbf{u}-\mathbf{v})\right\}+\mathbf{F}, \qquad (7.31)$$

and by inserting (7.30) into (7.27) and using (7.8), we deduce the following conservation of the angular momentum:

$$
\begin{aligned}
&\frac{\partial}{\partial t}\left\{\mathbf{x}\times\left(\mu\mathbf{u}+Q\mathbf{v}+\frac{1}{4\pi c}\mathbf{D}\times\mathbf{B}\right)\right\}+\frac{1}{4\pi G}\,\mathrm{div}_{\mathbf{x}}\left\{(\mathbf{x}\times\nabla_{\mathbf{x}}\Phi)\otimes\nabla_{\mathbf{x}}\Phi\right\}\\
&\quad+\frac{1}{8\pi G}\mathrm{curl}_{\mathbf{x}}\left\{|\nabla_{\mathbf{x}}\Phi|^2\,\mathbf{x}\right\}+\mathrm{div}_{\mathbf{x}}\left\{\mu\,(\mathbf{x}\times\mathbf{u})\otimes\mathbf{u}\right.\\
&\quad\left.+Q\,(\mathbf{x}\times\mathbf{v})\otimes\mathbf{v}+(\mathbf{x}\times\mathbf{v})\otimes\left(\frac{1}{4\pi c}\mathbf{D}\times\mathbf{B}\right)+\left(\mathbf{x}\times\left(\frac{1}{4\pi c}\mathbf{D}\times\mathbf{B}\right)\right)\otimes\mathbf{v}\right\}\\
&=\frac{\partial}{\partial t}\left\{\mathbf{x}\times\left(\mu(\mathbf{u}-\mathbf{v})+\frac{1}{4\pi c}\mathbf{D}\times\mathbf{B}\right)\right\}+\mathrm{div}_{\mathbf{x}}\left\{\mu\,(\mathbf{x}\times(\mathbf{u}-\mathbf{v}))\otimes(\mathbf{u}-\mathbf{v})\right\}\\
&+\mathrm{div}_{\mathbf{x}}\left\{(\mathbf{x}\times\mathbf{v})\otimes\left(\mu(\mathbf{u}-\mathbf{v})+\frac{1}{4\pi c}\mathbf{D}\times\mathbf{B}\right)+\left(\mathbf{x}\times\left(\mu(\mathbf{u}-\mathbf{v})+\frac{1}{4\pi c}\mathbf{D}\times\mathbf{B}\right)\right)\otimes\mathbf{v}\right\}\\
&\quad+\frac{\partial}{\partial t}\left((\mu+Q)\,\mathbf{x}\times\mathbf{v}\right)+\mathrm{div}_{\mathbf{x}}\left\{(\mu+Q)\,(\mathbf{x}\times\mathbf{v})\otimes\mathbf{v}\right\}\\
&\quad+\frac{1}{4\pi G}\,\mathrm{div}_{\mathbf{x}}\left\{(\mathbf{x}\times\nabla_{\mathbf{x}}\Phi)\otimes\nabla_{\mathbf{x}}\Phi\right\}+\frac{1}{8\pi G}\mathrm{curl}_{\mathbf{x}}\left\{|\nabla_{\mathbf{x}}\Phi|^2\,\mathbf{x}\right\}\\
&=\frac{1}{4\pi}\,\mathrm{div}_{\mathbf{x}}\left\{(\mathbf{x}\times\mathbf{D})\otimes\mathbf{D}+(\mathbf{x}\times\mathbf{B})\otimes\mathbf{B}\right\}+\mathrm{curl}_{\mathbf{x}}\left\{\frac{1}{2}\left(|\mathbf{D}|^2+|\mathbf{B}|^2\right)\mathbf{x}\right\}+\mathbf{x}\times\mathbf{F}.
\end{aligned}
\tag{7.32}
$$

Finally, by (7.28), (7.8) and (7.9), we have

$$
\begin{aligned}
&\frac{\partial\mathbf{v}}{\partial t}\cdot\left(\mu\,(\mathbf{u}-\mathbf{v})+\frac{1}{4\pi c}\mathbf{D}\times\mathbf{B}\right)=-\nabla_{\mathbf{x}}\left(\Phi+\frac{1}{2}\,|\mathbf{v}|^2\right)\cdot\left(\mu\,(\mathbf{u}-\mathbf{v})+\frac{1}{4\pi c}\mathbf{D}\times\mathbf{B}\right)=\\
&\quad\left(\Phi+\frac{1}{2}\,|\mathbf{v}|^2\right)\left(\mathrm{div}_{\mathbf{x}}\left\{\mu\,(\mathbf{u}-\mathbf{v})+\frac{1}{4\pi c}\mathbf{D}\times\mathbf{B}\right\}\right)\\
&\quad-\mathrm{div}_{\mathbf{x}}\left\{\left(\Phi+\frac{1}{2}\,|\mathbf{v}|^2\right)\left(\mu\,(\mathbf{u}-\mathbf{v})+\frac{1}{4\pi c}\mathbf{D}\times\mathbf{B}\right)\right\}=\\
&\quad-\left(\Phi+\frac{1}{2}\,|\mathbf{v}|^2\right)\left(\frac{\partial}{\partial t}(\mu+Q)+\mathrm{div}_{\mathbf{x}}\left\{(\mu+Q)\,\mathbf{v}\right\}\right)\\
&\quad-\mathrm{div}_{\mathbf{x}}\left\{\left(\Phi+\frac{1}{2}\,|\mathbf{v}|^2\right)\left(\mu\,(\mathbf{u}-\mathbf{v})+\frac{1}{4\pi c}\mathbf{D}\times\mathbf{B}\right)\right\}\\
&=-\left(\Phi+\frac{1}{2}\,|\mathbf{v}|^2\right)\left(\frac{\partial}{\partial t}(\mu+Q)\right)+(\mu+Q)\,\nabla_{\mathbf{x}}\left(\Phi+\frac{1}{2}\,|\mathbf{v}|^2\right)\cdot\mathbf{v}\\
&-\mathrm{div}_{\mathbf{x}}\left\{\left(\Phi+\frac{1}{2}\,|\mathbf{v}|^2\right)(\mu+Q)\,\mathbf{v}\right\}-\mathrm{div}_{\mathbf{x}}\left\{\left(\Phi+\frac{1}{2}\,|\mathbf{v}|^2\right)\left(\mu\,(\mathbf{u}-\mathbf{v})+\frac{1}{4\pi c}\mathbf{D}\times\mathbf{B}\right)\right\}\\
&=-\frac{1}{4\pi G}\Phi\left(\frac{\partial}{\partial t}(\Delta_{\mathbf{x}}\Phi)\right)-\left(\frac{1}{2}\,|\mathbf{v}|^2\right)\left(\frac{\partial}{\partial t}(\mu+Q)\right)-(\mu+Q)\,\frac{\partial\mathbf{v}}{\partial t}\cdot\mathbf{v}\\
&\quad-\mathrm{div}_{\mathbf{x}}\left\{\left(\Phi+\frac{1}{2}\,|\mathbf{v}|^2\right)(\mu+Q)\,\mathbf{v}\right\}\\
&\quad-\mathrm{div}_{\mathbf{x}}\left\{\left(\Phi+\frac{1}{2}\,|\mathbf{v}|^2\right)\left(\mu\,(\mathbf{u}-\mathbf{v})+\frac{1}{4\pi c}\mathbf{D}\times\mathbf{B}\right)\right\}\\
&=\frac{1}{8\pi G}\frac{\partial}{\partial t}\left(|\nabla_{\mathbf{x}}\Phi|^2\right)-\frac{\partial}{\partial t}\left(\frac{1}{2}\,(\mu+Q)\,|\mathbf{v}|^2\right)\\
&-\frac{1}{4\pi G}\,\mathrm{div}_{\mathbf{x}}\left\{\Phi\frac{\partial}{\partial t}(\nabla_{\mathbf{x}}\Phi)\right\}-\mathrm{div}_{\mathbf{x}}\left\{\left(\Phi+\frac{1}{2}\,|\mathbf{v}|^2\right)\left(\mu\mathbf{u}+Q\mathbf{v}+\frac{1}{4\pi c}\mathbf{D}\times\mathbf{B}\right)\right\}
\end{aligned}
\tag{7.33}
$$

Then, by inserting (7.33) into (7.23), we deuce

$$
\begin{aligned}
&\frac{\partial}{\partial t}\left\{\frac{\mu}{2}|\mathbf{u}-\mathbf{v}|^2+\frac{|\mathbf{D}|^2+|\mathbf{B}|^2}{8\pi}+\mathbf{v}\cdot\left(\mu(\mathbf{u}-\mathbf{v})+\frac{1}{4\pi c}\mathbf{D}\times\mathbf{B}\right)\right\}\\
&\quad+\operatorname{div}_{\mathbf{x}}\left\{\left(\frac{\mu}{2}|\mathbf{u}-\mathbf{v}|^2+\left(\frac{|\mathbf{D}|^2+|\mathbf{B}|^2}{8\pi}\right)\right)\mathbf{v}\right\}+\operatorname{div}_{\mathbf{x}}\left\{\frac{\mu}{2}|\mathbf{u}-\mathbf{v}|^2(\mathbf{u}-\mathbf{v})+\frac{c}{4\pi}\mathbf{D}\times\mathbf{B}\right\}\\
&\qquad=\frac{1}{8\pi G}\frac{\partial}{\partial t}\left(|\nabla_{\mathbf{x}}\Phi|^2\right)-\frac{\partial}{\partial t}\left(\frac{1}{2}(\mu+Q)|\mathbf{v}|^2\right)\\
&\quad-\frac{1}{4\pi G}\operatorname{div}_{\mathbf{x}}\left\{\Phi\frac{\partial}{\partial t}(\nabla_{\mathbf{x}}\Phi)\right\}-\operatorname{div}_{\mathbf{x}}\left\{\left(\Phi+\frac{1}{2}|\mathbf{v}|^2\right)\left(\mu\mathbf{u}+Q\mathbf{v}+\frac{1}{4\pi c}\mathbf{D}\times\mathbf{B}\right)\right\}\\
&\qquad-\operatorname{div}_{\mathbf{x}}\left\{\left(\left(\mu(\mathbf{u}-\mathbf{v})+\frac{1}{4\pi c}\mathbf{D}\times\mathbf{B}\right)\cdot\mathbf{v}\right)\mathbf{v}\right\}\\
&\quad-\operatorname{div}_{\mathbf{x}}\left\{\mu((\mathbf{u}-\mathbf{v})\cdot\mathbf{v})(\mathbf{u}-\mathbf{v})\right\}+\frac{1}{4\pi}\operatorname{div}_{\mathbf{x}}\left\{\left(\mathbf{D}\otimes\mathbf{D}+\mathbf{B}\otimes\mathbf{B}-\frac{1}{2}\left(|\mathbf{D}|^2+|\mathbf{B}|^2\right)I\right)\cdot\mathbf{v}\right\}\\
&\qquad+\mathbf{u}\cdot\mathbf{F}. \qquad (7.34)
\end{aligned}
$$

Then, using (2.1) and the last two equalities in (7.11), we rewrite (7.34) in the form of the following conservation of the energy:

$$
\begin{aligned}
&\frac{\partial}{\partial t}\left\{\frac{\mu}{2}|\mathbf{u}|^2+\frac{Q}{2}|\mathbf{v}|^2+\frac{\mathbf{D}\cdot\mathbf{E}+\mathbf{B}\cdot\mathbf{H}}{8\pi}-\frac{1}{8\pi G}|\nabla_{\mathbf{x}}\Phi|^2\right\}\\
&\quad+\operatorname{div}_{\mathbf{x}}\left\{\frac{\mu}{2}|\mathbf{u}|^2\mathbf{u}+\frac{Q}{2}|\mathbf{v}|^2\mathbf{v}+\left(\frac{\mathbf{D}\cdot\mathbf{E}+\mathbf{B}\cdot\mathbf{H}}{8\pi}\right)\mathbf{v}+\frac{1}{8\pi c}|\mathbf{v}|^2(\mathbf{D}\times\mathbf{B})\right\}\\
&+\operatorname{div}_{\mathbf{x}}\left\{\frac{c}{4\pi}\mathbf{D}\times\mathbf{B}\right\}=-\frac{1}{4\pi G}\operatorname{div}_{\mathbf{x}}\left\{\Phi\frac{\partial}{\partial t}(\nabla_{\mathbf{x}}\Phi)\right\}-\operatorname{div}_{\mathbf{x}}\left\{\Phi\left(\mu\mathbf{u}+Q\mathbf{v}+\frac{1}{4\pi c}\mathbf{D}\times\mathbf{B}\right)\right\}\\
&\quad+\frac{1}{4\pi}\operatorname{div}_{\mathbf{x}}\left\{\left(\mathbf{D}\otimes\mathbf{D}+\mathbf{B}\otimes\mathbf{B}-\frac{1}{2}\left(|\mathbf{D}|^2+|\mathbf{B}|^2\right)I\right)\cdot\mathbf{v}\right\}+\mathbf{u}\cdot\mathbf{F}. \qquad (7.35)
\end{aligned}
$$

As a consequence of (7.31), (7.32) and (7.35), we infer that we have the following proposition:

Proposition 7.1 *Consider the Maxwell equation for the vacuum in the form* (7.11) *and the Second Law of Newton for the moving continuum in the form* (7.13). *Next, assume that, in some Cartesian coordinate system* $(*)$, *we observe the gravitational law in the form of* (7.8), (7.9) *and* (7.3). *Then, in the system* $(*)$, *we have the following conservation laws of the linear and angular momentums and energy:*

$$
\begin{aligned}
&\frac{\partial}{\partial t}\left(\mu\mathbf{u}+Q\mathbf{v}+\frac{1}{4\pi c}\mathbf{D}\times\mathbf{B}\right)=\\
&\qquad-\operatorname{div}_{\mathbf{x}}\left\{\mu\mathbf{u}\otimes\mathbf{u}+Q\mathbf{v}\otimes\mathbf{v}+\left(\frac{1}{4\pi c}\mathbf{D}\times\mathbf{B}\right)\otimes\mathbf{v}+\mathbf{v}\otimes\left(\frac{1}{4\pi c}\mathbf{D}\times\mathbf{B}\right)\right\}\\
&\quad+\frac{1}{4\pi}\operatorname{div}_{\mathbf{x}}\left\{\mathbf{D}\otimes\mathbf{D}+\mathbf{B}\otimes\mathbf{B}-\frac{1}{2}\left(|\mathbf{D}|^2+|\mathbf{B}|^2\right)I-\frac{1}{G}\nabla_{\mathbf{x}}\Phi\otimes\nabla_{\mathbf{x}}\Phi+\frac{1}{2G}|\nabla_{\mathbf{x}}\Phi|^2 I\right\}\\
&\qquad+\mathbf{F}, \qquad (7.36)
\end{aligned}
$$

$$\frac{\partial}{\partial t}\left(\mathbf{x}\times(\mu\mathbf{u})+\mathbf{x}\times(Q\mathbf{v})+\mathbf{x}\times\left(\frac{1}{4\pi c}\mathbf{D}\times\mathbf{B}\right)\right)=-\operatorname{div}_{\mathbf{x}}\Big\{\mu(\mathbf{x}\times\mathbf{u})\otimes\mathbf{u}$$
$$+Q(\mathbf{x}\times\mathbf{v})\otimes\mathbf{v}+\left(\mathbf{x}\times\left(\frac{1}{4\pi c}\mathbf{D}\times\mathbf{B}\right)\right)\otimes\mathbf{v}+(\mathbf{x}\times\mathbf{v})\otimes\left(\frac{1}{4\pi c}\mathbf{D}\times\mathbf{B}\right)\Big\}$$
$$+\frac{1}{4\pi}\operatorname{div}_{\mathbf{x}}\left\{(\mathbf{x}\times\mathbf{D})\otimes\mathbf{D}+(\mathbf{x}\times\mathbf{B})\otimes\mathbf{B}-\frac{1}{G}(\mathbf{x}\times\nabla_{\mathbf{x}}\Phi)\otimes\nabla_{\mathbf{x}}\Phi\right\}$$
$$+\frac{1}{8\pi}curl_{\mathbf{x}}\left\{\left(|\mathbf{D}|^2+|\mathbf{B}|^2-\frac{1}{G}|\nabla_{\mathbf{x}}\Phi|^2\right)\mathbf{x}\right\}+\mathbf{x}\times\mathbf{F},\quad(7.37)$$

and

$$\frac{\partial}{\partial t}\left(\frac{\mu|\mathbf{u}|^2}{2}+\frac{Q}{2}|\mathbf{v}|^2+\frac{\mathbf{D}\cdot\mathbf{E}+\mathbf{B}\cdot\mathbf{H}}{8\pi}-\frac{1}{8\pi G}|\nabla_{\mathbf{x}}\Phi|^2\right)=$$
$$=-\operatorname{div}_{\mathbf{x}}\left\{\left(\frac{\mu|\mathbf{u}|^2}{2}\right)\mathbf{u}+\left(\frac{Q|\mathbf{v}|^2}{2}\right)\mathbf{v}+\frac{1}{2}|\mathbf{v}|^2\left(\frac{1}{4\pi c}\mathbf{D}\times\mathbf{B}\right)+\left(\frac{\mathbf{D}\cdot\mathbf{E}+\mathbf{B}\cdot\mathbf{H}}{8\pi}\right)\mathbf{v}\right\}$$
$$+\frac{1}{4\pi}\operatorname{div}_{\mathbf{x}}\left\{(\mathbf{D}\otimes\mathbf{D}+\mathbf{B}\otimes\mathbf{B})\cdot\mathbf{v}-\frac{1}{2}\left(|\mathbf{D}|^2+|\mathbf{B}|^2\right)\mathbf{v}-c\mathbf{D}\times\mathbf{B}\right\}$$
$$-\operatorname{div}_{\mathbf{x}}\left\{\Phi\left(\mu\mathbf{u}+Q\mathbf{v}+\frac{1}{4\pi c}\mathbf{D}\times\mathbf{B}\right)\right\}-\frac{1}{4\pi G}\operatorname{div}_{\mathbf{x}}\left\{\Phi\frac{\partial}{\partial t}(\nabla_{\mathbf{x}}\Phi)\right\}+\mathbf{F}\cdot\mathbf{u}.\quad(7.38)$$

Chapter 8
Lagrangian of the Unified Gravitational–Electromagnetic Field

8.1 Lagrangian in the Case of Newtonian Gravity

Given the distribution of inertial mass density of some continuum medium $\mu := \mu(\mathbf{x}, t)$, the field of velocities of this medium $\mathbf{u} := \mathbf{u}(\mathbf{x}, t)$, the charge density $\rho := \rho(\mathbf{x}, t)$ and the current density $\mathbf{j} := \mathbf{j}(\mathbf{x}, t)$, consider a Lagrangian density L defined by

$$
\begin{aligned}
&L\left(\mathbf{A}, \Psi, \mathbf{v}, \Phi, \mathbf{p}, \mathbf{x}, t\right) := \\
&\frac{1}{8\pi}\left|-\nabla_{\mathbf{x}}\Psi - \frac{1}{c}\frac{\partial \mathbf{A}}{\partial t} + \frac{1}{c}\mathbf{v} \times \operatorname{curl}_{\mathbf{x}}\mathbf{A}\right|^2 - \frac{1}{8\pi}\left|\operatorname{curl}_{\mathbf{x}}\mathbf{A}\right|^2 - \left(\rho\Psi - \frac{1}{c}\mathbf{A}\cdot\mathbf{j}\right) \\
&+ \frac{\mu}{2}\left|\mathbf{u} - \mathbf{v}\right|^2 + \frac{1}{2}\left(d_{\mathbf{x}}\mathbf{v} + \{d_{\mathbf{x}}\mathbf{v}\}^T\right) : \left(d_{\mathbf{x}}\mathbf{p} + \{d_{\mathbf{x}}\mathbf{p}\}^T\right) - 2\left(\operatorname{div}_{\mathbf{x}}\mathbf{v}\right)\left(\operatorname{div}_{\mathbf{x}}\mathbf{p}\right) \\
&+ \frac{1}{4\pi G}\left(\operatorname{div}_{\mathbf{x}}\mathbf{v}\right)\left(\frac{\partial \Phi}{\partial t} + \mathbf{v}\cdot\nabla_{\mathbf{x}}\Phi\right) + \frac{1}{4\pi G}\Phi\left(\operatorname{div}_{\mathbf{x}}\mathbf{v}\right)^2 - \frac{\Phi}{16\pi G}\left|d_{\mathbf{x}}\mathbf{v} + \{d_{\mathbf{x}}\mathbf{v}\}^T\right|^2 \\
&+ \frac{1}{8\pi G}\left|\nabla_{\mathbf{x}}\Phi\right|^2, \qquad (8.1)
\end{aligned}
$$

where Φ is an ancillary proper scalar field and $\mathbf{p}$ is an ancillary proper vector field. Then L is invariant under the change of inertial or non-inertial Cartesian coordinate system. We investigate critical points of the functional

$$
J = \int_0^T \int_{\mathbb{R}^3} L\left(\mathbf{A}, \Psi, \mathbf{v}, \Phi, \mathbf{p}, \mathbf{x}, t\right) d\mathbf{x}dt. \qquad (8.2)
$$

We denote

A. Poliakovsky, *The Laws of Gravity and Electromagnetism*,
https://doi.org/10.1007/978-3-031-61407-1_8

$$\begin{cases} \mathbf{D} = -\nabla_{\mathbf{x}}\Psi - \frac{1}{c}\frac{\partial \mathbf{A}}{\partial t} + \frac{1}{c}\mathbf{v} \times \text{curl}_{\mathbf{x}}\mathbf{A} \\ \mathbf{B} = \text{curl}_{\mathbf{x}}\mathbf{A} \\ \mathbf{E} = -\nabla_{\mathbf{x}}\Psi - \frac{1}{c}\frac{\partial \mathbf{A}}{\partial t} = \mathbf{D} - \frac{1}{c}\mathbf{v} \times \mathbf{B} \\ \mathbf{H} = \text{curl}_{\mathbf{x}}\mathbf{A} + \frac{1}{c}\mathbf{v} \times \left(-\nabla_{\mathbf{x}}\Psi - \frac{1}{c}\frac{\partial \mathbf{A}}{\partial t} + \frac{1}{c}\mathbf{v} \times \text{curl}_{\mathbf{x}}\mathbf{A}\right) = \mathbf{B} + \frac{1}{c}\mathbf{v} \times \mathbf{D}. \end{cases} \tag{8.3}$$

Then, by (8.3), we have

$$\begin{cases} \text{curl}_{\mathbf{x}}\mathbf{E} + \frac{1}{c}\frac{\partial \mathbf{B}}{\partial t} = 0 \\ \text{div}_{\mathbf{x}}\,\mathbf{B} = 0. \end{cases} \tag{8.4}$$

Moreover, by (8.1), (2.5) and (2.11), we have

$$\frac{\delta L}{\delta \mathbf{p}} = -\,\text{div}_{\mathbf{x}}\left(d_{\mathbf{x}}\mathbf{v} + \{d_{\mathbf{x}}\mathbf{v}\}^T\right) + 2\nabla_{\mathbf{x}}\left(\text{div}_{\mathbf{x}}\,\mathbf{v}\right) = \text{curl}_{\mathbf{x}}\left(\text{curl}_{\mathbf{x}}\mathbf{v}\right) = 0, \tag{8.5}$$

$$\frac{\delta L}{\delta \Phi} = -\frac{1}{4\pi G}\left(\frac{\partial}{\partial t}\{\text{div}_{\mathbf{x}}\,\mathbf{v}\} + \mathbf{v}\cdot\nabla_{\mathbf{x}}\left(\text{div}_{\mathbf{x}}\,\mathbf{v}\right) + \frac{1}{4}\left|d_{\mathbf{x}}\mathbf{v} + \{d_{\mathbf{x}}\mathbf{v}\}^T\right|^2\right) - \frac{1}{4\pi G}\Delta_{\mathbf{x}}\Phi = 0, \tag{8.6}$$

$$\frac{\delta L}{\delta \mathbf{v}} = -\left(\mu\mathbf{u} - \mu\mathbf{v} + \frac{1}{4\pi c}\mathbf{D}\times\mathbf{B}\right) - \text{div}_{\mathbf{x}}\left(d_{\mathbf{x}}\mathbf{p} + \{d_{\mathbf{x}}\mathbf{p}\}^T\right) + 2\nabla_{\mathbf{x}}\left(\text{div}_{\mathbf{x}}\,\mathbf{p}\right)$$
$$+\frac{1}{4\pi G}\,\text{div}_{\mathbf{x}}\left\{\left(d_{\mathbf{x}}\mathbf{v} + \{d_{\mathbf{x}}\mathbf{v}\}^T\right)\Phi\right\} - \frac{1}{2\pi G}\nabla_{\mathbf{x}}\left(\Phi\left(\text{div}_{\mathbf{x}}\,\mathbf{v}\right)\right) - \frac{1}{4\pi G}\nabla_{\mathbf{x}}\left(\frac{\partial \Phi}{\partial t} + \mathbf{v}\cdot\nabla_{\mathbf{x}}\Phi\right)$$
$$+\frac{1}{4\pi G}\left(\text{div}_{\mathbf{x}}\,\mathbf{v}\right)\nabla_{\mathbf{x}}\Phi =$$
$$-\left(\mu\mathbf{u} - \mu\mathbf{v} + \frac{1}{4\pi c}\mathbf{D}\times\mathbf{B}\right) + \text{curl}_{\mathbf{x}}\left(\text{curl}_{\mathbf{x}}\mathbf{p}\right) - \frac{1}{4\pi G}\Phi\,\text{curl}_{\mathbf{x}}\left(\text{curl}_{\mathbf{x}}\mathbf{v}\right)$$
$$-\frac{1}{4\pi G}\left(\frac{\partial}{\partial t}\left(\nabla_{\mathbf{x}}\Phi\right) - \text{curl}_{\mathbf{x}}\left(\mathbf{v}\times\nabla_{\mathbf{x}}\Phi\right) + \left(\Delta_{\mathbf{x}}\Phi\right)\mathbf{v}\right) = 0, \tag{8.7}$$

$$\frac{\delta L}{\delta \Psi} = \frac{1}{4\pi}\text{div}_{\mathbf{x}}\,\mathbf{D} - \rho = 0, \tag{8.8}$$

and

$$\frac{\delta L}{\delta \mathbf{A}} =$$
$$\frac{1}{c}\mathbf{j} + \frac{1}{4\pi c}\frac{\partial \mathbf{D}}{\partial t} - \frac{1}{4\pi}\text{curl}_{\mathbf{x}}\mathbf{B} - \frac{1}{4\pi c}\text{curl}_{\mathbf{x}}\left(\mathbf{v}\times\mathbf{D}\right) = \frac{1}{c}\mathbf{j} + \frac{1}{4\pi c}\frac{\partial \mathbf{D}}{\partial t} - \frac{1}{4\pi}\text{curl}_{\mathbf{x}}\mathbf{H} = 0. \tag{8.9}$$

So,

$$\begin{cases} \text{curl}_{\mathbf{x}}\mathbf{H} = \frac{4\pi}{c}\mathbf{j} + \frac{1}{c}\frac{\partial \mathbf{D}}{\partial t} \\ \text{div}_{\mathbf{x}}\,\mathbf{D} = 4\pi\rho \\ \text{curl}_{\mathbf{x}}\mathbf{E} + \frac{1}{c}\frac{\partial \mathbf{B}}{\partial t} = 0 \\ \text{div}_{\mathbf{x}}\,\mathbf{B} = 0 \\ \mathbf{E} = \mathbf{D} - \frac{1}{c}\mathbf{v} \times \mathbf{B} \\ \mathbf{H} = \mathbf{B} + \frac{1}{c}\mathbf{v} \times \mathbf{D} \\ \text{curl}_{\mathbf{x}}\,(\text{curl}_{\mathbf{x}}\mathbf{v}) = 0 \\ \frac{\partial}{\partial t}\{\text{div}_{\mathbf{x}}\,\mathbf{v}\} + \mathbf{v}\cdot\nabla_{\mathbf{x}}\,(\text{div}_{\mathbf{x}}\,\mathbf{v}) + \frac{1}{4}\left|d_{\mathbf{x}}\mathbf{v} + \{d_{\mathbf{x}}\mathbf{v}\}^T\right|^2 = -\Delta_{\mathbf{x}}\Phi \\ \left(\mu\mathbf{u} - \mu\mathbf{v} + \frac{1}{4\pi c}\mathbf{D}\times\mathbf{B}\right) = \\ \text{curl}_{\mathbf{x}}\,(\text{curl}_{\mathbf{x}}\mathbf{p}) - \frac{1}{4\pi G}\left(\frac{\partial}{\partial t}\,(\nabla_{\mathbf{x}}\Phi) - \text{curl}_{\mathbf{x}}\,(\mathbf{v}\times\nabla_{\mathbf{x}}\Phi) + (\Delta_{\mathbf{x}}\Phi)\,\mathbf{v}\right). \end{cases} \tag{8.10}$$

In particular, using continuum equation $\partial_t\mu + \text{div}_{\mathbf{x}}\,(\mu\mathbf{u}) = 0$ from the last equality in (8.10), we deduce

$$\frac{\partial}{\partial t}\left(\frac{1}{4\pi G}\Delta_{\mathbf{x}}\Phi - \mu\right) + \text{div}_{\mathbf{x}}\left\{\left(\frac{1}{4\pi G}\Delta_{\mathbf{x}}\Phi - \mu\right)\mathbf{v}\right\} = -\,\text{div}_{\mathbf{x}}\left\{\frac{1}{4\pi c}\mathbf{D}\times\mathbf{B}\right\}.$$

Thus, denoting $Q = \Delta_{\mathbf{x}}\Phi/4\pi G - \mu$, we deduce

$$\begin{cases} \text{curl}_{\mathbf{x}}\mathbf{H} = \frac{4\pi}{c}\mathbf{j} + \frac{1}{c}\frac{\partial \mathbf{D}}{\partial t} \\ \text{div}_{\mathbf{x}}\,\mathbf{D} = 4\pi\rho \\ \text{curl}_{\mathbf{x}}\mathbf{E} + \frac{1}{c}\frac{\partial \mathbf{B}}{\partial t} = 0 \\ \text{div}_{\mathbf{x}}\,\mathbf{B} = 0 \\ \mathbf{E} = \mathbf{D} - \frac{1}{c}\mathbf{v} \times \mathbf{B} \\ \mathbf{H} = \mathbf{B} + \frac{1}{c}\mathbf{v} \times \mathbf{D} \\ \text{curl}_{\mathbf{x}}\,(\text{curl}_{\mathbf{x}}\mathbf{v}) = 0 \\ \frac{\partial}{\partial t}\{\text{div}_{\mathbf{x}}\,\mathbf{v}\} + \mathbf{v}\cdot\nabla_{\mathbf{x}}\,(\text{div}_{\mathbf{x}}\,\mathbf{v}) + \frac{1}{4}\left|d_{\mathbf{x}}\mathbf{v} + \{d_{\mathbf{x}}\mathbf{v}\}^T\right|^2 = -4\pi G(\mu + Q) \\ \frac{\partial Q}{\partial t} + \text{div}_{\mathbf{x}}\,(Q\mathbf{v}) = -\,\text{div}_{\mathbf{x}}\left\{\frac{1}{4\pi c}\mathbf{D}\times\mathbf{B}\right\}. \end{cases} \tag{8.11}$$

8.2 Lagrangian in the Case of Some Possible Alternative Model of the Gravity

Consider $\mathbf{k}$ to be the vectorial potential of the inertia, which is a generally trivial speed-like vector field, assumed to be fixed in every fixed inertial or non-inertial Cartesian coordinate system (see Definition 4.2). Given the distribution of inertial mass density of some continuum medium $\mu := \mu(\mathbf{x}, t)$, the field of velocities of this medium $\mathbf{u} := \mathbf{u}(\mathbf{x}, t)$, the charge density $\rho := \rho(\mathbf{x}, t)$ and the current density $\mathbf{j} := \mathbf{j}(\mathbf{x}, t)$, consider a Lagrangian density L defined by

$$
\begin{aligned}
&L(\mathbf{A}, \Psi, \mathbf{v}, \Phi_0, \mathbf{h}, \mathbf{x}, t) := \\
&\frac{1}{8\pi}\left|-\nabla_{\mathbf{x}}\Psi - \frac{1}{c}\frac{\partial \mathbf{A}}{\partial t} + \frac{1}{c}\mathbf{v}\times \operatorname{curl}_{\mathbf{x}}\mathbf{A}\right|^2 - \frac{1}{8\pi}\left|\operatorname{curl}_{\mathbf{x}}\mathbf{A}\right|^2 - \left(\rho\Psi - \frac{1}{c}\mathbf{A}\cdot\mathbf{j}\right) + \frac{\mu}{2}|\mathbf{u}-\mathbf{v}|^2 \\
&\quad - \frac{c^2}{8\pi G}\left|-\nabla_{\mathbf{x}}\Phi_0 - \frac{1}{c}\frac{\partial \mathbf{h}}{\partial t} + \frac{1}{c}\mathbf{v}\times\operatorname{curl}_{\mathbf{x}}\mathbf{h} - \frac{1}{c}\nabla_{\mathbf{x}}(\mathbf{h}\cdot\mathbf{v})\right|^2 + \frac{c^2}{8\pi G}\left|\operatorname{curl}_{\mathbf{x}}\mathbf{h}\right|^2 \\
&\quad + \frac{c^2\beta}{4\pi G}\left(\frac{1}{4}\left|d_{\mathbf{x}}\mathbf{v} + \{d_{\mathbf{x}}\mathbf{v}\}^T\right|^2 - \left|\operatorname{div}_{\mathbf{x}}\mathbf{v}\right|^2\right), \qquad (8.12)
\end{aligned}
$$

where

$$
\mathbf{h} = \mathbf{v} - \mathbf{k} \quad \text{and} \quad \Phi_0 = -\frac{1}{2c}|\mathbf{v}-\mathbf{k}|^2, \qquad (8.13)
$$

and $\beta \in \mathbb{R}$ is some constant. In other words, we have

$$
\begin{aligned}
&L(\mathbf{A}, \Psi, \mathbf{v}, \Phi_0, \mathbf{h}, \mathbf{x}, t) = L_1(\mathbf{A}, \Psi, \mathbf{v}, \mathbf{x}, t) := \\
&\frac{1}{8\pi}\left|-\nabla_{\mathbf{x}}\Psi - \frac{1}{c}\frac{\partial \mathbf{A}}{\partial t} + \frac{1}{c}\mathbf{v}\times \operatorname{curl}_{\mathbf{x}}\mathbf{A}\right|^2 - \frac{1}{8\pi}\left|\operatorname{curl}_{\mathbf{x}}\mathbf{A}\right|^2 - \left(\rho\Psi - \frac{1}{c}\mathbf{A}\cdot\mathbf{j}\right) + \frac{\mu}{2}|\mathbf{u}-\mathbf{v}|^2 \\
&- \frac{c^2}{8\pi G}\left|\frac{1}{c}\nabla_{\mathbf{x}}\left(\frac{1}{2}|\mathbf{v}-\mathbf{k}|^2\right) - \frac{1}{c}\frac{\partial}{\partial t}(\mathbf{v}-\mathbf{k}) + \frac{1}{c}\mathbf{v}\times\operatorname{curl}_{\mathbf{x}}(\mathbf{v}-\mathbf{k}) - \frac{1}{c}\nabla_{\mathbf{x}}((\mathbf{v}-\mathbf{k})\cdot\mathbf{v})\right|^2 \\
&\quad + \frac{c^2}{8\pi G}\left|\operatorname{curl}_{\mathbf{x}}(\mathbf{v}-\mathbf{k})\right|^2 + \frac{c^2\beta}{4\pi G}\left(\frac{1}{4}\left|d_{\mathbf{x}}\mathbf{v} + \{d_{\mathbf{x}}\mathbf{v}\}^T\right|^2 - \left|\operatorname{div}_{\mathbf{x}}\mathbf{v}\right|^2\right). \qquad (8.14)
\end{aligned}
$$

In particular, in the inertial coordinate system where $d_{\mathbf{x}}\mathbf{k} = 0$ and $\partial_t\mathbf{k} = 0$, we have

$$
\begin{aligned}
&L(\mathbf{A}, \Psi, \mathbf{v}, \Phi_0, \mathbf{h}, \mathbf{x}, t) = L_1(\mathbf{A}, \Psi, \mathbf{v}, \mathbf{x}, t) := \\
&\frac{1}{8\pi}\left|-\nabla_{\mathbf{x}}\Psi - \frac{1}{c}\frac{\partial \mathbf{A}}{\partial t} + \frac{1}{c}\mathbf{v}\times \operatorname{curl}_{\mathbf{x}}\mathbf{A}\right|^2 - \frac{1}{8\pi}\left|\operatorname{curl}_{\mathbf{x}}\mathbf{A}\right|^2 - \left(\rho\Psi - \frac{1}{c}\mathbf{A}\cdot\mathbf{j}\right) + \frac{\mu}{2}|\mathbf{u}-\mathbf{v}|^2 \\
&\quad - \frac{c^2}{8\pi G}\left|-\frac{1}{c}\frac{\partial \mathbf{v}}{\partial t} + \frac{1}{c}\mathbf{v}\times\operatorname{curl}_{\mathbf{x}}\mathbf{v} - \frac{1}{c}\nabla_{\mathbf{x}}\left(\frac{1}{2}|\mathbf{v}|^2\right)\right|^2 + \frac{c^2}{8\pi G}\left|\operatorname{curl}_{\mathbf{x}}\mathbf{v}\right|^2 \\
&\quad + \frac{c^2\beta}{4\pi G}\left(\frac{1}{4}\left|d_{\mathbf{x}}\mathbf{v} + \{d_{\mathbf{x}}\mathbf{v}\}^T\right|^2 - \left|\operatorname{div}_{\mathbf{x}}\mathbf{v}\right|^2\right), \qquad (8.15)
\end{aligned}
$$

Note here the advantage of inertial coordinate systems, where L and L_1 are completely independent of the vectorial potential of the inertia $\mathbf{k}$. Furthermore, by (2.15), we rewrite (8.15) as

$$
\begin{aligned}
&L(\mathbf{A}, \Psi, \mathbf{v}, \Phi_0, \mathbf{h}, \mathbf{x}, t) = L_1(\mathbf{A}, \Psi, \mathbf{v}, \mathbf{x}, t) := \\
&\frac{1}{8\pi}\left|-\nabla_{\mathbf{x}}\Psi - \frac{1}{c}\frac{\partial \mathbf{A}}{\partial t} + \frac{1}{c}\mathbf{v}\times \operatorname{curl}_{\mathbf{x}}\mathbf{A}\right|^2 - \frac{1}{8\pi}\left|\operatorname{curl}_{\mathbf{x}}\mathbf{A}\right|^2 - \left(\rho\Psi - \frac{1}{c}\mathbf{A}\cdot\mathbf{j}\right) \\
&\quad + \frac{\mu}{2}|\mathbf{u}-\mathbf{v}|^2 - \frac{c^2}{8\pi G}\left|-\frac{1}{c}\left(\frac{\partial \mathbf{v}}{\partial t} + d_{\mathbf{x}}\mathbf{v}\cdot\mathbf{v}\right)\right|^2
\end{aligned}
$$

$$+\frac{c^2}{8\pi G}|\text{curl}_{\mathbf{x}}\mathbf{v}|^2+\frac{c^2\beta}{4\pi G}\left(\frac{1}{4}\left|d_{\mathbf{x}}\mathbf{v}+\{d_{\mathbf{x}}\mathbf{v}\}^T\right|^2-|\text{div}_{\mathbf{x}}\,\mathbf{v}|^2\right). \quad (8.16)$$

Then, using Proposition 3.1 by (8.12), (8.13) and (8.14), we deduce that L and L_1 are invariant under the change of inertial or non-inertial Cartesian coordinate system, provided that $\mathbf{h}$ and $\mathbf{A}$ are proper vector fields, $\mathbf{v}$ is a speed-like vector field and Φ_0 and $\Psi_0 := \Psi - \frac{1}{c}\mathbf{A}\cdot\mathbf{v}$ are proper scalar fields.

We investigate critical points of the functional

$$J=\int_0^T\int_{\mathbb{R}^3}L_1\left(\mathbf{A},\Psi,\mathbf{v},\mathbf{x},t\right)d\mathbf{x}dt. \quad (8.17)$$

We denote

$$\begin{cases}\Psi_0=\Psi-\frac{1}{c}\mathbf{A}\cdot\mathbf{v}\\ \mathbf{D}=-\nabla_{\mathbf{x}}\Psi-\frac{1}{c}\frac{\partial\mathbf{A}}{\partial t}+\frac{1}{c}\mathbf{v}\times\text{curl}_{\mathbf{x}}\mathbf{A}\\ =-\nabla_{\mathbf{x}}\Psi_0-\frac{1}{c}\frac{\partial\mathbf{A}}{\partial t}+\frac{1}{c}\mathbf{v}\times\text{curl}_{\mathbf{x}}\mathbf{A}-\frac{1}{c}\nabla_{\mathbf{x}}\left(\mathbf{A}\cdot\mathbf{v}\right)\\ \mathbf{B}=\text{curl}_{\mathbf{x}}\mathbf{A}\\ \mathbf{E}=-\nabla_{\mathbf{x}}\Psi-\frac{1}{c}\frac{\partial\mathbf{A}}{\partial t}=\mathbf{D}-\frac{1}{c}\mathbf{v}\times\mathbf{B}\\ \mathbf{H}=\text{curl}_{\mathbf{x}}\mathbf{A}+\frac{1}{c}\mathbf{v}\times\left(-\nabla_{\mathbf{x}}\Psi-\frac{1}{c}\frac{\partial\mathbf{A}}{\partial t}+\frac{1}{c}\mathbf{v}\times\text{curl}_{\mathbf{x}}\mathbf{A}\right)=\mathbf{B}+\frac{1}{c}\mathbf{v}\times\mathbf{D}\end{cases} \quad (8.18)$$

and

$$\begin{cases}\mathbf{R}=-\nabla_{\mathbf{x}}\Phi_0-\frac{1}{c}\frac{\partial\mathbf{h}}{\partial t}+\frac{1}{c}\mathbf{v}\times\text{curl}_{\mathbf{x}}\mathbf{h}-\frac{1}{c}\nabla_{\mathbf{x}}\left(\mathbf{h}\cdot\mathbf{v}\right)\\ \mathbf{Q}=\text{curl}_{\mathbf{x}}\mathbf{h},\end{cases} \quad (8.19)$$

where $\mathbf{h}$ is a proper vector field and Φ_0 is a proper scalar field that are given by (8.13). In other words,

$$\begin{cases}\mathbf{R}=\frac{1}{c}\nabla_{\mathbf{x}}\left(\frac{1}{2}\left|\mathbf{v}-\mathbf{k}\right|^2\right)-\frac{1}{c}\frac{\partial}{\partial t}\left(\mathbf{v}-\mathbf{k}\right)+\frac{1}{c}\mathbf{v}\times\text{curl}_{\mathbf{x}}\left(\mathbf{v}-\mathbf{k}\right)-\frac{1}{c}\nabla_{\mathbf{x}}\left(\left(\mathbf{v}-\mathbf{k}\right)\cdot\mathbf{v}\right)\\ \mathbf{Q}=\text{curl}_{\mathbf{x}}(\mathbf{v}-\mathbf{k}),\end{cases} \quad (8.20)$$

and in inertial coordinate system where $d_{\mathbf{x}}\mathbf{k}=0$ and $\partial_t\mathbf{k}=0$, we also have

$$\begin{cases}\mathbf{R}=-\frac{1}{c}\left(\frac{\partial\mathbf{v}}{\partial t}+d_{\mathbf{x}}\mathbf{v}\cdot\mathbf{v}\right)\\ \mathbf{Q}=\text{curl}_{\mathbf{x}}\mathbf{v}.\end{cases} \quad (8.21)$$

As before, by (8.18) and (8.19) and Proposition 3.1 we infer that both $\mathbf{D}$, $\mathbf{B}$ and $\mathbf{R}$, $\mathbf{Q}$ are proper vector fields. Next, by (8.18), we have

$$\begin{cases}\text{curl}_{\mathbf{x}}\mathbf{D}+\frac{1}{c}\frac{\partial\mathbf{B}}{\partial t}-\frac{1}{c}\text{curl}_{\mathbf{x}}\left(\mathbf{v}\times\mathbf{B}\right)=\text{curl}_{\mathbf{x}}\mathbf{E}+\frac{1}{c}\frac{\partial\mathbf{B}}{\partial t}=0\\ \text{div}_{\mathbf{x}}\,\mathbf{B}=0,\end{cases} \quad (8.22)$$

and by (8.19), we have

$$\begin{cases} \mathrm{curl}_{\mathbf{x}}\mathbf{R} + \frac{1}{c}\frac{\partial \mathbf{Q}}{\partial t} - \frac{1}{c}\,\mathrm{curl}_{\mathbf{x}}\,(\mathbf{v}\times\mathbf{Q}) = 0 \\ \mathrm{div}_{\mathbf{x}}\,\mathbf{Q} = 0. \end{cases} \tag{8.23}$$

Furthermore, by (8.20) and (2.15), we deduce

$$\mathbf{R} = \frac{1}{c}\nabla_{\mathbf{x}}\left(\frac{1}{2}|\mathbf{v}-\mathbf{k}|^2\right) - \frac{1}{c}\{d_{\mathbf{x}}\mathbf{v}\}^T\cdot(\mathbf{v}-\mathbf{k}) - \frac{1}{c}\frac{\partial}{\partial t}(\mathbf{v}-\mathbf{k}) - \frac{1}{c}\{d_{\mathbf{x}}(\mathbf{v}-\mathbf{k})\}^T\cdot\mathbf{v}$$
$$+\frac{1}{c}\mathbf{v}\times\mathrm{curl}_{\mathbf{x}}(\mathbf{v}-\mathbf{k}) = -\frac{1}{c}\{d_{\mathbf{x}}\mathbf{k}\}^T\cdot(\mathbf{v}-\mathbf{k}) - \frac{1}{c}\frac{\partial}{\partial t}(\mathbf{v}-\mathbf{k}) - \frac{1}{c}d_{\mathbf{x}}(\mathbf{v}-\mathbf{k})\cdot\mathbf{v}$$
$$\text{and}\quad \mathbf{Q} = \mathrm{curl}_{\mathbf{x}}(\mathbf{v}-\mathbf{k}), \tag{8.24}$$

and thus, since $d_{\mathbf{x}}\mathbf{k} + \{d_{\mathbf{x}}\mathbf{k}\}^T = 0$, $\mathrm{curl}_{\mathbf{x}}(\mathrm{curl}_{\mathbf{x}}\mathbf{k}) = 0$ and $\mathrm{div}_{\mathbf{x}}\,\mathbf{k} = 0$, we infer

$$\mathrm{div}_{\mathbf{x}}\,\mathbf{R} = -\frac{1}{c}(\mathbf{v}-\mathbf{k})\cdot\Delta_{\mathbf{x}}\mathbf{k} - \frac{1}{c}\mathrm{tr}\left(\{d_{\mathbf{x}}\mathbf{k}\}^T\cdot d_{\mathbf{x}}(\mathbf{v}-\mathbf{k})\right)$$
$$-\frac{1}{c}\frac{\partial}{\partial t}(\mathrm{div}_{\mathbf{x}}\,\mathbf{v}) - \frac{1}{c}d_{\mathbf{x}}(\mathrm{div}_{\mathbf{x}}\,\mathbf{v})\cdot\mathbf{v} - \frac{1}{c}\mathrm{tr}\left(d_{\mathbf{x}}(\mathbf{v}-\mathbf{k})\cdot d_{\mathbf{x}}\mathbf{v}\right)$$
$$= -\frac{1}{c}\frac{\partial}{\partial t}(\mathrm{div}_{\mathbf{x}}\,\mathbf{v}) - \frac{1}{c}d_{\mathbf{x}}(\mathrm{div}_{\mathbf{x}}\,\mathbf{v})\cdot\mathbf{v} - \frac{1}{c}\mathrm{tr}\left(d_{\mathbf{x}}(\mathbf{v}-\mathbf{k})\cdot d_{\mathbf{x}}(\mathbf{v}-\mathbf{k})\right)$$
$$= -\frac{1}{c}\frac{\partial}{\partial t}(\mathrm{div}_{\mathbf{x}}\,\mathbf{v}) - \frac{1}{c}d_{\mathbf{x}}(\mathrm{div}_{\mathbf{x}}\,\mathbf{v})\cdot\mathbf{v} - \frac{1}{4c}\left|d_{\mathbf{x}}(\mathbf{v}-\mathbf{k}) + \{d_{\mathbf{x}}(\mathbf{v}-\mathbf{k})\}^T\right|^2$$
$$+\frac{1}{4c}\left|d_{\mathbf{x}}(\mathbf{v}-\mathbf{k}) - \{d_{\mathbf{x}}(\mathbf{v}-\mathbf{k})\}^T\right|^2 =$$
$$-\frac{1}{c}\frac{\partial}{\partial t}(\mathrm{div}_{\mathbf{x}}\,\mathbf{v}) - \frac{1}{c}d_{\mathbf{x}}(\mathrm{div}_{\mathbf{x}}\,\mathbf{v})\cdot\mathbf{v} - \frac{1}{4c}\left|d_{\mathbf{x}}\mathbf{v} + \{d_{\mathbf{x}}\mathbf{v}\}^T\right|^2 + \frac{1}{2c}|\mathbf{Q}|^2$$
$$\text{and}\quad \mathrm{curl}_{\mathbf{x}}\mathbf{Q} = \mathrm{curl}_{\mathbf{x}}(\mathrm{curl}_{\mathbf{x}}\mathbf{v}). \tag{8.25}$$

So,

$$\frac{1}{c}\left(\frac{\partial}{\partial t}(\mathrm{div}_{\mathbf{x}}\,\mathbf{v}) + d_{\mathbf{x}}(\mathrm{div}_{\mathbf{x}}\,\mathbf{v})\cdot\mathbf{v} + \frac{1}{4}\left|d_{\mathbf{x}}\mathbf{v} + \{d_{\mathbf{x}}\mathbf{v}\}^T\right|^2\right) = \frac{1}{2c}|\mathbf{Q}|^2 - \mathrm{div}_{\mathbf{x}}\,\mathbf{R}$$
$$\text{and}\quad \mathrm{curl}_{\mathbf{x}}(\mathrm{curl}_{\mathbf{x}}\mathbf{v}) = \mathrm{curl}_{\mathbf{x}}\mathbf{Q}. \tag{8.26}$$

In other words,

$$\frac{1}{c}\left(\frac{\partial}{\partial t}(\mathrm{div}_{\mathbf{x}}\,\mathbf{v}) + \mathrm{div}_{\mathbf{x}}\{(\mathrm{div}_{\mathbf{x}}\,\mathbf{v})\,\mathbf{v}\} + \frac{1}{4}\left|d_{\mathbf{x}}\mathbf{v} + \{d_{\mathbf{x}}\mathbf{v}\}^T\right|^2 - |\mathrm{div}_{\mathbf{x}}\,\mathbf{v}|^2\right)$$
$$= \frac{1}{2c}|\mathbf{Q}|^2 - \mathrm{div}_{\mathbf{x}}\,\mathbf{R} \quad\text{and}\quad \mathrm{curl}_{\mathbf{x}}(\mathrm{curl}_{\mathbf{x}}\mathbf{v}) = \mathrm{curl}_{\mathbf{x}}\mathbf{Q}. \tag{8.27}$$

Moreover, by (8.12), and (2.5), we have

$$\frac{\delta L}{\delta \mathbf{h}} = \frac{c^2}{4\pi G}\text{curl}_{\mathbf{x}}\mathbf{Q} - \frac{c}{4\pi G}\left(\frac{\partial \mathbf{R}}{\partial t} - \text{curl}_{\mathbf{x}}\left\{\mathbf{v}\times\mathbf{R}\right\} + (\text{div}_{\mathbf{x}}\,\mathbf{R})\,\mathbf{v}\right), \tag{8.28}$$

$$\frac{\delta L}{\delta \Phi_0} = -\frac{c^2}{4\pi G}(\text{div}_{\mathbf{x}}\,\mathbf{R}) \tag{8.29}$$

and

$$\frac{\delta L}{\delta \mathbf{v}} = \\ -\left(\mu\mathbf{u} - \mu\mathbf{v} + \frac{1}{4\pi c}\mathbf{D}\times\mathbf{B} - \frac{c}{4\pi G}\mathbf{R}\times\mathbf{Q}\right) + \frac{c^2\beta}{4\pi G}\text{curl}_{\mathbf{x}}(\text{curl}_{\mathbf{x}}\mathbf{v}) - \frac{c}{4\pi G}(\text{div}_{\mathbf{x}}\,\mathbf{R})\,\mathbf{h}. \tag{8.30}$$

Thus, since by (8.13), we have

$$L_1(\mathbf{A}, \Psi, \mathbf{v}, \mathbf{x}, t) = L\left(\mathbf{A}, \Psi, \mathbf{v}, -\frac{1}{2c}|\mathbf{v}-\mathbf{k}|^2, (\mathbf{v}-\mathbf{k}), \mathbf{x}, t\right) = L(\mathbf{A}, \Psi, \mathbf{v}, \Phi_0, \mathbf{h}, \mathbf{x}, t), \tag{8.31}$$

by Chain rule, we have

$$\frac{\delta L_1}{\delta \mathbf{v}}(\mathbf{A}, \Psi, \mathbf{v}, \mathbf{x}, t) = \frac{\delta L}{\delta \mathbf{v}} + \frac{\delta L}{\delta \mathbf{h}} - \frac{1}{c}\frac{\delta L}{\delta \Phi_0}(\mathbf{v}-\mathbf{k}). \tag{8.32}$$

Therefore, using (8.28), (8.29), (8.30) in (8.32), we deduce

$$\left(\mu\mathbf{u} - \mu\mathbf{v} + \frac{1}{4\pi c}\mathbf{D}\times\mathbf{B} - \frac{c}{4\pi G}\mathbf{R}\times\mathbf{Q}\right) = \\ \frac{c^2\beta}{4\pi G}\text{curl}_{\mathbf{x}}(\text{curl}_{\mathbf{x}}\mathbf{v}) + \frac{c^2}{4\pi G}\text{curl}_{\mathbf{x}}\mathbf{Q} - \frac{c}{4\pi G}\left(\frac{\partial \mathbf{R}}{\partial t} - \text{curl}_{\mathbf{x}}\left\{\mathbf{v}\times\mathbf{R}\right\} + (\text{div}_{\mathbf{x}}\,\mathbf{R})\,\mathbf{v}\right). \tag{8.33}$$

Moreover,

$$\frac{\delta L}{\delta \Psi} = \frac{1}{4\pi}\text{div}_{\mathbf{x}}\,\mathbf{D} - \rho = 0, \tag{8.34}$$

and

$$\frac{\delta L}{\delta \mathbf{A}} = \frac{1}{c}\mathbf{j} + \frac{1}{4\pi c}\frac{\partial \mathbf{D}}{\partial t} - \frac{1}{4\pi}\text{curl}_{\mathbf{x}}\mathbf{B} - \frac{1}{4\pi c}\text{curl}_{\mathbf{x}}(\mathbf{v}\times\mathbf{D}) = \frac{1}{c}\mathbf{j} + \frac{1}{4\pi c}\frac{\partial \mathbf{D}}{\partial t} - \frac{1}{4\pi}\text{curl}_{\mathbf{x}}\mathbf{H} = 0. \tag{8.35}$$

So,

$$
\begin{cases}
\mathrm{curl}_{\mathbf{x}}\mathbf{B} = \frac{4\pi}{c}\mathbf{j} + \frac{1}{c}\frac{\partial \mathbf{D}}{\partial t} - \frac{1}{c}\mathrm{curl}_{\mathbf{x}}\,(\mathbf{v}\times\mathbf{D})\\
\mathrm{div}_{\mathbf{x}}\,\mathbf{D} = 4\pi\rho\\
\mathrm{curl}_{\mathbf{x}}\mathbf{D} + \frac{1}{c}\frac{\partial \mathbf{B}}{\partial t} - \frac{1}{c}\,\mathrm{curl}_{\mathbf{x}}\,(\mathbf{v}\times\mathbf{B}) = 0\\
\mathrm{div}_{\mathbf{x}}\,\mathbf{B} = 0\\
\mathrm{curl}_{\mathbf{x}}\mathbf{R} + \frac{1}{c}\frac{\partial \mathbf{Q}}{\partial t} - \frac{1}{c}\mathrm{curl}_{\mathbf{x}}\,(\mathbf{v}\times\mathbf{Q}) = 0\\
\mathrm{div}_{\mathbf{x}}\,\mathbf{Q} = 0\\
\frac{1}{c}\left(\frac{\partial}{\partial t}\,(\mathrm{div}_{\mathbf{x}}\,\mathbf{v}) + \mathbf{v}\cdot\nabla_{\mathbf{x}}\,(\mathrm{div}_{\mathbf{x}}\,\mathbf{v}) + \frac{1}{4}\left|d_{\mathbf{x}}\mathbf{v} + \{d_{\mathbf{x}}\mathbf{v}\}^T\right|^2\right) = \frac{1}{2c}\,|\mathbf{Q}|^2 - \mathrm{div}_{\mathbf{x}}\,\mathbf{R}\\
\mathrm{curl}_{\mathbf{x}}(\mathrm{curl}_{\mathbf{x}}\mathbf{v}) = \mathrm{curl}_{\mathbf{x}}\mathbf{Q}\\
\frac{4\pi G}{c^2}\left(\mu\mathbf{u} - \mu\mathbf{v} + \frac{1}{4\pi c}\mathbf{D}\times\mathbf{B} - \frac{c}{4\pi G}\mathbf{R}\times\mathbf{Q}\right) =\\
(1+\beta)\,\mathrm{curl}_{\mathbf{x}}\mathbf{Q} - \frac{1}{c}\left(\frac{\partial \mathbf{R}}{\partial t} - \mathrm{curl}_{\mathbf{x}}\,\{\mathbf{v}\times\mathbf{R}\} + (\mathrm{div}_{\mathbf{x}}\,\mathbf{R})\,\mathbf{v}\right),
\end{cases}
\tag{8.36}
$$

and by (8.21) in the inertial frame, we have

$$
\begin{cases}
\mathrm{curl}_{\mathbf{x}}\mathbf{B} = \frac{4\pi}{c}\mathbf{j} + \frac{1}{c}\frac{\partial \mathbf{D}}{\partial t} - \frac{1}{c}\mathrm{curl}_{\mathbf{x}}\,(\mathbf{v}\times\mathbf{D})\\
\mathrm{div}_{\mathbf{x}}\,\mathbf{D} = 4\pi\rho\\
\mathrm{curl}_{\mathbf{x}}\mathbf{D} + \frac{1}{c}\frac{\partial \mathbf{B}}{\partial t} - \frac{1}{c}\,\mathrm{curl}_{\mathbf{x}}\,(\mathbf{v}\times\mathbf{B}) = 0\\
\mathrm{div}_{\mathbf{x}}\,\mathbf{B} = 0\\
\mathbf{R} = -\frac{1}{c}\left(\frac{\partial \mathbf{v}}{\partial t} + d_{\mathbf{x}}\mathbf{v}\cdot\mathbf{v}\right)\\
\mathbf{Q} = \mathrm{curl}_{\mathbf{x}}\mathbf{v}\\
\frac{4\pi G}{c^2}\left(\mu\mathbf{u} - \mu\mathbf{v} + \frac{1}{4\pi c}\mathbf{D}\times\mathbf{B} - \frac{c}{4\pi G}\mathbf{R}\times\mathbf{Q}\right) =\\
(1+\beta)\,\mathrm{curl}_{\mathbf{x}}\mathbf{Q} - \frac{1}{c}\left(\frac{\partial \mathbf{R}}{\partial t} - \mathrm{curl}_{\mathbf{x}}\,\{\mathbf{v}\times\mathbf{R}\} + (\mathrm{div}_{\mathbf{x}}\,\mathbf{R})\,\mathbf{v}\right).
\end{cases}
\tag{8.37}
$$

Furthermore, taking $\mathrm{div}_{\mathbf{x}}$ of the both sides of the last equality in (8.36) and using continuum equation $\partial_t\mu + \mathrm{div}_{\mathbf{x}}\,(\mu\mathbf{u}) = 0$ we deduce

$$
\begin{aligned}
-\left(\partial_t\mu + \mathrm{div}_{\mathbf{x}}\,(\mu\mathbf{v})\right) + \mathrm{div}_{\mathbf{x}}\left\{\frac{1}{4\pi c}\mathbf{D}\times\mathbf{B} - \frac{c}{4\pi G}\mathbf{R}\times\mathbf{Q}\right\} &=\\
\mathrm{div}_{\mathbf{x}}\left(\mu\mathbf{u} - \mu\mathbf{v} + \frac{1}{4\pi c}\mathbf{D}\times\mathbf{B} - \frac{c}{4\pi G}\mathbf{R}\times\mathbf{Q}\right) &=\\
-\frac{c}{4\pi G}\left(\frac{\partial}{\partial t}\,(\mathrm{div}_{\mathbf{x}}\,\mathbf{R}) + \mathrm{div}_{\mathbf{x}}\,\{(\mathrm{div}_{\mathbf{x}}\,\mathbf{R})\,\mathbf{v}\}\right)&.
\end{aligned}
\tag{8.38}
$$

Therefore, considering the proper scalar quantity Q_0, that we call the field mass, which satisfies

$$
Q_0 := -\mu + \frac{c}{4\pi G}\,\mathrm{div}_{\mathbf{x}}\,\mathbf{R},
\tag{8.39}
$$

by (8.38), we deduce

$$
\frac{\partial Q_0}{\partial t} + \mathrm{div}_{\mathbf{x}}\,(Q_0\mathbf{v}) = -\,\mathrm{div}_{\mathbf{x}}\left\{\frac{1}{4\pi c}\mathbf{D}\times\mathbf{B} - \frac{c}{4\pi G}\mathbf{R}\times\mathbf{Q}\right\}.
\tag{8.40}
$$

Thus, we rewrite (8.36) as

$$\begin{cases} \mathrm{curl}_{\mathbf{x}}\mathbf{B} = \frac{4\pi}{c}\left(\mathbf{j} - \rho\mathbf{v}\right) + \frac{1}{c}\frac{\partial \mathbf{D}}{\partial t} - \frac{1}{c}\mathrm{curl}_{\mathbf{x}}\left(\mathbf{v}\times\mathbf{D}\right) + \frac{1}{c}\left(\mathrm{div}_{\mathbf{x}}\mathbf{D}\right)\mathbf{v}, \\ \mathrm{div}_{\mathbf{x}}\mathbf{D} = 4\pi\rho, \\ \mathrm{curl}_{\mathbf{x}}\mathbf{D} + \frac{1}{c}\frac{\partial \mathbf{B}}{\partial t} - \frac{1}{c}\mathrm{curl}_{\mathbf{x}}\left(\mathbf{v}\times\mathbf{B}\right) = 0, \\ \mathrm{div}_{\mathbf{x}}\mathbf{B} = 0, \\ \mathrm{curl}_{\mathbf{x}}\mathbf{R} + \frac{1}{c}\frac{\partial \mathbf{Q}}{\partial t} - \frac{1}{c}\mathrm{curl}_{\mathbf{x}}\left(\mathbf{v}\times\mathbf{Q}\right) = 0, \\ \mathrm{div}_{\mathbf{x}}\mathbf{Q} = 0, \\ \frac{4\pi G}{c^2}\left(\mu\mathbf{u} - \mu\mathbf{v} + \frac{1}{4\pi c}\mathbf{D}\times\mathbf{B} - \frac{c}{4\pi G}\mathbf{R}\times\mathbf{Q}\right) = \\ (1+\beta)\,\mathrm{curl}_{\mathbf{x}}\mathbf{Q} - \frac{1}{c}\left(\frac{\partial \mathbf{R}}{\partial t} - \mathrm{curl}_{\mathbf{x}}\left\{\mathbf{v}\times\mathbf{R}\right\} + \left(\mathrm{div}_{\mathbf{x}}\mathbf{R}\right)\mathbf{v}\right), \\ \mathrm{div}_{\mathbf{x}}\mathbf{R} = \frac{4\pi G}{c}\left(\mu + Q_0\right), \\ \frac{1}{c}\left(\frac{\partial}{\partial t}\left(\mathrm{div}_{\mathbf{x}}\mathbf{v}\right) + \mathbf{v}\cdot\nabla_{\mathbf{x}}\left(\mathrm{div}_{\mathbf{x}}\mathbf{v}\right) + \frac{1}{4}\left|d_{\mathbf{x}}\mathbf{v} + \{d_{\mathbf{x}}\mathbf{v}\}^T\right|^2\right) = \frac{1}{2c}\left|\mathbf{Q}\right|^2 - \mathrm{div}_{\mathbf{x}}\mathbf{R}, \\ \mathrm{curl}_{\mathbf{x}}(\mathrm{curl}_{\mathbf{x}}\mathbf{v}) = \mathrm{curl}_{\mathbf{x}}\mathbf{Q}, \\ \frac{\partial Q_0}{\partial t} + \mathrm{div}_{\mathbf{x}}\left(Q_0\mathbf{v}\right) = -\,\mathrm{div}_{\mathbf{x}}\left\{\frac{1}{4\pi c}\mathbf{D}\times\mathbf{B} - \frac{c}{4\pi G}\mathbf{R}\times\mathbf{Q}\right\}, \end{cases} \tag{8.41}$$

and we rewrite (8.37) in the inertial frame as

$$\begin{cases} \mathrm{curl}_{\mathbf{x}}\mathbf{B} = \frac{4\pi}{c}\mathbf{j} + \frac{1}{c}\frac{\partial \mathbf{D}}{\partial t} - \frac{1}{c}\mathrm{curl}_{\mathbf{x}}\left(\mathbf{v}\times\mathbf{D}\right), \\ \mathrm{div}_{\mathbf{x}}\mathbf{D} = 4\pi\rho, \\ \mathrm{curl}_{\mathbf{x}}\mathbf{D} + \frac{1}{c}\frac{\partial \mathbf{B}}{\partial t} - \frac{1}{c}\mathrm{curl}_{\mathbf{x}}\left(\mathbf{v}\times\mathbf{B}\right) = 0, \\ \mathrm{div}_{\mathbf{x}}\mathbf{B} = 0, \\ \mathbf{R} = -\frac{1}{c}\left(\frac{\partial \mathbf{v}}{\partial t} + d_{\mathbf{x}}\mathbf{v}\cdot\mathbf{v}\right), \\ \mathbf{Q} = \mathrm{curl}_{\mathbf{x}}\mathbf{v}, \\ \frac{4\pi G}{c^2}\left(\mu\mathbf{u} - \mu\mathbf{v} + \frac{1}{4\pi c}\mathbf{D}\times\mathbf{B} - \frac{c}{4\pi G}\mathbf{R}\times\mathbf{Q}\right) = \\ (1+\beta)\,\mathrm{curl}_{\mathbf{x}}\mathbf{Q} - \frac{1}{c}\left(\frac{\partial \mathbf{R}}{\partial t} - \mathrm{curl}_{\mathbf{x}}\left\{\mathbf{v}\times\mathbf{R}\right\} + \left(\mathrm{div}_{\mathbf{x}}\mathbf{R}\right)\mathbf{v}\right), \\ \mathrm{div}_{\mathbf{x}}\mathbf{R} = \frac{4\pi G}{c}\left(\mu + Q_0\right), \\ \frac{\partial Q_0}{\partial t} + \mathrm{div}_{\mathbf{x}}\left(Q_0\mathbf{v}\right) = -\,\mathrm{div}_{\mathbf{x}}\left\{\frac{1}{4\pi c}\mathbf{D}\times\mathbf{B} - \frac{c}{4\pi G}\mathbf{R}\times\mathbf{Q}\right\}. \end{cases} \tag{8.42}$$

As before, by Proposition 3.1, we deduce that (8.41) is invariant under the change of inertial or non-inertial Cartesian coordinate systems. Moreover, (8.42) is invariant under the change of inertial Cartesian coordinate systems. Note also that, both, (8.41) in an arbitrary inertial or non-inertial Cartesian coordinate system and (8.42) in an arbitrary inertial Cartesian coordinate system, are completely independent on the vectorial potential of the inertia **k**.

Finally, note that, in the case of large constant $|\beta| \gg 1$, we have $\mathbf{Q} \to 0$ in (8.41) and thus, the gravity Eq. (8.41) reduce to the equations of the Newtonian-type gravity in the form of (8.11). In that case, the gravity field propagates with the infinite speed. On the other hand, in the case of vanishing constant $\beta = 0$, the form of equations for **R** and **Q** in (8.41) is completely the same as the form of the Maxwell equations for **D** and **B** in (8.41), except of the different meaning of "charges" and "currents" in these two sets of equations. In that case, the electromagnetic and the gravity fields propagate with the same speed. However, in the mixed case of constant $\beta \sim 1$, the electromagnetic and the gravity fields propagate with different finite speeds.

Remark 8.1 One can wonder: what should be possible values of the vectorial gravitational potential $\mathbf{v}$ in the proximity of the Earth or another massive body? In Remark 4.4, we answered this question in the case of the Newtonian-type gravity, given by (4.49). In order to answer this question in the case of more general laws of the gravity, given by (8.41) with arbitrary constant β, consider two Cartesian coordinate systems: non-rotating system $(*)$ with the center that coincides with the center of masses of the Earth and inertial system $(**)$ related to some external cosmic bodies. Assume that the center of masses of the Earth has place $\mathbf{R}(t')$ and velocity $\mathbf{W}(t') := \frac{d\mathbf{R}}{dt'}(t')$ in the coordinate system $(**)$. Thus, the change in coordinate system $(*)$ to coordinate system $(**)$ is given by

$$\begin{cases} \mathbf{x}' = \mathbf{x} + \mathbf{R}(t), \\ t' = t, \end{cases} \tag{8.43}$$

and the vectorial gravitational potential $\mathbf{v}$, being a speed like vector field, transforms as

$$\mathbf{v}' = \mathbf{v} + \mathbf{W}(t). \tag{8.44}$$

Next, since the system $(**)$ is inertial, consistently with (8.42) in the system $(**)$ we have

$$\begin{cases} \operatorname{curl}_{\mathbf{x}'} \mathbf{B}' = \frac{4\pi}{c}\left(\mathbf{j}' - \rho'\mathbf{v}'\right) + \frac{1}{c}\frac{\partial \mathbf{D}'}{\partial t'} - \frac{1}{c}\operatorname{curl}_{\mathbf{x}'}\left(\mathbf{v}' \times \mathbf{D}'\right) + \frac{1}{c}\left(\operatorname{div}_{\mathbf{x}'} \mathbf{D}'\right)\mathbf{v}', \\ \operatorname{div}_{\mathbf{x}} \mathbf{D}' = 4\pi\rho', \\ \operatorname{curl}_{\mathbf{x}'} \mathbf{D}' + \frac{1}{c}\frac{\partial \mathbf{B}'}{\partial t'} - \frac{1}{c}\operatorname{curl}_{\mathbf{x}'}\left(\mathbf{v}' \times \mathbf{B}'\right) = 0, \\ \operatorname{div}_{\mathbf{x}'} \mathbf{B}' = 0, \\ \mathbf{R}' = -\frac{1}{c}\left(\frac{\partial \mathbf{v}'}{\partial t'} + d_{\mathbf{x}'}\mathbf{v}' \cdot \mathbf{v}'\right), \\ \mathbf{Q}' = \operatorname{curl}_{\mathbf{x}'} \mathbf{v}', \\ \operatorname{curl}_{\mathbf{x}'} \mathbf{R}' + \frac{1}{c}\frac{\partial \mathbf{Q}'}{\partial t'} - \frac{1}{c}\operatorname{curl}_{\mathbf{x}'}\left(\mathbf{v}' \times \mathbf{Q}'\right) = 0, \\ \operatorname{div}_{\mathbf{x}'} \mathbf{Q}' = 0, \\ \frac{4\pi G}{c^2}\left(\mu'\mathbf{u}' - \mu'\mathbf{v}' + \frac{1}{4\pi c}\mathbf{D}' \times \mathbf{B}' - \frac{c}{4\pi G}\mathbf{R}' \times \mathbf{Q}'\right) = \\ (1+\beta)\operatorname{curl}_{\mathbf{x}'} \mathbf{Q}' - \frac{1}{c}\left(\frac{\partial \mathbf{R}'}{\partial t'} - \operatorname{curl}_{\mathbf{x}'}\left\{\mathbf{v}' \times \mathbf{R}'\right\} + \left(\operatorname{div}_{\mathbf{x}} \mathbf{R}'\right)\mathbf{v}'\right), \\ \operatorname{div}_{\mathbf{x}'} \mathbf{R}' = \frac{4\pi G}{c}(\mu' + Q'_0), \\ \frac{\partial Q'_0}{\partial t'} + \operatorname{div}_{\mathbf{x}'}\left(Q'_0\mathbf{v}'\right) = -\operatorname{div}_{\mathbf{x}'}\left\{\frac{1}{4\pi c}\mathbf{D}' \times \mathbf{B}' - \frac{c}{4\pi G}\mathbf{R}' \times \mathbf{Q}'\right\}, \end{cases} \tag{8.45}$$

and by (8.41) in the system $(*)$, we have

$$\begin{cases} \mathrm{curl}_{\mathbf{x}}\mathbf{B} = \frac{4\pi}{c}\left(\mathbf{j} - \rho\mathbf{v}\right) + \frac{1}{c}\frac{\partial \mathbf{D}}{\partial t} - \frac{1}{c}\mathrm{curl}_{\mathbf{x}}\left(\mathbf{v}\times\mathbf{D}\right) + \frac{1}{c}\left(\mathrm{div}_{\mathbf{x}}\,\mathbf{D}\right)\mathbf{v},\\ \mathrm{div}_{\mathbf{x}}\,\mathbf{D} = 4\pi\rho,\\ \mathrm{curl}_{\mathbf{x}}\mathbf{D} + \frac{1}{c}\frac{\partial \mathbf{B}}{\partial t} - \frac{1}{c}\,\mathrm{curl}_{\mathbf{x}}\left(\mathbf{v}\times\mathbf{B}\right) = 0,\\ \mathrm{div}_{\mathbf{x}}\,\mathbf{B} = 0,\\ \mathrm{curl}_{\mathbf{x}}\mathbf{R} + \frac{1}{c}\frac{\partial \mathbf{Q}}{\partial t} - \frac{1}{c}\mathrm{curl}_{\mathbf{x}}\left(\mathbf{v}\times\mathbf{Q}\right) = 0,\\ \mathrm{div}_{\mathbf{x}}\,\mathbf{Q} = 0,\\ \frac{4\pi G}{c^2}\left(\mu\mathbf{u} - \mu\mathbf{v} + \frac{1}{4\pi c}\mathbf{D}\times\mathbf{B} - \frac{c}{4\pi G}\mathbf{R}\times\mathbf{Q}\right) =\\ (1+\beta)\,\mathrm{curl}_{\mathbf{x}}\mathbf{Q} - \frac{1}{c}\left(\frac{\partial \mathbf{R}}{\partial t} - \mathrm{curl}_{\mathbf{x}}\left\{\mathbf{v}\times\mathbf{R}\right\} + \left(\mathrm{div}_{\mathbf{x}}\,\mathbf{R}\right)\mathbf{v}\right),\\ \mathrm{div}_{\mathbf{x}}\,\mathbf{R} = \frac{4\pi G}{c}\left(\mu + Q_0\right),\\ \frac{1}{c}\left(\frac{\partial}{\partial t}\left(\mathrm{div}_{\mathbf{x}}\,\mathbf{v}\right) + \mathbf{v}\cdot\nabla_{\mathbf{x}}\left(\mathrm{div}_{\mathbf{x}}\,\mathbf{v}\right) + \frac{1}{4}\left|d_{\mathbf{x}}\mathbf{v} + \left\{d_{\mathbf{x}}\mathbf{v}\right\}^T\right|^2\right) = \frac{1}{2c}\left|\mathbf{Q}\right|^2 - \mathrm{div}_{\mathbf{x}}\,\mathbf{R},\\ \mathrm{curl}_{\mathbf{x}}(\mathrm{curl}_{\mathbf{x}}\mathbf{v}) = \mathrm{curl}_{\mathbf{x}}\mathbf{Q},\\ \frac{\partial Q_0}{\partial t} + \mathrm{div}_{\mathbf{x}}\left(Q_0\mathbf{v}\right) = -\,\mathrm{div}_{\mathbf{x}}\left\{\frac{1}{4\pi c}\mathbf{D}\times\mathbf{B} - \frac{c}{4\pi G}\mathbf{R}\times\mathbf{Q}\right\}. \end{cases} \tag{8.46}$$

On the other hand, by (8.43) and (8.44) and using Proposition 3.1, we deduce

$$\mathrm{curl}_{\mathbf{x}}\mathbf{v} = \mathrm{curl}_{\mathbf{x}'}\mathbf{v}', \tag{8.47}$$

and

$$\frac{d\mathbf{W}}{\partial t}(t) + \frac{\partial \mathbf{v}}{\partial t} + d_{\mathbf{x}}\mathbf{v}\cdot\mathbf{v} = \frac{d\mathbf{W}}{\partial t'}(t') + \frac{\partial \mathbf{v}}{\partial t'} + d_{\mathbf{x}'}\mathbf{v}\cdot\mathbf{W}(t') + d_{\mathbf{x}'}\mathbf{v}\cdot\mathbf{v} = \frac{\partial \mathbf{v}'}{\partial t'} + d_{\mathbf{x}'}\mathbf{v}'\cdot\mathbf{v}'. \tag{8.48}$$

Thus, since **R** is a proper field, by (8.45), (8.47) and (8.48), we deduce

$$\mathrm{curl}_{\mathbf{x}}\mathbf{v} = \mathbf{Q}' = \mathbf{Q}, \tag{8.49}$$

and

$$\frac{d\mathbf{W}}{\partial t}(t) + \frac{\partial \mathbf{v}}{\partial t} + d_{\mathbf{x}}\mathbf{v}\cdot\mathbf{v} = -c\,\mathbf{R}' = -c\,\mathbf{R}. \tag{8.50}$$

Therefore, by (8.46), (8.49) and (8.50) in the system (∗), we have

$$\begin{cases} \operatorname{curl}_{\mathbf{x}}\mathbf{B} = \frac{4\pi}{c}\mathbf{j} + \frac{1}{c}\frac{\partial \mathbf{D}}{\partial t} - \frac{1}{c}\operatorname{curl}_{\mathbf{x}}(\mathbf{v}\times\mathbf{D}), \\ \operatorname{div}_{\mathbf{x}}\mathbf{D} = 4\pi\rho, \\ \operatorname{curl}_{\mathbf{x}}\mathbf{D} + \frac{1}{c}\frac{\partial \mathbf{B}}{\partial t} - \frac{1}{c}\operatorname{curl}_{\mathbf{x}}(\mathbf{v}\times\mathbf{B}) = 0, \\ \operatorname{div}_{\mathbf{x}}\mathbf{B} = 0, \\ \frac{1}{c}\frac{d\mathbf{W}}{\partial t}(t) + \mathbf{R} = -\frac{1}{c}\left(\frac{\partial \mathbf{v}}{\partial t} + d_{\mathbf{x}}\mathbf{v}\cdot\mathbf{v}\right), \\ \mathbf{Q} = \operatorname{curl}_{\mathbf{x}}\mathbf{v}, \\ \frac{4\pi G}{c^2}\left(\mu\mathbf{u} - \mu\mathbf{v} + \frac{1}{4\pi c}\mathbf{D}\times\mathbf{B} - \frac{c}{4\pi G}\mathbf{R}\times\mathbf{Q}\right) = \\ (1+\beta)\operatorname{curl}_{\mathbf{x}}\mathbf{Q} - \frac{1}{c}\left(\frac{\partial \mathbf{R}}{\partial t} - \operatorname{curl}_{\mathbf{x}}\{\mathbf{v}\times\mathbf{R}\} + (\operatorname{div}_{\mathbf{x}}\mathbf{R})\,\mathbf{v}\right), \\ \operatorname{div}_{\mathbf{x}}\mathbf{R} = \frac{4\pi G}{c}(\mu + Q_0), \\ \frac{\partial Q_0}{\partial t} + \operatorname{div}_{\mathbf{x}}(Q_0\mathbf{v}) = -\operatorname{div}_{\mathbf{x}}\left\{\frac{1}{4\pi c}\mathbf{D}\times\mathbf{B} - \frac{c}{4\pi G}\mathbf{R}\times\mathbf{Q}\right\}. \end{cases} \tag{8.51}$$

On the other hand, since the system (∗∗) is inertial, the quantity $\frac{d\mathbf{W}}{\partial t}(t)$, being generated by the gravitational field from the far bodies, is insignificant with respect to the quantity $c\mathbf{R}$ in the scale compatible to the Earth size. Thus, we rewrite (8.51) as

$$\begin{cases} \operatorname{curl}_{\mathbf{x}}\mathbf{B} = \frac{4\pi}{c}\mathbf{j} + \frac{1}{c}\frac{\partial \mathbf{D}}{\partial t} - \frac{1}{c}\operatorname{curl}_{\mathbf{x}}(\mathbf{v}\times\mathbf{D}), \\ \operatorname{div}_{\mathbf{x}}\mathbf{D} = 4\pi\rho, \\ \operatorname{curl}_{\mathbf{x}}\mathbf{D} + \frac{1}{c}\frac{\partial \mathbf{B}}{\partial t} - \frac{1}{c}\operatorname{curl}_{\mathbf{x}}(\mathbf{v}\times\mathbf{B}) = 0, \\ \operatorname{div}_{\mathbf{x}}\mathbf{B} = 0, \\ \mathbf{R} = -\frac{1}{c}\left(\frac{\partial \mathbf{v}}{\partial t} + d_{\mathbf{x}}\mathbf{v}\cdot\mathbf{v}\right), \\ \mathbf{Q} = \operatorname{curl}_{\mathbf{x}}\mathbf{v}, \\ \frac{4\pi G}{c^2}\left(\mu\mathbf{u} - \mu\mathbf{v} + \frac{1}{4\pi c}\mathbf{D}\times\mathbf{B} - \frac{c}{4\pi G}\mathbf{R}\times\mathbf{Q}\right) = \\ (1+\beta)\operatorname{curl}_{\mathbf{x}}\mathbf{Q} - \frac{1}{c}\left(\frac{\partial \mathbf{R}}{\partial t} - \operatorname{curl}_{\mathbf{x}}\{\mathbf{v}\times\mathbf{R}\} + (\operatorname{div}_{\mathbf{x}}\mathbf{R})\,\mathbf{v}\right), \\ \operatorname{div}_{\mathbf{x}}\mathbf{R} = \frac{4\pi G}{c}(\mu + Q_0), \\ \frac{\partial Q_0}{\partial t} + \operatorname{div}_{\mathbf{x}}(Q_0\mathbf{v}) = -\operatorname{div}_{\mathbf{x}}\left\{\frac{1}{4\pi c}\mathbf{D}\times\mathbf{B} - \frac{c}{4\pi G}\mathbf{R}\times\mathbf{Q}\right\}. \end{cases} \tag{8.52}$$

Next, we can neglect all the far cosmic body masses expect of the Earth itself and thus we can consider

$$\mu(\mathbf{x}, t) = \mu_1(|\mathbf{x}|) \quad \text{and} \quad \mathbf{u}(\mathbf{x}, t) = 0 \qquad \forall\,(\mathbf{x}, t), \tag{8.53}$$

where $\mu_1 := \mu_1(|\mathbf{x}|)$ is the inertial mass density of the Earth which is assumed to be a radial function such that

$$\mu_1(|\mathbf{x}|) = 0 \quad \text{if} \quad |\mathbf{x}| > r_0, \tag{8.54}$$

where r_0 is the Earth radius. Moreover, we can neglect all the electromagnetical masses and thus we simplify the equations for the Gravity in (8.52) as

$$\begin{cases} \mathbf{R} = -\frac{1}{c}\left(\frac{\partial \mathbf{v}}{\partial t} + d_{\mathbf{x}}\mathbf{v}\cdot\mathbf{v}\right), \\ \mathbf{Q} = \text{curl}_{\mathbf{x}}\mathbf{v}, \\ \frac{4\pi G}{c^2}\left(-\mu_1\left(|\mathbf{x}|\right)\mathbf{v} - \frac{c}{4\pi G}\mathbf{R}\times\mathbf{Q}\right) = \\ (1+\beta)\,\text{curl}_{\mathbf{x}}\mathbf{Q} - \frac{1}{c}\left(\frac{\partial \mathbf{R}}{\partial t} - \text{curl}_{\mathbf{x}}\left\{\mathbf{v}\times\mathbf{R}\right\} + \left(\text{div}_{\mathbf{x}}\,\mathbf{R}\right)\mathbf{v}\right), \\ \text{div}_{\mathbf{x}}\,\mathbf{R} = \frac{4\pi G}{c}\left(\mu_1\left(|\mathbf{x}|\right) + Q_0\right), \\ \frac{\partial Q_0}{\partial t} + \text{div}_{\mathbf{x}}\left(Q_0\mathbf{v}\right) = \text{div}_{\mathbf{x}}\left\{\frac{c}{4\pi G}\mathbf{R}\times\mathbf{Q}\right\}. \end{cases} \tag{8.55}$$

Being in the system (∗) which is stationary with respect to the center of the Earth we look for stationary (i.e. time-independent) solutions of (8.55). Thus, (8.55) implies

$$\begin{cases} \mathbf{R} = -\frac{1}{c}\, d_{\mathbf{x}}\mathbf{v}\cdot\mathbf{v}, \\ \mathbf{Q} = \text{curl}_{\mathbf{x}}\mathbf{v}, \\ \frac{4\pi G}{c^2}\left(-\mu_1\left(|\mathbf{x}|\right)\mathbf{v} - \frac{c}{4\pi G}\mathbf{R}\times\mathbf{Q}\right) = \\ (1+\beta)\,\text{curl}_{\mathbf{x}}\mathbf{Q} - \frac{1}{c}\left(-\text{curl}_{\mathbf{x}}\left\{\mathbf{v}\times\mathbf{R}\right\} + \left(\text{div}_{\mathbf{x}}\,\mathbf{R}\right)\mathbf{v}\right), \\ \text{div}_{\mathbf{x}}\,\mathbf{R} = \frac{4\pi G}{c}\left(\mu_1\left(|\mathbf{x}|\right) + Q_0\right), \\ \text{div}_{\mathbf{x}}\left(Q_0\mathbf{v}\right) = \text{div}_{\mathbf{x}}\left\{\frac{c}{4\pi G}\mathbf{R}\times\mathbf{Q}\right\}. \end{cases} \tag{8.56}$$

On the other hand, by the symmetry considerations of the problem, we look for the solution of (8.56) that satisfies $\mathbf{v}(\mathbf{x}) = \nabla_{\mathbf{x}} Z_0\left(|\mathbf{x}|\right)$ where, again by the symmetry of the problem, the scalar function $Z_0\left(|\mathbf{x}|\right)$ should be radial. In particular, by (8.56), we obtain

$$\mathbf{Q} = \text{curl}_{\mathbf{x}}\mathbf{v} = 0 \tag{8.57}$$

and thus we simplify (8.56) as

$$\begin{cases} \mathbf{R} = -\frac{1}{c}\, d_{\mathbf{x}}\mathbf{v}\cdot\mathbf{v} = -\frac{1}{c}\nabla_{\mathbf{x}}\left(\frac{1}{2}|\mathbf{v}|^2\right), \\ \mathbf{Q} = \text{curl}_{\mathbf{x}}\mathbf{v} = 0, \\ \frac{4\pi G}{c^2}\mu_1\left(|\mathbf{x}|\right)\mathbf{v} = \frac{1}{c}\left(-\text{curl}_{\mathbf{x}}\left\{\mathbf{v}\times\mathbf{R}\right\} + \left(\text{div}_{\mathbf{x}}\,\mathbf{R}\right)\mathbf{v}\right), \\ \text{div}_{\mathbf{x}}\,\mathbf{R} = \frac{4\pi G}{c}\mu_1\left(|\mathbf{x}|\right), \\ \text{curl}_{\mathbf{x}}\mathbf{R} = 0, \\ Q_0 = 0. \end{cases} \tag{8.58}$$

In particular, since $\mathbf{v} = \nabla_{\mathbf{x}} Z_0\left(|\mathbf{x}|\right)$ and $\mathbf{R} = -\frac{1}{c}\nabla_{\mathbf{x}}\left(\frac{1}{2}|\mathbf{v}|^2\right)$ are both gradients of radial functions, we have $\mathbf{v}\times\mathbf{R} = 0$ and thus, we further simplify (8.58) as

$$\begin{cases} \mathbf{R} = -\frac{1}{c}\nabla_{\mathbf{x}}\left(\frac{1}{2}|\mathbf{v}|^2\right), \\ \mathbf{Q} = \text{curl}_{\mathbf{x}}\mathbf{v} = 0, \\ \text{div}_{\mathbf{x}}\,\mathbf{R} = \frac{4\pi G}{c}\mu_1\left(|\mathbf{x}|\right), \\ \text{curl}_{\mathbf{x}}\mathbf{R} = 0, \\ Q_0 = 0. \end{cases} \tag{8.59}$$

However, (8.59) is equivalent to the following:

$$\begin{cases} \Delta_{\mathbf{x}}\left(-\frac{1}{2}|\mathbf{v}|^2\right) = 4\pi G\,\mu_1(|\mathbf{x}|)\,, \\ \mathrm{curl}_{\mathbf{x}}\mathbf{v} = 0, \\ \mathbf{R} = -\frac{1}{c}\nabla_{\mathbf{x}}\left(\frac{1}{2}|\mathbf{v}|^2\right), \\ \mathbf{Q} = 0, \\ Q_0 = 0. \end{cases} \tag{8.60}$$

Therefore, denoting

$$\Phi_1 := -\frac{1}{2}|\mathbf{v}|^2\,, \tag{8.61}$$

we rewrite (8.60) as

$$\begin{cases} \Delta_{\mathbf{x}}\Phi_1 = 4\pi G\,\mu_1(|\mathbf{x}|)\,, \\ \Phi_1 := -\frac{1}{2}|\mathbf{v}|^2\,, \\ \mathrm{curl}_{\mathbf{x}}\mathbf{v} = 0, \\ \mathbf{R} = -\frac{1}{c}\nabla_{\mathbf{x}}\left(\frac{1}{2}|\mathbf{v}|^2\right), \\ \mathbf{Q} = 0, \\ Q_0 = 0, \end{cases} \tag{8.62}$$

where the radial scalar field $\Phi_1 := \Phi_1(|\mathbf{x}|)$, coincides with the usual Newtonian potential of the Earth, and outside of the Earth surface, it is known that $\Phi_1(\mathbf{x}) = -\frac{Gm_0}{|\mathbf{x}|}$, where m_0 is the total mass of the Earth (we have $m_0 = \iiint_{|\mathbf{x}|\le r_0}\mu_1(|\mathbf{x}|)\,d\mathbf{x}$). Thus, since there exists a scalar radial field $Z_0(\mathbf{x})$ such that $\mathbf{v}(\mathbf{x}) = \nabla_{\mathbf{x}}Z_0(|\mathbf{x}|)$ by (8.62), we obtain

$$\left|\frac{dZ_0}{d(|\mathbf{x}|)}(|\mathbf{x}|)\right| = \sqrt{-2\Phi_1(\mathbf{x})}, \tag{8.63}$$

that implies either

$$\mathbf{v}(\mathbf{x}) = \frac{\sqrt{-2\Phi_1(|\mathbf{x}|)}}{|\mathbf{x}|}\mathbf{x} \tag{8.64}$$

or

$$\mathbf{v}(\mathbf{x}) = -\frac{\sqrt{-2\Phi_1(|\mathbf{x}|)}}{|\mathbf{x}|}\mathbf{x}, \tag{8.65}$$

exactly as in the case of the usual Newtonian gravity. In particular, on the Earth surface, we have

$$|\mathbf{v}| = \sqrt{\frac{2Gm_0}{r_0}}, \tag{8.66}$$

where r_0 is the Earth radius and m_0 is the Earth mass, i.e. the absolute value of the vectorial gravitational potential on the Earth surface approximately equals to the escape velocity and its direction is normal to the Earth, either downward or upward.

Chapter 9
Covariant Formulation of the Physical Laws in the Four-Dimensional Space-Time

9.1 Four Vectors, Four Covectors and Tensors in the Four-Dimensional Space-Time

First of all, we would like to remind the definitions of the vectors, covectors and covariant and contravariant tensors of second order in $\mathbb{R}^4$.

Definition 9.1 Given $\mathcal{S}$, that is a certain subgroup of the group of all smooth non-degenerate invertible transformations from $\mathbb{R}^4$ onto $\mathbb{R}^4$ having the form

$$\begin{cases} x'^0 = f^{(0)}(x^0, x^1, x^2, x^3), \\ x'^1 = f^{(1)}(x^0, x^1, x^2, x^3), \\ x'^2 = f^{(2)}(x^0, x^1, x^2, x^3), \\ x'^3 = f^{(3)}(x^0, x^1, x^2, x^3), \end{cases} \tag{9.1}$$

we say that a one-component field $a := a(x^0, x^1, x^2, x^3)$ is a scalar field on the group $\mathcal{S}$, if under the coordinate transformation in the group $\mathcal{S}$ of the form (9.1) this field transforms as

$$a' = a. \tag{9.2}$$

Next we say that a four-component field (a^0, a^1, a^2, a^3) is a four-vector field on the group $\mathcal{S}$, if under the coordinate transformation in the group $\mathcal{S}$ of the form (9.1) every of four components of this field transforms as

$$a'^j = \sum_{k=0}^{3} \frac{\partial f^{(j)}}{\partial x^k} a^k \qquad \forall j = 0, 1, 2, 3. \tag{9.3}$$

A. Poliakovsky, *The Laws of Gravity and Electromagnetism*,
https://doi.org/10.1007/978-3-031-61407-1_9

Next we say that a four-component field (a_0, a_1, a_2, a_3) is a four-covector field on the group $\mathcal{S}$, if under the coordinate transformation in the group $\mathcal{S}$ of the form (9.1) every of four components of this field transforms as

$$a_j = \sum_{k=0}^{3} \frac{\partial f^{(k)}}{\partial x^j} a'_k \qquad \forall j = 0, 1, 2, 3. \tag{9.4}$$

Furthermore, we say that a 16-component field $\{a_{mn}\}_{m,n=0,1,2,3}$ is a two times covariant tensor field on the group $\mathcal{S}$, if under the coordinate transformation in the group $\mathcal{S}$ of the form (9.1) every of 16 components of this field transforms as

$$a_{mn} = \sum_{j=0}^{3} \sum_{k=0}^{3} \frac{\partial f^{(k)}}{\partial x^m} \frac{\partial f^{(j)}}{\partial x^n} a'_{kj} \qquad \forall m, n = 0, 1, 2, 3. \tag{9.5}$$

Next we say that a 16-component field $\{a^{mn}\}_{m,n=0,1,2,3}$ is a two times contravariant tensor field on the group $\mathcal{S}$, if under the coordinate transformation in the group $\mathcal{S}$ of the form (9.1) every of 16 components of this field transforms as

$$a'^{mn} = \sum_{j=0}^{3} \sum_{k=0}^{3} \frac{\partial f^{(m)}}{\partial x^k} \frac{\partial f^{(n)}}{\partial x^j} a^{kj} \qquad \forall m, n = 0, 1, 2, 3. \tag{9.6}$$

Then it is well known that for every two four vectors (a^0, a^1, a^2, a^3) and (b^0, b^1, b^2, b^3) on $\mathcal{S}$, the 16-component field $\{c^{mn}\}_{m,n=0,1,2,3}$, defined in every coordinate system by

$$c^{mn} := a^m b^n \qquad \forall m, n = 0, 1, 2, 3, \tag{9.7}$$

is a two times contravariant tensor on $\mathcal{S}$. Moreover, for every two four covectors (a_0, a_1, a_2, a_3) and (b_0, b_1, b_2, b_3) on $\mathcal{S}$, the 16-component field $\{c_{mn}\}_{m,n=0,1,2,3}$, defined in every coordinate system by

$$c_{mn} := a_m b_n \qquad \forall m, n = 0, 1, 2, 3 \tag{9.8}$$

is a two times covariant tensor on $\mathcal{S}$. It is also well known that if $\{a^{mn}\}_{m,n=0,1,2,3}$ is a two times contravariant tensor field on the group $\mathcal{S}$ and if a 16-component field $\{b_{mn}\}_{m,n=0,1,2,3}$ satisfies

$$\sum_{k=0}^{3} a^{mk} b_{kn} = \begin{cases} 1 & \text{if} \quad m = n \\ 0 & \text{if} \quad m \neq n \end{cases} \qquad \forall m, n = 0, 1, 2, 3, \tag{9.9}$$

then $\{b_{mn}\}_{m,n=0,1,2,3}$ is a two times covariant tensor on $\mathcal{S}$. Next it is well known that, given a four covector (a_0, a_1, a_2, a_3) a four vector (b^0, b^1, b^2, b^3), a two times covariant tensor $\{c_{mn}\}_{m,n=0,1,2,3}$ and a two times contravariant tensor $\{d^{mn}\}_{m,n=0,1,2,3}$ on the group $\mathcal{S}$, the quantities

$$\sum_{k=0}^{3} a_k b^k \quad \text{and} \quad \sum_{m=0}^{3}\sum_{n=0}^{3} c_{mn} d^{mn} \tag{9.10}$$

are scalars on $\mathcal{S}$, the four-component fields defined by

$$\left\{\sum_{k=0}^{3} d^{mk} a_k\right\}_{m=0,1,2,3} \quad \text{and} \quad \left\{\sum_{k=0}^{3} c_{mk} b^k\right\}_{m=0,1,2,3} \tag{9.11}$$

are four vector and four covector on $\mathcal{S}$, and moreover 16-component fields $\{\hat{c}^{mn}\}_{m,n=0,1,2,3}$ and $\{\hat{d}_{mn}\}_{m,n=0,1,2,3}$ defined by

$$\hat{c}^{mn} := \sum_{k=0}^{3}\sum_{j=0}^{3} d^{mj} d^{nk} c_{jk} \quad \text{and} \quad \hat{d}_{mn} := \sum_{j=0}^{3}\sum_{k=0}^{3} c_{mj} c_{nk} d^{jk} \qquad \forall\, m, n = 0, 1, 2, 3 \tag{9.12}$$

are two times contravariant and two times covariant tensors on $\mathcal{S}$. Next, it is also well known that given a two times covariant tensor $\{c_{mn}\}_{m,n=0,1,2,3}$ and a two times contravariant tensor $\{d^{mn}\}_{m,n=0,1,2,3}$ on the group $\mathcal{S}$ the 16-component fields $\{c_{nm}\}_{m,n=0,1,2,3}$ and $\{d^{nm}\}_{m,n=0,1,2,3}$ are also two times covariant and two times contravariant tensors on $\mathcal{S}$. Finally, it is well known that, if $a := a(x^0, x^1, x^2, x^3)$ is a scalar field on the group $\mathcal{S}$, then the four-component field (w_0, w_1, w_2, w_3) defined by

$$w_j := \frac{\partial a}{\partial x^j} \qquad \forall\, j = 0, 1, 2, 3 \tag{9.13}$$

is a <u>four-covector</u> field on the group $\mathcal{S}$.

Next consider the four-dimensional space-time $\mathbb{R}^4$, such that for every point in space $\mathbf{x} = (x_1, x_2, x_3) \in \mathbb{R}^3$ and every instant of time t we correspond the point $(x^0, x^1, x^2, x^3) \in \mathbb{R}^4$ that has the form:

$$(x^0, x^1, x^2, x^3) := (ct, x_1, x_2, x_3) = (ct, \mathbf{x})\,, \tag{9.14}$$

where c is the universal constant in Maxwell equations for vacuum. In this space we denote by $\mathcal{S}_0$, the subgroup of the group of smooth non-degenerate invertible mappings, containing transformations of the form

$$\begin{cases} x'^0 = x^0 \\ x'^j = \sum\limits_{k=1}^{3} A_{jk}\left(\frac{x^0}{c}\right) x_k + z_j\left(\frac{x^0}{c}\right) \quad \forall j = 1, 2, 3, \end{cases} \tag{9.15}$$

where

$$\left\{A_{jk}(t)\right\}_{j,k=1,2,3} = A(t) : \mathbb{R} \to SO(3)$$

is a rotation, smoothly dependent on t and

$$(z_1(t), z_2(t), z_3(t)) = \mathbf{z}(t) : \mathbb{R} \to \mathbb{R}^3$$

also smoothly dependent on t. Then in the terms of time t and three-dimensional space, we rewrite (9.15) as

$$\begin{cases} \mathbf{x}' = A(t) \cdot \mathbf{x} + \mathbf{z}(t), \\ t' = t, \end{cases} \tag{9.16}$$

where $A(t) \in SO(3)$ is a rotation. I.e. the group $\mathcal{S}_0$ represents all transformations of Cartesian non-inertial coordinate systems in the non-relativistic space-time. It can be easily checked by trivial calculations that $\mathcal{S}_0$ is indeed a group, i.e. for every two transformations $f, g \in \mathcal{S}_0$ the composition $g \circ f$ and the inverse transformation $f^{(-1)}$ are also contained in $\mathcal{S}_0$, that means that they also have a form of (9.15). Next assume that a four covector (a_0, a_1, a_2, a_3) and a four vector (b^0, b^1, b^2, b^3) on the group $\mathcal{S}_0$ are given. Then by inserting (9.15) into (9.3) and (9.4) we obtain the following laws of transformations under the acting in the group $\mathcal{S}_0$:

$$\begin{cases} a_0 = a'_0 + \sum_{k=1}^{3} \frac{1}{c} \left(\sum_{j=1}^{3} \frac{dA_{kj}}{dt} \left(\frac{x^0}{c}\right) x_j + \frac{dz_k}{dt} \left(\frac{x^0}{c}\right) \right) a'_k \\ a_j = \sum_{k=1}^{3} A_{kj} \left(\frac{x^0}{c}\right) a'_k \qquad \forall j = 1, 2, 3, \end{cases} \tag{9.17}$$

and

$$\begin{cases} b'^0 = b^0 \\ b'^j = \frac{1}{c} \left(\sum_{k=1}^{3} \frac{dA_{jk}}{dt} \left(\frac{x^0}{c}\right) x_k + \frac{dz_j}{dt} \left(\frac{x^0}{c}\right) \right) b^0 + \sum_{k=1}^{3} A_{jk} \left(\frac{x^0}{c}\right) b^k \qquad \forall j = 1, 2, 3. \end{cases} \tag{9.18}$$

In particular, since $A(t) \in SO(3)$ and thus

$$\sum_{j=1}^{3} A_{mj}(t) A_{nj}(t) = \begin{cases} 1 & \text{if} \quad m = n \\ 0 & \text{if} \quad m \neq n \end{cases} \qquad \forall m, n = 1, 2, 3, \tag{9.19}$$

by (9.17) we deduce

$$\begin{cases} a'_0 = a_0 - \sum_{k=1}^{3} \frac{1}{c} \left(\sum_{j=1}^{3} \frac{dA_{kj}}{dt} \left(\frac{x^0}{c}\right) x_j + \frac{dz_k}{dt} \left(\frac{x^0}{c}\right) \right) \left(\sum_{j=1}^{3} A_{kj} \left(\frac{x^0}{c}\right) a_j \right) \\ a'_k = \sum_{j=1}^{3} A_{kj} \left(\frac{x^0}{c}\right) a_j \qquad \forall k = 1, 2, 3. \end{cases} \tag{9.20}$$

So, by (9.20) and (9.18), we obtained the following laws of transformation of four covectors and four vectors in the group $\mathcal{S}_0$, i.e. under the change of non-inertial Cartesian coordinate systems:

$$\begin{cases} a'_0 = a_0 - \sum_{k=1}^{3} \frac{1}{c}\left(\sum_{j=1}^{3} \frac{dA_{kj}}{dt}\left(\frac{x^0}{c}\right) x_j + \frac{dz_k}{dt}\left(\frac{x^0}{c}\right)\right)\left(\sum_{j=1}^{3} A_{kj}\left(\frac{x^0}{c}\right) a_j\right) \\ a'_k = \sum_{j=1}^{3} A_{kj}\left(\frac{x^0}{c}\right) a_j \qquad \forall k = 1, 2, 3, \end{cases} \tag{9.21}$$

and

$$\begin{cases} b'^0 = b^0 \\ b'^j = \frac{1}{c}\left(\sum_{k=1}^{3} \frac{dA_{jk}}{dt}\left(\frac{x^0}{c}\right) x_k + \frac{dz_j}{dt}\left(\frac{x^0}{c}\right)\right) b^0 + \sum_{k=1}^{3} A_{jk}\left(\frac{x^0}{c}\right) b^k \qquad \forall j = 1, 2, 3. \end{cases} \tag{9.22}$$

Therefore, if we denote the four vector (b^0, b^1, b^2, b^3) and the four covector (a_0, a_1, a_2, a_3) on the group $\mathcal{S}_0$ as

$$\begin{cases} (b^0, b^1, b^2, b^3) = \left(\sigma, \frac{1}{c}\mathbf{b}\right) \quad \text{where} \quad \sigma := b^0 \quad \text{and} \quad \mathbf{b} := c(b^1, b^2, b^3) \in \mathbb{R}^3, \\ (a_0, a_1, a_2, a_3) = (\psi, -\mathbf{a}) \quad \text{where} \quad \psi := a_0 \quad \text{and} \quad \mathbf{a} := -(a_1, a_2, a_3) \in \mathbb{R}^3, \end{cases} \tag{9.23}$$

then by (9.21) and (9.22) in the terms of time t and three-dimensional space $\mathbf{x}$, we obtain the following laws of transformations of σ, $\mathbf{b}$, ψ and $\mathbf{a}$ under the change of non-inertial Cartesian coordinate system:

$$\begin{cases} \sigma' = \sigma \\ \mathbf{b}' = A(t) \cdot \mathbf{b} + \left(\frac{dA}{dt}(t) \cdot \mathbf{x} + \frac{d\mathbf{z}}{dt}(t)\right)\sigma, \end{cases} \tag{9.24}$$

and

$$\begin{cases} \psi' = \psi + \frac{1}{c}\left(\frac{dA}{dt}(t) \cdot \mathbf{x} + \frac{d\mathbf{z}}{dt}(t)\right) \cdot (A(t) \cdot \mathbf{a}) \\ \mathbf{a}' = A(t) \cdot \mathbf{a}. \end{cases} \tag{9.25}$$

In particular, if $\sigma := b^0$ is the first coordinate of an arbitrary four vector (b^0, b^1, b^2, b^3) on the group $\mathcal{S}_0$, then σ is a proper scalar field in the frames of Definition 3.1. Moreover, if $\mathbf{a} := -(a_1, a_2, a_3)$, where a_1, a_2, a_3 are the last three coordinates of an arbitrary four covector (a_0, a_1, a_2, a_3) on the group $\mathcal{S}_0$, then $\mathbf{a}$ is a proper vector field in the frames of Definition 3.1.

Next, since by Definition 3.1 every three-dimensional speed-like vector field $\mathbf{u}$ transforms under the change of non-inertial Cartesian coordinate system as

$$\mathbf{u}' = A(t) \cdot \mathbf{u} + \frac{dA}{dt}(t) \cdot \mathbf{x} + \frac{d\mathbf{z}}{dt}(t), \tag{9.26}$$

by comparing (9.26) with (9.24) we deduce that for every speed-like vector field **u** the four-component field (u^0, u^1, u^2, u^3) defined by

$$(u^0, u^1, u^2, u^3) := \left(1, \frac{1}{c}\mathbf{u}\right) \quad \text{where} \quad u^0 = 1 \quad \text{and} \quad (u^1, u^2, u^3) = \frac{1}{c}\mathbf{u} \in \mathbb{R}^3 \tag{9.27}$$

is a four-vector field on the group $\mathcal{S}_0$. We call such four vectors by the name vectors of type 1. In particular, if **u** is the velocity field, then the quantity defined by (9.27) is a a four-vector field on the group $\mathcal{S}_0$ that we call the four-dimensional speed. Regarding the field of velocity **u** we also can give a different argumentation that the four-component field (u^0, u^1, u^2, u^3) defined by (9.27) is a four-vector field on the group $\mathcal{S}_0$: indeed it is well known from Tensor Analysis that if $\left(x^0(s), x^1(s), x^2(s), x^3(s)\right)$ is a curve in $\mathbb{R}^4$, parameterized by some scalar parameter s, then the four-component field $\left(\frac{dx^0}{ds}(s), \frac{dx^1}{ds}(s), \frac{dx^2}{ds}(s), \frac{dx^3}{ds}(s)\right)$ is a four-vector field on an arbitrary group $\mathcal{S}$ and, in particular, on the group $\mathcal{S}_0$. Thus, if $\mathbf{r}(t) = (r_1(t), r_2(t), r_3(t))$ is a three-dimensional trajectory of the motion of some particle, parameterized by the global time t, then if we consider a curve $\frac{1}{c}(ct, r_1(t), r_2(t), r_3(t))$ in $\mathbb{R}^4$, parameterized by the global time t, then the four-component field:

$$\left(1, \frac{1}{c}\frac{d\mathbf{r}}{dt}(t)\right) := \left(1, \frac{1}{c}\frac{dr_1}{dt}(t), \frac{1}{c}\frac{dr_2}{dt}(t), \frac{1}{c}\frac{dr_3}{dt}(t)\right) \tag{9.28}$$

is a four-vector field on the group $\mathcal{S}_0$.

Similarly, if **v** is the vectorial gravitational potential, then since **v** is a speed-like vector field, the four-component field (v^0, v^1, v^2, v^3) defined by

$$(v^0, v^1, v^2, v^3) := \left(1, \frac{1}{c}\mathbf{v}\right) \quad \text{where} \quad v^0 = 1 \quad \text{and} \quad (v^1, v^2, v^3) = \frac{1}{c}\mathbf{v}, \tag{9.29}$$

is also a four-vector field on the group $\mathcal{S}_0$ that we call the four-dimensional gravitational potential. In the same way, if **k** is the vectorial potential of inertia, defined by Definition 4.2, then since **k** is a speed-like vector field, the four-component field (k^0, k^1, k^2, k^3) defined by

$$(k^0, k^1, k^2, k^3) := \left(1, \frac{1}{c}\mathbf{k}\right) \quad \text{where} \quad k^0 = 1 \quad \text{and} \quad (k^1, k^2, k^3) = \frac{1}{c}\mathbf{k} \tag{9.30}$$

is also a four-vector field on the group $\mathcal{S}_0$ that we call the four-dimensional potential of inertia.

Moreover, by (9.24), for every speed-like vector field **u** and every proper scalar field σ the four-component field (b^0, b^1, b^2, b^3) defined by

$$(b^0, b^1, b^2, b^3) := \left(\sigma, \frac{\sigma}{c}\mathbf{u}\right) \quad \text{where} \quad b^0 = \sigma \quad \text{and} \quad (b^1, b^2, b^3) = \frac{\sigma}{c}\mathbf{u} \tag{9.31}$$

is also a four-vector field on the group $\mathcal{S}_0$. In particular, if we consider the field of four-dimensional moment of a particle (p^0, p^1, p^2, p^3) defined by

$$(p^0, p^1, p^2, p^3) := \left(m, \frac{1}{c}(m\mathbf{u})\right) \quad \text{where} \quad p^0 = m \quad \text{and} \quad (p^1, p^2, p^3) = \frac{1}{c}(m\mathbf{u}), \tag{9.32}$$

where m is the mass of the particle and $\mathbf{u}$ is the velocity of the particle, then (p^0, p^1, p^2, p^3) is also a four vector on the group $\mathcal{S}_0$. Moreover, by comparing (5.17) and (5.19) with (9.24) we deduce that if we consider the field of four-dimensional electric current (j^0, j^1, j^2, j^3) defined by

$$(j^0, j^1, j^2, j^3) := \left(\rho, \frac{1}{c}\mathbf{j}\right) \quad \text{where} \quad j^0 = \rho \quad \text{and} \quad (j^1, j^2, j^3) = \frac{1}{c}\mathbf{j}, \tag{9.33}$$

where ρ is the electric charge density and $\mathbf{j}$ is the electric current density, then (j^0, j^1, j^2, j^3) is also a four vector on the group $\mathcal{S}_0$.

On the other hand, for every proper three-dimensional vector field $\mathbf{G}$ that satisfies due to Definition 3.1:

$$\mathbf{G}' = A(t) \cdot \mathbf{G}, \tag{9.34}$$

by comparing (9.34) with (9.24) we deduce that the four-component field (G^0, G^1, G^2, G^3) defined by

$$(G^0, G^1, G^2, G^3) := (0, \mathbf{G}) \quad \text{where} \quad G^0 = 0 \quad \text{and} \quad (G^1, G^2, G^3) = \mathbf{G} \tag{9.35}$$

is also a four-vector field on the group $\mathcal{S}_0$. We call such four vectors by the name vectors of type 0.

Next, since by (5.54) the scalar electromagnetic potential Ψ and the vector electromagnetic potential $\mathbf{A}$, under the change of non-inertial Cartesian coordinate system transform as

$$\begin{cases} \Psi' = \Psi + \frac{1}{c}\left(\frac{dA}{dt}(t) \cdot \mathbf{x} + \frac{d\mathbf{z}}{dt}(t)\right) \cdot (A(t) \cdot \mathbf{A}) \\ \mathbf{A}' = A(t) \cdot \mathbf{A}, \end{cases} \tag{9.36}$$

by comparing (9.36) with (9.25) we deduce that the four-component field (A_0, A_1, A_2, A_3) defined as

$$(A_0, A_1, A_2, A_3) = (\Psi, -\mathbf{A}) \quad \text{where} \quad A_0 = \Psi \quad \text{and} \quad (A_1, A_2, A_3) = -\mathbf{A} \tag{9.37}$$

is a four-covector field on the group $\mathcal{S}_0$. We call this four-covector field by the name four-dimensional electromagnetic potential. Next, since (A_0, A_1, A_2, A_3) is a four-covector field on the group $\mathcal{S}_0$, then it is well known from the tensor analysis that the 16-component field $\{F_{ij}\}_{0 \le i,j \le 3}$ defined in every non-inertial Cartesian coordinate system by

$$F_{ij} := \frac{\partial A_j}{\partial x^i} - \frac{\partial A_i}{\partial x^j} \qquad \forall i, j = 0, 1, 2, 3 \tag{9.38}$$

is an antisymmetric two times covariant tensor field on the group $\mathcal{S}_0$, which we call the covariant tensor of the electromagnetic field. In particular, by inserting (9.37) and (9.14) into (9.38) we deduce

$$\begin{cases} F_{00} = 0 \\ F_{0j} = -F_{j0} = -\frac{1}{c}\frac{\partial(-A_j)}{\partial t} - \frac{\partial \Psi}{\partial x^j} & \forall j = 1,2,3 \\ F_{jj} = 0 \quad \forall j = 1,2,3 \\ F_{ij} = -F_{ji} = \frac{\partial(-A_i)}{\partial x^j} - \frac{\partial(-A_j)}{\partial x^i} & \forall i \neq j = 1,2,3\,. \end{cases} \tag{9.39}$$

Thus if as in (5.38) we denote:

$$\begin{cases} \mathbf{B} := \mathrm{curl}_{\mathbf{x}}\mathbf{A}, \\ \mathbf{E} := -\nabla_{\mathbf{x}}\Psi - \frac{1}{c}\frac{\partial \mathbf{A}}{\partial t}, \end{cases} \tag{9.40}$$

then denoting $\mathbf{E} := (E_1, E_2, E_3)$ and $\mathbf{B} := (B_1, B_2, B_3)$, by (9.40) we rewrite (9.39) as

$$\begin{cases} F_{00} = 0 \\ F_{0j} = -F_{j0} = E_j \quad \forall j = 1,2,3 \\ F_{jj} = 0 \quad \forall j = 1,2,3 \\ F_{12} = -F_{21} = -B_3 \\ F_{13} = -F_{31} = B_2 \\ F_{23} = -F_{32} = -B_1\,. \end{cases} \tag{9.41}$$

Next assume that $T := \left\{T_{ij}\right\}_{i,j=1,2,3} \in \mathbb{R}^{3\times 3}$ is a 9-component proper matrix-valued field, which, being a proper matrix field, by Definition 3.1 satisfies:

$$T' = A(t)\cdot T\cdot A^T(t) = A(t)\cdot T\cdot \{A(t)\}^{-1}\,. \tag{9.42}$$

Next consider a 16-component field $\{\mathcal{T}^{ij}\}_{0\le i,j\le 3}$ defined in every non-inertial Cartesian coordinate system by

$$\begin{cases} \mathcal{T}^{00} = 0 \\ \mathcal{T}^{0j} = \mathcal{T}^{j0} = 0 \quad \forall j = 1,2,3 \\ \mathcal{T}^{ij} := T_{ij} \quad \forall i,j = 1,2,3\,. \end{cases} \tag{9.43}$$

Then by inserting (9.15) and (9.42) into (9.6), we can prove that the field $\{\mathcal{T}^{ij}\}_{0\le i,j\le 3}$ defined by (9.43) is a two times <u>contravariant</u> tensor field on the group $\mathcal{S}_0$. Indeed, by (9.6) for every two times contravariant tensor field $\{a^{ij}\}_{0\le i,j\le 3}$ we have

$$a'^{mn} = \frac{\partial f^{(m)}}{\partial x^0}\frac{\partial f^{(n)}}{\partial x^0}a^{00} + \sum_{k=1}^{3}\frac{\partial f^{(m)}}{\partial x^k}\frac{\partial f^{(n)}}{\partial x^0}a^{k0} + \sum_{j=1}^{3}\frac{\partial f^{(m)}}{\partial x^0}\frac{\partial f^{(n)}}{\partial x^j}a^{0j}$$
$$+ \sum_{j=1}^{3}\sum_{k=1}^{3}\frac{\partial f^{(m)}}{\partial x^k}\frac{\partial f^{(n)}}{\partial x^j}a^{kj} \quad \forall m, n = 0, 1, 2, 3. \tag{9.44}$$

Then, since by (9.15) we have $\frac{\partial f^{(0)}}{\partial x^0} = 1$, $\frac{\partial f^{(0)}}{\partial x^k} = 0\ \forall k = 1, 2, 3$ and $\frac{\partial f^{(m)}}{\partial x^k} = A_{mk}\left(\frac{x^0}{c}\right)$ $\forall k, m = 1, 2, 3$, in the case where $a^{00} = 0$ and $a^{0j} = a^{j0} = 0\ \forall j = 1, 2, 3$ we rewrite (9.44) as

$$\begin{cases} a'^{00} = 0 \\ a'^{j0} = a'^{0j} = 0 \quad \forall j = 1, 2, 3, \\ a'^{mn} = \sum_{j=1}^{3}\sum_{k=1}^{3} A_{mk}\left(\frac{x^0}{c}\right) A_{nj}\left(\frac{x^0}{c}\right) a^{kj} \quad \forall m, n = 1, 2, 3, \end{cases} \tag{9.45}$$

that is compatible with (9.43) and (9.42).

In particular, if we consider the 9-component matrix field I that defined in every Cartesian coordinate system as $I := \left\{\delta_{ij}\right\}_{1,j=1,2,3} \in \mathbb{R}^{3\times 3}$, where

$$\delta_{ij} := \begin{cases} 1 & \text{if } i = j \\ 0 & \text{if } i \neq j, \end{cases} \tag{9.46}$$

which is a proper matrix field, since

$$I = A(t) \cdot I \cdot \{A(t)\}^{-1}, \tag{9.47}$$

then the 16-component field $\{\Theta^{ij}\}_{0\leq i,j\leq 3}$ defined in every non-inertial Cartesian coordinate system by

$$\begin{cases} \Theta^{00} = 0 \\ \Theta^{0j} = \Theta^{j0} = 0 \quad \forall j = 1, 2, 3 \\ \Theta^{ij} := \delta_{ij} \quad \forall i, j = 1, 2, 3 \end{cases} \tag{9.48}$$

is a two times contravariant tensor field on the group $\mathcal{S}_0$, and moreover this tensor is symmetric. We call $\{\Theta^{ij}\}_{0\leq i,j\leq 3}$ the contravariant tensor of the three-dimensional geometry.

Next, the scalar field $\tau := \tau(x^0, x^1, x^2, x^3)$, defined in every Cartesian coordinate system as

$$\tau := \frac{x^0}{c} = t, \tag{9.49}$$

is a scalar on the group $\mathcal{S}_0$. Here t is the global non-relativistic time. Moreover, by (9.13) and (9.49), the four-component field (v_0, v_1, v_2, v_3) defined by

$$v_0 := c\frac{\partial\tau}{\partial x^0}(x^0, x^1, x^2, x^3) = 1 \quad \text{and} \quad v_j := c\frac{\partial\tau}{\partial x^j}(x^0, x^1, x^2, x^3) = 0 \qquad \forall\, j = 1, 2, 3 \tag{9.50}$$

is a four-covector field on the group $\mathcal{S}_0$.

Finally, consider a motion of a classical particle with inertial mass m, charge σ, place $\mathbf{r}(t)$ and velocity $\mathbf{u}(t) = \mathbf{r}'(t)$ in the outer gravitational field with the vectorial gravitational potential $\mathbf{v}(\mathbf{x}, t)$, the outer electromagnetic field with vectorial and scalar potentials $\mathbf{A}(\mathbf{x}, t)$ and $\Psi(\mathbf{x}, t)$, and additional conservative field with scalar potential $V(\mathbf{x}, t)$ ruled by a Lagrangian (1.69):

$$L_0\left(\frac{d\mathbf{r}}{dt}, \mathbf{r}, t\right) := \frac{m}{2}\left|\frac{d\mathbf{r}}{dt} - \mathbf{v}(\mathbf{r}, t)\right|^2 - \sigma\left(\Psi(\mathbf{r}, t) - \frac{1}{c}\mathbf{A}(\mathbf{r}, t)\cdot\frac{d\mathbf{r}}{dt}\right) + V(\mathbf{r}, t). \tag{9.51}$$

Then L_0 is a scalar on the group $\mathcal{S}_0$. Moreover, consider the generalized momentum of the particle m by (1.72):

$$\mathbf{P} := \nabla_{\mathbf{r}'} L_0\left(\mathbf{r}', \mathbf{r}, t\right) = m\frac{d\mathbf{r}}{dt} - m\mathbf{v}(\mathbf{r}, t) + \frac{\sigma}{c}\mathbf{A}(\mathbf{r}, t), \tag{9.52}$$

consider a Hamiltonian

$$H_0\left(\mathbf{P}, \mathbf{r}, t\right) := \mathbf{P}\cdot\frac{d\mathbf{r}}{dt} - L_0\left(\frac{d\mathbf{r}}{dt}, \mathbf{r}, t\right), \tag{9.53}$$

which by (1.74) satisfies

$$H_0\left(\mathbf{P}, \mathbf{r}, t\right) = \mathbf{P}\cdot\mathbf{v}(\mathbf{r}, t) + \frac{1}{2m}\left|\mathbf{P} - \frac{\sigma}{c}\mathbf{A}(\mathbf{r}, t)\right|^2 + \sigma\left(\Psi(\mathbf{r}, t) - \frac{1}{c}\mathbf{A}(\mathbf{r}, t)\cdot\mathbf{v}(\mathbf{r}, t)\right) - V\left(\mathbf{r}, t\right), \tag{9.54}$$

and furthermore define the four-dimensional generalized momentum (P_0, P_1, P_2, P_3) as

$$(P_0, P_1, P_2, P_3) := \left(\frac{1}{c}H_0, -\mathbf{P}\right) \quad \text{where} \quad P_0 = \frac{1}{c}H_0 \quad \text{and} \quad (P_1, P_2, P_3) = -\mathbf{P}. \tag{9.55}$$

Then, since by (9.54) and (9.52), under the change of non-inertial Cartesian coordinate system H_0 and $\mathbf{P}$ transform as

$$\begin{cases} H_0' = H_0 + \left(\frac{dA}{dt}(t)\cdot \mathbf{x} + \frac{d\mathbf{z}}{dt}(t)\right)\cdot (A(t)\cdot \mathbf{P}) \\ \mathbf{P}' = A(t)\cdot \mathbf{P}, \end{cases} \tag{9.56}$$

by comparing (9.56) with (9.25) we deduce that the four-dimensional momentum (P_0, P_1, P_2, P_3) is a four-covector on the group $\mathcal{S}_0$.

9.2 Pseudo-metric Tensors of the Four-Dimensional Space-Time

Consider $\{g^{ij}\}_{0\le i,j\le 3}$ to be a two times contravariant tensor field on the group $\mathcal{S}_0$, defined by

$$g^{ij} := v^i v^j - \Theta^{ij} \qquad \forall i, j = 0, 1, 2, 3, \tag{9.57}$$

where $\{\Theta^{ij}\}_{0\le i,j\le 3}$ is the contravariant tensor of the three-dimensional geometry, defined by (9.48) and being a two times contravariant tensor, and (v^0, v^1, v^2, v^3) is the four-dimensional gravitational potential, defined by (9.29) and being a four vector. Then by (9.7) we obtain that $\{g^{ij}\}_{0\le i,j\le 3}$ is indeed a two times contravariant tensor field on the group $\mathcal{S}_0$, and moreover this tensor is symmetric. Moreover, by (9.48) and (9.29) we have

$$\begin{cases} g^{00} = 1 \\ g^{ij} = -\delta_{ij} + \frac{v^i v^j}{c^2} \quad \forall 1 \le i, j \le 3 \\ g^{0j} = g^{j0} = \frac{v^j}{c} \quad \forall 1 \le j \le 3, \end{cases} \tag{9.58}$$

where $\mathbf{v} = (v^1, v^2, v^3)$ is the three-dimensional vectorial gravitational potential. We call the tensor $\{g^{ij}\}_{0\le i,j\le 3}$ the contravariant pseudo-metric tensor of the four-dimensional space-time. Next consider a 16-component field $\{g_{ij}\}_{0\le i,j\le 3}$ defined by

$$\begin{cases} g_{00} = 1 - \frac{|\mathbf{v}|^2}{c^2} \\ g_{ij} = -\delta_{ij} \quad \forall 1 \le i, j \le 3 \\ g_{0j} = g_{j0} = \frac{v^j}{c} \quad \forall 1 \le j \le 3. \end{cases} \tag{9.59}$$

Then

$$\sum_{k=0}^{3} g_{0k} g^{k0} = g_{00} g^{00} + \sum_{k=1}^{3} g_{0k} g^{k0} = 1 - \frac{|\mathbf{v}|^2}{c^2} + \frac{|\mathbf{v}|^2}{c^2} = 1,$$

$$\sum_{k=0}^{3} g_{ik} g^{kj} = g_{i0} g^{0j} + \sum_{k=1}^{3} g_{ik} g^{kj} = \frac{v^i v^j}{c^2} + \delta_{ij} - \frac{v^i v^j}{c^2} = \delta_{ij} \quad \forall 1 \le i, j \le 3,$$

and

$$\sum_{k=0}^{3} g_{ik}g^{k0} = g_{i0}g^{00} + \sum_{k=1}^{3} g_{ik}g^{k0} = \frac{v^i}{c} - \frac{v^i}{c} = 0 \quad \forall 1 \le i \le 3,$$

$$\sum_{k=0}^{3} g_{0k}g^{kj} = g_{00}g^{0j} + \sum_{k=1}^{3} g_{0k}g^{kj} = \left(1 - \frac{|\mathbf{v}|^2}{c^2}\right)\frac{v^j}{c} - \sum_{k=1}^{3} \frac{v^k}{c}\left(\delta_{kj} - \frac{v^k v^j}{c^2}\right) = 0$$
$$\forall 1 \le j \le 3,$$

where $\{g^{ij}\}_{0\le i,j\le 3}$ is the contravariant pseudo-metric tensor of the four-dimensional space-time, defined by (9.58). So,

$$\sum_{k=0}^{3} g^{ik}g_{kj} = \begin{cases} 1 & \text{if } i = j \\ 0 & \text{if } i \neq j \end{cases} \qquad \forall i, j = 0, 1, 2, 3. \tag{9.60}$$

Therefore, by comparing (9.60) and (9.9) we deduce that $\{g_{ij}\}_{i,j=0,1,2,3}$ is a two times covariant tensor on the group $\mathcal{S}_0$, and moreover this tensor is symmetric. We call the tensor $\{g_{ij}\}_{0\le i,j\le 3}$ covariant pseudo-metric tensor of the four-dimensional space-time. Using (9.60) we also obtain that the pseudo-metric tensors $\{g_{ij}\}_{i,j=0,1,2,3}$ and $\{g^{ij}\}_{0\le i,j\le 3}$ are non-degenerate and they are inverse of each other. Moreover, it can be easily calculated that if we consider the 4×4-matrix:

$$G = \{g_{ij}\}_{0\le i,j\le 3}, \tag{9.61}$$

then

$$\det G = -1. \tag{9.62}$$

Thus, with the covariant and contravariant pseudo-metric tensors we can lower and lift indexes of arbitrary tensors. In particular, given a four covector (a_0, a_1, a_2, a_3) and a four vector (b^0, b^1, b^2, b^3) on the group $\mathcal{S}_0$ we can define the corresponding lifted four vector (a^0, a^1, a^2, a^3) and the corresponded lowered four covector (b_0, b_1, b_2, b_3) by

$$(a^0, a^1, a^2, a^3) := \left\{\sum_{k=0}^{3} g^{mk}a_k\right\}_{m=0,1,2,3}$$
$$\text{and} \quad (b_0, b_1, b_2, b_3) := \left\{\sum_{k=0}^{3} g_{mk}b^k\right\}_{m=0,1,2,3}. \tag{9.63}$$

Then by (9.58), (9.59) and (9.63), we have

$$a^0 = a_0 + \sum_{k=1}^{3} \frac{1}{c} v^k a_k \quad \text{and} \quad a^m = -a_m + \frac{1}{c}\left(a_0 + \sum_{k=1}^{3} \frac{1}{c} v^k a_k\right) v^m \qquad \forall m = 1, 2, 3, \tag{9.64}$$

and

$$b_0 = \left(1 - \frac{|\mathbf{v}|^2}{c^2}\right) b^0 + \sum_{k=1}^{3} \frac{1}{c} v^k b^k = b^0 - \sum_{k=1}^{3} \frac{1}{c} v^k \left(-b^k + \frac{1}{c} b^0 v^k\right)$$
$$\text{and} \quad b_m = -b^m + \frac{1}{c} b^0 v^m \qquad \forall m = 1, 2, 3. \tag{9.65}$$

We also can rewrite (9.64) and (9.65) as

$$a^0 = a_0 + \sum_{k=1}^{3} \frac{1}{c} v^k a_k \quad \text{and} \quad a^m = -a_m + \frac{1}{c} a^0 v^m \qquad \forall m = 1, 2, 3, \tag{9.66}$$

and

$$b_m = -b^m + \frac{1}{c} b^0 v^m \quad \text{and} \quad b_0 = b^0 - \sum_{k=1}^{3} \frac{1}{c} v^k b_k \qquad \forall m = 1, 2, 3. \tag{9.67}$$

In particular, if we consider the scalar field Λ on the group $\mathcal{S}_0$ defined by

$$\Lambda := b^0 a_0 + \sum_{k=1}^{3} b^k a_k \tag{9.68}$$

then by inserting (9.66) and (9.67) into (9.68) we deduce

$$\Lambda = b^0 \left(a^0 - \sum_{k=1}^{3} \frac{1}{c} v^k a_k\right) + \sum_{k=1}^{3} \left(-b_k + \frac{1}{c} b^0 v^k\right) a_k = b^0 a^0 - \sum_{k=1}^{3} b_k a_k. \tag{9.69}$$

So,

$$\Lambda = b^0 a_0 + \sum_{k=1}^{3} b^k a_k = b^0 a^0 - \sum_{k=1}^{3} b_k a_k. \tag{9.70}$$

Next, if for every speed-like vector field $\mathbf{u}$ we consider the four-vector field (u^0, u^1, u^2, u^3) defined by (9.27) as

$$(u^0, u^1, u^2, u^3) := \left(1, \frac{1}{c}\mathbf{u}\right) \quad \text{where} \quad u^0 = 1 \text{ and } (u^1, u^2, u^3) = \frac{1}{c}\mathbf{u} \in \mathbb{R}^3, \tag{9.71}$$

then by (9.67) the corresponding lowered four-covector field (u_0, u_1, u_2, u_3) satisfies

$$(u_0, u_1, u_2, u_3) := \left(1 + \frac{1}{c^2}(\mathbf{u} - \mathbf{v}) \cdot \mathbf{v}, -\frac{1}{c}(\mathbf{u} - \mathbf{v})\right) \quad \text{where}$$
$$u_0 = 1 + \frac{1}{c^2}(\mathbf{u} - \mathbf{v}) \cdot \mathbf{v} \text{ and } (u_1, u_2, u_3) = -\frac{1}{c}(\mathbf{u} - \mathbf{v}) \in \mathbb{R}^3. \tag{9.72}$$

Moreover, in the case where (u^0, u^1, u^2, u^3) is a four-dimensional speed, we call the corresponding lowered four-covector field (u_0, u_1, u_2, u_3) by the name four-dimensional cospeed. In particular, if we consider the four-dimensional gravitational potential (v^0, v^1, v^2, v^3) defined by (9.29):

$$(v^0, v^1, v^2, v^3) := \left(1, \frac{1}{c}\mathbf{v}\right) \quad \text{where} \quad v^0 = 1 \text{ and } (v^1, v^2, v^3) = \frac{1}{c}\mathbf{v}, \tag{9.73}$$

then by (9.72) we obtain that the corresponding lowered four-covector field (v_0, v_1, v_2, v_3), that we call the four covector of gravitational potential, satisfies

$$(v_0, v_1, v_2, v_3) := (1, 0, 0, 0) \quad \text{where} \quad v_0 = 1 \text{ and } (v_1, v_2, v_3) = 0 := (0, 0, 0). \tag{9.74}$$

Note that the four covector of gravitational potential, defined by (9.74), coincides with the four covector defined by (9.50) as the gradient of the scalar of global time. Next, by (9.73) and (9.74) we clearly have

$$c^2 \left(\sum_{j=0}^{3}\sum_{k=0}^{3} g^{jk} \frac{\partial \tau}{\partial x^j}\frac{\partial \tau}{\partial x^k}\right) = \sum_{j=0}^{3}\sum_{k=0}^{3} g^{jk} v_j v_k = \sum_{j=0}^{3}\sum_{k=0}^{3} g_{jk} v^j v^k = \sum_{j=0}^{3} v^j v_j = 1, \tag{9.75}$$

where τ is the scalar of the global time on the group $\mathcal{S}_0$, defined by (9.49). Finally, we clearly have

$$\sum_{k=0}^{3} \Theta^{mk} \frac{\partial \tau}{\partial x^k} = \sum_{k=0}^{3} \Theta^{mk} v_k = 0 \qquad \forall m = 0, 1, 2, 3, \tag{9.76}$$

where Θ^{ij} is the contravariant tensor of the three-dimensional geometry, defined by (9.48).

More generally, if for every speed-like vector field $\mathbf{u}$ and every proper scalar field σ we consider the four-vector field (b^0, b^1, b^2, b^3) on the group $\mathcal{S}_0$ defined by (9.31) as

$$(b^0, b^1, b^2, b^3) := \left(\sigma, \frac{\sigma}{c}\mathbf{u}\right) \quad \text{where} \quad b^0 = \sigma \ \text{ and } \ (b^1, b^2, b^3) = \frac{\sigma}{c}\mathbf{u}, \tag{9.77}$$

then by (9.67) the corresponding lowered four-covector field (b_0, b_1, b_2, b_3) satisfies

$$(b_0, b_1, b_2, b_3) := \left(\sigma\left(1 + \frac{1}{c^2}(\mathbf{u} - \mathbf{v}) \cdot \mathbf{v}\right), -\frac{\sigma}{c}(\mathbf{u} - \mathbf{v})\right) \quad \text{where}$$
$$b_0 = \sigma\left(1 + \frac{1}{c^2}(\mathbf{u} - \mathbf{v}) \cdot \mathbf{v}\right) \ \text{ and } \ (b_1, b_2, b_3) = -\frac{\sigma}{c}(\mathbf{u} - \mathbf{v}). \tag{9.78}$$

In particular, if we consider the field of four vector of the moment of a particle (p^0, p^1, p^2, p^3) defined by (9.32) as

$$(p^0, p^1, p^2, p^3) := \left(m, \frac{1}{c}(m\mathbf{u})\right) \quad \text{where} \quad p^0 = m \ \text{ and } \ (p^1, p^2, p^3) = \frac{1}{c}(m\mathbf{u}), \tag{9.79}$$

where m is the mass of the particle and $\mathbf{u}$ is the velocity of the particle, then the corresponding lowered four-covector field (p_0, p_1, p_2, p_3), which we call the four covector of momentum, satisfies

$$(p_0, p_1, p_2, p_3) := \left(m\left(1 + \frac{1}{c^2}(\mathbf{u} - \mathbf{v}) \cdot \mathbf{v}\right), -\frac{m}{c}(\mathbf{u} - \mathbf{v})\right) \quad \text{where}$$
$$p_0 = m\left(1 + \frac{1}{c^2}(\mathbf{u} - \mathbf{v}) \cdot \mathbf{v}\right) \ \text{ and } \ (p_1, p_2, p_3) = -\frac{m}{c}(\mathbf{u} - \mathbf{v}). \tag{9.80}$$

In particular, the scalar field J_0 defined by

$$J_0 := -\frac{c^2}{2m}\left(p^0 p_0 + \sum_{k=1}^{3} p^k p_k\right), \tag{9.81}$$

by (9.70), (9.79) and (9.80) satisfies

$$J_0 = \frac{mc^2}{2}\left(\frac{1}{c^2}|\mathbf{u} - \mathbf{v}|^2 - 1\right) = \frac{m}{2}|\mathbf{u} - \mathbf{v}|^2 - \frac{mc^2}{2}. \tag{9.82}$$

Moreover, if we consider the four-dimensional electric current (j^0, j^1, j^2, j^3) defined by (9.33) as

$$(j^0, j^1, j^2, j^3) := \left(\rho, \frac{1}{c}\mathbf{j}\right) \quad \text{where} \quad j^0 = \rho \ \text{ and } \ (j^1, j^2, j^3) = \frac{1}{c}\mathbf{j}, \tag{9.83}$$

where ρ is the electric charge density and $\mathbf{j}$ is the electric current density, then the corresponding lowered four-covector field (j_0, j_1, j_2, j_3), which we call the four covector of current, satisfies

$$(j_0, j_1, j_2, j_3) := \left(\rho + \frac{1}{c^2}(\mathbf{j} - \rho\mathbf{v}) \cdot \mathbf{v}, -\frac{1}{c}(\mathbf{j} - \rho\mathbf{v})\right) \quad \text{where}$$
$$j_0 = \rho + \frac{1}{c^2}(\mathbf{j} - \rho\mathbf{v}) \cdot \mathbf{v} \text{ and } (j_1, j_2, j_3) = -\frac{1}{c}(\mathbf{j} - \rho\mathbf{v}). \tag{9.84}$$

Finally, if Ψ is the scalar electromagnetic potential and $\mathbf{A}$ is the vector electromagnetic potential and we consider the four-covector field of four-dimensional electromagnetic potential (A_0, A_1, A_2, A_3), defined by (9.37) as

$$(A_0, A_1, A_2, A_3) = (\Psi, -\mathbf{A}) \quad \text{where} \quad A_0 = \Psi \text{ and } (A_1, A_2, A_3) = -\mathbf{A}, \tag{9.85}$$

then by inserting (9.85) into (9.66) we deduce that the corresponding lifted four-vector field (A^0, A^1, A^2, A^3), which we call the four vector of electromagnetic potential, satisfies

$$A^0 = \Psi - \frac{1}{c}\mathbf{v} \cdot \mathbf{A} \text{ and } (A^1, A^2, A^3) = \mathbf{A} + \frac{1}{c}\left(\Psi - \frac{1}{c}\mathbf{v} \cdot \mathbf{A}\right)\mathbf{v}. \tag{9.86}$$

On the other hand, the proper scalar electromagnetic potential Ψ_0 was defined by (5.40) as

$$\Psi_0 := \Psi - \frac{1}{c}\mathbf{A} \cdot \mathbf{v}. \tag{9.87}$$

Thus we rewrite (9.86) as

$$A^0 = \Psi_0 \text{ and } (A^1, A^2, A^3) = \mathbf{A} + \frac{1}{c}\Psi_0\mathbf{v}. \tag{9.88}$$

Next given a two times covariant tensor $\{c_{mn}\}_{m,n=0,1,2,3}$ on the group $\mathcal{S}_0$ by (9.12) we consider two times contravariant tensor on $\mathcal{S}_0$: $\{c^{mn}\}_{m,n=0,1,2,3}$ defined by

$$c^{mn} := \sum_{k=0}^{3}\sum_{j=0}^{3} g^{mj} g^{nk} c_{jk} \quad \forall m, n = 0, 1, 2, 3. \tag{9.89}$$

We rewrite (9.89) as

$$c^{mn} = g^{m0} g^{n0} c_{00} + \sum_{k=1}^{3} g^{m0} g^{nk} c_{0k} + \sum_{j=1}^{3} g^{mj} g^{n0} c_{j0} + \sum_{k=1}^{3}\sum_{j=1}^{3} g^{mj} g^{nk} c_{jk}$$
$$\forall m, n = 0, 1, 2, 3. \quad (9.90)$$

In particular, by inserting (9.58) and (9.59) into (9.90), we deduce

$$\begin{cases} c^{00} = c_{00} + \sum_{k=1}^{3} \frac{v^k}{c} c_{0k} + \sum_{j=1}^{3} \frac{v^j}{c} c_{j0} + \sum_{k=1}^{3}\sum_{j=1}^{3} \frac{v^j}{c}\frac{v^k}{c} c_{jk} \\ c^{m0} = \frac{v^m}{c} c^{00} - c_{m0} - \sum_{k=1}^{3} \frac{v^k}{c} c_{mk} \quad \forall m = 1, 2, 3, \\ c^{0n} = \frac{v^n}{c} c^{00} - c_{0n} - \sum_{j=1}^{3} \frac{v^j}{c} c_{jn} \quad \forall n = 1, 2, 3, \\ c^{mn} = \frac{v^m}{c}\frac{v^n}{c} c^{00} - \sum_{k=1}^{3} \frac{v^n}{c}\frac{v^k}{c} c_{mk} - \sum_{j=1}^{3} \frac{v_m}{c}\frac{v^j}{c} c_{jn} - \frac{v^m}{c} c_{0n} - \frac{v^n}{c} c_{m0} + c_{mn} \\ \quad \forall m, n = 1, 2, 3. \end{cases} \quad (9.91)$$

We rewrite (9.91) as

$$\begin{cases} c^{00} = c_{00} + \sum_{k=1}^{3} \frac{v^k}{c} c_{0k} + \sum_{j=1}^{3} \frac{v^j}{c} c_{j0} + \sum_{k=1}^{3}\sum_{j=1}^{3} \frac{v^j}{c}\frac{v^k}{c} c_{jk} \\ c^{m0} = \frac{v^m}{c} c^{00} - c_{m0} - \sum_{k=1}^{3} \frac{v^k}{c} c_{mk} \quad \forall m = 1, 2, 3, \\ c^{0n} = \frac{v^n}{c} c^{00} - c_{0n} - \sum_{j=1}^{3} \frac{v^j}{c} c_{jn} \quad \forall n = 1, 2, 3, \\ c^{mn} = \frac{v^m}{c} c^{0n} + \frac{v^n}{c} c^{m0} - \frac{v^m}{c}\frac{v^n}{c} c^{00} + c_{mn} \quad \forall m, n = 1, 2, 3. \end{cases} \quad (9.92)$$

In particular, if the tensor $\{c_{mn}\}_{m,n=0,1,2,3}$ is antisymmetric, i.e. $c_{mn} = -c_{nm}\ \forall m, n = 0, 1, 2, 3$, then we simplify (9.92) as

$$\begin{cases} c^{00} = 0 \\ c^{mm} = 0 \quad \forall m = 1, 2, 3, \\ c^{0m} = -c^{m0} = -c_{0m} + \sum_{k=1}^{3} \frac{v^k}{c} c_{mk} \quad \forall m = 1, 2, 3, \\ c^{mn} = \frac{v^m}{c} c^{0n} - \frac{v^n}{c} c^{0m} + c_{mn} \quad \forall m, n = 1, 2, 3. \end{cases} \quad (9.93)$$

In particular, if $\{F_{ij}\}_{0\le i,j\le 3}$ is the antisymmetric two times covariant tensor field of the electromagnetic field on the group $\mathcal{S}_0$, which by (9.41) satisfies:

$$\begin{cases} F_{00} = 0 \\ F_{0j} = -F_{j0} = E_j \quad \forall j = 1, 2, 3 \\ F_{jj} = 0 \quad \forall j = 1, 2, 3 \\ F_{12} = -F_{21} = -B_3 \\ F_{13} = -F_{31} = B_2 \\ F_{23} = -F_{32} = -B_1 , \end{cases} \quad (9.94)$$

where $\mathbf{E} := (E_1, E_2, E_3)$ and $\mathbf{B} := (B_1, B_2, B_3)$, then by inserting (9.94) into (9.93) we deduce

$$\begin{cases} F^{00} = 0 \\ F^{jj} = 0 \quad \forall\, j = 1, 2, 3, \\ F^{01} = -F^{10} = -F_{01} + \frac{v^2}{c} F_{12} + \frac{v^3}{c} F_{13} = -\left(E_1 + \frac{1}{c}\left(v^2 B_3 - v^3 B_2\right)\right) \\ F^{02} = -F^{20} = -F_{02} + \frac{v^1}{c} F_{21} + \frac{v^3}{c} F_{23} = -\left(E_2 + \frac{1}{c}\left(v^3 B_1 - v^1 B_3\right)\right) \\ F^{03} = -F^{30} = -F_{03} + \frac{v^1}{c} F_{31} + \frac{v^2}{c} F_{32} = -\left(E_3 + \frac{1}{c}\left(v^1 B_2 - v^2 B_1\right)\right) \\ F^{12} = -F^{21} = \frac{v^1}{c} F^{02} - \frac{v^2}{c} F^{01} + F_{12} = -\left(B_3 + \frac{1}{c}\left(v^1 F^{20} - v^2 F^{10}\right)\right) \\ F^{13} = -F^{31} = \frac{v^1}{c} F^{03} - \frac{v^3}{c} F^{01} + F_{13} = B_2 + \frac{1}{c}\left(v^3 F^{10} - v^1 F^{30}\right) \\ F^{23} = -F^{32} = \frac{v^2}{c} F^{03} - \frac{v^3}{c} F^{02} + F_{23} = -\left(B_1 + \frac{1}{c}\left(v^2 F^{30} - v^3 F^{20}\right)\right). \end{cases} \tag{9.95}$$

Thus, as before in (5.10), denoting

$$\begin{cases} \mathbf{D} := \mathbf{E} + \frac{1}{c}\mathbf{v} \times \mathbf{B} \\ \mathbf{H} := \mathbf{B} + \frac{1}{c}\mathbf{v} \times \mathbf{D}, \end{cases} \tag{9.96}$$

and denoting $\mathbf{D} := (D_1, D_2, D_3)$ and $\mathbf{H} := (H_1, H_2, H_3)$ we rewrite (9.95) as

$$\begin{cases} F^{00} = 0 \\ F^{0j} = -F^{j0} = -D_j \quad \forall\, j = 1, 2, 3, \\ F^{jj} = 0 \quad \forall\, j = 1, 2, 3, \\ F^{12} = -F^{21} = -H_3 \\ F^{13} = -F^{31} = H_2 \\ F^{23} = -F^{32} = -H_1. \end{cases} \tag{9.97}$$

In particular, by (9.94) and (9.97), using (9.96) we deduce that the scalar field on the group $\mathcal{S}_0$: L_e, defined as

$$L_e := \sum_{j=0}^{3} \sum_{k=0}^{3} F^{jk} F_{jk}, \tag{9.98}$$

satisfies

$$L_e = F^{00} F_{00} + \sum_{k=1}^{3} F^{0k} F_{0k} + \sum_{j=1}^{3} F^{j0} F_{j0} + \sum_{j=1}^{3} \sum_{k=1}^{3} F^{jk} F_{jk} = -2\mathbf{E} \cdot \mathbf{D} + 2\mathbf{B} \cdot \mathbf{H} =$$
$$-2\left(\left(\mathbf{D} - \frac{1}{c}\mathbf{v} \times \mathbf{B}\right) \cdot \mathbf{D} - \mathbf{B} \cdot \left(\mathbf{B} + \frac{1}{c}\mathbf{v} \times \mathbf{D}\right)\right) = -2\left(|\mathbf{D}|^2 - |\mathbf{B}|^2\right). \tag{9.99}$$

9.3 Maxwell Equations in Covariant Formulation

It is well known from tensor analysis that if $\{S^{ij}\}_{0\leq i,j\leq 3}$ is the antisymmetric two times contravariant tensor and if $\{\xi_{ij}\}_{0\leq i,j\leq 3}$ is a symmetric two times covariant and non-degenerate tensor, both on the certain group $\mathcal{S}$, then the four-component field $\{\theta_k\}_{0\leq k\leq 3}$ defined by

$$\theta_k := \sum_{j=0}^{3}\frac{\partial S^{kj}}{\partial x^j}+\sum_{j=0}^{3}\frac{S^{kj}}{\sqrt{|\det\xi|}}\frac{\partial}{\partial x^j}\left(\sqrt{|\det\xi|}\right) \qquad \forall k=0,1,2,3, \tag{9.100}$$

is a four vector on $\mathcal{S}$. Here ξ is a 4×4-matrix defined by

$$\xi = \{\xi_{ij}\}_{0\leq i,j\leq 3}. \tag{9.101}$$

In particular, if we consider the 4×4-matrix G defined by (9.61) as

$$G = \{g_{ij}\}_{0\leq i,j\leq 3}, \tag{9.102}$$

which satisfies (9.62) in every Cartesian coordinate system, i.e.

$$\det G = -1. \tag{9.103}$$

Then for the lifted contravariant tensor of the electromagnetic field $\{F^{ij}\}_{0\leq i,j\leq 3}$ on the group $\mathcal{S}_0$, considered in (9.97), as in (9.100) we can define the four-vector field:

$$\left\{\sum_{j=0}^{3}\frac{\partial F^{kj}}{\partial x^j}+\sum_{j=0}^{3}\frac{F^{kj}}{\sqrt{|\det G|}}\frac{\partial}{\partial x^j}\left(\sqrt{|\det G|}\right)\right\}_{0\leq k\leq 3} = \left\{\sum_{j=0}^{3}\frac{\partial F^{kj}}{\partial x^j}\right\}_{0\leq k\leq 3} \tag{9.104}$$

on the group $\mathcal{S}_0$. Note here that we denoted the matrix $G = \{g_{ij}\}_{0\leq i,j\leq 3}$ by the same letter as the Gravitational Constant G. However, there is no ambiguity, since in the second case G is a constant scalar and in the first case G is a matrix. Moreover, we will use the matrix notation $G = \{g_{ij}\}_{0\leq i,j\leq 3}$ only in the expressions containing term $\det G$. Then by (9.97), denoting

$$(x^0,x^1,x^2,x^3) := (ct,x_1,x_2,x_3) = (ct,\mathbf{x}),$$

we deduce

$$\begin{cases}\sum_{j=0}^{3}\frac{\partial F^{0j}}{\partial x^j} = -\mathrm{div}_{\mathbf{x}}\mathbf{D}\\ \sum_{j=0}^{3}\frac{\partial F^{1j}}{\partial x^j} = \frac{1}{c}\frac{\partial D_1}{\partial t}-\left(\frac{\partial H_3}{\partial x_2}-\frac{\partial H_2}{\partial x_3}\right)\\ \sum_{j=0}^{3}\frac{\partial F^{2j}}{\partial x^j} = \frac{1}{c}\frac{\partial D_2}{\partial t}-\left(\frac{\partial H_1}{\partial x_3}-\frac{\partial H_3}{\partial x_1}\right)\\ \sum_{j=0}^{3}\frac{\partial F^{3j}}{\partial x^j} = \frac{1}{c}\frac{\partial D_3}{\partial t}-\left(\frac{\partial H_2}{\partial x_1}-\frac{\partial H_1}{\partial x_2}\right).\end{cases} \tag{9.105}$$

I.e.:

$$\left(\sum_{j=0}^{3}\frac{\partial F^{0j}}{\partial x^j},\sum_{j=0}^{3}\frac{\partial F^{1j}}{\partial x^j},\sum_{j=0}^{3}\frac{\partial F^{2j}}{\partial x^j},\sum_{j=0}^{3}\frac{\partial F^{3j}}{\partial x^j}\right)=\left(-\mathrm{div}_{\mathbf{x}}\mathbf{D},\left(\frac{1}{c}\frac{\partial \mathbf{D}}{\partial t}-\mathrm{curl}_{\mathbf{x}}\mathbf{H}\right)\right). \tag{9.106}$$

Therefore, by (9.106), the first pair of Maxwell Equations in (5.10):

$$\begin{cases}\mathrm{curl}_{\mathbf{x}}\mathbf{H}=\frac{4\pi}{c}\mathbf{j}+\frac{1}{c}\frac{\partial \mathbf{D}}{\partial t}\\ \mathrm{div}_{\mathbf{x}}\mathbf{D}=4\pi\rho\end{cases} \tag{9.107}$$

is equivalent to the following equations:

$$\left(\sum_{j=0}^{3}\frac{\partial F^{0j}}{\partial x^j},\sum_{j=0}^{3}\frac{\partial F^{1j}}{\partial x^j},\sum_{j=0}^{3}\frac{\partial F^{2j}}{\partial x^j},\sum_{j=0}^{3}\frac{\partial F^{3j}}{\partial x^j}\right)=-4\pi(j^0,j^1,j^2,j^3), \tag{9.108}$$

where (j^0, j^1, j^2, j^3) is the four vector of electric current on the group $\mathcal{S}_0$ defined by (9.33) as

$$(j^0,j^1,j^2,j^3):=\left(\rho,\frac{1}{c}\mathbf{j}\right). \tag{9.109}$$

Note that in both sides of equation (9.108) we have four vectors and thus (9.108) is a covariant form of (9.107). On the other hand, the second pair of Maxwell equations in (5.10):

$$\begin{cases}\mathrm{curl}_{\mathbf{x}}\mathbf{E}+\frac{1}{c}\frac{\partial \mathbf{B}}{\partial t}=0\\ \mathrm{div}_{\mathbf{x}}\mathbf{B}=0\end{cases} \tag{9.110}$$

is equivalent to (9.40), i.e. to the following:

$$\begin{cases}\mathbf{B}=\mathrm{curl}_{\mathbf{x}}\mathbf{A},\\ \mathbf{E}=-\nabla_{\mathbf{x}}\Psi-\frac{1}{c}\frac{\partial \mathbf{A}}{\partial t}.\end{cases} \tag{9.111}$$

On the other hand, as before, by (9.41) we can rewrite (9.111) in the form of (9.38):

$$F_{ij}=\frac{\partial A_j}{\partial x^i}-\frac{\partial A_i}{\partial x^j}\qquad\forall i,j=0,1,2,3, \tag{9.112}$$

where (A_0, A_1, A_2, A_3) is the four covector of the electromagnetic potential on the group $\mathcal{S}_0$ defined by (9.37) as

$$(A_0,A_1,A_2,A_3)=(\Psi,-\mathbf{A}). \tag{9.113}$$

Note that in both sides of equation (9.112) we have two time covariant tensors, and thus (9.112) is a covariant form of (9.110). Finally, by (9.95), the relations between $(\mathbf{E}, \mathbf{B})$ and $(\mathbf{D}, \mathbf{H})$ in (9.96):

$$\begin{cases} \mathbf{D} = \mathbf{E} + \frac{1}{c}\mathbf{v} \times \mathbf{B} \\ \mathbf{H} = \mathbf{B} + \frac{1}{c}\mathbf{v} \times \mathbf{D} \end{cases} \tag{9.114}$$

are equivalent to the following covariant equations:

$$F^{mn} := \sum_{k=0}^{3}\sum_{j=0}^{3} g^{mj} g^{nk} F_{jk} \qquad \forall m, n = 0, 1, 2, 3. \tag{9.115}$$

Thus by (9.112), (9.115) and (9.108) together, we deduce that the full system of Maxwell equations in (5.10):

$$\begin{cases} \mathrm{curl}_{\mathbf{x}}\mathbf{H} = \frac{4\pi}{c}\mathbf{j} + \frac{1}{c}\frac{\partial \mathbf{D}}{\partial t} \\ \mathrm{div}_{\mathbf{x}}\mathbf{D} = 4\pi\rho \\ \mathrm{curl}_{\mathbf{x}}\mathbf{E} + \frac{1}{c}\frac{\partial \mathbf{B}}{\partial t} = 0 \\ \mathrm{div}_{\mathbf{x}}\mathbf{B} = 0 \\ \mathbf{E} = \mathbf{D} - \frac{1}{c}\mathbf{v} \times \mathbf{B} \\ \mathbf{H} = \mathbf{B} + \frac{1}{c}\mathbf{v} \times \mathbf{D} \end{cases} \tag{9.116}$$

is equivalent to the following covariant equations:

$$\sum_{j=0}^{3} \frac{\partial}{\partial x^j}\left(\sum_{m=0}^{3}\sum_{n=0}^{3} g^{km} g^{jn}\left(\frac{\partial A_n}{\partial x^m} - \frac{\partial A_m}{\partial x^n}\right)\right) = -4\pi j^k \qquad \forall k = 0, 1, 2, 3. \tag{9.117}$$

Note that equations (9.117) are fully analogous to the covariant formulation of Maxwell equations in Special Relativity and the only difference is the choice of the pseudo-metric tensor $\{g^{ij}\}_{0\le i,j\le 3}$ (Note that for the Special Relativity case we also have $\det G = -1$). As for the cases of the General Relativity, the covariant formulation of Maxwell equations is still similar to (9.117); however, in addition to the different choice of the pseudo-metric tensor $\{g^{ij}\}_{0\le i,j\le 3}$ we also have $\det G \neq Const.$ and thus for the full analogy equations (9.117) should be rewritten in the enlarged form, due to (9.100) and (9.104):

$$\sum_{j=0}^{3}\frac{\partial}{\partial x^j}\left(\sum_{m=0}^{3}\sum_{n=0}^{3}g^{km}g^{jn}\left(\frac{\partial A_n}{\partial x^m}-\frac{\partial A_m}{\partial x^n}\right)\right)+$$
$$\sum_{j=0}^{3}\frac{1}{\sqrt{|\det G|}}\frac{\partial}{\partial x^j}\left(\sqrt{|\det G|}\right)\left(\sum_{m=0}^{3}\sum_{n=0}^{3}g^{km}g^{jn}\left(\frac{\partial A_n}{\partial x^m}-\frac{\partial A_m}{\partial x^n}\right)\right)=-4\pi j^k$$
$$\forall k=0,1,2,3. \quad (9.118)$$

Note also that we can rewrite (9.118) as

$$\sum_{j=0}^{3}\frac{\partial}{\partial x^j}\left(\sum_{m=0}^{3}\sum_{n=0}^{3}\sqrt{|\det G|}\,g^{km}g^{jn}\left(\frac{\partial A_n}{\partial x^m}-\frac{\partial A_m}{\partial x^n}\right)\right)=-4\pi\sqrt{|\det G|}\,j^k$$
$$\forall k=0,1,2,3. \quad (9.119)$$

Next by (9.98) and (9.99) we have

$$\frac{1}{2}|\mathbf{D}|^2-\frac{1}{2}|\mathbf{B}|^2=-\sum_{j=0}^{3}\sum_{k=0}^{3}\frac{1}{4}F^{jk}F_{jk}. \quad (9.120)$$

Therefore, by (9.109), (9.113) and (9.120), we can rewrite the density of the Lagrangian of the electromagnetic field, defined in (5.62) as

$$L_1(\mathbf{A},\Psi,\mathbf{x},t):=\frac{1}{4\pi}\left(\frac{1}{2}|\mathbf{D}|^2-\frac{1}{2}|\mathbf{B}|^2-4\pi\left(\rho\Psi-\frac{1}{c}\mathbf{A}\cdot\mathbf{j}\right)\right), \quad (9.121)$$

in the equivalent covariant form:

$$L_1=\frac{1}{4\pi}\left(-\sum_{n=0}^{3}\sum_{k=0}^{3}\frac{1}{4}F^{nk}F_{nk}-\sum_{k=0}^{3}4\pi j^k A_k\right)=$$
$$\frac{1}{4\pi}\left(-\sum_{n=0}^{3}\sum_{k=0}^{3}\sum_{m=0}^{3}\sum_{p=0}^{3}\frac{1}{4}g^{mn}g^{pk}\left(\frac{\partial A_p}{\partial x^m}-\frac{\partial A_m}{\partial x^p}\right)\left(\frac{\partial A_k}{\partial x^n}-\frac{\partial A_n}{\partial x^k}\right)-\sum_{k=0}^{3}4\pi j^k A_k\right). \quad (9.122)$$

The density of Lagrangian in (9.122) is also fully analogous to the covariant formulation of the Lagrangian density of the electromagnetic field in Special and General Relativity and the only difference is the choice of the pseudo-metric tensor $\{g^{ij}\}_{0\le i,j\le 3}$.

9.4 Covariant Form of Lagrangian of Motion of a Classical Charged Particle

Given a classical charged particle with inertial mass m, charge σ, three-dimensional place $\mathbf{r}(t)$ and three-dimensional velocity $\frac{d\mathbf{r}}{dt}$ in the outer gravitational field with three-dimensional vectorial potential $\mathbf{v}(\mathbf{x}, t)$, the outer electromagnetical field with three-dimensional vectorial potential $\mathbf{A}(\mathbf{x}, t)$ and scalar potential $\Psi(\mathbf{x}, t)$, consider a usual Lagrangian that is a particular case of (6.1):

$$L_0\left(\frac{d\mathbf{r}}{dt}, t\right) := \left\{\frac{m}{2}\left|\frac{d\mathbf{r}}{dt} - \mathbf{v}(\mathbf{r}, t)\right|^2 - \sigma\left(\Psi(\mathbf{r}, t) - \frac{1}{c}\mathbf{A}(\mathbf{r}, t) \cdot \frac{d\mathbf{r}}{dt}\right)\right\}. \quad (9.123)$$

Then, since we are interesting in critical points of the functional

$$J_0 = \int_0^T L_0\left(\frac{d\mathbf{r}}{dt}, \mathbf{r}, t\right) dt, \quad (9.124)$$

adding a constant does not changes the physical meaning of the Lagrangian and we can rewrite (9.123) as

$$L_0'\left(\frac{d\mathbf{r}}{dt}, t\right) := \left\{\left(\frac{m}{2}\left|\frac{d\mathbf{r}}{dt} - \mathbf{v}(\mathbf{r}, t)\right|^2 - \frac{mc^2}{2}\right) - \sigma\left(\Psi(\mathbf{r}, t) - \frac{1}{c}\mathbf{A}(\mathbf{r}, t) \cdot \frac{d\mathbf{r}}{dt}\right)\right\}, \quad (9.125)$$

and (9.124) as

$$J_0' := J_0 - \frac{Tmc^2}{2} = \int_0^T L_0'\left(\frac{d\mathbf{r}}{dt}, \mathbf{r}, t\right) dt =$$
$$\int_0^T \left\{\left(\frac{m}{2}\left|\frac{d\mathbf{r}}{dt} - \mathbf{v}(\mathbf{r}, t)\right|^2 - \frac{mc^2}{2}\right) - \sigma\left(\Psi(\mathbf{r}, t) - \frac{1}{c}\mathbf{A}(\mathbf{r}, t) \cdot \frac{d\mathbf{r}}{dt}\right)\right\} dt. \quad (9.126)$$

Next consider the four-vector field of the momentum on the group $\mathcal{S}_0$: $\left(p^0(t), p^1(t), p^2(t), p^3(t)\right)$, defined by (9.28) and (9.32) as

$$\left(p^0(t), p^1(t), p^2(t), p^3(t)\right) := \left(m, \frac{m}{c}\frac{d\mathbf{r}}{dt}(t)\right) = \left(m, \frac{m}{c}\frac{dr_1}{dt}(t), \frac{m}{c}\frac{dr_2}{dt}(t), \frac{m}{c}\frac{dr_3}{dt}(t)\right). \quad (9.127)$$

Then by (9.81) and (9.82) we have

$$\frac{mc^2}{2}\left(\frac{1}{c^2}\left|\frac{d\mathbf{r}}{dt}-\mathbf{v}(\mathbf{r},t)\right|^2-1\right)=\frac{m}{2}\left|\frac{d\mathbf{r}}{dt}-\mathbf{v}(\mathbf{r},t)\right|^2-\frac{mc^2}{2}$$
$$=-\frac{c^2}{2m}\left(\sum_{k=0}^{3}p^k p_k\right)=-\frac{mc^2}{2}\left(\sum_{j=0}^{3}\sum_{k=0}^{3}g_{jk}(\mathbf{r},t)\frac{p^j}{m}\frac{p^k}{m}\right). \tag{9.128}$$

On the other hand if we consider the four covector of the electromagnetic potential on the group $\mathcal{S}_0$: (A_0, A_1, A_2, A_3), defined by (9.37) as

$$(A_0, A_1, A_2, A_3) = (\Psi, -\mathbf{A}), \tag{9.129}$$

then we can write

$$\sigma\left(\Psi(\mathbf{r},t)-\frac{1}{c}\mathbf{A}(\mathbf{r},t)\cdot\frac{d\mathbf{r}}{dt}\right)=\sum_{k=0}^{3}\sigma A_k(\mathbf{r},t)\frac{p^k}{m}. \tag{9.130}$$

Thus by (9.128) and (9.130) we rewrite (9.126) in a covariant form:

$$J_0' = \int_0^T L_0'\left(\frac{d\mathbf{r}}{dt},\mathbf{r},t\right)dt =$$
$$\int_0^T\left\{-\frac{mc^2}{2}\left(\sum_{j=0}^{3}\sum_{k=0}^{3}g_{jk}(\mathbf{r},t)\frac{p^j}{m}\frac{p^k}{m}\right)-\sum_{k=0}^{3}\sigma A_k(\mathbf{r},t)\frac{p^k}{m}\right\}dt. \tag{9.131}$$

Thus if we consider the four-dimensional space-time trajectory of the particle:

$$\left(\chi^0(t),\chi^1(t),\chi^2(t),\chi^3(t)\right)=\left(t,\frac{1}{c}r_1(t),\frac{1}{c}r_2(t),\frac{1}{c}r_3(t)\right), \tag{9.132}$$

then we rewrite (9.131) as

$$J_0'=\int_0^T\left\{-\frac{mc^2}{2}\left(\sum_{j=0}^{3}\sum_{k=0}^{3}g_{jk}(\chi(t))\frac{d\chi^j}{dt}\frac{d\chi^k}{dt}\right)-\sum_{k=0}^{3}\sigma A_k(\chi(t))\frac{d\chi^k}{dt}\right\}dt. \tag{9.133}$$

Moreover, $\left(\frac{d\chi^0}{dt},\frac{d\chi^1}{dt},\frac{d\chi^2}{dt},\frac{d\chi^3}{dt}\right)$ is a four vector on the group $\mathcal{S}_0$ and the global non-relativistic time t is the scalar on the group $\mathcal{S}_0$.

Next we also can consider a more general Lagrangian then (9.133): given a function $\mathcal{G}(\tau) : \mathbb{R} \to \mathbb{R}$ define

$$J_{\mathcal{G}}(\chi) = \int_0^T \left\{ -mc^2\, \mathcal{G}\left(\sum_{j=0}^{3}\sum_{k=0}^{3} g_{jk}(\chi(t)) \frac{d\chi^j}{dt}\frac{d\chi^k}{dt} \right) - \sum_{k=0}^{3} \sigma A_k(\chi(t)) \frac{d\chi^k}{dt} \right\} dt. \quad (9.134)$$

Clearly, (9.134) is written in covariant form, and, in particular, (9.134) is invariant under the change of non-inertial Cartesian coordinate systems. In particular, for $\mathcal{G}(\tau) := \frac{1}{2}\tau$, we obtain (9.133).

Another important particular case is the following choice: $\mathcal{G}(\tau) := \sqrt{\tau}$. Then we deduce

$$J_{rl}(\chi) = \int_0^T \left\{ -mc^2 \sqrt{\left(\sum_{j=0}^{3}\sum_{k=0}^{3} g_{jk}(\chi(t)) \frac{d\chi^j}{dt}\frac{d\chi^k}{dt} \right)} - \sum_{k=0}^{3} \sigma A_k(\chi(t)) \frac{d\chi^k}{dt} \right\} dt, \quad (9.135)$$

that is in somewhat analogous to the relativistic Lagrangian of the motion of charged particle. Due to (9.132) we rewrite (9.135) in a three-dimensional form as

$$J_{rl}(\mathbf{r}) = \int_0^T \left\{ -mc^2 \sqrt{1 - \frac{1}{c^2}\left| \frac{d\mathbf{r}}{dt} - \mathbf{v}(\mathbf{r},t) \right|^2} - \sigma\left(\Psi(\mathbf{r},t) - \frac{1}{c}\mathbf{A}(\mathbf{r},t)\cdot\frac{d\mathbf{r}}{dt} \right) \right\} dt. \quad (9.136)$$

Thus in the case

$$\frac{1}{c^2}\left| \frac{d\mathbf{r}}{dt} - \mathbf{v}(\mathbf{r},t) \right|^2 \ll 1,$$

up to additive constant, (9.136) becomes to be (9.124), where L_0 is given by (9.123). Note that the Lagrangian in (9.135) has the following advantage with respect to (9.133): if we parameterize the curve in (9.132) by some arbitrary parameter s with increasing mapping $t \leftrightarrow s$, which, however, can differ from the global time t, then changing variables of integration in (9.135) from t to s gives

$$J_{rl}(\chi) = \int_a^b \left\{ -mc^2 \sqrt{\left(\sum_{j=0}^{3}\sum_{k=0}^{3} g_{jk}(\chi(s)) \frac{d\chi^j}{ds}\frac{d\chi^k}{ds} \right)} - \sum_{k=0}^{3} \sigma A_k(\chi(s)) \frac{d\chi^k}{ds} \right\} ds, \quad (9.137)$$

that has exactly the same form as (9.135), however s in (9.137) can be arbitrary parameter of the curve with increasing mapping $t \leftrightarrow s$.

Finally, we would like to note that if the motion of some particle is ruled by the relativistic-like Lagrangian in (9.136), then, although the absolute value of the velocity of the particle $\left|\frac{d\mathbf{r}}{dt}\right|$ can be arbitrary large, the absolute value of the difference between the velocity of the particle and the local gravitational potential cannot exceed the value c, i.e.:

$$|\mathbf{u}(t) - \mathbf{v}(\mathbf{r},t)| := \left| \frac{d\mathbf{r}}{dt} - \mathbf{v}(\mathbf{r},t) \right| < c \qquad \forall t, \quad (9.138)$$

provided that (9.138) is satisfied in some initial instant of time. Note also that the quantity in the left-hand side of (9.138) is invariant under the change of inertial or non-inertial Cartesian coordinate system.

9.5 Kinematic Pseudo-Metric Tensors of Inertia

Consider $\{J^{ij}\}_{0\leq i,j\leq 3}$ to be a two times contravariant tensor field on the group $\mathcal{S}_0$, defined by

$$J^{ij} := k^i k^j - \Theta^{ij} \qquad \forall i, j = 0, 1, 2, 3, \tag{9.139}$$

where $\{\Theta^{ij}\}_{0\leq i,j\leq 3}$ is the contravariant tensor of the three-dimensional geometry, defined by (9.48) and being a two times contravariant tensor, and (k^0, k^1, k^2, k^3) is the four-dimensional potential of inertia, defined by (9.30) and being a four vector. Then by (9.7) we obtain that $\{J^{ij}\}_{0\leq i,j\leq 3}$ is indeed a two times contravariant tensor field on the group $\mathcal{S}_0$, and moreover this tensor is symmetric. Moreover, by (9.48) and (9.30), we have

$$\begin{cases} J^{00} = 1 \\ J^{ij} = -\delta_{ij} + \frac{k^i k^j}{c^2} \quad \forall 1 \leq i, j \leq 3 \\ J^{0j} = J^{j0} = \frac{k^j}{c} \quad \forall 1 \leq j \leq 3, \end{cases} \tag{9.140}$$

where $\mathbf{k} = (k^1, k^2, k^3)$ is the three-dimensional vectorial potential of inertia. We call the tensor $\{J^{ij}\}_{0\leq i,j\leq 3}$ the contravariant kinematic pseudo-metric tensor of inertia. Next consider a 16-component field $\{J_{ij}\}_{0\leq i,j\leq 3}$ defined by

$$\begin{cases} J_{00} = 1 - \frac{|\mathbf{k}|^2}{c^2} \\ J_{ij} = -\delta_{ij} \quad \forall 1 \leq i, j \leq 3 \\ J_{0j} = J_{j0} = \frac{k^j}{c} \quad \forall 1 \leq j \leq 3. \end{cases} \tag{9.141}$$

Then, as before, we have

$$\sum_{m=0}^{3} J_{0m} J^{m0} = J_{00} J^{00} + \sum_{m=1}^{3} J_{0m} J^{m0} = 1 - \frac{|\mathbf{k}|^2}{c^2} + \frac{|\mathbf{k}|^2}{c^2} = 1,$$

$$\sum_{m=0}^{3} J_{im} J^{mj} = J_{i0} J^{0j} + \sum_{m=1}^{3} J_{im} J^{mj} = \frac{k^i k^j}{c^2} + \delta_{ij} - \frac{k^i k^j}{c^2} = \delta_{ij} \quad \forall 1 \leq i, j \leq 3,$$

and

$$\sum_{m=0}^{3} J_{im} J^{m0} = J_{i0} J^{00} + \sum_{m=1}^{3} J_{im} J^{m0} = \frac{k^i}{c} - \frac{k^i}{c} = 0 \ \ \forall 1 \le i \le 3,$$

$$\sum_{m=0}^{3} J_{0m} J^{mj} = J_{00} J^{0j} + \sum_{m=1}^{3} J_{0m} J^{mj} = \left(1 - \frac{|\mathbf{k}|^2}{c^2}\right) \frac{k^j}{c} - \sum_{m=1}^{3} \frac{k^m}{c} \left(\delta_{mj} - \frac{k^m k^j}{c^2}\right) = 0$$
$$\forall 1 \le j \le 3,$$

where $\{J^{ij}\}_{0 \le i,j \le 3}$ is the contravariant kinematic pseudo-metric tensors of inertia, defined by (9.140). So,

$$\sum_{m=0}^{3} J^{im} J_{mj} = \begin{cases} 1 & \text{if } \ i = j \\ 0 & \text{if } \ i \neq j \end{cases} \quad \forall i, j = 0, 1, 2, 3. \tag{9.142}$$

Therefore, by comparing (9.142) and (9.9) we deduce that $\{J_{ij}\}_{i,j=0,1,2,3}$ is a two times covariant tensor on the group $\mathcal{S}_0$, and moreover this tensor is symmetric. We call the tensor $\{J_{ij}\}_{0 \le i,j \le 3}$ covariant kinematic pseudo-metric tensors of inertia. Using (9.142) we also obtain that the pseudo-metric tensors $\{J_{ij}\}_{i,j=0,1,2,3}$ and $\{J^{ij}\}_{0 \le i,j \le 3}$ are non-degenerate and they are inverse of each other. Moreover, as before, it can be easily calculated that if we consider the 4×4-matrix:

$$J = \{J_{ij}\}_{0 \le i,j \le 3}, \tag{9.143}$$

then

$$\det J = -1. \tag{9.144}$$

In particular, by (9.144) and (9.62) we deduce

$$\det G = \det J, \tag{9.145}$$

where 4×4-matrix G is given by

$$G = \{g_{ij}\}_{0 \le i,j \le 3}, \tag{9.146}$$

with the covariant pseudo-metric tensor of the four-dimensional space-time $\{g_{ij}\}_{0 \le i,j \le 3}$, given by (9.59). Next, obviously we have

$$g^{ij} = J^{ij} + \left(v^i v^j - k^i k^j\right) \qquad \forall i, j = 0, 1, 2, 3, \tag{9.147}$$

where J^{ij} is the contravariant kinematic pseudo-metric tensor of inertia, given by (9.139), and g^{ij} is the contravariant pseudo-metric tensor of the four-dimensional space-time, given by (9.57). In particular, in the case of the absence of the genuine gravity where $\mathbf{v} = \mathbf{k}$ we have

$$g^{ij} = J^{ij} \qquad \forall i, j = 0, 1, 2, 3, \tag{9.148}$$

or, equivalently,

$$g_{ij} = J_{ij} \qquad \forall i, j = 0, 1, 2, 3. \tag{9.149}$$

Furthermore, by (9.140), we have

$$\sum_{m=0}^{3} J^{jm} \left(c \frac{\partial \tau}{\partial x^m} \right) = k^j \qquad \forall j = 0, 1, 2, 3, \tag{9.150}$$

where (k^0, k^1, k^2, k^3) is the four-dimensional potential of inertia, defined by (9.30) and τ is the scalar of the global time on the group $\mathcal{S}_0$, defined by (9.49). Thus since

$$\sum_{j=0}^{3} k^j \left(c \frac{\partial \tau}{\partial x^j} \right) = 1, \tag{9.151}$$

we deduce the following eikonal-type equation

$$c^2 \left(\sum_{j=0}^{3} \sum_{m=0}^{3} J^{jm} \frac{\partial \tau}{\partial x^j} \frac{\partial \tau}{\partial x^m} \right) = 1, \tag{9.152}$$

which is similar to (9.75), despite the different pseudo-metrics. Moreover, by (9.150) and (9.142) we infer

$$\sum_{m=0}^{3} J_{jm} k^m = c \frac{\partial \tau}{\partial x^j} \qquad \forall j = 0, 1, 2, 3. \tag{9.153}$$

Next, as before, note that the Kinematic pseudo-metric tensors of inertia $\{J^{ij}\}_{0 \le i, j \le 3}$ and $\{J_{ij}\}_{0 \le i, j \le 3}$ depend only on the coordinate system in the space-time and are completely independent on the physical matter or physical fields filling this space. In contrast, the pseudo-metric tensors of the four-dimensional space-time $\{g^{ij}\}_{0 \le i, j \le 3}$ and $\{g_{ij}\}_{0 \le i, j \le 3}$, given by (9.58) and (9.59), depend essentially on the surrounding physical matter (in the model of the Newtonian gravity through gravitational masses). Furthermore, note that in the absence of the genuine gravity $\{g^{ij}\}_{0 \le i, j \le 3}$ and $\{g_{ij}\}_{0 \le i, j \le 3}$ coincide with $\{J^{ij}\}_{0 \le i, j \le 3}$ and $\{J_{ij}\}_{0 \le i, j \le 3}$, respectively. In the model of the Newtonian gravity this happens away of essential gravitational masses. In particular, $\{g^{ij}\}_{0 \le i, j \le 3}$ and $\{g_{ij}\}_{0 \le i, j \le 3}$ tend to $\{J^{ij}\}_{0 \le i, j \le 3}$ and $\{J_{ij}\}_{0 \le i, j \le 3}$, respectively, as $|\mathbf{x}| \to +\infty$. Finally note that since in every inertial Cartesian coordinate system $\mathbf{k}$ is a constant we deduce that in every such a system $\{J^{ij}\}_{0 \le i, j \le 3}$ and $\{J_{ij}\}_{0 \le i, j \le 3}$

are constant in the four-dimensional space-time. So, in contrast to $\{g^{ij}\}_{0\le i,j\le 3}$ and $\{g_{ij}\}_{0\le i,j\le 3}$, the pseudo-metrics $\{J^{ij}\}_{0\le i,j\le 3}$ and $\{J_{ij}\}_{0\le i,j\le 3}$ are completely flat. In particular, if we consider the Christoffel symbols of the kinematic pseudo-metric tensors of inertia $\{J_{ij}\}_{0\le i,j\le 3}$ defined by

$$\begin{cases} \left\{\Gamma_{i,mn}\right\}_J := \frac{1}{2}\left(\frac{\partial J_{im}}{\partial x_n} + \frac{\partial J_{in}}{\partial x_m} - \frac{\partial J_{mn}}{\partial x_i}\right) \\ \left\{\Gamma^i_{mn}\right\}_J := \sum_{j=0}^{3} J^{ij}\left\{\Gamma_{j,mn}\right\}_J \end{cases} \quad \forall i, m, n = 0, 1, 2, 3, \tag{9.154}$$

then, since in every inertial Cartesian coordinate system $\{J_{ij}\}_{0\le i,j\le 3}$ is constant in the four-dimensional space-time, we deduce that in every such coordinate system, we have

$$\left\{\Gamma^i_{mn}\right\}_J = \left\{\Gamma_{i,mn}\right\}_J = 0 \quad \forall i, m, n = 0, 1, 2, 3. \tag{9.155}$$

Thus, if we consider the two times covariant tensor of the covariant derivative to the covector of the gradient to the scalar of the global time τ from (9.49), denoted by $\left\{\delta_j\left(\frac{\partial\tau}{\partial x_i}\right)\right\}_J$, and defined by

$$\begin{aligned} \left\{\delta_j\left(\frac{\partial\tau}{\partial x_i}\right)\right\}_J = \left\{\delta_i\left(\frac{\partial\tau}{\partial x_j}\right)\right\}_J &:= \frac{\partial^2\tau}{\partial x_i\partial x_j} - \sum_{m=0}^{3}\left\{\Gamma^m_{ij}\right\}_J\frac{\partial\tau}{\partial x_m} \\ &= \frac{\partial^2\tau}{\partial x_i\partial x_j} - \sum_{m=0}^{3}\left\{\Gamma_{m,ij}\right\}_J\frac{k^m}{c} \quad \forall i, j = 0, 1, 2, 3, \end{aligned} \tag{9.156}$$

then by (9.50) and (9.155) we prove the following identity, first in every inertial Cartesian coordinate system, and then by the covariance of this identity, also in every non-inertial Cartesian coordinate system:

$$\left\{\delta_j\left(\frac{\partial\tau}{\partial x_i}\right)\right\}_J = 0 \quad \forall i, j = 0, 1, 2, 3. \tag{9.157}$$

Note here that we put brackets $\{\cdot\}_J$ for the Christoffel symbols and covariant derivative with respect to kinematical pseudo-metrics $\{J^{ij}\}_{0\le i,j\le 3}$ and $\{J_{ij}\}_{0\le i,j\le 3}$ in order to distinguish them from Christoffel symbols and covariant derivative with respect to dynamical pseudo-metrics $\{g^{ij}\}_{0\le i,j\le 3}$ and $\{g_{ij}\}_{0\le i,j\le 3}$ that we will write without any brackets.

Furthermore, since $\mathbf{k}$ is generally trivial speed-like vector field, by the definition, there exists a unique, up to equivalence, "preferable" inertial Cartesian coordinate system ($\{0\}$) where we have

$$\mathbf{k} = 0, \tag{9.158}$$

and inserting it into (9.140) and (9.141) we deduce that in system ($\{0\}$) we have

$$\begin{cases} J^{00} = 1 \\ J^{ij} = -\delta_{ij} \quad \forall 1 \leq i, j \leq 3 \\ J^{0j} = J^{j0} = 0 \quad \forall 1 \leq j \leq 3, \end{cases} \tag{9.159}$$

and

$$\begin{cases} J_{00} = 1 \\ J_{ij} = -\delta_{ij} \quad \forall 1 \leq i, j \leq 3 \\ J_{0j} = J_{j0} = 0 \quad \forall 1 \leq j \leq 3, \end{cases} \tag{9.160}$$

which is the same as the basic pseudo-metrics in the Special Relativity.

Next, we define the dynamical four covector of genuine gravity (s_0, s_1, s_2, s_3) by the following:

$$s_j = \frac{1}{2}\left(\sum_{m=0}^{3} g_{jm}k^m - \sum_{m=0}^{3} J_{jm}v^m\right) \quad \forall j = 0, 1, 2, 3, \tag{9.161}$$

where (k^0, k^1, k^2, k^3) is the four-dimensional potential of inertia, defined by (9.30), (v^0, v^1, v^2, v^3) is the four-dimensional gravitational potential, defined by (9.29), $\{g_{ij}\}_{0\leq i,j\leq 3}$ defined by (9.59) and $\{J_{ij}\}_{0\leq i,j\leq 3}$ defined by (9.141). Then, by inserting (9.30), (9.29), (9.59) and (9.141) into (9.161) we obtain

$$s_0 = \frac{1}{2}\left(\sum_{m=0}^{3} g_{0m}k^m - \sum_{m=0}^{3} J_{0m}v^m\right) =$$
$$\frac{1}{2}\left(g_{00}k^0 - J_{00}v^0\right) + \frac{1}{2}\left(\sum_{m=1}^{3} g_{0m}k^m - \sum_{m=1}^{3} J_{0m}v^m\right) =$$
$$\frac{1}{2}\left(\left(1 - \frac{|\mathbf{v}|^2}{c^2}\right) - \left(1 - \frac{|\mathbf{k}|^2}{c^2}\right)\right) + \frac{1}{2}\left(\sum_{m=1}^{3} v^m k^m - \sum_{m=1}^{3} k^m v^m\right) = \frac{1}{2c^2}|\mathbf{k}|^2 - \frac{1}{2c^2}|\mathbf{v}|^2 =$$
$$-\frac{1}{2c^2}(\mathbf{v}-\mathbf{k})\cdot(\mathbf{k}+\mathbf{v}) = -\frac{1}{2c^2}|\mathbf{v}-\mathbf{k}|^2 - \frac{1}{c^2}\mathbf{k}\cdot(\mathbf{v}-\mathbf{k}) = \frac{1}{2c^2}|\mathbf{v}-\mathbf{k}|^2 - \frac{1}{c^2}\mathbf{v}\cdot(\mathbf{v}-\mathbf{k}), \tag{9.162}$$

and

$$s_j = \frac{1}{2}\left(\sum_{m=0}^{3} g_{jm}k^m - \sum_{m=0}^{3} J_{jm}v^m\right) =$$
$$\frac{1}{2}\left(g_{j0}k^0 - J_{j0}v^0\right) + \frac{1}{2}\left(\sum_{m=1}^{3} g_{jm}k^m - \sum_{m=1}^{3} J_{jm}v^m\right) =$$
$$\frac{1}{2}\left(v^j k^0 - k^j v^0\right) + \frac{1}{2}\left(\sum_{m=1}^{3} (-\delta_{jm})\, k^m - \sum_{m=1}^{3} (-\delta_{jm})\, v^m\right) = v^j - k^j \quad \forall j = 1, 2, 3. \tag{9.163}$$

So, by (9.162) and (9.163) we deduce that

$$(s_0, s_1, s_2, s_3) = \left(-\frac{1}{2c^2}|\mathbf{h}|^2 - \frac{1}{c^2}\mathbf{k}\cdot\mathbf{h}\,,\ \frac{1}{c}\mathbf{h}\right) = \left(\frac{1}{2c^2}|\mathbf{h}|^2 - \frac{1}{c^2}\mathbf{v}\cdot\mathbf{h}\,,\ \frac{1}{c}\mathbf{h}\right) \quad \text{where}$$
$$s_0 = -\frac{1}{2c^2}|\mathbf{h}|^2 - \frac{1}{c^2}\mathbf{k}\cdot\mathbf{h} = \frac{1}{2c^2}|\mathbf{h}|^2 - \frac{1}{c^2}\mathbf{v}\cdot\mathbf{h} \quad \text{and} \quad (s_1, s_2, s_3) = \frac{1}{c}\mathbf{h}, \tag{9.164}$$

where $\mathbf{h}$ is the three-dimensional proper vector field of the vectorial potential of genuine gravity defined as in (4.101) by

$$\mathbf{h} := \mathbf{v} - \mathbf{k}. \tag{9.165}$$

In particular, by (9.59), (9.141), (9.164) and (9.50) together we deduce

$$g_{ij} = J_{ij} + \left(s_i\left(c\frac{\partial\tau}{\partial x^j}\right) + \left(c\frac{\partial\tau}{\partial x^i}\right)s_j\right) \qquad \forall\, i, j = 0, 1, 2, 3. \tag{9.166}$$

Moreover, by (9.164) (9.48) and (9.139) we obtain

$$v^j - k^j = \sum_{m=0}^{3}\Theta^{jm}s_m = \sum_{m=0}^{3}\left(k^j k^m - J^{jm}\right)s_m \qquad \forall\, j = 0, 1, 2, 3. \tag{9.167}$$

9.6 Physical Laws in Non-Cartesian or Curvilinear Coordinate Systems in the Space-Time

Let $\mathcal{S}$ be the group of all smooth non-degenerate invertible transformations from $\mathbb{R}^4$ onto $\mathbb{R}^4$ having the form (9.1):

$$\begin{cases} x'^0 = f^{(0)}(x^0, x^1, x^2, x^3),\\ x'^1 = f^{(1)}(x^0, x^1, x^2, x^3),\\ x'^2 = f^{(2)}(x^0, x^1, x^2, x^3),\\ x'^3 = f^{(3)}(x^0, x^1, x^2, x^3), \end{cases} \tag{9.168}$$

and let $\mathcal{S}_0$ be a subgroup of transformations of the form (9.15). Then, it is clear, that given any object that is a scalar, four vector, four covector, two times covariant tensor or two times contravariant tensor on the group $\mathcal{S}_0$, defined in every Cartesian non-inertial coordinate system, we can uniquely extend the definition of this object, in such a way that it will be defined also in every curvilinear coordinate systems in $\mathbb{R}^4$ and will be, respectively, a scalar, four vector, four covector, two times covariant tensor or two times contravariant tensor on the wider group $\mathcal{S}$. Thus all the physical laws that have a covariant form preserve their form also in transformations of the form

(9.168), i.e. in curvilinear coordinate systems. In particular, the Maxwell equations in every curvilinear coordinate system have the form of (9.118) or equivalently of (9.119):

$$\sum_{j=0}^{3}\frac{\partial}{\partial x^j}\left(\sum_{m=0}^{3}\sum_{n=0}^{3}g^{km}g^{jn}\left(\frac{\partial A_n}{\partial x^m}-\frac{\partial A_m}{\partial x^n}\right)\right)+$$
$$\sum_{j=0}^{3}\frac{1}{\sqrt{|\det G|}}\frac{\partial}{\partial x^j}\left(\sqrt{|\det G|}\right)\left(\sum_{m=0}^{3}\sum_{n=0}^{3}g^{km}g^{jn}\left(\frac{\partial A_n}{\partial x^m}-\frac{\partial A_m}{\partial x^n}\right)\right)=-4\pi j^k$$
$$\forall k=0,1,2,3, \quad (9.169)$$

or equivalently:

$$\sum_{j=0}^{3}\frac{\partial}{\partial x^j}\left(\sum_{m=0}^{3}\sum_{n=0}^{3}\sqrt{|\det G|}\,g^{km}g^{jn}\left(\frac{\partial A_n}{\partial x^m}-\frac{\partial A_m}{\partial x^n}\right)\right)=-4\pi\sqrt{|\det G|}\,j^k$$
$$\forall k=0,1,2,3. \quad (9.170)$$

Here $\{A_k\}_{k=0,1,2,3}$ is the four covector of the electromagnetic potential, $\{j^k\}_{k=0,1,2,3}$ is the four vector of the current and $G:=\{g_{kj}\}_{k,j=0,1,2,3}$, $\{g^{kj}\}_{k,j=0,1,2,3}$ are pseudo-metric covariant and contravariant tensors. Note that in curvilinear coordinate system we can have $\det G\neq Const$ and thus we need to consider the enlarged form (9.118) instead of (9.117). Moreover, the density of the Lagrangian of the electromagnetic field in every curvilinear coordinate system in $\mathbb{R}^4$ also has a form of (9.122):

$$L_1=\frac{1}{4\pi}\left(-\sum_{n=0}^{3}\sum_{k=0}^{3}\frac{1}{4}F^{nk}F_{nk}-\sum_{k=0}^{3}4\pi j^kA_k\right)=$$
$$\frac{1}{4\pi}\left(-\sum_{n=0}^{3}\sum_{k=0}^{3}\sum_{m=0}^{3}\sum_{p=0}^{3}\frac{1}{4}g^{mn}g^{pk}\left(\frac{\partial A_p}{\partial x^m}-\frac{\partial A_m}{\partial x^p}\right)\left(\frac{\partial A_k}{\partial x^n}-\frac{\partial A_n}{\partial x^k}\right)-\sum_{k=0}^{3}4\pi j^kA_k\right),$$
$$(9.171)$$

where

$$F_{ij}:=\frac{\partial A_j}{\partial x^i}-\frac{\partial A_i}{\partial x^j}\quad\forall i,j=0,1,2,3. \quad (9.172)$$

Next the general Lagrangian of motion of the charged particle in the gravitational and electromagnetic field (9.134) preserves its form in every curvilinear coordinate system:

$$J_{\mathcal{G}}(\chi)=\int_0^T\left\{-mc^2\,\mathcal{G}\left(\sum_{j=0}^{3}\sum_{k=0}^{3}g_{jk}(\chi(t))\frac{d\chi^j}{dt}\frac{d\chi^k}{dt}\right)-\sum_{k=0}^{3}\sigma A_k(\chi(t))\frac{d\chi^k}{dt}\right\}dt, \quad (9.173)$$

where t is the global time, which is a scalar on the group $\mathcal{S}$,

$$\left(\chi^0(t), \chi^1(t), \chi^2(t), \chi^3(t)\right) := \left(\frac{1}{c}x^0(t), \frac{1}{c}x_1(t), \frac{1}{c}x_2(t), \frac{1}{c}x_3(t)\right), \tag{9.174}$$

and $\left(x^0(t), x^1(t), x^2(t), x^3(t)\right) \in \mathbb{R}^4$ is a four-dimensional space-time trajectory of the particle, parameterized by the global time.

Note that if we denote by t the scalar of global time, then in a general curvilinear coordinate system the coordinate x^0 can differ from ct, and the equality $x^0 = ct$ valid, in general, only in Cartesian inertial or non-inertial coordinate systems. However, since the equality in (9.75) has a covariant form, the scalar of the global time t satisfies the following eikonal-type equation in every curvilinear coordinate system:

$$\sum_{j=0}^{3}\sum_{m=0}^{3} g^{jm}(x^0, x^1, x^2, x^3)\,\frac{\partial t}{\partial x^j}(x^0, x^1, x^2, x^3)\,\frac{\partial t}{\partial x^m}(x^0, x^1, x^2, x^3) = \frac{1}{c^2}. \tag{9.175}$$

Moreover, since the equality in (9.75) also has a covariant form, the following identity is valid in every curvilinear coordinate system:

$$\sum_{m=0}^{3} \Theta^{jm}\,\frac{\partial t}{\partial x^m} = 0 \quad \forall\, j = 0, 1, 2, 3, \tag{9.176}$$

where Θ^{ij} is the contravariant tensor of the three-dimensional geometry that has the form (9.48) only in Cartesian inertial or non-inertial coordinate systems.

Next, in the particular case of the relativistic-like Lagrangian where $\mathcal{G}(\tau) := \sqrt{\tau}$, the Lagrangian in (9.137) also preserves their form in every curvilinear coordinate system:

$$J_{rl}(\chi) = \int_a^b \left\{ -mc^2 \sqrt{\left(\sum_{j=0}^{3}\sum_{k=0}^{3} g_{jk}(\chi(s))\,\frac{d\chi^j}{ds}\,\frac{d\chi^k}{ds}\right)} - \sum_{k=0}^{3} \sigma A_k(\chi(s))\,\frac{d\chi^k}{ds} \right\} ds, \tag{9.177}$$

where s is the arbitrary parameter of the trajectory with increasing mapping $t \leftrightarrow s$:

$$\left(\chi^0(s), \chi^1(s), \chi^2(s), \chi^3(s)\right) := \left(\frac{1}{c}x^0(s), \frac{1}{c}x_1(s), \frac{1}{c}x_2(s), \frac{1}{c}x_3(s)\right). \tag{9.178}$$

In particular, in the case $\frac{\partial \chi^0}{\partial t} > 0$ we can take $s := \chi^0$ in (9.177).

Finally, we would like to note the following fact: since in the absence of essential gravitational masses we have $g_{ij} = J_{ij}$, there exists a unique, up to equivalence, "preferable" inertial coordinate system where $\mathbf{v} = \mathbf{k} = 0$ everywhere. In this particular system by (9.160) we have

$$\begin{cases} g_{00} = 1 \\ g_{ij} = -\delta_{ij} \quad \forall 1 \le i, j \le 3 \\ g_{0j} = g_{j0} = 0 \quad \forall 1 \le j \le 3, \end{cases} \tag{9.179}$$

and thus the Maxwell equations are the same as in the Special Relativity. Moreover, in this system, the Lagrangian of the motion of the particle of the form (9.177) is also the same as in the Special Relativity. Thus, since Maxwell equations (9.169) and the Lagrangian of the motion of particles (9.177) preserve their form in every Cartesian, non-Cartesian or curvilinear coordinate system of the group $\mathcal{S}$, they stay the same as in Special Relativity also in the case of every Cartesian, non-Cartesian or curvilinear coordinate system. Thus in the particular case of $\mathcal{G}(\tau) := \sqrt{\tau}$ in (9.173) and in the absence of essential gravitational masses, the unique formal mathematical difference between our model and the Special Relativity is that in the frames of our model we consider the Galilean transformations as transformations of the change of inertial Cartesian coordinate systems, up to equivalence, and (3.1) as transformations of the change of non-inertial Cartesian coordinate systems; however, the Lorenz transformations lead to inertial non-Cartesian coordinate systems (see Definition 9.3). In contrast, in the Special Relativity the fundamental role of the Lorenz transformations, i.e. the transformations that preserve the form (9.179) of the pseudo-metric tensor, is postulated as the role of transformations of the change of inertial Cartesian coordinate systems, and, at the same time, the Galilean transformations lead to inertial non-Cartesian coordinate systems, and transformations (3.1) lead to non-inertial non-Cartesian coordinate systems.

9.6.1 Some General Covariant Identities in Non-Cartesian or Curvilinear Coordinate Systems

Consider the contravariant kinematic pseudo-metric tensor of inertia $\{J^{ij}\}_{0 \le i,j \le 3}$ on the group $\mathcal{S}$, defined by

$$J^{ij} := k^i k^j - \Theta^{ij} \qquad \forall i, j = 0, 1, 2, 3, \tag{9.180}$$

where $\{\Theta^{ij}\}_{0 \le i,j \le 3}$ is the contravariant tensor of the three-dimensional geometry on the group $\mathcal{S}$ (the same as in (9.176)) defined in Cartesian coordinate systems by (9.48), and (k^0, k^1, k^2, k^3) is the four vector of the potential of inertia on the group $\mathcal{S}$, defined in Cartesian coordinate systems by (9.30). Then, as before, the reverse to $\{J^{ij}\}_{0 \le i,j \le 3}$ covariant tensor on the group $\mathcal{S}$ is denoted as $\{J_{ij}\}_{0 \le i,j \le 3}$ and satisfies

$$\sum_{m=0}^{3} J^{im} J_{mj} = \begin{cases} 1 & \text{if } i = j \\ 0 & \text{if } i \neq j \end{cases} \qquad \forall i, j = 0, 1, 2, 3 \tag{9.181}$$

(as in (9.142)).

Definition 9.2 We say that a given general coordinate system $(*)$ is Cartesian if the contravariant tensor of the three-dimensional geometry $\{\Theta^{ij}\}_{0 \leq i,j \leq 3}$ has the following simple form in the system $(*)$ (the same as in (9.48)):

$$\begin{cases} \Theta^{00} = 0 \\ \Theta^{0j} = \Theta^{j0} = 0 \qquad \forall j = 1, 2, 3 \\ \Theta^{ij} := \delta_{ij} \qquad \forall i, j = 1, 2, 3, \end{cases} \tag{9.182}$$

and at the same time in the system $(*)$ we have

$$c \frac{\partial t}{\partial x^0}(x^0, x^1, x^2, x^3) = 1, \tag{9.183}$$

where t is the scalar of global time (the same as in (9.176)). In particular, by (9.176), (9.182) and (9.183) all together in the system $(*)$ we have

$$c\left(\frac{\partial t}{\partial x^0}(x^0, x^1, x^2, x^3), \frac{\partial t}{\partial x^1}(x^0, x^1, x^2, x^3), \frac{\partial t}{\partial x^2}(x^0, x^1, x^2, x^3), \frac{\partial t}{\partial x^3}(x^0, x^1, x^2, x^3)\right) = (1, 0, 0, 0). \tag{9.184}$$

Then, it easily can be shown that given a Cartesian coordinate system $(*)$ and a general coordinate system $(**)$, the system $(**)$ is Cartesian if and only if, up to a constant shift of the time, the change of coordinates from system $(*)$ to system $(**)$ is given by (9.15) or equivalently by (9.16):

$$\begin{cases} \mathbf{x}' = A\left(\frac{x_0}{c}\right) \cdot \mathbf{x} + \mathbf{z}\left(\frac{x_0}{c}\right) = A(t) \cdot \mathbf{x} + \mathbf{z}(t), \\ x_0' = x_0 = t, \end{cases} \tag{9.185}$$

where $A(t) \in SO(3)$ is a rotation.

Definition 9.3 Similar to Cartesian coordinate systems, we say that a given non-Cartesian or curvilinear coordinate system $(*)$ is inertial if the four vector of the potential of inertia (k^0, k^1, k^2, k^3), defined in Cartesian systems by (9.30), is constant in $\mathbb{R}^4$. Then, it easily can be shown that given an inertial coordinate system $(*)$ and a general coordinate system $(**)$, the system $(**)$ is inertial if and only if the change of coordinates from system $(*)$ to system $(**)$ is given by a linear transformation in $\mathbb{R}^4$. In particular, a general coordinate system $(***)$ is inertial if and only if the tensors $\{J^{ij}\}_{0 \leq i,j \leq 3}$ and $\{J_{ij}\}_{0 \leq i,j \leq 3}$ are constant in the four-dimensional space-time in the given system $(***)$. Moreover, given an inertial Cartesian coordinate system

$(*)$ and a general coordinate system $(**)$, the system $(**)$ is inertial and, at the same time, is Cartesian if and only if, up to a constant shift of time and, up to equivalence of coordinate systems, the change of coordinates from system $(*)$ to system $(**)$ is given by the Galilean transformation.

In the general non-Cartesian or curvilinear coordinate system, we obviously have the following list of general covariant identities:

-
$$\det G = \det J \tag{9.186}$$

  (see (9.145)), where 4×4-matrix J is given by

$$J = \{J_{ij}\}_{0 \le i,j \le 3}, \tag{9.187}$$

  and 4×4-matrix G is given by

$$G = \{g_{ij}\}_{0 \le i,j \le 3}, \tag{9.188}$$

  with the covariant pseudo-metric tensor of the four-dimensional space-time $\{g_{ij}\}_{0 \le i,j \le 3}$.

-
$$\left\{\delta_j\left(\frac{\partial t}{\partial x_i}\right)\right\}_J = 0 \quad \forall i, j = 0, 1, 2, 3 \tag{9.189}$$

  (see (9.157)), where

$$\left\{\delta_j\left(\frac{\partial t}{\partial x_i}\right)\right\}_J = \left\{\delta_i\left(\frac{\partial t}{\partial x_j}\right)\right\}_J := \frac{\partial^2 t}{\partial x^i \partial x^j} - \sum_{m=0}^{3} \left\{\Gamma^m_{ij}\right\}_J \frac{\partial t}{\partial x^m} \quad \forall i, j = 0, 1, 2, 3, \tag{9.190}$$

  with

$$\begin{cases} \left\{\Gamma_{i,mn}\right\}_J := \frac{1}{2}\left(\frac{\partial J_{im}}{\partial x_n} + \frac{\partial J_{in}}{\partial x_m} - \frac{\partial J_{mn}}{\partial x_i}\right) \\ \left\{\Gamma^i_{mn}\right\}_J := \sum_{j=0}^{3} J^{ij} \left\{\Gamma_{j,mn}\right\}_J \end{cases} \quad \forall i, m, n = 0, 1, 2, 3. \tag{9.191}$$

-
$$\sum_{m=0}^{3} J^{jm}\left(c\frac{\partial t}{\partial x^m}\right) = k^j \quad \forall j = 0, 1, 2, 3, \tag{9.192}$$

  and

$$\sum_{m=0}^{3} g^{jm}\left(c\frac{\partial t}{\partial x^m}\right) = v^j \quad \forall j = 0, 1, 2, 3. \tag{9.193}$$

- $$\sum_{j=0}^{3} k^j \left(c \frac{\partial t}{\partial x^j} \right) = 1 = \sum_{j=0}^{3} v^j \left(c \frac{\partial t}{\partial x^j} \right), \tag{9.194}$$

- $$c^2 \left(\sum_{j=0}^{3} \sum_{m=0}^{3} J^{jm} \frac{\partial t}{\partial x^j} \frac{\partial t}{\partial x^m} \right) = 1 = c^2 \left(\sum_{j=0}^{3} \sum_{m=0}^{3} g^{jm} \frac{\partial t}{\partial x^j} \frac{\partial t}{\partial x^m} \right). \tag{9.195}$$

- $$\sum_{m=0}^{3} J_{jm} k^m = c \frac{\partial t}{\partial x^j} = \sum_{m=0}^{3} g_{jm} v^m \qquad \forall j = 0, 1, 2, 3. \tag{9.196}$$

- $$\sum_{m=0}^{3} \Theta^{jm} \frac{\partial t}{\partial x^m} = 0 \qquad \forall j = 0, 1, 2, 3. \tag{9.197}$$

- $$J^{ij} = k^i k^j - \Theta^{ij} \qquad \forall i, j = 0, 1, 2, 3, \tag{9.198}$$

and

$$g^{ij} = v^i v^j - \Theta^{ij} \qquad \forall i, j = 0, 1, 2, 3, \tag{9.199}$$

that implies, in particular,

$$g^{ij} = \left(v^i v^j - k^i k^j \right) + J^{ij} \qquad \forall i, j = 0, 1, 2, 3. \tag{9.200}$$

- $$\sum_{m=0}^{3} J^{im} J_{mj} = \sum_{m=0}^{3} g^{im} g_{mj} = \begin{cases} 1 & \text{if } i = j \\ 0 & \text{if } i \neq j \end{cases} \qquad \forall i, j = 0, 1, 2, 3, \tag{9.201}$$

that implies together with (9.196) and (9.195) the following:

$$\sum_{j=0}^{3} \sum_{m=0}^{3} J_{jm} k^j k^m = 1 = \sum_{j=0}^{3} \sum_{m=0}^{3} g_{jm} v^j v^m. \tag{9.202}$$

Moreover, in the general non-Cartesian or curvilinear coordinate system, as before, we can define the dynamical four covector of genuine gravity (s_0, s_1, s_2, s_3) by

$$s_j = \frac{1}{2} \left(\sum_{m=0}^{3} g_{jm} k^m - \sum_{m=0}^{3} J_{jm} v^m \right) \qquad \forall j = 0, 1, 2, 3 \tag{9.203}$$

(see (9.161)). Then we have the following covariant identities:

$$g_{ij} = J_{ij} + \left(s_i \left(c \frac{\partial t}{\partial x^j} \right) + \left(c \frac{\partial t}{\partial x^i} \right) s_j \right) \quad \forall\, i, j = 0, 1, 2, 3 \tag{9.204}$$

(see (9.166)), and

$$v^j - k^j = \sum_{m=0}^{3} \Theta^{jm} s_m = \sum_{m=0}^{3} \left(k^j k^m - J^{jm} \right) s_m \quad \forall\, j = 0, 1, 2, 3 \tag{9.205}$$

(see (9.167)).

Definition 9.4 We say that a given general coordinate system $(*)$ is <u>Lorentzian</u> if in system $(*)$ kinematic pseudo-metric tensors of inertia $\{J^{ij}\}_{0 \le i,j \le 3}$ and $\{J_{ij}\}_{0 \le i,j \le 3}$ have the following simple form (the same as in (9.159) and (9.160)):

$$\begin{cases} J^{00} = 1 \\ J^{ij} = -\delta_{ij} \quad \forall 1 \le i, j \le 3 \\ J^{0j} = J^{j0} = 0 \quad \forall 1 \le j \le 3, \end{cases} \tag{9.206}$$

and

$$\begin{cases} J_{00} = 1 \\ J_{ij} = -\delta_{ij} \quad \forall 1 \le i, j \le 3 \\ J_{0j} = J_{j0} = 0 \quad \forall 1 \le j \le 3. \end{cases} \tag{9.207}$$

Thus, since in every Lorentzian coordinate system $\{J^{ij}\}_{0 \le i,j \le 3}$ and $\{J_{ij}\}_{0 \le i,j \le 3}$ are constant in $\mathbb{R}^4$, as it was explained in the end of Definition 9.3 every Lorentzian coordinate system is necessary <u>inertial</u>, however it is not necessarily Cartesian. In particular, a unique, up to equivalence, "preferable" inertial Cartesian coordinate system ($\{0\}$), where the three-dimensional vector potential of inertia satisfies $\mathbf{k} = 0$, is a unique, up to equivalence, coordinate system which is Cartesian and Lorentzian <u>simultaneously</u>. On the other hand, there exists a plenty of Lorentzian coordinate systems. In particular, it is well known that, given a Lorentzian coordinate system $(*)$ and a general coordinate system $(**)$, the system $(**)$ is also Lorentzian, if and only if the change of coordinates from system $(*)$ to system $(**)$ is given by the usual <u>Lorentz</u> transformation, up to certain trivial transformations. In every Lorentzian coordinate system the four vector of the potential of inertia (k^1, k^2, k^3, k^4) is <u>constant</u> which is by (9.202) and (9.207) always satisfies

$$\left(k^0\right)^2 - \left(\left(k^1\right)^2 + \left(k^2\right)^2 + \left(k^3\right)^2\right) = 1. \tag{9.208}$$

In fact, in a general Lorentzian coordinate system, the four vector (k^1, k^2, k^3, k^4) can be arbitrary constant four vector, satisfying (9.208). We remind that in the unique

Cartesian Lorentzian coordinate system, we have $(k^1, k^2, k^3, k^4) = (1, 0, 0, 0)$. Next by (9.196) together with (9.206) we have

$$\left(\frac{\partial t}{\partial x^0}, \frac{\partial t}{\partial x^1}, \frac{\partial t}{\partial x^2}, \frac{\partial t}{\partial x^3}\right) = \frac{1}{c}\left(k^0, -k^1, -k^2, -k^3\right), \tag{9.209}$$

and so, up to a constant shift in time, in every Lorentzian coordinate system we have

$$t = \frac{1}{c}k^0x^0 - \frac{1}{c}\left(k^1x^1 + k^2x^2 + k^3x^3\right). \tag{9.210}$$

Note that in the unique Cartesian Lorentzian coordinate system we have $t = \frac{x^0}{c}$.

9.7 Certain Coordinate System for Stationary Radial Field and the Schwarzschild Metric

Assume that for a given part of the space in some inertial or non-inertial Cartesian coordinate system $(*)$ the gravitational field is stationary and radially symmetric that means that the vectorial gravitational potential $\mathbf{v} = (v_1, v_2, v_3)$ is independent on time variable t and having the form

$$\mathbf{v}(\mathbf{x}) = g\left(|\mathbf{x}|\right)\frac{\mathbf{x}}{|\mathbf{x}|} \qquad \forall \mathbf{x}, \tag{9.211}$$

for some scalar function $g(s) : \mathbb{R} \to \mathbb{R}$. Next, given some differentiable function $\Theta(\mathbf{x}) : \mathbb{R}^3 \to \mathbb{R}$, consider the change of variables in the four-dimensional space-time $\mathbb{R}^4$:

$$\begin{cases} x'^0 = x^0 + \frac{\Theta((x^1, x^2, x^3))}{c} \\ x'^j = x^j \quad \forall j = 1, 2, 3 \end{cases} \tag{9.212}$$

that transforms the Cartesian coordinate system $(*)$ to the curvilinear coordinate system $(**)$ in the four-dimensional space-time $\mathbb{R}^4$. Then in terms of the three-dimensional space and one-dimensional time:

$$(x^0, x^1, x^2, x^3) := (ct, x_1, x_2, x_3) = (ct, \mathbf{x}), \tag{9.213}$$

we rewrite (9.212) as

$$\begin{cases} t' = t + \frac{\Theta(\mathbf{x})}{c^2} \\ \mathbf{x}' = \mathbf{x}. \end{cases} \tag{9.214}$$

Note again, that since the new coordinate system $(**)$ in $\mathbb{R}^4$ is <u>curvilinear</u>, the time-like coordinate t' in coordinate system $(**)$ <u>differ</u> from the proper scalar of the global time.

Next if we define a matrix

$$A = \left\{a_j^i\right\}_{0\le i,j\le 3} \in \mathbb{R}^{4\times 4} = \left\{\frac{\partial x'^i}{\partial x^j}\right\}_{0\le i,j\le 3} \in \mathbb{R}^{4\times 4}, \tag{9.215}$$

then

$$\begin{cases} a_0^0 = 1 \\ a_j^i = \delta_{ij} \quad \forall 1 \le i, j \le 3 \\ a_j^0 = \frac{1}{c}\frac{\partial \Theta}{\partial x^j} \quad \forall 1 \le j \le 3 \\ a_0^j = 0 \quad \forall 1 \le j \le 3. \end{cases} \tag{9.216}$$

Next consider the contravariant pseudo-metric tensor of the four-dimensional space-time $\{g^{ij}\}_{0\le i,j\le 3}$ that due to (9.58) has the form of

$$\begin{cases} g^{00} = 1 \\ g^{ij} = -\delta_{ij} + \frac{v^i v^j}{c^2} \quad \forall 1 \le i, j \le 3 \\ g^{0j} = g^{j0} = \frac{v^j}{c} \quad \forall 1 \le j \le 3, \end{cases} \tag{9.217}$$

in the Cartesian coordinate system $(*)$. We would like to find the form $\{g'^{ij}\}_{0\le i,j\le 3}$ of this tensor in the curvilinear coordinate system $(**)$. Then by (9.6) we have

$$g'^{mn} = \sum_{i=0}^{3}\sum_{j=0}^{3}\left(a_i^m g^{ij} a_j^n\right) = \sum_{i=0}^{3} a_i^m \left(\sum_{j=0}^{3} g^{ij} a_j^n\right) \qquad \forall 0 \le m, n \le 3. \tag{9.218}$$

I.e.

$$g'^{mn} = a_0^m \left(g^{00} a_0^n + \sum_{j=1}^{3} w^{0j} a_j^n\right) + \sum_{i=1}^{3} a_i^m \left(g^{i0} a_0^n + \sum_{j=1}^{3} g^{ij} a_j^n\right) \qquad \forall 0 \le m, n \le 3. \tag{9.219}$$

In particular, by (9.216) and (9.219), we obtain

$$\begin{aligned} g'^{00} &= a_0^0 \left(g^{00} a_0^0 + \sum_{j=1}^{3} g^{0j} a_j^0\right) + \sum_{i=1}^{3} a_i^0 \left(g^{i0} a_0^0 + \sum_{j=1}^{3} g^{ij} a_j^0\right) \\ &= a_0^0 \left(a_0^0 + \sum_{j=1}^{3} 2\frac{v^j}{c} a_j^0\right) + \left(\sum_{j=1}^{3} a_j^0 \frac{v^j}{c}\right)^2 - \sum_{j=1}^{3}\left(a_j^0\right)^2 = \left(a_0^0 + \sum_{j=1}^{3}\frac{v^j}{c} a_j^0\right)^2 - \sum_{j=1}^{3}\left(a_j^0\right)^2, \end{aligned} \tag{9.220}$$

$$g'^{0n} = g'^{n0} = a_0^0 \left(g^{00} a_0^n + \sum_{j=1}^{3} g^{0j} a_j^n \right) + \sum_{i=1}^{3} a_i^0 \left(g^{i0} a_0^n + \sum_{j=1}^{3} g^{ij} a_j^n \right)$$
$$= a_0^0 \frac{v^n}{c} - a_n^0 + \sum_{i=1}^{3} a_i^0 \frac{v^i}{c} \frac{v^n}{c} \qquad \forall 1 \le n \le 3, \tag{9.221}$$

and

$$g'^{mn} = a_0^m \left(g^{00} a_0^n + \sum_{j=1}^{3} g^{0j} a_j^n \right) + \sum_{i=1}^{3} a_i^m \left(g^{i0} a_0^n + \sum_{j=1}^{3} g^{ij} a_j^n \right) = \frac{v^m}{c} \frac{v^n}{c} - \delta_{mn}$$
$$\forall 1 \le m, n \le 3. \tag{9.222}$$

Thus by three equations together with (9.216) we deduce

$$\begin{cases} g'^{00} = \left(1 + \sum_{j=1}^{3} \frac{1}{c^2} v^j \frac{\partial \Theta}{\partial x^j}\right)^2 - \sum_{j=1}^{3} \frac{1}{c^2} \left(\frac{\partial \Theta}{\partial x^j}\right)^2 \\ g'^{0n} = g'^{n0} = \frac{v^n}{c} \left(1 + \sum_{j=1}^{3} \frac{1}{c^2} v^j \frac{\partial \Theta}{\partial x^j}\right) - \frac{1}{c} \frac{\partial \Theta}{\partial x^n} \qquad \forall 1 \le n \le 3, \\ g'^{mn} = \frac{v^m}{c} \frac{v^n}{c} - \delta_{mn} \qquad \forall 1 \le m, n \le 3. \end{cases} \tag{9.223}$$

Next if $\mathbf{v}$ satisfies (9.211) then choosing the function Θ to be defined as

$$\Theta(\mathbf{x}) = \xi\left(|\mathbf{x}|\right) \quad \forall \mathbf{x}, \qquad \text{where} \quad \frac{d\xi}{ds}(s) = \frac{g(s)}{1 - \frac{g^2(s)}{c^2}} \qquad \forall s, \tag{9.224}$$

we find that:

$$v_n = \left(1 + \sum_{j=1}^{3} \frac{1}{c^2} v^j \frac{\partial \Theta}{\partial x^j}\right)^{-1} \frac{\partial \Theta}{\partial x^n} \qquad \forall 1 \le n \le 3, \tag{9.225}$$

i.e.

$$v_n \left(1 + \sum_{j=1}^{3} \frac{1}{c^2} v^j \frac{\partial \Theta}{\partial x^j}\right) = \frac{\partial \Theta}{\partial x^n} \qquad \forall 1 \le n \le 3. \tag{9.226}$$

Then we rewrite (9.223) as

$$\begin{cases} g'^{00} = \left(1 - \frac{|\mathbf{v}|^2}{c^2}\right) \left(1 + \sum_{j=1}^{3} \frac{1}{c^2} v^j \frac{\partial \Theta}{\partial x^j}\right)^2, \\ g'^{0n} = g'^{n0} = 0 \qquad \forall 1 \le n \le 3, \\ g'^{mn} = \frac{v^m}{c} \frac{v^n}{c} - \delta_{mn} \qquad \forall 1 \le m, n \le 3. \end{cases} \tag{9.227}$$

On the other hand, by (9.226) we have

$$|\mathbf{v}|^2 = \left(1 - \frac{|\mathbf{v}|^2}{c^2}\right)\left(\sum_{j=1}^{3} v^j \frac{\partial \Theta}{\partial x^j}\right). \tag{9.228}$$

We rewrite (9.228) as

$$1 = \left(1 - \frac{|\mathbf{v}|^2}{c^2}\right)\left(1 + \sum_{j=1}^{3} \frac{1}{c^2} v^j \frac{\partial \Theta}{\partial x^j}\right) \tag{9.229}$$

Therefore, by (9.227) and (9.229) we deduce

$$\begin{cases} g'^{00} = \left(1 - \frac{|\mathbf{v}|^2}{c^2}\right)^{-1}, \\ g'^{0n} = g'^{n0} = 0 \quad \forall 1 \le n \le 3, \\ g'^{mn} = \frac{v^m}{c}\frac{v^n}{c} - \delta_{mn} \quad \forall 1 \le m, n \le 3. \end{cases} \tag{9.230}$$

Next we find that the <u>covariant</u> pseudo-metric tensor $\{g'_{ij}\}_{0 \le i,j \le 3}$ in the curvilinear coordinate system (∗∗) has the following form:

$$\begin{cases} g'_{00} = \left(1 - \frac{|\mathbf{v}|^2}{c^2}\right), \\ g'_{0n} = g'_{n0} = 0 \quad \forall 1 \le n \le 3, \\ g'_{mn} = -\left(\left(1 - \frac{|\mathbf{v}|^2}{c^2}\right)^{-1} \frac{v^m}{c}\frac{v^n}{c} + \delta_{mn}\right) \quad \forall 1 \le m, n \le 3. \end{cases} \tag{9.231}$$

Indeed, if $\{g'_{ij}\}_{0 \le i,j \le 3}$ is defined by (9.231), then by (9.230) we have

$$\sum_{k=0}^{3} g'_{0k} g'^{k0} = g'_{00} g'^{00} + \sum_{k=1}^{3} g'_{0k} g'^{k0} = 1,$$

$$\sum_{k=0}^{3} g'_{ik} g'^{kj} = g'_{i0} g'^{0j} + \sum_{k=1}^{3} g'_{ik} g'^{kj} =$$
$$\sum_{k=1}^{3} \left(\left(1 - \frac{|\mathbf{v}|^2}{c^2}\right)^{-1} \frac{v^i}{c}\frac{v^k}{c} + \delta_{ik}\right)\left(\delta_{kj} - \frac{v^k}{c}\frac{v^j}{c}\right)$$
$$= \delta_{ij} - \left(1 - \frac{|\mathbf{v}|^2}{c^2}\right)^{-1} \frac{|\mathbf{v}|^2}{c^2}\frac{v^i}{c}\frac{v^j}{c} - \frac{v^i}{c}\frac{v^j}{c} + \left(1 - \frac{|\mathbf{v}|^2}{c^2}\right)^{-1} \frac{v^i}{c}\frac{v^j}{c} = \delta_{ij} \quad \forall 1 \le i, j \le 3,$$

and

$$\sum_{k=0}^{3} g'_{ik} g'^{k0} = g'_{i0} g'^{00} + \sum_{k=1}^{3} g'_{ik} g'^{k0} = 0 \quad \forall 1 \leq i \leq 3,$$

$$\sum_{k=0}^{3} g'_{0k} g'^{kj} = g'_{00} g'^{0j} + \sum_{k=1}^{3} g'_{0k} g'^{kj} = 0 \quad \forall 1 \leq j \leq 3.$$

So

$$\sum_{k=0}^{3} g'_{ik} g'^{kj} = \delta_{ij} \quad \forall 0 \leq i, j \leq 3,$$

and thus equalities (9.231) indeed define the covariant form of the pseudo-metric tensor. So by (9.231) we have

$$\begin{cases} g'_{00} = \left(1 - \frac{|\mathbf{v}|^2}{c^2}\right), \\ g'_{0n} = g'_{n0} = 0 \qquad \forall 1 \leq n \leq 3, \\ g'_{mn} = -\left(\left(1 - \frac{|\mathbf{v}|^2}{c^2}\right)^{-1} \frac{v^m}{c} \frac{v^n}{c} + \delta_{mn}\right) \qquad \forall 1 \leq m, n \leq 3. \end{cases} \tag{9.232}$$

In particular, the quadratic form, induced by the covariant form of the pseudo-metric tensor $\{g'_{ij}\}_{0 \leq i, j \leq 3}$ in the curvilinear coordinate system (**), that defined on the tangent vectors $\left(dx'^0, dx'^1, dx'^2, dx'^3\right) \in \mathbb{R}^4$ where $d\mathbf{x}' := (dx'^1, dx'^2, dx'^3)$ has the following form:

$$\begin{aligned} \sum_{i=0}^{3}\sum_{j=0}^{3} g'_{ij} dx'^i dx'^j &= \left(1 - \frac{|\mathbf{v}|^2}{c^2}\right) dx_0'^2 - \left(|d\mathbf{x}'|^2 + \frac{1}{c^2}\left(1 - \frac{|\mathbf{v}|^2}{c^2}\right)^{-1} \left|\mathbf{v} \cdot d\mathbf{x}'\right|^2\right) \\ &= \left(1 - \frac{|\mathbf{v}|^2}{c^2}\right) dx_0'^2 \\ &- \left(\left(|d\mathbf{x}'|^2 - \left|\frac{\mathbf{v}}{|\mathbf{v}|} \cdot d\mathbf{x}'\right|^2\right) + \left(1 - \frac{|\mathbf{v}|^2}{c^2}\right)^{-1} \frac{|\mathbf{v}|^2}{c^2} \left|\frac{\mathbf{v}}{|\mathbf{v}|} \cdot d\mathbf{x}'\right|^2 + \left|\frac{\mathbf{v}}{|\mathbf{v}|} \cdot d\mathbf{x}'\right|^2\right) \\ &= \left(1 - \frac{|\mathbf{v}|^2}{c^2}\right) dx_0'^2 - \left(\left(1 - \frac{|\mathbf{v}|^2}{c^2}\right)^{-1} \left|\frac{\mathbf{v}}{|\mathbf{v}|} \cdot d\mathbf{x}'\right|^2 + \left(|d\mathbf{x}'|^2 - \left|\frac{\mathbf{v}}{|\mathbf{v}|} \cdot d\mathbf{x}'\right|^2\right)\right). \end{aligned} \tag{9.233}$$

Thus taking into account (9.214) and (9.211) we rewrite (9.233) as

$$\sum_{i=0}^{3}\sum_{j=0}^{3} g'_{ij}dx'^i dx'^j =$$

$$\left(1-\frac{\left|\mathbf{v}(\mathbf{x}')\right|^2}{c^2}\right)dx_0'^2-\left(\left(1-\frac{\left|\mathbf{v}(\mathbf{x}')\right|^2}{c^2}\right)^{-1}\left|\frac{\mathbf{x}'}{|\mathbf{x}'|}\cdot d\mathbf{x}'\right|^2+\left(|d\mathbf{x}'|^2-\left|\frac{\mathbf{x}'}{|\mathbf{x}'|}\cdot d\mathbf{x}'\right|^2\right)\right). \tag{9.234}$$

Next, up to the end of this section, assume that our Cartesian coordinate system $(*)$ is non-rotating and our gravitational field is formed by the spherical symmetric massive body of mass m_0 and radius R_0 like the Earth, the Sun et.al. with the center at the point 0. Then, as we get either in (4.84) and (4.85) of Remark 4.4 in the case of the Newtonian gravity, or in (8.64) and (8.65) of Sect. 8.2 in the case of the more general gravity model, given by (8.41) with arbitrary constant β, we have either

$$\mathbf{v}(\mathbf{x}) = \frac{\sqrt{-2\Phi_1(|\mathbf{x}|)}}{|\mathbf{x}|}\mathbf{x}, \tag{9.235}$$

or

$$\mathbf{v}(\mathbf{x}) = -\frac{\sqrt{-2\Phi_1(|\mathbf{x}|)}}{|\mathbf{x}|}\mathbf{x}, \tag{9.236}$$

where Φ_1 is the classical Newtonian potential of our massive body m_0 that satisfies

$$\Phi_1(\mathbf{x}) = -\frac{Gm_0}{|\mathbf{x}|} \tag{9.237}$$

outside of the body surface. Thus, in particular,

$$|\mathbf{v}(\mathbf{x})|^2 = -2\Phi_1(|\mathbf{x}|), \tag{9.238}$$

and outside of the massive body surface we have

$$|\mathbf{v}(\mathbf{x})|^2 = \frac{2Gm_0}{|\mathbf{x}|}. \tag{9.239}$$

Both (9.235) and (9.236) are particular cases of (9.211), with

$$g(s) = \pm\sqrt{-2\Phi_1(s)}, \tag{9.240}$$

and, in particular, outside of the massive body surface we have

$$g(|x|) = \pm\sqrt{\frac{2Gm_0}{|\mathbf{x}|}}. \tag{9.241}$$

Thus defining the function $\Theta(\mathbf{x})$ as in (9.224) that always can be done in the case $\frac{2Gm_0}{R_0} < c^2$, we can define the change of variables from coordinate system $(*)$ to the curvilinear coordinate system $(**)$ in the four-dimensional space-time $\mathbb{R}^4$ as in (9.214):

$$\begin{cases} t' = t + \frac{\Theta(\mathbf{x})}{c^2} \\ \mathbf{x}' = \mathbf{x}. \end{cases} \tag{9.242}$$

Then by inserting (9.235) or (9.236) into (9.232) we deduce the form of the covariant pseudo-metric tensor in the curvilinear coordinate system $(**)$:

$$\begin{cases} g'_{00} = \left(1 + \frac{2\Phi_1(|\mathbf{x}'|)}{c^2}\right), \\ g'_{0n} = g'_{n0} = 0 \quad \forall 1 \le n \le 3, \\ g'_{mn} = \left(\left(1 + \frac{2\Phi_1(|\mathbf{x}'|)}{c^2}\right)^{-1} \frac{2\Phi_1(|\mathbf{x}'|)}{c^2} \frac{x'_m}{|\mathbf{x}'|} \frac{x'_n}{|\mathbf{x}'|} - \delta_{mn}\right) \quad \forall 1 \le m, n \le 3. \end{cases} \tag{9.243}$$

Moreover, by (9.234) we have

$$\sum_{i=0}^{3}\sum_{j=0}^{3} g'_{ij} dx'^i dx'^j = \left(1 + \frac{2\Phi_1(|\mathbf{x}'|)}{c^2}\right) dx_0'^2 - \left(\left(1 + \frac{2\Phi_1(|\mathbf{x}'|)}{c^2}\right)^{-1} \left|\frac{\mathbf{x}'}{|\mathbf{x}'|} \cdot d\mathbf{x}'\right|^2 + \left(|d\mathbf{x}'|^2 - \left|\frac{\mathbf{x}'}{|\mathbf{x}'|} \cdot d\mathbf{x}'\right|^2\right)\right). \tag{9.244}$$

In particular, outside of the massive body surface, i.e. when $|x'| > R_0$ we rewrite (9.243) and (9.244) as

$$\begin{cases} g'_{00} = \left(1 - \frac{2Gm_0}{c^2|\mathbf{x}'|}\right), \\ g'_{0n} = g'_{n0} = 0 \quad \forall 1 \le n \le 3, \\ g'_{mn} = -\left(\left(1 - \frac{2Gm_0}{c^2|\mathbf{x}'|}\right)^{-1} \frac{2Gm_0}{c^2|\mathbf{x}'|} \frac{x'_m}{|\mathbf{x}'|} \frac{x'_n}{|\mathbf{x}'|} + \delta_{mn}\right) \quad \forall 1 \le m, n \le 3, \end{cases} \tag{9.245}$$

and

$$\sum_{i=0}^{3}\sum_{j=0}^{3} g'_{ij} dx'^i dx'^j = \left(1 - \frac{2Gm_0}{c^2|\mathbf{x}'|}\right) dx_0'^2 - \left(\left(1 - \frac{2Gm_0}{c^2|\mathbf{x}'|}\right)^{-1} \left|\frac{\mathbf{x}'}{|\mathbf{x}'|} \cdot d\mathbf{x}'\right|^2 + \left(|d\mathbf{x}'|^2 - \left|\frac{\mathbf{x}'}{|\mathbf{x}'|} \cdot d\mathbf{x}'\right|^2\right)\right). \tag{9.246}$$

Therefore, we get that in coordinate system $(**)$, outside of the massive body, the covariant pseudo-metric tensor in (9.245) and (9.246) exactly the same as the well-known Schwarzschild metric from the General Relativity. Indeed in the spherical coordinates in $\mathbb{R}^3$ we rewrite (9.246) as

$$\sum_{i=0}^{3}\sum_{j=0}^{3} g'_{ij}dx'^i dx'^j =$$
$$\left(1-\frac{2Gm_0}{c^2r'}\right)dx_0'^2 - \left(\left(1-\frac{2Gm_0}{c^2r'}\right)^{-1}(dr')^2 + (r')^2\left((d\theta')^2 + \sin^2(\theta')(d\varphi')^2\right)\right), \tag{9.247}$$

and this is exactly the classical Schwarzschild metric.

In particular, if we consider the monochromatic electromagnetic wave of frequency ω of the form $e^{i\omega t}U(\mathbf{x})$ in the coordinate system $(*)$, then by (9.242) in the coordinate system $(**)$ the form of this light is $e^{i\omega t'}U'(\mathbf{x}')$ where $U'(\mathbf{x}') = U(\mathbf{x}')e^{-i\omega\frac{\Theta(\mathbf{x}')}{c^2}}$, i.e. the electromagnetic wave in the coordinate system $(**)$ is also monochromatic of the same frequency ω. Thus all the optical effects that we find in the frames of our model coincide with the effects considered in the frames of General Relativity for the Schwarzschild metric. In particular, the Michelson–Morley experiment and all Sagnac-type effects will lead to the same result in the frame of our model like in the case of the General Relativity. Moreover, since the Maxwell equations in both models have the same tensor form, all the electromagnetic effects, where the time does not appear explicitly, will be the same. Similarly, the curvature of the light path in the Sun's gravitational field will be the same in both models. Finally, in the particular case of $\mathcal{G}(\tau) = \sqrt{\tau}$ in (9.173), i.e. in the case of the relativistic-like Lagrangian of the motion in (9.136) all the mechanical effects will be the same in the frame of our model like in the case of the General Relativity for the Schwarzschild metric, provided that the time does not appear explicitly in this effects. In particular, the movement of the Mercury planet in the Sun's gravitational field will be the same in both models, provided we take into account the relativistic-like Lagrangian of the motion as in (9.136).

Finally, note that the similar solution as in (9.246) or (9.247) is valid also for the general laws of the gravity, given by either (9.263) or (9.295), where in all places we take M_0 instead of m_0 and $\Phi_1(\mathbf{x}) = -\frac{GM_0}{|\mathbf{x}|}$ instead of (9.237), with M_0 being the total effective gravitational mass of the Earth, defined as in (9.317) (see Sect. 9.8.2 for the details).

9.8 General Lagrangian of the Gravitational Field, Related to the Lagrangian in (9.134)

9.8.1 The Case of the Newtonian-Type Gravity

Given known the distribution of inertial mass density of some continuum medium $\mu := \mu(\mathbf{x}, t)$, the field of velocities of this medium $\mathbf{u} := \mathbf{u}(\mathbf{x}, t)$, the charge density $\rho := \rho(\mathbf{x}, t)$ and the current density $\mathbf{j} := \mathbf{j}(\mathbf{x}, t)$ in a Cartesian coordinate system, consider a general Lagrangian density L of the unified gravitational-electromagnetic field that generalizes the Lagrangian density defined by (8.1) and is consistent with the Lagrangian of the motion of particles of the general form (9.134):

$$
\begin{aligned}
&L\left(\mathbf{A}, \Psi, \mathbf{v}, \Phi, \mathbf{p}, \mathbf{x}, t\right) := \\
&\frac{1}{8\pi}\left|-\nabla_{\mathbf{x}}\Psi - \frac{1}{c}\frac{\partial \mathbf{A}}{\partial t} + \frac{1}{c}\mathbf{v}\times \operatorname{curl}_{\mathbf{x}}\mathbf{A}\right|^2 - \frac{1}{8\pi}\left|\operatorname{curl}_{\mathbf{x}}\mathbf{A}\right|^2 - \left(\rho\Psi - \frac{1}{c}\mathbf{A}\cdot\mathbf{j}\right) \\
&-\mu c^2\mathcal{G}\left(1 - \frac{1}{c^2}\left|\mathbf{u}-\mathbf{v}\right|^2\right) + \frac{1}{2}\left(d_{\mathbf{x}}\mathbf{v} + \{d_{\mathbf{x}}\mathbf{v}\}^T\right) : \left(d_{\mathbf{x}}\mathbf{p} + \{d_{\mathbf{x}}\mathbf{p}\}^T\right) - 2\left(\operatorname{div}_{\mathbf{x}}\mathbf{v}\right)\left(\operatorname{div}_{\mathbf{x}}\mathbf{p}\right) \\
&+\frac{1}{4\pi G}\left(\operatorname{div}_{\mathbf{x}}\mathbf{v}\right)\left(\frac{\partial\Phi}{\partial t} + \mathbf{v}\cdot\nabla_{\mathbf{x}}\Phi\right) + \frac{1}{4\pi G}\Phi\left(\operatorname{div}_{\mathbf{x}}\mathbf{v}\right)^2 - \frac{\Phi}{16\pi G}\left|d_{\mathbf{x}}\mathbf{v} + \{d_{\mathbf{x}}\mathbf{v}\}^T\right|^2 \\
&+\frac{1}{8\pi G}\left|\nabla_{\mathbf{x}}\Phi\right|^2, \qquad (9.248)
\end{aligned}
$$

where $\mathcal{G}(s) : \mathbb{R} \to \mathbb{R}$ is a given function, Φ is some ancillary proper scalar field and $\mathbf{p}$ is some ancillary proper vector field. Then L is invariant under the change of inertial or non-inertial Cartesian coordinate system of the form (9.16). Then denoting the function

$$
g(s) := -c^2\mathcal{G}\left(1 - \frac{2s}{c^2}\right) \quad \forall s \qquad (9.249)
$$

we rewrite (9.248) as

$$
\begin{aligned}
&L\left(\mathbf{A}, \Psi, \mathbf{v}, \Phi, \mathbf{p}, \mathbf{x}, t\right) := \\
&\frac{1}{8\pi}\left|-\nabla_{\mathbf{x}}\Psi - \frac{1}{c}\frac{\partial \mathbf{A}}{\partial t} + \frac{1}{c}\mathbf{v}\times \operatorname{curl}_{\mathbf{x}}\mathbf{A}\right|^2 - \frac{1}{8\pi}\left|\operatorname{curl}_{\mathbf{x}}\mathbf{A}\right|^2 - \left(\rho\Psi - \frac{1}{c}\mathbf{A}\cdot\mathbf{j}\right) \\
&+\mu g\left(\frac{1}{2}\left|\mathbf{u}-\mathbf{v}\right|^2\right) + \frac{1}{2}\left(d_{\mathbf{x}}\mathbf{v} + \{d_{\mathbf{x}}\mathbf{v}\}^T\right) : \left(d_{\mathbf{x}}\mathbf{p} + \{d_{\mathbf{x}}\mathbf{p}\}^T\right) - 2\left(\operatorname{div}_{\mathbf{x}}\mathbf{v}\right)\left(\operatorname{div}_{\mathbf{x}}\mathbf{p}\right) \\
&+\frac{1}{4\pi G}\left(\operatorname{div}_{\mathbf{x}}\mathbf{v}\right)\left(\frac{\partial\Phi}{\partial t} + \mathbf{v}\cdot\nabla_{\mathbf{x}}\Phi\right) + \frac{1}{4\pi G}\Phi\left(\operatorname{div}_{\mathbf{x}}\mathbf{v}\right)^2 - \frac{\Phi}{16\pi G}\left|d_{\mathbf{x}}\mathbf{v} + \{d_{\mathbf{x}}\mathbf{v}\}^T\right|^2 \\
&+\frac{1}{8\pi G}\left|\nabla_{\mathbf{x}}\Phi\right|^2. \qquad (9.250)
\end{aligned}
$$

We point out to the most important choices of function $\mathcal{G}(s)$: fully non-relativistic choice $\mathcal{G}(s) = \frac{s}{2}$ and correspondingly $g(s) = \left(s - \frac{c^2}{2}\right)$; and relativistic-like choice

$\mathcal{G}(s) = \sqrt{s}$ and correspondingly $g(s) := -c^2\sqrt{1-\frac{2s}{c^2}}$. Note also that in the first case we have $\frac{dg}{ds}(s) = 1$ and in the second case $\frac{dg}{ds}(s) = \left(1-\frac{2s}{c^2}\right)^{-\frac{1}{2}} \approx 1$, where the last equation is valid in the case where $2s \ll c^2$.

We investigate stationary points of the functional

$$J = \int_0^T \int_{\mathbb{R}^3} L\left(\mathbf{A}, \Psi, \mathbf{v}, \Phi, \mathbf{p}, \mathbf{x}, t\right) d\mathbf{x}dt. \tag{9.251}$$

We denote

$$\begin{cases} \mathbf{D} = -\nabla_{\mathbf{x}}\Psi - \frac{1}{c}\frac{\partial \mathbf{A}}{\partial t} + \frac{1}{c}\mathbf{v}\times \operatorname{curl}_{\mathbf{x}}\mathbf{A} \\ \mathbf{B} = \operatorname{curl}_{\mathbf{x}}\mathbf{A} \\ \mathbf{E} = -\nabla_{\mathbf{x}}\Psi - \frac{1}{c}\frac{\partial \mathbf{A}}{\partial t} = \mathbf{D} - \frac{1}{c}\mathbf{v}\times\mathbf{B} \\ \mathbf{H} = \operatorname{curl}_{\mathbf{x}}\mathbf{A} + \frac{1}{c}\mathbf{v}\times\left(-\nabla_{\mathbf{x}}\Psi - \frac{1}{c}\frac{\partial \mathbf{A}}{\partial t} + \frac{1}{c}\mathbf{v}\times \operatorname{curl}_{\mathbf{x}}\mathbf{A}\right) = \mathbf{B} + \frac{1}{c}\mathbf{v}\times\mathbf{D}. \end{cases} \tag{9.252}$$

Then by (9.252) we have

$$\begin{cases} \operatorname{curl}_{\mathbf{x}}\mathbf{E} + \frac{1}{c}\frac{\partial \mathbf{B}}{\partial t} = 0 \\ \operatorname{div}_{\mathbf{x}}\mathbf{B} = 0. \end{cases} \tag{9.253}$$

Moreover by (9.250), (2.10) and (2.11) we have

$$\frac{\delta L}{\delta \mathbf{p}} = -\operatorname{div}_{\mathbf{x}}\left(d_{\mathbf{x}}\mathbf{v} + \{d_{\mathbf{x}}\mathbf{v}\}^T\right) + 2\nabla_{\mathbf{x}}\left(\operatorname{div}_{\mathbf{x}}\mathbf{v}\right) = \operatorname{curl}_{\mathbf{x}}\left(\operatorname{curl}_{\mathbf{x}}\mathbf{v}\right) = 0, \tag{9.254}$$

$$\frac{\delta L}{\delta \Phi} = \\ -\frac{1}{4\pi G}\left(\frac{\partial}{\partial t}\{\operatorname{div}_{\mathbf{x}}\mathbf{v}\} + \mathbf{v}\cdot\nabla_{\mathbf{x}}\left(\operatorname{div}_{\mathbf{x}}\mathbf{v}\right) + \frac{1}{4}\left|d_{\mathbf{x}}\mathbf{v} + \{d_{\mathbf{x}}\mathbf{v}\}^T\right|^2\right) - \frac{1}{4\pi G}\Delta_{\mathbf{x}}\Phi = 0, \tag{9.255}$$

$$\begin{aligned} \frac{\delta L}{\delta \mathbf{v}} = &-\left(\mu g'\left(\frac{1}{2}|\mathbf{u}-\mathbf{v}|^2\right)(\mathbf{u}-\mathbf{v}) + \frac{1}{4\pi c}\mathbf{D}\times\mathbf{B}\right) \\ &- \operatorname{div}_{\mathbf{x}}\left(d_{\mathbf{x}}\mathbf{p} + \{d_{\mathbf{x}}\mathbf{p}\}^T\right) + 2\nabla_{\mathbf{x}}\left(\operatorname{div}_{\mathbf{x}}\mathbf{p}\right) + \frac{1}{4\pi G}\operatorname{div}_{\mathbf{x}}\left\{\left(d_{\mathbf{x}}\mathbf{v} + \{d_{\mathbf{x}}\mathbf{v}\}^T\right)\Phi\right\} \\ &- \frac{1}{2\pi G}\nabla_{\mathbf{x}}\left(\Phi\left(\operatorname{div}_{\mathbf{x}}\mathbf{v}\right)\right) - \frac{1}{4\pi G}\nabla_{\mathbf{x}}\left(\frac{\partial \Phi}{\partial t} + \mathbf{v}\cdot\nabla_{\mathbf{x}}\Phi\right) + \frac{1}{4\pi G}\left(\operatorname{div}_{\mathbf{x}}\mathbf{v}\right)\nabla_{\mathbf{x}}\Phi = \\ &-\left(\mu g'\left(\frac{1}{2}|\mathbf{u}-\mathbf{v}|^2\right)(\mathbf{u}-\mathbf{v}) + \frac{1}{4\pi c}\mathbf{D}\times\mathbf{B}\right) + \operatorname{curl}_{\mathbf{x}}\left(\operatorname{curl}_{\mathbf{x}}\mathbf{p}\right) \\ &- \frac{1}{4\pi G}\Phi\operatorname{curl}_{\mathbf{x}}\left(\operatorname{curl}_{\mathbf{x}}\mathbf{v}\right) - \frac{1}{4\pi G}\left(\frac{\partial}{\partial t}\left(\nabla_{\mathbf{x}}\Phi\right) - \operatorname{curl}_{\mathbf{x}}\left(\mathbf{v}\times\nabla_{\mathbf{x}}\Phi\right) + \left(\Delta_{\mathbf{x}}\Phi\right)\mathbf{v}\right) = 0, \end{aligned} \tag{9.256}$$

$$\frac{\delta L}{\delta \Psi} = \frac{1}{4\pi}\mathrm{div}_{\mathbf{x}}\mathbf{D} - \rho = 0, \tag{9.257}$$

and

$$\frac{\delta L}{\delta \mathbf{A}} = \frac{1}{c}\mathbf{j} + \frac{1}{4\pi c}\frac{\partial \mathbf{D}}{\partial t} - \frac{1}{4\pi}\mathrm{curl}_{\mathbf{x}}\mathbf{B} - \frac{1}{4\pi c}\mathrm{curl}_{\mathbf{x}}\left(\mathbf{v}\times\mathbf{D}\right)$$
$$= \frac{1}{c}\mathbf{j} + \frac{1}{4\pi c}\frac{\partial \mathbf{D}}{\partial t} - \frac{1}{4\pi}\mathrm{curl}_{\mathbf{x}}\mathbf{H} = 0. \tag{9.258}$$

So

$$\begin{cases} \mathrm{curl}_{\mathbf{x}}\mathbf{H} = \frac{4\pi}{c}\mathbf{j} + \frac{1}{c}\frac{\partial \mathbf{D}}{\partial t} \\ \mathrm{div}_{\mathbf{x}}\mathbf{D} = 4\pi\rho \\ \mathrm{curl}_{\mathbf{x}}\mathbf{E} + \frac{1}{c}\frac{\partial \mathbf{B}}{\partial t} = 0 \\ \mathrm{div}_{\mathbf{x}}\mathbf{B} = 0 \\ \mathbf{E} = \mathbf{D} - \frac{1}{c}\mathbf{v}\times\mathbf{B} \\ \mathbf{H} = \mathbf{B} + \frac{1}{c}\mathbf{v}\times\mathbf{D} \\ \mathrm{curl}_{\mathbf{x}}\left(\mathrm{curl}_{\mathbf{x}}\mathbf{v}\right) = 0 \\ \frac{\partial}{\partial t}\left\{\mathrm{div}_{\mathbf{x}}\mathbf{v}\right\} + \mathbf{v}\cdot\nabla_{\mathbf{x}}\left(\mathrm{div}_{\mathbf{x}}\mathbf{v}\right) + \frac{1}{4}\left|d_{\mathbf{x}}\mathbf{v} + \{d_{\mathbf{x}}\mathbf{v}\}^T\right|^2 = -\Delta_{\mathbf{x}}\Phi \\ \left(\mu g'\left(\frac{1}{2}\left|\mathbf{u}-\mathbf{v}\right|^2\right)\left(\mathbf{u}-\mathbf{v}\right) + \frac{1}{4\pi c}\mathbf{D}\times\mathbf{B}\right) \\ = \mathrm{curl}_{\mathbf{x}}\left(\mathrm{curl}_{\mathbf{x}}\mathbf{p}\right) - \frac{1}{4\pi G}\left(\frac{\partial}{\partial t}\left(\nabla_{\mathbf{x}}\Phi\right) - \mathrm{curl}_{\mathbf{x}}\left(\mathbf{v}\times\nabla_{\mathbf{x}}\Phi\right) + \left(\Delta_{\mathbf{x}}\Phi\right)\mathbf{v}\right). \end{cases} \tag{9.259}$$

Next consider the equations of the gravitational-electromagnetic field in the form (9.259). Then defining the gravitational mass

$$M := \frac{1}{4\pi G}\Delta_{\mathbf{x}}\Phi \tag{9.260}$$

we obtain

$$\begin{cases} \mathrm{curl}_{\mathbf{x}}\mathbf{H} = \frac{4\pi}{c}\mathbf{j} + \frac{1}{c}\frac{\partial \mathbf{D}}{\partial t}, \\ \mathrm{div}_{\mathbf{x}}\mathbf{D} = 4\pi\rho, \\ \mathrm{curl}_{\mathbf{x}}\mathbf{E} + \frac{1}{c}\frac{\partial \mathbf{B}}{\partial t} = 0, \\ \mathrm{div}_{\mathbf{x}}\mathbf{B} = 0, \\ \mathbf{E} = \mathbf{D} - \frac{1}{c}\mathbf{v}\times\mathbf{B}, \\ \mathbf{H} = \mathbf{B} + \frac{1}{c}\mathbf{v}\times\mathbf{D}, \\ \frac{\partial}{\partial t}\left\{\mathrm{div}_{\mathbf{x}}\mathbf{v}\right\} + \mathbf{v}\cdot\nabla_{\mathbf{x}}\left(\mathrm{div}_{\mathbf{x}}\mathbf{v}\right) + \frac{1}{4}\left|d_{\mathbf{x}}\mathbf{v} + \{d_{\mathbf{x}}\mathbf{v}\}^T\right|^2 = -4\pi G M, \\ \mathrm{curl}_{\mathbf{x}}\left(\mathrm{curl}_{\mathbf{x}}\mathbf{v}\right) = 0, \\ \frac{\partial M}{\partial t} + \mathrm{div}_{\mathbf{x}}\left\{M\mathbf{v} + \mu g'\left(\frac{1}{2}\left|\mathbf{u}-\mathbf{v}\right|^2\right)\left(\mathbf{u}-\mathbf{v}\right) + \frac{1}{4\pi c}\mathbf{D}\times\mathbf{B}\right\} = 0. \end{cases} \tag{9.261}$$

Using the continuum equation

$$\frac{\partial \mu}{\partial t} + \operatorname{div}_{\mathbf{x}}(\mu \mathbf{u}) = 0, \tag{9.262}$$

we rewrite (9.261) as

$$\begin{cases} \operatorname{curl}_{\mathbf{x}} \mathbf{H} = \frac{4\pi}{c}\mathbf{j} + \frac{1}{c}\frac{\partial \mathbf{D}}{\partial t}, \\ \operatorname{div}_{\mathbf{x}} \mathbf{D} = 4\pi\rho, \\ \operatorname{curl}_{\mathbf{x}} \mathbf{E} + \frac{1}{c}\frac{\partial \mathbf{B}}{\partial t} = 0, \\ \operatorname{div}_{\mathbf{x}} \mathbf{B} = 0, \\ \mathbf{E} = \mathbf{D} - \frac{1}{c}\mathbf{v} \times \mathbf{B}, \\ \mathbf{H} = \mathbf{B} + \frac{1}{c}\mathbf{v} \times \mathbf{D}, \\ \frac{\partial}{\partial t}\{\operatorname{div}_{\mathbf{x}} \mathbf{v}\} + \mathbf{v} \cdot \nabla_{\mathbf{x}}(\operatorname{div}_{\mathbf{x}} \mathbf{v}) + \frac{1}{4}\left|d_{\mathbf{x}}\mathbf{v} + \{d_{\mathbf{x}}\mathbf{v}\}^T\right|^2 = -4\pi G M, \\ \operatorname{curl}_{\mathbf{x}}(\operatorname{curl}_{\mathbf{x}} \mathbf{v}) = 0, \\ \frac{\partial}{\partial t}(M - \mu) + \operatorname{div}_{\mathbf{x}}\{(M - \mu)\mathbf{v}\} = \\ -\operatorname{div}_{\mathbf{x}}\left\{\mu\left(g'\left(\frac{1}{2}|\mathbf{u} - \mathbf{v}|^2\right) - 1\right)(\mathbf{u} - \mathbf{v}) + \frac{1}{4\pi c}\mathbf{D} \times \mathbf{B}\right\}. \end{cases} \tag{9.263}$$

Note again for the last equation in (9.263) that: in the fully non-relativistic case we have $g'(s) = 1$ and in the relativistic-like case we have $g'(s) = \left(1 - \frac{2s}{c^2}\right)^{-\frac{1}{2}} \approx 1$, where the last equation is valid in the case where $2s \ll c^2$.

9.8.2 *The Case of Some Possible Alternative Model of the Gravity*

Consider **k** to be the vectorial potential of the inertia, which is a generally trivial speed-like vector field, assumed to be fixed in every fixed inertial or non-inertial Cartesian coordinate system (see Definition 4.2). Given known the distribution of inertial mass density of some continuum medium $\mu := \mu(\mathbf{x}, t)$, the field of velocities of this medium $\mathbf{u} := \mathbf{u}(\mathbf{x}, t)$, the charge density $\rho := \rho(\mathbf{x}, t)$ and the current density $\mathbf{j} := \mathbf{j}(\mathbf{x}, t)$ consider a Lagrangian density L defined by

$$\begin{aligned} L(\mathbf{A}, \Psi, \mathbf{v}, \Phi_0, \mathbf{h}, \mathbf{x}, t) := &\frac{1}{8\pi}\left|-\nabla_{\mathbf{x}}\Psi - \frac{1}{c}\frac{\partial \mathbf{A}}{\partial t} + \frac{1}{c}\mathbf{v} \times \operatorname{curl}_{\mathbf{x}} \mathbf{A}\right|^2 \\ &- \frac{1}{8\pi}|\operatorname{curl}_{\mathbf{x}} \mathbf{A}|^2 - \left(\rho\Psi - \frac{1}{c}\mathbf{A} \cdot \mathbf{j}\right) - \mu c^2 \mathcal{G}\left(1 - \frac{1}{c^2}|\mathbf{u} - \mathbf{v}|^2\right) \\ &- \frac{c^2}{8\pi G}\left|-\nabla_{\mathbf{x}}\Phi_0 - \frac{1}{c}\frac{\partial \mathbf{h}}{\partial t} + \frac{1}{c}\mathbf{v} \times \operatorname{curl}_{\mathbf{x}} \mathbf{h} - \frac{1}{c}\nabla_{\mathbf{x}}(\mathbf{h} \cdot \mathbf{v})\right|^2 + \frac{c^2}{8\pi G}|\operatorname{curl}_{\mathbf{x}} \mathbf{h}|^2 \\ &+ \frac{c^2\beta}{4\pi G}\left(\frac{1}{4}\left|d_{\mathbf{x}}\mathbf{v} + \{d_{\mathbf{x}}\mathbf{v}\}^T\right|^2 - |\operatorname{div}_{\mathbf{x}} \mathbf{v}|^2\right), \end{aligned} \tag{9.264}$$

where

$$\mathbf{h} = \mathbf{v} - \mathbf{k} \quad \text{and} \quad \Phi_0 = -\frac{1}{2c}\left|\mathbf{v} - \mathbf{k}\right|^2, \tag{9.265}$$

$\mathcal{G}(s) : \mathbb{R} \to \mathbb{R}$ is a given function, and $\beta \in \mathbb{R}$ is some constant. Then, as before, denoting the function

$$g(s) := -c^2\mathcal{G}\left(1 - \frac{2s}{c^2}\right) \qquad \forall s \tag{9.266}$$

we rewrite (9.264) as

$$\begin{aligned} L\left(\mathbf{A}, \Psi, \mathbf{v}, \Phi_0, \mathbf{h}, \mathbf{x}, t\right) := \frac{1}{8\pi}\left|-\nabla_{\mathbf{x}}\Psi - \frac{1}{c}\frac{\partial \mathbf{A}}{\partial t} + \frac{1}{c}\mathbf{v} \times \mathrm{curl}_{\mathbf{x}}\mathbf{A}\right|^2 \\ - \frac{1}{8\pi}\left|\mathrm{curl}_{\mathbf{x}}\mathbf{A}\right|^2 - \left(\rho\Psi - \frac{1}{c}\mathbf{A}\cdot\mathbf{j}\right) + \mu g\left(\frac{1}{2}\left|\mathbf{u} - \mathbf{v}\right|^2\right) \\ - \frac{c^2}{8\pi G}\left|-\nabla_{\mathbf{x}}\Phi_0 - \frac{1}{c}\frac{\partial \mathbf{h}}{\partial t} + \frac{1}{c}\mathbf{v} \times \mathrm{curl}_{\mathbf{x}}\mathbf{h} - \frac{1}{c}\nabla_{\mathbf{x}}\left(\mathbf{h}\cdot\mathbf{v}\right)\right|^2 + \frac{c^2}{8\pi G}\left|\mathrm{curl}_{\mathbf{x}}\mathbf{h}\right|^2 \\ + \frac{c^2\beta}{4\pi G}\left(\frac{1}{4}\left|d_{\mathbf{x}}\mathbf{v} + \{d_{\mathbf{x}}\mathbf{v}\}^T\right|^2 - \left|\mathrm{div}_{\mathbf{x}}\mathbf{v}\right|^2\right). \end{aligned} \tag{9.267}$$

In other words, we have

$$\begin{aligned} L\left(\mathbf{A}, \Psi, \mathbf{v}, \Phi_0, \mathbf{h}, \mathbf{x}, t\right) = L_1\left(\mathbf{A}, \Psi, \mathbf{v}, \mathbf{x}, t\right) := \frac{1}{8\pi}\left|-\nabla_{\mathbf{x}}\Psi - \frac{1}{c}\frac{\partial \mathbf{A}}{\partial t} + \frac{1}{c}\mathbf{v} \times \mathrm{curl}_{\mathbf{x}}\mathbf{A}\right|^2 \\ - \frac{1}{8\pi}\left|\mathrm{curl}_{\mathbf{x}}\mathbf{A}\right|^2 - \left(\rho\Psi - \frac{1}{c}\mathbf{A}\cdot\mathbf{j}\right) + \mu g\left(\frac{1}{2}\left|\mathbf{u} - \mathbf{v}\right|^2\right) \\ - \frac{c^2}{8\pi G}\left|\frac{1}{c}\nabla_{\mathbf{x}}\left(\frac{1}{2}\left|\mathbf{v} - \mathbf{k}\right|^2\right) - \frac{1}{c}\frac{\partial}{\partial t}\left(\mathbf{v} - \mathbf{k}\right) + \frac{1}{c}\mathbf{v} \times \mathrm{curl}_{\mathbf{x}}\left(\mathbf{v} - \mathbf{k}\right) - \frac{1}{c}\nabla_{\mathbf{x}}\left(\left(\mathbf{v} - \mathbf{k}\right)\cdot\mathbf{v}\right)\right|^2 \\ + \frac{c^2}{8\pi G}\left|\mathrm{curl}_{\mathbf{x}}\left(\mathbf{v} - \mathbf{k}\right)\right|^2 + \frac{c^2\beta}{4\pi G}\left(\frac{1}{4}\left|d_{\mathbf{x}}\mathbf{v} + \{d_{\mathbf{x}}\mathbf{v}\}^T\right|^2 - \left|\mathrm{div}_{\mathbf{x}}\mathbf{v}\right|^2\right). \end{aligned} \tag{9.268}$$

In particular, in the inertial coordinate system where $d_{\mathbf{x}}\mathbf{k} = 0$ and $\partial_t\mathbf{k} = 0$, we have

$$\begin{aligned} L\left(\mathbf{A}, \Psi, \mathbf{v}, \Phi_0, \mathbf{h}, \mathbf{x}, t\right) = L_1\left(\mathbf{A}, \Psi, \mathbf{v}, \mathbf{x}, t\right) := \frac{1}{8\pi}\left|-\nabla_{\mathbf{x}}\Psi - \frac{1}{c}\frac{\partial \mathbf{A}}{\partial t} + \frac{1}{c}\mathbf{v} \times \mathrm{curl}_{\mathbf{x}}\mathbf{A}\right|^2 \\ - \frac{1}{8\pi}\left|\mathrm{curl}_{\mathbf{x}}\mathbf{A}\right|^2 - \left(\rho\Psi - \frac{1}{c}\mathbf{A}\cdot\mathbf{j}\right) + \mu g\left(\frac{1}{2}\left|\mathbf{u} - \mathbf{v}\right|^2\right) \\ - \frac{c^2}{8\pi G}\left|-\frac{1}{c}\frac{\partial \mathbf{v}}{\partial t} + \frac{1}{c}\mathbf{v} \times \mathrm{curl}_{\mathbf{x}}\mathbf{v} - \frac{1}{c}\nabla_{\mathbf{x}}\left(\frac{1}{2}\left|\mathbf{v}\right|^2\right)\right|^2 + \frac{c^2}{8\pi G}\left|\mathrm{curl}_{\mathbf{x}}\mathbf{v}\right|^2 \\ + \frac{c^2\beta}{4\pi G}\left(\frac{1}{4}\left|d_{\mathbf{x}}\mathbf{v} + \{d_{\mathbf{x}}\mathbf{v}\}^T\right|^2 - \left|\mathrm{div}_{\mathbf{x}}\mathbf{v}\right|^2\right). \end{aligned} \tag{9.269}$$

Note here the advantage of inertial coordinate systems, where L and L_1 are completely independent on the vectorial potential of the inertia $\mathbf{k}$. Furthermore, by (2.15) we rewrite (9.269) as

$$L(\mathbf{A},\Psi,\mathbf{v},\Phi_0,\mathbf{h},\mathbf{x},t)=L_1(\mathbf{A},\Psi,\mathbf{v},\mathbf{x},t):=\frac{1}{8\pi}\left|-\nabla_{\mathbf{x}}\Psi-\frac{1}{c}\frac{\partial\mathbf{A}}{\partial t}+\frac{1}{c}\mathbf{v}\times\operatorname{curl}_{\mathbf{x}}\mathbf{A}\right|^2$$
$$-\frac{1}{8\pi}|\operatorname{curl}_{\mathbf{x}}\mathbf{A}|^2-\left(\rho\Psi-\frac{1}{c}\mathbf{A}\cdot\mathbf{j}\right)+\mu g\left(\frac{1}{2}|\mathbf{u}-\mathbf{v}|^2\right)-\frac{c^2}{8\pi G}\left|-\frac{1}{c}\left(\frac{\partial\mathbf{v}}{\partial t}+d_{\mathbf{x}}\mathbf{v}\cdot\mathbf{v}\right)\right|^2$$
$$+\frac{c^2}{8\pi G}|\operatorname{curl}_{\mathbf{x}}\mathbf{v}|^2+\frac{c^2\beta}{4\pi G}\left(\frac{1}{4}\left|d_{\mathbf{x}}\mathbf{v}+\{d_{\mathbf{x}}\mathbf{v}\}^T\right|^2-|\operatorname{div}_{\mathbf{x}}\mathbf{v}|^2\right). \quad (9.270)$$

Then, using Proposition 3.1 by (9.267), (9.265) and (9.268) we deduce that L and L_1 are invariant under the change of inertial or non-inertial Cartesian coordinate system, provided that $\mathbf{h}$ and $\mathbf{A}$ are proper vector fields, $\mathbf{v}$ is a speed-like vector field and Φ_0 and $\Psi_0:=\Psi-\frac{1}{c}\mathbf{A}\cdot\mathbf{v}$ are proper scalar fields.

We point out, again, to the most important choices of function $\mathcal{G}(s)$ in (9.264): fully non-relativistic choice $\mathcal{G}(s)=\frac{s}{2}$ and correspondingly $g(s)=\left(s-\frac{c^2}{2}\right)$; and relativistic-like choice $\mathcal{G}(s)=\sqrt{s}$ and correspondingly $g(s):=-c^2\sqrt{1-\frac{2s}{c^2}}$. Note also that in the first case we have $\frac{dg}{ds}(s)=1$ and in the second case $\frac{dg}{ds}(s)=\left(1-\frac{2s}{c^2}\right)^{-\frac{1}{2}}\approx 1$, where the last equation is valid in the case where $2s\ll c^2$.

We investigate critical points of the functional

$$J=\int_0^T\int_{\mathbb{R}^3}L_1(\mathbf{A},\Psi,\mathbf{v},\mathbf{x},t)\,d\mathbf{x}dt. \quad (9.271)$$

We denote

$$\begin{cases}\Psi_0=\Psi-\frac{1}{c}\mathbf{A}\cdot\mathbf{v}\\ \mathbf{D}=-\nabla_{\mathbf{x}}\Psi-\frac{1}{c}\frac{\partial\mathbf{A}}{\partial t}+\frac{1}{c}\mathbf{v}\times\operatorname{curl}_{\mathbf{x}}\mathbf{A}\\ \quad=-\nabla_{\mathbf{x}}\Psi_0-\frac{1}{c}\frac{\partial\mathbf{A}}{\partial t}+\frac{1}{c}\mathbf{v}\times\operatorname{curl}_{\mathbf{x}}\mathbf{A}-\frac{1}{c}\nabla_{\mathbf{x}}(\mathbf{A}\cdot\mathbf{v})\\ \mathbf{B}=\operatorname{curl}_{\mathbf{x}}\mathbf{A}\\ \mathbf{E}=-\nabla_{\mathbf{x}}\Psi-\frac{1}{c}\frac{\partial\mathbf{A}}{\partial t}=\mathbf{D}-\frac{1}{c}\mathbf{v}\times\mathbf{B}\\ \mathbf{H}=\operatorname{curl}_{\mathbf{x}}\mathbf{A}+\frac{1}{c}\mathbf{v}\times\left(-\nabla_{\mathbf{x}}\Psi-\frac{1}{c}\frac{\partial\mathbf{A}}{\partial t}+\frac{1}{c}\mathbf{v}\times\operatorname{curl}_{\mathbf{x}}\mathbf{A}\right)=\mathbf{B}+\frac{1}{c}\mathbf{v}\times\mathbf{D}\end{cases} \quad (9.272)$$

and

$$\begin{cases}\mathbf{R}=-\nabla_{\mathbf{x}}\Phi_0-\frac{1}{c}\frac{\partial\mathbf{h}}{\partial t}+\frac{1}{c}\mathbf{v}\times\operatorname{curl}_{\mathbf{x}}\mathbf{h}-\frac{1}{c}\nabla_{\mathbf{x}}(\mathbf{h}\cdot\mathbf{v})\\ \mathbf{Q}=\operatorname{curl}_{\mathbf{x}}\mathbf{h},\end{cases} \quad (9.273)$$

where $\mathbf{h}$ is a proper vector field and Φ_0 is a proper scalar field that are given by (9.265). In other words,

$$\begin{cases}\mathbf{R}=\frac{1}{c}\nabla_{\mathbf{x}}\left(\frac{1}{2}\left|\mathbf{v}-\mathbf{k}\right|^2\right)-\frac{1}{c}\frac{\partial}{\partial t}\left(\mathbf{v}-\mathbf{k}\right)+\frac{1}{c}\mathbf{v}\times\text{curl}_{\mathbf{x}}\left(\mathbf{v}-\mathbf{k}\right)-\frac{1}{c}\nabla_{\mathbf{x}}\left(\left(\mathbf{v}-\mathbf{k}\right)\cdot\mathbf{v}\right)\\ \mathbf{Q}=\text{curl}_{\mathbf{x}}(\mathbf{v}-\mathbf{k}),\end{cases} \tag{9.274}$$

and in inertial coordinate system where $d_{\mathbf{x}}\mathbf{k}=0$ and $\partial_t\mathbf{k}=0$ we also have

$$\begin{cases}\mathbf{R}=-\frac{1}{c}\left(\frac{\partial\mathbf{v}}{\partial t}+d_{\mathbf{x}}\mathbf{v}\cdot\mathbf{v}\right)\\ \mathbf{Q}=\text{curl}_{\mathbf{x}}\mathbf{v}.\end{cases} \tag{9.275}$$

As before, by (9.272) and (9.273) and Proposition 3.1 we infer that both **D**, **B** and **R**, **Q** are proper vector fields. Next, by (9.272) we have

$$\begin{cases}\text{curl}_{\mathbf{x}}\mathbf{D}+\frac{1}{c}\frac{\partial\mathbf{B}}{\partial t}-\frac{1}{c}\text{curl}_{\mathbf{x}}\left(\mathbf{v}\times\mathbf{B}\right)=\text{curl}_{\mathbf{x}}\mathbf{E}+\frac{1}{c}\frac{\partial\mathbf{B}}{\partial t}=0\\ \text{div}_{\mathbf{x}}\mathbf{B}=0,\end{cases} \tag{9.276}$$

and by (9.273) we have

$$\begin{cases}\text{curl}_{\mathbf{x}}\mathbf{R}+\frac{1}{c}\frac{\partial\mathbf{Q}}{\partial t}-\frac{1}{c}\text{curl}_{\mathbf{x}}\left(\mathbf{v}\times\mathbf{Q}\right)=0\\ \text{div}_{\mathbf{x}}\mathbf{Q}=0.\end{cases} \tag{9.277}$$

Furthermore, by (9.274) and (2.15) we deduce

$$\mathbf{R}=\frac{1}{c}\nabla_{\mathbf{x}}\left(\frac{1}{2}\left|\mathbf{v}-\mathbf{k}\right|^2\right)-\frac{1}{c}\left\{d_{\mathbf{x}}\mathbf{v}\right\}^T\cdot(\mathbf{v}-\mathbf{k})-\frac{1}{c}\frac{\partial}{\partial t}(\mathbf{v}-\mathbf{k})-\frac{1}{c}\left\{d_{\mathbf{x}}\left(\mathbf{v}-\mathbf{k}\right)\right\}^T\cdot\mathbf{v}$$
$$+\frac{1}{c}\mathbf{v}\times\text{curl}_{\mathbf{x}}\left(\mathbf{v}-\mathbf{k}\right)=-\frac{1}{c}\left\{d_{\mathbf{x}}\mathbf{k}\right\}^T\cdot(\mathbf{v}-\mathbf{k})-\frac{1}{c}\frac{\partial}{\partial t}(\mathbf{v}-\mathbf{k})-\frac{1}{c}d_{\mathbf{x}}\left(\mathbf{v}-\mathbf{k}\right)\cdot\mathbf{v}$$
$$\text{and}\quad\mathbf{Q}=\text{curl}_{\mathbf{x}}(\mathbf{v}-\mathbf{k}), \tag{9.278}$$

and thus, since $d_{\mathbf{x}}\mathbf{k}+\{d_{\mathbf{x}}\mathbf{k}\}^T=0$, $\text{curl}_{\mathbf{x}}(\text{curl}_{\mathbf{x}}\mathbf{k})=0$ and $\text{div}_{\mathbf{x}}\mathbf{k}=0$ we infer

$$\text{div}_{\mathbf{x}}\mathbf{R}=-\frac{1}{c}(\mathbf{v}-\mathbf{k})\cdot\Delta_{\mathbf{x}}\mathbf{k}-\frac{1}{c}\text{tr}\left(\{d_{\mathbf{x}}\mathbf{k}\}^T\cdot d_{\mathbf{x}}\left(\mathbf{v}-\mathbf{k}\right)\right)-\frac{1}{c}\frac{\partial}{\partial t}\left(\text{div}_{\mathbf{x}}\mathbf{v}\right)$$
$$-\frac{1}{c}d_{\mathbf{x}}\left(\text{div}_{\mathbf{x}}\mathbf{v}\right)\cdot\mathbf{v}-\frac{1}{c}\text{tr}\left(d_{\mathbf{x}}\left(\mathbf{v}-\mathbf{k}\right)\cdot d_{\mathbf{x}}\mathbf{v}\right)=-\frac{1}{c}\frac{\partial}{\partial t}\left(\text{div}_{\mathbf{x}}\mathbf{v}\right)-\frac{1}{c}d_{\mathbf{x}}\left(\text{div}_{\mathbf{x}}\mathbf{v}\right)\cdot\mathbf{v}$$
$$-\frac{1}{c}\text{tr}\left(d_{\mathbf{x}}\left(\mathbf{v}-\mathbf{k}\right)\cdot d_{\mathbf{x}}(\mathbf{v}-\mathbf{k})\right)=-\frac{1}{c}\frac{\partial}{\partial t}\left(\text{div}_{\mathbf{x}}\mathbf{v}\right)-\frac{1}{c}d_{\mathbf{x}}\left(\text{div}_{\mathbf{x}}\mathbf{v}\right)\cdot\mathbf{v}$$
$$-\frac{1}{4c}\left|d_{\mathbf{x}}(\mathbf{v}-\mathbf{k})+\{d_{\mathbf{x}}(\mathbf{v}-\mathbf{k})\}^T\right|^2+\frac{1}{4c}\left|d_{\mathbf{x}}(\mathbf{v}-\mathbf{k})-\{d_{\mathbf{x}}(\mathbf{v}-\mathbf{k})\}^T\right|^2$$
$$=-\frac{1}{c}\frac{\partial}{\partial t}\left(\text{div}_{\mathbf{x}}\mathbf{v}\right)-\frac{1}{c}d_{\mathbf{x}}\left(\text{div}_{\mathbf{x}}\mathbf{v}\right)\cdot\mathbf{v}-\frac{1}{4c}\left|d_{\mathbf{x}}\mathbf{v}+\{d_{\mathbf{x}}\mathbf{v}\}^T\right|^2+\frac{1}{2c}\left|\mathbf{Q}\right|^2$$
$$\text{and}\quad\text{curl}_{\mathbf{x}}\mathbf{Q}=\text{curl}_{\mathbf{x}}(\text{curl}_{\mathbf{x}}\mathbf{v}). \tag{9.279}$$

So,

$$\frac{1}{c}\left(\frac{\partial}{\partial t}\left(\mathrm{div}_{\mathbf{x}}\mathbf{v}\right)+d_{\mathbf{x}}\left(\mathrm{div}_{\mathbf{x}}\mathbf{v}\right)\cdot\mathbf{v}+\frac{1}{4}\left|d_{\mathbf{x}}\mathbf{v}+\{d_{\mathbf{x}}\mathbf{v}\}^T\right|^2\right)=\frac{1}{2c}\left|\mathbf{Q}\right|^2-\mathrm{div}_{\mathbf{x}}\mathbf{R}$$
$$\text{and}\quad \mathrm{curl}_{\mathbf{x}}(\mathrm{curl}_{\mathbf{x}}\mathbf{v})=\mathrm{curl}_{\mathbf{x}}\mathbf{Q}. \quad (9.280)$$

In other words,

$$\frac{1}{c}\left(\frac{\partial}{\partial t}\left(\mathrm{div}_{\mathbf{x}}\mathbf{v}\right)+\mathrm{div}_{\mathbf{x}}\left\{\left(\mathrm{div}_{\mathbf{x}}\mathbf{v}\right)\mathbf{v}\right\}+\frac{1}{4}\left|d_{\mathbf{x}}\mathbf{v}+\{d_{\mathbf{x}}\mathbf{v}\}^T\right|^2-\left|\mathrm{div}_{\mathbf{x}}\mathbf{v}\right|^2\right)$$
$$=\frac{1}{2c}\left|\mathbf{Q}\right|^2-\mathrm{div}_{\mathbf{x}}\mathbf{R}$$
$$\text{and}\quad \mathrm{curl}_{\mathbf{x}}(\mathrm{curl}_{\mathbf{x}}\mathbf{v})=\mathrm{curl}_{\mathbf{x}}\mathbf{Q}. \quad (9.281)$$

Moreover, by (9.267) and (2.5) we have

$$\frac{\delta L}{\delta \mathbf{h}}=\frac{c^2}{4\pi G}\mathrm{curl}_{\mathbf{x}}\mathbf{Q}-\frac{c}{4\pi G}\left(\frac{\partial \mathbf{R}}{\partial t}-\mathrm{curl}_{\mathbf{x}}\left\{\mathbf{v}\times\mathbf{R}\right\}+\left(\mathrm{div}_{\mathbf{x}}\mathbf{R}\right)\mathbf{v}\right), \quad (9.282)$$

$$\frac{\delta L}{\delta \Phi_0}=-\frac{c^2}{4\pi G}\left(\mathrm{div}_{\mathbf{x}}\mathbf{R}\right) \quad (9.283)$$

and

$$\frac{\delta L}{\delta \mathbf{v}}=-\left(\mu g'\left(\frac{1}{2}\left|\mathbf{u}-\mathbf{v}\right|^2\right)(\mathbf{u}-\mathbf{v})+\frac{1}{4\pi c}\mathbf{D}\times\mathbf{B}-\frac{c}{4\pi G}\mathbf{R}\times\mathbf{Q}\right)$$
$$+\frac{c^2\beta}{4\pi G}\mathrm{curl}_{\mathbf{x}}(\mathrm{curl}_{\mathbf{x}}\mathbf{v})-\frac{c}{4\pi G}\left(\mathrm{div}_{\mathbf{x}}\mathbf{R}\right)\mathbf{h}. \quad (9.284)$$

Thus, since by (9.265) we have

$$L_1\left(\mathbf{A},\Psi,\mathbf{v},\mathbf{x},t\right)=L\left(\mathbf{A},\Psi,\mathbf{v},-\frac{1}{2c}\left|\mathbf{v}-\mathbf{k}\right|^2,(\mathbf{v}-\mathbf{k}),\mathbf{x},t\right)=L\left(\mathbf{A},\Psi,\mathbf{v},\Phi_0,\mathbf{h},\mathbf{x},t\right), \quad (9.285)$$

by chain rule we have

$$\frac{\delta L_1}{\delta \mathbf{v}}\left(\mathbf{A},\Psi,\mathbf{v},\mathbf{x},t\right)=\frac{\delta L}{\delta \mathbf{v}}+\frac{\delta L}{\delta \mathbf{h}}-\frac{1}{c}\frac{\delta L}{\delta \Phi_0}(\mathbf{v}-\mathbf{k}). \quad (9.286)$$

Therefore, using (9.282), (9.283) (9.284) in (9.286) we deduce

$$\left(\mu g'\left(\frac{1}{2}|\mathbf{u}-\mathbf{v}|^2\right)(\mathbf{u}-\mathbf{v})+\frac{1}{4\pi c}\mathbf{D}\times\mathbf{B}-\frac{c}{4\pi G}\mathbf{R}\times\mathbf{Q}\right)=$$
$$\frac{c^2\beta}{4\pi G}\,\mathrm{curl}_{\mathbf{x}}(\mathrm{curl}_{\mathbf{x}}\mathbf{v})+\frac{c^2}{4\pi G}\mathrm{curl}_{\mathbf{x}}\mathbf{Q}-\frac{c}{4\pi G}\left(\frac{\partial \mathbf{R}}{\partial t}-\mathrm{curl}_{\mathbf{x}}\left\{\mathbf{v}\times\mathbf{R}\right\}+(\mathrm{div}_{\mathbf{x}}\mathbf{R})\,\mathbf{v}\right). \tag{9.287}$$

Moreover,

$$\frac{\delta L}{\delta \Psi}=\frac{1}{4\pi}\mathrm{div}_{\mathbf{x}}\mathbf{D}-\rho=0, \tag{9.288}$$

and

$$\frac{\delta L}{\delta \mathbf{A}}=\frac{1}{c}\mathbf{j}+\frac{1}{4\pi c}\frac{\partial \mathbf{D}}{\partial t}-\frac{1}{4\pi}\mathrm{curl}_{\mathbf{x}}\mathbf{B}-\frac{1}{4\pi c}\mathrm{curl}_{\mathbf{x}}\left(\mathbf{v}\times\mathbf{D}\right)$$
$$=\frac{1}{c}\mathbf{j}+\frac{1}{4\pi c}\frac{\partial \mathbf{D}}{\partial t}-\frac{1}{4\pi}\mathrm{curl}_{\mathbf{x}}\mathbf{H}=0. \tag{9.289}$$

So

$$\begin{cases}\mathrm{curl}_{\mathbf{x}}\mathbf{B}=\frac{4\pi}{c}\mathbf{j}+\frac{1}{c}\frac{\partial \mathbf{D}}{\partial t}-\frac{1}{c}\mathrm{curl}_{\mathbf{x}}\left(\mathbf{v}\times\mathbf{D}\right)\\ \mathrm{div}_{\mathbf{x}}\mathbf{D}=4\pi\rho\\ \mathrm{curl}_{\mathbf{x}}\mathbf{D}+\frac{1}{c}\frac{\partial \mathbf{B}}{\partial t}-\frac{1}{c}\,\mathrm{curl}_{\mathbf{x}}\left(\mathbf{v}\times\mathbf{B}\right)=0\\ \mathrm{div}_{\mathbf{x}}\mathbf{B}=0\\ \mathrm{curl}_{\mathbf{x}}\mathbf{R}+\frac{1}{c}\frac{\partial \mathbf{Q}}{\partial t}-\frac{1}{c}\mathrm{curl}_{\mathbf{x}}\left(\mathbf{v}\times\mathbf{Q}\right)=0\\ \mathrm{div}_{\mathbf{x}}\mathbf{Q}=0\\ \frac{1}{c}\left(\frac{\partial}{\partial t}\left(\mathrm{div}_{\mathbf{x}}\mathbf{v}\right)+\mathbf{v}\cdot\nabla_{\mathbf{x}}\left(\mathrm{div}_{\mathbf{x}}\mathbf{v}\right)+\frac{1}{4}\left|d_{\mathbf{x}}\mathbf{v}+\{d_{\mathbf{x}}\mathbf{v}\}^T\right|^2\right)=\frac{1}{2c}|\mathbf{Q}|^2-\mathrm{div}_{\mathbf{x}}\mathbf{R}\\ \mathrm{curl}_{\mathbf{x}}(\mathrm{curl}_{\mathbf{x}}\mathbf{v})=\mathrm{curl}_{\mathbf{x}}\mathbf{Q}\\ \frac{4\pi G}{c^2}\left(\mu g'\left(\frac{1}{2}|\mathbf{u}-\mathbf{v}|^2\right)(\mathbf{u}-\mathbf{v})+\frac{1}{4\pi c}\mathbf{D}\times\mathbf{B}-\frac{c}{4\pi G}\mathbf{R}\times\mathbf{Q}\right)=\\ (1+\beta)\,\mathrm{curl}_{\mathbf{x}}\mathbf{Q}-\frac{1}{c}\left(\frac{\partial \mathbf{R}}{\partial t}-\mathrm{curl}_{\mathbf{x}}\left\{\mathbf{v}\times\mathbf{R}\right\}+(\mathrm{div}_{\mathbf{x}}\mathbf{R})\,\mathbf{v}\right),\end{cases} \tag{9.290}$$

and by (9.275) in the inertial frame we have

$$\begin{cases}\mathrm{curl}_{\mathbf{x}}\mathbf{B}=\frac{4\pi}{c}\mathbf{j}+\frac{1}{c}\frac{\partial \mathbf{D}}{\partial t}-\frac{1}{c}\mathrm{curl}_{\mathbf{x}}\left(\mathbf{v}\times\mathbf{D}\right)\\ \mathrm{div}_{\mathbf{x}}\mathbf{D}=4\pi\rho\\ \mathrm{curl}_{\mathbf{x}}\mathbf{D}+\frac{1}{c}\frac{\partial \mathbf{B}}{\partial t}-\frac{1}{c}\,\mathrm{curl}_{\mathbf{x}}\left(\mathbf{v}\times\mathbf{B}\right)=0\\ \mathrm{div}_{\mathbf{x}}\mathbf{B}=0\\ \mathbf{R}=-\frac{1}{c}\left(\frac{\partial \mathbf{v}}{\partial t}+d_{\mathbf{x}}\mathbf{v}\cdot\mathbf{v}\right)\\ \mathbf{Q}=\mathrm{curl}_{\mathbf{x}}\mathbf{v}\\ \frac{4\pi G}{c^2}\left(\mu g'\left(\frac{1}{2}|\mathbf{u}-\mathbf{v}|^2\right)(\mathbf{u}-\mathbf{v})+\frac{1}{4\pi c}\mathbf{D}\times\mathbf{B}-\frac{c}{4\pi G}\mathbf{R}\times\mathbf{Q}\right)=\\ (1+\beta)\,\mathrm{curl}_{\mathbf{x}}\mathbf{Q}-\frac{1}{c}\left(\frac{\partial \mathbf{R}}{\partial t}-\mathrm{curl}_{\mathbf{x}}\left\{\mathbf{v}\times\mathbf{R}\right\}+(\mathrm{div}_{\mathbf{x}}\mathbf{R})\,\mathbf{v}\right).\end{cases} \tag{9.291}$$

Furthermore, taking $\mathrm{div}_{\mathbf{x}}$ of the both sides of the last equality in (9.290) and using continuum equation $\partial_t \mu + \mathrm{div}_{\mathbf{x}}(\mu \mathbf{u}) = 0$ we deduce

$$\begin{aligned}
&-(\partial_t \mu + \mathrm{div}_{\mathbf{x}}(\mu \mathbf{v})) \\
&+ \mathrm{div}_{\mathbf{x}} \left\{ \mu \left(g'\left(\frac{1}{2}|\mathbf{u}-\mathbf{v}|^2\right) - 1 \right)(\mathbf{u}-\mathbf{v}) + \frac{1}{4\pi c}\mathbf{D}\times\mathbf{B} - \frac{c}{4\pi G}\mathbf{R}\times\mathbf{Q} \right\} = \\
&\mathrm{div}_{\mathbf{x}} \left(\mu g'\left(\frac{1}{2}|\mathbf{u}-\mathbf{v}|^2\right)(\mathbf{u}-\mathbf{v}) + \frac{1}{4\pi c}\mathbf{D}\times\mathbf{B} - \frac{c}{4\pi G}\mathbf{R}\times\mathbf{Q} \right) = \\
&-\frac{c}{4\pi G}\left(\frac{\partial}{\partial t}(\mathrm{div}_{\mathbf{x}}\mathbf{R}) + \mathrm{div}_{\mathbf{x}}\{(\mathrm{div}_{\mathbf{x}}\mathbf{R})\,\mathbf{v}\} \right). \qquad (9.292)
\end{aligned}$$

Therefore, considering the proper scalar quantity Q_0, that we call the field mass, which satisfies

$$Q_0 := -\mu + \frac{c}{4\pi G}\mathrm{div}_{\mathbf{x}}\mathbf{R}, \qquad (9.293)$$

by (9.292) we deduce

$$\begin{aligned}
&\frac{\partial Q_0}{\partial t} + \mathrm{div}_{\mathbf{x}}(Q_0\mathbf{v}) = \\
&-\mathrm{div}_{\mathbf{x}} \left\{ \mu \left(g'\left(\frac{1}{2}|\mathbf{u}-\mathbf{v}|^2\right) - 1 \right)(\mathbf{u}-\mathbf{v}) + \frac{1}{4\pi c}\mathbf{D}\times\mathbf{B} - \frac{c}{4\pi G}\mathbf{R}\times\mathbf{Q} \right\}.
\end{aligned} \qquad (9.294)$$

Thus, we rewrite (9.290) as

$$\begin{cases}
\mathrm{curl}_{\mathbf{x}}\mathbf{B} = \frac{4\pi}{c}(\mathbf{j}-\rho\mathbf{v}) + \frac{1}{c}\frac{\partial \mathbf{D}}{\partial t} - \frac{1}{c}\mathrm{curl}_{\mathbf{x}}(\mathbf{v}\times\mathbf{D}) + \frac{1}{c}(\mathrm{div}_{\mathbf{x}}\mathbf{D})\,\mathbf{v}, \\
\mathrm{div}_{\mathbf{x}}\mathbf{D} = 4\pi\rho, \\
\mathrm{curl}_{\mathbf{x}}\mathbf{D} + \frac{1}{c}\frac{\partial \mathbf{B}}{\partial t} - \frac{1}{c}\,\mathrm{curl}_{\mathbf{x}}(\mathbf{v}\times\mathbf{B}) = 0, \\
\mathrm{div}_{\mathbf{x}}\mathbf{B} = 0, \\
\mathrm{curl}_{\mathbf{x}}\mathbf{R} + \frac{1}{c}\frac{\partial \mathbf{Q}}{\partial t} - \frac{1}{c}\mathrm{curl}_{\mathbf{x}}(\mathbf{v}\times\mathbf{Q}) = 0, \\
\mathrm{div}_{\mathbf{x}}\mathbf{Q} = 0, \\
\frac{4\pi G}{c^2}\left(\mu g'\left(\frac{1}{2}|\mathbf{u}-\mathbf{v}|^2\right)(\mathbf{u}-\mathbf{v}) + \frac{1}{4\pi c}\mathbf{D}\times\mathbf{B} - \frac{c}{4\pi G}\mathbf{R}\times\mathbf{Q}\right) = \\
(1+\beta)\,\mathrm{curl}_{\mathbf{x}}\mathbf{Q} - \frac{1}{c}\left(\frac{\partial \mathbf{R}}{\partial t} - \mathrm{curl}_{\mathbf{x}}\{\mathbf{v}\times\mathbf{R}\} + (\mathrm{div}_{\mathbf{x}}\mathbf{R})\,\mathbf{v}\right), \\
\mathrm{div}_{\mathbf{x}}\mathbf{R} = \frac{4\pi G}{c}(\mu + Q_0), \\
\frac{1}{c}\left(\frac{\partial}{\partial t}(\mathrm{div}_{\mathbf{x}}\mathbf{v}) + \mathbf{v}\cdot\nabla_{\mathbf{x}}(\mathrm{div}_{\mathbf{x}}\mathbf{v}) + \frac{1}{4}\left|d_{\mathbf{x}}\mathbf{v} + \{d_{\mathbf{x}}\mathbf{v}\}^T\right|^2\right) = \frac{1}{2c}|\mathbf{Q}|^2 - \mathrm{div}_{\mathbf{x}}\mathbf{R}, \\
\mathrm{curl}_{\mathbf{x}}(\mathrm{curl}_{\mathbf{x}}\mathbf{v}) = \mathrm{curl}_{\mathbf{x}}\mathbf{Q}, \\
\frac{\partial Q_0}{\partial t} + \mathrm{div}_{\mathbf{x}}(Q_0\mathbf{v}) = \\
-\mathrm{div}_{\mathbf{x}}\left\{\mu\left(g'\left(\frac{1}{2}|\mathbf{u}-\mathbf{v}|^2\right) - 1\right)(\mathbf{u}-\mathbf{v}) + \frac{1}{4\pi c}\mathbf{D}\times\mathbf{B} - \frac{c}{4\pi G}\mathbf{R}\times\mathbf{Q}\right\},
\end{cases} \qquad (9.295)$$

and we rewrite (9.291) in the inertial frame as

$$\begin{cases}\mathrm{curl}_{\mathbf{x}}\mathbf{B} = \frac{4\pi}{c}\mathbf{j} + \frac{1}{c}\frac{\partial \mathbf{D}}{\partial t} - \frac{1}{c}\mathrm{curl}_{\mathbf{x}}\left(\mathbf{v}\times\mathbf{D}\right),\\ \mathrm{div}_{\mathbf{x}}\mathbf{D} = 4\pi\rho,\\ \mathrm{curl}_{\mathbf{x}}\mathbf{D} + \frac{1}{c}\frac{\partial \mathbf{B}}{\partial t} - \frac{1}{c}\mathrm{curl}_{\mathbf{x}}\left(\mathbf{v}\times\mathbf{B}\right) = 0,\\ \mathrm{div}_{\mathbf{x}}\mathbf{B} = 0,\\ \mathbf{R} = -\frac{1}{c}\left(\frac{\partial \mathbf{v}}{\partial t} + d_{\mathbf{x}}\mathbf{v}\cdot\mathbf{v}\right),\\ \mathbf{Q} = \mathrm{curl}_{\mathbf{x}}\mathbf{v},\\ \frac{4\pi G}{c^2}\left(\mu g'\left(\frac{1}{2}\left|\mathbf{u}-\mathbf{v}\right|^2\right)(\mathbf{u}-\mathbf{v}) + \frac{1}{4\pi c}\mathbf{D}\times\mathbf{B} - \frac{c}{4\pi G}\mathbf{R}\times\mathbf{Q}\right) =\\ (1+\beta)\,\mathrm{curl}_{\mathbf{x}}\mathbf{Q} - \frac{1}{c}\left(\frac{\partial \mathbf{R}}{\partial t} - \mathrm{curl}_{\mathbf{x}}\left\{\mathbf{v}\times\mathbf{R}\right\} + (\mathrm{div}_{\mathbf{x}}\mathbf{R})\,\mathbf{v}\right),\\ \mathrm{div}_{\mathbf{x}}\mathbf{R} = \frac{4\pi G}{c}(\mu + Q_0),\\ \frac{\partial Q_0}{\partial t} + \mathrm{div}_{\mathbf{x}}\left(Q_0\mathbf{v}\right) =\\ -\mathrm{div}_{\mathbf{x}}\left\{\mu\left(g'\left(\frac{1}{2}\left|\mathbf{u}-\mathbf{v}\right|^2\right) - 1\right)(\mathbf{u}-\mathbf{v}) + \frac{1}{4\pi c}\mathbf{D}\times\mathbf{B} - \frac{c}{4\pi G}\mathbf{R}\times\mathbf{Q}\right\}.\end{cases} \tag{9.296}$$

As before, by Proposition 3.1 we deduce that (9.295) is invariant under the change of inertial or non-inertial Cartesian coordinate systems. Moreover, (9.296) is invariant under the change of inertial Cartesian coordinate systems. Note also that, both, (9.295) in an arbitrary inertial or non-inertial Cartesian coordinate system and (9.296) in an arbitrary inertial Cartesian coordinate system, are completely independent on the vectorial potential of the inertia **k**. Note again for the last equation in (9.295) or (9.296) that: in the fully non-relativistic case we have $g'(s) = 1$ and in the relativistic-like case we have $g'(s) = \left(1 - \frac{2s}{c^2}\right)^{-\frac{1}{2}} \approx 1$, where the last equation is valid in the case where $2s \ll c^2$.

Finally, as before, note that in the case of large constant $|\beta| \gg 1$ we have $\mathbf{Q} \to 0$ in (9.295), and thus the gravity equations (9.295) reduce to the equations of the Newtonian-type gravity in the form of (9.263). In that case the gravity field propagates with the infinite speed. On the other hand, in the case of vanishing constant $\beta = 0$ the form of equations for $\mathbf{R}$ and $\mathbf{Q}$ in (9.295) is completely the same as the form of the Maxwell equations for $\mathbf{D}$ and $\mathbf{B}$ in (9.295), except of the different meaning of "charges" and "currents" in these two sets of equations. In that case the electromagnetic and the gravity fields propagate with the same speed. However, in the mixed case of constant $\beta \sim 1$, the electromagnetic and the gravity fields propagate with different finite speeds.

Remark 9.5 One can wonder: what should be possible values of the vectorial gravitational potential $\mathbf{v}$ in the proximity of the Earth or another massive body? In Remark 4.4 we answered this question in the case of the Newtonian-type gravity, given by (4.49). Moreover, in Remark 8.1 we answered this question in the case of more general laws of the gravity, given by (8.41) with arbitrary constant β. In order to answer this question in the case of more general laws of the gravity, given by either (9.263) or (9.295) with arbitrary constant β, consider two Cartesian coordinate systems: non-rotating system $(*)$ with the center that coincides with the center of masses of the Earth and inertial system $(**)$ related to some external cosmic bodies. Assume that the center of masses of the Earth has place $\mathbf{R}(t')$ and velocity $\mathbf{W}(t') := \frac{d\mathbf{R}}{dt'}(t')$ in the coordinate system $(**)$. Thus the change of coordinate system $(*)$ to coordinate system $(**)$ is given by

$$\begin{cases} \mathbf{x}' = \mathbf{x} + \mathbf{R}(t), \\ t' = t, \end{cases} \tag{9.297}$$

and the vectorial gravitational potential $\mathbf{v}$, being a speed like vector field, transforms as

$$\mathbf{v}' = \mathbf{v} + \mathbf{W}(t). \tag{9.298}$$

Next, since the system $(**)$ is inertial, consistently with (9.296) in the system $(**)$ we have

$$\begin{cases} curl_{\mathbf{x}'}\mathbf{B}' = \frac{4\pi}{c}\left(\mathbf{j}' - \rho'\mathbf{v}'\right) + \frac{1}{c}\frac{\partial \mathbf{D}'}{\partial t'} - \frac{1}{c}curl_{\mathbf{x}'}\left(\mathbf{v}' \times \mathbf{D}'\right) + \frac{1}{c}\left(div_{\mathbf{x}'}\mathbf{D}'\right)\mathbf{v}', \\ div_{\mathbf{x}}\mathbf{D}' = 4\pi\rho', \\ curl_{\mathbf{x}'}\mathbf{D}' + \frac{1}{c}\frac{\partial \mathbf{B}'}{\partial t'} - \frac{1}{c}\,curl_{\mathbf{x}'}\left(\mathbf{v}' \times \mathbf{B}'\right) = 0, \\ div_{\mathbf{x}'}\mathbf{B}' = 0, \\ \mathbf{R}' = -\frac{1}{c}\left(\frac{\partial \mathbf{v}'}{\partial t'} + d_{\mathbf{x}'}\mathbf{v}' \cdot \mathbf{v}'\right), \\ \mathbf{Q}' = curl_{\mathbf{x}'}\mathbf{v}', \\ curl_{\mathbf{x}'}\mathbf{R}' + \frac{1}{c}\frac{\partial \mathbf{Q}'}{\partial t'} - \frac{1}{c}curl_{\mathbf{x}'}\left(\mathbf{v}' \times \mathbf{Q}'\right) = 0, \\ div_{\mathbf{x}'}\mathbf{Q}' = 0, \\ \frac{4\pi G}{c^2}\left(\mu' g'\left(\frac{1}{2}\left|\mathbf{u}' - \mathbf{v}'\right|^2\right)\left(\mathbf{u}' - \mathbf{v}'\right) + \frac{1}{4\pi c}\mathbf{D}' \times \mathbf{B}' - \frac{c}{4\pi G}\mathbf{R}' \times \mathbf{Q}'\right) = \\ (1+\beta)\,curl_{\mathbf{x}'}\mathbf{Q}' - \frac{1}{c}\left(\frac{\partial \mathbf{R}'}{\partial t'} - curl_{\mathbf{x}'}\left\{\mathbf{v}' \times \mathbf{R}'\right\} + \left(div_{\mathbf{x}}\mathbf{R}'\right)\mathbf{v}'\right), \\ div_{\mathbf{x}'}\mathbf{R}' = \frac{4\pi G}{c}(\mu' + Q_0'), \\ \frac{\partial Q_0'}{\partial t'} + div_{\mathbf{x}'}\left(Q_0'\mathbf{v}'\right) = \\ -div_{\mathbf{x}'}\left\{\mu'\left(g'\left(\frac{1}{2}\left|\mathbf{u}' - \mathbf{v}'\right|^2\right) - 1\right)\left(\mathbf{u}' - \mathbf{v}'\right) + \frac{1}{4\pi c}\mathbf{D}' \times \mathbf{B}' - \frac{c}{4\pi G}\mathbf{R}' \times \mathbf{Q}'\right\}, \end{cases} \tag{9.299}$$

and by (9.295) in the system $(*)$ we have

$$\begin{cases} curl_{\mathbf{x}}\mathbf{B} = \frac{4\pi}{c}\left(\mathbf{j} - \rho\mathbf{v}\right) + \frac{1}{c}\frac{\partial \mathbf{D}}{\partial t} - \frac{1}{c}curl_{\mathbf{x}}\left(\mathbf{v} \times \mathbf{D}\right) + \frac{1}{c}\left(div_{\mathbf{x}}\mathbf{D}\right)\mathbf{v}, \\ div_{\mathbf{x}}\mathbf{D} = 4\pi\rho, \\ curl_{\mathbf{x}}\mathbf{D} + \frac{1}{c}\frac{\partial \mathbf{B}}{\partial t} - \frac{1}{c}\,curl_{\mathbf{x}}\left(\mathbf{v} \times \mathbf{B}\right) = 0, \\ div_{\mathbf{x}}\mathbf{B} = 0, \\ curl_{\mathbf{x}}\mathbf{R} + \frac{1}{c}\frac{\partial \mathbf{Q}}{\partial t} - \frac{1}{c}curl_{\mathbf{x}}\left(\mathbf{v} \times \mathbf{Q}\right) = 0, \\ div_{\mathbf{x}}\mathbf{Q} = 0, \\ \frac{4\pi G}{c^2}\left(\mu g'\left(\frac{1}{2}\left|\mathbf{u} - \mathbf{v}\right|^2\right)\left(\mathbf{u} - \mathbf{v}\right) + \frac{1}{4\pi c}\mathbf{D} \times \mathbf{B} - \frac{c}{4\pi G}\mathbf{R} \times \mathbf{Q}\right) = \\ (1+\beta)\,curl_{\mathbf{x}}\mathbf{Q} - \frac{1}{c}\left(\frac{\partial \mathbf{R}}{\partial t} - curl_{\mathbf{x}}\left\{\mathbf{v} \times \mathbf{R}\right\} + \left(div_{\mathbf{x}}\mathbf{R}\right)\mathbf{v}\right), \\ div_{\mathbf{x}}\mathbf{R} = \frac{4\pi G}{c}(\mu + Q_0), \\ \frac{1}{c}\left(\frac{\partial}{\partial t}\left(div_{\mathbf{x}}\mathbf{v}\right) + \mathbf{v} \cdot \nabla_{\mathbf{x}}\left(div_{\mathbf{x}}\mathbf{v}\right) + \frac{1}{4}\left|d_{\mathbf{x}}\mathbf{v} + \left\{d_{\mathbf{x}}\mathbf{v}\right\}^T\right|^2\right) = \frac{1}{2c}\left|\mathbf{Q}\right|^2 - div_{\mathbf{x}}\mathbf{R}, \\ curl_{\mathbf{x}}\left(curl_{\mathbf{x}}\mathbf{v}\right) = curl_{\mathbf{x}}\mathbf{Q}, \\ \frac{\partial Q_0}{\partial t} + div_{\mathbf{x}}\left(Q_0\mathbf{v}\right) = \\ -div_{\mathbf{x}}\left\{\mu\left(g'\left(\frac{1}{2}\left|\mathbf{u} - \mathbf{v}\right|^2\right) - 1\right)\left(\mathbf{u} - \mathbf{v}\right) + \frac{1}{4\pi c}\mathbf{D} \times \mathbf{B} - \frac{c}{4\pi G}\mathbf{R} \times \mathbf{Q}\right\}. \end{cases} \tag{9.300}$$

On the other hand, by (9.297) and (9.298) and using Proposition 3.1 we deduce

$$\mathrm{curl}_{\mathbf{x}}\mathbf{v} = \mathrm{curl}_{\mathbf{x}'}\mathbf{v}', \tag{9.301}$$

and

$$\frac{d\mathbf{W}}{\partial t}(t) + \frac{\partial \mathbf{v}}{\partial t} + d_{\mathbf{x}}\mathbf{v}\cdot\mathbf{v} = \frac{d\mathbf{W}}{\partial t'}(t') + \frac{\partial \mathbf{v}}{\partial t'} + d_{\mathbf{x}'}\mathbf{v}\cdot\mathbf{W}(t') + d_{\mathbf{x}'}\mathbf{v}\cdot\mathbf{v} = \frac{\partial \mathbf{v}'}{\partial t'} + d_{\mathbf{x}'}\mathbf{v}'\cdot\mathbf{v}'. \tag{9.302}$$

Thus since **R** is a proper field, by (9.299), (9.301) and (9.302) we deduce

$$\mathrm{curl}_{\mathbf{x}}\mathbf{v} = \mathbf{Q}' = \mathbf{Q}, \tag{9.303}$$

and

$$\frac{d\mathbf{W}}{\partial t}(t) + \frac{\partial \mathbf{v}}{\partial t} + d_{\mathbf{x}}\mathbf{v}\cdot\mathbf{v} = -c\,\mathbf{R}' = -c\,\mathbf{R}. \tag{9.304}$$

Therefore, by (9.300), (9.303) and (9.304) in the system (∗) we have

$$\begin{cases} \mathrm{curl}_{\mathbf{x}}\mathbf{B} = \frac{4\pi}{c}\mathbf{j} + \frac{1}{c}\frac{\partial \mathbf{D}}{\partial t} - \frac{1}{c}\mathrm{curl}_{\mathbf{x}}\left(\mathbf{v}\times\mathbf{D}\right), \\ \mathrm{div}_{\mathbf{x}}\mathbf{D} = 4\pi\rho, \\ \mathrm{curl}_{\mathbf{x}}\mathbf{D} + \frac{1}{c}\frac{\partial \mathbf{B}}{\partial t} - \frac{1}{c}\,\mathrm{curl}_{\mathbf{x}}\left(\mathbf{v}\times\mathbf{B}\right) = 0, \\ \mathrm{div}_{\mathbf{x}}\mathbf{B} = 0, \\ \frac{1}{c}\frac{d\mathbf{W}}{\partial t}(t) + \mathbf{R} = -\frac{1}{c}\left(\frac{\partial \mathbf{v}}{\partial t} + d_{\mathbf{x}}\mathbf{v}\cdot\mathbf{v}\right), \\ \mathbf{Q} = \mathrm{curl}_{\mathbf{x}}\mathbf{v}, \\ \frac{4\pi G}{c^2}\left(\mu g'\left(\frac{1}{2}\left|\mathbf{u}-\mathbf{v}\right|^2\right)(\mathbf{u}-\mathbf{v}) + \frac{1}{4\pi c}\mathbf{D}\times\mathbf{B} - \frac{c}{4\pi G}\mathbf{R}\times\mathbf{Q}\right) = \\ (1+\beta)\,\mathrm{curl}_{\mathbf{x}}\mathbf{Q} - \frac{1}{c}\left(\frac{\partial \mathbf{R}}{\partial t} - \mathrm{curl}_{\mathbf{x}}\left\{\mathbf{v}\times\mathbf{R}\right\} + (\mathrm{div}_{\mathbf{x}}\mathbf{R})\,\mathbf{v}\right), \\ \mathrm{div}_{\mathbf{x}}\mathbf{R} = \frac{4\pi G}{c}(\mu + Q_0), \\ \frac{\partial Q_0}{\partial t} + \mathrm{div}_{\mathbf{x}}\left(Q_0\mathbf{v}\right) = \\ -\mathrm{div}_{\mathbf{x}}\left\{\mu\left(g'\left(\frac{1}{2}\left|\mathbf{u}-\mathbf{v}\right|^2\right) - 1\right)(\mathbf{u}-\mathbf{v}) + \frac{1}{4\pi c}\mathbf{D}\times\mathbf{B} - \frac{c}{4\pi G}\mathbf{R}\times\mathbf{Q}\right\}. \end{cases} \tag{9.305}$$

On the other hand, since the system (∗∗) is inertial, the quantity $\frac{d\mathbf{W}}{\partial t}(t)$, being generated by the gravitational field from the far bodies, is insignificant with respect to the quantity $c\mathbf{R}$ in the scale compatible to the Earth size. Thus, we rewrite (9.305) as

$$\begin{cases} \text{curl}_{\mathbf{x}}\mathbf{B} = \frac{4\pi}{c}\mathbf{j} + \frac{1}{c}\frac{\partial \mathbf{D}}{\partial t} - \frac{1}{c}\text{curl}_{\mathbf{x}}\left(\mathbf{v} \times \mathbf{D}\right), \\ \text{div}_{\mathbf{x}}\mathbf{D} = 4\pi\rho, \\ \text{curl}_{\mathbf{x}}\mathbf{D} + \frac{1}{c}\frac{\partial \mathbf{B}}{\partial t} - \frac{1}{c}\,\text{curl}_{\mathbf{x}}\left(\mathbf{v} \times \mathbf{B}\right) = 0, \\ \text{div}_{\mathbf{x}}\mathbf{B} = 0, \\ \mathbf{R} = -\frac{1}{c}\left(\frac{\partial \mathbf{v}}{\partial t} + d_{\mathbf{x}}\mathbf{v}\cdot\mathbf{v}\right), \\ \mathbf{Q} = \text{curl}_{\mathbf{x}}\mathbf{v}, \\ \frac{4\pi G}{c^2}\left(\mu g'\left(\frac{1}{2}\left|\mathbf{u}-\mathbf{v}\right|^2\right)(\mathbf{u}-\mathbf{v}) + \frac{1}{4\pi c}\mathbf{D}\times\mathbf{B} - \frac{c}{4\pi G}\mathbf{R}\times\mathbf{Q}\right) = \\ (1+\beta)\,\text{curl}_{\mathbf{x}}\mathbf{Q} - \frac{1}{c}\left(\frac{\partial \mathbf{R}}{\partial t} - \text{curl}_{\mathbf{x}}\left\{\mathbf{v}\times\mathbf{R}\right\} + (\text{div}_{\mathbf{x}}\mathbf{R})\,\mathbf{v}\right), \\ \text{div}_{\mathbf{x}}\mathbf{R} = \frac{4\pi G}{c}(\mu + Q_0), \\ \frac{\partial Q_0}{\partial t} + \text{div}_{\mathbf{x}}\left(Q_0\mathbf{v}\right) = \\ -\text{div}_{\mathbf{x}}\left\{\mu\left(g'\left(\frac{1}{2}\left|\mathbf{u}-\mathbf{v}\right|^2\right) - 1\right)(\mathbf{u}-\mathbf{v}) + \frac{1}{4\pi c}\mathbf{D}\times\mathbf{B} - \frac{c}{4\pi G}\mathbf{R}\times\mathbf{Q}\right\}. \end{cases} \tag{9.306}$$

Next, we can neglect all the far cosmic body masses expect of the Earth itself and thus we can consider

$$\mu(\mathbf{x}, t) = \mu_1\left(|\mathbf{x}|\right) \quad \text{and} \quad \mathbf{u}(\mathbf{x}, t) = 0 \qquad \forall\,(\mathbf{x}, t), \tag{9.307}$$

where $\mu_1 := \mu_1\left(|\mathbf{x}|\right)$ is the inertial mass density of the Earth which is assumed to be a radial function such that

$$\mu_1\left(|\mathbf{x}|\right) = 0 \qquad \text{if} \quad |\mathbf{x}| > r_0, \tag{9.308}$$

where r_0 is the Earth radius. Moreover, we can neglect all the electromagnetical masses and thus we simplify the equations for the gravity in (8.52) as

$$\begin{cases} \mathbf{R} = -\frac{1}{c}\left(\frac{\partial \mathbf{v}}{\partial t} + d_{\mathbf{x}}\mathbf{v}\cdot\mathbf{v}\right), \\ \mathbf{Q} = \text{curl}_{\mathbf{x}}\mathbf{v}, \\ \frac{4\pi G}{c^2}\left(-\mu_1\left(|\mathbf{x}|\right) g'\left(\frac{1}{2}\left|\mathbf{v}\right|^2\right)\mathbf{v} - \frac{c}{4\pi G}\mathbf{R}\times\mathbf{Q}\right) = \\ (1+\beta)\,\text{curl}_{\mathbf{x}}\mathbf{Q} - \frac{1}{c}\left(\frac{\partial \mathbf{R}}{\partial t} - \text{curl}_{\mathbf{x}}\left\{\mathbf{v}\times\mathbf{R}\right\} + (\text{div}_{\mathbf{x}}\mathbf{R})\,\mathbf{v}\right), \\ \text{div}_{\mathbf{x}}\mathbf{R} = \frac{4\pi G}{c}\left(\mu_1\left(|\mathbf{x}|\right) + Q_0\right), \\ \frac{\partial Q_0}{\partial t} + \text{div}_{\mathbf{x}}\left(Q_0\mathbf{v}\right) = \text{div}_{\mathbf{x}}\left\{\mu_1\left(|\mathbf{x}|\right)\left(g'\left(\frac{1}{2}\left|\mathbf{v}\right|^2\right) - 1\right)\mathbf{v} + \frac{c}{4\pi G}\mathbf{R}\times\mathbf{Q}\right\}. \end{cases} \tag{9.309}$$

Being in the system (∗) which is stationary with respect to the center of the Earth we look for stationary (i.e. time independent) solutions of (9.309). Thus (9.309) implies

$$
\begin{cases}
\mathbf{R} = -\frac{1}{c}\, d_{\mathbf{x}}\mathbf{v}\cdot\mathbf{v},\\
\mathbf{Q} = \text{curl}_{\mathbf{x}}\mathbf{v},\\
\frac{4\pi G}{c^2}\left(-\mu_1\left(|\mathbf{x}|\right) g'\left(\frac{1}{2}|\mathbf{v}|^2\right)\mathbf{v} - \frac{c}{4\pi G}\mathbf{R}\times\mathbf{Q}\right) =\\
(1+\beta)\,\text{curl}_{\mathbf{x}}\mathbf{Q} - \frac{1}{c}\left(-\text{curl}_{\mathbf{x}}\left\{\mathbf{v}\times\mathbf{R}\right\} + \left(\text{div}_{\mathbf{x}}\mathbf{R}\right)\mathbf{v}\right),\\
\text{div}_{\mathbf{x}}\mathbf{R} = \frac{4\pi G}{c}\left(\mu_1\left(|\mathbf{x}|\right) + Q_0\right),\\
\text{div}_{\mathbf{x}}\left(Q_0\mathbf{v}\right) = \text{div}_{\mathbf{x}}\left\{\mu_1\left(|\mathbf{x}|\right)\left(g'\left(\frac{1}{2}|\mathbf{v}|^2\right) - 1\right)\mathbf{v} + \frac{c}{4\pi G}\mathbf{R}\times\mathbf{Q}\right\}.
\end{cases}
\tag{9.310}
$$

On the other hand, by the symmetry considerations of the problem we look for the solution of (9.310) that satisfies $\mathbf{v}(\mathbf{x}) = \nabla_{\mathbf{x}} Z_0\left(|\mathbf{x}|\right)$ where, again by the symmetry of the problem, the scalar function $Z_0\left(|\mathbf{x}|\right)$ should be radial. In particular, by (9.310) we obtain

$$
\mathbf{Q} = \text{curl}_{\mathbf{x}}\mathbf{v} = 0 \tag{9.311}
$$

and thus we simplify (9.310) as

$$
\begin{cases}
\mathbf{R} = -\frac{1}{c}\, d_{\mathbf{x}}\mathbf{v}\cdot\mathbf{v} = -\frac{1}{c}\nabla_{\mathbf{x}}\left(\frac{1}{2}|\mathbf{v}|^2\right),\\
\mathbf{Q} = \text{curl}_{\mathbf{x}}\mathbf{v} = 0,\\
\frac{4\pi G}{c^2}\mu_1\left(|\mathbf{x}|\right) g'\left(\frac{1}{2}|\mathbf{v}|^2\right)\mathbf{v} = \frac{1}{c}\left(-\text{curl}_{\mathbf{x}}\left\{\mathbf{v}\times\mathbf{R}\right\} + \left(\text{div}_{\mathbf{x}}\mathbf{R}\right)\mathbf{v}\right),\\
\text{div}_{\mathbf{x}}\mathbf{R} = \frac{4\pi G}{c}\mu_1\left(|\mathbf{x}|\right)\, g'\left(\frac{1}{2}|\mathbf{v}|^2\right),\\
\text{curl}_{\mathbf{x}}\mathbf{R} = 0,\\
Q_0 = \mu_1\left(|\mathbf{x}|\right)\left(g'\left(\frac{1}{2}|\mathbf{v}|^2\right) - 1\right).
\end{cases}
\tag{9.312}
$$

In particular, since $\mathbf{v} = \nabla_{\mathbf{x}} Z_0\left(|\mathbf{x}|\right)$ and $\mathbf{R} = -\frac{1}{c}\nabla_{\mathbf{x}}\left(\frac{1}{2}|\mathbf{v}|^2\right)$ are both gradients of radial functions, we have $\mathbf{v}\times\mathbf{R} = 0$, and thus we further simplify (9.312) as

$$
\begin{cases}
\mathbf{R} = -\frac{1}{c}\nabla_{\mathbf{x}}\left(\frac{1}{2}|\mathbf{v}|^2\right),\\
\mathbf{Q} = \text{curl}_{\mathbf{x}}\mathbf{v} = 0,\\
\text{div}_{\mathbf{x}}\mathbf{R} = \frac{4\pi G}{c}\mu_1\left(|\mathbf{x}|\right)\, g'\left(\frac{1}{2}|\mathbf{v}|^2\right),\\
\text{curl}_{\mathbf{x}}\mathbf{R} = 0,\\
Q_0 = \mu_1\left(|\mathbf{x}|\right)\left(g'\left(\frac{1}{2}|\mathbf{v}|^2\right) - 1\right).
\end{cases}
\tag{9.313}
$$

However, (9.313) is equivalent to the following:

$$
\begin{cases}
\Delta_{\mathbf{x}}\left(-\frac{1}{2}|\mathbf{v}|^2\right) = 4\pi G\,\mu_1\left(|\mathbf{x}|\right)\, g'\left(\frac{1}{2}|\mathbf{v}|^2\right),\\
\text{curl}_{\mathbf{x}}\mathbf{v} = 0,\\
\mathbf{R} = -\frac{1}{c}\nabla_{\mathbf{x}}\left(\frac{1}{2}|\mathbf{v}|^2\right),\\
\mathbf{Q} = 0,\\
Q_0 = \mu_1\left(|\mathbf{x}|\right)\left(g'\left(\frac{1}{2}|\mathbf{v}|^2\right) - 1\right).
\end{cases}
\tag{9.314}
$$

Therefore, denoting

$$\Phi_1 := -\frac{1}{2}|\mathbf{v}|^2 \tag{9.315}$$

we rewrite (9.314) as

$$\begin{cases} \Delta_{\mathbf{x}}\Phi_1 = 4\pi G\,\mu_1(|\mathbf{x}|)\,g'(-\Phi_1)\,,\\ \Phi_1 := -\frac{1}{2}|\mathbf{v}|^2\,,\\ \mathrm{curl}_{\mathbf{x}}\mathbf{v} = 0,\\ \mathbf{R} = -\frac{1}{c}\nabla_{\mathbf{x}}\left(\frac{1}{2}|\mathbf{v}|^2\right),\\ \mathbf{Q} = 0,\\ Q_0 = \mu_1(|\mathbf{x}|)\left(g'(-\Phi_1) - 1\right), \end{cases} \tag{9.316}$$

where the scalar field $\Phi_1 = \Phi(|\mathbf{x}|)$ is radial, and outside of the Earth surface it coincides with the following Newtonian potential of the Earth: $\Phi_1(\mathbf{x}) = -\frac{GM_0}{|\mathbf{x}|}$, where M_0 is the total effective gravitational mass of the Earth, defined as

$$M_0 = \iiint_{|\mathbf{x}|\le r_0} \mu_1(|\mathbf{x}|)\,g'(-\Phi_1(|\mathbf{x}|))\,d\mathbf{x}$$
$$= \iiint_{|\mathbf{x}|\le r_0} \mu_1(|\mathbf{x}|)\,g'\left(\frac{1}{2}|\nabla_{\mathbf{x}}Z_0(|\mathbf{x}|)|^2\right)d\mathbf{x}, \tag{9.317}$$

(for the inertial mass of the Earth m_0 we have $m_0 = \iiint_{|\mathbf{x}|\le r_0}\mu_1(|\mathbf{x}|)\,d\mathbf{x}$, and thus, in the relativistic-like case, where $g'(s) = \left(1 - \frac{2s}{c^2}\right)^{-\frac{1}{2}} > 1$ we have $M_0 > m_0$). Thus, since there exists a scalar radial field $Z_0(|\mathbf{x}|)$ such that $\mathbf{v}(\mathbf{x}) = \nabla_{\mathbf{x}}Z_0(|\mathbf{x}|)$ by (9.316) we obtain

$$\left|\frac{dZ_0}{d(|\mathbf{x}|)}(|\mathbf{x}|)\right| = \sqrt{-2\Phi_1(\mathbf{x})}, \tag{9.318}$$

that implies either

$$\mathbf{v}(\mathbf{x}) = \frac{\sqrt{-2\Phi_1(|\mathbf{x}|)}}{|\mathbf{x}|}\mathbf{x}, \tag{9.319}$$

or

$$\mathbf{v}(\mathbf{x}) = -\frac{\sqrt{-2\Phi_1(|\mathbf{x}|)}}{|\mathbf{x}|}\mathbf{x}, \tag{9.320}$$

exactly as in the case of the usual Newtonian gravity. In particular, on the Earth surface we have

$$|\mathbf{v}| = \sqrt{\frac{2GM_0}{r_0}}, \tag{9.321}$$

where r_0 is the Earth radius and M_0 is the total effective gravitational mass of the Earth, i.e. the absolute value of the vectorial gravitational potential on the Earth surface approximately equals to the escape velocity and its direction is normal to the Earth, either downward or upward.

Finally, note that the same solution as in (9.316) and (9.319) or (9.320) is valid also for the laws of the Newtonian-type gravity given by (9.263). We can get this either by a direct calculations or by passing to the limit $\beta \to +\infty$ in (9.295).

9.9 Covariant Formulation of the Laws of Gravity in Curvilinear Coordinate Systems

9.9.1 The Case of the Newtonian-Type Gravity

In this section, we find a equivalent form of the Lagrangian density of the gravitational-electromagnetic field in the case of the Newtonian-type gravity having the general form (9.250):

$$
\begin{aligned}
L(\mathbf{A}, \Psi, \mathbf{v}, \Phi, \mathbf{p}, \mathbf{x}, t) :=\\
&\frac{1}{8\pi}\left|-\nabla_{\mathbf{x}}\Psi - \frac{1}{c}\frac{\partial \mathbf{A}}{\partial t} + \frac{1}{c}\mathbf{v}\times \operatorname{curl}_{\mathbf{x}}\mathbf{A}\right|^2 - \frac{1}{8\pi}|\operatorname{curl}_{\mathbf{x}}\mathbf{A}|^2 - \left(\rho\Psi - \frac{1}{c}\mathbf{A}\cdot\mathbf{j}\right)\\
&+\mu g\left(\frac{1}{2}|\mathbf{u}-\mathbf{v}|^2\right) + \frac{1}{2}\left(d_{\mathbf{x}}\mathbf{v} + \{d_{\mathbf{x}}\mathbf{v}\}^T\right) : \left(d_{\mathbf{x}}\mathbf{p} + \{d_{\mathbf{x}}\mathbf{p}\}^T\right) - 2\left(\operatorname{div}_{\mathbf{x}}\mathbf{v}\right)\left(\operatorname{div}_{\mathbf{x}}\mathbf{p}\right)\\
&+\frac{1}{4\pi G}\left(\operatorname{div}_{\mathbf{x}}\mathbf{v}\right)\left(\frac{\partial\Phi}{\partial t} + \mathbf{v}\cdot\nabla_{\mathbf{x}}\Phi\right) + \frac{1}{4\pi G}\Phi\left(\operatorname{div}_{\mathbf{x}}\mathbf{v}\right)^2 - \frac{\Phi}{16\pi G}\left|d_{\mathbf{x}}\mathbf{v} + \{d_{\mathbf{x}}\mathbf{v}\}^T\right|^2\\
&+\frac{1}{8\pi G}|\nabla_{\mathbf{x}}\Phi|^2. \qquad (9.322)
\end{aligned}
$$

Our purpose is to make the equivalent form of this Lagrangian density to be covariant and valid in every curvilinear coordinate system.

Assume first, that we deal with a Cartesian inertial or non-inertial coordinate system. Then consider a three-dimensional vectorial gravitational potential $\mathbf{v} = (v^1, v^2, v^3)$ and consider the covariant pseudo-metric tensor $G = \{g_{ij}\}_{0\le i,j\le 3}$ defined by (9.59):

$$
\begin{cases}
g_{00} = 1 - \frac{|\mathbf{v}|^2}{c^2}\\
g_{ij} = -\delta_{ij} \quad \forall 1 \le i, j \le 3\\
g_{0j} = g_{j0} = \frac{v^j}{c} \quad \forall 1 \le j \le 3.
\end{cases}
\qquad (9.323)
$$

Next consider the contravariant pseudo-metric tensor $\tilde{G} = \{g^{ij}\}_{0\le i,j\le 3}$ defined by (9.58):

$$\begin{cases} g^{00} = 1 \\ g^{ij} = -\delta_{ij} + \frac{v^i v^j}{c^2} & \forall 1 \le i, j \le 3 \\ g^{0j} = g^{j0} = \frac{v^j}{c} & \forall 1 \le j \le 3. \end{cases} \tag{9.324}$$

Next consider the Christoffel symbols:

$$\begin{cases} \Gamma_{i,kn} := \frac{1}{2}\left(\frac{\partial g_{ik}}{\partial x_n} + \frac{\partial g_{in}}{\partial x_k} - \frac{\partial g_{kn}}{\partial x_i}\right) \\ \Gamma^i_{kn} := \sum_{j=0}^3 g^{ij}\Gamma_{j,kn} \end{cases} \quad \forall i, k, n = 0, 1, 2, 3, \tag{9.325}$$

where $\mathbf{x} = (x_1, x_2, x_3)$, $x_0 = ct$ and the point in the four-dimensional space-time is denoted as $(x^0, x^1, x^2, x^3) := (ct, \mathbf{x}) = (x_0, x_1, x_2, x_3)$. In particular by (9.323) and by the first equation in (9.325) we obtain

$$\begin{cases} \Gamma_{0,00} = -\frac{1}{2c^3}\frac{\partial(|\mathbf{v}|^2)}{\partial t} \\ \Gamma_{0,k0} = \Gamma_{0,0k} = -\frac{1}{2c^2}\frac{\partial(|\mathbf{v}|^2)}{\partial x_k} & \forall k = 1, 2, 3, \\ \Gamma_{0,kn} = \frac{1}{2c}\left(\frac{\partial v^k}{\partial x_n} + \frac{\partial v^n}{\partial x_k}\right) & \forall k, n = 1, 2, 3 \\ \Gamma_{i,00} = \frac{1}{c^2}\left(\frac{\partial v^i}{\partial t} + \frac{1}{2}\frac{\partial(|\mathbf{v}|^2)}{\partial x_i}\right) & \forall i = 1, 2, 3, \\ \Gamma_{i,k0} = \Gamma_{i,0k} = \frac{1}{2c}\left(\frac{\partial v^i}{\partial x_k} - \frac{\partial v^k}{\partial x_i}\right) & \forall i, k = 1, 2, 3, \\ \Gamma_{i,kn} = 0 & \forall i, k, n = 1, 2, 3. \end{cases} \tag{9.326}$$

Next given a four-covector field (S_0, S_1, S_2, S_3) on the group $\mathcal{S}_0$ and the corresponding lifted four vector (S^0, S^1, S^2, S^3) given by

$$(S^0, S^1, S^2, S^3) := \left\{\sum_{k=0}^{3} g^{mk} S_k\right\}_{m=0,1,2,3}, \tag{9.327}$$

by (9.58), (9.59) and (9.63) we have

$$S^0 = S_0 + \sum_{k=1}^{3}\frac{1}{c}v^k S_k \quad \text{and} \quad S^m = -S_m + \frac{1}{c}\left(S_0 + \sum_{k=1}^{3}\frac{1}{c}v^k S_k\right)v^m \quad \forall m = 1, 2, 3. \tag{9.328}$$

On the other hand, if we denote

$$\begin{cases} \phi := S^0 \\ h_j := -S_j & \forall 1 \le j \le 3 \quad \text{and} \quad \mathbf{h} := (h_1, h_2, h_3), \end{cases} \tag{9.329}$$

then, as we get above in Sect. 9.2, ϕ is a proper scalar field and $\mathbf{h}$ is a proper vector field and by (9.328) we can write

$$\begin{cases} S_0 = \phi + \sum_{k=1}^{3} \frac{1}{c} v^k h_k = \phi + \frac{1}{c} \mathbf{v} \cdot \mathbf{h} \\ S_j = -h_j \quad \forall 1 \le j \le 3, \end{cases} \tag{9.330}$$

and

$$\begin{cases} S^0 = \phi \\ S^j = \phi \frac{v^j}{c} + h_j \quad \forall 1 \le j \le 3. \end{cases} \tag{9.331}$$

Next consider $Z_{ij} = \delta_j S_i$, where δ_j means the covariant derivative with respect to the dynamical pseudo-metrics $G = \{g_{ij}\}_{0 \le i, j \le 3}$. Then

$$Z_{ij} := \delta_j S_i := \frac{\partial S_i}{\partial x_j} - \sum_{k=0}^{3} \Gamma_{ij}^k S_k \quad \forall 0 \le i, j \le 3. \tag{9.332}$$

It is well known from the tensor analysis that, given a covector field S_j on the group $\mathcal{S}_0$, Z_{ij}, defined by (9.332), is a two times covariant tensor field on the group $\mathcal{S}_0$. On the other hand, we can write by (9.332) and by the second equality in (9.325):

$$Z_{ij} := \delta_j S_i := \frac{\partial S_i}{\partial x_j} - \sum_{k=0}^{3} \sum_{m=0}^{3} g^{km} \Gamma_{m,ij} S_k = \frac{\partial S_i}{\partial x_j} - \sum_{m=0}^{3} \Gamma_{m,ij} S^m \quad \forall 0 \le i, j \le 3. \tag{9.333}$$

Thus by inserting (9.330) and (9.331) into (9.333) we deduce

$$\begin{aligned} Z_{ij} := \delta_j S_i &= \frac{\partial S_i}{\partial x_j} - \Gamma_{0,ij} S^0 - \sum_{m=1}^{3} \Gamma_{m,ij} S^m \\ &= \frac{\partial S_i}{\partial x_j} - \Gamma_{0,ij} \phi - \sum_{m=1}^{3} \Gamma_{m,ij} h_m - \sum_{m=1}^{3} \Gamma_{m,ij} \frac{\phi v^m}{c} \quad \forall 0 \le i, j \le 3. \end{aligned} \tag{9.334}$$

In particular by inserting (9.330) and (9.326) into (9.334) we deduce

$$\begin{aligned} Z_{00} := \delta_0 S_0 &= \frac{\partial S_0}{\partial x_0} - \Gamma_{0,00} \phi - \sum_{m=1}^{3} \Gamma_{m,00} h_m - \sum_{m=1}^{3} \Gamma_{m,00} \frac{\phi v^m}{c} = \\ &\frac{1}{c} \frac{\partial}{\partial t} \left(\phi + \frac{1}{c} \mathbf{v} \cdot \mathbf{h} \right) + \frac{\phi}{2c^3} \frac{\partial (|\mathbf{v}|^2)}{\partial t} - \sum_{m=1}^{3} \frac{1}{c^2} \left(\frac{\partial v^m}{\partial t} + \frac{1}{2} \frac{\partial (|\mathbf{v}|^2)}{\partial x_m} \right) h_m \\ &- \sum_{m=1}^{3} \frac{1}{c^2} \left(\frac{\partial v^m}{\partial t} + \frac{1}{2} \frac{\partial (|\mathbf{v}|^2)}{\partial x_m} \right) \frac{\phi v^m}{c} \\ &= \frac{1}{c} \frac{\partial \phi}{\partial t} + \frac{1}{c^2} \mathbf{v} \cdot \frac{\partial \mathbf{h}}{\partial t} - \sum_{m=1}^{3} \frac{1}{2c^2} \left(\frac{\partial (|\mathbf{v}|^2)}{\partial x_m} \right) \left(h_m + \frac{\phi v^m}{c} \right) \end{aligned} \tag{9.335}$$

$$Z_{0j} := \delta_j S_0 = \frac{\partial S_0}{\partial x_j} - \Gamma_{0,0j}\,\phi - \sum_{m=1}^{3} \Gamma_{m,0j}\,h_m - \sum_{m=1}^{3} \Gamma_{m,0j}\,\frac{\phi\, v^m}{c} =$$
$$\frac{\partial}{\partial x_j}\left(\phi + \frac{1}{c}\mathbf{v}\cdot\mathbf{h}\right) + \frac{\phi}{2c^2}\frac{\partial(|\mathbf{v}|^2)}{\partial x_j} - \sum_{m=1}^{3} \frac{1}{2c}\left(\frac{\partial v^m}{\partial x^j} - \frac{\partial v^j}{\partial x_m}\right) h_m$$
$$- \sum_{m=1}^{3} \frac{1}{2c}\left(\frac{\partial v^m}{\partial x_j} - \frac{\partial v^j}{\partial x_m}\right)\frac{\phi\, v^m}{c} = \frac{\partial \phi}{\partial x_j} + \sum_{m=1}^{3} \frac{v^m}{c}\frac{\partial h_m}{\partial x_j}$$
$$+ \sum_{m=1}^{3} \frac{1}{2c}\left(\frac{\partial v^j}{\partial x_m} + \frac{\partial v^m}{\partial x_j}\right) h_m + \sum_{m=1}^{3} \frac{1}{2c}\left(\frac{\partial v^j}{\partial x_m} + \frac{\partial v^m}{\partial x_j}\right)\frac{\phi\, v^m}{c} \qquad \forall\, 1 \le j \le 3, \tag{9.336}$$

$$Z_{i0} := \delta_0 S_i = \frac{\partial S_i}{\partial x_0} - \Gamma_{0,i0}\,\phi - \sum_{m=1}^{3} \Gamma_{m,i0}\,h_m - \sum_{m=1}^{3} \Gamma_{m,i0}\,\frac{\phi\, v^m}{c} = -\frac{1}{c}\frac{\partial h_i}{\partial t}$$
$$+ \frac{\phi}{2c^2}\frac{\partial(|\mathbf{v}|^2)}{\partial x_i} - \sum_{m=1}^{3} \frac{1}{2c}\left(\frac{\partial v^m}{\partial x_i} - \frac{\partial v^i}{\partial x_m}\right) h_m - \sum_{m=1}^{3} \frac{1}{2c}\left(\frac{\partial v^m}{\partial x_i} - \frac{\partial v^i}{\partial x_m}\right)\frac{\phi\, v^m}{c} =$$
$$-\frac{1}{c}\frac{\partial h_i}{\partial t} + \sum_{m=1}^{3} \frac{1}{2c}\left(\frac{\partial v^i}{\partial x_m} - \frac{\partial v^m}{\partial x_i}\right) h_m + \sum_{m=1}^{3} \frac{1}{2c}\left(\frac{\partial v^m}{\partial x_i} + \frac{\partial v^i}{\partial x_m}\right)\frac{\phi\, v^m}{c}$$
$$\forall\, 1 \le i \le 3, \tag{9.337}$$

and

$$Z_{ij} := \delta_j S_i = \frac{\partial S_i}{\partial x_j} - \Gamma_{0,ij}\,\phi - \sum_{m=1}^{3} \Gamma_{m,ij}\,h_m - \sum_{m=1}^{3} \Gamma_{m,ij}\,\frac{\phi\, v^m}{c}$$
$$= -\left(\frac{\partial h_i}{\partial x_j} + \frac{\phi}{2c}\left(\frac{\partial v^i}{\partial x_j} + \frac{\partial v^j}{\partial x_i}\right)\right) \qquad \forall\, 1 \le i, j \le 3. \tag{9.338}$$

Therefore, if we denote the symmetric two times covariant tensor $\hat{Z}_{ij}$ defined by

$$\hat{Z}_{ij} := Z_{ij} + Z_{ji} = \delta_j S_i + \delta_i S_j = \frac{\partial S_i}{\partial x_j} + \frac{\partial S_j}{\partial x_i} - \sum_{k=0}^{3} 2\Gamma_{ij}^{k} S_k \qquad \forall\, 0 \le i, j \le 3, \tag{9.339}$$

then by (9.335), (9.336), (9.337) and (9.338) we have

$$\hat{Z}_{00} = \frac{2}{c}\frac{\partial \phi}{\partial t} + \sum_{m=1}^{3} \frac{2v^m}{c^2}\frac{\partial h_m}{\partial t} - \sum_{m=1}^{3} \frac{1}{c^2}\left(\frac{\partial(|\mathbf{v}|^2)}{\partial x_m}\right)\left(h_m + \frac{\phi\, v^m}{c}\right)$$
$$= \frac{2}{c}\left(\frac{\partial \phi}{\partial t} + \sum_{m=1}^{3} v^m \frac{\partial \phi}{\partial x_m}\right)$$
$$- \sum_{m=1}^{3} \frac{2v^m}{c}\left(\frac{\partial \phi}{\partial x_m} - \frac{1}{c}\frac{\partial h_m}{\partial t} - \sum_{k=1}^{3}\frac{v^k}{c}\frac{\partial h_m}{\partial x_k} + \sum_{k=1}^{3}\frac{h_k}{c}\frac{\partial v^m}{\partial x_k}\right)$$
$$- \sum_{m=1}^{3} \frac{v^m}{c}\left(\sum_{k=1}^{3}\frac{v_k}{c}\left(\left(\frac{\partial h_k}{\partial x_m} + \frac{\partial h_m}{\partial x_k}\right) + \left(\frac{\partial v^k}{\partial x_m} + \frac{\partial v^m}{\partial x_k}\right)\frac{\phi}{c}\right)\right) \tag{9.340}$$

$$\hat{Z}_{0j} = \hat{Z}_{j0} =$$
$$\frac{\partial \phi}{\partial x_j} + \sum_{m=1}^{3}\frac{v^m}{c}\frac{\partial h_m}{\partial x_j} + \sum_{m=1}^{3}\frac{1}{2c}\left(\frac{\partial v^j}{\partial x_m} + \frac{\partial v^m}{\partial x_j}\right)h_m + \sum_{m=1}^{3}\frac{1}{2c}\left(\frac{\partial v^j}{\partial x_m} + \frac{\partial v^m}{\partial x_j}\right)\frac{\phi\, v^m}{c}$$
$$- \frac{1}{c}\frac{\partial h_j}{\partial t} + \sum_{m=1}^{3}\frac{1}{2c}\left(\frac{\partial v^j}{\partial x_m} - \frac{\partial v^m}{\partial x_j}\right)h_m + \sum_{m=1}^{3}\frac{1}{2c}\left(\frac{\partial v^m}{\partial x_j} + \frac{\partial v^j}{\partial x_m}\right)\frac{\phi\, v^m}{c} =$$
$$\frac{\partial \phi}{\partial x_j} - \frac{1}{c}\frac{\partial h_j}{\partial t} - \sum_{m=1}^{3}\frac{v^m}{c}\frac{\partial h_j}{\partial x_m} + \sum_{m=1}^{3}\frac{h_m}{c}\frac{\partial v^j}{\partial x_m}$$
$$+ \sum_{m=1}^{3}\frac{v^m}{c}\left(\left(\frac{\partial h_j}{\partial x_m} + \frac{\partial h_m}{\partial x_j}\right) + \left(\frac{\partial v^j}{\partial x_m} + \frac{\partial v^m}{\partial x_j}\right)\frac{\phi}{c}\right) \quad \forall 1 \le j \le 3, \tag{9.341}$$

and

$$\hat{Z}_{ij} = \hat{Z}_{ji} = -\left(\left(\frac{\partial h_i}{\partial x_j} + \frac{\partial h_j}{\partial x_i}\right) + \frac{\phi}{c}\left(\frac{\partial v^i}{\partial x_j} + \frac{\partial v^j}{\partial x_i}\right)\right) \quad \forall 1 \le i, j \le 3. \tag{9.342}$$

Therefore, by (9.342), (9.341) and (9.340) we obtain

$$\begin{cases} \hat{Z}_{ij} = \hat{Z}_{ji} = -\left(\left(\frac{\partial h_i}{\partial x_j} + \frac{\partial h_j}{\partial x_i}\right) + \frac{\phi}{c}\left(\frac{\partial v^i}{\partial x_j} + \frac{\partial v^j}{\partial x_i}\right)\right) \quad \forall 1 \le i, j \le 3, \\ \hat{Z}_{0j} = \hat{Z}_{j0} = \left(\frac{\partial \phi}{\partial x_j} - \frac{1}{c}\frac{\partial h_j}{\partial t} - \sum_{m=1}^{3}\frac{v^m}{c}\frac{\partial h_j}{\partial x_m} + \sum_{m=1}^{3}\frac{h_m}{c}\frac{\partial v^j}{\partial x_m}\right) \\ -\sum_{m=1}^{3}\frac{v^m}{c}\hat{Z}_{mj} \quad \forall 1 \le j \le 3 \\ \hat{Z}_{00} = \frac{2}{c}\left(\frac{\partial \phi}{\partial t} + \sum_{m=1}^{3} v^m \frac{\partial \phi}{\partial x_m}\right) - \sum_{m=1}^{3}\frac{2v^m}{c}\left(\hat{Z}_{0m} + \sum_{k=1}^{3}\frac{v^k}{c}\hat{Z}_{mk}\right) \\ +\sum_{m=1}^{3}\sum_{k=1}^{3}\frac{v^m}{c}\frac{v^k}{c}\hat{Z}_{mk}. \end{cases} \tag{9.343}$$

In particular, by (9.343) and (9.324) we deduce

$$\sum_{i=0}^{3}\sum_{j=0}^{3} g^{ij}\hat{Z}_{ij} = g^{00}\hat{Z}_{00} + \sum_{j=1}^{3} 2g^{0j}\hat{Z}_{0j} + \sum_{i=1}^{3}\sum_{j=1}^{3} g^{ij}\hat{Z}_{ij}$$

$$= \hat{Z}_{00} + \sum_{j=1}^{3}\frac{2v^j}{c}\hat{Z}_{0j} + \sum_{i=1}^{3}\sum_{j=1}^{3}\frac{v^i}{c}\frac{v^j}{c}\hat{Z}_{ij} - \sum_{i=1}^{3}\hat{Z}_{ii} =$$

$$\left(\frac{2}{c}\left(\frac{\partial\phi}{\partial t} + \sum_{m=1}^{3} v^m\frac{\partial\phi}{\partial x_m}\right) - \sum_{m=1}^{3}\frac{2v^m}{c}\left(\hat{Z}_{0m} + \sum_{k=1}^{3}\frac{v^k}{c}\hat{Z}_{mk}\right) + \sum_{m=1}^{3}\sum_{k=1}^{3}\frac{v^m}{c}\frac{v^k}{c}\hat{Z}_{mk}\right)$$

$$+ \sum_{j=1}^{3}\frac{2v^j}{c}\hat{Z}_{0j} + \sum_{i=1}^{3}\sum_{j=1}^{3}\frac{v^i}{c}\frac{v^j}{c}\hat{Z}_{ij} + \sum_{i=1}^{3} 2\left(\frac{\partial h_i}{\partial x_i} + \frac{\phi}{c}\frac{\partial v^i}{\partial x_i}\right)$$

$$= 2\left(\frac{1}{c}\left(\frac{\partial\phi}{\partial t} + \mathrm{div}_{\mathbf{x}}\{\phi\mathbf{v}\}\right) + \mathrm{div}_{\mathbf{x}}\mathbf{h}\right). \qquad (9.344)$$

In particular, if we consider the four-dimensional gravitational potential (v^0, v^1, v^2, v^3) defined by (9.29) as

$$(v^0, v^1, v^2, v^3) := \left(1, \frac{1}{c}\mathbf{v}\right), \qquad (9.345)$$

and the corresponding lowered four-covector field (v_0, v_1, v_2, v_3), that we called the four covector of gravitational potential:

$$(v_0, v_1, v_2, v_3) := (1, 0, 0, 0), \qquad (9.346)$$

then denoting

$$\hat{Y}_{ij} := \delta_j v_i + \delta_i v_j \qquad \forall 0 \le i, j \le 3, \qquad (9.347)$$

as a particular case of (9.343) and (9.344) with $\phi = 1$ and $\mathbf{h} = 0$, we deduce

$$\begin{cases} \hat{Y}_{ij} = \hat{Y}_{ji} = -\frac{1}{c}\left(\frac{\partial v^i}{\partial x_j} + \frac{\partial v^j}{\partial x_i}\right) & \forall 1 \le i, j \le 3, \\ \hat{Y}_{0j} = \hat{Y}_{j0} = -\sum_{m=1}^{3}\frac{v^m}{c}\hat{Y}_{mj} & \forall 1 \le j \le 3 \\ \hat{Y}_{00} = -\sum_{m=1}^{3}\frac{2v^m}{c}\left(\hat{Y}_{0m} + \sum_{k=1}^{3}\frac{v^k}{c}\hat{Y}_{mk}\right) + \sum_{m=1}^{3}\sum_{k=1}^{3}\frac{v^m}{c}\frac{v^k}{c}\hat{Y}_{mk}, & \end{cases} \qquad (9.348)$$

and

$$\sum_{i=0}^{3}\sum_{j=0}^{3} g^{ij}\hat{Y}_{ij} = \frac{2}{c}\left(\mathrm{div}_{\mathbf{x}}\mathbf{v}\right). \qquad (9.349)$$

Moreover, denoting the two times contravariant lifted tensor:

$$\hat{Y}^{mn} := \sum_{i=0}^{3}\sum_{j=0}^{3} g^{mj} g^{in} \hat{Y}_{ij} \quad \forall 0 \le m, n \le 3, \tag{9.350}$$

since then

$$\hat{Y}^{mn} = g^{m0} g^{0n} \hat{Y}_{ij} + \sum_{i=1}^{3} g^{m0} g^{in} \hat{Y}_{i0} + \sum_{j=1}^{3} g^{mj} g^{0n} \hat{Y}_{0j} + \sum_{i=1}^{3}\sum_{j=1}^{3} g^{mj} g^{in} \hat{Y}_{ij} \quad \forall 0 \le m, n \le 3, \tag{9.351}$$

by (9.348) and (9.324) we also deduce

$$\hat{Y}^{00} = g^{00} g^{00} \hat{Y}_{00} + \sum_{i=1}^{3} g^{00} g^{i0} \hat{Y}_{i0} + \sum_{j=1}^{3} g^{0j} g^{00} \hat{Y}_{0j} + \sum_{i=1}^{3}\sum_{j=1}^{3} g^{0j} g^{i0} \hat{Y}_{ij} = \hat{Y}_{00} + \sum_{j=1}^{3} \frac{2v_j}{c} \hat{Y}_{0j} + \sum_{i=1}^{3}\sum_{j=1}^{3} \frac{v^i}{c}\frac{v^j}{c} \hat{Y}_{ij} = 0, \tag{9.352}$$

$$\begin{aligned}\hat{Y}^{0n} = \hat{Y}^{n0} &= g^{00} g^{0n} \hat{Y}_{00} + \sum_{i=1}^{3} g^{00} g^{in} \hat{Y}_{i0} + \sum_{j=1}^{3} g^{0j} g^{0n} \hat{Z}_{0j} + \sum_{i=1}^{3}\sum_{j=1}^{3} g^{0j} g^{in} \hat{Y}_{ij} \\ &= \frac{v^n}{c} \hat{Y}_{00} + \sum_{i=1}^{3} \left(-\delta_{in} + \frac{v^i v^n}{c^2}\right) \hat{Y}_{i0} + \sum_{j=1}^{3} \frac{v^j}{c}\frac{v^n}{c} \hat{Y}_{0j} + \sum_{i=1}^{3}\sum_{j=1}^{3} \frac{v^j}{c} \left(-\delta_{in} + \frac{v^i v^n}{c^2}\right) \hat{Y}_{ij} \\ &= -\left(\hat{Y}_{n0} + \sum_{j=1}^{3} \frac{v^j}{c} \hat{Y}_{nj}\right) = 0 \quad \forall 1 \le n \le 3,\end{aligned} \tag{9.353}$$

and

$$\begin{aligned}\hat{Y}^{mn} &= g^{m0} g^{0n} \hat{Y}_{00} + \sum_{i=1}^{3} g^{m0} g^{in} \hat{Y}_{i0} + \sum_{j=1}^{3} g^{mj} g^{0n} \hat{Y}_{0j} + \sum_{i=1}^{3}\sum_{j=1}^{3} g^{mj} g^{in} \hat{Y}_{ij} \\ &= \frac{v^m v^n}{c^2} \hat{Y}_{00} + \sum_{i=1}^{3} \frac{v^m}{c} \left(-\delta_{in} + \frac{v^i v^n}{c^2}\right) \hat{Y}_{i0} + \sum_{j=1}^{3} \left(-\delta_{mj} + \frac{v^m v^j}{c^2}\right) \frac{v^n}{c} \hat{Y}_{0j} \\ &+ \sum_{i=1}^{3}\sum_{j=1}^{3} \left(-\delta_{mj} + \frac{v^m v^j}{c^2}\right)\left(-\delta_{in} + \frac{v^i v^n}{c^2}\right) \hat{Y}_{ij} = -\frac{v^m}{c}\left(\hat{Y}_{n0} + \sum_{j=1}^{3} \frac{v^j}{c} \hat{Y}_{nj}\right) \\ &- \frac{v^n}{c}\left(\hat{Y}_{0m} + \sum_{i=1}^{3} \frac{v^i}{c} \hat{Y}_{im}\right) + \hat{Y}_{mn} = -\frac{1}{c}\left(\frac{\partial v^m}{\partial x_n} + \frac{\partial v^n}{\partial x_m}\right) \quad \forall m, n = 1, 2, 3.\end{aligned} \tag{9.354}$$

As a consequence of (9.352), (9.353) and (9.354) we have

$$\begin{cases} \hat{Y}^{00} = 0, \\ \hat{Y}^{0n} = \hat{Y}^{n0} = 0 \quad \forall 1 \leq n \leq 3, \\ \hat{Y}^{mn} = -\frac{1}{c}\left(\frac{\partial v^m}{\partial x_n} + \frac{\partial v^n}{\partial x_m}\right) \quad \forall 1 \leq m, n \leq 3. \end{cases} \tag{9.355}$$

Therefore, in particular, by (9.343) and (9.355) we obtain

$$\begin{aligned} \sum_{i=0}^{3}\sum_{j=0}^{3} \hat{Z}_{ij}\hat{Y}^{ij} &= \sum_{i=1}^{3}\sum_{j=1}^{3} \frac{1}{c}\left(\left(\frac{\partial h_i}{\partial x_j} + \frac{\partial h_j}{\partial x_i}\right) + \frac{\phi}{c}\left(\frac{\partial v^i}{\partial x_j} + \frac{\partial v^j}{\partial x_i}\right)\right)\left(\frac{\partial v^i}{\partial x_j} + \frac{\partial v^j}{\partial x_i}\right) \\ &= \frac{1}{c}\left(d_{\mathbf{x}}\mathbf{v} + \{d_{\mathbf{x}}\mathbf{v}\}^T\right) : \left(d_{\mathbf{x}}\mathbf{h} + \{d_{\mathbf{x}}\mathbf{h}\}^T\right) + \frac{\phi}{c^2}\left|d_{\mathbf{x}}\mathbf{v} + \{d_{\mathbf{x}}\mathbf{v}\}^T\right|^2. \end{aligned} \tag{9.356}$$

On the other hand, by (9.344) and (9.349) we have

$$\begin{aligned} \left(\sum_{i=0}^{3}\sum_{j=0}^{3} g^{ij}\hat{Y}_{ij}\right)\left(\sum_{i=0}^{3}\sum_{j=0}^{3} g^{ij}\hat{Z}_{ij}\right) &= \frac{4}{c}\left(\operatorname{div}_{\mathbf{x}}\mathbf{v}\right)\left(\frac{1}{c}\left(\frac{\partial \phi}{\partial t} + \operatorname{div}_{\mathbf{x}}\{\phi\mathbf{v}\}\right) + \operatorname{div}_{\mathbf{x}}\mathbf{h}\right) \\ &= \frac{4}{c}\left(\operatorname{div}_{\mathbf{x}}\mathbf{h}\right)\left(\operatorname{div}_{\mathbf{x}}\mathbf{v}\right) + \frac{4\phi}{c^2}\left(\operatorname{div}_{\mathbf{x}}\mathbf{v}\right)^2 + \frac{4}{c^2}\left(\operatorname{div}_{\mathbf{x}}\mathbf{v}\right)\left(\frac{\partial \phi}{\partial t} + \mathbf{v}\cdot\nabla_{\mathbf{x}}\phi\right). \end{aligned} \tag{9.357}$$

Thus by (9.356) and (9.357) we deduce

$$\begin{aligned} &\left(\sum_{i=0}^{3}\sum_{j=0}^{3} g^{ij}\hat{Y}_{ij}\right)\left(\sum_{i=0}^{3}\sum_{j=0}^{3} g^{ij}\hat{Z}_{ij}\right) - \sum_{i=0}^{3}\sum_{j=0}^{3} \hat{Z}_{ij}\hat{Y}^{ij} \\ &= \frac{4}{c}\left(\left(\operatorname{div}_{\mathbf{x}}\mathbf{h}\right)\left(\operatorname{div}_{\mathbf{x}}\mathbf{v}\right) - \frac{1}{4}\left(d_{\mathbf{x}}\mathbf{v} + \{d_{\mathbf{x}}\mathbf{v}\}^T\right) : \left(d_{\mathbf{x}}\mathbf{h} + \{d_{\mathbf{x}}\mathbf{h}\}^T\right)\right) \\ &+ \frac{4}{c^2}\left(\left(\operatorname{div}_{\mathbf{x}}\mathbf{v}\right)\left(\frac{\partial \phi}{\partial t} + \mathbf{v}\cdot\nabla_{\mathbf{x}}\phi\right) + \phi\left(\left(\operatorname{div}_{\mathbf{x}}\mathbf{v}\right)^2 - \frac{1}{4}\left|d_{\mathbf{x}}\mathbf{v} + \{d_{\mathbf{x}}\mathbf{v}\}^T\right|^2\right)\right). \end{aligned} \tag{9.358}$$

Note again that by (9.346) and (9.50) the four-covector of gravitational potential is a gradient of the global time multiplied by the constant c:

$$(v_0, v_1, v_2, v_3) := (1, 0, 0, 0) = c\left(\frac{\partial t}{\partial x^0}, \frac{\partial t}{\partial x^1}, \frac{\partial t}{\partial x^2}, \frac{\partial t}{\partial x^1}\right). \tag{9.359}$$

Next by (9.329) and (9.346) we deduce that the covariant scalar $\sum_{k=0}^{3} v_k S^k$ satisfies in Cartesian coordinate system:

$$\sum_{k=0}^{3} v_k S^k = \phi. \tag{9.360}$$

Therefore by (9.13) we deduce that the four-component field (d_0, d_1, d_2, d_3) defined by

$$d_j := \frac{\partial}{\partial x^j}\left(\sum_{k=0}^{3} v_k S^k\right) \quad \forall j = 0, 1, 2, 3 \tag{9.361}$$

is a four covector. Then the following quantity

$$\sum_{m=0}^{3}\sum_{n=0}^{3} \Theta^{mn} d_m d_n = \sum_{m=0}^{3}\sum_{n=0}^{3} \Theta^{mn} \frac{\partial}{\partial x^m}\left(\sum_{k=0}^{3} v_k S^k\right) \frac{\partial}{\partial x^n}\left(\sum_{k=0}^{3} v_k S^k\right) \tag{9.362}$$

is a covariant scalar, where $\{\Theta^{ij}\}_{0\leq i,j\leq 3}$ is the contravariant tensor of the three-dimensional geometry that satisfies (9.48) in every non-inertial Cartesian coordinate system:

$$\begin{cases} \Theta^{00} = 0 \\ \Theta^{0j} = \Theta^{j0} = 0 \quad \forall j = 1, 2, 3 \\ \Theta^{ij} := \delta_{ij} \quad \forall i, j = 1, 2, 3. \end{cases} \tag{9.363}$$

Moreover, by (9.57) we have

$$g^{ij} := v^i v^j - \Theta^{ij} \quad \forall i, j = 0, 1, 2, 3. \tag{9.364}$$

On the other hand, inserting (9.360) and (9.363) into (9.362) we deduce that in the Cartesian coordinate system we have

$$\begin{aligned} &\sum_{m=0}^{3}\sum_{n=0}^{3} \Theta^{mn} \frac{\partial}{\partial x^m}\left(\sum_{j=0}^{3}\sum_{k=0}^{3} g^{kj} v_k S_j\right) \frac{\partial}{\partial x^n}\left(\sum_{j=0}^{3}\sum_{k=0}^{3} g^{kj} v_k S_j\right) \\ &= \sum_{m=0}^{3}\sum_{n=0}^{3} \Theta^{mn} \frac{\partial}{\partial x^m}\left(\sum_{k=0}^{3} v_k S^k\right) \frac{\partial}{\partial x^n}\left(\sum_{k=0}^{3} v_k S^k\right) = |\nabla_{\mathbf{x}}\phi|^2. \end{aligned} \tag{9.365}$$

Therefore, by (9.358) and (9.365) we deduce that

$$\sum_{m=0}^{3}\sum_{n=0}^{3}\frac{32\pi G}{c^4}\Theta^{mn}\frac{\partial}{\partial x^m}\left(\sum_{j=0}^{3}\sum_{k=0}^{3}g^{kj}v_kS_j\right)\frac{\partial}{\partial x^n}\left(\sum_{j=0}^{3}\sum_{k=0}^{3}g^{kj}v_kS_j\right)$$
$$+\left(\sum_{i=0}^{3}\sum_{j=0}^{3}g^{ij}\left(\delta_j v_i+\delta_i v_j\right)\right)\left(\sum_{i=0}^{3}\sum_{j=0}^{3}g^{ij}\left(\delta_j S_i+\delta_i S_j\right)\right)$$
$$-\sum_{k=0}^{3}\sum_{j=0}^{3}\sum_{m=0}^{3}\sum_{n=0}^{3}\left(\delta_j S_k+\delta_k S_j\right)g^{km}g^{jn}\left(\delta_m v_n+\delta_n v_m\right)$$
$$=\frac{4}{c}\left((\mathrm{div}_{\mathbf{x}}\mathbf{h})\,(\mathrm{div}_{\mathbf{x}}\mathbf{v})-\frac{1}{4}\left(d_{\mathbf{x}}\mathbf{v}+\{d_{\mathbf{x}}\mathbf{v}\}^T\right):\left(d_{\mathbf{x}}\mathbf{h}+\{d_{\mathbf{x}}\mathbf{h}\}^T\right)\right)$$
$$+\frac{4}{c^2}\Bigg((\mathrm{div}_{\mathbf{x}}\mathbf{v})\left(\frac{\partial\phi}{\partial t}+\mathbf{v}\cdot\nabla_{\mathbf{x}}\phi\right)+\phi\left((\mathrm{div}_{\mathbf{x}}\mathbf{v})^2-\frac{1}{4}\left|d_{\mathbf{x}}\mathbf{v}+\{d_{\mathbf{x}}\mathbf{v}\}^T\right|^2\right)$$
$$+\frac{8\pi G}{c^2}\left|\nabla_{\mathbf{x}}\phi\right|^2\Bigg), \tag{9.366}$$

where the left-hand side of (9.366) is written in a covariant form. Therefore, for the choice

$$\phi=\frac{c^2}{16\pi G}\Phi,\quad\text{and}\quad\mathbf{h}=-\frac{c}{2}\mathbf{p}:=-\frac{c}{2}(p_1,p_2,p_3)\,, \tag{9.367}$$

we can rewrite (9.366) in Cartesian coordinate system as

$$\frac{1}{2}\left(d_{\mathbf{x}}\mathbf{v}+\{d_{\mathbf{x}}\mathbf{v}\}^T\right):\left(d_{\mathbf{x}}\mathbf{p}+\{d_{\mathbf{x}}\mathbf{p}\}^T\right)-2\,(\mathrm{div}_{\mathbf{x}}\mathbf{v})\,(\mathrm{div}_{\mathbf{x}}\mathbf{p})$$
$$+\frac{1}{4\pi G}(\mathrm{div}_{\mathbf{x}}\mathbf{v})\left(\frac{\partial\Phi}{\partial t}+\mathbf{v}\cdot\nabla_{\mathbf{x}}\Phi\right)+\frac{1}{4\pi G}\Phi\,(\mathrm{div}_{\mathbf{x}}\mathbf{v})^2-\frac{\Phi}{16\pi G}\left|d_{\mathbf{x}}\mathbf{v}+\{d_{\mathbf{x}}\mathbf{v}\}^T\right|^2$$
$$+\frac{1}{8\pi G}\left|\nabla_{\mathbf{x}}\Phi\right|^2=\sum_{m=0}^{3}\sum_{n=0}^{3}\frac{32\pi G}{c^4}\Theta^{mn}\frac{\partial}{\partial x^m}\left(\sum_{j=0}^{3}\sum_{k=0}^{3}g^{kj}v_kS_j\right)\frac{\partial}{\partial x^n}\left(\sum_{j=0}^{3}\sum_{k=0}^{3}g^{kj}v_kS_j\right)$$
$$+\left(\sum_{i=0}^{3}\sum_{j=0}^{3}g^{ij}\left(\delta_j v_i+\delta_i v_j\right)\right)\left(\sum_{i=0}^{3}\sum_{j=0}^{3}g^{ij}\left(\delta_j S_i+\delta_i S_j\right)\right)$$
$$-\sum_{k=0}^{3}\sum_{j=0}^{3}\sum_{m=0}^{3}\sum_{n=0}^{3}\left(\delta_j S_k+\delta_k S_j\right)g^{km}g^{jn}\left(\delta_m v_n+\delta_n v_m\right), \tag{9.368}$$

where the right-hand side is written in the covariant form which is valid in every curvilinear coordinate system and, due to (9.330), (9.331) and (9.367) in a Cartesian coordinate system we have

$$\begin{cases}S_0=\frac{c^2}{16\pi G}\Phi-\frac{1}{2}\mathbf{v}\cdot\mathbf{p}\\ S_j=\frac{c}{2}p_j\qquad\forall 1\le j\le 3\end{cases} \tag{9.369}$$

and

$$\begin{cases} S^0 = \frac{c^2}{16\pi G}\Phi \\ S^j = \frac{c^2}{16\pi G}\Phi\frac{v^j}{c} - \frac{c}{2}p_j \quad \forall 1 \le j \le 3. \end{cases} \tag{9.370}$$

Next, given a system of n particles with inertial masses $m_1, \ldots, m_n$, charges $\sigma_1, \ldots, \sigma_n$, places $\mathbf{r}_1(t), \ldots, \mathbf{r}_n(t)$ and velocities $\frac{d\mathbf{r}_1}{dt}(t), \ldots, \frac{d\mathbf{r}_n}{dt}(t)$ in the Cartesian coordinate system, the usual definitions of the charge density, current density and the mass density of this system are the following:

$$\begin{cases} \rho(\mathbf{x}, t) := \sum_{j=1}^{n} \sigma_j\, \delta^{(3)}\left(\mathbf{x} - \mathbf{r}_j(t)\right), \\ \mathbf{j}(\mathbf{x}, t) := \sum_{j=1}^{n} \sigma_j \frac{d\mathbf{r}_j}{dt}(t)\, \delta^{(3)}\left(\mathbf{x} - \mathbf{r}_j(t)\right), \\ \mu(\mathbf{x}, t) := \sum_{j=1}^{n} m_j\, \delta^{(3)}\left(\mathbf{x} - \mathbf{r}_j(t)\right), \end{cases} \tag{9.371}$$

where $\delta^{(3)}$ is the usual Dirac-delta distribution (generalized function) in $\mathbb{R}^3$. Then denoting by $\delta^{(4)}$ the Dirac-delta distribution in $\mathbb{R}^4$ and by $\delta^{(1)}$ the Dirac-delta distribution in $\mathbb{R}$, since we have

$$\delta^{(4)}\left(\left(x^0, x^1, x^2, x^3\right) - \left(a^0, a^1, a^2, a^3\right)\right) = \delta^{(3)}\left(\mathbf{x} - \mathbf{a}\right)\delta^{(1)}\left(x^0 - a^0\right)$$
$$\forall \left(x^0, x^1, x^2, x^3\right) = \left(x^0, \mathbf{x}\right) \in \mathbb{R}^4, \quad \forall \left(a^0, a^1, a^2, a^3\right) = \left(a^0, \mathbf{a}\right) \in \mathbb{R}^4,$$
$$\text{and} \quad \delta^{(3)}\left(\mathbf{x} - \mathbf{a}\right) = \frac{1}{c^3}\delta^{(3)}\left(\frac{1}{c}\left(\mathbf{x} - \mathbf{a}\right)\right), \tag{9.372}$$

we rewrite (9.371) as

$$\begin{cases} \rho(\mathbf{x}, t) = \frac{1}{c^3}\int_{\mathbb{R}}\left(\sum_{j=1}^{n} \sigma_j\, \delta^{(4)}\left(\left(x^0, x^1, x^2, x^3\right) - \left(\chi_j^0(t), \chi_j^1(t), \chi_j^2(t), \chi_j^3(t)\right)\right)\right) dx^0, \\ \mathbf{j}(\mathbf{x}, t) = \\ \frac{1}{c^3}\int_{\mathbb{R}}\left(\sum_{j=1}^{n} \sigma_j \frac{d\mathbf{r}_j}{dx^0}(x^0)\, \delta^{(4)}\left(\left(x^0, x^1, x^2, x^3\right) - \left(\chi_j^0(t), \chi_j^1(t), \chi_j^2(t), \chi_j^3(t)\right)\right)\right) dx^0 \\ \mu(\mathbf{x}, t) = \frac{1}{c^3}\int_{\mathbb{R}}\left(\sum_{j=1}^{n} m_j\, \delta^{(4)}\left(\left(x^0, x^1, x^2, x^3\right) - \left(\chi_j^0(t), \chi_j^1(t), \chi_j^2(t), \chi_j^3(t)\right)\right)\right) dx^0, \end{cases} \tag{9.373}$$

where we denote

$$\left(x^0, x^1, x^2, x^3\right) := \left(t, \frac{1}{c}\mathbf{x}\right) \tag{9.374}$$

and $\left(\chi_j^0(t), \chi_j^1(t), \chi_j^2(t), \chi_j^3(t)\right) \in \mathbb{R}^4$ is a four-dimensional space-time trajectory of the jth particle, parameterized by the global time, which is defined by the following:

$$\left(\chi_j^0(t), \chi_j^1(t), \chi_j^2(t), \chi_j^3(t)\right) := \left(t, \frac{1}{c}\mathbf{r}_j(t)\right). \tag{9.375}$$

Thus, if we denote by G the 4×4-matrix $G := \{g_{ij}\}_{0\le i,j\le 3}$. Then, since the matrix G satisfies $\det G = -1$ in every Cartesian coordinate system, we can rewrite the first two equations in (9.373) as

$$\left(j^0, j^1, j^2, j^3\right) := \left(\rho, \frac{1}{c}\mathbf{j}\right)(\mathbf{x}, t) =$$
$$\sum_{j=1}^{n}\int_{\mathbb{R}} \frac{\sigma_j\left(\frac{d\chi_j^0}{dx^0}, \frac{d\chi_j^1}{dx^0}, \frac{d\chi_j^2}{dx^0}, \frac{d\chi_j^0}{dx^0}\right)(x^0)}{c^3\sqrt{\left|\det G\left(x^0, x^1, x^2, x^3\right)\right|}}\,\delta^{(4)}\Bigg(\left(x^0, x^1, x^2, x^3\right)$$
$$-\left(\chi_j^0(t), \chi_j^1(t), \chi_j^2(t), \chi_j^3(t)\right)\Bigg)dx^0. \tag{9.376}$$

Note here that we denoted the matrix $G = \{g_{ij}\}_{0\le i,j\le 3}$ by the same letter as the gravitational constant G. However, there is no ambiguity, since in the second case G is a constant scalar and in the first case G is a matrix. Moreover, we will use the matrix notation $G = \{g_{ij}\}_{0\le i,j\le 3}$ only in the expressions containing term $\det G$.

Similarly, we can write

$$\mu(\mathbf{x}, t)\sqrt{1 - \frac{1}{c^2}\left|\mathbf{u}(\mathbf{x}, t) - \mathbf{v}(\mathbf{x}, t)\right|^2} =$$
$$\sum_{j=1}^{n}\int_{\mathbb{R}} \frac{m_j\sqrt{\left(\sum_{m=0}^{3}\sum_{k=0}^{3} g_{mk}\left(\left(\chi_j^0, \ldots \chi_j^3\right)\right)\frac{d\chi_j^m}{dx^0}\frac{d\chi_j^k}{dx^0}\right)(x^0)}}{c^3\sqrt{\left|\det G\left(x^0, \ldots x^3\right)\right|}}\,\delta^{(4)}\Bigg(\left(x^0, \ldots x^3\right)$$
$$-\left(\chi_j^0, \ldots \chi_j^3\right)(t)\Bigg)dx^0, \tag{9.377}$$

where $\mathbf{u}(\mathbf{x}, t)$ is the field of velocities of the given system of particles, considered as a continuum. Next we observe that the four-dimensional quantity in the right-hand side of (9.376):

$$\sum_{j=1}^{n}\int_{\mathbb{R}} \frac{\sigma_j\left(\frac{d\chi_j^0}{dx^0}, \frac{d\chi_j^1}{dx^0}, \frac{d\chi_j^2}{dx^0}, \frac{d\chi_j^0}{dx^0}\right)(x^0)}{c^3\sqrt{\left|\det G\left(x^0, x^1, x^2, x^3\right)\right|}}\,\delta^{(4)}\Bigg(\left(x^0, x^1, x^2, x^3\right)$$
$$-\left(\chi_j^0, \chi_j^1, \chi_j^2, \chi_j^3\right)\Bigg)dx^0 \tag{9.378}$$

can be defined also in curvilinear systems of coordinates, where $\frac{\partial t}{\partial x^0} > 0$, and it can be easily proved in a similar way as it is proved in the general relativity that this quantity, defined in every curvilinear coordinate system with $\frac{\partial t}{\partial x^0} > 0$, equals to some four vector. Similarly, the scalar quantity in the right-hand side of (9.377):

$$\sum_{j=1}^{n}\int_{\mathbb{R}} \frac{m_j\sqrt{\left(\sum_{m=0}^{3}\sum_{k=0}^{3} g_{mk}\left(\left(\chi_j^0,\dots\chi_j^3\right)\right)\frac{d\chi_j^m}{dx^0}\frac{d\chi_j^k}{dx^0}\right)(x^0)}}{c^3\sqrt{\left|\det G\left(x^0,\dots x^3\right)\right|}}\delta^{(4)}\Big(\left(x^0,\dots x^3\right) - \left(\chi_j^0,\dots\chi_j^3\right)\Big)dx^0 \tag{9.379}$$

can also be defined in curvilinear systems of coordinates and this quantity forms a covariant scalar under the change of curvilinear coordinate system. On the other hand, since (9.372) also valid in every curvilinear coordinate systems we rewrite (9.378) and (9.379) as

$$\begin{aligned}
&\sum_{j=1}^{n}\int_{\mathbb{R}} \frac{\sigma_j\left(\frac{d\chi_j^0}{dx^0},\frac{d\chi_j^1}{dx^0},\frac{d\chi_j^2}{dx^0},\frac{d\chi_j^0}{dx^0}\right)(x^0)}{c^3\sqrt{\left|\det G\left(x^0,x^1,x^2,x^3\right)\right|}}\delta^{(4)}\left(\left(x^0,x^1,x^2,x^3\right)-\left(\chi_j^0,\chi_j^1,\chi_j^2,\chi_j^3\right)\right)dx^0\\
&=\sum_{j=1}^{n}\frac{\sigma_j\left(\frac{d\chi_j^0}{d\chi_j^0},\frac{d\chi_j^1}{d\chi_j^0},\frac{d\chi_j^2}{d\chi_j^0},\frac{d\chi_j^0}{d\chi_j^0}\right)(\chi_j^0)}{c^3\sqrt{\left|\det G\left(\chi_j^0,x^1,x^2,x^3\right)\right|}}\delta^{(3)}\left(\left(x^1,x^2,x^3\right)-\left(\chi_j^1,\chi_j^2,\chi_j^3\right)(\chi_j^0)\right)\\
&=\left(\sum_{j=1}^{n}\frac{\sigma_j}{\sqrt{\left|\det G\left(\chi_j^0,x^1,x^2,x^3\right)\right|}}\frac{1}{c^3}\delta^{(3)}\left(\left(x^1,x^2,x^3\right)-\left(\chi_j^1,\chi_j^2,\chi_j^3\right)(\chi_j^0)\right),\right.\\
&\left.\sum_{j=1}^{n}\frac{\sigma_j\left(\frac{d\chi_j^1}{dx^0},\frac{d\chi_j^2}{dx^0},\frac{d\chi_j^0}{dx^0}\right)(\chi_j^0)}{\sqrt{\left|\det G\left(\chi_j^0,x^1,x^2,x^3\right)\right|}}\frac{1}{c^3}\delta^{(3)}\left(\left(x^1,x^2,x^3\right)-\left(\chi_j^1,\chi_j^2,\chi_j^3\right)(\chi_j^0)\right)\right)
\end{aligned} \tag{9.380}$$

and

$$\sum_{j=1}^{n}\int_{\mathbb{R}} \frac{m_j\sqrt{\left(\sum_{m=0}^{3}\sum_{k=0}^{3} g_{mk}\left(\left(\chi_j^0,\dots\chi_j^3\right)\right)\frac{d\chi_j^m}{dx^0}\frac{d\chi_j^k}{dx^0}\right)(x^0)}}{c^3\sqrt{\left|\det G\left(x^0,\dots x^3\right)\right|}}\delta^{(4)}\Big(\left(x^0,\dots x^3\right) - \left(\chi_j^0,\dots\chi_j^3\right)\Big)dx^0 =$$

$$\sum_{j=1}^{n} \frac{m_j \sqrt{\left(\sum_{m=0}^{3}\sum_{k=0}^{3} g_{mk}\left(\left(\chi_j^0,\dots\chi_j^3\right)\right)\frac{d\chi_j^m}{d\chi_j^0}\frac{d\chi_j^k}{d\chi_j^0}\right)}}{c^3\sqrt{\left|\det G\left(\chi_j^0, x^1,\dots x^3\right)\right|}}\delta^{(3)}\left(\left(x^1,\dots x^3\right) - \left(\chi_j^1,\dots\chi_j^3\right)\left(\chi_j^0\right)\right). \quad (9.381)$$

So, by (9.376) and (9.380) in every curvilinear coordinate system, where $\frac{\partial t}{\partial x^0} > 0$, we have

$$\left(j^0, j^1, j^2, j^3\right) := \frac{1}{\sqrt{|\det G|}}\left(\hat{\rho}, \frac{1}{c}\hat{\rho}\,\hat{\mathbf{u}}_{x^0}\right) \quad \text{where}$$

$$\hat{\rho} := \sum_{j=1}^{n} \sigma_j\, \delta^{(3)}\left(\hat{\mathbf{x}} - c\left(\chi_j^1, \chi_j^2, \chi_j^3\right)\left(\chi_j^0\right)\right)$$

is the local charge density, calculated in the curvilinear coordinate system, (9.382)

$\hat{\mathbf{x}} := (cx^1, cx^2, cx^3)$ and $\hat{\mathbf{u}}_{x^0}$ is the field of velocities of the system, calculated in a given curvilinear coordinate system by the differentiation of the last three coordinates of the particle: $\hat{\mathbf{r}} := (c\chi^1, c\chi^2, c\chi^3)$ by the coordinate χ^0 that can be considered as the local time instead of global time t. The quantity in (9.382) equals to a four vector, under the change of curvilinear coordinate systems. Similarly by (9.381) the following quantities, defined in every coordinate system where $\frac{\partial t}{\partial x^0} > 0$, equal to some four vector and some covariant scalar, respectively, under the change of curvilinear coordinate systems:

$$\frac{1}{\sqrt{|\det G|}}\left(\hat{\mu}, \frac{1}{c}\hat{\mu}\,\hat{\mathbf{u}}_{x^0}\right) \quad \text{and} \quad \frac{\hat{\mu}}{\sqrt{|\det G|}}\sqrt{\left(\sum_{m=0}^{3}\sum_{k=0}^{3} g_{mk}\,\hat{u}_{x^0}^m\,\hat{u}_{x^0}^k\right)} \quad \text{where}$$

$$\hat{\mu} := \sum_{j=1}^{n} m_j\, \delta^{(3)}\left(\hat{\mathbf{x}} - c\left(\chi_j^1, \chi_j^2, \chi_j^3\right)\left(\chi_j^0\right)\right)$$

is the local mass density, calculated in the curvilinear coordinate system, (9.383)

and $\left(\hat{u}^0, \hat{u}^1, \hat{u}^2, \hat{u}^3\right)_{x^0} = \left(1, \frac{1}{c}\hat{\mathbf{u}}_{x^0}\right)$ is the field of four-dimensional velocities of the system, calculated in a given curvilinear coordinate system by the differentiation of the four-dimensional coordinates of the particles by the first coordinate χ^0. Again note that although the quantities $\hat{\rho}$ and $\hat{\mu}$ are not covariant scalars and $\left(\hat{u}^0, \hat{u}^1, \hat{u}^2, \hat{u}^3\right)_{x^0}$ is not a four vector, the first quantity in (9.382) equals to a four vector and the two first quantities in (9.383) equal to a four vector and a covariant scalar, under the change of curvilinear coordinate systems. Moreover, clearly the four-dimensional speed $\left(u^0, u^1, u^2, u^3\right)_{,t}$, obtained in curvilinear coordinate system by the differentiation by the global time t, instead of the first local coordinate χ_0, indeed forms a four vector, and therefore the quantity, defined in every coordinate system where $\frac{\partial t}{\partial x^0} > 0$,

$$\frac{\hat{\mu}}{\sqrt{|\det G|}}\left(\sum_{m=0}^{3}\sum_{k=0}^{3} g_{mk}\,\hat{u}^m_{x^0}\,u^k_t\right) \quad \text{where } \hat{\mu} = \sum_{j=1}^{n} m_j\,\delta^{(3)}\left(\hat{\mathbf{x}} - c\left(\chi^1_j, \chi^2_j, \chi^3_j\right)(\chi^0_j)\right) \tag{9.384}$$

equals to a covariant scalar, under the change of curvilinear coordinate systems.

Next we can write the density of the Lagrangian of the electromagnetic field, defined in (5.62) in the equivalent form (9.122), where the right-hand side is written in a covariant form which is valid for every curvilinear coordinate system:

$$\frac{1}{4\pi}\left(\frac{1}{2}|\mathbf{D}|^2 - \frac{1}{2}|\mathbf{B}|^2 - 4\pi\left(\rho\Psi - \frac{1}{c}\mathbf{A}\cdot\mathbf{j}\right)\right) =$$

$$\frac{1}{4\pi}\left(-\sum_{n=0}^{3}\sum_{k=0}^{3}\sum_{m=0}^{3}\sum_{p=0}^{3}\frac{1}{4}g^{mn}g^{pk}\left(\frac{\partial A_p}{\partial x^m} - \frac{\partial A_m}{\partial x^p}\right)\left(\frac{\partial A_k}{\partial x^n} - \frac{\partial A_n}{\partial x^k}\right) - \sum_{k=0}^{3} 4\pi j^k A_k\right), \tag{9.385}$$

where (j^0, j^1, j^2, j^3) is the four vector of the current that satisfies (9.382) in every curvilinear coordinate system, where $\frac{\partial t}{\partial x^0} > 0$, and (A_0, A_1, A_2, A_3) is the four covector of the electromagnetic potential. Then by (9.385), (9.368) and (9.384) we rewrite the Lagrangian density in (9.322) in the Cartesian coordinate system as

$$\begin{aligned}
&\frac{1}{8\pi}\left|-\nabla_{\mathbf{x}}\Psi - \frac{1}{c}\frac{\partial \mathbf{A}}{\partial t} + \frac{1}{c}\mathbf{v}\times\mathrm{curl}_{\mathbf{x}}\mathbf{A}\right|^2 - \frac{1}{8\pi}|\mathrm{curl}_{\mathbf{x}}\mathbf{A}|^2 - \left(\rho\Psi - \frac{1}{c}\mathbf{A}\cdot\mathbf{j}\right)\\
&\quad + \mu g\left(\frac{1}{2}|\mathbf{u}-\mathbf{v}|^2\right) + \frac{1}{2}\left(d_{\mathbf{x}}\mathbf{v} + \{d_{\mathbf{x}}\mathbf{v}\}^T\right) : \left(d_{\mathbf{x}}\mathbf{p} + \{d_{\mathbf{x}}\mathbf{p}\}^T\right) - 2\,(\mathrm{div}_{\mathbf{x}}\mathbf{v})\,(\mathrm{div}_{\mathbf{x}}\mathbf{p})\\
&\quad + \frac{1}{4\pi G}(\mathrm{div}_{\mathbf{x}}\mathbf{v})\left(\frac{\partial\Phi}{\partial t} + \mathbf{v}\cdot\nabla_{\mathbf{x}}\Phi\right) + \frac{1}{4\pi G}\Phi\,(\mathrm{div}_{\mathbf{x}}\mathbf{v})^2\\
&\quad - \frac{\Phi}{16\pi G}\left|d_{\mathbf{x}}\mathbf{v} + \{d_{\mathbf{x}}\mathbf{v}\}^T\right|^2 + \frac{1}{8\pi G}|\nabla_{\mathbf{x}}\Phi|^2 = L_{ge} =\\
&\frac{1}{4\pi}\left(-\sum_{n=0}^{3}\sum_{k=0}^{3}\sum_{m=0}^{3}\sum_{p=0}^{3}\frac{1}{4}g^{mn}g^{pk}\left(\frac{\partial A_p}{\partial x^m} - \frac{\partial A_m}{\partial x^p}\right)\left(\frac{\partial A_k}{\partial x^n} - \frac{\partial A_n}{\partial x^k}\right) - \sum_{k=0}^{3}4\pi j^k A_k\right)\\
&+ \frac{\hat{\mu}}{\sqrt{|\det G|}}\left(\sum_{m=0}^{3}\sum_{k=0}^{3} g_{mk}\hat{u}^m_{x^0}u^k_t\right)\left(\sum_{m=0}^{3}\sum_{k=0}^{3} g_{mk}u^m_t u^k_t\right)^{-1} g\left(\frac{c^2}{2} - \sum_{m=0}^{3}\sum_{k=0}^{3}\frac{c^2}{2}g_{mk}u^m_t u^k_t\right)\\
&\quad + \sum_{m=0}^{3}\sum_{n=0}^{3}\frac{32\pi G}{c^4}\Theta^{mn}\frac{\partial}{\partial x^m}\left(\sum_{j=0}^{3}\sum_{k=0}^{3} g^{kj}v_k S_j\right)\frac{\partial}{\partial x^n}\left(\sum_{j=0}^{3}\sum_{k=0}^{3} g^{kj}v_k S_j\right)\\
&\quad + \left(\sum_{i=0}^{3}\sum_{j=0}^{3} g^{ij}\left(\delta_j v_i + \delta_i v_j\right)\right)\left(\sum_{i=0}^{3}\sum_{j=0}^{3} g^{ij}\left(\delta_j S_i + \delta_i S_j\right)\right)\\
&\quad - \sum_{k=0}^{3}\sum_{j=0}^{3}\sum_{m=0}^{3}\sum_{n=0}^{3}\left(\delta_j S_k + \delta_k S_j\right)g^{km}g^{jn}\left(\delta_m v_n + \delta_n v_m\right),
\end{aligned} \tag{9.386}$$

where the right-hand side is written in the covariant form and the second equality is valid in every curvilinear coordinate system with $\frac{\partial t}{\partial x^0} > 0$, (j^0, j^1, j^2, j^3) is the four vector of the current that satisfies (9.382), $\hat{\mu}$ is given by (9.384) and in a Cartesian coordinate system we have

$$\begin{cases} S_0 = \frac{c^2}{16\pi G}\Phi - \frac{1}{2}\mathbf{v}\cdot\mathbf{p} \\ S_j = \frac{c}{2}p_j \quad\quad \forall 1 \le j \le 3 \end{cases} \tag{9.387}$$

and

$$\begin{cases} S^0 = \frac{c^2}{16\pi G}\Phi \\ S^j = \frac{c^2}{16\pi G}\Phi\frac{v^j}{c} - \frac{c}{2}p_j \quad\quad \forall 1 \le j \le 3. \end{cases} \tag{9.388}$$

Moreover in the particular case of the relativistic-like choice $g(s) := -c^2\sqrt{1-\frac{2s}{c^2}}$, by (9.383) we can write an alternative to (9.386) as

$$\begin{aligned}
&\frac{1}{8\pi}\left|-\nabla_{\mathbf{x}}\Psi - \frac{1}{c}\frac{\partial \mathbf{A}}{\partial t} + \frac{1}{c}\mathbf{v}\times\mathrm{curl}_{\mathbf{x}}\mathbf{A}\right|^2 - \frac{1}{8\pi}\left|\mathrm{curl}_{\mathbf{x}}\mathbf{A}\right|^2 - \left(\rho\Psi - \frac{1}{c}\mathbf{A}\cdot\mathbf{j}\right) \\
&\quad + \mu g\left(\frac{1}{2}|\mathbf{u}-\mathbf{v}|^2\right) + \frac{1}{2}\left(d_{\mathbf{x}}\mathbf{v} + \{d_{\mathbf{x}}\mathbf{v}\}^T\right) : \left(d_{\mathbf{x}}\mathbf{p} + \{d_{\mathbf{x}}\mathbf{p}\}^T\right) - 2\left(\mathrm{div}_{\mathbf{x}}\mathbf{v}\right)\left(\mathrm{div}_{\mathbf{x}}\mathbf{p}\right) \\
&\quad + \frac{1}{4\pi G}\left(\mathrm{div}_{\mathbf{x}}\mathbf{v}\right)\left(\frac{\partial\Phi}{\partial t} + \mathbf{v}\cdot\nabla_{\mathbf{x}}\Phi\right) + \frac{1}{4\pi G}\Phi\left(\mathrm{div}_{\mathbf{x}}\mathbf{v}\right)^2 \\
&\quad - \frac{\Phi}{16\pi G}\left|d_{\mathbf{x}}\mathbf{v} + \{d_{\mathbf{x}}\mathbf{v}\}^T\right|^2 + \frac{1}{8\pi G}\left|\nabla_{\mathbf{x}}\Phi\right|^2 = L_{ge} = \\
&\frac{1}{4\pi}\left(-\sum_{n=0}^{3}\sum_{k=0}^{3}\sum_{m=0}^{3}\sum_{p=0}^{3}\frac{1}{4}g^{mn}g^{pk}\left(\frac{\partial A_p}{\partial x^m} - \frac{\partial A_m}{\partial x^p}\right)\left(\frac{\partial A_k}{\partial x^n} - \frac{\partial A_n}{\partial x^k}\right) - \sum_{k=0}^{3}4\pi j^k A_k\right) \\
&\quad - \frac{c^2\hat{\mu}}{\sqrt{|\det G|}}\sqrt{\left(\sum_{m=0}^{3}\sum_{k=0}^{3}g_{mk}\,\hat{u}^m_{x^0}\,\hat{u}^k_{x^0}\right)} \\
&\quad + \sum_{m=0}^{3}\sum_{n=0}^{3}\frac{32\pi G}{c^4}\Theta^{mn}\frac{\partial}{\partial x^m}\left(\sum_{j=0}^{3}\sum_{k=0}^{3}g^{kj}v_k S_j\right)\frac{\partial}{\partial x^n}\left(\sum_{j=0}^{3}\sum_{k=0}^{3}g^{kj}v_k S_j\right) \\
&\quad + \left(\sum_{i=0}^{3}\sum_{j=0}^{3}g^{ij}\left(\delta_j v_i + \delta_i v_j\right)\right)\left(\sum_{i=0}^{3}\sum_{j=0}^{3}g^{ij}\left(\delta_j S_i + \delta_i S_j\right)\right) \\
&\quad - \sum_{k=0}^{3}\sum_{j=0}^{3}\sum_{m=0}^{3}\sum_{n=0}^{3}\left(\delta_j S_k + \delta_k S_j\right)g^{km}g^{jn}\left(\delta_m v_n + \delta_n v_m\right),
\end{aligned} \tag{9.389}$$

where the right-hand side is written in the covariant form and the second equality is valid in every curvilinear coordinate system with $\frac{\partial t}{\partial x^0} > 0$. On the other hand, in the case of fully non-relativistic Lagrangian, where $g(s) = \left(s - \frac{c^2}{2}\right)$, we can write an alternative to (9.386) as

$$\frac{1}{8\pi}\left|-\nabla_{\mathbf{x}}\Psi - \frac{1}{c}\frac{\partial \mathbf{A}}{\partial t} + \frac{1}{c}\mathbf{v}\times \mathrm{curl}_{\mathbf{x}}\mathbf{A}\right|^2 - \frac{1}{8\pi}\left|\mathrm{curl}_{\mathbf{x}}\mathbf{A}\right|^2 - \left(\rho\Psi - \frac{1}{c}\mathbf{A}\cdot\mathbf{j}\right)$$
$$+\mu g\left(\frac{1}{2}\left|\mathbf{u}-\mathbf{v}\right|^2\right) + \frac{1}{2}\left(d_{\mathbf{x}}\mathbf{v} + \{d_{\mathbf{x}}\mathbf{v}\}^T\right) : \left(d_{\mathbf{x}}\mathbf{p} + \{d_{\mathbf{x}}\mathbf{p}\}^T\right) - 2\left(\mathrm{div}_{\mathbf{x}}\mathbf{v}\right)\left(\mathrm{div}_{\mathbf{x}}\mathbf{p}\right)$$
$$+\frac{1}{4\pi G}\left(\mathrm{div}_{\mathbf{x}}\mathbf{v}\right)\left(\frac{\partial\Phi}{\partial t} + \mathbf{v}\cdot\nabla_{\mathbf{x}}\Phi\right) + \frac{1}{4\pi G}\Phi\left(\mathrm{div}_{\mathbf{x}}\mathbf{v}\right)^2$$
$$-\frac{\Phi}{16\pi G}\left|d_{\mathbf{x}}\mathbf{v} + \{d_{\mathbf{x}}\mathbf{v}\}^T\right|^2 + \frac{1}{8\pi G}\left|\nabla_{\mathbf{x}}\Phi\right|^2 = L_{ge} =$$
$$\frac{1}{4\pi}\left(-\sum_{n=0}^{3}\sum_{k=0}^{3}\sum_{m=0}^{3}\sum_{p=0}^{3}\frac{1}{4}g^{mn}g^{pk}\left(\frac{\partial A_p}{\partial x^m} - \frac{\partial A_m}{\partial x^p}\right)\left(\frac{\partial A_k}{\partial x^n} - \frac{\partial A_n}{\partial x^k}\right) - \sum_{k=0}^{3}4\pi j^k A_k\right)$$
$$-\frac{c^2\hat{\mu}}{\sqrt{|\det G|}}\left(\sum_{m=0}^{3}\sum_{k=0}^{3}g_{mk}\,\hat{u}^m_{x^0}\,u^k_t\right)$$
$$+\sum_{m=0}^{3}\sum_{n=0}^{3}\frac{32\pi G}{c^4}\Theta^{mn}\frac{\partial}{\partial x^m}\left(\sum_{j=0}^{3}\sum_{k=0}^{3}g^{kj}v_k S_j\right)\frac{\partial}{\partial x^n}\left(\sum_{j=0}^{3}\sum_{k=0}^{3}g^{kj}v_k S_j\right)$$
$$+\left(\sum_{i=0}^{3}\sum_{j=0}^{3}g^{ij}\left(\delta_j v_i + \delta_i v_j\right)\right)\left(\sum_{i=0}^{3}\sum_{j=0}^{3}g^{ij}\left(\delta_j S_i + \delta_i S_j\right)\right)$$
$$-\sum_{k=0}^{3}\sum_{j=0}^{3}\sum_{m=0}^{3}\sum_{n=0}^{3}\left(\delta_j S_k + \delta_k S_j\right)g^{km}g^{jn}\left(\delta_m v_n + \delta_n v_m\right), \tag{9.390}$$

where again the right-hand side is written in the covariant form and the second equality is valid in every curvilinear coordinate system with $\frac{\partial t}{\partial x^0} > 0$.

Once we wrote the Lagrangian density $L_{ge} := L\left(S^k, v^k, A^k, x^k\right)_{k=0,\ldots 4}$ as a covariant scalar, under the changes of curvilinear coordinate systems, we can write a covariant Lagrangian as

$$J_{ge}(S^k, v^k, A^k) := \int_{(x^0,x^1,x^2,x^3)} L_{ge}\left(S^k, v^k, A^k, x^k\right)\sqrt{|\det G|}\,dx^0 dx^1 dx^2 dx^3. \tag{9.391}$$

Although we need a term $\sqrt{|\det G|}$ for the covariance of the Lagrangian in curvilinear coordinate systems, in the Cartesian coordinate systems we always have $\sqrt{|\det G|} = 1$.

Next note that the contravariant tensor of the three-dimensional geometry Θ^{ij} which satisfies (9.363) in non-inertial Cartesian coordinate systems and the scalar of the global time t are dependent on the geometry of the space-time only and are fully determined in a given curvilinear coordinate system by change of variables rules. In particular, the four covector of the gravitational potential (v_0, v_1, v_2, v_3) is fully determined in the given curvilinear coordinate system, since we have

$$v_k = c\frac{\partial t}{\partial x^k} \qquad \forall k = 0, 1, 2, 3. \tag{9.392}$$

Moreover, by (9.175) and (9.176) we have the following covariant identities which are valid in every curvilinear coordinate system:

$$\begin{cases} \sum_{k=0}^{3} \Theta^{mk} v_k = \sum_{k=0}^{3} c\,\Theta^{mk} \frac{\partial t}{\partial x^k} = 0 \qquad \forall m = 0, 1, 2, 3 \\ \sum_{k=0}^{3} \sum_{j=0}^{3} g^{kj} v_k v_j = \sum_{k=0}^{3} \sum_{j=0}^{3} c^2 g^{kj} \frac{\partial t}{\partial x^k} \frac{\partial t}{\partial x^j} = 1. \end{cases} \tag{9.393}$$

However, the four vector of the gravitational potential (v^0, v^1, v^2, v^3), the contravariant pseudo-metric tensor $g^{mn} = v^m v^n - \Theta^{mn}$ and thus also the covariant pseudo-metric tensor g_{mn} depend also on the physical properties of the matter. Without knowing the physical properties of the matter the four vector of the gravitational potential can be arbitrary vector (v^0, v^1, v^2, v^3) that satisfies:

$$\sum_{k=0}^{3} v_k v^k = \sum_{k=0}^{3} c\frac{\partial t}{\partial x^k} v^k = 1. \tag{9.394}$$

Indeed for an arbitrary four vector (v^0, v^1, v^2, v^3) that satisfies (9.394), denoting $g^{mn} := v^m v^n - \Theta^{mn}$, using (9.393) and (9.394) we clearly obtain the following consistency:

$$\sum_{j=0}^{3} g^{kj} v_j = \sum_{j=0}^{3} \left(v^k v^j - \Theta^{kj}\right) v_j = v^k \left(\sum_{j=0}^{3} v^j v_j\right) - \sum_{j=0}^{3} \Theta^{kj} v_j = v^k \qquad \forall k = 0, 1, 2, 3. \tag{9.395}$$

Thus we obtained that the four vector of the gravitational potential can be arbitrary four vector in (9.391) that satisfies the linear constraint (9.394) where the four covector v_k is prescribed. So the four vector (v^0, v^1, v^2, v^3) actually contains three independent scalar functions similarly as in Cartesian coordinate systems where we have $v^0 = 1$. On the other hand, the four vector S^k contains four independent scalar functions. Thus the Lagrangian in (9.391) depends on seven independent scalar functions characterizing the gravitational field and the four vector of electromagnetic potential, exactly as in Cartesian coordinate systems where we have four independent scalar functions that characterize the electromagnetic field: scalar Ψ and three-dimensional vector $\mathbf{A}$ and seven independent scalar functions that characterize the gravitational field: three are contained in the three-dimensional vectorial gravitational potential $\mathbf{v}$ and other four are the ancillary scalar field Φ and the ancillary three-dimensional vector field $\mathbf{p}$.

Finally note that, since our model of the Newtonian-type gravity in the case of fully non-relativistic choice (9.390) and in the absence of electromagnetic fields coincides with the classical Newtonian gravitation, as a particular case, we obtained a covariant formulation of the classical Newtonian gravity in curvilinear coordinate systems.

9.9.2 The Case of Some Alternative Model of the Gravity

Consider $\mathbf{k}$ to be the vectorial potential of the inertia, which is a generally trivial speed-like vector field, assumed to be fixed in every fixed inertial or non-inertial Cartesian coordinate system (see Definition 4.2). In this section, we find a equivalent form of the Lagrangian density of the gravitational-electromagnetic field in the case of the alternative model of the gravity having the general form (9.267) or (9.268):

$$\begin{aligned} L(\mathbf{A}, \Psi, \mathbf{v}, \Phi_0, \mathbf{h}, \mathbf{x}, t) := \frac{1}{8\pi}\left|-\nabla_{\mathbf{x}}\Psi - \frac{1}{c}\frac{\partial \mathbf{A}}{\partial t} + \frac{1}{c}\mathbf{v}\times \operatorname{curl}_{\mathbf{x}}\mathbf{A}\right|^2 \\ - \frac{1}{8\pi}|\operatorname{curl}_{\mathbf{x}}\mathbf{A}|^2 - \left(\rho\Psi - \frac{1}{c}\mathbf{A}\cdot\mathbf{j}\right) + \mu g\left(\frac{1}{2}|\mathbf{u}-\mathbf{v}|^2\right) \\ - \frac{c^2}{8\pi G}\left|-\nabla_{\mathbf{x}}\Phi_0 - \frac{1}{c}\frac{\partial \mathbf{h}}{\partial t} + \frac{1}{c}\mathbf{v}\times\operatorname{curl}_{\mathbf{x}}\mathbf{h} - \frac{1}{c}\nabla_{\mathbf{x}}(\mathbf{h}\cdot\mathbf{v})\right|^2 + \frac{c^2}{8\pi G}|\operatorname{curl}_{\mathbf{x}}\mathbf{h}|^2 \\ + \frac{c^2\beta}{4\pi G}\left(\frac{1}{4}\left|d_{\mathbf{x}}\mathbf{v} + \{d_{\mathbf{x}}\mathbf{v}\}^T\right|^2 - |\operatorname{div}_{\mathbf{x}}\mathbf{v}|^2\right), \end{aligned} \tag{9.396}$$

where

$$\mathbf{h} = \mathbf{v} - \mathbf{k} \quad \text{and} \quad \Phi_0 = -\frac{1}{2c}|\mathbf{v}-\mathbf{k}|^2, \tag{9.397}$$

so that we have

$$\begin{aligned} L(\mathbf{A}, \Psi, \mathbf{v}, \Phi_0, \mathbf{h}, \mathbf{x}, t) = L_1(\mathbf{A}, \Psi, \mathbf{v}, \mathbf{x}, t) := \frac{1}{8\pi}\left|-\nabla_{\mathbf{x}}\Psi - \frac{1}{c}\frac{\partial \mathbf{A}}{\partial t} + \frac{1}{c}\mathbf{v}\times \operatorname{curl}_{\mathbf{x}}\mathbf{A}\right|^2 \\ - \frac{1}{8\pi}|\operatorname{curl}_{\mathbf{x}}\mathbf{A}|^2 - \left(\rho\Psi - \frac{1}{c}\mathbf{A}\cdot\mathbf{j}\right) + \mu g\left(\frac{1}{2}|\mathbf{u}-\mathbf{v}|^2\right) \\ - \frac{c^2}{8\pi G}\left|\frac{1}{c}\nabla_{\mathbf{x}}\left(\frac{1}{2}|\mathbf{v}-\mathbf{k}|^2\right) - \frac{1}{c}\frac{\partial}{\partial t}(\mathbf{v}-\mathbf{k}) + \frac{1}{c}\mathbf{v}\times\operatorname{curl}_{\mathbf{x}}(\mathbf{v}-\mathbf{k}) - \frac{1}{c}\nabla_{\mathbf{x}}((\mathbf{v}-\mathbf{k})\cdot\mathbf{v})\right|^2 \\ + \frac{c^2}{8\pi G}|\operatorname{curl}_{\mathbf{x}}(\mathbf{v}-\mathbf{k})|^2 + \frac{c^2\beta}{4\pi G}\left(\frac{1}{4}\left|d_{\mathbf{x}}\mathbf{v} + \{d_{\mathbf{x}}\mathbf{v}\}^T\right|^2 - |\operatorname{div}_{\mathbf{x}}\mathbf{v}|^2\right). \end{aligned} \tag{9.398}$$

Our purpose is to make the equivalent form of this Lagrangian density to be covariant and valid in every curvilinear coordinate system. Assume first that we deal with a Cartesian inertial or non-inertial coordinate system. Consider again the three-dimensional vectorial gravitational potential $\mathbf{v} = (v^1, v^2, v^3)$ and consider the covariant pseudo-metric tensor $G = \{g_{ij}\}_{0\le i,j\le 3}$ defined by (9.323) and the contravariant pseudo-metric tensor $\tilde{G} = \{g^{ij}\}_{0\le i,j\le 3}$ defined by (9.324). Next, as before, we consider the four-dimensional gravitational potential (v^0, v^1, v^2, v^3) defined by (9.345) and the corresponding lowered four-covector field (v_0, v_1, v_2, v_3), that we called the four covector of gravitational potential, defined by (9.346) and, as in (9.347), denote

$$\hat{Y}_{ij} := \delta_j v_i + \delta_i v_j \qquad \forall 0 \le i, j \le 3, \tag{9.399}$$

where δ_j means the covariant derivative with respect to the dynamical pseudo-metrics $G = \{g_{ij}\}_{0\le i,j\le 3}$. Then by the particular case of (9.358) with $\hat{Z}_{ij} = \hat{Y}_{ij}$ we have

$$\left((\mathrm{div}_{\mathbf{x}}\mathbf{v})^2 - \frac{1}{4}\left|d_{\mathbf{x}}\mathbf{v} + \{d_{\mathbf{x}}\mathbf{v}\}^T\right|^2\right) = \frac{c^2}{4}\left(\left(\sum_{i=0}^{3}\sum_{j=0}^{3} g^{ij}\left(\delta_j v_i + \delta_i v_j\right)\right)^2 \right.$$
$$\left. - \sum_{k=0}^{3}\sum_{j=0}^{3}\sum_{m=0}^{3}\sum_{n=0}^{3}\left(\delta_j v_k + \delta_k v_j\right) g^{km} g^{jn}\left(\delta_m v_n + \delta_n v_m\right)\right). \tag{9.400}$$

Furthermore, consider the dynamical four covector of genuine gravity (s_0, s_1, s_2, s_3), defined as in (9.164) by

$$(s_0, s_1, s_2, s_3) = -\frac{1}{c}\left(\left(\Phi_0 + \frac{1}{c}\mathbf{v}\cdot\mathbf{h}\right), -\mathbf{h}\right) = \left(\frac{1}{2c^2}|\mathbf{h}|^2 - \frac{1}{c^2}\mathbf{v}\cdot\mathbf{h}\,,\ \frac{1}{c}\mathbf{h}\right) \quad \text{where}$$
$$s_0 = -\frac{1}{c}\left(\Phi_0 + \frac{1}{c}\mathbf{v}\cdot\mathbf{h}\right) = \frac{1}{2c^2}|\mathbf{h}|^2 - \frac{1}{c^2}\mathbf{v}\cdot\mathbf{h} \quad \text{and} \quad (s_1, s_2, s_3) = \frac{1}{c}\mathbf{h}, \tag{9.401}$$

where $\mathbf{h}$ and Φ_0 are given by (9.397). Next, as in (9.120) and (9.122), we have the following:

$$\frac{1}{4\pi}\left(\frac{1}{2}\left|-\nabla_{\mathbf{x}}\Psi - \frac{1}{c}\frac{\partial \mathbf{A}}{\partial t} + \frac{1}{c}\mathbf{v}\times \mathrm{curl}_{\mathbf{x}}\mathbf{A}\right|^2 - \frac{1}{2}\left|\mathrm{curl}_{\mathbf{x}}\mathbf{A}\right|^2 - 4\pi\left(\rho\Psi - \frac{1}{c}\mathbf{A}\cdot\mathbf{j}\right)\right) =$$
$$\frac{1}{4\pi}\left(-\sum_{n=0}^{3}\sum_{k=0}^{3}\sum_{m=0}^{3}\sum_{p=0}^{3}\frac{1}{4}g^{mn}g^{pk}\left(\frac{\partial A_p}{\partial x^m} - \frac{\partial A_m}{\partial x^p}\right)\left(\frac{\partial A_k}{\partial x^n} - \frac{\partial A_n}{\partial x^k}\right) - \sum_{k=0}^{3} 4\pi j^k A_k\right), \tag{9.402}$$

where (A_0, A_1, A_2, A_3) is the four covector of the four-dimensional electromagnetic potential, defined as in (9.37) by

$$(A_0, A_1, A_2, A_3) = (\Psi, -\mathbf{A}) \quad \text{where} \quad A_0 = \Psi \ \text{and} \ (A_1, A_2, A_3) = -\mathbf{A}, \tag{9.403}$$

and (j^0, j^1, j^2, j^3) is the four vector of the current, given by (9.382). Then, completely analogously, as we get in (9.120) and (9.122) the following part of (9.402):

$$\left(\frac{1}{2}\left|-\nabla_{\mathbf{x}}\Psi-\frac{1}{c}\frac{\partial\mathbf{A}}{\partial t}+\frac{1}{c}\mathbf{v}\times\mathrm{curl}_{\mathbf{x}}\mathbf{A}\right|^2-\frac{1}{2}\left|\mathrm{curl}_{\mathbf{x}}\mathbf{A}\right|^2\right)=$$
$$-\sum_{n=0}^{3}\sum_{k=0}^{3}\sum_{m=0}^{3}\sum_{p=0}^{3}\frac{1}{4}g^{mn}g^{pk}\left(\frac{\partial A_p}{\partial x^m}-\frac{\partial A_m}{\partial x^p}\right)\left(\frac{\partial A_k}{\partial x^n}-\frac{\partial A_n}{\partial x^k}\right), \quad (9.404)$$

we can get also:

$$\left(\frac{1}{2}\left|-\nabla_{\mathbf{x}}\Phi_0-\frac{1}{c}\frac{\partial\mathbf{h}}{\partial t}+\frac{1}{c}\mathbf{v}\times\mathrm{curl}_{\mathbf{x}}\mathbf{h}-\frac{1}{c}\nabla_{\mathbf{x}}(\mathbf{h}\cdot\mathbf{v})\right|^2-\frac{1}{2}\left|\mathrm{curl}_{\mathbf{x}}\mathbf{h}\right|^2\right)=$$
$$\left(\frac{1}{2}\left|-\nabla_{\mathbf{x}}\left(\Phi_0+\frac{1}{c}\mathbf{v}\cdot\mathbf{h}\right)-\frac{1}{c}\frac{\partial\mathbf{h}}{\partial t}+\frac{1}{c}\mathbf{v}\times\mathrm{curl}_{\mathbf{x}}\mathbf{h}\right|^2-\frac{1}{2}\left|\mathrm{curl}_{\mathbf{x}}\mathbf{h}\right|^2\right)=$$
$$-\sum_{n=0}^{3}\sum_{k=0}^{3}\sum_{m=0}^{3}\sum_{p=0}^{3}\frac{c^2}{4}g^{mn}g^{pk}\left(\frac{\partial s_p}{\partial x^m}-\frac{\partial s_m}{\partial x^p}\right)\left(\frac{\partial s_k}{\partial x^n}-\frac{\partial s_n}{\partial x^k}\right). \quad (9.405)$$

Then, similarly as it was done in (9.386), by (9.402), (9.405) and (9.400) we rewrite the Lagrangian density in (9.396) or (9.398) in the Cartesian coordinate system as

$$\frac{1}{8\pi}\left|-\nabla_{\mathbf{x}}\Psi-\frac{1}{c}\frac{\partial\mathbf{A}}{\partial t}+\frac{1}{c}\mathbf{v}\times\mathrm{curl}_{\mathbf{x}}\mathbf{A}\right|^2-\frac{1}{8\pi}\left|\mathrm{curl}_{\mathbf{x}}\mathbf{A}\right|^2-\left(\rho\Psi-\frac{1}{c}\mathbf{A}\cdot\mathbf{j}\right)$$
$$+\mu g\left(\frac{1}{2}\left|\mathbf{u}-\mathbf{v}\right|^2\right)-\frac{c^2}{8\pi G}\left|-\nabla_{\mathbf{x}}\Phi_0-\frac{1}{c}\frac{\partial\mathbf{h}}{\partial t}+\frac{1}{c}\mathbf{v}\times\mathrm{curl}_{\mathbf{x}}\mathbf{h}-\frac{1}{c}\nabla_{\mathbf{x}}(\mathbf{h}\cdot\mathbf{v})\right|^2$$
$$+\frac{c^2}{8\pi G}\left|\mathrm{curl}_{\mathbf{x}}\mathbf{h}\right|^2+\frac{c^2\beta}{4\pi G}\left(\frac{1}{4}\left|d_{\mathbf{x}}\mathbf{v}+\{d_{\mathbf{x}}\mathbf{v}\}^T\right|^2-\left|\mathrm{div}_{\mathbf{x}}\mathbf{v}\right|^2\right)=L_{ge}=$$
$$\frac{1}{4\pi}\left(-\sum_{n=0}^{3}\sum_{k=0}^{3}\sum_{m=0}^{3}\sum_{p=0}^{3}\frac{1}{4}g^{mn}g^{pk}\left(\frac{\partial A_p}{\partial x^m}-\frac{\partial A_m}{\partial x^p}\right)\left(\frac{\partial A_k}{\partial x^n}-\frac{\partial A_n}{\partial x^k}\right)-\sum_{k=0}^{3}4\pi j^k A_k\right)$$
$$+\frac{\hat{\mu}}{\sqrt{|\det G|}}\left(\sum_{m=0}^{3}\sum_{k=0}^{3}g_{mk}\hat{u}^m_{x0}u^k_t\right)\left(\sum_{m=0}^{3}\sum_{k=0}^{3}g_{mk}u^m_t u^k_t\right)^{-1}g\left(\frac{c^2}{2}-\sum_{m=0}^{3}\sum_{k=0}^{3}\frac{c^2}{2}g_{mk}u^m_t u^k_t\right)$$
$$+\frac{c^4}{4\pi G}\left(\sum_{n=0}^{3}\sum_{k=0}^{3}\sum_{m=0}^{3}\sum_{p=0}^{3}\frac{1}{4}g^{mn}g^{pk}\left(\frac{\partial s_p}{\partial x^m}-\frac{\partial s_m}{\partial x^p}\right)\left(\frac{\partial s_k}{\partial x^n}-\frac{\partial s_n}{\partial x^k}\right)\right)$$
$$-\frac{c^4\beta}{16\pi G}\left(\left(\sum_{i=0}^{3}\sum_{j=0}^{3}g^{ij}\left(\delta_j v_i+\delta_i v_j\right)\right)^2\right.$$
$$\left.-\sum_{k=0}^{3}\sum_{j=0}^{3}\sum_{m=0}^{3}\sum_{n=0}^{3}\left(\delta_j v_k+\delta_k v_j\right)g^{km}g^{jn}\left(\delta_m v_n+\delta_n v_m\right)\right), \quad (9.406)$$

where the right-hand side is written in the covariant form and the second equality is valid in every curvilinear coordinate system with $\frac{\partial t}{\partial x^0} > 0$, (j^0, j^1, j^2, j^3) is the four vector of the current, $\hat{\mu}$, u^j_t and $u^j_{x^0}$ are the same as in (9.384), and $\mathbf{h}$ and Φ_0 are given by

$$\mathbf{h} = \mathbf{v} - \mathbf{k} \quad \text{and} \quad \Phi_0 = -\frac{1}{2c}\left|\mathbf{v} - \mathbf{k}\right|^2. \tag{9.407}$$

Moreover, in the particular case of the relativistic-like choice $g(s) := -c^2\sqrt{1 - \frac{2s}{c^2}}$, by (9.383) we can write an alternative to (9.406) as

$$\begin{aligned}
&\frac{1}{8\pi}\left|-\nabla_{\mathbf{x}}\Psi - \frac{1}{c}\frac{\partial \mathbf{A}}{\partial t} + \frac{1}{c}\mathbf{v}\times \text{curl}_{\mathbf{x}}\mathbf{A}\right|^2 - \frac{1}{8\pi}\left|\text{curl}_{\mathbf{x}}\mathbf{A}\right|^2 - \left(\rho\Psi - \frac{1}{c}\mathbf{A}\cdot\mathbf{j}\right) \\
&\quad + \mu g\left(\frac{1}{2}\left|\mathbf{u}-\mathbf{v}\right|^2\right) - \frac{c^2}{8\pi G}\left|-\nabla_{\mathbf{x}}\Phi_0 - \frac{1}{c}\frac{\partial \mathbf{h}}{\partial t} + \frac{1}{c}\mathbf{v}\times\text{curl}_{\mathbf{x}}\mathbf{h} - \frac{1}{c}\nabla_{\mathbf{x}}\left(\mathbf{h}\cdot\mathbf{v}\right)\right|^2 \\
&\quad + \frac{c^2}{8\pi G}\left|\text{curl}_{\mathbf{x}}\mathbf{h}\right|^2 + \frac{c^2\beta}{4\pi G}\left(\frac{1}{4}\left|d_{\mathbf{x}}\mathbf{v} + \{d_{\mathbf{x}}\mathbf{v}\}^T\right|^2 - \left|\text{div}_{\mathbf{x}}\mathbf{v}\right|^2\right) = L_{ge} = \\
&\frac{1}{4\pi}\left(-\sum_{n=0}^{3}\sum_{k=0}^{3}\sum_{m=0}^{3}\sum_{p=0}^{3}\frac{1}{4}g^{mn}g^{pk}\left(\frac{\partial A_p}{\partial x^m} - \frac{\partial A_m}{\partial x^p}\right)\left(\frac{\partial A_k}{\partial x^n} - \frac{\partial A_n}{\partial x^k}\right) - \sum_{k=0}^{3}4\pi j^k A_k\right) \\
&\quad - \frac{c^2\hat{\mu}}{\sqrt{|\det G|}}\sqrt{\left(\sum_{m=0}^{3}\sum_{k=0}^{3}g_{mk}\,\hat{u}^m_{x^0}\,\hat{u}^k_{x^0}\right)} \\
&\quad + \frac{c^4}{4\pi G}\left(\sum_{n=0}^{3}\sum_{k=0}^{3}\sum_{m=0}^{3}\sum_{p=0}^{3}\frac{1}{4}g^{mn}g^{pk}\left(\frac{\partial s_p}{\partial x^m} - \frac{\partial s_m}{\partial x^p}\right)\left(\frac{\partial s_k}{\partial x^n} - \frac{\partial s_n}{\partial x^k}\right)\right) \\
&\quad - \frac{c^4\beta}{16\pi G}\left(\left(\sum_{i=0}^{3}\sum_{j=0}^{3}g^{ij}\left(\delta_j v_i + \delta_i v_j\right)\right)^2\right. \\
&\quad \left. - \sum_{k=0}^{3}\sum_{j=0}^{3}\sum_{m=0}^{3}\sum_{n=0}^{3}\left(\delta_j v_k + \delta_k v_j\right)g^{km}g^{jn}\left(\delta_m v_n + \delta_n v_m\right)\right),
\end{aligned} \tag{9.408}$$

where the right-hand side is written in the covariant form and the second equality is valid in every curvilinear coordinate system with $\frac{\partial t}{\partial x^0} > 0$. On the other hand, in the case of fully non-relativistic Lagrangian, where $g(s) = \left(s - \frac{c^2}{2}\right)$, we can write an alternative to (9.406) as

$$\frac{1}{8\pi}\left|-\nabla_{\mathbf{x}}\Psi-\frac{1}{c}\frac{\partial\mathbf{A}}{\partial t}+\frac{1}{c}\mathbf{v}\times\operatorname{curl}_{\mathbf{x}}\mathbf{A}\right|^2-\frac{1}{8\pi}\left|\operatorname{curl}_{\mathbf{x}}\mathbf{A}\right|^2-\left(\rho\Psi-\frac{1}{c}\mathbf{A}\cdot\mathbf{j}\right)$$
$$+\mu g\left(\frac{1}{2}\left|\mathbf{u}-\mathbf{v}\right|^2\right)-\frac{c^2}{8\pi G}\left|-\nabla_{\mathbf{x}}\Phi_0-\frac{1}{c}\frac{\partial\mathbf{h}}{\partial t}+\frac{1}{c}\mathbf{v}\times\operatorname{curl}_{\mathbf{x}}\mathbf{h}-\frac{1}{c}\nabla_{\mathbf{x}}\left(\mathbf{h}\cdot\mathbf{v}\right)\right|^2$$
$$+\frac{c^2}{8\pi G}\left|\operatorname{curl}_{\mathbf{x}}\mathbf{h}\right|^2+\frac{c^2\beta}{4\pi G}\left(\frac{1}{4}\left|d_{\mathbf{x}}\mathbf{v}+\{d_{\mathbf{x}}\mathbf{v}\}^T\right|^2-\left|\operatorname{div}_{\mathbf{x}}\mathbf{v}\right|^2\right)=L_{ge}=$$
$$\frac{1}{4\pi}\left(-\sum_{n=0}^{3}\sum_{k=0}^{3}\sum_{m=0}^{3}\sum_{p=0}^{3}\frac{1}{4}g^{mn}g^{pk}\left(\frac{\partial A_p}{\partial x^m}-\frac{\partial A_m}{\partial x^p}\right)\left(\frac{\partial A_k}{\partial x^n}-\frac{\partial A_n}{\partial x^k}\right)-\sum_{k=0}^{3}4\pi j^k A_k\right)$$
$$-\frac{c^2\hat{\mu}}{\sqrt{|\det G|}}\left(\sum_{m=0}^{3}\sum_{k=0}^{3}g_{mk}\,\hat{u}^m_{x0}\,u^k_t\right)$$
$$+\frac{c^4}{4\pi G}\left(\sum_{n=0}^{3}\sum_{k=0}^{3}\sum_{m=0}^{3}\sum_{p=0}^{3}\frac{1}{4}g^{mn}g^{pk}\left(\frac{\partial s_p}{\partial x^m}-\frac{\partial s_m}{\partial x^p}\right)\left(\frac{\partial s_k}{\partial x^n}-\frac{\partial s_n}{\partial x^k}\right)\right)$$
$$-\frac{c^4\beta}{16\pi G}\left(\left(\sum_{i=0}^{3}\sum_{j=0}^{3}g^{ij}\left(\delta_j v_i+\delta_i v_j\right)\right)^2\right.$$
$$\left.-\sum_{k=0}^{3}\sum_{j=0}^{3}\sum_{m=0}^{3}\sum_{n=0}^{3}\left(\delta_j v_k+\delta_k v_j\right)g^{km}g^{jn}\left(\delta_m v_n+\delta_n v_m\right)\right),\qquad(9.409)$$

where again the right-hand side is written in the covariant form and the second equality is valid in every curvilinear coordinate system with $\frac{\partial t}{\partial x^0}>0$.

Note that the four vector of the gravitational potential (v^0, v^1, v^2, v^3) and the four covector of the genuine gravity (s_0, s_1, s_2, s_3) are <u>not</u> independent arguments of L_{ge}. Moreover, in the general non-Cartesian or curvilinear coordinate system, knowing (v^0, v^1, v^2, v^3) and thus also knowing $\{g_{ij}\}_{0\le i,j\le 3}$, as before in (9.203), we can find (s_0, s_1, s_2, s_3) by

$$s_j=\frac{1}{2}\left(\sum_{m=0}^{3}g_{jm}k^m-\sum_{m=0}^{3}J_{jm}v^m\right)\quad\forall j=0,1,2,3,\qquad(9.410)$$

where (k^0, k^1, k^2, k^3) is the four vector of the potential of inertia and $\{J_{ij}\}_{0\le i,j\le 3}$ is the covariant kinematic pseudo-metric tensor of inertia that are prescribed in every Cartesian, non-Cartesian or curvilinear coordinate system. On the other hand, knowing (s_0, s_1, s_2, s_3), we can find $\{g_{ij}\}_{0\le i,j\le 3}$ and (v^0, v^1, v^2, v^3) by the following covariant identities:

$$g_{ij}=J_{ij}+\left(s_i\left(c\frac{\partial t}{\partial x^j}\right)+\left(c\frac{\partial t}{\partial x^i}\right)s_j\right)\quad\forall i,j=0,1,2,3\qquad(9.411)$$

(see (9.204)), and

$$v^j - k^j = \sum_{m=0}^{3} \Theta^{jm} s_m = \sum_{m=0}^{3} \left(k^j k^m - J^{jm}\right) s_m \quad \forall j = 0, 1, 2, 3 \tag{9.412}$$

(see (9.205)).

Once we wrote the Lagrangian density $L_{ge} := L_{ge1}\left(v^k, A^k, x^k\right)_{k=0,\dots 4} = L_{ge2}\left(s^k, A^k, x^k\right)_{k=0,\dots 4}$ as a covariant scalar, under the changes of curvilinear coordinate systems, we can write a covariant Lagrangian as

$$\begin{aligned} J_{ge1}(v^k, A^k) = J_{ge1}(s^k, A^k) := \\ \int_{(x^0,x^1,x^2,x^3)} L_{ge1}\left(v^k, A^k, x^k\right) \sqrt{|\det G|}\, dx^0 dx^1 dx^2 dx^3 \\ = \int_{(x^0,x^1,x^2,x^3)} L_{ge2}\left(s^k, A^k, x^k\right) \sqrt{|\det G|}\, dx^0 dx^1 dx^2 dx^3. \end{aligned} \tag{9.413}$$

Although we need a term $\sqrt{|\det G|}$ for the covariance of the Lagrangian in curvilinear coordinate systems, in the Cartesian coordinate systems we always have $\sqrt{|\det G|} = 1$.

Finally, note again that the four vector of the gravitational potential, as an independent argument, is restricted by (9.394):

$$\sum_{k=0}^{3} c \frac{\partial t}{\partial x^k} v^k = 1. \tag{9.414}$$

So the four vector (v^0, v^1, v^2, v^3) actually contains three independent scalar functions similarly as in Cartesian coordinate systems where we have $v^0 = 1$. On the other hand, if we consider (s_0, s_1, s_2, s_3) instead of (v^0, v^1, v^2, v^3), as an independent argument, then by (9.145) and (9.411) it is restricted by the following identity:

$$\det\left(\left\{J_{ij} + s_i\left(c\frac{\partial t}{\partial x^j}\right) + \left(c\frac{\partial t}{\partial x^i}\right) s_j\right\}_{0\le i,j\le 3}\right) = \det\left(\{J_{ij}\}_{0\le i,j\le 3}\right), \tag{9.415}$$

which is valid in every Cartesian, non-Cartesian or curvilinear coordinate system. So the four covector (s_0, s_1, s_2, s_3) also contains three independent scalar functions.

Chapter 10
Relativistic-Like Dirac Equation

10.1 Classical Relativistic-Like Lagrangian and Hamiltonian of the Motion of Particles

As in (9.136) consider the relativistic-like Lagrangian of the motion of the particle with mass m and charge σ in the outer gravitational and electromagnetic fields and additional field with potential $V(\mathbf{x}, t)$:

$$J_{rl}(\mathbf{r}) = \int_0^T L_0\left(\frac{d\mathbf{r}}{dt}, \mathbf{r}, t\right) dt :=$$
$$\int_0^T \left\{ -mc^2\sqrt{1 - \frac{1}{c^2}\left|\frac{d\mathbf{r}}{dt} - \mathbf{v}(\mathbf{r}, t)\right|^2} - \sigma\left(\Psi(\mathbf{r}, t) - \frac{1}{c}\mathbf{A}(\mathbf{r}, t)\cdot\frac{d\mathbf{r}}{dt}\right) + V(\mathbf{r}, t)\right\} dt. \tag{10.1}$$

Next define the generalized momentum of the particle by

$$\mathbf{P} := \nabla_{\mathbf{r}'} L_0(\mathbf{r}', \mathbf{r}, t) = m\left(1 - \frac{1}{c^2}\left|\frac{d\mathbf{r}}{dt} - \mathbf{v}(\mathbf{r}, t)\right|^2\right)^{-\frac{1}{2}}\left(\frac{d\mathbf{r}}{dt} - \mathbf{v}(\mathbf{r}, t)\right) + \frac{\sigma}{c}\mathbf{A}(\mathbf{r}, t). \tag{10.2}$$

Then

$$\left(1 - \frac{1}{c^2}\left|\frac{d\mathbf{r}}{dt} - \mathbf{v}(\mathbf{r}, t)\right|^2\right)^{-\frac{1}{2}}\left(\frac{d\mathbf{r}}{dt} - \mathbf{v}(\mathbf{r}, t)\right) = \left(\frac{1}{m}\mathbf{P} - \frac{\sigma}{mc}\mathbf{A}(\mathbf{r}, t)\right). \tag{10.3}$$

A. Poliakovsky, *The Laws of Gravity and Electromagnetism*,
https://doi.org/10.1007/978-3-031-61407-1_10

So

$$\frac{d\mathbf{r}}{dt} = \left(1 + \frac{1}{c^2}\left|\frac{1}{m}\mathbf{P} - \frac{\sigma}{mc}\mathbf{A}(\mathbf{r}, t)\right|^2\right)^{-\frac{1}{2}} \left(\frac{1}{m}\mathbf{P} - \frac{\sigma}{mc}\mathbf{A}(\mathbf{r}, t)\right) + \mathbf{v}(\mathbf{r}, t). \quad (10.4)$$

Thus if we consider a Hamiltonian

$$H_0\left(\mathbf{P}, \mathbf{r}, t\right) := \mathbf{P} \cdot \frac{d\mathbf{r}}{dt} - L_0\left(\frac{d\mathbf{r}}{dt}, \mathbf{r}, t\right) \quad (10.5)$$

then by (10.5), (10.1), (10.3) and (10.4) we have

$$\begin{aligned}
H_0\left(\mathbf{P}, \mathbf{r}, t\right) &= -V\left(\mathbf{r}, t\right) + \mathbf{P} \cdot \frac{d\mathbf{r}}{dt} \\
&\quad + \left(mc^2\left(1 - \frac{1}{c^2}\left|\frac{d\mathbf{r}}{dt} - \mathbf{v}(\mathbf{r}, t)\right|^2\right)^{\frac{1}{2}} + \sigma\left(\Psi(\mathbf{r}, t) - \frac{1}{c}\mathbf{A}(\mathbf{r}, t) \cdot \frac{d\mathbf{r}}{dt}\right)\right) \\
&= -V\left(\mathbf{r}, t\right) + mc^2\left(1 + \frac{1}{c^2}\left|\frac{1}{m}\mathbf{P} - \frac{\sigma}{mc}\mathbf{A}(\mathbf{r}, t)\right|^2\right)^{-\frac{1}{2}} \\
&\quad + \mathbf{P} \cdot \left(\left(1 + \frac{1}{c^2}\left|\frac{1}{m}\mathbf{P} - \frac{\sigma}{mc}\mathbf{A}(\mathbf{r}, t)\right|^2\right)^{-\frac{1}{2}} \left(\frac{1}{m}\mathbf{P} - \frac{\sigma}{mc}\mathbf{A}(\mathbf{r}, t)\right) + \mathbf{v}(\mathbf{r}, t)\right) + \sigma\Bigg(\Psi(\mathbf{r}, t) \\
&\quad - \frac{1}{c}\mathbf{A}(\mathbf{r}, t) \cdot \left(\left(1 + \frac{1}{c^2}\left|\frac{1}{m}\mathbf{P} - \frac{\sigma}{mc}\mathbf{A}(\mathbf{r}, t)\right|^2\right)^{-\frac{1}{2}} \left(\frac{1}{m}\mathbf{P} - \frac{\sigma}{mc}\mathbf{A}(\mathbf{r}, t)\right) + \mathbf{v}(\mathbf{r}, t)\right)\Bigg) \\
&= mc^2\left(1 + \frac{1}{c^2}\left|\frac{1}{m}\mathbf{P} - \frac{\sigma}{mc}\mathbf{A}(\mathbf{r}, t)\right|^2\right)^{\frac{1}{2}} + \sigma\left(\Psi(\mathbf{r}, t) - \frac{1}{c}\mathbf{v}(\mathbf{r}, t) \cdot \mathbf{A}(\mathbf{r}, t)\right) \\
&\quad - V\left(\mathbf{r}, t\right) + \mathbf{v}(\mathbf{r}, t) \cdot \mathbf{P}. \quad (10.6)
\end{aligned}$$

So, the relativistic-like Hamiltonian for macro-particles has the form:

$$\begin{aligned}
H_0\left(\mathbf{P}, \mathbf{r}, t\right) = mc^2\left(1 + \frac{1}{c^2}\left|\frac{1}{m}\mathbf{P} - \frac{\sigma}{mc}\mathbf{A}(\mathbf{r}, t)\right|^2\right)^{\frac{1}{2}} &+ \sigma\left(\Psi(\mathbf{r}, t) - \frac{1}{c}\mathbf{v}(\mathbf{r}, t) \cdot \mathbf{A}(\mathbf{r}, t)\right) \\
&- V\left(\mathbf{r}, t\right) + \mathbf{v}(\mathbf{r}, t) \cdot \mathbf{P}. \quad (10.7)
\end{aligned}$$

In particular, if similarly to (9.55) we define the four-dimensional generalized momentum (P_0, P_1, P_2, P_3) as

$$(P_0, P_1, P_2, P_3) := \left(\frac{1}{c}H_0, -\mathbf{P}\right) \quad \text{where} \quad P_0 = \frac{1}{c}H_0 \quad \text{and} \quad (P_1, P_2, P_3) = -\mathbf{P}. \quad (10.8)$$

Then, since by (10.2) and (10.7), under the change of non-inertial Cartesian coordinate system H_0 and $\mathbf{P}$ transform as

$$\begin{cases} H_0' = H_0 + \left(\frac{dA}{dt}(t)\cdot\mathbf{x} + \frac{d\mathbf{z}}{dt}(t)\right)\cdot(A(t)\cdot\mathbf{P}) \\ \mathbf{P}' = A(t)\cdot\mathbf{P}, \end{cases} \tag{10.9}$$

by comparing (10.9) with (9.25) we deduce that the four-dimensional momentum (P_0, P_1, P_2, P_3) is a four-covector on the group $\mathcal{S}_0$ that is the group of changes of Cartesian non-inertial coordinate systems.

10.2 Dirac Equation

Next consider the motion of a spin-half quantum relativistic-like micro-particle with inertial mass m and charge σ in the outer gravitational and electromagnetical field with characteristics $\mathbf{v}(\mathbf{x}, t)$, $\mathbf{A}(\mathbf{x}, t)$ and $\Psi(\mathbf{x}, t)$ and additional conservative field with potential $V(\mathbf{y}, t)$. The evolution equation for this particle is

$$i\hbar\frac{\partial\psi}{\partial t} = \hat{H}_0\cdot\psi, \tag{10.10}$$

where $\psi(\mathbf{x}, t) = (\psi_1(\mathbf{x}, t), \psi_2(\mathbf{x}, t)) \in \mathbb{C}^2\times\mathbb{C}^2$ is a four-component wave function and $\hat{H}_0$ is the Hamiltonian operator. Since the relativistic-like Hamiltonian for macro-particles has the form (10.7), analogously to the usual Dirac Hamiltonian operator, we built the Hermitian–Hamiltonian operator as $\hat{H}_0\cdot\psi = \left(\hat{H}_1\cdot\psi, \hat{H}_2\cdot\psi\right)$, where

$$\begin{aligned} \hat{H}_1\cdot\psi = {} & mc^2\psi_1 - c\,\mathbf{S}\cdot\left(i\hbar\nabla_{\mathbf{x}}\psi_2 + \frac{\sigma}{c}\mathbf{A}(\mathbf{x}, t)\psi_2\right) + \sigma\left(\Psi(\mathbf{x}, t) - \frac{1}{c}\mathbf{v}(\mathbf{x}, t)\cdot\mathbf{A}(\mathbf{x}_j, t)\right)\psi_1 \\ & - V(\mathbf{x}, t)\,\psi_1 - \frac{i\hbar}{2}\mathrm{div}_{\mathbf{x}}\{\psi_1\mathbf{v}(\mathbf{x}, t)\} - \frac{i\hbar}{2}\nabla_{\mathbf{x}}\psi_1\cdot\mathbf{v}(\mathbf{x}, t) + \frac{\hbar}{4}\mathbf{S}\cdot(\mathrm{curl}_{\mathbf{x}}\mathbf{v}(\mathbf{x}, t)\,\psi_1) \\ & - \frac{(g-1)\sigma\hbar}{2mc}\mathbf{S}\cdot(\mathrm{curl}_{\mathbf{x}}\mathbf{A}(\mathbf{x}, t)\,\psi_1), \end{aligned} \tag{10.11}$$

and

$$\begin{aligned} \hat{H}_2\cdot\psi = {} & -mc^2\psi_2 - c\,\mathbf{S}\cdot\left(i\hbar\nabla_{\mathbf{x}}\psi_1 + \frac{\sigma}{c}\mathbf{A}(\mathbf{x}, t)\psi_1\right) + \sigma\left(\Psi(\mathbf{x}, t) - \frac{1}{c}\mathbf{v}(\mathbf{x}, t)\cdot\mathbf{A}(\mathbf{x}, t)\right)\psi_2 \\ & - V(\mathbf{x}, t)\,\psi_2 - \frac{i\hbar}{2}\mathrm{div}_{\mathbf{x}}\{\psi_2\mathbf{v}(\mathbf{x}, t)\} - \frac{i\hbar}{2}\nabla_{\mathbf{x}}\psi_2\cdot\mathbf{v}(\mathbf{x}, t) + \frac{\hbar}{4}\mathbf{S}\cdot(\mathrm{curl}_{\mathbf{x}}\mathbf{v}(\mathbf{x}, t)\,\psi_2) \\ & - \frac{(g-1)\sigma\hbar}{2mc}\mathbf{S}\cdot(\mathrm{curl}_{\mathbf{x}}\mathbf{A}(\mathbf{x}, t)\,\psi_2), \end{aligned} \tag{10.12}$$

where $\mathbf{S} := (S_1, S_2, S_3)$ and

$$S_1 = \begin{pmatrix} 0 & 1 \\ 1 & 0 \end{pmatrix}, \quad S_2 = \begin{pmatrix} 0 & -i \\ i & 0 \end{pmatrix} \quad S_3 = \begin{pmatrix} 1 & 0 \\ 0 & -1 \end{pmatrix}$$

are Pauli matrices and g is a constant that depends on the type of the particle (for electron we have $g = 1$). As before for the Schrödinger–Pauli equation, we added an additional term to the Hamiltonian, namely $\frac{\hbar}{4}\mathbf{S} \cdot (\mathrm{curl}_{\mathbf{x}}\mathbf{v}(\mathbf{x}, t)\, \psi)$. Although in the case of the Newtonian-type gravity, this term vanishes in inertial coordinate systems, it provides, however, invariance of our Dirac-type equation, under the change of non-inertial coordinate systems as we will see below. Thus, we have the following two evolution equations that we call together Dirac system of equations:

$$\begin{aligned} i\hbar\frac{\partial \psi_1}{\partial t} &= mc^2\psi_1 - c\,\mathbf{S}\cdot\left(i\hbar\nabla_{\mathbf{x}}\psi_2 + \frac{\sigma}{c}\mathbf{A}(\mathbf{x},t)\psi_2\right) + \sigma\left(\Psi(\mathbf{x},t) - \frac{1}{c}\mathbf{v}(\mathbf{x},t)\cdot\mathbf{A}(\mathbf{x}_j,t)\right)\psi_1 \\ &\quad - V(\mathbf{x},t)\,\psi_1 - \frac{i\hbar}{2}\mathrm{div}_{\mathbf{x}}\left\{\psi_1\mathbf{v}(\mathbf{x},t)\right\} - \frac{i\hbar}{2}\nabla_{\mathbf{x}}\psi_1\cdot\mathbf{v}(\mathbf{x},t) + \frac{\hbar}{4}\mathbf{S}\cdot(\mathrm{curl}_{\mathbf{x}}\mathbf{v}(\mathbf{x},t)\,\psi_1) \\ &\quad - \frac{(g-1)\sigma\hbar}{2mc}\mathbf{S}\cdot(\mathrm{curl}_{\mathbf{x}}\mathbf{A}(\mathbf{x},t)\,\psi_1), \end{aligned} \tag{10.13}$$

and

$$\begin{aligned} i\hbar\frac{\partial \psi_2}{\partial t} &= \\ &- mc^2\psi_2 - c\,\mathbf{S}\cdot\left(i\hbar\nabla_{\mathbf{x}}\psi_1 + \frac{\sigma}{c}\mathbf{A}(\mathbf{x},t)\psi_1\right) + \sigma\left(\Psi(\mathbf{x},t) - \frac{1}{c}\mathbf{v}(\mathbf{x},t)\cdot\mathbf{A}(\mathbf{x},t)\right)\psi_2 \\ &- V(\mathbf{x},t)\,\psi_2 - \frac{i\hbar}{2}\mathrm{div}_{\mathbf{x}}\left\{\psi_2\mathbf{v}(\mathbf{x},t)\right\} - \frac{i\hbar}{2}\nabla_{\mathbf{x}}\psi_2\cdot\mathbf{v}(\mathbf{x},t) + \frac{\hbar}{4}\mathbf{S}\cdot(\mathrm{curl}_{\mathbf{x}}\mathbf{v}(\mathbf{x},t)\,\psi_2) \\ &- \frac{(g-1)\sigma\hbar}{2mc}\mathbf{S}\cdot(\mathrm{curl}_{\mathbf{x}}\mathbf{A}(\mathbf{x},t)\,\psi_2). \end{aligned} \tag{10.14}$$

Then we can rewrite Dirac equations as

$$\begin{aligned} i\hbar\left(\frac{\partial \psi_1}{\partial t} + \frac{1}{2}\mathrm{div}_{\mathbf{x}}\left\{\psi_1\mathbf{v}(\mathbf{x},t)\right\} + \frac{1}{2}\nabla_{\mathbf{x}}\psi_1\cdot\mathbf{v}(\mathbf{x},t)\right) &= \\ mc^2\psi_1 - c\,\mathbf{S}\cdot\left(i\hbar\nabla_{\mathbf{x}}\psi_2 + \frac{\sigma}{c}\mathbf{A}(\mathbf{x},t)\psi_2\right) & \\ + \sigma\left(\Psi(\mathbf{x},t) - \frac{1}{c}\mathbf{v}(\mathbf{x},t)\cdot\mathbf{A}(\mathbf{x},t)\right)\psi_1 - V(\mathbf{x},t)\,\psi_1 + \frac{\hbar}{4}\mathbf{S}\cdot(\mathrm{curl}_{\mathbf{x}}\mathbf{v}(\mathbf{x},t)\,\psi_1) & \\ - \frac{(g-1)\sigma\hbar}{2mc}\mathbf{S}\cdot(\mathrm{curl}_{\mathbf{x}}\mathbf{A}(\mathbf{x},t)\,\psi_1), & \end{aligned} \tag{10.15}$$

and

$$
\begin{aligned}
i\hbar\left(\frac{\partial\psi_2}{\partial t}+\frac{1}{2}\mathrm{div}_{\mathbf{x}}\left\{\psi_2\mathbf{v}(\mathbf{x},t)\right\}+\frac{1}{2}\nabla_{\mathbf{x}}\psi_2\cdot\mathbf{v}(\mathbf{x},t)\right)=&\\
-mc^2\psi_2-c\,\mathbf{S}\cdot\left(i\hbar\nabla_{\mathbf{x}}\psi_1+\frac{\sigma}{c}\mathbf{A}(\mathbf{x},t)\psi_1\right)&\\
+\sigma\left(\Psi(\mathbf{x},t)-\frac{1}{c}\mathbf{v}(\mathbf{x},t)\cdot\mathbf{A}(\mathbf{x},t)\right)\psi_2-V(\mathbf{x},t)\,\psi_2+\frac{\hbar}{4}\mathbf{S}\cdot(\mathrm{curl}_{\mathbf{x}}\mathbf{v}(\mathbf{x},t)\,\psi_2)&\\
-\frac{(g-1)\sigma\hbar}{2mc}\mathbf{S}\cdot(\mathrm{curl}_{\mathbf{x}}\mathbf{A}(\mathbf{x},t)\,\psi_2)\,.&\quad(10.16)
\end{aligned}
$$

Then similarly to the proof of Theorem 6.1 about the invariance of the Shrödinger–Pauli equation, we can prove the following Theorem for Dirac equations:

Theorem 10.1 *Consider that the change of some Cartesian coordinate system* $(*)$ *to another Cartesian coordinate system* $(**)$ *is given by*

$$
\begin{cases}\mathbf{x}'=A(t)\cdot\mathbf{x}+\mathbf{z}(t),\\ t'=t,\end{cases}\qquad(10.17)
$$

where $A(t)\in SO(3)$ *is a rotation. Next, assume that in the coordinate system* $(**)$ *we observe a validity of the Dirac equations of the form:*

$$
\begin{aligned}
i\hbar\left(\frac{\partial\psi_1'}{\partial t'}+\frac{1}{2}div_{\mathbf{x}'}\left\{\psi_1'\mathbf{v}'(\mathbf{x}',t')\right\}+\frac{1}{2}\nabla_{\mathbf{x}'}\psi_1'\cdot\mathbf{v}'(\mathbf{x}',t')\right)=&\\
m'c^2\psi_1'-c\,\mathbf{S}\cdot\left(i\hbar\nabla_{\mathbf{x}'}\psi_2'+\frac{\sigma'}{c}\mathbf{A}'(\mathbf{x}',t)\psi_2'\right)&\\
+\sigma'\left(\Psi'(\mathbf{x}',t')-\frac{1}{c}\mathbf{v}'(\mathbf{x}',t')\cdot\mathbf{A}'(\mathbf{x}',t')\right)\psi_1'-V'(\mathbf{x}',t')\,\psi_1'+\frac{\hbar}{4}\mathbf{S}\cdot(curl_{\mathbf{x}'}\mathbf{v}'(\mathbf{x}',t')\,\psi_1')&\\
-\frac{(g'-1)\sigma'\hbar}{2m'c}\mathbf{S}\cdot(curl_{\mathbf{x}'}\mathbf{A}'(\mathbf{x}',t)\,\psi_1')\,,&\quad(10.18)
\end{aligned}
$$

and

$$
\begin{aligned}
i\hbar\left(\frac{\partial\psi_2'}{\partial t'}+\frac{1}{2}div_{\mathbf{x}'}\left\{\psi_2'\mathbf{v}'(\mathbf{x}',t')\right\}+\frac{1}{2}\nabla_{\mathbf{x}'}\psi_2'\cdot\mathbf{v}'(\mathbf{x}',t')\right)=-m'c^2\psi_2'&\\
-c\,\mathbf{S}\cdot\left(i\hbar\nabla_{\mathbf{x}'}\psi_1'+\frac{\sigma'}{c}\mathbf{A}'(\mathbf{x}',t')\psi_1'\right)+\sigma'\left(\Psi'(\mathbf{x}',t')-\frac{1}{c}\mathbf{v}'(\mathbf{x}',t')\cdot\mathbf{A}'(\mathbf{x}',t')\right)\psi_2'&\\
-V'(\mathbf{x}',t')\,\psi_2'+\frac{\hbar}{4}\mathbf{S}\cdot(curl_{\mathbf{x}'}\mathbf{v}'(\mathbf{x}',t')\,\psi_2')&\\
-\frac{(g'-1)\sigma'\hbar}{2m'c}\mathbf{S}\cdot(curl_{\mathbf{x}'}\mathbf{A}'(\mathbf{x}',t)\,\psi_2')\,,&\quad(10.19)
\end{aligned}
$$

where $\psi = (\psi_1, \psi_2) \in \mathbb{C}^2 \times \mathbb{C}^2$ is a four-component wave function. Then in the coordinate system (∗) we have the validity of Dirac equations of the same as (10.18) and (10.19) form:

$$
\begin{aligned}
i\hbar\left(\frac{\partial\psi_1}{\partial t} + \frac{1}{2}div_{\mathbf{x}}\left\{\psi_1\mathbf{v}(\mathbf{x},t)\right\} + \frac{1}{2}\nabla_{\mathbf{x}}\psi_1\cdot\mathbf{v}(\mathbf{x},t)\right) = \\
mc^2\psi_1 - c\,\mathbf{S}\cdot\left(i\hbar\nabla_{\mathbf{x}}\psi_2 + \frac{\sigma}{c}\mathbf{A}(\mathbf{x},t)\psi_2\right) \\
+\sigma\left(\Psi(\mathbf{x},t) - \frac{1}{c}\mathbf{v}(\mathbf{x},t)\cdot\mathbf{A}(\mathbf{x},t)\right)\psi_1 - V(\mathbf{x},t)\,\psi_1 + \frac{\hbar}{4}\mathbf{S}\cdot\left(curl_{\mathbf{x}}\mathbf{v}(\mathbf{x},t)\,\psi_1\right) \\
-\frac{(g-1)\sigma\hbar}{2mc}\mathbf{S}\cdot\left(curl_{\mathbf{x}}\mathbf{A}(\mathbf{x},t)\,\psi_1\right),
\end{aligned}
\tag{10.20}
$$

and

$$
\begin{aligned}
i\hbar\left(\frac{\partial\psi_2}{\partial t} + \frac{1}{2}div_{\mathbf{x}}\left\{\psi_2\mathbf{v}(\mathbf{x},t)\right\} + \frac{1}{2}\nabla_{\mathbf{x}}\psi_2\cdot\mathbf{v}(\mathbf{x},t)\right) = \\
-mc^2\psi_2 - c\,\mathbf{S}\cdot\left(i\hbar\nabla_{\mathbf{x}}\psi_1 + \frac{\sigma}{c}\mathbf{A}(\mathbf{x},t)\psi_1\right) \\
+\sigma\left(\Psi(\mathbf{x},t) - \frac{1}{c}\mathbf{v}(\mathbf{x},t)\cdot\mathbf{A}(\mathbf{x},t)\right)\psi_2 - V(\mathbf{x},t)\,\psi_2 + \frac{\hbar}{4}\mathbf{S}\cdot\left(curl_{\mathbf{x}}\mathbf{v}(\mathbf{x},t)\,\psi_2\right) \\
-\frac{(g-1)\sigma\hbar}{2mc}\mathbf{S}\cdot\left(curl_{\mathbf{x}}\mathbf{A}(\mathbf{x},t)\,\psi_2\right),
\end{aligned}
\tag{10.21}
$$

provided that

$$
\begin{cases}
V' = V, \\
\sigma' = \sigma, \\
m' = m, \\
\mathbf{v}' = A(t)\cdot\mathbf{v} + \frac{dA}{dt}(t)\cdot\mathbf{x} + \frac{d\mathbf{z}}{dt}(t), \\
\mathbf{A}' = A(t)\cdot\mathbf{A}, \\
\Psi' - \mathbf{v}'\cdot\mathbf{A}' = \Psi - \mathbf{v}\cdot\mathbf{A}, \\
\psi_1' = U(t)\cdot\psi_1, \\
\psi_2' = U(t)\cdot\psi_2,
\end{cases}
\tag{10.22}
$$

where, as before, $U(t) \in SU(2)$ is characterized by

$$
U^*(t)\cdot\mathbf{S}\cdot U(t) = A(t)\cdot\mathbf{S},
\tag{10.23}
$$

that means

$$\Big(U^*(t)\cdot S_1\cdot U(t), U^*(t)\cdot S_2\cdot U(t), U^*(t)\cdot S_3\cdot U(t)\Big) =$$
$$\Big(a_{11}(t)S_1 + a_{12}(t)S_2 + a_{13}(t)S_3\,,\ a_{21}(t)S_1 + a_{22}(t)S_2 + a_{23}(t)S_3\,,$$
$$a_{31}(t)S_1 + a_{32}(t)S_2 + a_{33}(t)S_3\Big),$$

where $A(t) = \{a_{mk}(t)\}_{\{1\le m,k\le 3\}}$.

Next, in the case that our particle has a positive energy, define

$$(\phi_1,\phi_2) = \left(e^{-\frac{ic^2mt}{\hbar}}\psi_1, e^{-\frac{ic^2mt}{\hbar}}\psi_2\right).$$

Then we rewrite the Dirac equations (10.13) and (10.14) as

$$\begin{aligned} i\hbar\frac{\partial\phi_1}{\partial t} &= -c\,\mathbf{S}\cdot\left(i\hbar\nabla_{\mathbf{x}}\phi_2 + \frac{\sigma}{c}\mathbf{A}(\mathbf{x},t)\phi_2\right) + \sigma\left(\Psi(\mathbf{x},t) - \frac{1}{c}\mathbf{v}(\mathbf{x},t)\cdot\mathbf{A}(\mathbf{x},t)\right)\phi_1 \\ &- V(\mathbf{x},t)\,\phi_1 - \frac{i\hbar}{2}\mathrm{div}_{\mathbf{x}}\,\{\phi_1\mathbf{v}(\mathbf{x},t)\} - \frac{i\hbar}{2}\nabla_{\mathbf{x}}\phi_1\cdot\mathbf{v}(\mathbf{x},t) + \frac{\hbar}{4}\mathbf{S}\cdot(\mathrm{curl}_{\mathbf{x}}\mathbf{v}(\mathbf{x},t)\,\phi_1) \\ &- \frac{(g-1)\sigma\hbar}{2mc}\mathbf{S}\cdot(\mathrm{curl}_{\mathbf{x}}\mathbf{A}(\mathbf{x},t)\,\phi_1)\,, \end{aligned} \tag{10.24}$$

and

$$\begin{aligned} i\hbar\frac{\partial\phi_2}{\partial t} &= \\ &- 2mc^2\phi_2 - c\,\mathbf{S}\cdot\left(i\hbar\nabla_{\mathbf{x}}\phi_1 + \frac{\sigma}{c}\mathbf{A}(\mathbf{x},t)\phi_1\right) + \sigma\left(\Psi(\mathbf{x},t) - \frac{1}{c}\mathbf{v}(\mathbf{x},t)\cdot\mathbf{A}(\mathbf{x},t)\right)\phi_2 \\ &- V(\mathbf{x},t)\,\phi_2 - \frac{i\hbar}{2}\mathrm{div}_{\mathbf{x}}\,\{\phi_2\mathbf{v}(\mathbf{x},t)\} - \frac{i\hbar}{2}\nabla_{\mathbf{x}}\phi_2\cdot\mathbf{v}(\mathbf{x},t) + \frac{\hbar}{4}\mathbf{S}\cdot(\mathrm{curl}_{\mathbf{x}}\mathbf{v}(\mathbf{x},t)\,\phi_2) \\ &- \frac{(g-1)\sigma\hbar}{2mc}\mathbf{S}\cdot(\mathrm{curl}_{\mathbf{x}}\mathbf{A}(\mathbf{x},t)\,\phi_2)\,. \end{aligned} \tag{10.25}$$

Thus from (10.25), in the non-relativistic limit, we have

$$2mc^2\phi_2 \approx -c\,\mathbf{S}\cdot\left(i\hbar\nabla_{\mathbf{x}}\phi_1 + \frac{\sigma}{c}\mathbf{A}(\mathbf{x},t)\phi_1\right). \tag{10.26}$$

I.e.

$$\phi_2 \approx -\frac{1}{2cm}\mathbf{S}\cdot\left(i\hbar\nabla_{\mathbf{x}}\phi_1 + \frac{\sigma}{c}\mathbf{A}(\mathbf{x},t)\phi_1\right). \tag{10.27}$$

Thus inserting (10.27) into (10.24) gives

$$
\begin{aligned}
i\hbar\frac{\partial\phi_1}{\partial t} \approx \frac{1}{2m}\mathbf{S}\cdot\left(i\hbar\nabla_{\mathbf{x}}\left(\mathbf{S}\cdot\left(i\hbar\nabla_{\mathbf{x}}\phi_1+\frac{\sigma}{c}\mathbf{A}(\mathbf{x},t)\phi_1\right)\right)\right)\\
+\frac{1}{2m}\mathbf{S}\cdot\left(\frac{\sigma}{c}\mathbf{A}(\mathbf{x},t)\left(\mathbf{S}\cdot\left(i\hbar\nabla_{\mathbf{x}}\phi_1+\frac{\sigma}{c}\mathbf{A}(\mathbf{x},t)\phi_1\right)\right)\right)\\
+\sigma\left(\Psi(\mathbf{x},t)-\frac{1}{c}\mathbf{v}(\mathbf{x},t)\cdot\mathbf{A}(\mathbf{x},t)\right)\phi_1 - V(\mathbf{x},t)\,\phi_1-\frac{i\hbar}{2}\mathrm{div}_{\mathbf{x}}\left\{\phi_1\mathbf{v}(\mathbf{x},t)\right\}\\
-\frac{i\hbar}{2}\nabla_{\mathbf{x}}\phi_1\cdot\mathbf{v}(\mathbf{x},t)+\frac{\hbar}{4}\mathbf{S}\cdot\left(\mathrm{curl}_{\mathbf{x}}\mathbf{v}(\mathbf{x},t)\,\phi_1\right)-\frac{(g-1)\sigma\hbar}{2mc}\mathbf{S}\cdot\left(\mathrm{curl}_{\mathbf{x}}\mathbf{A}(\mathbf{x},t)\,\phi_1\right),
\end{aligned}
\tag{10.28}
$$

that we rewrite as a non-relativistic Shrödinger–Pauli equation that we studied above:

$$
\begin{aligned}
i\hbar\frac{\partial\phi_1}{\partial t} \approx\\
-\frac{\hbar^2}{2m}\Delta_{\mathbf{x}}\phi_1+\frac{i\hbar\sigma}{2mc}\mathrm{div}_{\mathbf{x}}\left\{\phi_1\mathbf{A}(\mathbf{x},t)\right\}+\frac{i\hbar\sigma}{2mc}\nabla_{\mathbf{x}}\phi_1\cdot\mathbf{A}(\mathbf{x},t)+\frac{\sigma^2}{2mc^2}\left|\mathbf{A}(\mathbf{x},t)\right|^2\phi_1\\
-\frac{g\sigma\hbar}{2mc}\mathbf{S}\cdot\left(\mathrm{curl}_{\mathbf{x}}\mathbf{A}(\mathbf{x},t)\,\phi_1\right)+\sigma\left(\Psi(\mathbf{x},t)-\frac{1}{c}\mathbf{v}(\mathbf{x},t)\cdot\mathbf{A}(\mathbf{x},t)\right)\phi_1-V(\mathbf{x},t)\,\phi_1\\
-\frac{i\hbar}{2}\mathrm{div}_{\mathbf{x}}\left\{\phi_1\mathbf{v}(\mathbf{x},t)\right\}-\frac{i\hbar}{2}\nabla_{\mathbf{x}}\phi_1\cdot\mathbf{v}(\mathbf{x},t)+\frac{\hbar}{4}\mathbf{S}\cdot\left(\mathrm{curl}_{\mathbf{x}}\mathbf{v}(\mathbf{x},t)\,\phi_1\right).
\end{aligned}
\tag{10.29}
$$

Next, consider a Lagrangian density L associated the motion of a spin-half quantum relativistic-like micro-particle with inertial mass m and charge σ in the outer gravitational and electromagnetical field with characteristics $\mathbf{v}(\mathbf{x},t)$, $\mathbf{A}(\mathbf{x},t)$ and $\Psi(\mathbf{x},t)$ and additional conservative field with potential $V(\mathbf{y},t)$:

$$
\begin{aligned}
L(\psi,\mathbf{x},t):=\frac{i\hbar}{2}\left(\left(\frac{\partial\psi_1}{\partial t}+\mathbf{v}\cdot\nabla_{\mathbf{x}}\psi_1\right)\cdot\bar{\psi}_1-\psi_1\cdot\left(\frac{\partial\bar{\psi}_1}{\partial t}+\mathbf{v}\cdot\nabla_{\mathbf{x}}\bar{\psi}_1\right)\right)\\
+\frac{c}{2}\left(\left(\mathbf{S}\cdot\left(i\hbar\nabla_{\mathbf{x}}\psi_2+\frac{\sigma}{c}\mathbf{A}(\mathbf{x},t)\psi_2\right)\right)\cdot\bar{\psi}_1-\psi_1\cdot\left(\mathbf{S}\cdot\left(i\hbar\nabla_{\mathbf{x}}\bar{\psi}_2-\frac{\sigma}{c}\mathbf{A}(\mathbf{x},t)\bar{\psi}_2\right)\right)\right)\\
+\frac{i\hbar}{2}\left(\left(\frac{\partial\psi_2}{\partial t}+\mathbf{v}\cdot\nabla_{\mathbf{x}}\psi_2\right)\cdot\bar{\psi}_2-\psi_2\cdot\left(\frac{\partial\bar{\psi}_2}{\partial t}+\mathbf{v}\cdot\nabla_{\mathbf{x}}\bar{\psi}_2\right)\right)\\
+\frac{c}{2}\left(\left(\mathbf{S}\cdot\left(i\hbar\nabla_{\mathbf{x}}\psi_1+\frac{\sigma}{c}\mathbf{A}(\mathbf{x},t)\psi_1\right)\right)\cdot\bar{\psi}_2-\psi_2\cdot\left(\mathbf{S}\cdot\left(i\hbar\nabla_{\mathbf{x}}\bar{\psi}_1-\frac{\sigma}{c}\mathbf{A}(\mathbf{x},t)\bar{\psi}_1\right)\right)\right)\\
-mc^2\left(\psi_1\cdot\bar{\psi}_1-\psi_2\cdot\bar{\psi}_2\right)-\sigma\left(\Psi-\frac{1}{c}\mathbf{v}\cdot\mathbf{A}\right)\left(\psi_1\cdot\bar{\psi}_1+\psi_2\cdot\bar{\psi}_2\right)\\
+V(\mathbf{x},t)\left(\psi_1\cdot\bar{\psi}_1+\psi_2\cdot\bar{\psi}_2\right)-\frac{\hbar}{4}\left(\mathbf{S}\cdot\left(\mathrm{curl}_{\mathbf{x}}\mathbf{v}\right)\psi_1\right)\cdot\bar{\psi}_1-\frac{\hbar}{4}\left(\mathbf{S}\cdot\left(\mathrm{curl}_{\mathbf{x}}\mathbf{v}\right)\psi_2\right)\cdot\bar{\psi}_2\\
+\frac{(g-1)\sigma\hbar}{2mc}\left(\mathbf{S}\cdot\left(\mathrm{curl}_{\mathbf{x}}\mathbf{A}\right)\psi_1\right)\cdot\bar{\psi}_1+\frac{(g-1)\sigma\hbar}{2mc}\left(\mathbf{S}\cdot\left(\mathrm{curl}_{\mathbf{x}}\mathbf{A}\right)\psi_2\right)\cdot\bar{\psi}_2,
\end{aligned}
\tag{10.30}
$$

where $\psi = (\psi_1, \psi_2) \in \mathbb{C}^2 \times \mathbb{C}^2$ is a four-component wave function. Then similarly to the proof of Theorem 10.1 we can prove that L is invariant under the change of inertial or non-inertial Cartesian coordinate system, given by (10.17), provided that we take into account (10.22). We investigate stationary points of the functional

$$J = \int_0^T \int_{\mathbb{R}^3} L(\psi, \mathbf{x}, t)\, d\mathbf{x}dt. \tag{10.31}$$

Then, by (10.30) we have

$$\begin{aligned}0 = \frac{\delta L}{\delta(\bar{\psi}_1)} &= i\hbar\left(\frac{\partial \psi_1}{\partial t} + \frac{1}{2}\mathbf{v}\cdot\nabla_{\mathbf{x}}\psi_1 + \frac{1}{2}\mathrm{div}_{\mathbf{x}}\{\psi_1\mathbf{v}\}\right)\\ &+ c\,\mathbf{S}\cdot\left(i\hbar\nabla_{\mathbf{x}}\psi_2 + \frac{\sigma}{c}\mathbf{A}(\mathbf{x},t)\psi_2\right) - mc^2\psi_1 - \sigma\left(\Psi - \frac{1}{c}\mathbf{v}\cdot\mathbf{A}\right)\psi_1\\ &- \frac{\hbar}{4}\mathbf{S}\cdot(\mathrm{curl}_{\mathbf{x}}\mathbf{v})\,\psi_1 + V(\mathbf{x},t)\psi_1 + \frac{(g-1)\sigma\hbar}{2mc}, \mathbf{S}\cdot(\mathrm{curl}_{\mathbf{x}}\mathbf{A})\,\psi_1 + V(\mathbf{x},t)\psi_1,\end{aligned} \tag{10.32}$$

$$\begin{aligned}0 = \frac{\delta L}{\delta(\bar{\psi}_2)} &= i\hbar\left(\frac{\partial \psi_2}{\partial t} + \frac{1}{2}\mathbf{v}\cdot\nabla_{\mathbf{x}}\psi_2 + \frac{1}{2}\mathrm{div}_{\mathbf{x}}\{\psi_2\mathbf{v}\}\right)\\ &+ c\,\mathbf{S}\cdot\left(i\hbar\nabla_{\mathbf{x}}\psi_1 + \frac{\sigma}{c}\mathbf{A}(\mathbf{x},t)\psi_1\right) + mc^2\psi_2 - \sigma\left(\Psi - \frac{1}{c}\mathbf{v}\cdot\mathbf{A}\right)\psi_2\\ &- \frac{\hbar}{4}\mathbf{S}\cdot(\mathrm{curl}_{\mathbf{x}}\mathbf{v})\,\psi_1 + V(\mathbf{x},t)\psi_2 + \frac{(g-1)\sigma\hbar}{2mc}, \mathbf{S}\cdot(\mathrm{curl}_{\mathbf{x}}\mathbf{A})\,\psi_2,\end{aligned} \tag{10.33}$$

$$\begin{aligned}0 = \frac{\delta L}{\delta(\psi_1)} &= (\bar{i})\hbar\left(\frac{\partial \bar{\psi}_1}{\partial t} + \frac{1}{2}\mathbf{v}\cdot\nabla_{\mathbf{x}}\bar{\psi}_1 + \frac{1}{2}\mathrm{div}_{\mathbf{x}}\{\bar{\psi}_1\mathbf{v}\}\right)\\ &+ c\,\mathbf{S}\cdot\left((\bar{i})\hbar\nabla_{\mathbf{x}}\bar{\psi}_2 + \frac{\sigma}{c}\mathbf{A}(\mathbf{x},t)\bar{\psi}_2\right) - mc^2\bar{\psi}_1 - \sigma\left(\Psi - \frac{1}{c}\mathbf{v}\cdot\mathbf{A}\right)\bar{\psi}_1\\ &- \frac{\hbar}{4}\mathbf{S}^T\cdot(\mathrm{curl}_{\mathbf{x}}\mathbf{v})\,\bar{\psi}_1 + V(\mathbf{x},t)\bar{\psi}_1 + \frac{(g-1)\sigma\hbar}{2mc}, \mathbf{S}^T\cdot(\mathrm{curl}_{\mathbf{x}}\mathbf{A})\,\bar{\psi}_1\end{aligned} \tag{10.34}$$

and

$$\begin{aligned}0 = \frac{\delta L}{\delta(\psi_2)} &= (\bar{i})\hbar\left(\frac{\partial \bar{\psi}_2}{\partial t} + \frac{1}{2}\mathbf{v}\cdot\nabla_{\mathbf{x}}\bar{\psi}_2 + \frac{1}{2}\mathrm{div}_{\mathbf{x}}\{\bar{\psi}_2\mathbf{v}\}\right)\\ &+ c\,\mathbf{S}\cdot\left((\bar{i})\hbar\nabla_{\mathbf{x}}\bar{\psi}_1 + \frac{\sigma}{c}\mathbf{A}(\mathbf{x},t)\bar{\psi}_1\right) + mc^2\bar{\psi}_2 - \sigma\left(\Psi - \frac{1}{c}\mathbf{v}\cdot\mathbf{A}\right)\bar{\psi}_2\\ &- \frac{\hbar}{4}\mathbf{S}^T\cdot(\mathrm{curl}_{\mathbf{x}}\mathbf{v})\,\bar{\psi}_2 + V(\mathbf{x},t)\bar{\psi}_2 + \frac{(g-1)\sigma\hbar}{2mc}, \mathbf{S}^T\cdot(\mathrm{curl}_{\mathbf{x}}\mathbf{A})\,\bar{\psi}_2.\end{aligned} \tag{10.35}$$

Note that (10.34) and (10.35) are just the complex conjugates of (10.32) and (10.33). So we get that the Euler–Lagrange equation for (10.30) coincide with Dirac equations in the form of (10.15) and (10.16).

10.3 Classical Relativistic-Like Lagrangian and Hamiltonian of a System of n Particles

As in (9.136) and (10.1) consider the relativistic-like Lagrangian of the motion of n relativistic-like particles with inertial masses $m_1, \ldots, m_n$ and the charges $\sigma_1, \ldots, \sigma_n$ in the outer gravitational and electromagnetical field with characteristics $\mathbf{v}(\mathbf{x}, t)$, $\mathbf{A}(\mathbf{x}, t)$ and $\Psi(\mathbf{x}, t)$ and additional conservative field with potential $V(\mathbf{y}_1, \ldots, \mathbf{y}_n, t)$:

$$J_{rl}(\mathbf{r}_1, \ldots, \mathbf{r}_n) = \int_0^T L_0\left(\frac{d\mathbf{r}_1}{dt}, \ldots, \frac{d\mathbf{r}_n}{dt}, \mathbf{r}_1, \ldots, \mathbf{r}_n, t\right) dt :=$$

$$\int_0^T \Bigg\{ -\sum_{j=1}^n m_j c^2 \sqrt{1 - \frac{1}{c^2}\left|\frac{d\mathbf{r}_j}{dt} - \mathbf{v}(\mathbf{r}_j, t)\right|^2} - \sum_{j=1}^n \sigma_j \left(\Psi(\mathbf{r}_j, t) - \frac{1}{c}\mathbf{A}(\mathbf{r}_j, t) \cdot \frac{d\mathbf{r}_j}{dt}\right)$$
$$+ V(\mathbf{r}_1, \ldots, \mathbf{r}_n, t) \Bigg\} dt. \quad (10.36)$$

Next define the generalized momentums of each particle by

$$\mathbf{P}_j := \nabla_{\mathbf{r}'_j} L_0\left(\mathbf{r}'_1, \ldots, \mathbf{r}'_n, \mathbf{r}_1, \ldots, \mathbf{r}_n, t\right)$$
$$= m_j \left(1 - \frac{1}{c^2}\left|\frac{d\mathbf{r}_j}{dt} - \mathbf{v}(\mathbf{r}_j, t)\right|^2\right)^{-\frac{1}{2}} \left(\frac{d\mathbf{r}_j}{dt} - \mathbf{v}(\mathbf{r}_j, t)\right) + \frac{\sigma_j}{c}\mathbf{A}(\mathbf{r}_j, t). \quad (10.37)$$

Then

$$\left(1 - \frac{1}{c^2}\left|\frac{d\mathbf{r}_j}{dt} - \mathbf{v}(\mathbf{r}_j, t)\right|^2\right)^{-\frac{1}{2}} \left(\frac{d\mathbf{r}_j}{dt} - \mathbf{v}(\mathbf{r}_j, t)\right) = \left(\frac{1}{m_j}\mathbf{P}_j - \frac{\sigma_j}{m_j c}\mathbf{A}(\mathbf{r}_j, t)\right). \quad (10.38)$$

So

$$\frac{d\mathbf{r}_j}{dt} = \left(1 + \frac{1}{c^2}\left|\frac{1}{m_j}\mathbf{P}_j - \frac{\sigma_j}{m_j c}\mathbf{A}(\mathbf{r}_j, t)\right|^2\right)^{-\frac{1}{2}} \left(\frac{1}{m_j}\mathbf{P}_j - \frac{\sigma_j}{m_j c}\mathbf{A}(\mathbf{r}_j, t)\right) + \mathbf{v}(\mathbf{r}_j, t). \quad (10.39)$$

Thus if we consider a Hamiltonian

$$H_0(\mathbf{P}_1, \ldots, \mathbf{P}_n, \mathbf{r}_1, \ldots, \mathbf{r}_n, t) := \sum_{j=1}^{n} \mathbf{P}_j \cdot \frac{d\mathbf{r}_j}{dt} - L_0\left(\frac{d\mathbf{r}_1}{dt}, \ldots, \frac{d\mathbf{r}_n}{dt}, \mathbf{r}_1, \ldots, \mathbf{r}_n, t\right), \tag{10.40}$$

then, as before in (10.6) and (6.15) by (10.40), (10.36), (10.38) and (10.39), we obtain that the relativistic-like Hamiltonian for macro-particles has the form:

$$H_0(\mathbf{P}_1, \ldots, \mathbf{P}_n, \mathbf{r}_1, \ldots, \mathbf{r}_n, t) = \sum_{j=1}^{n} m_j c^2 \left(1 + \frac{1}{c^2}\left|\frac{1}{m_j}\mathbf{P}_j - \frac{\sigma_j}{m_j c}\mathbf{A}(\mathbf{r}_j, t)\right|^2\right)^{\frac{1}{2}}$$
$$+ \sum_{j=1}^{n} \sigma_j \left(\Psi(\mathbf{r}_j, t) - \frac{1}{c}\mathbf{v}(\mathbf{r}_j, t) \cdot \mathbf{A}(\mathbf{r}_j, t)\right) - V(\mathbf{r}_1, \ldots, \mathbf{r}_n, t) + \sum_{j=1}^{n} \mathbf{v}(\mathbf{r}_j, t) \cdot \mathbf{P}_j. \tag{10.41}$$

10.4 Liouville's Equation for a System of n Relativistic-Like Classical Particles

Assume that the number of particles n in the system, ruled by the Hamiltonian (10.41), is large and we describe this system statistically. Then let $w(\mathbf{P}_1, \ldots, \mathbf{P}_n, \mathbf{r}_1, \ldots, \mathbf{r}_n, t) \to [0, +\infty)$ be the probability density of the system which satisfies the well-known classical Liouville equation of the form:

$$\frac{\partial w}{\partial t} + \sum_{j=1}^{n} \left(\mathrm{div}_{\mathbf{r}_j}\left\{w \nabla_{\mathbf{P}_j} H_0\right\} - \mathrm{div}_{\mathbf{P}_j}\left\{w \nabla_{\mathbf{r}_j} H_0\right\}\right) =$$
$$\frac{\partial w}{\partial t} + \sum_{j=1}^{n} \left(\nabla_{\mathbf{P}_j} H_0 \cdot \nabla_{\mathbf{r}_j} w - \nabla_{\mathbf{r}_j} H_0 \cdot \nabla_{\mathbf{P}_j} w\right) = 0. \tag{10.42}$$

Then since by (10.41) we have

$$H_0(\mathbf{P}_1, \ldots, \mathbf{P}_n, \mathbf{r}_1, \ldots, \mathbf{r}_n, t) =$$
$$\sum_{j=1}^{n} \mathbf{P}_j \cdot \mathbf{v}(\mathbf{r}_j, t) + \sum_{j=1}^{n} m_j c^2 \left(1 + \frac{1}{c^2}\left|\frac{1}{m_j}\mathbf{P}_j - \frac{\sigma_j}{m_j c}\mathbf{A}(\mathbf{r}_j, t)\right|^2\right)^{\frac{1}{2}}$$
$$+ \sum_{j=1}^{n} \sigma_j \left(\Psi(\mathbf{r}_j, t) - \frac{1}{c}\mathbf{A}(\mathbf{r}_j, t) \cdot \mathbf{v}(\mathbf{r}_j, t)\right) - V(\mathbf{r}_1, \ldots, \mathbf{r}_n, t), \tag{10.43}$$

and in particular,

$$\nabla_{\mathbf{P}_j} H_0 = \frac{1}{m_j}\left(1+\frac{1}{c^2}\left|\frac{1}{m_j}\mathbf{P}_j - \frac{\sigma_j}{m_j c}\mathbf{A}(\mathbf{r}_j,t)\right|^2\right)^{-\frac{1}{2}}\left(\mathbf{P}_j - \frac{\sigma_j}{c}\mathbf{A}(\mathbf{r}_j,t)\right) + \mathbf{v}(\mathbf{r}_j,t)$$

$$\text{and}\quad \nabla_{\mathbf{r}_j} H_0 = \left\{d_{\mathbf{x}}\mathbf{v}(\mathbf{r}_j,t)\right\}^T\cdot\mathbf{P}_j$$

$$-\frac{\sigma_j}{cm_j}\left(1+\frac{1}{c^2}\left|\frac{1}{m_j}\mathbf{P}_j - \frac{\sigma_j}{m_j c}\mathbf{A}(\mathbf{r}_j,t)\right|^2\right)^{-\frac{1}{2}}\left\{d_{\mathbf{x}}\mathbf{A}(\mathbf{r}_j,t)\right\}^T\cdot\left(\mathbf{P}_j - \frac{\sigma_j}{c}\mathbf{A}(\mathbf{r}_j,t)\right)$$

$$+\sigma_j\nabla_{\mathbf{x}}\left(\Psi(\mathbf{r}_j,t) - \frac{1}{c}\mathbf{A}(\mathbf{r}_j,t)\cdot\mathbf{v}(\mathbf{r}_j,t)\right) - \nabla_{\mathbf{y}_j} V\left(\mathbf{r}_1,\ldots,\mathbf{r}_n,t\right)\qquad\forall j=1,\ldots n, \tag{10.44}$$

inserting (10.44) into (10.42) gives

$$0 = \frac{\partial w}{\partial t} + \sum_{j=1}^{n}\nabla_{\mathbf{P}_j}H_0\cdot\nabla_{\mathbf{r}_j}w - \sum_{j=1}^{n}\nabla_{\mathbf{r}_j}H_0\cdot\nabla_{\mathbf{P}_j}w = \frac{\partial w}{\partial t} + \sum_{j=1}^{n}\mathbf{v}(\mathbf{r}_j,t)\cdot\nabla_{\mathbf{r}_j}w$$

$$+\sum_{j=1}^{n}\frac{1}{m_j}\left(1+\frac{1}{c^2}\left|\frac{1}{m_j}\mathbf{P}_j - \frac{\sigma_j}{m_j c}\mathbf{A}(\mathbf{r}_j,t)\right|^2\right)^{-\frac{1}{2}}\left(\mathbf{P}_j - \frac{\sigma_j}{c}\mathbf{A}(\mathbf{r}_j,t)\right)\cdot\nabla_{\mathbf{r}_j}w$$

$$-\sum_{j=1}^{n}\left(\left\{d_{\mathbf{x}}\mathbf{v}(\mathbf{r}_j,t)\right\}^T\cdot\mathbf{P}_j\right)\cdot\nabla_{\mathbf{P}_j}w$$

$$+\sum_{j=1}^{n}\left(\frac{\sigma_j}{cm_j}\left(1+\frac{1}{c^2}\left|\frac{1}{m_j}\mathbf{P}_j - \frac{\sigma_j}{m_j c}\mathbf{A}(\mathbf{r}_j,t)\right|^2\right)^{-\frac{1}{2}}\left\{d_{\mathbf{x}}\mathbf{A}(\mathbf{r}_j,t)\right\}^T\cdot\left(\mathbf{P}_j\right.\right.$$

$$\left.\left.-\frac{\sigma_j}{c}\mathbf{A}(\mathbf{r}_j,t)\right)\right)\cdot\nabla_{\mathbf{P}_j}w$$

$$-\sum_{j=1}^{n}\left(\sigma_j\nabla_{\mathbf{x}}\left(\Psi(\mathbf{r}_j,t) - \frac{1}{c}\mathbf{A}(\mathbf{r}_j,t)\cdot\mathbf{v}(\mathbf{r}_j,t)\right) - \nabla_{\mathbf{y}_j}V\left(\mathbf{r}_1,\ldots,\mathbf{r}_n,t\right)\right)\cdot\nabla_{\mathbf{P}_j}w$$

$$= \frac{\partial w}{\partial t} + \sum_{j=1}^{n}\mathbf{v}(\mathbf{r}_j,t)\cdot\nabla_{\mathbf{r}_j}w$$

$$+\sum_{j=1}^{n}\frac{1}{m_j}\left(1+\frac{1}{c^2}\left|\frac{1}{m_j}\mathbf{P}_j - \frac{\sigma_j}{m_j c}\mathbf{A}(\mathbf{r}_j,t)\right|^2\right)^{-\frac{1}{2}}\left(\mathbf{P}_j - \frac{\sigma_j}{c}\mathbf{A}(\mathbf{r}_j,t)\right)\cdot\nabla_{\mathbf{r}_j}w$$

$$+\sum_{j=1}^{n}\frac{1}{2}\left(\left(d_{\mathbf{x}}\mathbf{v}(\mathbf{r}_j,t) - \left\{d_{\mathbf{x}}\mathbf{v}(\mathbf{r}_j,t)\right\}^T\right)\cdot\mathbf{P}_j\right)\cdot\nabla_{\mathbf{P}_j}w$$

$$
\begin{aligned}
&-\sum_{j=1}^{n}\frac{1}{2}\left(\left(d_{\mathbf{x}}\mathbf{v}(\mathbf{r}_j,t)+\left\{d_{\mathbf{x}}\mathbf{v}(\mathbf{r}_j,t)\right\}^T\right)\cdot\mathbf{P}_j\right)\cdot\nabla_{\mathbf{P}_j}w\\
&+\sum_{j=1}^{n}\left(\frac{\sigma_j}{cm_j}\left(1+\frac{1}{c^2}\left|\frac{1}{m_j}\mathbf{P}_j-\frac{\sigma_j}{m_jc}\mathbf{A}(\mathbf{r}_j,t)\right|^2\right)^{-\frac{1}{2}}\left\{d_{\mathbf{x}}\mathbf{A}(\mathbf{r}_j,t)\right\}^T\cdot\left(\mathbf{P}_j\right.\right.\\
&\qquad\left.\left.-\frac{\sigma_j}{c}\mathbf{A}(\mathbf{r}_j,t)\right)\right)\cdot\nabla_{\mathbf{P}_j}w\\
&-\sum_{j=1}^{n}\left(\sigma_j\nabla_{\mathbf{x}}\left(\Psi(\mathbf{r}_j,t)-\frac{1}{c}\mathbf{A}(\mathbf{r}_j,t)\cdot\mathbf{v}(\mathbf{r}_j,t)\right)-\nabla_{\mathbf{y}_j}V\left(\mathbf{r}_1,\ldots,\mathbf{r}_n,t\right)\right)\cdot\nabla_{\mathbf{P}_j}w.
\end{aligned}
\tag{10.45}
$$

Thus, by (10.45), using (2.15), we rewrite Liouville's equation as

$$
\begin{aligned}
&\frac{\partial w}{\partial t}+\sum_{j=1}^{n}\mathbf{v}(\mathbf{r}_j,t)\cdot\nabla_{\mathbf{r}_j}w+\sum_{j=1}^{n}\frac{1}{2}\left(\left(\mathrm{curl}_{\mathbf{x}}\mathbf{v}(\mathbf{r}_j,t)\right)\times\mathbf{P}_j\right)\cdot\nabla_{\mathbf{P}_j}w\\
&-\sum_{j=1}^{n}\frac{1}{2}\left(\left(d_{\mathbf{x}}\mathbf{v}(\mathbf{r}_j,t)+\left\{d_{\mathbf{x}}\mathbf{v}(\mathbf{r}_j,t)\right\}^T\right)\cdot\mathbf{P}_j\right)\cdot\nabla_{\mathbf{P}_j}w\\
&+\sum_{j=1}^{n}\frac{1}{m_j}\left(1+\frac{1}{c^2}\left|\frac{1}{m_j}\mathbf{P}_j-\frac{\sigma_j}{m_jc}\mathbf{A}(\mathbf{r}_j,t)\right|^2\right)^{-\frac{1}{2}}\left(\mathbf{P}_j-\frac{\sigma_j}{c}\mathbf{A}(\mathbf{r}_j,t)\right)\cdot\nabla_{\mathbf{r}_j}w\\
&+\sum_{j=1}^{n}\left(\frac{\sigma_j}{cm_j}\left(1+\frac{1}{c^2}\left|\frac{1}{m_j}\mathbf{P}_j-\frac{\sigma_j}{m_jc}\mathbf{A}(\mathbf{r}_j,t)\right|^2\right)^{-\frac{1}{2}}\left\{d_{\mathbf{x}}\mathbf{A}(\mathbf{r}_j,t)\right\}^T\cdot\left(\mathbf{P}_j\right.\right.\\
&\qquad\left.\left.-\frac{\sigma_j}{c}\mathbf{A}(\mathbf{r}_j,t)\right)\right)\cdot\nabla_{\mathbf{P}_j}w\\
&-\sum_{j=1}^{n}\left(\sigma_j\nabla_{\mathbf{x}}\left(\Psi(\mathbf{r}_j,t)-\frac{1}{c}\mathbf{A}(\mathbf{r}_j,t)\cdot\mathbf{v}(\mathbf{r}_j,t)\right)-\nabla_{\mathbf{y}_j}V\left(\mathbf{r}_1,\ldots,\mathbf{r}_n,t\right)\right)\cdot\nabla_{\mathbf{P}_j}w=0.
\end{aligned}
\tag{10.46}
$$

Next if the change of some non-inertial Cartesian coordinate system $(*)$ to another Cartesian coordinate system $(**)$ is of the form (4.2):

$$
\begin{cases}
\mathbf{x}'=A(t)\cdot\mathbf{x}+\mathbf{z}(t),\\
t'=t,
\end{cases}
\tag{10.47}
$$

where $A(t)\in SO(3)$ is a rotation, then consistently with (10.47) and (10.37) we have the following change of variables $(\mathbf{P}_1,\ldots,\mathbf{P}_n,\mathbf{x}_1,\ldots,\mathbf{x}_n,t)\to(\mathbf{P}'_1,\ldots,\mathbf{P}'_n,\mathbf{x}'_1,\ldots,\mathbf{x}'_n,t')$:

$$\begin{cases} t' = t, \\ \mathbf{x}'_k = A(t) \cdot \mathbf{x}_k + \mathbf{z}(t) \qquad \forall k = 1, \dots, n, \\ \mathbf{P}'_k = A(t) \cdot \mathbf{P}_k \qquad \forall k = 1, \dots, n. \end{cases} \tag{10.48}$$

Thus, since consistently with (10.47) we have

$$\begin{cases} V'\left(\mathbf{x}'_1, \dots, \mathbf{x}'_n, t'\right) = V\left(\mathbf{x}_1, \dots, \mathbf{x}_n, t\right), \\ \sigma'_j = \sigma_j, \\ m'_j = m_j, \\ \mathbf{v}'(\mathbf{x}', t) = A(t) \cdot \mathbf{v}(\mathbf{x}, t) + \frac{dA}{dt}(t) \cdot \mathbf{x} + \frac{d\mathbf{z}}{dt}(t), \\ \mathbf{A}'(\mathbf{x}', t) = A(t) \cdot \mathbf{A}(\mathbf{x}, t), \\ \Psi'(\mathbf{x}', t) - \mathbf{v}'(\mathbf{x}', t) \cdot \mathbf{A}'(\mathbf{x}', t) = \Psi(\mathbf{x}, t) - \mathbf{v}(\mathbf{x}, t) \cdot \mathbf{A}(\mathbf{x}, t), \end{cases} \tag{10.49}$$

by (10.48) and (10.49) we deduce that the Liouville equation (10.46) is invariant under the the change of non-inertial Cartesian coordinate system of the form (10.47).

10.4.1 Thermodynamical Equilibrium in the Relativistic-Like Case

Next let $w\left(\mathbf{P}_1, \dots, \mathbf{P}_n, \mathbf{r}_1, \dots, \mathbf{r}_n, t\right) \to [0, +\infty)$ be the probability density of the system, ruled by the Hamiltonian H_0 from (10.43), in the case of thermodynamical equilibrium. Then in the case that $\frac{\partial H_0}{\partial t} = 0$ and the given equilibrium system rests macroscopically, i.e. it has the macroscopical velocity field zero: $\mathbf{u}(\mathbf{x}, t) = 0$ it is well known that

$$w\left(\mathbf{P}_1, \dots, \mathbf{P}_n, \mathbf{r}_1, \dots, \mathbf{r}_n, t\right) := \frac{1}{K(f, H_0)} f\left(H_0\left(\mathbf{P}_1, \dots, \mathbf{P}_n, \mathbf{r}_1, \dots, \mathbf{r}_n, t\right)\right), \tag{10.50}$$

with

$$K(f, H_0) := \int f\left(H_0\left(\mathbf{P}_1, \dots, \mathbf{P}_n, \mathbf{r}_1, \dots, \mathbf{r}_n, t\right)\right) d\mathbf{P}_1 \dots d\mathbf{P}_n d\mathbf{r}_1 \dots d\mathbf{r}_n, \tag{10.51}$$

and

- in the case of a system in thermostat, having the Kelvin temperature T, we have the canonical ensemble: $f(s) = e^{-\frac{s}{kT}}\ \forall s$, where k is the Boltzmann constant,
- in the case of a thermally isolated system with the average internal energy E, we have the micro-canonical ensemble: $f(s) = \delta(s - E)\ \forall s$.

We would like to find alternative forms of the above laws of thermodynamical equilibrium, which are invariant under the change of inertial or non-inertial Cartesian coordinate system. Then it is clear that if the given system rests in the old coordinate system, then it obviously has non-trivial macroscopical velocity field $\mathbf{u}(\mathbf{x}, t)$ in the new one. Moreover, $\mathbf{u}(\mathbf{x}, t)$ can depend on $\mathbf{x}$ and t, as it indeed happens in the case of a rotation of the new coordinate system with respect to the old one. On the other hand, the concept of thermodynamical equilibrium is clearly independent of the coordinate system.

In order to find the forms of the laws of thermodynamical equilibrium, which are indeed invariant under the change of inertial or non-inertial Cartesian coordinate system, we follow the steps as below. By (10.43) the Hamiltonian H_0 of our system has the following form:

$$H_0\left(\mathbf{P}_1, \ldots, \mathbf{P}_n, \mathbf{r}_1, \ldots, \mathbf{r}_n, t\right) = \sum_{j=1}^{n} \mathbf{P}_j \cdot \mathbf{v}(\mathbf{r}_j, t) + \sum_{j=1}^{n} m_j c^2 \left(1 + \frac{1}{c^2}\left|\frac{1}{m_j}\mathbf{P}_j - \frac{\sigma_j}{m_j c}\mathbf{A}(\mathbf{r}_j, t)\right|^2\right)^{\frac{1}{2}} + \sum_{j=1}^{n} \sigma_j \left(\Psi(\mathbf{r}_j, t) - \frac{1}{c}\mathbf{A}(\mathbf{r}_j, t) \cdot \mathbf{v}(\mathbf{r}_j, t)\right) - V\left(\mathbf{r}_1, \ldots, \mathbf{r}_n, t\right), \tag{10.52}$$

Next define an invariant quantity $H_{\mathbf{u}}$ as

$$H_{\mathbf{u}}\left(\mathbf{P}_1, \ldots, \mathbf{P}_n, \mathbf{r}_1, \ldots, \mathbf{r}_n, t\right) := H_0\left(\mathbf{P}_1, \ldots, \mathbf{P}_n, \mathbf{r}_1, \ldots, \mathbf{r}_n, t\right) - \sum_{j=1}^{n} \mathbf{P}_j \cdot \mathbf{u}(\mathbf{r}_j, t)$$
$$= \sum_{j=1}^{n} \mathbf{P}_j \cdot \left(\mathbf{v}(\mathbf{r}_j, t) - \mathbf{u}(\mathbf{r}_j, t)\right) + \sum_{j=1}^{n} m_j c^2 \left(1 + \frac{1}{c^2}\left|\frac{1}{m_j}\mathbf{P}_j - \frac{\sigma_j}{m_j c}\mathbf{A}(\mathbf{r}_j, t)\right|^2\right)^{\frac{1}{2}} + \sum_{j=1}^{n} \sigma_j \left(\Psi(\mathbf{r}_j, t) - \frac{1}{c}\mathbf{A}(\mathbf{r}_j, t) \cdot \mathbf{v}(\mathbf{r}_j, t)\right) - V\left(\mathbf{r}_1, \ldots, \mathbf{r}_n, t\right), \tag{10.53}$$

where $\mathbf{u}(\mathbf{x}, t)$ is the macroscopical velocity field of the given system of particles. Then, as before, it can be easily deduced that under the change of some non-inertial Cartesian coordinate system $(*)$ to another Cartesian coordinate system $(**)$ of the form (10.47) the quantity $H_{\mathbf{u}}$ transforms as

$$H'_{\mathbf{u}'} = H_{\mathbf{u}}, \tag{10.54}$$

provided that $\mathbf{u}$ is the speed-like vector field, i.e. under the above change we have

$$\mathbf{u}'(\mathbf{x}', t) = A(t) \cdot \mathbf{u}(\mathbf{x}, t) + \frac{dA}{dt}(t) \cdot \mathbf{x} + \frac{d\mathbf{z}}{dt}(t), \tag{10.55}$$

and we have (10.48) and (10.49). Next consider the canonical and micro-canonical ensembles as

$$w\left(\mathbf{P}_1,\ldots,\mathbf{P}_n,\mathbf{r}_1,\ldots,\mathbf{r}_n,t\right):=\frac{1}{K(f,H_{\mathbf{u}})}f\left(H_{\mathbf{u}}\left(\mathbf{P}_1,\ldots,\mathbf{P}_n,\mathbf{r}_1,\ldots,\mathbf{r}_n,t\right)\right), \tag{10.56}$$

where as before,

$$K(f,H_{\mathbf{u}}):=\int f\left(H_{\mathbf{u}}\left(\mathbf{P}_1,\ldots,\mathbf{P}_n,\mathbf{r}_1,\ldots,\mathbf{r}_n,t\right)\right)d\mathbf{P}_1\ldots d\mathbf{P}_n d\mathbf{r}_1\ldots d\mathbf{r}_n, \tag{10.57}$$

and

- in the case of a system in thermostat, having the Kelvin temperature T, we have $f(s)=e^{-\frac{s}{kT}}\ \forall s$,
- in the case of a thermally isolated system with the average internal energy E we have $f(s)=\delta(s-E)\ \forall s$.

Note here that by (10.53) for the finite integrability in (10.57) the vector field $\mathbf{u}(\mathbf{x},t)$ must satisfy

$$|\mathbf{u}(\mathbf{x},t)-\mathbf{v}(\mathbf{x},t)|<c \qquad \forall(\mathbf{x},t). \tag{10.58}$$

Next, by (10.54) we deduce that under the change of some non-inertial Cartesian coordinate system $(*)$ to another Cartesian coordinate system $(**)$ of the form (10.47) the quantity w transforms as

$$w'=w. \tag{10.59}$$

Moreover, in the case $\mathbf{u}=0$ (10.56) coincides with (10.50). However, we still need to derive the restrictions on the field $\mathbf{u}$ and the Hamiltonian H_0, providing that our system can indeed be found in the state of thermodynamical equilibrium. We remind that in the case $\mathbf{u}=0$ the appropriate restriction is $\frac{\partial H_0}{\partial t}=0$. In order to get these restrictions in the general case, we need to insert w in (10.56) into Liouville's equation in (10.45) having the form:

$$\frac{\partial w}{\partial t}+\sum_{j=1}^{n}\left(\nabla_{\mathbf{P}_j}H_0\cdot\nabla_{\mathbf{r}_j}w-\nabla_{\mathbf{r}_j}H_0\cdot\nabla_{\mathbf{P}_j}w\right)=0. \tag{10.60}$$

By (10.53) we have

$$\nabla_{\mathbf{P}_j}H_{\mathbf{u}}=\nabla_{\mathbf{P}_j}H_0-\mathbf{u}(\mathbf{r}_j,t)\quad\text{and}\quad\nabla_{\mathbf{r}_j}H_{\mathbf{u}}=\nabla_{\mathbf{r}_j}H_0-\left\{d_{\mathbf{x}}\mathbf{u}(\mathbf{r}_j,t)\right\}^T\cdot\mathbf{P}_j \qquad \forall j=1,\ldots n. \tag{10.61}$$

Next inserting (10.56) into (10.60) we deduce

$$\frac{\partial H_{\mathbf{u}}}{\partial t}+\sum_{j=1}^{n}\left(\nabla_{\mathbf{P}_j}H_0\cdot\nabla_{\mathbf{r}_j}H_{\mathbf{u}}-\nabla_{\mathbf{r}_j}H_0\cdot\nabla_{\mathbf{P}_j}H_{\mathbf{u}}\right)=0. \tag{10.62}$$

Thus inserting (10.61) into (10.62) we obtain

$$0 = \frac{\partial H_{\mathbf{u}}}{\partial t} + \sum_{j=1}^{n}\left(\left(\nabla_{\mathbf{P}_j} H_{\mathbf{u}} + \mathbf{u}(\mathbf{r}_j, t)\right)\cdot \nabla_{\mathbf{r}_j} H_{\mathbf{u}} - \left(\nabla_{\mathbf{r}_j} H_{\mathbf{u}} + \left\{d_{\mathbf{x}}\mathbf{u}(\mathbf{r}_j, t)\right\}^T \cdot \mathbf{P}_j\right)\cdot \nabla_{\mathbf{P}_j} H_{\mathbf{u}}\right)$$

$$= \frac{\partial H_{\mathbf{u}}}{\partial t} + \sum_{j=1}^{n} \mathbf{u}(\mathbf{r}_j, t)\cdot \nabla_{\mathbf{r}_j} H_{\mathbf{u}} - \sum_{j=1}^{n}\left(\left\{d_{\mathbf{x}}\mathbf{u}(\mathbf{r}_j, t)\right\}^T \cdot \mathbf{P}_j\right)\cdot \nabla_{\mathbf{P}_j} H_{\mathbf{u}} =$$

$$\frac{\partial H_{\mathbf{u}}}{\partial t} + \sum_{j=1}^{n} \mathbf{u}(\mathbf{r}_j, t)\cdot \nabla_{\mathbf{r}_j} H_{\mathbf{u}} + \sum_{j=1}^{n}\frac{1}{2}\left(\left(d_{\mathbf{x}}\mathbf{u}(\mathbf{r}_j, t) - \left\{d_{\mathbf{x}}\mathbf{u}(\mathbf{r}_j, t)\right\}^T\right)\cdot \mathbf{P}_j\right)\cdot \nabla_{\mathbf{P}_j} H_{\mathbf{u}}$$
$$- \sum_{j=1}^{n}\frac{1}{2}\left(\left(d_{\mathbf{x}}\mathbf{u}(\mathbf{r}_j, t) + \left\{d_{\mathbf{x}}\mathbf{u}(\mathbf{r}_j, t)\right\}^T\right)\cdot \mathbf{P}_j\right)\cdot \nabla_{\mathbf{P}_j} H_{\mathbf{u}}. \quad (10.63)$$

Thus, by (2.15), we rewrite (10.63) as

$$\frac{\partial H_{\mathbf{u}}}{\partial t} + \sum_{j=1}^{n} \mathbf{u}(\mathbf{r}_j, t)\cdot \nabla_{\mathbf{r}_j} H_{\mathbf{u}} + \sum_{j=1}^{n}\frac{1}{2}\left(\left(\mathrm{curl}_{\mathbf{x}}\mathbf{u}(\mathbf{r}_j, t)\right)\times \mathbf{P}_j\right)\cdot \nabla_{\mathbf{P}_j} H_{\mathbf{u}}$$
$$- \sum_{j=1}^{n}\frac{1}{2}\left(\left(d_{\mathbf{x}}\mathbf{u}(\mathbf{r}_j, t) + \left\{d_{\mathbf{x}}\mathbf{u}(\mathbf{r}_j, t)\right\}^T\right)\cdot \mathbf{P}_j\right)\cdot \nabla_{\mathbf{P}_j} H_{\mathbf{u}} = 0. \quad (10.64)$$

Equality (10.64) is the required restriction on the field $\mathbf{u}$ and the Hamiltonian H_0, providing that our system can indeed be found in state of thermodynamical equilibrium. In particular, if $\frac{\partial H_0}{\partial t} = 0$ and $\mathbf{u} = 0$ then (10.64) indeed holds. Moreover, as before in the case of Liouville's equation, we deduce that the equation (10.64) is invariant under the change of non-inertial Cartesian coordinate system of the form (10.47), provided that we have (10.55), (10.48) and (10.49). Finally, we still need to prove that if w is given by (10.56) then the vector field $\mathbf{u}(\mathbf{x}, t)$ is indeed the macroscopic (average) velocity field of every particle that we can find near the point $\mathbf{x}$ at the instant of time t. Indeed, we need to prove the following:

$$\mathbf{u}(\mathbf{x}, t) :=$$
$$\frac{m_j \int \left(\nabla_{\mathbf{P}_j} H_0\right) w\left(\mathbf{P}_1, \ldots, \mathbf{P}_n, \mathbf{r}_1, \ldots, \mathbf{r}_n, t\right)\delta(\mathbf{x} - \mathbf{r}_j)\, d\mathbf{P}_1, \ldots, d\mathbf{P}_n, d\mathbf{r}_1, \ldots, d\mathbf{r}_n}{\mu_j(\mathbf{x}, t)},$$
$$(10.65)$$

where

$$\mu_j(\mathbf{x},t) := \int m_j\, w\left(\mathbf{P}_1,\ldots,\mathbf{P}_n,\mathbf{r}_1,\ldots,\mathbf{r}_n,t\right)\delta(\mathbf{x}-\mathbf{r}_j)\,d\mathbf{P}_1,\ldots,d\mathbf{P}_n,d\mathbf{r}_1,\ldots,d\mathbf{r}_n. \tag{10.66}$$

On the other hand, by (10.61) we have

$$\nabla_{\mathbf{P}_j} H_0 = \nabla_{\mathbf{P}_j} H_{\mathbf{u}} + \mathbf{u}(\mathbf{r}_j,t) \qquad \forall j = 1,\ldots n. \tag{10.67}$$

Thus, by (10.67) and (10.56), we deduce

$$\begin{aligned}
&\int \left(\nabla_{\mathbf{P}_j} H_0\right) w\left(\mathbf{P}_1,\ldots,\mathbf{P}_n,\mathbf{r}_1,\ldots,\mathbf{r}_n,t\right)\delta(\mathbf{x}-\mathbf{r}_j)\,d\mathbf{P}_1,\ldots,d\mathbf{P}_n,d\mathbf{r}_1,\ldots,d\mathbf{r}_n = \\
&\frac{\int \left(\nabla_{\mathbf{P}_j} H_{\mathbf{u}}\right) f\left(H_{\mathbf{u}}\left(\mathbf{P}_1,\ldots,\mathbf{P}_n,\mathbf{r}_1,\ldots,\mathbf{r}_n,t\right)\right)\delta(\mathbf{x}-\mathbf{r}_j)\,d\mathbf{P}_1,\ldots,d\mathbf{P}_n,d\mathbf{r}_1,\ldots,d\mathbf{r}_n}{K(f,H_{\mathbf{u}})} \\
&\quad + \int \mathbf{u}(\mathbf{r}_j,t) w\left(\mathbf{P}_1,\ldots,\mathbf{P}_n,\mathbf{r}_1,\ldots,\mathbf{r}_n,t\right)\delta(\mathbf{x}-\mathbf{r}_j)\,d\mathbf{P}_1,\ldots,d\mathbf{P}_n,d\mathbf{r}_1,\ldots,d\mathbf{r}_n = \\
&\frac{\int \left(\nabla_{\mathbf{P}_j}\left(F\left(H_{\mathbf{u}}\left(\mathbf{P}_1,\ldots,\mathbf{P}_n,\mathbf{r}_1,\ldots,\mathbf{r}_n,t\right)\right)\right)\right)\delta(\mathbf{x}-\mathbf{r}_j)\,d\mathbf{P}_1,\ldots,d\mathbf{P}_n,d\mathbf{r}_1,\ldots,d\mathbf{r}_n}{K(f,H_{\mathbf{u}})} \\
&\qquad + \frac{\mu_j(\mathbf{x},t)}{m_j}\mathbf{u}(x,t) = 0 + \frac{\mu_j(\mathbf{x},t)}{m_j}\mathbf{u}(x,t),
\end{aligned} \tag{10.68}$$

where $F(s) := \int_0^s f(\tau)d\tau$. So, by (10.68) we indeed prove (10.65).

Next consider the following conditions:

$$\begin{cases}
\frac{\partial U}{\partial t}\left(\mathbf{x}_1,\ldots,\mathbf{x}_n,t\right) + \sum_{j=1}^n \mathbf{u}(\mathbf{x}_j,t)\cdot\nabla_{\mathbf{x}_j} U\left(\mathbf{x}_1,\ldots,\mathbf{x}_n,t\right) = 0, \\
\frac{\partial}{\partial t}\left(\mathbf{u}(\mathbf{x},t)-\mathbf{v}(\mathbf{x},t)\right) - \operatorname{curl}_{\mathbf{x}}\left(\mathbf{u}(\mathbf{x},t)\times\left(\mathbf{u}(\mathbf{x},t)-\mathbf{v}(\mathbf{x},t)\right)\right) \\
+\left(\operatorname{div}_{\mathbf{x}}\left(\mathbf{u}(\mathbf{x},t)-\mathbf{v}(\mathbf{x},t)\right)\right)\mathbf{u}(\mathbf{x},t) = 0, \\
\frac{\partial \mathbf{A}}{\partial t}(\mathbf{x},t) - \operatorname{curl}_{\mathbf{x}}\left(\mathbf{u}(\mathbf{x},t)\times\mathbf{A}(\mathbf{x},t)\right) + \left(\operatorname{div}_{\mathbf{x}}\mathbf{A}(\mathbf{x},t)\right)\mathbf{u}(\mathbf{x},t) = 0, \\
d_{\mathbf{x}}\mathbf{u}(\mathbf{x},t) + \{d_{\mathbf{x}}\mathbf{u}(\mathbf{x},t)\}^T = 0,
\end{cases} \tag{10.69}$$

where

$$\begin{aligned}
&U\left(\mathbf{x}_1,\ldots,\mathbf{x}_n,t\right) := \\
&\sum_{j=1}^n\left(\frac{\sigma_j^2}{2m_jc^2}\left|\mathbf{A}(\mathbf{x}_j,t)\right|^2 + \sigma_j\Psi(\mathbf{x}_j,t) - \frac{\sigma_j}{c}\mathbf{A}(\mathbf{x}_j,t)\cdot\mathbf{v}(\mathbf{x}_j,t)\right) - V\left(\mathbf{x}_1,\ldots,\mathbf{x}_n,t\right),
\end{aligned} \tag{10.70}$$

that are the same conditions as in (6.55), (6.56) and were the necessary conditions for the Thermodynamical equilibrium in the fully non-relativistic case. Moreover, as before, equations (10.69), (10.70) are invariant under the change of inertial or non-inertial Cartesian coordinate systems. We will prove now that, as in the fully non-relativistic case, conditions (10.69), (10.70) imply (10.63) or equivalently (10.64) also in the case of the relativistic-like Hamiltonian (10.52) and the quantity $H_{\mathbf{u}}$ in (10.53).

Indeed, by (10.69) and Proposition 3.3 there exists another Cartesian coordinate system $(**)$ such that under the change of coordinate system $(*)$ to another Cartesian coordinate system $(**)$, given by (3.54), we have

$$A(t) \cdot \mathbf{u}(\mathbf{x}, t) + \frac{dA}{dt}(t) \cdot \mathbf{x} + \frac{d\mathbf{z}}{dt}(t) = \mathbf{u}'(\mathbf{x}', t') = 0. \tag{10.71}$$

Thus, since (10.69), (10.70) are invariant under the change of inertial or non-inertial Cartesian coordinate systems, in system $(**)$ we have

$$\begin{cases} \frac{\partial U'}{\partial t'}\left(\mathbf{x}'_1, \ldots, \mathbf{x}'_n, t'\right) = 0, \\ \frac{\partial \mathbf{v}'}{\partial t'}(\mathbf{x}', t') = 0, \\ \frac{\partial \mathbf{A}'}{\partial t'}(\mathbf{x}', t') = 0 \\ \mathbf{u}'(\mathbf{x}', t') = 0, \end{cases} \tag{10.72}$$

where

$$U'\left(\mathbf{x}'_1, \ldots, \mathbf{x}'_n, t'\right) := \sum_{j=1}^{n} \left(\frac{(\sigma'_j)^2}{2m'_j c^2} \left|\mathbf{A}'(\mathbf{x}'_j, t')\right|^2 + \sigma'_j \Psi'(\mathbf{x}'_j, t') - \frac{\sigma'_j}{c} \mathbf{A}'(\mathbf{x}'_j, t') \cdot \mathbf{v}'(\mathbf{x}'_j, t') \right) - V'\left(\mathbf{x}'_1, \ldots, \mathbf{x}'_n, t'\right). \tag{10.73}$$

On the other hand, (10.72) is equivalent to

$$\frac{\partial H'_0}{\partial t'}\left(\mathbf{P}'_1, \ldots, \mathbf{P}'_n, \mathbf{r}'_1, \ldots, \mathbf{r}'_n, t'\right) = 0 \quad \text{and} \quad \mathbf{u}'(\mathbf{x}', t') = 0, \tag{10.74}$$

where H_0 is given by (10.52). Then (10.74) implies that we have the primed version of (10.64) and therefore, since (10.64) is invariant under the change of Cartesian coordinate systems obtain also the original version of (10.64). So, (10.69), (10.70) imply (10.64) or equivalently (10.63). Thus, we again obtain that the condition for the existence of the possibility to achieve the Thermodynamical equilibrium in the given system, ruled by the Hamiltonian $H_0\left(\mathbf{P}_1, \ldots, \mathbf{P}_n, \mathbf{r}_1, \ldots, \mathbf{r}_n, t\right)$ of the form (10.52), is equivalent to the existence of a Cartesian coordinate system $(**)$, where the average velocity of every particle vanishes and at the same time in the system $(**)$ the Hamiltonian is independent of the time variable explicitly.

10.5 Relativistic-Like Dirac Equation for a System of n Spin-Half Micro-particles

Consider the motion of a system of n spin-half relativistic-like stable micro-particles with inertial masses $m_1, \ldots, m_n$ and the charges $\sigma_1, \ldots, \sigma_n$ in the outer gravitational and electromagnetical field with characteristics $\mathbf{v}(\mathbf{x}, t)$, $\mathbf{A}(\mathbf{x}, t)$ and $\Psi(\mathbf{x}, t)$ and additional conservative field with potential $V(\mathbf{y}_1, \ldots, \mathbf{y}_n, t)$, taking into account the spin interaction. Then the system is characterized by 4^n-component complex wave function $\psi(\mathbf{x}_1, \ldots, \mathbf{x}_n, t) \in \mathbb{C}^{4^n}$ where by $\mathbb{C}^{4^n}$ we denote the tensor product of n copies of the space $\mathbb{C}^2 \otimes \mathbb{C}^2 = \mathbb{C}^4$:

$$\mathbb{C}^{4^n} := \left(\mathbb{C}^4\right) \otimes_1 \left(\mathbb{C}^4\right) \otimes_2 \left(\mathbb{C}^4\right) \ldots \otimes_{(n-1)} \left(\mathbb{C}^4\right), \tag{10.75}$$

and given $a = (a_1, a_2) \in \mathbb{C}^2$ and $b = (b_1, b_2) \in \mathbb{C}^2$ we identify their tensor product $a \otimes b \in \mathbb{C}^2 \otimes \mathbb{C}^2$ with the vector $(a_1, a_2, b_1, b_2) \in \mathbb{C}^4$. Then, as before in (10.11),(10.12) and in (6.192) consistently with the classical Hamiltonian (10.41), we built the Hermitian–Hamiltonian operator as

$$\begin{aligned}
\hat{H}_0 \cdot \psi = &\sum_{j=1}^{n} m_j c^2 D_4^j \cdot \psi - \sum_{j=1}^{n} c\,\mathbf{D}_j \cdot \left(i\hbar\nabla_{\mathbf{x}_j}\psi + \frac{\sigma_j}{c}\mathbf{A}(\mathbf{x}_j, t)\psi\right) \\
&+ \sum_{j=1}^{n} \sigma_j \left(\Psi(\mathbf{x}_j, t) - \frac{1}{c}\mathbf{v}(\mathbf{x}_j, t) \cdot \mathbf{A}(\mathbf{x}_j, t)\right)\psi - V(\mathbf{x}_1, \ldots, \mathbf{x}_n, t)\,\psi \\
- \sum_{j=1}^{n} \frac{i\hbar}{2}\mathrm{div}_{\mathbf{x}_j}\left\{\psi\mathbf{v}(\mathbf{x}_j, t)\right\} &- \sum_{j=1}^{n} \frac{i\hbar}{2}\nabla_{\mathbf{x}_j}\psi \cdot \mathbf{v}(\mathbf{x}_j, t) + \sum_{j=1}^{n} \frac{\hbar}{4}\mathbf{M}_j \cdot \left(\mathrm{curl}_{\mathbf{x}_j}\mathbf{v}(\mathbf{x}_j, t)\,\psi\right) \\
&- \sum_{j=1}^{n} \frac{(g_j - 1)\sigma_j\hbar}{2m_j c}\mathbf{M}_j \cdot \left(\mathrm{curl}_{\mathbf{x}_j}\mathbf{A}(\mathbf{x}_j, t)\,\psi\right),
\end{aligned} \tag{10.76}$$

where $\hat{H}_0$ is the Hamiltonian operator, for every $j = 1, \ldots, n$ g_j is a constant that depends on the type of the particle (for electron we have $g_j = 1$) and for every $j = 1, \ldots, n$ we denote

$$\mathbf{D}_j := (D_1^j, D_2^j, D_3^j) \quad \text{and} \quad \mathbf{M}_j := (M_1^j, M_2^j, M_3^j) \qquad \forall\, j = 1, 2, \ldots, n, \tag{10.77}$$

where for every $k = 1, 2, 3, 4$ and every $j = 1, 2, \ldots, n$: $D_k^j : \mathbb{C}^{4^n} \to \mathbb{C}^{4^n}$ is a linear operator on $\mathbb{C}^{4^n}$ (i.e. it is a $4^n \times 4^n$-complex matrix) defined by the following identities:

$$D_k^1 := (D_k) \otimes_1 (I^{4\times4}) \otimes_2 (I^{4\times4}) \ldots \otimes_{(n-1)} (I^{4\times4}), \quad \ldots$$
$$D_k^j = (I^{4\times4}) \otimes_1 (I^{4\times4}) \otimes_2 (I^{4\times4}) \ldots \otimes_{(j-1)} (D_k) \otimes_j (I^{4\times4}) \otimes_{(j+1)} (I^{4\times4}) \ldots \otimes_{(n-1)} (I^{4\times4}),$$
$$\ldots \text{ and } \quad D_k^n := (I^{4\times4}) \otimes_1 (I^{4\times4}) \otimes_2 (I^{4\times4}) \ldots \otimes_{(n-2)} (I^{4\times4}) \otimes_{(n-1)} (D_k), \tag{10.78}$$

and for every $k = 1, 2, 3$ and every $j = 1, 2, \ldots, n$: $M_k^j : \mathbb{C}^{4^n} \to \mathbb{C}^{4^n}$ is a linear operator on $\mathbb{C}^{4^n}$ (i.e. it is a $4^n \times 4^n$-complex matrix) defined by the following identities:

$$M_k^1 := (M_k) \otimes_1 \left(I^{4\times4}\right) \otimes_2 \left(I^{4\times4}\right) \ldots \otimes_{(n-1)} \left(I^{4\times4}\right), \quad \ldots \quad M_k^j :=$$
$$\left(I^{4\times4}\right) \otimes_1 \left(I^{4\times4}\right) \otimes_2 \left(I^{4\times4}\right) \ldots \otimes_{(j-1)} (M_k) \otimes_j \left(I^{4\times4}\right) \otimes_{(j+1)} \left(I^{4\times4}\right) \ldots \otimes_{(n-1)} \left(I^{4\times4}\right),$$
$$\ldots \text{ and } \quad M_k^n := \left(I^{4\times4}\right) \otimes_1 \left(I^{4\times4}\right) \otimes_2 \left(I^{4\times4}\right) \ldots \otimes_{(n-2)} \left(I^{4\times4}\right) \otimes_{(n-1)} (M_k). \tag{10.79}$$

Here D_k for $k = 1, 2, 3, 4$ are 4×4 Dirac α-matrixes defined as

$$D_1 = \begin{pmatrix} O^{2\times2} & S_1 \\ S_1 & O^{2\times2} \end{pmatrix} = \begin{pmatrix} 0&0&0&1 \\ 0&0&1&0 \\ 0&1&0&0 \\ 1&0&0&0 \end{pmatrix}, \quad D_2 = \begin{pmatrix} O^{2\times2} & S_2 \\ S_2 & O^{2\times2} \end{pmatrix} = \begin{pmatrix} 0&0&0&-i \\ 0&0&i&0 \\ 0&-i&0&0 \\ i&0&0&0 \end{pmatrix},$$
$$D_3 = \begin{pmatrix} O^{2\times2} & S_3 \\ S_3 & O^{2\times2} \end{pmatrix} = \begin{pmatrix} 0&0&1&0 \\ 0&0&0&-1 \\ 1&0&0&0 \\ 0&-1&0&0 \end{pmatrix}, \quad D_4 = \begin{pmatrix} I^{2\times2} & O^{2\times2} \\ O^{2\times2} & -I^{2\times2} \end{pmatrix} = \begin{pmatrix} 1&0&0&0 \\ 0&1&0&0 \\ 0&0&-1&0 \\ 0&0&0&-1 \end{pmatrix}, \tag{10.80}$$

and M_k for $k = 1, 2, 3, 4$ are 4×4 matrixes defined as

$$M_1 = \begin{pmatrix} S_1 & O^{2\times2} \\ O^{2\times2} & S_1 \end{pmatrix} = \begin{pmatrix} 0&1&0&0 \\ 1&0&0&0 \\ 0&0&0&1 \\ 0&0&1&0 \end{pmatrix}, \quad M_2 = \begin{pmatrix} S_2 & O^{2\times2} \\ O^{2\times2} & S_2 \end{pmatrix} = \begin{pmatrix} 0&-i&0&0 \\ i&0&0&0 \\ 0&0&0&-i \\ 0&0&i&0 \end{pmatrix},$$
$$M_3 = \begin{pmatrix} S_3 & O^{2\times2} \\ O^{2\times2} & S_3 \end{pmatrix} = \begin{pmatrix} 1&0&0&0 \\ 0&-1&0&0 \\ 0&0&1&0 \\ 0&0&0&-1 \end{pmatrix}, \tag{10.81}$$

where S_k for $k = 1, 2, 3$ are Pauli matrixes defined as

$$S_1 = \begin{pmatrix} 0&1 \\ 1&0 \end{pmatrix}, \quad S_2 = \begin{pmatrix} 0&-i \\ i&0 \end{pmatrix}, \quad S_3 = \begin{pmatrix} 1&0 \\ 0&-1 \end{pmatrix} \tag{10.82}$$

and the sign $\otimes$ in (10.78) and (10.79) means the tensor product of the matrices, i.e. for given two linear operators $A : \mathbb{C}^p \to \mathbb{C}^r$ and $B : \mathbb{C}^q \to \mathbb{C}^d$ their tensor product $A \otimes B$ is a linear operator from $\mathbb{C}^p \otimes \mathbb{C}^q$ to $\mathbb{C}^r \otimes \mathbb{C}^d$, defined by the identity:

$$(A \otimes B) \cdot (a \otimes b) = (A \cdot a) \otimes (B \cdot b) \qquad \forall a \in \mathbb{C}^p,\ \forall b \in \mathbb{C}^q. \tag{10.83}$$

Note that we added an additional term to the Hamiltonian, namely $\sum_{j=1}^{n} \frac{\hbar}{4}\mathbf{D}_j \cdot \left(\mathrm{curl}_{\mathbf{x}_j}\mathbf{v}(\mathbf{x}_j,t)\,\psi\right)$. In the case of the Newtonian-type gravity, this term vanishes in every non-rotating and, in particular, in every inertial coordinate system, however, it provides the invariance of the Dirac equation, under the change of non-inertial Cartesian coordinate system, as can be seen in the following Theorem 10.2. Thus the corresponding Dirac equation will be the following:

$$\begin{aligned}
i\hbar\frac{\partial\psi}{\partial t} = \hat{H}_0\cdot\psi = \sum_{j=1}^{n} m_j c^2 D_4^j\cdot\psi - \sum_{j=1}^{n} c\,\mathbf{D}_j\cdot\left(i\hbar\nabla_{\mathbf{x}_j}\psi + \frac{\sigma_j}{c}\mathbf{A}(\mathbf{x}_j,t)\psi\right)\\
+\sum_{j=1}^{n}\sigma_j\left(\Psi(\mathbf{x}_j,t) - \frac{1}{c}\mathbf{v}(\mathbf{x}_j,t)\cdot\mathbf{A}(\mathbf{x}_j,t)\right)\psi - V(\mathbf{x}_1,\dots,\mathbf{x}_n,t)\,\psi\\
-\sum_{j=1}^{n}\frac{i\hbar}{2}\mathrm{div}_{\mathbf{x}_j}\left\{\psi\mathbf{v}(\mathbf{x}_j,t)\right\} - \sum_{j=1}^{n}\frac{i\hbar}{2}\nabla_{\mathbf{x}_j}\psi\cdot\mathbf{v}(\mathbf{x}_j,t) + \sum_{j=1}^{n}\frac{\hbar}{4}\mathbf{M}_j\cdot\left(\mathrm{curl}_{\mathbf{x}_j}\mathbf{v}(\mathbf{x}_j,t)\,\psi\right)\\
-\sum_{j=1}^{n}\frac{(g_j-1)\sigma_j\hbar}{2m_jc}\mathbf{M}_j\cdot\left(\mathrm{curl}_{\mathbf{x}_j}\mathbf{A}(\mathbf{x}_j,t)\,\psi\right).
\end{aligned} \tag{10.84}$$

So

$$\begin{aligned}
i\hbar\left(\frac{\partial\psi}{\partial t} + \sum_{j=1}^{n}\mathbf{v}(\mathbf{x}_j,t)\cdot\nabla_{\mathbf{x}_j}\psi\right) + \sum_{j=1}^{n}\frac{i\hbar}{2}\left(\mathrm{div}_{\mathbf{x}_j}\mathbf{v}(\mathbf{x}_j,t)\right)\psi = \sum_{j=1}^{n} m_j c^2 D_4^j\cdot\psi\\
-\sum_{j=1}^{n} c\,\mathbf{D}_j\cdot\left(i\hbar\nabla_{\mathbf{x}_j}\psi + \frac{\sigma_j}{c}\mathbf{A}(\mathbf{x}_j,t)\psi\right) + \sum_{j=1}^{n}\sigma_j\left(\Psi(\mathbf{x}_j,t) - \frac{1}{c}\mathbf{v}(\mathbf{x}_j,t)\cdot\mathbf{A}(\mathbf{x}_j,t)\right)\psi\\
- V(\mathbf{x}_1,\dots,\mathbf{x}_n,t)\,\psi + \sum_{j=1}^{n}\frac{\hbar}{4}\mathbf{M}_j\cdot\left(\mathrm{curl}_{\mathbf{x}_j}\mathbf{v}(\mathbf{x}_j,t)\,\psi\right)\\
-\sum_{j=1}^{n}\frac{(g_j-1)\sigma_j\hbar}{2m_jc}\mathbf{M}_j\cdot\left(\mathrm{curl}_{\mathbf{x}_j}\mathbf{A}(\mathbf{x}_j,t)\,\psi\right).
\end{aligned} \tag{10.85}$$

We can rewrite the Dirac equation (10.85) in the following alternative form. First, define the linear operators $T_0 : \mathbb{C}^4 \to \mathbb{C}^2$ and $T_1 : \mathbb{C}^4 \to \mathbb{C}^2$ as

$$T_0 \cdot (a_1, a_2, b_1, b_2) = (a_1, a_2) \quad \text{and} \quad T_1 \cdot (a_1, a_2, b_1, b_2) = (b_1, b_2) \quad \forall (a_1, a_2, b_1, b_2) \in \mathbb{C}^4. \tag{10.86}$$

Then for every $k_1, \ldots, k_n = 0, 1$ define $\tilde{T}_{k_1,\ldots,k_n} : \mathbb{C}^{4^n} \to \mathbb{C}^{2^n}$ as

$$\tilde{T}_{k_1,\ldots,k_n} := \left(T_{k_1}\right) \otimes_1 \left(T_{k_2}\right) \otimes_2 \left(T_{k_3}\right) \ldots \otimes_{(n-1)} \left(T_{k_n}\right) \qquad \forall k_1, \ldots k_n \in \{0, 1\}, \tag{10.87}$$

and define $\psi_{k_1,\ldots,k_n} \in \mathbb{C}^{2^n}$ as

$$\psi_{k_1,\ldots,k_n}(\mathbf{x}_1, \ldots, \mathbf{x}_n, t) := \tilde{T}_{k_1,\ldots,k_n} \cdot \psi(\mathbf{x}_1, \ldots, \mathbf{x}_n, t) \qquad \forall k_1, \ldots k_n \in \{0, 1\}. \tag{10.88}$$

Then we rewrite (10.85) as

$$\begin{aligned}
&i\hbar \left(\frac{\partial}{\partial t} \psi_{k_1,\ldots,k_n} + \sum_{j=1}^{n} \mathbf{v}(\mathbf{x}_j, t) \cdot \nabla_{\mathbf{x}_j} \psi_{k_1,\ldots,k_n} \right) + \sum_{j=1}^{n} \frac{i\hbar}{2} \left(\operatorname{div}_{\mathbf{x}_j} \mathbf{v}(\mathbf{x}_j, t)\right) \psi_{k_1,\ldots,k_n} \\
&= \sum_{j=1}^{n} \frac{\hbar}{4} \mathbf{S}_j \cdot \left(\operatorname{curl}_{\mathbf{x}_j} \mathbf{v}(\mathbf{x}_j, t)\, \psi_{k_1,\ldots,k_n}\right) + \sum_{j=1}^{n} m_j c^2 (-1)^{2+k_j} \psi_{k_1,\ldots,k_n} \\
&- \sum_{j=1}^{n} c\, \mathbf{S}_j \cdot \left(i\hbar \nabla_{\mathbf{x}_j} \psi_{k_1,\ldots,k_{(j-1)},(1-k_j),k_{(j+1)},\ldots,k_n} + \frac{\sigma_j}{c} \mathbf{A}(\mathbf{x}_j, t)\, \psi_{k_1,\ldots,k_{(j-1)},(1-k_j),k_{(j+1)},\ldots,k_n} \right) \\
&+ \sum_{j=1}^{n} \sigma_j \left(\Psi(\mathbf{x}_j, t) - \frac{1}{c} \mathbf{v}(\mathbf{x}_j, t) \cdot \mathbf{A}(\mathbf{x}_j, t) \right) \psi_{k_1,\ldots,k_n} - V(\mathbf{x}_1, \ldots, \mathbf{x}_n, t)\, \psi_{k_1,\ldots,k_n} \\
&- \sum_{j=1}^{n} \frac{(g_j - 1)\sigma_j \hbar}{2m_j c} \mathbf{S}_j \cdot \left(\operatorname{curl}_{\mathbf{x}_j} \mathbf{A}(\mathbf{x}_j, t)\, \psi_{k_1,\ldots,k_n}\right) \qquad \forall k_1, \ldots k_n \in \{0, 1\}.
\end{aligned} \tag{10.89}$$

Next, in the similar way as the proof of Theorems 10.1 and 6.2 we can prove the following more general theorem about the invariance of the Dirac equation (10.85) under the change of inertial or non-inertial Cartesian coordinate system:

Theorem 10.2 *Consider that the change of some Cartesian coordinate system* $(*)$ *to another Cartesian coordinate system* $(**)$ *is given by*

$$\begin{cases} t' = t, \\ \mathbf{x}' = A(t) \cdot \mathbf{x} + \mathbf{z}(t), \\ \mathbf{x}'_k = A(t) \cdot \mathbf{x}_k + \mathbf{z}(t) \qquad \forall k = 1, \ldots, n, \end{cases} \tag{10.90}$$

where $A(t) \in SO(3)$ *is a rotation. Next, assume that in the coordinate system* $(**)$ *we observe a validity of the Dirac equation of the form:*

$$i\hbar\left(\frac{\partial\psi'}{\partial t'}+\sum_{j=1}^{n}\mathbf{v}'(\mathbf{x}'_j,t')\cdot\nabla_{\mathbf{x}'_j}\psi'\right)+\sum_{j=1}^{n}\frac{i\hbar}{2}\left(div_{\mathbf{x}'_j}\mathbf{v}'(\mathbf{x}'_j,t')\right)\psi'=$$
$$\sum_{j=1}^{n}m'_jc^2D_4^j\psi'-\sum_{j=1}^{n}c\,\mathbf{D}_j\cdot\left(i\hbar\nabla_{\mathbf{x}'_j}\psi'+\frac{\sigma'_j}{c}\mathbf{A}'(\mathbf{x}'_j,t')\psi'\right)$$
$$+\sum_{j=1}^{n}\sigma'_j\left(\Psi'(\mathbf{x}'_j,t')-\frac{1}{c}\mathbf{v}'(\mathbf{x}'_j,t')\cdot\mathbf{A}'(\mathbf{x}'_j,t')\right)\psi'-V'\left(\mathbf{x}'_1,\ldots,\mathbf{x}'_n,t'\right)\psi'$$
$$+\sum_{j=1}^{n}\frac{\hbar}{4}\mathbf{M}_j\cdot\left(curl_{\mathbf{x}'_j}\mathbf{v}'(\mathbf{x}'_j,t')\,\psi'\right)-\sum_{j=1}^{n}\frac{(g'_j-1)\sigma'_j\hbar}{2m'_jc}\mathbf{M}_j\cdot\left(curl_{\mathbf{x}'_j}\mathbf{A}'(\mathbf{x}'_j,t')\,\psi'\right) \tag{10.91}$$

where $\psi(\mathbf{x}_1,\ldots,\mathbf{x}_n,t)\in\mathbb{C}^{4^n}$ is a 4^n-component complex wave function defined above. Then in the coordinate system $(*)$ we have the validity of Dirac equation of the same as (10.91) form:

$$i\hbar\left(\frac{\partial\psi}{\partial t}+\sum_{j=1}^{n}\mathbf{v}(\mathbf{x}_j,t)\cdot\nabla_{\mathbf{x}_j}\psi\right)+\sum_{j=1}^{n}\frac{i\hbar}{2}\left(div_{\mathbf{x}_j}\mathbf{v}(\mathbf{x}_j,t)\right)\psi=$$
$$\sum_{j=1}^{n}m_jc^2D_4^j\psi-\sum_{j=1}^{n}c\,\mathbf{D}_j\cdot\left(i\hbar\nabla_{\mathbf{x}_j}\psi+\frac{\sigma_j}{c}\mathbf{A}(\mathbf{x}_j,t)\psi\right)$$
$$+\sum_{j=1}^{n}\sigma_j\left(\Psi(\mathbf{x}_j,t)-\frac{1}{c}\mathbf{v}(\mathbf{x}_j,t)\cdot\mathbf{A}(\mathbf{x}_j,t)\right)\psi-V\left(\mathbf{x}_1,\ldots,\mathbf{x}_n,t\right)\psi$$
$$+\sum_{j=1}^{n}\frac{\hbar}{4}\mathbf{M}_j\cdot\left(curl_{\mathbf{x}_j}\mathbf{v}(\mathbf{x}_j,t)\,\psi\right)-\sum_{j=1}^{n}\frac{(g_j-1)\sigma_j\hbar}{2mc}\mathbf{M}_j\cdot\left(curl_{\mathbf{x}_j}\mathbf{A}(\mathbf{x}_j,t)\,\psi\right), \tag{10.92}$$

provided that we have

$$\begin{cases} V'\left(\mathbf{x}'_1,\ldots,\mathbf{x}'_n,t'\right)=V\left(\mathbf{x}_1,\ldots,\mathbf{x}_n,t\right),\\ g'_j=g_j,\\ \sigma'_j=\sigma_j,\\ m'_j=m_j,\\ \mathbf{v}'(\mathbf{x}',t)=A(t)\cdot\mathbf{v}(\mathbf{x},t)+\frac{dA}{dt}(t)\cdot\mathbf{x}+\frac{d\mathbf{z}}{dt}(t),\\ \mathbf{A}'(\mathbf{x}',t)=A(t)\cdot\mathbf{A}(\mathbf{x},t),\\ \Psi'(\mathbf{x}',t)-\mathbf{v}'(\mathbf{x}',t)\cdot\mathbf{A}'(\mathbf{x}',t)=\Psi(\mathbf{x},t)-\mathbf{v}(\mathbf{x},t)\cdot\mathbf{A}(\mathbf{x},t),\\ \psi'\left(\mathbf{x}'_1,\ldots,\mathbf{x}'_n,t'\right)=\left(W(t)\otimes_1 W(t)\otimes_2 W(t)\ldots\otimes_{(n-1)}W(t)\right)\cdot\psi\left(\mathbf{x}_1,\ldots,\mathbf{x}_n,t\right), \end{cases} \tag{10.93}$$

with 4×4 *matrix* $W(t)$ *defined by*

$$W(t) := \begin{pmatrix} U(t) & O^{2\times 2} \\ O^{2\times 2} & U(t) \end{pmatrix} = U(t) \otimes I^{2\times 2}, \tag{10.94}$$

where, as before, $U(t) \in SU(2)$ *is characterized by*

$$U^*(t) \cdot \mathbf{S} \cdot U(t) = A(t) \cdot \mathbf{S}, \tag{10.95}$$

where $\mathbf{S} := (S_1, S_2, S_3)$ *with Pauli matrices defined by (10.82), that means*

$$\Big(U^*(t) \cdot S_1 \cdot U(t), U^*(t) \cdot S_2 \cdot U(t), U^*(t) \cdot S_3 \cdot U(t)\Big) =$$
$$\Big(a_{11}(t)S_1 + a_{12}(t)S_2 + a_{13}(t)S_3\,,\ a_{21}(t)S_1 + a_{22}(t)S_2 + a_{23}(t)S_3\,,$$
$$a_{31}(t)S_1 + a_{32}(t)S_2 + a_{33}(t)S_3\Big),$$

where $A(t) = \{a_{mk}(t)\}_{\{1\le m,k\le 3\}}$.

Next, in the case that the system of our particles has a positive energy, define

$$\phi := e^{-\frac{1}{\hbar}ic^2(\sum_{p=1}^n m_p)t}\psi \quad \text{and} \quad \phi_{k_1,\dots,k_n} := e^{-\frac{1}{\hbar}ic^2(\sum_{p=1}^n m_p)t}\psi_{k_1,\dots,k_n}$$
$$\forall k_1, \dots k_n \in \{0,1\}. \tag{10.96}$$

Then we rewrite the Dirac equations (10.89) as

$$\sum_{j=1}^n m_j c^2 \left(1 - (-1)^{2+k_j}\right) \phi_{k_1,\dots,k_n} + i\hbar \left(\frac{\partial}{\partial t}\phi_{k_1,\dots,k_n} + \sum_{j=1}^n \mathbf{v}(\mathbf{x}_j, t) \cdot \nabla_{\mathbf{x}_j}\phi_{k_1,\dots,k_n} \right)$$
$$+ \sum_{j=1}^n \frac{i\hbar}{2}\left(\mathrm{div}_{\mathbf{x}_j}\mathbf{v}(\mathbf{x}_j, t)\right)\phi_{k_1,\dots,k_n} = \sum_{j=1}^n \frac{\hbar}{4}\mathbf{S}_j \cdot \left(\mathrm{curl}_{\mathbf{x}_j}\mathbf{v}(\mathbf{x}_j, t)\,\phi_{k_1,\dots,k_n}\right)$$
$$- \sum_{j=1}^n \frac{(g_j - 1)\sigma_j\hbar}{2m_j c}\mathbf{S}_j \cdot \left(\mathrm{curl}_{\mathbf{x}_j}\mathbf{A}(\mathbf{x}_j, t)\,\phi_{k_1,\dots,k_n}\right)$$
$$- \sum_{j=1}^n c\,\mathbf{S}_j \cdot \left(i\hbar\nabla_{\mathbf{x}_j}\phi_{k_1,\dots,k_{(j-1)},(1-k_j),k_{(j+1)},\dots,k_n} + \frac{\sigma_j}{c}\mathbf{A}(\mathbf{x}_j, t)\,\phi_{k_1,\dots,k_{(j-1)},(1-k_j),k_{(j+1)},\dots,k_n} \right)$$
$$+ \sum_{j=1}^n \sigma_j \left(\Psi(\mathbf{x}_j, t) - \frac{1}{c}\mathbf{v}(\mathbf{x}_j, t) \cdot \mathbf{A}(\mathbf{x}_j, t) \right)\phi_{k_1,\dots,k_n} - V(\mathbf{x}_1, \dots, \mathbf{x}_n, t)\,\phi_{k_1,\dots,k_n}$$
$$\forall k_1, \dots k_n \in \{0,1\}. \tag{10.97}$$

In, particular,

$$
\begin{aligned}
& i\hbar \left(\frac{\partial}{\partial t} \phi_{0,\ldots,0} + \sum_{j=1}^{n} \mathbf{v}(\mathbf{x}_j, t) \cdot \nabla_{\mathbf{x}_j} \phi_{0,\ldots,0} \right) + \sum_{j=1}^{n} \frac{i\hbar}{2} \left(\mathrm{div}_{\mathbf{x}_j} \mathbf{v}(\mathbf{x}_j, t) \right) \phi_{0,\ldots,0} \\
&= \sum_{j=1}^{n} \frac{\hbar}{4} \mathbf{S}_j \cdot \left(\mathrm{curl}_{\mathbf{x}_j} \mathbf{v}(\mathbf{x}_j, t)\, \phi_{0,\ldots,0} \right) - \sum_{j=1}^{n} \frac{(g_j - 1)\sigma_j \hbar}{2m_j c} \mathbf{S}_j \cdot \left(\mathrm{curl}_{\mathbf{x}_j} \mathbf{A}(\mathbf{x}_j, t)\, \phi_{0,\ldots,0} \right) \\
&- \sum_{j=1}^{n} c\, \mathbf{S}_j \cdot \left(i\hbar \nabla_{\mathbf{x}_j} \phi_{0_1,\ldots,0_{(j-1)},1_j,0_{(j+1)},\ldots,0_n} + \frac{\sigma_j}{c} \mathbf{A}(\mathbf{x}_j, t)\, \phi_{0_1,\ldots,0_{(j-1)},1_j,0_{(j+1)},\ldots,0_n} \right) \\
&+ \sum_{j=1}^{n} \sigma_j \left(\Psi(\mathbf{x}_j, t) - \frac{1}{c} \mathbf{v}(\mathbf{x}_j, t) \cdot \mathbf{A}(\mathbf{x}_j, t) \right) \phi_{0,\ldots,0} - V(\mathbf{x}_1, \ldots, \mathbf{x}_n, t)\, \phi_{0,\ldots,0} \,.
\end{aligned}
\tag{10.98}
$$

On the other hand, if $(k_1 + \ldots + k_n) \geq 1$ then

$$
\sum_{j=1}^{n} m_j \left(1 - (-1)^{2+k_j}\right) \geq 2 \min \{m_1, \ldots, m_n\} > 0.
$$

Thus, as before in the proof of (10.27), for (10.97) where $(k_1 + \ldots + k_n) \geq 1$, in the non-relativistic limit we have

$$
\begin{aligned}
\phi_{k_1,\ldots,k_n} \approx & - \sum_{j=1}^{n} \frac{1}{c \left(\sum_{p=1}^{n} m_p \left(1 - (-1)^{2+k_p}\right) \right)} \Bigg\{ \\
& \mathbf{S}_j \cdot \left(i\hbar \nabla_{\mathbf{x}_j} \phi_{k_1,\ldots,k_{(j-1)},(1-k_j),k_{(j+1)},\ldots,k_n} + \frac{\sigma_j}{c} \mathbf{A}(\mathbf{x}_j, t)\, \phi_{k_1,\ldots,k_{(j-1)},(1-k_j),k_{(j+1)},\ldots,k_n} \right) \Bigg\} \\
& \forall k_1, \ldots k_n \in \{0, 1\}, \quad (k_1 + \ldots + k_n) \geq 1\,.
\end{aligned}
\tag{10.99}
$$

Then, using again the above-mentioned non-relativistic approximation, we further approximate (10.99) as

$$
\begin{aligned}
\phi_{0_1,\ldots,0_{(j-1)},1_j,0_{(j+1)},\ldots,0_n} &\approx -\frac{1}{2m_j c} \mathbf{S}_j \cdot \left(i\hbar \nabla_{\mathbf{x}_j} \phi_{0,\ldots,0} + \frac{\sigma_j}{c} \mathbf{A}(\mathbf{x}_j, t)\, \phi_{0,\ldots,0} \right) \\
& \forall j = 1, \ldots n \\
\text{and} \qquad & \phi_{k_1,\ldots,k_n} \approx 0 \quad \text{if} \quad k_1 + \ldots + k_n \geq 2.
\end{aligned}
\tag{10.100}
$$

Thus inserting (10.100) into (10.98) gives

$$
\begin{aligned}
& i\hbar\left(\frac{\partial}{\partial t}\phi_{0,\ldots,0}+\sum_{j=1}^{n}\mathbf{v}(\mathbf{x}_j,t)\cdot\nabla_{\mathbf{x}_j}\phi_{0,\ldots,0}\right)+\sum_{j=1}^{n}\frac{i\hbar}{2}\left(\operatorname{div}_{\mathbf{x}_j}\mathbf{v}(\mathbf{x}_j,t)\right)\phi_{0,\ldots,0}= \\
& \qquad \sum_{j=1}^{n}\frac{\hbar}{4}\mathbf{S}_j\cdot\left(\operatorname{curl}_{\mathbf{x}_j}\mathbf{v}(\mathbf{x}_j,t)\,\phi_{0,\ldots,0}\right) \\
& \quad +\sum_{j=1}^{n}\frac{1}{2m_j}\mathbf{S}_j\cdot\Bigg(i\hbar\nabla_{\mathbf{x}_j}\left\{\mathbf{S}_j\cdot\left(i\hbar\nabla_{\mathbf{x}_j}\phi_{0,\ldots,0}+\frac{\sigma_j}{c}\mathbf{A}(\mathbf{x}_j,t)\,\phi_{0,\ldots,0}\right)\right\} \\
& \qquad +\frac{\sigma_j}{c}\mathbf{A}(\mathbf{x}_j,t)\left\{\mathbf{S}_j\cdot\left(i\hbar\nabla_{\mathbf{x}_j}\phi_{0,\ldots,0}+\frac{\sigma_j}{c}\mathbf{A}(\mathbf{x}_j,t)\,\phi_{0,\ldots,0}\right)\right\}\Bigg) \\
& \qquad -\sum_{j=1}^{n}\frac{(g_j-1)\sigma_j\hbar}{2m_jc}\mathbf{S}_j\cdot\left(\operatorname{curl}_{\mathbf{x}_j}\mathbf{A}(\mathbf{x}_j,t)\,\phi_{0,\ldots,0}\right) \\
& +\sum_{j=1}^{n}\sigma_j\left(\Psi(\mathbf{x}_j,t)-\frac{1}{c}\mathbf{v}(\mathbf{x}_j,t)\cdot\mathbf{A}(\mathbf{x}_j,t)\right)\phi_{0,\ldots,0}-V\left(\mathbf{x}_1,\ldots,\mathbf{x}_n,t\right)\phi_{0,\ldots,0}\,,
\end{aligned}
\tag{10.101}
$$

that we rewrite as a non-relativistic Shrödinger–Pauli equation that we studied above in (6.197):

$$
\begin{aligned}
& i\hbar\left(\frac{\partial}{\partial t}\phi_{0,\ldots,0}+\sum_{j=1}^{n}\mathbf{v}(\mathbf{x}_j,t)\cdot\nabla_{\mathbf{x}_j}\phi_{0,\ldots,0}\right)+\sum_{j=1}^{n}\frac{i\hbar}{2}\left(\operatorname{div}_{\mathbf{x}_j}\mathbf{v}(\mathbf{x}_j,t)\right)\phi_{0,\ldots,0} \\
& \qquad =-\sum_{j=1}^{n}\frac{\hbar^2}{2m_j}\Delta_{\mathbf{x}_j}\phi_{0,\ldots,0}-V\left(\mathbf{x}_1,\ldots,\mathbf{x}_n,t\right)\phi_{0,\ldots,0} \\
& \quad +\sum_{j=1}^{n}\frac{i\hbar\sigma_j}{2m_jc}\operatorname{div}_{\mathbf{x}_j}\left\{\phi_{0,\ldots,0}\mathbf{A}(\mathbf{x}_j,t)\right\}+\sum_{j=1}^{n}\frac{i\hbar\sigma_j}{2m_jc}\mathbf{A}(\mathbf{x}_j,t)\cdot\nabla_{\mathbf{x}_j}\phi_{0,\ldots,0} \\
& \quad +\sum_{j=1}^{n}\left(\sigma_j\Psi(\mathbf{x}_j,t)-\frac{\sigma_j}{c}\mathbf{A}(\mathbf{x}_j,t)\cdot\mathbf{v}(\mathbf{x}_j,t)+\frac{\sigma_j^2}{2m_jc^2}\left|\mathbf{A}(\mathbf{x}_j,t)\right|^2\right)\phi_{0,\ldots,0} \\
& \quad -\sum_{j=1}^{n}\frac{g_j\sigma_j\hbar}{2m_jc}\mathbf{S}_j\cdot\left(\operatorname{curl}_{\mathbf{x}_j}\mathbf{A}(\mathbf{x}_j,t)\,\phi_{0,\ldots,0}\right)+\sum_{j=1}^{n}\frac{\hbar}{4}\mathbf{S}_j\cdot\left(\operatorname{curl}_{\mathbf{x}_j}\mathbf{v}(\mathbf{x}_j,t)\,\phi_{0,\ldots,0}\right).
\end{aligned}
\tag{10.102}
$$

Next, consider the relativistic-like Lagrangian of the motion of system of n spin-half quantum micro-particles with inertial masses $m_1, \ldots, m_n$ and the charges $\sigma_1, \ldots, \sigma_n$ in the outer gravitational and electromagnetical fields with characteristics $\mathbf{v}(\mathbf{x}, t)$, $\mathbf{A}(\mathbf{x}, t)$ and $\Psi(\mathbf{x}, t)$ and additional conservative field with potential $V(\mathbf{y}_1, \ldots, \mathbf{y}_n, t)$. Then consider a Lagrangian density L defined by

$$
\begin{aligned}
&L(\psi, \mathbf{A}, \Psi, \mathbf{v}, \mathbf{x}_1, \ldots, \mathbf{x}_n, t) := \\
&\frac{i\hbar}{2}\left(\left(\frac{\partial\psi}{\partial t} + \sum_{k=1}^{n}\mathbf{v}(\mathbf{x}_k, t)\cdot\nabla_{\mathbf{x}_k}\psi\right)\cdot\bar{\psi} - \psi\cdot\left(\frac{\partial\bar{\psi}}{\partial t} + \sum_{k=1}^{n}\mathbf{v}(\mathbf{x}_k, t)\cdot\nabla_{\mathbf{x}_k}\bar{\psi}\right)\right) \\
&+\sum_{k=1}^{n}\frac{c}{2}\left(\mathbf{D}_k\cdot\left(i\hbar\nabla_{\mathbf{x}_k}\psi + \frac{\sigma_k}{c}\mathbf{A}(\mathbf{x}_k, t)\psi\right)\right)\cdot\bar{\psi} - \psi\cdot\left(\mathbf{D}_k\cdot\left(i\hbar\nabla_{\mathbf{x}_k}\bar{\psi} - \frac{\sigma_k}{c}\mathbf{A}(\mathbf{x}_k, t)\bar{\psi}\right)\right) \\
&\qquad + V(\mathbf{x}_1, \ldots, \mathbf{x}_n, t)\,\psi\cdot\bar{\psi} - \sum_{k=1}^{n} m_k c^2\left(D_4^k\cdot\psi\right)\cdot\bar{\psi} \\
&\qquad - \sum_{k=1}^{n}\sigma_k\left(\Psi(\mathbf{x}_k, t) - \frac{1}{c}\mathbf{v}(\mathbf{x}_k, t)\cdot\mathbf{A}(\mathbf{x}_k, t)\right)\psi\cdot\bar{\psi} \\
&\qquad + \sum_{k=1}^{n}\frac{(g_k - 1)\sigma_k\hbar}{2m_k c}\left(\left(\mathbf{M}_k\cdot\operatorname{curl}_{\mathbf{x}_k}\mathbf{A}(\mathbf{x}_k, t)\right)\cdot\psi\right)\cdot\bar{\psi} \\
&\qquad\qquad - \sum_{k=1}^{n}\frac{\hbar}{4}\left(\left(\mathbf{M}_k\cdot\operatorname{curl}_{\mathbf{x}_k}\mathbf{v}(\mathbf{x}_k, t)\right)\cdot\psi\right)\cdot\bar{\psi},
\end{aligned}
\tag{10.103}
$$

where $\psi := \psi(\mathbf{x}_1, \ldots \mathbf{x}_n, t) \in \mathbb{C}^{4^n}$ is a wave function of the system. Then, as before, we can get that the density L is invariant under the change of inertial or non-inertial Cartesian coordinate systems of the form

$$
\begin{cases}
t' = t \\
\mathbf{x}' = A(t)\cdot\mathbf{x} + \mathbf{z}(t) \\
\mathbf{x}'_k = A(t)\cdot\mathbf{x}_k + \mathbf{z}(t) \qquad \forall k = 1, \ldots, n,
\end{cases}
$$

provided we take (10.93) into account. Next we investigate stationary points of the functional

$$
J(\psi) = \int_0^T\int_{(\mathbb{R}^3)^n} L(\psi, \mathbf{A}, \Psi, \mathbf{v}, \mathbf{x}_1, \ldots, \mathbf{x}_n, t)\, d\mathbf{x}_1 \ldots, d\mathbf{x}_n dt. \tag{10.104}
$$

Then

$$0 = \frac{\delta L}{\delta(\bar{\psi})} = i\hbar\left(\frac{\partial\psi}{\partial t} + \sum_{k=1}^{n}\frac{1}{2}\mathbf{v}(\mathbf{x}_k,t)\cdot\nabla_{\mathbf{x}_k}\psi + \sum_{k=1}^{n}\frac{1}{2}\mathrm{div}_{\mathbf{x}_k}\{\psi\mathbf{v}(\mathbf{x}_k,t)\}\right)$$
$$+\sum_{k=1}^{n}c\mathbf{D}_k\cdot\left(i\hbar\nabla_{\mathbf{x}_k}\psi + \frac{\sigma_k}{c}\mathbf{A}(\mathbf{x}_k,t)\psi\right) - \sum_{k=1}^{n}m_k c^2 D_4^k\cdot\psi + V(\mathbf{x}_1,\ldots,\mathbf{x}_n,t)\,\psi$$
$$-\sum_{k=1}^{n}\sigma_k\left(\Psi(\mathbf{x}_k,t) - \frac{1}{c}\mathbf{v}(\mathbf{x}_k,t)\cdot\mathbf{A}(\mathbf{x}_k,t)\right)\psi - \sum_{k=1}^{n}\frac{\hbar}{4}\left(\mathbf{M}_k\cdot\mathrm{curl}_{\mathbf{x}_k}\mathbf{v}(\mathbf{x}_k,t)\right)\cdot\psi$$
$$+\sum_{k=1}^{n}\frac{(g_k-1)\sigma_k\hbar}{2m_k c}\left(\mathbf{M}_k\cdot\mathrm{curl}_{\mathbf{x}_k}\mathbf{A}(\mathbf{x}_k,t)\right)\cdot\psi, \quad (10.105)$$

and

$$0 = \frac{\delta L}{\delta(\psi)} = (\bar{i})\hbar\left(\frac{\partial\bar{\psi}}{\partial t} + \sum_{k=1}^{n}\frac{1}{2}\mathbf{v}(\mathbf{x}_k,t)\cdot\nabla_{\mathbf{x}_k}\bar{\psi} + \sum_{k=1}^{n}\frac{1}{2}\mathrm{div}_{\mathbf{x}_k}\{\bar{\psi}\mathbf{v}(\mathbf{x}_k,t)\}\right)$$
$$+\sum_{k=1}^{n}c\mathbf{D}_k\cdot\left((\bar{i})\hbar\nabla_{\mathbf{x}_k}\bar{\psi} + \frac{\sigma_k}{c}\mathbf{A}(\mathbf{x}_k,t)\bar{\psi}\right) - \sum_{k=1}^{n}m_k c^2 D_4^k\cdot\bar{\psi} + V(\mathbf{x}_1,\ldots,\mathbf{x}_n,t)\,\bar{\psi}$$
$$-\sum_{k=1}^{n}\sigma_k\left(\Psi(\mathbf{x}_k,t) - \frac{1}{c}\mathbf{v}(\mathbf{x}_k,t)\cdot\mathbf{A}(\mathbf{x}_k,t)\right)\bar{\psi} - \sum_{k=1}^{n}\frac{\hbar}{4}\left(\mathbf{M}_k\cdot\mathrm{curl}_{\mathbf{x}_k}\mathbf{v}(\mathbf{x}_k,t)\right)\cdot\bar{\psi}$$
$$+\sum_{k=1}^{n}\frac{(g_k-1)\sigma_k\hbar}{2m_k c}\left(\mathbf{M}_k^T\cdot\mathrm{curl}_{\mathbf{x}_k}\mathbf{A}(\mathbf{x}_k,t)\right)\cdot\bar{\psi}. \quad (10.106)$$

Equation (10.106) is just a complex conjugate of equation (10.105). Thus the Euler–Lagrange for (10.104) coincides with the Dirac equation (10.85).

Finally, assume that the first and the second particles have the same mass $m_1 = m_2$ and the same charge $\sigma_1 = \sigma_2$, and moreover, assume that we have

$$V(\mathbf{x}_1,\mathbf{x}_2,\mathbf{x}_3,\ldots,\mathbf{x}_n,t) = V(\mathbf{x}_2,\mathbf{x}_1,\mathbf{x}_3,\ldots,\mathbf{x}_n,t) \qquad \forall\mathbf{x}_1,\mathbf{x}_2,\mathbf{x}_3,\ldots,\mathbf{x}_n\in\mathbb{R}^3,\quad \forall t.$$

In this case it can be easily deduced that if $\psi(\mathbf{x}_1,\mathbf{x}_2,\mathbf{x}_3,\ldots,\mathbf{x}_n,t)\in\mathbb{C}^{4^n}$ is a solution of (10.85), then $B_{1,2}\cdot\psi(\mathbf{x}_2,\mathbf{x}_1,\mathbf{x}_3,\ldots,\mathbf{x}_n,t)$ is also a solution of (10.85), where by $B_{1,2}:\mathbb{C}^{4^n}\to\mathbb{C}^{4^n}$ we denote the linear operator (matrix) defined as

$$B_{1,2}\cdot(b_1\otimes b_2\otimes b_3\otimes\ldots\otimes b_n) = (b_2\otimes b_1\otimes b_3\otimes\ldots\otimes b_n) \qquad \forall b_1,\ldots b_n\in\mathbb{C}^4. \quad (10.107)$$

Therefore, if $\psi(\mathbf{x}_1, \mathbf{x}_2, \mathbf{x}_3, \ldots, \mathbf{x}_n, t) \in \mathbb{C}^{4^n}$ is a solution of (10.85) then for every $t \geq 0$ we will have

$$B_{1,2} \cdot \psi(\mathbf{x}_2, \mathbf{x}_1, \mathbf{x}_3, \ldots, \mathbf{x}_n, t) = -\psi(\mathbf{x}_1, \mathbf{x}_2, \mathbf{x}_3, \ldots, \mathbf{x}_n, t) \quad \forall \mathbf{x}_1, \mathbf{x}_2, \mathbf{x}_3, \ldots, \mathbf{x}_n \in \mathbb{R}^3, \tag{10.108}$$

provided that (10.108) holds for the initial instant of time $t = 0$. So we have a consistency with the Pauli Exclusion Principle for two or more identical fermions.

10.6 Quantum Liouville's Equation for a Finite System of Spin-Half Relativistic-Like Particles

Consider the statistical description of the motion of a system of n spin-half relativistic-like stable micro-particles with inertial masses $m_1, \ldots, m_n$ and the charges $\sigma_1, \ldots, \sigma_n$ in the outer gravitational and electromagnetical field with characteristics $\mathbf{v}(\mathbf{x}, t)$, $\mathbf{A}(\mathbf{x}, t)$ and $\Psi(\mathbf{x}, t)$ and additional conservative field with potential $V(\mathbf{y}_1, \ldots, \mathbf{y}_n, t)$, taking into account the spin interaction. Then, as before, the Quantum Liouville's equation for this system of particles has the following form:

$$\begin{aligned} i\hbar\frac{\partial \xi}{\partial t}(\mathbf{x}_1, \ldots, \mathbf{x}_n, \mathbf{y}_1, \ldots, \mathbf{y}_n, t) = \\ \left(\hat{H}_0(\mathbf{x}_1, \ldots, \mathbf{x}_n, t) \otimes I\right) \cdot \xi(\mathbf{x}_1, \ldots, \mathbf{x}_n, \mathbf{y}_1, \ldots, \mathbf{y}_n, t) \\ - \left(I \otimes \hat{H}_0^*(\mathbf{y}_1, \ldots, \mathbf{y}_n, t)\right) \cdot \xi(\mathbf{x}_1, \ldots, \mathbf{x}_n, \mathbf{y}_1, \ldots, \mathbf{y}_n, t), \end{aligned} \tag{10.109}$$

where $\xi(\mathbf{x}_1, \ldots, \mathbf{x}_n, \mathbf{y}_1, \ldots, \mathbf{y}_n, t) \in \mathbb{C}^{4^n} \otimes \mathbb{C}^{4^n}$ is a density-matrix function, $\hat{H}_0(\mathbf{x}_1, \ldots, \mathbf{x}_n, t)$ is the Hamiltonian operator acting on the variables $(\mathbf{x}_1, \ldots, \mathbf{x}_n)$, $\hat{H}_0^*(\mathbf{y}_1, \ldots, \mathbf{y}_n, t)$ is the complex conjugate (not the Hermitian adjoint) to the Hamiltonian operator acting on the variables $(\mathbf{y}_1, \ldots, \mathbf{y}_n)$ and $\hat{H}_0(\mathbf{x}_1, \ldots, \mathbf{x}_n, t) \otimes I$ and $I \otimes \hat{H}_0^*(\mathbf{y}_1, \ldots, \mathbf{y}_n, t)$ are linear operators acting on functions $\xi(\mathbf{x}_1, \ldots, \mathbf{x}_n, \mathbf{y}_1, \ldots, \mathbf{y}_n, t) \in \mathbb{C}^{4^n} \otimes \mathbb{C}^{4^n}$, defined by

$$\begin{aligned} \left(\hat{H}_0(\mathbf{x}_1, \ldots, \mathbf{x}_n, t) \otimes I\right) \cdot (\psi_1(\mathbf{x}_1, \ldots, \mathbf{x}_n, t) \otimes \psi_2(\mathbf{y}_1, \ldots, \mathbf{y}_n, t)) = \\ \left(\hat{H}_0(\mathbf{x}_1, \ldots, \mathbf{x}_n, t) \cdot \psi_1(\mathbf{x}_1, \ldots, \mathbf{x}_n, t)\right) \otimes \psi_2(\mathbf{y}_1, \ldots, \mathbf{y}_n, t) \quad \text{and} \\ \left(I \otimes \hat{H}_0^*(\mathbf{y}_1, \ldots, \mathbf{y}_n, t)\right) \cdot (\psi_1(\mathbf{x}_1, \ldots, \mathbf{x}_n, t) \otimes \psi_2(\mathbf{y}_1, \ldots, \mathbf{y}_n, t)) = \\ \psi_1(\mathbf{x}_1, \ldots, \mathbf{x}_n, t) \otimes \left(\hat{H}_0^*(\mathbf{y}_1, \ldots, \mathbf{y}_n, t) \cdot \psi_2(\mathbf{y}_1, \ldots, \mathbf{y}_n, t)\right) \quad \forall \psi_1, \psi_2 \in \mathbb{C}^{4^n}. \end{aligned} \tag{10.110}$$

Since by (10.76) the Hamiltonian operator $\hat{H}_0$ has the forms:

$$
\begin{aligned}
&\hat{H}_0\left(\mathbf{x}_1, \ldots, \mathbf{x}_n, t\right) \cdot \psi\left(\mathbf{x}_1, \ldots, \mathbf{x}_n, t\right)=\sum_{j=1}^n m_j c^2 D_4^j \cdot \psi \\
&-\sum_{j=1}^n c\, \mathbf{D}_j \cdot\left(i \hbar \nabla_{\mathbf{x}_j} \psi+\frac{\sigma_j}{c} \mathbf{A}(\mathbf{x}_j, t) \psi\right)+\sum_{j=1}^n \sigma_j\left(\Psi(\mathbf{x}_j, t)-\frac{1}{c} \mathbf{v}(\mathbf{x}_j, t) \cdot \mathbf{A}(\mathbf{x}_j, t)\right) \psi \\
&\quad-V\left(\mathbf{x}_1, \ldots, \mathbf{x}_n, t\right) \psi-\sum_{j=1}^n \frac{i \hbar}{2} \operatorname{div}_{\mathbf{x}_j}\left\{\psi \mathbf{v}(\mathbf{x}_j, t)\right\}-\sum_{j=1}^n \frac{i \hbar}{2} \nabla_{\mathbf{x}_j} \psi \cdot \mathbf{v}(\mathbf{x}_j, t) \\
&\quad+\sum_{j=1}^n \frac{\hbar}{4} \mathbf{M}_j \cdot\left(\operatorname{curl}_{\mathbf{x}_j} \mathbf{v}(\mathbf{x}_j, t)\, \psi\right)-\sum_{j=1}^n \frac{(g_j-1) \sigma_j \hbar}{2 m_j c} \mathbf{M}_j \cdot\left(\operatorname{curl}_{\mathbf{x}_j} \mathbf{A}(\mathbf{x}_j, t)\, \psi\right),
\end{aligned}
\tag{10.111}
$$

and consistently with (10.111) we write

$$
\begin{aligned}
&\hat{H}_0^*\left(\mathbf{y}_1, \ldots, \mathbf{y}_n, t\right) \cdot \psi\left(\mathbf{y}_1, \ldots, \mathbf{y}_n, t\right)=\sum_{j=1}^n m_j c^2 D_4^j \cdot \psi \\
&-\sum_{j=1}^n c\, \bar{\mathbf{D}}_j \cdot\left(-i \hbar \nabla_{\mathbf{y}_j} \psi+\frac{\sigma_j}{c} \mathbf{A}(\mathbf{y}_j, t) \psi\right)+\sum_{j=1}^n \sigma_j\left(\Psi(\mathbf{y}_j, t)-\frac{1}{c} \mathbf{v}(\mathbf{y}_j, t) \cdot \mathbf{A}(\mathbf{y}_j, t)\right) \psi \\
&\quad-V\left(\mathbf{y}_1, \ldots, \mathbf{y}_n, t\right) \psi+\sum_{j=1}^n \frac{i \hbar}{2} \operatorname{div}_{\mathbf{y}_j}\left\{\psi \mathbf{v}(\mathbf{y}_j, t)\right\}+\sum_{j=1}^n \frac{i \hbar}{2} \nabla_{\mathbf{y}_j} \psi \cdot \mathbf{v}(\mathbf{y}_j, t) \\
&\quad+\sum_{j=1}^n \frac{\hbar}{4} \bar{\mathbf{M}}_j \cdot\left(\operatorname{curl}_{\mathbf{y}_j} \mathbf{v}(\mathbf{y}_j, t)\, \psi\right)-\sum_{j=1}^n \frac{(g_j-1) \sigma_j \hbar}{2 m_j c} \bar{\mathbf{M}}_j \cdot\left(\operatorname{curl}_{\mathbf{y}_j} \mathbf{A}(\mathbf{y}_j, t)\, \psi\right),
\end{aligned}
\tag{10.112}
$$

where $\bar{\mathbf{D}}_j$ and $\bar{\mathbf{M}}_j$ are the complex conjugates (not the Hermitian adjoint) of the operators $\mathbf{D}_j$ and $\mathbf{M}_j$. Thus we rewrite the corresponding Quantum Liouville's equation (10.109) as

$$
\begin{aligned}
&i \hbar\left(\frac{\partial \xi}{\partial t}+\sum_{j=1}^n \mathbf{v}(\mathbf{x}_j, t) \cdot \nabla_{\mathbf{x}_j} \xi+\sum_{j=1}^n \mathbf{v}(\mathbf{y}_j, t) \cdot \nabla_{\mathbf{y}_j} \xi\right) \\
&\qquad\qquad+\sum_{j=1}^n \frac{i \hbar}{2}\left(\operatorname{div}_{\mathbf{x}_j} \mathbf{v}(\mathbf{x}_j, t)+\operatorname{div}_{\mathbf{y}_j} \mathbf{v}(\mathbf{y}_j, t)\right) \xi \\
&\quad=\sum_{j=1}^n \frac{\hbar}{4}(\mathbf{M}_j \otimes I) \cdot\left(\operatorname{curl}_{\mathbf{x}_j} \mathbf{v}(\mathbf{x}_j, t)\, \xi\right)-\sum_{j=1}^n \frac{\hbar}{4}(I \otimes \bar{\mathbf{M}}_j) \cdot\left(\operatorname{curl}_{\mathbf{y}_j} \mathbf{v}(\mathbf{y}_j, t)\, \xi\right) \\
&+\sum_{j=1}^n m_j c^2\left((D_4^j \otimes I)-(I \otimes D_4^j)\right) \cdot \xi-\sum_{j=1}^n c\,(\mathbf{D}_j \otimes I) \cdot\left(i \hbar \nabla_{\mathbf{x}_j} \xi+\frac{\sigma_j}{c} \mathbf{A}(\mathbf{x}_j, t) \xi\right)
\end{aligned}
$$

$$
-\sum_{j=1}^{n} c\,(I\otimes\bar{\mathbf{D}}_j)\cdot\Big(i\hbar\nabla_{\mathbf{y}_j}\xi-\frac{\sigma_j}{c}\mathbf{A}(\mathbf{y}_j,t)\xi\Big)-\left(V\left(\mathbf{x}_1,\ldots,\mathbf{x}_n,t\right)-V\left(\mathbf{y}_1,\ldots,\mathbf{y}_n,t\right)\right)\xi
$$
$$
+\sum_{j=1}^{n}\Big(\sigma_j\Psi(\mathbf{x}_j,t)-\frac{\sigma_j}{c}\mathbf{A}(\mathbf{x}_j,t)\cdot\mathbf{v}(\mathbf{x}_j,t)\Big)\xi-\sum_{j=1}^{n}\Big(\sigma_j\Psi(\mathbf{y}_j,t)-\frac{\sigma_j}{c}\mathbf{A}(\mathbf{y}_j,t)\cdot\mathbf{v}(\mathbf{y}_j,t)\Big)\xi
$$
$$
-\sum_{j=1}^{n}\frac{(g_j-1)\sigma_j\hbar}{2mc}(\mathbf{M}_j\otimes I)\cdot\big(\mathrm{curl}_{\mathbf{x}_j}\mathbf{A}(\mathbf{x}_j,t)\,\xi\big)
$$
$$
+\sum_{j=1}^{n}\frac{(g_j-1)\sigma_j\hbar}{2mc}(I\otimes\bar{\mathbf{M}}_j)\cdot\big(\mathrm{curl}_{\mathbf{y}_j}\mathbf{A}(\mathbf{y}_j,t)\,\xi\big),\qquad(10.113)
$$

where, as before,

$$
\begin{aligned}
(D_4^j\otimes I)\cdot(a\otimes b)&=(D_4^j\cdot a)\otimes b \text{ and } (I\otimes D_4^j)\cdot(a\otimes b)=a\otimes(D_4^j\cdot b)\ \ \forall a,b\in\mathbb{C}^{4^n},\\
(\mathbf{D}_j\otimes I)\cdot(a\otimes b)&=(\mathbf{D}_j\cdot a)\otimes b \text{ and } (I\otimes\mathbf{D}_j)\cdot(a\otimes b)=a\otimes(\mathbf{D}_j\cdot b)\ \ \forall a,b\in\mathbb{C}^{4^n},\\
(\mathbf{M}_j\otimes I)\cdot(a\otimes b)&=(\mathbf{M}_j\cdot a)\otimes b \text{ and } (I\otimes\mathbf{M}_j)\cdot(a\otimes b)=a\otimes(\mathbf{M}_j\cdot b)\ \ \forall a,b\in\mathbb{C}^{4^n}.
\end{aligned}\qquad(10.114)
$$

Next consider a change of some non-inertial Cartesian coordinate system $(*)$ to another Cartesian coordinate system $(**)$ of the form (4.2):

$$
\begin{cases}\mathbf{x}'=A(t)\cdot\mathbf{x}+\mathbf{z}(t),\\ t'=t,\end{cases}\qquad(10.115)
$$

where $A(t)\in SO(3)$ is a rotation. Then, as before in Theorem 10.2, we deduce that Quantum Liouville's equation of the form (10.113) is invariant under the change of non-inertial Cartesian coordinate system, provided we have

$$
\begin{cases}
\mathbf{x}'_j=A(t)\cdot\mathbf{x}_j+\mathbf{z}(t)\quad\forall j=1,\ldots,n\\
\mathbf{y}'_j=A(t)\cdot\mathbf{y}_j+\mathbf{z}(t)\quad\forall j=1,\ldots,n\\
\xi'\left(\mathbf{x}'_1,\ldots,\mathbf{x}'_n,\mathbf{y}'_1,\ldots,\mathbf{y}'_n,t'\right)=\Big(\big(W(t)\otimes_1 W(t)\ldots\otimes_{(n-1)}W(t)\big)\\
\otimes\big(\bar{W}(t)\otimes_1\bar{W}(t)\ldots\otimes_{(n-1)}\bar{W}(t)\big)\Big)\cdot\xi\left(\mathbf{x}_1,\ldots,\mathbf{x}_n,\mathbf{y}_1,\ldots,\mathbf{y}_n,t\right)\\
V'\left(\mathbf{y}'_1,\ldots,\mathbf{y}'_n,t'\right)=V\left(\mathbf{y}_1,\ldots,\mathbf{y}_n,t\right)\\
\mathbf{A}'=A(t)\cdot\mathbf{A}\\
\Psi'-\frac{1}{c}\mathbf{A}'\cdot\mathbf{v}'=\Psi-\frac{1}{c}\mathbf{A}\cdot\mathbf{v},
\end{cases}\qquad(10.116)
$$

with 4×4 matrix $W(t)$ defined by

$$
W(t):=\begin{pmatrix}U(t)&O^{2\times2}\\ O^{2\times2}&U(t)\end{pmatrix}=U(t)\otimes I^{2\times2},\qquad(10.117)
$$

where, as before, $U(t) \in SU(2)$ is characterized by

$$U^*(t) \cdot \mathbf{S} \cdot U(t) = A(t) \cdot \mathbf{S}, \tag{10.118}$$

where $\mathbf{S} := (S_1, S_2, S_3)$ with Pauli matrices defined by (10.82), and $\bar{W}(t)$ is the complex conjugate (not the Hermitian adjoint) of the matrix $W(t)$.

Next assume that $\hat{A}(\mathbf{x}_1, \ldots, \mathbf{x}_n, t)$ is some Hermitian operator acting on the functions $\psi(\mathbf{x}_1, \ldots, \mathbf{x}_n, t) \in \mathbb{C}^{4^n}$ with respect to spatial variables $(\mathbf{x}_1, \ldots, \mathbf{x}_n)$. Then the average $\tilde{A}(t)$ of $\hat{A}$ on the density matrix ξ is defined by

$$\tilde{A}(t) = \frac{\int trace\left(\left(\hat{A}(\mathbf{x}_1, \ldots, \mathbf{x}_n, t) \otimes I\right) \cdot \xi\right)(\mathbf{z}_1, \ldots, \mathbf{z}_n, \mathbf{z}_1, \ldots, \mathbf{z}_n, t)\, d\mathbf{z}_1 \ldots d\mathbf{z}_n}{\int trace(\xi)(\mathbf{z}_1, \ldots, \mathbf{z}_n, \mathbf{z}_1, \ldots, \mathbf{z}_n, t)\, d\mathbf{z}_1 \ldots d\mathbf{z}_n}, \tag{10.119}$$

where for given $R \in \mathbb{C}^p \otimes \mathbb{C}^p$ $trace(R)$ is a linear functional from $\mathbb{C}^p \otimes \mathbb{C}^p$ to $\mathbb{C}$ defined by

$$trace(a \otimes b) = \sum_{k=1}^{p} a_k b_k \quad \forall a = (a_1, \ldots a_p) \in \mathbb{C}^p, \ \forall b = (b_1, \ldots b_p) \in \mathbb{C}^p. \tag{10.120}$$

On the other hand, using the fact that $\hat{A}$ is Hermitian, it can be easily deduced that in addition to (10.119) we have the following identity:

$$\begin{gathered}\int trace\left(\left(\hat{A}(\mathbf{x}_1, \ldots, \mathbf{x}_n, t) \otimes I\right) \cdot \xi\right)(\mathbf{z}_1, \ldots, \mathbf{z}_n, \mathbf{z}_1, \ldots, \mathbf{z}_n, t)\, d\mathbf{z}_1 \ldots d\mathbf{z}_n = \\ \int trace\left(\left(I \otimes \hat{A}^*(\mathbf{y}_1, \ldots, \mathbf{y}_n, t)\right) \cdot \xi\right)(\mathbf{z}_1, \ldots, \mathbf{z}_n, \mathbf{z}_1, \ldots, \mathbf{z}_n, t)\, d\mathbf{z}_1 \ldots d\mathbf{z}_n,\end{gathered} \tag{10.121}$$

where $\hat{A}^*(\mathbf{y}_1, \ldots, \mathbf{y}_n, t)$ is the complex conjugate (not the Hermitian adjoint) to the operator $\hat{A}$ acting on the variables $(\mathbf{y}_1, \ldots, \mathbf{y}_n)$ and so,

$$\tilde{A}(t) = \frac{\int trace\left(\left(I \otimes \hat{A}^*(\mathbf{y}_1, \ldots, \mathbf{y}_n, t)\right) \cdot \xi\right)(\mathbf{z}_1, \ldots, \mathbf{z}_n, \mathbf{z}_1, \ldots, \mathbf{z}_n, t)\, d\mathbf{z}_1 \ldots d\mathbf{z}_n}{\int trace(\xi)(\mathbf{z}_1, \ldots, \mathbf{z}_n, \mathbf{z}_1, \ldots, \mathbf{z}_n, t)\, d\mathbf{z}_1 \ldots d\mathbf{z}_n}. \tag{10.122}$$

In particular, by inserting (10.121) in the particular case $\hat{A} = \hat{H}_0$ into (10.109) we deduce

$$i\hbar \frac{\partial}{\partial t} \int trace(\xi)(\mathbf{z}_1, \ldots, \mathbf{z}_n, \mathbf{z}_1, \ldots, \mathbf{z}_n, t)\, d\mathbf{z}_1 \ldots d\mathbf{z}_n = 0, \tag{10.123}$$

and so,

$$\int trace\,(\xi)\,(\mathbf{z}_1,\ldots,\mathbf{z}_n,\mathbf{z}_1,\ldots,\mathbf{z}_n,t)\,d\mathbf{z}_1\ldots d\mathbf{z}_n$$
$$= \int trace\,(\xi)\,(\mathbf{z}_1,\ldots,\mathbf{z}_n,\mathbf{z}_1,\ldots,\mathbf{z}_n,0)\,d\mathbf{z}_1\ldots d\mathbf{z}_n. \quad (10.124)$$

Next, by (10.109) we have

$$i\hbar\frac{\partial\xi}{\partial t}(\mathbf{x}_1,\ldots,\mathbf{x}_n,\mathbf{y}_1,\ldots,\mathbf{y}_n,t) =$$
$$\left(\hat{H}_0(\mathbf{x}_1,\ldots,\mathbf{x}_n,t)\otimes I\right)\cdot\xi(\mathbf{x}_1,\ldots,\mathbf{x}_n,\mathbf{y}_1,\ldots,\mathbf{y}_n,t)$$
$$-\left(I\otimes\hat{H}_0^*(\mathbf{y}_1,\ldots,\mathbf{y}_n,t)\right)\cdot\xi(\mathbf{x}_1,\ldots,\mathbf{x}_n,\mathbf{y}_1,\ldots,\mathbf{y}_n,t). \quad (10.125)$$

Then, denote

$$\xi_1(\mathbf{x}_1,\ldots,\mathbf{x}_n,\mathbf{y}_1,\ldots,\mathbf{y}_n,t) := P\cdot\bar{\xi}(\mathbf{y}_1,\ldots,\mathbf{y}_n,\mathbf{x}_1,\ldots,\mathbf{x}_n,t), \quad (10.126)$$

where P is a linear operator (matrix), acting on vectors in $\mathbb{C}^{4^n}\otimes\mathbb{C}^{4^n}$ and satisfying

$$P\cdot(a\otimes b) = b\otimes a \quad \forall a\in\mathbb{C}^{4^n},\ \forall b\in\mathbb{C}^{4^n}, \quad (10.127)$$

and consider

$$\zeta(\mathbf{x}_1,\ldots,\mathbf{x}_n,\mathbf{y}_1,\ldots,\mathbf{y}_n,t) :=$$
$$\xi(\mathbf{x}_1,\ldots,\mathbf{x}_n,\mathbf{y}_1,\ldots,\mathbf{y}_n,t) - \xi_1(\mathbf{x}_1,\ldots,\mathbf{x}_n,\mathbf{y}_1,\ldots,\mathbf{y}_n,t). \quad (10.128)$$

Thus

$$-i\hbar P\cdot\frac{\partial\bar{\xi}}{\partial t}(\mathbf{y}_1,\ldots,\mathbf{y}_n,\mathbf{x}_1,\ldots,\mathbf{x}_n,t) =$$
$$\left(I\otimes\hat{H}_0^*(\mathbf{y}_1,\ldots,\mathbf{y}_n,t)\right)\cdot\left(P\cdot\bar{\xi}(\mathbf{y}_1,\ldots,\mathbf{y}_n,\mathbf{x}_1,\ldots,\mathbf{x}_n,t)\right)$$
$$-\left(\hat{H}_0(\mathbf{x}_1,\ldots,\mathbf{x}_n,t)\otimes I\right)\cdot\left(P\cdot\bar{\xi}(\mathbf{y}_1,\ldots,\mathbf{y}_n,\mathbf{x}_1,\ldots,\mathbf{x}_n,t)\right), \quad (10.129)$$

and then

$$i\hbar\frac{\partial\xi_1}{\partial t}(\mathbf{x}_1,\ldots,\mathbf{x}_n,\mathbf{y}_1,\ldots,\mathbf{y}_n,t) =$$
$$\left(\hat{H}_0(\mathbf{x}_1,\ldots,\mathbf{x}_n,t)\otimes I\right)\cdot\xi_1(\mathbf{x}_1,\ldots,\mathbf{x}_n,\mathbf{y}_1,\ldots,\mathbf{y}_n,t)$$
$$-\left(I\otimes\hat{H}_0^*(\mathbf{y}_1,\ldots,\mathbf{y}_n,t)\right)\cdot\xi_1(\mathbf{x}_1,\ldots,\mathbf{x}_n,\mathbf{y}_1,\ldots,\mathbf{y}_n,t). \quad (10.130)$$

Therefore,

$$i\hbar\frac{\partial\zeta}{\partial t}\left(\mathbf{x}_1,\ldots,\mathbf{x}_n,\mathbf{y}_1,\ldots,\mathbf{y}_n,t\right)=$$
$$\left(\hat{H}_0\left(\mathbf{x}_1,\ldots,\mathbf{x}_n,t\right)\otimes I\right)\cdot\zeta\left(\mathbf{x}_1,\ldots,\mathbf{x}_n,\mathbf{y}_1,\ldots,\mathbf{y}_n,t\right)$$
$$-\left(I\otimes\hat{H}_0^*\left(\mathbf{y}_1,\ldots,\mathbf{y}_n,t\right)\right)\cdot\zeta\left(\mathbf{x}_1,\ldots,\mathbf{x}_n,\mathbf{y}_1,\ldots,\mathbf{y}_n,t\right). \quad (10.131)$$

Thus if ζ satisfies initial condition $\zeta\left(\mathbf{x}_1,\ldots,\mathbf{x}_n,\mathbf{y}_1,\ldots,\mathbf{y}_n,0\right)=0$ then $\zeta\left(\mathbf{x}_1,\ldots,\mathbf{x}_n,\mathbf{y}_1,\ldots,\mathbf{y}_n,t\right)=0$ for every $t>0$. So

$$\xi\left(\mathbf{x}_1,\ldots,\mathbf{x}_n,\mathbf{y}_1,\ldots,\mathbf{y}_n,0\right)=P\cdot\bar{\xi}\left(\mathbf{y}_1,\ldots,\mathbf{y}_n,\mathbf{x}_1,\ldots,\mathbf{x}_n,0\right)\quad\text{implies}$$
$$\xi\left(\mathbf{x}_1,\ldots,\mathbf{x}_n,\mathbf{y}_1,\ldots,\mathbf{y}_n,t\right)=P\cdot\bar{\xi}\left(\mathbf{y}_1,\ldots,\mathbf{y}_n,\mathbf{x}_1,\ldots,\mathbf{x}_n,t\right)\quad\forall t\geq 0. \quad (10.132)$$

Finally, assume that the first and the second particles have the same mass $m_1=m_2$ and the same charge $\sigma_1=\sigma_2$, and moreover, assume that we have

$$V\left(\mathbf{x}_1,\mathbf{x}_2,\mathbf{x}_3,\ldots,\mathbf{x}_n,t\right)=V\left(\mathbf{x}_2,\mathbf{x}_1,\mathbf{x}_3,\ldots,\mathbf{x}_n,t\right)\qquad\forall\mathbf{x}_1,\mathbf{x}_2,\mathbf{x}_3,\ldots,\mathbf{x}_n\in\mathbb{R}^3,\quad\forall t.$$

In this case it can be easily deduced that if $\xi\left(\mathbf{x}_1,\mathbf{x}_2,\mathbf{x}_3,\ldots,\mathbf{x}_n,\mathbf{y}_1,\mathbf{y}_2,\mathbf{y}_3,\ldots,\mathbf{y}_n,t\right)\in\mathbb{C}^{4^n}\otimes\mathbb{C}^{4^n}$ is a solution of (10.125), then $C_{1,2}\cdot\xi(\mathbf{x}_2,\mathbf{x}_1,\mathbf{x}_3,\ldots,\mathbf{x}_n,\mathbf{y}_2,\mathbf{y}_1,\mathbf{y}_3,\ldots,\mathbf{y}_n,t)$ is also a solution of (10.125), where by $C_{1,2}:\mathbb{C}^{4^n}\otimes\mathbb{C}^{4^n}\to\mathbb{C}^{4^n}\otimes\mathbb{C}^{4^n}$ we denote the linear operator (matrix) defined as

$$C_{1,2}\cdot\left(\left(a_1\otimes a_2\otimes a_3\otimes\ldots\otimes a_n\right)\otimes\left(b_1\otimes b_2\otimes b_3\otimes\ldots\otimes b_n\right)\right)=$$
$$\left(\left(a_2\otimes a_1\otimes a_3\otimes\ldots\otimes a_n\right)\otimes\left(b_2\otimes b_1\otimes b_3\otimes\ldots\otimes b_n\right)\right)$$
$$\forall a_1,\ldots a_n,b_1,\ldots,b_n\in\mathbb{C}^4. \quad (10.133)$$

Therefore, if $\xi\left(\mathbf{x}_1,\mathbf{x}_2,\mathbf{x}_3,\ldots,\mathbf{x}_n,\mathbf{y}_1,\mathbf{y}_2,\mathbf{y}_3,\ldots,\mathbf{y}_n,t\right)\in\mathbb{C}^{4^n}\otimes\mathbb{C}^{4^n}$ is a solution of (10.125) then for every $t\geq 0$ we will have

$$C_{1,2}\cdot\xi(\mathbf{x}_2,\mathbf{x}_1,\mathbf{x}_3,\ldots,\mathbf{x}_n,\mathbf{y}_2,\mathbf{y}_1,\mathbf{y}_3,\ldots,\mathbf{y}_n,t)$$
$$=\xi\left(\mathbf{x}_1,\mathbf{x}_2,\mathbf{x}_3,\ldots,\mathbf{x}_n,\mathbf{y}_1,\mathbf{y}_2,\mathbf{y}_3,\ldots,\mathbf{y}_n,t\right)$$
$$\forall\mathbf{x}_1,\mathbf{x}_2,\mathbf{x}_3,\ldots,\mathbf{x}_n,\mathbf{y}_1,\mathbf{y}_2,\mathbf{y}_3,\ldots,\mathbf{y}_n\in\mathbb{R}^3, \quad (10.134)$$

provided that (10.134) holds for the initial instant of time $t=0$. So we have a consistency with the Pauli Exclusion Principle for two or more identical fermions.

Next given $\xi(\mathbf{x}_1, \ldots, \mathbf{x}_n, \mathbf{y}_1, \ldots, \mathbf{y}_n, t)$ that satisfies

$$\xi(\mathbf{x}_1, \ldots, \mathbf{x}_n, \mathbf{y}_1, \ldots, \mathbf{y}_n, t) = P \cdot \bar{\xi}(\mathbf{y}_1, \ldots, \mathbf{y}_n, \mathbf{x}_1, \ldots, \mathbf{x}_n, t) \quad \forall t \geq 0, \tag{10.135}$$

define the operator $\hat{R}_\xi$ by

$$\hat{R}_\xi(\mathbf{x}_1, \ldots, \mathbf{x}_n, t) \cdot \psi(\mathbf{x}_1, \ldots, \mathbf{x}_n, t) = \int B(\xi(\mathbf{x}_1, \ldots, \mathbf{x}_n, \mathbf{y}_1, \ldots, \mathbf{y}_n, t), \psi(\mathbf{y}_1, \ldots, \mathbf{y}_n, t))\, d\mathbf{y}_1 \ldots d\mathbf{y}_n, \tag{10.136}$$

where B is a bilinear mapping, acting on the pairs of vectors in $\left(\mathbb{C}^{4^n} \otimes \mathbb{C}^{4^n}\right) \times \mathbb{C}^{4^n}$, taking the values in $\mathbb{C}^{4^n}$ and having the form

$$B((a \otimes b), c) = (b \cdot c)a \quad \forall a \in \mathbb{C}^{4^n},\ \forall b \in \mathbb{C}^{4^n},\ \forall c \in \mathbb{C}^{4^n}. \tag{10.137}$$

Then the mapping $\xi \to \hat{R}_\xi$ is one-to-one. Moreover, by (10.135) $\hat{R}_\xi$ is an Hermitian operator. Finally by (10.125) we have

$$i\hbar \frac{\partial \hat{R}_\xi}{\partial t}(\mathbf{x}_1, \ldots, \mathbf{x}_n, t) = \hat{H}_0(\mathbf{x}_1, \ldots, \mathbf{x}_n, t) \cdot \hat{R}_\xi(\mathbf{x}_1, \ldots, \mathbf{x}_n, t) - \hat{R}_\xi(\mathbf{x}_1, \ldots, \mathbf{x}_n, t) \cdot \hat{H}_0(\mathbf{x}_1, \ldots, \mathbf{x}_n, t). \tag{10.138}$$

Equation (10.138) is equivalent to (10.125).

Next, clearly, ξ satisfies (10.134) if and only if we have the following relation:

$$\begin{gathered} B_{1,2} \cdot \psi(\mathbf{x}_2, \mathbf{x}_1, \mathbf{x}_3, \ldots, \mathbf{x}_n, t) = -\psi(\mathbf{x}_1, \mathbf{x}_2, \mathbf{x}_3, \ldots, \mathbf{x}_n, t) \ \forall \mathbf{x}_1, \ldots, \mathbf{x}_n \in \mathbb{R}^3 \\ \text{and} \quad \phi = \hat{R}_\xi \cdot \psi \\ \text{implies} \quad B_{1,2} \cdot \phi(\mathbf{x}_2, \mathbf{x}_1, \mathbf{x}_3, \ldots, \mathbf{x}_n, t) = -\phi(\mathbf{x}_1, \mathbf{x}_2, \mathbf{x}_3, \ldots, \mathbf{x}_n, t) \\ \forall \mathbf{x}_1, \ldots, \mathbf{x}_n \in \mathbb{R}^3, \end{gathered} \tag{10.139}$$

where by $B_{1,2} : \mathbb{C}^{4^n} \to \mathbb{C}^{4^n}$ we denote the linear operator (matrix) defined as in (10.107) by the following:

$$B_{1,2} \cdot (a_1 \otimes a_2 \otimes a_3 \otimes \ldots \otimes a_n) = (a_2 \otimes a_1 \otimes a_3 \otimes \ldots \otimes a_n) \qquad \forall a_1, \ldots a_n \in \mathbb{C}^4. \tag{10.140}$$

Finally, assume that $\hat{A}(\mathbf{x}_1, \ldots, \mathbf{x}_n, t)$ is some Hermitian operator acting on the functions $\psi(\mathbf{x}_1, \ldots, \mathbf{x}_n, t) \in \mathbb{C}^{4^n}$ with respect to spatial variables $(\mathbf{x}_1, \ldots, \mathbf{x}_n)$. Then, we remind that the average $\tilde{A}(t)$ of $\hat{A}$ on the density matrix ξ is defined by (10.119):

$$\tilde{A}(t) = \frac{\int trace\left(\left(\hat{A}\left(\mathbf{x}_1,\ldots,\mathbf{x}_n,t\right)\otimes I\right)\cdot\xi\right)\left(\mathbf{z}_1,\ldots,\mathbf{z}_n,\mathbf{z}_1,\ldots,\mathbf{z}_n,t\right)d\mathbf{z}_1\ldots d\mathbf{z}_n}{\int trace\left(\xi\right)\left(\mathbf{z}_1,\ldots,\mathbf{z}_n,\mathbf{z}_1,\ldots,\mathbf{z}_n,t\right)d\mathbf{z}_1\ldots d\mathbf{z}_n}. \tag{10.141}$$

On the other hand,

$$trace\left(\hat{A}\circ\hat{R}_\xi\right) = \int trace\left(\left(\hat{A}\left(\mathbf{x}_1,\ldots,\mathbf{x}_n,t\right)\otimes I\right)\cdot\xi\right)\left(\mathbf{z}_1,\ldots,\mathbf{z}_n,\mathbf{z}_1,\ldots,\mathbf{z}_n,t\right)d\mathbf{z}_1\ldots d\mathbf{z}_n, \tag{10.142}$$

and so,

$$\tilde{A}(t) = \frac{trace\left(\hat{A}\circ\hat{R}_\xi\right)}{trace\left(\hat{R}_\xi\right)} = \frac{trace\left(\hat{R}_\xi\circ\hat{A}\right)}{trace\left(\hat{R}_\xi\right)}, \tag{10.143}$$

where by the trace we mean the trace of an operator on a Hilbert space. Moreover, by (10.124) and (10.142), we have

$$trace\left(\hat{R}_\xi\right)(t) = trace\left(\hat{R}_\xi\right)(0). \tag{10.144}$$

10.6.1 Thermodynamical Equilibrium for Relativistic-Like Spin-Half Particles

Let $\xi\left(\mathbf{x}_1,\ldots,\mathbf{x}_n,\mathbf{y}_1,\ldots,\mathbf{y}_n,t\right)$ be the density matrix of the quantum system, ruled by the Hamiltonian operator $\hat{H}_0$ from (10.111), in the case of thermodynamical equilibrium and $\hat{R}_\xi$ is given by (10.136) for this ξ. Then in the case that $\frac{\partial\hat{H}_0}{\partial t} = 0$ and the given equilibrium system rests macroscopically, i.e. it has the macroscopical velocity field zero: $\mathbf{u}(\mathbf{x},t) = 0$ it is well known that

$$\hat{R}_\xi = \frac{1}{trace\left(f\circ\hat{H}_0\right)}f\circ\hat{H}_0, \tag{10.145}$$

where $f(z):\mathbb{C}\to\mathbb{C}$ is some holomorphic function defined as a sum of the power series

$$f(z) := \sum_{m=0}^{+\infty} a_m z^m, \tag{10.146}$$

with $a_m \in \mathbb{C}$, and the operator $f \circ \hat{H}_0$ is given by

$$f \circ \hat{H}_0 := \sum_{m=0}^{+\infty} a_m \hat{H}_0^m. \tag{10.147}$$

We remind that in the case of a non-relativistic system in thermostat, having the Kelvin temperature T, we considered the canonical ensemble:

$$f(s) = e^{-\frac{s}{kT}} = \sum_{m=0}^{+\infty} \frac{1}{m!}\left(-\frac{s}{kT}\right)^m \quad \forall s \in \mathbb{C}, \tag{10.148}$$

where k is the Boltzmann constant and by $f \circ \hat{H}_0$ we denoted the following operator:

$$f \circ \hat{H}_0 = \sum_{m=0}^{+\infty} \frac{1}{m!}\left(-\frac{1}{kT}\hat{H}_0\right)^m \quad \forall s \in \mathbb{C}. \tag{10.149}$$

However, in the non-relativistic case the Hamiltonian operator $\hat{H}_0$ was bounded from below and then the trace of $f \circ \hat{H}_0$ was finite. On the other hand, in the case of the Dirac system, the infinitely small negative energies are possible and then the trace of $f \circ \hat{H}_0$ could be infinite. Thus, we need to assume in addition that $f(s)$ decays rapidly as $s \to \pm\infty$. For example, the possible choice is the following:

$$f(s) = e^{-\frac{s^2}{2\left(\sum_{j=1}^n m_j\right)c^2kT}} = \sum_{m=0}^{+\infty} \frac{1}{m!}\left(-\frac{s^2}{2\left(\sum_{j=1}^n m_j\right)c^2kT}\right)^m \quad \forall s \in \mathbb{C}, \tag{10.150}$$

where k is the Boltzmann constant and by $f \circ \hat{H}_0$ we denote the following operator

$$f \circ \hat{H}_0 = \sum_{m=0}^{+\infty} \frac{1}{m!}\left(-\frac{1}{2\left(\sum_{j=1}^n m_j\right)c^2kT}\hat{H}_0^2\right)^m \quad \forall s \in \mathbb{C}. \tag{10.151}$$

We would like to find an alternative form of the above law of thermodynamical equilibrium, which is invariant under the change of inertial or non-inertial Cartesian coordinate system. Then it is clear that if the given system rests in the old coordinate system, then it obviously has non-trivial macroscopical velocity field $\mathbf{u}(\mathbf{x}, t)$ in the new one. Moreover, $\mathbf{u}(\mathbf{x}, t)$ can depend on $\mathbf{x}$ and t, as it indeed happens in the case of a rotation of the new coordinate system with respect to the old one. On the other hand the concept of thermodynamical equilibrium is clearly independent of the coordinate system.

In order to find the forms of the law of thermodynamical equilibrium, which are indeed invariant under the change of inertial or non-inertial Cartesian coordinate system we follow the steps as below. Given a speed-like vector field $\mathbf{u} = \mathbf{u}(\mathbf{x}, t)$, define

$$\hat{H}_{\mathbf{u}}(\mathbf{x}_1, \ldots, \mathbf{x}_n, t) :=$$
$$\hat{H}_0(\mathbf{x}_1, \ldots, \mathbf{x}_n, t) - \frac{1}{2}\left(\sum_{j=1}^{n} \hat{\mathbf{P}}_j \cdot \mathbf{u}(\mathbf{x}_j, t) + \mathbf{u}(\mathbf{x}_j, t) \cdot \hat{\mathbf{P}}_j\right) - \sum_{j=1}^{n} \frac{\hbar}{4} \mathbf{M}_j \cdot (\mathrm{curl}_{\mathbf{x}_j} \mathbf{u}(\mathbf{x}_j, t)), \tag{10.152}$$

where $\hat{\mathbf{P}}_j$ is operator defined as

$$\hat{\mathbf{P}}_j \cdot \psi(\mathbf{x}_1, \ldots, \mathbf{x}_n) = -i\hbar\nabla_{\mathbf{x}_j}\psi(\mathbf{x}_1, \ldots, \mathbf{x}_n). \tag{10.153}$$

On the other hand, by (10.111) the Hamiltonian operator $\hat{H}_0$ has the forms:

$$\hat{H}_0(\mathbf{x}_1, \ldots, \mathbf{x}_n, t) \cdot \psi(\mathbf{x}_1, \ldots, \mathbf{x}_n, t) =$$
$$\sum_{j=1}^{n} m_j c^2 D_4^j \cdot \psi - \sum_{j=1}^{n} c\,\mathbf{D}_j \cdot \left(i\hbar\nabla_{\mathbf{x}_j}\psi + \frac{\sigma_j}{c}\mathbf{A}(\mathbf{x}_j, t)\psi\right)$$
$$+ \sum_{j=1}^{n} \sigma_j \left(\Psi(\mathbf{x}_j, t) - \frac{1}{c}\mathbf{v}(\mathbf{x}_j, t) \cdot \mathbf{A}(\mathbf{x}_j, t)\right)\psi - V(\mathbf{x}_1, \ldots, \mathbf{x}_n, t)\,\psi$$
$$- \sum_{j=1}^{n} \frac{i\hbar}{2} \mathrm{div}_{\mathbf{x}_j}\left\{\psi \mathbf{v}(\mathbf{x}_j, t)\right\} - \sum_{j=1}^{n} \frac{i\hbar}{2} \nabla_{\mathbf{x}_j}\psi \cdot \mathbf{v}(\mathbf{x}_j, t)$$
$$+ \sum_{j=1}^{n} \frac{\hbar}{4} \mathbf{M}_j \cdot \left(\mathrm{curl}_{\mathbf{x}_j} \mathbf{v}(\mathbf{x}_j, t)\,\psi\right) - \sum_{j=1}^{n} \frac{(g_j - 1)\sigma_j \hbar}{2m_j c} \mathbf{M}_j \cdot \left(\mathrm{curl}_{\mathbf{x}_j} \mathbf{A}(\mathbf{x}_j, t)\,\psi\right). \tag{10.154}$$

Then by (10.154) and (10.152) we have

$$\hat{H}_{\mathbf{u}}(\mathbf{x}_1, \ldots, \mathbf{x}_n, t) \cdot \psi(\mathbf{x}_1, \ldots, \mathbf{x}_n, t) =$$
$$\sum_{j=1}^{n} m_j c^2 D_4^j \cdot \psi - \sum_{j=1}^{n} c\,\mathbf{D}_j \cdot \left(i\hbar\nabla_{\mathbf{x}_j}\psi + \frac{\sigma_j}{c}\mathbf{A}(\mathbf{x}_j, t)\psi\right)$$
$$+ \sum_{j=1}^{n} \sigma_j \left(\Psi(\mathbf{x}_j, t) - \frac{1}{c}\mathbf{v}(\mathbf{x}_j, t) \cdot \mathbf{A}(\mathbf{x}_j, t)\right)\psi - V(\mathbf{x}_1, \ldots, \mathbf{x}_n, t)\,\psi$$
$$- \sum_{j=1}^{n} \frac{i\hbar}{2} \mathrm{div}_{\mathbf{x}_j}\left\{\psi\left(\mathbf{v}(\mathbf{x}_j, t) - \mathbf{u}(\mathbf{x}_j, t)\right)\right\} - \sum_{j=1}^{n} \frac{i\hbar}{2} \nabla_{\mathbf{x}_j}\psi \cdot \left(\mathbf{v}(\mathbf{x}_j, t) - \mathbf{u}(\mathbf{x}_j, t)\right)$$
$$+ \sum_{j=1}^{n} \frac{\hbar}{4} \mathbf{M}_j \cdot \left(\mathrm{curl}_{\mathbf{x}_j} \left(\mathbf{v}(\mathbf{x}_j, t) - \mathbf{u}(\mathbf{x}_j, t)\right)\psi\right) - \sum_{j=1}^{n} \frac{(g_j - 1)\sigma_j \hbar}{2m_j c} \mathbf{M}_j \cdot \left(\mathrm{curl}_{\mathbf{x}_j} \mathbf{A}(\mathbf{x}_j, t)\,\psi\right). \tag{10.155}$$

Thus, as before, we can prove that (10.155) is invariant under the change of inertial or non-inertial Cartesian coordinate systems, i.e.

$$\hat{H}'_{\mathbf{u}'}\left(\mathbf{x}'_1,\ldots,\mathbf{x}'_n,t'\right)\cdot\psi'\left(\mathbf{x}'_1,\ldots,\mathbf{x}'_n,t'\right)=\phi'\left(\mathbf{x}'_1,\ldots,\mathbf{x}'_n,t'\right)\quad\text{if and only if}$$
$$\hat{H}_{\mathbf{u}}\left(\mathbf{x}_1,\ldots,\mathbf{x}_n,t\right)\cdot\psi\left(\mathbf{x}_1,\ldots,\mathbf{x}_n,t\right)=\phi\left(\mathbf{x}_1,\ldots,\mathbf{x}_n,t\right),\tag{10.156}$$

provided that, as before in (10.93), we have

$$\begin{cases}V'\left(\mathbf{x}'_1,\ldots,\mathbf{x}'_n,t'\right) = V\left(\mathbf{x}_1,\ldots,\mathbf{x}_n,t\right),\\ \sigma'_j = \sigma_j,\\ m'_j = m_j,\\ g'_j = g_j,\\ \mathbf{v}'(\mathbf{x}',t) = A(t)\cdot\mathbf{v}(\mathbf{x},t)+\frac{dA}{dt}(t)\cdot\mathbf{x}+\frac{d\mathbf{z}}{dt}(t),\\ \mathbf{u}'(\mathbf{x}',t) = A(t)\cdot\mathbf{u}(\mathbf{x},t)+\frac{dA}{dt}(t)\cdot\mathbf{x}+\frac{d\mathbf{z}}{dt}(t),\\ \mathbf{A}'(\mathbf{x}',t) = A(t)\cdot\mathbf{A}(\mathbf{x},t),\\ \Psi'(\mathbf{x}',t)-\mathbf{v}'(\mathbf{x}',t)\cdot\mathbf{A}'(\mathbf{x}',t) = \Psi(\mathbf{x},t)-\mathbf{v}(\mathbf{x},t)\cdot\mathbf{A}(\mathbf{x},t),\\ \psi'\left(\mathbf{x}'_1,\ldots,\mathbf{x}'_n,t'\right) = \left(W(t)\otimes_1 W(t)\otimes_2 W(t)\ldots\otimes_{(n-1)} W(t)\right)\cdot\psi\left(\mathbf{x}_1,\ldots,\mathbf{x}_n,t\right),\\ \phi'\left(\mathbf{x}'_1,\ldots,\mathbf{x}'_n,t'\right) = \left(W(t)\otimes_1 W(t)\otimes_2 W(t)\ldots\otimes_{(n-1)} W(t)\right)\cdot\phi\left(\mathbf{x}_1,\ldots,\mathbf{x}_n,t\right).\end{cases}\tag{10.157}$$

Next consider

$$\hat{R}_{\xi}=\frac{1}{trace\left(f\circ\hat{H}_{\mathbf{u}}\right)}f\circ\hat{H}_{\mathbf{u}},\tag{10.158}$$

where $f(z):\mathbb{C}\to\mathbb{C}$ is some holomorphic function defined as a sum of the power series

$$f(z):=\sum_{m=0}^{+\infty}a_m z^m,\tag{10.159}$$

with $a_m\in\mathbb{C}$, such that $f(s)$ decays rapidly as $s\to\pm\infty$. For example, we can take $f(s)$ be defined by (10.150). Moreover, consider the operator $f\circ\hat{H}_{\mathbf{u}}$, given by

$$f\circ\hat{H}_{\mathbf{u}}:=\sum_{m=0}^{+\infty}a_m\hat{H}^m_{\mathbf{u}}.\tag{10.160}$$

We would like to note here that by (10.155) in the case that vector field $\mathbf{u}(\mathbf{x},t)$ satisfies

$$|\mathbf{u}(\mathbf{x},t)-\mathbf{v}(\mathbf{x},t)|<c\qquad\forall\,(\mathbf{x},t),\tag{10.161}$$

since $f(s)$ decays rapidly as $s\to\pm\infty$, there exists a density matrix ξ_1, such that the operator $\hat{R}_{\xi_1}$, given by (10.136), equals to the operator $f\circ\hat{H}_{\mathbf{u}}$ and thus (10.158) indeed has sense. Next by (10.156) and (10.160) the operator $\hat{R}_{\xi}$ in the left-hand side of (10.158) is invariant under the change of inertial or non-inertial Cartesian coordinate systems, i.e.

$$\hat{R}_{\xi'}\left(\mathbf{x}'_1,\ldots,\mathbf{x}'_n,t'\right)\cdot\psi'\left(\mathbf{x}'_1,\ldots,\mathbf{x}'_n,t'\right)=\phi'\left(\mathbf{x}'_1,\ldots,\mathbf{x}'_n,t'\right)\quad\text{if and only if}$$
$$\hat{R}_{\xi}\left(\mathbf{x}_1,\ldots,\mathbf{x}_n,t\right)\cdot\psi\left(\mathbf{x}_1,\ldots,\mathbf{x}_n,t\right)=\phi\left(\mathbf{x}_1,\ldots,\mathbf{x}_n,t\right),\tag{10.162}$$

provided that, as before, we have (10.157) and

$$\xi'\left(\mathbf{x}'_1,\ldots,\mathbf{x}'_n,\mathbf{y}'_1,\ldots,\mathbf{y}'_n,t'\right)=\Big(\left(W(t)\otimes_1 W(t)\ldots\otimes_{(n-1)}W(t)\right)$$
$$\otimes\left(\bar{W}(t)\otimes_1\bar{W}(t)\ldots\otimes_{(n-1)}\bar{W}(t)\right)\Big)\cdot\xi\left(\mathbf{x}_1,\ldots,\mathbf{x}_n,\mathbf{y}_1,\ldots,\mathbf{y}_n,t\right).\tag{10.163}$$

Moreover, in the case $\mathbf{u}=0$ (10.158) coincides with (10.145). However, we still need to derive the restrictions on the field $\mathbf{u}$ and the Hamiltonian operator $\hat{H}_0$, providing that our system can indeed be found in the state of thermodynamical equilibrium. We remind that in the case $\mathbf{u}=0$ the appropriate restriction is $\frac{\partial\hat{H}_0}{\partial t}=0$. In order to get these restrictions in the general case, we need to insert $\hat{R}_\xi$ in (10.158) into the equation in (10.138) which is equivalent to the quantum Liouville equation.

Therefore, assume that the vector field $\mathbf{u}$ satisfies (10.69) with (10.70), i.e.

$$\begin{cases}\frac{\partial U}{\partial t}\left(\mathbf{x}_1,\ldots,\mathbf{x}_n,t\right)+\sum_{j=1}^n\mathbf{u}(\mathbf{x}_j,t)\cdot\nabla_{\mathbf{x}_j}U\left(\mathbf{x}_1,\ldots,\mathbf{x}_n,t\right)=0,\\ \frac{\partial}{\partial t}\left(\mathbf{u}(\mathbf{x},t)-\mathbf{v}(\mathbf{x},t)\right)-\mathrm{curl}_{\mathbf{x}}\left(\mathbf{u}(\mathbf{x},t)\times\left(\mathbf{u}(\mathbf{x},t)-\mathbf{v}(\mathbf{x},t)\right)\right)\\ +\left(\mathrm{div}_{\mathbf{x}}\left(\mathbf{u}(\mathbf{x},t)-\mathbf{v}(\mathbf{x},t)\right)\right)\mathbf{u}(\mathbf{x},t)=0,\\ \frac{\partial\mathbf{A}}{\partial t}(\mathbf{x},t)-\mathrm{curl}_{\mathbf{x}}\left(\mathbf{u}(\mathbf{x},t)\times\mathbf{A}(\mathbf{x},t)\right)+\left(\mathrm{div}_{\mathbf{x}}\mathbf{A}(\mathbf{x},t)\right)\mathbf{u}(\mathbf{x},t)=0,\\ d_{\mathbf{x}}\mathbf{u}(\mathbf{x},t)+\{d_{\mathbf{x}}\mathbf{u}(\mathbf{x},t)\}^T=0,\end{cases}\tag{10.164}$$

where

$$U\left(\mathbf{x}_1,\ldots,\mathbf{x}_n,t\right):=$$
$$\sum_{j=1}^n\left(\frac{\sigma_j^2}{2m_jc^2}\left|\mathbf{A}(\mathbf{x}_j,t)\right|^2+\sigma_j\Psi(\mathbf{x}_j,t)-\frac{\sigma_j}{c}\mathbf{A}(\mathbf{x}_j,t)\cdot\mathbf{v}(\mathbf{x}_j,t)\right)-V\left(\mathbf{x}_1,\ldots,\mathbf{x}_n,t\right).\tag{10.165}$$

Then, by Proposition 10.1 below we have

$$i\hbar\frac{\partial}{\partial t}\left(f\circ\hat{H}_{\mathbf{u}}\right)=\hat{H}_0\cdot\left(f\circ\hat{H}_{\mathbf{u}}\right)-\left(f\circ\hat{H}_{\mathbf{u}}\right)\cdot\hat{H}_0.\tag{10.166}$$

Then by (10.166) together with (10.144) we deduce that

$$i\hbar\frac{\partial}{\partial t}\left(\hat{R}_\xi\right) = \hat{H}_0 \cdot \left(\hat{R}_\xi\right) - \left(\hat{R}_\xi\right) \cdot \hat{H}_0, \tag{10.167}$$

where $\hat{R}_\xi$ is the operator in the left-hand side of (10.158). So we indeed get (10.138) in the case of (10.164), (10.165). Moreover, (10.164), (10.165) and (10.161) are invariant under the change of inertial or non-inertial coordinate system.

Proposition 10.1 *Assume that the speed-like vector field* $\mathbf{u}$ *satisfies (10.164) and (10.165). Next assume that the holomorphic function* $f(z) : \mathbb{C} \to \mathbb{C}$ *defined as a sum of the power series*

$$f(z) := \sum_{m=0}^{+\infty} a_m z^m, \tag{10.168}$$

with $a_m \in \mathbb{C}$*, is such that for the operator* $f \circ \hat{H}_{\mathbf{u}}$*, given by*

$$f \circ \hat{H}_{\mathbf{u}} := \sum_{m=0}^{+\infty} a_m \hat{H}_{\mathbf{u}}^m, \tag{10.169}$$

where the operator $\hat{H}_{\mathbf{u}}$ *is given by (10.155), there exists a density matrix* $\xi(\mathbf{x}_1, \ldots, \mathbf{x}_n, \mathbf{y}_1, \ldots, \mathbf{y}_n, t)$*, such that*

$$\left(f \circ \hat{H}_{\mathbf{u}}\right)(\mathbf{x}_1, \ldots, \mathbf{x}_n, t) = \hat{R}_\xi(\mathbf{x}_1, \ldots, \mathbf{x}_n, t), \tag{10.170}$$

where $\hat{R}_\xi$ *is given by (10.136). Then the operator* $f \circ \hat{H}_{\mathbf{u}}$ *satisfies*

$$i\hbar\frac{\partial}{\partial t}\left(f \circ \hat{H}_{\mathbf{u}}\right) = \hat{H}_0 \cdot \left(f \circ \hat{H}_{\mathbf{u}}\right) - \left(f \circ \hat{H}_{\mathbf{u}}\right) \cdot \hat{H}_0, \tag{10.171}$$

or equivalently

$$\begin{aligned} i\hbar\frac{\partial \xi}{\partial t}(\mathbf{x}_1, \ldots, \mathbf{x}_n, \mathbf{y}_1, \ldots, \mathbf{y}_n, t) = \\ \left(\hat{H}_0(\mathbf{x}_1, \ldots, \mathbf{x}_n, t) \otimes I\right) \cdot \xi(\mathbf{x}_1, \ldots, \mathbf{x}_n, \mathbf{y}_1, \ldots, \mathbf{y}_n, t) \\ - \left(I \otimes \hat{H}_0^*(\mathbf{y}_1, \ldots, \mathbf{y}_n, t)\right) \cdot \xi(\mathbf{x}_1, \ldots, \mathbf{x}_n, \mathbf{y}_1, \ldots, \mathbf{y}_n, t). \end{aligned} \tag{10.172}$$

Moreover, $f \circ \hat{H}_{\mathbf{u}}$ is invariant under the change of inertial or non-inertial Cartesian coordinate systems, i.e.

$$\left(f \circ \hat{H}'_{\mathbf{u}'}\right)(\mathbf{x}'_1, \ldots, \mathbf{x}'_n, t') \cdot \psi'(\mathbf{x}'_1, \ldots, \mathbf{x}'_n, t') = \phi'(\mathbf{x}'_1, \ldots, \mathbf{x}'_n, t') \quad \textit{if and only if}$$
$$\left(f \circ \hat{H}_{\mathbf{u}}\right)(\mathbf{x}_1, \ldots, \mathbf{x}_n, t) \cdot \psi(\mathbf{x}_1, \ldots, \mathbf{x}_n, t) = \phi(\mathbf{x}_1, \ldots, \mathbf{x}_n, t), \qquad (10.173)$$

provided that we have (10.157). Finally, if we assume that the first and the second particles have the same mass $m_1 = m_2$ and the same charge $\sigma_1 = \sigma_2$, and moreover, if we assume that

$$V(\mathbf{x}_1, \mathbf{x}_2, \mathbf{x}_3, \ldots, \mathbf{x}_n, t) = V(\mathbf{x}_2, \mathbf{x}_1, \mathbf{x}_3, \ldots, \mathbf{x}_n, t) \qquad \forall \mathbf{x}_1, \mathbf{x}_2, \mathbf{x}_3, \ldots, \mathbf{x}_n \in \mathbb{R}^3, \quad \forall t,$$

then

$$B_{1,2} \cdot \psi(\mathbf{x}_2, \mathbf{x}_1, \mathbf{x}_3, \ldots, \mathbf{x}_n, t) = -\psi(\mathbf{x}_1, \mathbf{x}_2, \mathbf{x}_3, \ldots, \mathbf{x}_n, t) \;\; \forall \mathbf{x}_1, \ldots, \mathbf{x}_n \in \mathbb{R}^3$$
$$\textit{and} \;\; \phi = \left(f \circ \hat{H}_{\mathbf{u}}\right) \cdot \psi$$
$$\textit{implies} \;\; B_{1,2} \cdot \phi(\mathbf{x}_2, \mathbf{x}_1, \mathbf{x}_3, \ldots, \mathbf{x}_n, t) = -\phi(\mathbf{x}_1, \mathbf{x}_2, \mathbf{x}_3, \ldots, \mathbf{x}_n, t)$$
$$\forall \mathbf{x}_1, \ldots, \mathbf{x}_n \in \mathbb{R}^3, \qquad (10.174)$$

where, as before, by $B_{1,2} : \mathbb{C}^{4^n} \to \mathbb{C}^{4^n}$ we denote the linear operator (matrix) defined as in (10.107) by the following:

$$B_{1,2} \cdot (a_1 \otimes a_2 \otimes a_3 \otimes \ldots \otimes a_n) = (a_2 \otimes a_1 \otimes a_3 \otimes \ldots \otimes a_n) \qquad \forall a_1, \ldots a_n \in \mathbb{C}^4. \qquad (10.175)$$

Proof Again by (10.164) and Proposition 3.3 there exists another Cartesian coordinate system $(**)$ such that under the change of coordinate system $(*)$ to another Cartesian coordinate system $(**)$, given by (3.54), we have

$$A(t) \cdot \mathbf{u}(\mathbf{x}, t) + \frac{dA}{dt}(t) \cdot \mathbf{x} + \frac{d\mathbf{z}}{dt}(t) = \mathbf{u}'(\mathbf{x}', t') = 0. \qquad (10.176)$$

Thus, since (10.164), (10.165) are invariant under the change of inertial or non-inertial Cartesian coordinate systems, as before, in system $(**)$ we have

$$\begin{cases} \frac{\partial U'}{\partial t'}(\mathbf{x}'_1, \ldots, \mathbf{x}'_n, t') = 0, \\ \frac{\partial \mathbf{v}'}{\partial t'}(\mathbf{x}', t') = 0, \\ \frac{\partial \mathbf{A}'}{\partial t'}(\mathbf{x}', t') = 0, \\ \mathbf{u}'(\mathbf{x}', t') = 0, \end{cases} \qquad (10.177)$$

where

$$U'\left(\mathbf{x}'_1,\ldots,\mathbf{x}'_n,t'\right) := \sum_{j=1}^{n}\left(\frac{(\sigma'_j)^2}{2m'_jc^2}\left|\mathbf{A}'(\mathbf{x}'_j,t')\right|^2 + \sigma'_j\Psi'(\mathbf{x}'_j,t') - \frac{\sigma'_j}{c}\mathbf{A}'(\mathbf{x}'_j,t')\cdot\mathbf{v}'(\mathbf{x}'_j,t')\right) - V'\left(\mathbf{x}'_1,\ldots,\mathbf{x}'_n,t'\right). \quad (10.178)$$

On the other hand, (10.177) is equivalent to

$$\frac{\partial \hat{H}'_0}{\partial t'}\left(\mathbf{x}'_1,\ldots,\mathbf{x}'_n,t'\right) = 0 \quad \text{and} \quad \mathbf{u}'(\mathbf{x}',t') = 0, \quad (10.179)$$

in system $(**)$. Thus, by (10.179), in system $(**)$ we deduce

$$\hat{H}'_0\cdot\hat{H}'_{\mathbf{u}'} - \hat{H}'_{\mathbf{u}'}\cdot\hat{H}'_0 = \hat{H}'_0\cdot\hat{H}'_0 - \hat{H}'_0\cdot\hat{H}'_0 = 0 = i\hbar\frac{\partial \hat{H}'_0}{\partial t'} = i\hbar\frac{\partial \hat{H}'_{\mathbf{u}'}}{\partial t'}. \quad (10.180)$$

So we get

$$i\hbar\frac{\partial \hat{H}'_{\mathbf{u}'}}{\partial t'} = \hat{H}'_0\cdot\hat{H}'_{\mathbf{u}'} - \hat{H}'_{\mathbf{u}'}\cdot\hat{H}'_0. \quad (10.181)$$

Therefore, given the holomorphic function $f(z):\mathbb{C}\to\mathbb{C}$ defined as a sum of the power series

$$f(z) := \sum_{m=0}^{+\infty} a_m z^m, \quad (10.182)$$

with $a_m \in \mathbb{C}$, if we define the operator $f\circ\hat{H}_{\mathbf{u}}$ as

$$f\circ\hat{H}_{\mathbf{u}} := \sum_{m=0}^{+\infty} a_m \hat{H}_{\mathbf{u}}^m, \quad (10.183)$$

by (10.181) and Proposition 6.1 we deduce

$$i\hbar\frac{\partial}{\partial t'}\left(f\circ\hat{H}'_{\mathbf{u}'}\right) = \hat{H}'_0\cdot\left(f\circ\hat{H}'_{\mathbf{u}'}\right) - \left(f\circ\hat{H}'_{\mathbf{u}'}\right)\cdot\hat{H}'_0. \quad (10.184)$$

Moreover, by (10.156) and (10.183), we can easily prove that $f\circ\hat{H}_{\mathbf{u}}$ is invariant under the change of inertial or non-inertial Cartesian coordinate systems, i.e.

$$\left(f\circ\hat{H}'_{\mathbf{u}'}\right)\left(\mathbf{x}'_1,\ldots,\mathbf{x}'_n,t'\right)\cdot\psi'\left(\mathbf{x}'_1,\ldots,\mathbf{x}'_n,t'\right) = \phi'\left(\mathbf{x}'_1,\ldots,\mathbf{x}'_n,t'\right) \quad \text{if and only if}$$
$$\left(f\circ\hat{H}_{\mathbf{u}}\right)\left(\mathbf{x}_1,\ldots,\mathbf{x}_n,t\right)\cdot\psi\left(\mathbf{x}_1,\ldots,\mathbf{x}_n,t\right) = \phi\left(\mathbf{x}_1,\ldots,\mathbf{x}_n,t\right), \quad (10.185)$$

provided that, as before, we have (10.157).

Next assume that the holomorphic function f in (10.182) is such that there exists a density matrix $\xi\left(\mathbf{x}_1,\ldots,\mathbf{x}_n,\mathbf{y}_1,\ldots,\mathbf{y}_n,t\right)$ satisfying

$$\left(f\circ\hat{H}_{\mathbf{u}}\right)\left(\mathbf{x}_1,\ldots,\mathbf{x}_n,t\right)=\hat{R}_{\xi}\left(\mathbf{x}_1,\ldots,\mathbf{x}_n,t\right), \tag{10.186}$$

where $\hat{R}_{\xi}$ is given by (10.136). Then, by (10.185) for the density matrix $\xi'\left(\mathbf{x}'_1,\ldots,\mathbf{x}'_n,\mathbf{y}'_1,\ldots,\mathbf{y}'_n,t'\right)$ we have

$$\left(f\circ\hat{H}'_{\mathbf{u}'}\right)\left(\mathbf{x}'_1,\ldots,\mathbf{x}'_n,t'\right)=\hat{R}_{\xi'}\left(\mathbf{x}'_1,\ldots,\mathbf{x}'_n,t'\right), \tag{10.187}$$

provided that

$$\xi'\left(\mathbf{x}'_1,\ldots,\mathbf{x}'_n,\mathbf{y}'_1,\ldots,\mathbf{y}'_n,t'\right)=\Bigg(\left(W(t)\otimes_1 W(t)\ldots\otimes_{(n-1)} W(t)\right)$$
$$\otimes\left(\bar{W}(t)\otimes_1\bar{W}(t)\ldots\otimes_{(n-1)}\bar{W}(t)\right)\Bigg)\cdot\xi\left(\mathbf{x}_1,\ldots,\mathbf{x}_n,\mathbf{y}_1,\ldots,\mathbf{y}_n,t\right). \tag{10.188}$$

Therefore, since we obtained before that equation (10.184) is equivalent to the primed version of (10.125) and at the same time equation (10.125) is invariant under the change of non-inertial Cartesian coordinate system, provided we have (10.188), with the help of (10.186), as before, we deduce

$$i\hbar\frac{\partial}{\partial t}\left(f\circ\hat{H}_{\mathbf{u}}\right)=\hat{H}_0\cdot\left(f\circ\hat{H}_{\mathbf{u}}\right)-\left(f\circ\hat{H}_{\mathbf{u}}\right)\cdot\hat{H}_0 \tag{10.189}$$

in an arbitrary coordinate system. So we get (10.171) or equivalently (10.172).

Finally, assume that the first and the second particles have the same mass $m_1=m_2$ and the same charge $\sigma_1=\sigma_2$, and moreover, assume that we have

$$V\left(\mathbf{x}_1,\mathbf{x}_2,\mathbf{x}_3,\ldots,\mathbf{x}_n,t\right)=V\left(\mathbf{x}_2,\mathbf{x}_1,\mathbf{x}_3,\ldots,\mathbf{x}_n,t\right)\qquad\forall\mathbf{x}_1,\mathbf{x}_2,\mathbf{x}_3,\ldots,\mathbf{x}_n\in\mathbb{R}^3,\quad\forall t.$$

In this case clearly by (10.155) we deduce the following relation:

$$B_{1,2}\cdot\psi(\mathbf{x}_2,\mathbf{x}_1,\mathbf{x}_3,\ldots,\mathbf{x}_n,t)=-\psi(\mathbf{x}_1,\mathbf{x}_2,\mathbf{x}_3,\ldots,\mathbf{x}_n,t)\ \ \forall\mathbf{x}_1,\ldots,\mathbf{x}_n\in\mathbb{R}^3$$
$$\text{and}\quad\phi=\hat{H}_{\mathbf{u}}\cdot\psi$$
$$\text{implies}\quad B_{1,2}\cdot\phi(\mathbf{x}_2,\mathbf{x}_1,\mathbf{x}_3,\ldots,\mathbf{x}_n,t)=-\phi(\mathbf{x}_1,\mathbf{x}_2,\mathbf{x}_3,\ldots,\mathbf{x}_n,t)$$
$$\forall\mathbf{x}_1,\ldots,\mathbf{x}_n\in\mathbb{R}^3, \tag{10.190}$$

where, as before, by $B_{1,2}:\mathbb{C}^{4^n}\to\mathbb{C}^{4^n}$ we denote the linear operator (matrix) defined as in (10.107) by the following:

$$B_{1,2} \cdot (a_1 \otimes a_2 \otimes a_3 \otimes \ldots \otimes a_n) = (a_2 \otimes a_1 \otimes a_3 \otimes \ldots \otimes a_n) \qquad \forall a_1, \ldots a_n \in \mathbb{C}^4. \tag{10.191}$$

Therefore, by (10.183) and (10.190) we obtain

$$\begin{aligned} B_{1,2} \cdot \psi(\mathbf{x}_2, \mathbf{x}_1, \mathbf{x}_3, \ldots, \mathbf{x}_n, t) = -\psi(\mathbf{x}_1, \mathbf{x}_2, \mathbf{x}_3, \ldots, \mathbf{x}_n, t) \;\; \forall \mathbf{x}_1, \ldots, \mathbf{x}_n \in \mathbb{R}^3 \\ \text{and} \quad \phi = \left(f \circ \hat{H}_{\mathbf{u}}\right) \cdot \psi \\ \text{implies} \quad B_{1,2} \cdot \phi(\mathbf{x}_2, \mathbf{x}_1, \mathbf{x}_3, \ldots, \mathbf{x}_n, t) = -\phi(\mathbf{x}_1, \mathbf{x}_2, \mathbf{x}_3, \ldots, \mathbf{x}_n, t) \\ \forall \mathbf{x}_1, \ldots, \mathbf{x}_n \in \mathbb{R}^3, \end{aligned} \tag{10.192}$$

consistently with (10.139). □

10.7 Spinless Relativistic-Like Particles: The Klein–Gordon Equation

Consider the motion of a system of n stable relativistic-like spinless quantum micro-particles with inertial masses $m_1, \ldots, m_n$ and the charges $\sigma_1, \ldots, \sigma_n$ in the outer gravitational and electromagnetical field with characteristics $\mathbf{v}(\mathbf{x}, t)$, $\mathbf{A}(\mathbf{x}, t)$ and $\Psi(\mathbf{x}, t)$ and additional conservative field with potential $V(\mathbf{y}_1, \ldots, \mathbf{y}_n, t)$. The Klein–Gordon equation for this system of particles is the following:

$$\begin{aligned} &\frac{\hbar^2}{2c^2\left(\sum_{k=1}^n m_k\right)} \frac{\partial}{\partial t}\left(\frac{\partial \psi}{\partial t} + \sum_{k=1}^n \mathbf{v}(\mathbf{x}_k, t) \cdot \nabla_{\mathbf{x}_k} \psi + \sum_{k=1}^n \frac{i\sigma_k}{\hbar} \Psi_0(\mathbf{x}_k, t)\psi\right) \\ &+ \sum_{j=1}^n \frac{\hbar^2}{2c^2\left(\sum_{k=1}^n m_k\right)} \operatorname{div}_{\mathbf{x}_j}\left\{\mathbf{v}(\mathbf{x}_j, t)\left(\frac{\partial \psi}{\partial t} + \sum_{k=1}^n \mathbf{v}(\mathbf{x}_k, t) \cdot \nabla_{\mathbf{x}_k} \psi + \sum_{k=1}^n \frac{i\sigma_k}{\hbar} \Psi_0(\mathbf{x}_k, t)\psi\right)\right\} \\ &+ \frac{\hbar^2}{2c^2\left(\sum_{k=1}^n m_k\right)} \left(\sum_{k=1}^n \frac{i\sigma_k}{\hbar} \Psi_0(\mathbf{x}_k, t)\right)\left(\frac{\partial \psi}{\partial t} + \sum_{k=1}^n \mathbf{v}(\mathbf{x}_k, t) \cdot \nabla_{\mathbf{x}_k} \psi + \sum_{k=1}^n \frac{i\sigma_k}{\hbar} \Psi_0(\mathbf{x}_k, t)\psi\right) \\ &+ \frac{c^2\left(\sum_{k=1}^n m_k\right)}{2}\psi - \sum_{j=1}^n \frac{\hbar^2}{2m_j} \Delta_{\mathbf{x}_j} \psi - V(\mathbf{x}_1, \ldots, \mathbf{x}_n, t)\psi + \sum_{j=1}^n \frac{i\hbar\sigma_j}{2m_j c} \operatorname{div}_{\mathbf{x}_j}\{\psi \mathbf{A}(\mathbf{x}_j, t)\} \\ &+ \sum_{j=1}^n \frac{i\hbar\sigma_j}{2m_j c} \mathbf{A}(\mathbf{x}_j, t) \cdot \nabla_{\mathbf{x}_j} \psi + \sum_{j=1}^n \left(\frac{\sigma_j^2}{2m_j c^2} \left|\mathbf{A}(\mathbf{x}_j, t)\right|^2\right)\psi = 0, \end{aligned} \tag{10.193}$$

where, as before,

$$\Psi_0(\mathbf{x}, t) := \Psi(\mathbf{x}, t) - \frac{1}{c}\mathbf{v}(\mathbf{x}, t) \cdot \mathbf{A}(\mathbf{x}, t) \tag{10.194}$$

is the proper electrical potential and $\psi = \psi(\mathbf{x}_1, \ldots, \mathbf{x}_n, t) \in \mathbb{C}$ is the scalar wave function. Next consider a change of some non-inertial Cartesian coordinate system $(*)$ to another Cartesian coordinate system $(**)$ of the form:

$$\begin{cases} \mathbf{x}' = A(t) \cdot \mathbf{x} + \mathbf{z}(t), \\ t' = t, \end{cases} \tag{10.195}$$

where $A(t) \in SO(3)$ is a rotation. Then, we deduce that the Klein–Gordon equation of the form (10.193) is invariant under the change of non-inertial Cartesian coordinate system, provided that under (10.195) we have

$$\begin{cases} \psi' = \psi \\ V' = V \\ \mathbf{A}' = A(t) \cdot \mathbf{A} \\ \Psi_0' := \Psi' - \frac{1}{c}\mathbf{A}' \cdot \mathbf{v}' = \Psi - \frac{1}{c}\mathbf{A} \cdot \mathbf{v} := \Psi_0. \end{cases} \tag{10.196}$$

Next defining

$$\psi_1(\mathbf{x}_1, \ldots, \mathbf{x}_n, t) := e^{\frac{ic^2\left(\sum_{k=1}^n m_k\right)t}{\hbar}} \psi(\mathbf{x}_1, \ldots, \mathbf{x}_n, t) \tag{10.197}$$

we rewrite (10.193) as

$$\begin{aligned}
&\frac{c^2\left(\sum_{k=1}^n m_k\right)}{2}\psi_1 \\
&+ \frac{\hbar^2}{2c^2\left(\sum_{k=1}^n m_k\right)} e^{\frac{ic^2\left(\sum_{k=1}^n m_k\right)t}{\hbar}} \frac{\partial}{\partial t}\left(\left(\frac{\partial \psi_1}{\partial t} + \sum_{k=1}^n \mathbf{v}(\mathbf{x}_k, t) \cdot \nabla_{\mathbf{x}_k}\psi_1\right) e^{-\frac{ic^2\left(\sum_{k=1}^n m_k\right)t}{\hbar}}\right) \\
&+ \frac{\hbar^2 e^{\frac{ic^2\left(\sum_{k=1}^n m_k\right)t}{\hbar}}}{2c^2\left(\sum_{k=1}^n m_k\right)} \frac{\partial}{\partial t}\left(\left(\sum_{k=1}^n \frac{i\sigma_k}{\hbar}\Psi_0(\mathbf{x}_k, t)\psi_1 - \frac{ic^2\left(\sum_{k=1}^n m_k\right)}{\hbar}\psi_1\right) e^{-\frac{ic^2\left(\sum_{k=1}^n m_k\right)t}{\hbar}}\right) \\
&+ \sum_{j=1}^n \frac{\hbar^2 \operatorname{div}_{\mathbf{x}_j}\left\{\mathbf{v}(\mathbf{x}_j, t)\left(\frac{\partial \psi_1}{\partial t} + \sum_{k=1}^n \mathbf{v}(\mathbf{x}_k, t) \cdot \nabla_{\mathbf{x}_k}\psi_1 + \sum_{k=1}^n \frac{i\sigma_k}{\hbar}\Psi_0(\mathbf{x}_k, t)\psi_1\right)\right\}}{2c^2\left(\sum_{k=1}^n m_k\right)} \\
&+ \frac{\hbar^2\left(\sum_{k=1}^n \frac{i\sigma_k}{\hbar}\Psi_0(\mathbf{x}_k, t)\right)}{2c^2\left(\sum_{k=1}^n m_k\right)}\left(\frac{\partial \psi_1}{\partial t} + \sum_{k=1}^n \mathbf{v}(\mathbf{x}_k, t) \cdot \nabla_{\mathbf{x}_k}\psi_1 + \sum_{k=1}^n \frac{i\sigma_k}{\hbar}\Psi_0(\mathbf{x}_k, t)\psi_1\right)
\end{aligned}$$

$$
-\sum_{j=1}^{n} \frac{\hbar^2}{2c^2\left(\sum_{k=1}^{n} m_k\right)} \mathrm{div}_{\mathbf{x}_j}\left\{\mathbf{v}(\mathbf{x}_j,t)\left(\frac{ic^2\left(\sum_{k=1}^{n} m_k\right)}{\hbar}\psi_1\right)\right\}
$$

$$
-\frac{\hbar^2}{2c^2\left(\sum_{k=1}^{n} m_k\right)}\left(\sum_{k=1}^{n}\frac{i\sigma_k}{\hbar}\Psi_0(\mathbf{x}_k,t)\right)\left(\frac{ic^2\left(\sum_{k=1}^{n} m_k\right)}{\hbar}\psi_1\right)-\sum_{j=1}^{n}\frac{\hbar^2}{2m_j}\Delta_{\mathbf{x}_j}\psi_1
$$

$$
-V(\mathbf{x}_1,\ldots,\mathbf{x}_n,t)\,\psi_1+\sum_{j=1}^{n}\frac{i\hbar\sigma_j}{2m_jc}\mathrm{div}_{\mathbf{x}_j}\left\{\psi_1\mathbf{A}(\mathbf{x}_j,t)\right\}
$$

$$
+\sum_{j=1}^{n}\frac{i\hbar\sigma_j}{2m_jc}\mathbf{A}(\mathbf{x}_j,t)\cdot\nabla_{\mathbf{x}_j}\psi_1+\sum_{j=1}^{n}\left(\frac{\sigma_j^2}{2m_jc^2}\left|\mathbf{A}(\mathbf{x}_j,t)\right|^2\right)\psi_1=0, \quad (10.198)
$$

that we further rewrite as

$$
\frac{\hbar^2}{2c^2\left(\sum_{k=1}^{n} m_k\right)}\frac{\partial}{\partial t}\left(\frac{\partial\psi_1}{\partial t}+\sum_{k=1}^{n}\mathbf{v}(\mathbf{x}_k,t)\cdot\nabla_{\mathbf{x}_k}\psi_1+\sum_{k=1}^{n}\frac{i\sigma_k}{\hbar}\Psi_0(\mathbf{x}_k,t)\psi_1\right)
$$

$$
+\sum_{j=1}^{n}\frac{\hbar^2\mathrm{div}_{\mathbf{x}_j}\left\{\mathbf{v}(\mathbf{x}_j,t)\left(\frac{\partial\psi_1}{\partial t}+\sum_{k=1}^{n}\mathbf{v}(\mathbf{x}_k,t)\cdot\nabla_{\mathbf{x}_k}\psi_1+\sum_{k=1}^{n}\frac{i\sigma_k}{\hbar}\Psi_0(\mathbf{x}_k,t)\psi_1\right)\right\}}{2c^2\left(\sum_{k=1}^{n} m_k\right)}
$$

$$
+\frac{\hbar^2\left(\sum_{k=1}^{n}\frac{i\sigma_k}{\hbar}\Psi_0(\mathbf{x}_k,t)\right)}{2c^2\left(\sum_{k=1}^{n} m_k\right)}\left(\frac{\partial\psi_1}{\partial t}+\sum_{k=1}^{n}\mathbf{v}(\mathbf{x}_k,t)\cdot\nabla_{\mathbf{x}_k}\psi_1+\sum_{k=1}^{n}\frac{i\sigma_k}{\hbar}\Psi_0(\mathbf{x}_k,t)\psi_1\right)
$$

$$
-i\hbar\left(\frac{\partial\psi_1}{\partial t}+\sum_{k=1}^{n}\frac{1}{2}\mathbf{v}(\mathbf{x}_k,t)\cdot\nabla_{\mathbf{x}_k}\psi_1+\sum_{k=1}^{n}\frac{1}{2}\mathrm{div}_{\mathbf{x}_k}\left\{\psi_1\mathbf{v}(\mathbf{x}_k,t)\right\}+\sum_{k=1}^{n}\frac{i\sigma_k}{\hbar}\Psi_0(\mathbf{x}_k,t)\psi_1\right)
$$

$$
-\sum_{j=1}^{n}\frac{\hbar^2}{2m_j}\Delta_{\mathbf{x}_j}\psi_1-V(\mathbf{x}_1,\ldots,\mathbf{x}_n,t)\,\psi_1+\sum_{j=1}^{n}\frac{i\hbar\sigma_j}{2m_jc}\mathrm{div}_{\mathbf{x}_j}\left\{\psi_1\mathbf{A}(\mathbf{x}_j,t)\right\}
$$

$$
+\sum_{j=1}^{n}\frac{i\hbar\sigma_j}{2m_jc}\mathbf{A}(\mathbf{x}_j,t)\cdot\nabla_{\mathbf{x}_j}\psi_1+\sum_{j=1}^{n}\left(\frac{\sigma_j^2}{2m_jc^2}\left|\mathbf{A}(\mathbf{x}_j,t)\right|^2\right)\psi_1=0. \quad (10.199)
$$

Thus by (10.199) in the non-relativistic limit we obtain

$$i\hbar\left(\frac{\partial\psi_1}{\partial t}+\sum_{j=1}^{n}\mathbf{v}(\mathbf{x}_j,t)\cdot\nabla_{\mathbf{x}_j}\psi_1\right)+\sum_{j=1}^{n}\frac{i\hbar}{2}\left(\operatorname{div}_{\mathbf{x}_j}\mathbf{v}(\mathbf{x}_j,t)\right)\psi_1\approx-\sum_{j=1}^{n}\frac{\hbar^2}{2m_j}\Delta_{\mathbf{x}_j}\psi_1$$
$$-V(\mathbf{x}_1,\ldots,\mathbf{x}_n,t)\,\psi_1+\sum_{j=1}^{n}\frac{i\hbar\sigma_j}{2m_jc}\operatorname{div}_{\mathbf{x}_j}\left\{\psi_1\mathbf{A}(\mathbf{x}_j,t)\right\}+\sum_{j=1}^{n}\frac{i\hbar\sigma_j}{2m_jc}\mathbf{A}(\mathbf{x}_j,t)\cdot\nabla_{\mathbf{x}_j}\psi_1$$
$$+\sum_{j=1}^{n}\left(\sigma_j\Psi(\mathbf{x}_j,t)-\frac{\sigma_j}{c}\mathbf{A}(\mathbf{x}_j,t)\cdot\mathbf{v}(\mathbf{x}_j,t)+\frac{\sigma_j^2}{2m_jc^2}\left|\mathbf{A}(\mathbf{x}_j,t)\right|^2\right)\psi_1, \quad (10.200)$$

that coincides with the Schrödinger equation of the form (6.197).

Next, again consider the motion and interaction of system of n stable quantum micro-particles having inertial masses $m_1,\ldots,m_n$ and the charges $\sigma_1,\ldots,\sigma_n$ with the known gravitational and electromagnetical field with potentials $\mathbf{v}(\mathbf{x},t)$, $\mathbf{A}(\mathbf{x},t)$ and $\Psi(\mathbf{x},t)$ and additional conservative field with potential $V(\mathbf{y}_1,\ldots,\mathbf{y}_n,t)$. Then consider a Lagrangian density L defined by

$$L_2(\psi,\mathbf{A},\Psi,\mathbf{v},\mathbf{x}_1,\ldots,\mathbf{x}_n,t):=-\frac{c^2\left(\sum_{k=1}^{n}m_k\right)}{2}\psi\cdot\bar{\psi}-\sum_{k=1}^{n}\frac{\hbar^2}{2m_k}\nabla_{\mathbf{x}_k}\psi\cdot\nabla_{\mathbf{x}_k}\bar{\psi}$$
$$-\sum_{k=1}^{n}\frac{\hbar\sigma_k i}{2m_kc}\left(\nabla_{\mathbf{x}_k}\psi\cdot\bar{\psi}-\psi\cdot\nabla_{\mathbf{x}_k}\bar{\psi}\right)\cdot\mathbf{A}(\mathbf{x}_k,t)-\sum_{k=1}^{n}\frac{\sigma_k^2}{2m_kc^2}\left|\mathbf{A}(\mathbf{x}_k,t)\right|^2\psi\cdot\bar{\psi}$$
$$+V(\mathbf{x}_1,\ldots,\mathbf{x}_n,t)\,\psi\cdot\bar{\psi}$$
$$+\frac{\hbar^2}{2c^2\left(\sum_{k=1}^{n}m_k\right)}\left(\frac{\partial\psi}{\partial t}+\sum_{k=1}^{n}\mathbf{v}(\mathbf{x}_k,t)\cdot\nabla_{\mathbf{x}_k}\psi+\sum_{k=1}^{n}\frac{i\sigma_k}{\hbar}\Psi_0(\mathbf{x}_k,t)\psi\right)\times$$
$$\times\left(\frac{\partial\bar{\psi}}{\partial t}+\sum_{k=1}^{n}\mathbf{v}(\mathbf{x}_k,t)\cdot\nabla_{\mathbf{x}_k}\bar{\psi}-\sum_{k=1}^{n}\frac{i\sigma_k}{\hbar}\Psi_0(\mathbf{x}_k,t)\bar{\psi}\right), \quad (10.201)$$

where $\psi(\mathbf{x}_1,\ldots,\mathbf{x}_n,t)\in\mathbb{C}$ is a wave function of the system. Then, as before, it can be proven that L is invariant under the change of inertial or non-inertial Cartesian coordinate systems of the form

$$\begin{cases}t'=t\\ \mathbf{x}'_k=A(t)\cdot\mathbf{x}_k+\mathbf{z}(t) & \forall k=1,\ldots,n,\end{cases}$$

provided that $\psi'=\psi$. We investigate stationary points of the functional

$$J=\int_0^T\int_{(\mathbb{R}^3)^n}L(\psi,\mathbf{A},\Psi,\mathbf{v},\mathbf{x}_1,\ldots,\mathbf{x}_n,t)\,d\mathbf{x}_1\ldots,d\mathbf{x}_ndt. \quad (10.202)$$

Then,

$$0=\frac{\delta L_2}{\delta(\bar{\psi})}=-\frac{\hbar^2}{2c^2\left(\sum\limits_{k=1}^n m_k\right)}\frac{\partial}{\partial t}\left(\frac{\partial\psi}{\partial t}+\sum_{k=1}^n \mathbf{v}(\mathbf{x}_k,t)\cdot\nabla_{\mathbf{x}_k}\psi+\sum_{k=1}^n\frac{i\sigma_k}{\hbar}\Psi_0(\mathbf{x}_k,t)\psi\right)$$
$$-\sum_{j=1}^n\frac{\hbar^2\mathrm{div}_{\mathbf{x}_j}\left\{\mathbf{v}(\mathbf{x}_j,t)\left(\frac{\partial\psi}{\partial t}+\sum\limits_{k=1}^n \mathbf{v}(\mathbf{x}_k,t)\cdot\nabla_{\mathbf{x}_k}\psi+\sum\limits_{k=1}^n\frac{i\sigma_k}{\hbar}\Psi_0(\mathbf{x}_k,t)\psi\right)\right\}}{2c^2\left(\sum\limits_{k=1}^n m_k\right)}$$
$$-\frac{\hbar^2\left(\sum\limits_{k=1}^n\frac{i\sigma_k}{\hbar}\Psi_0(\mathbf{x}_k,t)\right)}{2c^2\left(\sum\limits_{k=1}^n m_k\right)}\left(\frac{\partial\psi}{\partial t}+\sum_{k=1}^n \mathbf{v}(\mathbf{x}_k,t)\cdot\nabla_{\mathbf{x}_k}\psi+\sum_{k=1}^n\frac{i\sigma_k}{\hbar}\Psi_0(\mathbf{x}_k,t)\psi\right)$$
$$-\frac{c^2\left(\sum\limits_{k=1}^n m_k\right)}{2}\psi+\sum_{k=1}^n\frac{\hbar^2}{2m_k}\Delta_{\mathbf{x}_k}\psi-\sum_{k=1}^n\frac{\hbar\sigma_k i}{2m_k c}\left(\mathbf{A}(\mathbf{x}_k,t)\cdot\nabla_{\mathbf{x}_k}\psi+\mathrm{div}_{\mathbf{x}_k}\{\psi\mathbf{A}(\mathbf{x}_k,t)\}\right)$$
$$-\sum_{k=1}^n\frac{\sigma_k^2}{2m_k c^2}|\mathbf{A}(\mathbf{x}_k,t)|^2\psi+V(\mathbf{x}_1,\ldots,\mathbf{x}_n,t)\,\psi,\quad(10.203)$$

and

$$0=\frac{\delta L_2}{\delta(\psi)}=-\frac{\hbar^2}{2c^2\left(\sum\limits_{k=1}^n m_k\right)}\frac{\partial}{\partial t}\left(\frac{\partial\bar{\psi}}{\partial t}+\sum_{k=1}^n \mathbf{v}(\mathbf{x}_k,t)\cdot\nabla_{\mathbf{x}_k}\bar{\psi}+\sum_{k=1}^n\frac{(\bar{i})\sigma_k}{\hbar}\Psi_0(\mathbf{x}_k,t)\bar{\psi}\right)$$
$$-\sum_{j=1}^n\frac{\hbar^2\mathrm{div}_{\mathbf{x}_j}\left\{\mathbf{v}(\mathbf{x}_j,t)\left(\frac{\partial\bar{\psi}}{\partial t}+\sum\limits_{k=1}^n \mathbf{v}(\mathbf{x}_k,t)\cdot\nabla_{\mathbf{x}_k}\bar{\psi}+\sum\limits_{k=1}^n\frac{(\bar{i})\sigma_k}{\hbar}\Psi_0(\mathbf{x}_k,t)\bar{\psi}\right)\right\}}{2c^2\left(\sum\limits_{k=1}^n m_k\right)}$$
$$-\frac{\hbar^2\left(\sum\limits_{k=1}^n\frac{(\bar{i})\sigma_k}{\hbar}\Psi_0(\mathbf{x}_k,t)\right)}{2c^2\left(\sum\limits_{k=1}^n m_k\right)}\left(\frac{\partial\bar{\psi}}{\partial t}+\sum_{k=1}^n \mathbf{v}(\mathbf{x}_k,t)\cdot\nabla_{\mathbf{x}_k}\bar{\psi}+\sum_{k=1}^n\frac{(\bar{i})\sigma_k}{\hbar}\Psi_0(\mathbf{x}_k,t)\bar{\psi}\right)$$
$$-\frac{c^2\left(\sum\limits_{k=1}^n m_k\right)}{2}\bar{\psi}+\sum_{k=1}^n\frac{\hbar^2}{2m_k}\Delta_{\mathbf{x}_k}\bar{\psi}-\sum_{k=1}^n\frac{\hbar\sigma_k(\bar{i})}{2m_k c}\left(\mathbf{A}(\mathbf{x}_k,t)\cdot\nabla_{\mathbf{x}_k}\bar{\psi}+\mathrm{div}_{\mathbf{x}_k}\left\{\bar{\psi}\mathbf{A}(\mathbf{x}_k,t)\right\}\right)$$
$$-\sum_{k=1}^n\frac{\sigma_k^2}{2m_k c^2}|\mathbf{A}(\mathbf{x}_k,t)|^2\bar{\psi}+V(\mathbf{x}_1,\ldots,\mathbf{x}_n,t)\,\bar{\psi},\quad(10.204)$$

where the last equality is just the complex conjugate of (10.203). So we get that the Euler–Lagrange equation for (10.202) coincides with the Klein–Gordon equation of the form (10.193).

Finally, assume that the first and the second particles have the same mass $m_1 = m_2$ and the same charge $\sigma_1 = \sigma_2$, and moreover, assume that we have

$$V(\mathbf{x}_1, \mathbf{x}_2, \mathbf{x}_3, \ldots, \mathbf{x}_n, t) = V(\mathbf{x}_2, \mathbf{x}_1, \mathbf{x}_3, \ldots, \mathbf{x}_n, t) \qquad \forall \mathbf{x}_1, \mathbf{x}_2, \mathbf{x}_3, \ldots, \mathbf{x}_n \in \mathbb{R}^3, \quad \forall t.$$

In this case it can be easily deduced that if $\psi(\mathbf{x}_1, \mathbf{x}_2, \mathbf{x}_3, \ldots, \mathbf{x}_n, t)$ is a solution of (10.193), then $\psi(\mathbf{x}_2, \mathbf{x}_1, \mathbf{x}_3, \ldots, \mathbf{x}_n, t)$ is also a solution of (10.193), Therefore, if $\psi(\mathbf{x}_1, \mathbf{x}_2, \mathbf{x}_3, \ldots, \mathbf{x}_n, t)$ is a solution of (10.193), then for every $t \geq 0$ we will have

$$\begin{aligned}
&\psi(\mathbf{x}_2, \mathbf{x}_1, \mathbf{x}_3, \ldots, \mathbf{x}_n, t) = \psi(\mathbf{x}_1, \mathbf{x}_2, \mathbf{x}_3, \ldots, \mathbf{x}_n, t) \quad \text{and} \\
&\frac{\partial \psi}{\partial t}(\mathbf{x}_2, \mathbf{x}_1, \mathbf{x}_3, \ldots, \mathbf{x}_n, t) = \frac{\partial \psi}{\partial t}(\mathbf{x}_1, \mathbf{x}_2, \mathbf{x}_3, \ldots, \mathbf{x}_n, t) \quad \forall \mathbf{x}_1, \mathbf{x}_2, \mathbf{x}_3, \ldots, \mathbf{x}_n \in \mathbb{R}^3,
\end{aligned} \tag{10.205}$$

provided that (10.205) holds for the initial instant of time $t = 0$. So we have a consistency with the principles of identity for two or more identical bosons, in the case of spinless particles.

Chapter 11
Thermodynamics of a Moving Continuum Medium

Again, consistently with (4.11), consider, in some Cartesian coordinate system (∗), the Second Law of Newton for the moving continuum medium with the inertial mass density μ, the field of average (macroscopic) velocities $\mathbf{u}$, the charge density ρ and the electric current density $\mathbf{j}$:

$$\mu\left(\frac{\partial \mathbf{u}}{\partial t}+d_{\mathbf{x}}\mathbf{u}\cdot\mathbf{u}\right)=\frac{\partial}{\partial t}(\mu\mathbf{u})+\operatorname{div}_{\mathbf{x}}\{\mu\mathbf{u}\otimes\mathbf{u}\}=$$
$$-\mu\mathbf{u}\times\operatorname{curl}_{\mathbf{x}}\mathbf{v}+\mu\left(\partial_t\mathbf{v}+\nabla_{\mathbf{x}}\frac{1}{2}|\mathbf{v}|^2\right)+\rho\mathbf{E}+\frac{1}{c}\mathbf{j}\times\mathbf{B}+\operatorname{div}_{\mathbf{x}}\mathcal{T}=$$
$$=-\mu(\mathbf{u}-\mathbf{v})\times\operatorname{curl}_{\mathbf{x}}\mathbf{v}+\mu\left(\partial_t\mathbf{v}+d_{\mathbf{x}}\mathbf{v}\cdot\mathbf{v}\right)+\rho\mathbf{E}+\frac{1}{c}\mathbf{j}\times\mathbf{B}+\operatorname{div}_{\mathbf{x}}\mathcal{T}. \tag{11.1}$$

Here $\rho\mathbf{E}+\frac{1}{c}\mathbf{j}\times\mathbf{B}$ is the volume density of the Lorentz force where $\mathbf{E}$ and $\mathbf{B}$ are outer electric and magnetic fields, assumed to be changing smoothly and almost constant in the microscopic level, $\mathbf{v}$ is a vectorial gravitational potential also assumed to be changing smoothly and almost constant in the microscopic level, and $\mathcal{T}\in\mathbb{R}^{3\times 3}$ is the symmetric Cauchy stress tensor of the continuum medium. Moreover, the mass density μ, clearly satisfies the continuum equation:

$$\frac{\partial\mu}{\partial t}+\operatorname{div}_{\mathbf{x}}(\mu\mathbf{u})=0. \tag{11.2}$$

In particular, multiplying (11.1) by $\mathbf{u}$ and using (11.2), we deduce the equality of the balance of the kinetic energy:

A. Poliakovsky, *The Laws of Gravity and Electromagnetism*,
https://doi.org/10.1007/978-3-031-61407-1_11

$$\frac{\partial}{\partial t}\left(\frac{\mu}{2}|\mathbf{u}|^2\right)+\mathrm{div}_{\mathbf{x}}\left\{\left(\frac{\mu}{2}|\mathbf{u}|^2\right)\mathbf{u}\right\}=\mu\left(\frac{\partial}{\partial t}\left(\frac{1}{2}|\mathbf{u}|^2\right)+\mathbf{u}\cdot\nabla_{\mathbf{x}}\left(\frac{1}{2}|\mathbf{u}|^2\right)\right)=$$
$$\mu\left(\partial_t\mathbf{v}+\nabla_{\mathbf{x}}\frac{1}{2}|\mathbf{v}|^2\right)\cdot\mathbf{u}+\left(\rho\mathbf{E}+\frac{1}{c}\mathbf{j}\times\mathbf{B}\right)\cdot\mathbf{u}+(\mathrm{div}_{\mathbf{x}}\mathcal{T})\cdot\mathbf{u}=$$
$$\mu\left(\partial_t\mathbf{v}+\nabla_{\mathbf{x}}\frac{1}{2}|\mathbf{v}|^2\right)\cdot\mathbf{u}+\rho\mathbf{E}\cdot\mathbf{u}-\frac{1}{c}(\mathbf{u}\times\mathbf{B})\cdot\mathbf{j}+(\mathrm{div}_{\mathbf{x}}\mathcal{T})\cdot\mathbf{u}. \quad (11.3)$$

Next, it is well known (see [4]) that the First Law of Thermodynamics of this moving medium has the following form:

$$\frac{\partial E}{\partial t}+\mathrm{div}_{\mathbf{x}}\{E\mathbf{u}\}=\mu\left(\frac{\partial}{\partial t}\left(\frac{E}{\mu}\right)+\mathbf{u}\cdot\nabla_{\mathbf{x}}\left(\frac{E}{\mu}\right)\right)$$
$$=\frac{1}{2}\left(d_{\mathbf{x}}\mathbf{u}+\{d_{\mathbf{x}}\mathbf{u}\}^T\right):\mathcal{T}-\mathrm{div}_{\mathbf{x}}\mathbf{q}+(\mathbf{j}-\rho\mathbf{u})\cdot\left(\mathbf{E}+\frac{1}{c}\mathbf{u}\times\mathbf{B}\right). \quad (11.4)$$

Here E is the volume density of the internal energy (energy per unit volume) and consistently $\frac{E}{\mu}$ is the internal energy per unit mass, $\mathbf{q}$ is the heat flux and

$$(\mathbf{j}-\rho\mathbf{u})\cdot\left(\mathbf{E}+\frac{1}{c}\mathbf{u}\times\mathbf{B}\right)$$

is the Joules heat term. In particular, adding (11.4) with (11.3) and using the symmetry of $\mathcal{T}$, we deduce the following equality of the balance of the energy:

$$\frac{\partial}{\partial t}\left(\frac{\mu}{2}|\mathbf{u}|^2+E\right)+\mathrm{div}_{\mathbf{x}}\left\{\left(\frac{\mu}{2}|\mathbf{u}|^2+E\right)\mathbf{u}\right\}$$
$$=\mu\left(\frac{\partial}{\partial t}\left(\frac{1}{2}|\mathbf{u}|^2+\frac{E}{\mu}\right)+\mathbf{u}\cdot\nabla_{\mathbf{x}}\left(\frac{1}{2}|\mathbf{u}|^2+\frac{E}{\mu}\right)\right)$$
$$=\mathrm{div}_{\mathbf{x}}(\mathcal{T}\cdot\mathbf{u})-\mathrm{div}_{\mathbf{x}}\mathbf{q}+\mu\left(\partial_t\mathbf{v}+\nabla_{\mathbf{x}}\frac{1}{2}|\mathbf{v}|^2\right)\cdot\mathbf{u}+\mathbf{E}\cdot\mathbf{j}. \quad (11.5)$$

Next, the Second Law of Thermodynamics states that

$$T\left(\frac{\partial S}{\partial t}+\mathrm{div}_{\mathbf{x}}\{S\mathbf{u}\}\right)=T\mu\left(\frac{\partial}{\partial t}\left(\frac{S}{\mu}\right)+\mathbf{u}\cdot\nabla_{\mathbf{x}}\left(\frac{S}{\mu}\right)\right)\geq-\mathrm{div}_{\mathbf{x}}\mathbf{q}. \quad (11.6)$$

Here $T:=T(\mathbf{x},t)$ is the Kelvin temperature field and S is the volume density of the entropy (entropy per unit volume) and consistently $\frac{S}{\mu}$ is the entropy per unit mass. Moreover, we have the equality in (11.6) in the case of reversible or quasi-reversible process. In the latter case, we rewrite the First Law (11.4) and the Second Law (11.6) together as

$$\frac{\partial E}{\partial t} + \mathrm{div}_{\mathbf{x}}\{E\mathbf{u}\} =$$
$$T\left(\frac{\partial S}{\partial t} + \mathrm{div}_{\mathbf{x}}\{S\mathbf{u}\}\right) + \frac{1}{2}\left(d_{\mathbf{x}}\mathbf{u} + \{d_{\mathbf{x}}\mathbf{u}\}^T\right) : \mathcal{T} + (\mathbf{j} - \rho\mathbf{u}) \cdot \left(\mathbf{E} + \frac{1}{c}\mathbf{u} \times \mathbf{B}\right)$$
$$= T\mu\left(\frac{\partial}{\partial t}\left(\frac{S}{\mu}\right) + \mathbf{u} \cdot \nabla_{\mathbf{x}}\left(\frac{S}{\mu}\right)\right) + \frac{1}{2}\left(d_{\mathbf{x}}\mathbf{u} + \{d_{\mathbf{x}}\mathbf{u}\}^T\right) : \mathcal{T} + (\mathbf{j} - \rho\mathbf{u}) \cdot \left(\mathbf{E} + \frac{1}{c}\mathbf{u} \times \mathbf{B}\right)$$
$$= \mu\left(\frac{\partial}{\partial t}\left(\frac{E}{\mu}\right) + \mathbf{u} \cdot \nabla_{\mathbf{x}}\left(\frac{E}{\mu}\right)\right). \tag{11.7}$$

In particular, if the stress tensor has the following particular form

$$\mathcal{T} = -p\, I, \tag{11.8}$$

where p is the scalar pressure and $I := Id \in \mathbb{R}^{3\times 3}$ is the identity matrix, then, since by (11.8) and (11.2), we have

$$\frac{1}{2}\left(d_{\mathbf{x}}\mathbf{u} + \{d_{\mathbf{x}}\mathbf{u}\}^T\right) : \mathcal{T} = -p\,(\mathrm{div}_{\mathbf{x}}\,\mathbf{u}) = -p\mu\left(-\frac{1}{\mu^2}\right)\left(\frac{\partial \mu}{\partial t} + \mathbf{u} \cdot \nabla_{\mathbf{x}}\mu\right)$$
$$= -p\mu\left(\frac{\partial}{\partial t}\left(\frac{1}{\mu}\right) + \mathbf{u} \cdot \nabla_{\mathbf{x}}\left(\frac{1}{\mu}\right)\right), \tag{11.9}$$

in the latter case, we rewrite the First Law of Thermodynamics (11.4) as

$$\frac{\partial}{\partial t}\left(\frac{E}{\mu}\right) + \mathbf{u} \cdot \nabla_{\mathbf{x}}\left(\frac{E}{\mu}\right) =$$
$$-p\left(\frac{\partial}{\partial t}\left(\frac{1}{\mu}\right) + \mathbf{u} \cdot \nabla_{\mathbf{x}}\left(\frac{1}{\mu}\right)\right) - \frac{1}{\mu}\,\mathrm{div}_{\mathbf{x}}\,\mathbf{q} + \frac{1}{\mu}(\mathbf{j} - \rho\mathbf{u}) \cdot \left(\mathbf{E} + \frac{1}{c}\mathbf{u} \times \mathbf{B}\right), \tag{11.10}$$

and in the case of quasi-reversible process, we rewrite (11.7) as

$$\frac{\partial}{\partial t}\left(\frac{E}{\mu}\right) + \mathbf{u} \cdot \nabla_{\mathbf{x}}\left(\frac{E}{\mu}\right) = -p\left(\frac{\partial}{\partial t}\left(\frac{1}{\mu}\right) + \mathbf{u} \cdot \nabla_{\mathbf{x}}\left(\frac{1}{\mu}\right)\right) + T\left(\frac{\partial}{\partial t}\left(\frac{S}{\mu}\right) + \mathbf{u} \cdot \nabla_{\mathbf{x}}\left(\frac{S}{\mu}\right)\right)$$
$$+ \frac{1}{\mu}(\mathbf{j} - \rho\mathbf{u}) \cdot \left(\mathbf{E} + \frac{1}{c}\mathbf{u} \times \mathbf{B}\right), \tag{11.11}$$

where clearly $\frac{1}{\mu}$ is the volume per unit mass. Moreover, the following inequality always holds

$$\mathbf{q} \cdot \nabla_{\mathbf{x}} T \leq 0. \tag{11.12}$$

In particular, by inserting (11.12) into (11.6), we deduce

$$\frac{\partial \mathcal{S}}{\partial t} + \mathrm{div}_{\mathbf{x}}\{\mathcal{S}\mathbf{u}\} = \mu\left(\frac{\partial}{\partial t}\left(\frac{\mathcal{S}}{\mu}\right) + \mathbf{u}\cdot\nabla_{\mathbf{x}}\left(\frac{\mathcal{S}}{\mu}\right)\right) \geq -\frac{1}{T}\,\mathrm{div}_{\mathbf{x}}\,\mathbf{q} \geq$$
$$-\frac{1}{T}\,\mathrm{div}_{\mathbf{x}}\,\mathbf{q} + \frac{1}{T^2}\nabla_{\mathbf{x}}T\cdot\mathbf{q} = -\,\mathrm{div}_{\mathbf{x}}\left(\frac{1}{T}\mathbf{q}\right), \tag{11.13}$$

i.e.

$$\frac{\partial \mathcal{S}}{\partial t} + \mathrm{div}_{\mathbf{x}}\{\mathcal{S}\mathbf{u}\} = \mu\left(\frac{\partial}{\partial t}\left(\frac{\mathcal{S}}{\mu}\right) + \mathbf{u}\cdot\nabla_{\mathbf{x}}\left(\frac{\mathcal{S}}{\mu}\right)\right) \geq -\,\mathrm{div}_{\mathbf{x}}\left(\frac{1}{T}\mathbf{q}\right). \tag{11.14}$$

Note that, by (11.14), the total entropy of an arbitrary moving thermally isolated system is a non-decreasing function of time. Indeed, integrating (11.14) on the region $\Omega(t)$, occupied by our system at the instant of time t, and using the divergence theorem, gives

$$\frac{d}{dt}\left(\int_{\Omega(t)} \mathcal{S}(\mathbf{x},t)\,d\mathbf{x}\right) = \int_{\Omega(t)} \frac{\partial \mathcal{S}}{\partial t}(\mathbf{x},t)\,d\mathbf{x} + \int_{\partial\Omega(t)} \mathcal{S}(\mathbf{x},t)\,\mathbf{u}(\mathbf{x},t)\cdot\mathbf{n}(\mathbf{x},t)\,ds_{\mathbf{x}}$$
$$= \int_{\Omega(t)} \mu\left(\frac{\partial}{\partial t}\left(\frac{\mathcal{S}}{\mu}\right) + \mathbf{u}\cdot\nabla_{\mathbf{x}}\left(\frac{\mathcal{S}}{\mu}\right)\right)d\mathbf{x} \geq -\int_{\partial\Omega(t)} \frac{1}{T(\mathbf{x},t)}\,\mathbf{q}(\mathbf{x},t)\cdot\mathbf{n}(\mathbf{x},t)\,ds_{\mathbf{x}}, \tag{11.15}$$

where $\mathbf{n} := \mathbf{n}(\mathbf{x},t)$ is an outward normal to the boundary $\partial\Omega(t)$ of the region $\Omega(t)$. On the other hand, as our system is thermally isolated, the heat flux $\mathbf{q}$ on the boundary of $\Omega(t)$ vanishes. So, the right-hand side of inequality (11.15) is negligible and thus the total entropy of the system is a non-decreasing function of time.

Finally, we remind the approximate Fourier's law (which is consistent with (11.12)):

$$\mathbf{q} = -\chi\nabla_{\mathbf{x}}T, \tag{11.16}$$

where χ is some positive material coefficient (not necessary a constant). The generalization of (11.16) to anisotropic mediums is the following:

$$\mathbf{q} = -\mathcal{R}\cdot\nabla_{\mathbf{x}}T, \tag{11.17}$$

where $\mathcal{R} \in \mathbb{R}^{3\times 3}$ is a proper matrix valued field, such that

$$\mathbf{a}\cdot(\mathcal{R}\cdot\mathbf{a}) \geq 0 \qquad \forall\,\mathbf{a} \in \mathbb{R}^3. \tag{11.18}$$

Next, consider the change in certain non-inertial Cartesian coordinate system $(*)$ to another Cartesian coordinate system $(**)$ as

$$\begin{cases} \mathbf{x}' = A(t) \cdot \mathbf{x} + \mathbf{z}(t), \\ t' = t, \end{cases} \tag{11.19}$$

where $A(t) \in SO(3)$ is a rotation. Then by Proposition 3.1, we easily deduce that the Laws in (11.4), (11.6), (11.7), (11.8), (11.10), (11.11), (11.12), (11.16), (11.14) and (11.18) are invariant under the change in a non-inertial Cartesian coordinate system given by (11.19), provided that under (11.19), we have

$$\begin{cases} \mu' = \mu, \\ E' = E, \\ \mathcal{S}' = \mathcal{S}, \\ T' = T, \\ \mathbf{q}' = A(t) \cdot \mathbf{q}, \\ \mathcal{T}' = A(t) \cdot \mathcal{T} \cdot A^T(t) \\ p' = p, \\ \chi' = \chi, \\ \mathcal{R}' = A(t) \cdot \mathcal{R} \cdot A^T(t), \\ \rho' = \rho, \\ \mathbf{u}' = A(t) \cdot \mathbf{u} + \frac{dA}{dt}(t) \cdot \mathbf{x} + \frac{d\mathbf{z}}{dt}(t), \\ \mathbf{v}' = A(t) \cdot \mathbf{v} + \frac{dA}{dt}(t) \cdot \mathbf{x} + \frac{d\mathbf{z}}{dt}(t), \\ \mathbf{j}' = A(t) \cdot \mathbf{j} + \rho \frac{dA}{dt}(t) \cdot \mathbf{x} + \rho \frac{d\mathbf{z}}{dt}(t). \end{cases} \tag{11.20}$$

11.1 Some Special Cases of Continuum Mediums

11.1.1 The Case of an Inviscid Fluid/Gas

In the case of an inviscid fluid or gas, equality (11.8) indeed holds. As a consequence, equality (11.10) holds, and moreover, in the case of quasi-reversible process, equality (11.11) also holds. Moreover, in the case of a classical ideal gas, the following state equality is well known:

$$p = \frac{\mu}{m_0} k T, \tag{11.21}$$

where m_0 is the mass of the single molecule of the given gas and k is the Boltzmann constant. Finally, for the ideal gas, we have the following expression for the volume density of the internal energy E:

$$E = \frac{\mu}{m_0} C_0 k T, \tag{11.22}$$

where $C_0 > 0$ is a constant that depends on the kind of the gas (for the monatomic gas we have $C_0 = \frac{3}{2}$). On the other hand, in the case of incompressible fluid, we have

$$\operatorname{div}_{\mathbf{x}} \mathbf{u} = 0, \tag{11.23}$$

and the pressure p is unspecified.

Then, as before, we easily deduce that the Laws in (11.21), (11.22) and (11.23) are invariant under the change in a non-inertial Cartesian coordinate system given by (11.19), provided that under (11.19) we have (11.20).

11.1.2 The Simplest Viscous Fluid/Gas

In the case of the simplest viscous fluid or gas, we have the following equality, that substitutes (11.8):

$$\mathcal{T} = -p\, I + \left(\alpha \left(d_{\mathbf{x}}\mathbf{u} + \{d_{\mathbf{x}}\mathbf{u}\}^T\right) + \beta \left(\operatorname{div}_{\mathbf{x}} \mathbf{u}\right) I\right), \tag{11.24}$$

where $\alpha \geq 0$ and β are some material coefficients. Then, as in (11.10) and (11.11), by (11.24), we rewrite the First Law of Thermodynamics (11.4) as

$$\begin{aligned} \frac{\partial}{\partial t}\left(\frac{E}{\mu}\right) + \mathbf{u} \cdot \nabla_{\mathbf{x}} \left(\frac{E}{\mu}\right) &= \frac{\alpha}{2} \left|d_{\mathbf{x}}\mathbf{u} + \{d_{\mathbf{x}}\mathbf{u}\}^T\right|^2 + \frac{\beta}{2} \left|\operatorname{div}_{\mathbf{x}} \mathbf{u}\right|^2 \\ - p\left(\frac{\partial}{\partial t}\left(\frac{1}{\mu}\right) + \mathbf{u} \cdot \nabla_{\mathbf{x}} \left(\frac{1}{\mu}\right)\right) &- \frac{1}{\mu} \operatorname{div}_{\mathbf{x}} \mathbf{q} + \frac{1}{\mu} (\mathbf{j} - \rho\mathbf{u}) \cdot \left(\mathbf{E} + \frac{1}{c}\mathbf{u} \times \mathbf{B}\right), \end{aligned} \tag{11.25}$$

and in the case of quasi-reversible process we rewrite (11.7) as

$$\begin{aligned} \frac{\partial}{\partial t}\left(\frac{E}{\mu}\right) + \mathbf{u} \cdot \nabla_{\mathbf{x}} \left(\frac{E}{\mu}\right) &= \frac{\alpha}{2} \left|d_{\mathbf{x}}\mathbf{u} + \{d_{\mathbf{x}}\mathbf{u}\}^T\right|^2 + \frac{\beta}{2} \left|\operatorname{div}_{\mathbf{x}} \mathbf{u}\right|^2 \\ &- p\left(\frac{\partial}{\partial t}\left(\frac{1}{\mu}\right) + \mathbf{u} \cdot \nabla_{\mathbf{x}} \left(\frac{1}{\mu}\right)\right) + T\left(\frac{\partial}{\partial t}\left(\frac{S}{\mu}\right) + \mathbf{u} \cdot \nabla_{\mathbf{x}} \left(\frac{S}{\mu}\right)\right) \\ &+ \frac{1}{\mu} (\mathbf{j} - \rho\mathbf{u}) \cdot \left(\mathbf{E} + \frac{1}{c}\mathbf{u} \times \mathbf{B}\right), \end{aligned} \tag{11.26}$$

Moreover, in the case of a classical ideal gas, the equalities (11.21) and (11.22) are still valid in the case of viscous flow. On the other hand, in the case of incompressible viscous fluid, we have (11.23) and the pressure p is also unspecified.

Then, as before, by Proposition 3.1, we easily deduce that the Laws in (11.24), (11.25), (11.26), (11.21), (11.22) and (11.23) are invariant under the change in a non-inertial Cartesian coordinate system given by (11.19), provided that under (11.19) we have (11.20).

Finally, in the case of an anisotropic Newtonian fluid, we have the following generalization of the viscosity law: (11.24):

$$\mathcal{T} = -p\,I + \Theta \cdot \left(d_{\mathbf{x}}\mathbf{u} + \{d_{\mathbf{x}}\mathbf{u}\}^T\right), \tag{11.27}$$

where $\Theta := \Theta(\mathbf{x}, t)$ is a linear operator (9×9 matrix) from $\mathbb{R}^{3\times 3}$ to $\mathbb{R}^{3\times 3}$, such that

$$(\Theta \cdot G)^T = \Theta \cdot G \qquad \forall G \in \mathbb{R}^{3\times 3} \quad \text{such that} \quad G^T = G.$$

Then, using Lemma 3.5, by (11.20), we easily deduce that the law in (11.27) is invariant under the change in a non-inertial Cartesian coordinate system given by (11.19), provided that, under (11.19), we have

$$\begin{cases} \mathcal{T}' = A(t) \cdot \mathcal{T} \cdot A^T(t), \\ \Theta' = (A(t) \otimes A(t)) \cdot \Theta \cdot \left(A^T(t) \otimes A^T(t)\right), \end{cases} \tag{11.28}$$

where the sign $\otimes$ in (11.28) means the tensor product of the matrices, i.e. for given two linear operators (matrices) $A \in \mathbb{R}^{3\times 3}$ and $B \in \mathbb{R}^{3\times 3}$ their tensor product $A \otimes B$ is a linear operator from $\mathbb{R}^{3\times 3}$ to $\mathbb{R}^{3\times 3}$, defined by the identity:

$$(A \otimes B) \cdot (\mathbf{a} \otimes \mathbf{b}) = (A \cdot \mathbf{a}) \otimes (B \cdot \mathbf{b}) \qquad \forall \mathbf{a} \in \mathbb{R}^3,\ \forall \mathbf{b} \in \mathbb{R}^3. \tag{11.29}$$

11.2 Lagrangian Coordinates and the Simplest Models of Elastic Bodies

In some Cartesian coordinate system, consider a motion of some continuum medium occupying a region $\Omega \subset \mathbb{R}^3$ at some fixed instant of time $t = t_0$ and having the velocity field $\mathbf{u} := \mathbf{u}(\mathbf{x}, t)$. Next let $\mathbf{r}(t, \mathbf{y}) : \mathbb{R} \times \Omega \to \mathbb{R}^3$ be a solution of the following initial value problem for an ordinary differential equation:

$$\begin{cases} \frac{\partial \mathbf{r}}{\partial t}(t, \mathbf{y}) = \mathbf{u}\left(\mathbf{r}(t, \mathbf{y}), t\right) \qquad \forall t \in \mathbb{R},\ \ \forall \mathbf{y} \in \Omega \\ \mathbf{r}(t_0, \mathbf{y}) = \mathbf{y} \qquad \forall \mathbf{y} \in \Omega. \end{cases} \tag{11.30}$$

Then, clearly, $\mathbf{x} = \mathbf{r}(t, \mathbf{y})$ stands for the spatial coordinates at the instant of time t of the parcel of continuum, having initial coordinates $\mathbf{y}$. Moreover, by (11.30) the matrix-valued function $d_{\mathbf{y}}\mathbf{r}(t, \mathbf{y})$ satisfies the following initial value problem for a matrix ordinary differential equation:

$$\begin{cases} \frac{\partial}{\partial t}\left(d_{\mathbf{y}}\mathbf{r}(t,\mathbf{y})\right) = \left(d_{\mathbf{x}}\mathbf{u}\left(\mathbf{r}(t,\mathbf{y}),t\right)\right)\cdot d_{\mathbf{y}}\mathbf{r}(t,\mathbf{y}) & \forall t \in \mathbb{R}, \ \ \forall \mathbf{y} \in \Omega \\ d_{\mathbf{y}}\mathbf{r}(t_0,\mathbf{y}) = I \quad \mathbf{y} \in \Omega. \end{cases} \tag{11.31}$$

In particular, using the Liouville Theorem in the theory of linear ordinary differential systems by (11.31), we deduce that $det\left\{d_{\mathbf{y}}\mathbf{r}(t,\mathbf{y})\right\}$ satisfies

$$det\left\{d_{\mathbf{y}}\mathbf{r}(t,\mathbf{y})\right\} = e^{\int_{t_0}^{t}(\mathrm{div}_{\mathbf{x}}\,\mathbf{u}(\mathbf{r}(\tau,\mathbf{y}),\tau))d\tau} \qquad \forall t \in \mathbb{R}, \ \ \forall \mathbf{y} \in \Omega. \tag{11.32}$$

Thus, by (11.32), we deduce that $det\left\{d_{\mathbf{y}}\mathbf{r}(t,\mathbf{y})\right\} \neq 0$ for every instant of time t and so, for the given instant of time t, the mapping $\mathbf{x} = \mathbf{r}(t,\mathbf{y})$ is locally invertible, i.e. the equation $\mathbf{x} = \mathbf{r}(t,\mathbf{y})$ can be resolved in $\mathbf{y}$. Thus, there exists a regular mapping $\mathbf{y} = \mathbf{f}(\mathbf{x},t)$, such that

$$\mathbf{f}\left(\mathbf{r}(t,\mathbf{y}),t\right) = \mathbf{y} \ \ \forall \mathbf{y} \in \Omega \qquad \text{and} \qquad \mathbf{r}\left(t,\mathbf{f}(\mathbf{x},t)\right) = \mathbf{x}. \tag{11.33}$$

Then clearly $\mathbf{y} = \mathbf{f}(\mathbf{x},t)$ stands for the initial coordinates at the time t_0 of the parcel of continuum, having coordinates $\mathbf{x}$ at the instant of time t. Next differentiating the first equation in (11.33) by t due to the chain rule and using (11.30), we deduce

$$\begin{aligned} 0 &= \frac{\partial \mathbf{f}}{\partial t}\left(\mathbf{r}(t,\mathbf{y}),t\right) + d_{\mathbf{x}}\mathbf{f}\left(\mathbf{r}(t,\mathbf{y}),t\right)\cdot\frac{\partial \mathbf{r}}{\partial t}(t,\mathbf{y}) \\ &= \frac{\partial \mathbf{f}}{\partial t}\left(\mathbf{r}(t,\mathbf{y}),t\right) + d_{\mathbf{x}}\mathbf{f}\left(\mathbf{r}(t,\mathbf{y}),t\right)\cdot\mathbf{u}\left(\mathbf{r}(t,\mathbf{y}),t\right) \quad \forall \mathbf{y} \in \Omega. \end{aligned} \tag{11.34}$$

So, $\mathbf{f}(\mathbf{x},t)$ satisfies

$$\begin{cases} \frac{\partial \mathbf{f}}{\partial t}\left(\mathbf{x},t\right) + d_{\mathbf{x}}\mathbf{f}\left(\mathbf{x},t\right)\cdot\mathbf{u}\left(\mathbf{x},t\right) = 0 \\ \mathbf{f}(\mathbf{x},t_0) = \mathbf{x}. \end{cases} \tag{11.35}$$

The Cartesian coordinates $(\mathbf{x},t)$ are called the Eulerian coordinates of the continuum medium. In contrast, change in variables

$$\begin{cases} t = t \\ \mathbf{y} = \mathbf{f}\left(\mathbf{x},t\right) \end{cases} \tag{11.36}$$

to the new coordinates $(\mathbf{y},t)$ of the space-time leads to generally non-Cartesian curvilinear coordinates that called the Lagrangian or the reference coordinates of the continuum medium. Next note that, in the case of a given scalar field in the Eulerian coordinates $\Theta = \Theta(\mathbf{x},t)$ and the corresponding field in Lagrangian coordinates $\Theta_1(\mathbf{y},t) := \Theta\left(\mathbf{r}(t,\mathbf{y}),t\right)$, due to the chain rule and using (11.30), we have

$$\frac{\partial \Theta_1}{\partial t}(\mathbf{y},t) = \frac{\partial \Theta}{\partial t}(\mathbf{r}(t,\mathbf{y}),t) + \nabla_{\mathbf{x}}\Theta(\mathbf{r}(t,\mathbf{y}),t)\cdot\frac{\partial \mathbf{r}}{\partial t}(t,\mathbf{y})$$
$$= \frac{\partial \Theta}{\partial t}(\mathbf{r}(t,\mathbf{y}),t) + \mathbf{u}(\mathbf{r}(t,\mathbf{y}),t)\cdot\nabla_{\mathbf{x}}\Theta(\mathbf{r}(t,\mathbf{y}),t) \quad \forall \mathbf{y}\in\Omega. \quad (11.37)$$

So, $\frac{\partial \Theta_1}{\partial t}(\mathbf{y},t)$ in the Lagrangian coordinates correspond to the following expression in the Eulerian coordinates:

$$\frac{\partial \Theta_1}{\partial t}(\mathbf{y},t) = \frac{\partial \Theta}{\partial t}(\mathbf{x},t) + \mathbf{u}(\mathbf{x},t)\cdot\nabla_{\mathbf{x}}\Theta(\mathbf{x},t). \quad (11.38)$$

Thus, by (11.38), we also obtain that, in the case of a given vector field in the Eulerian coordinates $\mathbf{g} = \mathbf{g}(\mathbf{x},t)$ and the corresponding field in Lagrangian coordinates $\mathbf{g}_1(\mathbf{y},t) := \mathbf{g}(\mathbf{x},t) = \mathbf{g}(\mathbf{r}(t,\mathbf{y}),t)$, $\frac{\partial \mathbf{g}_1}{\partial t}(\mathbf{y},t)$ in the Lagrangian coordinates corresponds to the following expression in the Eulerian coordinates:

$$\frac{\partial \mathbf{g}_1}{\partial t}(\mathbf{y},t) = \frac{\partial \mathbf{g}}{\partial t}(\mathbf{x},t) + d_{\mathbf{x}}\mathbf{g}(\mathbf{x},t)\cdot\mathbf{u}(\mathbf{x},t). \quad (11.39)$$

Next, as before, assume that the change in some non-inertial Cartesian system (∗) of Eulerian coordinates to another Cartesian system (∗∗) of Eulerian coordinates is of the form:

$$\begin{cases} \mathbf{x}' = A(t)\cdot\mathbf{x} + \mathbf{z}(t), \\ t' = t, \end{cases} \quad (11.40)$$

where $A(t) \in SO(3)$. We would like to derive that the law of transformation of the Lagrangian coordinates $(\mathbf{y},t) \to (\mathbf{y}',t')$, consistent with (11.40), is the following:

$$\begin{cases} \mathbf{y}' = A(t_0)\cdot\mathbf{y} + \mathbf{z}(t_0), \\ t' = t, \end{cases} \quad (11.41)$$

Indeed consistently with (11.40), we have

$$\mathbf{u}'(\mathbf{x}',t) = A(t)\cdot\mathbf{u}(\mathbf{x},t) + \frac{dA}{dt}(t)\cdot\mathbf{x} + \frac{d\mathbf{z}}{dt}(t). \quad (11.42)$$

Thus, if we define $\mathbf{r}'(t',\mathbf{y}') : \mathbb{R}\times\mathbb{R}^3 \to \mathbb{R}^3$ as

$$\mathbf{r}'(t',\mathbf{y}') = \mathbf{r}'(t, A(t_0)\cdot\mathbf{y} + \mathbf{z}(t_0)) := A(t)\cdot\mathbf{r}(t,\mathbf{y}) + \mathbf{z}(t), \quad (11.43)$$

then, by (11.30), we obtain firstly

$$\mathbf{r}'(t_0,\mathbf{y}') = \mathbf{r}'(t_0, A(t_0)\cdot\mathbf{y} + \mathbf{z}(t_0)) = A(t_0)\cdot\mathbf{r}(t_0,\mathbf{y}) + \mathbf{z}(t_0) = A(t_0)\cdot\mathbf{y} + \mathbf{z}(t_0) = \mathbf{y}', \quad (11.44)$$

and secondly with the help of (11.43) and (11.42):

$$\frac{\partial \mathbf{r}'}{\partial t'}(t', \mathbf{y}') = \frac{\partial \mathbf{r}'}{\partial t}(t, A(t_0) \cdot \mathbf{y} + \mathbf{z}(t_0)) = A(t) \cdot \frac{\partial \mathbf{r}}{\partial t}(t, \mathbf{y}) + \frac{dA}{dt}(t) \cdot \mathbf{r}(t, \mathbf{y}) + \frac{d\mathbf{z}}{dt}(t)$$
$$= A(t) \cdot \mathbf{u}(\mathbf{r}(t, \mathbf{y}), t) + \frac{dA}{dt}(t) \cdot \mathbf{r}(t, \mathbf{y}) + \frac{d\mathbf{z}}{dt}(t) = \mathbf{u}'\left(\mathbf{r}'(t', \mathbf{y}'), t'\right). \quad (11.45)$$

So, by (11.44) and (11.45) consistently with (11.30), we have

$$\begin{cases} \frac{\partial \mathbf{r}'}{\partial t'}(t', \mathbf{y}') = \mathbf{u}'\left(\mathbf{r}'(t', \mathbf{y}'), t'\right) \\ \mathbf{r}'(t_0, \mathbf{y}') = \mathbf{y}'. \end{cases} \quad (11.46)$$

On the other hand, by (11.43) for inverse mappings, we have

$$\mathbf{f}'(\mathbf{x}', t') = \mathbf{f}'\left(A(t) \cdot \mathbf{x} + \mathbf{z}(t), t\right) = A(t_0) \cdot \mathbf{f}(\mathbf{x}, t) + \mathbf{z}(t_0). \quad (11.47)$$

11.2.1 Strain Tensor and Hooke's Law

Next, it is well known that the finite strain tensor $\mathcal{E}(\mathbf{x}, t) \in \mathbb{R}^{3\times 3}$ of an elastic continuum medium in the Cartesian Eulerian coordinates $(\mathbf{x}, t)$ has the form:

$$\mathcal{E}(\mathbf{x}, t) := \frac{1}{2}\left(I - \{d_{\mathbf{x}}\mathbf{f}(\mathbf{x}, t)\}^T \cdot d_{\mathbf{x}}\mathbf{f}(\mathbf{x}, t)\right), \quad (11.48)$$

where the mapping $\mathbf{f}(\mathbf{x}, t)$ is given by (11.33) and satisfies (11.35). Then, by (11.40) and (11.47), we deduce that under (11.40) the matrix-valued field $\mathcal{E}$ transforms as

$$\mathcal{E}'(\mathbf{x}', t) = A(t) \cdot \mathcal{E}(\mathbf{x}, t) \cdot A^T(t), \quad (11.49)$$

i.e. $\mathcal{E}$ is a proper matrix field as it was defined in Definition 3.1.

Remark 11.1 It is quite clear that the finite strain tensor, defined by (11.48) depends essentially on the choice of initial instant of time t_0. Thus, the initial time t_0 for (11.48) is always chosen (possibly fictitiously) in such a way that our elastic body is relaxed at time t_0.

Next, in the case of the simplest elastic body, we have the following Hooke's law that is similar to (11.24):

$$\mathcal{T}(\mathbf{x}, t) = \left(\alpha\mathcal{E}(\mathbf{x}, t) + \beta\left(\operatorname{tr}\{\mathcal{E}(\mathbf{x}, t)\}\right) I\right), \quad (11.50)$$

where α and β are some material coefficients, which are not necessary constant. Here $\mathcal{T}(\mathbf{x}, t)$ is the Cauchy stress tensor appearing in the equations of the motion of the medium (11.1) in the Eulerian coordinates $(\mathbf{x}, t)$ and $\mathcal{E}(\mathbf{x}, t)$ is the finite strain tensor defined by (11.48). Then, by (11.49), we easily deduce that Hooke's law (11.50) is invariant under the change in a non-inertial Cartesian coordinate system given by (11.40), provided that, as usual, under (11.40), we have

$$\begin{cases} \alpha' = \alpha, \\ \beta' = \beta, \\ \mathcal{T}' = A(t) \cdot \mathcal{T} \cdot A^T(t). \end{cases} \tag{11.51}$$

Finally, in the case of the anisotropic elastic body, we have the following generalization of Hooke's law (11.50):

$$\mathcal{T}(\mathbf{x}, t) = \Lambda(\mathbf{x}, t) \cdot \mathcal{E}(\mathbf{x}, t), \tag{11.52}$$

where $\Lambda := \Lambda(\mathbf{x}, t)$ is a linear operator (9×9 matrix) from $\mathbb{R}^{3\times 3}$ to $\mathbb{R}^{3\times 3}$, such that

$$(\Lambda \cdot G)^T = \Lambda \cdot G \qquad \forall G \in \mathbb{R}^{3\times 3} \quad \text{such that} \quad G^T = G.$$

Then, as before, using Lemma 3.5, by (11.49), we easily deduce that the law in (11.52) is invariant under the change in a non-inertial Cartesian coordinate system given by (11.40), provided that, under (11.40), we have

$$\begin{cases} \mathcal{T}' = A(t) \cdot \mathcal{T} \cdot A^T(t), \\ \Lambda' = (A(t) \otimes A(t)) \cdot \Lambda \cdot \left(A^T(t) \otimes A^T(t)\right), \end{cases} \tag{11.53}$$

where the sign $\otimes$ in (11.53) means the tensor product of the matrices, i.e. for given two linear operators (matrices) $A \in \mathbb{R}^{3\times 3}$ and $B \in \mathbb{R}^{3\times 3}$, their tensor product $A \otimes B$ is a linear operator from $\mathbb{R}^{3\times 3}$ to $\mathbb{R}^{3\times 3}$, defined by the identity:

$$(A \otimes B) \cdot (\mathbf{a} \otimes \mathbf{b}) = (A \cdot \mathbf{a}) \otimes (B \cdot \mathbf{b}) \qquad \forall \mathbf{a} \in \mathbb{R}^3, \ \forall \mathbf{b} \in \mathbb{R}^3. \tag{11.54}$$

11.2.2 The Equations of the Motion of the General Continuum Medium in Lagrangian Coordinates

The equation of the motion of the continuum in the Cartesian Eulerian coordinates $(\mathbf{x}, t)$ has the form (11.1), i.e:

$$\mu\left(\frac{\partial \mathbf{u}}{\partial t} + d_{\mathbf{x}}\mathbf{u} \cdot \mathbf{u}\right) = -\mu \mathbf{u} \times \mathrm{curl}_{\mathbf{x}}\mathbf{v} + \mu\left(\partial_t \mathbf{v} + \nabla_{\mathbf{x}}\frac{1}{2}|\mathbf{v}|^2\right) + \rho\mathbf{E} + \frac{1}{c}\mathbf{j} \times \mathbf{B} + \mathrm{div}_{\mathbf{x}}\mathcal{T} =$$
$$= -\mu(\mathbf{u} - \mathbf{v}) \times \mathrm{curl}_{\mathbf{x}}\mathbf{v} + \mu\left(\partial_t \mathbf{v} + d_{\mathbf{x}}\mathbf{v} \cdot \mathbf{v}\right) + \rho\mathbf{E} + \frac{1}{c}\mathbf{j} \times \mathbf{B} + \mathrm{div}_{\mathbf{x}}\mathcal{T}, \tag{11.55}$$

and the continuum equation has the form (11.2), i.e:

$$\frac{\partial \mu}{\partial t} + \mathrm{div}_{\mathbf{x}}(\mu \mathbf{u}) = 0. \tag{11.56}$$

We would like to get the analogues of these equations in the Lagrangian coordinates $(\mathbf{y}, t)$. First of all observe that, since the inverse change in coordinates is given by the formula:

$$\begin{cases} t = t \\ \mathbf{x} = \mathbf{r}(t, \mathbf{y}), \end{cases} \tag{11.57}$$

then clearly, by the formula of the Change of Variable of the Integration, we deduce

$$\mu(\mathbf{y}, t_0) = \mu\left(\mathbf{r}(t, \mathbf{y}), t\right) \left|det\{d_{\mathbf{y}}\mathbf{r}(t, \mathbf{y})\}\right|, \tag{11.58}$$

i.e:

$$\mu\left(\mathbf{r}(t, \mathbf{y}), t\right) = \mu(\mathbf{y}, t_0) \left|det\{d_{\mathbf{y}}\mathbf{r}(t, \mathbf{y})\}\right|^{-1} \quad \forall t, \ \forall \mathbf{y} \in \Omega. \tag{11.59}$$

In particular, by (11.33), (11.59) in Eulerian coordinates reads as

$$\mu\left(\mathbf{x}, t\right) = \mu(\mathbf{x}, t_0) \left|det\{d_{\mathbf{x}}\mathbf{f}(\mathbf{x}, t)\}\right|. \tag{11.60}$$

On the other hand, consistently with (11.38), the analog of continuum equation (11.56) in Lagrangian coordinates $(\mathbf{y}, t)$ is the following:

$$\frac{\partial}{\partial t}\left(\mu\left(\mathbf{r}(t, \mathbf{y}), t\right)\right) + \left(\operatorname{div}_{\mathbf{x}} \mathbf{u}\left(\mathbf{r}(t, \mathbf{y}), t\right)\right)\left(\mu\left(\mathbf{r}(t, \mathbf{y}), t\right)\right) = 0. \tag{11.61}$$

However, we can easily deduce that equality (11.32) together with (11.59) implies (11.61). Thus, equality (11.59) indeed substitutes the continuum equation (11.56) in the Lagrangian coordinates $(\mathbf{y}, t)$.

Next, the following Calculus fact is well known in Continuum Mechanics: if, given some matrix-valued field

$$\mathcal{T}(\mathbf{x}, t) \in \mathbb{R}^{3\times 3}, \tag{11.62}$$

we denote

$$F(\mathbf{y}, t) := d_{\mathbf{y}}\mathbf{r}(t, \mathbf{y}) \in \mathbb{R}^{3\times 3}, \tag{11.63}$$

and

$$\mathcal{R}(\mathbf{y}, t) := \mathcal{T}\left(\mathbf{r}(t, \mathbf{y}), t\right) \cdot \{F^{-1}(\mathbf{y}, t)\}^T \left(\det F(\mathbf{y}, t)\right), \tag{11.64}$$

then we must have

$$\operatorname{div}_{\mathbf{y}} \mathcal{R}(\mathbf{y}, t) = \left(\operatorname{div}_{\mathbf{x}} \mathcal{T}\left(\mathbf{r}(t, \mathbf{y}), t\right)\right)\left(\det F(\mathbf{y}, t)\right). \tag{11.65}$$

In particular, if $\mathcal{T}(\mathbf{x}, t)$ is the Cauchy stress tensor, then $\mathcal{R}(\mathbf{y}, t)$ defined by (11.64) is called the first Piola–Kirchhoff stress tensor. Then, by (11.65), (11.59), (11.30) and (11.39), we finally rewrite (11.55) in Lagrangian coordinates as

$$\mu(\mathbf{y}, t_0)\frac{\partial^2 \mathbf{r}}{\partial t^2}(t, \mathbf{y}) = \mathrm{div}_{\mathbf{y}}\, \mathcal{R}(\mathbf{y}, t)$$
$$- \mu(\mathbf{y}, t_0)\frac{\partial \mathbf{r}}{\partial t}(t, \mathbf{y}) \times \mathrm{curl}_{\mathbf{x}} \mathbf{v}\,(\mathbf{r}(t, \mathbf{y}), t) + \mu(\mathbf{y}, t_0)\left(\partial_t \mathbf{v}\,(\mathbf{r}(t, \mathbf{y}), t) + \nabla_{\mathbf{x}} \frac{1}{2}\,|\mathbf{v}\,(\mathbf{r}(t, \mathbf{y}), t)|^2\right)$$
$$+ \frac{\mu(\mathbf{y}, t_0)}{\mu\,(\mathbf{r}(t, \mathbf{y}), t)}\left(\rho\,(\mathbf{r}(t, \mathbf{y}), t)\,\mathbf{E}\,(\mathbf{r}(t, \mathbf{y}), t) + \frac{1}{c}\mathbf{j}\,(\mathbf{r}(t, \mathbf{y}), t) \times \mathbf{B}\,(\mathbf{r}(t, \mathbf{y}), t)\right). \quad (11.66)$$

11.3 Propagation of Sound Wave in the Moving Inviscid Gas or Fluid

Again, consistently with (11.1), consider, in some Cartesian coordinate system $(*)$, the Second Law of Newton for the moving continuum medium with the inertial mass density μ, the field of average (macroscopic) velocities $\mathbf{u}$, the charge density ρ and the electric current density $\mathbf{j}$:

$$\mu\left(\frac{\partial \mathbf{u}}{\partial t} + d_{\mathbf{x}}\mathbf{u} \cdot \mathbf{u}\right) = \frac{\partial}{\partial t}(\mu\mathbf{u}) + \mathrm{div}_{\mathbf{x}}\,\{\mu\mathbf{u} \otimes \mathbf{u}\} =$$
$$- \mu\mathbf{u} \times \mathrm{curl}_{\mathbf{x}}\mathbf{v} + \mu\left(\partial_t \mathbf{v} + \nabla_{\mathbf{x}}\frac{1}{2}|\mathbf{v}|^2\right) + \rho\mathbf{E} + \frac{1}{c}\mathbf{j} \times \mathbf{B} + \mathrm{div}_{\mathbf{x}}\,\mathcal{T} =$$
$$= -\mu(\mathbf{u} - \mathbf{v}) \times \mathrm{curl}_{\mathbf{x}}\mathbf{v} + \mu\,(\partial_t \mathbf{v} + d_{\mathbf{x}}\mathbf{v} \cdot \mathbf{v}) + \rho\mathbf{E} + \frac{1}{c}\mathbf{j} \times \mathbf{B} + \mathrm{div}_{\mathbf{x}}\,\mathcal{T}. \quad (11.67)$$

Here $\rho\mathbf{E} + \frac{1}{c}\mathbf{j} \times \mathbf{B}$ is the volume density of the Lorentz force where $\mathbf{E}$ and $\mathbf{B}$ are outer electric and magnetic fields, assumed to be changing smoothly and almost constant in the microscopic level, $\mathbf{v}$ is a vectorial gravitational potential also assumed to be changing smoothly and almost constant in the microscopic level, and $\mathcal{T} \in \mathbb{R}^{3\times 3}$ is the symmetric Cauchy stress tensor of the continuum medium. Moreover, the mass density μ, clearly satisfies the continuum equation (11.2):

$$\frac{\partial \mu}{\partial t} + \mathrm{div}_{\mathbf{x}}\,(\mu\mathbf{u}) = 0. \quad (11.68)$$

Next, by (11.4) the First Law of Thermodynamics of this moving medium has the following form:

$$\frac{\partial E}{\partial t} + \mathrm{div}_{\mathbf{x}}\,\{E\mathbf{u}\} = \mu\left(\frac{\partial}{\partial t}\left(\frac{E}{\mu}\right) + \mathbf{u} \cdot \nabla_{\mathbf{x}}\left(\frac{E}{\mu}\right)\right)$$
$$= \frac{1}{2}\left(d_{\mathbf{x}}\mathbf{u} + \{d_{\mathbf{x}}\mathbf{u}\}^T\right) : \mathcal{T} - \mathrm{div}_{\mathbf{x}}\,\mathbf{q} + (\mathbf{j} - \rho\mathbf{u}) \cdot \left(\mathbf{E} + \frac{1}{c}\mathbf{u} \times \mathbf{B}\right). \quad (11.69)$$

Here E is the volume density of the internal energy (energy per unit volume) and consistently $\frac{E}{\mu}$ is the internal energy per unit mass and $\mathbf{q}$ is the heat flux. Next, in the case of inviscid fluid or gas (11.8) holds

$$\mathcal{T} = -pI. \tag{11.70}$$

Thus, as before, we rewrite (11.67) as

$$\begin{aligned}\mu\left(\frac{\partial \mathbf{u}}{\partial t} + d_{\mathbf{x}}\mathbf{u}\cdot\mathbf{u}\right) &= \frac{\partial}{\partial t}(\mu\mathbf{u}) + \operatorname{div}_{\mathbf{x}}\{\mu\mathbf{u}\otimes\mathbf{u}\} = \\ &- \mu\mathbf{u}\times\operatorname{curl}_{\mathbf{x}}\mathbf{v} + \mu\left(\partial_t\mathbf{v} + \nabla_{\mathbf{x}}\frac{1}{2}|\mathbf{v}|^2\right) + \rho\mathbf{E} + \frac{1}{c}\mathbf{j}\times\mathbf{B} - \nabla_{\mathbf{x}}p = \\ &= -\mu(\mathbf{u}-\mathbf{v})\times\operatorname{curl}_{\mathbf{x}}\mathbf{v} + \mu\left(\partial_t\mathbf{v} + d_{\mathbf{x}}\mathbf{v}\cdot\mathbf{v}\right) + \rho\mathbf{E} + \frac{1}{c}\mathbf{j}\times\mathbf{B} - \nabla_{\mathbf{x}}p,\end{aligned} \tag{11.71}$$

and we rewrite (11.69) as

$$\begin{aligned}\frac{\partial E}{\partial t} + \operatorname{div}_{\mathbf{x}}\{E\mathbf{u}\} &= \mu\left(\frac{\partial}{\partial t}\left(\frac{E}{\mu}\right) + \mathbf{u}\cdot\nabla_{\mathbf{x}}\left(\frac{E}{\mu}\right)\right) \\ &= -\left(\operatorname{div}_{\mathbf{x}}\mathbf{u}\right)p - \operatorname{div}_{\mathbf{x}}\mathbf{q} + (\mathbf{j}-\rho\mathbf{u})\cdot\left(\mathbf{E} + \frac{1}{c}\mathbf{u}\times\mathbf{B}\right).\end{aligned} \tag{11.72}$$

Moreover, the internal energy per unit mass is a function of the density μ, pressure p and Kelvin temperature T:

$$\frac{E}{\mu} = U(\mu, p, T). \tag{11.73}$$

There is also a state equation of the form:

$$T = g(\mu, p). \tag{11.74}$$

We remember that, in the case of the simplest ideal gas, (11.73) takes particular form of (11.22):

$$\frac{E}{\mu} = \frac{C_0}{m_0} k\,T, \tag{11.75}$$

and (11.74) takes particular form of (11.21):

$$T = \frac{m_0}{k}\frac{p}{\mu}, \tag{11.76}$$

where m_0 is the mass of the single molecule of the given gas and k is the Boltzmann constant and $C_0 > 0$ is a constant that depends on the kind of the gas (for the monatomic gas, we have $C_0 = \frac{3}{2}$). So, by inserting (11.74) into (11.73), we have

$$\frac{E}{\mu} = U\left(\mu, p, g\left(\mu, p\right)\right) := F\left(\mu, p\right). \tag{11.77}$$

In particular, in the case where (11.75) and (11.76) are valid, we have

$$\frac{E}{\mu} = F\left(\mu, p\right) := C_0 \frac{p}{\mu}. \tag{11.78}$$

Next, we assume that

$$\mathbf{u} = \mathbf{u}_0 + \mathbf{u}_1, \quad \mu = \mu_0 + \mu_1, \quad p = p_0 + p_1, \tag{11.79}$$

where $\mathbf{u}_0(\mathbf{x}, t)$, $\mu_0(\mathbf{x}, t)$, $p_0(\mathbf{x}, t)$ are the averages of $\mathbf{u}(\mathbf{x}, t)$, $\mu(\mathbf{x}, t)$, $p(\mathbf{x}, t)$ on small spatial and temporal intervals, surrounding the point $(\mathbf{x}, t)$. Although we assume these intervals of space and time to be very small, we also assume them to be quite macroscopic. We call $\mathbf{u}_1$, μ_1, p_1 the oscillating parts of $\mathbf{u}$, μ, p and we assume that they are small with respect to the averages $\mathbf{u}_0$, μ_0, p_0, i.e. we have

$$|\mathbf{u}_1| \ll |\mathbf{u}_0|, \quad |\mu_1| \ll |\mu_0|, \quad |p_1| \ll |p_0|. \tag{11.80}$$

However, we assume that $\mathbf{u}_1$, μ_1, p_1 are highly oscillate and thus they changes spatially and temporary much faster than the averages $\mathbf{u}_0$, μ_0, p_0 and the fields $\mathbf{v}$, $\mathbf{E}$, $\mathbf{B}$, ρ, $\mathbf{j}$, i.e. we have

$$\begin{gathered}
\frac{\left|d_{\mathbf{x}}\left(\rho\mathbf{E} + \frac{1}{c}\mathbf{j}\times\mathbf{B}\right)\right|}{\left|\rho\mathbf{E} + \frac{1}{c}\mathbf{j}\times\mathbf{B}\right|} + \frac{|d_{\mathbf{x}}\mathbf{v}|}{|\mathbf{v}|} + \frac{|d_{\mathbf{x}}\mu_0|}{|\mu_0|} + \frac{|d_{\mathbf{x}}p_0|}{|p_0|} + \frac{|d_{\mathbf{x}}\mathbf{u}_0|}{|\mathbf{u}_0|} \\
\ll \min\left\{\frac{|d_{\mathbf{x}}\mathbf{u}_1|}{|\mathbf{u}_1|}, \frac{|d_{\mathbf{x}}\mu_1|}{|\mu_1|}, \frac{|d_{\mathbf{x}}p_1|}{|p_1|}\right\}, \\
\frac{\left|\partial_t\left(\rho\mathbf{E} + \frac{1}{c}\mathbf{j}\times\mathbf{B}\right)\right|}{\left|\rho\mathbf{E} + \frac{1}{c}\mathbf{j}\times\mathbf{B}\right|} + \frac{|\partial_t\mathbf{v}|}{|\mathbf{v}|} + \frac{|\partial_t\mu_0|}{|\mu_0|} + \frac{|\partial_t p_0|}{|p_0|} + \frac{|\partial_t\mathbf{u}_0|}{|\mathbf{u}_0|} \\
\ll \min\left\{\frac{|\partial_t\mathbf{u}_1|}{|\mathbf{u}_1|}, \frac{|\partial_t\mu_1|}{|\mu_1|}, \frac{|\partial_t p_1|}{|p_1|}\right\}, \\
\frac{|\mathbf{u}_1|}{|\mathbf{u}_0|} \ll \min\left\{\frac{|d_{\mathbf{x}}\mathbf{u}_1|}{|d_{\mathbf{x}}\mathbf{u}_0|}, \frac{|d_{\mathbf{x}}\mathbf{u}_1|}{|d_{\mathbf{x}}\mathbf{v}|}\right\}, \qquad \frac{|\mu_1|}{|\mu_0|} + \frac{|p_1|}{|p_0|} \ll \min\left\{\frac{|d_{\mathbf{x}}p_1|}{|d_{\mathbf{x}}p_0|}, \frac{|d_{\mathbf{x}}\mu_1|}{|d_{\mathbf{x}}\mu_0|}\right\} \\
\text{and} \qquad \frac{|\mu_1|}{|\mu_0|} \ll \frac{|d_{\mathbf{x}}\mathbf{u}_1||\mathbf{u}_0||\mu_0|}{\left|\rho\mathbf{E} + \frac{1}{c}\mathbf{j}\times\mathbf{B}\right|}.
\end{gathered} \tag{11.81}$$

Finally, we assume that the fields $\mathbf{v}$, $\mathbf{E}$, $\mathbf{B}$, ρ, $\mathbf{j}$ and $(\mathrm{div}_{\mathbf{x}}\, \mathbf{q})$ change slowly with respect to the oscillations of $\mathbf{u}_1$, μ_1, p_1 and thus we assume that $\mathbf{v}$, $\mathbf{E}$, $\mathbf{B}$, ρ, $\mathbf{j}$ and $(\mathrm{div}_{\mathbf{x}}\, \mathbf{q})$ can be replaced by their spatial and temporal averages. Note that μ, p, μ_0, p_0, μ_1, p_1 behave like proper scalar fields and $\mathbf{u}$, $\mathbf{u}_0$ behave like speed-like vector fields under the change in Cartesian coordinate systems. Thus, since $\mathbf{u}_1 = \mathbf{u} - \mathbf{u}_0$, we deduce that $\mathbf{u}_1$ behaves like a proper vector field under the change in Cartesian coordinate systems. Finally, note that obviously the averages of $\mathbf{u}_1$, μ_1, p_1 vanish.

Next, we would like to approximate (11.71). In the rough level of approximation, we could just replace $\mathbf{u}$, μ, p by $\mathbf{u}_0$, μ_0, p_0. However, we would like to use a more delicate approximation, taking into the account the first-order terms with $\mathbf{u}_1$, μ_1, p_1. Then, by inserting (11.79) into (11.71) and using (11.80), we conclude

$$\begin{aligned}
&\mu\left(\frac{\partial \mathbf{u}}{\partial t} + d_{\mathbf{x}}\mathbf{u}\cdot\mathbf{u}\right) \approx \\
&\mu_0\left(\frac{\partial \mathbf{u}_1}{\partial t} + d_{\mathbf{x}}\mathbf{u}_0\cdot\mathbf{u}_1 + d_{\mathbf{x}}\mathbf{u}_1\cdot\mathbf{u}_0\right) + \mu_1\left(\frac{\partial \mathbf{u}_0}{\partial t} + d_{\mathbf{x}}\mathbf{u}_0\cdot\mathbf{u}_0\right) + \mu_0\left(\frac{\partial \mathbf{u}_0}{\partial t} + d_{\mathbf{x}}\mathbf{u}_0\cdot\mathbf{u}_0\right) \\
&\qquad \approx -\mu_0\mathbf{u}_0\times \mathrm{curl}_{\mathbf{x}}\mathbf{v} + \mu_0\left(\partial_t\mathbf{v} + \nabla_{\mathbf{x}}\frac{1}{2}|\mathbf{v}|^2\right) + \rho\mathbf{E} + \frac{1}{c}\mathbf{j}\times\mathbf{B} - \nabla_{\mathbf{x}}p_0 \\
&\qquad + \mu_1\left(\partial_t\mathbf{v} + \nabla_{\mathbf{x}}\frac{1}{2}|\mathbf{v}|^2\right) - \mu_1\mathbf{u}_0\times\mathrm{curl}_{\mathbf{x}}\mathbf{v} - \mu_0\mathbf{u}_1\times\mathrm{curl}_{\mathbf{x}}\mathbf{v} - \nabla_{\mathbf{x}}p_1. \qquad (11.82)
\end{aligned}$$

On the other hand, by inserting (11.79) into (11.68) and using (11.80), we have

$$\begin{aligned}
&0 = \left(\frac{\partial\mu}{\partial t} + \mathbf{u}\cdot\nabla_{\mathbf{x}}\mu\right) + \mu\ \mathrm{div}_{\mathbf{x}}\,\mathbf{u} \approx \\
&\qquad \left(\frac{\partial\mu_0}{\partial t} + \mathbf{u}_0\cdot\nabla_{\mathbf{x}}\mu_0\right) + \left(\frac{\partial\mu_1}{\partial t} + \mathbf{u}_0\cdot\nabla_{\mathbf{x}}\mu_1 + \mathbf{u}_1\cdot\nabla_{\mathbf{x}}\mu_0\right) \\
&\qquad\qquad + \mu_0\ \mathrm{div}_{\mathbf{x}}\,\mathbf{u}_0 + \mu_0\ \mathrm{div}_{\mathbf{x}}\,\mathbf{u}_1 + \mu_1\ \mathrm{div}_{\mathbf{x}}\,\mathbf{u}_0 \approx 0. \qquad (11.83)
\end{aligned}$$

In particular, averaging of (11.82) gives

$$\begin{aligned}
&\mu_0\left(\frac{\partial \mathbf{u}_0}{\partial t} + d_{\mathbf{x}}\mathbf{u}_0\cdot\mathbf{u}_0\right) \approx \\
&-\mu_0\mathbf{u}_0\times\mathrm{curl}_{\mathbf{x}}\mathbf{v} + \mu_0\left(\partial_t\mathbf{v} + \nabla_{\mathbf{x}}\frac{1}{2}|\mathbf{v}|^2\right) + \rho\mathbf{E} + \frac{1}{c}\mathbf{j}\times\mathbf{B} - \nabla_{\mathbf{x}}p_0, \qquad (11.84)
\end{aligned}$$

and averaging of (11.83) gives

$$\left(\frac{\partial\mu_0}{\partial t} + \mathbf{u}_0\cdot\nabla_{\mathbf{x}}\mu_0\right) + \mu_0\ \mathrm{div}_{\mathbf{x}}\,\mathbf{u}_0 \approx 0. \qquad (11.85)$$

Here we used the fact that $\mathbf{v}$, $\mathbf{E}$, $\mathbf{B}$, ρ, $\mathbf{j}$ can be replaced by their spatial and temporal averages. Thus, subtracting (11.84) from the unaveraged (11.82) gives

$$\mu_0\left(\frac{\partial \mathbf{u}_1}{\partial t}+d_{\mathbf{x}}\mathbf{u}_0\cdot\mathbf{u}_1+d_{\mathbf{x}}\mathbf{u}_1\cdot\mathbf{u}_0\right)+\mu_1\left(\frac{\partial \mathbf{u}_0}{\partial t}+d_{\mathbf{x}}\mathbf{u}_0\cdot\mathbf{u}_0\right)\approx$$
$$+\mu_1\left(\partial_t\mathbf{v}+\nabla_{\mathbf{x}}\frac{1}{2}|\mathbf{v}|^2\right)-\mu_1\mathbf{u}_0\times\operatorname{curl}_{\mathbf{x}}\mathbf{v}-\mu_0\mathbf{u}_1\times\operatorname{curl}_{\mathbf{x}}\mathbf{v}-\nabla_{\mathbf{x}}p_1, \quad (11.86)$$

and thus, by (11.86) and (11.84), we have

$$\mu_0\left(\frac{\partial \mathbf{u}_1}{\partial t}+d_{\mathbf{x}}\mathbf{u}_0\cdot\mathbf{u}_1+d_{\mathbf{x}}\mathbf{u}_1\cdot\mathbf{u}_0\right)\approx$$
$$-\mu_0\mathbf{u}_1\times\operatorname{curl}_{\mathbf{x}}\mathbf{v}-\nabla_{\mathbf{x}}p_1+\frac{\mu_1}{\mu_0}\nabla_{\mathbf{x}}p_0-\frac{\mu_1}{\mu_0}\left(\rho\mathbf{E}+\frac{1}{c}\mathbf{j}\times\mathbf{B}\right). \quad (11.87)$$

Furthermore, subtracting (11.85) from the unaveraged (11.83) gives

$$\left(\frac{\partial \mu_1}{\partial t}+\mathbf{u}_0\cdot\nabla_{\mathbf{x}}\mu_1+\mathbf{u}_1\cdot\nabla_{\mathbf{x}}\mu_0\right)+\mu_0\operatorname{div}_{\mathbf{x}}\mathbf{u}_1+\mu_1\operatorname{div}_{\mathbf{x}}\mathbf{u}_0\approx 0. \quad (11.88)$$

Therefore, using (11.81) we further approximate (11.87) as

$$\mu_0\left(\frac{\partial \mathbf{u}_1}{\partial t}+d_{\mathbf{x}}\mathbf{u}_1\cdot\mathbf{u}_0\right)\approx-\nabla_{\mathbf{x}}p_1. \quad (11.89)$$

Thus, again using (11.81), we approximate (11.89) as

$$\mu_0\left(\frac{\partial \mathbf{u}_1}{\partial t}+(\operatorname{div}_{\mathbf{x}}\mathbf{u}_0)\,\mathbf{u}_1+d_{\mathbf{x}}\mathbf{u}_1\cdot\mathbf{u}_0-d_{\mathbf{x}}\mathbf{u}_0\cdot\mathbf{u}_1\right)\approx-\nabla_{\mathbf{x}}p_1. \quad (11.90)$$

Then, by (2.11) we rewrite (11.90) as

$$\mu_0\left(\frac{\partial \mathbf{u}_1}{\partial t}-\operatorname{curl}_{\mathbf{x}}\{\mathbf{u}_0\times\mathbf{u}_1\}+(\operatorname{div}_{\mathbf{x}}\mathbf{u}_1)\,\mathbf{u}_0\right)\approx-\nabla_{\mathbf{x}}p_1. \quad (11.91)$$

Moreover, using (11.81), we further approximate (11.88) as

$$\left(\frac{\partial \mu_1}{\partial t}+\mathbf{u}_0\cdot\nabla_{\mathbf{x}}\mu_1\right)+\mu_0\operatorname{div}_{\mathbf{x}}\mathbf{u}_1\approx 0, \quad (11.92)$$

and thus,

$$\frac{1}{\mu_0}\left(\frac{\partial \mu_1}{\partial t}+\mathbf{u}_0\cdot\nabla_{\mathbf{x}}\mu_1\right)+\operatorname{div}_{\mathbf{x}}\mathbf{u}_1\approx 0. \quad (11.93)$$

In particular, taking the divergence of both sides of (11.91) and using again (11.81), gives

$$\mu_0\left(\frac{\partial}{\partial t}(\operatorname{div}_{\mathbf{x}}\mathbf{u}_1)+\operatorname{div}_{\mathbf{x}}\{(\operatorname{div}_{\mathbf{x}}\mathbf{u}_1)\,\mathbf{u}_0\}\right)\approx-\Delta_{\mathbf{x}}p_1. \quad (11.94)$$

Therefore, inserting (11.93) into (11.94) and using again (11.81), we deduce

$$\frac{\partial}{\partial t}\left(\frac{\partial \mu_1}{\partial t}+\mathbf{u}_0\cdot\nabla_{\mathbf{x}}\mu_1\right)+\mathrm{div}_{\mathbf{x}}\left\{\left(\frac{\partial \mu_1}{\partial t}+\mathbf{u}_0\cdot\nabla_{\mathbf{x}}\mu_1\right)\mathbf{u}_0\right\}\approx\Delta_{\mathbf{x}}p_1. \tag{11.95}$$

Note that, as before, it can be easily proved that (11.95) and (11.91) are invariant under the change in inertial or non-inertial Cartesian coordinate system. Thus, (11.95) and (11.91) are still valid if (11.80) and (11.81) hold only in some specific (possibly artificial) inertial or non-inertial Cartesian coordinate system.

Next, we proceed in two cases:

(i) In the case of the simple models of inviscid gas:
(ii) In the case of inviscid barotropic fluid:

In the case of inviscid gas, approximating (11.72), by taking into account up to the first order terms with $\mathbf{u}_1$, μ_1, p_1 and using (11.77), we infer

$$\begin{aligned}
&\mu\left(\frac{\partial}{\partial t}\left(\frac{E}{\mu}\right)+\mathbf{u}\cdot\nabla_{\mathbf{x}}\left(\frac{E}{\mu}\right)\right)= \\
&(\mu_0+\mu_1)\left(\frac{\partial}{\partial t}(F(\mu_0+\mu_1,p_0+p_1))+(\mathbf{u}_0+\mathbf{u}_1)\cdot\nabla_{\mathbf{x}}(F(\mu_0+\mu_1,p_0+p_1))\right) \\
&\approx\mu_1\left(\frac{\partial}{\partial t}(F(\mu_0,p_0))+\mathbf{u}_0\cdot\nabla_{\mathbf{x}}(F(\mu_0,p_0))\right) \\
&+\mu_0\bigg(\frac{\partial}{\partial t}(F(\mu_0+\mu_1,p_0+p_1))+\mathbf{u}_0\cdot\nabla_{\mathbf{x}}(F(\mu_0+\mu_1,p_0+p_1)) \\
&+\mathbf{u}_1\cdot\nabla_{\mathbf{x}}(F(\mu_0,p_0))\bigg)\approx\mu_0\left(\frac{\partial}{\partial t}(F(\mu_0,p_0))+\mathbf{u}_0\cdot\nabla_{\mathbf{x}}(F(\mu_0,p_0))\right) \\
&+\mu_1\left(\frac{\partial}{\partial t}(F(\mu_0,p_0))+\mathbf{u}_0\cdot\nabla_{\mathbf{x}}(F(\mu_0,p_0))\right)+\mu_0\mathbf{u}_1\cdot\nabla_{\mathbf{x}}(F(\mu_0,p_0)) \\
&+\mu_0\bigg(\frac{\partial}{\partial t}\left(\mu_1\frac{\partial F}{\partial\mu}(\mu_0,p_0)+p_1\frac{\partial F}{\partial p}(\mu_0,p_0)\right) \\
&+\mathbf{u}_0\cdot\nabla_{\mathbf{x}}\left(\mu_1\frac{\partial F}{\partial\mu}(\mu_0,p_0)+p_1\frac{\partial F}{\partial p}(\mu_0,p_0)\right)\bigg) \\
&\approx-(\mathrm{div}_{\mathbf{x}}\,\mathbf{u}_1)\,p_0-(\mathrm{div}_{\mathbf{x}}\,\mathbf{u}_0)\,p_1-(\mathrm{div}_{\mathbf{x}}\,\mathbf{u}_0)\,p_0-\mathrm{div}_{\mathbf{x}}\,\mathbf{q}+(\mathbf{j}-\rho\mathbf{u})\cdot\left(\mathbf{E}+\frac{1}{c}\mathbf{u}\times\mathbf{B}\right).
\end{aligned} \tag{11.96}$$

In particular, averaging of (11.96) gives

$$\begin{aligned}
&\mu_0\left(\frac{\partial}{\partial t}(F(\mu_0,p_0))+\mathbf{u}_0\cdot\nabla_{\mathbf{x}}(F(\mu_0,p_0))\right) \\
&\approx-(\mathrm{div}_{\mathbf{x}}\,\mathbf{u}_0)\,p_0-\mathrm{div}_{\mathbf{x}}\,\mathbf{q}+(\mathbf{j}-\rho\mathbf{u})\cdot\left(\mathbf{E}+\frac{1}{c}\mathbf{u}\times\mathbf{B}\right).
\end{aligned} \tag{11.97}$$

Here we again used the fact that that $\mathbf{v}$, $\mathbf{E}$, $\mathbf{B}$, ρ, $\mathbf{j}$ and $\operatorname{div}_{\mathbf{x}} \mathbf{q}$ can be replaced by their spatial and temporal averages. Thus, subtracting (11.97) from the unaveraged (11.96) gives

$$\begin{aligned} &-(\operatorname{div}_{\mathbf{x}} \mathbf{u}_1)\, p_0 - (\operatorname{div}_{\mathbf{x}} \mathbf{u}_0)\, p_1 \approx \\ &\quad \mu_1 \left(\frac{\partial}{\partial t} \left(F(\mu_0, p_0)\right) + \mathbf{u}_0 \cdot \nabla_{\mathbf{x}} \left(F(\mu_0, p_0)\right) \right) + \mu_0 \mathbf{u}_1 \cdot \nabla_{\mathbf{x}} \left(F(\mu_0, p_0)\right) \\ &\qquad + \mu_0 \left(\frac{\partial}{\partial t} \left(\mu_1 \frac{\partial F}{\partial \mu}(\mu_0, p_0) + p_1 \frac{\partial F}{\partial p}(\mu_0, p_0) \right) \right. \\ &\qquad\qquad \left. + \mathbf{u}_0 \cdot \nabla_{\mathbf{x}} \left(\mu_1 \frac{\partial F}{\partial \mu}(\mu_0, p_0) + p_1 \frac{\partial F}{\partial p}(\mu_0, p_0) \right) \right). \end{aligned} \tag{11.98}$$

Therefore, using (11.81), we further approximate (11.98) as

$$\begin{aligned} \mu_0 \frac{\partial F}{\partial \mu}(\mu_0, p_0) \left(\frac{\partial \mu_1}{\partial t} + \mathbf{u}_0 \cdot \nabla_{\mathbf{x}} \mu_1 \right) + \mu_0 \frac{\partial F}{\partial p}(\mu_0, p_0) \left(\frac{\partial p}{\partial t} + \mathbf{u}_0 \cdot \nabla_{\mathbf{x}} p_1 \right) \\ \approx -(\operatorname{div}_{\mathbf{x}} \mathbf{u}_1)\, p_0. \end{aligned} \tag{11.99}$$

Thus, inserting (11.93) into (11.99) gives

$$\begin{aligned} \mu_0 \frac{\partial F}{\partial \mu}(\mu_0, p_0) \left(\frac{\partial \mu_1}{\partial t} + \mathbf{u}_0 \cdot \nabla_{\mathbf{x}} \mu_1 \right) + \mu_0 \frac{\partial F}{\partial p}(\mu_0, p_0) \left(\frac{\partial p}{\partial t} + \mathbf{u}_0 \cdot \nabla_{\mathbf{x}} p_1 \right) \\ \approx \frac{p_0}{\mu_0} \left(\frac{\partial \mu_1}{\partial t} + \mathbf{u}_0 \cdot \nabla_{\mathbf{x}} \mu_1 \right). \end{aligned} \tag{11.100}$$

So,

$$\begin{aligned} \left(\mu_0 \frac{\partial F}{\partial \mu}(\mu_0, p_0) - \frac{p_0}{\mu_0} \right) \left(\frac{\partial \mu_1}{\partial t} + \mathbf{u}_0 \cdot \nabla_{\mathbf{x}} \mu_1 \right) \\ + \mu_0 \frac{\partial F}{\partial p}(\mu_0, p_0) \left(\frac{\partial p}{\partial t} + \mathbf{u}_0 \cdot \nabla_{\mathbf{x}} p_1 \right) \approx 0. \end{aligned} \tag{11.101}$$

Therefore, denoting the scalar field $g(\mathbf{x}, t)$ defined by

$$g := \mu_0 \frac{\partial F}{\partial p}(\mu_0, p_0)\, p_1 - \left(\frac{p_0}{\mu_0} - \mu_0 \frac{\partial F}{\partial \mu}(\mu_0, p_0) \right) \mu_1, \tag{11.102}$$

inserting it into (11.101) and using (11.81) again, we deduce

$$\frac{\partial g}{\partial t} + \mathbf{u}_0 \cdot \nabla_{\mathbf{x}} g \approx 0. \tag{11.103}$$

In particular, we deduce

$$\text{If} \quad g(\mathbf{x}, 0) \approx 0 \quad \text{then} \quad g(\mathbf{x}, t) \approx 0 \quad \forall t. \tag{11.104}$$

In this case, by (11.102), we deduce

$$\mu_1 \approx \left(\frac{p_0}{\mu_0} - \mu_0 \frac{\partial F}{\partial \mu} (\mu_0, p_0) \right)^{-1} \mu_0 \frac{\partial F}{\partial p} (\mu_0, p_0)\, p_1. \tag{11.105}$$

Therefore, inserting (11.105) into (11.95) and using again (11.81), we finally obtain the wave equation for the oscillating part of the pressure p_1 of the form:

$$\frac{\partial}{\partial t} \left\{ \frac{1}{c_0^2} \left(\frac{\partial p_1}{\partial t} + \mathbf{u}_0 \cdot \nabla_{\mathbf{x}} p_1 \right) \right\} + \operatorname{div}_{\mathbf{x}} \left\{ \frac{1}{c_0^2} \left(\frac{\partial p_1}{\partial t} + \mathbf{u}_0 \cdot \nabla_{\mathbf{x}} p_1 \right) \mathbf{u}_0 \right\} \approx \Delta_{\mathbf{x}} p_1, \tag{11.106}$$

where we denote

$$c_0 := \sqrt{\frac{\left(\frac{p_0}{\mu_0} - \mu_0 \frac{\partial F}{\partial \mu} (\mu_0, p_0) \right)}{\mu_0 \frac{\partial F}{\partial p} (\mu_0, p_0)}}. \tag{11.107}$$

Moreover, the oscillating parts of the density μ_1 and the velocity $\mathbf{u}_1$ can be found from (11.105) and (11.91), respectively, i.e. we have

$$\mu_1 \approx \left(\frac{p_0}{\mu_0} - \mu_0 \frac{\partial F}{\partial \mu} (\mu_0, p_0) \right)^{-1} \mu_0 \frac{\partial F}{\partial p} (\mu_0, p_0)\, p_1, \tag{11.108}$$

and

$$\mu_0 \left(\frac{\partial \mathbf{u}_1}{\partial t} - \operatorname{curl}_{\mathbf{x}} \{\mathbf{u}_0 \times \mathbf{u}_1\} + (\operatorname{div}_{\mathbf{x}} \mathbf{u}_1)\, \mathbf{u}_0 \right) \approx -\nabla_{\mathbf{x}} p_1. \tag{11.109}$$

Note that, as before, it can be easily proved that (11.106), (11.108) and (11.109) are invariant under the change in inertial or non-inertial Cartesian coordinate system. Thus, as before, (11.106), (11.108) and (11.109) are still valid if (11.80) and (11.81) hold only in some specific (possibly artificial) inertial or non-inertial Cartesian coordinate system.

In particular, in the case of the simplest ideal gas where (11.75) and (11.76) are valid, by (11.78), we have

$$\frac{E}{\mu} = F(\mu, p) := C_0 \frac{p}{\mu}, \tag{11.110}$$

and therefore, inserting (11.110) into (11.107) gives

$$c_0 := \sqrt{\frac{(1 + C_0)\, p_0}{C_0 \mu_0}}. \tag{11.111}$$

Moreover, in this case (11.105) reads as

$$\mu_1 \approx \frac{C_0}{1 + C_0} \frac{\mu_0}{p_0}\, p_1. \tag{11.112}$$

On the other hand, in the case of barotropic fluid, the pressure p is a function of the density μ only, i.e.

$$p = \mathcal{R}(\mu). \tag{11.113}$$

Thus, inserting (11.79) into (11.113) and using (11.80), we deduce

$$p_1 \approx \frac{d\mathcal{R}}{d\mu}(\mu_0)\, \mu_1. \tag{11.114}$$

Therefore, inserting (11.114) into (11.95) and using again (11.81), we finally obtain the analogous to (11.106) wave equation for the oscillating part of the pressure p_1 of the form:

$$\frac{\partial}{\partial t}\left\{\frac{1}{c_0^2}\left(\frac{\partial p_1}{\partial t} + \mathbf{u}_0 \cdot \nabla_{\mathbf{x}} p_1\right)\right\} + \mathrm{div}_{\mathbf{x}}\left\{\frac{1}{c_0^2}\left(\frac{\partial p_1}{\partial t} + \mathbf{u}_0 \cdot \nabla_{\mathbf{x}} p_1\right)\mathbf{u}_0\right\} \approx \Delta_{\mathbf{x}} p_1, \tag{11.115}$$

where we denote

$$c_0 := \sqrt{\frac{d\mathcal{R}}{d\mu}(\mu_0)}. \tag{11.116}$$

Moreover, the oscillating parts of the density μ_1 and the velocity $\mathbf{u}_1$ can be found from (11.114) and (11.91), respectively, i.e. we have

$$p_1 \approx \frac{d\mathcal{R}}{d\mu}(\mu_0)\, \mu_1, \tag{11.117}$$

and

$$\mu_0 \left(\frac{\partial \mathbf{u}_1}{\partial t} - \mathrm{curl}_{\mathbf{x}} \{\mathbf{u}_0 \times \mathbf{u}_1\} + (\mathrm{div}_{\mathbf{x}}\, \mathbf{u}_1)\, \mathbf{u}_0\right) \approx -\nabla_{\mathbf{x}} p_1. \tag{11.118}$$

Again note that (11.115), (11.117) and (11.118) are invariant under the change in inertial or non-inertial Cartesian coordinate system. Thus, as before, (11.115), (11.117) and (11.118) are still valid if (11.80) and (11.81) hold only in some specific (possibly artificial) inertial or non-inertial Cartesian coordinate system.

11.4 Propagation of Waves in the Moving Elastic Body

Consistently with the general equation of motion of a continuum medium (11.1), consider the motion of an elastic body:

$$\mu\left(\frac{\partial \mathbf{u}}{\partial t}+d_{\mathbf{x}}\mathbf{u}\cdot\mathbf{u}\right)=$$
$$-\mu\mathbf{u}\times\operatorname{curl}_{\mathbf{x}}\mathbf{v}+\mu\left(\partial_t\mathbf{v}+\nabla_{\mathbf{x}}\frac{1}{2}|\mathbf{v}|^2\right)+\rho\mathbf{E}+\frac{1}{c}\mathbf{j}\times\mathbf{B}+\operatorname{div}_{\mathbf{x}}\left\{\alpha\mathcal{E}(\mathbf{x},t)+\beta\left(\operatorname{tr}\left\{\mathcal{E}(\mathbf{x},t)\right\}\right)I\right\}, \tag{11.119}$$

where

$$\mathcal{E}(\mathbf{x},t):=\frac{1}{2}\left(I-\{d_{\mathbf{x}}\mathbf{f}(\mathbf{x},t)\}^T\cdot d_{\mathbf{x}}\mathbf{f}(\mathbf{x},t)\right), \tag{11.120}$$

consistently with (11.48) and (11.50), with

$$\begin{cases}\frac{\partial\mathbf{f}}{\partial t}(\mathbf{x},t)+d_{\mathbf{x}}\mathbf{f}(\mathbf{x},t)\cdot\mathbf{u}(\mathbf{x},t)=0\\ \mathbf{f}(\mathbf{x},t_0)=\mathbf{x},\end{cases} \tag{11.121}$$

consistently with (11.35), where the initial instant of time t_0 is chosen (possibly fictitiously) in such a way that our elastic body is relaxed at time t_0. Moreover, by (11.60), we have

$$\mu(\mathbf{x},t)=\mu(\mathbf{x},t_0)\left|det\{d_{\mathbf{x}}\mathbf{f}(\mathbf{x},t)\}\right|. \tag{11.122}$$

Next we assume that

$$\mathbf{u}(\mathbf{x},t)=\mathbf{u}_0(\mathbf{x},t)+\mathbf{u}_1(\mathbf{x},t),\quad \mu(\mathbf{x},t)=\mu_0(\mathbf{x},t)+\mu_1(\mathbf{x},t),$$
$$\mathbf{f}(\mathbf{x},t)=\mathbf{f}_0(\mathbf{x},t)+\mathbf{f}_1(\mathbf{x},t)\quad\text{and}\quad\mathcal{E}(\mathbf{x},t)=\mathcal{E}_0(\mathbf{x},t)+\mathcal{E}_1(\mathbf{x},t), \tag{11.123}$$

where $\mathbf{u}_0(\mathbf{x},t),\mu_0(\mathbf{x},t),\mathbf{f}_0(\mathbf{x},t),\mathcal{E}_0(\mathbf{x},t)$ are the averages of $\mathbf{u}(\mathbf{x},t),\mu(\mathbf{x},t),\mathbf{f}(\mathbf{x},t),\mathcal{E}(\mathbf{x},t)$ on small spatial and temporal intervals, surrounding the point $(\mathbf{x},t)$. Although we assume these intervals of space and time to be very small, we also assume them to be quite macroscopic. We call $\mathbf{u}_1,\mu_1,\mathbf{f}_1,\mathcal{E}_1$ the oscillating parts of $\mathbf{u},\mu,\mathbf{f},\mathcal{E}$ and we assume that they are small with respect to the averages $\mathbf{u}_0,\mu_0,\mathbf{f}_0,\mathcal{E}_0$, i.e. we have

$$|\mathbf{u}_1|\ll|\mathbf{u}_0|,\quad|\mu_1|\ll|\mu_0|,\quad|\mathbf{f}_1|\ll|\mathbf{f}_0|. \tag{11.124}$$

However, we assume that $\mathbf{u}_1$, μ_1, $\mathbf{f}_1$ are highly oscillate and thus they change spatially and temporary much faster than the averages $\mathbf{u}_0$, μ_0, $\mathbf{f}_0$ and the fields $\mathbf{v}$, $\mathbf{E}$, $\mathbf{B}$, ρ, $\mathbf{j}$, i.e. we have

$$\frac{|d_{\mathbf{x}}\alpha|}{|\alpha|}+\frac{|d_{\mathbf{x}}\beta|}{|\beta|}+\frac{\left|d_{\mathbf{x}}\left(\rho\mathbf{E}+\frac{1}{c}\mathbf{j}\times\mathbf{B}\right)\right|}{\left|\rho\mathbf{E}+\frac{1}{c}\mathbf{j}\times\mathbf{B}\right|}+\frac{|d_{\mathbf{x}}\mathbf{v}|}{|\mathbf{v}|}+\frac{|d_{\mathbf{x}}\mu_0|}{|\mu_0|}+\frac{|d_{\mathbf{x}}\mathbf{f}_0|}{|\mathbf{f}_0|}+\frac{|d_{\mathbf{x}}\mathbf{u}_0|}{|\mathbf{u}_0|}$$
$$\ll \min\left\{\frac{|d_{\mathbf{x}}\mathbf{u}_1|}{|\mathbf{u}_1|},\frac{|d_{\mathbf{x}}\mu_1|}{|\mu_1|},\frac{|d_{\mathbf{x}}\mathbf{f}_1|}{|\mathbf{f}_1|}\right\},$$
$$\frac{\left|\partial_t\left(\rho\mathbf{E}+\frac{1}{c}\mathbf{j}\times\mathbf{B}\right)\right|}{\left|\rho\mathbf{E}+\frac{1}{c}\mathbf{j}\times\mathbf{B}\right|}+\frac{|\partial_t\mathbf{v}|}{|\mathbf{v}|}+\frac{|\partial_t\mu_0|}{|\mu_0|}+\frac{|\partial_t\mathbf{f}_0|}{|\mathbf{f}_0|}+\frac{|\partial_t\mathbf{u}_0|}{|\mathbf{u}_0|}$$
$$\ll \min\left\{\frac{|\partial_t\mathbf{u}_1|}{|\mathbf{u}_1|},\frac{|\partial_t\mu_1|}{|\mu_1|},\frac{|\partial_t\mathbf{f}_1|}{|\mathbf{f}_1|}\right\},$$
$$\frac{|d^2_{\mathbf{x}}\mathbf{f}_0|}{|d_{\mathbf{x}}\mathbf{f}_0|}\ll\frac{|d^2_{\mathbf{x}}\mathbf{f}_1|}{|d_{\mathbf{x}}\mathbf{f}_1|}\qquad\frac{|\mathbf{u}_1|}{|\mathbf{u}_0|}\ll\min\left\{\frac{|d_{\mathbf{x}}\mathbf{u}_1|}{|d_{\mathbf{x}}\mathbf{u}_0|},\frac{|d_{\mathbf{x}}\mathbf{u}_1|}{|d_{\mathbf{x}}\mathbf{v}|}\right\},$$
$$\text{and}\qquad\frac{|\mu_1|}{|\mu_0|}\ll\frac{|d_{\mathbf{x}}\mathbf{u}_1||\mathbf{u}_0||\mu_0|}{\left|\rho\mathbf{E}+\frac{1}{c}\mathbf{j}\times\mathbf{B}\right|}.\tag{11.125}$$

Finally, we assume that the fields $\mathbf{v}, \mathbf{E}, \mathbf{B}, \rho, \mathbf{j}$ change slowly with respect to the oscillations of $\mathbf{u}_1, \mu_1, \mathbf{f}_1$ and thus we assume that $\mathbf{v}, \mathbf{E}, \mathbf{B}, \rho, \mathbf{j}$ can be replaced by their spatial and temporal averages. Note that μ, μ_0, μ_1 behave like proper scalar fields, $\mathcal{E}, \mathcal{E}_0, \mathcal{E}_1$ behave like proper matrix fields and $\mathbf{u}, \mathbf{u}_0$ behave like speed-like vector fields under the change in Cartesian coordinate systems. Thus, since $\mathbf{u}_1 = \mathbf{u} - \mathbf{u}_0$, we deduce that $\mathbf{u}_1$ behaves like a proper vector field under the change in Cartesian coordinate systems. Furthermore, using (11.47), we deduce that the vector field $\mathbf{g}_1(\mathbf{x}, t)$ defined by the following:

$$\mathbf{g}_1(\mathbf{x},t) := \left(d_{\mathbf{x}}\mathbf{f}_0(\mathbf{x},t)\right)^{-1}\cdot\mathbf{f}_1(\mathbf{x},t),\tag{11.126}$$

behaves like a proper vector field. Finally, note that obviously the averages of $\mathbf{u}_1, \mu_1, \mathbf{f}_1, \mathcal{E}_1$ vanish.

Next, we would like to approximate (11.121) and (11.119). In the rough level of approximation, we could just replace $\mathbf{u}, \mu, \mathbf{f}$ by $\mathbf{u}_0, \mu_0, \mathbf{f}_0$. However, we would like to use a more delicate approximation, taking into the account the first-order terms with $\mathbf{u}_1, \mu_1, \mathbf{f}_1$. Then, as before, by (11.121) and (11.123) and using (11.124), we have

$$\frac{\partial\mathbf{f}_0}{\partial t}(\mathbf{x},t)+\frac{\partial\mathbf{f}_1}{\partial t}(\mathbf{x},t)+d_{\mathbf{x}}\mathbf{f}_0(\mathbf{x},t)\cdot\mathbf{u}_0(\mathbf{x},t)+d_{\mathbf{x}}\mathbf{f}_0(\mathbf{x},t)\cdot\mathbf{u}_1(\mathbf{x},t)+d_{\mathbf{x}}\mathbf{f}_1(\mathbf{x},t)\cdot\mathbf{u}_0(\mathbf{x},t)\approx 0.\tag{11.127}$$

Thus, averaging of (11.127) gives

$$\frac{\partial\mathbf{f}_0}{\partial t}(\mathbf{x},t)+d_{\mathbf{x}}\mathbf{f}_0(\mathbf{x},t)\cdot\mathbf{u}_0(\mathbf{x},t)\approx 0.\tag{11.128}$$

Therefore, subtracting (11.128) from unaveraged (11.127), we obtain

$$\frac{\partial\mathbf{f}_1}{\partial t}(\mathbf{x},t)+d_{\mathbf{x}}\mathbf{f}_1(\mathbf{x},t)\cdot\mathbf{u}_0(\mathbf{x},t)+d_{\mathbf{x}}\mathbf{f}_0(\mathbf{x},t)\cdot\mathbf{u}_1(\mathbf{x},t)\approx 0.\tag{11.129}$$

Then, by (11.129) and (11.125), we deduce

$$\frac{\partial}{\partial t}\left((d_{\mathbf{x}}\mathbf{f}_0(\mathbf{x},t))^{-1}\cdot\mathbf{f}_1(\mathbf{x},t)\right)+d_{\mathbf{x}}\left((d_{\mathbf{x}}\mathbf{f}_0(\mathbf{x},t))^{-1}\cdot\mathbf{f}_1(\mathbf{x},t)\right)\cdot\mathbf{u}_0(\mathbf{x},t)\approx-\mathbf{u}_1(\mathbf{x},t), \quad (11.130)$$

i.e.

$$\frac{\partial\mathbf{g}_1}{\partial t}+d_{\mathbf{x}}\mathbf{g}_1\cdot\mathbf{u}_0\approx-\mathbf{u}_1, \quad (11.131)$$

where $\mathbf{g}_1(\mathbf{x},t)$ is given by (11.126). Note that, by (11.125), (11.131) can be written in the alternative approximate form

$$\frac{\partial\mathbf{g}_1}{\partial t}+(\mathrm{div}_{\mathbf{x}}\,\mathbf{u}_0)\,\mathbf{g}_1+d_{\mathbf{x}}\mathbf{g}_1\cdot\mathbf{u}_0-d_{\mathbf{x}}\mathbf{u}_0\cdot\mathbf{g}_1\approx-\mathbf{u}_1, \quad (11.132)$$

and thus, by (11.132) and (2.11), we have

$$\frac{\partial\mathbf{g}_1}{\partial t}-\mathrm{curl}_{\mathbf{x}}\,\{\mathbf{u}_0\times\mathbf{g}_1\}+(\mathrm{div}_{\mathbf{x}}\,\mathbf{g}_1)\,\mathbf{u}_0\approx-\mathbf{u}_1. \quad (11.133)$$

On the other hand, as before, by (11.123) and using (11.124), we rewrite (11.119) as

$$\begin{aligned}
&\mu_0\left(\frac{\partial\mathbf{u}_0}{\partial t}+d_{\mathbf{x}}\mathbf{u}_0\cdot\mathbf{u}_0\right)+\mu_1\left(\frac{\partial\mathbf{u}_0}{\partial t}+d_{\mathbf{x}}\mathbf{u}_0\cdot\mathbf{u}_0\right)\\
&\qquad+\mu_0\left(\frac{\partial\mathbf{u}_1}{\partial t}+d_{\mathbf{x}}\mathbf{u}_1\cdot\mathbf{u}_0+d_{\mathbf{x}}\mathbf{u}_0\cdot\mathbf{u}_1\right)\\
&\approx-\mu_0\mathbf{u}_0\times\mathrm{curl}_{\mathbf{x}}\mathbf{v}-\mu_1\mathbf{u}_0\times\mathrm{curl}_{\mathbf{x}}\mathbf{v}-\mu_0\mathbf{u}_1\times\mathrm{curl}_{\mathbf{x}}\mathbf{v}+\mu_0\left(\partial_t\mathbf{v}+\nabla_{\mathbf{x}}\frac{1}{2}|\mathbf{v}|^2\right)\\
&\qquad+\mu_1\left(\partial_t\mathbf{v}+\nabla_{\mathbf{x}}\frac{1}{2}|\mathbf{v}|^2\right)+\rho\mathbf{E}+\frac{1}{c}\mathbf{j}\times\mathbf{B}\\
&\qquad+\mathrm{div}_{\mathbf{x}}\,\{\alpha\mathcal{E}_0+\beta\,(tr\,\mathcal{E}_0)\,I\}+\mathrm{div}_{\mathbf{x}}\,\{\alpha\mathcal{E}_1+\beta\,(tr\,\mathcal{E}_1)\,I\}. \quad (11.134)
\end{aligned}$$

Thus, as before, averaging (11.134) gives

$$\begin{aligned}
\mu_0\left(\frac{\partial\mathbf{u}_0}{\partial t}+d_{\mathbf{x}}\mathbf{u}_0\cdot\mathbf{u}_0\right)&\approx-\mu_0\mathbf{u}_0\times\mathrm{curl}_{\mathbf{x}}\mathbf{v}+\mu_0\left(\partial_t\mathbf{v}+\nabla_{\mathbf{x}}\frac{1}{2}|\mathbf{v}|^2\right)+\rho\mathbf{E}+\frac{1}{c}\mathbf{j}\times\mathbf{B}\\
&\qquad+\mathrm{div}_{\mathbf{x}}\,\{\alpha\mathcal{E}_0+\beta\,(tr\,\mathcal{E}_0)\,I\}. \quad (11.135)
\end{aligned}$$

Therefore, subtracting (11.135) from unaveraged (11.134), we obtain

$$\begin{aligned}
&\mu_1\left(\frac{\partial\mathbf{u}_0}{\partial t}+d_{\mathbf{x}}\mathbf{u}_0\cdot\mathbf{u}_0\right)+\mu_0\left(\frac{\partial\mathbf{u}_1}{\partial t}+d_{\mathbf{x}}\mathbf{u}_1\cdot\mathbf{u}_0+d_{\mathbf{x}}\mathbf{u}_0\cdot\mathbf{u}_1\right)\approx\\
&\quad-\mu_1\mathbf{u}_0\times\mathrm{curl}_{\mathbf{x}}\mathbf{v}-\mu_0\mathbf{u}_1\times\mathrm{curl}_{\mathbf{x}}\mathbf{v}+\mu_1\left(\partial_t\mathbf{v}+\nabla_{\mathbf{x}}\frac{1}{2}|\mathbf{v}|^2\right)+\mathrm{div}_{\mathbf{x}}\,\{\alpha\mathcal{E}_1+\beta\,(tr\,\mathcal{E}_1)\,I\}.
\end{aligned}$$

(11.136)

Thus, by (11.136) and (11.135), we infer

$$\mu_0\left(\frac{\partial \mathbf{u}_1}{\partial t} + d_{\mathbf{x}}\mathbf{u}_1 \cdot \mathbf{u}_0 + d_{\mathbf{x}}\mathbf{u}_0 \cdot \mathbf{u}_1\right) \approx -\mu_0 \mathbf{u}_1 \times \mathrm{curl}_{\mathbf{x}}\mathbf{v} + \mathrm{div}_{\mathbf{x}}\left\{\alpha\mathcal{E}_1 + \beta\,(tr, \mathcal{E}_1)\,I\right\}$$
$$- \frac{\mu_1}{\mu_0}\,\mathrm{div}_{\mathbf{x}}\left\{\alpha\mathcal{E}_0 + \beta\,(tr\,\mathcal{E}_0)\,I\right\} - \frac{\mu_1}{\mu_0}\left(\rho\mathbf{E} + \frac{1}{c}\mathbf{j} \times \mathbf{B}\right). \quad (11.137)$$

Therefore, using (11.125), we further approximate (11.137) as

$$\mu_0\left(\frac{\partial \mathbf{u}_1}{\partial t} + d_{\mathbf{x}}\mathbf{u}_1 \cdot \mathbf{u}_0\right) \approx \mathrm{div}_{\mathbf{x}}\left\{\alpha\mathcal{E}_1 + \beta\,(tr\,\mathcal{E}_1)\,I\right\} - \frac{\mu_1}{\mu_0}\,\mathrm{div}_{\mathbf{x}}\left\{\alpha\mathcal{E}_0 + \beta\,(tr\,\mathcal{E}_0)\,I\right\}. \quad (11.138)$$

On the other hand, by (11.122), (11.123), (11.120) and (11.125), we deduce

$$\left|\frac{\mu_1}{\mu_0}\,\mathrm{div}_{\mathbf{x}}\left\{\alpha\mathcal{E}_0 + \beta\,(tr\,\mathcal{E}_0)\,I\right\}\right| \ll \left|\mathrm{div}_{\mathbf{x}}\left\{\alpha\mathcal{E}_1 + \beta\,(tr\,\mathcal{E}_1)\,I\right\}\right|. \quad (11.139)$$

Thus, by (11.139), we further approximate (11.138) as

$$\mu_0\left(\frac{\partial \mathbf{u}_1}{\partial t} + d_{\mathbf{x}}\mathbf{u}_1 \cdot \mathbf{u}_0\right) \approx \mathrm{div}_{\mathbf{x}}\left\{\alpha\mathcal{E}_1 + \beta\,(tr\,\mathcal{E}_1)\,I\right\}. \quad (11.140)$$

Therefore, by inserting (11.131) into (11.140), we deduce

$$\mu_0\left(\frac{\partial}{\partial t}\left(\frac{\partial \mathbf{g}_1}{\partial t} + d_{\mathbf{x}}\mathbf{g}_1 \cdot \mathbf{u}_0\right) + d_{\mathbf{x}}\left(\frac{\partial \mathbf{g}_1}{\partial t} + d_{\mathbf{x}}\mathbf{g}_1 \cdot \mathbf{u}_0\right) \cdot \mathbf{u}_0\right)$$
$$\approx -\,\mathrm{div}_{\mathbf{x}}\left\{\alpha\mathcal{E}_1 + \beta\,(\mathrm{tr}\,\{\mathcal{E}_1\})\,I\right\}, \quad (11.141)$$

where $\mathbf{g}_1(\mathbf{x}, t)$ is given by (11.126). On the other hand, by (11.120), (11.123), (11.124) and (11.125), we have

$$\mathcal{E}_0(\mathbf{x}, t) \approx \frac{1}{2}\left(I - \{d_{\mathbf{x}}\mathbf{f}_0(\mathbf{x}, t)\}^T \cdot d_{\mathbf{x}}\mathbf{f}_0(\mathbf{x}, t)\right), \quad (11.142)$$

and

$$\mathcal{E}_1(\mathbf{x}, t) \approx -\frac{1}{2}\left(\{d_{\mathbf{x}}\mathbf{f}_1(\mathbf{x}, t)\}^T \cdot d_{\mathbf{x}}\mathbf{f}_0(\mathbf{x}, t) + \{d_{\mathbf{x}}\mathbf{f}_0(\mathbf{x}, t)\}^T \cdot d_{\mathbf{x}}\mathbf{f}_1(\mathbf{x}, t)\right) \approx$$
$$-\frac{1}{2}\{d_{\mathbf{x}}\mathbf{f}_0(\mathbf{x}, t)\}^T \cdot d_{\mathbf{x}}\mathbf{f}_0(\mathbf{x}, t) \cdot d_{\mathbf{x}}\left((d_{\mathbf{x}}\mathbf{f}_0(\mathbf{x}, t))^{-1} \cdot \mathbf{f}_1(\mathbf{x}, t)\right)$$
$$-\frac{1}{2}\left\{d_{\mathbf{x}}\left((d_{\mathbf{x}}\mathbf{f}_0(\mathbf{x}, t))^{-1} \cdot \mathbf{f}_1(\mathbf{x}, t)\right)\right\}^T \cdot \{d_{\mathbf{x}}\mathbf{f}_0(\mathbf{x}, t)\}^T \cdot d_{\mathbf{x}}\mathbf{f}_0(\mathbf{x}, t) =$$
$$\mathcal{E}_0(\mathbf{x}, t) \cdot d_{\mathbf{x}}\mathbf{g}_1(\mathbf{x}, t) + \{d_{\mathbf{x}}\mathbf{g}_1(\mathbf{x}, t)\}^T \cdot \mathcal{E}_0(\mathbf{x}, t) - \frac{1}{2}\left(d_{\mathbf{x}}\mathbf{g}_1(\mathbf{x}, t) + \{d_{\mathbf{x}}\mathbf{g}_1(\mathbf{x}, t)\}^T\right), \quad (11.143)$$

where the last equality of (11.143) follows from (11.126) and (11.142). So,

$$\mathcal{E}_1(\mathbf{x},t) \approx$$
$$\mathcal{E}_0(\mathbf{x},t)\cdot d_{\mathbf{x}}\mathbf{g}_1(\mathbf{x},t)+\{d_{\mathbf{x}}\mathbf{g}_1(\mathbf{x},t)\}^T\cdot\mathcal{E}_0(\mathbf{x},t)-\frac{1}{2}\left(d_{\mathbf{x}}\mathbf{g}_1(\mathbf{x},t)+\{d_{\mathbf{x}}\mathbf{g}_1(\mathbf{x},t)\}^T\right). \tag{11.144}$$

Thus, inserting (11.144) into (11.141) and using (11.125) finally gives

$$\begin{aligned}\mu_0\left(\frac{\partial}{\partial t}\left(\frac{\partial\mathbf{g}_1}{\partial t}+d_{\mathbf{x}}\mathbf{g}_1\cdot\mathbf{u}_0\right)+d_{\mathbf{x}}\left(\frac{\partial\mathbf{g}_1}{\partial t}+d_{\mathbf{x}}\mathbf{g}_1\cdot\mathbf{u}_0\right)\cdot\mathbf{u}_0\right)&\approx\\ -\alpha\,\mathrm{div}_{\mathbf{x}}\{\mathcal{E}_0\cdot d_{\mathbf{x}}\mathbf{g}_1\}-\alpha\,\mathrm{div}_{\mathbf{x}}\left\{\{d_{\mathbf{x}}\mathbf{g}_1\}^T\cdot\mathcal{E}_0\right\}&-\beta\,\nabla_{\mathbf{x}}\left(\mathrm{tr}\left(\mathcal{E}_0\cdot d_{\mathbf{x}}\mathbf{g}_1\right)\right)\\ -\beta\,\nabla_{\mathbf{x}}\left(\mathrm{tr}\left(\{d_{\mathbf{x}}\mathbf{g}_1\}^T\cdot\mathcal{E}_0\right)\right)+\frac{\alpha}{2}\,\Delta_{\mathbf{x}}\mathbf{g}_1&+\left(\frac{\alpha}{2}+\beta\right)\nabla_{\mathbf{x}}\left(\mathrm{div}_{\mathbf{x}}\,\mathbf{g}_1\right).\end{aligned} \tag{11.145}$$

Note again that similarly as we proved (11.133), i.e.:

$$\frac{\partial\mathbf{g}_1}{\partial t}-\mathrm{curl}_{\mathbf{x}}\{\mathbf{u}_0\times\mathbf{g}_1\}+(\mathrm{div}_{\mathbf{x}}\,\mathbf{g}_1)\,\mathbf{u}_0\approx-\mathbf{u}_1, \tag{11.146}$$

we can derive the approximation of (11.145) as

$$\begin{aligned}&\mu_0\frac{\partial}{\partial t}\left(\frac{\partial\mathbf{g}_1}{\partial t}-\mathrm{curl}_{\mathbf{x}}\{\mathbf{u}_0\times\mathbf{g}_1\}+(\mathrm{div}_{\mathbf{x}}\,\mathbf{g}_1)\,\mathbf{u}_0\right)\\ &\quad-\mu_0\,\mathrm{curl}_{\mathbf{x}}\left\{\mathbf{u}_0\times\left(\frac{\partial\mathbf{g}_1}{\partial t}-\mathrm{curl}_{\mathbf{x}}\{\mathbf{u}_0\times\mathbf{g}_1\}+(\mathrm{div}_{\mathbf{x}}\,\mathbf{g}_1)\,\mathbf{u}_0\right)\right\}\\ &\quad+\left(\mathrm{div}_{\mathbf{x}}\left\{\frac{\partial\mathbf{g}_1}{\partial t}-\mathrm{curl}_{\mathbf{x}}\{\mathbf{u}_0\times\mathbf{g}_1\}+(\mathrm{div}_{\mathbf{x}}\,\mathbf{g}_1)\,\mathbf{u}_0\right\}\right)\mathbf{u}_0\approx\\ &\frac{\alpha}{2}\,\Delta_{\mathbf{x}}\mathbf{g}_1+\left(\frac{\alpha}{2}+\beta\right)\nabla_{\mathbf{x}}\left(\mathrm{div}_{\mathbf{x}}\,\mathbf{g}_1\right)-\alpha\,\mathrm{div}_{\mathbf{x}}\{\mathcal{E}_0\cdot d_{\mathbf{x}}\mathbf{g}_1\}-\alpha\,\mathrm{div}_{\mathbf{x}}\left\{\{d_{\mathbf{x}}\mathbf{g}_1\}^T\cdot\mathcal{E}_0\right\}\\ &\qquad-\beta\,\nabla_{\mathbf{x}}\left(\mathrm{tr}\left(\mathcal{E}_0\cdot d_{\mathbf{x}}\mathbf{g}_1\right)\right)-\beta\,\nabla_{\mathbf{x}}\left(\mathrm{tr}\left(\{d_{\mathbf{x}}\mathbf{g}_1\}^T\cdot\mathcal{E}_0\right)\right).\end{aligned} \tag{11.147}$$

In particular, if our elastic body is nearly rigid, we have

$$\mathcal{E}_0\approx 0, \tag{11.148}$$

and we simplify (11.145) as

$$\begin{aligned}\mu_0\left(\frac{\partial}{\partial t}\left(\frac{\partial\mathbf{g}_1}{\partial t}+d_{\mathbf{x}}\mathbf{g}_1\cdot\mathbf{u}_0\right)+d_{\mathbf{x}}\left(\frac{\partial\mathbf{g}_1}{\partial t}+d_{\mathbf{x}}\mathbf{g}_1\cdot\mathbf{u}_0\right)\cdot\mathbf{u}_0\right)&\\ \approx\frac{\alpha}{2}\,\Delta_{\mathbf{x}}\mathbf{g}_1+\left(\frac{\alpha}{2}+\beta\right)\nabla_{\mathbf{x}}\left(\mathrm{div}_{\mathbf{x}}\,\mathbf{g}_1\right),&\end{aligned} \tag{11.149}$$

and (11.144) as

$$\mathcal{E}_1 \approx -\frac{1}{2}\left(d_{\mathbf{x}}\mathbf{g}_1 + \{d_{\mathbf{x}}\mathbf{g}_1\}^T\right). \tag{11.150}$$

Moreover, we rewrite (11.147) as

$$\begin{aligned}
&\mu_0 \frac{\partial}{\partial t}\left(\frac{\partial \mathbf{g}_1}{\partial t} - \text{curl}_{\mathbf{x}}\{\mathbf{u}_0 \times \mathbf{g}_1\} + (\text{div}_{\mathbf{x}}\,\mathbf{g}_1)\,\mathbf{u}_0\right) \\
&\quad - \mu_0\,\text{curl}_{\mathbf{x}}\left\{\mathbf{u}_0 \times \left(\frac{\partial \mathbf{g}_1}{\partial t} - \text{curl}_{\mathbf{x}}\{\mathbf{u}_0 \times \mathbf{g}_1\} + (\text{div}_{\mathbf{x}}\,\mathbf{g}_1)\,\mathbf{u}_0\right)\right\} \\
&\quad + \mu_0\left(\text{div}_{\mathbf{x}}\left\{\frac{\partial \mathbf{g}_1}{\partial t} - \text{curl}_{\mathbf{x}}\{\mathbf{u}_0 \times \mathbf{g}_1\} + (\text{div}_{\mathbf{x}}\,\mathbf{g}_1)\,\mathbf{u}_0\right\}\right)\mathbf{u}_0 \\
&\qquad \approx \frac{\alpha}{2}\Delta_{\mathbf{x}}\mathbf{g}_1 + \left(\frac{\alpha}{2} + \beta\right)\nabla_{\mathbf{x}}\left(\text{div}_{\mathbf{x}}\,\mathbf{g}_1\right).
\end{aligned} \tag{11.151}$$

Next, note that, since μ, μ_0, μ_1 are proper scalar fields, $\mathcal{E}$, $\mathcal{E}_0$, $\mathcal{E}_1$ are proper matrix fields, $\mathbf{u}$, $\mathbf{u}_0$ are speed-like vector fields, $\mathbf{u}_1$ is a proper vector field and the vector field $\mathbf{g}_1(\mathbf{x}, t)$ defined by (11.126) is a proper vector field, as before, it can be easily proved that (11.147), (11.146), (11.142), (11.144), (11.151) (11.148) and (11.150) are invariant under the change in inertial or non-inertial Cartesian coordinate system. Thus, (11.147), (11.146), (11.142), (11.144) and in the case of nearly rigid elastic body also (11.151) (11.148) and (11.150) are still valid if (11.124) and (11.125) hold only in some specific (possibly artificial) inertial or non-inertial Cartesian coordinate system.

Next, taking the divergence of both sides of (11.151) and using (11.125) gives

$$\begin{aligned}
&\frac{\partial}{\partial t}\left\{\mu_0\left(\frac{\partial}{\partial t}(\text{div}_{\mathbf{x}}\,\mathbf{g}_1) + \mathbf{u}_0 \cdot \nabla_{\mathbf{x}}(\text{div}_{\mathbf{x}}\,\mathbf{g}_1)\right)\right\} \\
&\quad + \text{div}_{\mathbf{x}}\left\{\mu_0\left(\frac{\partial}{\partial t}(\text{div}_{\mathbf{x}}\,\mathbf{g}_1) + \mathbf{u}_0 \cdot \nabla_{\mathbf{x}}(\text{div}_{\mathbf{x}}\,\mathbf{g}_1)\right)\mathbf{u}_0\right\} \approx (\alpha + \beta)\,\Delta_{\mathbf{x}}(\text{div}_{\mathbf{x}}\,\mathbf{g}_1),
\end{aligned} \tag{11.152}$$

i.e.:

$$\frac{\partial}{\partial t}\left\{\frac{1}{c_{01}^2}\left(\frac{\partial h_1}{\partial t} + \mathbf{u}_0 \cdot \nabla_{\mathbf{x}} h_1\right)\right\} + \text{div}_{\mathbf{x}}\left\{\frac{1}{c_{01}^2}\left(\frac{\partial h_1}{\partial t} + \mathbf{u}_0 \cdot \nabla_{\mathbf{x}} h_1\right)\mathbf{u}_0\right\} \approx \Delta_{\mathbf{x}} h_1, \tag{11.153}$$

where

$$h_1(\mathbf{x}, t) := \text{div}_{\mathbf{x}}\,\mathbf{g}_1(\mathbf{x}, t) \quad \text{and} \quad c_{01} := \sqrt{\frac{(\alpha + \beta)}{\mu_0}}. \tag{11.154}$$

On the other hand, taking the curl of both sides of (11.149) and using (11.125) gives

$$\frac{\partial}{\partial t}\left\{\mu_0\left(\frac{\partial}{\partial t}(\mathrm{curl}_{\mathbf{x}}\,\mathbf{g}_1)+d_{\mathbf{x}}(\mathrm{curl}_{\mathbf{x}}\,\mathbf{g}_1)\cdot\mathbf{u}_0\right)\right\}$$
$$+\,\mathrm{div}_{\mathbf{x}}\left\{\mu_0\left(\frac{\partial}{\partial t}(\mathrm{curl}_{\mathbf{x}}\,\mathbf{g}_1)+d_{\mathbf{x}}(\mathrm{curl}_{\mathbf{x}}\,\mathbf{g}_1)\cdot\mathbf{u}_0\right)\otimes\mathbf{u}_0\right\}\approx\frac{\alpha}{2}\Delta_{\mathbf{x}}(\mathrm{curl}_{\mathbf{x}}\,\mathbf{g}_1),\tag{11.155}$$

i.e.

$$\frac{\partial}{\partial t}\left\{\frac{1}{c_{02}^2}\left(\frac{\partial\mathbf{h}_2}{\partial t}+d_{\mathbf{x}}\mathbf{h}_2\cdot\mathbf{u}_0\right)\right\}+\mathrm{div}_{\mathbf{x}}\left\{\frac{1}{c_{02}^2}\left(\frac{\partial\mathbf{h}_2}{\partial t}+d_{\mathbf{x}}\mathbf{h}_2\cdot\mathbf{u}_0\right)\otimes\mathbf{u}_0\right\}\approx\Delta_{\mathbf{x}}\mathbf{h}_2,\tag{11.156}$$

where

$$\mathbf{h}_2(\mathbf{x},t):=\mathrm{curl}_{\mathbf{x}}\,\mathbf{g}_1(\mathbf{x},t)\quad\text{and}\quad c_{02}:=\sqrt{\frac{\alpha}{2\mu_0}}.\tag{11.157}$$

Again, using (11.125) we can rewrite (11.156) as

$$\frac{1}{c_{02}^2}\frac{\partial}{\partial t}\left(\frac{\partial\mathbf{h}_2}{\partial t}-\mathbf{u}_0\times\mathrm{curl}_{\mathbf{x}}\mathbf{h}_2+\nabla_{\mathbf{x}}(\mathbf{h}_2\cdot\mathbf{u}_0)\right)$$
$$-\frac{1}{c_{02}^2}\mathrm{curl}_{\mathbf{x}}\left\{\mathbf{u}_0\times\left(\frac{\partial\mathbf{h}_2}{\partial t}-\mathbf{u}_0\times\mathrm{curl}_{\mathbf{x}}\mathbf{h}_2+\nabla_{\mathbf{x}}(\mathbf{h}_2\cdot\mathbf{u}_0)\right)\right\}$$
$$+\frac{1}{c_{02}^2}\left(\mathrm{div}_{\mathbf{x}}\left(\frac{\partial\mathbf{h}_2}{\partial t}-\mathbf{u}_0\times\mathrm{curl}_{\mathbf{x}}\mathbf{h}_2+\nabla_{\mathbf{x}}(\mathbf{h}_2\cdot\mathbf{u}_0)\right)\right)\mathbf{u}_0\approx\Delta_{\mathbf{x}}\mathbf{h}_2.\tag{11.158}$$

Again note that, using Proposition 3.1, we deduce that equations (11.153) and (11.158) are invariant under the change in inertial or non-inertial Cartesian coordinate system, provided that $\mathbf{h}_2$ is a proper vector field and h_1 is a proper scalar field. Thus, in the case of nearly rigid elastic body (11.153) and (11.158) are still valid if (11.124) and (11.125) hold only in some specific (possibly artificial) inertial or non-inertial Cartesian coordinate system. Finally note that, although the equation (11.156) is invariant under the Galilean Transformation, it is not invariant under the more general change in non-inertial Cartesian coordinate system. However, (11.156) is more convenient than (11.158), since every of the three scalar components of the vector field $\mathbf{h}_2$ in (11.156) satisfies three decoupled wave equations of the same type. On the other hand, if we consider some three proper vector fields $\mathbf{e}_1:=\mathbf{e}_1(\mathbf{x},t)$, $\mathbf{e}_2:=\mathbf{e}_2(\mathbf{x},t)$, and $\mathbf{e}_3:=\mathbf{e}_3(\mathbf{x},t)$, which are mutually orthogonal to each other and satisfy the following approximation analogous to (11.125):

$$\frac{|d_{\mathbf{x}}\mathbf{e}_1|+c_{02}|\partial_t\mathbf{e}_1|}{|\mathbf{e}_1|}+\frac{|d_{\mathbf{x}}\mathbf{e}_2|+c_{02}|\partial_t\mathbf{e}_2|}{|\mathbf{e}_2|}+\frac{|d_{\mathbf{x}}\mathbf{e}_3|+c_{02}|\partial_t\mathbf{e}_3|}{|\mathbf{e}_3|}\ll\frac{|d_{\mathbf{x}}\mathbf{h}_2|+c_{02}|\partial_t\mathbf{h}_2|}{|\mathbf{h}_2|}.\tag{11.159}$$

i.e. the field $\mathbf{e}_k$ vary in space and time much weaker than $\mathbf{h}_2$, then we may write the alternative to (11.156) approximate equations in the form of three decoupled scalar wave equations of the same type:

$$\frac{\partial}{\partial t}\left\{\frac{1}{c_{02}^2}\left(\frac{\partial}{\partial t}(\mathbf{e}_k\cdot\mathbf{h}_2)+\mathbf{u}_0\cdot\nabla_{\mathbf{x}}(\mathbf{e}_k\cdot\mathbf{h}_2)\right)\right\}$$
$$+\operatorname{div}_{\mathbf{x}}\left\{\frac{1}{c_{02}^2}\left(\frac{\partial}{\partial t}(\mathbf{e}_k\cdot\mathbf{h}_2)+\mathbf{u}_0\cdot\nabla_{\mathbf{x}}(\mathbf{e}_k\cdot\mathbf{h}_2)\right)\mathbf{u}_0\right\}\approx\Delta_{\mathbf{x}}(\mathbf{e}_k\cdot\mathbf{h}_2)\quad\forall k=1,2,3. \tag{11.160}$$

Then, clearly, the new alternative approximate Eq. (11.160) together with (11.153) are indeed invariant under the more general change in non-inertial Cartesian coordinate system.

As a final corollary of the above in the case of nearly rigid body, we have (11.160) and (11.153), i.e.:

$$\frac{\partial}{\partial t}\left\{\frac{1}{c_{01}^2}\left(\frac{\partial h_1}{\partial t}+\mathbf{u}_0\cdot\nabla_{\mathbf{x}}h_1\right)\right\}+\operatorname{div}_{\mathbf{x}}\left\{\frac{1}{c_{01}^2}\left(\frac{\partial h_1}{\partial t}+\mathbf{u}_0\cdot\nabla_{\mathbf{x}}h_1\right)\mathbf{u}_0\right\}\approx\Delta_{\mathbf{x}}h_1, \tag{11.161}$$

and

$$\frac{\partial}{\partial t}\left\{\frac{1}{c_{02}^2}\left(\frac{\partial}{\partial t}(\mathbf{e}_k\cdot\mathbf{h}_2)+\mathbf{u}_0\cdot\nabla_{\mathbf{x}}(\mathbf{e}_k\cdot\mathbf{h}_2)\right)\right\}$$
$$+\operatorname{div}_{\mathbf{x}}\left\{\frac{1}{c_{02}^2}\left(\frac{\partial}{\partial t}(\mathbf{e}_k\cdot\mathbf{h}_2)+\mathbf{u}_0\cdot\nabla_{\mathbf{x}}(\mathbf{e}_k\cdot\mathbf{h}_2)\right)\mathbf{u}_0\right\}\approx\Delta_{\mathbf{x}}(\mathbf{e}_k\cdot\mathbf{h}_2)\quad\forall k=1,2,3, \tag{11.162}$$

where

$$h_1(\mathbf{x},t):=\operatorname{div}_{\mathbf{x}}\mathbf{g}_1(\mathbf{x},t)\quad\text{and}\quad c_{01}:=\sqrt{\frac{(\alpha+\beta)}{\mu_0}}. \tag{11.163}$$

and

$$\mathbf{h}_2(\mathbf{x},t):=\operatorname{curl}_{\mathbf{x}}\mathbf{g}_1(\mathbf{x},t)\quad\text{and}\quad c_{02}:=\sqrt{\frac{\alpha}{2\mu_0}}. \tag{11.164}$$

Moreover, we have (11.150), i.e.:

$$\mathcal{E}_1\approx-\frac{1}{2}\left(d_{\mathbf{x}}\mathbf{g}_1+\{d_{\mathbf{x}}\mathbf{g}_1\}^T\right), \tag{11.165}$$

and (11.146), i.e.:

$$\frac{\partial\mathbf{g}_1}{\partial t}-\operatorname{curl}_{\mathbf{x}}\{\mathbf{u}_0\times\mathbf{g}_1\}+(\operatorname{div}_{\mathbf{x}}\mathbf{g}_1)\,\mathbf{u}_0\approx-\mathbf{u}_1, \tag{11.166}$$

and the above six equations are invariant under the change in inertial or non-inertial Cartesian coordinate system.

As we can easily see, the divergence of $\mathbf{g}_l$ and any of the scalar components of the curl of $\mathbf{g}_l$ satisfy the invariant wave equation of the same type as (11.106) or (11.115). However, as we can see from (11.163) and (11.164) the characteristic parameter c_{0l} in the wave equation for the divergence part of $\mathbf{g}_l$ (11.161) differ from the characteristic parameter c_{02} in the wave equation for the curl part of $\mathbf{g}_l$ (11.162), i.e. the divergence part and the curl parts of $\mathbf{g}_l$ propagate as two different waves with the different speeds. As it can be easily seen, in the case of the flat waves in the resting body, the divergence part of $\mathbf{g}_l$ (which is curl-free) propagate as a longitudinal wave, similarly to the sound in fluid or gas, and at the same time, the curl part of $\mathbf{g}_l$ (which is divergence-free) propagate as a transverse wave. Moreover, the longitudinal and transverse waves in an elastic body propagate with two different speeds.

Chapter 12
Maxwell Equations in the Presence of Dielectrics and/or Magnetics

12.1 Polarization and Magnetization of a Totally Neutral System of Point Charges

Proposition 12.1 *Consider an arbitrary moving point with place* $\mathbf{r}(t)$ *and velocity* $\frac{d\mathbf{r}}{dt}(t)$ *and let* $\mathbf{u}(\mathbf{x}, t)$ *be a speed-like vector field. Next, consider a totally neutral system of n point charges* $\sigma_1, \ldots, \sigma_n$*, with places* $\mathbf{r}_1(t), \ldots, \mathbf{r}_n(t)$ *and velocities* $\frac{d\mathbf{r}_1}{dt}(t), \ldots, \frac{d\mathbf{r}_n}{dt}(t)$ *satisfying*

$$\sum_{k=1}^{n} \sigma_k = 0. \tag{12.1}$$

Then, denoting, as usual, the charge and the current densities as

$$\rho(\mathbf{x}, t) := \sum_{k=1}^{n} \sigma_k \delta\left(\mathbf{x} - \mathbf{r}_k(t)\right), \quad \text{and} \quad \mathbf{j}(\mathbf{x}, t) := \sum_{k=1}^{n} \sigma_k \frac{d\mathbf{r}_k}{dt}(t)\, \delta\left(\mathbf{x} - \mathbf{r}_k(t)\right), \tag{12.2}$$

we have

$$\begin{cases} \rho(\mathbf{x}, t) = -\operatorname{div}_{\mathbf{x}} \mathbf{P}(\mathbf{x}, t) \\ \mathbf{j}(x, t) = \frac{\partial \mathbf{P}}{\partial t}(\mathbf{x}, t) - curl_{\mathbf{x}} \left\{\mathbf{u}\left(\mathbf{x}, t\right) \times \mathbf{P}(\mathbf{x}, t)\right\} + c\, curl_{\mathbf{x}} \mathbf{M}_{\mathbf{u}}(\mathbf{x}, t), \end{cases} \tag{12.3}$$

or equivalently

$$\begin{cases} \rho(\mathbf{x}, t) = -\operatorname{div}_{\mathbf{x}} \mathbf{P}(\mathbf{x}, t) \\ \mathbf{j}(x, t) = \frac{\partial \mathbf{P}}{\partial t}(\mathbf{x}, t) + c\, curl_{\mathbf{x}} \mathbf{M}(\mathbf{x}, t), \end{cases} \tag{12.4}$$

A. Poliakovsky, *The Laws of Gravity and Electromagnetism*,
https://doi.org/10.1007/978-3-031-61407-1_12

where $\mathbf{P}(\mathbf{x},t)$, $\mathbf{M}(\mathbf{x},t)$ *and* $\mathbf{M}_{\mathbf{u}}(\mathbf{x},t)$ *are the Polarization and the Magnetization of the given neutral system defined as*

$$\mathbf{P}(\mathbf{x},t) := \sum_{k=1}^{n} \sigma_k \mathbf{l}_k(t) \int_0^1 \delta\left(\mathbf{x} - (\mathbf{r}(t) + s\mathbf{l}_k(t))\right) ds,$$

$$\mathbf{M}(\mathbf{x},t) :=$$

$$-\frac{1}{c}\left(\sum_{k=1}^{n} \sigma_k \int_0^1 \left(\left(\frac{d\mathbf{r}}{dt}(t) + s\frac{d\mathbf{l}_k}{dt}(t)\right) \times \mathbf{l}_k(t)\, \delta\left(\mathbf{x} - (\mathbf{r}(t) + s\mathbf{l}_k(t))\right)\right) ds\right)$$

$$\text{and} \qquad \mathbf{M}_{\mathbf{u}}(\mathbf{x},t) := \mathbf{M}(\mathbf{x},t) + \frac{1}{c}\mathbf{u}(\mathbf{x},t) \times \mathbf{P}(\mathbf{x},t)$$

$$= -\sum_{k=1}^{n} \frac{\sigma_k}{c} \int_0^1 \left(\left(\left(\frac{d\mathbf{r}}{dt}(t) + s\frac{d\mathbf{l}_k}{dt}(t)\right)\right.\right.$$

$$\left.\left. - \mathbf{u}\left(\mathbf{r}(t) + s\mathbf{l}_k(t), t\right)\right) \times \mathbf{l}_k(t)\, \delta\left(\mathbf{x} - (\mathbf{r}(t) + s\mathbf{l}_k(t))\right)\right) ds, \tag{12.5}$$

where we denoted

$$\mathbf{l}_k(t) := \mathbf{r}_k(t) - \mathbf{r}(t) \qquad \forall k = 1, 2, \ldots, n. \tag{12.6}$$

Next, under the change of Cartesian coordinate system $(*)$ *to another Cartesian coordinate system* $(**)$ *of the form*

$$\begin{cases} \mathbf{x}' = A(t) \cdot \mathbf{x} + \mathbf{z}(t), \\ t' = t, \end{cases}$$

where $A(t) \in SO(3)$ *is a rotation, we have*

$$\begin{cases} \mathbf{P}'(\mathbf{x}',t') = A(t) \cdot \mathbf{P}(\mathbf{x},t), \\ \mathbf{M}'_{\mathbf{u}'}(\mathbf{x}',t') = A(t) \cdot \mathbf{M}_{\mathbf{u}}(\mathbf{x},t), \\ \mathbf{M}' = A(t) \cdot \mathbf{M} - \frac{1}{c}\left(\frac{dA}{dt}(t) \cdot \mathbf{x} + \frac{d\mathbf{z}}{dt}(t)\right) \times \left(A(t) \cdot \mathbf{P}\right), \\ \rho'(\mathbf{x}',t') = \rho(\mathbf{x},t), \\ \mathbf{j}'(\mathbf{x}',t') = A(t) \cdot \mathbf{j}(\mathbf{x},t) + \rho(\mathbf{x},t)\, \frac{dA}{dt}(t) \cdot \mathbf{x} + \rho(\mathbf{x},t)\, \frac{d\mathbf{z}}{dt}(t), \end{cases} \tag{12.7}$$

provided that

$$\mathbf{u}'(\mathbf{x}',t') = A(t) \cdot \mathbf{u}(\mathbf{x},t) + \frac{dA}{dt}(t) \cdot \mathbf{x} + \frac{d\mathbf{z}}{dt}(t). \tag{12.8}$$

Moreover, identities in (12.3) *are invariant under the change of Cartesian coordinate system systems, provided we have* (12.7) *and* (12.8). *Finally, if all the quantities*

$\left|\mathbf{l}_k(t)\right|$ *in* (12.6) *are infinitely small, then we have the following approximate equalities for* $\mathbf{P}(\mathbf{x},t)$, $\mathbf{M}(\mathbf{x},t)$ *and* $\mathbf{M}_{\mathbf{u}}(\mathbf{x},t)$ *in* (12.5)*:*

$$\mathbf{P}(\mathbf{x},t) \approx \sum_{k=1}^{n} \sigma_k \mathbf{l}_k(t)\,\delta\left(\mathbf{x}-\mathbf{r}(t)\right),$$

$$\mathbf{M}(\mathbf{x},t) \approx -\frac{1}{c}\left(\sum_{k=1}^{n} \sigma_k \frac{d\mathbf{r}}{dt}(t)\times \mathbf{l}_k(t)\,\delta\left(\mathbf{x}-\mathbf{r}(t)\right)\right) \qquad \textit{and}$$

$$\mathbf{M}_{\mathbf{u}}(\mathbf{x},t) \approx -\frac{1}{c}\left(\sum_{k=1}^{n} \sigma_k \left(\frac{d\mathbf{r}}{dt}(t)-\mathbf{u}\left(\mathbf{r}(t),t\right)\right)\times \mathbf{l}_k(t)\,\delta\left(\mathbf{x}-\mathbf{r}(t)\right)\right)$$
$$\approx \mathbf{M}(\mathbf{x},t)+\frac{1}{c}\mathbf{u}(\mathbf{x},t)\times\mathbf{P}(\mathbf{x},t) \qquad (12.9)$$

and moreover, if $\mathbf{P}$, $\mathbf{M}$ *and* $\mathbf{M}_{\mathbf{u}}$ *are given by* (12.9) *rather than* (12.5), *then, under the change of Cartesian coordinate system, the first three equations in* (12.7) *are still valid.*

Proof Let

$$\mathbf{P}(\mathbf{x},t) := \sum_{k=1}^{n} \sigma_k \mathbf{l}_k(t)\int_0^1 \delta\left(\mathbf{x}-\left(\mathbf{r}(t)+s\mathbf{l}_k(t)\right)\right)ds,$$
$$\mathbf{M}(\mathbf{x},t) :=$$
$$-\frac{1}{c}\left(\sum_{k=1}^{n}\sigma_k\int_0^1\left(\left(\frac{d\mathbf{r}}{dt}(t)+s\frac{d\mathbf{l}_k}{dt}(t)\right)\times\mathbf{l}_k(t)\,\delta\left(\mathbf{x}-\left(\mathbf{r}(t)+s\mathbf{l}_k(t)\right)\right)\right)ds\right)$$
$$\text{and} \qquad \mathbf{M}_{\mathbf{u}}(\mathbf{x},t) :=$$
$$-\sum_{k=1}^{n}\frac{\sigma_k}{c}\int_0^1\left(\left(\left(\frac{d\mathbf{r}}{dt}(t)+s\frac{d\mathbf{l}_k}{dt}(t)\right)\right.\right.$$
$$\left.\left.-\,\mathbf{u}\left(\mathbf{r}(t)+s\mathbf{l}_k(t),t\right)\right)\times\mathbf{l}_k(t)\,\delta\left(\mathbf{x}-\left(\mathbf{r}(t)+s\mathbf{l}_k(t)\right)\right)\right)ds. \qquad (12.10)$$

Then, by definition in (12.10), we obviously have

$$\mathbf{M}_{\mathbf{u}}(\mathbf{x},t)=\mathbf{M}(\mathbf{x},t)+\frac{1}{c}\mathbf{u}(\mathbf{x},t)\times\mathbf{P}(\mathbf{x},t). \qquad (12.11)$$

Next, by (12.2) and (12.1), we have

$$\begin{aligned}\rho(\mathbf{x},t) &:= \sum_{k=1}^{n} \sigma_k \delta\left(\mathbf{x} - (\mathbf{r}(t) + \mathbf{l}_k(t))\right) \\ &= \sum_{k=1}^{n} \sigma_k \left(\delta\left(\mathbf{x} - (\mathbf{r}(t) + \mathbf{l}_k(t))\right) - \delta\left(\mathbf{x} - \mathbf{r}(t)\right)\right) \\ &= -\sum_{k=1}^{n} \sigma_k \int_0^1 \mathbf{l}_k(t) \cdot \nabla_{\mathbf{x}} \delta\left(\mathbf{x} - (\mathbf{r}(t) + s\mathbf{l}_k(t))\right) ds \\ &= -\operatorname{div}_{\mathbf{x}} \left(\sum_{k=1}^{n} \sigma_k \mathbf{l}_k(t) \int_0^1 \delta\left(\mathbf{x} - (\mathbf{r}(t) + s\mathbf{l}_k(t))\right) ds\right). \end{aligned} \tag{12.12}$$

Thus,

$$\rho(\mathbf{x},t) = -\operatorname{div}_{\mathbf{x}} \mathbf{P}(\mathbf{x},t), \tag{12.13}$$

where

$$\mathbf{P}(\mathbf{x},t) = \sum_{k=1}^{n} \sigma_k \mathbf{l}_k(t) \int_0^1 \delta\left(\mathbf{x} - (\mathbf{r}(t) + s\mathbf{l}_k(t))\right) ds. \tag{12.14}$$

Moreover, if all the quantities $\left|\mathbf{l}_k(t)\right|$ are infinitely small, then, by (12.14), we obviously have the following approximate equality:

$$\mathbf{P}(\mathbf{x},t) \approx \sum_{k=1}^{n} \sigma_k \mathbf{l}_k(t)\, \delta\left(\mathbf{x} - \mathbf{r}(t)\right). \tag{12.15}$$

Next, by (12.2) and (12.1), we obtain

$$\begin{aligned}\mathbf{j}(\mathbf{x},t) &:= \sum_{k=1}^{n} \sigma_k \left(\frac{d\mathbf{r}}{dt}(t) + \frac{d\mathbf{l}_k}{dt}(t)\right) \delta\left(\mathbf{x} - (\mathbf{r}(t) + \mathbf{l}_k(t))\right) \\ &= \sum_{k=1}^{n} \sigma_k \left(\left(\frac{d\mathbf{r}}{dt}(t) + \frac{d\mathbf{l}_k}{dt}(t)\right) \delta\left(\mathbf{x} - (\mathbf{r}(t) + \mathbf{l}_k(t))\right) - \frac{d\mathbf{r}}{dt}(t)\delta\left(\mathbf{x} - \mathbf{r}(t)\right)\right) \\ &= \sum_{k=1}^{n} \sigma_k \int_0^1 \frac{\partial}{\partial s}\left(\left(\frac{d\mathbf{r}}{dt}(t) + s\frac{d\mathbf{l}_k}{dt}(t)\right) \delta\left(\mathbf{x} - (\mathbf{r}(t) + s\mathbf{l}_k(t))\right)\right) ds \\ &= \sum_{k=1}^{n} \sigma_k \int_0^1 \left(\frac{d\mathbf{l}_k}{dt}(t)\delta\left(\mathbf{x} - (\mathbf{r}(t) + s\mathbf{l}_k(t))\right)\right. \\ &\quad \left. - \left(\frac{d\mathbf{r}}{dt}(t) + s\frac{d\mathbf{l}_k}{dt}(t)\right) \left(\mathbf{l}_k(t) \cdot \nabla_{\mathbf{x}} \delta\left(\mathbf{x} - (\mathbf{r}(t) + s\mathbf{l}_k(t))\right)\right)\right) ds \end{aligned}$$

$$
\begin{aligned}
&= \sum_{k=1}^{n} \sigma_k \int_0^1 \left(\frac{d\mathbf{l}_k}{dt}(t)\delta\left(\mathbf{x} - (\mathbf{r}(t) + s\mathbf{l}_k(t))\right) \right.\\
&\quad \left. - \mathbf{l}_k(t) \left(\left(\frac{d\mathbf{r}}{dt}(t) + s\frac{d\mathbf{l}_k}{dt}(t) \right) \cdot \nabla_{\mathbf{x}}\delta\left(\mathbf{x} - (\mathbf{r}(t) + s\mathbf{l}_k(t))\right) \right) \right) ds \\
&+ \sum_{k=1}^{n} \sigma_k \int_0^1 \left(\mathbf{l}_k(t) \left(\left(\frac{d\mathbf{r}}{dt}(t) + s\frac{d\mathbf{l}_k}{dt}(t) \right) \cdot \nabla_{\mathbf{x}}\delta\left(\mathbf{x} - (\mathbf{r}(t) + s\mathbf{l}_k(t))\right) \right) \right) ds \\
&- \sum_{k=1}^{n} \sigma_k \int_0^1 \left(\left(\frac{d\mathbf{r}}{dt}(t) + s\frac{d\mathbf{l}_k}{dt}(t) \right) \left(\mathbf{l}_k(t) \cdot \nabla_{\mathbf{x}}\delta\left(\mathbf{x} - (\mathbf{r}(t) + s\mathbf{l}_k(t))\right) \right) \right) ds \\
&= \frac{\partial}{\partial t} \left(\sum_{k=1}^{n} \sigma_k \mathbf{l}_k(t) \int_0^1 \delta\left(\mathbf{x} - (\mathbf{r}(t) + s\mathbf{l}_k(t))\right) ds \right) \\
&+ \operatorname{div}_{\mathbf{x}} \left\{ \sum_{k=1}^{n} \sigma_k \int_0^1 \left(\mathbf{l}_k(t) \otimes \left(\frac{d\mathbf{r}}{dt}(t) + s\frac{d\mathbf{l}_k}{dt}(t) \right) \delta\left(\mathbf{x} - (\mathbf{r}(t) + s\mathbf{l}_k(t))\right) \right) ds \right\} \\
&- \operatorname{div}_{\mathbf{x}} \left\{ \sum_{k=1}^{n} \sigma_k \int_0^1 \left(\left(\frac{d\mathbf{r}}{dt}(t) + s\frac{d\mathbf{l}_k}{dt}(t) \right) \otimes \mathbf{l}_k(t)\, \delta\left(\mathbf{x} - (\mathbf{r}(t) + s\mathbf{l}_k(t))\right) \right) ds \right\}.
\end{aligned}
\tag{12.16}
$$

Then, using (2.12), by (12.16), we deduce

$$
\begin{aligned}
\mathbf{j}(\mathbf{x}, t) &= \frac{\partial}{\partial t} \left(\sum_{k=1}^{n} \sigma_k \mathbf{l}_k(t) \int_0^1 \delta\left(\mathbf{x} - (\mathbf{r}(t) + s\mathbf{l}_k(t))\right) ds \right) \\
&- \operatorname{curl}_{\mathbf{x}} \left\{ \sum_{k=1}^{n} \sigma_k \int_0^1 \left(\left(\frac{d\mathbf{r}}{dt}(t) + s\frac{d\mathbf{l}_k}{dt}(t) \right) \times \mathbf{l}_k(t)\, \delta\left(\mathbf{x} - (\mathbf{r}(t) + s\mathbf{l}_k(t))\right) \right) ds \right\}.
\end{aligned}
\tag{12.17}
$$

If $\mathbf{u}(\mathbf{x}, t)$ is the speed like vector field, then, by (12.17), we have

$$
\begin{aligned}
\mathbf{j}(\mathbf{x}, t) &= \frac{\partial}{\partial t} \left(\sum_{k=1}^{n} \sigma_k \mathbf{l}_k(t) \int_0^1 \delta\left(\mathbf{x} - (\mathbf{r}(t) + s\mathbf{l}_k(t))\right) ds \right) \\
&- \operatorname{curl}_{\mathbf{x}} \left\{ \sum_{k=1}^{n} \sigma_k \int_0^1 \left(\mathbf{u}\left(\mathbf{r}(t) + s\mathbf{l}_k(t), t\right) \times \mathbf{l}_k(t)\, \delta\left(\mathbf{x} - (\mathbf{r}(t) + s\mathbf{l}_k(t))\right) \right) ds \right\} \\
&\qquad - \operatorname{curl}_{\mathbf{x}} \left\{ \sum_{k=1}^{n} \sigma_k \int_0^1 \left(\left(\left(\frac{d\mathbf{r}}{dt}(t) + s\frac{d\mathbf{l}_k}{dt}(t) \right) \right. \right. \right. \\
&\quad \left. \left. \left. - \mathbf{u}\left(\mathbf{r}(t) + s\mathbf{l}_k(t), t\right) \right) \times \mathbf{l}_k(t)\, \delta\left(\mathbf{x} - (\mathbf{r}(t) + s\mathbf{l}_k(t))\right) \right) ds \right\}
\end{aligned}
$$

$$= \frac{\partial}{\partial t}\left(\sum_{k=1}^{n} \sigma_k \mathbf{l}_k(t) \int_0^1 \delta\left(\mathbf{x} - (\mathbf{r}(t) + s\mathbf{l}_k(t))\right) ds\right)$$

$$- \operatorname{curl}_{\mathbf{x}} \left\{\mathbf{u}(\mathbf{x}, t) \times \left(\sum_{k=1}^{n} \sigma_k \mathbf{l}_k(t) \int_0^1 \delta\left(\mathbf{x} - (\mathbf{r}(t) + s\mathbf{l}_k(t))\right) ds\right)\right\}$$

$$- \operatorname{curl}_{\mathbf{x}} \left\{ \sum_{k=1}^{n} \sigma_k \int_0^1 \left(\left(\left(\frac{d\mathbf{r}}{dt}(t) + s\frac{d\mathbf{l}_k}{dt}(t)\right)\right.\right.\right.$$

$$\left.\left.\left. - \mathbf{u}\left(\mathbf{r}(t) + s\mathbf{l}_k(t), t\right)\right) \times \mathbf{l}_k(t)\, \delta\left(\mathbf{x} - (\mathbf{r}(t) + s\mathbf{l}_k(t))\right)\right) ds\right\}. \quad (12.18)$$

Thus, denoting by (12.12) and (12.10), we have

$$\rho(\mathbf{x}, t) = -\operatorname{div}_{\mathbf{x}} \mathbf{P}(\mathbf{x}, t), \quad (12.19)$$

and by (12.18) and (12.10), we have

$$\mathbf{j}(x, t) = \frac{\partial \mathbf{P}}{\partial t}(\mathbf{x}, t) - \operatorname{curl}_{\mathbf{x}} \left\{\mathbf{u}(\mathbf{x}, t) \times \mathbf{P}(\mathbf{x}, t)\right\} + c \operatorname{curl}_{\mathbf{x}} \mathbf{M}_{\mathbf{u}}(\mathbf{x}, t). \quad (12.20)$$

Furthermore, by (12.20) and (12.11), we have

$$\mathbf{j}(x, t) = \frac{\partial \mathbf{P}}{\partial t}(\mathbf{x}, t) + c \operatorname{curl}_{\mathbf{x}} \mathbf{M}(\mathbf{x}, t). \quad (12.21)$$

Moreover, if all the quantities $\left|\mathbf{l}_k(t)\right|$ are infinitely small, then, by (12.10), we obviously have the following approximate equalities:

$$\mathbf{P}(\mathbf{x}, t) \approx \sum_{k=1}^{n} \sigma_k \mathbf{l}_k(t)\, \delta\left(\mathbf{x} - \mathbf{r}(t)\right),$$

$$\mathbf{M}(\mathbf{x}, t) \approx -\frac{1}{c}\left(\sum_{k=1}^{n} \sigma_k \frac{d\mathbf{r}}{dt}(t) \times \mathbf{l}_k(t)\, \delta\left(\mathbf{x} - \mathbf{r}(t)\right)\right) \quad \text{and}$$

$$\mathbf{M}_{\mathbf{u}}(\mathbf{x}, t) \approx -\frac{1}{c}\left(\sum_{k=1}^{n} \sigma_k \left(\frac{d\mathbf{r}}{dt}(t) - \mathbf{u}(\mathbf{r}(t), t)\right) \times \mathbf{l}_k(t)\, \delta\left(\mathbf{x} - \mathbf{r}(t)\right)\right)$$

$$\approx \mathbf{M}(\mathbf{x}, t) + \frac{1}{c}\mathbf{u}(\mathbf{x}, t) \times \mathbf{P}(\mathbf{x}, t). \quad (12.22)$$

Next, consider the change of Cartesian coordinate system $(*)$ to another Cartesian coordinate system $(**)$ of the form

$$\begin{cases} \mathbf{x}' = A(t) \cdot \mathbf{x} + \mathbf{z}(t), \\ t' = t, \end{cases}$$

where $A(t) \in SO(3)$ is a rotation. Then obviously we have

$$\begin{cases} \mathbf{u}'(\mathbf{x}', t') = A(t) \cdot \mathbf{u}(\mathbf{x}, t) + \frac{dA}{dt}(t) \cdot \mathbf{x} + \frac{d\mathbf{z}}{dt}(t) \\ \rho'(\mathbf{x}', t') = \rho(\mathbf{x}, t), \\ \mathbf{j}'(\mathbf{x}', t') = A(t) \cdot \mathbf{j}(\mathbf{x}, t) + \rho(\mathbf{x}, t) \frac{dA}{dt}(t) \cdot \mathbf{x} + \rho(\mathbf{x}, t) \frac{d\mathbf{z}}{dt}(t), \\ \mathbf{r}'(t') = A(t) \cdot \mathbf{r}(t) + \mathbf{z}(t), \\ \frac{d\mathbf{r}'}{dt'}(t') = A(t) \cdot \frac{d\mathbf{r}}{dt}(t) + \frac{dA}{dt}(t) \cdot \mathbf{r}(t) + \frac{d\mathbf{z}}{dt}(t), \\ \mathbf{r}'_k(t') = A(t) \cdot \mathbf{r}_k(t) + \mathbf{z}(t) \qquad \forall k = 1, 2, \ldots, n, \\ \frac{d\mathbf{r}'_k}{dt'}(t') = A(t) \cdot \frac{d\mathbf{r}_k}{dt}(t) + \frac{dA}{dt}(t) \cdot \mathbf{r}_k(t) + \frac{d\mathbf{z}}{dt}(t) \qquad \forall k = 1, 2, \ldots, n, \end{cases} \tag{12.23}$$

and thus also,

$$\begin{cases} \mathbf{l}'_k(t') = A(t) \cdot \mathbf{l}_k(t) \qquad \forall k = 1, 2, \ldots, n, \\ \frac{d\mathbf{l}'_k}{dt'}(t') = A(t) \cdot \frac{d\mathbf{l}_k}{dt}(t) + \frac{dA}{dt}(t) \cdot \mathbf{l}_k(t) \qquad \forall k = 1, 2, \ldots, n, \\ \left(\mathbf{r}'(t') + s\mathbf{l}'_k(t')\right) = A(t) \cdot (\mathbf{r}(t) + s\mathbf{l}_k(t)) + \mathbf{z}(t) \qquad \forall k = 1, 2, \ldots, n \ \forall s, \\ \left(\frac{d\mathbf{r}'}{dt'}(t') + s\frac{d\mathbf{l}'_k}{dt'}(t')\right) = A(t) \cdot \left(\frac{d\mathbf{r}}{dt}(t) + s\frac{d\mathbf{l}_k}{dt}(t)\right) + \frac{dA}{dt}(t) \cdot (\mathbf{r}(t) + s\mathbf{l}_k(t)) + \frac{d\mathbf{z}}{dt}(t), \\ \left(\frac{d\mathbf{r}'}{dt'}(t') + s\frac{d\mathbf{l}'_k}{dt'}(t') - \mathbf{u}'\left(\mathbf{r}'(t') + s\mathbf{l}'_k(t'), t'\right)\right) \\ = A(t) \cdot \left(\frac{d\mathbf{r}}{dt}(t) + s\frac{d\mathbf{l}_k}{dt}(t) - \mathbf{u}\left(\mathbf{r}(t) + s\mathbf{l}_k(t), t\right)\right). \end{cases} \tag{12.24}$$

Therefore, as before, by (12.23) and (12.24), $\mathbf{P}(\mathbf{x}, t)$ and $\mathbf{M}_{\mathbf{u}}(\mathbf{x}, t)$, defined either by (12.10) or by (12.22), clearly satisfy the first two equations in (12.7). Moreover, by the first two equations in (12.7) together with (12.11) and the first equation in (12.23), we deduce the third equation in (12.7). Finally, as before, identities in either (12.3) or (12.4) are invariant under the change of Cartesian coordinate systems, provided we have (12.7) and (12.8). □

Next, consider a complex system containing m totally neutral simpler systems, such that for every $k = 1, 2, \ldots, m$ every simple system consists of n_k point charges $\sigma_1^{(k)}, \ldots, \sigma_{n_k}^{(k)}$, with places $\mathbf{r}_1^{(k)}(t), \ldots, \mathbf{r}_{n_k}^{(k)}(t)$ which are very close to the place $\mathbf{r}^{(k)}(t)$ and velocities $\frac{d\mathbf{r}_1^{(k)}}{dt}(t), \ldots, \frac{d\mathbf{r}_{n_k}^{(k)}}{dt}(t)$ which are very close to the velocity $\frac{d\mathbf{r}^{(k)}}{dt}(t)$, satisfying

$$\sum_{j=1}^{n_k} \sigma_j^{(k)} = 0 \qquad \forall k = 1, 2, \ldots, m. \tag{12.25}$$

and such that all the quantities $\left|\mathbf{l}_j^{(k)}(t)\right|$ defined by

$$\mathbf{l}_j^{(k)}(t) := \mathbf{r}_j^{(k)}(t) - \mathbf{r}^{(k)}(t) \qquad \forall k = 1, 2, \ldots, m, \ \forall j = 1, 2, \ldots, n_k, \tag{12.26}$$

are infinitely small. As an example of such a complex system, we can take a system containing m point dipoles or m totally neutral molecules. Then, consistently with (12.9), we can write the Polarization and the Magnetization of the system as

$$\mathbf{P}(\mathbf{x},t) \approx \sum_{k=1}^{m}\sum_{j=1}^{n_k} \sigma_j^{(k)} \mathbf{l}_j^{(k)}(t)\,\delta\left(\mathbf{x}-\mathbf{r}^{(k)}(t)\right) \qquad \text{and}$$

$$\mathbf{M}(\mathbf{x},t) \approx -\frac{1}{c}\left(\sum_{k=1}^{m}\sum_{j=1}^{n_k} \sigma_j^{(k)} \frac{d\mathbf{r}^{(k)}}{dt}(t) \times \mathbf{l}_j^{(k)}(t)\,\delta\left(\mathbf{x}-\mathbf{r}^{(k)}(t)\right)\right). \tag{12.27}$$

Remark 12.1 In the frames of Quantum Physics, one can show that there is an additional different term in the expression of the Magnetization $\mathbf{M}(\mathbf{x},t)$, arisen from the spin of charged particles. However, it is impossible to describe this term in the frames of Classical Mechanics.

12.2 General Setting of Maxwell Equations in the Medium

Consider system (5.37) in some inertial or non-inertial Cartesian coordinate system inside a dielectric and/or magnetic medium:

$$\begin{cases} \operatorname{curl}_{\mathbf{x}}\mathbf{H}_0 = \frac{4\pi}{c}\left(\mathbf{j}+\mathbf{j}_{mp}\right) + \frac{1}{c}\frac{\partial \mathbf{D}_0}{\partial t} & \text{for } (\mathbf{x},t)\in\mathbb{R}^3\times[t_0,+\infty), \\ \operatorname{div}_{\mathbf{x}}\mathbf{D}_0 = 4\pi\left(\rho+\rho_p\right) & \text{for } (\mathbf{x},t)\in\mathbb{R}^3\times[t_0,+\infty), \\ \operatorname{curl}_{\mathbf{x}}\mathbf{E} + \frac{1}{c}\frac{\partial \mathbf{B}}{\partial t} = 0 & \text{for } (\mathbf{x},t)\in\mathbb{R}^3\times[t_0,+\infty), \\ \operatorname{div}_{\mathbf{x}}\mathbf{B} = 0 & \text{for } (\mathbf{x},t)\in\mathbb{R}^3\times[t_0,+\infty), \end{cases} \tag{12.28}$$

where $\mathbf{E}$ is the electric field, $\mathbf{B}$ is the magnetic field, $\mathbf{v} := \mathbf{v}(\mathbf{x},t)$ is the vectorial gravitational potential, ρ is the average (macroscopic) charge density, ρ_p is the density of the charge of polarization, $\mathbf{j}$ is the average (macroscopic) current density, $\mathbf{j}_{mp}$ is the density of the current of polarization and magnetization and

$$\mathbf{D}_0 := \mathbf{E} + \frac{1}{c}\mathbf{v}\times\mathbf{B} \quad \text{and} \quad \mathbf{H}_0 := \mathbf{B} + \frac{1}{c}\mathbf{v}\times\mathbf{D}_0. \tag{12.29}$$

Consistently with Sect. 12.1, it is well known from the Lorentz theory that in the case of a moving dielectric/magnetic medium

$$\rho_p = -\operatorname{div}_{\mathbf{x}}\mathbf{P} \quad \text{and} \quad \mathbf{j}_{mp} = \frac{\partial \mathbf{P}}{\partial t} + c\operatorname{curl}_{\mathbf{x}}\mathbf{M}, \tag{12.30}$$

where $\mathbf{P} : \mathbb{R}^3\times[t_0,+\infty)\to\mathbb{R}^3$ is the field of polarization, $\mathbf{M} : \mathbb{R}^3\times[t_0,+\infty)\to\mathbb{R}^3$ is the field of magnetization and $\mathbf{u} := \mathbf{u}(\mathbf{x},t)$ is the field of velocities of the dielectric medium, see Eq. (12.4) in Proposition 12.1 for the proof of (12.30). Thus, if we consider

$$\mathbf{D} := \mathbf{D}_0 + 4\pi\mathbf{P} = \mathbf{E} + \frac{1}{c}\mathbf{v}\times\mathbf{B} + 4\pi\mathbf{P} \tag{12.31}$$

and

$$\mathbf{H} := \mathbf{H}_0 - 4\pi\mathbf{M} = \mathbf{B} + \frac{1}{c}\mathbf{v}\times\mathbf{D}_0 - 4\pi\mathbf{M} = \mathbf{B} + \frac{1}{c}\mathbf{v}\times\mathbf{E} + \frac{1}{c}\mathbf{v}\times\left(\frac{1}{c}\mathbf{v}\times\mathbf{B}\right) - 4\pi\mathbf{M}, \quad (12.32)$$

we obtain the usual Maxwell equations of the form:

$$\begin{cases} \mathrm{curl}_{\mathbf{x}}\mathbf{H} = \frac{4\pi}{c}\mathbf{j} + \frac{1}{c}\frac{\partial\mathbf{D}}{\partial t} \quad \text{for } (\mathbf{x},t)\in\mathbb{R}^3\times[t_0,+\infty), \\ \mathrm{div}_{\mathbf{x}}\,\mathbf{D} = 4\pi\rho \quad \text{for } (\mathbf{x},t)\in\mathbb{R}^3\times[t_0,+\infty), \\ \mathrm{curl}_{\mathbf{x}}\mathbf{E} + \frac{1}{c}\frac{\partial\mathbf{B}}{\partial t} = 0 \quad \text{for } (\mathbf{x},t)\in\mathbb{R}^3\times[t_0,+\infty), \\ \mathrm{div}_{\mathbf{x}}\,\mathbf{B} = 0 \quad \text{for } (\mathbf{x},t)\in\mathbb{R}^3\times[t_0,+\infty), \end{cases} \quad (12.33)$$

We call **D** by the electric displacement field and **H** by the **H**-magnetic field in a medium.

12.3 Change of Non-inertial Coordinate System

Consider the change of certain non-inertial Cartesian coordinate system $(*)$ to another Cartesian coordinate system $(**)$ of the form

$$\begin{cases} \mathbf{x}' = A(t)\cdot\mathbf{x} + \mathbf{z}(t), \\ t' = t, \end{cases}$$

where $A(t) \in SO(3)$ is a rotation. Then, as before in (5.36), denoting $\mathbf{w}(t) = \mathbf{z}'(t)$, we have the following relations between the physical quantities in coordinate systems $(*)$ and $(**)$:

$$\begin{cases} \mathbf{E}' = A(t)\cdot\mathbf{E} - \frac{1}{c}\left(A'(t)\cdot\mathbf{x} + \mathbf{w}(t)\right)\times\left(A(t)\cdot\mathbf{B}\right), \\ \mathbf{B}' = A(t)\cdot\mathbf{B}, \\ \mathbf{D}'_0 = A(t)\cdot\mathbf{D}_0, \\ \mathbf{H}'_0 = A(t)\cdot\mathbf{H}_0 + \frac{1}{c}\left(A'(t)\cdot\mathbf{x} + \mathbf{w}(t)\right)\times\left(A(t)\cdot\mathbf{D}_0\right), \\ \mathbf{u}' = A(t)\cdot\mathbf{u} + A'(t)\cdot\mathbf{x} + \mathbf{w}(t), \\ \mathbf{P}' = A(t)\cdot\mathbf{P}, \\ \mathbf{M}' = A(t)\cdot\mathbf{M} - \frac{1}{c}\left(A'(t)\cdot\mathbf{x} + \mathbf{w}(t)\right)\times\left(A(t)\cdot\mathbf{P}\right), \end{cases} \quad (12.34)$$

see Sect. 12.1 for the justification of the last two equalities in (12.34). Plugging it into (12.31) and (12.32), we deduce

$$\mathbf{D}' := \mathbf{D}'_0 + 4\pi\mathbf{P}' = A(t)\cdot(\mathbf{D}_0 + 4\pi\mathbf{P}) = A(t)\cdot\mathbf{D} \quad (12.35)$$

and

$$
\begin{aligned}
\mathbf{H}' := \mathbf{H}_0' - 4\pi\mathbf{M}' = A(t)\cdot\mathbf{H}_0 + \frac{1}{c}\left(A'(t)\cdot\mathbf{x} + \mathbf{w}(t)\right) \times \left(A(t)\cdot\mathbf{D}_0\right) \\
- 4\pi A(t)\cdot\mathbf{M} + \frac{4\pi}{c}\left(A'(t)\cdot\mathbf{x} + \mathbf{w}(t)\right) \times \left(A(t)\cdot\mathbf{P}\right) \\
= A(t)\cdot(\mathbf{H}_0 - 4\pi\mathbf{M}) + \frac{1}{c}\left(A'(t)\cdot\mathbf{x} + \mathbf{w}(t)\right) \times \left(A(t)\cdot(\mathbf{D}_0 + 4\pi\mathbf{P})\right) \\
= A(t)\cdot\mathbf{H} + \frac{1}{c}\left(A'(t)\cdot\mathbf{x} + \mathbf{w}(t)\right) \times \left(A(t)\cdot\mathbf{D}\right).
\end{aligned}
\tag{12.36}
$$

So the expressions of transformations under the change of non-inertial Cartesian coordinate system in a dielectric/magnetic medium exactly the same as in the vacuum, i.e. having the form of

$$
\begin{cases}
\mathbf{D}' = A(t)\cdot\mathbf{D} \\
\mathbf{B}' = A(t)\cdot\mathbf{B} \\
\mathbf{E}' = A(t)\cdot\mathbf{E} - \frac{1}{c}\left(A'(t)\cdot\mathbf{x} + \mathbf{w}(t)\right) \times \left(A(t)\cdot\mathbf{B}\right) \\
\mathbf{H}' = A(t)\cdot\mathbf{H} + \frac{1}{c}\left(A'(t)\cdot\mathbf{x} + \mathbf{w}(t)\right) \times \left(A(t)\cdot\mathbf{D}\right).
\end{cases}
\tag{12.37}
$$

Moreover, as before, the Maxwell equations in the medium of the form (12.33) stays invariant under the change of inertial or non-inertial Cartesian coordinate system, provided we have (12.37) and

$$
\mathbf{j}' = A(t)\cdot\mathbf{j} + \rho\, A'(t)\cdot\mathbf{x} + \rho\,\mathbf{w}(t). \tag{12.38}
$$

12.4 Case of Simplest Dielectrics/Magnetics

It is well known that, in the case of simplest resting homogenous isotropic nonmagnetic dielectric medium, we have

$$
\begin{cases}
\mathbf{P} = \frac{n^2-1}{4\pi}\,\mathbf{E}, \\
\mathbf{M} = 0,
\end{cases}
\tag{12.39}
$$

where n is a material coefficient (not necessary constant), called refraction index. Moreover, It is well known that, in the case of simplest resting homogenous isotropic dielectric medium with certain magnetic properties, we rewrite (12.39) as

$$
\begin{cases}
\mathbf{P} = \frac{n^2-1}{4\pi}\,\mathbf{E} - \kappa\mathbf{D}, \\
\mathbf{M} = \kappa\mathbf{H},
\end{cases}
\tag{12.40}
$$

where n is the refraction index and κ is an additional material coefficient (not necessary constant). Either (12.39) or (12.40) are valid only for points and instants of time where the velocity of the medium $\mathbf{u}$ vanishes. In the case of nonmagnetic dielectric, we just put $\kappa = 0$ and (12.40) becomes to be the same as (12.39).

Next, in the case of simplest moving homogenous isotropic nonmagnetic dielectric medium, we can generalize (12.39) as

$$\begin{cases} \mathbf{P} = \frac{n^2-1}{4\pi}\left(\mathbf{E} + \frac{1}{c}\mathbf{u}\times\mathbf{B}\right), \\ \mathbf{M} = -\frac{1}{c}\frac{n^2-1}{4\pi}\mathbf{u}\times\left(\mathbf{E} + \frac{1}{c}\mathbf{u}\times\mathbf{B}\right) = -\frac{1}{c}\mathbf{u}\times\mathbf{P}, \end{cases} \tag{12.41}$$

where n is a material coefficient (not necessary constant), called refraction index and $\mathbf{u}$ is the velocity of the medium. Moreover, in the case of simplest moving homogenous isotropic dielectric medium with certain magnetic properties we can generalize (12.40) as

$$\begin{cases} \mathbf{P} = \frac{n^2-1}{4\pi}\left(\mathbf{E} + \frac{1}{c}\mathbf{u}\times\mathbf{B}\right) - \kappa\mathbf{D}, \\ \mathbf{M} = -\frac{1}{c}\frac{n^2-1}{4\pi}\mathbf{u}\times\left(\mathbf{E} + \frac{1}{c}\mathbf{u}\times\mathbf{B}\right) + \kappa\mathbf{H} = -\frac{1}{c}\mathbf{u}\times\mathbf{P} + \kappa\left(\mathbf{H} - \frac{1}{c}\mathbf{u}\times\mathbf{D}\right), \end{cases} \tag{12.42}$$

where n is the refraction index and κ is an additional material coefficient (not necessary constant). In the case of nonmagnetic dielectric we just put $\kappa = 0$ and (12.42) becomes to be the same as (12.41). Then, using (12.34) and (12.37), it can be easily seen that the laws in either (12.41) or (12.42) are invariant under the changes of inertial or non-inertial Cartesian coordinate system. Alternatively, one can assume either (12.39) and (12.40) for the case $\mathbf{u} = 0$ and to postulate the invariance under the changes of inertial or non-inertial Cartesian coordinate system. Then, using (12.34) and (12.37), one can definitely deduce either (12.41) or (12.42) in the case $\mathbf{u} \neq 0$, similarly as it was done in Sect. 5.1, for establishing the relations $\mathbf{D} \sim \mathbf{E}$ and $\mathbf{H} \sim \mathbf{B}$ in the case $\mathbf{v} \neq 0$.

Next, plugging (12.42) into (12.31) and (12.32) gives,

$$\mathbf{D} := \mathbf{E} + \frac{1}{c}\mathbf{v}\times\mathbf{B} + 4\pi\left(\frac{n^2-1}{4\pi}\left(\mathbf{E} + \frac{1}{c}\mathbf{u}\times\mathbf{B}\right) - \kappa\mathbf{D}\right) \tag{12.43}$$

and

$$\mathbf{H} := \mathbf{B} + \frac{1}{c}\mathbf{v}\times\mathbf{E} + \frac{1}{c}\mathbf{v}\times\left(\frac{1}{c}\mathbf{v}\times\mathbf{B}\right) - 4\pi\left(-\frac{1}{c}\frac{n^2-1}{4\pi}\mathbf{u}\times\left(\mathbf{E} + \frac{1}{c}\mathbf{u}\times\mathbf{B}\right) + \kappa\mathbf{H}\right). \tag{12.44}$$

We rewrite (12.43) as

$$\mathbf{E} = \frac{1+4\pi\kappa}{n^2}\mathbf{D} - \frac{1}{c}\left(\frac{1}{n^2}\mathbf{v} + \left(1 - \frac{1}{n^2}\right)\mathbf{u}\right)\times\mathbf{B}, \tag{12.45}$$

and, by (12.43) and (12.45), we rewrite (12.44) as

$$
\begin{aligned}
(1+4\pi\kappa)\,\mathbf{H} &= \mathbf{B} + \frac{1}{c}\,\mathbf{v}\times\mathbf{E} + \frac{1}{c}\left(n^2-1\right)\mathbf{u}\times\mathbf{E} + \frac{1}{c}\,\mathbf{v}\times\left(\frac{1}{c}\,\mathbf{v}\times\mathbf{B}\right) \\
&+ \frac{1}{c}\left(n^2-1\right)\mathbf{u}\times\left(\frac{1}{c}\,\mathbf{u}\times\mathbf{B}\right) = \mathbf{B} + \frac{1}{c}\,(1+4\pi\kappa)\left(\frac{1}{n^2}\mathbf{v} + \left(1-\frac{1}{n^2}\right)\mathbf{u}\right)\times\mathbf{D} \\
&- \frac{1}{c}\,\mathbf{v}\times\left(\frac{1}{c}\left(\frac{1}{n^2}\mathbf{v} + \left(1-\frac{1}{n^2}\right)\mathbf{u}\right)\times\mathbf{B}\right) \\
&- \frac{1}{c}\left(n^2-1\right)\mathbf{u}\times\left(\frac{1}{c}\left(\frac{1}{n^2}\mathbf{v} + \left(1-\frac{1}{n^2}\right)\mathbf{u}\right)\times\mathbf{B}\right) \\
&+ \frac{1}{c}\,\mathbf{v}\times\left(\frac{1}{c}\,\mathbf{v}\times\mathbf{B}\right) + \frac{1}{c}\left(n^2-1\right)\mathbf{u}\times\left(\frac{1}{c}\,\mathbf{u}\times\mathbf{B}\right) = \mathbf{B} \\
&+ \frac{1}{c}\,(1+4\pi\kappa)\left(\frac{1}{n^2}\mathbf{v} + \left(1-\frac{1}{n^2}\right)\mathbf{u}\right)\times\mathbf{D} + \frac{1}{c^2}\left(1-\frac{1}{n^2}\right)(\mathbf{u}-\mathbf{v})\times((\mathbf{u}-\mathbf{v})\times\mathbf{B}),
\end{aligned}
\tag{12.46}
$$

so that

$$
\begin{aligned}
\mathbf{H} = \frac{1}{1+4\pi\kappa}\mathbf{B} &+ \frac{1}{c}\left(\frac{1}{n^2}\mathbf{v} + \left(1-\frac{1}{n^2}\right)\mathbf{u}\right)\times\mathbf{D} \\
&+ \frac{1}{c^2}\,\frac{1}{1+4\pi\kappa}\left(1-\frac{1}{n^2}\right)(\mathbf{u}-\mathbf{v})\times((\mathbf{u}-\mathbf{v})\times\mathbf{B}).
\end{aligned}
\tag{12.47}
$$

Thus, denoting

$$
\kappa_0 = \frac{1}{1+4\pi\kappa} \quad \text{and} \quad \gamma_0 = \frac{1+4\pi\kappa}{n^2} = \frac{1}{\kappa_0 n^2} \quad \text{so that} \quad n = \frac{1}{\sqrt{\gamma_0\kappa_0}}
\tag{12.48}
$$

and defining the speed-like vector field

$$
\tilde{\mathbf{u}} := \left(\frac{1}{n^2}\mathbf{v} + \left(1-\frac{1}{n^2}\right)\mathbf{u}\right) = \left(\gamma_0\kappa_0\mathbf{v} + (1-\gamma_0\kappa_0)\,\mathbf{u}\right),
\tag{12.49}
$$

by (12.45) and (12.47), we deduce

$$
\mathbf{E} = \gamma_0\mathbf{D} - \frac{1}{c}\,\tilde{\mathbf{u}}\times\mathbf{B},
\tag{12.50}
$$

and

$$
\mathbf{H} = \kappa_0\mathbf{B} + \frac{1}{c}\,\tilde{\mathbf{u}}\times\mathbf{D} + \kappa_0\frac{(1-\gamma_0\kappa_0)}{c^2}\,(\mathbf{u}-\mathbf{v})\times((\mathbf{u}-\mathbf{v})\times\mathbf{B}),
\tag{12.51}
$$

where we call γ_0 and κ_0 dielectric and magnetic permeability of the medium. Thus, by (12.33), (12.49) (12.50) and (12.51), we have

$$\begin{cases} \operatorname{curl}_{\mathbf{x}}\mathbf{H} = \frac{4\pi}{c}\mathbf{j} + \frac{1}{c}\frac{\partial \mathbf{D}}{\partial t}, \\ \operatorname{div}_{\mathbf{x}} \mathbf{D} = 4\pi\rho, \\ \operatorname{curl}_{\mathbf{x}}\mathbf{E} + \frac{1}{c}\frac{\partial \mathbf{B}}{\partial t} = 0, \\ \operatorname{div}_{\mathbf{x}} \mathbf{B} = 0, \\ \mathbf{E} = \gamma_0\mathbf{D} - \frac{1}{c}\tilde{\mathbf{u}} \times \mathbf{B}, \\ \mathbf{H} = \kappa_0\mathbf{B} + \frac{1}{c}\tilde{\mathbf{u}} \times \mathbf{D} + \frac{\kappa_0(1-\gamma_0\kappa_0)}{c^2}(\mathbf{u}-\mathbf{v}) \times ((\mathbf{u}-\mathbf{v}) \times \mathbf{B}), \\ \tilde{\mathbf{u}} := (\gamma_0\kappa_0\mathbf{v} + (1-\gamma_0\kappa_0)\,\mathbf{u}), \end{cases} \tag{12.52}$$

where $\tilde{\mathbf{u}}$ is a speed-like vector field that we call the optical displacement of the moving medium. Note that for the case $\gamma_0 = 1$ and $\kappa_0 = 1$, the system (12.52) is exactly the same as the corresponding system in the vacuum. The equations in (12.52) take much simpler forms in the case where the quantity

$$\frac{|\kappa_0||1-\gamma_0\kappa_0| \cdot |\mathbf{u}-\mathbf{v}|^2}{c^2} \ll 1 \tag{12.53}$$

is negligible, that happens if either the absolute value of the difference between the medium velocity and vectorial gravitational potential is much less then the constant c or $\gamma_0\kappa_0 = \frac{1}{n^2}$ is close to the value 1. Indeed, in this case, instead of (12.50) and (12.51), we obtain the following relations:

$$\mathbf{E} = \gamma_0\mathbf{D} - \frac{1}{c}\tilde{\mathbf{u}} \times \mathbf{B}, \tag{12.54}$$

$$\mathbf{H} = \kappa_0\mathbf{B} + \frac{1}{c}\tilde{\mathbf{u}} \times \mathbf{D}. \tag{12.55}$$

As a consequence, we obtain the full system of Maxwell equations in the medium:

$$\begin{cases} \operatorname{curl}_{\mathbf{x}}\mathbf{H} = \frac{4\pi}{c}\mathbf{j} + \frac{1}{c}\frac{\partial \mathbf{D}}{\partial t}, \\ \operatorname{div}_{\mathbf{x}} \mathbf{D} = 4\pi\rho, \\ \operatorname{curl}_{\mathbf{x}}\mathbf{E} + \frac{1}{c}\frac{\partial \mathbf{B}}{\partial t} = 0, \\ \operatorname{div}_{\mathbf{x}} \mathbf{B} = 0, \\ \mathbf{E} = \gamma_0\mathbf{D} - \frac{1}{c}\tilde{\mathbf{u}} \times \mathbf{B}, \\ \mathbf{H} = \kappa_0\mathbf{B} + \frac{1}{c}\tilde{\mathbf{u}} \times \mathbf{D}, \\ \tilde{\mathbf{u}} = (\gamma_0\kappa_0\mathbf{v} + (1-\gamma_0\kappa_0)\,\mathbf{u}), \end{cases} \tag{12.56}$$

where $\tilde{\mathbf{u}}$ is the speed-like vector field and γ_0 and κ_0 are dielectric and magnetic permeability of the medium. Note that (12.56) is analogous to the system of Maxwell equations in the vacuum and it is also invariant under the change of inertial or non-inertial Cartesian coordinate system, provided that, under this transformation, we have (12.37).

12.5 Case of Anisotropic Dielectrics/Magnetics

It is well known that, in the case the simplest anisotropic dielectrics and/or magnetics, we have the following generalization of (12.42):

$$\begin{cases} \mathbf{P} = \Gamma \cdot \left(\mathbf{E} + \frac{1}{c}\mathbf{u} \times \mathbf{B}\right) - \Upsilon \cdot \mathbf{D}, \\ \mathbf{M} = -\frac{1}{c}\mathbf{u} \times \mathbf{P} + \Upsilon \cdot \left(\mathbf{H} - \frac{1}{c}\mathbf{u} \times \mathbf{D}\right), \end{cases} \tag{12.57}$$

where $\Gamma \in \mathbb{R}^{3\times 3}$ and $\Upsilon \in \mathbb{R}^{3\times 3}$ are matrix-valued fields. Then, using (12.34), it can be easily seen that the laws in (12.57) are invariant under the changes of inertial or non-inertial Cartesian coordinate system, provided that Γ and Υ are proper matrix fields (see Definition 3.1).

12.6 Ohm's Law in a Conducting Medium

It is well known that Ohm's Law in a conducting medium has the form

$$\mathbf{j} - \rho\mathbf{u} = \varepsilon \left(\mathbf{E} + \frac{1}{c}\mathbf{u} \times \mathbf{B}\right), \tag{12.58}$$

where $\mathbf{u}$ is the velocity of the medium and ε is a material coefficient. As before, using (12.34), it can be easily seen that Ohm's Law is invariant under the changes of inertial or non-inertial Cartesian coordinate system.

Furthermore, it is well known that, in the case of the strong magnetic field, the modification of Ohm's Law including the Hall effect has the following form:

$$\mathbf{j} - \rho\mathbf{u} = \varepsilon \left(\mathbf{E} + \frac{1}{c}\mathbf{u} \times \mathbf{B}\right) - \varsigma\,(\mathbf{j} - \rho\mathbf{u}) \times \mathbf{B}, \tag{12.59}$$

where ς is a material coefficient. Then, as before, using (12.34), it can be easily seen that the generalized Ohm's Law (12.59), including the Hall effect, is invariant under the changes of inertial or non-inertial Cartesian coordinate system.

Next, it is well known that, in the case of the anisotropic conducting medium, Ohm's Law has the following form: generalizing (12.58):

$$\mathbf{j} - \rho\mathbf{u} = \Xi \cdot \left(\mathbf{E} + \frac{1}{c}\mathbf{u} \times \mathbf{B}\right), \tag{12.60}$$

where $\Xi \in \mathbb{R}^{3\times 3}$ is a matrix-valued field. As before, using (12.34), it can be easily seen that (12.60) is invariant under the changes of inertial or non-inertial Cartesian coordinate system, provided that Ξ is a proper matrix field.

Finally, the generalization of (12.59) to anisotropic mediums is the following:

$$\mathbf{j} - \rho \mathbf{u} = \Xi \cdot \left(\mathbf{E} + \frac{1}{c} \mathbf{u} \times \mathbf{B} \right) - \Pi \cdot ((\mathbf{j} - \rho \mathbf{u}) \times \mathbf{B}), \tag{12.61}$$

where $\Xi \in \mathbb{R}^{3\times 3}$ and $\Pi \in \mathbb{R}^{3\times 3}$ are matrix-valued fields. As before, using (12.34), it can be easily seen that (12.61) is invariant under the changes of inertial or non-inertial Cartesian coordinate system, provided that Ξ and Π are proper matrix fields.

12.7 Optical Dispersion in Moving Mediums

Remind that if U is a real-valued field, then we can write the field U as a Furier's transform on the time variable:

$$U(\mathbf{x}, t) = Re\left\{ 2 \int_0^{+\infty} \hat{U}(\mathbf{x}, \omega) e^{i\omega t} d\omega \right\}$$
$$\text{where} \quad \hat{U}(\mathbf{x}, \omega) := \frac{1}{2\pi} \int_{-\infty}^{+\infty} U(\mathbf{x}, t) e^{-i\omega t} dt. \tag{12.62}$$

Thus, since **E**, **B**, **D** and **H** are real vectors, we can write them as a Furier's transform on the time variable:

$$\mathbf{E}(\mathbf{x}, t) = Re\left\{ 2 \int_0^{+\infty} \hat{\mathbf{E}}(\omega, \mathbf{x}) e^{i\omega t} d\omega \right\}$$
$$\text{where} \quad \hat{\mathbf{E}}(\omega, \mathbf{x}) := \frac{1}{2\pi} \int_{-\infty}^{+\infty} \mathbf{E}(\mathbf{x}, t) e^{-i\omega t} dt, \tag{12.63}$$

$$\mathbf{B}(\mathbf{x}, t) = Re\left\{ 2 \int_0^{+\infty} \hat{\mathbf{B}}(\omega, \mathbf{x}) e^{i\omega t} d\omega \right\}$$
$$\text{where} \quad \hat{\mathbf{B}}(\omega, \mathbf{x}) := \frac{1}{2\pi} \int_{-\infty}^{+\infty} \mathbf{B}(\mathbf{x}, t) e^{-i\omega t} dt, \tag{12.64}$$

$$\mathbf{D}(\mathbf{x}, t) = Re\left\{ 2 \int_0^{+\infty} \hat{\mathbf{D}}(\omega, \mathbf{x}) e^{i\omega t} d\omega \right\}$$
$$\text{where} \quad \hat{\mathbf{D}}(\omega, \mathbf{x}) := \frac{1}{2\pi} \int_{-\infty}^{+\infty} \mathbf{D}(\mathbf{x}, t) e^{-i\omega t} dt, \tag{12.65}$$

and

$$\mathbf{H}(\mathbf{x},t) = Re\left\{2\int_0^{+\infty} \hat{\mathbf{H}}(\omega,\mathbf{x})e^{i\omega t}d\omega\right\}$$

$$\text{where } \hat{\mathbf{H}}(\omega,\mathbf{x}) := \frac{1}{2\pi}\int_{-\infty}^{+\infty} \mathbf{H}(\mathbf{x},t)e^{-i\omega t}dt, \tag{12.66}$$

where given $\mathbf{a} = (a_1, a_2, a_3) \in \mathbb{C}^3$ we denote $Re\{\mathbf{a}\} = (Re\{a_1\}, Re\{a_2\}, Re\{a_3\}) \in \mathbb{R}^3$. Then, it is well known that, in the case of the simplest optical dispersion in the resting medium with $\mathbf{u} = 0$, we have the following generalization of (12.40):

$$\begin{cases} \mathbf{P}(\mathbf{x},t) = Re\left\{2\int_0^{+\infty} \frac{n^2(\omega,\mathbf{x},t)-1}{4\pi}\hat{\mathbf{E}}(\omega,\mathbf{x})e^{i\omega t}d\omega\right\} \\ -Re\left\{2\int_0^{+\infty} \kappa(\omega,\mathbf{x},t)\hat{\mathbf{D}}(\omega,\mathbf{x})e^{i\omega t}d\omega\right\}, \\ \mathbf{M}(\mathbf{x},t) = Re\left\{2\int_0^{+\infty} \kappa(\omega,\mathbf{x},t)\hat{\mathbf{H}}(\omega,\mathbf{x})e^{i\omega t}d\omega\right\}, \end{cases} \tag{12.67}$$

where $\hat{\mathbf{E}}(\omega,\mathbf{x})$, $\hat{\mathbf{B}}(\omega,\mathbf{x})$, $\hat{\mathbf{D}}(\omega,\mathbf{x})$ and $\hat{\mathbf{H}}(\omega,\mathbf{x})$ are Furier's transforms by time of $\mathbf{E}(\mathbf{x},t)$, $\mathbf{B}(\mathbf{x},t)$, $\mathbf{D}(\mathbf{x},t)$ and $\mathbf{H}(\mathbf{x},t)$, given by the second equalities of (12.63), (12.64), (12.65) and (12.66) and the quantities $n(\omega,\mathbf{x},t)$ and $\kappa(\omega,\mathbf{x},t)$ in (12.67) are assumed to be complex in general and assumed to depend on ω in addition to the dependence on $(\mathbf{x},t)$.

We would like to obtain the law being an analog of (12.67) in the case of moving medium (i.e. $\mathbf{u} \neq 0$), which is invariant under the change of inertial or non-inertial Cartesian coordinate system. In addition to the invariance under the change of Cartesian coordinate systems and the equivalence to (12.67) in the particular case $\mathbf{u} = 0$, we also need to assume that in the case of an arbitrary moving transparent medium without dispersion our law is equivalent to (12.42).

Then, in some Cartesian coordinate system, consider a motion of some continuum medium occupying a region $\Omega \subset \mathbb{R}^3$ at some fixed instant of time $t = t_0$ and having the velocity field $\mathbf{u} := \mathbf{u}(\mathbf{x},t)$. Next, as before, let $\mathbf{r}(t,\mathbf{y}) : \mathbb{R}\times\Omega \to \mathbb{R}^3$ be a solution of (11.30), i.e. it satisfies the following initial value problem for an ordinary differential equation:

$$\begin{cases} \frac{\partial \mathbf{r}}{\partial t}(t,\mathbf{y}) = \mathbf{u}(\mathbf{r}(t,\mathbf{y}),t) & \forall t\in\mathbb{R},\ \forall \mathbf{y}\in\Omega \\ \mathbf{r}(t_0,\mathbf{y}) = \mathbf{y} & \forall \mathbf{y}\in\Omega. \end{cases} \tag{12.68}$$

Then, clearly, $\mathbf{x} = \mathbf{r}(t,\mathbf{y})$ stands for the spatial coordinates at the instant of time t of the parcel of continuum, having initial coordinates $\mathbf{y}$. Thus, by (11.32), we deduce that $det\left\{d_{\mathbf{y}}\mathbf{r}(t,\mathbf{y})\right\} \neq 0$ for every instant of time t and so, for the given instant of time t, the mapping $\mathbf{x} = \mathbf{r}(t,\mathbf{y})$ is locally invertible, i.e. the equation $\mathbf{x} = \mathbf{r}(t,\mathbf{y})$ can be resolved in $\mathbf{y}$. Thus, there exists a regular mapping $\mathbf{y} = \mathbf{f}(\mathbf{x},t)$, such that

$$\mathbf{f}(\mathbf{r}(t,\mathbf{y}),t) = \mathbf{y}\ \ \forall \mathbf{y}\in\Omega \quad \text{and} \quad \mathbf{r}(t,\mathbf{f}(\mathbf{x},t)) = \mathbf{x}. \tag{12.69}$$

Then, clearly, $\mathbf{y} = \mathbf{f}(\mathbf{x}, t)$ stands for the initial coordinates at the time t_0 of the parcel of continuum, having coordinates $\mathbf{x}$ at the instant of time t. Next $\mathbf{f}(\mathbf{x}, t)$ satisfies

$$\begin{cases} \frac{\partial \mathbf{f}}{\partial t}(\mathbf{x}, t) + d_{\mathbf{x}}\mathbf{f}(\mathbf{x}, t) \cdot \mathbf{u}(\mathbf{x}, t) = 0 \\ \mathbf{f}(\mathbf{x}, t_0) = \mathbf{x}. \end{cases} \tag{12.70}$$

Next, consider

$$\begin{cases} \mathbf{E}^*(t, \mathbf{y}) := \left\{d_{\mathbf{y}}\mathbf{r}(t, \mathbf{y})\right\}^T \cdot \left(\mathbf{E}(\mathbf{r}(t, \mathbf{y}), t) + \frac{1}{c}\mathbf{u}(\mathbf{r}(t, \mathbf{y}), t) \times \mathbf{B}(\mathbf{r}(t, \mathbf{y}), t)\right) \\ \quad \forall t \in \mathbb{R}, \quad \forall \mathbf{y} \in \Omega, \\ \mathbf{B}^*(t, \mathbf{y}) := \left\{d_{\mathbf{y}}\mathbf{r}(t, \mathbf{y})\right\}^T \cdot \mathbf{B}(\mathbf{r}(t, \mathbf{y}), t) \qquad \forall t \in \mathbb{R}, \quad \forall \mathbf{y} \in \Omega, \\ \mathbf{D}^*(t, \mathbf{y}) := \left\{d_{\mathbf{y}}\mathbf{r}(t, \mathbf{y})\right\}^T \cdot \mathbf{D}(\mathbf{r}(t, \mathbf{y}), t) \qquad \forall t \in \mathbb{R}, \quad \forall \mathbf{y} \in \Omega, \\ \mathbf{H}^*(t, \mathbf{y}) := \left\{d_{\mathbf{y}}\mathbf{r}(t, \mathbf{y})\right\}^T \cdot \left(\mathbf{H}(\mathbf{r}(t, \mathbf{y}), t) - \frac{1}{c}\mathbf{u}(\mathbf{r}(t, \mathbf{y}), t) \times \mathbf{D}(\mathbf{r}(t, \mathbf{y}), t)\right) \\ \quad \forall t \in \mathbb{R}, \quad \forall \mathbf{y} \in \Omega, \end{cases} \tag{12.71}$$

where $\mathbf{r}(t, \mathbf{y})$ is given by (12.68). Then, since $\mathbf{E}^*$, $\mathbf{B}^*$, $\mathbf{D}^*$ and $\mathbf{H}^*$ are real vectors, we can write them as a Furier's transform on the time variable:

$$\mathbf{E}^*(t, \mathbf{y}) = Re\left\{2\int_0^{+\infty} \hat{\mathbf{E}}^*(\tilde{\omega}, \mathbf{y})e^{i\tilde{\omega}t}d\tilde{\omega}\right\}$$
$$\text{where} \quad \hat{\mathbf{E}}^*(\tilde{\omega}, \mathbf{y}) := \frac{1}{2\pi}\int_{-\infty}^{+\infty} \mathbf{E}^*(t, \mathbf{y})e^{-i\tilde{\omega}t}dt, \tag{12.72}$$

$$\mathbf{B}^*(t, \mathbf{y}) = Re\left\{2\int_0^{+\infty} \hat{\mathbf{B}}^*(\tilde{\omega}, \mathbf{y})e^{i\tilde{\omega}t}d\tilde{\omega}\right\}$$
$$\text{where} \quad \hat{\mathbf{B}}^*(\tilde{\omega}, \mathbf{y}) := \frac{1}{2\pi}\int_{-\infty}^{+\infty} \mathbf{B}^*(t, \mathbf{y})e^{-i\tilde{\omega}t}dt, \tag{12.73}$$

$$\mathbf{D}^*(t, \mathbf{y}) = Re\left\{2\int_0^{+\infty} \hat{\mathbf{D}}^*(\tilde{\omega}, \mathbf{y})e^{i\tilde{\omega}t}d\tilde{\omega}\right\}$$
$$\text{where} \quad \hat{\mathbf{D}}^*(\tilde{\omega}, \mathbf{y}) := \frac{1}{2\pi}\int_{-\infty}^{+\infty} \mathbf{D}^*(t, \mathbf{y})e^{-i\tilde{\omega}t}dt, \tag{12.74}$$

and

$$\mathbf{H}^*(t, \mathbf{y}) = Re\left\{2\int_0^{+\infty} \hat{\mathbf{H}}^*(\tilde{\omega}, \mathbf{y})e^{i\tilde{\omega}t}d\tilde{\omega}\right\}$$
$$\text{where} \quad \hat{\mathbf{H}}^*(\tilde{\omega}, \mathbf{y}) := \frac{1}{2\pi}\int_{-\infty}^{+\infty} \mathbf{H}^*(t, \mathbf{y})e^{-i\tilde{\omega}t}dt. \tag{12.75}$$

Then we write the generalization of (12.67) as

$$\begin{aligned}
&\mathbf{P}\left(\mathbf{r}(t,\mathbf{y}),t\right)\\
&= Re\left\{2\int_0^{+\infty}\frac{n^2\left(\tilde{\omega},\mathbf{r}(t,\mathbf{y}),t\right)-1}{4\pi}\left(\left(\left\{d_{\mathbf{y}}\mathbf{r}\left(t,\mathbf{y}\right)\right\}^T\right)^{-1}\cdot\hat{\mathbf{E}}^*\left(\omega,\mathbf{x}\right)\right)e^{i\omega t}d\omega\right\}\\
&\quad - Re\left\{2\int_0^{+\infty}\kappa\left(\tilde{\omega},\mathbf{r}(t,\mathbf{y}),t\right)\left(\left(\left\{d_{\mathbf{y}}\mathbf{r}\left(t,\mathbf{y}\right)\right\}^T\right)^{-1}\cdot\hat{\mathbf{D}}^*\left(\omega,\mathbf{x}\right)\right)e^{i\omega t}d\omega\right\}\\
&\text{and}\qquad \mathbf{M}\left(\mathbf{r}(t,\mathbf{y}),t\right)+\frac{1}{c}\mathbf{u}\left(\mathbf{r}(t,\mathbf{y}),t\right)\times\mathbf{P}\left(\mathbf{r}(t,\mathbf{y}),t\right)\\
&= Re\left\{2\int_0^{+\infty}\kappa\left(\tilde{\omega},\mathbf{r}(t,\mathbf{y}),t\right)\left(\left(\left\{d_{\mathbf{y}}\mathbf{r}\left(t,\mathbf{y}\right)\right\}^T\right)^{-1}\cdot\hat{\mathbf{H}}^*(\tilde{\omega},\mathbf{y})\right)e^{i\tilde{\omega}t}d\tilde{\omega}\right\},
\end{aligned}\tag{12.76}$$

i.e.

$$\begin{aligned}
\mathbf{P}\left(\mathbf{x},t\right) &= Re\left\{2\int_0^{+\infty}\frac{n^2\left(\tilde{\omega},\mathbf{x},t\right)-1}{4\pi}\left(\left\{d_{\mathbf{x}}\mathbf{f}\left(\mathbf{x},t\right)\right\}^T\cdot\hat{\mathbf{E}}^*\left(\tilde{\omega},\mathbf{f}\left(\mathbf{x},t\right)\right)\right)e^{i\tilde{\omega}t}d\tilde{\omega}\right\}\\
&\quad - Re\left\{2\int_0^{+\infty}\kappa\left(\tilde{\omega},\mathbf{x},t\right)\left(\left\{d_{\mathbf{x}}\mathbf{f}\left(\mathbf{x},t\right)\right\}^T\cdot\hat{\mathbf{D}}^*\left(\tilde{\omega},\mathbf{f}\left(\mathbf{x},t\right)\right)\right)e^{i\tilde{\omega}t}d\tilde{\omega}\right\}\qquad\text{and}\\
&\mathbf{M}\left(\mathbf{x},t\right)+\frac{1}{c}\mathbf{u}(\mathbf{x},t)\times\mathbf{P}(\mathbf{x},t)\\
&= Re\left\{2\int_0^{+\infty}\kappa\left(\tilde{\omega},\mathbf{x},t\right)\left(\left\{d_{\mathbf{x}}\mathbf{f}\left(\mathbf{x},t\right)\right\}^T\cdot\hat{\mathbf{H}}^*\left(\tilde{\omega},\mathbf{f}\left(\mathbf{x},t\right)\right)\right)e^{i\tilde{\omega}t}d\tilde{\omega}\right\}.
\end{aligned}\tag{12.77}$$

The quantities $n\left(\tilde{\omega},\mathbf{x},t\right)$ and $\kappa\left(\tilde{\omega},\mathbf{x},t\right)$ in (12.76) and (12.77) are assumed to be complex in general and assumed to depend on $\tilde{\omega}$ in addition to the dependence on $(\mathbf{x},t)$.

In particular, in the case $\mathbf{u}=0$, we clearly have $\mathbf{r}(t,\mathbf{y})=\mathbf{y}$, $\mathbf{f}\left(\mathbf{x},t\right)=\mathbf{x}$ and $\mathbf{E}^*(t,\mathbf{x})=\mathbf{E}(\mathbf{x},t)$, $\mathbf{B}^*(t,\mathbf{x})=\mathbf{B}(\mathbf{x},t)$, $\mathbf{D}^*(t,\mathbf{x})=\mathbf{D}(\mathbf{x},t)$, $\mathbf{H}^*(t,\mathbf{x})=\mathbf{H}(\mathbf{x},t)$ and therefore, (12.76) and (12.77) coincide with (12.67).

On the other hand, in the case of an arbitrary moving transparent medium without dispersion i.e. in the case where we assume that $n:=n\left(\mathbf{x},t\right)$ and $\kappa:=\kappa\left(\mathbf{x},t\right)$ are independent on the argument $\tilde{\omega}$, and moreover, we assume them to be real, by the properties of Furier's transform, we rewrite (12.76) as

$$\begin{aligned}
\mathbf{P}\left(\mathbf{r}(t,\mathbf{y}),t\right) &= \frac{n^2\left(\mathbf{r}(t,\mathbf{y}),t\right)-1}{4\pi}\left(\left\{d_{\mathbf{y}}\mathbf{r}\left(t,\mathbf{y}\right)\right\}^T\right)^{-1}\cdot Re\left\{2\int_0^{+\infty}\hat{\mathbf{E}}^*(\tilde{\omega},\mathbf{y})e^{i\tilde{\omega}t}d\tilde{\omega}\right\}\\
&\quad - \kappa\left(\mathbf{r}(t,\mathbf{y}),t\right)\left(\left\{d_{\mathbf{y}}\mathbf{r}\left(t,\mathbf{y}\right)\right\}^T\right)^{-1}\cdot Re\left\{2\int_0^{+\infty}\hat{\mathbf{D}}^*(\tilde{\omega},\mathbf{y})e^{i\tilde{\omega}t}d\tilde{\omega}\right\}\\
&= \gamma\left(\mathbf{r}(t,\mathbf{y}),t\right)\left(\left\{d_{\mathbf{y}}\mathbf{r}\left(t,\mathbf{y}\right)\right\}^T\right)^{-1}\cdot\mathbf{E}^*(t,\mathbf{y})-\kappa\left(\mathbf{r}(t,\mathbf{y}),t\right)\left(\left\{d_{\mathbf{y}}\mathbf{r}\left(t,\mathbf{y}\right)\right\}^T\right)^{-1}\cdot\mathbf{D}^*(t,\mathbf{y})
\end{aligned}$$

and
$$\mathbf{M}\left(\mathbf{r}(t,\mathbf{y}),t\right)+\frac{1}{c}\mathbf{u}\left(\mathbf{r}(t,\mathbf{y}),t\right)\times\mathbf{P}\left(\mathbf{r}(t,\mathbf{y}),t\right)$$
$$=\kappa\left(\mathbf{r}(t,\mathbf{y}),t\right)\left(\left\{d_{\mathbf{y}}\mathbf{r}\left(t,\mathbf{y}\right)\right\}^T\right)^{-1}\cdot Re\left\{2\int_0^{+\infty}\hat{\mathbf{H}}^*\left(\tilde{\omega},\mathbf{y}\right)e^{i\tilde{\omega}t}d\tilde{\omega}\right\}$$
$$=\kappa\left(\mathbf{r}(t,\mathbf{y}),t\right)\left(\left\{d_{\mathbf{y}}\mathbf{r}\left(t,\mathbf{y}\right)\right\}^T\right)^{-1}\cdot\mathbf{H}^*(t,\mathbf{y}). \quad (12.78)$$

Thus, inserting (12.71) into (12.78), we deduce

$$\mathbf{P}\left(\mathbf{r}(t,\mathbf{y}),t\right)=\frac{n^2\left(\mathbf{r}(t,\mathbf{y}),t\right)-1}{4\pi}\left(\mathbf{E}\left(\mathbf{r}(t,\mathbf{y}),t\right)+\frac{1}{c}\,\mathbf{u}\left(\mathbf{r}(t,\mathbf{y}),t\right)\times\mathbf{B}\left(\mathbf{r}(t,\mathbf{y}),t\right)\right)$$
$$-\kappa\left(\mathbf{r}(t,\mathbf{y}),t\right)\mathbf{D}\left(\mathbf{r}(t,\mathbf{y}),t\right)\qquad\text{and}$$
$$\mathbf{M}\left(\mathbf{r}(t,\mathbf{y}),t\right)+\frac{1}{c}\mathbf{u}\left(\mathbf{r}(t,\mathbf{y}),t\right)\times\mathbf{P}\left(\mathbf{r}(t,\mathbf{y}),t\right)=$$
$$\kappa\left(\mathbf{r}(t,\mathbf{y}),t\right)\left(\mathbf{H}\left(\mathbf{r}(t,\mathbf{y}),t\right)-\frac{1}{c}\,\mathbf{u}\left(\mathbf{r}(t,\mathbf{y}),t\right)\times\mathbf{D}\left(\mathbf{r}(t,\mathbf{y}),t\right)\right). \quad (12.79)$$

I.e.

$$\begin{cases}\mathbf{P}(\mathbf{x},t)=\frac{n^2(\mathbf{x},t)-1}{4\pi}\left(\mathbf{E}(\mathbf{x},t)+\frac{1}{c}\,\mathbf{u}(\mathbf{x},t)\times\mathbf{B}(\mathbf{x},t)\right)-\kappa(\mathbf{x},t)\,\mathbf{D}(\mathbf{x},t),\\ \mathbf{M}(\mathbf{x},t)=\kappa(\mathbf{x},t)\left(\mathbf{H}(\mathbf{x},t)-\frac{1}{c}\,\mathbf{u}(\mathbf{x},t)\times\mathbf{D}(\mathbf{x},t)\right)-\frac{1}{c}\mathbf{u}\left(\mathbf{x},t\right)\times\mathbf{P}\left(\mathbf{x},t\right),\end{cases} \quad (12.80)$$

consistently with (12.42).

Next, as before, assume that the change of some non-inertial Cartesian system (∗) of coordinates to another Cartesian system (∗∗) of coordinates is of the form:

$$\begin{cases}\mathbf{x}'=A(t)\cdot\mathbf{x}+\mathbf{z}(t),\\ t'=t,\end{cases} \quad (12.81)$$

where $A(t)\in SO(3)$ is a rotation. Then, by (11.41) the law of transformation of the Lagrangian coordinates $(\mathbf{y},t)\to(\mathbf{y}',t')$, consistent with (12.81), is the following:

$$\begin{cases}\mathbf{y}'=A(t_0)\cdot\mathbf{y}+\mathbf{z}(t_0),\\ t'=t.\end{cases} \quad (12.82)$$

Moreover, by (11.43), we have

$$\mathbf{r}'(t',\mathbf{y}')=\mathbf{r}'\left(t,A(t_0)\cdot\mathbf{y}+\mathbf{z}(t_0)\right)=A(t)\cdot\mathbf{r}(t,\mathbf{y})+\mathbf{z}(t), \quad (12.83)$$

and by (11.47) for inverse mappings, we have

$$\mathbf{f}'(\mathbf{x}',t')=\mathbf{f}'\left(A(t)\cdot\mathbf{x}+\mathbf{z}(t),t\right)=A(t_0)\cdot\mathbf{f}(\mathbf{x},t)+\mathbf{z}(t_0). \quad (12.84)$$

In particular,

$$d_{\mathbf{y}'}\mathbf{r}'(t',\mathbf{y}') = A(t)\cdot d_{\mathbf{y}}\mathbf{r}(t,\mathbf{y})\cdot\{A(t_0)\}^{-1} = A(t)\cdot d_{\mathbf{y}}\mathbf{r}(t,\mathbf{y})\cdot A^T(t_0), \tag{12.85}$$

and

$$d_{\mathbf{x}'}\mathbf{f}'(\mathbf{x}',t') = A(t_0)\cdot d_{\mathbf{x}}\mathbf{f}(\mathbf{x},t)\cdot\{A(t)\}^{-1} = A(t_0)\cdot d_{\mathbf{x}}\mathbf{f}(\mathbf{x},t)\cdot A^T(t). \tag{12.86}$$

However, by (12.37) and the fifth equation in (12.34), we have

$$\begin{cases}\mathbf{D}'(\mathbf{x}',t') = A(t)\cdot\mathbf{D}(\mathbf{x},t)\\ \mathbf{B}'(\mathbf{x}',t') = A(t)\cdot\mathbf{B}(\mathbf{x},t)\\ \left(\mathbf{E}'(\mathbf{x}',t') + \frac{1}{c}\mathbf{u}'(\mathbf{x}',t')\times\mathbf{B}'(\mathbf{x}',t')\right) = A(t)\cdot\left(\mathbf{E}(x,t) + \frac{1}{c}\mathbf{u}(x,t)\times\mathbf{B}(x,t)\right)\\ \left(\mathbf{H}'(\mathbf{x}',t') - \frac{1}{c}\mathbf{u}'(\mathbf{x}',t')\times\mathbf{D}'(\mathbf{x}',t')\right) = A(t)\cdot\left(\mathbf{H}(x,t) - \frac{1}{c}\mathbf{u}(x,t)\times\mathbf{D}(x,t)\right).\end{cases} \tag{12.87}$$

In, particular, by inserting (12.82), (12.83) and (12.85) into (12.71) and using (12.87), we deduce

$$\begin{cases}\mathbf{E}^{*\prime}(t',\mathbf{y}') = A(t_0)\cdot\mathbf{E}^*(t,\mathbf{y})\\ \mathbf{B}^{*\prime}(t',\mathbf{y}') = A(t_0)\cdot\mathbf{B}^*(t,\mathbf{y})\\ \mathbf{D}^{*\prime}(t',\mathbf{y}') = A(t_0)\cdot\mathbf{D}^*(t,\mathbf{y})\\ \mathbf{H}^{*\prime}(t',\mathbf{y}') = A(t_0)\cdot\mathbf{H}^*(t,\mathbf{y}).\end{cases} \tag{12.88}$$

Thus inserting (12.88) into (12.72) and (12.73) gives

$$\begin{cases}\hat{\mathbf{E}}^{*\prime}(\tilde{\omega}',\mathbf{y}') = A(t_0)\cdot\hat{\mathbf{E}}^*(\tilde{\omega},\mathbf{y})\\ \hat{\mathbf{B}}^{*\prime}(\tilde{\omega}',\mathbf{y}') = A(t_0)\cdot\hat{\mathbf{B}}^*(\tilde{\omega},\mathbf{y})\\ \hat{\mathbf{D}}^{*\prime}(\tilde{\omega}',\mathbf{y}') = A(t_0)\cdot\hat{\mathbf{D}}^*(\tilde{\omega},\mathbf{y})\\ \hat{\mathbf{H}}^{*\prime}(\tilde{\omega}',\mathbf{y}') = A(t_0)\cdot\hat{\mathbf{H}}^*(\tilde{\omega},\mathbf{y}),\end{cases} \tag{12.89}$$

provided that we have

$$\tilde{\omega}' = \tilde{\omega}. \tag{12.90}$$

Then, inserting (12.84) and (12.86) into (12.89), we deduce

$$\begin{cases}\left\{d_{\mathbf{x}'}\mathbf{f}'(\mathbf{x}',t')\right\}^T\cdot\hat{\mathbf{E}}^{*\prime}\left(\tilde{\omega}',\mathbf{f}'(\mathbf{x}',t')\right) = A(t)\cdot\left(\{d_{\mathbf{x}}\mathbf{f}(\mathbf{x},t)\}^T\cdot\hat{\mathbf{E}}^*(\tilde{\omega},\mathbf{f}(\mathbf{x},t))\right),\\ \left\{d_{\mathbf{x}'}\mathbf{f}'(\mathbf{x}',t')\right\}^T\cdot\hat{\mathbf{B}}^{*\prime}\left(\tilde{\omega}',\mathbf{f}'(\mathbf{x}',t')\right) = A(t)\cdot\left(\{d_{\mathbf{x}}\mathbf{f}(\mathbf{x},t)\}^T\cdot\hat{\mathbf{B}}^*(\tilde{\omega},\mathbf{f}(\mathbf{x},t))\right)\\ \left\{d_{\mathbf{x}'}\mathbf{f}'(\mathbf{x}',t')\right\}^T\cdot\hat{\mathbf{D}}^{*\prime}\left(\tilde{\omega}',\mathbf{f}'(\mathbf{x}',t')\right) = A(t)\cdot\left(\{d_{\mathbf{x}}\mathbf{f}(\mathbf{x},t)\}^T\cdot\hat{\mathbf{D}}^*(\tilde{\omega},\mathbf{f}(\mathbf{x},t))\right),\\ \left\{d_{\mathbf{x}'}\mathbf{f}'(\mathbf{x}',t')\right\}^T\cdot\hat{\mathbf{H}}^{*\prime}\left(\tilde{\omega}',\mathbf{f}'(\mathbf{x}',t')\right) = A(t)\cdot\left(\{d_{\mathbf{x}}\mathbf{f}(\mathbf{x},t)\}^T\cdot\hat{\mathbf{H}}^*(\tilde{\omega},\mathbf{f}(\mathbf{x},t))\right).\end{cases} \tag{12.91}$$

Thus, inserting (12.91) into (12.77), we obtain

$$\begin{cases} \mathbf{P}'\left(\mathbf{x}', t'\right) = A(t) \cdot \mathbf{P}\left(\mathbf{x}, t\right) \\ \mathbf{M}'\left(\mathbf{x}', t'\right) = A(t) \cdot \mathbf{M}\left(\mathbf{x}, t\right) - \frac{1}{c}\left(A'(t) \cdot \mathbf{x} + \mathbf{w}(t)\right) \times \left(A(t) \cdot \mathbf{P}\right), \end{cases} \tag{12.92}$$

provided that we have

$$\tilde{\omega}' = \tilde{\omega}, \quad n'\left(\tilde{\omega}', \mathbf{x}', t'\right) = n\left(\tilde{\omega}, \mathbf{x}, t\right) \quad \text{and} \quad \kappa'\left(\tilde{\omega}', \mathbf{x}', t'\right) = \kappa\left(\tilde{\omega}, \mathbf{x}, t\right). \tag{12.93}$$

On the other hand, (12.92) is fully consistent with the last two equalities in (12.34) and therefore, the dispersion law, in the forms of (12.76) or equivalently (12.77), is invariant under the change of inertial or non-inertial Cartesian coordinate system, provided we have (12.93).

Next, we will prove that the right-hand side of (12.76) or (12.77) is independent on the initial instant of time t_0 in (12.68). Indeed let $t_1 \in \mathbb{R}$ be another choice of initial instant of time, let $\mathbf{r}(t, \mathbf{y})$ be given by (12.68)

$$\begin{cases} \frac{\partial \mathbf{r}}{\partial t}(t, \mathbf{y}) = \mathbf{u}\left(\mathbf{r}(t, \mathbf{y}), t\right) \\ \mathbf{r}(t_0, \mathbf{y}) = \mathbf{y}, \end{cases} \tag{12.94}$$

and let $\mathbf{r}_1(t, \mathbf{y})$ be given by the following:

$$\begin{cases} \frac{\partial \mathbf{r}_1}{\partial t}(t, \mathbf{y}) = \mathbf{u}\left(\mathbf{r}_1(t, \mathbf{y}), t\right) \\ \mathbf{r}_1(t_1, \mathbf{y}) = \mathbf{y}. \end{cases} \tag{12.95}$$

Then considering $\mathbf{r}_2(t, \mathbf{y})$ defined by

$$\mathbf{r}_2(t, \mathbf{y}) = \mathbf{r}\left(t, \mathbf{h}(\mathbf{y})\right) \quad \text{where} \quad \mathbf{h}(\mathbf{y}) := \mathbf{r}_1(t_0, \mathbf{y}), \tag{12.96}$$

gives

$$\begin{cases} \frac{\partial \mathbf{r}_2}{\partial t}(t, \mathbf{y}) = \mathbf{u}\left(\mathbf{r}_2(t, \mathbf{y}), t\right) \\ \mathbf{r}_2(t_0, \mathbf{y}) = \mathbf{r}\left(t_0, \mathbf{r}_1(t_0, \mathbf{y})\right) = \mathbf{r}_1(t_0, \mathbf{y}). \end{cases} \tag{12.97}$$

Thus, by (12.97), using the uniqueness of solution to the ordinary differential equation, we deduce

$$\mathbf{r}_1(t, \mathbf{y}) = \mathbf{r}_2(t, \mathbf{y}) = \mathbf{r}\left(t, \mathbf{h}(\mathbf{y})\right) \qquad \forall t, \mathbf{y}. \tag{12.98}$$

Thus considering,

$$\begin{cases} \mathbf{D}_1^*(t,\mathbf{y}) := \left\{d_\mathbf{y}\mathbf{r}_1\left(t,\mathbf{y}\right)\right\}^T \cdot \mathbf{D}\left(\mathbf{r}_1(t,\mathbf{y}),t\right) \\ \mathbf{B}_1^*(t,\mathbf{y}) := \left\{d_\mathbf{y}\mathbf{r}_1\left(t,\mathbf{y}\right)\right\}^T \cdot \mathbf{B}\left(\mathbf{r}_1(t,\mathbf{y}),t\right) \\ \mathbf{E}_1^*(t,\mathbf{y}) := \left\{d_\mathbf{y}\mathbf{r}_1\left(t,\mathbf{y}\right)\right\}^T \cdot \left(\mathbf{E}\left(\mathbf{r}_1(t,\mathbf{y}),t\right) + \frac{1}{c}\mathbf{u}\left(\mathbf{r}_1(t,\mathbf{y}),t\right)\times\mathbf{B}\left(\mathbf{r}_1(t,\mathbf{y}),t\right)\right) \\ \mathbf{H}_1^*(t,\mathbf{y}) := \left\{d_\mathbf{y}\mathbf{r}_1\left(t,\mathbf{y}\right)\right\}^T \cdot \left(\mathbf{H}\left(\mathbf{r}_1(t,\mathbf{y}),t\right) - \frac{1}{c}\mathbf{u}\left(\mathbf{r}_1(t,\mathbf{y}),t\right)\times\mathbf{D}\left(\mathbf{r}_1(t,\mathbf{y}),t\right)\right), \end{cases} \tag{12.99}$$

where $\mathbf{r}_1(t,\mathbf{y})$ is given by (12.95), by (12.98), we deduce

$$\mathbf{E}_1^*(t,\mathbf{y}) = \left\{d_\mathbf{y}\mathbf{h}\left(\mathbf{y}\right)\right\}^T \cdot \mathbf{E}^*\left(t,\mathbf{h}(\mathbf{y})\right), \qquad \mathbf{B}_1^*(t,\mathbf{y}) = \left\{d_\mathbf{y}\mathbf{h}\left(\mathbf{y}\right)\right\}^T \cdot \mathbf{B}^*\left(t,\mathbf{h}(\mathbf{y})\right),$$
$$\mathbf{D}_1^*(t,\mathbf{y}) = \left\{d_\mathbf{y}\mathbf{h}\left(\mathbf{y}\right)\right\}^T \cdot \mathbf{D}^*\left(t,\mathbf{h}(\mathbf{y})\right) \quad \text{and} \quad \mathbf{H}_1^*(t,\mathbf{y}) = \left\{d_\mathbf{y}\mathbf{h}\left(\mathbf{y}\right)\right\}^T \cdot \mathbf{H}^*\left(t,\mathbf{h}(\mathbf{y})\right). \tag{12.100}$$

Then defining

$$\hat{\mathbf{E}}_1^*(\tilde{\omega},\mathbf{y}) := \frac{1}{2\pi}\int_{-\infty}^{+\infty} \mathbf{E}_1^*(t,\mathbf{y})e^{-i\tilde{\omega}t}dt,$$
$$\hat{\mathbf{D}}_1^*(\tilde{\omega},\mathbf{y}) := \frac{1}{2\pi}\int_{-\infty}^{+\infty} \mathbf{D}_1^*(t,\mathbf{y})e^{-i\tilde{\omega}t}dt,$$
$$\hat{\mathbf{B}}_1^*(\tilde{\omega},\mathbf{y}) := \frac{1}{2\pi}\int_{-\infty}^{+\infty} \mathbf{B}_1^*(t,\mathbf{y})e^{-i\tilde{\omega}t}dt$$
$$\text{and} \quad \hat{\mathbf{H}}_1^*(\tilde{\omega},\mathbf{y}) := \frac{1}{2\pi}\int_{-\infty}^{+\infty} \mathbf{H}_1^*(t,\mathbf{y})e^{-i\tilde{\omega}t}dt, \tag{12.101}$$

we deduce

$$\hat{\mathbf{E}}_1^*(\tilde{\omega},\mathbf{y}) = \left\{d_\mathbf{y}\mathbf{h}\left(\mathbf{y}\right)\right\}^T \cdot \hat{\mathbf{E}}^*\left(\tilde{\omega},\mathbf{h}(\mathbf{y})\right),$$
$$\hat{\mathbf{B}}_1^*(\tilde{\omega},\mathbf{y}) = \left\{d_\mathbf{y}\mathbf{h}\left(\mathbf{y}\right)\right\}^T \cdot \hat{\mathbf{B}}^*\left(\tilde{\omega},\mathbf{h}(\mathbf{y})\right),$$
$$\hat{\mathbf{D}}_1^*(\tilde{\omega},\mathbf{y}) = \left\{d_\mathbf{y}\mathbf{h}\left(\mathbf{y}\right)\right\}^T \cdot \hat{\mathbf{D}}^*\left(\tilde{\omega},\mathbf{h}(\mathbf{y})\right)$$
$$\text{and} \quad \hat{\mathbf{H}}_1^*(\tilde{\omega},\mathbf{y}) = \left\{d_\mathbf{y}\mathbf{h}\left(\mathbf{y}\right)\right\}^T \cdot \hat{\mathbf{H}}^*\left(\tilde{\omega},\mathbf{h}(\mathbf{y})\right). \tag{12.102}$$

Thus, we write the analog of (12.76) as

$$\mathbf{P}\left(\mathbf{r}_1(t,\mathbf{y}),t\right) = \mathbf{P}\left(\mathbf{r}\left(t,\mathbf{h}(\mathbf{y})\right),t\right) = 2\int_0^{+\infty} Re\Bigg\{$$
$$\frac{n^2\left(\tilde{\omega},\mathbf{r}\left(t,\mathbf{h}(\mathbf{y})\right),t\right)-1}{4\pi}\left(\left\{d_\mathbf{y}\mathbf{r}\left(t,\mathbf{h}(\mathbf{y})\right)\cdot d_\mathbf{y}\mathbf{h}(\mathbf{y})\right\}^T\right)^{-1}\cdot\left\{d_\mathbf{y}\mathbf{h}\left(\mathbf{y}\right)\right\}^T\cdot\hat{\mathbf{E}}^*\left(\tilde{\omega},\mathbf{h}(y)\right)e^{i\tilde{\omega}t}$$
$$\Bigg\}d\tilde{\omega} - 2\int_0^{+\infty} Re\Bigg\{$$
$$\kappa\left(\tilde{\omega},\mathbf{r}\left(t,\mathbf{h}(\mathbf{y})\right),t\right)\left(\left\{d_\mathbf{y}\mathbf{r}\left(t,\mathbf{h}(\mathbf{y})\right)\cdot d_\mathbf{y}\mathbf{h}(\mathbf{y})\right\}^T\right)^{-1}\cdot\left\{d_\mathbf{y}\mathbf{h}\left(\mathbf{y}\right)\right\}^T\cdot\hat{\mathbf{D}}^*\left(\tilde{\omega},\mathbf{h}(y)\right)e^{i\tilde{\omega}t}$$

$$\Big\}d\tilde{\omega} = Re\left\{2\int_0^{+\infty}\frac{n^2(\tilde{\omega},\mathbf{r}_1(t,\mathbf{y}),t)-1}{4\pi}\left(\left\{d_{\mathbf{y}}\mathbf{r}_1(t,\mathbf{y})\right\}^T\right)^{-1}\cdot\hat{\mathbf{E}}_1^*(\tilde{\omega},\mathbf{y})e^{i\tilde{\omega}t}d\tilde{\omega}\right\}$$

$$-Re\left\{2\int_0^{+\infty}\kappa(\tilde{\omega},\mathbf{r}_1(t,\mathbf{y}),t)\left(\left\{d_{\mathbf{y}}\mathbf{r}_1(t,\mathbf{y})\right\}^T\right)^{-1}\cdot\hat{\mathbf{D}}_1^*(\tilde{\omega},\mathbf{y})e^{i\tilde{\omega}t}d\tilde{\omega}\right\}, \quad (12.103)$$

and

$$\mathbf{M}(\mathbf{r}_1(t,\mathbf{y}),t)+\frac{1}{c}\mathbf{u}(\mathbf{r}_1(t,\mathbf{y}),t)\times\mathbf{P}(\mathbf{r}_1(t,\mathbf{y}),t)$$
$$=\mathbf{M}(\mathbf{r}(t,\mathbf{h}(\mathbf{y})),t)+\frac{1}{c}\mathbf{u}(\mathbf{r}(t,\mathbf{h}(\mathbf{y})),t)\times\mathbf{P}(\mathbf{r}(t,\mathbf{h}(\mathbf{y})),t)=2\int_0^{+\infty}Re\Big\{$$
$$\kappa(\tilde{\omega},\mathbf{r}(t,\mathbf{h}(\mathbf{y})),t)\left(\left\{d_{\mathbf{y}}\mathbf{r}(t,\mathbf{h}(\mathbf{y}))\cdot d_{\mathbf{y}}\mathbf{h}(\mathbf{y})\right\}^T\right)^{-1}\cdot\left\{d_{\mathbf{y}}\mathbf{h}(\mathbf{y})\right\}^T\cdot\hat{\mathbf{H}}^*(\tilde{\omega},\mathbf{h}(y))e^{i\tilde{\omega}t}$$
$$\Big\}d\tilde{\omega}=Re\left\{2\int_0^{+\infty}\kappa(\tilde{\omega},\mathbf{r}_1(t,\mathbf{y}),t)\left(\left\{d_{\mathbf{y}}\mathbf{r}_1(t,\mathbf{y})\right\}^T\right)^{-1}\cdot\hat{\mathbf{H}}_1^*(\tilde{\omega},\mathbf{y})e^{i\tilde{\omega}t}d\tilde{\omega}\right\}. \quad (12.104)$$

So we proved that the right-hand side of (12.76) or (12.77) is independent on the initial instant of time t_0.

Finally, we can write the following possible alternative to the dispersion law in the form of (12.71) and (12.76) or (12.77). We could consider,

$$\begin{cases}\mathbf{E}^*(t,\mathbf{y}):=\left\{d_{\mathbf{y}}\mathbf{r}(t,\mathbf{y})\right\}^{-1}\cdot\left(\mathbf{E}(\mathbf{r}(t,\mathbf{y}),t)+\frac{1}{c}\mathbf{u}(\mathbf{r}(t,\mathbf{y}),t)\times\mathbf{B}(\mathbf{r}(t,\mathbf{y}),t)\right)\\ \quad\forall t\in\mathbb{R},\ \ \forall\mathbf{y}\in\Omega,\\ \mathbf{B}^*(t,\mathbf{y}):=\left\{d_{\mathbf{y}}\mathbf{r}(t,\mathbf{y})\right\}^{-1}\cdot\mathbf{B}(\mathbf{r}(t,\mathbf{y}),t)\qquad\forall t\in\mathbb{R},\ \ \forall\mathbf{y}\in\Omega,\\ \mathbf{D}^*(t,\mathbf{y}):=\left\{d_{\mathbf{y}}\mathbf{r}(t,\mathbf{y})\right\}^{-1}\cdot\mathbf{D}(\mathbf{r}(t,\mathbf{y}),t)\qquad\forall t\in\mathbb{R},\ \ \forall\mathbf{y}\in\Omega,\\ \mathbf{H}^*(t,\mathbf{y}):=\left\{d_{\mathbf{y}}\mathbf{r}(t,\mathbf{y})\right\}^{-1}\cdot\left(\mathbf{H}(\mathbf{r}(t,\mathbf{y}),t)-\frac{1}{c}\mathbf{u}(\mathbf{r}(t,\mathbf{y}),t)\times\mathbf{D}(\mathbf{r}(t,\mathbf{y}),t)\right)\\ \quad\forall t\in\mathbb{R},\ \ \forall\mathbf{y}\in\Omega,\end{cases} \quad (12.105)$$

where $\mathbf{r}(t,\mathbf{y})$ is given by (12.68). Then, since $\mathbf{E}^*$, $\mathbf{D}^*$, $\mathbf{B}^*$ and $\mathbf{H}^*$ are real vectors, we could write them as a Furier's transform on the time variable:

$$\mathbf{E}^*(t,\mathbf{y})=Re\left\{2\int_0^{+\infty}\hat{\mathbf{E}}^*(\tilde{\omega},\mathbf{y})e^{i\tilde{\omega}t}d\tilde{\omega}\right\}$$
$$\text{where}\quad\hat{\mathbf{E}}^*(\tilde{\omega},\mathbf{y}):=\frac{1}{2\pi}\int_{-\infty}^{+\infty}\mathbf{E}^*(t,\mathbf{y})e^{-i\tilde{\omega}t}dt, \quad (12.106)$$

$$\mathbf{B}^*(t,\mathbf{y}) = Re\left\{2\int_0^{+\infty} \hat{\mathbf{B}}^*(\tilde{\omega},\mathbf{y})e^{i\tilde{\omega}t}d\tilde{\omega}\right\}$$
$$\text{where}\quad \hat{\mathbf{B}}^*(\tilde{\omega},\mathbf{y}) := \frac{1}{2\pi}\int_{-\infty}^{+\infty} \mathbf{B}^*(t,\mathbf{y})e^{-i\tilde{\omega}t}dt, \qquad (12.107)$$

$$\mathbf{D}^*(t,\mathbf{y}) = Re\left\{2\int_0^{+\infty} \hat{\mathbf{D}}^*(\tilde{\omega},\mathbf{y})e^{i\tilde{\omega}t}d\tilde{\omega}\right\}$$
$$\text{where}\quad \hat{\mathbf{D}}^*(\tilde{\omega},\mathbf{y}) := \frac{1}{2\pi}\int_{-\infty}^{+\infty} \mathbf{D}^*(t,\mathbf{y})e^{-i\tilde{\omega}t}dt, \qquad (12.108)$$

and

$$\mathbf{H}^*(t,\mathbf{y}) = Re\left\{2\int_0^{+\infty} \hat{\mathbf{H}}^*(\tilde{\omega},\mathbf{y})e^{i\tilde{\omega}t}d\tilde{\omega}\right\}$$
$$\text{where}\quad \hat{\mathbf{H}}^*(\tilde{\omega},\mathbf{y}) := \frac{1}{2\pi}\int_{-\infty}^{+\infty} \mathbf{H}^*(t,\mathbf{y})e^{-i\tilde{\omega}t}dt. \qquad (12.109)$$

Then, alternatively to (12.71) and (12.76), we could consider

$$\begin{aligned}
\mathbf{P}(\mathbf{r}(t,\mathbf{y}),t) &= Re\left\{2\int_0^{+\infty} \frac{n^2(\tilde{\omega},\mathbf{r}(t,\mathbf{y}),t)-1}{4\pi}\left(d_{\mathbf{y}}\mathbf{r}(t,\mathbf{y})\cdot\hat{\mathbf{E}}^*(\omega,\mathbf{x})\right)e^{i\omega t}d\omega\right\}\\
&\quad - Re\left\{2\int_0^{+\infty} \kappa(\tilde{\omega},\mathbf{r}(t,\mathbf{y}),t)\left(d_{\mathbf{y}}\mathbf{r}(t,\mathbf{y})\cdot\hat{\mathbf{D}}^*(\omega,\mathbf{x})\right)e^{i\omega t}d\omega\right\}\\
&\text{and}\quad \mathbf{M}(\mathbf{r}(t,\mathbf{y}),t) + \frac{1}{c}\mathbf{u}(\mathbf{r}(t,\mathbf{y}),t)\times\mathbf{P}(\mathbf{r}(t,\mathbf{y}),t)\\
&= Re\left\{2\int_0^{+\infty} \kappa(\tilde{\omega},\mathbf{r}(t,\mathbf{y}),t)\left(d_{\mathbf{y}}\mathbf{r}(t,\mathbf{y})\cdot\hat{\mathbf{H}}^*(\tilde{\omega},\mathbf{y})\right)e^{i\tilde{\omega}t}d\tilde{\omega}\right\}, \qquad (12.110)
\end{aligned}$$

i.e.

$$\begin{aligned}
\mathbf{P}(\mathbf{x},t) &= Re\left\{2\int_0^{+\infty} \frac{n^2(\tilde{\omega},\mathbf{x},t)-1}{4\pi}\left(\{d_{\mathbf{x}}\mathbf{f}(\mathbf{x},t)\}^{-1}\cdot\hat{\mathbf{E}}^*(\tilde{\omega},\mathbf{f}(\mathbf{x},t))\right)e^{i\tilde{\omega}t}d\tilde{\omega}\right\}\\
&\quad - Re\left\{2\int_0^{+\infty} \kappa(\tilde{\omega},\mathbf{x},t)\left(\{d_{\mathbf{x}}\mathbf{f}(\mathbf{x},t)\}^{-1}\cdot\hat{\mathbf{D}}^*(\tilde{\omega},\mathbf{f}(\mathbf{x},t))\right)e^{i\tilde{\omega}t}d\tilde{\omega}\right\} \quad \text{and}\\
&\mathbf{M}(\mathbf{x},t) + \frac{1}{c}\mathbf{u}(\mathbf{x},t)\times\mathbf{P}(\mathbf{x},t)\\
&= Re\left\{2\int_0^{+\infty} \kappa(\tilde{\omega},\mathbf{x},t)\left(\{d_{\mathbf{x}}\mathbf{f}(\mathbf{x},t)\}^{-1}\cdot\hat{\mathbf{H}}^*(\tilde{\omega},\mathbf{f}(\mathbf{x},t))\right)e^{i\tilde{\omega}t}d\tilde{\omega}\right\}. \qquad (12.111)
\end{aligned}$$

Then, exactly in the same way we already proved for (12.71) and (12.76) or (12.77), we can also prove that the above alternative (12.105) and (12.110) or (12.111) is invariant under the change of inertial or non-inertial Cartesian coordinate system. Furthermore, as before, the above alternative is equivalent to (12.67) in the particular case $\mathbf{u} = 0$. Moreover, in the case of an arbitrary moving transparent medium without dispersion the alternative law is equivalent to (12.42). Finally, as before, (12.105) and (12.110) or (12.111) is also independent on the initial instant of time t_0.

We are unable to see any clear advantage of the dispersion law in the form of (12.71) and (12.76) or (12.77) with respect to the law (12.105) and (12.110) or (12.111). We just note that the above two forms of the dispersion law coincide in the case where our medium moves as a rigid body.

12.7.1 Optical Dispersion in Anisotropic Moving Mediums

The generalization of (12.77) to anisotropic mediums is the following

$$\begin{aligned}\mathbf{P}(\mathbf{x}, t) = Re\left\{2\int_0^{+\infty} \Gamma(\tilde{\omega}, \mathbf{x}, t)\cdot\Big(\{d_{\mathbf{x}}\mathbf{f}(\mathbf{x}, t)\}^T\cdot\hat{\mathbf{E}}^*(\tilde{\omega}, \mathbf{f}(\mathbf{x}, t))\Big)e^{i\tilde{\omega}t}d\tilde{\omega}\right\}\\ - Re\left\{2\int_0^{+\infty} \Upsilon(\tilde{\omega}, \mathbf{x}, t)\cdot\Big(\{d_{\mathbf{x}}\mathbf{f}(\mathbf{x}, t)\}^T\cdot\hat{\mathbf{D}}^*(\tilde{\omega}, \mathbf{f}(\mathbf{x}, t))\Big)e^{i\tilde{\omega}t}d\tilde{\omega}\right\} \quad \text{and}\\ \mathbf{M}(\mathbf{x}, t) + \frac{1}{c}\mathbf{u}(\mathbf{x}, t)\times\mathbf{P}(\mathbf{x}, t)\\ = Re\left\{2\int_0^{+\infty} \Upsilon(\tilde{\omega}, \mathbf{x}, t)\cdot\Big(\{d_{\mathbf{x}}\mathbf{f}(\mathbf{x}, t)\}^T\cdot\hat{\mathbf{H}}^*(\tilde{\omega}, \mathbf{f}(\mathbf{x}, t))\Big)e^{i\tilde{\omega}t}d\tilde{\omega}\right\}. \quad (12.112)\end{aligned}$$

Here $\mathbf{E}^*$, $\mathbf{D}^*$, $\mathbf{B}^*$ and $\mathbf{H}^*$ are given by (12.71) with $\hat{\mathbf{E}}^*$, $\hat{\mathbf{D}}^*$, $\hat{\mathbf{B}}^*$ and $\hat{\mathbf{H}}^*$ being their Furier's transforms on the time variable and $\Gamma := \Gamma(\tilde{\omega}, \mathbf{x}, t) \in \mathbb{C}^{3\times 3}$ and $\Upsilon := \Upsilon(\tilde{\omega}, \mathbf{x}, t) \in \mathbb{C}^{3\times 3}$ are complex matrices describing the dispersion. Then, (12.112) is invariant under the change of inertial or non-inertial coordinate system, provided that under (12.81), we have

$$\begin{aligned}\tilde{\omega}' = \tilde{\omega}, \quad \Gamma'\left(\tilde{\omega}', \mathbf{x}', t'\right) = A(t)\cdot\Gamma(\tilde{\omega}, \mathbf{x}, t)\cdot A^T(t)\\ \text{and} \quad \Upsilon'\left(\tilde{\omega}', \mathbf{x}', t'\right) = A(t)\cdot\Upsilon(\tilde{\omega}, \mathbf{x}, t)\cdot A^T(t). \quad (12.113)\end{aligned}$$

Next, in the case of an anisotropic transparent medium without dispersion i.e. in the case where we assume that $\Gamma := \Gamma(\mathbf{x}, t)$ and $\Upsilon := \Upsilon(\mathbf{x}, t)$ are independent on the argument $\tilde{\omega}$ and moreover we assume them to be real matrices, (12.112) coincides with (12.57). Finally, as before, the right-hand side of (12.112) is independent on the initial instant of time t_0 in (12.68).

On the other hand, the generalization of (12.111) to anisotropic mediums is the following

$$\mathbf{P}(\mathbf{x},t) = Re\left\{2\int_0^{+\infty}\Gamma(\tilde{\omega},\mathbf{x},t)\cdot\Big(\{d_{\mathbf{x}}\mathbf{f}(\mathbf{x},t)\}^{-1}\cdot\hat{\mathbf{E}}^*(\tilde{\omega},\mathbf{f}(\mathbf{x},t))\Big)e^{i\tilde{\omega}t}d\tilde{\omega}\right\} - Re\left\{2\int_0^{+\infty}\Upsilon(\tilde{\omega},\mathbf{x},t)\cdot\Big(\{d_{\mathbf{x}}\mathbf{f}(\mathbf{x},t)\}^{-1}\cdot\hat{\mathbf{D}}^*(\tilde{\omega},\mathbf{f}(\mathbf{x},t))\Big)e^{i\tilde{\omega}t}d\tilde{\omega}\right\} \quad \text{and}$$

$$\mathbf{M}(\mathbf{x},t) + \frac{1}{c}\mathbf{u}(\mathbf{x},t)\times\mathbf{P}(\mathbf{x},t) = Re\left\{2\int_0^{+\infty}\Upsilon(\tilde{\omega},\mathbf{x},t)\cdot\Big(\{d_{\mathbf{x}}\mathbf{f}(\mathbf{x},t)\}^{-1}\cdot\hat{\mathbf{H}}^*(\tilde{\omega},\mathbf{f}(\mathbf{x},t))\Big)e^{i\tilde{\omega}t}d\tilde{\omega}\right\}. \tag{12.114}$$

Now $\mathbf{E}^*$ and $\mathbf{B}^*$ are given by (12.105) with $\hat{\mathbf{E}}^*$ and $\hat{\mathbf{B}}^*$ being their Furier's transforms on the time variable and $\Gamma := \Gamma(\tilde{\omega},\mathbf{x},t) \in \mathbb{C}^{3\times3}$ and $\Upsilon := \Upsilon(\tilde{\omega},\mathbf{x},t) \in \mathbb{C}^{3\times3}$ are, as before, complex matrices describing the dispersion. Then again, (12.114) is invariant under the change of inertial or non-inertial coordinate system, provided that under (12.81), we have (12.113). Moreover, in the case of an anisotropic transparent medium without dispersion i.e. in the case where we assume that $\Gamma := \Gamma(\mathbf{x},t)$ and $\Upsilon := \Upsilon(\mathbf{x},t)$ are independent on the argument $\tilde{\omega}$ and moreover we assume them to be <u>real</u> matrices, (12.114) again coincides with (12.57). Finally, as before, the right-hand side of (12.114) is independent on the initial instant of time t_0 in (12.68).

Chapter 13
Some Further Consequences of Maxwell Equations in a Medium or Vacuum

13.1 General Case

Again consider the system of Maxwell equations in the vacuum or in a medium, without dispersion, having the form (12.56):

$$\begin{cases} \text{curl}_{\mathbf{x}}\mathbf{H} = \frac{4\pi}{c}\mathbf{j} + \frac{1}{c}\frac{\partial \mathbf{D}}{\partial t}, \\ \text{div}_{\mathbf{x}}\mathbf{D} = 4\pi\rho, \\ \text{curl}_{\mathbf{x}}\mathbf{E} + \frac{1}{c}\frac{\partial \mathbf{B}}{\partial t} = 0, \\ \text{div}_{\mathbf{x}}\mathbf{B} = 0 \\ \mathbf{E} = \gamma_0\mathbf{D} - \frac{1}{c}\tilde{\mathbf{u}} \times \mathbf{B} \\ \mathbf{H} = \kappa_0\mathbf{B} + \frac{1}{c}\tilde{\mathbf{u}} \times \mathbf{D}, \\ \tilde{\mathbf{u}} = (\gamma_0\kappa_0\mathbf{v} + (1 - \gamma_0\kappa_0)\mathbf{u}), \end{cases} \tag{13.1}$$

where $\gamma_0 \neq 0$ and $\kappa_0 \neq 0$ are material coefficients, $\mathbf{v}$ is the vectorial gravitational potential, $\mathbf{u}$ is the medium velocity and $\tilde{\mathbf{u}} = (\gamma_0\kappa_0\mathbf{v} + (1 - \gamma_0\kappa_0)\mathbf{u})$ is the speed-like vector field. Remind that in the case of the vacuum we have $\gamma_0 = \kappa_0 = 1$, $\tilde{\mathbf{u}} = \mathbf{v}$ and Eq. (13.1) are precise (in the frames of our model). Otherwise, in the case $\kappa_0\gamma_0 \neq 1$ Eq. (13.1) is just an approximation that is good enough for the case:

$$\frac{|\kappa_0||1 - \kappa_0\gamma_0| \cdot |\mathbf{u} - \mathbf{v}|^2}{c^2} \ll 1. \tag{13.2}$$

Next again by the third and the fourth equations in (13.1) we can write

$$\begin{cases} \mathbf{B} = \text{curl}_{\mathbf{x}}\mathbf{A}, \\ \mathbf{E} = -\nabla_{\mathbf{x}}\Psi - \frac{1}{c}\frac{\partial \mathbf{A}}{\partial t}, \end{cases} \tag{13.3}$$

A. Poliakovsky, *The Laws of Gravity and Electromagnetism*,
https://doi.org/10.1007/978-3-031-61407-1_13

where Ψ and $\mathbf{A}$ are the usual scalar and the vectorial electromagnetic potentials. Then by (13.3) and (13.1) we have

$$\begin{cases}\mathbf{B} = \text{curl}_\mathbf{x}\mathbf{A}\\ \mathbf{E} = -\nabla_\mathbf{x}\Psi - \frac{1}{c}\frac{\partial \mathbf{A}}{\partial t}\\ \mathbf{D} = -\frac{1}{\gamma_0}\nabla_\mathbf{x}\Psi - \frac{1}{\gamma_0 c}\frac{\partial \mathbf{A}}{\partial t} + \frac{1}{c\gamma_0}\tilde{\mathbf{u}}\times\text{curl}_\mathbf{x}\mathbf{A}\\ \mathbf{H} = \kappa_0\,\text{curl}_\mathbf{x}\mathbf{A} + \frac{1}{c}\tilde{\mathbf{u}}\times\left(-\frac{1}{\gamma_0}\nabla_\mathbf{x}\Psi - \frac{1}{\gamma_0 c}\frac{\partial \mathbf{A}}{\partial t} + \frac{1}{\gamma_0 c}\tilde{\mathbf{u}}\times\text{curl}_\mathbf{x}\mathbf{A}\right).\end{cases} \tag{13.4}$$

Next we remind the definition of the proper scalar electromagnetic potential:

$$\Psi_0 := \Psi - \frac{1}{c}\mathbf{A}\cdot\mathbf{v}, \tag{13.5}$$

and remind also that $\mathbf{A}$ is a proper vector field and Ψ_0 is a proper scalar field. Then in the case of the medium we also define an additional scalar electromagnetic potential:

$$\Psi_1 := \Psi - \frac{1}{c}\mathbf{A}\cdot\tilde{\mathbf{u}}. \tag{13.6}$$

Then, since $\mathbf{A}$ is a proper vector field, we deduce that Ψ_1 is also a proper scalar field. Moreover, in the case of the vacuum or more generally in the case where $\gamma_0\kappa_0 \approx 1$ we have $\Psi_1 = \Psi_0$. Thus by (13.6) we rewrite (13.4) as

$$\begin{cases}\mathbf{B} = \text{curl}_\mathbf{x}\mathbf{A}\\ \mathbf{E} = -\nabla_\mathbf{x}\Psi_1 - \frac{1}{c}\frac{\partial \mathbf{A}}{\partial t} - \frac{1}{c}\nabla_\mathbf{x}(\mathbf{A}\cdot\tilde{\mathbf{u}})\\ \mathbf{D} = -\frac{1}{\gamma_0}\nabla_\mathbf{x}\Psi_1 - \frac{1}{\gamma_0 c}\left(\frac{\partial \mathbf{A}}{\partial t} - \tilde{\mathbf{u}}\times\text{curl}_\mathbf{x}\mathbf{A} + \nabla_\mathbf{x}(\mathbf{A}\cdot\tilde{\mathbf{u}})\right)\\ \mathbf{H} = \kappa_0\,\text{curl}_\mathbf{x}\mathbf{A} - \frac{1}{c}\tilde{\mathbf{u}}\times\left(\frac{1}{\gamma_0}\nabla_\mathbf{x}\Psi_1 + \frac{1}{\gamma_0 c}\left(\frac{\partial \mathbf{A}}{\partial t} - \tilde{\mathbf{u}}\times\text{curl}_\mathbf{x}\mathbf{A} + \nabla_\mathbf{x}(\mathbf{A}\cdot\tilde{\mathbf{u}})\right)\right).\end{cases} \tag{13.7}$$

Using Proposition 3.1 we rewrite the third equation in (13.7) as

$$\mathbf{D} = -\frac{1}{\gamma_0}\nabla_\mathbf{x}\Psi_1 - \frac{1}{\gamma_0 c}\left(\frac{\partial \mathbf{A}}{\partial t} - \text{curl}_\mathbf{x}(\tilde{\mathbf{u}}\times\mathbf{A}) + (\text{div}_\mathbf{x}\mathbf{A})\,\tilde{\mathbf{u}} + \left(d_\mathbf{x}\tilde{\mathbf{u}} + \{d_\mathbf{x}\tilde{\mathbf{u}}\}^T\right)\cdot\mathbf{A} - (\text{div}_\mathbf{x}\tilde{\mathbf{u}})\,\mathbf{A}\right). \tag{13.8}$$

Then by (13.8), (13.7) and (13.1) we have

$$\text{div}_\mathbf{x}\left\{\frac{1}{\gamma_0 c}\left(\frac{\partial \mathbf{A}}{\partial t} - \tilde{\mathbf{u}}\times\text{curl}_\mathbf{x}\mathbf{A} + \nabla_\mathbf{x}(\mathbf{A}\cdot\tilde{\mathbf{u}})\right)\right\} + \text{div}_\mathbf{x}\left\{\frac{1}{\gamma_0}\nabla_\mathbf{x}\Psi_1\right\} = -4\pi\rho \tag{13.9}$$

and

$$\operatorname{curl}_{\mathbf{x}}\left\{\kappa_0 \operatorname{curl}_{\mathbf{x}}\mathbf{A} - \frac{1}{\gamma_0 c}\tilde{\mathbf{u}} \times \left(\nabla_{\mathbf{x}}\Psi_1 + \frac{1}{c}\left(\frac{\partial \mathbf{A}}{\partial t} - \tilde{\mathbf{u}} \times \operatorname{curl}_{\mathbf{x}}\mathbf{A} + \nabla_{\mathbf{x}}(\mathbf{A}\cdot\tilde{\mathbf{u}})\right)\right)\right\}$$
$$= \frac{4\pi}{c}\mathbf{j} + \frac{\partial}{\partial t}\left\{\frac{1}{\gamma_0 c}\left(-\nabla_{\mathbf{x}}\Psi_1 - \frac{1}{c}\left(\frac{\partial \mathbf{A}}{\partial t} - \tilde{\mathbf{u}} \times \operatorname{curl}_{\mathbf{x}}\mathbf{A} + \nabla_{\mathbf{x}}(\mathbf{A}\cdot\tilde{\mathbf{u}})\right)\right)\right\}. \quad (13.10)$$

In particular, by (13.9) and (13.10) we infer

$$\frac{1}{\kappa_0}\operatorname{curl}_{\mathbf{x}}\{\kappa_0 \operatorname{curl}_{\mathbf{x}}\mathbf{A}\} = \frac{4\pi}{\kappa_0 c}(\mathbf{j} - \rho\tilde{\mathbf{u}})$$
$$-\frac{1}{\kappa_0}\left(\frac{\partial}{\partial t}\left\{\frac{1}{\gamma_0 c}\nabla_{\mathbf{x}}\Psi_1\right\} - \operatorname{curl}_{\mathbf{x}}\left\{\tilde{\mathbf{u}} \times \left\{\frac{1}{\gamma_0 c}\nabla_{\mathbf{x}}\Psi_1\right\}\right\} + \left(\operatorname{div}_{\mathbf{x}}\left\{\frac{1}{\gamma_0 c}\nabla_{\mathbf{x}}\Psi_1\right\}\right)\tilde{\mathbf{u}}\right)$$
$$-\frac{1}{\kappa_0}\frac{\partial}{\partial t}\left\{\frac{1}{\gamma_0 c^2}\left(\frac{\partial \mathbf{A}}{\partial t} - \tilde{\mathbf{u}} \times \operatorname{curl}_{\mathbf{x}}\mathbf{A} + \nabla_{\mathbf{x}}(\mathbf{A}\cdot\tilde{\mathbf{u}})\right)\right\}$$
$$+\frac{1}{\kappa_0}\operatorname{curl}_{\mathbf{x}}\left\{\tilde{\mathbf{u}} \times \left(\frac{1}{\gamma_0 c^2}\left(\frac{\partial \mathbf{A}}{\partial t} - \tilde{\mathbf{u}} \times \operatorname{curl}_{\mathbf{x}}\mathbf{A} + \nabla_{\mathbf{x}}(\mathbf{A}\cdot\tilde{\mathbf{u}})\right)\right)\right\}$$
$$-\frac{1}{\kappa_0}\left(\operatorname{div}_{\mathbf{x}}\left\{\frac{1}{\gamma_0 c^2}\left(\frac{\partial \mathbf{A}}{\partial t} - \tilde{\mathbf{u}} \times \operatorname{curl}_{\mathbf{x}}\mathbf{A} + \nabla_{\mathbf{x}}(\mathbf{A}\cdot\tilde{\mathbf{u}})\right)\right\}\right)\tilde{\mathbf{u}}. \quad (13.11)$$

Then, by (2.11), (2.14) and (2.15) we rewrite (13.9) and (13.11) as

$$\operatorname{div}_{\mathbf{x}}\left\{\frac{1}{\gamma_0 c}\left(\frac{\partial \mathbf{A}}{\partial t} - \operatorname{curl}_{\mathbf{x}}\{\tilde{\mathbf{u}} \times \mathbf{A}\} + (\operatorname{div}_{\mathbf{x}}\mathbf{A})\,\tilde{\mathbf{u}}\right)\right\} + \operatorname{div}_{\mathbf{x}}\left\{\frac{1}{\gamma_0}\nabla_{\mathbf{x}}\Psi_1\right\}$$
$$+\operatorname{div}_{\mathbf{x}}\left\{\frac{1}{\gamma_0 c}\left(d_{\mathbf{x}}\tilde{\mathbf{u}} + \{d_{\mathbf{x}}\tilde{\mathbf{u}}\}^T\right)\cdot\mathbf{A} - \frac{1}{\gamma_0 c}(\operatorname{div}_{\mathbf{x}}\tilde{\mathbf{u}})\,\mathbf{A}\right\} = -4\pi\rho \quad (13.12)$$

and

$$\frac{1}{\kappa_0}\operatorname{curl}_{\mathbf{x}}\{\kappa_0 \operatorname{curl}_{\mathbf{x}}\mathbf{A}\} - \frac{4\pi}{\kappa_0 c}(\mathbf{j} - \rho\tilde{\mathbf{u}})$$
$$+\frac{1}{\kappa_0}\left(\frac{\partial}{\partial t}\left\{\frac{1}{\gamma_0 c}\nabla_{\mathbf{x}}\Psi_1\right\} - \tilde{\mathbf{u}} \times \operatorname{curl}_{\mathbf{x}}\left\{\frac{1}{\gamma_0 c}\nabla_{\mathbf{x}}\Psi_1\right\} + \nabla_{\mathbf{x}}\left(\left\{\frac{1}{\gamma_0 c}\nabla_{\mathbf{x}}\Psi_1\right\}\cdot\tilde{\mathbf{u}}\right)\right)$$
$$-\frac{1}{\kappa_0}\left(\left(d_{\mathbf{x}}\tilde{\mathbf{u}} + \{d_{\mathbf{x}}\tilde{\mathbf{u}}\}^T\right)\cdot\left\{\frac{1}{\gamma_0 c}\nabla_{\mathbf{x}}\Psi_1\right\} - (\operatorname{div}_{\mathbf{x}}\tilde{\mathbf{u}})\left\{\frac{1}{\gamma_0 c}\nabla_{\mathbf{x}}\Psi_1\right\}\right)$$
$$= -\frac{1}{\kappa_0}\frac{\partial}{\partial t}\left\{\frac{1}{\gamma_0 c^2}\left(\frac{\partial \mathbf{A}}{\partial t} - \tilde{\mathbf{u}} \times \operatorname{curl}_{\mathbf{x}}\mathbf{A} + \nabla_{\mathbf{x}}(\mathbf{A}\cdot\tilde{\mathbf{u}})\right)\right\}$$
$$-\frac{1}{\kappa_0}\operatorname{div}_{\mathbf{x}}\left\{\left(\frac{1}{\gamma_0 c^2}\left(\frac{\partial \mathbf{A}}{\partial t} - \tilde{\mathbf{u}} \times \operatorname{curl}_{\mathbf{x}}\mathbf{A} + \nabla_{\mathbf{x}}(\mathbf{A}\cdot\tilde{\mathbf{u}})\right)\right)\otimes\tilde{\mathbf{u}}\right\}$$
$$+\frac{1}{\kappa_0}\left(d_{\mathbf{x}}\tilde{\mathbf{u}}\cdot\left\{\frac{1}{\gamma_0 c^2}\left(\frac{\partial \mathbf{A}}{\partial t} - \tilde{\mathbf{u}} \times \operatorname{curl}_{\mathbf{x}}\mathbf{A} + \nabla_{\mathbf{x}}(\mathbf{A}\cdot\tilde{\mathbf{u}})\right)\right\}\right)$$
$$= -\frac{1}{\kappa_0}\frac{\partial}{\partial t}\left\{\frac{1}{\gamma_0 c^2}\left(\frac{\partial \mathbf{A}}{\partial t} + d_{\mathbf{x}}\mathbf{A}\cdot\tilde{\mathbf{u}} + \{d_{\mathbf{x}}\tilde{\mathbf{u}}\}^T\cdot\mathbf{A}\right)\right\}$$

$$- \frac{1}{\kappa_0} \operatorname{div}_{\mathbf{x}} \left\{ \left(\frac{1}{\gamma_0 c^2} \left(\frac{\partial \mathbf{A}}{\partial t} + d_{\mathbf{x}} \mathbf{A} \cdot \tilde{\mathbf{u}} + \{d_{\mathbf{x}} \tilde{\mathbf{u}}\}^T \cdot \mathbf{A} \right) \right) \otimes \tilde{\mathbf{u}} \right\}$$
$$+ \frac{1}{\kappa_0} \left(d_{\mathbf{x}} \tilde{\mathbf{u}} \cdot \left\{ \frac{1}{\gamma_0 c^2} \left(\frac{\partial \mathbf{A}}{\partial t} + d_{\mathbf{x}} \mathbf{A} \cdot \tilde{\mathbf{u}} + \{d_{\mathbf{x}} \tilde{\mathbf{u}}\}^T \cdot \mathbf{A} \right) \right\} \right). \quad (13.13)$$

Next, up to the end of this section we study Eq. (13.1) or equivalently (13.12), (13.13) in domains where we assume that the coefficients $\gamma_0 \neq 0$ and $\kappa_0 \neq 0$ vary sufficiently slow on the place and time and thus their spatial and temporal derivatives are negligible with respect to other terms. Then, we rewrite (13.12) and (13.13) as

$$\left(\frac{\partial}{\partial t} \left\{ \frac{1}{c} \operatorname{div}_{\mathbf{x}} \mathbf{A} \right\} + \operatorname{div}_{\mathbf{x}} \left\{ \frac{1}{c} (\operatorname{div}_{\mathbf{x}} \mathbf{A}) \tilde{\mathbf{u}} \right\} \right) + \Delta_{\mathbf{x}} \Psi_1$$
$$+ \operatorname{div}_{\mathbf{x}} \left\{ \frac{1}{c} \left(d_{\mathbf{x}} \tilde{\mathbf{u}} + \{d_{\mathbf{x}} \tilde{\mathbf{u}}\}^T \right) \cdot \mathbf{A} - \frac{1}{c} (\operatorname{div}_{\mathbf{x}} \tilde{\mathbf{u}}) \mathbf{A} \right\} = -4\pi \gamma_0 \rho \quad (13.14)$$

and

$$- \Delta_{\mathbf{x}} \mathbf{A} = \frac{4\pi}{\kappa_0 c} (\mathbf{j} - \rho \tilde{\mathbf{u}}) - \nabla_{\mathbf{x}} \left(\frac{1}{\kappa_0 \gamma_0 c} \left(\frac{\partial \Psi_1}{\partial t} + \tilde{\mathbf{u}} \cdot \nabla_{\mathbf{x}} \Psi_1 \right) + (\operatorname{div}_{\mathbf{x}} \mathbf{A}) \right)$$
$$+ \left(\left(d_{\mathbf{x}} \tilde{\mathbf{u}} + \{d_{\mathbf{x}} \tilde{\mathbf{u}}\}^T \right) \cdot \left\{ \frac{1}{\kappa_0 \gamma_0 c} \nabla_{\mathbf{x}} \Psi_1 \right\} - (\operatorname{div}_{\mathbf{x}} \tilde{\mathbf{u}}) \left\{ \frac{1}{\kappa_0 \gamma_0 c} \nabla_{\mathbf{x}} \Psi_1 \right\} \right)$$
$$- \frac{\partial}{\partial t} \left\{ \frac{1}{\gamma_0 \kappa_0 c^2} \left(\frac{\partial \mathbf{A}}{\partial t} + d_{\mathbf{x}} \mathbf{A} \cdot \tilde{\mathbf{u}} + \{d_{\mathbf{x}} \tilde{\mathbf{u}}\}^T \cdot \mathbf{A} \right) \right\}$$
$$- \operatorname{div}_{\mathbf{x}} \left\{ \left(\frac{1}{\gamma_0 \kappa_0 c^2} \left(\frac{\partial \mathbf{A}}{\partial t} + d_{\mathbf{x}} \mathbf{A} \cdot \tilde{\mathbf{u}} + \{d_{\mathbf{x}} \tilde{\mathbf{u}}\}^T \cdot \mathbf{A} \right) \right) \otimes \tilde{\mathbf{u}} \right\}$$
$$+ \left(d_{\mathbf{x}} \tilde{\mathbf{u}} \cdot \left\{ \frac{1}{\gamma_0 \kappa_0 c^2} \left(\frac{\partial \mathbf{A}}{\partial t} + d_{\mathbf{x}} \mathbf{A} \cdot \tilde{\mathbf{u}} + \{d_{\mathbf{x}} \tilde{\mathbf{u}}\}^T \cdot \mathbf{A} \right) \right\} \right). \quad (13.15)$$

Alternatively, we rewrite (13.15) as

$$- \Delta_{\mathbf{x}} \mathbf{A} = \frac{4\pi}{\kappa_0 c} (\mathbf{j} - \rho \tilde{\mathbf{u}}) - \nabla_{\mathbf{x}} (\operatorname{div}_{\mathbf{x}} \mathbf{A})$$
$$- \frac{1}{\gamma_0 \kappa_0 c} \left(\frac{\partial}{\partial t} \{\nabla_{\mathbf{x}} \Psi_1\} - \operatorname{curl}_{\mathbf{x}} \{\tilde{\mathbf{u}} \times \nabla_{\mathbf{x}} \Psi_1\} + (\Delta_{\mathbf{x}} \Psi_1) \tilde{\mathbf{u}} \right)$$
$$- \frac{\partial}{\partial t} \left\{ \frac{1}{\gamma_0 \kappa_0 c^2} \left(\frac{\partial \mathbf{A}}{\partial t} + d_{\mathbf{x}} \mathbf{A} \cdot \tilde{\mathbf{u}} + \{d_{\mathbf{x}} \tilde{\mathbf{u}}\}^T \cdot \mathbf{A} \right) \right\}$$
$$- \operatorname{div}_{\mathbf{x}} \left\{ \left(\frac{1}{\gamma_0 \kappa_0 c^2} \left(\frac{\partial \mathbf{A}}{\partial t} + d_{\mathbf{x}} \mathbf{A} \cdot \tilde{\mathbf{u}} + \{d_{\mathbf{x}} \tilde{\mathbf{u}}\}^T \cdot \mathbf{A} \right) \right) \otimes \tilde{\mathbf{u}} \right\}$$
$$+ \left(d_{\mathbf{x}} \tilde{\mathbf{u}} \cdot \left\{ \frac{1}{\gamma_0 \kappa_0 c^2} \left(\frac{\partial \mathbf{A}}{\partial t} + d_{\mathbf{x}} \mathbf{A} \cdot \tilde{\mathbf{u}} + \{d_{\mathbf{x}} \tilde{\mathbf{u}}\}^T \cdot \mathbf{A} \right) \right\} \right). \quad (13.16)$$

Next if we assume the following calibration of the potentials:

$$\operatorname{div}_{\mathbf{x}} \mathbf{A} = 0, \tag{13.17}$$

then by (13.17), (13.14), (13.16) and (2.11) we have

$$-\Delta_{\mathbf{x}}\Psi_1 = 4\pi\gamma_0\rho + \operatorname{div}_{\mathbf{x}}\left\{\frac{1}{c}\left(d_{\mathbf{x}}\tilde{\mathbf{u}} + \{d_{\mathbf{x}}\tilde{\mathbf{u}}\}^T\right)\cdot\mathbf{A} - \frac{1}{c}\left(\operatorname{div}_{\mathbf{x}}\tilde{\mathbf{u}}\right)\mathbf{A}\right\}, \tag{13.18}$$

and

$$\begin{aligned}
-\Delta_{\mathbf{x}}\mathbf{A} = \frac{4\pi}{\kappa_0 c}\Bigg(\mathbf{j} &- \frac{1}{4\pi\gamma_0}\left(\frac{\partial}{\partial t}\{\nabla_{\mathbf{x}}\Psi_1\} - \operatorname{curl}_{\mathbf{x}}\{\tilde{\mathbf{u}}\times\nabla_{\mathbf{x}}\Psi_1\}\right)\Bigg)\\
&+ \frac{1}{\gamma_0\kappa_0 c^2}\left(\operatorname{div}_{\mathbf{x}}\left\{\left(d_{\mathbf{x}}\tilde{\mathbf{u}} + \{d_{\mathbf{x}}\tilde{\mathbf{u}}\}^T\right)\cdot\mathbf{A} - \left(\operatorname{div}_{\mathbf{x}}\tilde{\mathbf{u}}\right)\mathbf{A}\right\}\right)\tilde{\mathbf{u}}\\
&- \frac{\partial}{\partial t}\left\{\frac{1}{\gamma_0\kappa_0 c^2}\left(\frac{\partial\mathbf{A}}{\partial t} + d_{\mathbf{x}}\mathbf{A}\cdot\tilde{\mathbf{u}} + \{d_{\mathbf{x}}\tilde{\mathbf{u}}\}^T\cdot\mathbf{A}\right)\right\}\\
&- \operatorname{div}_{\mathbf{x}}\left\{\left(\frac{1}{\gamma_0\kappa_0 c^2}\left(\frac{\partial\mathbf{A}}{\partial t} + d_{\mathbf{x}}\mathbf{A}\cdot\tilde{\mathbf{u}} + \{d_{\mathbf{x}}\tilde{\mathbf{u}}\}^T\cdot\mathbf{A}\right)\right)\otimes\tilde{\mathbf{u}}\right\}\\
&+ \left(d_{\mathbf{x}}\tilde{\mathbf{u}}\cdot\left\{\frac{1}{\gamma_0\kappa_0 c^2}\left(\frac{\partial\mathbf{A}}{\partial t} + d_{\mathbf{x}}\mathbf{A}\cdot\tilde{\mathbf{u}} + \{d_{\mathbf{x}}\tilde{\mathbf{u}}\}^T\cdot\mathbf{A}\right)\right\}\right).
\end{aligned} \tag{13.19}$$

On the other hand, if we assume the following alternative calibration of the potentials:

$$\frac{1}{\kappa_0\gamma_0 c}\left(\frac{\partial\Psi_1}{\partial t} + \tilde{\mathbf{u}}\cdot\nabla_{\mathbf{x}}\Psi_1\right) + \operatorname{div}_{\mathbf{x}}\mathbf{A} = 0, \tag{13.20}$$

then by (13.14), (13.15) we have

$$\begin{aligned}
&\left(\frac{\partial}{\partial t}\left\{\frac{1}{\kappa_0\gamma_0 c^2}\left(\frac{\partial\Psi_1}{\partial t} + \tilde{\mathbf{u}}\cdot\nabla_{\mathbf{x}}\Psi_1\right)\right\} + \operatorname{div}_{\mathbf{x}}\left\{\frac{1}{\kappa_0\gamma_0 c^2}\left(\frac{\partial\Psi_1}{\partial t} + \tilde{\mathbf{u}}\cdot\nabla_{\mathbf{x}}\Psi_1\right)\tilde{\mathbf{u}}\right\}\right)\\
&\quad - \Delta_{\mathbf{x}}\Psi_1 = 4\pi\gamma_0\rho + \operatorname{div}_{\mathbf{x}}\left\{\frac{1}{c}\left(d_{\mathbf{x}}\tilde{\mathbf{u}} + \{d_{\mathbf{x}}\tilde{\mathbf{u}}\}^T\right)\cdot\mathbf{A} - \frac{1}{c}\left(\operatorname{div}_{\mathbf{x}}\tilde{\mathbf{u}}\right)\mathbf{A}\right\},
\end{aligned} \tag{13.21}$$

and

$$
\begin{aligned}
-\Delta_{\mathbf{x}}\mathbf{A} = &\frac{4\pi}{\kappa_0 c}\left(\mathbf{j}-\rho\tilde{\mathbf{u}}\right)\\
&+\Bigg(\left(d_{\mathbf{x}}\tilde{\mathbf{u}}+\{d_{\mathbf{x}}\tilde{\mathbf{u}}\}^T\right)\cdot\left\{\frac{1}{\kappa_0\gamma_0 c}\nabla_{\mathbf{x}}\Psi_1\right\}-(\mathrm{div}_{\mathbf{x}}\,\tilde{\mathbf{u}})\left\{\frac{1}{\kappa_0\gamma_0 c}\nabla_{\mathbf{x}}\Psi_1\right\}\Bigg)\\
&-\frac{\partial}{\partial t}\left\{\frac{1}{\gamma_0\kappa_0 c^2}\left(\frac{\partial \mathbf{A}}{\partial t}+d_{\mathbf{x}}\mathbf{A}\cdot\tilde{\mathbf{u}}+\{d_{\mathbf{x}}\tilde{\mathbf{u}}\}^T\cdot\mathbf{A}\right)\right\}\\
&-\mathrm{div}_{\mathbf{x}}\left\{\left(\frac{1}{\gamma_0\kappa_0 c^2}\left(\frac{\partial \mathbf{A}}{\partial t}+d_{\mathbf{x}}\mathbf{A}\cdot\tilde{\mathbf{u}}+\{d_{\mathbf{x}}\tilde{\mathbf{u}}\}^T\cdot\mathbf{A}\right)\right)\otimes\tilde{\mathbf{u}}\right\}\\
&+\left(d_{\mathbf{x}}\tilde{\mathbf{u}}\cdot\left\{\frac{1}{\gamma_0\kappa_0 c^2}\left(\frac{\partial \mathbf{A}}{\partial t}+d_{\mathbf{x}}\mathbf{A}\cdot\tilde{\mathbf{u}}+\{d_{\mathbf{x}}\tilde{\mathbf{u}}\}^T\cdot\mathbf{A}\right)\right\}\right). \qquad (13.22)
\end{aligned}
$$

Next, from now we assume that $\tilde{\mathbf{u}}$ varies sufficiently slowly in space and time variables, so that the following approximation is valid:

$$
\frac{\left|d_{\mathbf{x}}\tilde{\mathbf{u}}+\{d_{\mathbf{x}}\tilde{\mathbf{u}}\}^T\right|}{c} \ll \frac{|\nabla_{\mathbf{x}}\Psi_1|}{|\mathbf{A}|+|\Psi_1|}+\frac{|\gamma_0\kappa_0||\Delta_{\mathbf{x}}\mathbf{A}|}{|d_{\mathbf{x}}\mathbf{A}|+|\nabla_{\mathbf{x}}\Psi_1|}. \qquad (13.23)
$$

In particular, if in some Cartesian coordinate system we have both

$$
\frac{\left|d_{\mathbf{x}}\tilde{\mathbf{u}}+\{d_{\mathbf{x}}\tilde{\mathbf{u}}\}^T\right|^2}{|\tilde{\mathbf{u}}|^2} \ll \left(\frac{|\nabla_{\mathbf{x}}\Psi_1|}{|\mathbf{A}|+|\Psi_1|}\right)^2+\left(\frac{|\gamma_0\kappa_0||\Delta_{\mathbf{x}}\mathbf{A}|}{|d_{\mathbf{x}}\mathbf{A}|+|\nabla_{\mathbf{x}}\Psi_1|}\right)^2, \qquad (13.24)
$$

and

$$
\frac{|\tilde{\mathbf{u}}|^2}{c^2} \ll 1, \qquad (13.25)
$$

then (13.23) indeed holds! Furthermore, taking into the account (13.23), under the calibration (13.17), we rewrite (13.18) and (13.19) as

$$
-\Delta_{\mathbf{x}}\Psi_1 \approx 4\pi\gamma_0\rho, \qquad (13.26)
$$

and

$$
\begin{aligned}
-\Delta_{\mathbf{x}}\mathbf{A} \approx &\frac{4\pi}{\kappa_0 c}\left(\mathbf{j}-\frac{1}{4\pi\gamma_0}\left(\frac{\partial}{\partial t}\{\nabla_{\mathbf{x}}\Psi_1\}-\mathrm{curl}_{\mathbf{x}}\{\tilde{\mathbf{u}}\times\nabla_{\mathbf{x}}\Psi_1\}\right)\right)\\
&-\frac{\partial}{\partial t}\left\{\frac{1}{\gamma_0\kappa_0 c^2}\left(\frac{\partial \mathbf{A}}{\partial t}+d_{\mathbf{x}}\mathbf{A}\cdot\tilde{\mathbf{u}}+\{d_{\mathbf{x}}\tilde{\mathbf{u}}\}^T\cdot\mathbf{A}\right)\right\}\\
&-\mathrm{div}_{\mathbf{x}}\left\{\left(\frac{1}{\gamma_0\kappa_0 c^2}\left(\frac{\partial \mathbf{A}}{\partial t}+d_{\mathbf{x}}\mathbf{A}\cdot\tilde{\mathbf{u}}+\{d_{\mathbf{x}}\tilde{\mathbf{u}}\}^T\cdot\mathbf{A}\right)\right)\otimes\tilde{\mathbf{u}}\right\}\\
&+\left(d_{\mathbf{x}}\tilde{\mathbf{u}}\cdot\left\{\frac{1}{\gamma_0\kappa_0 c^2}\left(\frac{\partial \mathbf{A}}{\partial t}+d_{\mathbf{x}}\mathbf{A}\cdot\tilde{\mathbf{u}}+\{d_{\mathbf{x}}\tilde{\mathbf{u}}\}^T\cdot\mathbf{A}\right)\right\}\right). \qquad (13.27)
\end{aligned}
$$

Note that, using Proposition 3.1 we deduce that the approximate Eqs. (13.26) and (13.27) are still invariant under the change of inertial or non-inertial Cartesian coordinate system, provided that **A** is a proper vector field and Ψ_1 is a proper scalar field.

On the other hand, taking into the account (13.23), under the calibration (13.20), we rewrite (13.21) and (13.22) as

$$\frac{1}{\kappa_0\gamma_0 c^2}\left(\frac{\partial}{\partial t}\left(\frac{\partial\Psi_1}{\partial t}+\tilde{\mathbf{u}}\cdot\nabla_{\mathbf{x}}\Psi_1\right)+\operatorname{div}_{\mathbf{x}}\left\{\left(\frac{\partial\Psi_1}{\partial t}+\tilde{\mathbf{u}}\cdot\nabla_{\mathbf{x}}\Psi_1\right)\tilde{\mathbf{u}}\right\}\right)-\Delta_{\mathbf{x}}\Psi_1\approx 4\pi\gamma_0\rho. \tag{13.28}$$

and

$$\begin{aligned}-\Delta_{\mathbf{x}}\mathbf{A}\approx\frac{4\pi}{\kappa_0 c}(\mathbf{j}-\rho\tilde{\mathbf{u}})-\frac{\partial}{\partial t}\left\{\frac{1}{\gamma_0\kappa_0 c^2}\left(\frac{\partial\mathbf{A}}{\partial t}+d_{\mathbf{x}}\mathbf{A}\cdot\tilde{\mathbf{u}}+\{d_{\mathbf{x}}\tilde{\mathbf{u}}\}^T\cdot\mathbf{A}\right)\right\}\\ -\operatorname{div}_{\mathbf{x}}\left\{\left(\frac{1}{\gamma_0\kappa_0 c^2}\left(\frac{\partial\mathbf{A}}{\partial t}+d_{\mathbf{x}}\mathbf{A}\cdot\tilde{\mathbf{u}}+\{d_{\mathbf{x}}\tilde{\mathbf{u}}\}^T\cdot\mathbf{A}\right)\right)\otimes\tilde{\mathbf{u}}\right\}\\ +\left(d_{\mathbf{x}}\tilde{\mathbf{u}}\cdot\left\{\frac{1}{\gamma_0\kappa_0 c^2}\left(\frac{\partial\mathbf{A}}{\partial t}+d_{\mathbf{x}}\mathbf{A}\cdot\tilde{\mathbf{u}}+\{d_{\mathbf{x}}\tilde{\mathbf{u}}\}^T\cdot\mathbf{A}\right)\right\}\right).\end{aligned} \tag{13.29}$$

Again note that, using Proposition 3.1 we deduce that the approximate Eqs. (13.28) and (13.29) are still invariant under the change of inertial or non-inertial Cartesian coordinate system, provided that **A** is a proper vector field and Ψ_1 is a proper scalar field.

In particular, assume that in some Cartesian coordinate system (∗) we have the following stronger than (13.23) approximation:

$$\frac{|d_{\mathbf{x}}\tilde{\mathbf{u}}|}{c}\ll\frac{|\gamma_0\kappa_0||\Delta_{\mathbf{x}}\mathbf{A}|+\frac{1}{c}\left|d_{\mathbf{x}}\left\{\frac{\partial\mathbf{A}}{\partial t}\right\}\right|+\frac{1}{c^2}\left|\frac{\partial^2\mathbf{A}}{\partial t^2}\right|+|\gamma_0\kappa_0||\nabla^2_{\mathbf{x}}\Psi_1|+\frac{1}{c}\left|\frac{\partial}{\partial t}(\nabla_{\mathbf{x}}\Psi_1)\right|+\frac{1}{c^2}\left|\frac{\partial^2\Psi_1}{\partial t^2}\right|}{|d_{\mathbf{x}}\mathbf{A}|+\frac{1}{c}\left|\frac{\partial\mathbf{A}}{\partial t}\right|+|\nabla_{\mathbf{x}}\Psi_1|+\frac{1}{c}\left|\frac{\partial\Psi_1}{\partial t}\right|}, \tag{13.30}$$

i.e. the field $\tilde{\mathbf{u}}$ changes in space much slower than $(\Psi_1, \mathbf{A})$. Estimation (13.30) holds especially good for the electromagnetic waves of high frequency, for example, for the visible light. However, (13.30) is still well for almost every electromagnetic field we meet in the common life, except probably the magnetic field of the Earth. Then, by (13.28), (13.29) and (13.30) we can write the further approximating equations:

$$\left(\frac{\partial}{\partial t}\left\{\frac{1}{c_0^2}\left(\frac{\partial\Psi_1}{\partial t}+\tilde{\mathbf{u}}\cdot\nabla_{\mathbf{x}}\Psi_1\right)\right\}+\operatorname{div}_{\mathbf{x}}\left\{\frac{1}{c_0^2}\left(\frac{\partial\Psi_1}{\partial t}+\tilde{\mathbf{u}}\cdot\nabla_{\mathbf{x}}\Psi_1\right)\tilde{\mathbf{u}}\right\}\right)-\Delta_{\mathbf{x}}\Psi_1\approx 4\pi\gamma_0\rho, \tag{13.31}$$

and

$$\left(\frac{\partial}{\partial t}\left\{\frac{1}{c_0^2}\left(\frac{\partial \mathbf{A}}{\partial t}+d_{\mathbf{x}}\mathbf{A}\cdot\tilde{\mathbf{u}}\right)\right\}+\mathrm{div}_{\mathbf{x}}\left\{\frac{1}{c_0^2}\left(\frac{\partial \mathbf{A}}{\partial t}+d_{\mathbf{x}}\mathbf{A}\cdot\tilde{\mathbf{u}}\right)\otimes\tilde{\mathbf{u}}\right\}\right)$$
$$-\Delta_{\mathbf{x}}\mathbf{A}\approx\frac{4\pi}{\kappa_0 c}\left(\mathbf{j}-\rho\tilde{\mathbf{u}}\right), \quad (13.32)$$

where the scalar quantity c_0, defined by

$$c_0=c\sqrt{\kappa_0\gamma_0}, \quad (13.33)$$

is called speed of light in the medium. Note that, although the approximate Eqs. (13.31) and (13.32) are invariant under the Galilean transformation, they are not invariant under the more general change of non-inertial Cartesian coordinate system. However, (13.31) and (13.32) are more convenient than (13.28) and (13.29), since the scalar potential Ψ_1 and each of the three scalar components of the vector potential $\mathbf{A}$ in (13.31) and (13.32) satisfy four decoupled equations of the same type that differ only by the right parts. On the other hand, if we consider some three proper vector fields $\mathbf{e}_1:=\mathbf{e}_1(\mathbf{x},t)$, $\mathbf{e}_2:=\mathbf{e}_2(\mathbf{x},t)$, and $\mathbf{e}_3:=\mathbf{e}_3(\mathbf{x},t)$, which are mutually orthogonal to each other and satisfy the following approximation:

$$\frac{|d_{\mathbf{x}}\mathbf{e}_1|+c_0|\partial_t\mathbf{e}_1|}{|\mathbf{e}_1|}+\frac{|d_{\mathbf{x}}\mathbf{e}_2|+c_0|\partial_t\mathbf{e}_2|}{|\mathbf{e}_2|}+\frac{|d_{\mathbf{x}}\mathbf{e}_3|+c_0|\partial_t\mathbf{e}_3|}{|\mathbf{e}_3|}$$
$$\ll\frac{|\nabla_{\mathbf{x}}\Psi_1|+c_0|\partial_t\Psi_1|+|d_{\mathbf{x}}\mathbf{A}|+c_0|\partial_t\mathbf{A}|}{|\Psi_1|+|\mathbf{A}|} \quad (13.34)$$

in the given before coordinate system $(*)$, i.e. the field $\mathbf{e}_k$ varies in space and time much weaker than $(\Psi_1,\mathbf{A})$, then we may write the alternative to (13.32) and (13.31) approximate equations in the form of four decoupled scalar invariant wave equations of the same type:

$$\frac{\partial}{\partial t}\left\{\frac{1}{c_0^2}\left(\frac{\partial}{\partial t}\left(\mathbf{e}_k\cdot\mathbf{A}\right)+\tilde{\mathbf{u}}\cdot\nabla_{\mathbf{x}}\left(\mathbf{e}_k\cdot\mathbf{A}\right)\right)\right\}$$
$$+\mathrm{div}_{\mathbf{x}}\left\{\frac{1}{c_0^2}\left(\frac{\partial}{\partial t}\left(\mathbf{e}_k\cdot\mathbf{A}\right)+\tilde{\mathbf{u}}\cdot\nabla_{\mathbf{x}}\left(\mathbf{e}_k\cdot\mathbf{A}\right)\right)\tilde{\mathbf{u}}\right\}$$
$$-\Delta_{\mathbf{x}}\left(\mathbf{e}_k\cdot\mathbf{A}\right)\approx\frac{4\pi}{\kappa_0 c}\left(\mathbf{j}-\rho\tilde{\mathbf{u}}\right)\cdot\mathbf{e}_k \qquad \forall k=1,2,3, \quad (13.35)$$

and

$$\left(\frac{\partial}{\partial t}\left\{\frac{1}{c_0^2}\left(\frac{\partial \Psi_1}{\partial t}+\tilde{\mathbf{u}}\cdot\nabla_{\mathbf{x}}\Psi_1\right)\right\}+\mathrm{div}_{\mathbf{x}}\left\{\frac{1}{c_0^2}\left(\frac{\partial \Psi_1}{\partial t}+\tilde{\mathbf{u}}\cdot\nabla_{\mathbf{x}}\Psi_1\right)\tilde{\mathbf{u}}\right\}\right)$$
$$-\Delta_{\mathbf{x}}\Psi_1\approx 4\pi\gamma_0\rho. \quad (13.36)$$

Then, clearly, the new alternative approximate Eqs. (13.35), (13.36) are indeed invariant under the more general change of non-inertial Cartesian coordinate system. So we can use approximate Eqs. (13.35) and (13.36) in the arbitrary Cartesian coordinate system (∗) even if (13.30) and (13.34) are not satisfied in the system (∗), provided that (13.30) and (13.34) are satisfied in another Cartesian system (∗∗).

In the absence of charges and currents (for example, for electromagnetic waves) Eqs. (13.31) and (13.32) become

$$\left(\frac{\partial}{\partial t}\left\{\frac{1}{c_0^2}\left(\frac{\partial \Psi_1}{\partial t}+\tilde{\mathbf{u}}\cdot\nabla_{\mathbf{x}}\Psi_1\right)\right\}+\operatorname{div}_{\mathbf{x}}\left\{\frac{1}{c_0^2}\left(\frac{\partial \Psi_1}{\partial t}+\tilde{\mathbf{u}}\cdot\nabla_{\mathbf{x}}\Psi_1\right)\tilde{\mathbf{u}}\right\}\right) - \Delta_{\mathbf{x}}\Psi_1 = 0, \tag{13.37}$$

and

$$\left(\left\{\frac{1}{c_0^2}\frac{\partial}{\partial t}\left(\frac{\partial \mathbf{A}}{\partial t}+d_{\mathbf{x}}\mathbf{A}\cdot\tilde{\mathbf{u}}\right)\right\}+\operatorname{div}_{\mathbf{x}}\left\{\frac{1}{c_0^2}\left(\frac{\partial \mathbf{A}}{\partial t}+d_{\mathbf{x}}\mathbf{A}\cdot\tilde{\mathbf{u}}\right)\otimes\tilde{\mathbf{u}}\right\}\right) - \Delta_{\mathbf{x}}\mathbf{A} = 0, \tag{13.38}$$

and Eqs. (13.35), (13.36) become

$$\frac{\partial}{\partial t}\left\{\frac{1}{c_0^2}\left(\frac{\partial}{\partial t}(\mathbf{e}_k\cdot\mathbf{A})+\tilde{\mathbf{u}}\cdot\nabla_{\mathbf{x}}(\mathbf{e}_k\cdot\mathbf{A})\right)\right\} + \operatorname{div}_{\mathbf{x}}\left\{\frac{1}{c_0^2}\left(\frac{\partial}{\partial t}(\mathbf{e}_k\cdot\mathbf{A})+\tilde{\mathbf{u}}\cdot\nabla_{\mathbf{x}}(\mathbf{e}_k\cdot\mathbf{A})\right)\tilde{\mathbf{u}}\right\} - \Delta_{\mathbf{x}}(\mathbf{e}_k\cdot\mathbf{A}) \approx 0 \quad \forall k=1,2,3, \tag{13.39}$$

and

$$\left(\frac{\partial}{\partial t}\left\{\frac{1}{c_0^2}\left(\frac{\partial \Psi_1}{\partial t}+\tilde{\mathbf{u}}\cdot\nabla_{\mathbf{x}}\Psi_1\right)\right\}+\operatorname{div}_{\mathbf{x}}\left\{\frac{1}{c_0^2}\left(\frac{\partial \Psi_1}{\partial t}+\tilde{\mathbf{u}}\cdot\nabla_{\mathbf{x}}\Psi_1\right)\tilde{\mathbf{u}}\right\}\right) - \Delta_{\mathbf{x}}\Psi_1 \approx 0. \tag{13.40}$$

Therefore, by (13.7), differentiating (13.37) and (13.38) or (13.39) and (13.40) and further usage of (13.30) and (13.34) gives that if the scalar field $U := U(\mathbf{x}, t)$ is one of any three scalar components of each of the fields **E**, **B**, **D** or **H**, then U satisfies the following approximate scalar equation of the wave type:

$$\left(\frac{\partial}{\partial t}\left\{\frac{1}{c_0^2}\left(\frac{\partial U}{\partial t}+\tilde{\mathbf{u}}\cdot\nabla_{\mathbf{x}}U\right)\right\}+\operatorname{div}_{\mathbf{x}}\left\{\frac{1}{c_0^2}\left(\frac{\partial U}{\partial t}+\tilde{\mathbf{u}}\cdot\nabla_{\mathbf{x}}U\right)\tilde{\mathbf{u}}\right\}\right) - \Delta_{\mathbf{x}}U \approx 0, \tag{13.41}$$

where

$$\tilde{\mathbf{u}} = \left(\gamma_0\kappa_0\mathbf{v} + (1-\gamma_0\kappa_0)\,\mathbf{u}\right). \tag{13.42}$$

Note that the wave equation (13.41) is of the same type as (11.106), (11.115), (11.161) or (11.162), with the only difference that the field $\tilde{\mathbf{u}}$ appear in (13.41) instead of the field $\mathbf{u}_0$ in (11.106), (11.115). (11.161) or (11.162).

13.2 The Case of Quasistationary Electromagnetic Fields

Assume that in the given inertial or non-inertial Cartesian coordinate system $(*)$ the field $\tilde{\mathbf{u}}$ is weak, meaning that at any instant on every point:

$$\frac{1}{\kappa_0\gamma_0}\frac{|\tilde{\mathbf{u}}|^2}{c^2} \ll 1. \tag{13.43}$$

Here $\tilde{\mathbf{u}} = (\gamma_0\kappa_0\mathbf{v} + (1-\gamma_0\kappa_0)\mathbf{u})$ is the speed-like vector field, where $\mathbf{v}$ is a vectorial gravitational potential in the system $(*)$ and $\mathbf{u}$ is the medium velocity. Furthermore, consider quasistationary electromagnetic fields. This means the following: assume that the changes in time of the physical characteristics of the electromagnetic fields become essential after certain interval of time T_e and the changes in space of the physical characteristics of the fields become essential in the spatial landscape L_e. Then we assume that

$$(\kappa_0\gamma_0)\frac{c^2T_e^2}{L_e^2} \gg 1. \tag{13.44}$$

Next assume that we are under the calibration (13.17). Then by (13.43) and (13.44) we rewrite (13.18) and (13.19) as

$$-\Delta_{\mathbf{x}}\Psi_1 = 4\pi\gamma_0\rho + \frac{1}{c}\,\mathrm{div}_{\mathbf{x}}\left\{\left(d_{\mathbf{x}}\tilde{\mathbf{u}} + \{d_{\mathbf{x}}\tilde{\mathbf{u}}\}^T\right)\cdot\mathbf{A} - (\mathrm{div}_{\mathbf{x}}\,\tilde{\mathbf{u}})\,\mathbf{A}\right\}, \tag{13.45}$$

and

$$-\Delta_{\mathbf{x}}\mathbf{A} \approx \frac{4\pi}{\kappa_0 c}(\mathbf{j}-\rho\tilde{\mathbf{u}}) - \frac{1}{\kappa_0\gamma_0 c}\left(\frac{\partial}{\partial t}(\nabla_{\mathbf{x}}\Psi_1) - \mathrm{curl}_{\mathbf{x}}(\tilde{\mathbf{u}}\times\nabla_{\mathbf{x}}\Psi_1) + (\Delta_{\mathbf{x}}\Psi_1)\,\tilde{\mathbf{u}}\right). \tag{13.46}$$

Moreover, by (13.43) and (13.44) we can perform further approximation of (13.46) and we get

$$\begin{aligned}-\Delta_{\mathbf{x}}\mathbf{A} &\approx \frac{4\pi}{\kappa_0 c}(\mathbf{j}-\rho\tilde{\mathbf{u}}) - \frac{1}{\kappa_0\gamma_0 c}\left(\frac{\partial}{\partial t}(\nabla_{\mathbf{x}}\Psi_1) - \mathrm{curl}_{\mathbf{x}}(\tilde{\mathbf{u}}\times\nabla_{\mathbf{x}}\Psi_1) + (\Delta_{\mathbf{x}}\Psi_1)\,\tilde{\mathbf{u}}\right)\\ &\approx \frac{4\pi}{\kappa_0 c}\mathbf{j} - \frac{1}{\kappa_0 c}\left(\frac{\partial}{\partial t}(\nabla_{\mathbf{x}}\psi_0) - \mathrm{curl}_{\mathbf{x}}(\mathbf{v}\times\nabla_{\mathbf{x}}\psi_0)\right),\end{aligned} \tag{13.47}$$

where $\psi_0(\mathbf{x}, t)$ is the classical Coulomb's potential which satisfies

$$-\Delta_{\mathbf{x}}\psi_0 = 4\pi\rho. \tag{13.48}$$

So we rewrite (13.45) and (13.47) as

$$\begin{cases} -\Delta_{\mathbf{x}}\mathbf{A} \approx \frac{4\pi}{\kappa_0 c}\widetilde{\mathbf{j}}, \\ -\Delta_{\mathbf{x}}\Psi_1 = 4\pi\gamma_0\rho + \frac{1}{c}\operatorname{div}_{\mathbf{x}}\left\{\left(d_{\mathbf{x}}\tilde{\mathbf{u}} + \{d_{\mathbf{x}}\tilde{\mathbf{u}}\}^T\right)\cdot\mathbf{A} - (\operatorname{div}_{\mathbf{x}}\tilde{\mathbf{u}})\,\mathbf{A}\right\}, \end{cases} \tag{13.49}$$

where we set the reduced current:

$$\begin{cases} \widetilde{\mathbf{j}} := \mathbf{j} - \frac{1}{4\pi}\frac{\partial}{\partial t}(\nabla_{\mathbf{x}}\psi_0) + \frac{1}{4\pi}\operatorname{curl}_{\mathbf{x}}(\tilde{\mathbf{u}}\times\nabla_{\mathbf{x}}\psi_0), \\ -\Delta_{\mathbf{x}}\psi_0 = 4\pi\rho. \end{cases} \tag{13.50}$$

Note that by the continuum equation of the conservation of charges:

$$\frac{\partial\rho}{\partial t} + \operatorname{div}_{\mathbf{x}}\mathbf{j} = 0, \tag{13.51}$$

the reduced current clearly satisfies

$$\operatorname{div}_{\mathbf{x}}\widetilde{\mathbf{j}} = 0. \tag{13.52}$$

Moreover, by (13.50), we clearly have

$$\widetilde{\mathbf{j}} := (\mathbf{j} - \rho\tilde{\mathbf{u}}) - \frac{1}{4\pi}\left(\frac{\partial}{\partial t}(\nabla_{\mathbf{x}}\psi_0) - \operatorname{curl}_{\mathbf{x}}(\tilde{\mathbf{u}}\times\nabla_{\mathbf{x}}\psi_0) + (\operatorname{div}_{\mathbf{x}}\{\nabla_{\mathbf{x}}\psi_0\})\,\tilde{\mathbf{u}}\right), \tag{13.53}$$

and thus, by (13.53), using Proposition 3.1 we deduce that $\widetilde{\mathbf{j}}$ is a proper vector field. Moreover, the approximate vectorial electromagnetic potential $\mathbf{A}$ from (13.49) clearly satisfies

$$\operatorname{div}_{\mathbf{x}}\mathbf{A} = 0. \tag{13.54}$$

Next, since by (13.6) we have

$$\Psi_1 := \Psi - \frac{1}{c}\mathbf{A}\cdot\tilde{\mathbf{u}}, \tag{13.55}$$

and, since by (13.54), (2.11) and (2.15) we have

$$\begin{aligned} &\operatorname{div}_{\mathbf{x}}\left\{\left(d_{\mathbf{x}}\tilde{\mathbf{u}} + \{d_{\mathbf{x}}\tilde{\mathbf{u}}\}^T\right)\cdot\mathbf{A} - (\operatorname{div}_{\mathbf{x}}\tilde{\mathbf{u}})\,\mathbf{A}\right\} - \Delta_{\mathbf{x}}(\mathbf{A}\cdot\tilde{\mathbf{u}}) = \\ &\quad\operatorname{div}_{\mathbf{x}}\left\{\left(d_{\mathbf{x}}\tilde{\mathbf{u}} + \{d_{\mathbf{x}}\tilde{\mathbf{u}}\}^T\right)\cdot\mathbf{A} - (\operatorname{div}_{\mathbf{x}}\tilde{\mathbf{u}})\,\mathbf{A} - \nabla_{\mathbf{x}}(\mathbf{A}\cdot\tilde{\mathbf{u}})\right\} = \\ &\operatorname{div}_{\mathbf{x}}\{d_{\mathbf{x}}\tilde{\mathbf{u}}\cdot\mathbf{A} - d_{\mathbf{x}}\mathbf{A}\cdot\tilde{\mathbf{u}} + (\operatorname{div}_{\mathbf{x}}\mathbf{A})\,\tilde{\mathbf{u}} - (\operatorname{div}_{\mathbf{x}}\tilde{\mathbf{u}})\,\mathbf{A} - \tilde{\mathbf{u}}\times\operatorname{curl}_{\mathbf{x}}\mathbf{A}\} \\ &= \operatorname{div}_{\mathbf{x}}\{\operatorname{curl}_{\mathbf{x}}(\tilde{\mathbf{u}}\times\mathbf{A}) - \tilde{\mathbf{u}}\times\operatorname{curl}_{\mathbf{x}}\mathbf{A}\} = -\operatorname{div}_{\mathbf{x}}\{\tilde{\mathbf{u}}\times\operatorname{curl}_{\mathbf{x}}\mathbf{A}\}, \end{aligned} \tag{13.56}$$

we rewrite (13.49) as

$$\begin{cases} -\Delta_{\mathbf{x}}\mathbf{A} \approx \frac{4\pi}{\kappa_0 c}\widetilde{\mathbf{j}}, \\ -\Delta_{\mathbf{x}}\Psi = 4\pi\gamma_0\rho - \frac{1}{c}\operatorname{div}_{\mathbf{x}}\left(\tilde{\mathbf{u}} \times \operatorname{curl}_{\mathbf{x}}\mathbf{A}\right), \end{cases} \tag{13.57}$$

where

$$\begin{cases} \widetilde{\mathbf{j}} := \mathbf{j} - \frac{1}{4\pi}\frac{\partial}{\partial t}\left(\nabla_{\mathbf{x}}\psi_0\right) + \frac{1}{4\pi}\operatorname{curl}_{\mathbf{x}}\left(\tilde{\mathbf{u}} \times \nabla_{\mathbf{x}}\psi_0\right), \\ -\Delta_{\mathbf{x}}\psi_0 = 4\pi\rho. \end{cases} \tag{13.58}$$

So in order to find the scalar and the vectorial electromagnetic potentials we just need to solve Laplace equations. Knowing the approximate electromagnetic potentials by (13.4) we can find the approximations of the electromagnetic fields:

$$\begin{cases} \mathbf{B} = \operatorname{curl}_{\mathbf{x}}\mathbf{A} \\ \mathbf{E} = -\nabla_{\mathbf{x}}\Psi - \frac{1}{c}\frac{\partial \mathbf{A}}{\partial t} \\ \mathbf{D} = -\frac{1}{\gamma_0}\nabla_{\mathbf{x}}\Psi - \frac{1}{\gamma_0 c}\frac{\partial \mathbf{A}}{\partial t} + \frac{1}{c\gamma_0}\tilde{\mathbf{u}} \times \operatorname{curl}_{\mathbf{x}}\mathbf{A} \\ \mathbf{H} = \kappa_0 \operatorname{curl}_{\mathbf{x}}\mathbf{A} + \frac{1}{c}\tilde{\mathbf{u}} \times \left(-\frac{1}{\gamma_0}\nabla_{\mathbf{x}}\Psi - \frac{1}{\gamma_0 c}\frac{\partial \mathbf{A}}{\partial t} + \frac{1}{\gamma_0 c}\tilde{\mathbf{u}} \times \operatorname{curl}_{\mathbf{x}}\mathbf{A}\right), \end{cases} \tag{13.59}$$

where Ψ and $\mathbf{A}$ are given by (13.57). Note also that, since $\widetilde{\mathbf{j}}$ is a proper vector field, by Proposition 3.1 we deduce that Eq. (13.49) and thus also Eq. (13.57) are invariant under the change of non-inertial Cartesian coordinate system, provided that $\mathbf{A}$ is a proper vector field and $\Psi_1 = \Psi - \frac{1}{c}\mathbf{A} \cdot \tilde{\mathbf{u}}$ is a proper scalar field. So the approximate solutions in the case of quasistationary fields in a weak gravitational field satisfy the same transformation as the exact solutions of Maxwell Equations. Therefore, if in coordinate system $(*)$ we can use the approximate equations, given by (13.57) and (13.59), then we can use the similar approximation also in coordinate system $(**)$, even in the case when in system $(**)$ (13.43) or (13.44) is not satisfied.

Remark 13.1 The solutions of (13.57) and (13.59) satisfy the following equations:

$$\begin{cases} \operatorname{curl}_{\mathbf{x}}\left(\kappa_0\mathbf{B} + \frac{1}{c}\tilde{\mathbf{u}} \times (-\nabla_{\mathbf{x}}\psi_0)\right) = \frac{4\pi}{c}\mathbf{j} + \frac{1}{c}\frac{\partial(-\nabla_{\mathbf{x}}\psi_0)}{\partial t}, \\ \operatorname{div}_{\mathbf{x}}\mathbf{D} = 4\pi\rho, \\ \operatorname{curl}_{\mathbf{x}}\mathbf{E} + \frac{1}{c}\frac{\partial \mathbf{B}}{\partial t} = 0, \\ \operatorname{div}_{\mathbf{x}}\mathbf{B} = 0 \\ \mathbf{E} = \gamma_0\mathbf{D} - \frac{1}{c}\tilde{\mathbf{u}} \times \mathbf{B} \\ \mathbf{H} = \kappa_0\mathbf{B} + \frac{1}{c}\tilde{\mathbf{u}} \times \mathbf{D}, \\ \tilde{\mathbf{u}} = \left(\gamma_0\kappa_0\mathbf{v} + (1 - \gamma_0\kappa_0)\mathbf{u}\right), \end{cases} \tag{13.60}$$

where ψ_0 was defined by (13.48). Equation (13.60) differ from the original Maxwell equations (13.1) only by neglecting the divergence-free part of the vector field $\mathbf{D}$ on the first equation.

Next, assume that, in addition to the validity of approximation (13.43) and (13.44), the approximation (13.30) also holds. Then we further approximate (13.49) as

$$\begin{cases} -\Delta_{\mathbf{x}}\Psi_1 = 4\pi\gamma_0\rho, \\ -\Delta_{\mathbf{x}}\mathbf{A} \approx \frac{4\pi}{\kappa_0 c}\mathbf{j} - \frac{1}{\kappa_0\gamma_0 c}\left(\frac{\partial}{\partial t}\left(\nabla_{\mathbf{x}}\Psi_1\right) - \mathrm{curl}_{\mathbf{x}}\left(\tilde{\mathbf{u}}\times\nabla_{\mathbf{x}}\Psi_1\right)\right) \\ \Psi = \Psi_1 + \frac{1}{c}\mathbf{A}\cdot\tilde{\mathbf{u}}. \end{cases} \tag{13.61}$$

Moreover, as before, we deduce that Eq. (13.61) is also invariant under the change of non-inertial Cartesian coordinate system. Therefore, as before, if in coordinate system $(*)$ we can use the approximation equations, given by (13.61) then we can use the similar equations also in coordinate system $(**)$, even in the case when in system $(**)$ (13.43), (13.44) or (13.30) is not satisfied.

Finally, assume that we are under the alternative calibration (13.20). Then by (13.43) and (13.44) we rewrite (13.21) and (13.22) as

$$-\Delta_{\mathbf{x}}\Psi_1 \approx 4\pi\gamma_0\rho + \frac{1}{c}\,\mathrm{div}_{\mathbf{x}}\left\{\left(d_{\mathbf{x}}\tilde{\mathbf{u}} + \{d_{\mathbf{x}}\tilde{\mathbf{u}}\}^T\right)\cdot\mathbf{A} - \left(\mathrm{div}_{\mathbf{x}}\,\tilde{\mathbf{u}}\right)\mathbf{A}\right\}, \tag{13.62}$$

and

$$-\Delta_{\mathbf{x}}\mathbf{A} \approx \frac{4\pi}{\kappa_0 c}\left(\mathbf{j} - \rho\tilde{\mathbf{u}}\right) + \frac{1}{\kappa_0\gamma_0 c}\left(\left(d_{\mathbf{x}}\tilde{\mathbf{u}} + \{d_{\mathbf{x}}\tilde{\mathbf{u}}\}^T\right)\cdot\nabla_{\mathbf{x}}\Psi_1 - \left(\mathrm{div}_{\mathbf{x}}\,\tilde{\mathbf{u}}\right)\nabla_{\mathbf{x}}\Psi_1\right). \tag{13.63}$$

Thus if we assume that in addition to the approximation (13.43) and (13.44) the approximation (13.30) also holds, we further approximate (13.62) and (13.63) as

$$\begin{cases} -\Delta_{\mathbf{x}}\Psi_1 \approx 4\pi\gamma_0\rho, \\ -\Delta_{\mathbf{x}}\mathbf{A} \approx \frac{4\pi}{\kappa_0 c}\left(\mathbf{j} - \rho\tilde{\mathbf{u}}\right) \\ \Psi = \Psi_1 + \frac{1}{c}\mathbf{A}\cdot\tilde{\mathbf{u}}. \end{cases} \tag{13.64}$$

Moreover, as before, we deduce that Eq. (13.64) is also invariant under the change of non-inertial Cartesian coordinate system. Therefore, as before, if in coordinate system $(*)$ we can use the approximation equations, given by (13.64) then we can use the similar equations also in coordinate system $(**)$, even in the case when in system $(**)$ (13.43), (13.44) or (13.30) is not satisfied.

13.2.1 Hertz's Notation for Quasistationary Electromagnetic Fields

Again consider in some domain the full system of Maxwell equations in the vacuum or in a medium, without dispersion, having the form (13.1):

$$\begin{cases} \operatorname{curl}_{\mathbf{x}}\mathbf{H} = \frac{4\pi}{c}\mathbf{j} + \frac{1}{c}\frac{\partial \mathbf{D}}{\partial t}, \\ \operatorname{div}_{\mathbf{x}}\mathbf{D} = 4\pi\rho, \\ \operatorname{curl}_{\mathbf{x}}\mathbf{E} + \frac{1}{c}\frac{\partial \mathbf{B}}{\partial t} = 0, \\ \operatorname{div}_{\mathbf{x}}\mathbf{B} = 0 \\ \mathbf{E} = \gamma_0\mathbf{D} - \frac{1}{c}\tilde{\mathbf{u}} \times \mathbf{B} \\ \mathbf{H} = \kappa_0\mathbf{B} + \frac{1}{c}\tilde{\mathbf{u}} \times \mathbf{D}, \\ \tilde{\mathbf{u}} = (\gamma_0\kappa_0\mathbf{v} + (1-\gamma_0\kappa_0)\mathbf{u}). \end{cases} \tag{13.65}$$

Next define the electromagnetic fields in Hertz's notation:

$$\begin{cases} \tilde{\mathbf{E}} := \mathbf{E} + \frac{1}{c}\mathbf{u} \times \mathbf{B}, \\ \tilde{\mathbf{D}} := \frac{1}{\gamma_0}\tilde{\mathbf{E}} = \frac{1}{\gamma_0}\left(\mathbf{E} + \frac{1}{c}\mathbf{u} \times \mathbf{B}\right), \\ \tilde{\mathbf{B}} := \frac{1}{\kappa_0}\left(\mathbf{H} - \frac{1}{c}\mathbf{u} \times \mathbf{D}\right), \\ \tilde{\mathbf{H}} := \kappa_0\tilde{\mathbf{B}} = \left(\mathbf{H} - \frac{1}{c}\mathbf{u} \times \mathbf{D}\right). \end{cases} \tag{13.66}$$

Note here that $\mathbf{u}$ is the velocity field in the medium, thus since this field is absent in vacuum we use the convention that for the vacuum we have $\mathbf{u} := \mathbf{v}$ (and in addition $\tilde{\mathbf{u}} = \mathbf{v}$). In particular, by (13.66) and (13.65) we have

$$\begin{cases} \tilde{\mathbf{E}} = \gamma_0\mathbf{D} + \frac{1}{c}(\mathbf{u} - \tilde{\mathbf{u}}) \times \mathbf{B} = \gamma_0\left(\mathbf{D} + \frac{\kappa_0}{c}(\mathbf{u} - \mathbf{v}) \times \mathbf{B}\right), \\ \tilde{\mathbf{D}} = \mathbf{D} + \frac{\kappa_0}{c}(\mathbf{u} - \mathbf{v}) \times \mathbf{B}, \\ \tilde{\mathbf{B}} = \frac{1}{\kappa_0}\left(\kappa_0\mathbf{B} - \frac{1}{c}(\mathbf{u} - \tilde{\mathbf{u}}) \times \mathbf{D}\right) = \mathbf{B} - \frac{\gamma_0}{c}(\mathbf{u} - \mathbf{v}) \times \mathbf{D}, \\ \tilde{\mathbf{H}} = \kappa_0\mathbf{B} - \frac{\gamma_0\kappa_0}{c}(\mathbf{u} - \mathbf{v}) \times \mathbf{D}. \end{cases} \tag{13.67}$$

Therefore, since $\mathbf{D}$, $\mathbf{B}$, $(\mathbf{u} - \mathbf{v})$ are all proper vector fields, we deduce that $\tilde{\mathbf{E}}$, $\tilde{\mathbf{D}}$, $\tilde{\mathbf{B}}$, $\tilde{\mathbf{H}}$ are all proper vector fields. Furthermore, by (13.66) and (13.67) we rewrite the first four equations in (13.65) as

$$\begin{cases} \operatorname{curl}_{\mathbf{x}}\tilde{\mathbf{H}} = \frac{4\pi}{c}\mathbf{j} + \frac{1}{c}\frac{\partial \mathbf{D}}{\partial t} - \frac{1}{c}\operatorname{curl}_{\mathbf{x}}(\mathbf{u} \times \mathbf{D}), \\ \operatorname{div}_{\mathbf{x}}\tilde{\mathbf{D}} = 4\pi\rho + \frac{1}{c}\operatorname{div}_{\mathbf{x}}(\kappa_0(\mathbf{u} - \mathbf{v}) \times \mathbf{B}), \\ \operatorname{curl}_{\mathbf{x}}\tilde{\mathbf{E}} = -\left(\frac{1}{c}\frac{\partial \mathbf{B}}{\partial t} - \frac{1}{c}\operatorname{curl}_{\mathbf{x}}(\mathbf{u} \times \mathbf{B})\right), \\ \operatorname{div}_{\mathbf{x}}\tilde{\mathbf{B}} = -\frac{1}{c}\operatorname{div}_{\mathbf{x}}(\gamma_0(\mathbf{u} - \mathbf{v}) \times \mathbf{D}). \end{cases} \tag{13.68}$$

Then by inserting (2.5) into (13.68) we deduce

$$\begin{cases} \mathrm{curl}_{\mathbf{x}}\tilde{\mathbf{H}} = \frac{4\pi}{c}\mathbf{j} + \frac{1}{c}\frac{\partial \mathbf{D}}{\partial t} - \frac{1}{c}\,\mathrm{curl}_{\mathbf{x}}\,(\mathbf{u}\times\mathbf{D})\,, \\ \mathrm{div}_{\mathbf{x}}\,\tilde{\mathbf{D}} = 4\pi\rho - \frac{1}{c}\,(\mathbf{u}-\mathbf{v})\cdot\mathrm{curl}_{\mathbf{x}}\,(\kappa_0\mathbf{B}) + \frac{\kappa_0}{c}\,\mathbf{B}\cdot\mathrm{curl}_{\mathbf{x}}\,((\mathbf{u}-\mathbf{v}))\,, \\ \mathrm{curl}_{\mathbf{x}}\tilde{\mathbf{E}} = -\left(\frac{1}{c}\frac{\partial \mathbf{B}}{\partial t} - \frac{1}{c}\,\mathrm{curl}_{\mathbf{x}}\,(\mathbf{u}\times\mathbf{B})\right), \\ \mathrm{div}_{\mathbf{x}}\,\tilde{\mathbf{B}} = \frac{1}{c}\,(\mathbf{u}-\mathbf{v})\cdot\mathrm{curl}_{\mathbf{x}}\,(\gamma_0\mathbf{D}) - \frac{\gamma_0}{c}\,\mathbf{D}\cdot\mathrm{curl}_{\mathbf{x}}\,((\mathbf{u}-\mathbf{v}))\,. \end{cases} \tag{13.69}$$

Therefore, inserting (13.65) into the second and the forth equations in (13.69) we deduce

$$\begin{cases} \mathrm{curl}_{\mathbf{x}}\tilde{\mathbf{H}} = \frac{4\pi}{c}\mathbf{j} + \frac{1}{c}\frac{\partial \mathbf{D}}{\partial t} - \frac{1}{c}\,\mathrm{curl}_{\mathbf{x}}\,(\mathbf{u}\times\mathbf{D})\,, \\ \mathrm{div}_{\mathbf{x}}\,\tilde{\mathbf{D}} = 4\pi\left(\rho - \frac{1}{c^2}\,(\mathbf{u}-\mathbf{v})\cdot\mathbf{j}\right) - \frac{1}{c^2}\,(\mathbf{u}-\mathbf{v})\cdot\left(\frac{\partial \mathbf{D}}{\partial t} - \mathrm{curl}_{\mathbf{x}}\,(\tilde{\mathbf{u}}\times\mathbf{D})\right) \\ +\frac{\kappa_0}{c}\,\mathbf{B}\cdot\mathrm{curl}_{\mathbf{x}}\,((\mathbf{u}-\mathbf{v}))\,, \\ \mathrm{curl}_{\mathbf{x}}\tilde{\mathbf{E}} = -\left(\frac{1}{c}\frac{\partial \mathbf{B}}{\partial t} - \frac{1}{c}\,\mathrm{curl}_{\mathbf{x}}\,(\mathbf{u}\times\mathbf{B})\right), \\ \mathrm{div}_{\mathbf{x}}\,\tilde{\mathbf{B}} = -\frac{1}{c^2}\,(\mathbf{u}-\mathbf{v})\cdot\left(\left(\frac{\partial \mathbf{B}}{\partial t} - \mathrm{curl}_{\mathbf{x}}\,(\tilde{\mathbf{u}}\times\mathbf{B})\right)\right) - \frac{\gamma_0}{c}\,\mathbf{D}\cdot\mathrm{curl}_{\mathbf{x}}\,((\mathbf{u}-\mathbf{v}))\,. \end{cases} \tag{13.70}$$

Next, analogously with (13.43), assume that in the given inertial or non-inertial Cartesian coordinate system (∗) the fields **u**, **v**, $\tilde{\mathbf{u}}$ are weak, meaning that at any instant on every point:

$$\frac{1}{\kappa_0\gamma_0}\left(\frac{|\mathbf{u}|^2}{c^2} + \frac{|\mathbf{v}|^2}{c^2} + \frac{|\tilde{\mathbf{u}}|^2}{c^2}\right) \ll 1, \tag{13.71}$$

and consider quasistationary electromagnetic fields, meaning the following: assume that the changes in time of the physical characteristics of the electromagnetic fields become essential after certain interval of time T_e and the changes in space of the physical characteristics of the fields become essential in the spatial landscape L_e. Then, analogously with (13.44), we assume that

$$(\kappa_0\gamma_0)\frac{c^2 T_e^2}{L_e^2} \gg 1. \tag{13.72}$$

Moreover, assume that we have the following approximation analogous to (13.30): if the changes in space of the physical characteristics of the electromagnetic fields become essential in the spatial landscape L_e and the changes in space of the fields **u**, **v**, $\tilde{\mathbf{u}}$ become essential in the spatial landscape L_u, then we assume

$$L_e \ll L_u, \tag{13.73}$$

i.e. the fields $\mathbf{u}$, $\mathbf{v}$, $\tilde{\mathbf{u}}$ change in space much slower than the electromagnetic fields. Then by (13.71), (13.72) and (13.73), using the second and the third equations in (13.67), we approximate (13.70) as

$$\begin{cases}\operatorname{curl}_{\mathbf{x}}\tilde{\mathbf{H}} \approx \frac{4\pi}{c}\mathbf{j} + \frac{1}{c}\frac{\partial \tilde{\mathbf{D}}}{\partial t} - \frac{1}{c}\operatorname{curl}_{\mathbf{x}}\left(\mathbf{u}\times\tilde{\mathbf{D}}\right),\\ \operatorname{div}_{\mathbf{x}}\tilde{\mathbf{D}} \approx 4\pi\left(\rho - \frac{1}{c^2}(\mathbf{u}-\mathbf{v})\cdot(\mathbf{j}-\rho\mathbf{u})\right),\\ \operatorname{curl}_{\mathbf{x}}\tilde{\mathbf{E}} \approx -\left(\frac{1}{c}\frac{\partial \tilde{\mathbf{B}}}{\partial t} - \frac{1}{c}\operatorname{curl}_{\mathbf{x}}\left(\mathbf{u}\times\tilde{\mathbf{B}}\right)\right),\\ \operatorname{div}_{\mathbf{x}}\tilde{\mathbf{B}} \approx 0.\end{cases} \tag{13.74}$$

Moreover, denoting

$$\begin{cases}\tilde{\rho}^* := \rho - \frac{1}{c^2}(\mathbf{u}-\mathbf{v})\cdot(\mathbf{j}-\rho\mathbf{u}),\\ \tilde{\mathbf{j}}^* = (\mathbf{j}-\rho\mathbf{u}) + \tilde{\rho}^*\mathbf{u},\end{cases} \tag{13.75}$$

again by (13.71), (13.72) and (13.73), using the equality of the conservation of the charge $\frac{\partial\rho}{\partial t} + \operatorname{div}_{\mathbf{x}}\mathbf{j} = 0$ we deduce

$$\begin{cases}\mathbf{j} \approx \tilde{\mathbf{j}}^*,\\ \frac{\partial\tilde{\rho}^*}{\partial t} + \operatorname{div}_{\mathbf{x}}\tilde{\mathbf{j}}^* \approx 0.\end{cases} \tag{13.76}$$

Moreover, by (13.75) we easily obtain that $\tilde{\rho}^*$ is a proper scalar and $\left(\tilde{\mathbf{j}}^* - \tilde{\rho}^*\mathbf{u}\right)$ is a proper vector. Finally, by (13.71), (13.67) and (13.66) we derive the following approximate equations:

$$\begin{cases}\mathbf{B} \approx \tilde{\mathbf{B}} + \frac{1}{\kappa_0}\frac{1}{c}(\mathbf{u}-\mathbf{v})\times\tilde{\mathbf{E}},\\ \mathbf{E} = \tilde{\mathbf{E}} - \frac{1}{c}\mathbf{u}\times\mathbf{B}.\end{cases} \tag{13.77}$$

So, by (13.74), (13.75), (13.76) and (13.66) we deduce the following approximate Maxwell equations in Hertz's form and write them together with (13.77) as the following two sets of equations:

$$\begin{cases}\operatorname{curl}_{\mathbf{x}}\left(\kappa_0\tilde{\mathbf{B}}\right) \approx \frac{4\pi}{c}\tilde{\mathbf{j}}^* + \frac{1}{c}\frac{\partial}{\partial t}\left(\frac{1}{\gamma_0}\tilde{\mathbf{E}}\right) - \frac{1}{c}\operatorname{curl}_{\mathbf{x}}\left(\mathbf{u}\times\frac{1}{\gamma_0}\tilde{\mathbf{E}}\right),\\ \operatorname{div}_{\mathbf{x}}\left(\frac{1}{\gamma_0}\tilde{\mathbf{E}}\right) \approx 4\pi\tilde{\rho}^*,\\ \operatorname{curl}_{\mathbf{x}}\tilde{\mathbf{E}} \approx -\left(\frac{1}{c}\frac{\partial\tilde{\mathbf{B}}}{\partial t} - \frac{1}{c}\operatorname{curl}_{\mathbf{x}}\left(\mathbf{u}\times\tilde{\mathbf{B}}\right)\right),\\ \operatorname{div}_{\mathbf{x}}\tilde{\mathbf{B}} \approx 0,\\ \frac{\partial\tilde{\rho}^*}{\partial t} + \operatorname{div}_{\mathbf{x}}\tilde{\mathbf{j}}^* \approx 0,\end{cases} \tag{13.78}$$

and

$$\begin{cases} \rho = \tilde{\rho}^* + \frac{1}{c^2}\frac{1}{\kappa_0}\left(\mathbf{u}-\mathbf{v}\right)\cdot\left(\tilde{\mathbf{j}}^* - \tilde{\rho}^*\mathbf{u}\right), \\ \mathbf{j} = \left(\tilde{\mathbf{j}}^* - \tilde{\rho}^*\mathbf{u}\right) + \rho\mathbf{u}, \\ \mathbf{B} \approx \tilde{\mathbf{B}} + \frac{1}{c}\left(\mathbf{u}-\mathbf{v}\right)\times\tilde{\mathbf{E}} \\ \mathbf{E} = \tilde{\mathbf{E}} - \frac{1}{c}\mathbf{u}\times\mathbf{B} \end{cases} \tag{13.79}$$

that are valid, however, only for quasistationary electromagnetic fields. Next, since $\tilde{\mathbf{E}}, \tilde{\mathbf{D}}, \tilde{\mathbf{B}}, \tilde{\mathbf{H}}, \mathbf{B}$, $\left(\mathbf{E} + \frac{1}{c}\mathbf{u}\times\mathbf{B}\right)$ and $\left(\tilde{\mathbf{j}}^* - \tilde{\rho}^*\mathbf{u}\right)$, $(\mathbf{j}-\rho\mathbf{u})$, $(\mathbf{u}-\mathbf{v})$ are all proper vector fields and $\tilde{\rho}^*$, ρ are proper scalars, as before we deduce that all equations in (13.78) and (13.79) are invariant under the change of inertial or non-inertial Cartesian coordinate system. Therefore, if in coordinate system $(*)$ we can use the approximate equations, given by (13.78) and (13.79), then we can use the similar approximation also in coordinate system $(**)$, even in the case when in system $(**)$ (13.71), (13.72) or (13.73) is not satisfied. Next, by (12.58) we adjoint (13.78) and (13.79) with Ohm's law that in our notations has the following simple form:

$$\tilde{\mathbf{j}}^* - \tilde{\rho}^*\,\mathbf{u} = \varepsilon\tilde{\mathbf{E}}, \tag{13.80}$$

and the Joules heat term, appearing in the First Law of Thermodynamics (11.4), in our notations, has the following simple form:

$$(\mathbf{j}-\rho\mathbf{u})\cdot\left(\mathbf{E} + \frac{1}{c}\mathbf{u}\times\mathbf{B}\right) = \left(\tilde{\mathbf{j}}^* - \tilde{\rho}^*\,\mathbf{u}\right)\cdot\tilde{\mathbf{E}}. \tag{13.81}$$

Finally, note that the system (13.78) is completely analogous to the system (13.65), where the field $\frac{1}{\gamma_0}\tilde{\mathbf{E}}$ substitutes the field $\mathbf{D}$; the field $\tilde{\mathbf{B}}$ substitutes the field $\mathbf{B}$; $\tilde{\rho}^*, \tilde{\mathbf{j}}^*$ substitute $\rho, \mathbf{j}$; and with the real velocity of the medium $\mathbf{u}$ instead of the field $\tilde{\mathbf{u}} = (\gamma_0\kappa_0\mathbf{v} + (1-\gamma_0\kappa_0)\mathbf{u})$. On the other hand, in the case of the real medium (not vacuum), the equations in (13.78) are completely independent on the vectorial gravitation potential $\mathbf{v}$. Since the systems (13.78) and (13.65) are similar they both obey wave-type solutions and thus one can wonder here: is it possible that (13.78) is approximately equivalent to (13.65) also for highly oscillating in time electromagnetic fields and/or for wave-type solutions in the absence of charges and currents, for example, in the case of the visible light? Unfortunately, the answer for the last question should be negative, indeed as it can be seen, the term $(1-\gamma_0\kappa_0)\mathbf{u}$ (consistently with the equality $\tilde{\mathbf{u}} = (\gamma_0\kappa_0\mathbf{v} + (1-\gamma_0\kappa_0)\mathbf{u}))$ appears in the Fizeau experiment for the light (see Sect. 13.3.6) rather than the term $\mathbf{u}$. The main reason for it is that (13.72) is not valid for highly oscillating in time electromagnetic fields. Moreover, in the case of the complete absence of charges and currents in the whole space (13.72) is not valid also for any nontrivial wave-type solution, even with moderate frequency. So (13.78) and (13.79) are valid, only for quasistationary electromagnetic fields, generated by charges and currents moderately changing in time, for example, in the case of estimation of DC or AC chains.

Next, as before, again by the third and the fourth equations in (13.78) we can write

$$\begin{cases} \tilde{\mathbf{B}} \approx \mathrm{curl}_{\mathbf{x}} \tilde{\mathbf{A}}^*, \\ \tilde{\mathbf{E}} \approx -\nabla_{\mathbf{x}} \tilde{\Psi}^* - \frac{1}{c}\frac{\partial \tilde{\mathbf{A}}^*}{\partial t} + \frac{1}{c}\mathbf{u} \times \left(\mathrm{curl}_{\mathbf{x}} \tilde{\mathbf{A}}^*\right), \end{cases} \tag{13.82}$$

where $\tilde{\Psi}^*$ and $\tilde{\mathbf{A}}^*$ are the scalar and the vectorial electromagnetic potentials in Hertz's notation. Next, analogously to (13.6) we define the proper scalar electromagnetic potential in Hertz's notation:

$$\tilde{\Psi}_1^* := \tilde{\Psi}^* - \frac{1}{c}\tilde{\mathbf{A}}^* \cdot \mathbf{u}. \tag{13.83}$$

As before, the electromagnetic potentials in Hertz's notation are not uniquely defined and thus we need to choose a calibration. For definiteness, we can take $\tilde{\mathbf{A}}^*$ to satisfy

$$\mathrm{div}_{\mathbf{x}} \tilde{\mathbf{A}}^* = 0. \tag{13.84}$$

Then, as before, $\tilde{\mathbf{A}}^*$ is a proper vector field and $\tilde{\Psi}_1^*$ is a proper scalar field. It is clear that any other choice of electromagnetic potentials with a different calibration, as before, can be obtained by

$$\begin{cases} \tilde{\Psi}^* \rightarrow \tilde{\Psi}^* + \frac{1}{c}\frac{\partial w}{\partial t} \\ \tilde{\mathbf{A}}^* \rightarrow \tilde{\mathbf{A}}^* - \nabla_{\mathbf{x}} w \\ \tilde{\Psi}_1^* \rightarrow \tilde{\Psi}_1^* + \frac{1}{c}\left(\frac{\partial w}{\partial t} + \mathbf{u} \cdot \nabla_{\mathbf{x}} w\right), \end{cases} \tag{13.85}$$

where $w := w(\mathbf{x}, t)$ is an arbitrary proper scalar field. Moreover, as before, under such a change of calibration, $\tilde{\mathbf{A}}^*$ is still a proper vector field and $\tilde{\Psi}_1^*$ is still a proper scalar field, provided the scalar field w is proper. In particular, the following alternative to (13.84) calibration of the potentials can be chosen:

$$\frac{1}{\kappa_0 \gamma_0 c}\left(\frac{\partial \tilde{\Psi}_1^*}{\partial t} + \mathbf{u} \cdot \nabla_{\mathbf{x}} \tilde{\Psi}_1^*\right) + \mathrm{div}_{\mathbf{x}} \tilde{\mathbf{A}}^* = 0. \tag{13.86}$$

Next, inserting (13.83) into (13.82) gives

$$\begin{cases} \tilde{\mathbf{B}} \approx \mathrm{curl}_{\mathbf{x}} \tilde{\mathbf{A}}^*, \\ \tilde{\mathbf{E}} \approx -\nabla_{\mathbf{x}} \tilde{\Psi}_1^* - \frac{1}{c}\left(\frac{\partial \tilde{\mathbf{A}}^*}{\partial t} - \mathbf{u} \times \left(\mathrm{curl}_{\mathbf{x}} \tilde{\mathbf{A}}^*\right) + \nabla_{\mathbf{x}}\left(\tilde{\mathbf{A}}^* \cdot \mathbf{u}\right)\right). \end{cases} \tag{13.87}$$

Therefore, by (3.8) and (3.9) in Proposition 3.1 we rewrite (13.87) as

$$\begin{cases} \tilde{\mathbf{B}} \approx \mathrm{curl}_{\mathbf{x}} \tilde{\mathbf{A}}^*, \\ \tilde{\mathbf{E}} \approx -\nabla_{\mathbf{x}} \tilde{\Psi}_1^* - \frac{1}{c}\Big(\frac{\partial \tilde{\mathbf{A}}^*}{\partial t} - \mathrm{curl}_{\mathbf{x}} \left(\mathbf{u} \times \tilde{\mathbf{A}}^* \right) + \left(\mathrm{div}_{\mathbf{x}} \tilde{\mathbf{A}}^* \right) \mathbf{u} \\ - \left(\mathrm{div}_{\mathbf{x}} \mathbf{u} \right) \tilde{\mathbf{A}}^* + \left(d_{\mathbf{x}} \mathbf{u} + \{ d_{\mathbf{x}} \mathbf{u} \}^T \right) \cdot \tilde{\mathbf{A}}^* \Big). \end{cases} \tag{13.88}$$

Thus, by (13.73) we approximate (13.88) as

$$\begin{cases} \tilde{\mathbf{B}} \approx \mathrm{curl}_{\mathbf{x}} \tilde{\mathbf{A}}^*, \\ \tilde{\mathbf{E}} \approx -\nabla_{\mathbf{x}} \tilde{\Psi}_1^* - \frac{1}{c}\left(\frac{\partial \tilde{\mathbf{A}}^*}{\partial t} - \mathrm{curl}_{\mathbf{x}} \left(\mathbf{u} \times \tilde{\mathbf{A}}^* \right) + \left(\mathrm{div}_{\mathbf{x}} \tilde{\mathbf{A}}^* \right) \mathbf{u} \right). \end{cases} \tag{13.89}$$

Furthermore, inserting (13.87) into (13.78) and using (13.71), (13.72) and (13.73) we deduce

$$\begin{cases} \mathrm{curl}_{\mathbf{x}} \left(\kappa_0 \tilde{\mathbf{B}} \right) \approx \frac{4\pi}{c} \left(\tilde{\mathbf{j}}^* - \tilde{\rho}^* \mathbf{u} \right) - \frac{1}{c}\Big(\frac{\partial}{\partial t} \left(\frac{1}{\gamma_0} \nabla_{\mathbf{x}} \tilde{\Psi}_1^* \right) \\ - \mathrm{curl}_{\mathbf{x}} \left(\mathbf{u} \times \frac{1}{\gamma_0} \nabla_{\mathbf{x}} \tilde{\Psi}_1^* \right) + \left(\mathrm{div}_{\mathbf{x}} \left\{ \frac{1}{\gamma_0} \nabla_{\mathbf{x}} \tilde{\Psi}_1^* \right\} \right) \mathbf{u} \Big), \\ \mathrm{div}_{\mathbf{x}} \left(\frac{1}{\gamma_0} \tilde{\mathbf{E}} \right) \approx 4\pi \tilde{\rho}^*, \\ \frac{\partial \tilde{\rho}^*}{\partial t} + \mathrm{div}_{\mathbf{x}} \tilde{\mathbf{j}}^* \approx 0, \\ \tilde{\mathbf{E}} \approx -\nabla_{\mathbf{x}} \tilde{\Psi}_1^* - \frac{1}{c}\left(\frac{\partial \tilde{\mathbf{A}}^*}{\partial t} - \mathrm{curl}_{\mathbf{x}} \left(\mathbf{u} \times \tilde{\mathbf{A}}^* \right) + \left(\mathrm{div}_{\mathbf{x}} \tilde{\mathbf{A}}^* \right) \mathbf{u} \right), \\ \tilde{\mathbf{B}} \approx \mathrm{curl}_{\mathbf{x}} \tilde{\mathbf{A}}^*. \end{cases} \tag{13.90}$$

Then again all equations in the new system (13.90) are invariant under the change of inertial or non-inertial Cartesian coordinate system.

Furthermore, if we assume that in the domain of the study the coefficients $\gamma_0 \neq 0$ and $\kappa_0 \neq 0$ vary sufficiently slow on the place and time and thus their spatial and temporal derivatives are negligible with respect to other terms, then again by (3.8) and (3.9) in Proposition 3.1 and by (13.73), (13.90) can be approximately rewritten as

$$\begin{cases} \kappa_0 \, \mathrm{curl}_{\mathbf{x}} \tilde{\mathbf{B}} \approx \frac{4\pi}{c} \left(\tilde{\mathbf{j}}^* - \tilde{\rho}^* \mathbf{u} \right) - \nabla_{\mathbf{x}} \left\{ \frac{1}{c} \frac{1}{\gamma_0} \left(\frac{\partial \tilde{\Psi}_1^*}{\partial t} + \mathbf{u} \cdot \nabla_{\mathbf{x}} \tilde{\Psi}_1^* \right) \right\}, \\ \mathrm{div}_{\mathbf{x}} \tilde{\mathbf{E}} \approx 4\pi \gamma_0 \tilde{\rho}^*, \\ \frac{\partial \tilde{\rho}^*}{\partial t} + \mathrm{div}_{\mathbf{x}} \tilde{\mathbf{j}}^* \approx 0, \\ \tilde{\mathbf{E}} \approx -\nabla_{\mathbf{x}} \tilde{\Psi}_1^* - \frac{1}{c}\left(\frac{\partial \tilde{\mathbf{A}}^*}{\partial t} - \mathrm{curl}_{\mathbf{x}} \left(\mathbf{u} \times \tilde{\mathbf{A}}^* \right) + \left(\mathrm{div}_{\mathbf{x}} \tilde{\mathbf{A}}^* \right) \mathbf{u} \right), \\ \tilde{\mathbf{B}} \approx \mathrm{curl}_{\mathbf{x}} \tilde{\mathbf{A}}^*. \end{cases} \tag{13.91}$$

Furthermore, inserting the last equation of (13.91) into the first one and using (2.10), we rewrite (13.91) as

$$\begin{cases} -\Delta_{\mathbf{x}}\tilde{\mathbf{A}}^* \approx \frac{4\pi}{\kappa_0 c}\left(\tilde{\mathbf{j}}^* - \tilde{\rho}^*\mathbf{u}\right) - \nabla_{\mathbf{x}}\left\{\mathrm{div}_{\mathbf{x}}\,\tilde{\mathbf{A}}^* + \frac{1}{c}\frac{1}{\gamma_0\kappa_0}\left(\frac{\partial\tilde{\Psi}_1^*}{\partial t} + \mathbf{u}\cdot\nabla_{\mathbf{x}}\tilde{\Psi}_1^*\right)\right\}, \\ \mathrm{div}_{\mathbf{x}}\,\tilde{\mathbf{E}} \approx 4\pi\gamma_0\tilde{\rho}^*, \\ \frac{\partial\tilde{\rho}^*}{\partial t} + \mathrm{div}_{\mathbf{x}}\,\tilde{\mathbf{j}}^* \approx 0, \\ \tilde{\mathbf{E}} \approx -\nabla_{\mathbf{x}}\tilde{\Psi}_1^* - \frac{1}{c}\left(\frac{\partial\tilde{\mathbf{A}}^*}{\partial t} - \mathrm{curl}_{\mathbf{x}}\left(\mathbf{u}\times\tilde{\mathbf{A}}^*\right) + \left(\mathrm{div}_{\mathbf{x}}\,\tilde{\mathbf{A}}^*\right)\mathbf{u}\right), \\ \tilde{\mathbf{B}} \approx \mathrm{curl}_{\mathbf{x}}\tilde{\mathbf{A}}^*. \end{cases} \tag{13.92}$$

On the other hand, taking the divergence of both sides of the second equation in (13.89) gives

$$\mathrm{div}_{\mathbf{x}}\,\tilde{\mathbf{E}} + \Delta_{\mathbf{x}}\tilde{\Psi}_1 \approx -\frac{1}{c}\left(\frac{\partial}{\partial t}\left(\mathrm{div}_{\mathbf{x}}\,\tilde{\mathbf{A}}^*\right) + \mathrm{div}_{\mathbf{x}}\left\{\left(\mathrm{div}_{\mathbf{x}}\,\tilde{\mathbf{A}}^*\right)\mathbf{u}\right\}\right). \tag{13.93}$$

Inserting (13.93) into (13.92) gives

$$\begin{cases} -\Delta_{\mathbf{x}}\tilde{\mathbf{A}}^* \approx \frac{4\pi}{\kappa_0 c}\left(\tilde{\mathbf{j}}^* - \tilde{\rho}^*\mathbf{u}\right) - \nabla_{\mathbf{x}}\left\{\mathrm{div}_{\mathbf{x}}\,\tilde{\mathbf{A}}^* + \frac{1}{c}\frac{1}{\gamma_0\kappa_0}\left(\frac{\partial\tilde{\Psi}_1^*}{\partial t} + \mathbf{u}\cdot\nabla_{\mathbf{x}}\tilde{\Psi}_1^*\right)\right\}, \\ -\Delta_{\mathbf{x}}\tilde{\Psi}_1 \approx 4\pi\gamma_0\tilde{\rho}^* + \frac{1}{c}\left(\frac{\partial}{\partial t}\left(\mathrm{div}_{\mathbf{x}}\,\tilde{\mathbf{A}}^*\right) + \mathrm{div}_{\mathbf{x}}\left\{\left(\mathrm{div}_{\mathbf{x}}\,\tilde{\mathbf{A}}^*\right)\mathbf{u}\right\}\right), \\ \frac{\partial\tilde{\rho}^*}{\partial t} + \mathrm{div}_{\mathbf{x}}\,\tilde{\mathbf{j}}^* \approx 0, \\ \tilde{\mathbf{E}} \approx -\nabla_{\mathbf{x}}\tilde{\Psi}_1^* - \frac{1}{c}\left(\frac{\partial\tilde{\mathbf{A}}^*}{\partial t} - \mathrm{curl}_{\mathbf{x}}\left(\mathbf{u}\times\tilde{\mathbf{A}}^*\right) + \left(\mathrm{div}_{\mathbf{x}}\,\tilde{\mathbf{A}}^*\right)\mathbf{u}\right), \\ \tilde{\mathbf{B}} \approx \mathrm{curl}_{\mathbf{x}}\tilde{\mathbf{A}}^*. \end{cases} \tag{13.94}$$

Therefore, by inserting (13.87) into (13.94) we deduce

$$\begin{cases} -\Delta_{\mathbf{x}}\tilde{\mathbf{A}}^* \approx \frac{4\pi}{\kappa_0 c}\left(\tilde{\mathbf{j}}^* - \tilde{\rho}^*\mathbf{u}\right) - \nabla_{\mathbf{x}}\left\{\mathrm{div}_{\mathbf{x}}\,\tilde{\mathbf{A}}^* + \frac{1}{c}\frac{1}{\gamma_0\kappa_0}\left(\frac{\partial\tilde{\Psi}_1^*}{\partial t} + \mathbf{u}\cdot\nabla_{\mathbf{x}}\tilde{\Psi}_1^*\right)\right\}, \\ -\Delta_{\mathbf{x}}\tilde{\Psi}_1 \approx 4\pi\gamma_0\tilde{\rho}^* + \frac{1}{c}\left(\frac{\partial}{\partial t}\left(\mathrm{div}_{\mathbf{x}}\,\tilde{\mathbf{A}}^*\right) + \mathrm{div}_{\mathbf{x}}\left\{\left(\mathrm{div}_{\mathbf{x}}\,\tilde{\mathbf{A}}^*\right)\mathbf{u}\right\}\right), \\ \frac{\partial\tilde{\rho}^*}{\partial t} + \mathrm{div}_{\mathbf{x}}\,\tilde{\mathbf{j}}^* \approx 0, \\ \tilde{\mathbf{E}} \approx -\nabla_{\mathbf{x}}\tilde{\Psi}_1^* - \frac{1}{c}\left(\frac{\partial\tilde{\mathbf{A}}^*}{\partial t} - \mathbf{u}\times\left(\mathrm{curl}_{\mathbf{x}}\tilde{\mathbf{A}}^*\right) + \nabla_{\mathbf{x}}\left(\tilde{\mathbf{A}}^*\cdot\mathbf{u}\right)\right), \\ \tilde{\mathbf{B}} \approx \mathrm{curl}_{\mathbf{x}}\tilde{\mathbf{A}}^*. \end{cases} \tag{13.95}$$

Thus, in the case of calibration (13.84) by (3.8) and (3.9) in Proposition 3.1, using (13.73), (13.95) can be rewritten as

$$\begin{cases} -\Delta_{\mathbf{x}}\tilde{\Psi}_1 \approx 4\pi\gamma_0\tilde{\rho}^*, \\ -\Delta_{\mathbf{x}}\tilde{\mathbf{A}}^* \approx \frac{4\pi}{\kappa_0 c}\left(\tilde{\mathbf{j}}^* - \tilde{\rho}^*\mathbf{u}\right) - \frac{1}{c}\frac{1}{\gamma_0\kappa_0}\left(\frac{\partial}{\partial t}\left(\nabla_{\mathbf{x}}\tilde{\Psi}_1^*\right) - \mathrm{curl}_{\mathbf{x}}\left(\mathbf{u}\times\nabla_{\mathbf{x}}\tilde{\Psi}_1^*\right) + \left(\Delta_{\mathbf{x}}\tilde{\Psi}_1^*\right)\mathbf{u}\right), \\ \frac{\partial\tilde{\rho}^*}{\partial t} + \mathrm{div}_{\mathbf{x}}\,\tilde{\mathbf{j}}^* \approx 0, \\ \tilde{\mathbf{E}} \approx -\nabla_{\mathbf{x}}\tilde{\Psi}_1^* - \frac{1}{c}\left(\frac{\partial\tilde{\mathbf{A}}^*}{\partial t} - \mathbf{u}\times\left(\mathrm{curl}_{\mathbf{x}}\tilde{\mathbf{A}}^*\right) + \nabla_{\mathbf{x}}\left(\tilde{\mathbf{A}}^*\cdot\mathbf{u}\right)\right), \\ \tilde{\mathbf{B}} \approx \mathrm{curl}_{\mathbf{x}}\tilde{\mathbf{A}}^*. \end{cases} \tag{13.96}$$

On the other hand, under the alternative calibration (13.86), again using (13.71), (13.72) and (13.73) in the second equation of (13.95), we approximately rewrite (13.95) as

$$\begin{cases} -\Delta_{\mathbf{x}}\tilde{\mathbf{A}}^* \approx \frac{4\pi}{\kappa_0 c}\left(\tilde{\mathbf{j}}^* - \tilde{\rho}^*\mathbf{u}\right), \\ -\Delta_{\mathbf{x}}\tilde{\Psi}_1 \approx 4\pi\gamma_0\tilde{\rho}^*, \\ \frac{\partial\tilde{\rho}^*}{\partial t} + \operatorname{div}_{\mathbf{x}}\tilde{\mathbf{j}}^* \approx 0, \\ \tilde{\mathbf{E}} \approx -\nabla_{\mathbf{x}}\tilde{\Psi}_1^* - \frac{1}{c}\left(\frac{\partial\tilde{\mathbf{A}}^*}{\partial t} - \mathbf{u}\times\left(\operatorname{curl}_{\mathbf{x}}\tilde{\mathbf{A}}^*\right) + \nabla_{\mathbf{x}}\left(\tilde{\mathbf{A}}^*\cdot\mathbf{u}\right)\right), \\ \tilde{\mathbf{B}} \approx \operatorname{curl}_{\mathbf{x}}\tilde{\mathbf{A}}^*. \end{cases} \tag{13.97}$$

Furthermore, defining the conduction currents:

$$\begin{cases} \mathbf{j}_{cond} := \mathbf{j} - \rho\mathbf{u}, \\ \tilde{\mathbf{j}}^*{}_{cond} := \tilde{\mathbf{j}}^* - \tilde{\rho}^*\mathbf{u}, \end{cases} \tag{13.98}$$

we obviously have

$$\begin{cases} \mathbf{j}_{cond} = \tilde{\mathbf{j}}^*{}_{cond}, \\ \frac{\partial\rho}{\partial t} = \operatorname{div}_{\mathbf{x}}\{\rho\mathbf{u}\} = -\operatorname{div}_{\mathbf{x}}\{\mathbf{j}_{cond}\}, \\ \frac{\partial\tilde{\rho}^*}{\partial t} + \operatorname{div}_{\mathbf{x}}\{\tilde{\rho}^*\mathbf{u}\} \approx -\operatorname{div}_{\mathbf{x}}\left\{\tilde{\mathbf{j}}^*{}_{cond}\right\}, \end{cases} \tag{13.99}$$

and, moreover, obviously $\mathbf{j}_{cond} = \tilde{\mathbf{j}}^*{}_{cond}$ is a <u>proper</u> vector field. Thus, by (13.98) and (13.99) we rewrite (13.96) as

$$\begin{cases} -\Delta_{\mathbf{x}}\tilde{\Psi}_1 \approx 4\pi\gamma_0\tilde{\rho}^*, \\ -\Delta_{\mathbf{x}}\tilde{\mathbf{A}}^* \approx \frac{4\pi}{\kappa_0 c}\tilde{\mathbf{j}}^*{}_{cond} - \frac{1}{c}\frac{1}{\gamma_0\kappa_0}\left(\frac{\partial}{\partial t}\left(\nabla_{\mathbf{x}}\tilde{\Psi}_1^*\right) - \operatorname{curl}_{\mathbf{x}}\left(\mathbf{u}\times\nabla_{\mathbf{x}}\tilde{\Psi}_1^*\right) + \left(\Delta_{\mathbf{x}}\tilde{\Psi}_1^*\right)\mathbf{u}\right), \\ \frac{\partial\tilde{\rho}^*}{\partial t} + \operatorname{div}_{\mathbf{x}}\{\tilde{\rho}^*\mathbf{u}\} \approx -\operatorname{div}_{\mathbf{x}}\left\{\tilde{\mathbf{j}}^*{}_{cond}\right\}, \\ \tilde{\mathbf{E}} \approx -\nabla_{\mathbf{x}}\tilde{\Psi}_1^* - \frac{1}{c}\left(\frac{\partial\tilde{\mathbf{A}}^*}{\partial t} - \mathbf{u}\times\left(\operatorname{curl}_{\mathbf{x}}\tilde{\mathbf{A}}^*\right) + \nabla_{\mathbf{x}}\left(\tilde{\mathbf{A}}^*\cdot\mathbf{u}\right)\right), \\ \tilde{\mathbf{B}} \approx \operatorname{curl}_{\mathbf{x}}\tilde{\mathbf{A}}^*. \end{cases} \tag{13.100}$$

On the other hand, by (13.98) and (13.99) we rewrite (13.97), which was obtained under the alternative calibration, as

$$\begin{cases} -\Delta_{\mathbf{x}}\tilde{\mathbf{A}}^* \approx \frac{4\pi}{\kappa_0 c}\tilde{\mathbf{j}}^*{}_{cond}, \\ -\Delta_{\mathbf{x}}\tilde{\Psi}_1 \approx 4\pi\gamma_0\tilde{\rho}^*, \\ \frac{\partial\tilde{\rho}^*}{\partial t} + \operatorname{div}_{\mathbf{x}}\{\tilde{\rho}^*\mathbf{u}\} \approx -\operatorname{div}_{\mathbf{x}}\left\{\tilde{\mathbf{j}}^*{}_{cond}\right\}, \\ \tilde{\mathbf{E}} \approx -\nabla_{\mathbf{x}}\tilde{\Psi}_1^* - \frac{1}{c}\left(\frac{\partial\tilde{\mathbf{A}}^*}{\partial t} - \mathbf{u}\times\left(\operatorname{curl}_{\mathbf{x}}\tilde{\mathbf{A}}^*\right) + \nabla_{\mathbf{x}}\left(\tilde{\mathbf{A}}^*\cdot\mathbf{u}\right)\right), \\ \tilde{\mathbf{B}} \approx \operatorname{curl}_{\mathbf{x}}\tilde{\mathbf{A}}^*. \end{cases} \tag{13.101}$$

In both given cases of calibration in (13.100) or (13.101) we just need to solve two Laplace equations in order to find $\tilde{\Psi}_1$ and $\tilde{\mathbf{A}}^*$ and then we find $\tilde{\mathbf{E}}$ and $\tilde{\mathbf{B}}$ by differentiation. Moreover, in either the case of (13.100) or the case of (13.101) we have

$$\operatorname{div}_{\mathbf{x}} \left\{ \tilde{\mathbf{E}} + \nabla_{\mathbf{x}} \tilde{\Psi}_1^* \right\} \approx 0 \tag{13.102}$$

and Ohm's Law in the form of (13.80):

$$\tilde{\mathbf{j}^*}_{cond} = \varepsilon \tilde{\mathbf{E}}. \tag{13.103}$$

Furthermore, the Joules heat term has the following simple form:

$$\tilde{\mathbf{j}^*}_{cond} \cdot \tilde{\mathbf{E}}. \tag{13.104}$$

Moreover, in both cases of (13.100) or (13.101), by the last two equations there we obviously have

$$\begin{cases} \operatorname{curl}_{\mathbf{x}} \tilde{\mathbf{E}} \approx -\left(\frac{1}{c} \frac{\partial \tilde{\mathbf{B}}}{\partial t} - \frac{1}{c} \operatorname{curl}_{\mathbf{x}} \left(\mathbf{u} \times \tilde{\mathbf{B}} \right) \right), \\ \operatorname{div}_{\mathbf{x}} \tilde{\mathbf{B}} \approx 0. \end{cases} \tag{13.105}$$

Next, in order to find the real electromagnetic fields in the usual notation we have (13.79), i.e. the following:

$$\begin{cases} \mathbf{B} \approx \tilde{\mathbf{B}} + \frac{1}{c} (\mathbf{u} - \mathbf{v}) \times \tilde{\mathbf{E}}, \\ \mathbf{E} = \tilde{\mathbf{E}} - \frac{1}{c} \mathbf{u} \times \mathbf{B}, \\ \rho = \tilde{\rho}^* + \frac{1}{c^2} (\mathbf{u} - \mathbf{v}) \cdot \tilde{\mathbf{j}^*}_{cond}, \\ \mathbf{j}_{cond} = \tilde{\mathbf{j}^*}_{cond}, \\ \mathbf{j} = \tilde{\mathbf{j}^*}_{cond} + \rho \mathbf{u}. \end{cases} \tag{13.106}$$

Finally, as before, we deduce that all equations in either (13.100) or (13.101), and, in addition, (13.102), (13.103), (13.105) and (13.106) are invariant under the change of inertial or non-inertial Cartesian coordinate system. Therefore, if in coordinate system (∗) we can use the approximate equations, given by either (13.100) or (13.101), and (13.102), (13.103), (13.105) and (13.106), then we can use the similar approximation also in coordinate system (∗∗), even in the case when in system (∗∗) (13.71), (13.72) or (13.73) is not satisfied.

Next, note that either (13.100) or (13.101), and (13.102), (13.103) and (13.105) are completely independent on the vectorial gravitational potential $\mathbf{v}$. Moreover, note that the first two equations in (13.101) are completely independent also on the velocity field of the medium $\mathbf{u}$. Furthermore, if $\Omega := \Omega(t) \subset \mathbb{R}^3$ is a three-dimensional domain, moving with velocity field $\mathbf{u}$ together with the given medium, then, by the third equation of either (13.100) or (13.101), using part **(iii)** of Proposition 3.7 and the Divergence Integral Theorem of the Calculus we deduce

$$\frac{d}{dt}\left(\iiint \tilde{\rho}^* \, d\Omega(t)\right) = \iiint \left(\frac{\partial \tilde{\rho}^*}{\partial t} + \mathrm{div}_{\mathbf{x}}\left\{\tilde{\rho}^* \mathbf{u}\right\}\right) d\Omega(t) \approx$$
$$- \iiint \left(\mathrm{div}_{\mathbf{x}}\left\{\tilde{\mathbf{j}}^*{}_{cond}\right\}\right) d\Omega(t) = -\iint \tilde{\mathbf{j}}^*{}_{cond} \cdot \mathbf{n}\, d\left(\partial\Omega(t)\right), \quad (13.107)$$

where $\partial\Omega(t)$ is the two-dimensional boundary of $\Omega(t)$, oriented by outer unit normal $\mathbf{n}$. Moreover, analogously, by the second equation in (13.99) we deduce

$$\frac{d}{dt}\left(\iiint \rho \, d\Omega(t)\right) = \iiint \left(\frac{\partial \rho}{\partial t} + \mathrm{div}_{\mathbf{x}}\left\{\rho \mathbf{u}\right\}\right) d\Omega(t) =$$
$$- \iiint \left(\mathrm{div}_{\mathbf{x}}\left\{\mathbf{j}_{cond}\right\}\right) d\Omega(t) = -\iint \mathbf{j}_{cond} \cdot \mathbf{n}\, d\left(\partial\Omega(t)\right). \quad (13.108)$$

So, we have

$$\frac{d}{dt}\left(\iiint \rho \, d\Omega(t)\right) = -\iint \mathbf{j}_{cond} \cdot \mathbf{n}\, d\left(\partial\Omega(t)\right) =$$
$$- \iint \tilde{\mathbf{j}}^*{}_{cond} \cdot \mathbf{n}\, d\left(\partial\Omega(t)\right) \approx \frac{d}{dt}\left(\iiint \tilde{\rho}^* \, d\Omega(t)\right). \quad (13.109)$$

Thus, denoting

$$\begin{cases} Q_{\Omega(t)}(t) := \iiint \rho \, d\Omega(t), \quad \tilde{Q}^*_{\Omega(t)}(t) := \iiint \tilde{\rho}^* \, d\Omega(t) \quad \text{and} \\ I_{\partial\Omega(t)}(t) := \iint \mathbf{j}_{cond} \cdot \mathbf{n} = \iint \tilde{\mathbf{j}}^*{}_{cond} \cdot \mathbf{n}\, d\left(\partial\Omega(t)\right) := \tilde{I}^*_{\partial\Omega(t)}(t), \end{cases} \quad (13.110)$$

which are the total charge inside the domain $\Omega(t)$, the total charge* inside the domain $\Omega(t)$ and the corresponding total current outward the boundary of $\Omega(t)$, by (13.109) we have

$$\frac{d}{dt}\left(Q_{\Omega(t)}(t)\right) = -I_{\partial\Omega(t)}(t) = -\tilde{I}^*_{\partial\Omega(t)}(t) \approx \frac{d}{dt}\left(\tilde{Q}^*_{\Omega(t)}(t)\right). \quad (13.111)$$

On the other hand, if $\gamma := \gamma(t) \subset \mathbb{R}^3$ is a one-dimensional curve oriented by the unit tangent vector $\mathbf{t} := \mathbf{t}(\mathbf{x}, t)$, having the starting and the ending points $\mathbf{r}_{begin}(t)$, $\mathbf{r}_{end}(t)$ and moving with velocity field $\mathbf{u}$ together with the given medium, then, by the fourth equation in either (13.100) or (13.101) using part **(ii)** of Proposition 3.7, we obtain

$$\int \tilde{\mathbf{E}} \cdot \mathbf{t}\, d\gamma(t) \approx -\int \nabla_{\mathbf{x}} \tilde{\Psi}^*_1 \cdot \mathbf{t}\, d\gamma(t)$$
$$- \frac{1}{c}\int \left(\frac{\partial \tilde{\mathbf{A}}^*}{\partial t} - \mathbf{u} \times \mathrm{curl}_{\mathbf{x}} \tilde{\mathbf{A}}^* + \nabla_{\mathbf{x}}\left(\mathbf{u} \cdot \tilde{\mathbf{A}}^*\right)\right) \cdot \mathbf{t}\, d\gamma(t)$$
$$= \tilde{\Psi}^*_1\left(\mathbf{r}_{begin}(t), t\right) - \tilde{\Psi}^*_1\left(\mathbf{r}_{end}(t), t\right) - \frac{1}{c}\frac{d}{dt}\left(\int \tilde{\mathbf{A}}^* \cdot \mathbf{t}\, d\gamma(t)\right),$$

and thus

$$\int \tilde{\mathbf{E}} \cdot \mathbf{t}\, d\gamma(t) \approx \tilde{\Psi}_1^* \left(\mathbf{r}_{begin}(t), t\right) - \tilde{\Psi}_1^* \left(\mathbf{r}_{end}(t), t\right) - \frac{1}{c}\frac{d}{dt}\left(\int \tilde{\mathbf{A}}^* \cdot \mathbf{t}\, d\gamma(t)\right). \tag{13.112}$$

In particular, in the case where $\gamma(t) := \partial\mathcal{S}(t)$ is a boundary of a two-dimensional surface $\mathcal{S} := \mathcal{S}(t) \subset \mathbb{R}^3$, oriented by the unit normal $\mathbf{n} := \mathbf{n}(\mathbf{x}, t)$, since in the later case we have $\mathbf{r}_{begin}(t) = \mathbf{r}_{end}(t)$, using the last equation in either (13.100) or (13.101) and the Stokes theorem, we rewrite (13.112) as

$$\int \tilde{\mathbf{E}} \cdot \mathbf{t}\, d\gamma(t) \approx -\frac{1}{c}\frac{d}{dt}\left(\iint \tilde{\mathbf{B}} \cdot \mathbf{n}\, d\mathcal{S}(t)\right) = -\frac{1}{c}\frac{d}{dt}\left(\iint \operatorname{curl}_{\mathbf{x}}\tilde{\mathbf{A}}^* \cdot \mathbf{n}\, d\mathcal{S}(t)\right). \tag{13.113}$$

Finally, the obtained relations in (13.102), (13.103), (13.104), (13.111) with (13.110), (13.112), (13.113) and the first two equations in (13.101) are completely independent on the velocity field $\mathbf{u}$ (and on the vectorial gravitational potential $\mathbf{v}$). However, the mentioned relations in (13.102), (13.103), (13.111) with (13.110), (13.112), (13.113) and the first two equations in (13.101) are together sufficient for estimation of DC or AC linear chains and, in particular, they are sufficient to obtain Kirchhoff's current and voltage laws, Faraday's law of induction and the law of capacity. Thus, the estimation of DC or quasistationary AC currents in the case of moving electrical chains and/or in the case of nontrivial gravitation is completely analogous to that estimation for the resting chains without influence of any gravitation!

13.3 Geometric Optics Inside a Moving Medium and/or in the Presence of Gravitational Field

13.3.1 Preliminary Calculations

Assume that in some inertial or non-inertial Cartesian coordinate system a real-valued scalar field $U := U(\mathbf{x}, t)$, characterizing some wave, satisfies the following wave equation:

$$\frac{\partial}{\partial t}\left\{\frac{1}{c_0^2}\left(\frac{\partial U}{\partial t} + \tilde{\mathbf{u}} \cdot \nabla_{\mathbf{x}} U\right)\right\} + \operatorname{div}_{\mathbf{x}}\left\{\frac{1}{c_0^2}\left(\frac{\partial U}{\partial t} + \tilde{\mathbf{u}} \cdot \nabla_{\mathbf{x}} U\right)\tilde{\mathbf{u}}\right\} - \Delta_{\mathbf{x}} U = 0, \tag{13.114}$$

where $\tilde{\mathbf{u}} := \tilde{\mathbf{u}}(\mathbf{x}, t)$ is some moderately changing (in space and in time) speed-like vector field and $c_0 := c_0(\mathbf{x}, t) > 0$ is a moderately changing (in space and in time) scalar quantity that we call wave propagation speed. Note that (13.114) coincides with (13.41) and thus, in particular, U can represent one of any scalar components of the electromagnetic field in the medium without dispersion, i.e. when (12.42) is valid. In this case, $\tilde{\mathbf{u}}$ is the linear combination of the velocity of the medium $\mathbf{u}$ and the vectorial gravitational potential $\mathbf{v}$. Note also that (11.106) or (11.115) also coincides with (13.114) and thus, in particular, U could represent the oscillating part of the pressure p_1 in the sound wave. However, in the later case, $\tilde{\mathbf{u}}$ is equal to the averaged (macroscopical) velocity of the fluid/gas $\mathbf{u}_0$. Moreover, (11.161) and (11.162) also coincide with (13.114) and thus, in particular, U could represent either longitudinal or transverse wave in an elastic body. In the later case, $\tilde{\mathbf{u}}$ is also equal to the averaged (macroscopical) velocity of the elastic medium $\mathbf{u}_0$.

Next if we assume that the fields $\tilde{\mathbf{u}}$ and c_0 are independent on the time variable, since U is a real-valued field, then we can write the field U as a Fourier's transform on the time variable:

$$U(\mathbf{x}, t) = Re\left\{2\int_0^{+\infty} \hat{U}(\mathbf{x}, \omega)e^{i\omega t}d\omega\right\}$$
$$\text{where}\quad \hat{U}(\mathbf{x}, \omega) := \frac{1}{2\pi}\int_{-\infty}^{+\infty} U(\mathbf{x}, t)e^{-i\omega t}dt. \tag{13.115}$$

Moreover, by (13.114) we obtain that Furier's transform $\hat{U}(\mathbf{x}, \omega)$ satisfies

$$i\omega\frac{1}{c_0^2}\left(i\omega\hat{U} + \tilde{\mathbf{u}}\cdot\nabla_{\mathbf{x}}\hat{U}\right) + \mathrm{div}_{\mathbf{x}}\left\{\frac{1}{c_0^2}\left(i\omega\hat{U} + \tilde{\mathbf{u}}\cdot\nabla_{\mathbf{x}}\hat{U}\right)\tilde{\mathbf{u}}\right\} - \Delta_{\mathbf{x}}\hat{U} = 0. \tag{13.116}$$

Thus by (13.116), for every given ω the monochromatic wave-type function

$$U_\omega(\mathbf{x}, t) := \hat{U}(\mathbf{x}, \omega)e^{i\omega t} \tag{13.117}$$

is a complex solution of

$$\frac{\partial}{\partial t}\left\{\frac{1}{c_0^2}\left(\frac{\partial U_\omega}{\partial t} + \tilde{\mathbf{u}}\cdot\nabla_{\mathbf{x}}U_\omega\right)\right\} + \mathrm{div}_{\mathbf{x}}\left\{\frac{1}{c_0^2}\left(\frac{\partial U_\omega}{\partial t} + \tilde{\mathbf{u}}\cdot\nabla_{\mathbf{x}}U_\omega\right)\tilde{\mathbf{u}}\right\} - \Delta_{\mathbf{x}}U_\omega = 0. \tag{13.118}$$

Note that Eq. (13.118) coincides with (13.114). Moreover, by (13.115) a general real solution of (13.114) can be represented as the real part of a superposition of monochromatic waves of type $U_\omega = f(\mathbf{x}, \omega)e^{i\omega t}$ that satisfy (13.118) for every ω. Finally if we consider a complex-valued function $\mathcal{U}(\mathbf{x}, t)$, defined by

$$\mathcal{U}(\mathbf{x}, t) := 2\int_0^{+\infty} \hat{U}(\mathbf{x}, \omega)e^{i\omega t}d\omega, \tag{13.119}$$

then $Re\,\mathcal{U} = U$ and $\mathcal{U}$ is a complex solution of (13.114) (i.e. not only the real part of $\mathcal{U}$ but also the imaginary part solves (13.114)).

Next assume that a scalar complex field $U := U(\mathbf{x}, t)$ satisfies (13.114). In particular, U can be a monochromatic solution of (13.118). Furthermore, we represent the complex field U as

$$U(\mathbf{x}, t) = A(\mathbf{x}, t)e^{iT(\mathbf{x},t)}, \tag{13.120}$$

where $A := A(\mathbf{x}, t)$ is a complex scalar field and $T := T(\mathbf{x}, t)$ is a real scalar field. Then define

$$\omega := \left\langle \left| \frac{\partial T}{\partial t} \right| \right\rangle, \tag{13.121}$$

where the sign $\langle\cdot\rangle$ means the spatial and temporal averaging. Next define k_0 and a scalar field $S := S(\mathbf{x}, t)$ by

$$k_0 := \frac{\omega}{c} \quad \text{and} \quad S(\mathbf{x}, t) = \frac{1}{k_0} T(\mathbf{x}, t), \tag{13.122}$$

where c is a constant in the Maxwell equations for the vacuum. So we clearly have

$$\langle \left| \frac{\partial S}{\partial t} \right| \rangle = c. \tag{13.123}$$

Moreover,

$$U(\mathbf{x}, t) = A(\mathbf{x}, t)e^{ik_0 S(\mathbf{x},t)}. \tag{13.124}$$

Then we are in position to insert it into the wave equation (13.114) of the form

$$\left(\frac{\partial}{\partial t} \left\{ \frac{1}{c_0^2} \left(\frac{\partial U}{\partial t} + \tilde{\mathbf{u}} \cdot \nabla_{\mathbf{x}} U \right) \right\} + \operatorname{div}_{\mathbf{x}} \left\{ \frac{1}{c_0^2} \left(\frac{\partial U}{\partial t} + \tilde{\mathbf{u}} \cdot \nabla_{\mathbf{x}} U \right) \tilde{\mathbf{u}} \right\} \right) - \Delta_{\mathbf{x}} U = 0. \tag{13.125}$$

Thus, since by (13.124) we have

$$\begin{cases} \nabla_{\mathbf{x}} U = e^{ik_0 S} \left(\nabla_{\mathbf{x}} A + ik_0 A \nabla_{\mathbf{x}} S \right) \\ \frac{\partial U}{\partial t} + \tilde{\mathbf{u}} \cdot \nabla_{\mathbf{x}} U = e^{ik_0 S} \left(\left(\frac{\partial A}{\partial t} + \tilde{\mathbf{u}} \cdot \nabla_{\mathbf{x}} A \right) + ik_0 A \left(\frac{\partial S}{\partial t} + \tilde{\mathbf{u}} \cdot \nabla_{\mathbf{x}} S \right) \right), \end{cases} \tag{13.126}$$

and furthermore

$$\begin{aligned} &\Delta_{\mathbf{x}} U = \\ &e^{ik_0 S} \left(\Delta_{\mathbf{x}} A + ik_0 \nabla_{\mathbf{x}} A \cdot \nabla_{\mathbf{x}} S + ik_0 A \Delta_{\mathbf{x}} S \right) + ik_0 e^{ik_0 S} \nabla_{\mathbf{x}} S \cdot \left(\nabla_{\mathbf{x}} A + ik_0 A \nabla_{\mathbf{x}} S \right) \\ &\qquad = e^{ik_0 S} \left(\Delta_{\mathbf{x}} A + 2ik_0 \nabla_{\mathbf{x}} A \cdot \nabla_{\mathbf{x}} S + ik_0 A \Delta_{\mathbf{x}} S - k_0^2 A \left| \nabla_{\mathbf{x}} S \right|^2 \right), \end{aligned} \tag{13.127}$$

and

$$\begin{aligned}
&\frac{\partial}{\partial t}\left\{\frac{1}{c_0^2}\left(\frac{\partial U}{\partial t}+\tilde{\mathbf{u}}\cdot\nabla_{\mathbf{x}}U\right)\right\}+\operatorname{div}_{\mathbf{x}}\left\{\frac{1}{c_0^2}\left(\frac{\partial U}{\partial t}+\tilde{\mathbf{u}}\cdot\nabla_{\mathbf{x}}U\right)\tilde{\mathbf{u}}\right\}\\
&\quad=e^{ik_0S}\left(\frac{\partial}{\partial t}\left\{\frac{1}{c_0^2}\left(\frac{\partial A}{\partial t}+\tilde{\mathbf{u}}\cdot\nabla_{\mathbf{x}}A\right)\right\}+\operatorname{div}_{\mathbf{x}}\left\{\frac{1}{c_0^2}\left(\frac{\partial A}{\partial t}+\tilde{\mathbf{u}}\cdot\nabla_{\mathbf{x}}A\right)\tilde{\mathbf{u}}\right\}\right)\\
&\qquad\qquad+ik_0e^{ik_0S}\frac{1}{c_0^2}\left(\frac{\partial A}{\partial t}+\tilde{\mathbf{u}}\cdot\nabla_{\mathbf{x}}A\right)\left(\frac{\partial S}{\partial t}+\tilde{\mathbf{u}}\cdot\nabla_{\mathbf{x}}S\right)\\
&\quad+ik_0e^{ik_0S}A\left(\frac{\partial}{\partial t}\left\{\frac{1}{c_0^2}\left(\frac{\partial S}{\partial t}+\tilde{\mathbf{u}}\cdot\nabla_{\mathbf{x}}S\right)\right\}+\operatorname{div}_{\mathbf{x}}\left\{\frac{1}{c_0^2}\left(\frac{\partial S}{\partial t}+\tilde{\mathbf{u}}\cdot\nabla_{\mathbf{x}}S\right)\tilde{\mathbf{u}}\right\}\right)\\
&\quad+ik_0e^{ik_0S}\frac{1}{c_0^2}\left(\frac{\partial S}{\partial t}+\tilde{\mathbf{u}}\cdot\nabla_{\mathbf{x}}S\right)\left(\left(\frac{\partial A}{\partial t}+\tilde{\mathbf{u}}\cdot\nabla_{\mathbf{x}}A\right)+ik_0A\left(\frac{\partial S}{\partial t}+\tilde{\mathbf{u}}\cdot\nabla_{\mathbf{x}}S\right)\right)\\
&\quad=e^{ik_0S}\left(\frac{\partial}{\partial t}\left\{\frac{1}{c_0^2}\left(\frac{\partial A}{\partial t}+\tilde{\mathbf{u}}\cdot\nabla_{\mathbf{x}}A\right)\right\}+\operatorname{div}_{\mathbf{x}}\left\{\frac{1}{c_0^2}\left(\frac{\partial A}{\partial t}+\tilde{\mathbf{u}}\cdot\nabla_{\mathbf{x}}A\right)\tilde{\mathbf{u}}\right\}\right)\\
&\qquad\qquad+2ik_0\frac{1}{c_0^2}e^{ik_0S}\left(\frac{\partial A}{\partial t}+\tilde{\mathbf{u}}\cdot\nabla_{\mathbf{x}}A\right)\left(\frac{\partial S}{\partial t}+\tilde{\mathbf{u}}\cdot\nabla_{\mathbf{x}}S\right)\\
&\quad+ik_0e^{ik_0S}A\left(\frac{\partial}{\partial t}\left\{\frac{1}{c_0^2}\left(\frac{\partial S}{\partial t}+\tilde{\mathbf{u}}\cdot\nabla_{\mathbf{x}}S\right)\right\}+\operatorname{div}_{\mathbf{x}}\left\{\frac{1}{c_0^2}\left(\frac{\partial S}{\partial t}+\tilde{\mathbf{u}}\cdot\nabla_{\mathbf{x}}S\right)\tilde{\mathbf{u}}\right\}\right)\\
&\qquad\qquad\qquad-k_0^2\frac{1}{c_0^2}e^{ik_0S}A\left(\frac{\partial S}{\partial t}+\tilde{\mathbf{u}}\cdot\nabla_{\mathbf{x}}S\right)^2,
\end{aligned}\tag{13.128}$$

inserting (13.127) and (13.128) into (13.125) we deduce

$$\begin{aligned}
&k_0^2\left(|\nabla_{\mathbf{x}}S|^2-\frac{1}{c_0^2}\left(\frac{\partial S}{\partial t}+\tilde{\mathbf{u}}\cdot\nabla_{\mathbf{x}}S\right)^2\right)A\\
&\quad+\left(\frac{\partial}{\partial t}\left\{\frac{1}{c_0^2}\left(\frac{\partial A}{\partial t}+\tilde{\mathbf{u}}\cdot\nabla_{\mathbf{x}}A\right)\right\}+\operatorname{div}_{\mathbf{x}}\left\{\frac{1}{c_0^2}\left(\frac{\partial A}{\partial t}+\tilde{\mathbf{u}}\cdot\nabla_{\mathbf{x}}A\right)\tilde{\mathbf{u}}\right\}\right)-\Delta_{\mathbf{x}}A\\
&\quad+i\kappa_0A\left(\frac{\partial}{\partial t}\left\{\frac{1}{c_0^2}\left(\frac{\partial S}{\partial t}+\tilde{\mathbf{u}}\cdot\nabla_{\mathbf{x}}S\right)\right\}+\operatorname{div}_{\mathbf{x}}\left\{\frac{1}{c_0^2}\left(\frac{\partial S}{\partial t}+\tilde{\mathbf{u}}\cdot\nabla_{\mathbf{x}}S\right)\tilde{\mathbf{u}}\right\}-\Delta_{\mathbf{x}}S\right)\\
&\quad+2i\kappa_0\left(\frac{1}{c_0^2}\left(\frac{\partial S}{\partial t}+\tilde{\mathbf{u}}\cdot\nabla_{\mathbf{x}}S\right)\left(\frac{\partial A}{\partial t}+\tilde{\mathbf{u}}\cdot\nabla_{\mathbf{x}}A\right)-\nabla_{\mathbf{x}}A\cdot\nabla_{\mathbf{x}}S\right)=0.
\end{aligned}\tag{13.129}$$

Finally, denoting

$$\tilde{\omega}(\mathbf{x},t):=\kappa_0\left(\frac{\partial S}{\partial t}(\mathbf{x},t)+\tilde{\mathbf{u}}(\mathbf{x},t)\cdot\nabla_{\mathbf{x}}S(\mathbf{x},t)\right),\tag{13.130}$$

$$\mathbf{k}(\mathbf{x},t):=c_0(\mathbf{x},t)\left(\frac{\partial S}{\partial t}(\mathbf{x},t)+\tilde{\mathbf{u}}(\mathbf{x},t)\cdot\nabla_{\mathbf{x}}S(\mathbf{x},t)\right)^{-1}\nabla_{\mathbf{x}}S(\mathbf{x},t),\tag{13.131}$$

$$\mathbf{h}(\mathbf{x},t) := \tilde{\mathbf{u}}(\mathbf{x},t) - c_0^2(\mathbf{x},t)\left(\frac{\partial S}{\partial t}(\mathbf{x},t) + \tilde{\mathbf{u}}(\mathbf{x},t)\cdot\nabla_{\mathbf{x}}S(\mathbf{x},t)\right)^{-1}\nabla_{\mathbf{x}}S(\mathbf{x},t)$$
$$= \tilde{\mathbf{u}}(\mathbf{x},t) - c_0(\mathbf{x},t)\mathbf{k}(\mathbf{x},t), \quad (13.132)$$

and

$$\begin{aligned} G(\mathbf{x},t) &:= \left(\frac{c_0^2}{2\left(\frac{\partial S}{\partial t} + \tilde{\mathbf{u}}\cdot\nabla_{\mathbf{x}}S\right)}\left(\Delta_{\mathbf{x}}S - \frac{\partial}{\partial t}\left\{\frac{1}{c_0^2}\left(\frac{\partial S}{\partial t} + \tilde{\mathbf{u}}\cdot\nabla_{\mathbf{x}}S\right)\right\}\right.\right. \\ &\qquad \left.\left. - \operatorname{div}_{\mathbf{x}}\left\{\frac{1}{c_0^2}\left(\frac{\partial S}{\partial t} + \tilde{\mathbf{u}}\cdot\nabla_{\mathbf{x}}S\right)\tilde{\mathbf{u}}\right\}\right)\right)(\mathbf{x},t) \\ &= \left(\frac{c_0^2}{2\tilde{\omega}}\left(\operatorname{div}_{\mathbf{x}}\left\{\frac{\tilde{\omega}}{c_0}\mathbf{k}\right\} - \frac{\partial}{\partial t}\left\{\frac{\tilde{\omega}}{c_0^2}\right\} - \operatorname{div}_{\mathbf{x}}\left\{\frac{\tilde{\omega}}{c_0^2}\tilde{\mathbf{u}}\right\}\right)\right)(\mathbf{x},t) \\ &= -\left(\frac{c_0^2}{2\tilde{\omega}}\left(\frac{\partial}{\partial t}\left\{\frac{\tilde{\omega}}{c_0^2}\right\} + \mathbf{h}\cdot\nabla_{\mathbf{x}}\left\{\frac{\tilde{\omega}}{c_0^2}\right\}\right)\right)(\mathbf{x},t) - \frac{1}{2}\left(\operatorname{div}_{\mathbf{x}}\mathbf{h}(\mathbf{x},t)\right), \end{aligned} \quad (13.133)$$

by inserting (13.130), (13.131), (13.132) and (13.133) into (13.129) we clearly have

$$\begin{aligned} &\frac{\tilde{\omega}}{2}\left(|\mathbf{k}|^2 - 1\right)A \\ &+ \frac{c_0^2}{2\tilde{\omega}}\left\{\frac{\partial}{\partial t}\left\{\frac{1}{c_0^2}\left(\frac{\partial A}{\partial t} + \tilde{\mathbf{u}}\cdot\nabla_{\mathbf{x}}A\right)\right\} + \operatorname{div}_{\mathbf{x}}\left\{\frac{1}{c_0^2}\left(\frac{\partial A}{\partial t} + \tilde{\mathbf{u}}\cdot\nabla_{\mathbf{x}}A\right)\tilde{\mathbf{u}}\right\} - \Delta_{\mathbf{x}}A\right\} \\ &\qquad + i\left(\frac{\partial A}{\partial t} + \mathbf{h}\cdot\nabla_{\mathbf{x}}A - G\,A\right) = 0. \end{aligned} \quad (13.134)$$

In particular, if we assume for the moment that

$$\frac{1}{c_0^2}\left(\frac{\partial S}{\partial t} + \tilde{\mathbf{u}}\cdot\nabla_{\mathbf{x}}S\right)^2 = |\nabla_{\mathbf{x}}S|^2, \quad (13.135)$$

i.e.

$$|\mathbf{k}|^2 = 1, \quad (13.136)$$

then we rewrite (13.134) as

$$\begin{aligned} &\frac{c_0^2}{2\tilde{\omega}}\left\{\frac{\partial}{\partial t}\left\{\frac{1}{c_0^2}\left(\frac{\partial A}{\partial t} + \tilde{\mathbf{u}}\cdot\nabla_{\mathbf{x}}A\right)\right\} + \operatorname{div}_{\mathbf{x}}\left\{\frac{1}{c_0^2}\left(\frac{\partial A}{\partial t} + \tilde{\mathbf{u}}\cdot\nabla_{\mathbf{x}}A\right)\tilde{\mathbf{u}}\right\} - \Delta_{\mathbf{x}}A\right\} \\ &\qquad + i\left(\frac{\partial A}{\partial t} + \mathbf{h}\cdot\nabla_{\mathbf{x}}A - G\,A\right) = 0. \end{aligned} \quad (13.137)$$

Moreover, in the later case, we obviously have

$$\frac{\tilde{\omega}^2}{c_0^2 k_0^2} = \frac{1}{c_0^2}\left(\frac{\partial S}{\partial t} + \tilde{\mathbf{u}} \cdot \nabla_{\mathbf{x}} S\right)^2 = |\nabla_{\mathbf{x}} S|^2, \qquad |\mathbf{k}|^2 = 1$$

$$\text{or equivalently} \qquad |\mathbf{h} - \tilde{\mathbf{u}}|^2 = c_0^2, \tag{13.138}$$

$$\frac{\partial S}{\partial t} + \tilde{\mathbf{u}} \cdot \nabla_{\mathbf{x}} S = c_0 \mathbf{k} \cdot \nabla_{\mathbf{x}} S \qquad \text{or equivalently} \qquad \frac{\partial S}{\partial t} + \mathbf{h} \cdot \nabla_{\mathbf{x}} S = 0, \tag{13.139}$$

$$\nabla_{\mathbf{x}} S = (\mathbf{k}(\mathbf{x}, t) \cdot \nabla_{\mathbf{x}} S)\, \mathbf{k}, \tag{13.140}$$

$$\frac{c_0}{\tilde{\omega}} \mathbf{k} = \frac{1}{k_0} |\nabla_{\mathbf{x}} S|^{-2}\, \nabla_{\mathbf{x}} S = \frac{c_0^2 k_0}{\tilde{\omega}^2} \nabla_{\mathbf{x}} S, \tag{13.141}$$

and

$$\frac{1}{k_0} \nabla_{\mathbf{x}} \tilde{\omega} = \frac{\partial}{\partial t} \{\nabla_{\mathbf{x}} S\} + \nabla_{\mathbf{x}}^2 S \cdot \tilde{\mathbf{u}} + \{d_{\mathbf{x}} \tilde{\mathbf{u}}\}^T \cdot \nabla_{\mathbf{x}} S. \tag{13.142}$$

In particular,

$$\frac{2}{k_0} \nabla_{\mathbf{x}} S \cdot \nabla_{\mathbf{x}} \tilde{\omega} = \frac{\partial}{\partial t} \left\{|\nabla_{\mathbf{x}} S|^2\right\} + \tilde{\mathbf{u}} \cdot \nabla_{\mathbf{x}} \left\{|\nabla_{\mathbf{x}} S|^2\right\} + \left(\nabla_{\mathbf{x}} S \cdot \left(\{d_{\mathbf{x}} \tilde{\mathbf{u}}\}^T + d_{\mathbf{x}} \tilde{\mathbf{u}}\right)\right) \cdot \nabla_{\mathbf{x}} S, \tag{13.143}$$

so that

$$c_0 \frac{1}{c_0^2} \mathbf{k} \cdot \nabla_{\mathbf{x}} \left\{\tilde{\omega}^2\right\} = \frac{\partial}{\partial t} \left\{\frac{\tilde{\omega}^2}{c_0^2}\right\} + \tilde{\mathbf{u}} \cdot \nabla_{\mathbf{x}} \left\{\frac{\tilde{\omega}^2}{c_0^2}\right\} + \frac{\tilde{\omega}^2}{c_0^2} \left(\mathbf{k} \cdot \left(\{d_{\mathbf{x}} \tilde{\mathbf{u}}\}^T + d_{\mathbf{x}} \tilde{\mathbf{u}}\right)\right) \cdot \mathbf{k}. \tag{13.144}$$

I.e.

$$\left(\frac{\tilde{\omega}^2}{c_0^2}\right)^{-1} \left(\frac{\partial}{\partial t} \left\{\frac{\tilde{\omega}^2}{c_0^2}\right\} + \mathbf{h} \cdot \nabla_{\mathbf{x}} \left\{\frac{\tilde{\omega}^2}{c_0^2}\right\}\right) = \frac{1}{c_0^2} (c_0 \mathbf{k}) \cdot \nabla_{\mathbf{x}} \left\{c_0^2\right\} - \left(\mathbf{k} \cdot \left(\{d_{\mathbf{x}} \tilde{\mathbf{u}}\}^T + d_{\mathbf{x}} \tilde{\mathbf{u}}\right)\right) \cdot \mathbf{k}. \tag{13.145}$$

We rewrite it as

$$\frac{1}{\tilde{\omega}} \left(\frac{\partial \tilde{\omega}}{\partial t} + \mathbf{h} \cdot \nabla_{\mathbf{x}} \tilde{\omega}\right) = \frac{1}{c_0} \left(\frac{\partial c_0}{\partial t} + \tilde{\mathbf{u}} \cdot \nabla_{\mathbf{x}} c_0\right) - \frac{1}{2} \left(\mathbf{k} \cdot \left(\{d_{\mathbf{x}} \tilde{\mathbf{u}}\}^T + d_{\mathbf{x}} \tilde{\mathbf{u}}\right)\right) \cdot \mathbf{k}, \tag{13.146}$$

provided, we have (13.135).

13.3.2 Derivation of the Eikonal Equation

In particular, in the case both S and A are real, since the zero complex number has both real and imaginary parts equal to zero, by (13.129) we have

$$k_0^2\left(|\nabla_{\mathbf{x}}S|^2-\frac{1}{c_0^2}\left(\frac{\partial S}{\partial t}+\tilde{\mathbf{u}}\cdot\nabla_{\mathbf{x}}S\right)^2\right)A$$
$$+\frac{\partial}{\partial t}\left\{\frac{1}{c_0^2}\left(\frac{\partial A}{\partial t}+\tilde{\mathbf{u}}\cdot\nabla_{\mathbf{x}}A\right)\right\}+\operatorname{div}_{\mathbf{x}}\left\{\frac{1}{c_0^2}\left(\frac{\partial A}{\partial t}+\tilde{\mathbf{u}}\cdot\nabla_{\mathbf{x}}A\right)\tilde{\mathbf{u}}\right\}-\Delta_{\mathbf{x}}A=0, \tag{13.147}$$

and

$$A\left(\frac{\partial}{\partial t}\left\{\frac{1}{c_0^2}\left(\frac{\partial S}{\partial t}+\tilde{\mathbf{u}}\cdot\nabla_{\mathbf{x}}S\right)\right\}+\operatorname{div}_{\mathbf{x}}\left\{\frac{1}{c_0^2}\left(\frac{\partial S}{\partial t}+\tilde{\mathbf{u}}\cdot\nabla_{\mathbf{x}}S\right)\tilde{\mathbf{u}}\right\}-\Delta_{\mathbf{x}}S\right)$$
$$+2\left(\frac{1}{c_0^2}\left(\frac{\partial S}{\partial t}+\tilde{\mathbf{u}}\cdot\nabla_{\mathbf{x}}S\right)\left(\frac{\partial A}{\partial t}+\tilde{\mathbf{u}}\cdot\nabla_{\mathbf{x}}A\right)-\nabla_{\mathbf{x}}A\cdot\nabla_{\mathbf{x}}S\right)=0. \tag{13.148}$$

Next assume either the delicate or the rough geometric optics approximation that is good for the electromagnetic wave of high frequency, for example, for the visible light. The delicate geometric optics approximation means the following: assume that the changes in time of c_0, $\tilde{\mathbf{u}}$, A and S become essential after certain interval of time T_e and the changes in space of c_0, A and S become essential in the spatial landscape L_e. Then we assume that

$$k_0^2c_0^2T_e^2\gg 1\quad\text{and}\quad k_0^2L_e^2\gg 1. \tag{13.149}$$

On the other hand, the rough geometric optics approximation (stronger than (13.149)) means the following:

$$k_0c_0T_e\gg 1\quad\text{and}\quad k_0L_e\gg 1. \tag{13.150}$$

Moreover, we assume that the order of c_0 is less or equal to the order of c. In particular, estimation (13.149) implies

$$\frac{\left|d_{\mathbf{x}}\tilde{\mathbf{u}}+\{d_{\mathbf{x}}\tilde{\mathbf{u}}\}^T\right|^2}{c_0^2}+\frac{\left(|\nabla_{\mathbf{x}}A|+\frac{1}{c_0}\left|\frac{\partial A}{\partial t}+\tilde{\mathbf{u}}\cdot\nabla_{\mathbf{x}}A\right|\right)^2}{|A|^2}+\frac{\left(|\nabla_{\mathbf{x}}c_0|+\frac{1}{c_0}\left|\frac{\partial c_0}{\partial t}+\tilde{\mathbf{u}}\cdot\nabla_{\mathbf{x}}c_0\right|\right)^2}{|c_0|^2}$$
$$+\frac{\left|\nabla_{\mathbf{x}}^2S\right|^2+\frac{1}{c_0^2}\left|\nabla_{\mathbf{x}}\left(\frac{\partial S}{\partial t}+\tilde{\mathbf{u}}\cdot\nabla_{\mathbf{x}}S\right)\right|^2}{|\nabla_{\mathbf{x}}S|^2}\ll k_0^2\quad\text{and}\quad\frac{\left|\frac{\partial S}{\partial t}+\tilde{\mathbf{u}}\cdot\nabla_{\mathbf{x}}S-c\right|^2}{c^2}\ll 1. \tag{13.151}$$

Thus, using (13.149), we approximate (13.147) as

$$\frac{1}{c_0^2}\left(\frac{\partial S}{\partial t}+\tilde{\mathbf{u}}\cdot\nabla_{\mathbf{x}}S\right)^2=|\nabla_{\mathbf{x}}S|^2\,. \tag{13.152}$$

Moreover, without any use in either (13.149) or (13.150) we rewrite (13.148) as

$$\left(\frac{\partial}{\partial t}\left\{\frac{1}{c_0^2}\left(\frac{\partial S}{\partial t}+\tilde{\mathbf{u}}\cdot\nabla_{\mathbf{x}}S\right)\right\}+\operatorname{div}_{\mathbf{x}}\left\{\frac{1}{c_0^2}\left(\frac{\partial S}{\partial t}+\tilde{\mathbf{u}}\cdot\nabla_{\mathbf{x}}S\right)\tilde{\mathbf{u}}\right\}-(\Delta_{\mathbf{x}}S)\right)A$$
$$+\frac{2}{c_0^2}\left(\frac{\partial S}{\partial t}+\tilde{\mathbf{u}}\cdot\nabla_{\mathbf{x}}S\right)\left(\frac{\partial A}{\partial t}+\tilde{\mathbf{u}}\cdot\nabla_{\mathbf{x}}A\right)-2\nabla_{\mathbf{x}}S\cdot\nabla_{\mathbf{x}}A=0. \tag{13.153}$$

Equality (13.152) is called the eikonal equation and equality (13.153) is called the equation of the beam propagation. Then, as before, we deduce that Eq. (13.152) is invariant under the change of non-inertial Cartesian coordinate system, provided that under such change we have $S'=S$. Moreover, (13.153) is also invariant under the change of non-inertial Cartesian coordinate system, in the case that under such change we have $A'=A$, provided that $S'=S$. Furthermore, note that although we established eikonal equation (13.152) in the delicate geometric optics approximation (13.149), since by (13.124) we have

$$U(\mathbf{x},t)=A(\mathbf{x},t)e^{ik_0S(\mathbf{x},t)}, \tag{13.154}$$

then, considering the approximations of A and S as in (13.153), (13.152) and putting them into (13.154) we actually establish $U(\mathbf{x},t)$ only in the rough geometric optics approximation (13.150).

Finally, as before, denoting

$$\tilde{\omega}(\mathbf{x},t):=\kappa_0\left(\frac{\partial S}{\partial t}(\mathbf{x},t)+\tilde{\mathbf{u}}(\mathbf{x},t)\cdot\nabla_{\mathbf{x}}S(\mathbf{x},t)\right), \tag{13.155}$$

$$\mathbf{k}(\mathbf{x},t):=c_0(\mathbf{x},t)\left(\frac{\partial S}{\partial t}(\mathbf{x},t)+\tilde{\mathbf{u}}(\mathbf{x},t)\cdot\nabla_{\mathbf{x}}S(\mathbf{x},t)\right)^{-1}\nabla_{\mathbf{x}}S(\mathbf{x},t), \tag{13.156}$$

$$\mathbf{h}(\mathbf{x},t):=\tilde{\mathbf{u}}(\mathbf{x},t)-c_0^2(\mathbf{x},t)\left(\frac{\partial S}{\partial t}(\mathbf{x},t)+\tilde{\mathbf{u}}(\mathbf{x},t)\cdot\nabla_{\mathbf{x}}S(\mathbf{x},t)\right)^{-1}\nabla_{\mathbf{x}}S(\mathbf{x},t)$$
$$=\tilde{\mathbf{u}}(\mathbf{x},t)-c_0(\mathbf{x},t)\mathbf{k}(\mathbf{x},t), \tag{13.157}$$

and

$$G(\mathbf{x},t) :=$$

$$\left(\frac{c_0^2}{2\left(\frac{\partial S}{\partial t} + \tilde{\mathbf{u}} \cdot \nabla_{\mathbf{x}} S\right)} \left(\Delta_{\mathbf{x}} S - \frac{\partial}{\partial t} \left\{ \frac{1}{c_0^2} \left(\frac{\partial S}{\partial t} + \tilde{\mathbf{u}} \cdot \nabla_{\mathbf{x}} S \right) \right\} \right.\right.$$
$$\left.\left. - \operatorname{div}_{\mathbf{x}} \left\{ \frac{1}{c_0^2} \left(\frac{\partial S}{\partial t} + \tilde{\mathbf{u}} \cdot \nabla_{\mathbf{x}} S \right) \tilde{\mathbf{u}} \right\} \right) \right) (\mathbf{x},t)$$
$$= \left(\frac{c_0^2}{2\tilde{\omega}} \left(\operatorname{div}_{\mathbf{x}} \left\{ \frac{\tilde{\omega}}{c_0} \mathbf{k} \right\} - \frac{\partial}{\partial t} \left\{ \frac{\tilde{\omega}}{c_0^2} \right\} - \operatorname{div}_{\mathbf{x}} \left\{ \frac{\tilde{\omega}}{c_0^2} \tilde{\mathbf{u}} \right\} \right) \right) (\mathbf{x},t)$$
$$= - \left(\frac{c_0^2}{2\tilde{\omega}} \left(\frac{\partial}{\partial t} \left\{ \frac{\tilde{\omega}}{c_0^2} \right\} + \mathbf{h} \cdot \nabla_{\mathbf{x}} \left\{ \frac{\tilde{\omega}}{c_0^2} \right\} \right) \right) (\mathbf{x},t) - \frac{1}{2} \left(\operatorname{div}_{\mathbf{x}} \mathbf{h}(\mathbf{x},t) \right), \quad (13.158)$$

we rewrite (13.153) and (13.152) as

$$\frac{\partial A}{\partial t}(\mathbf{x},t) + \mathbf{h}(\mathbf{x},t) \cdot \nabla_{\mathbf{x}} A(\mathbf{x},t) - G(\mathbf{x},t) A(\mathbf{x},t) = 0, \quad (13.159)$$

and

$$|\mathbf{k}(\mathbf{x},t)|^2 = 1 \quad \text{or equivalently} \quad |\mathbf{h}(\mathbf{x},t) - \tilde{\mathbf{u}}(\mathbf{x},t)|^2 = c_0^2(\mathbf{x},t). \quad (13.160)$$

Moreover, as before, we obviously have

$$\frac{\tilde{\omega}^2}{c_0^2 k_0^2} = \frac{1}{c_0^2} \left(\frac{\partial S}{\partial t} + \tilde{\mathbf{u}} \cdot \nabla_{\mathbf{x}} S \right)^2 = |\nabla_{\mathbf{x}} S|^2, \qquad |\mathbf{k}|^2 = 1$$
$$\text{or equivalently} \quad |\mathbf{h} - \tilde{\mathbf{u}}|^2 = c_0^2, \quad (13.161)$$

$$\frac{\partial S}{\partial t} + \tilde{\mathbf{u}} \cdot \nabla_{\mathbf{x}} S = c_0 \mathbf{k} \cdot \nabla_{\mathbf{x}} S \quad \text{or equivalently} \quad \frac{\partial S}{\partial t} + \mathbf{h} \cdot \nabla_{\mathbf{x}} S = 0, \quad (13.162)$$

$$\nabla_{\mathbf{x}} S = \left(\mathbf{k}(\mathbf{x},t) \cdot \nabla_{\mathbf{x}} S \right) \mathbf{k}, \quad (13.163)$$

$$\frac{c_0}{\tilde{\omega}} \mathbf{k} = \frac{1}{k_0} |\nabla_{\mathbf{x}} S|^{-2} \nabla_{\mathbf{x}} S = \frac{c_0^2 k_0}{\tilde{\omega}^2} \nabla_{\mathbf{x}} S, \quad (13.164)$$

$$\frac{1}{k_0} \nabla_{\mathbf{x}} \tilde{\omega} = \frac{\partial}{\partial t} \{\nabla_{\mathbf{x}} S\} + \nabla_{\mathbf{x}}^2 S \cdot \tilde{\mathbf{u}} + \{d_{\mathbf{x}} \tilde{\mathbf{u}}\}^T \cdot \nabla_{\mathbf{x}} S, \quad (13.165)$$

and

$$\frac{1}{\tilde{\omega}} \left(\frac{\partial \tilde{\omega}}{\partial t} + \mathbf{h} \cdot \nabla_{\mathbf{x}} \tilde{\omega} \right) = \frac{1}{c_0} \left(\frac{\partial c_0}{\partial t} + \tilde{\mathbf{u}} \cdot \nabla_{\mathbf{x}} c_0 \right) - \frac{1}{2} \left(\mathbf{k} \cdot \left(\{d_{\mathbf{x}} \tilde{\mathbf{u}}\}^T + d_{\mathbf{x}} \tilde{\mathbf{u}} \right) \right) \cdot \mathbf{k}. \quad (13.166)$$

Next if we consider an arbitrary characteristic curve $\mathbf{r}(t) : [t_0, b] \to \mathbb{R}^3$ of Eq. (13.159), parameterized by the time variable t, defined as a solution of ordinary differential equation:

$$\begin{cases} \frac{d\mathbf{r}}{dt}(t) = \mathbf{h}\left(\mathbf{r}(t), t\right) \\ \mathbf{r}(t_0) = \mathbf{x}_0, \end{cases} \tag{13.167}$$

where $\mathbf{h}$ was defined in (13.157), then, as before, by (13.159), (13.167) and the chain rule we have

$$\frac{d}{dt}\left(A\left(\mathbf{r}(t), t\right)\right) = \nabla_{\mathbf{x}} A\left(\mathbf{r}(t), t\right) \cdot \frac{d\mathbf{r}}{dt}(t) + \frac{\partial A}{\partial t}\left(\mathbf{r}(t), t\right) = G\left(\mathbf{r}(t), t\right) A\left(\mathbf{r}(t), t\right), \tag{13.168}$$

where G was defined in (13.158). Then (13.168) implies

$$A\left(\mathbf{r}(t), t\right) = A\left(\mathbf{x}_0, t_0\right) e^{\int_{t_0}^{t} G(\mathbf{r}(\tau), \tau) d\tau} \quad \forall t \in [t_0, b]. \tag{13.169}$$

In particular,

$$\begin{aligned} &A\left(\mathbf{x}_0, t_0\right) = 0 \;\text{ implies }\; A\left(\mathbf{r}(t), t\right) = 0 \;\; \forall t \in [t_0, b], \\ &\qquad \text{and }\; A\left(\mathbf{x}_0, t_0\right) \neq 0 \;\text{ implies }\; A\left(\mathbf{r}(t), t\right) \neq 0 \;\; \forall t \in [t_0, b]. \end{aligned} \tag{13.170}$$

Therefore, by (13.170) we deduce that the curve that satisfies (13.167) coincides with the ray of light that passes through the point $\mathbf{x}_0$ at the instant of time t_0. So, equality (13.167) is the equation of a ray and the vector field $\mathbf{h}$ defined for every $\mathbf{x}$ by (13.157) is the direction of the propagation of the ray that passes through point $\mathbf{x}$ at the instant of time t. On the other hand, as before, we can easily prove that the vector field defined in every inertial or non-inertial coordinate system by (13.157) is a speed-like vector field. Moreover, by (13.161) the following implication holds:

$$\tilde{\mathbf{u}} = 0 \;\text{ implies }\; |\mathbf{h}| = c_0. \tag{13.171}$$

Finally, by chain rule, for the curve that satisfies (13.167) we have

$$\begin{aligned} \frac{d}{dt}\left(S\left(\mathbf{r}(t), t\right)\right) &= \frac{\partial S}{\partial t}\left(\mathbf{r}(t), t\right) + \nabla_{\mathbf{x}} S\left(\mathbf{r}(t), t\right) \cdot \frac{d\mathbf{r}}{dt}(t) \\ &= \frac{\partial S}{\partial t}\left(\mathbf{r}(t), t\right) + \nabla_{\mathbf{x}} S\left(\mathbf{r}(t), t\right) \cdot \mathbf{h}\left(\mathbf{r}(t), t\right). \end{aligned} \tag{13.172}$$

Thus, by (13.162) and (13.172) we deduce

$$\frac{d}{dt}\left(S\left(\mathbf{r}(t), t\right)\right) = 0 \quad \forall t \geq t_0, \tag{13.173}$$

and so

$$S(\mathbf{r}(t), t) = S(\mathbf{x}_0, t_0) \qquad \forall t \geq t_0. \tag{13.174}$$

By all these facts, vector field $\mathbf{h}(\mathbf{x}, t)$ defined by (13.157) can be considered as the vector of the velocity (speed) of the wave at the point $\mathbf{x}$ at the instant of time t. Moreover, by (13.161) we have

$$|\mathbf{h} - \tilde{\mathbf{u}}|^2 = c_0^2. \tag{13.175}$$

Remark 13.2 In contrast to the proof of (13.152), we do not use any of the geometric optics approximations (13.149) or (13.150) in the proof of (13.153) and (13.159). So, (13.170) is still valid without assumption of the geometric optics approximations (13.149) or (13.150), and thus the vector field $\mathbf{h}$, defined by (13.157), is the direction of the propagation of the ray that passes through point $\mathbf{x}$ also in the general case. However, without assumptions of the geometric optics approximation we cannot derive (13.152), (13.171), (13.174) and (13.175) anymore.

13.3.3 The Case of the Monochromatic Wave

Next, up to the end of this section, consider the case of monochromatic wave of the constant frequency $\nu = \frac{\omega}{2\pi}$ where the fields $\tilde{\mathbf{u}}$ and c_0 are independent on the time variable, i.e. the case of (13.120) where we have

$$\begin{cases} \frac{\partial T}{\partial t} = \omega \\ \frac{\partial A}{\partial t} = 0 \\ \frac{\partial \tilde{\mathbf{u}}}{\partial t} = 0 \\ \frac{\partial c_0}{\partial t} = 0. \end{cases} \tag{13.176}$$

Moreover, we assume the rough geometric optics approximation (13.150). We also assume that either our medium has no dispersion (i.e. in the case of an electromagnetic wave (12.42) is valid), or the velocity of the medium is negligible, i.e. $\mathbf{u} = 0$ (and so in the case of an electromagnetic wave we have $\tilde{\mathbf{u}} = \mathbf{v}$) and our medium is transparent for the given frequency $\nu = \frac{\omega}{2\pi}$. Then, by (13.121) and (13.122) we rewrite (13.176) as

$$\begin{cases} \frac{\partial S}{\partial t} = c \\ \frac{\partial A}{\partial t} = 0. \end{cases} \tag{13.177}$$

Thus $\nabla_{\mathbf{x}} S$ is independent on t, and moreover we rewrite (13.152) and (13.153) as

$$\frac{c^2}{c_0^2}\left(1+\frac{1}{c}\tilde{\mathbf{u}}\cdot\nabla_{\mathbf{x}} S\right)^2 = |\nabla_{\mathbf{x}} S|^2, \tag{13.178}$$

and

$$2\left(\nabla_{\mathbf{x}} S - \frac{c}{c_0}\left(1+\frac{1}{c}\left(\tilde{\mathbf{u}}\cdot\nabla_{\mathbf{x}} S\right)\right)\frac{\tilde{\mathbf{u}}}{c_0}\right)\cdot\nabla_{\mathbf{x}} A$$
$$= \left(\operatorname{div}_{\mathbf{x}}\left\{\frac{1}{c_0^2}\left(c+\tilde{\mathbf{u}}\cdot\nabla_{\mathbf{x}} S\right)\tilde{\mathbf{u}}\right\} - \left(\Delta_{\mathbf{x}} S\right)\right)A. \tag{13.179}$$

In particular, in the case of the region of the space where the following approximation is valid:

$$\frac{|\tilde{\mathbf{u}}|^2}{c_0^2} \ll 1, \tag{13.180}$$

up to order $O\left(\frac{|\tilde{\mathbf{u}}|^2}{c_0^2}\right)$, we rewrite (13.178) as

$$\left|\frac{c\tilde{\mathbf{u}}}{c_0^2} - \nabla_{\mathbf{x}} S\right|^2 = \frac{c^2}{c_0^2}, \tag{13.181}$$

and (13.179) as

$$\left(\frac{c\tilde{\mathbf{u}}}{c_0^2} - \nabla_{\mathbf{x}} S\right)\cdot\nabla_{\mathbf{x}} A + \frac{1}{2}\left(\operatorname{div}_{\mathbf{x}}\left\{\frac{c}{c_0^2}\tilde{\mathbf{u}}\right\} - \Delta_{\mathbf{x}} S\right)A = 0. \tag{13.182}$$

The eikonal equation (13.181) and equation of the beam propagation (13.182) are two basic equations of propagation of monochromatic light in the geometric optics approximation inside a moving medium or/and in the presence of nontrivial gravitational field, provided that the field $\tilde{\mathbf{u}}$ satisfies (13.180).

Next if we consider an arbitrary characteristic curve $\mathbf{r}(s) : [a, b] \to \mathbb{R}^3$ of Eq. (13.182) defined as a solution of ordinary differential equation

$$\begin{cases} \frac{d\mathbf{r}}{ds}(s) = \frac{c}{c_0^2(\mathbf{r}(s))}\tilde{\mathbf{u}}\left(\mathbf{r}(s)\right) - \nabla_{\mathbf{x}} S\left(\mathbf{r}(s)\right) \\ \mathbf{r}(a) = \mathbf{x}_0, \end{cases} \tag{13.183}$$

then, as before, by (13.182) and (13.183) we have

$$\frac{d}{ds}\left(A\left(\mathbf{r}(s)\right)\right) = \nabla_{\mathbf{x}} A\left(\mathbf{r}(s)\right) \cdot \frac{d\mathbf{r}}{ds}(s)$$
$$= \frac{1}{2}\left(\Delta_{\mathbf{x}} S\left(\mathbf{r}(s)\right) - \mathrm{div}_{\mathbf{x}}\left\{\frac{c}{c_0^2\left(\mathbf{r}(s)\right)}\tilde{\mathbf{u}}\left(\mathbf{r}(s)\right)\right\}\right) A\left(\mathbf{r}(s)\right) \quad (13.184)$$

that implies

$$A\left(\mathbf{r}(s)\right) = A\left(\mathbf{x}_0\right) e^{\frac{1}{2}\int_a^s \left(\Delta_{\mathbf{x}} S(\mathbf{r}(\tau)) - \mathrm{div}_{\mathbf{x}}\left\{\frac{c}{c_0^2(\mathbf{r}(\tau))}\tilde{\mathbf{u}}(\mathbf{r}(\tau))\right\}\right) d\tau} \quad \forall s \in [a, b]. \quad (13.185)$$

In particular,

$$A\left(\mathbf{x}_0\right) = 0 \text{ implies } A\left(\mathbf{r}(s)\right) = 0 \quad \forall s \in [a, b], \quad \text{and} \quad A\left(\mathbf{x}_0\right) \neq 0$$
$$\text{implies } A\left(\mathbf{r}(s)\right) \neq 0 \quad \forall s \in [a, b]. \quad (13.186)$$

Therefore, by (13.186) we deduce that in the case of (13.180) the curve that satisfies (13.183) coincides with the ray of light that passes through the point $\mathbf{x}_0$. So in the case of (13.180), equality (13.183) is the equation of a ray and the vector field $\mathbf{h}_0$, defined for every $\mathbf{x}$ by

$$\mathbf{h}_0(\mathbf{x}) := \frac{c}{c_0^2(\mathbf{x})}\tilde{\mathbf{u}}(\mathbf{x}) - \nabla_{\mathbf{x}} S(\mathbf{x}) \approx \frac{c}{c_0^2(\mathbf{x})}\mathbf{h}(\mathbf{x}) \quad (13.187)$$

is the direction of the propagation of the ray that passes through point $\mathbf{x}$. Moreover, by (13.181) $\mathbf{h}_0$ satisfies

$$|\mathbf{h}_0|^2 = \frac{c^2}{c_0^2}. \quad (13.188)$$

Remark 13.3 As before in Remark 13.2, in contrast to the proof of (13.152), (13.178) or (13.181), we do not use the geometric optics approximations (13.149) or (13.150) in the proof of (13.148), (13.153), (13.179) and (13.182). We just need the estimation (13.180) for the proof of (13.182). So, (13.186) is still valid without assumption of the geometric optics approximations (13.149) or (13.150), and thus the vector field $\mathbf{h}_0$, defined by (13.187), is the direction of the propagation of the ray that passes through point $\mathbf{x}$ also in the general case, provided the estimation (13.180) holds. However, without assumption of the geometric optics approximation we cannot derive (13.188) anymore.

Next, under the approximation (13.180) consider a curve $\mathbf{r}(s) : [a, b] \to \mathbb{R}^3$ with endpoints $\mathbf{r}(a) = N$ and $\mathbf{r}(b) = M$. Then integrating the square root of both sides of (13.181) over the curve $\mathbf{r}(s)$ we deduce

$$\int_a^b \left|\frac{c}{c_0^2\left(\mathbf{r}(s)\right)}\tilde{\mathbf{u}}\left(\mathbf{r}(s)\right) - \nabla_{\mathbf{x}} S\left(\mathbf{r}(s)\right)\right| \left|\mathbf{r}'(s)\right| ds = \int_a^b \frac{c}{c_0\left(\mathbf{r}(s)\right)}\left|\mathbf{r}'(s)\right| ds. \quad (13.189)$$

Thus, in particular,

$$\int_a^b \left(\frac{c}{c_0^2(\mathbf{r}(s))}\tilde{\mathbf{u}}(\mathbf{r}(s)) - \nabla_{\mathbf{x}} S(\mathbf{r}(s))\right) \cdot \mathbf{r}'(s)ds \le \int_a^b \frac{c}{c_0(\mathbf{r}(s))}\left|\mathbf{r}'(s)\right| ds, \tag{13.190}$$

i.e.

$$(-S(M)) - (-S(N)) \le \int_a^b \frac{c}{c_0(\mathbf{r}(s))}\left|\mathbf{r}'(s)\right| ds - \int_a^b \frac{c}{c_0^2(\mathbf{r}(s))}\tilde{\mathbf{u}}(\mathbf{r}(s)) \cdot \mathbf{r}'(s)ds. \tag{13.191}$$

Moreover, if

$$\frac{d\mathbf{r}}{ds}(s) = \sigma(s)\mathbf{h}_0(\mathbf{r}(s)) := \sigma(s)\left(\frac{c}{c_0^2(\mathbf{r}(s))}\tilde{\mathbf{u}}(\mathbf{r}(s)) - \nabla_{\mathbf{x}} S(\mathbf{r}(s))\right), \tag{13.192}$$

for some nonnegative scalar factor $\sigma = \sigma(s)$ then by (13.192) we rewrite (13.189) as

$$(-S(M)) - (-S(N)) = \int_a^b \frac{c}{c_0(\mathbf{r}(s))}\left|\mathbf{r}'(s)\right| ds - \int_a^b \frac{c}{c_0^2(\mathbf{r}(s))}\tilde{\mathbf{u}}(\mathbf{r}(s)) \cdot \mathbf{r}'(s)ds. \tag{13.193}$$

Thus, by comparing (13.183) with (13.192) and using (13.191) and (13.193), we deduce that if we assume that the light travel from the point N to the point M across the curve $\tilde{\mathbf{r}}(s) : [a, b] \to \mathbb{R}^3$ such that $\mathbf{r}(\tilde{a}) = N$ and $\tilde{\mathbf{r}}(b) = M$, then

$$(-S(M)) - (-S(N)) = \int_a^b \frac{c}{c_0(\tilde{\mathbf{r}}(s))}\left|\tilde{\mathbf{r}}'(s)\right| ds - \int_a^b \frac{c}{c_0^2(\tilde{\mathbf{r}}(s))}\tilde{\mathbf{u}}(\tilde{\mathbf{r}}(s)) \cdot \tilde{\mathbf{r}}'(s)ds, \tag{13.194}$$

and for every other curve $\mathbf{r}(s) : [a, b] \to \mathbb{R}^3$ with endpoints $\mathbf{r}(a) = N$ and $\mathbf{r}(b) = M$ we have

$$\int_a^b \frac{c}{c_0(\mathbf{r}(s))}\left|\mathbf{r}'(s)\right| ds - \int_a^b \frac{c}{c_0^2(\mathbf{r}(s))}\tilde{\mathbf{u}}(\mathbf{r}(s)) \cdot \mathbf{r}'(s)ds \ge$$
$$\int_a^b \frac{c}{c_0(\tilde{\mathbf{r}}(s))}\left|\tilde{\mathbf{r}}'(s)\right| ds - \int_a^b \frac{c}{c_0^2(\tilde{\mathbf{r}}(s))}\tilde{\mathbf{u}}(\tilde{\mathbf{r}}(s)) \cdot \tilde{\mathbf{r}}'(s)ds. \tag{13.195}$$

So we obtain the following Fermat principle:

Proposition 13.1 *Assume geometric optics approximation together with* (13.180). *Then the light that travels from point N to point M chooses the path* $\mathbf{r}(s) : [a, b] \to \mathbb{R}^3$ *with endpoints* $\mathbf{r}(a) = N$ *and* $\mathbf{r}(b) = M$ *which minimizes the quantity:*

$$J(\mathbf{r}(\cdot)) := \int_a^b n(\mathbf{r}(s))\left|\mathbf{r}'(s)\right| ds - \int_a^b \frac{1}{c} n^2(\mathbf{r}(s))\,\tilde{\mathbf{u}}(\mathbf{r}(s)) \cdot \mathbf{r}'(s)ds, \tag{13.196}$$

where we set the refraction index:

$$n(\mathbf{x}) := \frac{c}{c_0(\mathbf{x})}. \tag{13.197}$$

Moreover, if $\mathbf{r}(s) : [a, b] \to \mathbb{R}^3$ *with endpoints* $\mathbf{r}(a) = N$ *and* $\mathbf{r}(b) = M$ *is the real path of the light, then:*

$$(-S(M)) - (-S(N)) = \int_a^b n(\mathbf{r}(s)) \left|\mathbf{r}'(s)\right| ds - \int_a^b \frac{1}{c} n^2(\mathbf{r}(s)) \tilde{\mathbf{u}}(\mathbf{r}(s)) \cdot \mathbf{r}'(s) ds. \tag{13.198}$$

In particular, by Proposition 13.1 the path of travel of the light satisfies the Euler–Lagrange equation for the functional $J(\mathbf{r}(\cdot))$:

$$\begin{aligned} &\frac{d}{ds}\left(n(\mathbf{r}(s)) \frac{1}{|\mathbf{r}'(s)|}\mathbf{r}'(s) - \frac{1}{c}n^2(\mathbf{r}(s))\tilde{\mathbf{u}}(\mathbf{r}(s))\right) = \left|\mathbf{r}'(s)\right| \nabla_{\mathbf{x}} n(\mathbf{r}(s)) \\ &\quad - \frac{2}{c}\left(\tilde{\mathbf{u}}(\mathbf{r}(s)) \cdot \mathbf{r}'(s)\right) n(\mathbf{r}(s)) \nabla_{\mathbf{x}} n(\mathbf{r}(s)) - \frac{1}{c}n^2(\mathbf{r}(s)) \left\{d_{\mathbf{x}}\tilde{\mathbf{u}}(\mathbf{r}(s))\right\}^T \cdot \mathbf{r}'(s), \end{aligned} \tag{13.199}$$

that we rewrite as

$$\begin{aligned} &\frac{1}{|\mathbf{r}'(s)|}\frac{d}{ds}\left(n(\mathbf{r}(s)) \frac{1}{|\mathbf{r}'(s)|}\mathbf{r}'(s)\right) = \\ &\quad \nabla_{\mathbf{x}} n(\mathbf{r}(s)) + \frac{1}{c}n^2(\mathbf{r}(s)) \left(d_{\mathbf{x}}\tilde{\mathbf{u}}(\mathbf{r}(s)) - \left\{d_{\mathbf{x}}\tilde{\mathbf{u}}(\mathbf{r}(s))\right\}^T\right) \cdot \left(\frac{1}{|\mathbf{r}'(s)|}\mathbf{r}'(s)\right) \\ &+ \frac{2}{c}n(\mathbf{r}(s)) \left\{\tilde{\mathbf{u}}(\mathbf{r}(s)) \otimes \nabla_{\mathbf{x}} n(\mathbf{r}(s)) - \nabla_{\mathbf{x}} n(\mathbf{r}(s)) \otimes \tilde{\mathbf{u}}(\mathbf{r}(s))\right\} \cdot \left(\frac{1}{|\mathbf{r}'(s)|}\mathbf{r}'(s)\right). \end{aligned} \tag{13.200}$$

Therefore by (2.15) and (13.200) we deduce the differential equation of the path of light:

$$\begin{aligned} &\frac{d}{d\lambda}\left(n(\mathbf{r}) \frac{d\mathbf{r}}{d\lambda}\right) = \frac{1}{c}n^2(\mathbf{r}) \left(\mathrm{curl}_{\mathbf{x}}\tilde{\mathbf{u}}(\mathbf{r})\right) \times \frac{d\mathbf{r}}{d\lambda} \\ &\quad + \nabla_{\mathbf{x}} n(\mathbf{r}) + \frac{2}{c}n(\mathbf{r}) \left\{\tilde{\mathbf{u}}(\mathbf{r}) \otimes \nabla_{\mathbf{x}} n(\mathbf{r}) - \nabla_{\mathbf{x}} n(\mathbf{r}) \otimes \tilde{\mathbf{u}}(\mathbf{r})\right\} \cdot \frac{d\mathbf{r}}{d\lambda}, \end{aligned} \tag{13.201}$$

where

$$\lambda := \int_a^s \left|\mathbf{r}'(\tau)\right| d\tau \tag{13.202}$$

is the natural parameter of the curve.

Next, assume that the wave we consider has an electromagnetic nature. Then by (13.33) and (13.42) we have

$$c_0 = c\sqrt{\kappa_0 \gamma_0} \quad \text{and} \quad \tilde{\mathbf{u}} = (\gamma_0 \kappa_0 \mathbf{v} + (1 - \gamma_0 \kappa_0)\mathbf{u}) \,, \tag{13.203}$$

where $\mathbf{u}$ is the medium velocity and $\mathbf{v}$ is the local vectorial gravitational potential. Moreover, assume that we consider light traveling in some region either filled with the resting medium of constant dielectric permeability γ_0 and magnetic permeability κ_0 or in the vacuum. Then by (13.203) and (13.197) we have

$$n = \frac{1}{\sqrt{\kappa_0 \gamma_0}} \quad \text{is a constant,} \quad \text{and} \quad \tilde{\mathbf{u}} = \gamma_0 \kappa_0 \mathbf{v}. \tag{13.204}$$

Then by (13.204) we rewrite (13.201) as

$$\frac{d^2\mathbf{r}}{d\lambda^2} = \frac{1}{c}\sqrt{\gamma_0 \kappa_0}\,(\text{curl}_{\mathbf{x}}\mathbf{v}\,(\mathbf{r})) \times \frac{d\mathbf{r}}{d\lambda} = \frac{1}{c}\frac{1}{n}\,(\text{curl}_{\mathbf{x}}\mathbf{v}\,(\mathbf{r})) \times \frac{d\mathbf{r}}{d\lambda}. \tag{13.205}$$

In particular, if our coordinate system is inertial, or more generally non-rotating, then $\text{curl}_{\mathbf{x}}\mathbf{v} = 0$ and we deduce that the path of the light from the point N to the point M is the direct line connecting these points, provided we take in the account estimation (13.180).

On the other hand, if our system is rotating, then, since $\mathbf{v}$ is a speed-like vector field, we clearly deduce

$$\text{curl}_{\mathbf{x}}\mathbf{v} = -2\mathbf{w}, \tag{13.206}$$

where $\mathbf{w}$ is the vector of the angular speed of rotation of our coordinate system. Thus by inserting (13.206) into (13.205) we deduce

$$\frac{d^2\mathbf{r}}{d\lambda^2} = -\frac{2}{c}\sqrt{\gamma_0 \kappa_0}\,\mathbf{w} \times \frac{d\mathbf{r}}{d\lambda}. \tag{13.207}$$

In particular, by (13.207) if we consider that $\mathbf{w} = (0, 0, w)$ and $\mathbf{r} = (x, y, z)$, then there exist three dimensionless constants C_1, C_2 and C_3 such that

$$\begin{cases} \frac{dx}{d\lambda} = -C_1 \sin\left(\frac{2w}{c}\sqrt{\gamma_0 \kappa_0}\,\lambda\right) + C_2 \cos\left(\frac{2w}{c}\sqrt{\gamma_0 \kappa_0}\,\lambda\right) \\ \frac{dy}{d\lambda} = -C_1 \cos\left(\frac{2w}{c}\sqrt{\gamma_0 \kappa_0}\,\lambda\right) - C_2 \sin\left(\frac{2w}{c}\sqrt{\gamma_0 \kappa_0}\,\lambda\right) \\ \frac{dz}{d\lambda} = C_3, \end{cases} \tag{13.208}$$

and moreover, since λ is a natural parameter, the constants satisfy

$$C_1^2 + C_2^2 + C_3^2 = 1. \tag{13.209}$$

Then by (13.208) there exist three additional constants D_1, D_2 and D_3 such that

$$\begin{cases} x(\lambda) = C_1 \frac{c\sqrt{\gamma_0\kappa_0}}{2w}\left(\cos\left(\frac{2w}{c}\sqrt{\gamma_0\kappa_0}\,\lambda\right) - 1\right) + C_2 \frac{c\sqrt{\gamma_0\kappa_0}}{2w}\sin\left(\frac{2w}{c}\sqrt{\gamma_0\kappa_0}\,\lambda\right) + D_1 \\ y(\lambda) = -C_1 \frac{c\sqrt{\gamma_0\kappa_0}}{2w}\sin\left(\frac{2w}{c}\sqrt{\gamma_0\kappa_0}\,\lambda\right) + C_2 \frac{c\sqrt{\gamma_0\kappa_0}}{2w}\left(\cos\left(\frac{2w}{c}\sqrt{\gamma_0\kappa_0}\,\lambda\right) - 1\right) + D_2 \\ z(\lambda) = C_3\lambda + D_3. \end{cases} \tag{13.210}$$

So, the curve in (13.210) is the trajectory of the light in the rotating coordinate system, provided we assume (13.180). In particular, by (13.210) and (13.208) we have

$$\begin{cases} x(0) = D_1, \quad y(0) = D_2, \quad z(0) = D_3, \\ \frac{dx}{d\lambda}(0) = C_2, \quad \frac{dy}{d\lambda}(0) = -C_1, \quad \frac{dz}{d\lambda}(0) = C_3. \end{cases} \tag{13.211}$$

The constants $C_1, C_2, C_3, D_1, D_2, D_3$ can be determined either by the initial data (13.211) or by the beginning and the ending points N and M of the curve.

13.3.4 The Laws of Reflection and Refraction

Next consider a monochromatic wave of the frequency $\nu = \omega/(2\pi)$ characterized by

$$U(\mathbf{x}, t) = A(\mathbf{x})e^{ik_0 S(\mathbf{x},t)}, \quad \text{where} \quad k_0 = \frac{\omega}{c} \quad \text{and} \quad \frac{\partial S}{\partial t} = c, \tag{13.212}$$

and consistently with (13.187) consider a direction field:

$$\mathbf{h}_0(\mathbf{x}) = \frac{c}{c_0^2(\mathbf{x})}\tilde{\mathbf{u}}(\mathbf{x}) - \nabla_{\mathbf{x}} S(\mathbf{x}) \approx \frac{c}{c_0^2(\mathbf{x})}\mathbf{h}(\mathbf{x}). \tag{13.213}$$

Furthermore, assume that this wave undergoes reflection and/or refraction on the stationary (time independent) surface $\mathcal{T}$ with the outcoming unit normal $\mathbf{n}$, separating two regions characterized, respectively, by $c_0 = c_0^{(1)}$ and $\tilde{\mathbf{u}} = \tilde{\mathbf{u}}_1$ and by $c_0^{(2)}$ and $\tilde{\mathbf{u}}_2$, with the formation of the reflected wave (of the same frequency), characterized by

$$U_1(\mathbf{x}, t) = A_1(\mathbf{x})e^{ik_0 S_1(\mathbf{x},t)}, \quad \text{where} \quad \frac{\partial S_1}{\partial t} = c, \tag{13.214}$$

and by a direction field:

$$\mathbf{h}_1(\mathbf{x}) = \frac{c}{c_0^2(\mathbf{x})}\tilde{\mathbf{u}}(\mathbf{x}) - \nabla_{\mathbf{x}} S_1(\mathbf{x}), \tag{13.215}$$

and formation of the refracted wave (of the same frequency), characterized by

$$U_2(\mathbf{x}, t) = A_2(\mathbf{x})e^{ik_0 S_2(\mathbf{x},t)}, \quad \text{where} \quad \frac{\partial S_2}{\partial t} = c. \tag{13.216}$$

and by a direction field:

$$\mathbf{h}_2(\mathbf{x}) = \frac{c}{\left(c_0^{(2)}(\mathbf{x})\right)^2}\tilde{\mathbf{u}_2}(\mathbf{x}) - \nabla_{\mathbf{x}} S_2(\mathbf{x}). \tag{13.217}$$

Then the boundary conditions of U, U_1 and U_2 depend on the physical meaning of these fields. However, one of the necessary conditions should be that

$$S_1(\mathbf{x}, t) = S_2(\mathbf{x}, t) + C_2 = S(\mathbf{x}, t) \qquad \forall \mathbf{x} \in \mathcal{T}, \tag{13.218}$$

where C_2 is a real constant. In particular, (13.218) implies

$$\nabla_{\mathbf{x}} S_1 - (\mathbf{n} \cdot \nabla_{\mathbf{x}} S_1)\,\mathbf{n} = \nabla_{\mathbf{x}} S_2 - (\mathbf{n} \cdot \nabla_{\mathbf{x}} S_2)\,\mathbf{n} = \nabla_{\mathbf{x}} S - (\mathbf{n} \cdot \nabla_{\mathbf{x}} S)\,\mathbf{n} \qquad \forall \mathbf{x} \in \mathcal{T}. \tag{13.219}$$

In particular, for every point on the surface $\mathcal{T}$ vectors $\nabla_{\mathbf{x}} S_1$ and $\nabla_{\mathbf{x}} S_2$ lie in the plane formed by vectors $\mathbf{n}$ and $\nabla_{\mathbf{x}} S$. Moreover, by (13.213), (13.215) and (13.219) we have

$$\mathbf{h}_1 - (\mathbf{n} \cdot \mathbf{h}_1)\,\mathbf{n} = \mathbf{h}_0 - (\mathbf{n} \cdot \mathbf{h}_0)\,\mathbf{n} \qquad \forall \mathbf{x} \in \mathcal{T}, \tag{13.220}$$

and, in particular, for every point on the surface $\mathcal{T}$ vector $\mathbf{h}_1$ lies in the plane formed by vectors $\mathbf{n}$ and $\mathbf{h}_0$. Next, assume that the approximate equations in (13.181) and (13.182) are valid in every of two regions on the both sides of $\mathcal{T}$. Then by (13.188) we have

$$|\mathbf{h}_1| = |\mathbf{h}_0| = \frac{c}{c_0}. \tag{13.221}$$

Then, since $\mathbf{h}_1 \neq \mathbf{h}_0$, by (13.220) and (13.221) we deduce

$$\mathbf{n} \cdot \mathbf{h}_1 = -\mathbf{n} \cdot \mathbf{h}_0 \qquad \forall \mathbf{x} \in \mathcal{T}. \tag{13.222}$$

So, by (13.221) and (13.222) we obtain the law of reflection: vector $\mathbf{h}_1$ lies in the plane formed by vectors $\mathbf{n}$ and $\mathbf{h}_0$, and we have

$$\theta\,(\mathbf{h}_0, -\mathbf{n}) = \theta_1\,(\mathbf{h}_1, \mathbf{n})\,, \tag{13.223}$$

where $\theta\,(\mathbf{h}_0, -\mathbf{n})$ is the angle between the incoming ray direction $\mathbf{h}_0$ and the incoming normal to the surface $-\mathbf{n}$ and $\theta_1\,(\mathbf{h}_1, \mathbf{n})$ is the angle between the reflected ray direction $\mathbf{h}_1$ and the outcoming normal $\mathbf{n}$.

Next assume that the wave we consider in (13.212) has an electromagnetic nature. Then by (13.203) we have

$$c_0 = c\sqrt{\kappa_0 \gamma_0} \quad \text{and} \quad \tilde{\mathbf{u}} = (\gamma_0 \kappa_0 \mathbf{v} + (1 - \gamma_0 \kappa_0)\mathbf{u}), \tag{13.224}$$

where $\mathbf{u}$ is the medium velocity and $\mathbf{v}$ is the local vectorial gravitational potential. Similarly, on the second side of surface $\mathcal{T}$ we have

$$c_0^{(2)} = c\sqrt{\kappa_0^{(2)} \gamma_0^{(2)}} \quad \text{and} \quad \tilde{\mathbf{u}}_2 = \left(\gamma_0^{(2)} \kappa_0^{(2)} \mathbf{v} + (1 - \gamma_0^{(2)})\kappa_0^{(2)} \mathbf{u}^{(2)}\right), \tag{13.225}$$

where $\mathbf{u}^{(2)}$ is the medium velocity on the second side of surface $\mathcal{T}$. Furthermore, assume that the medium rests on the both sides of surface $\mathcal{T}$. Then in this particular case we rewrite (13.224) and (13.225) as

$$c_0 = c\sqrt{\kappa_0 \gamma_0} \quad \text{and} \quad \tilde{\mathbf{u}} = \gamma_0 \kappa_0 \mathbf{v}, \tag{13.226}$$

and

$$c_0^{(2)} = c\sqrt{\kappa_0^{(2)} \gamma_0^{(2)}} \quad \text{and} \quad \tilde{\mathbf{u}}_2 = \gamma_0^{(2)} \kappa_0^{(2)} \mathbf{v}, \tag{13.227}$$

Then, in particular, by (13.226) and (13.227) we deduce

$$\frac{c}{\left(c_0^{(2)}\right)^2} \tilde{\mathbf{u}}_2 = \frac{c}{c_0^2} \tilde{\mathbf{u}} = \frac{1}{c} \mathbf{v}. \tag{13.228}$$

Thus, by inserting (13.213) and (13.228) into (13.219), we deduce

$$\mathbf{h}_2 - (\mathbf{n} \cdot \mathbf{h}_2)\,\mathbf{n} = \mathbf{h}_0 - (\mathbf{n} \cdot \mathbf{h}_0)\,\mathbf{n} \qquad \forall \mathbf{x} \in \mathcal{T}, \tag{13.229}$$

and, in particular, for every point on the surface $\mathcal{T}$ vector $\mathbf{h}_2$ lies in the plane formed by vectors $\mathbf{n}$ and $\mathbf{h}_0$. On the other hand, by (13.188) we have

$$|\mathbf{h}_0| = \frac{c}{c_0} \quad \text{and} \quad |\mathbf{h}_2| = \frac{c}{c_0^{(2)}}. \tag{13.230}$$

So, by (13.229) and (13.230), we have Snell's law of refraction: vector $\mathbf{h}_2$ lies in the plane formed by vectors $\mathbf{n}$ and $\mathbf{h}_0$, and we have

$$n \sin\left(\theta\left(\mathbf{h}_0, \mathbf{n}\right)\right) = n_2 \sin\left(\theta_2\left(\mathbf{h}_2, \mathbf{n}\right)\right), \tag{13.231}$$

where $\theta\left(\mathbf{h}_0, \mathbf{n}\right)$ is the angle between the incoming ray direction $\mathbf{h}_0$ and the normal to the surface $\mathbf{n}$, $\theta_2\left(\mathbf{h}_2, \mathbf{n}\right)$ is the angle between the refracted ray direction $\mathbf{h}_2$ and the normal $\mathbf{n}$ and as in (13.197) we set refraction indexes:

$$n := \frac{c}{c_0} \quad \text{and} \quad n_2 := \frac{c}{c_0^{(2)}}. \tag{13.232}$$

13.3.5 Sagnac Effect

Assume again the monochromatic electromagnetic wave of the frequency $\nu = \omega/(2\pi)$ characterized by

$$U(\mathbf{x},t) = A(\mathbf{x},t)e^{iT(\mathbf{x},t)} = A(\mathbf{x},t)e^{ik_0 S(\mathbf{x},t)}, \quad \text{where} \quad k_0 = \frac{\omega}{c} \quad \text{and} \quad \frac{\partial S}{\partial t} = c. \tag{13.233}$$

Then by (13.203) we have

$$c_0 = c\sqrt{\kappa_0\gamma_0} \quad \text{and} \quad \tilde{\mathbf{u}} = (\gamma_0\kappa_0\mathbf{v} + (1-\gamma_0\kappa_0)\mathbf{u})\,, \tag{13.234}$$

where $\mathbf{u}$ is the medium velocity and $\mathbf{v}$ is the local vectorial gravitational potential. Moreover, assume again that we consider light traveling in some region either filled with the resting medium of constant dielectric permeability γ_0 and magnetic permeability κ_0 or in the vacuum. Then by (13.234) and (13.197) we have

$$n = \frac{1}{\sqrt{\kappa_0\gamma_0}} \quad \text{is a constant,} \quad \text{and} \quad \tilde{\mathbf{u}} = \gamma_0\kappa_0\mathbf{v}. \tag{13.235}$$

Next, assume that the light travels from point N to point M across the curve $\mathbf{r}(s) : [a,b] \to \mathbb{R}^3$ with endpoints $\mathbf{r}(a) = N$ and $\mathbf{r}(b) = M$ undergoing possibly certain number of reflections from mirrors during its travel. Then by (13.193), (13.235) and (13.218) we have

$$\delta(-S) := (-S(M^-)) - (-S(N^+)) = \frac{1}{\sqrt{\kappa_0\gamma_0}}\int_a^b |\mathbf{r}'(s)|\,ds - \frac{1}{c}\int_a^b \mathbf{v}\,(\mathbf{r}(s))\cdot\mathbf{r}'(s)ds. \tag{13.236}$$

In particular, if we assume that $M = N$, i.e. our curve is closed, and moreover our curve is the boundary of some surface $\mathcal{S}_0$, then by Stokes theorem we have

$$\begin{aligned}\delta(-S) = (-S(M^-)) - (-S(M^+)) &= \frac{1}{\sqrt{\kappa_0\gamma_0}}\int_a^b |\mathbf{r}'(s)|\,ds - \frac{1}{c}\iint (\mathrm{curl}_{\mathbf{x}}\mathbf{v})\cdot\mathbf{n}\,d\mathcal{S}_0 \\ &= \frac{1}{\sqrt{\kappa_0\gamma_0}}|\partial\mathcal{S}_0| - \frac{1}{c}\iint (\mathrm{curl}_{\mathbf{x}}\mathbf{v})\cdot\mathbf{n}\,d\mathcal{S}_0,\end{aligned} \tag{13.237}$$

where $\mathbf{n}$ is the unit normal to the surface. In particular, if our coordinate system is inertial, or more generally non-rotating, then $\mathrm{curl}_{\mathbf{x}}\mathbf{v} = 0$ and by (13.237) we deduce

$$\delta(-S) = \frac{1}{\sqrt{\kappa_0\gamma_0}}|\partial\mathcal{S}_0|\,. \tag{13.238}$$

On the other hand, if our system is rotating, then as in (13.206) we clearly deduce

$$\text{curl}_{\mathbf{x}}\mathbf{v} = -2\mathbf{w}, \tag{13.239}$$

where **w** is the vector of the angular speed of rotation of our coordinate system. Then by (13.239) and (13.237) we deduce

$$\delta(-S) = \frac{1}{\sqrt{\kappa_0\gamma_0}}\,|\partial\mathcal{S}_0| + \frac{2}{c}\iint \mathbf{w}\cdot\mathbf{n}\,d\mathcal{S}_0. \tag{13.240}$$

In particular, if the surface $\mathcal{S}_0$ is a part of some plain then we rewrite (13.240) as

$$\delta(-S) = \frac{1}{\sqrt{\kappa_0\gamma_0}}\,|\partial\mathcal{S}_0| + \frac{2}{c}\,(\mathbf{w}\cdot\mathbf{n})\,|\mathcal{S}_0|\,. \tag{13.241}$$

On the other hand, if the light travels across the same curve in the opposite direction, then we must have

$$\delta(-S^-) = \frac{1}{\sqrt{\kappa_0\gamma_0}}\,|\partial\mathcal{S}_0| - \frac{2}{c}\,(\mathbf{w}\cdot\mathbf{n})\,|\mathcal{S}_0|\,. \tag{13.242}$$

Thus, by taking the difference in two cases and using (13.233), we deduce

$$\left(\delta(-T) - \delta(-T^-)\right) = k_0\left(\delta(-S) - \delta(-S^-)\right) = \frac{4\omega}{c^2}\,(\mathbf{w}\cdot\mathbf{n})\,|\mathcal{S}_0|\,. \tag{13.243}$$

Here, γ_0 and κ_0 are the dielectric and the magnetic permeability of the medium, T is given in (13.233), $|\mathcal{S}_0|$ is the area of the flat surface bounded by the closed path of the light, **n** is the unit normal to the surface, ω is the frequency of the light and **w** is the angular speed vector of the rotation of our coordinate system.

13.3.6 Fizeau Experiment

Assume again the monochromatic electromagnetic wave of the frequency $\nu = \omega/(2\pi)$ characterized by

$$U(\mathbf{x},t) = A(\mathbf{x},t)e^{iT(\mathbf{x},t)} = A(\mathbf{x},t)e^{ik_0 S(\mathbf{x},t)}, \quad \text{where } k_0 = \frac{\omega}{c} \text{ and } \frac{\partial S}{\partial t} = c. \tag{13.244}$$

Then by (13.203) we have

$$c_0 = c\sqrt{\kappa_0\gamma_0} \quad \text{and} \quad \tilde{\mathbf{u}} = (\gamma_0\kappa_0\mathbf{v} + (1-\gamma_0\kappa_0)\mathbf{u})\,, \tag{13.245}$$

where **u** is the medium velocity and **v** is the local vectorial gravitational potential. Moreover, assume that we consider light traveling in some region filled with the moving medium of constant dielectric permeability γ_0 and magnetic permeability κ_0. Then by (13.245) and (13.197) we have

$$n = \frac{c}{c_0} = \frac{1}{\sqrt{\kappa_0 \gamma_0}} \quad \text{is a constant, and} \quad \tilde{\mathbf{u}} = \frac{1}{n^2}\mathbf{v} + \left(1 - \frac{1}{n^2}\right)\mathbf{u}. \tag{13.246}$$

Next, assume that the light travels from point N to point M across the curve $\mathbf{r}(s) : [a, b] \to \mathbb{R}^3$ with endpoints $\mathbf{r}(a) = N$ and $\mathbf{r}(b) = M$ undergoing possibly certain number of reflections from mirrors during its travel. Then, as before, by (13.193), (13.246) and (13.218) we have

$$\delta(-S) := \left(-S(M^-)\right) - \left(-S(N^+)\right) =$$
$$n\int_a^b \left|\mathbf{r}'(s)\right| ds - \frac{1}{c}\int_a^b \mathbf{v}\left(\mathbf{r}(s)\right)\cdot\mathbf{r}'(s)ds - \frac{n^2}{c}\left(1 - \frac{1}{n^2}\right)\int_a^b \mathbf{u}\left(\mathbf{r}(s)\right)\cdot\mathbf{r}'(s)ds. \tag{13.247}$$

Next assume that, either our curve is perpendicular to the direction of the vectorial gravitational potential $\mathbf{v}$, that happens, for example, if our path of the light is tangent to the Earth surface, or assume that our curve is closed, i.e. $M = N$, and moreover our coordinate system is inertial, or more generally non-rotating. In particular, if we assume that $M = N$, i.e. our curve is closed, and moreover our coordinate system is inertial, or more generally non-rotating, then, as before, by Stokes theorem we have

$$\int_a^b \mathbf{v}\left(\mathbf{r}(s)\right)\cdot\mathbf{r}'(s)ds = 0. \tag{13.248}$$

On the other hand, in the case that our curve is perpendicular to the direction of the vectorial gravitational potential $\mathbf{v}$, (13.248) also trivially follows. Therefore, by inserting (13.248) into (13.247) in both cases we obtain:

$$\delta(-S) = \left(-S(M^-)\right) - \left(-S(N^+)\right) =$$
$$n\int_a^b \left|\mathbf{r}'(s)\right| ds - \frac{n^2}{c}\left(1 - \frac{1}{n^2}\right)\int_a^b \mathbf{u}\left(\mathbf{r}(s)\right)\cdot\mathbf{r}'(s)ds. \tag{13.249}$$

Then by (13.249) and (13.244) we deduce

$$\delta(-T) := \left(-T(M^-)\right) - \left(-T(N^+)\right) = k_0\delta(-S)$$
$$= \frac{n\omega}{c}\int_a^b \left|\mathbf{r}'(s)\right| ds - \frac{n^2\omega}{c^2}\left(1 - \frac{1}{n^2}\right)\int_a^b \mathbf{u}\left(\mathbf{r}(s)\right)\cdot\mathbf{r}'(s)ds$$
$$\frac{n^2\omega}{c^2}\left(c_0\int_a^b \left|\mathbf{r}'(s)\right| ds - \left(1 - \frac{1}{n^2}\right)\int_a^b \mathbf{u}\left(\mathbf{r}(s)\right)\cdot\mathbf{r}'(s)ds\right). \tag{13.250}$$

In particular, if the absolute value $|\mathbf{u}(\mathbf{r}(s))|$ is a constant across the curve and if the angle between $\mathbf{r}'(s)$ and $\mathbf{u}(\mathbf{r}(s))$ is a constant across the curve and equals to the value θ then denoting the length of the path by L:

$$L := \int_a^b \left|\mathbf{r}'(s)\right| ds, \tag{13.251}$$

by (13.250) we deduce

$$\delta(-T) = k_0\delta(-S) = \frac{\omega L n^2}{c^2}\left(c_0 - \left(1 - \frac{1}{n^2}\right)|\mathbf{u}|\cos(\theta)\right). \tag{13.252}$$

Thus, if the direction of $\mathbf{u}$ coincides with the direction of the light, i.e. $\theta = 0$ then

$$\delta(-T) = k_0\delta(-S) = \frac{\omega L n^2}{c^2}\left(c_0 - \left(1 - \frac{1}{n^2}\right)|\mathbf{u}|\right) \approx \frac{\omega L}{\left(c_0 + \left(1 - \frac{1}{n^2}\right)|\mathbf{u}|\right)}. \tag{13.253}$$

On the other hand, if the direction of $\mathbf{u}$ is opposite to the direction of the light, i.e. $\theta = \pi$ then

$$\delta(-T) = k_0\delta(-S) = \frac{\omega L n^2}{c^2}\left(c_0 + \left(1 - \frac{1}{n^2}\right)|\mathbf{u}|\right) \approx \frac{\omega L}{\left(c_0 - \left(1 - \frac{1}{n^2}\right)|\mathbf{u}|\right)}. \tag{13.254}$$

So, in the frames of our model we explain the results of the Fizeau experiment.

13.3.7 Fermat Principle of Geometric Optics in the Case When We Cannot Neglect Effects of Order $O\left(\frac{|\tilde{\mathbf{u}}|^2}{c_0^2}\right)$

Consider again a monochromatic wave of the frequency $\nu = \omega/(2\pi)$ characterized by

$$U(\mathbf{x}, t) = A(\mathbf{x}, t)e^{ik_0 S(\mathbf{x},t)}, \quad \text{where } k_0 = \frac{\omega}{c} \quad \text{and} \quad \frac{\partial S}{\partial t} = c. \tag{13.255}$$

As before, assume the validity of the rough geometric optics approximation. However, assume that we cannot consider anymore the case of the approximation, given by (13.180), i.e. we assume that we cannot neglect anymore effects of order $O\left(\frac{|\tilde{\mathbf{u}}|^2}{c_0^2}\right)$. This happens, for example, in the case of the Michelson–Morley experiment. Thus instead of (13.181) and (13.182) we need to deal with (13.178) and (13.179). On the other hand, by (13.178) we deduce

$$\left|\nabla_{\mathbf{x}} S - \frac{c}{c_0}\left(1 + \frac{1}{c}\left(\tilde{\mathbf{u}} \cdot \nabla_{\mathbf{x}} S\right)\right)\frac{\tilde{\mathbf{u}}}{c_0}\right|^2 = \frac{c^2}{c_0^2}\left(1 + \frac{1}{c^2}\left|\nabla_{\mathbf{x}} S\right|^2 |\tilde{\mathbf{u}}|^2 - \frac{1}{c^2}\left|\tilde{\mathbf{u}} \cdot \nabla_{\mathbf{x}} S\right|^2\right)$$
$$= \frac{c^2}{c_0^2}\left(1 + \frac{1}{c^2}\left|\nabla_{\mathbf{x}} S - \frac{c}{c_0}\left(1 + \frac{1}{c}\left(\tilde{\mathbf{u}} \cdot \nabla_{\mathbf{x}} S\right)\right)\frac{\tilde{\mathbf{u}}}{c_0}\right|^2 |\tilde{\mathbf{u}}|^2\right.$$
$$\left. - \frac{1}{c^2}\left|\tilde{\mathbf{u}} \cdot \left(\nabla_{\mathbf{x}} S - \frac{c}{c_0}\left(1 + \frac{1}{c}\left(\tilde{\mathbf{u}} \cdot \nabla_{\mathbf{x}} S\right)\right)\frac{\tilde{\mathbf{u}}}{c_0}\right)\right|^2\right). \tag{13.256}$$

Then we rewrite (13.179) and (13.256) as

$$\mathbf{h}_0 \cdot \nabla_{\mathbf{x}} A = \frac{1}{2}\left(\left(\Delta_{\mathbf{x}} S\right) - \operatorname{div}_{\mathbf{x}}\left\{\frac{1}{c_0^2}\left(\tilde{\mathbf{u}} \cdot \nabla_{\mathbf{x}} S\right)\tilde{\mathbf{u}}\right\}\right) A. \tag{13.257}$$

and

$$\left(1 - \frac{|\tilde{\mathbf{u}}|^2}{c_0^2}\right)\left|\mathbf{h} - \frac{1}{|\tilde{\mathbf{u}}|^2}\left(\tilde{\mathbf{u}} \cdot \mathbf{h}\right)\tilde{\mathbf{u}}\right|^2 + \left|\frac{1}{|\tilde{\mathbf{u}}|}\tilde{\mathbf{u}} \cdot \mathbf{h}\right|^2$$
$$= |\mathbf{h}|^2\left(1 - \frac{|\tilde{\mathbf{u}}|^2}{c_0^2}\right) + \frac{1}{c_0^2}\left|\tilde{\mathbf{u}} \cdot \mathbf{h}\right|^2 = \frac{c^2}{c_0^2}, \tag{13.258}$$

where the vector field $\mathbf{h}_0$ defined for every $\mathbf{x}$ by

$$\mathbf{h}_0(\mathbf{x}) := \frac{c}{c_0^2(\mathbf{x})}\left(1 + \frac{1}{c}\left(\tilde{\mathbf{u}(\mathbf{x})} \cdot \nabla_{\mathbf{x}} S(\mathbf{x})\right)\right)\tilde{\mathbf{u}}(\mathbf{x}) - \nabla_{\mathbf{x}} S(\mathbf{x})$$
$$= \frac{c}{c_0^2(\mathbf{x})}\left(1 + \frac{1}{c}\left(\tilde{\mathbf{u}(\mathbf{x})} \cdot \nabla_{\mathbf{x}} S(\mathbf{x})\right)\right)\mathbf{h}(\mathbf{x}) \tag{13.259}$$

is called the direction of propagation of the ray that passes through point $\mathbf{x}$. We clarify this name below. The eikonal equation (13.258) and equation of the beam propagation (13.257) are two basic equations of propagation of monochromatic light in the geometric optics approximation inside a moving medium or/and in the presence of nontrivial gravitational field.

Next if we consider an arbitrary characteristic curve $\mathbf{r}(s) : [a, b] \to \mathbb{R}^3$ of Eq. (13.257) defined as a solution of ordinary differential equation

$$\begin{cases} \frac{d\mathbf{r}}{ds}(s) = \mathbf{h}_0\left(\mathbf{r}(s)\right) \\ \mathbf{r}(a) = \mathbf{x}_0, \end{cases} \tag{13.260}$$

then by (13.257) and (13.260) we have

$$\frac{d}{ds}\left(A\left(\mathbf{r}(s)\right)\right) = \nabla_{\mathbf{x}} A\left(\mathbf{r}(s)\right) \cdot \frac{d\mathbf{r}}{ds}(s) = \frac{1}{2} g\left(\mathbf{r}(s)\right) A\left(\mathbf{r}(s)\right), \tag{13.261}$$

where we denote

$$g(\mathbf{x}) := \Delta_{\mathbf{x}} S(\mathbf{x}) - \frac{1}{c_0^2(\mathbf{x})} \left(\left(\nabla_{\mathbf{x}}^2 S(\mathbf{x}) \cdot \tilde{\mathbf{u}}(\mathbf{x}) \right) \cdot \tilde{\mathbf{u}}(\mathbf{x}) \right). \tag{13.262}$$

Then (13.261) implies

$$A\left(\mathbf{r}(s)\right) = A\left(\mathbf{x}_0\right) e^{\frac{1}{2}\int_a^s g(\mathbf{r}(\tau))d\tau} \qquad \forall s \in [a, b]. \tag{13.263}$$

In particular,

$$A\left(\mathbf{x}_0\right) = 0 \text{ implies } A\left(\mathbf{r}(s)\right) = 0 \quad \forall s \in [a, b], \quad \text{and} \quad A\left(\mathbf{x}_0\right) \neq 0 \\ \text{implies } A\left(\mathbf{r}(s)\right) \neq 0 \quad \forall s \in [a, b]. \tag{13.264}$$

Therefore, by (13.264) we deduce that the curve that satisfies (13.260) coincides with the ray of light that passes through the point $\mathbf{x}_0$. So (13.260) is the equation of a ray and the vector field $\mathbf{h}_0$ defined for every $\mathbf{x}$ by (13.259) is indeed the direction of propagation of the ray that passes through point $\mathbf{x}$.

Remark 13.4 As before in Remark 13.3, in contrast to the proof of (13.178) or (13.256), we do not use the geometric optics approximation (13.149) in the proof of (13.179) and (13.257). So, (13.264) is still valid without assumption of the geometric optics approximation (13.149), and thus the vector field $\mathbf{h}_0$, defined by (13.259), is the direction of the propagation of the ray that passes through point $\mathbf{x}$ also in the general case. However, without assumption of the geometric optics approximation we cannot derive (13.258) anymore.

Next, by (13.259) we have

$$\left(1 - \frac{|\tilde{\mathbf{u}}|^2}{c_0^2}\right)^{-1} \left(\frac{1}{|\tilde{\mathbf{u}}|}\tilde{\mathbf{u}} \cdot \mathbf{h}_0\right) = \frac{c}{c_0^2}\left(1 - \frac{|\tilde{\mathbf{u}}|^2}{c_0^2}\right)^{-1} |\tilde{\mathbf{u}}| - \left(\frac{1}{|\tilde{\mathbf{u}}|}\tilde{\mathbf{u}} \cdot \nabla_{\mathbf{x}} S\right), \tag{13.265}$$

and

$$\mathbf{h}_0 - \frac{1}{|\tilde{\mathbf{u}}|^2}\left(\tilde{\mathbf{u}} \cdot \mathbf{h}_0\right)\tilde{\mathbf{u}} = -\left(\nabla_{\mathbf{x}} S - \frac{1}{|\tilde{\mathbf{u}}|^2}\left(\tilde{\mathbf{u}} \cdot \nabla_{\mathbf{x}} S\right)\tilde{\mathbf{u}}\right). \tag{13.266}$$

On the other hand by (13.258) we have

$$\left|\mathbf{h}_0 - \frac{1}{|\tilde{\mathbf{u}}|^2}\left(\tilde{\mathbf{u}} \cdot \mathbf{h}_0\right)\tilde{\mathbf{u}}\right|^2 + \left(1 - \frac{|\tilde{\mathbf{u}}|^2}{c_0^2}\right)^{-1} \left|\frac{1}{|\tilde{\mathbf{u}}|}\tilde{\mathbf{u}} \cdot \mathbf{h}_0\right|^2 = \frac{c^2}{c_0^2}\left(1 - \frac{|\tilde{\mathbf{u}}|^2}{c_0^2}\right)^{-1}. \tag{13.267}$$

Therefore if we consider a curve $\mathbf{r}(s) : [a, b] \to \mathbb{R}^3$ with endpoints $\mathbf{r}(a) = N$ and $\mathbf{r}(b) = M$, then integrating the square root of both sides of (13.267) over the curve $\mathbf{r}(s)$ we deduce

$$\int_a^b \left(\left| \mathbf{h}_0(\mathbf{r}(s)) - (\tilde{\mathbf{u}}(\mathbf{r}(s)) \cdot \mathbf{h}_0(\mathbf{r}(s))) \frac{\tilde{\mathbf{u}}(\mathbf{r}(s))}{|\tilde{\mathbf{u}}(\mathbf{r}(s))|^2} \right|^2 \right.$$
$$\left. + \left(1 - \frac{|\tilde{\mathbf{u}}(\mathbf{r}(s))|^2}{c_0^2(\mathbf{r}(s))}\right)^{-1} \left| \frac{\tilde{\mathbf{u}}(\mathbf{r}(s))}{|\tilde{\mathbf{u}}(\mathbf{r}(s))|} \cdot \mathbf{h}_0(\mathbf{r}(s)) \right|^2 \right)^{\frac{1}{2}} \left(\left| \mathbf{r}'(s) - (\tilde{\mathbf{u}}(\mathbf{r}(s)) \cdot \mathbf{r}'(s)) \frac{\tilde{\mathbf{u}}(\mathbf{r}(s))}{|\tilde{\mathbf{u}}(\mathbf{r}(s))|^2} \right|^2 \right.$$
$$\left. + \left(1 - \frac{|\tilde{\mathbf{u}}(\mathbf{r}(s))|^2}{c_0^2(\mathbf{r}(s))}\right)^{-1} \left| \frac{\tilde{\mathbf{u}}(\mathbf{r}(s))}{|\tilde{\mathbf{u}}(\mathbf{r}(s))|} \cdot \mathbf{r}'(s) \right|^2 \right)^{\frac{1}{2}} ds =$$
$$\int_a^b \frac{c}{c_0(\mathbf{r}(s))} \left(1 - \frac{|\tilde{\mathbf{u}}(\mathbf{r}(s))|^2}{c_0^2(\mathbf{r}(s))}\right)^{-\frac{1}{2}} \left(\left| \mathbf{r}'(s) - (\tilde{\mathbf{u}}(\mathbf{r}(s)) \cdot \mathbf{r}'(s)) \frac{\tilde{\mathbf{u}}(\mathbf{r}(s))}{|\tilde{\mathbf{u}}(\mathbf{r}(s))|^2} \right|^2 \right.$$
$$\left. + \left(1 - \frac{|\tilde{\mathbf{u}}(\mathbf{r}(s))|^2}{c_0^2(\mathbf{r}(s))}\right)^{-1} \left| \frac{\tilde{\mathbf{u}}(\mathbf{r}(s))}{|\tilde{\mathbf{u}}(\mathbf{r}(s))|} \cdot \mathbf{r}'(s) \right|^2 \right)^{\frac{1}{2}} ds. \quad (13.268)$$

Thus, in particular, by inserting (13.265) and (13.266) into (13.268) and using inequality $\mathbf{a} \cdot \mathbf{b} \leq |\mathbf{a}||\mathbf{b}|$, we deduce

$$\int_a^b \frac{c}{c_0^2(\mathbf{r}(s))} \left(1 - \frac{|\tilde{\mathbf{u}}(\mathbf{r}(s))|^2}{c_0^2}\right)^{-1} \tilde{\mathbf{u}}(\mathbf{r}(s)) \cdot \mathbf{r}'(s) ds - \int_a^b \nabla_{\mathbf{x}} S(\mathbf{r}(s)) \cdot \mathbf{r}'(s) ds$$
$$\leq \int_a^b \frac{c}{c_0(\mathbf{r}(s))} \left(1 - \frac{|\tilde{\mathbf{u}}(\mathbf{r}(s))|^2}{c_0^2(\mathbf{r}(s))}\right)^{-\frac{1}{2}} \left(\left| \mathbf{r}'(s) - (\tilde{\mathbf{u}}(\mathbf{r}(s)) \cdot \mathbf{r}'(s)) \frac{\tilde{\mathbf{u}}(\mathbf{r}(s))}{|\tilde{\mathbf{u}}(\mathbf{r}(s))|^2} \right|^2 \right.$$
$$\left. + \left(1 - \frac{|\tilde{\mathbf{u}}(\mathbf{r}(s))|^2}{c_0^2(\mathbf{r}(s))}\right)^{-1} \left| \frac{\tilde{\mathbf{u}}(\mathbf{r}(s))}{|\tilde{\mathbf{u}}(\mathbf{r}(s))|} \cdot \mathbf{r}'(s) \right|^2 \right)^{\frac{1}{2}} ds, \quad (13.269)$$

i.e.

$$(-S(M)) - (-S(N)) \leq -\int_a^b \frac{c}{c_0^2(\mathbf{r}(s))} \left(1 - \frac{|\tilde{\mathbf{u}}(\mathbf{r}(s))|^2}{c_0^2(\mathbf{r}(s))}\right)^{-1} \tilde{\mathbf{u}}(\mathbf{r}(s)) \cdot \mathbf{r}'(s) ds$$
$$+ \int_a^b \frac{c}{c_0(\mathbf{r}(s))} \left(1 - \frac{|\tilde{\mathbf{u}}(\mathbf{r}(s))|^2}{c_0^2(\mathbf{r}(s))}\right)^{-\frac{1}{2}} \left(\left| \mathbf{r}'(s) - (\tilde{\mathbf{u}}(\mathbf{r}(s)) \cdot \mathbf{r}'(s)) \frac{\tilde{\mathbf{u}}(\mathbf{r}(s))}{|\tilde{\mathbf{u}}(\mathbf{r}(s))|^2} \right|^2 \right.$$
$$\left. + \left(1 - \frac{|\tilde{\mathbf{u}}(\mathbf{r}(s))|^2}{c_0^2(\mathbf{r}(s))}\right)^{-1} \left| \frac{\tilde{\mathbf{u}}(\mathbf{r}(s))}{|\tilde{\mathbf{u}}(\mathbf{r}(s))|} \cdot \mathbf{r}'(s) \right|^2 \right)^{\frac{1}{2}} ds. \quad (13.270)$$

Moreover, if

$$\frac{d\mathbf{r}}{ds}(s) = \sigma(s) \mathbf{h}_0(\mathbf{r}(s)), \quad (13.271)$$

for some nonnegative scalar factor $\sigma = \sigma(s)$ then by (13.271), exactly in the same way as we get (13.269), we rewrite (13.268) as

$$(-S(M)) - (-S(N)) = -\int_a^b \frac{c}{c_0^2(\mathbf{r}(s))}\left(1 - \frac{|\tilde{\mathbf{u}}(\mathbf{r}(s))|^2}{c_0^2(\mathbf{r}(s))}\right)^{-1} \tilde{\mathbf{u}}(\mathbf{r}(s)) \cdot \mathbf{r}'(s)ds$$
$$+\int_a^b \frac{c}{c_0(\mathbf{r}(s))}\left(1 - \frac{|\tilde{\mathbf{u}}(\mathbf{r}(s))|^2}{c_0^2(\mathbf{r}(s))}\right)^{-\frac{1}{2}} \left(\left|\mathbf{r}'(s) - \left(\tilde{\mathbf{u}}(\mathbf{r}(s)) \cdot \mathbf{r}'(s)\right)\frac{\tilde{\mathbf{u}}(\mathbf{r}(s))}{|\tilde{\mathbf{u}}(\mathbf{r}(s))|^2}\right|^2\right.$$
$$\left.+\left(1 - \frac{|\tilde{\mathbf{u}}(\mathbf{r}(s))|^2}{c_0^2(\mathbf{r}(s))}\right)^{-1}\left|\frac{\tilde{\mathbf{u}}(\mathbf{r}(s))}{|\tilde{\mathbf{u}}(\mathbf{r}(s))|} \cdot \mathbf{r}'(s)\right|^2\right)^{\frac{1}{2}} ds. \quad (13.272)$$

Thus, by comparing (13.260) with (13.271) and using (13.270) and (13.272), we deduce that if we assume that the light travel from the point N to the point M across the curve $\tilde{\mathbf{r}}(s) : [a, b] \to \mathbb{R}^3$ such that $\tilde{\mathbf{r}(a)} = N$ and $\tilde{\mathbf{r}}(b) = M$, then

$$(-S(M)) - (-S(N)) = -\int_a^b \frac{c}{c_0^2(\tilde{\mathbf{r}}(s))}\left(1 - \frac{|\tilde{\mathbf{u}}(\tilde{\mathbf{r}}(s))|^2}{c_0^2(\tilde{\mathbf{r}}(s))}\right)^{-1} \tilde{\mathbf{u}}(\tilde{\mathbf{r}}(s)) \cdot \mathbf{r}'(s)ds$$
$$+\int_a^b \frac{c}{c_0(\tilde{\mathbf{r}}(s))}\left(1 - \frac{|\tilde{\mathbf{u}}(\tilde{\mathbf{r}}(s))|^2}{c_0^2(\tilde{\mathbf{r}}(s))}\right)^{-\frac{1}{2}} \left(\left|\tilde{\mathbf{r}}'(s) - \left(\tilde{\mathbf{u}}(\tilde{\mathbf{r}}(s)) \cdot \tilde{\mathbf{r}}'(s)\right)\frac{\tilde{\mathbf{u}}(\tilde{\mathbf{r}}(s))}{|\tilde{\mathbf{u}}(\tilde{\mathbf{r}}(s))|^2}\right|^2\right.$$
$$\left.+\left(1 - \frac{|\tilde{\mathbf{u}}(\tilde{\mathbf{r}}(s))|^2}{c_0^2(\tilde{\mathbf{r}}(s))}\right)^{-1}\left|\frac{\tilde{\mathbf{u}}(\tilde{\mathbf{r}}(s))}{|\tilde{\mathbf{u}}(\tilde{\mathbf{r}}(s))|} \cdot \tilde{\mathbf{r}}'(s)\right|^2\right)^{\frac{1}{2}} ds, \quad (13.273)$$

and for every other curve $\mathbf{r}(s) : [a, b] \to \mathbb{R}^3$ with endpoints $\mathbf{r}(a) = N$ and $\mathbf{r}(b) = M$ we have

$$-\int_a^b \frac{c}{c_0^2(\mathbf{r}(s))}\left(1 - \frac{|\tilde{\mathbf{u}}(\mathbf{r}(s))|^2}{c_0^2(\mathbf{r}(s))}\right)^{-1} \tilde{\mathbf{u}}(\mathbf{r}(s)) \cdot \mathbf{r}'(s)ds$$
$$+\int_a^b \frac{c}{c_0(\mathbf{r}(s))}\left(1 - \frac{|\tilde{\mathbf{u}}(\mathbf{r}(s))|^2}{c_0^2(\mathbf{r}(s))}\right)^{-\frac{1}{2}} \left(\left|\mathbf{r}'(s) - \left(\tilde{\mathbf{u}}(\mathbf{r}(s)) \cdot \mathbf{r}'(s)\right)\frac{\tilde{\mathbf{u}}(\mathbf{r}(s))}{|\tilde{\mathbf{u}}(\mathbf{r}(s))|^2}\right|^2\right.$$
$$\left.+\left(1 - \frac{|\tilde{\mathbf{u}}(\mathbf{r}(s))|^2}{c_0^2(\mathbf{r}(s))}\right)^{-1}\left|\frac{\tilde{\mathbf{u}}(\mathbf{r}(s))}{|\tilde{\mathbf{u}}(\mathbf{r}(s))|} \cdot \mathbf{r}'(s)\right|^2\right)^{\frac{1}{2}} ds$$
$$\geq -\int_a^b \frac{c}{c_0^2(\tilde{\mathbf{r}}(s))}\left(1 - \frac{|\tilde{\mathbf{u}}(\tilde{\mathbf{r}}(s))|^2}{c_0^2(\tilde{\mathbf{r}}(s))}\right)^{-1} \tilde{\mathbf{u}}(\tilde{\mathbf{r}}(s)) \cdot \mathbf{r}'(s)ds$$

$$+\int_a^b \frac{c}{c_0(\tilde{\mathbf{r}}(s))}\left(1-\frac{|\tilde{\mathbf{u}}(\tilde{\mathbf{r}}(s))|^2}{c_0^2(\tilde{\mathbf{r}}(s))}\right)^{-\frac{1}{2}}\left(\left|\tilde{\mathbf{r}}'(s)-\left(\tilde{\mathbf{u}}(\tilde{\mathbf{r}}(s))\cdot\tilde{\mathbf{r}}'(s)\right)\frac{\tilde{\mathbf{u}}(\tilde{\mathbf{r}}(s))}{|\tilde{\mathbf{u}}(\tilde{\mathbf{r}}(s))|^2}\right|^2\right.$$
$$\left.+\left(1-\frac{|\tilde{\mathbf{u}}(\tilde{\mathbf{r}}(s))|^2}{c_0^2(\tilde{\mathbf{r}}(s))}\right)^{-1}\left|\frac{\tilde{\mathbf{u}}(\tilde{\mathbf{r}}(s))}{|\tilde{\mathbf{u}}(\tilde{\mathbf{r}}(s))|}\cdot\tilde{\mathbf{r}}'(s)\right|^2\right)^{\frac{1}{2}} ds. \quad (13.274)$$

So we obtain the following Fermat principle:

Proposition 13.2 *Assume geometric optics approximation. Then the light that travels from point N to point M chooses the path $\mathbf{r}(s):[a,b]\to\mathbb{R}^3$ with endpoints $\mathbf{r}(a)=N$ and $\mathbf{r}(b)=M$ which minimizes the quantity:*

$$J(\mathbf{r}(\cdot)) := -\int_a^b \frac{1}{c}n^2(\mathbf{r}(s))\left(1-\frac{|\tilde{\mathbf{u}}(\mathbf{r}(s))|^2}{c_0^2(\mathbf{r}(s))}\right)^{-1}\tilde{\mathbf{u}}(\mathbf{r}(s))\cdot\mathbf{r}'(s)ds$$
$$+\int_a^b n(\mathbf{r}(s))\left(1-\frac{|\tilde{\mathbf{u}}(\mathbf{r}(s))|^2}{c_0^2(\mathbf{r}(s))}\right)^{-1}\cdot$$
$$\cdot\sqrt{\left(1-\frac{|\tilde{\mathbf{u}}(\mathbf{r}(s))|^2}{c_0^2(\mathbf{r}(s))}\right)\left|\mathbf{r}'(s)-\left(\tilde{\mathbf{u}}(\mathbf{r}(s))\cdot\mathbf{r}'(s)\right)\frac{\tilde{\mathbf{u}}(\mathbf{r}(s))}{|\tilde{\mathbf{u}}(\mathbf{r}(s))|^2}\right|^2+\left|\frac{\tilde{\mathbf{u}}(\mathbf{r}(s))}{|\tilde{\mathbf{u}}(\mathbf{r}(s))|}\cdot\mathbf{r}'(s)\right|^2}\, ds,$$
$$(13.275)$$

where we set refraction index:

$$n(\mathbf{x}) := \frac{c}{c_0(\mathbf{x})}. \quad (13.276)$$

13.3.8 The Case of the Non-monochromatic Wave or/and Moving Domains of Propagation of Light

Next, assume that the wave is not monochromatic or/and the fields $\tilde{\mathbf{u}}$ and c_0 depend on the time variable or/and we consider the case of moving domains of propagation of light (in particular moving surfaces of reflection/refraction). In other words, we cannot assume (13.176) or (13.177) anymore. However, we do assume the rough geometric optics approximation (13.150). We also assume that our medium has no dispersion. Then, due to (13.124) we have

$$U(\mathbf{x},t) = A(\mathbf{x},t)e^{ik_0 S(\mathbf{x},t)}, \quad (13.277)$$

and by (13.121) and (13.122) we have

$$\langle \left|\frac{\partial S}{\partial t}\right| \rangle = c, \tag{13.278}$$

where the sign $\langle \cdot \rangle$ means the spatial and temporal averaging. Then, due to (13.152) we have the eikonal-type equation:

$$\frac{1}{c_0^2}\left(\frac{\partial S}{\partial t} + \tilde{\mathbf{u}} \cdot \nabla_{\mathbf{x}} S\right)^2 = |\nabla_{\mathbf{x}} S|^2, \tag{13.279}$$

and we rewrite the equation of the propagation of the beam (13.153):

$$\left(\frac{\partial}{\partial t}\left\{\frac{1}{c_0^2}\left(\frac{\partial S}{\partial t} + \tilde{\mathbf{u}} \cdot \nabla_{\mathbf{x}} S\right)\right\} + \operatorname{div}_{\mathbf{x}}\left\{\frac{1}{c_0^2}\left(\frac{\partial S}{\partial t} + \tilde{\mathbf{u}} \cdot \nabla_{\mathbf{x}} S\right)\tilde{\mathbf{u}}\right\} - (\Delta_{\mathbf{x}} S)\right) A$$
$$+ \frac{2}{c_0^2}\left(\frac{\partial S}{\partial t} + \tilde{\mathbf{u}} \cdot \nabla_{\mathbf{x}} S\right)\left(\frac{\partial A}{\partial t} + \tilde{\mathbf{u}} \cdot \nabla_{\mathbf{x}} A\right) - 2\nabla_{\mathbf{x}} S \cdot \nabla_{\mathbf{x}} A = 0. \tag{13.280}$$

Then, as before, denoting

$$\mathbf{h}(\mathbf{x}, t) := \tilde{\mathbf{u}}(\mathbf{x}, t) - c_0^2(\mathbf{x}, t)\left(\frac{\partial S}{\partial t}(\mathbf{x}, t) + \tilde{\mathbf{u}}(\mathbf{x}, t) \cdot \nabla_{\mathbf{x}} S(\mathbf{x}, t)\right)^{-1} \nabla_{\mathbf{x}} S(\mathbf{x}, t), \tag{13.281}$$

and

$$G(\mathbf{x}, t) := \frac{c_0^2}{2\left(\frac{\partial S}{\partial t} + \tilde{\mathbf{u}} \cdot \nabla_{\mathbf{x}} S\right)}\left(\Delta_{\mathbf{x}} S\right.$$
$$\left.- \left(\frac{\partial}{\partial t}\left\{\frac{1}{c_0^2}\left(\frac{\partial S}{\partial t} + \tilde{\mathbf{u}} \cdot \nabla_{\mathbf{x}} S\right)\right\} + \operatorname{div}_{\mathbf{x}}\left\{\frac{1}{c_0^2}\left(\frac{\partial S}{\partial t} + \tilde{\mathbf{u}} \cdot \nabla_{\mathbf{x}} S\right)\tilde{\mathbf{u}}\right\}\right)\right), \tag{13.282}$$

by inserting (13.281) and (13.282) into (13.280) we clearly have

$$\frac{\partial A}{\partial t}(\mathbf{x}, t) + \mathbf{h}(\mathbf{x}, t) \cdot \nabla_{\mathbf{x}} A(\mathbf{x}, t) = G(\mathbf{x}, t) A(\mathbf{x}, t). \tag{13.283}$$

Next consider a curve $\mathbf{r}(t) : [t_0, b] \to \mathbb{R}^3$, parameterized by the time variable t, defined as a solution of ordinary differential equation:

$$\begin{cases} \frac{d\mathbf{r}}{dt}(t) = \mathbf{h}\left(\mathbf{r}(t), t\right) \\ \mathbf{r}(t_0) = \mathbf{x}_0, \end{cases} \tag{13.284}$$

here $\mathbf{h}$ was defined in (13.281). Note that the equations of the ray propagation of the form (13.284) are not always convenient since $\mathbf{h}$ in the right-hand side of (13.167) depend on the partial derivatives of the quantity S, which we do not know a priori unless we solve (13.279). In the following proposition, we present the alternative form for the equations of the ray propagation being the second-order ordinary differential equations which does not contain the quantity S or its partial derivatives.

Proposition 13.3 *Consider a smooth solution of* (13.279). *Next let* $\mathbf{r}(t) : [t_0, b] \to \mathbb{R}^3$ *be any curve parameterized by the time variable* t*, defined as a solution* (13.284). *Then,* $\mathbf{r}(t)$ *satisfies the following second-order ordinary differential equations of the ray propagation:*

$$\begin{aligned}\frac{d}{dt}\left(\frac{1}{c_0\left(\mathbf{r}(t),t\right)}\left(\frac{d\mathbf{r}}{dt}(t)-\tilde{\mathbf{u}}\left(\mathbf{r}(t),t\right)\right)\right)=&\\ -\nabla_{\mathbf{x}}c_0(\mathbf{r},t)-\frac{1}{c_0(\mathbf{r},t)}\left\{d_{\mathbf{x}}\tilde{\mathbf{u}}(\mathbf{r},t)\right\}^T\cdot\left(\frac{d\mathbf{r}}{dt}-\tilde{\mathbf{u}}(\mathbf{r},t)\right)&\\ +\frac{1}{c_0^2(\mathbf{r},t)}\left(\left(\frac{d\mathbf{r}}{dt}-\tilde{\mathbf{u}}(\mathbf{r},t)\right)\cdot\nabla_{\mathbf{x}}c_0(\mathbf{r},t)\right)\left(\frac{d\mathbf{r}}{dt}-\tilde{\mathbf{u}}(\mathbf{r},t)\right)&\\ +\frac{\left(\left(\left(\left\{d_{\mathbf{x}}\tilde{\mathbf{u}}(\mathbf{r},t)\right\}^T+d_{\mathbf{x}}\tilde{\mathbf{u}}(\mathbf{r},t)\right)\cdot\left(\frac{d\mathbf{r}}{dt}-\tilde{\mathbf{u}}(\mathbf{r},t)\right)\right)\cdot\left(\frac{d\mathbf{r}}{dt}-\tilde{\mathbf{u}}(\mathbf{r},t)\right)\right)\left(\frac{d\mathbf{r}}{dt}-\tilde{\mathbf{u}}(\mathbf{r},t)\right)}{2c_0^3(\mathbf{r},t)}&.\end{aligned} \tag{13.285}$$

Moreover, the proper scalar quantity $\tilde{\omega}(t) : [t_0, b] \to \mathbb{R}$*, defined by*

$$\tilde{\omega}(t)=\frac{\partial S}{\partial t}\left(\mathbf{r}(t),t\right)+\tilde{\mathbf{u}}\left(\mathbf{r}(t),t\right)\cdot\nabla_{\mathbf{x}}S\left(\mathbf{r}(t),t\right) \tag{13.286}$$

satisfies the following ordinary differential equation:

$$\begin{aligned}\frac{d\tilde{\omega}}{dt}(t)=\frac{1}{c_0(\mathbf{r},t)}\left(\frac{\partial c_0}{\partial t}(\mathbf{r},t)+\tilde{\mathbf{u}}(\mathbf{r},t)\cdot\nabla_{\mathbf{x}}c_0(\mathbf{r},t)\right)\tilde{\omega}(t)&\\ -\frac{\left(\left(\left(\left\{d_{\mathbf{x}}\tilde{\mathbf{u}}(\mathbf{r},t)\right\}^T+d_{\mathbf{x}}\tilde{\mathbf{u}}(\mathbf{r},t)\right)\cdot\left(\frac{d\mathbf{r}}{dt}(t)-\tilde{\mathbf{u}}(\mathbf{r},t)\right)\right)\cdot\left(\frac{d\mathbf{r}}{dt}(t)-\tilde{\mathbf{u}}(\mathbf{r},t)\right)\right)\tilde{\omega}(t)}{2c_0^2(\mathbf{r},t)}&.\end{aligned} \tag{13.287}$$

Proof Consider $\mathbf{r}(t) : [t_0, b] \to \mathbb{R}^3$ to be an arbitrary solution of ordinary differential equation (13.167), i.e.

$$\begin{cases}\frac{d\mathbf{r}}{dt}(t)=\mathbf{h}\left(\mathbf{r}(t),t\right)\\ \mathbf{r}(t_0)=\mathbf{x}_0,\end{cases} \tag{13.288}$$

where $\mathbf{h}$ was defined in (13.281). In particular, by (13.288) and (13.175) we have

$$\left|\frac{d\mathbf{r}}{dt}(t)-\tilde{\mathbf{u}}(\mathbf{r},t)\right|^2=c_0^2(\mathbf{r},t), \tag{13.289}$$

and by (13.288) and (13.281) we have

$$\nabla_{\mathbf{x}} S(\mathbf{r},t) = -\frac{1}{c_0^2(\mathbf{r},t)}\left(\frac{\partial S}{\partial t}(\mathbf{r},t) + \tilde{\mathbf{u}}(\mathbf{r},t)\cdot\nabla_{\mathbf{x}} S(\mathbf{r},t)\right)\left(\frac{d\mathbf{r}}{dt}(t) - \tilde{\mathbf{u}}(\mathbf{r},t)\right). \tag{13.290}$$

Then, by (13.290) and (13.289) obviously we have

$$\frac{d\mathbf{r}}{dt}(t) - \tilde{\mathbf{u}}(\mathbf{r},t) = -c_0^2(\mathbf{r},t)\left(\frac{\partial S}{\partial t}(\mathbf{r},t) + \tilde{\mathbf{u}}(\mathbf{r},t)\cdot\nabla_{\mathbf{x}} S(\mathbf{r},t)\right)^{-1}\nabla_{\mathbf{x}} S(\mathbf{r},t), \tag{13.291}$$

$$\frac{\partial S}{\partial t}(\mathbf{r},t) + \tilde{\mathbf{u}}(\mathbf{r},t)\cdot\nabla_{\mathbf{x}} S(\mathbf{r},t) = -\left(\frac{d\mathbf{r}}{dt} - \tilde{\mathbf{u}}(\mathbf{r},t)\right)\cdot\nabla_{\mathbf{x}} S(\mathbf{r},t), \tag{13.292}$$

and thus

$$\frac{d\mathbf{r}}{dt}(t) - \tilde{\mathbf{u}}(\mathbf{r},t) = c_0^2(\mathbf{r},t)\left(\left(\frac{d\mathbf{r}}{dt} - \tilde{\mathbf{u}}(\mathbf{r},t)\right)\cdot\nabla_{\mathbf{x}} S(\mathbf{r},t)\right)^{-1}\nabla_{\mathbf{x}} S(\mathbf{r},t). \tag{13.293}$$

On the other hand, by (13.279) and the chain rule we have

$$\begin{aligned}\frac{1}{c_0^2}\left(\frac{\partial S}{\partial t} + \tilde{\mathbf{u}}\cdot\nabla_{\mathbf{x}} S\right)\nabla_{\mathbf{x}}\left(\frac{\partial S}{\partial t} + \tilde{\mathbf{u}}\cdot\nabla_{\mathbf{x}} S\right) - \frac{1}{c_0^3}\left(\frac{\partial S}{\partial t} + \tilde{\mathbf{u}}\cdot\nabla_{\mathbf{x}} S\right)^2\nabla_{\mathbf{x}} c_0 \\ = \nabla_{\mathbf{x}}^2 S(\mathbf{r},t)\cdot\nabla_{\mathbf{x}} S(\mathbf{r},t).\end{aligned} \tag{13.294}$$

Thus,

$$\begin{aligned}\left(\frac{\partial S}{\partial t} + \tilde{\mathbf{u}}\cdot\nabla_{\mathbf{x}} S\right)\nabla_{\mathbf{x}}\left(\frac{\partial S}{\partial t} + \tilde{\mathbf{u}}\cdot\nabla_{\mathbf{x}} S\right) - \frac{1}{c_0}\left(\frac{\partial S}{\partial t} + \tilde{\mathbf{u}}\cdot\nabla_{\mathbf{x}} S\right)^2\nabla_{\mathbf{x}} c_0 \\ = c_0^2\nabla_{\mathbf{x}}^2 S(\mathbf{r},t)\cdot\nabla_{\mathbf{x}} S(\mathbf{r},t).\end{aligned} \tag{13.295}$$

So,

$$\begin{aligned}\frac{\partial}{\partial t}(\nabla_{\mathbf{x}} S) + \nabla_{\mathbf{x}}^2 S\cdot\tilde{\mathbf{u}} + \{d_{\mathbf{x}}\tilde{\mathbf{u}}\}^T\cdot\nabla_{\mathbf{x}} S - \frac{1}{c_0}\left(\frac{\partial S}{\partial t} + \tilde{\mathbf{u}}\cdot\nabla_{\mathbf{x}} S\right)\nabla_{\mathbf{x}} c_0 \\ = c_0^2\left(\frac{\partial S}{\partial t} + \tilde{\mathbf{u}}\cdot\nabla_{\mathbf{x}} S\right)^{-1}\nabla_{\mathbf{x}}^2 S(\mathbf{r},t)\cdot\nabla_{\mathbf{x}} S(\mathbf{r},t).\end{aligned} \tag{13.296}$$

Therefore,

$$\frac{\partial}{\partial t}(\nabla_{\mathbf{x}} S) + \nabla^2_{\mathbf{x}} S \cdot \tilde{\mathbf{u}} - c_0^2 \left(\frac{\partial S}{\partial t} + \tilde{\mathbf{u}} \cdot \nabla_{\mathbf{x}} S \right)^{-1} \nabla^2_{\mathbf{x}} S(\mathbf{r}, t) \cdot \nabla_{\mathbf{x}} S(\mathbf{r}, t) =$$
$$\frac{1}{c_0} \left(\frac{\partial S}{\partial t} + \tilde{\mathbf{u}} \cdot \nabla_{\mathbf{x}} S \right) \nabla_{\mathbf{x}} c_0 - \{d_{\mathbf{x}} \tilde{\mathbf{u}}\}^T \cdot \nabla_{\mathbf{x}} S. \quad (13.297)$$

Next, by (13.297) and (13.291) we infer

$$\nabla^2_{\mathbf{x}} S(\mathbf{r}, t) \cdot \left(\frac{d\mathbf{r}}{dt} - \tilde{\mathbf{u}}(\mathbf{r}, t) \right) + \frac{\partial}{\partial t}(\nabla_{\mathbf{x}} S)(\mathbf{r}, t) + \nabla^2_{\mathbf{x}} S(\mathbf{r}, t) \cdot \tilde{\mathbf{u}}(\mathbf{r}, t) =$$
$$\frac{1}{c_0(\mathbf{r}, t)} \left(\frac{\partial S}{\partial t}(\mathbf{r}, t) + \tilde{\mathbf{u}}(\mathbf{r}, t) \cdot \nabla_{\mathbf{x}} S(\mathbf{r}, t) \right) \nabla_{\mathbf{x}} c_0(\mathbf{r}, t) - \{d_{\mathbf{x}} \tilde{\mathbf{u}}(\mathbf{r}, t)\}^T \cdot \nabla_{\mathbf{x}} S(\mathbf{r}, t). \quad (13.298)$$

On the other hand, by (13.292) and the chain rule we have

$$\frac{d}{dt} \left(\frac{\partial S}{\partial t}(\mathbf{r}(t), t) + \tilde{\mathbf{u}}(\mathbf{r}(t), t) \cdot \nabla_{\mathbf{x}} S(\mathbf{r}(t), t) \right) =$$
$$-\frac{d}{dt} \left(\left(\frac{d\mathbf{r}}{dt}(t) - \tilde{\mathbf{u}}(\mathbf{r}t), t) \right) \cdot \nabla_{\mathbf{x}} S(\mathbf{r}(t), t) \right) =$$
$$-\frac{d}{dt} \left(\frac{d\mathbf{r}}{dt}(t) - \tilde{\mathbf{u}}(\mathbf{r}(t), t) \right) \cdot \nabla_{\mathbf{x}} S(\mathbf{r}, t)$$
$$-\left(\frac{d\mathbf{r}}{dt} - \tilde{\mathbf{u}}(\mathbf{r}, t) \right) \cdot \left(\nabla^2_{\mathbf{x}} S(\mathbf{r}, t) \cdot \frac{d\mathbf{r}}{dt} + \frac{\partial}{\partial t}(\nabla_{\mathbf{x}} S)(\mathbf{r}, t) \right) =$$
$$-\frac{d}{dt} \left(\frac{d\mathbf{r}}{dt}(t) - \tilde{\mathbf{u}}(\mathbf{r}(t), t) \right) \cdot \nabla_{\mathbf{x}} S(\mathbf{r}, t)$$
$$-\left(\frac{d\mathbf{r}}{dt} - \tilde{\mathbf{u}}(\mathbf{r}, t) \right) \cdot \left(\nabla^2_{\mathbf{x}} S(\mathbf{r}, t) \cdot \left(\frac{d\mathbf{r}}{dt} - \tilde{\mathbf{u}}(\mathbf{r}, t) \right) \right.$$
$$\left. + \frac{\partial}{\partial t}(\nabla_{\mathbf{x}} S)(\mathbf{r}, t) + \nabla^2_{\mathbf{x}} S(\mathbf{r}, t) \cdot \tilde{\mathbf{u}}(\mathbf{r}, t) \right). \quad (13.299)$$

Thus, inserting (13.290) and (13.298) into (13.299) we deduce

$$
\begin{aligned}
&\frac{d}{dt}\left(\frac{\partial S}{\partial t}(\mathbf{r}(t),t)+\tilde{\mathbf{u}}(\mathbf{r}(t),t)\cdot\nabla_{\mathbf{x}}S(\mathbf{r}(t),t)\right)=\\
&\frac{1}{c_0^2(\mathbf{r},t)}\left(\frac{\partial S}{\partial t}(\mathbf{r},t)+\tilde{\mathbf{u}}(\mathbf{r},t)\cdot\nabla_{\mathbf{x}}S(\mathbf{r},t)\right)\left(\frac{d\mathbf{r}}{dt}-\tilde{\mathbf{u}}(\mathbf{r},t)\right)\cdot\frac{d}{dt}\left(\frac{d\mathbf{r}}{dt}(t)-\tilde{\mathbf{u}}(\mathbf{r}(t),t)\right)\\
&\quad-\left(\frac{d\mathbf{r}}{dt}-\tilde{\mathbf{u}}(\mathbf{r},t)\right)\cdot\left(\frac{1}{c_0(\mathbf{r},t)}\left(\frac{\partial S}{\partial t}(\mathbf{r},t)+\tilde{\mathbf{u}}(\mathbf{r},t)\cdot\nabla_{\mathbf{x}}S(\mathbf{r},t)\right)\nabla_{\mathbf{x}}c_0(\mathbf{r},t)\right.\\
&\qquad\left.-\{d_{\mathbf{x}}\tilde{\mathbf{u}}(\mathbf{r},t)\}^T\cdot\nabla_{\mathbf{x}}S(\mathbf{r},t)\right)=\\
&\frac{1}{2c_0^2(\mathbf{r},t)}\left(\frac{\partial S}{\partial t}(\mathbf{r},t)+\tilde{\mathbf{u}}(\mathbf{r},t)\cdot\nabla_{\mathbf{x}}S(\mathbf{r},t)\right)\frac{d}{dt}\left(\left|\frac{d\mathbf{r}}{dt}(t)-\tilde{\mathbf{u}}(\mathbf{r}(t),t)\right|^2\right)\\
&\quad-\left(\frac{d\mathbf{r}}{dt}-\tilde{\mathbf{u}}(\mathbf{r},t)\right)\cdot\left(\frac{1}{c_0(\mathbf{r},t)}\left(\frac{\partial S}{\partial t}(\mathbf{r},t)+\tilde{\mathbf{u}}(\mathbf{r},t)\cdot\nabla_{\mathbf{x}}S(\mathbf{r},t)\right)\nabla_{\mathbf{x}}c_0(\mathbf{r},t)\right.\\
&\qquad\left.-\{d_{\mathbf{x}}\tilde{\mathbf{u}}(\mathbf{r},t)\}^T\cdot\nabla_{\mathbf{x}}S(\mathbf{r},t)\right). \qquad (13.300)
\end{aligned}
$$

Thus, inserting (13.289) into (13.300) we deduce

$$
\begin{aligned}
&\frac{d}{dt}\left(\frac{\partial S}{\partial t}(\mathbf{r}(t),t)+\tilde{\mathbf{u}}(\mathbf{r}(t),t)\cdot\nabla_{\mathbf{x}}S(\mathbf{r}(t),t)\right)\\
&\quad=\frac{1}{2c_0^2(\mathbf{r},t)}\left(\frac{\partial S}{\partial t}(\mathbf{r},t)+\tilde{\mathbf{u}}(\mathbf{r},t)\cdot\nabla_{\mathbf{x}}S(\mathbf{r},t)\right)\frac{d}{dt}\left(c_0^2(\mathbf{r}(t),t)\right)\\
&\quad-\left(\frac{d\mathbf{r}}{dt}-\tilde{\mathbf{u}}(\mathbf{r},t)\right)\cdot\left(\frac{1}{c_0(\mathbf{r},t)}\left(\frac{\partial S}{\partial t}(\mathbf{r},t)+\tilde{\mathbf{u}}(\mathbf{r},t)\cdot\nabla_{\mathbf{x}}S(\mathbf{r},t)\right)\nabla_{\mathbf{x}}c_0(\mathbf{r},t)\right.\\
&\qquad\left.-\{d_{\mathbf{x}}\tilde{\mathbf{u}}(\mathbf{r},t)\}^T\cdot\nabla_{\mathbf{x}}S(\mathbf{r},t)\right)=\\
&\frac{1}{c_0(\mathbf{r},t)}\left(\frac{\partial S}{\partial t}(\mathbf{r},t)+\tilde{\mathbf{u}}(\mathbf{r},t)\cdot\nabla_{\mathbf{x}}S(\mathbf{r},t)\right)\left(\nabla_{\mathbf{x}}c_0(\mathbf{r},t)\cdot\left(\frac{d\mathbf{r}}{dt}-\tilde{\mathbf{u}}(\mathbf{r},t)\right)\right.\\
&\qquad\left.+\frac{\partial c_0}{\partial t}(\mathbf{r},t)+\tilde{\mathbf{u}}(\mathbf{r},t)\cdot\nabla_{\mathbf{x}}c_0(\mathbf{r},t)\right)\\
&\quad-\left(\frac{d\mathbf{r}}{dt}-\tilde{\mathbf{u}}(\mathbf{r},t)\right)\cdot\left(\frac{1}{c_0(\mathbf{r},t)}\left(\frac{\partial S}{\partial t}(\mathbf{r},t)+\tilde{\mathbf{u}}(\mathbf{r},t)\cdot\nabla_{\mathbf{x}}S(\mathbf{r},t)\right)\nabla_{\mathbf{x}}c_0(\mathbf{r},t)\right.\\
&\qquad\left.-\{d_{\mathbf{x}}\tilde{\mathbf{u}}(\mathbf{r},t)\}^T\cdot\nabla_{\mathbf{x}}S(\mathbf{r},t)\right). \qquad (13.301)
\end{aligned}
$$

Then inserting (13.290) into (13.301) we infer

$$
\begin{aligned}
&\frac{d}{dt}\left(\frac{\partial S}{\partial t}\left(\mathbf{r}(t),t\right)+\tilde{\mathbf{u}}\left(\mathbf{r}(t),t\right)\cdot\nabla_{\mathbf{x}}S\left(\mathbf{r}(t),t\right)\right)=\\
&\quad\frac{1}{c_0(\mathbf{r},t)}\left(\frac{\partial S}{\partial t}(\mathbf{r},t)+\tilde{\mathbf{u}}(\mathbf{r},t)\cdot\nabla_{\mathbf{x}}S(\mathbf{r},t)\right)\left(\nabla_{\mathbf{x}}c_0(\mathbf{r},t)\cdot\left(\frac{d\mathbf{r}}{dt}-\tilde{\mathbf{u}}(\mathbf{r},t)\right)\right.\\
&\qquad\qquad\left.+\frac{\partial c_0}{\partial t}(\mathbf{r},t)+\tilde{\mathbf{u}}(\mathbf{r},t)\cdot\nabla_{\mathbf{x}}c_0(\mathbf{r},t)\right)\\
&\qquad\qquad-\frac{1}{c_0^2(\mathbf{r},t)}\left(\frac{\partial S}{\partial t}(\mathbf{r},t)+\tilde{\mathbf{u}}(\mathbf{r},t)\cdot\nabla_{\mathbf{x}}S(\mathbf{r},t)\right)\times\\
&\times\left(\frac{d\mathbf{r}}{dt}-\tilde{\mathbf{u}}(\mathbf{r},t)\right)\cdot\left(c_0(\mathbf{r},t)\nabla_{\mathbf{x}}c_0(\mathbf{r},t)+\{d_{\mathbf{x}}\tilde{\mathbf{u}}(\mathbf{r},t)\}^T\cdot\left(\frac{d\mathbf{r}}{dt}(t)-\tilde{\mathbf{u}}(\mathbf{r},t)\right)\right).
\end{aligned}
\tag{13.302}
$$

So

$$
\begin{aligned}
&\frac{d}{dt}\left(\frac{\partial S}{\partial t}\left(\mathbf{r}(t),t\right)+\tilde{\mathbf{u}}\left(\mathbf{r}(t),t\right)\cdot\nabla_{\mathbf{x}}S\left(\mathbf{r}(t),t\right)\right)=\\
&\quad\frac{1}{c_0(\mathbf{r},t)}\left(\frac{\partial S}{\partial t}(\mathbf{r},t)+\tilde{\mathbf{u}}(\mathbf{r},t)\cdot\nabla_{\mathbf{x}}S(\mathbf{r},t)\right)\left(\frac{\partial c_0}{\partial t}(\mathbf{r},t)+\tilde{\mathbf{u}}(\mathbf{r},t)\cdot\nabla_{\mathbf{x}}c_0(\mathbf{r},t)\right)\\
&\qquad\qquad-\frac{1}{c_0^2(\mathbf{r},t)}\left(\frac{\partial S}{\partial t}(\mathbf{r},t)\right.\\
&\quad\left.+\tilde{\mathbf{u}}(\mathbf{r},t)\cdot\nabla_{\mathbf{x}}S(\mathbf{r},t)\right)\left(\left(\{d_{\mathbf{x}}\tilde{\mathbf{u}}(\mathbf{r},t)\}^T\cdot\left(\frac{d\mathbf{r}}{dt}-\tilde{\mathbf{u}}(\mathbf{r},t)\right)\right)\cdot\left(\frac{d\mathbf{r}}{dt}-\tilde{\mathbf{u}}(\mathbf{r},t)\right)\right)\\
&\quad=\frac{1}{c_0(\mathbf{r},t)}\left(\frac{\partial S}{\partial t}(\mathbf{r},t)+\tilde{\mathbf{u}}(\mathbf{r},t)\cdot\nabla_{\mathbf{x}}S(\mathbf{r},t)\right)\left(\frac{\partial c_0}{\partial t}(\mathbf{r},t)+\tilde{\mathbf{u}}(\mathbf{r},t)\cdot\nabla_{\mathbf{x}}c_0(\mathbf{r},t)\right)\\
&\qquad\qquad-\frac{1}{2c_0^2(\mathbf{r},t)}\left(\frac{\partial S}{\partial t}(\mathbf{r},t)+\tilde{\mathbf{u}}(\mathbf{r},t)\cdot\nabla_{\mathbf{x}}S(\mathbf{r},t)\right)\times\\
&\times\left(\left(\left(\{d_{\mathbf{x}}\tilde{\mathbf{u}}(\mathbf{r},t)\}^T+d_{\mathbf{x}}\tilde{\mathbf{u}}(\mathbf{r},t)\right)\cdot\left(\frac{d\mathbf{r}}{dt}-\tilde{\mathbf{u}}(\mathbf{r},t)\right)\right)\cdot\left(\frac{d\mathbf{r}}{dt}-\tilde{\mathbf{u}}(\mathbf{r},t)\right)\right).
\end{aligned}
\tag{13.303}
$$

So we deduce (13.287). Moreover, we can rewrite it as

$$
\begin{aligned}
&\left(\frac{\partial S}{\partial t}(\mathbf{r},t)+\tilde{\mathbf{u}}(\mathbf{r},t)\cdot\nabla_{\mathbf{x}}S(\mathbf{r},t)\right)^{-1}\frac{d}{dt}\left(\frac{\partial S}{\partial t}\left(\mathbf{r}(t),t\right)+\tilde{\mathbf{u}}\left(\mathbf{r}(t),t\right)\cdot\nabla_{\mathbf{x}}S\left(\mathbf{r}(t),t\right)\right)\\
&\qquad=\frac{1}{c_0(\mathbf{r},t)}\left(\frac{\partial c_0}{\partial t}(\mathbf{r},t)+\tilde{\mathbf{u}}(\mathbf{r},t)\cdot\nabla_{\mathbf{x}}c_0(\mathbf{r},t)\right)\\
&-\frac{1}{2c_0^2(\mathbf{r},t)}\left(\left(\left(\{d_{\mathbf{x}}\tilde{\mathbf{u}}(\mathbf{r},t)\}^T+d_{\mathbf{x}}\tilde{\mathbf{u}}(\mathbf{r},t)\right)\cdot\left(\frac{d\mathbf{r}}{dt}-\tilde{\mathbf{u}}(\mathbf{r},t)\right)\right)\cdot\left(\frac{d\mathbf{r}}{dt}-\tilde{\mathbf{u}}(\mathbf{r},t)\right)\right).
\end{aligned}
\tag{13.304}
$$

On the other hand, by (13.291) we have

$$
\begin{aligned}
&\frac{d}{dt}\left(\frac{1}{c_0^2\left(\mathbf{r}(t),t\right)}\left(\frac{d\mathbf{r}}{dt}(t)-\tilde{\mathbf{u}}\left(\mathbf{r}(t),t\right)\right)\right)=\\
&-\left(\frac{\partial S}{\partial t}(\mathbf{r},t)+\tilde{\mathbf{u}}(\mathbf{r},t)\cdot\nabla_{\mathbf{x}}S(\mathbf{r},t)\right)^{-1}\left(\nabla^2_{\mathbf{x}}S(\mathbf{r},t)\cdot\frac{d\mathbf{r}}{dt}+\frac{\partial}{\partial t}\left(\nabla_{\mathbf{x}}S\right)(\mathbf{r},t)\right)+\left(\frac{\partial S}{\partial t}(\mathbf{r},t)\right.\\
&\left.+\tilde{\mathbf{u}}(\mathbf{r},t)\cdot\nabla_{\mathbf{x}}S(\mathbf{r},t)\right)^{-2}\left(\frac{d}{dt}\left(\frac{\partial S}{\partial t}\left(\mathbf{r}(t),t\right)+\tilde{\mathbf{u}}\left(\mathbf{r}(t),t\right)\cdot\nabla_{\mathbf{x}}S\left(\mathbf{r}(t),t\right)\right)\right)\nabla_{\mathbf{x}}S(\mathbf{r},t)\\
&=-\left(\frac{\partial S}{\partial t}(\mathbf{r},t)+\tilde{\mathbf{u}}(\mathbf{r},t)\cdot\nabla_{\mathbf{x}}S(\mathbf{r},t)\right)^{-1}\times\\
&\times\left(\nabla^2_{\mathbf{x}}S(\mathbf{r},t)\cdot\left(\frac{d\mathbf{r}}{dt}-\tilde{\mathbf{u}}(\mathbf{r},t)\right)+\frac{\partial}{\partial t}\left(\nabla_{\mathbf{x}}S\right)(\mathbf{r},t)+\nabla^2_{\mathbf{x}}S(\mathbf{r},t)\cdot\tilde{\mathbf{u}}(\mathbf{r},t)\right)+\left(\frac{\partial S}{\partial t}(\mathbf{r},t)\right.\\
&\left.+\tilde{\mathbf{u}}(\mathbf{r},t)\cdot\nabla_{\mathbf{x}}S(\mathbf{r},t)\right)^{-2}\left(\frac{d}{dt}\left(\frac{\partial S}{\partial t}\left(\mathbf{r}(t),t\right)+\tilde{\mathbf{u}}\left(\mathbf{r}(t),t\right)\cdot\nabla_{\mathbf{x}}S\left(\mathbf{r}(t),t\right)\right)\right)\nabla_{\mathbf{x}}S(\mathbf{r},t).
\end{aligned}
\tag{13.305}
$$

Thus, by inserting (13.298) into (13.305) we deduce

$$
\begin{aligned}
&\frac{d}{dt}\left(\frac{1}{c_0^2\left(\mathbf{r}(t),t\right)}\left(\frac{d\mathbf{r}}{dt}(t)-\tilde{\mathbf{u}}\left(\mathbf{r}(t),t\right)\right)\right)=-\left(\frac{\partial S}{\partial t}(\mathbf{r},t)+\tilde{\mathbf{u}}(\mathbf{r},t)\cdot\nabla_{\mathbf{x}}S(\mathbf{r},t)\right)^{-1}\times\\
&\times\left(\frac{1}{c_0(\mathbf{r},t)}\left(\frac{\partial S}{\partial t}(\mathbf{r},t)+\tilde{\mathbf{u}}(\mathbf{r},t)\cdot\nabla_{\mathbf{x}}S(\mathbf{r},t)\right)\nabla_{\mathbf{x}}c_0(\mathbf{r},t)-\left\{d_{\mathbf{x}}\tilde{\mathbf{u}}(\mathbf{r},t)\right\}^T\cdot\nabla_{\mathbf{x}}S(\mathbf{r},t)\right)\\
&\qquad+\left(\frac{\partial S}{\partial t}(\mathbf{r},t)\right.\\
&\left.+\tilde{\mathbf{u}}(\mathbf{r},t)\cdot\nabla_{\mathbf{x}}S(\mathbf{r},t)\right)^{-2}\left(\frac{d}{dt}\left(\frac{\partial S}{\partial t}\left(\mathbf{r}(t),t\right)+\tilde{\mathbf{u}}\left(\mathbf{r}(t),t\right)\cdot\nabla_{\mathbf{x}}S\left(\mathbf{r}(t),t\right)\right)\right)\nabla_{\mathbf{x}}S(\mathbf{r},t)\\
&=-\left(\frac{1}{c_0(\mathbf{r},t)}\nabla_{\mathbf{x}}c_0(\mathbf{r},t)-\left(\frac{\partial S}{\partial t}(\mathbf{r},t)+\tilde{\mathbf{u}}(\mathbf{r},t)\cdot\nabla_{\mathbf{x}}S(\mathbf{r},t)\right)^{-1}\left\{d_{\mathbf{x}}\tilde{\mathbf{u}}(\mathbf{r},t)\right\}^T\cdot\nabla_{\mathbf{x}}S(\mathbf{r},t)\right)\\
&\qquad+\left(\frac{\partial S}{\partial t}(\mathbf{r},t)\right.\\
&\left.+\tilde{\mathbf{u}}(\mathbf{r},t)\cdot\nabla_{\mathbf{x}}S(\mathbf{r},t)\right)^{-2}\left(\frac{d}{dt}\left(\frac{\partial S}{\partial t}\left(\mathbf{r}(t),t\right)+\tilde{\mathbf{u}}\left(\mathbf{r}(t),t\right)\cdot\nabla_{\mathbf{x}}S\left(\mathbf{r}(t),t\right)\right)\right)\nabla_{\mathbf{x}}S(\mathbf{r},t).
\end{aligned}
\tag{13.306}
$$

Thus inserting (13.304) into (13.306) gives

$$\frac{d}{dt}\left(\frac{1}{c_0^2\left(\mathbf{r}(t),t\right)}\left(\frac{d\mathbf{r}}{dt}(t)-\tilde{\mathbf{u}}\left(\mathbf{r}(t),t\right)\right)\right)=$$
$$-\frac{1}{c_0(\mathbf{r},t)}\nabla_{\mathbf{x}}c_0(\mathbf{r},t)+\left(\frac{\partial S}{\partial t}(\mathbf{r},t)+\tilde{\mathbf{u}}(\mathbf{r},t)\cdot\nabla_{\mathbf{x}}S(\mathbf{r},t)\right)^{-1}\{d_{\mathbf{x}}\tilde{\mathbf{u}}(\mathbf{r},t)\}^T\cdot\nabla_{\mathbf{x}}S(\mathbf{r},t)$$
$$+\frac{1}{c_0(\mathbf{r},t)}\left(\frac{\partial c_0}{\partial t}(\mathbf{r},t)+\tilde{\mathbf{u}}(\mathbf{r},t)\cdot\nabla_{\mathbf{x}}c_0(\mathbf{r},t)\right)\left(\frac{\partial S}{\partial t}(\mathbf{r},t)+\tilde{\mathbf{u}}(\mathbf{r},t)\cdot\nabla_{\mathbf{x}}S(\mathbf{r},t)\right)^{-1}\nabla_{\mathbf{x}}S(\mathbf{r},t)$$
$$-\frac{1}{2c_0^2(\mathbf{r},t)}\left(\left(\left(\{d_{\mathbf{x}}\tilde{\mathbf{u}}(\mathbf{r},t)\}^T+d_{\mathbf{x}}\tilde{\mathbf{u}}(\mathbf{r},t)\right)\cdot\left(\frac{d\mathbf{r}}{dt}-\tilde{\mathbf{u}}(\mathbf{r},t)\right)\right)\cdot\left(\frac{d\mathbf{r}}{dt}-\tilde{\mathbf{u}}(\mathbf{r},t)\right)\right)\times$$
$$\times\left(\frac{\partial S}{\partial t}(\mathbf{r},t)+\tilde{\mathbf{u}}(\mathbf{r},t)\cdot\nabla_{\mathbf{x}}S(\mathbf{r},t)\right)^{-1}\nabla_{\mathbf{x}}S(\mathbf{r},t). \quad (13.307)$$

Therefore, inserting (13.291) into (13.307) we finally deduce

$$\frac{d}{dt}\left(\frac{1}{c_0^2\left(\mathbf{r}(t),t\right)}\left(\frac{d\mathbf{r}}{dt}(t)-\tilde{\mathbf{u}}\left(\mathbf{r}(t),t\right)\right)\right)=$$
$$-\frac{1}{c_0(\mathbf{r},t)}\nabla_{\mathbf{x}}c_0(\mathbf{r},t)-\frac{1}{c_0^2(\mathbf{r},t)}\{d_{\mathbf{x}}\tilde{\mathbf{u}}(\mathbf{r},t)\}^T\cdot\left(\frac{d\mathbf{r}}{dt}-\tilde{\mathbf{u}}(\mathbf{r},t)\right)$$
$$-\frac{1}{c_0^3(\mathbf{r},t)}\left(\frac{\partial c_0}{\partial t}(\mathbf{r},t)+\tilde{\mathbf{u}}(\mathbf{r},t)\cdot\nabla_{\mathbf{x}}c_0(\mathbf{r},t)\right)\left(\frac{d\mathbf{r}}{dt}-\tilde{\mathbf{u}}(\mathbf{r},t)\right)$$
$$+\frac{1}{2c_0^4(\mathbf{r},t)}\left(\left(\left(\{d_{\mathbf{x}}\tilde{\mathbf{u}}(\mathbf{r},t)\}^T\right.\right.\right.$$
$$\left.\left.\left.+d_{\mathbf{x}}\tilde{\mathbf{u}}(\mathbf{r},t)\right)\cdot\left(\frac{d\mathbf{r}}{dt}-\tilde{\mathbf{u}}(\mathbf{r},t)\right)\right)\cdot\left(\frac{d\mathbf{r}}{dt}-\tilde{\mathbf{u}}(\mathbf{r},t)\right)\right)\left(\frac{d\mathbf{r}}{dt}-\tilde{\mathbf{u}}(\mathbf{r},t)\right) \quad (13.308)$$

that we can finally rewrite in the alternative form of (13.285). □

Note that, by (13.285) we obviously deduce

$$\frac{1}{2}\frac{d}{dt}\left(\left|\frac{1}{c_0\left(\mathbf{r}(t),t\right)}\left(\frac{d\mathbf{r}}{dt}(t)-\tilde{\mathbf{u}}\left(\mathbf{r}(t),t\right)\right)\right|^2-1\right)=$$
$$\frac{1}{c_0(\mathbf{r},t)}\left(\left(\frac{d\mathbf{r}}{dt}-\tilde{\mathbf{u}}(\mathbf{r},t)\right)\cdot\nabla_{\mathbf{x}}c_0(\mathbf{r},t)\right)\left(\left|\frac{1}{c_0\left(\mathbf{r}(t),t\right)}\left(\frac{d\mathbf{r}}{dt}(t)-\tilde{\mathbf{u}}\left(\mathbf{r}(t),t\right)\right)\right|^2-1\right)$$
$$+\frac{1}{2c_0^2(\mathbf{r},t)}\left(\left(\left(\{d_{\mathbf{x}}\tilde{\mathbf{u}}(\mathbf{r},t)\}^T+d_{\mathbf{x}}\tilde{\mathbf{u}}(\mathbf{r},t)\right)\cdot\left(\frac{d\mathbf{r}}{dt}-\tilde{\mathbf{u}}(\mathbf{r},t)\right)\right)\cdot\left(\frac{d\mathbf{r}}{dt}-\tilde{\mathbf{u}}(\mathbf{r},t)\right)\right)\times$$
$$\times\left(\left|\frac{1}{c_0\left(\mathbf{r}(t),t\right)}\left(\frac{d\mathbf{r}}{dt}(t)-\tilde{\mathbf{u}}\left(\mathbf{r}(t),t\right)\right)\right|^2-1\right). \quad (13.309)$$

Therefore, if $\mathbf{r}(t)$ satisfies (13.285) then by (13.309) we obtain that the following equality for the initial instant of time t_0:

$$\left|\frac{1}{c_0\left(\mathbf{r}(t_0), t_0\right)}\left(\frac{d\mathbf{r}}{dt}(t_0) - \tilde{\mathbf{u}}\left(\mathbf{r}(t_0), t_0\right)\right)\right|^2 - 1 = 0$$

implies

$$\left|\frac{1}{c_0\left(\mathbf{r}(t), t\right)}\left(\frac{d\mathbf{r}}{dt}(t) - \tilde{\mathbf{u}}\left(\mathbf{r}(t), t\right)\right)\right|^2 - 1 = 0$$

for every instant of time. So we get (13.289), i.e. we have a consistency with (13.175) and (13.167).

Next note that, as before, we can easily deduce that equations of the ray propagation either of the form (13.167) or of the form (13.285) are invariant under the change of non-inertial Cartesian coordinate system. Moreover, (13.287) is also invariant under the change of non-inertial Cartesian coordinate system.

Next consider a wave characterized by

$$U(\mathbf{x}, t) = A(\mathbf{x}, t)e^{ik_0 S(\mathbf{x},t)}, \tag{13.310}$$

and consistently with (13.281) consider a velocity field of the wave:

$$\mathbf{h}(\mathbf{x}, t) := \tilde{\mathbf{u}}(\mathbf{x}, t) - c_0^2(\mathbf{x}, t)\left(\frac{\partial S}{\partial t}(\mathbf{x}, t) + \tilde{\mathbf{u}}(\mathbf{x}, t)\cdot\nabla_{\mathbf{x}} S(\mathbf{x}, t)\right)^{-1}\nabla_{\mathbf{x}} S(\mathbf{x}, t). \tag{13.311}$$

Furthermore, assume that the wave we consider undergoes reflection and/or refraction on the time-dependent surface $\mathcal{T}$ having the outcoming three-dimensional unit normal $\mathbf{n}(\mathbf{x}, t)$ and the motion velocity field $\mathbf{w}_{\mathcal{T}}(\mathbf{x}, t)$, separating two regions characterized, respectively, by $c_0 = c_0^{(1)}$ and $\tilde{\mathbf{u}} = \tilde{\mathbf{u}}_1$ and by $c_0^{(2)}$ and $\tilde{\mathbf{u}}_2$, with the formation of the reflected wave, characterized by

$$U_1(\mathbf{x}, t) = A_1(\mathbf{x}, t)e^{ik_0 S_1(\mathbf{x},t)}, \tag{13.312}$$

and by the velocity field:

$$\mathbf{h}_1(\mathbf{x}) = \tilde{\mathbf{u}}_1(\mathbf{x}, t) - c_0^2(\mathbf{x}, t)\left(\frac{\partial S_1}{\partial t}(\mathbf{x}, t) + \tilde{\mathbf{u}}_1(\mathbf{x}, t)\cdot\nabla_{\mathbf{x}} S_1(\mathbf{x}, t)\right)^{-1}\nabla_{\mathbf{x}} S_1(\mathbf{x}, t), \tag{13.313}$$

and formation of the refracted wave characterized by

$$U_2(\mathbf{x}, t) = A_2(\mathbf{x}, t)e^{ik_0 S_2(\mathbf{x},t)}, \tag{13.314}$$

and by the velocity field:

$$\mathbf{h}_2(\mathbf{x}) = \tilde{\mathbf{u}}_2(\mathbf{x}, t) - (c_0^{(2)})^2(\mathbf{x}, t)\left(\frac{\partial S_2}{\partial t}(\mathbf{x}, t) + \tilde{\mathbf{u}}_2(\mathbf{x}, t)\cdot\nabla_{\mathbf{x}} S_2(\mathbf{x}, t)\right)^{-1}\nabla_{\mathbf{x}} S_2(\mathbf{x}, t). \tag{13.315}$$

Then the boundary conditions of U, U_1 and U_2 depend on the physical meaning of these fields. However, one of the necessary conditions should be that

$$S_1(\mathbf{x}, t) = S_2(\mathbf{x}, t) + C_2 = S(\mathbf{x}, t) \qquad \forall \mathbf{x} \in \mathcal{T},\ \forall t, \tag{13.316}$$

where C_2 is a real constant. In particular (13.316) implies

$$\nabla_{\mathbf{x}} S_1 - (\mathbf{n}\cdot\nabla_{\mathbf{x}} S_1)\,\mathbf{n} = \nabla_{\mathbf{x}} S_2 - (\mathbf{n}\cdot\nabla_{\mathbf{x}} S_2)\,\mathbf{n} = \nabla_{\mathbf{x}} S - (\mathbf{n}\cdot\nabla_{\mathbf{x}} S)\,\mathbf{n} \qquad \forall \mathbf{x} \in \mathcal{T},\ \forall t. \tag{13.317}$$

In particular, for every point on the surface $\mathcal{T}$ vectors $\nabla_{\mathbf{x}} S_1$ and $\nabla_{\mathbf{x}} S_2$ lie in the plane formed by vectors $\mathbf{n}$ and $\nabla_{\mathbf{x}} S$. Moreover, by (13.316) we also have

$$\frac{\partial S}{\partial t} + \mathbf{w}_{\mathcal{T}}\cdot\nabla_{\mathbf{x}} S = \frac{\partial S_1}{\partial t} + \mathbf{w}_{\mathcal{T}}\cdot\nabla_{\mathbf{x}} S_1 = \frac{\partial S_2}{\partial t} + \mathbf{w}_{\mathcal{T}}\cdot\nabla_{\mathbf{x}} S_2 \qquad \forall \mathbf{x} \in \mathcal{T},\ \forall t. \tag{13.318}$$

Finally, by (13.175) we have

$$|\mathbf{h} - \tilde{\mathbf{u}}|^2 = c_0^2, \quad |\mathbf{h}_1 - \tilde{\mathbf{u}}|^2 = c_0^2 \quad \text{and} \quad |\mathbf{h}_2 - \tilde{\mathbf{u}}_2|^2 = (c_0^{(2)})^2. \tag{13.319}$$

In particular, by (13.317) and (13.318) we have

$$\begin{aligned}
&\nabla_{\mathbf{x}} S_1 - \frac{1}{c_0^2}\left(\frac{\partial S_1}{\partial t} + \mathbf{w}_{\mathcal{T}}\cdot\nabla_{\mathbf{x}} S_1\right)(\tilde{\mathbf{u}} - \mathbf{w}_{\mathcal{T}}) \\
&\quad - \left(\mathbf{n}\cdot\left(\nabla_{\mathbf{x}} S_1 - \frac{1}{c_0^2}\left(\frac{\partial S_1}{\partial t} + \mathbf{w}_{\mathcal{T}}\cdot\nabla_{\mathbf{x}} S_1\right)(\tilde{\mathbf{u}} - \mathbf{w}_{\mathcal{T}})\right)\right)\mathbf{n} \\
&\quad = \left(\nabla_{\mathbf{x}} S - \frac{1}{c_0^2}\left(\frac{\partial S}{\partial t} + \mathbf{w}_{\mathcal{T}}\cdot\nabla_{\mathbf{x}} S\right)(\tilde{\mathbf{u}} - \mathbf{w}_{\mathcal{T}})\right) \\
&\quad - \left(\mathbf{n}\cdot\left(\nabla_{\mathbf{x}} S - \frac{1}{c_0^2}\left(\frac{\partial S}{\partial t} + \mathbf{w}_{\mathcal{T}}\cdot\nabla_{\mathbf{x}} S\right)(\tilde{\mathbf{u}} - \mathbf{w}_{\mathcal{T}})\right)\right)\mathbf{n} \\
&\quad = \left(\nabla_{\mathbf{x}} S_2 - \frac{1}{c_0^2}\left(\frac{\partial S_2}{\partial t} + \mathbf{w}_{\mathcal{T}}\cdot\nabla_{\mathbf{x}} S_2\right)(\tilde{\mathbf{u}} - \mathbf{w}_{\mathcal{T}})\right) \\
&\quad - \left(\mathbf{n}\cdot\left(\nabla_{\mathbf{x}} S_2 - \frac{1}{c_0^2}\left(\frac{\partial S_2}{\partial t} + \mathbf{w}_{\mathcal{T}}\cdot\nabla_{\mathbf{x}} S_2\right)(\tilde{\mathbf{u}} - \mathbf{w}_{\mathcal{T}})\right)\right)\mathbf{n} \qquad \forall \mathbf{x} \in \mathcal{T},\ \forall t.
\end{aligned} \tag{13.320}$$

Next, assume that the following approximation is valid on the both sides of the surface $\mathcal{T}$:

$$\frac{|\mathbf{w}_{\mathcal{T}}|^2}{c_0^2 + (c_0^{(2)})^2} \ll 1, \quad \frac{|\tilde{\mathbf{u}}|^2}{c_0^2} \ll 1 \quad \text{and} \quad \frac{|\tilde{\mathbf{u}}_2|^2}{(c_0^{(2)})^2} \ll 1. \tag{13.321}$$

Then, by (13.321) we approximate (13.279) as

$$\frac{1}{c_0^2}\left(\frac{\partial S}{\partial t}\right)^2 \approx \left|\nabla_{\mathbf{x}} S - \frac{1}{c_0^2}\left(\frac{\partial S}{\partial t} + \tilde{\mathbf{u}} \cdot \nabla_{\mathbf{x}} S\right)\tilde{\mathbf{u}}\right|^2. \tag{13.322}$$

Thus by (13.321) we further approximate (13.322) as

$$\frac{1}{c_0^2}\left(\frac{\partial S}{\partial t} + \mathbf{w}_{\mathcal{T}} \cdot \nabla_{\mathbf{x}} S\right)^2 \approx \left|\nabla_{\mathbf{x}} S - \frac{1}{c_0^2}\left(\frac{\partial S}{\partial t} + \tilde{\mathbf{u}} \cdot \nabla_{\mathbf{x}} S\right)(\tilde{\mathbf{u}} - \mathbf{w}_{\mathcal{T}})\right|^2. \tag{13.323}$$

Then by (13.281) we finally rewrite (13.323) as

$$\left(\frac{\partial S}{\partial t} + \mathbf{w}_{\mathcal{T}} \cdot \nabla_{\mathbf{x}} S\right)^2 \approx \left|\frac{1}{c_0}\left(\frac{\partial S}{\partial t} + \tilde{\mathbf{u}} \cdot \nabla_{\mathbf{x}} S\right)(\mathbf{h} - \mathbf{w}_{\mathcal{T}})\right|^2. \tag{13.324}$$

Similarly we obtain

$$\left(\frac{\partial S_1}{\partial t} + \mathbf{w}_{\mathcal{T}} \cdot \nabla_{\mathbf{x}} S_1\right)^2 \approx \left|\frac{1}{c_0}\left(\frac{\partial S_1}{\partial t} + \tilde{\mathbf{u}} \cdot \nabla_{\mathbf{x}} S_1\right)(\mathbf{h}_1 - \mathbf{w}_{\mathcal{T}})\right|^2, \tag{13.325}$$

and

$$\left(\frac{\partial S_2}{\partial t} + \mathbf{w}_{\mathcal{T}} \cdot \nabla_{\mathbf{x}} S_2\right)^2 \approx \left|\frac{1}{c_0^{(2)}}\left(\frac{\partial S_2}{\partial t} + \tilde{\mathbf{u}}_2 \cdot \nabla_{\mathbf{x}} S_2\right)(\mathbf{h}_2 - \mathbf{w}_{\mathcal{T}})\right|^2. \tag{13.326}$$

In particular, by (13.324), (13.325), (13.326) and (13.318) we deduce

$$\begin{aligned}\frac{(c_0^{(2)})^2}{c_0^2}&\left|\frac{c_0}{(c_0^{(2)})^2}\left(\frac{\partial S_2}{\partial t} + \tilde{\mathbf{u}}_2 \cdot \nabla_{\mathbf{x}} S_2\right)(\mathbf{h}_2 - \mathbf{w}_{\mathcal{T}})\right|^2 \approx \\ &\left|\frac{1}{c_0}\left(\frac{\partial S_1}{\partial t} + \tilde{\mathbf{u}} \cdot \nabla_{\mathbf{x}} S_1\right)(\mathbf{h}_1 - \mathbf{w}_{\mathcal{T}})\right|^2 \approx \left|\frac{1}{c_0}\left(\frac{\partial S}{\partial t} + \tilde{\mathbf{u}} \cdot \nabla_{\mathbf{x}} S\right)(\mathbf{h} - \mathbf{w}_{\mathcal{T}})\right|^2.\end{aligned} \tag{13.327}$$

Moreover, by (13.321) we approximate (13.320) as

$$\nabla_{\mathbf{x}} S_1 - \frac{1}{c_0^2}\left(\frac{\partial S_1}{\partial t} + \tilde{\mathbf{u}} \cdot \nabla_{\mathbf{x}} S_1\right)(\tilde{\mathbf{u}} - \mathbf{w}_{\mathcal{T}})$$
$$- \left(\mathbf{n} \cdot \left(\nabla_{\mathbf{x}} S_1 - \frac{1}{c_0^2}\left(\frac{\partial S_1}{\partial t} + \tilde{\mathbf{u}} \cdot \nabla_{\mathbf{x}} S_1\right)(\tilde{\mathbf{u}} - \mathbf{w}_{\mathcal{T}})\right)\right)\mathbf{n} =$$
$$\left(\nabla_{\mathbf{x}} S - \frac{1}{c_0^2}\left(\frac{\partial S}{\partial t} + \tilde{\mathbf{u}} \cdot \nabla_{\mathbf{x}} S\right)(\tilde{\mathbf{u}} - \mathbf{w}_{\mathcal{T}})\right)$$
$$- \left(\mathbf{n} \cdot \left(\nabla_{\mathbf{x}} S - \frac{1}{c_0^2}\left(\frac{\partial S}{\partial t} + \tilde{\mathbf{u}} \cdot \nabla_{\mathbf{x}} S\right)(\tilde{\mathbf{u}} - \mathbf{w}_{\mathcal{T}})\right)\right)\mathbf{n}$$
$$= \left(\nabla_{\mathbf{x}} S_2 - \frac{1}{(c_0^{(2)})^2}\left(\frac{\partial S_2}{\partial t} + \tilde{\mathbf{u}}_2 \cdot \nabla_{\mathbf{x}} S_2\right)(\tilde{\mathbf{u}}_2 - \mathbf{w}_{\mathcal{T}})\right)$$
$$- \left(\mathbf{n} \cdot \left(\nabla_{\mathbf{x}} S_2 - \frac{1}{(c_0^{(2)})^2}\left(\frac{\partial S_2}{\partial t} + \tilde{\mathbf{u}}_2 \cdot \nabla_{\mathbf{x}} S_2\right)(\tilde{\mathbf{u}}_2 - \mathbf{w}_{\mathcal{T}})\right)\right)\mathbf{n}$$
$$+ \frac{1}{c_0^2}\left(\frac{\partial S_2}{\partial t} + \tilde{\mathbf{u}}_2 \cdot \nabla_{\mathbf{x}} S_2\right)\left(\frac{c_0^2}{(c_0^{(2)})^2}\tilde{\mathbf{u}}_2 - \tilde{\mathbf{u}} - \left(\frac{c_0^2}{(c_0^{(2)})^2} - 1\right)\mathbf{w}_{\mathcal{T}}\right)$$
$$- \frac{1}{c_0^2}\left(\frac{\partial S_2}{\partial t} + \tilde{\mathbf{u}}_2 \cdot \nabla_{\mathbf{x}} S_2\right)\left(\mathbf{n} \cdot \left(\frac{c_0^2}{(c_0^{(2)})^2}\tilde{\mathbf{u}}_2 - \tilde{\mathbf{u}} - \left(\frac{c_0^2}{(c_0^{(2)})^2} - 1\right)\mathbf{w}_{\mathcal{T}}\right)\right)\mathbf{n}$$
$$\forall \mathbf{x} \in \mathcal{T},\ \forall t, \quad (13.328)$$

and then by (13.281) rewrite it as

$$\frac{1}{c_0}\left(\frac{\partial S_1}{\partial t} + \tilde{\mathbf{u}} \cdot \nabla_{\mathbf{x}} S_1\right)(\mathbf{h}_1 - \mathbf{w}_{\mathcal{T}}) - \frac{1}{c_0}\left(\frac{\partial S_1}{\partial t} + \tilde{\mathbf{u}} \cdot \nabla_{\mathbf{x}} S_1\right)(\mathbf{n} \cdot (\mathbf{h}_1 - \mathbf{w}_{\mathcal{T}}))\,\mathbf{n} =$$
$$\frac{1}{c_0}\left(\frac{\partial S}{\partial t} + \tilde{\mathbf{u}} \cdot \nabla_{\mathbf{x}} S\right)(\mathbf{h} - \mathbf{w}_{\mathcal{T}}) - \frac{1}{c_0}\left(\frac{\partial S}{\partial t} + \tilde{\mathbf{u}} \cdot \nabla_{\mathbf{x}} S\right)(\mathbf{n} \cdot (\mathbf{h} - \mathbf{w}_{\mathcal{T}}))\,\mathbf{n} =$$
$$\frac{c_0}{(c_0^{(2)})^2}\left(\frac{\partial S_2}{\partial t} + \tilde{\mathbf{u}}_2 \cdot \nabla_{\mathbf{x}} S_2\right)(\mathbf{h}_2 - \mathbf{w}_{\mathcal{T}})$$
$$- \frac{c_0}{(c_0^{(2)})^2}\left(\frac{\partial S_2}{\partial t} + \tilde{\mathbf{u}}_2 \cdot \nabla_{\mathbf{x}} S_2\right)(\mathbf{n} \cdot (\mathbf{h}_2 - \mathbf{w}_{\mathcal{T}}))\,\mathbf{n}$$
$$- \frac{1}{c_0}\left(\frac{\partial S_2}{\partial t} + \tilde{\mathbf{u}}_2 \cdot \nabla_{\mathbf{x}} S_2\right)\left(\frac{c_0^2}{(c_0^{(2)})^2}\tilde{\mathbf{u}}_2 - \tilde{\mathbf{u}} - \left(\frac{c_0^2}{(c_0^{(2)})^2} - 1\right)\mathbf{w}_{\mathcal{T}}\right)$$
$$+ \frac{1}{c_0}\left(\frac{\partial S_2}{\partial t} + \tilde{\mathbf{u}}_2 \cdot \nabla_{\mathbf{x}} S_2\right)\left(\mathbf{n} \cdot \left(\frac{c_0^2}{(c_0^{(2)})^2}\tilde{\mathbf{u}}_2 - \tilde{\mathbf{u}} - \left(\frac{c_0^2}{(c_0^{(2)})^2} - 1\right)\mathbf{w}_{\mathcal{T}}\right)\right)\mathbf{n}$$
$$\forall \mathbf{x} \in \mathcal{T},\ \forall t. \quad (13.329)$$

Then, since the directions of vectors **h** and $\mathbf{h}_1$ should be different, by (13.329) and (13.327) we deduce

$$\left(\frac{\partial S_1}{\partial t}+\tilde{\mathbf{u}}\cdot\nabla_{\mathbf{x}}S_1\right)\left(\mathbf{n}\cdot(\mathbf{h}_1-\mathbf{w}_{\mathcal{T}})\right)=-\left(\frac{\partial S}{\partial t}+\tilde{\mathbf{u}}\cdot\nabla_{\mathbf{x}}S\right)\left(\mathbf{n}\cdot(\mathbf{h}-\mathbf{w}_{\mathcal{T}})\right)\quad\forall\mathbf{x}\in\mathcal{T}. \tag{13.330}$$

So, by (13.329) and (13.330) we obtain the law of reflection: vector $(\mathbf{h}_1-\mathbf{w}_{\mathcal{T}})$ lies in the plane formed by vectors **n** and $(\mathbf{h}-\mathbf{w}_{\mathcal{T}})$, and we have

$$\theta\left((\mathbf{h}-\mathbf{w}_{\mathcal{T}}),-\mathbf{n}\right)=\theta_1\left((\mathbf{h}_1-\mathbf{w}_{\mathcal{T}}),\mathbf{n}\right), \tag{13.331}$$

where $\theta\left((\mathbf{h}-\mathbf{w}_{\mathcal{T}}),-\mathbf{n}\right)$ is the angle between the vector of the relative velocity of the incoming ray, relative to the surface of reflection, $(\mathbf{h}-\mathbf{w}_{\mathcal{T}})$ and the incoming normal to the surface $-\mathbf{n}$ and $\theta_1\left((\mathbf{h}_1-\mathbf{w}_{\mathcal{T}}),\mathbf{n}\right)$ is the angle between the vector of the relative velocity of the reflected ray, relative to the surface of reflection, $(\mathbf{h}_1-\mathbf{w}_{\mathcal{T}})$ and the outcoming normal **n**. Note that, since **h** and $\mathbf{w}_{\mathcal{T}}$ are both speed-like vector fields then $(\mathbf{h}-\mathbf{w}_{\mathcal{T}})$ is a proper vector field and thus the above law of reflection together with (13.331) is invariant under the change of inertial or non-inertial Cartesian coordinate systems. In particular, if (13.321) holds for some Cartesian coordinate system, then we can use this law also in other coordinate systems where (13.321) does not hold. Therefore, for the validity of the above law of reflection, we may assume the following relation instead of (13.321):

$$\frac{|\tilde{\mathbf{u}}-\mathbf{w}_{\mathcal{T}}|^2}{c_0^2}\ll 1. \tag{13.332}$$

Next, assume that the wave we consider in (13.310) has an electromagnetic nature. Then by (13.203) we have

$$c_0=c\sqrt{\kappa_0\gamma_0}\quad\text{and}\quad\tilde{\mathbf{u}}=\left(\gamma_0\kappa_0\mathbf{v}+(1-\gamma_0\kappa_0)\mathbf{u}\right), \tag{13.333}$$

where **u** is the medium velocity and **v** is the local vectorial gravitational potential. Similarly, on the second side of surface $\mathcal{T}$, we have

$$c_0^{(2)}=c\sqrt{\kappa_0^{(2)}\gamma_0^{(2)}}\quad\text{and}\quad\tilde{\mathbf{u}}_2=\left(\gamma_0^{(2)}\kappa_0^{(2)}\mathbf{v}+(1-\gamma_0^{(2)}\kappa_0^{(2)})\mathbf{u}_2\right), \tag{13.334}$$

where $\mathbf{u}_2$ is the medium velocity on the second side of surface $\mathcal{T}$. Then we rewrite the first equalities in (13.333) and (13.334) as

$$c_0=c\sqrt{\kappa_0\gamma_0}\quad\text{and}\quad c_0^{(2)}=c\sqrt{\kappa_0^{(2)}\gamma_0^{(2)}}. \tag{13.335}$$

Then, in particular, by (13.335) we deduce

$$\frac{c_0^2}{(c_0^{(2)})^2}=\frac{\gamma_0\kappa_0}{\gamma_0^{(2)}\kappa_0^{(2)}}. \tag{13.336}$$

Thus by (13.336), (13.333) and (13.334) we deduce

$$\left(\frac{c_0^2}{(c_0^{(2)})^2}\tilde{\mathbf{u}}_2 - \tilde{\mathbf{u}} - \left(\frac{c_0^2}{(c_0^{(2)})^2} - 1\right)\mathbf{w}_{\mathcal{T}}\right) = \left(\frac{\gamma_0\kappa_0}{\gamma_0^{(2)}\kappa_0^{(2)}}\left(\gamma_0^{(2)}\kappa_0^{(2)}\mathbf{v} + (1 - \gamma_0^{(2)}\kappa_0^{(2)})\mathbf{u}_2\right)\right.$$
$$\left. - (\gamma_0\kappa_0\mathbf{v} + (1 - \gamma_0\kappa_0)\mathbf{u}) - \left(\frac{\gamma_0\kappa_0}{\gamma_0^{(2)}\kappa_0^{(2)}} - 1\right)\mathbf{w}_{\mathcal{T}}\right)$$
$$= \left(\frac{\gamma_0\kappa_0}{\gamma_0^{(2)}\kappa_0^{(2)}} - 1\right)(\mathbf{u}_2 - \mathbf{w}_{\mathcal{T}}) + (1 - \gamma_0\kappa_0)(\mathbf{u}_2 - \mathbf{u}). \quad (13.337)$$

On the other hand, since $\mathbf{u}$ and $\mathbf{u}_2$ are velocities of the medium on both sides of the surface $\mathcal{T}$ and $\mathbf{w}_{\mathcal{T}}$ is the velocity of the surface, we have equalities of these three vectors on the surface:

$$\mathbf{u}_2(\mathbf{x}, t) = \mathbf{u}(\mathbf{x}, t) = \mathbf{w}_{\mathcal{T}}(\mathbf{x}, t) \qquad \forall \mathbf{x} \in \mathcal{T},\ \forall t. \quad (13.338)$$

Thus with the help of (13.337) and (13.338) we infer

$$\frac{1}{c_0}\left(\frac{\partial S_2}{\partial t} + \tilde{\mathbf{u}}_2 \cdot \nabla_{\mathbf{x}} S_2\right)\left(\frac{c_0^2}{(c_0^{(2)})^2}\tilde{\mathbf{u}}_2 - \tilde{\mathbf{u}} - \left(\frac{c_0^2}{(c_0^{(2)})^2} - 1\right)\mathbf{w}_{\mathcal{T}}\right)$$
$$- \frac{1}{c_0}\left(\frac{\partial S_2}{\partial t} + \tilde{\mathbf{u}}_2 \cdot \nabla_{\mathbf{x}} S_2\right)\left(\mathbf{n} \cdot \left(\frac{c_0^2}{(c_0^{(2)})^2}\tilde{\mathbf{u}}_2 - \tilde{\mathbf{u}} - \left(\frac{c_0^2}{(c_0^{(2)})^2} - 1\right)\mathbf{w}_{\mathcal{T}}\right)\right)\mathbf{n} = 0$$
$$\forall \mathbf{x} \in \mathcal{T},\ \forall t. \quad (13.339)$$

Note that if there is a vacuum on the one side of surface $\mathcal{T}$ then we can obtain (13.339) from (13.337) with even easier proof. Therefore, using (13.339) by (13.329) we deduce

$$\frac{1}{c_0}\left(\frac{\partial S_1}{\partial t} + \tilde{\mathbf{u}} \cdot \nabla_{\mathbf{x}} S_1\right)(\mathbf{h}_1 - \mathbf{w}_{\mathcal{T}}) - \frac{1}{c_0}\left(\frac{\partial S_1}{\partial t} + \tilde{\mathbf{u}} \cdot \nabla_{\mathbf{x}} S_1\right)(\mathbf{n} \cdot (\mathbf{h}_1 - \mathbf{w}_{\mathcal{T}}))\,\mathbf{n} =$$
$$\frac{c_0}{(c_0^{(2)})^2}\left(\frac{\partial S_2}{\partial t} + \tilde{\mathbf{u}}_2 \cdot \nabla_{\mathbf{x}} S_2\right)(\mathbf{h}_2 - \mathbf{w}_{\mathcal{T}})$$
$$- \frac{c_0}{(c_0^{(2)})^2}\left(\frac{\partial S_2}{\partial t} + \tilde{\mathbf{u}}_2 \cdot \nabla_{\mathbf{x}} S_2\right)(\mathbf{n} \cdot (\mathbf{h}_2 - \mathbf{w}_{\mathcal{T}}))\,\mathbf{n} \qquad \forall \mathbf{x} \in \mathcal{T},\ \forall t, \quad (13.340)$$

and, in particular, for every point on the surface $\mathcal{T}$ vector $(\mathbf{h}_2 - \mathbf{w}_{\mathcal{T}})$ lies in the plane formed by vectors $\mathbf{n}$ and $(\mathbf{h} - \mathbf{w}_{\mathcal{T}})$. On the other hand, by (13.327) we have

$$\frac{(c_0^{(2)})^2}{c_0^2}\left|\frac{c_0}{(c_0^{(2)})^2}\left(\frac{\partial S_2}{\partial t}+\tilde{\mathbf{u}}_2\cdot\nabla_{\mathbf{x}}S_2\right)(\mathbf{h}_2-\mathbf{w}_{\mathcal{T}})\right|^2$$
$$\approx\left|\frac{1}{c_0}\left(\frac{\partial S_1}{\partial t}+\tilde{\mathbf{u}}\cdot\nabla_{\mathbf{x}}S_1\right)(\mathbf{h}_1-\mathbf{w}_{\mathcal{T}})\right|^2. \quad (13.341)$$

Thus by (13.340) and (13.341), exactly as in the proof of (13.231), we finally deduce that we have the following Snell's law of refraction: $(\mathbf{h}_2-\mathbf{w}_{\mathcal{T}})$ lies in the plane formed by vectors $\mathbf{n}$ and $(\mathbf{h}-\mathbf{w}_{\mathcal{T}})$ and we have

$$n\sin\left(\theta\left((\mathbf{h}-\mathbf{w}_{\mathcal{T}}),\mathbf{n}\right)\right)=n_2\sin\left(\theta_2\left((\mathbf{h}_2-\mathbf{w}_{\mathcal{T}}),\mathbf{n}\right)\right) \quad (13.342)$$

where $\theta\left((\mathbf{h}-\mathbf{w}_{\mathcal{T}}),\mathbf{n}\right)$ is the vector of the relative velocity of the incoming ray, relative to the surface of refraction, $(\mathbf{h}-\mathbf{w}_{\mathcal{T}})$ and the normal to the surface $\mathbf{n}$, $\theta_2\left((\mathbf{h}_2-\mathbf{w}_{\mathcal{T}}),\mathbf{n}\right)$ is the vector of the relative velocity of the refracted ray, relative to the surface of refraction, $(\mathbf{h}_2-\mathbf{w}_{\mathcal{T}})$ and the normal $\mathbf{n}$ and as in (13.197) we set refraction indexes:

$$n:=\frac{c}{c_0}\quad\text{and}\quad n_2:=\frac{c}{c_0^{(2)}}. \quad (13.343)$$

Note that as the above law of reflection Snell's law together with (13.342) is invariant under the change of inertial or non-inertial Cartesian coordinate systems. In particular, if (13.321) holds for some Cartesian coordinate system, then we can use this law also in other coordinate systems where (13.321) does not hold. Therefore, as before, for the validity of the above Snell's law we may assume the following relation instead of (13.321):

$$\frac{|\tilde{\mathbf{u}}-\mathbf{w}_{\mathcal{T}}|^2}{c_0^2}\ll 1\quad\text{and}\quad\frac{|\tilde{\mathbf{u}}_2-\mathbf{w}_{\mathcal{T}}|^2}{(c_0^{(2)})^2}\ll 1. \quad (13.344)$$

13.3.9 Polarization of the Light Inside a Moving Medium and/or in the Presence of Gravitational Field

Again consider the system of Maxwell equations in the vacuum or in a medium, having the form (12.56) in the absence of macroscopic charges and/or currents:

$$\begin{cases}\operatorname{curl}_{\mathbf{x}}\mathbf{H} = \frac{1}{c}\frac{\partial \mathbf{D}}{\partial t},\\ \operatorname{div}_{\mathbf{x}}\mathbf{D} = 0,\\ \operatorname{curl}_{\mathbf{x}}\mathbf{E} + \frac{1}{c}\frac{\partial \mathbf{B}}{\partial t} = 0,\\ \operatorname{div}_{\mathbf{x}}\mathbf{B} = 0,\\ \mathbf{E} = \gamma_0\mathbf{D} - \frac{1}{c}\tilde{\mathbf{u}}\times\mathbf{B},\\ \mathbf{H} = \kappa_0\mathbf{B} + \frac{1}{c}\tilde{\mathbf{u}}\times\mathbf{D},\\ \tilde{\mathbf{u}} = (\gamma_0\kappa_0\mathbf{v} + (1-\gamma_0\kappa_0)\mathbf{u}),\end{cases} \tag{13.345}$$

and consider the case of monochromatic wave of the constant frequency $\nu = \frac{\omega}{2\pi}$ where the fields $\tilde{\mathbf{u}}$, γ_0 and κ_0 are independent on the time variable. We also again assume that either our medium has no dispersion or $\mathbf{u} = 0$ (and so $\tilde{\mathbf{u}} = \mathbf{v}$) and the medium is transparent for the given frequency $\nu = \frac{\omega}{2\pi}$. Next assume the rough geometric optics approximation (13.150) (stronger than (13.149)) that means the following: assume that the changes in time of the basic characteristics of the electromagnetic field become essential after certain interval of time T_e and the changes in space of the basic characteristics of the electromagnetic field become essential in the spatial landscape L_e. Then we assume that

$$k_0c_0T_e \gg 1 \quad \text{and} \quad k_0L_e \gg 1, \tag{13.346}$$

where

$$k_0 = \frac{\omega}{c}. \tag{13.347}$$

We also assume (13.180):

$$\frac{|\tilde{\mathbf{u}}|^2}{c_0^2} \ll 1. \tag{13.348}$$

Then consistently with (13.124) we can write

$$\begin{cases}\mathbf{D}(\mathbf{x},t) = \Xi_1\cdot\mathbf{D}_a(\mathbf{x})e^{ik_0S(\mathbf{x},t)}\\ \mathbf{B}(\mathbf{x},t) = \Xi_2\cdot\mathbf{B}_a(\mathbf{x})e^{ik_0S(\mathbf{x},t)}\\ \mathbf{E}(\mathbf{x},t) = \Xi_3\cdot\mathbf{E}_a(\mathbf{x})e^{ik_0S(\mathbf{x},t)}\\ \mathbf{H}(\mathbf{x},t) = \Xi_4\cdot\mathbf{H}_a(\mathbf{x})e^{ik_0S(\mathbf{x},t)}\\ \mathbf{E} = \gamma_0\mathbf{D} - \frac{1}{c}\tilde{\mathbf{u}}\times\mathbf{B}\\ \mathbf{H} = \kappa_0\mathbf{B} + \frac{1}{c}\tilde{\mathbf{u}}\times\mathbf{D}.\end{cases} \tag{13.349}$$

Here $\mathbf{D}_a(\mathbf{x}), \mathbf{B}_a(\mathbf{x}), \mathbf{E}_a(\mathbf{x}), \mathbf{H}_a(\mathbf{x}) : \mathbb{R}^3 \to \mathbb{R}^3$ are real vector fields, independent on the time variable, $S(\mathbf{x}, t)$ is a real function such that $\frac{\partial S}{\partial t} = c$ and $\Xi_1, \Xi_2, \Xi_3, \Xi_4 \in \mathbb{C}^{3\times 3}$ are constant complex diagonal matrices of the form:

$$\Xi_k = \begin{pmatrix} e^{i\theta_{1k}} & 0 & 0 \\ 0 & e^{i\theta_{2k}} & 0 \\ 0 & 0 & e^{i\theta_{3k}} \end{pmatrix} \qquad \forall k = 1, 2, 3, 4, \tag{13.350}$$

where $\theta_{1k}, \theta_{2k}, \theta_{3k}$ are real constants. Then denoting:

$$\begin{cases} \mathbf{D}_l(\mathbf{x}) = \Xi_1 \cdot \mathbf{D}_a(\mathbf{x}) \\ \mathbf{B}_l(\mathbf{x}) = \Xi_2 \cdot \mathbf{B}_a(\mathbf{x}) \\ \mathbf{E}_l(\mathbf{x}) = \Xi_3 \cdot \mathbf{E}_a(\mathbf{x}) \\ \mathbf{H}_l(\mathbf{x}) = \Xi_4 \cdot \mathbf{H}_a(\mathbf{x}), \end{cases} \tag{13.351}$$

we deduce

$$\begin{cases} \mathbf{D}(\mathbf{x}, t) = \mathbf{D}_l(\mathbf{x}) e^{ik_0 S(\mathbf{x},t)} \\ \mathbf{B}(\mathbf{x}, t) = \mathbf{B}_l(\mathbf{x}) e^{ik_0 S(\mathbf{x},t)} \\ \mathbf{E}(\mathbf{x}, t) = \mathbf{E}_l(\mathbf{x}) e^{ik_0 S(\mathbf{x},t)} \\ \mathbf{H}(\mathbf{x}, t) = \mathbf{H}_l(\mathbf{x}) e^{ik_0 S(\mathbf{x},t)} \\ \mathbf{E}_l = \gamma_0 \mathbf{D}_l - \frac{1}{c} \tilde{\mathbf{u}} \times \mathbf{B}_l \\ \mathbf{H}_l = \kappa_0 \mathbf{B}_l + \frac{1}{c} \tilde{\mathbf{u}} \times \mathbf{D}_l. \end{cases} \tag{13.352}$$

Note that $\mathbf{D}_l, \mathbf{B}_l, \mathbf{E}_l, \mathbf{H}_l$ are complex. Then, consistently with (13.181), S satisfies the eikonal equation:

$$\left| \frac{c\tilde{\mathbf{u}}}{c_0^2} - \nabla_{\mathbf{x}} S \right|^2 = \frac{c^2}{c_0^2}, \tag{13.353}$$

and consistently with (13.182) if $\mathbf{A}_l$ denotes either one of the vectors $\mathbf{D}_a, \mathbf{B}_a, \mathbf{E}_a, \mathbf{H}_a$ or one of the vectors $\mathbf{D}_l, \mathbf{B}_l, \mathbf{E}_l, \mathbf{H}_l$ then:

$$\{d_{\mathbf{x}} \mathbf{A}_l\}^T \cdot \left(\frac{c\tilde{\mathbf{u}}}{c_0^2} - \nabla_{\mathbf{x}} S \right) + \frac{1}{2} \left(-\Delta_{\mathbf{x}} S \right) \mathbf{A}_l = 0. \tag{13.354}$$

Moreover, consistently with (13.176), (13.177) and (13.203) we have

$$\begin{cases} c_0 = c\sqrt{\kappa_0 \gamma_0} \\ \frac{\partial S}{\partial t} = c \\ \frac{\partial \mathbf{A}_l}{\partial t} = 0, \\ \frac{\partial \tilde{\mathbf{u}}}{\partial t} = 0 \\ \frac{\partial c_0}{\partial t} = 0, \end{cases} \tag{13.355}$$

and consistently with (13.187), the vector field $\mathbf{h}_0$ defined for every $\mathbf{x}$ by

$$\mathbf{h}_0(\mathbf{x}) := \frac{c}{c_0^2(\mathbf{x})} \tilde{\mathbf{u}}(\mathbf{x}) - \nabla_{\mathbf{x}} S(\mathbf{x}) \approx \frac{c}{c_0^2(\mathbf{x})} \mathbf{h}(\mathbf{x}) \tag{13.356}$$

is the direction of the propagation of the ray that passes through point $\mathbf{x}$.

Next, by inserting (13.352) into (13.345), using the fact that $\frac{\partial S}{\partial t} = c$ and using the rough approximation (13.346) together with (2.7) we obtain

$$\begin{cases} \mathbf{D}_1 \approx \nabla_{\mathbf{x}} S \times \mathbf{H}_1, \\ \nabla_{\mathbf{x}} S \cdot \mathbf{D}_1 \approx 0, \\ \mathbf{B}_1 \approx -\nabla_{\mathbf{x}} S \times \mathbf{E}_1, \\ \nabla_{\mathbf{x}} S \cdot \mathbf{B}_1 \approx 0. \end{cases} \tag{13.357}$$

Thus, by (13.357) and the last two equations in (13.352), using (2.2) we obtain

$$\mathbf{D}_1 \approx \nabla_{\mathbf{x}} S \times \left(\kappa_0 \mathbf{B}_1 + \frac{1}{c} \tilde{\mathbf{u}} \times \mathbf{D}_1 \right) = \kappa_0 \nabla_{\mathbf{x}} S \times \mathbf{B}_1 - \frac{1}{c} (\tilde{\mathbf{u}} \cdot \nabla_{\mathbf{x}} S) \mathbf{D}_1 \quad \text{and}$$
$$\mathbf{B}_1 \approx -\nabla_{\mathbf{x}} S \times \left(\gamma_0 \mathbf{D}_1 - \frac{1}{c} \tilde{\mathbf{u}} \times \mathbf{B}_1 \right) = -\gamma_0 \nabla_{\mathbf{x}} S \times \mathbf{D}_1 - \frac{1}{c} (\tilde{\mathbf{u}} \cdot \nabla_{\mathbf{x}} S) \mathbf{B}_1 . \tag{13.358}$$

So

$$\mathbf{D}_1 \approx \frac{\kappa_0}{\left(1 + \frac{1}{c}\tilde{\mathbf{u}} \cdot \nabla_{\mathbf{x}} S\right)} \nabla_{\mathbf{x}} S \times \mathbf{B}_1 \quad \text{and} \quad \mathbf{B}_1 \approx -\frac{\gamma_0}{\left(1 + \frac{1}{c}\tilde{\mathbf{u}} \cdot \nabla_{\mathbf{x}} S\right)} \nabla_{\mathbf{x}} S \times \mathbf{D}_1 . \tag{13.359}$$

Thus by inserting (13.359) and (13.357) into (13.352) we infer

$$\begin{cases} \mathbf{B} \approx \frac{\gamma_0}{(1+\frac{1}{c}\tilde{\mathbf{u}}\cdot\nabla_{\mathbf{x}} S)} (-\nabla_{\mathbf{x}} S) \times \mathbf{D} \\ \mathbf{D} \approx -\frac{\kappa_0}{(1+\frac{1}{c}\tilde{\mathbf{u}}\cdot\nabla_{\mathbf{x}} S)} (-\nabla_{\mathbf{x}} S) \times \mathbf{B} \\ (-\nabla_{\mathbf{x}} S) \cdot \mathbf{D} \approx 0 \\ (-\nabla_{\mathbf{x}} S) \cdot \mathbf{B} \approx 0. \end{cases} \tag{13.360}$$

So the vectors $(-\nabla_{\mathbf{x}} S)$, $\mathbf{D}$ and $\mathbf{B}$ form together rightly orientated orthogonal system of vectors.

Next by (13.359) and the last two equations in (13.352) we deduce

$$\mathbf{E}_1 \approx \left(\frac{\gamma_0 \kappa_0}{\left(1 + \frac{1}{c}\tilde{\mathbf{u}} \cdot \nabla_{\mathbf{x}} S\right)} \nabla_{\mathbf{x}} S - \frac{1}{c} \tilde{\mathbf{u}} \right) \times \mathbf{B}_1$$
$$\text{and} \quad \mathbf{H}_1 \approx -\left(\frac{\gamma_0 \kappa_0}{\left(1 + \frac{1}{c}\tilde{\mathbf{u}} \cdot \nabla_{\mathbf{x}} S\right)} \nabla_{\mathbf{x}} S - \frac{1}{c} \tilde{\mathbf{u}} \right) \times \mathbf{D}_1, \tag{13.361}$$

and, in particular,

$$\left(\frac{\gamma_0\kappa_0}{\left(1+\frac{1}{c}\tilde{\mathbf{u}}\cdot\nabla_{\mathbf{x}}S\right)}\nabla_{\mathbf{x}}S-\frac{1}{c}\tilde{\mathbf{u}}\right)\cdot\mathbf{E}_1\approx 0$$
$$\text{and}\quad\left(\frac{\gamma_0\kappa_0}{\left(1+\frac{1}{c}\tilde{\mathbf{u}}\cdot\nabla_{\mathbf{x}}S\right)}\nabla_{\mathbf{x}}S-\frac{1}{c}\tilde{\mathbf{u}}\right)\cdot\mathbf{H}_1\approx 0. \tag{13.362}$$

Then, by (13.361) and by the last two equations in (13.352), using (13.348) we deduce

$$\mathbf{E}_1\approx\frac{1}{\kappa_0}\left(\frac{\gamma_0\kappa_0}{\left(1+\frac{1}{c}\tilde{\mathbf{u}}\cdot\nabla_{\mathbf{x}}S\right)}\nabla_{\mathbf{x}}S-\frac{1}{c}\tilde{\mathbf{u}}\right)\times\left(\mathbf{H}_1-\frac{1}{c\gamma_0}\tilde{\mathbf{u}}\times\mathbf{E}_1\right)\quad\text{and}$$
$$\mathbf{H}_1\approx-\frac{1}{\gamma_0}\left(\frac{\gamma_0\kappa_0}{\left(1+\frac{1}{c}\tilde{\mathbf{u}}\cdot\nabla_{\mathbf{x}}S\right)}\nabla_{\mathbf{x}}S-\frac{1}{c}\tilde{\mathbf{u}}\right)\times\left(\mathbf{E}_1+\frac{1}{c\kappa_0}\tilde{\mathbf{u}}\times\mathbf{H}_1\right). \tag{13.363}$$

So, by (2.2) and (13.362), using (13.348), we rewrite (13.363) as

$$\frac{1}{\left(1+\frac{1}{c}\tilde{\mathbf{u}}\cdot\nabla_{\mathbf{x}}S\right)}\mathbf{E}_1\approx\frac{1}{\kappa_0}\left(\frac{\gamma_0\kappa_0}{\left(1+\frac{1}{c}\tilde{\mathbf{u}}\cdot\nabla_{\mathbf{x}}S\right)}\nabla_{\mathbf{x}}S-\frac{1}{c}\tilde{\mathbf{u}}\right)\times\mathbf{H}_1\quad\text{and}$$
$$\frac{1}{\left(1+\frac{1}{c}\tilde{\mathbf{u}}\cdot\nabla_{\mathbf{x}}S\right)}\mathbf{H}_1\approx-\frac{1}{\gamma_0}\left(\frac{\gamma_0\kappa_0}{\left(1+\frac{1}{c}\tilde{\mathbf{u}}\cdot\nabla_{\mathbf{x}}S\right)}\nabla_{\mathbf{x}}S-\frac{1}{c}\tilde{\mathbf{u}}\right)\times\mathbf{E}_1. \tag{13.364}$$

Then, using (13.348) and (13.355), we finally rewrite (13.364) as

$$\mathbf{E}_1\approx-\gamma_0\left(\frac{c}{c_0^2}\tilde{\mathbf{u}}-\nabla_{\mathbf{x}}S\right)\times\mathbf{H}_1\quad\text{and}\quad\mathbf{H}_1\approx\kappa_0\left(\frac{c}{c_0^2}\tilde{\mathbf{u}}-\nabla_{\mathbf{x}}S\right)\times\mathbf{E}_1, \tag{13.365}$$

and, in particular,

$$\left(\frac{c}{c_0^2}\tilde{\mathbf{u}}-\nabla_{\mathbf{x}}S\right)\cdot\mathbf{E}_1\approx 0\quad\text{and}\quad\left(\frac{c}{c_0^2}\tilde{\mathbf{u}}-\nabla_{\mathbf{x}}S\right)\cdot\mathbf{H}_1\approx 0. \tag{13.366}$$

Thus, by inserting (13.365) and (13.366) into (13.352), and using (13.356) we finally deduce

$$\begin{cases}\mathbf{H}\approx\kappa_0\mathbf{h}_0\times\mathbf{E}\\ \mathbf{E}\approx-\gamma_0\mathbf{h}_0\times\mathbf{H}\\ \mathbf{h}_0\cdot\mathbf{E}\approx 0\\ \mathbf{h}_0\cdot\mathbf{H}\approx 0.\end{cases} \tag{13.367}$$

So the vectors $\mathbf{h}_0$, $\mathbf{E}$ and $\mathbf{H}$ form together rightly orientated orthogonal system of vectors. We remind here again that $\mathbf{h}_0$ is the direction of the propagation of the ray.

13.3.10 More Delicate Approximation of the Wave Equation

Again consider the scalar wave equation of the form

$$\left(\frac{\partial}{\partial t}\left\{\frac{1}{c_0^2}\left(\frac{\partial U}{\partial t}+\tilde{\mathbf{u}}\cdot\nabla_{\mathbf{x}}U\right)\right\}+\operatorname{div}_{\mathbf{x}}\left\{\frac{1}{c_0^2}\left(\frac{\partial U}{\partial t}+\tilde{\mathbf{u}}\cdot\nabla_{\mathbf{x}}U\right)\tilde{\mathbf{u}}\right\}\right)-\Delta_{\mathbf{x}}U=0. \tag{13.368}$$

Above, we established $U(\mathbf{x}, t)$ only in the rough geometric optics approximation (13.150). Since we have (13.154) in the rough geometric optics approximation and $S(\mathbf{x}, t)$ satisfies the following eikonal equation:

$$\frac{1}{c_0^2}\left(\frac{\partial S}{\partial t}+\tilde{\mathbf{u}}\cdot\nabla_{\mathbf{x}}S\right)^2=|\nabla_{\mathbf{x}}S|^2. \tag{13.369}$$

In the case of more precise delicate geometric optics approximation, in order to get better approximation for $U(\mathbf{x}, t)$ we can use the following alternative consideration, slightly different than we did above: from now we consider $U(\mathbf{x}, t)$ be of the form

$$U(\mathbf{x}, t)=A(\mathbf{x}, t)e^{ik_0 S(\mathbf{x},t)}, \tag{13.370}$$

similarly to (13.154). However, we consider that S in (13.370) satisfied (13.369) precisely, although we consider the amplitude A to be complex rather than real! So we include the correction term of $\kappa_0 S$ into the complex amplitude A! Thus either (13.153) or (13.159) is not valid anymore, since it was derived under the assumption of real amplitude A. However, either (13.152) or (13.160) still holds. I.e. we have (13.369), which assumed to hold precisely. Moreover, by (13.129) together with (13.369) we have

$$\begin{aligned}&\left(\frac{\partial}{\partial t}\left\{\frac{1}{c_0^2}\left(\frac{\partial A}{\partial t}+\tilde{\mathbf{u}}\cdot\nabla_{\mathbf{x}}A\right)\right\}+\operatorname{div}_{\mathbf{x}}\left\{\frac{1}{c_0^2}\left(\frac{\partial A}{\partial t}+\tilde{\mathbf{u}}\cdot\nabla_{\mathbf{x}}A\right)\tilde{\mathbf{u}}\right\}\right)-\Delta_{\mathbf{x}}A\\&+i\kappa_0 A\left(\frac{\partial}{\partial t}\left\{\frac{1}{c_0^2}\left(\frac{\partial S}{\partial t}+\tilde{\mathbf{u}}\cdot\nabla_{\mathbf{x}}S\right)\right\}+\operatorname{div}_{\mathbf{x}}\left\{\frac{1}{c_0^2}\left(\frac{\partial S}{\partial t}+\tilde{\mathbf{u}}\cdot\nabla_{\mathbf{x}}S\right)\tilde{\mathbf{u}}\right\}-\Delta_{\mathbf{x}}S\right)\\&+2i\kappa_0\left(\frac{1}{c_0^2}\left(\frac{\partial S}{\partial t}+\tilde{\mathbf{u}}\cdot\nabla_{\mathbf{x}}S\right)\left(\frac{\partial A}{\partial t}+\tilde{\mathbf{u}}\cdot\nabla_{\mathbf{x}}A\right)-\nabla_{\mathbf{x}}A\cdot\nabla_{\mathbf{x}}S\right)=0.\end{aligned} \tag{13.371}$$

Then denoting, as before, the proper scalar field:

$$\tilde{\omega}(\mathbf{x}, t):=\kappa_0\left(\frac{\partial S}{\partial t}(\mathbf{x}, t)+\tilde{\mathbf{u}}(\mathbf{x}, t)\cdot\nabla_{\mathbf{x}}S(\mathbf{x}, t)\right), \tag{13.372}$$

the proper vector field:

$$\mathbf{k}(\mathbf{x},t) := c_0(\mathbf{x},t)\left(\frac{\partial S}{\partial t}(\mathbf{x},t) + \tilde{\mathbf{u}}(\mathbf{x},t)\cdot\nabla_{\mathbf{x}}S(\mathbf{x},t)\right)^{-1}\nabla_{\mathbf{x}}S(\mathbf{x},t), \tag{13.373}$$

the speed-like vector field:

$$\begin{aligned}\mathbf{h}(\mathbf{x},t) &:= \tilde{\mathbf{u}}(\mathbf{x},t) - c_0^2(\mathbf{x},t)\left(\frac{\partial S}{\partial t}(\mathbf{x},t) + \tilde{\mathbf{u}}(\mathbf{x},t)\cdot\nabla_{\mathbf{x}}S(\mathbf{x},t)\right)^{-1}\nabla_{\mathbf{x}}S(\mathbf{x},t)\\ &= \tilde{\mathbf{u}}(\mathbf{x},t) - c_0(\mathbf{x},t)\mathbf{k}(\mathbf{x},t),\end{aligned} \tag{13.374}$$

and the proper scalar field:

$$\begin{aligned}G\,(\mathbf{x},t) &:= \left(\frac{c_0^2}{2\left(\frac{\partial S}{\partial t} + \tilde{\mathbf{u}}\cdot\nabla_{\mathbf{x}}S\right)}\left(\Delta_{\mathbf{x}}S\right.\right.\\ &\left.\left. - \frac{\partial}{\partial t}\left\{\frac{1}{c_0^2}\left(\frac{\partial S}{\partial t} + \tilde{\mathbf{u}}\cdot\nabla_{\mathbf{x}}S\right)\right\} - \operatorname{div}_{\mathbf{x}}\left\{\frac{1}{c_0^2}\left(\frac{\partial S}{\partial t} + \tilde{\mathbf{u}}\cdot\nabla_{\mathbf{x}}S\right)\tilde{\mathbf{u}}\right\}\right)\right)(\mathbf{x},t)\\ &= \left(\frac{c_0^2}{2\tilde{\omega}}\left(\operatorname{div}_{\mathbf{x}}\left\{\frac{\tilde{\omega}}{c_0}\mathbf{k}\right\} - \frac{\partial}{\partial t}\left\{\frac{\tilde{\omega}}{c_0^2}\right\} - \operatorname{div}_{\mathbf{x}}\left\{\frac{\tilde{\omega}}{c_0^2}\tilde{\mathbf{u}}\right\}\right)\right)(\mathbf{x},t)\\ &= -\left(\frac{c_0^2}{2\tilde{\omega}}\left(\frac{\partial}{\partial t}\left\{\frac{\tilde{\omega}}{c_0^2}\right\} + \mathbf{h}\cdot\nabla_{\mathbf{x}}\left\{\frac{\tilde{\omega}}{c_0^2}\right\}\right)\right)(\mathbf{x},t) - \frac{1}{2}\left(\operatorname{div}_{\mathbf{x}}\mathbf{h}(\mathbf{x},t)\right),\end{aligned} \tag{13.375}$$

we clearly rewrite (13.371) as

$$\begin{aligned}\frac{c_0^2}{2\tilde{\omega}}\left\{\frac{\partial}{\partial t}\left\{\frac{1}{c_0^2}\left(\frac{\partial A}{\partial t} + \tilde{\mathbf{u}}\cdot\nabla_{\mathbf{x}}A\right)\right\} + \operatorname{div}_{\mathbf{x}}\left\{\frac{1}{c_0^2}\left(\frac{\partial A}{\partial t} + \tilde{\mathbf{u}}\cdot\nabla_{\mathbf{x}}A\right)\tilde{\mathbf{u}}\right\} - \Delta_{\mathbf{x}}A\right\}\\ + i\left(\frac{\partial A}{\partial t} + \mathbf{h}\cdot\nabla_{\mathbf{x}}A - G\,A\right) = 0,\end{aligned} \tag{13.376}$$

and (13.369), as

$$|\mathbf{k}(\mathbf{x},t)|^2 = 1 \quad \text{or equivalently} \quad |\mathbf{h}(\mathbf{x},t) - \tilde{\mathbf{u}}(\mathbf{x},t)|^2 = c_0^2(\mathbf{x},t). \tag{13.377}$$

Moreover, as before, we obviously have

$$\begin{aligned}\frac{\tilde{\omega}^2}{c_0^2k_0^2} = \frac{1}{c_0^2}\left(\frac{\partial S}{\partial t} + \tilde{\mathbf{u}}\cdot\nabla_{\mathbf{x}}S\right)^2 = |\nabla_{\mathbf{x}}S|^2\,, \qquad |\mathbf{k}|^2 = 1\\ \text{or equivalently} \qquad |\mathbf{h} - \tilde{\mathbf{u}}|^2 = c_0^2,\end{aligned} \tag{13.378}$$

$$\frac{\partial S}{\partial t} + \tilde{\mathbf{u}}\cdot\nabla_{\mathbf{x}}S = c_0\mathbf{k}\cdot\nabla_{\mathbf{x}}S \qquad \text{or equivalently} \qquad \frac{\partial S}{\partial t} + \mathbf{h}\cdot\nabla_{\mathbf{x}}S = 0, \tag{13.379}$$

$$\nabla_{\mathbf{x}} S = (\mathbf{k}(\mathbf{x}, t) \cdot \nabla_{\mathbf{x}} S)\, \mathbf{k}, \tag{13.380}$$

$$\frac{c_0}{\tilde{\omega}} \mathbf{k} = \frac{1}{k_0} |\nabla_{\mathbf{x}} S|^{-2} \nabla_{\mathbf{x}} S = \frac{c_0^2 k_0}{\tilde{\omega}^2} \nabla_{\mathbf{x}} S, \tag{13.381}$$

$$\frac{1}{k_0} \nabla_{\mathbf{x}} \tilde{\omega} = \frac{\partial}{\partial t} \{\nabla_{\mathbf{x}} S\} + \nabla_{\mathbf{x}}^2 S \cdot \tilde{\mathbf{u}} + \{d_{\mathbf{x}} \tilde{\mathbf{u}}\}^T \cdot \nabla_{\mathbf{x}} S, \tag{13.382}$$

and

$$\frac{1}{\tilde{\omega}} \left(\frac{\partial \tilde{\omega}}{\partial t} + \mathbf{h} \cdot \nabla_{\mathbf{x}} \tilde{\omega} \right) = \frac{1}{c_0} \left(\frac{\partial c_0}{\partial t} + \tilde{\mathbf{u}} \cdot \nabla_{\mathbf{x}} c_0 \right) - \frac{1}{2} \left(\mathbf{k} \cdot \left(\{d_{\mathbf{x}} \tilde{\mathbf{u}}\}^T + d_{\mathbf{x}} \tilde{\mathbf{u}} \right) \right) \cdot \mathbf{k}. \tag{13.383}$$

Furthermore, from now assume the case of delicate geometric optics approximation. Remind that the delicate geometric optics approximation means the following: assume that the changes in time of c_0, $\tilde{\mathbf{u}}$, A, and S become essential after certain interval of time T_e and the changes in space of c_0, A and S become essential in the spatial landscape L_e. Then we assume that

$$k_0^2 c_0^2 T_e^2 \gg 1 \quad \text{and} \quad k_0^2 L_e^2 \gg 1. \tag{13.384}$$

Moreover, as before, we assume that the order of c_0 is less or equal to the order of c. Then, estimation (13.384) implies

$$\frac{\left| d_{\mathbf{x}} \tilde{\mathbf{u}} + \{d_{\mathbf{x}} \tilde{\mathbf{u}}\}^T \right|^2}{c_0^2} + \frac{\left(|\nabla_{\mathbf{x}} A| + \frac{1}{c_0} \left| \frac{\partial A}{\partial t} + \tilde{\mathbf{u}} \cdot \nabla_{\mathbf{x}} A \right| \right)^2}{|A|^2} + \frac{\left(|\nabla_{\mathbf{x}} c_0| + \frac{1}{c_0} \left| \frac{\partial c_0}{\partial t} + \tilde{\mathbf{u}} \cdot \nabla_{\mathbf{x}} c_0 \right| \right)^2}{|c_0|^2}$$
$$+ \frac{\left| \nabla_{\mathbf{x}}^2 S \right|^2 + \frac{1}{c_0^2} \left| \nabla_{\mathbf{x}} \left(\frac{\partial S}{\partial t} + \tilde{\mathbf{u}} \cdot \nabla_{\mathbf{x}} S \right) \right|^2}{|\nabla_{\mathbf{x}} S|^2} \ll k_0^2 \quad \text{and} \quad \frac{\left| \frac{\partial S}{\partial t} + \tilde{\mathbf{u}} \cdot \nabla_{\mathbf{x}} S - c \right|^2}{c^2} \ll 1, \tag{13.385}$$

so that we have

$$\frac{\left| d_{\mathbf{x}} \tilde{\mathbf{u}} + \{d_{\mathbf{x}} \tilde{\mathbf{u}}\}^T \right|^2}{c_0^2} + \frac{\left(|\nabla_{\mathbf{x}} A| + \frac{1}{c_0} \left| \frac{\partial A}{\partial t} + \tilde{\mathbf{u}} \cdot \nabla_{\mathbf{x}} A \right| \right)^2}{|A|^2} + \frac{\left(|\nabla_{\mathbf{x}} c_0| + \frac{1}{c_0} \left| \frac{\partial c_0}{\partial t} + \tilde{\mathbf{u}} \cdot \nabla_{\mathbf{x}} c_0 \right| \right)^2}{|c_0|^2}$$
$$+ \frac{\left| \nabla_{\mathbf{x}}^2 S \right|^2 + \frac{1}{c_0^2} \left| \nabla_{\mathbf{x}} \left(\frac{\partial S}{\partial t} + \tilde{\mathbf{u}} \cdot \nabla_{\mathbf{x}} S \right) \right|^2}{|\nabla_{\mathbf{x}} S|^2} \ll \frac{\tilde{\omega}^2}{c^2} \leq \frac{\tilde{\omega}^2}{c_0^2}. \tag{13.386}$$

Moreover, by (13.386) together with (13.383) we also have

$$\frac{|\operatorname{div}_{\mathbf{x}} \mathbf{h}|^2}{c_0^2} + \frac{\left(|d_{\mathbf{x}}\mathbf{k}| + \frac{1}{c_0}\left|\frac{\partial \mathbf{k}}{\partial t} + d_{\mathbf{x}}\mathbf{k}\cdot\tilde{\mathbf{u}} - d_{\mathbf{x}}\tilde{\mathbf{u}}\cdot\mathbf{k}\right|\right)^2}{|\mathbf{k}|^2} + \frac{\left(|\nabla_{\mathbf{x}}\tilde{\omega}| + \frac{1}{c_0}\left|\frac{\partial \tilde{\omega}}{\partial t} + \tilde{\mathbf{u}}\cdot\nabla_{\mathbf{x}}\tilde{\omega}\right|\right)^2}{|\tilde{\omega}|^2} \ll \frac{\tilde{\omega}^2}{c_0^2}. \tag{13.387}$$

Therefore, denoting

$$H := \frac{\partial A}{\partial t} + \mathbf{h}\cdot\nabla_{\mathbf{x}}A - G\,A = \frac{\partial A}{\partial t} + \tilde{\mathbf{u}}\cdot\nabla_{\mathbf{x}}A - c_0\mathbf{k}\cdot\nabla_{\mathbf{x}}A - G\,A \tag{13.388}$$

we rewrite (13.376) as

$$\frac{c_0^2}{2\tilde{\omega}}\left\{\frac{\partial}{\partial t}\left(\frac{1}{c_0^2}H\right) + \operatorname{div}_{\mathbf{x}}\left(\frac{1}{c_0^2}H\tilde{\mathbf{u}}\right)\right\} + \frac{c_0^2}{2\tilde{\omega}}\left\{\frac{\partial}{\partial t}\left\{\frac{1}{c_0^2}\left(c_0\mathbf{k}\cdot\nabla_{\mathbf{x}}A + G\,A\right)\right\} + \operatorname{div}_{\mathbf{x}}\left\{\frac{1}{c_0^2}\left(c_0\mathbf{k}\cdot\nabla_{\mathbf{x}}A + G\,A\right)\tilde{\mathbf{u}}\right\}\right\} - \frac{c_0^2}{2\tilde{\omega}}\left(\Delta_{\mathbf{x}}A\right) = -iH\,. \tag{13.389}$$

However, by (13.386) and (13.387) we have

$$\frac{\left(|\nabla_{\mathbf{x}}H| + \frac{1}{c_0}\left|\frac{\partial H}{\partial t} + \tilde{\mathbf{u}}\cdot\nabla_{\mathbf{x}}H\right|\right)^2}{|H|^2} \ll \frac{\tilde{\omega}^2}{c_0^2}. \tag{13.390}$$

In particular, by (13.389), (13.386), (13.387), and (13.390), we deduce

$$H^2 \ll |c_0\mathbf{k}\cdot\nabla_{\mathbf{x}}A + G\,A|^2\,. \tag{13.391}$$

Thus, by (13.386), (13.387), (13.390) and (13.391) we approximate (13.389) as

$$\frac{c_0^2}{2\tilde{\omega}}\left\{\frac{\partial}{\partial t}\left\{\frac{1}{c_0^2}\left(c_0\mathbf{k}\cdot\nabla_{\mathbf{x}}A + G\,A\right)\right\} + \operatorname{div}_{\mathbf{x}}\left\{\frac{1}{c_0^2}\left(c_0\mathbf{k}\cdot\nabla_{\mathbf{x}}A + G\,A\right)\tilde{\mathbf{u}}\right\}\right\} - \frac{c_0^2}{2\tilde{\omega}}\left(\Delta_{\mathbf{x}}A\right) \approx -iH = -i\left\{\frac{\partial A}{\partial t} + \mathbf{h}\cdot\nabla_{\mathbf{x}}A + \frac{1}{2}\left(\operatorname{div}_{\mathbf{x}}\mathbf{h}\right)A\right\} - iA\left(\frac{c_0}{\sqrt{\tilde{\omega}}}\left(\frac{\partial}{\partial t}\left\{\frac{\sqrt{\tilde{\omega}}}{c_0}\right\} + \mathbf{h}\cdot\nabla_{\mathbf{x}}\left\{\frac{\sqrt{\tilde{\omega}}}{c_0}\right\}\right)\right). \tag{13.392}$$

We rewrite it as

$$
\begin{aligned}
&\frac{c_0^2}{2\tilde{\omega}}\nabla_{\mathbf{x}}A\cdot\left\{\frac{\partial}{\partial t}\left\{\frac{1}{c_0}\mathbf{k}\right\}+\mathrm{div}_{\mathbf{x}}\left\{\frac{1}{c_0}\mathbf{k}\otimes\tilde{\mathbf{u}}\right\}-d_{\mathbf{x}}\tilde{\mathbf{u}}\cdot\left(\frac{1}{c_0}\mathbf{k}\right)\right\}\\
&\quad+\frac{1}{2\tilde{\omega}}c_0\mathbf{k}\cdot\nabla_{\mathbf{x}}\left(\frac{\partial A}{\partial t}+\tilde{\mathbf{u}}\cdot\nabla_{\mathbf{x}}A\right)+\frac{1}{2\tilde{\omega}}G\left(\frac{\partial A}{\partial t}+\tilde{\mathbf{u}}\cdot\nabla_{\mathbf{x}}A\right)\\
&\quad+\frac{c_0^2}{2\tilde{\omega}}A\left(\frac{\partial}{\partial t}\left\{\frac{1}{c_0^2}G\right\}+\mathrm{div}_{\mathbf{x}}\left\{\frac{1}{c_0^2}G\tilde{\mathbf{u}}\right\}\right)-\frac{c_0^2}{2\tilde{\omega}}(\Delta_{\mathbf{x}}A)\approx -iH=\\
&\qquad -i\left\{\frac{\partial A}{\partial t}+\mathbf{h}\cdot\nabla_{\mathbf{x}}A+\frac{1}{2}(\mathrm{div}_{\mathbf{x}}\mathbf{h})A\right\}\\
&\qquad -iA\left(\frac{c_0}{\sqrt{\tilde{\omega}}}\left(\frac{\partial}{\partial t}\left\{\frac{\sqrt{\tilde{\omega}}}{c_0}\right\}+\mathbf{h}\cdot\nabla_{\mathbf{x}}\left\{\frac{\sqrt{\tilde{\omega}}}{c_0}\right\}\right)\right). \qquad (13.393)
\end{aligned}
$$

Therefore,

$$
\begin{aligned}
&\frac{c_0^2}{2\tilde{\omega}}\nabla_{\mathbf{x}}A\cdot\left\{\frac{\partial}{\partial t}\left\{\frac{1}{c_0}\mathbf{k}\right\}+\mathrm{div}_{\mathbf{x}}\left\{\frac{1}{c_0}\mathbf{k}\otimes\tilde{\mathbf{u}}\right\}-d_{\mathbf{x}}\tilde{\mathbf{u}}\cdot\left(\frac{1}{c_0}\mathbf{k}\right)\right\}+\frac{1}{2\tilde{\omega}}c_0\mathbf{k}\cdot\nabla_{\mathbf{x}}H\\
&\quad+\frac{1}{2\tilde{\omega}}G\,H+\frac{1}{2\tilde{\omega}}c_0\mathbf{k}\cdot\nabla_{\mathbf{x}}\left(c_0\mathbf{k}\cdot\nabla_{\mathbf{x}}A+G\,A\right)+\frac{1}{2\tilde{\omega}}G\left(c_0\mathbf{k}\cdot\nabla_{\mathbf{x}}A+G\,A\right)\\
&\quad+\frac{c_0^2}{2\tilde{\omega}}A\left(\frac{\partial}{\partial t}\left\{\frac{1}{c_0^2}G\right\}+\mathrm{div}_{\mathbf{x}}\left\{\frac{1}{c_0^2}G\tilde{\mathbf{u}}\right\}\right)-\frac{c_0^2}{2\tilde{\omega}}(\Delta_{\mathbf{x}}A)\approx -iH=\\
&\qquad -i\left\{\frac{\partial A}{\partial t}+\mathbf{h}\cdot\nabla_{\mathbf{x}}A+\frac{1}{2}(\mathrm{div}_{\mathbf{x}}\mathbf{h})A\right\}\\
&\qquad -iA\left(\frac{c_0}{\sqrt{\tilde{\omega}}}\left(\frac{\partial}{\partial t}\left\{\frac{\sqrt{\tilde{\omega}}}{c_0}\right\}+\mathbf{h}\cdot\nabla_{\mathbf{x}}\left\{\frac{\sqrt{\tilde{\omega}}}{c_0}\right\}\right)\right). \qquad (13.394)
\end{aligned}
$$

Thus, by (13.387), (13.390) and (13.391) we further approximate (13.394) as

$$
\begin{aligned}
&\frac{c_0^2}{2\tilde{\omega}}\nabla_{\mathbf{x}}A\cdot\left\{\frac{\partial}{\partial t}\left\{\frac{1}{c_0}\mathbf{k}\right\}+\mathrm{div}_{\mathbf{x}}\left\{\frac{1}{c_0}\mathbf{k}\otimes\tilde{\mathbf{u}}\right\}-d_{\mathbf{x}}\tilde{\mathbf{u}}\cdot\left(\frac{1}{c_0}\mathbf{k}\right)\right\}\\
&\quad+\frac{1}{2\tilde{\omega}}c_0\mathbf{k}\cdot\nabla_{\mathbf{x}}\left(c_0\mathbf{k}\cdot\nabla_{\mathbf{x}}A+G\,A\right)+\frac{1}{2\tilde{\omega}}G\left(c_0\mathbf{k}\cdot\nabla_{\mathbf{x}}A+G\,A\right)\\
&\quad+\frac{c_0^2}{2\tilde{\omega}}A\left(\frac{\partial}{\partial t}\left\{\frac{1}{c_0^2}G\right\}+\mathrm{div}_{\mathbf{x}}\left\{\frac{1}{c_0^2}G\tilde{\mathbf{u}}\right\}\right)-\frac{c_0^2}{2\tilde{\omega}}(\Delta_{\mathbf{x}}A)\approx\\
&\qquad -i\left\{\frac{\partial A}{\partial t}+\mathbf{h}\cdot\nabla_{\mathbf{x}}A+\frac{1}{2}(\mathrm{div}_{\mathbf{x}}\mathbf{h})A\right\}\\
&\qquad -iA\left(\frac{c_0}{\sqrt{\tilde{\omega}}}\left(\frac{\partial}{\partial t}\left\{\frac{\sqrt{\tilde{\omega}}}{c_0}\right\}+\mathbf{h}\cdot\nabla_{\mathbf{x}}\left\{\frac{\sqrt{\tilde{\omega}}}{c_0}\right\}\right)\right). \qquad (13.395)
\end{aligned}
$$

We rewrite the last equation as

$$\frac{c_0^2}{2\tilde{\omega}}\nabla_{\mathbf{x}}A\cdot\left\{\frac{\partial}{\partial t}\left\{\frac{1}{c_0}\mathbf{k}\right\}+\mathrm{div}_{\mathbf{x}}\left\{\frac{1}{c_0}\mathbf{k}\otimes\tilde{\mathbf{u}}\right\}-d_{\mathbf{x}}\tilde{\mathbf{u}}\cdot\left(\frac{1}{c_0}\mathbf{k}\right)\right\}$$
$$+\frac{c_0^2}{2\tilde{\omega}}\left(\frac{1}{c_0}\mathbf{k}\right)\cdot\nabla_{\mathbf{x}}\left(c_0\mathbf{k}\cdot\nabla_{\mathbf{x}}A\right)+\frac{1}{\tilde{\omega}}G\left(c_0\mathbf{k}\cdot\nabla_{\mathbf{x}}A\right)$$
$$+\frac{c_0^2}{2\tilde{\omega}}A\left(\frac{\partial}{\partial t}\left\{\frac{1}{c_0^2}G\right\}+\mathrm{div}_{\mathbf{x}}\left\{\frac{1}{c_0^2}G\tilde{\mathbf{u}}\right\}+\frac{1}{c_0^2}(c_0\mathbf{k})\cdot\nabla_{\mathbf{x}}G+\frac{1}{c_0^2}G^2\right)-\frac{c_0^2}{2\tilde{\omega}}\left(\Delta_{\mathbf{x}}A\right)$$
$$\approx -i\left\{\frac{\partial A}{\partial t}+\mathbf{h}\cdot\nabla_{\mathbf{x}}A+\frac{1}{2}\left(\mathrm{div}_{\mathbf{x}}\,\mathbf{h}\right)A\right\}$$
$$-iA\left(\frac{c_0}{\sqrt{\tilde{\omega}}}\left(\frac{\partial}{\partial t}\left\{\frac{\sqrt{\tilde{\omega}}}{c_0}\right\}+\mathbf{h}\cdot\nabla_{\mathbf{x}}\left\{\frac{\sqrt{\tilde{\omega}}}{c_0}\right\}\right)\right), \qquad (13.396)$$

and further as

$$\frac{c_0^2}{2\tilde{\omega}}\nabla_{\mathbf{x}}A\cdot\left\{\frac{\partial}{\partial t}\left\{\frac{1}{c_0}\mathbf{k}\right\}+\mathrm{div}_{\mathbf{x}}\left\{\frac{1}{c_0}\mathbf{k}\otimes\tilde{\mathbf{u}}\right\}-d_{\mathbf{x}}\tilde{\mathbf{u}}\cdot\left(\frac{1}{c_0}\mathbf{k}\right)\right\}$$
$$+\frac{c_0^2}{2\tilde{\omega}}\left(\frac{1}{c_0}\mathbf{k}\cdot\nabla_{\mathbf{x}}c_0\right)\left(\mathbf{k}\cdot\nabla_{\mathbf{x}}A\right)+\frac{c_0^2}{2\tilde{\omega}}\nabla_{\mathbf{x}}\left(\mathbf{k}\cdot\nabla_{\mathbf{x}}A\right)\cdot\mathbf{k}+\frac{1}{\tilde{\omega}}G\left(c_0\mathbf{k}\cdot\nabla_{\mathbf{x}}A\right)$$
$$+\frac{c_0^2}{2\tilde{\omega}}A\left(\frac{\partial}{\partial t}\left\{\frac{1}{c_0^2}G\right\}+\mathrm{div}_{\mathbf{x}}\left\{\frac{1}{c_0^2}G\tilde{\mathbf{u}}\right\}+\frac{1}{c_0^2}(c_0\mathbf{k})\cdot\nabla_{\mathbf{x}}G+\frac{1}{c_0^2}G^2\right)-\frac{c_0^2}{2\tilde{\omega}}\left(\Delta_{\mathbf{x}}A\right)$$
$$\approx -i\left\{\frac{\partial A}{\partial t}+\mathbf{h}\cdot\nabla_{\mathbf{x}}A+\frac{1}{2}\left(\mathrm{div}_{\mathbf{x}}\,\mathbf{h}\right)A\right\}$$
$$-iA\left(\frac{c_0}{\sqrt{\tilde{\omega}}}\left(\frac{\partial}{\partial t}\left\{\frac{\sqrt{\tilde{\omega}}}{c_0}\right\}+\mathbf{h}\cdot\nabla_{\mathbf{x}}\left\{\frac{\sqrt{\tilde{\omega}}}{c_0}\right\}\right)\right). \qquad (13.397)$$

Therefore, we infer

$$\frac{c_0^2}{2\tilde{\omega}}\nabla_{\mathbf{x}}A\cdot\left\{\frac{\partial}{\partial t}\left\{\frac{1}{c_0}\mathbf{k}\right\}+\mathrm{div}_{\mathbf{x}}\left\{\frac{1}{c_0}\mathbf{k}\otimes\tilde{\mathbf{u}}\right\}-d_{\mathbf{x}}\tilde{\mathbf{u}}\cdot\left(\frac{1}{c_0}\mathbf{k}\right)\right\}$$
$$+\frac{c_0^2}{2\tilde{\omega}}\left(\frac{1}{c_0}\mathbf{k}\cdot\nabla_{\mathbf{x}}c_0\right)\left(\mathbf{k}\cdot\nabla_{\mathbf{x}}A\right)-\frac{c_0^2}{2\tilde{\omega}}\left(\mathbf{k}\cdot\nabla_{\mathbf{x}}A\right)\left(\mathrm{div}_{\mathbf{x}}\,\mathbf{k}\right)+\frac{1}{\tilde{\omega}}G\left(c_0\mathbf{k}\cdot\nabla_{\mathbf{x}}A\right)$$
$$+\frac{c_0^2}{2\tilde{\omega}}A\left(\frac{\partial}{\partial t}\left\{\frac{1}{c_0^2}G\right\}+\mathrm{div}_{\mathbf{x}}\left\{\frac{1}{c_0^2}G\tilde{\mathbf{u}}\right\}+\frac{1}{c_0^2}(c_0\mathbf{k})\cdot\nabla_{\mathbf{x}}G+\frac{1}{c_0^2}G^2\right)$$
$$-\frac{c_0^2}{2\tilde{\omega}}\left(\mathrm{div}_{\mathbf{x}}\left\{\nabla_{\mathbf{x}}A-\left(\mathbf{k}\cdot\nabla_{\mathbf{x}}A\right)\mathbf{k}\right\}\right)\approx -i\left\{\frac{\partial A}{\partial t}+\mathbf{h}\cdot\nabla_{\mathbf{x}}A+\frac{1}{2}\left(\mathrm{div}_{\mathbf{x}}\,\mathbf{h}\right)A\right\}$$
$$-iA\left(\frac{c_0}{\sqrt{\tilde{\omega}}}\left(\frac{\partial}{\partial t}\left\{\frac{\sqrt{\tilde{\omega}}}{c_0}\right\}+\mathbf{h}\cdot\nabla_{\mathbf{x}}\left\{\frac{\sqrt{\tilde{\omega}}}{c_0}\right\}\right)\right). \qquad (13.398)$$

So,

$$\frac{c_0^2}{2\tilde{\omega}}\nabla_{\mathbf{x}}A\cdot\left\{\frac{\partial}{\partial t}\left\{\frac{1}{c_0}\mathbf{k}\right\}+\operatorname{div}_{\mathbf{x}}\left\{\frac{1}{c_0}\mathbf{k}\otimes\tilde{\mathbf{u}}\right\}-d_{\mathbf{x}}\tilde{\mathbf{u}}\cdot\left(\frac{1}{c_0}\mathbf{k}\right)-\left(\operatorname{div}_{\mathbf{x}}\left\{\frac{1}{c_0}\mathbf{k}\right\}\right)(c_0\mathbf{k})\right.$$
$$\left.+2G\left(\frac{1}{c_0}\mathbf{k}\right)\right\}+\frac{c_0^2}{2\tilde{\omega}}A\left(\frac{\partial}{\partial t}\left\{\frac{1}{c_0^2}G\right\}+\operatorname{div}_{\mathbf{x}}\left\{\frac{1}{c_0^2}G\tilde{\mathbf{u}}\right\}+\frac{1}{c_0^2}(c_0\mathbf{k})\cdot\nabla_{\mathbf{x}}G+\frac{1}{c_0^2}G^2\right)$$
$$-\frac{c_0^2}{2\tilde{\omega}}\left(\operatorname{div}_{\mathbf{x}}\left\{\nabla_{\mathbf{x}}A-(\mathbf{k}\cdot\nabla_{\mathbf{x}}A)\,\mathbf{k}\right\}\right)\approx -i\left\{\frac{\partial A}{\partial t}+\mathbf{h}\cdot\nabla_{\mathbf{x}}A+\frac{1}{2}\left(\operatorname{div}_{\mathbf{x}}\mathbf{h}\right)A\right\}$$
$$-iA\left(\frac{c_0}{\sqrt{\tilde{\omega}}}\left(\frac{\partial}{\partial t}\left\{\frac{\sqrt{\tilde{\omega}}}{c_0}\right\}+\mathbf{h}\cdot\nabla_{\mathbf{x}}\left\{\frac{\sqrt{\tilde{\omega}}}{c_0}\right\}\right)\right). \tag{13.399}$$

On the other hand by (13.381) we have

$$\frac{\tilde{\omega}}{k_0c_0}\mathbf{k}=\nabla_{\mathbf{x}}S. \tag{13.400}$$

Thus,

$$\frac{\tilde{\omega}}{k_0}\left\{\frac{\partial}{\partial t}\left\{\frac{1}{c_0}\mathbf{k}\right\}+\operatorname{div}_{\mathbf{x}}\left\{\frac{1}{c_0}\mathbf{k}\otimes\tilde{\mathbf{u}}\right\}-d_{\mathbf{x}}\tilde{\mathbf{u}}\cdot\left(\frac{1}{c_0}\mathbf{k}\right)-\left(\operatorname{div}_{\mathbf{x}}\left\{\frac{1}{c_0}\mathbf{k}\right\}\right)(c_0\mathbf{k})\right.$$
$$\left.+2G\left(\frac{1}{c_0}\mathbf{k}\right)\right\}=$$
$$\left\{\frac{\partial}{\partial t}\{\nabla_{\mathbf{x}}S\}+\operatorname{div}_{\mathbf{x}}\{\nabla_{\mathbf{x}}S\otimes\tilde{\mathbf{u}}\}-d_{\mathbf{x}}\tilde{\mathbf{u}}\cdot(\nabla_{\mathbf{x}}S)-(\operatorname{div}_{\mathbf{x}}\{\nabla_{\mathbf{x}}S\})(c_0\mathbf{k})\right.$$
$$\left.+2G(\nabla_{\mathbf{x}}S)\right\}-\frac{1}{c_0}\left\{\frac{\partial}{\partial t}\left\{\frac{\tilde{\omega}}{k_0}\right\}+\tilde{\mathbf{u}}\cdot\nabla_{\mathbf{x}}\left\{\frac{\tilde{\omega}}{k_0}\right\}-(c_0\mathbf{k})\cdot\nabla_{\mathbf{x}}\left\{\frac{\tilde{\omega}}{k_0}\right\}\right\}\mathbf{k}=$$
$$\nabla_{\mathbf{x}}\left(\frac{\partial S}{\partial t}+\tilde{\mathbf{u}}\cdot\nabla_{\mathbf{x}}S\right)+(\operatorname{div}_{\mathbf{x}}\{\tilde{\mathbf{u}}\})\nabla_{\mathbf{x}}S-\left(d_{\mathbf{x}}\tilde{\mathbf{u}}+\{d_{\mathbf{x}}\tilde{\mathbf{u}}\}^T\right)\cdot(\nabla_{\mathbf{x}}S)$$
$$-(\operatorname{div}_{\mathbf{x}}\{\nabla_{\mathbf{x}}S\})(c_0\mathbf{k})+2G(\nabla_{\mathbf{x}}S)-\frac{1}{c_0}\left\{\frac{\partial}{\partial t}\left\{\frac{\tilde{\omega}}{k_0}\right\}+\mathbf{h}\cdot\nabla_{\mathbf{x}}\left\{\frac{\tilde{\omega}}{k_0}\right\}\right\}\mathbf{k}. \tag{13.401}$$

I.e.

$$\tilde{\omega}\left\{\frac{\partial}{\partial t}\left\{\frac{1}{c_0}\mathbf{k}\right\}+\operatorname{div}_{\mathbf{x}}\left\{\frac{1}{c_0}\mathbf{k}\otimes\tilde{\mathbf{u}}\right\}-d_{\mathbf{x}}\tilde{\mathbf{u}}\cdot\left(\frac{1}{c_0}\mathbf{k}\right)-\left(\operatorname{div}_{\mathbf{x}}\left\{\frac{1}{c_0}\mathbf{k}\right\}\right)(c_0\mathbf{k})\right.$$
$$\left.+2G\left(\frac{1}{c_0}\mathbf{k}\right)\right\}=\nabla_{\mathbf{x}}\tilde{\omega}+(\operatorname{div}_{\mathbf{x}}\{\tilde{\mathbf{u}}\})\left(\frac{\tilde{\omega}}{c_0}\mathbf{k}\right)-\left(d_{\mathbf{x}}\tilde{\mathbf{u}}+\{d_{\mathbf{x}}\tilde{\mathbf{u}}\}^T\right)\cdot\left(\frac{\tilde{\omega}}{c_0}\mathbf{k}\right)$$
$$-\left(\operatorname{div}_{\mathbf{x}}\left\{\frac{\tilde{\omega}}{c_0}\mathbf{k}\right\}\right)(c_0\mathbf{k})+2G\left(\frac{\tilde{\omega}}{c_0}\mathbf{k}\right)-\frac{1}{c_0}\left\{\frac{\partial\tilde{\omega}}{\partial t}+\mathbf{h}\cdot\nabla_{\mathbf{x}}\tilde{\omega}\right\}\mathbf{k}=\nabla_{\mathbf{x}}\tilde{\omega}-(\mathbf{k}\cdot\nabla_{\mathbf{x}}\tilde{\omega})\,\mathbf{k}$$

$$
\begin{aligned}
&+\tilde{\omega}\left(\operatorname{div}_{\mathbf{x}}\{\tilde{\mathbf{u}}\}\right)\left(\frac{1}{c_0}\mathbf{k}\right)-\tilde{\omega}\left(d_{\mathbf{x}}\tilde{\mathbf{u}}+\{d_{\mathbf{x}}\tilde{\mathbf{u}}\}^T\right)\cdot\left(\frac{1}{c_0}\mathbf{k}\right)-\tilde{\omega}\left(c_0^2\operatorname{div}_{\mathbf{x}}\left\{\frac{1}{c_0}\mathbf{k}\right\}\right)\left(\frac{1}{c_0}\mathbf{k}\right)\\
&\quad+2G\left(\frac{\tilde{\omega}}{c_0}\mathbf{k}\right)-\frac{1}{c_0}\left\{\frac{\partial\tilde{\omega}}{\partial t}+\mathbf{h}\cdot\nabla_{\mathbf{x}}\tilde{\omega}\right\}\mathbf{k}=\nabla_{\mathbf{x}}\tilde{\omega}-\left(\mathbf{k}\cdot\nabla_{\mathbf{x}}\tilde{\omega}\right)\mathbf{k}\\
&+\tilde{\omega}\left(\operatorname{div}_{\mathbf{x}}\mathbf{h}\right)\left(\frac{1}{c_0}\mathbf{k}\right)-\tilde{\omega}\left(d_{\mathbf{x}}\tilde{\mathbf{u}}+\{d_{\mathbf{x}}\tilde{\mathbf{u}}\}^T\right)\cdot\left(\frac{1}{c_0}\mathbf{k}\right)+2\tilde{\omega}\left(\mathbf{k}\cdot\nabla_{\mathbf{x}}c_0\right)\left(\frac{1}{c_0}\mathbf{k}\right)\\
&\quad+2G\left(\frac{\tilde{\omega}}{c_0}\mathbf{k}\right)-\frac{1}{c_0}\left\{\frac{\partial\tilde{\omega}}{\partial t}+\mathbf{h}\cdot\nabla_{\mathbf{x}}\tilde{\omega}\right\}\mathbf{k}.
\end{aligned}
\tag{13.402}
$$

However, we remind that by (13.375) we have

$$
2G=-\left(\frac{c_0^2}{\tilde{\omega}}\left(\frac{\partial}{\partial t}\left\{\frac{\tilde{\omega}}{c_0^2}\right\}+\mathbf{h}\cdot\nabla_{\mathbf{x}}\left\{\frac{\tilde{\omega}}{c_0^2}\right\}\right)\right)-\left(\operatorname{div}_{\mathbf{x}}\mathbf{h}\right).
\tag{13.403}
$$

Thus, inserting (13.403) into (13.402) gives

$$
\begin{aligned}
&\tilde{\omega}\left\{\frac{\partial}{\partial t}\left\{\frac{1}{c_0}\mathbf{k}\right\}+\operatorname{div}_{\mathbf{x}}\left\{\frac{1}{c_0}\mathbf{k}\otimes\tilde{\mathbf{u}}\right\}-d_{\mathbf{x}}\tilde{\mathbf{u}}\cdot\left(\frac{1}{c_0}\mathbf{k}\right)-\left(\operatorname{div}_{\mathbf{x}}\left\{\frac{1}{c_0}\mathbf{k}\right\}\right)\left(c_0\mathbf{k}\right)\right.\\
&\quad\left.+2G\left(\frac{1}{c_0}\mathbf{k}\right)\right\}=\nabla_{\mathbf{x}}\tilde{\omega}-\left(\mathbf{k}\cdot\nabla_{\mathbf{x}}\tilde{\omega}\right)\mathbf{k}-\tilde{\omega}\left(d_{\mathbf{x}}\tilde{\mathbf{u}}+\{d_{\mathbf{x}}\tilde{\mathbf{u}}\}^T\right)\cdot\left(\frac{1}{c_0}\mathbf{k}\right)\\
&\quad+2\tilde{\omega}\left(\mathbf{k}\cdot\nabla_{\mathbf{x}}c_0\right)\left(\frac{1}{c_0}\mathbf{k}\right)-\left(c_0\left(\frac{\partial}{\partial t}\left\{\frac{\tilde{\omega}}{c_0^2}\right\}+\mathbf{h}\cdot\nabla_{\mathbf{x}}\left\{\frac{\tilde{\omega}}{c_0^2}\right\}\right)\right)\mathbf{k}\\
&\quad-\frac{1}{c_0}\left\{\frac{\partial\tilde{\omega}}{\partial t}+\mathbf{h}\cdot\nabla_{\mathbf{x}}\tilde{\omega}\right\}\mathbf{k}=\nabla_{\mathbf{x}}\tilde{\omega}-\left(\mathbf{k}\cdot\nabla_{\mathbf{x}}\tilde{\omega}\right)\mathbf{k}\\
&\quad-\tilde{\omega}\left(d_{\mathbf{x}}\tilde{\mathbf{u}}+\{d_{\mathbf{x}}\tilde{\mathbf{u}}\}^T\right)\cdot\left(\frac{1}{c_0}\mathbf{k}\right)+2\tilde{\omega}\left(\mathbf{k}\cdot\nabla_{\mathbf{x}}c_0\right)\left(\frac{1}{c_0}\mathbf{k}\right)\\
&\quad+2\tilde{\omega}\left(\frac{1}{c_0}\left(\frac{\partial c_0}{\partial t}+\mathbf{h}\cdot\nabla_{\mathbf{x}}c_0\right)\right)\left(\frac{1}{c_0}\mathbf{k}\right)-\frac{2}{c_0}\left\{\frac{\partial\tilde{\omega}}{\partial t}+\mathbf{h}\cdot\nabla_{\mathbf{x}}\tilde{\omega}\right\}\mathbf{k}.
\end{aligned}
\tag{13.404}
$$

So,

$$
\begin{aligned}
&\tilde{\omega}\left\{\frac{\partial}{\partial t}\left\{\frac{1}{c_0}\mathbf{k}\right\}+\operatorname{div}_{\mathbf{x}}\left\{\frac{1}{c_0}\mathbf{k}\otimes\tilde{\mathbf{u}}\right\}-d_{\mathbf{x}}\tilde{\mathbf{u}}\cdot\left(\frac{1}{c_0}\mathbf{k}\right)-\left(\operatorname{div}_{\mathbf{x}}\left\{\frac{1}{c_0}\mathbf{k}\right\}\right)\left(c_0\mathbf{k}\right)\right.\\
&\quad\left.+2G\left(\frac{1}{c_0}\mathbf{k}\right)\right\}=\nabla_{\mathbf{x}}\tilde{\omega}-\left(\mathbf{k}\cdot\nabla_{\mathbf{x}}\tilde{\omega}\right)\mathbf{k}-\frac{2}{c_0}\left\{\frac{\partial\tilde{\omega}}{\partial t}+\mathbf{h}\cdot\nabla_{\mathbf{x}}\tilde{\omega}\right\}\mathbf{k}\\
&-\tilde{\omega}\left(d_{\mathbf{x}}\tilde{\mathbf{u}}+\{d_{\mathbf{x}}\tilde{\mathbf{u}}\}^T\right)\cdot\left(\frac{1}{c_0}\mathbf{k}\right)+2\tilde{\omega}\left(\frac{1}{c_0}\left(\frac{\partial c_0}{\partial t}+\tilde{\mathbf{u}}\cdot\nabla_{\mathbf{x}}c_0\right)\right)\left(\frac{1}{c_0}\mathbf{k}\right).
\end{aligned}
\tag{13.405}
$$

However, by (13.383) we have

$$
\frac{1}{\tilde{\omega}}\left(\frac{\partial\tilde{\omega}}{\partial t}+\mathbf{h}\cdot\nabla_{\mathbf{x}}\tilde{\omega}\right)=\frac{1}{c_0}\left(\frac{\partial c_0}{\partial t}+\tilde{\mathbf{u}}\cdot\nabla_{\mathbf{x}}c_0\right)-\frac{1}{2}\left(\mathbf{k}\cdot\left(\{d_{\mathbf{x}}\tilde{\mathbf{u}}\}^T+d_{\mathbf{x}}\tilde{\mathbf{u}}\right)\right)\cdot\mathbf{k}.
\tag{13.406}
$$

Thus, inserting (13.406) into (13.405) gives

$$\begin{aligned}
\tilde{\omega}\Bigg\{ \frac{\partial}{\partial t}\left\{\frac{1}{c_0}\mathbf{k}\right\} + \operatorname{div}_{\mathbf{x}}\left\{\frac{1}{c_0}\mathbf{k}\otimes\tilde{\mathbf{u}}\right\} - d_{\mathbf{x}}\tilde{\mathbf{u}}\cdot\left(\frac{1}{c_0}\mathbf{k}\right) - \left(\operatorname{div}_{\mathbf{x}}\left\{\frac{1}{c_0}\mathbf{k}\right\}\right)(c_0\mathbf{k}) \\
+ 2G\left(\frac{1}{c_0}\mathbf{k}\right)\Bigg\} = \nabla_{\mathbf{x}}\tilde{\omega} - (\mathbf{k}\cdot\nabla_{\mathbf{x}}\tilde{\omega})\,\mathbf{k} - \frac{\tilde{\omega}}{c_0}\left(d_{\mathbf{x}}\tilde{\mathbf{u}} + \{d_{\mathbf{x}}\tilde{\mathbf{u}}\}^T\right)\cdot\mathbf{k} \\
+ \frac{\tilde{\omega}}{c_0}\left(\left(\mathbf{k}\cdot\left(\{d_{\mathbf{x}}\tilde{\mathbf{u}}\}^T + d_{\mathbf{x}}\tilde{\mathbf{u}}\right)\right)\cdot\mathbf{k}\right)\mathbf{k}.
\end{aligned} \tag{13.407}$$

Thus, by (13.407) we rewrite (13.399) as

$$\begin{aligned}
\frac{c_0^2}{2\tilde{\omega}^2}\nabla_{\mathbf{x}}A\cdot\Bigg(\nabla_{\mathbf{x}}\tilde{\omega} - (\mathbf{k}\cdot\nabla_{\mathbf{x}}\tilde{\omega})\,\mathbf{k} - \frac{\tilde{\omega}}{c_0}\left(d_{\mathbf{x}}\tilde{\mathbf{u}} + \{d_{\mathbf{x}}\tilde{\mathbf{u}}\}^T\right)\cdot\mathbf{k} \\
+ \frac{\tilde{\omega}}{c_0}\left(\left(\mathbf{k}\cdot\left(\{d_{\mathbf{x}}\tilde{\mathbf{u}}\}^T + d_{\mathbf{x}}\tilde{\mathbf{u}}\right)\right)\cdot\mathbf{k}\right)\mathbf{k}\Bigg) \\
+ \frac{c_0^2}{2\tilde{\omega}}A\left(\frac{\partial}{\partial t}\left\{\frac{1}{c_0^2}G\right\} + \operatorname{div}_{\mathbf{x}}\left\{\frac{1}{c_0^2}G\tilde{\mathbf{u}}\right\} + \frac{1}{c_0^2}(c_0\mathbf{k})\cdot\nabla_{\mathbf{x}}G + \frac{1}{c_0^2}G^2\right) \\
- \frac{c_0^2}{2\tilde{\omega}}\left(\operatorname{div}_{\mathbf{x}}\left\{\nabla_{\mathbf{x}}A - (\mathbf{k}\cdot\nabla_{\mathbf{x}}A)\,\mathbf{k}\right\}\right) \approx \\
- i\left\{\frac{\partial A}{\partial t} + \mathbf{h}\cdot\nabla_{\mathbf{x}}A + \frac{1}{2}\left(\operatorname{div}_{\mathbf{x}}\mathbf{h}\right)A\right\} \\
- iA\left(\frac{c_0}{\sqrt{\tilde{\omega}}}\left(\frac{\partial}{\partial t}\left\{\frac{\sqrt{\tilde{\omega}}}{c_0}\right\} + \mathbf{h}\cdot\nabla_{\mathbf{x}}\left\{\frac{\sqrt{\tilde{\omega}}}{c_0}\right\}\right)\right),
\end{aligned} \tag{13.408}$$

We rewrite it as

$$\begin{aligned}
-\frac{c_0}{2\tilde{\omega}}\left(\nabla_{\mathbf{x}}A - (\mathbf{k}\cdot\nabla_{\mathbf{x}}A)\,\mathbf{k}\right)\cdot\Bigg(\left(d_{\mathbf{x}}\tilde{\mathbf{u}} + \{d_{\mathbf{x}}\tilde{\mathbf{u}}\}^T\right)\cdot\mathbf{k} \\
- \left(\left(\mathbf{k}\cdot\left(\{d_{\mathbf{x}}\tilde{\mathbf{u}}\}^T + d_{\mathbf{x}}\tilde{\mathbf{u}}\right)\right)\cdot\mathbf{k}\right)\mathbf{k}\Bigg) \\
+ \frac{c_0^2}{2\tilde{\omega}}A\left(\frac{\partial}{\partial t}\left\{\frac{1}{c_0^2}G\right\} + \operatorname{div}_{\mathbf{x}}\left\{\frac{1}{c_0^2}G\tilde{\mathbf{u}}\right\} + \frac{1}{c_0^2}(c_0\mathbf{k})\cdot\nabla_{\mathbf{x}}G + \frac{1}{c_0^2}G^2\right) \\
- \frac{c_0^2}{2}\left(\operatorname{div}_{\mathbf{x}}\left\{\frac{1}{\tilde{\omega}}\left(\nabla_{\mathbf{x}}A - (\mathbf{k}\cdot\nabla_{\mathbf{x}}A)\,\mathbf{k}\right)\right\}\right) \\
\approx -i\left\{\frac{\partial A}{\partial t} + \mathbf{h}\cdot\nabla_{\mathbf{x}}A + \frac{1}{2}\left(\operatorname{div}_{\mathbf{x}}\mathbf{h}\right)A\right\} \\
- iA\left(\frac{c_0}{\sqrt{\tilde{\omega}}}\left(\frac{\partial}{\partial t}\left\{\frac{\sqrt{\tilde{\omega}}}{c_0}\right\} + \mathbf{h}\cdot\nabla_{\mathbf{x}}\left\{\frac{\sqrt{\tilde{\omega}}}{c_0}\right\}\right)\right).
\end{aligned} \tag{13.409}$$

By (13.409) we have

$$
\begin{aligned}
&-\frac{c_0}{2\tilde{\omega}}\left(\nabla_{\mathbf{x}}A-(\mathbf{k}\cdot\nabla_{\mathbf{x}}A)\,\mathbf{k}\right)\cdot\Bigg(\left(d_{\mathbf{x}}\tilde{\mathbf{u}}+\{d_{\mathbf{x}}\tilde{\mathbf{u}}\}^T\right)\cdot\mathbf{k}\\
&\qquad\qquad-\left(\left(\mathbf{k}\cdot\left(\{d_{\mathbf{x}}\tilde{\mathbf{u}}\}^T+d_{\mathbf{x}}\tilde{\mathbf{u}}\right)\right)\cdot\mathbf{k}\right)\mathbf{k}\Bigg)\\
&\quad+\frac{c_0^2}{2\tilde{\omega}}A\left(\frac{\partial}{\partial t}\left\{\frac{1}{c_0^2}G\right\}+\operatorname{div}_{\mathbf{x}}\left\{\frac{1}{c_0^2}G\tilde{\mathbf{u}}\right\}+\frac{1}{c_0^2}(c_0\mathbf{k})\cdot\nabla_{\mathbf{x}}G+\frac{1}{c_0^2}G^2\right)\\
&\qquad-\frac{c_0^2}{2}\left(\operatorname{div}_{\mathbf{x}}\left\{\frac{1}{\tilde{\omega}}\left(\nabla_{\mathbf{x}}A-(\mathbf{k}\cdot\nabla_{\mathbf{x}}A)\,\mathbf{k}\right)\right\}\right)\\
&\qquad\approx -i\left\{\frac{\partial A}{\partial t}+\mathbf{h}\cdot\nabla_{\mathbf{x}}A+\frac{1}{2}\left(\operatorname{div}_{\mathbf{x}}\mathbf{h}\right)A\right\}\\
&\qquad-iA\frac{1}{2\tilde{\omega}}\left(\frac{\partial\tilde{\omega}}{\partial t}+\mathbf{h}\cdot\nabla_{\mathbf{x}}\tilde{\omega}\right)+iA\frac{1}{c_0}\left(\frac{\partial c_0}{\partial t}+\mathbf{h}\cdot\nabla_{\mathbf{x}}c_0\right).
\end{aligned}
\tag{13.410}
$$

On the other hand, by (13.383) we have

$$
\frac{1}{2\tilde{\omega}}\left(\frac{\partial\tilde{\omega}}{\partial t}+\mathbf{h}\cdot\nabla_{\mathbf{x}}\tilde{\omega}\right)=\frac{1}{2c_0}\left(\frac{\partial c_0}{\partial t}+\tilde{\mathbf{u}}\cdot\nabla_{\mathbf{x}}c_0\right)-\frac{1}{4}\left(\mathbf{k}\cdot\left(\{d_{\mathbf{x}}\tilde{\mathbf{u}}\}^T+d_{\mathbf{x}}\tilde{\mathbf{u}}\right)\right)\cdot\mathbf{k}.
\tag{13.411}
$$

Therefore, inserting (13.411) into (13.410) gives

$$
\begin{aligned}
&-\frac{c_0}{2\tilde{\omega}}\left(\nabla_{\mathbf{x}}A-(\mathbf{k}\cdot\nabla_{\mathbf{x}}A)\,\mathbf{k}\right)\cdot\Bigg(\left(d_{\mathbf{x}}\tilde{\mathbf{u}}+\{d_{\mathbf{x}}\tilde{\mathbf{u}}\}^T\right)\cdot\mathbf{k}\\
&\qquad\qquad-\left(\left(\mathbf{k}\cdot\left(\{d_{\mathbf{x}}\tilde{\mathbf{u}}\}^T+d_{\mathbf{x}}\tilde{\mathbf{u}}\right)\right)\cdot\mathbf{k}\right)\mathbf{k}\Bigg)\\
&\quad+\frac{c_0^2}{2\tilde{\omega}}A\left(\frac{\partial}{\partial t}\left\{\frac{1}{c_0^2}G\right\}+\operatorname{div}_{\mathbf{x}}\left\{\frac{1}{c_0^2}G\tilde{\mathbf{u}}\right\}+\frac{1}{c_0^2}(c_0\mathbf{k})\cdot\nabla_{\mathbf{x}}G+\frac{1}{c_0^2}G^2\right)\\
&-\frac{c_0^2}{2}\left(\operatorname{div}_{\mathbf{x}}\left\{\frac{1}{\tilde{\omega}}\left(\nabla_{\mathbf{x}}A-(\mathbf{k}\cdot\nabla_{\mathbf{x}}A)\,\mathbf{k}\right)\right\}\right)\approx -i\left\{\frac{\partial A}{\partial t}+\mathbf{h}\cdot\nabla_{\mathbf{x}}A+\frac{1}{2}\left(\operatorname{div}_{\mathbf{x}}\mathbf{h}\right)A\right\}\\
&-iA\frac{1}{2c_0}\left(c_0\mathbf{k}\cdot\nabla_{\mathbf{x}}c_0\right)+iA\frac{1}{2c_0}\left(\frac{\partial c_0}{\partial t}+\mathbf{h}\cdot\nabla_{\mathbf{x}}c_0\right)+\frac{iA}{4}\left(\mathbf{k}\cdot\left(\{d_{\mathbf{x}}\tilde{\mathbf{u}}\}^T+d_{\mathbf{x}}\tilde{\mathbf{u}}\right)\right)\cdot\mathbf{k}.
\end{aligned}
\tag{13.412}
$$

Moreover, as before, we can easily deduce that (13.412) invariant under the change of non-inertial Cartesian coordinate system in the case that under such change we have $A' = A$ and $S' = S$.

In particular, if in some Cartesian coordinate system we assume time-independent settings of the problem so that $\frac{\partial S}{\partial t} = c$, $\frac{\partial c_0}{\partial t} = 0$ and $\frac{\partial \tilde{\mathbf{u}}}{\partial t} = 0$, then we also have $\frac{\partial \mathbf{k}}{\partial t} =$

$\frac{\partial \mathbf{h}}{\partial t} = 0$ and $\frac{\partial \tilde{\omega}}{\partial t} = \frac{\partial G}{\partial t} = 0$, and therefore time-independent solutions $A := A(\mathbf{x})$ of (13.412) are admitted and they satisfy

$$
\begin{aligned}
-\frac{c_0}{2\tilde{\omega}}\left(\nabla_{\mathbf{x}}A - (\mathbf{k}\cdot\nabla_{\mathbf{x}}A)\,\mathbf{k}\right)\cdot\Bigg(\left(d_{\mathbf{x}}\tilde{\mathbf{u}} + \{d_{\mathbf{x}}\tilde{\mathbf{u}}\}^T\right)\cdot\mathbf{k}\\
-\left(\left(\mathbf{k}\cdot\left(\{d_{\mathbf{x}}\tilde{\mathbf{u}}\}^T + d_{\mathbf{x}}\tilde{\mathbf{u}}\right)\right)\cdot\mathbf{k}\right)\mathbf{k}\Bigg)\\
+\frac{c_0^2}{2\tilde{\omega}}A\left(\operatorname{div}_{\mathbf{x}}\left\{\frac{1}{c_0^2}G\tilde{\mathbf{u}}\right\} + \frac{1}{c_0^2}(c_0\mathbf{k})\cdot\nabla_{\mathbf{x}}G + \frac{1}{c_0^2}G^2\right)\\
-\frac{c_0^2}{2}\left(\operatorname{div}_{\mathbf{x}}\left\{\frac{1}{\tilde{\omega}}\left(\nabla_{\mathbf{x}}A - (\mathbf{k}\cdot\nabla_{\mathbf{x}}A)\,\mathbf{k}\right)\right\}\right) \approx -i\left\{\mathbf{h}\cdot\nabla_{\mathbf{x}}A + \frac{1}{2}\left(\operatorname{div}_{\mathbf{x}}\mathbf{h}\right)A\right\}\\
-iA\frac{1}{2c_0}\left(c_0\mathbf{k}\cdot\nabla_{\mathbf{x}}c_0\right) + iA\frac{1}{2c_0}\left(\mathbf{h}\cdot\nabla_{\mathbf{x}}c_0\right) + \frac{iA}{4}\left(\mathbf{k}\cdot\left(\{d_{\mathbf{x}}\tilde{\mathbf{u}}\}^T + d_{\mathbf{x}}\tilde{\mathbf{u}}\right)\right)\cdot\mathbf{k}.
\end{aligned}
\tag{13.413}
$$

Finally, from now we assume that $\tilde{\mathbf{u}}$ and c_0 vary sufficiently slowly in space and time variables, so that the following approximations are valid:

$$
\frac{|\nabla_{\mathbf{x}}c_0| + \frac{1}{c_0}\left|\frac{\partial c_0}{\partial t} + \tilde{\mathbf{u}}\cdot\nabla_{\mathbf{x}}c_0\right|}{c_0} + \frac{\left|d_{\mathbf{x}}\tilde{\mathbf{u}} + \{d_{\mathbf{x}}\tilde{\mathbf{u}}\}^T\right|}{c_0} \ll \frac{\left(|\nabla_{\mathbf{x}}A| + \frac{1}{c_0}\left|\frac{\partial A}{\partial t} + \tilde{\mathbf{u}}\cdot\nabla_{\mathbf{x}}A\right|\right)}{|A|}.
\tag{13.414}
$$

In particular, if in some Cartesian coordinate system, we have

$$
\begin{cases}
\frac{\left|d_{\mathbf{x}}\tilde{\mathbf{u}} + \{d_{\mathbf{x}}\tilde{\mathbf{u}}\}^T\right|^2}{|\tilde{\mathbf{u}}|^2} \ll \frac{\left(|\nabla_{\mathbf{x}}A| + \frac{1}{c_0}\left|\frac{\partial A}{\partial t} + \tilde{\mathbf{u}}\cdot\nabla_{\mathbf{x}}A\right|\right)^2}{|A|^2}\\
\frac{|\tilde{\mathbf{u}}|^2}{c_0^2} \ll 1\\
\frac{|\nabla_{\mathbf{x}}c_0| + \frac{1}{c_0}\left|\frac{\partial c_0}{\partial t} + \tilde{\mathbf{u}}\cdot\nabla_{\mathbf{x}}c_0\right|}{c_0} \ll \frac{\left(|\nabla_{\mathbf{x}}A| + \frac{1}{c_0}\left|\frac{\partial A}{\partial t} + \tilde{\mathbf{u}}\cdot\nabla_{\mathbf{x}}A\right|\right)}{|A|},
\end{cases}
\tag{13.415}
$$

then (13.414) indeed holds! Then, by (13.414) we rewrite and simplify (13.412) as

$$
\begin{aligned}
\frac{1}{2\tilde{\omega}}A\left(\frac{\partial G}{\partial t} + \operatorname{div}_{\mathbf{x}}\{G\tilde{\mathbf{u}}\} + (c_0\mathbf{k})\cdot\nabla_{\mathbf{x}}G + G^2\right)\\
-\frac{1}{2}\left(\operatorname{div}_{\mathbf{x}}\left\{\frac{c_0^2}{\tilde{\omega}}\left(\nabla_{\mathbf{x}}A - (\mathbf{k}\cdot\nabla_{\mathbf{x}}A)\,\mathbf{k}\right)\right\}\right) \approx -i\left\{\frac{\partial A}{\partial t} + \mathbf{h}\cdot\nabla_{\mathbf{x}}A + \frac{1}{2}\left(\operatorname{div}_{\mathbf{x}}\mathbf{h}\right)A\right\}.
\end{aligned}
\tag{13.416}
$$

Moreover, as before, we can easily deduce that (13.416) invariant under the change of non-inertial Cartesian coordinate system in the case that under such change we have $A' = A$ and $S' = S$.

Furthermore, in time-independent settings we rewrite (13.413) as

$$\frac{1}{2\tilde{\omega}}A\left(\mathrm{div}_{\mathbf{x}}\left\{G\tilde{\mathbf{u}}\right\}+(c_0\mathbf{k})\cdot\nabla_{\mathbf{x}}G+G^2\right)$$
$$-\frac{1}{2}\left(\mathrm{div}_{\mathbf{x}}\left\{\frac{c_0^2}{\tilde{\omega}}\left(\nabla_{\mathbf{x}}A-(\mathbf{k}\cdot\nabla_{\mathbf{x}}A)\,\mathbf{k}\right)\right\}\right)\approx -i\left\{\mathbf{h}\cdot\nabla_{\mathbf{x}}A+\frac{1}{2}\left(\mathrm{div}_{\mathbf{x}}\,\mathbf{h}\right)A\right\}. \tag{13.417}$$

In particular, in the case $\tilde{\mathbf{u}}=0$ and constant $\mathbf{k}$, c_0 and $\tilde{\omega}$, (13.417) simplifies as the paraxial approximation of the Helmholtz equation, with respect to direction $(-\mathbf{k})$, which is common in optics:

$$\frac{c_0}{2\tilde{\omega}}\left(\mathrm{div}_{\mathbf{x}}\left\{\left(\nabla_{\mathbf{x}}A-(\mathbf{h}\cdot\nabla_{\mathbf{x}}A)\,\mathbf{h}\right)\right\}\right)\approx -i\mathbf{k}\cdot\nabla_{\mathbf{x}}A. \tag{13.418}$$

Next, note that the time-dependent Eq. (13.416) is very similar to the Schrödinger equation and similarly can be written in the form

$$i\hbar\frac{\partial A}{\partial t}=\hat{H}\cdot A, \tag{13.419}$$

where

$$\hat{H}\cdot A:=\frac{\hbar}{2}\left(\mathrm{div}_{\mathbf{x}}\left\{\frac{c_0^2}{\tilde{\omega}}\left(\nabla_{\mathbf{x}}A-(\mathbf{k}\cdot\nabla_{\mathbf{x}}A)\,\mathbf{k}\right)\right\}\right)-i\hbar\left\{\mathbf{h}\cdot\nabla_{\mathbf{x}}A+\frac{1}{2}\left(\mathrm{div}_{\mathbf{x}}\,\mathbf{h}\right)A\right\}$$
$$-\frac{\hbar}{2\tilde{\omega}}\left(\frac{\partial G}{\partial t}+\mathrm{div}_{\mathbf{x}}\left\{G\tilde{\mathbf{u}}\right\}+(c_0\mathbf{k})\cdot\nabla_{\mathbf{x}}G+G^2\right)A \tag{13.420}$$

is a self-adjoint linear operator on the complex Hilbert space.

Finally, solving either (13.412) or (13.416) and inserting the solution into (13.370) we established $U(\mathbf{x},t)$ in (13.368) in the delicate geometric optics approximation (13.149).

13.3.11 Wave Optics Inside a Moving Medium and/or in the Presence of Gravitational Field

Assume that in some inertial or non-inertial Cartesian coordinate system a complex-valued scalar field $U:=U(\mathbf{x},t)$, characterizing some wave, satisfies the wave equation (13.114):

$$\frac{\partial}{\partial t}\left\{\frac{1}{c_0^2}\left(\frac{\partial U}{\partial t}+\tilde{\mathbf{u}}\cdot\nabla_{\mathbf{x}}U\right)\right\}+\operatorname{div}_{\mathbf{x}}\left\{\frac{1}{c_0^2}\left(\frac{\partial U}{\partial t}+\tilde{\mathbf{u}}\cdot\nabla_{\mathbf{x}}U\right)\tilde{\mathbf{u}}\right\}-\Delta_{\mathbf{x}}U=0, \tag{13.421}$$

where, as before, $\tilde{\mathbf{u}} := \tilde{\mathbf{u}}(\mathbf{x}, t)$ is some moderately changing (in space and in time) speed-like vector field and $c_0 := c_0(\mathbf{x}, t) > 0$ is a moderately changing scalar quantity that we call wave propagation speed. In particular, if we assume that the fields $\tilde{\mathbf{u}}$ and c_0 are independent on the time variable, and we consider the case of a monochromatic wave of the constant frequency $\nu = \frac{\omega}{2\pi}$:

$$U(\mathbf{x}, t) := \mathcal{U}(\mathbf{x})e^{i\omega t} \qquad \forall(\mathbf{x}, t), \tag{13.422}$$

where $\mathcal{U} := \mathcal{U}(\mathbf{x})$ is a complex-valued scalar field, independent on time, then inserting (13.422) into (13.421) gives the following time-independent equation:

$$\begin{aligned}&\frac{i\omega}{c_0^2(\mathbf{x})}\left(i\omega\mathcal{U}(\mathbf{x})+\tilde{\mathbf{u}}(\mathbf{x})\cdot\nabla_{\mathbf{x}}\mathcal{U}(\mathbf{x})\right)\\&\quad+\operatorname{div}_{\mathbf{x}}\left\{\frac{1}{c_0^2(\mathbf{x})}\left(i\omega\mathcal{U}(\mathbf{x})+\tilde{\mathbf{u}}(\mathbf{x})\cdot\nabla_{\mathbf{x}}\mathcal{U}(\mathbf{x})\right)\tilde{\mathbf{u}}(\mathbf{x})\right\}-\Delta_{\mathbf{x}}\mathcal{U}(\mathbf{x})=0.\end{aligned} \tag{13.423}$$

Next, suppose that, although we cannot consider the geometric optics approximation (13.149), we can consider, however, the following assumption:

$$\tilde{\mathbf{u}}(\mathbf{x}, t) = \nabla_{\mathbf{x}}\tilde{Z}(\mathbf{x}, t) \qquad \forall(\mathbf{x}, t), \tag{13.424}$$

where $\tilde{Z} := \tilde{Z}(\mathbf{x}, t)$ is some real-valued scalar function, which satisfies

$$\frac{1}{c_0^2}\left|\frac{\partial\tilde{Z}}{\partial t}\right|+\frac{1}{c_0^2}\left|\nabla_{\mathbf{x}}\tilde{Z}\right|^2 \ll 1. \tag{13.425}$$

Note that (13.424) is valid in the particular case of the electromagnetic waves, propagating in vacuum, in the inertial or more generally non-rotating Cartesian coordinate system (where we indeed have $\tilde{\mathbf{u}} = \mathbf{v}$ and $\operatorname{curl}_{\mathbf{x}}\mathbf{v} = 0$). On the other hand, (13.425) means that $\frac{|\tilde{\mathbf{u}}|^2}{c_0^2} \ll 1$ and $\tilde{\mathbf{u}}$ is either independent on time or depends on time slowly. Moreover, assume that c_0 is a constant in the given region.

Next we do the change of variables $(\mathbf{x}, t) \to (\mathbf{z}, \tau)$ as

$$\begin{cases}\mathbf{z} := \mathbf{x},\\ \tau := t + \frac{1}{c_0^2}\tilde{Z}(\mathbf{x}, t),\end{cases} \tag{13.426}$$

so that, we can find a complex-valued function $V := V(\mathbf{z}, \tau)$ depending on $\mathbf{z} \in \mathbb{R}^3$ and a real τ, which satisfies

$$U(\mathbf{x},t) := V\left(\mathbf{x}, t + \frac{1}{c_0^2}\tilde{Z}(\mathbf{x},t)\right). \tag{13.427}$$

Next we differentiate (13.427), assuming that c_0 vary very slowly and we can consider it to be a constant in the differentiation. Thus, by the chain rule we deduce

$$\begin{cases} \nabla_{\mathbf{x}} U(\mathbf{x},t) = \nabla_{\mathbf{z}} V\left(\mathbf{x}, t + \frac{1}{c_0^2}\tilde{Z}(\mathbf{x},t)\right) + \frac{1}{c_0^2}\frac{\partial V}{\partial \tau}\left(\mathbf{x}, t + \frac{1}{c_0^2}\tilde{Z}(\mathbf{x},t)\right)\nabla_{\mathbf{x}}\tilde{Z}(\mathbf{x},t) \\ \frac{\partial U}{\partial t}(\mathbf{x},t) = \frac{\partial V}{\partial \tau}\left(\mathbf{x}, t + \tilde{Z}(\mathbf{x},t)\right)\left(1 + \frac{1}{c_0^2}\frac{\partial \tilde{Z}}{\partial t}(\mathbf{x},t)\right). \end{cases} \tag{13.428}$$

Therefore, again by the chain rule we infer

$$\begin{aligned} \nabla_{\mathbf{x}}^2 U(\mathbf{x},t) = \nabla_{\mathbf{z}}^2 V\left(\mathbf{x}, t + \frac{1}{c_0^2}\tilde{Z}(\mathbf{x},t)\right) + \frac{1}{c_0^2}\nabla_{\mathbf{z}}\left(\frac{\partial V}{\partial \tau}\right)\left(\mathbf{x}, t + \frac{1}{c_0^2}\tilde{Z}(\mathbf{x},t)\right) \otimes \nabla_{\mathbf{x}}\tilde{Z}(\mathbf{x},t) \\ + \frac{1}{c_0^2}\nabla_{\mathbf{x}}\tilde{Z}(\mathbf{x},t) \otimes \nabla_{\mathbf{z}}\left(\frac{\partial V}{\partial \tau}\right)\left(\mathbf{x}, t + \frac{1}{c_0^2}\tilde{Z}(\mathbf{x},t)\right) \\ + \frac{1}{c_0^4}\frac{\partial^2 V}{\partial \tau^2}\left(\mathbf{x}, t + \frac{1}{c_0^2}\tilde{Z}(\mathbf{x},t)\right)\nabla_{\mathbf{x}}\tilde{Z}(\mathbf{x},t) \otimes \nabla_{\mathbf{x}}\tilde{Z}(\mathbf{x},t) \\ + \frac{1}{c_0^2}\frac{\partial V}{\partial \tau}\left(\mathbf{x}, t + \frac{1}{c_0^2}\tilde{Z}(\mathbf{x},t)\right)\nabla_{\mathbf{x}}^2\tilde{Z}(\mathbf{x},t). \end{aligned} \tag{13.429}$$

In particular,

$$\begin{aligned} \Delta_{\mathbf{x}} U(\mathbf{x},t) = \Delta_{\mathbf{z}} V\left(\mathbf{x}, t + \frac{1}{c_0^2}\tilde{Z}(\mathbf{x},t)\right) + \frac{2}{c_0^2}\nabla_{\mathbf{z}}\left(\frac{\partial V}{\partial \tau}\right)\left(\mathbf{x}, t + \frac{1}{c_0^2}\tilde{Z}(\mathbf{x},t)\right) \cdot \nabla_{\mathbf{x}}\tilde{Z}(\mathbf{x},t) \\ + \frac{1}{c_0^4}\frac{\partial^2 V}{\partial \tau^2}\left(\mathbf{x}, t + \frac{1}{c_0^2}\tilde{Z}(\mathbf{x},t)\right)\left|\nabla_{\mathbf{x}}\tilde{Z}(\mathbf{x},t)\right|^2 + \frac{1}{c_0^2}\frac{\partial V}{\partial \tau}\left(\mathbf{x}, t + \frac{1}{c_0^2}\tilde{Z}(\mathbf{x},t)\right)\Delta_{\mathbf{x}}\tilde{Z}(\mathbf{x},t). \end{aligned} \tag{13.430}$$

Moreover, again by the chain rule we have

$$\begin{aligned} \nabla_{\mathbf{x}}\left(\frac{\partial U}{\partial t}\right)(\mathbf{x},t) = \left(1 + \frac{1}{c_0^2}\frac{\partial \tilde{Z}}{\partial t}(\mathbf{x},t)\right)\nabla_{\mathbf{z}}\left(\frac{\partial V}{\partial \tau}\right)\left(\mathbf{x}, t + \frac{1}{c_0^2}\tilde{Z}(\mathbf{x},t)\right) \\ + \frac{1}{c_0^2}\left(1 + \frac{1}{c_0^2}\frac{\partial \tilde{Z}}{\partial t}(\mathbf{x},t)\right)\frac{\partial^2 V}{\partial \tau^2}\left(\mathbf{x}, t + \frac{1}{c_0^2}\tilde{Z}(\mathbf{x},t)\right)\nabla_{\mathbf{x}}\tilde{Z}(\mathbf{x},t) \\ + \frac{1}{c_0^2}\frac{\partial V}{\partial \tau}\left(\mathbf{x}, t + \frac{1}{c_0^2}\tilde{Z}(\mathbf{x},t)\right)\nabla_{\mathbf{x}}\left(\frac{\partial \tilde{Z}}{\partial t}(\mathbf{x},t)\right), \end{aligned} \tag{13.431}$$

and

$$\frac{\partial^2 U}{\partial t^2}(\mathbf{x},t) = \frac{\partial^2 V}{\partial \tau^2}\left(\mathbf{x}, t + \frac{1}{c_0^2}\tilde{Z}(\mathbf{x},t)\right)\left(1 + \frac{1}{c_0^2}\frac{\partial \tilde{Z}}{\partial t}(\mathbf{x},t)\right)^2 + \frac{1}{c_0^2}\frac{\partial V}{\partial \tau}\left(\mathbf{x}, t + \frac{1}{c_0^2}\tilde{Z}(\mathbf{x},t)\right)\frac{\partial^2 \tilde{Z}}{\partial t^2}(\mathbf{x},t). \quad (13.432)$$

Then, by (13.430), (13.431) and (13.432) together with (13.425) we have

$$\Delta_{\mathbf{x}} U(\mathbf{x},t) = \Delta_{\mathbf{z}} V\left(\mathbf{x}, t + \frac{1}{c_0^2}\tilde{Z}(\mathbf{x},t)\right) + \frac{2}{c_0^2}\nabla_{\mathbf{z}}\left(\frac{\partial V}{\partial \tau}\right)\left(\mathbf{x}, t + \frac{1}{c_0^2}\tilde{Z}(\mathbf{x},t)\right)\cdot \nabla_{\mathbf{x}}\tilde{Z}(\mathbf{x},t) + \frac{1}{c_0^2}\frac{\partial V}{\partial \tau}\left(\mathbf{x}, t + \frac{1}{c_0^2}\tilde{Z}(\mathbf{x},t)\right)\Delta_{\mathbf{x}}\tilde{Z}(\mathbf{x},t) + \frac{1}{c_0^4}\frac{\partial^2 V}{\partial \tau^2}\left(\mathbf{x}, t + \frac{1}{c_0^2}\tilde{Z}(\mathbf{x},t)\right)\left|\nabla_{\mathbf{x}}\tilde{Z}(\mathbf{x},t)\right|^2, \quad (13.433)$$

$$\nabla_{\mathbf{x}}\left(\frac{\partial U}{\partial t}\right)(\mathbf{x},t) \approx \nabla_{\mathbf{z}}\left(\frac{\partial V}{\partial \tau}\right)\left(\mathbf{x}, t + \frac{1}{c_0^2}\tilde{Z}(\mathbf{x},t)\right) + \frac{1}{c_0^2}\frac{\partial^2 V}{\partial \tau^2}\left(\mathbf{x}, t + \frac{1}{c_0^2}\tilde{Z}(\mathbf{x},t)\right)\nabla_{\mathbf{x}}\tilde{Z}(\mathbf{x},t) + \frac{1}{c_0^2}\frac{\partial V}{\partial \tau}\left(\mathbf{x}, t + \frac{1}{c_0^2}\tilde{Z}(\mathbf{x},t)\right)\nabla_{\mathbf{x}}\left(\frac{\partial \tilde{Z}}{\partial t}(\mathbf{x},t)\right), \quad (13.434)$$

and

$$\frac{\partial^2 U}{\partial t^2}(\mathbf{x},t) \approx, \frac{\partial^2 V}{\partial \tau^2}\left(\mathbf{x}, t + \frac{1}{c_0^2}\tilde{Z}(\mathbf{x},t)\right) + \frac{1}{c_0^2}\frac{\partial V}{\partial \tau}\left(\mathbf{x}, t + \frac{1}{c_0^2}\tilde{Z}(\mathbf{x},t)\right)\frac{\partial^2 \tilde{Z}}{\partial t^2}(\mathbf{x},t). \quad (13.435)$$

Moreover, by (13.428) together with (13.425) we have

$$\begin{cases} \nabla_{\mathbf{x}} U(\mathbf{x},t) = \nabla_{\mathbf{z}} V\left(\mathbf{x}, t + \frac{1}{c_0^2}\tilde{Z}(\mathbf{x},t)\right) + \frac{1}{c_0^2}\frac{\partial V}{\partial \tau}\left(\mathbf{x}, t + \frac{1}{c_0^2}\tilde{Z}(\mathbf{x},t)\right)\nabla_{\mathbf{x}}\tilde{Z}(\mathbf{x},t) \\ \frac{\partial U}{\partial t}(\mathbf{x},t) \approx \frac{\partial V}{\partial \tau}\left(\mathbf{x}, t + \frac{1}{c_0^2}\tilde{Z}(\mathbf{x},t)\right). \end{cases} \quad (13.436)$$

Thus, by (13.424) and (13.425) using (13.433), (13.434), (13.435) and (13.436) we deduce

$$
\frac{1}{c_0^2}\left(\frac{\partial}{\partial t}\left(\frac{\partial U}{\partial t}+\tilde{\mathbf{u}}\cdot\nabla_{\mathbf{x}}U\right)+\operatorname{div}_{\mathbf{x}}\left\{\left(\frac{\partial U}{\partial t}+\tilde{\mathbf{u}}\cdot\nabla_{\mathbf{x}}U\right)\tilde{\mathbf{u}}\right\}\right)-\Delta_{\mathbf{x}}U\approx
$$

$$
\frac{1}{c_0^2}\left(\frac{\partial^2 U}{\partial t^2}+2\tilde{\mathbf{u}}\cdot\nabla_{\mathbf{x}}\left(\frac{\partial U}{\partial t}\right)+(\operatorname{div}_{\mathbf{x}}\tilde{\mathbf{u}})\frac{\partial U}{\partial t}\right)-\Delta_{\mathbf{x}}U
$$

$$
\approx\frac{1}{c_0^2}\frac{\partial^2 V}{\partial\tau^2}\left(\mathbf{x},t+\frac{1}{c_0^2}\tilde{Z}(\mathbf{x},t)\right)-\Delta_{\mathbf{z}}V\left(\mathbf{x},t+\frac{1}{c_0^2}\tilde{Z}(\mathbf{x},t)\right)
$$

$$
+\frac{1}{c_0^4}\frac{\partial V}{\partial\tau}\left(\mathbf{x},t+\frac{1}{c_0^2}\tilde{Z}(\mathbf{x},t)\right)\frac{\partial^2\tilde{Z}}{\partial t^2}(\mathbf{x},t)
$$

$$
+\frac{2}{c_0^4}\frac{\partial V}{\partial\tau}\left(\mathbf{x},t+\frac{1}{c_0^2}\tilde{Z}(\mathbf{x},t)\right)\nabla_{\mathbf{x}}\left(\frac{\partial\tilde{Z}}{\partial t}(\mathbf{x},t)\right)\cdot\nabla_{\mathbf{x}}\tilde{Z}(\mathbf{x},t)
$$

$$
+\frac{1}{c_0^4}\frac{\partial^2 V}{\partial\tau^2}\left(\mathbf{x},t+\frac{1}{c_0^2}\tilde{Z}(\mathbf{x},t)\right)\left|\nabla_{\mathbf{x}}\tilde{Z}(\mathbf{x},t)\right|^2\approx
$$

$$
\frac{1}{c_0^2}\frac{\partial^2 V}{\partial\tau^2}\left(\mathbf{x},t+\frac{1}{c_0^2}\tilde{Z}(\mathbf{x},t)\right)-\Delta_{\mathbf{z}}V\left(\mathbf{x},t+\frac{1}{c_0^2}\tilde{Z}(\mathbf{x},t)\right)=\frac{1}{c_0^2}\frac{\partial^2 V}{\partial\tau^2}(\mathbf{z},\tau)-\Delta_{\mathbf{z}}V(\mathbf{z},\tau)\,.
\tag{13.437}
$$

Therefore, inserting (13.437) into (13.421) we deduce that the function $V(\mathbf{z},\tau)$ solves the standard type of the wave equation of the form:

$$
\frac{1}{c_0^2}\frac{\partial^2 V}{\partial\tau^2}(\mathbf{z},\tau)-\Delta_{\mathbf{z}}V(\mathbf{z},\tau)\approx 0.
\tag{13.438}
$$

Fortunately, the plenty of tools for analytical resolution of the standard wave equation (13.438) is well known. In particular, if we assume that the fields $\tilde{\mathbf{u}}$ and c_0 are independent of the time variable, and we consider the case of a monochromatic wave of the constant frequency $\nu=\frac{\omega}{2\pi}$, as in (13.422), then, we can choose $\tilde{Z}$ to be independent of time, and moreover inserting (13.422) into (13.427) gives

$$
\mathcal{U}(\mathbf{x})e^{i\omega t}:=V\left(\mathbf{x},t+\frac{1}{c_0^2}\tilde{Z}(\mathbf{x})\right),
\tag{13.439}
$$

and so

$$
V(\mathbf{z},\tau)=\mathcal{V}(\mathbf{z})\,e^{i\omega\tau}\quad\text{and}\quad\mathcal{U}(\mathbf{x}):=\mathcal{V}(\mathbf{x})\,e^{\frac{i\omega}{c_0^2}\tilde{Z}(\mathbf{x})},
\tag{13.440}
$$

where $\mathcal{V}:=\mathcal{V}(\mathbf{z})$ is a complex-valued scalar field, independent of time. Moreover, inserting (13.440) into (13.438) gives that $\mathcal{V}(\mathbf{x})$ solves the Helmholtz equation of the form:

$$
\frac{\omega^2}{c_0^2(\mathbf{x})}\mathcal{V}(\mathbf{x})+\Delta_{\mathbf{x}}\mathcal{V}(\mathbf{x})=0,
\tag{13.441}
$$

and

$$U(\mathbf{x},t) = \mathcal{V}(\mathbf{x})\, e^{i\omega\left(t+\frac{1}{c_0^2}\tilde{Z}(\mathbf{x})\right)}. \tag{13.442}$$

Thus, if, as before in (13.122), we denote $k_0 := \frac{\omega}{c}$, where c is a constant in the Maxwell equations for the vacuum, and express $U(\mathbf{x},t)$ as in (13.124) by the following:

$$U(\mathbf{x},t) = A(\mathbf{x},t)e^{ik_0 S(\mathbf{x},t)}, \tag{13.443}$$

where $A := A(\mathbf{x},t)$ and $S := S(\mathbf{x},t)$ are real scalar fields, then by (13.442) we deduce

$$U(\mathbf{x},t) = A(\mathbf{x},t)e^{ik_0 S(\mathbf{x},t)}, \quad \text{where} \quad A(\mathbf{x},t) = B(\mathbf{x}) \quad \text{and} \quad S(\mathbf{x},t) = ct + P(\mathbf{x}) + \frac{c}{c_0^2}\tilde{Z}(\mathbf{x}), \tag{13.444}$$

with

$$B(\mathbf{x})e^{ik_0 P(\mathbf{x})} := \mathcal{V}(\mathbf{x}), \tag{13.445}$$

where $B := B(\mathbf{x})$ and $P := P(\mathbf{x})$ are real scalar fields and $\mathcal{V}(\mathbf{x})$ is a complex solution of (13.441). In particular, if the vector field $\mathbf{h}$ is the direction of the propagation of the ray that passes through point $\mathbf{x}$, defined for every $\mathbf{x}$, as in (13.187) by

$$\mathbf{h}_0(\mathbf{x}) := \frac{c}{c_0^2}\tilde{\mathbf{u}}(\mathbf{x}) - \nabla_{\mathbf{x}} S(\mathbf{x}), \tag{13.446}$$

then by inserting (13.444) and (13.424) into (13.446) gives

$$\mathbf{h}_0(\mathbf{x}) = -\nabla_{\mathbf{x}} P(\mathbf{x}). \tag{13.447}$$

As $P(\mathbf{x})$ being the phase of $\mathcal{V}(\mathbf{x})$ is independent of the field $\tilde{\mathbf{u}}$ we obtain that the direction of the ray $\mathbf{h}_0$, in the general case of nontrivial field $\tilde{\mathbf{u}}$, is the same as in the case $\tilde{\mathbf{u}}(\mathbf{x}) = 0$. So the direction of the ray propagation $\mathbf{h}_0$ is not affected by the nontrivial motion of the medium and the nontrivial vectorial gravitational potential, provided we assume (13.424) and (13.425). Moreover, if we denote by $U(\mathbf{x},t) = A(\mathbf{x},t)e^{ik_0 S(\mathbf{x},t)}$ the monochromatic wave solution of the problem in the general case of nontrivial field $\tilde{\mathbf{u}}$ and by $U_0(\mathbf{x},t) = A_0(\mathbf{x},t)e^{ik_0 S_0(\mathbf{x},t)}$ the monochromatic wave solution of the corresponding simpler problem in the case of the trivial field $\tilde{\mathbf{u}} = 0$, then we have

$$U(\mathbf{x},t) = U_0(\mathbf{x},t)e^{\frac{i\omega}{c_0^2}\tilde{Z}(\mathbf{x})}, \quad A(\mathbf{x},t) = A_0(\mathbf{x},t) \quad \text{and} \quad S(\mathbf{x},t) = S_0(\mathbf{x},t) + \frac{c}{c_0^2}\tilde{Z}(\mathbf{x}), \tag{13.448}$$

i.e. in the case, where (13.424) and (13.425) hold, the wave solution for the general case of nontrivial $\tilde{\mathbf{u}}$ can be obtained from the corresponding solution in the simplest trivial case by simple adding the term $\frac{\omega}{c_0^2}\tilde{Z}(\mathbf{x})$ to the phase. In particular, all the pictures of the interference or the diffraction for given nontrivial $\tilde{\mathbf{u}}$ will be the same as in the trivial case, provided we assume (13.424) and (13.425). Finally, by simple adding the phase $\frac{\omega}{c_0^2}\tilde{Z}(\mathbf{x})$ we obtain the particular monochromatic solutions of (13.421) from the following well-known particular monochromatic solutions of the simplest wave equation (i.e. for the case $\tilde{\mathbf{u}} = 0$):

- plane wave,
- spherical wave,
- the approximate solution in the form of the Gaussian beam.

Chapter 14
Appendix

Consider the equations:

$$\begin{cases} \mathrm{curl}_{\mathbf{x}}\mathbf{H} = \frac{4\pi}{c}\mathbf{j} + \frac{1}{c}\frac{\partial \mathbf{D}}{\partial t}, \\ \mathrm{div}_{\mathbf{x}}\,\mathbf{D} = 4\pi\rho, \\ \mathrm{curl}_{\mathbf{x}}\mathbf{E} + \frac{1}{c}\frac{\partial \mathbf{B}}{\partial t} = 0, \\ \mathrm{div}_{\mathbf{x}}\,\mathbf{B} = 0, \\ \mathbf{E} = \mathbf{D} - \frac{1}{c}\mathbf{v}\times\mathbf{B}, \\ \mathbf{H} = \mathbf{B} + \frac{1}{c}\mathbf{v}\times\mathbf{D}. \end{cases} \tag{14.1}$$

Lemma 14.1 *Let* **D**, **B**, **E**, **H**, ρ, **j**, **v** *be solutions of* (14.1). *Then*

$$\frac{\partial}{\partial t}\left(\frac{|\mathbf{D}|^2+|\mathbf{B}|^2}{8\pi}\right) + \mathrm{div}_{\mathbf{x}}\left\{\left(\frac{|\mathbf{D}|^2+|\mathbf{B}|^2}{8\pi}\right)\mathbf{v}\right\} =$$
$$\frac{1}{4\pi}\mathrm{div}_{\mathbf{x}}\left\{(\mathbf{D}\otimes\mathbf{D}+\mathbf{B}\otimes\mathbf{B})\cdot\mathbf{v} - \frac{1}{2}\left(|\mathbf{D}|^2+|\mathbf{B}|^2\right)\mathbf{v} - c\mathbf{D}\times\mathbf{B}\right\}$$
$$-\left\{\frac{1}{4\pi}\left(\mathrm{div}_{\mathbf{x}}\left\{\mathbf{D}\otimes\mathbf{D}+\mathbf{B}\otimes\mathbf{B} - \frac{1}{2}\left(|\mathbf{D}|^2+|\mathbf{B}|^2\right)I\right\}\right) - \left(\rho\mathbf{E} + \frac{1}{c}\mathbf{j}\times\mathbf{B}\right)\right\}\cdot\mathbf{v} - \mathbf{j}\cdot\mathbf{E}, \tag{14.2}$$

where I is the identity matrix.

Proof By (14.1) and (2.5), we infer

$$\frac{1}{2c}\frac{\partial}{\partial t}\left(|\mathbf{D}|^2+|\mathbf{B}|^2\right)$$
$$= \frac{1}{c}\frac{\partial \mathbf{D}}{\partial t}\cdot\mathbf{D} + \frac{1}{c}\frac{\partial \mathbf{B}}{\partial t}\cdot\mathbf{B} = \left(\mathrm{curl}_{\mathbf{x}}\mathbf{H} - \frac{4\pi}{c}\mathbf{j}\right)\cdot\mathbf{D} - (\mathrm{curl}_{\mathbf{x}}\mathbf{E})\cdot\mathbf{B}$$
$$= \left\{\mathrm{curl}_{\mathbf{x}}\left(\mathbf{B} + \frac{1}{c}\mathbf{v}\times\mathbf{D}\right)\right\}\cdot\mathbf{D} - \left\{\mathrm{curl}_{\mathbf{x}}\left(\mathbf{D} - \frac{1}{c}\mathbf{v}\times\mathbf{B}\right)\right\}\cdot\mathbf{B} - \frac{4\pi}{c}\mathbf{j}\cdot\mathbf{D} =$$

A. Poliakovsky, *The Laws of Gravity and Electromagnetism*,
https://doi.org/10.1007/978-3-031-61407-1_14

$$\frac{1}{c}\mathbf{D}\cdot \mathrm{curl}_{\mathbf{x}}(\mathbf{v}\times\mathbf{D})+\frac{1}{c}\mathbf{B}\cdot \mathrm{curl}_{\mathbf{x}}(\mathbf{v}\times\mathbf{B})+\mathbf{D}\cdot \mathrm{curl}_{\mathbf{x}}\mathbf{B}-\mathbf{B}\cdot \mathrm{curl}_{\mathbf{x}}\mathbf{D}-\frac{4\pi}{c}\mathbf{j}\cdot\mathbf{D}$$
$$=\frac{1}{c}\mathbf{D}\cdot \mathrm{curl}_{\mathbf{x}}(\mathbf{v}\times\mathbf{D})+\frac{1}{c}\mathbf{B}\cdot \mathrm{curl}_{\mathbf{x}}(\mathbf{v}\times\mathbf{B})-\mathrm{div}_{\mathbf{x}}(\mathbf{D}\times\mathbf{B})-\frac{4\pi}{c}\mathbf{j}\cdot\mathbf{D}. \quad (14.3)$$

On the other hand, by (2.11) and (14.1), we obtain

$$\begin{aligned}
&\frac{1}{c}\mathbf{D}\cdot \mathrm{curl}_{\mathbf{x}}(\mathbf{v}\times\mathbf{D})+\frac{1}{c}\mathbf{B}\cdot \mathrm{curl}_{\mathbf{x}}(\mathbf{v}\times\mathbf{B})= \\
&\quad \frac{1}{c}(\mathrm{div}_{\mathbf{x}}\,\mathbf{D})\,\mathbf{v}\cdot\mathbf{D}-\frac{1}{c}(\mathrm{div}_{\mathbf{x}}\,\mathbf{v})|\mathbf{D}|^2+\frac{1}{c}\mathbf{D}\cdot\{(d_{\mathbf{x}}\mathbf{v})\cdot\mathbf{D}\}-\frac{1}{2c}\mathbf{v}\cdot\nabla_{\mathbf{x}}|\mathbf{D}|^2 \\
&\quad +(\mathrm{div}_{\mathbf{x}}\,\mathbf{B})\frac{1}{c}\,\mathbf{v}\cdot\mathbf{B}-(\mathrm{div}_{\mathbf{x}}\,\mathbf{v})\frac{1}{c}|\mathbf{B}|^2+\frac{1}{c}\,\mathbf{B}\cdot\{(d_{\mathbf{x}}\mathbf{v})\cdot\mathbf{B}\}-\frac{1}{2c}\mathbf{v}\cdot\nabla_{\mathbf{x}}|\mathbf{B}|^2 \\
&\quad =\frac{4\pi\rho}{c}\mathbf{v}\cdot\mathbf{D}-(\mathrm{div}_{\mathbf{x}}\,\mathbf{v})\frac{1}{c}\left(|\mathbf{D}|^2+|\mathbf{B}|^2\right)+\frac{1}{c}\,\mathbf{B}\cdot\{(d_{\mathbf{x}}\mathbf{v})\cdot\mathbf{B}\} \\
&\qquad +\frac{1}{c}\mathbf{D}\cdot\{(d_{\mathbf{x}}\mathbf{v})\cdot\mathbf{D}\}-\frac{1}{2c}\left\{\mathbf{v}\cdot\nabla_{\mathbf{x}}\left(|\mathbf{D}|^2+|\mathbf{B}|^2\right)\right\} \\
&\quad =\frac{4\pi\rho}{c}\mathbf{v}\cdot\mathbf{D}-\frac{1}{c}\left(\mathrm{div}_{\mathbf{x}}\left\{\mathbf{D}\otimes\mathbf{D}+\mathbf{B}\otimes\mathbf{B}-\frac{1}{2}\left(|\mathbf{D}|^2+|\mathbf{B}|^2\right)I\right\}\right)\cdot\mathbf{v} \\
&\qquad +\frac{1}{c}\,\mathrm{div}_{\mathbf{x}}\left\{(\mathbf{D}\otimes\mathbf{D}+\mathbf{B}\otimes\mathbf{B})\cdot\mathbf{v}-\left(|\mathbf{D}|^2+|\mathbf{B}|^2\right)\mathbf{v}\right\}. \quad (14.4)
\end{aligned}$$

Therefore, by (14.3) and (14.4) we obtain

$$\begin{aligned}
&\frac{1}{2c}\frac{\partial}{\partial t}\left(|\mathbf{D}|^2+|\mathbf{B}|^2\right)+\frac{1}{2c}\,\mathrm{div}_{\mathbf{x}}\left\{\left(|\mathbf{D}|^2+|\mathbf{B}|^2\right)\mathbf{v}\right\}= \\
&\quad \frac{1}{c}\,\mathrm{div}_{\mathbf{x}}\left\{(\mathbf{D}\otimes\mathbf{D}+\mathbf{B}\otimes\mathbf{B})\cdot\mathbf{v}-\frac{1}{2}\left(|\mathbf{D}|^2+|\mathbf{B}|^2\right)\mathbf{v}-c\mathbf{D}\times\mathbf{B}\right\} \\
&\quad -\frac{1}{c}\left(\mathrm{div}_{\mathbf{x}}\left\{\mathbf{D}\otimes\mathbf{D}+\mathbf{B}\otimes\mathbf{B}-\frac{1}{2}\left(|\mathbf{D}|^2+|\mathbf{B}|^2\right)I\right\}\right)\cdot\mathbf{v}-\frac{4\pi}{c}(\mathbf{j}-\rho\mathbf{v})\cdot\mathbf{D}.
\end{aligned} \quad (14.5)$$

Thus, since

$$(\mathbf{j}-\rho\mathbf{v})\cdot\mathbf{D}=(\mathbf{j}-\rho\mathbf{v})\cdot\left(\mathbf{E}+\frac{1}{c}\,\mathbf{v}\times\mathbf{B}\right)=\mathbf{j}\cdot\mathbf{E}-\mathbf{v}\cdot\left(\rho\mathbf{E}+\frac{1}{c}\,\mathbf{j}\times\mathbf{B}\right), \quad (14.6)$$

we rewrite (14.5) in the form (14.2). □

Lemma 14.2 *Let* $\mathbf{D}, \mathbf{B}, \mathbf{E}, \mathbf{H}, \rho, \mathbf{j}, \mathbf{v}$ *be solutions of* (14.1). *Then*

$$\frac{\partial}{\partial t}\left(\frac{1}{4\pi c}\mathbf{D}\times\mathbf{B}\right)+\mathrm{div}_{\mathbf{x}}\left\{\left(\frac{1}{4\pi c}\mathbf{D}\times\mathbf{B}\right)\otimes\mathbf{v}\right\}=-(d_{\mathbf{x}}\mathbf{v})^T\cdot\left(\frac{1}{4\pi c}\mathbf{D}\times\mathbf{B}\right)$$

$$+\frac{1}{4\pi}\operatorname{div}_{\mathbf{x}}\left\{\mathbf{D}\otimes\mathbf{D}+\mathbf{B}\otimes\mathbf{B}-\frac{1}{2}\left(|\mathbf{D}|^2+|\mathbf{B}|^2\right)I\right\}-\left(\rho\mathbf{E}+\frac{1}{c}\mathbf{j}\times\mathbf{B}\right). \tag{14.7}$$

Proof By (14.1), we have

$$\begin{aligned}\frac{\partial}{\partial t}\left(\frac{1}{c}\mathbf{D}\times\mathbf{B}\right)&=\frac{1}{c}\frac{\partial\mathbf{D}}{\partial t}\times\mathbf{B}+\mathbf{D}\times\frac{1}{c}\frac{\partial\mathbf{B}}{\partial t}=\left(\operatorname{curl}_{\mathbf{x}}\mathbf{H}-\frac{4\pi}{c}\mathbf{j}\right)\times\mathbf{B}-\mathbf{D}\times\operatorname{curl}_{\mathbf{x}}\mathbf{E}\\&=\operatorname{curl}_{\mathbf{x}}\left(\mathbf{B}+\frac{1}{c}\mathbf{v}\times\mathbf{D}\right)\times\mathbf{B}-\mathbf{D}\times\operatorname{curl}_{\mathbf{x}}\left(\mathbf{D}-\frac{1}{c}\mathbf{v}\times\mathbf{B}\right)-\frac{4\pi}{c}\mathbf{j}\times\mathbf{B}=\\&\frac{1}{c}\mathbf{D}\times\operatorname{curl}_{\mathbf{x}}(\mathbf{v}\times\mathbf{B})+\frac{1}{c}\operatorname{curl}_{\mathbf{x}}(\mathbf{v}\times\mathbf{D})\times\mathbf{B}-\mathbf{D}\times\operatorname{curl}_{\mathbf{x}}\mathbf{D}-\mathbf{B}\times\operatorname{curl}_{\mathbf{x}}\mathbf{B}-\frac{4\pi}{c}\mathbf{j}\times\mathbf{B}.\end{aligned} \tag{14.8}$$

On the other hand, by (2.11) and (14.1), we obtain

$$\begin{aligned}&\frac{1}{c}\mathbf{D}\times\operatorname{curl}_{\mathbf{x}}(\mathbf{v}\times\mathbf{B})+\frac{1}{c}\operatorname{curl}_{\mathbf{x}}(\mathbf{v}\times\mathbf{D})\times\mathbf{B}=\\&(\operatorname{div}_{\mathbf{x}}\mathbf{B})\frac{1}{c}\mathbf{D}\times\mathbf{v}-(\operatorname{div}_{\mathbf{x}}\mathbf{v})\frac{1}{c}\mathbf{D}\times\mathbf{B}+\frac{1}{c}\mathbf{D}\times\{(d_{\mathbf{x}}\mathbf{v})\cdot\mathbf{B}\}-\frac{1}{c}\mathbf{D}\times\{(d_{\mathbf{x}}\mathbf{B})\cdot\mathbf{v}\}\\&+\frac{1}{c}(\operatorname{div}_{\mathbf{x}}\mathbf{D})\,\mathbf{v}\times\mathbf{B}-\frac{1}{c}(\operatorname{div}_{\mathbf{x}}\mathbf{v})\,\mathbf{D}\times\mathbf{B}+\frac{1}{c}\{(d_{\mathbf{x}}\mathbf{v})\cdot\mathbf{D}\}\times\mathbf{B}-\frac{1}{c}\{(d_{\mathbf{x}}\mathbf{D})\cdot\mathbf{v}\}\times\mathbf{B}\\&\qquad=\frac{1}{c}\mathbf{D}\times\{(d_{\mathbf{x}}\mathbf{v})\cdot\mathbf{B}\}+\frac{1}{c}\{(d_{\mathbf{x}}\mathbf{v})\cdot\mathbf{D}\}\times\mathbf{B}\\&\qquad-2(\operatorname{div}_{\mathbf{x}}\mathbf{v})\frac{1}{c}\mathbf{D}\times\mathbf{B}-\frac{1}{c}\{d_{\mathbf{x}}(\mathbf{D}\times\mathbf{B})\}\cdot\mathbf{v}+\frac{4\pi\rho}{c}\mathbf{v}\times\mathbf{B}=\\&\qquad\frac{1}{c}\mathbf{D}\times\{(d_{\mathbf{x}}\mathbf{v})\cdot\mathbf{B}\}+\frac{1}{c}\{(d_{\mathbf{x}}\mathbf{v})\cdot\mathbf{D}\}\times\mathbf{B}\\&\qquad-(\operatorname{div}_{\mathbf{x}}\mathbf{v})\frac{1}{c}\mathbf{D}\times\mathbf{B}-\frac{1}{c}\operatorname{div}_{\mathbf{x}}\{(\mathbf{D}\times\mathbf{B})\otimes\mathbf{v}\}+\frac{4\pi\rho}{c}\mathbf{v}\times\mathbf{B},\end{aligned} \tag{14.9}$$

and by (2.16) and (14.1), we deduce

$$\begin{aligned}-\mathbf{D}\times\operatorname{curl}_{\mathbf{x}}\mathbf{D}-\mathbf{B}\times\operatorname{curl}_{\mathbf{x}}\mathbf{B}&=(d_{\mathbf{x}}\mathbf{D})\cdot\mathbf{D}-\frac{1}{2}\nabla_{\mathbf{x}}|\mathbf{D}|^2+(d_{\mathbf{x}}\mathbf{B})\cdot\mathbf{B}-\frac{1}{2}\nabla_{\mathbf{x}}|\mathbf{B}|^2\\&=\operatorname{div}_{\mathbf{x}}\left\{\mathbf{D}\otimes\mathbf{D}+\mathbf{B}\otimes\mathbf{B}-\frac{1}{2}\left(|\mathbf{D}|^2+|\mathbf{B}|^2\right)I\right\}-4\pi\rho\mathbf{D},\end{aligned} \tag{14.10}$$

where $I\in\mathbb{R}^{3\times3}$ is the unit matrix (identity linear operator). Thus, plugging (14.9) and (14.10) into (14.8) and using (2.3), we obtain

$$\begin{aligned}&\frac{\partial}{\partial t}\left(\frac{1}{c}\mathbf{D}\times\mathbf{B}\right)+\operatorname{div}_{\mathbf{x}}\left\{\left(\frac{1}{c}\mathbf{D}\times\mathbf{B}\right)\otimes\mathbf{v}\right\}=\\&\qquad\frac{1}{c}\mathbf{D}\times\{(d_{\mathbf{x}}\mathbf{v})\cdot\mathbf{B}\}+\frac{1}{c}\{(d_{\mathbf{x}}\mathbf{v})\cdot\mathbf{D}\}\times\mathbf{B}-(\operatorname{div}_{\mathbf{x}}\mathbf{v})\frac{1}{c}\mathbf{D}\times\mathbf{B}\end{aligned}$$

$$
\begin{aligned}
&+ \operatorname{div}_{\mathbf{x}} \left\{ \mathbf{D} \otimes \mathbf{D} + \mathbf{B} \otimes \mathbf{B} - \frac{1}{2} \left(|\mathbf{D}|^2 + |\mathbf{B}|^2 \right) I \right\} - 4\pi\rho\mathbf{D} - \frac{4\pi}{c} (\mathbf{j} - \rho\mathbf{v}) \times \mathbf{B} \\
&= -\frac{1}{c} (d_{\mathbf{x}}\mathbf{v})^T \cdot (\mathbf{D} \times \mathbf{B}) + \operatorname{div}_{\mathbf{x}} \left\{ \mathbf{D} \otimes \mathbf{D} + \mathbf{B} \otimes \mathbf{B} - \frac{1}{2} \left(|\mathbf{D}|^2 + |\mathbf{B}|^2 \right) I \right\} \\
&\quad - 4\pi\rho\mathbf{E} - \frac{4\pi}{c} \mathbf{j} \times \mathbf{B} = \frac{1}{c} \left\{ d_{\mathbf{x}} (\mathbf{D} \times \mathbf{B}) \right\}^T \cdot \mathbf{v} \\
&\quad + \operatorname{div}_{\mathbf{x}} \left\{ \mathbf{D} \otimes \mathbf{D} + \mathbf{B} \otimes \mathbf{B} - \frac{1}{2} \left(|\mathbf{D}|^2 + |\mathbf{B}|^2 + \frac{2}{c} \mathbf{v} \cdot (\mathbf{D} \times \mathbf{B}) \right) I \right\} \\
&\quad - 4\pi\rho\mathbf{E} - \frac{4\pi}{c} \mathbf{j} \times \mathbf{B}. \qquad (14.11)
\end{aligned}
$$

So, we finally deduce (14.7). □

References

1. P. Cornille, *Advansed Electromagnetism and Vacuum Physics*, World Scientific series in Contemporary Chemical Physics, vol. **21**.
2. L.D. Landau, E.M. Lifshitz, *Quantum Mechanics: Non-Relativistic Theory. Vol. 3* (3rd ed.) (1977), Pergamon Press.
3. A. Poliakovsky, *Lorentzian geometrical structures with global time, Gravity and Electrodynamics*, Springer, Cham, 2023, Book doi: https://doi.org/10.1007/978-3-031-23762-1
4. L.I. Sedov, *Mechanics of Continuous Media: Volume 1*, World Scientific (1997), Series in Theoretical and Applied Mechanics, vol. **4**.

A. Poliakovsky, *The Laws of Gravity and Electromagnetism*,
https://doi.org/10.1007/978-3-031-61407-1

The manufacturer's authorised representative in the EU is Springer Nature Customer Service Centre GmbH, Europaplatz 3, 69115 Heidelberg, Germany. If you have any concerns regarding our products, please contact ProductSafety@springernature.com

Printed and bound by CPI Group (UK) Ltd, Croydon, CR0 4YY
01/10/2026
02225909-0019